Radiation from Medical Procedures in the Pathogenesis of Cancer and Ischemic Heart Disease:

Dose–Response Studies with Physicians per 100,000 Population

John W. Gofman, M.D., Ph. D.
Professor Emeritus, Molecular and Cell Biology
University of California, Berkeley

Edited by Egan O'Connor

● Hypothesis–1: Medical radiation is a highly important cause (probably the principal cause) of cancer mortality in the United States during the Twentieth Century. Medical radiation means, primarily, exposure by xrays (including fluoroscopy and CT scans).

● Hypothesis–2: Medical radiation, received even at very low and moderate doses, is an important cause of death from Ischemic Heart Disease; the probable mechanism is radiation-induction of mutations in the coronary arteries, resulting in dysfunctional clones (mini-tumors) of smooth muscle cells.

First Edition: 1999

C.N.R. Book Division

Committee for Nuclear Responsibility, Inc.

Post Office Box 421993

San Francisco, California 94142

U.S.A.

The Committee for Nuclear Responsibility, Inc. (CNR) is a non-profit educational group organized in 1971 to provide independent analyses of the health effects and sources of ionizing radiation. Research in this field is not commercially viable. Most radiation research, analysis, and publications are funded, directly or indirectly, by governments and other sponsors whose activities cause exposure to ionizing radiation. By contrast, CNR's work is independent of sponsorship from such sources. The work is supported by persons who value "second opinions."

This book, RADIATION FROM MEDICAL PROCEDURES IN THE PATHOGENESIS OF CANCER AND ISCHEMIC HEART DISEASE, begins where the previous CNR study ended.

Library of Congress Cataloging-in-Publication Data

Gofman, John W.
  Radiation from medical procedures in the pathogenesis of cancer and ischemic heart disease: dose-response studies with physicians per 100,000 population / John W. Gofman ; edited by Egan O'Connor.-- 1st ed.
    p. ; cm.
  Includes bibliographical references and index.
  ISBN 0-932682-97-9 (hardcover) -- ISBN 0-932682-98-7 (soft cover)
  1. Radiation carcinogenesis. 2. Coronary heart disease--Etiology. 3. Ionizing radiation--Health aspects. 4. Radiology, Medical--Health aspects. 5. Cancer--United States--Epidemiology. 6. Coronary heart disease--United States--Epidemiology. I. O'Connor, Egan. II. Committee for Nuclear Responsibility, Inc. III. Title.
  [DNLM: 1. Neoplasms--mortality. 2. Neoplasms--etiology. 3. Myocardial Ischemia--mortality. 4. Myocardial Ischemia--etiology. 5. Radiography--adverse effects. 6. Dose-Response Relationship, Radiation. 7. United States--epidemiology. QZ 200 G612ra 1999]
  RC268.55 .G628 1999
  616.99'4071--dc21
                                                                99-045096

The Library of Congress Catalog Number (LCCN) is 99-045096.
This book is manufactured in the United States of America, at Consolidated Printers in Berkeley CA. It is available from the publisher, from online bookstores, and from library distributors.

Published by:
Committee for Nuclear Responsibility. Tel/fax 415-776-8299. Email: cnr123@webtv.net
Hardcover: ISBN 0-932682-97-9. $35.00.   Paperback: ISBN 0-932682-98-7. $27.00.

    Paper in both editions is pH neutral. There are 699 pages per book, and page-size is 8.5 x 11 inches upright (21.6 x 27.9 cm).
    Both the hardcover and softcover editions are sturdy and durable. The hardcover has a glossy laminated casing (no jacket); the textblock is Smyth-sewn. Librarians can obtain additional details about the binding from the publisher.

Related books by John W. Gofman:

What We Do Know about Heart Attacks. 1957.
Dietary Prevention and Treatment of Heart Disease. Co-authors: Alex V. Nichols and E. Virginia Dobbin. 1958. LCCN 58-10072.
Coronary Heart Disease. 1959. LCCN 58-14073.
Radiation and Human Health. 1981. LCCN 80-26484.
Xrays: Health Effects of Common Exams. Co-author: Egan O'Connor. 1985. LCCN 84-23527.
Radiation-Induced Cancer from Low-Dose Exposure: An Independent Analysis. 1990. LCCN 89-62431.
Chernobyl Accident: Radiation Consequences for This and Future Generations. 1994. ISBN 5-339-00869-X (Vysheishaya Shkola Publishing House, Minsk, Belarus).
Preventing Breast Cancer: The Story of a Major, Proven, Preventable Cause of This Disease. 1995/96. LCCN 94-69129 (1st edition, 1995). LCCN 96-2453 (2nd edition, 1996).

Acknowledgments

The publisher takes great pleasure in thanking the following individuals and foundations for their support of CNR's publishing program on the health effects of exposure to ionizing radiation:

June Allen, M.Ed.
Phillip M. Allen, M.D., Ph.D.
Rodney Baird, in memoriam
Charles Bloomstein
David C. Bradley, M.D.
Paul F. Burmeister
Columbia Foundation
C.S. Fund
Richard Dorwin Davis, in memoriam
Leo and Kay Drey
Genevieve Ellis Estes, in memoriam
Robin M. Fleck, M.D.
Friendship Fund
Franklin L. Gage
John B. Gilpin
Richard and Rhoda Goldman Fund
The Grodzins Fund
Anne D. Hahn, in memoriam
Hahn Family Foundation
Betty Rhodes Latner
Robert Lowitz
James P. McGinley, M.D.
Mary R. Morgan
David T. Ratcliffe
Right Livelihood Award Foundation
Caroline H. Robinson Social Service Fund
Lawrence G. Tesler
May C. Tooker

SECTION ONE
Orientation, Materials, Methods

SECTION TWO
Cancer Mortality: Evidence that Medical Radiation Became a Principal Cause by 1940

SECTION THREE
NonCancer NonIHD Mortality: Inverse Dose-Response with Medical Radiation

SECTION FOUR
Ischemic Heart Disease: Evidence that Medical Radiation Became an Important Cause by 1950

To denote emphasis, this monograph uses CAPITALIZED words (instead of italics).

The Author's History
by Egan O'Connor

John William Gofman is Professor Emeritus of Molecular and Cell Biology, University of California at Berkeley, CA 94720-5706. He is also on the faculty at the University of California Medical School at San Francisco (UCSF). His life's work is divisible into three main areas, which converge for the first time in this monograph. Some of the earlier work is cited in the monograph's Reference List.

● (1) While a graduate student at U.C. Berkeley, Gofman earned his Ph.D. (1943) in nuclear/physical chemistry, with his dissertation on the discovery of Pa-232, U-232, Pa-233, and U-233, the proof that U-233 is fissionable by slow and fast neutrons, and discovery of the 4n + 1 radioactive series. His faculty advisor was Glenn T. Seaborg (who became Chairman of the Atomic Energy Commission, 1961-1971). Seaborg, Gofman, and Raymond W. Stoughton share Patent #3,123,535 on the slow and fast neutron fissionability of uranium-233, with its application to production of nuclear power or nuclear weapons. The work is recounted in Seaborg's book "Nuclear Milestones" (1972).

Post-doctorally, Gofman continued research related to the first atomic bombs ––– particularly the chemistry of plutonium, at a time when the world's total supply was less than 0.25 milligram. He shares patents #2,671,251 and #2,912,302 on two processes for separating plutonium from the uranium and fission products of irradiated nuclear fuel. "We all were pushing the envelope in those years, and in the process, we learned the habit of observing details very closely."

● (2) After the plutonium work, Gofman completed medical school (1946) at UCSF, where the faculty and his classmates selected him to receive the annual Gold-Headed Cane Award for having the qualities of "a true physician."

In 1947, following his internship in Internal Medicine, Gofman joined the faculty at U.C. Berkeley (Division of Medical Physics), where he began his research on lipoproteins and Coronary Heart Disease at the Donner Laboratory. At the time, only two types of blood lipoproteins were known: Alpha and beta. By devising special flotation techniques with the ultracentrifuge, he and Frank T. Lindgren and co-workers at the Donner Lab began to reveal (1949-1950) the great diversity of very-low-density, intermediate-density, low-density, and high-density lipoproteins (VLDL, IDL, LDL, HDL) which truly exist in the bloodstream.

Their work on the chemistry of lipoproteins (e.g., the cholesterol-rich and triglyceride-rich varieties), and on dietary experiments, and on epidemiologic studies, soon produced evidence that high blood levels of the LDL, IDL, and VLDL lipoproteins are a risk-factor for Coronary Heart Disease.

In 1954, Gofman received the Modern Medicine Award for outstanding contributions to heart disease research. In 1965, he received the Lyman Duff Lectureship Award of the American Heart Association, for his research in atherosclerosis and Coronary Heart Disease. In 1972, he shared the Stouffer Prize for outstanding contributions to research in arteriosclerosis. In 1974, the American College of Cardiology selected him as one of twenty-five leading researchers in cardiology of the past quarter-century.

● (3) Meanwhile, in the early 1960s, the Atomic Energy Commission (AEC) asked Gofman to establish a Biomedical Research Division at the AEC's Livermore National Laboratory, for the purpose of evaluating the health effects of all types of nuclear activities. From 1963-1965, Gofman served as the division's first director and concurrently as an Associate Director of the full laboratory. Then he stepped down from the administrative activities in order to have more time for his own laboratory research on Cancer and chromosomes (the Boveri Hypothesis), on radiation-induced chromosomal mutations and genomic instability, and for his analytical work on the epidemiologic data from the Japanese atomic-bomb survivors and other irradiated human populations.

By 1969, Gofman and a Livermore colleague, Dr. Arthur R. Tamplin, had concluded that human exposure to ionizing radiation was much more serious than previously recognized. Because of this finding, Gofman and Tamplin spoke out publicly against two AEC programs which they had previously accepted. One was Project Plowshare, a program to explode hundreds or thousands of underground nuclear bombs in the Rocky Mountains in order to liberate (radioactive) natural gas, and to use nuclear explosives also to excavate harbors and canals. The second was the plan to license about 1,000 commercial nuclear power plants (USA) as quickly as possible. In 1970, Gofman and Tamplin proposed a 5-year moratorium on that activity.

The AEC was not pleased. Seaborg recounts some of the heated conversations among the Commissioners in his book "The Atomic Energy Commission under Nixon: Adjusting to Troubled Times" (1993). By 1973, Livermore de-funded Gofman's laboratory research on chromosomes and Cancer. He returned to teaching full-time at U.C. Berkeley, until choosing an early and active "retirement" in order to concentrate fully on pro-bono research into human health-effects from radiation.

His 1981, 1985, 1990, 1994, and 1995/96 books present a series of findings. His 1990 book includes his proof, "by any reasonable standard of biomedical proof," that there is no threshold level (no harmless dose) of ionizing radiation with respect to radiation mutagenesis and carcinogenesis ––– a conclusion supported in 1995 by a government-funded radiation committee. His 1995/96 book provides evidence that medical radiation is a necessary co-actor in about 75% of the recent and current Breast Cancer incidence (USA) ––– a conclusion doubted but not at all refuted by several peer-reviewers.

John W. Gofman is the son of David and Sarah Gofman ––– who immigrated to the USA from czarist Russia in about 1905. JWG was born in Cleveland, Ohio, in September 1918.

INTRODUCTION

Overview, and Some Practical Implications of This Work

Part 1. Practical Implications of Hypotheses One and Two
Part 2. Differing Origins of the Two Hypotheses
Part 3. Some Rather Dazzling Results to Examine
Part 4. Why Our Findings Do Not Challenge the Importance of Other Causes of Cancer and IHD
Part 5. How to Reconcile High Fractional Causations by Xrays, Smoking, Diet

● **Part 1. Practical Implications of Hypotheses One and Two**

During the 1990s, approximately 23% of the U.S. deaths have been caused by Cancer, and 22% by Ischemic Heart Disease (also called Coronary Heart Disease, and Coronary Artery Disease).

Would anyone NOT welcome a simple, safe, and painless way either to postpone many cases of such diseases or to prevent many cases from occurring at all? The findings in this book, combined with already-published wisdom from some mainstream radiologists and radiologic physicists, identify such a way --- with certainty for Cancer, and with great likelihood for Ischemic Heart Disease (IHD).

The word "practical" is featured above, because prevention of these two diseases has always been our chief reason for investigating their causes. The evidence assembled and analyzed in this monograph identifies medical radiation as a very important cause of both diseases. The work is organized around two hypotheses.

1a. Statement of Hypothesis-1 (Cancer) and Hypothesis-2 (IHD)

● Hypothesis-1 is this: Medical radiation is a highly important cause (probably the principal cause) of cancer mortality in the United States during the Twentieth Century. (Hypothesis-1 is about causation, so it is silent about radiation-therapy used after a Cancer has been diagnosed.)

We are well aware of a belief that medical radiation causes only a very low percentage of cancer mortality. That belief rests on a few estimates whose input-data are highly unreliable and sometimes inherently irrelevant, for the reasons presented in Chapters 1, 2, and 67 (Part 5). By contrast, the evidence in this book strongly supports Hypothesis-1. We are confident --- for the reasons listed in Chapter 1 --- that our findings are far more credible, scientifically, than the low estimates. Also we are confident, for reasons stated in Part 5, that our findings do not conflict with estimates that more than half of the cancer rate is a result of smoking and poor diet.

● Hypothesis-2 is this: Medical radiation, received even at very low and moderate doses, is an important cause of Ischemic Heart Disease (IHD); the probable mechanism is radiation-induction of mutations in the coronary arteries, resulting in dysfunctional clones (mini-tumors) of smooth muscle cells. (Here at the outset, we can prevent some confusion about Hypothesis-2 by stating that (a) it was discovered decades ago that medical radiation at very high doses can damage the heart and its vessels, and that (b) the kinds of damage reported from very high-dose radiation seldom resemble the lesions of Ischemic Heart Disease --- details in Appendix J.)

Chapter 45 presents a Unified Model of Atherogenesis and Acute IHD Events which is consistent with the evidence in this book, is consistent with the findings (first by Earl Benditt in 1973) of monoclonal cells in atherosclerotic plaques, is consistent with well-established knowledge about atherogenic lipoproteins and other non-xray causes of fatal IHD, and is consistent with recent findings about the weaker connection than expected between degree of arterial stenosis and the fatal rupturing of specific atherosclerotic plaques.

1b. What Constitutes "Medical Radiation"?

Because not all readers will "arrive" here from the same fields, or with the same backgrounds, or with English as the native language, this book defines various terms and concepts in the fields of radiation, Cancer, Ischemic Heart Disease, and dose-response analysis. Definitions can be located with the combined Index and Glossary.

By medical radiation, Hypotheses One and Two mean primarily but not exclusively xrays (including fluoroscopy and CT scans).

There is no doubt that medical radiation can both be a cause of Cancer and also be used to treat Cancer. Cancerous activities are done by living cells, whose cancerous behavior can result from radiation-induced mutations of numerous types --- types which do not kill or sterilize the cells. When radiation is used for treatment of Cancer, it is used in very high doses which do enough damage to kill or sterilize cells. Clearly, dead or non-dividing cells cannot behave like cancer cells.

1c. Practical Implications of Hypotheses One and Two

The validity of Hypotheses One and Two is a question with major implications for future health, in the USA and elsewhere. Validity means that medical professionals and other humans have, already at hand, an opportunity which is guaranteed to achieve large reductions in FUTURE mortality-rates from Cancer and which is very likely to achieve similar reductions in Ischemic Heart Disease, in countries where medical radiation is widely in use.

Knowledgeable "mainstream" experts in radiology and radiologic physics have shown that xray dosage, from nontherapeutic diagnostic and interventional radiology in current medicine, could readily be cut by a factor of two or more (Chapter 1, Box 3) --- while still obtaining all the benefits of such radiology and without eliminating a single procedure (specifics in Chapters 1 and 2). Example: While radiographers have reduced the xray dose per mammographic examination by more than 10-fold, use of mammography has risen dramatically. The result of dose-reduction has certainly not been less mammography --- but rather, less-risky mammography.

Beyond diagnostic radiology, there is extensive and growing use of xray fluoroscopy, nondiagnostically, during placement of catheters and during surgical procedures. There is no doubt that dosage could be reduced many-fold during such procedures (Chapter 1, Box 3; Chapter 2, Part 3).

● Part 2. Differing Origins of the Two Hypotheses

How we happened to arrive at Hypothesis-1 is related in Chapter 2, Part 9. It deserves emphasis that Hypothesis-1 is not "Medical radiation can induce Cancer." Induction of Cancer in humans by ionizing radiation, including xrays, was proven long ago (Chapter 2, Part 4). The proof is so solid that it is accepted even by industries and professions which irradiate people.

Hypothesis-1 is that MEDICAL radiation causes a very LARGE part of the nation's cancer problem. This book was undertaken in order to test, modify, or discard Hypothesis-1. In the process, the work also provides a bonus: Some of the most powerful evidence ever assembled CONFIRMING that ionizing radiation is a potent cause of virtually all types of human cancer.

By contrast, ionizing radiation was NOT a proven cause of Ischemic Heart Disease when Hypothesis-2 came into existence. Hypothesis-2 "fell out of the data" which we assembled in order to test Hypothesis-1. This book presents the first powerful evidence that ionizing radiation IS a cause of Ischemic Heart Disease --- a very important cause.

● Part 3. Some Rather Dazzling Results to Examine

In approximately 50 years of biomedical research, we have rarely seen support for an hypothesis (Hypothesis-1), and indication for a new hypothesis (Hypothesis-2), "fall out of data" so strongly as they do in this monograph. Such events have to be taken seriously by objective analysts.

Even though the evidence is uncomplicated and the logic is straightforward, this book is long because we have the unusual policy of showing the steps which connect the raw data with the conclusions. For readers who want to know only the "bottom line," we provide an Abstract and Executive Summary (Chapter 1).

● Part 4. Why Our Findings Do Not Challenge the Importance of Other Causes of Cancer and IHD

Both Cancer and Ischemic Heart Disease are well established as multi-cause diseases. There is convincing evidence that several different causes increase the death-rate from Cancer, and likewise, that several different causes increase the death-rate from IHD. Moreover, it is safe to say that multiple causes generally (perhaps always) contribute to a SINGLE CASE of fatal IHD, and to a SINGLE CASE of fatal Cancer. The case would not occur when it does, without co-action by multiple causes.

The concept of NECESSARY co-actors is an old one. For instance, in the famous 1964 "Surgeon General's Report" on cigarette smoking as a cause of Lung Cancer, the authors wrote (p.31): "It is recognized that often the co-existence of several factors is required for the occurrence of a disease, and that one of the factors may play a dominant role; that is, without it, the other factors (such as genetic susceptibility) seldom lead to the occurrence of the disease."

The assumption, of more than one cause per case of Cancer, arises from various lines of evidence. For example, the rate of Breast Cancer is higher in women who inherit one mutated copy of a "Breast Cancer Gene" than in women without that inheritance, but that inheritance certainly does not guarantee the development of Breast Cancer in every breast-cell --- even though every breast-cell contains the mutation. One or more additional causes are necessary in order to turn even one of those breast-cells into a Cancer.

The concept, that more than ONE cause is necessary to produce a case of Cancer, is embraced by the widely accepted initiator-promoter model of Cancer. In that model, inherited or acquired carcinogenic mutations require help from a "promoter" --- for example, a hormone or infectious agent. The concept of mutually dependent co-actors is also inherent in the widely accepted multi-mutation multi-step models of carcinogenesis --- i.e., Cancer "is typically a multi-step process resulting from an accumulation of as many as 10 genetic changes in a single cell" (p.471 in Understanding Genetics: A Molecular Approach, Norman V. Rothwell; Wiley-Liss Publishers, 1993).

By definition, absence of a NECESSARY co-actor prevents the result. When two or more co-actors each have a required role, in producing a particular case of disease, then the absence of any ONE of them will prevent the case. We would regard such co-actors as equally important.

Thus, neither Hypothesis-1 nor Hypothesis-2 challenges the very important roles, already established, for various nonradiation causes of Cancer and IHD. When we propose that medical radiation is a highly important cause of Cancer and IHD mortality, we mean that in the ABSENCE of medical radiation, many or most of the cases would not have occurred when they did. While medical radiation has not been the ONLY factor contributing to such cases, we mean that it has been a NECESSARY co-actor in such cases. Discussion of co-action continues in Chapter 6, Part 6.

● Part 5. How to Reconcile High Fractional Causations by Xrays, Smoking, Diet

Fractional Causation refers to the fraction of the cancer mortality rate which would be absent (prevented) in the absence of a specified carcinogen --- which is medical radiation, in this monograph. Therefore, Fractional Causation is the fraction or percentage of the cancer mortality rate attributable to medical radiation --- or caused by medical radiation, in ordinary parlance.

A related term, widely in use, is "radiation-induced Cancer." The term is a brief and convenient way to refer to cancer cases which would have been absent in the absence of exposure to ionizing radiation. It does not mean that radiation is necessarily the ONLY cause contributing to cases of radiation-induced Cancer. Similarly, when people refer to "occupationally-induced Cancer," they do not mean that occupation is the ONLY cause contributing to such cases. They refer to cases which would have been absent in the absence of occupational exposure to carcinogens.

An Illustration of 100 Cancer Cases Resulting from Co-Action

Suppose that the evidence in this book indicates that Fractional Causation by medical radiation, of the national cancer death-rate, is 90% in a certain decade. Because of co-action, such a finding would NOT leave only 10% for all other causes combined --- as we will illustrate here with some hypothetical values. We will limit our illustration to only four carcinogens: Xrays, smoking, poor diet, and particular inherited mutations. For brevity, we exclude other workplace, at-home, and environmental carcinogens. Then, we arbitrarily specify that the total cancer death-rate per year is 100 cases per 100,000 population and that these 100 cases are the result of co-action as follows. Our First List (illustrative):

- 40 cases by co-action of xrays + smoking + poor diet.
- 25 cases by co-action of xrays + poor diet + inherited mutations.
- 25 cases by co-action of xrays + smoking + inherited mutations.
- 10 cases by co-action of smoking + poor diet + inherited mutations.

The meaning of the first row, above, is that xrays, smoking, and poor diet each make a NECESSARY contribution to each case of Cancer in the first row. In the absence of any ONE of the necessary co-actors, the 40 cases in the first row could not occur. That is the meaning of "necessary." The meaning is similar for all four rows of hypothetical values.

A Second List, also adding up to 100 cases, would have very different implications if it were: 90 cases caused by xrays acting ALONE, 4 cases caused by a dietary factor acting alone, 3 cases caused by smoking acting alone, and 3 cases caused by an inherited mutation acting alone. In both lists, the sum of cases = 100 cases, but every case in the First List is the result of more than one cause per case, whereas every case in the Second List is the result of only one cause per case (no co-action in the Second List).

The Illustrative Fractional Causations by Xrays, Diet, Smoking, and Inherited Mutations

Out of the mixture of cases in the First List, we will explore how many cases could be prevented if we could remove just ONE cause, while the other causes remain as they were. Xrays are a required co-actor in (40 + 25 + 25), or 90 cases per 100 total cases. Because absence of a required co-actor prevents the result, 90% of the cancer death-rate would be absent, in the absence of exposure to medical radiation. Fractional Causation = 90% by medical radiation.

Next, we put radiation back into the mixture, and we remove just "poor diet." In our supposition, it is a required co-actor in (40 + 25 + 10), or 75 cases per 100 total cases. Because absence of a required co-actor prevents the result, 75% of the cancer death-rate would be absent, in the absence of poor diet in this illustration. Fractional Causation = 75% by poor diet. In our hypothetical illustration, Fractional Causation = 75% by smoking and 60% by inherited mutations. It is obvious that a HIGH Fractional Causation by xrays does not require a LOW Fractional Causation by any other cause of Cancer.

Because Fractional Causation means the fraction or percentage of the death-rate which would be absent (prevented) by the absence of a specified co-actor, ADDITION of the separate Fractional Causations produces nonsense (a total greater than 100%). Such addition would be equivalent to counting the same cases of absent Cancer more than once.

Our warning against adding Fractional Causations applies to a statement in the 1999 report of the National Research Council's sixth Committee on the Biological Effects of Ionizing Radiation (the BEIR-6 Report, from the National Academy Press, 1999). The BEIR-6 Committee, referring to evidence of co-action between smoking and exposure to radon (and radon's decay-products), states that "Some lung-cancer cases reflect the joint effect of the two agents and are in principle preventable by removing either agent" (BEIR-6, p.33). Although Fractional Causation of such cases is 100% by radon and 100% by smoking, addition of the two Fractional Causations would clearly count each prevented case twice.

Implications of Co-Action for Progress in Preventing Cancer and IHD

When more than one cause is REQUIRED per case of Cancer or Ischemic Heart Disease, it means that reducing exposure to a single necessary carcinogen or atherogen reduces the impact of all its partners. If one can identify a single agent which is a necessary co-actor in a high fraction of cases of Cancer and Ischemic Heart Disease, one can make real progress in preventing these diseases by reducing exposure to that cause. The evidence uncovered in this book strongly indicates that medical radiation is such an agent.

>>>>>>>>>>

ABSTRACT

Radiation from Medical Procedures in the Pathogenesis of Cancer and
Ischemic Heart Disease:
Dose-Response Studies with Physicians per 100,000 Population.

John W. Gofman, M.D., Ph.D. 1999. 699 pages. LCCN 99-045096.
Hardcover: ISBN 0-932682-97-9. Softcover: ISBN 0-932682-98-7.
Committee for Nuclear Responsibility Books, San Francisco.

ORIENTATION:

For decades, xrays and other classes of ionizing radiation have been a proven cause, in vivo
and/or in vitro, of virtually all types of mutation --- especially structural chromosomal mutations (such
as deletions, translocations, and rings), for which the doubling-dose by xrays is extremely low.
Additionally, xrays are an established cause of in vitro genomic instability.

This monograph looks at the impact of medical radiation --- primarily from xrays, including
fluoroscopy and CT scans --- upon mortality-rates from both Cancer and Ischemic (Coronary) Heart
Disease, from mid-century to 1990. The evidence in this book strongly indicates that medical radiation
has become a necessary co-actor (but not the only necessary co-actor) in causing over 50% of the
death-rates from Cancer and Ischemic Heart Disease (IHD) --- a finding which is consistent with
participation of non-xray causes as necessary co-actors in the same cases (Introduction). In
multi-cause diseases such as Cancer and IHD, more than one necessary co-actor per fatal case is very
likely. Absence of any necessary co-actor, by definition, prevents such cases. The concept, of cases
due to medical radiation, means cases which would be absent in the absence of medical radiation.

PURPOSE:

Xrays have been a well-established cause of human Cancer for decades. This monograph was
undertaken (a) to quantify what share of U.S. age-adjusted cancer mortality, for each gender, is caused
by medical radiation, and (b) to check on the author's 1995 finding, based on completely different data,
that exposure to medical radiation accounts for about 75% of Breast Cancer incidence in the USA. In
the process of evaluating cancer mortality vs. noncancer mortality for this monograph, it became
obvious that the impact of medical radiation upon death-rates specifically from Ischemic Heart Disease
also demanded evaluation.

MATERIALS AND METHODS:

This study is based on mortality rates among 130-250 million persons --- namely, the entire
United States population, 1940-1990. Age-adjusted cancer mortality rates (MortRates) per 100,000
population are available by gender for each of the Nine Census Divisions (USA), for the 1940-1990
decades, from Vital Statistics. Such rates for noncancer mortality rates also are available. For
Ischemic Heart Disease, such rates are available starting in 1950, which means that NonCancer
NonIHD MortRates, by Census Divisions, are available starting in 1950.

For reasons presented in Chapter 2 (Parts 2+3), there are no reliable estimates of average per
capita population dose, accumulated from medical radiation, currently or in the past. Also not
available, for reasons presented in Chapter 2 (Part 7c), are reliable estimates of cancer-risk per
unit of dose from medical xrays. This monograph avoids these two types of uncertainty by using the
number of physicians per 100,000 population (PhysPop) as a reasonable approximation of the
RELATIVE magnitude of exposure from medical radiation in the Nine Census Divisions. The ranking
of averaged PhysPop values by Census Divisions, over the 1940-1990 period, is remarkably stable.

MortRates are regressed upon PhysPop values, by Census Divisions, to determine the presence
and direction of any dose-response. When a significant positive dose-response exists, the line of best
fit is extended to the y-axis, where the intercept's value indicates what the MortRate would have been

for that disease, if there had been NO physicians per 100,000 population in a Census Division. The national MortRate for the disease under study, minus the intercept's value, provides a reasonable estimate of the share of that national MortRate which is due to medical radiation (i.e., the share which would be absent in the absence of medical radiation). Confidence limits are provided in Chapter 22, Box 1.

RESULTS:

Cancer and IHD MortRates each have very significant positive correlations with PhysPop, for males and females separately. By contrast, NonCancer NonIHD MortRates have a significant negative correlation with PhysPop. The following groups of Cancer were studied: All-Cancers-Combined, Breast Cancers, Digestive-System Cancers, Urinary-System Cancers, Genital Cancers, Buccal/Pharynx Cancers, Respiratory-System Cancers, Difference-Cancers (All-Except-Respiratory). Only female Genital Cancers failed to have a significant positive dose-response with PhysPop. The percentages, of the death-rates from Cancer and IHD caused by medical radiation (i.e., the shares which would be absent, in the absence of medical radiation), are shown in Box 1 of Chapter 1. For example:

	Year	Percent	Year	Percent
● All-Cancers-Combined, m	1940	90%	1988	74%
● All-Cancers-Combined, f	1940	58%	1988	50%
● All-Cancer-Except-Genital, f	1940	75%	1980	66%
● Breast Cancer, f	1940	~ 100%	1990	83%
● Ischemic Heart Disease, m	1950	79%	1993	63%
● Ischemic Heart Disease, f	1950	97%	1993	78%

The growing impact of cigarette-smoking (Chapters 48, 49) almost certainly explains why the shares from medical radiation in 1980-1993 are somewhat lower than in 1940-50.

CONCLUSIONS:

Since its introduction in 1896, medical radiation has become a necessary co-actor in most fatal cases of Cancer and Ischemic Heart Disease (IHD).

It is proposed that, for radiation-induced IHD, the probable mechanism is radiation-induction of mutations in the coronary arteries, resulting in dysfunctional clones (mini-tumors) of smooth muscle cells. A Unified Model of Atherogenesis and Acute IHD Events is presented (Chapter 45), which is consistent with the findings in this book, is consistent with the findings (first by Earl Benditt in 1973) of monoclonal cells in atherosclerotic plaques, is consistent with well-established knowledge about atherogenic lipoproteins and other non-xray causes of fatal IHD, and is consistent with recent findings about the weaker connection than expected between degree of arterial stenosis and the fatal rupturing of specific atherosclerotic plaques.

The evidence in this monograph has major implications for prevention of Cancer and IHD. This monograph points to demonstrations, by others, of proven ways to reduce dose-levels of nontherapeutic medical radiation by 50% or considerably more, without eliminating a single diagnostic or interventional radiologic procedure and without degrading the information provided by medical radiation. Reduction of exposure to medical radiation can and will reduce mortality rates from both Cancer and Ischemic Heart Disease.

CHAPTER 1

Executive Summary of This Book

Part 1. Orientation: What Is Old, and What Is New
Part 2. Some Key Facts about Xrays and Ionizing Radiation in General
Part 3. No Doubt about Benefits from Medical Radiation
Part 4. Role of Medical Radiation in Causing Cancer and IHD, Past and Present
Part 5. Our Method for Calculating Fractional Causation
Part 6. Eight Features Which Confer High Credibility on the Findings
Part 7. Our Unified Model of Atherogenesis, and NonXray Co-Actors in IHD
Part 8. A Personal Word: The Xray Deserves Its Honored Place in Health
Part 9. Every Benefit of Medical Radiation: Same Procedures, Lower Dose-Levels
Part 10. An Immense Opportunity: All Benefit, No Risk

   Boxes, Figures, and Tables, in that (alphabetical) order, are located
   in this book at the ends of the corresponding chapters.

Box 1. Final Summary for Fractional Causation, by Medical Radiation, of Cancer and IHD.
Box 2. Comparison of Dose-Response at Mid-Century: NonCancer NonIHD, Cancer, IHD.
Box 3. Known Procedures Which Reduce Dosage from Medical Xrays.
Figure 1-A: All-Cancers-Combined: Dose-Response between PhysPop and MortRates.
Figure 1-B: Ischemic Heart Disease: Dose-Response between PhysPop and MortRates.
Figure 1-C: NonCancer NonIHD Deaths: Dose-Response between PhysPop and MortRates.

● Part 1. Orientation: What Is Old, and What Is New

The evidence presented in this book strongly indicates that over 50% of the death-rate from Cancer today, and over 60% of the death-rate from Ischemic Heart Disease today, are xray-induced as defined and explained in Part 5 of the Introduction. The finding means that xrays (including fluoroscopy and CT scans) have become a necessary co-actor --- but not the only necessary co-actor --- in causing most of the death-rate from Cancer and from Ischemic Heart Disease (also called Coronary Heart Disease, and Coronary Artery Disease). In multi-cause diseases such as Cancer and Ischemic Heart Disease, more than one necessary co-actor per fatal case is very likely. Absence of any necessary co-actor, by definition, prevents such cases. The concept of xray-induced cases means cases which would be absent in the absence of exposure to xrays.

Xrays and other classes of ionizing radiation have been, for decades, a proven cause of virtually all types of mutations --- especially structural chromosomal mutations (such as deletions, translocations, and rings), for which the doubling dose by xrays is extremely low. Additionally, xrays are an established cause of genomic instability, often a characteristic of the most aggressive Cancers.

Not surprisingly, a host of epidemiologic studies have firmly established that xrays and other classes of ionizing radiation are a cause of most varieties of human Cancer. This monograph presents (a) the first compelling evidence that xrays are a cause also of Ischemic Heart Disease (IHD) --- a very important cause --- and presents (b) a Unified Model of Atherogenesis and Acute IHD Events (Part 7 of this chapter).

We have a high level of confidence that our findings, about the important causal role of medical radiation in both Cancer and IHD, are correct. Part 6 of this chapter identifies the features of the work which produce this confidence.

Part 9 of this chapter points to demonstrations, by others, of proven ways to reduce dose-levels of nontherapeutic medical radiation by 50% or considerably more, without eliminating a single diagnostic or interventional radiologic procedure and without degrading the information provided by medical radiation.

Reduction of exposure to medical radiation can and will reduce mortality rates --- from Cancer with certainty, and with very great probability from Ischemic Heart Disease too.

● **Part 2.   Some Key Facts about Xrays and Ionizing Radiation in General**

Most physicians and other people appreciate the imaging capability of the xray, but --- through no fault of their own --- they are taught very little about the biological action of those xrays which never reach the film or other image-receptor. Part 2 provides some information about xrays and ionizing radiation in general. These facts are well supported in the peer-reviewed biomedical literature, in our text, and in our Reference List.

2a.   Capacity to Commit Mayhem among the Genetic Molecules

The biological damage from a medical xray procedure does not come directly from the xray photons. The damage comes from electrons, which those photons "kick" out of their normal atomic orbits within human tissues. Endowed with biologically unnatural energy by the photons, such electrons leave their atomic orbits and travel with high speed and high energy through their "home" cells and neighboring cells. Each such electron gradually slows down, as it unloads portions of its biologically unnatural energy, at irregular intervals, onto various biological molecules along its primary track (path).

The molecular victims include, of course, chromosomal DNA, and the structural proteins of chromosomes, and water. Even though each energy-deposit transfers only a portion of the total energy of a high-speed high-energy electron, the single deposits very often have energies far exceeding any energy-transfer which occurs in a natural biochemical reaction. Such energy-deposits are more like grenades and small bombs (Chapter 2, Part 4a). None of this is in dispute.

2b.   The Free-Radical Fallacy

There is no doubt that, along the path of each high-speed high-energy electron described above, the energy-deposits produce various species of free radicals. Nonetheless, it is a demonstrated fallacy (Appendix-C) to assume equivalence between the biological potency of xrays and the biological potency of the free radicals which are routinely produced by a cell's own natural metabolism.

The uniquely violent and concentrated energy-transfers, resulting from xrays, are simply absent in a cell's natural biochemistry. As a result of these "grenades" and "small bombs," both strands of opposing DNA can experience a level of mayhem far exceeding the damage which metabolic free-radicals (and most other chemical species) generally inflict upon a comparable segment of the DNA double helix.

2c.   Ionizing Radiation: A Uniquely Potent Mutagen

The extra level of mayhem is what makes xrays (and other types of ionizing radiation) uniquely potent mutagens. Cells can not correctly repair every type of complex genetic damage, induced by ionizing radiation, and sometimes cells can not repair such damage at all (evidence discussed in Appendix-B and Appendix-C). Not all mutated cells die, of course. If they all died, there would be very little Cancer and no inherited afflictions. Indeed, certain mutations confer a proliferative advantage on the mutated cells. Exposure to xrays is a proven cause of genomic instability --- a characteristic of many of the most aggressive Cancers (Chapter 2, Part 4b, and Appendix-D).

Unlike some other mutagens, xrays have access to the genetic molecules of every internal organ, if the organ is within the xray beam. Within such organs, even a single high-speed high-energy electron, set into motion by an xray photon, has a chance (far from a certainty) of inducing the types of damage which defy repair. That is why there is no risk-free (no safe) dose-level (Appendix-B).

There is widespread agreement that, by its very nature, ionizing radiation at any dose-level can induce particularly complex injuries to the genetic molecules. There is growing mainstream acknowledgment that cellular repair processes are fallible, or entirely absent, for various complex injuries to the genetic molecules (Appendix-B and Appendix-C).

2d.   The Very Low Doubling-Dose for Xray-Induced Chromosomal Mutations

The inability of human cells, to repair correctly every type of radiation-induced chromosomal damage, has been demonstrated in nuclear workers (who received their extra low-dose radiation at minimal dose-rates) and in numerous studies of xray-irradiated human cells at low doses. Besides demonstrating non-repair or imperfect repair, such studies have established that xrays have an extremely low doubling-dose for structural chromosomal mutations. (The doubling dose of an effect is the dose which adds a frequency equal to the pre-existing frequency of that effect.)

For instance, the doubling-dose for the dicentric mutation is in the dose range delivered by some common xray procedures, such as CT scans and fluoroscopy --- i.e., in the dose range of 2 to 20 rads (references in Chapter 2, Part 4b). The rad is a dose-unit which is identical to the centi-gray (Appendix-A). We, and many others, prefer the simpler name: Rad.

Xrays are capable of causing virtually every known kind of mutation --- from the very common types to the very complex types, from deletions of single nucleotides, to chromosomal deletions of every size and position, and chromosomal re-arrangements of every type. When such mutations are not cell-lethal, they endure and accumulate with each additional exposure to xrays or other ionizing radiation (Chapter 2, Part 8c; and Appendix-B, Part 2d).

2e. Medical Xrays as a Proven Cause of Human Cancer

Ionizing radiation is firmly established by epidemiologic evidence as a proven cause of almost every major type of human Cancer (Chapter 2, Part 4c). Some of the strongest evidence comes from the study of medical patients exposed to xrays --- even at minimal dose-levels per exposure (Appendix-B, Part 2d). Mounting mainstream evidence indicates that medical xrays are 2 to 4 times more mutagenic than high-energy beta and gamma rays, per rad of exposure (Chapter 2, Part 7.)

● Part 3. No Doubt about Benefits from Medical Radiation

Radiation was introduced into medicine almost immediately after discovery of the xray (by Wilhelm Roentgen) in 1895.

There is simply no doubt that the use of radiation in medicine has many benefits. The findings in this book provide no argument against medical radiation. The findings do provide a powerful argument for acquiring all the benefits of medical radiation with the use of much lower doses of radiation, in both diagnostic and interventional radiology. (Interventional radiology refers primarily, but not exclusively, to the use of fluoroscopy to acquire information during surgery and during placement of catheters, needles, and other devices.)

Within the professions of radiology and radiologic physics, there are mainstream experts who have shown how the dosage of xrays in current practice could be cut by 50%, or by considerably more, in diagnostic and interventional radiology --- without any loss of information and without eliminating a single procedure (discussion in Part 9, below). Among the current leaders in dose-reduction education are Joel Gray, Ph.D. (recently retired from the Mayo Clinic's Department of Radiology in Rochester, Minnesota) and Fred Mettler, M.D. (Chief of Radiology, University of New Mexico School of Medicine in Albuquerque, New Mexico).

● Part 4. Role of Medical Radiation in Causing Cancer and IHD, Past and Present

This monograph has produced evidence with regard to two hypotheses.

● - Hypothesis-1: Medical radiation is a highly important cause (probably the principal cause) of cancer mortality in the United States during the Twentieth Century. Medical radiation means, primarily but not exclusively, exposure by xrays --- including fluoroscopy and CT scans. (Hypothesis-1 is about causation of Cancer, so it is silent about radiation-therapy used after a Cancer has been diagnosed.)

● - Hypothesis-2: Medical radiation, received even at very low and moderate doses, is an important cause of death from Ischemic Heart Disease (IHD); the probable mechanism is radiation-induced mutations in the coronary arteries, resulting in dysfunctional clones (mini-tumors) of smooth muscle cells. (The kinds of damage to the heart and its vessels, observed from very high-dose radiation and reported for decades, seldom resemble the lesions of IHD --- details in Appendix J.)

4a. These Hypotheses in Terms of Multi-Cause Diseases

Cancer and Ischemic Heart Disease are well established as multi-cause diseases. The concept, that more than one necessary co-actor is required per case, has already been discussed in Parts 4 and 5 of the Introduction. In efforts to prevent these multi-cause diseases, reduction or removal of any necessary co-actor is a central goal. The evidence in this book is that medical radiation has become a necessary co-actor in a high fraction of the U.S. mortality rates from BOTH diseases. Fortunately, dosage from medical radiation is demonstrably reducible without eliminating a single procedure.

4b. Fractional Causation: Percentage of Death-Rates due to Medical Radiation

The tabulation below shows the percentages, of the age-adjusted death rates (m=male, f=female) from Cancer and IHD, due to medical radiation at mid-century and at the most recent year for which we have data. Box 1 at the end of this chapter shows percentages for several specific types of Cancer. Percentages for each intervening decade are shown in the appropriate chapters and assembled in Chapter 66.

When an entry of $\sim$ 100% occurs, such a finding is fully consistent with the fact that these diseases occurred before the introduction of radiation into medicine, over a century ago. Other mutagens (including radiation exposure from nature itself) have been operative both before and after the introduction of medical radiation. A finding, of about 100% of the death-rate due to medical radiation in 1940, means that by 1940, a very low fraction of such deaths would have occurred without medical radiation as a co-actor.

	Year	Percent	Year	Percent
• All-Cancers-Combined, m	1940	90%	1988	74%
• All-Cancers-Combined, f	1940	58%	1988	50%
• Breast Cancer, f	1940	$\sim$ 100%	1990	83%
• All-Cancer-Except-Genital, f	1940	75%	1980	66%
• Ischemic Heart Disease, m	1950	79%	1993	63%
• Ischemic Heart Disease, f	1950	97%	1993	78%

The growing impact of cigarette smoking (Chapters 48, 49) almost certainly explains why the shares from medical radiation in 1980-1993 are somewhat lower than in 1940-1950.

A percentage such as 90% due to medical radiation (Fractional Causation by medical radiation = 0.90) means that about 90% of the death-rate would have been absent in the absence of medical radiation. Circumstantial evidence is strong that nonxray agents ALSO were necessary co-actors in these same deaths. Thus, Fractional Causation of 90% by medical radiation certainly does not leave "just 10%" for all other causes combined, as already illustrated in Part 5 of the Introduction.

Fractional Causation, of a year-specific mortality rate (MortRate) by medical radiation, refers to whatever rate occurs in that year, and says nothing about whether the MortRate has been rising or falling over time. Indeed, changes over time, in the types and concentrations of non-xray co-actors to which populations are exposed, can cause cancer MortRates simultaneously to rise for some organs, fall for other organs, and remain constant for still other organs (discussion in Chapter 67, Part 2).

The results in this book amply support Hypothesis-1 and the first part of Hypothesis-2. While the central estimates of Fractional Causation are statistically the most likely to be correct, of course the actual percentages could be either higher or lower. We note that percentages VERY much lower than the central estimates would support each hypothesis, too.

• Part 5. Our Method for Calculating Fractional Causation

When increments, in the death-rate from a disease, are proportional to increments in exposure to an identified cause, a linear dose-response exists between the causal agent and increments in the death-rate.

The evidence in this monograph repeatedly reveals a positive and tight linear dose-response, between dose from medical radiation and mortality rates from Cancer (discussion in Chapter 5, Part 5d). By "tight," we mean highly reliable (statistically). As we will explain, no group in our database escapes entirely from exposure to medical radiation. In order to estimate what the cancer mortality rates would be in the ABSENCE of medical radiation, we use the basic technique of linear regression analysis (Part 5c, below). After that basic step, it is not at all complicated to calculate Fractional Causation due to medical radiation (Part 5g, below).

5a. The Database for Age-Adjusted Mortality Rates (MortRates)

We acquired the age-adjusted cancer MortRates per 100,000 population in each of the Nine Census Divisions of the USA, from 1940 onward --- separately for males and females, and for all races combined (no exclusions). Such data are published by the U.S. Government (details in Chapter 4). For most types of Cancer, our data end in 1988-1990 (some end in 1980).

Also we acquired the comparable age-adjusted MortRates for All NonCancer Causes of Death --- as well as for selected individual causes (such as IHD, Stroke, Diabetes Mellitus, Influenza and Pneumonia, Accidents, etc.) --- in each of the Nine Census Divisions.

These MortRates, by Census Divisions, are the dependent variables (the responses) in our dose-response studies. Because the MortRates are age-adjusted, the Census Divisions are matched with each other for age.

5b. The Database for Dose: Physicians per 100,000 Population

During the 1985-1990 period, the number of diagnostic medical xray examinations performed per year in the USA was approximately 200 million, excluding 100 million dental xray examinations and 6.8 million diagnostic nuclear medicine examinations. The source of these estimates (the 1993 Report of UNSCEAR, the United Nations Scientific Committee on Atomic Radiation, p.229, p.275) warns that 200 million could be an underestimate by up to sixty percent.

Not only is the number of annual examinations quite uncertain, but the average doses per examination --- in actual practice, not measured with a dummy during ideal practice --- vary sometimes by many-fold from one facility to another, even for patients of the same size. The variation by facility has been established by a few on-site surveys of selected facilities, because measurement and recording of xray doses are not required for actual procedures (Part 9, below).

Fluoroscopy is a major source of xray dosage, because the xray beam stays "on" during fluoroscopy. Such doses are rarely measured. When fluoroscopic xrays are used during common diagnostic examinations, the total dose delivered varies with the operator. When fluoroscopic xrays are used during surgery and other nondiagnostic procedures, the total dose delivered varies both with the operator and the particular circumstances.

The uncertain number of procedures and the very uncertain doses per procedure combine to cause profound uncertainty about current average per capita population dose from medical radiation (Chapter 2, Part 3). Dose estimates for past decades are even MORE uncertain (Chapter 2, Part 2).

An Additional Gap in Knowledge: Risk-per-Rad Estimates

In most of the studies which produce estimates of cancer-risk per rad of xray dose, it is far from certain which participants received which xray doses over their lifetimes, because such doses were neither measured nor recorded. When a few participants are (unintentionally) assigned a wrong dose-estimate, the error can substantially alter the resulting risk-per-rad estimates. This contributes to the great uncertainty about the true risk-per-rad from xrays (Chapter 2, Part 7c). The uncertainty is no secret. For example, the fifth Committee on the Biological Effects of Ionizing Radiation stated in its 1990 report (National Academy Press, at pp.46-47): "A number of low-dose studies have reported risks that are substantially in excess of those estimated in the present report ... Although such studies do not provide sufficient statistical precision to contribute to the risk estimation procedure per se, they do raise legitimate questions about the validity of the currently accepted estimates."

A Solution to These Gaps in Knowledge

Medical radiation procedures are initiated by a physician, even if someone else actually performs the procedure. It is very reasonable to think that the more physicians there are per 100,000 population, the more radiation procedures per 100,000 population will be ordered. Thus, we arrive at the premise that average radiation dose, received per capita of population in a specific Census Division from medical procedures during a specific year, is approximately proportional to the number of physicians per 100,000 population in that same Census Division during that same year.

This common-sense premise is well supported in the 1988 and 1993 reports of the United Nations Scientific Committee on Atomic Radiation (details in our Chapter 3, Part 1a), and is supported specifically for the USA by data in a 1989 report from the National Council on Radiation Protection and Measurements (details in Chapter 3, Part 1a).

"PhysPop" Values in the Nine Census Divisions, over Many Decades

We use the abbreviation, "PhysPop," for the quantity "Physicians per 100,000 Population." A PhysPop value of 134 means 134 Physicians per 100,000 population, for the specified year and place.

PhysPop values for various calendar years have been compiled and published for each state by the American Medical Association over many decades (details in Chapter 3). It is a routine matter to combine such data appropriately, in order to obtain PhysPop values for the Nine Census Divisions (details in Chapter 3). Because substantial DIFFERENCES in PhysPop values exist among the Nine Census Divisions, it has been possible for us to do dose-response studies, with PhysPop values in each Census Division as surrogates for average per capita dose from medical radiation in each corresponding Census Division.

Of course, dose is cumulative (i.e., radiation-induced mutations are cumulative). Moreover, in a population of mixed ages (newborn to very advanced ages), the cancer-response to ionizing radiation is spread out over at least four to five decades (Chapter 2, Part 8). Thus, the age-adjusted cancer MortRates in any single year --- say 1990 --- incorporate cases which are due to radiation received in 1940, 1950, 1960, 1970, etc. It happens that, during the 1921-1990 period, the rank order of the Census Divisions --- by the size of their PhysPop values --- has been remarkably stable (details in Chapter 3, Box 1; see also Chapter 47, Table 47-A). Thus, PhysPop values are well-suited to be surrogates for the RELATIVE size of average ACCUMULATED per capita dose from medical radiation, among the Nine Census Divisions.

5c. Illustrative Regression (Input and Output), for All Cancers Combined

Linear regression analysis is a branch of mathematics which, among other things, evaluates how well correlated are sets of paired values. In our dose-response studies, there are always nine pairs of values, because there are Nine Census Divisions --- each having its own age-adjusted MortRate (the y-variable) and its own PhysPop value (the x-variable). On the lefthand side of the next page, we show the input data for a regression whose output is shown on the righthand side.

In the output, two quantities measure the goodness (strength) of the correlation: The R-squared value, and the ratio of the X-coefficient divided by its Standard Error (X-Coef/S.E.).

● An R-squared value of 1.00 is perfection. An R-squared value of 0.70 is very good. Those who are familiar with the correlation coefficient, R, will recognize that R-squared values are lower than the corresponding R-values (for instance, when R = 0.83666, R-squared = 0.70; when R = 0.94868, R-squared = 0.90).

● A ratio of (X-Coef/S.E.) of about 2.0 generally indicates a statistically significant correlation. A ratio of 4.0 is a tight correlation. A ratio above 4.0 is very tight. The ratio describes the reliability of the slope in a line of best fit.

In Part 5d, the male 1940 MortRates per 100,000 population, for All-Cancers-Combined, are regressed upon the 1940 PhysPop values (which represent accumulated doses from earlier years of medical radiation). The regression reveals a spectacularly tight correlation: R-squared = 0.9508.

5d. Figure 1-A: Graph of the 1940 PhysPop-Cancer Dose-Response (Males, Females)

The regression output (below) provides all the information necessary to calculate and to graph the line of best fit for the nine pairs of real-world observations (listed below). Chapter 6, Part 3, shows how. The resulting graph is presented in the upper half of Figure 1-A, at the end of this chapter. The nine boxy symbols in Figure 1-A represent the nine pairs of actual observations from the x,y columns below. For example, the box farthest to the right represents the pair with the highest PhysPop value: The Mid-Atlantic pair.

Census Division	1940 PhysPop x	1940 All-Ca y	All-Cancer MortRates 1940 (males) vs. PhysPop 1940 Regression Output:	
Pacific	159.72	122.9	Constant	11.5484
New England	161.55	135.5	Std Err of Y Est	5.4727
West North Central	123.14	110.9	R Squared	0.9508
Mid-Atlantic	169.76	140.9	No. of Observations	9
East North Central	133.36	119.6	Degrees of Freedom	7
Mountain	119.89	99.8		
West South Central	103.94	86.9	X Coefficient(s)	0.7557
East South Central	85.83	73.6	Std Err of Coef.	0.0650
South Atlantic	100.74	88.9	X-Coef/S.E. =	11.6275

Figure 1-A also presents the comparable graph for females (borrowed from Chapter 7). It was prepared after regressing the female 1940 MortRates per 100,000 population, for All-Cancers-Combined, upon the 1940 PhysPop values (which represent accumulated doses from earlier years of medical radiation).

5e. The Dose-Response Findings for Specific Sets of Cancer

In addition to All-Cancers, we examined the dose-response for various sets of Cancers. With only one exception (female Genital Cancers), all the regression analyses revealed strong POSITIVE correlations between PhysPop and the 1940 Cancer MortRates, by Census Divisions. A summary of their R-squared values is in Column D of Box 1, after the text of this chapter.

5f. NonCancer Causes of Death: IHD Separates Itself from Other Causes

Before exploring the post-1940 decades, we asked, "Do the same strong positive correlations exist for noncancer causes of death?"

They definitely do not. When we studied All Causes Except Cancer (Chapter 24), we found a nonsignificant NEGATIVE relationship between PhysPop and MortRates. Curiosity drove us also to study SPECIFIC noncancer MortRates in 1940 versus PhysPop. Almost all regression analyses revealed negative relationships between PhysPop and noncancer MortRates. There is a summary of those findings in the upper part of Box 2, at the end of this chapter. A negative X-coefficient means a downward slope.

Strong POSITIVE Correlation between PhysPop and 1950 IHD MortRates

We arrived late at regressing Ischemic Heart Disease (IHD) MortRates on PhysPop, by Census Divisions, because there are no MortRate data for IHD until 1950. When we finally regressed the 1950 MortRates for IHD on PhysPop, we were astonished by the results (Chapters 40 and 41). What fell out of the data are very strong POSITIVE correlations with PhysPop --- which are graphed as Figure 1-B at the end of this chapter.

- Male IHD MortRates vs. PhysPop: R-sq = 0.95 and Xcoef/SE = 11.25.
- Female IHD MortRates vs. PhysPop: R-sq = 0.87 and Xcoef/SE = 6.75.

Such spectacular correlations do not happen by accident. They "demand" an explanation. The resemblance to the positive dose-response for Cancer is self-evident. These two diseases unambiguously sort THEMSELVES out from NonCancer NonIHD causes of death, with respect to medical radiation (PhysPop). The positive dose-response between PhysPop and Cancer is no surprise, because xrays are a proven cause of Cancer. For IHD, the findings above invoke the Law of Minimum Hypotheses: Medical radiation is a cause of Ischemic Heart Disease, too. Our Unified Model of Atherogenesis (Part 7, below) proposes HOW radiation-induced dysfunctional clones of smooth muscle cells, in the coronary arteries, may interact with atherogenic lipoproteins to explain the strong positive correlations presented above.

Strong NEGATIVE Correlation between PhysPop and 1950 NonCancer NonIHD MortRates

When BOTH Cancer and IHD are removed from Causes of Death, the correlation between PhysPop and MortRates for the remaining Causes of Death (NonCancer NonIHD) is not only NEGATIVE, but it also is statistically significant. That relationship is depicted in Figure 1-C --- borrowed from Chapter 25. The contrast is dramatic, between Figure 1-C and the two preceding figures. Box 2, at the end of this chapter, presents the findings for specific NonCancer Non IHD causes of death.

5g. From Positive Dose-Response to Fractional Causation: The Calculation

The observed PhysPop values and the observed MortRates, by Census Divisions, reveal a positive, linear dose-response of great strength between medical radiation and the mid-century MortRates for Cancer and (separately) for Ischemic Heart Disease.

In order to estimate what SHARE of the National MortRates for these diseases was due to medical radiation, we use the regression output to identify what the MortRates for each disease would have been at that time, if the population had received NO medical radiation. The Constant is the value of the y-variable (the MortRate) when the x-variable (PhysPop) is zero. Obviously, if there had been no physicians per 100,000 population, there would have been no medical radiation. On our graphs, the Constant is the value of y where the line of best fit intercepts the vertical y-axis.

Example from Part 5d, above: In the regression output, the Constant = 11.5 --- matching the y-intercept in the upper graph of Figure 1-A. From Chapter 6, Table 6-B, we have the datum that the 1940 NATIONAL age-adjusted male MortRate from All Cancers Combined was 115.0 fatal Cancers per 100,000 male population. Of these 115.0 cases, only 11.5 cases would have occurred if there had been no medical radiation. The number of fatal cases (per 100,000 population) in which medical radiation was a required co-actor was (115.0 minus 11.5), or 103.5 cases. And the Fractional Causation by medical radiation was 103.5 / 115.0, or 0.90 --- 90%.

This is the manner in which Fractional Causation by medical radiation is estimated, both for Cancer and for IHD MortRates, throughout this book. For the decades beyond mid-century, one adjustment was required (and executed in plain view) for the impact of cigarette smoking, an important co-actor whose intensity was not matched across the Nine Census Divisions (Chapter 48).

Returning to the example from Part 5d, we want to estimate the Upper and Lower 90% Confidence Limits on the Fractional Causation by medical radiation of the male 1940 National All-Cancer MortRate. These limits are, respectively, 99% and 75%. These limits are derived from the reliability of the slope of the line of best fit, because its slope (the X-coefficient) determines the value of the y-intercept (the Constant). The regression output in Part 5d provides the required values: The X-coefficient is 0.7557 units of y per unit of x, with a Standard Error of 0.0650. Calculation of the Confidence Limits is first demonstrated in Chapter 6, Part 4.

● Part 6. Eight Features Which Confer High Credibility on the Findings

This monograph presents evidence that medical radiation is an important cause of both fatal Cancer and fatal Ischemic Heart Disease in the USA. There are eight features of our findings which endow us with high confidence that the findings are correct, and so we call those features to the attention of readers:

● First, the findings occur from data which were collected long ago for other purposes --- namely the collection of Vital Statistics from each state on the causes of death per 100,000 population, and the collection of information from each state on the number of physicians per 100,000 population (PhysPop values). Thus, these databases are free from any conceivable bias with respect to Hypothesis-1 or Hypothesis-2. This is no small matter. The first obligation of objective analysts is to be able to assure themselves and the public that the raw data which they employ are trustworthy and neutral with respect to the topic.

● Second, the findings occur from an enormous database: The entire U.S. population. (132 million in 1940; 247 million in 1990). It is hard to imagine a larger prospective study than one which "enrolls" the entire U.S. population in its nine dose-cohorts (Chapter 22, Part 4). All other things being equal, the larger the database, the more reliable are the results.

● Third, the findings occur without dependence on permanently uncertain dose-estimates in medical rads and without dependence on unsettled estimates of cancer-risk per medical rad (Part 5b, above). Instead, the RELATIVE sizes of medical doses, proportional to PhysPop values in the Nine Census Divisions, directly reveal the magnitude of Fractional Causation, by medical radiation, of the death-rates from Cancer and from Ischemic Heart Disease. This aspect of the method itself is a source of enormous credibility for the results.

● Fourth, the findings are not the product of elaborate statistical manuevers and adjustments occurring, beyond realistic review, in a computer. While statistical operations are an essential part of epidemiology, we regard findings in the biomedical literature as unreliable, if they are the product of layer upon layer of such operations. In this monograph, we have confined ourselves to one layer of statistical operation: The basic linear regression with just one independent variable. (Every step in our findings --- from the raw data to the estimated values of Fractional Causation by medical radiation --- has been presented in the open.)

● Fifth, the mid-century dose-responses between PhysPop and the MortRates for Cancer and for Ischemic Heart Disease are extremely strong. There is nothing marginal about the findings. They are almost spectacular in their strength. Even without linear regression, it would be clear from Figures 1-A and 1-B that the nine real-world observations (the boxy symbols) cluster very closely around a straight and upward line. The nearly perfect correlations provide a solid foundation for confidence in the resulting estimates of Fractional Causation by medical radiation, both for Cancer and for Ischemic Heart Disease.

● Sixth, MortRates from diseases in GENERAL very definitely do not share a strong positive correlation with PhysPop values. On the contrary. PhysPop discriminates among diseases. Figure 1-C displays the significant NEGATIVE correlation between PhysPop and all NonCancer NonIHD Causes of Death at mid-century --- and the negative correlation persists through subsequent decades (Chapter 25, Box 1).

Box 2 summarizes the findings for specific as well as combined NonCancer NonIHD Causes of Death, and contrasts them with the findings for All-Cancers, specific Cancers, and IHD.

A mountain of powerful evidence is summarized on that single page. The real-world observations clearly show that Cancer and Ischemic Heart Disease belong together, and not with the other diseases, with respect to PhysPop. These observations "demand" an explanation, which is supplied by the proportionality between PhysPop and average accumulated per capita dose from medical radiation.

Figure 1-A has a ready explanation, based on two undisputed facts: 1) Physicians cause exposure to medical radiation, and 2) Radiation is a proven cause of Cancer. Figure 1-B also has an explanation which is tied to real-world evidence: 1) Physicians cause exposure to medical radiation; 2) Radiation is a proven cause of mutations of virtually every sort; and 3) Some evidence exists, prior to this monograph, that acquired mutations ARE co-actors in atherogenesis (Chapter 44, Parts 8 and 9). In contrast to the evidence-based explanations above, various speculations are possible (Chapter 68). For example, perhaps physicians do something additional (besides causing exposure to radiation) which causes both Cancer and Ischemic Heart Disease. If that speculation seems credible, then clearly the National Institutes of Health should give top priority to IDENTIFYING what the physicians do.

● Seventh, the conclusion, that medical radiation is a major cause of both fatal Cancer and fatal Ischemic Heart Disease, very reasonably explains the tight positive correlations between PhysPop and the MortRates for Cancer and for IHD (and the absence of such correlations for NonCancer NonIHD MortRates), while various alternative proposals fall short (Chapter 68). Moreover, the conclusion does not produce conflicts with well-established facts (Introduction, and Chapters 46 and 67). Indeed, the conclusion helps to explain some of them (Chapter 46).

● Eighth, this monograph --- although employing completely independent data and methods from our 1995/96 monograph about Breast Cancer --- nonetheless produces remarkably similar estimates of the Fractional Causation of recent Breast Cancer rates by medical radiation (Chapter 67, Part 5c).

● **Part 7. Our Unified Model of Atherogenesis, and NonXray Co-Actors in IHD**

As noted above, this monograph's real-world evidence clearly shows that Cancer and Ischemic Heart Disease belong together, and not with the other causes of death, with respect to PhysPop. The positive dose-response between PhysPop and Cancer is certainly not strange. Cancer is the single cause of DEATH already well-proven (prior to this monograph) to be inducible by ionizing radiation --- and average population exposure to ionizing radiation from medical procedures is approximately proportional to PhysPop.

The surprise is our unambiguous finding of a tight positive correlation between PhysPop and IHD MortRates, a result which indicates strongly that Ischemic Heart Disease also is inducible by medical radiation. With respect to "surprise," a reminder is appropriate: The kinds of damage to the heart and its vessels, observed from very high-dose radiation and reported for decades, seldom resemble the lesions of IHD --- details in Appendix-J.

Our monograph is essentially the first, large prospective study on induction of fatal Ischemic Heart Disease by medical radiation. The results are stunning in their strength. Such strong dose-response relationships do not occur by accident.

7a. Earl Benditt's Work on Monoclonality in Atherosclerotic Plaques

We might be less surprised, by the strong positive dose-response between medical radiation and IHD MortRates, if we (and others) had paid more attention to a different type of evidence, available since 1973. We mean evidence supporting a role for mutagens in atherosclerosis. Such evidence came into existence at the University of Washington School of Medicine, Department of Pathology, when Earl Benditt and colleagues found evidence of monoclonality in atherosclerotic plaques in 1973 --- findings which have been replicated several times (Chapter 44, Parts 8 + 9). The fact, that ionizing radiation is a uniquely potent mutagen, provides the foundation for the second part of Hypothesis-2 --- our Unified Model of Atherogenesis (Part 7c, below).

7b. A Reality-Check, for Consistency in Our Findings

Our dose-response evidence, that medical radiation is an important cause of both Cancer and Ischemic Heart Disease, elicits a "prediction." The MortRates for the two diseases should show a persistent positive correlation with each OTHER, by Census Divisions, over time --- and should simultaneously show a distinctly DIFFERENT relationship with MortRates for NonCancer NonIHD Causes of Death, which are NOT inducible by ionizing radiation. The expectation is well met, as we show in Appendix-N.

7c. Our Unified Model of Atherogenesis and Acute IHD Events

Our Unified Model of Atherogenesis and Acute IHD Events (Chapter 45) combines the evidence in this book, that medical radiation has an important causal role in mortality from Ischemic Heart Disease, with the abundant evidence elsewhere that certain lipoproteins in the bloodstream also have an important causal role in mortality from Ischemic Heart Disease (Chapter 44, Parts 3,4,5,6,7).

Our view (shared by many others) is that the plasma lipoproteins have no physiologic function in the intimal layer of the coronary arteries, and that under normal circumstances, their rate of entry and exit from the intimal layer is in balance. We propose that what disrupts this lifelong egress of lipoproteins from the intima --- with the disruption occurring only at specific locations --- are mutations acquired from medical radiation and from other mutagens.

In our Unified Model, some mutations acquired by smooth muscle cells render such cells dysfunctional AND give such cells a proliferative advantage --- so that they gradually replace competent smooth muscle cells at a localized patch of artery (a mini-tumor). And this patch of cells, unable to process lipoproteins correctly, becomes the site of chronic inflammation, resulting in construction of an atherosclerotic plaque --- whose fibrous cap is sometimes too fragile to contain the highly thrombogenic lipid-core within the plaque. The Unified Model is described in more detail in Chapter 45. Then Chapter 46 describes how the model helps to explain, or is consistent with, established observations --- including the existence of many additional co-actors in the causation of mortality from Ischemic Heart Disease.

● Part 8. A Personal Word: The Xray Deserves Its Honored Place in Health

The finding, that radiation from medical procedures is a major cause of both Cancer and Ischemic Heart Disease, does NOT argue against the use of xrays, CT scans, fluoroscopy, and radioisotopes in diagnostic and interventional radiology. Such uses also make very POSITIVE contributions to health. We deeply respect those contributions, and the men and women who achieve them.

This author is most definitely not "anti-xray" or "radio-phobic." As a graduate student in physical chemistry, I worked very intimately with radiation, in the quest for the first three atomic-bombs. Subsequently, in medical school, I considered becoming a radiologist. In the late 1940s, I did nuclear medicine with patients having a variety of hematological disorders. In the 1960s, I did chemical elemental analysis of human blood by xray spectroscopy. In the early 1970s, our group at the Livermore National Laboratory induced genomic instability in human cells with gamma rays.

In short, I fully appreciate the benefits and insights (in medicine and other fields) which ionizing radiation makes possible.

But no one HONORS the xray by treating it casually or by failing to acknowledge that it is a uniquely potent mutagen. One honors the xray by taking it seriously. While doses from diagnostic and interventional radiology are very low RELATIVE TO DOSES USED FOR CANCER THERAPY, diagnostic and interventional xray doses today are far from negligible (some examples in Chapter 2, Part 7e). The widely used CT scans, and the common diagnostic examinations which use fluoroscopy, and interventional fluoroscopy (e.g., during surgery), deliver some of the largest nontherapeutic doses of xrays. In 1993, the United Nations Scientific Committee on the Effects of Atomic Radiation warned, appropriately, in its Annual Report:

"Although the doses from diagnostic xray examinations are generally relatively low, the magnitude of the practice makes for a significant radiological impact" (UNSCEAR 1993, p.228/40). In the USA until about 1970, fetal irradiation occurred during ∼ 1 pregnancy per 14 (Chapter 2, Part 2d).

● Part 9. Every Benefit of Medical Radiation: Same Procedures, Lower Dose-Levels

The fact that ionizing radiation is a uniquely potent mutagen, and the finding that radiation from medical procedures is a major cause of both Cancer and Ischemic Heart Disease, clearly indicate that it would be appropriate in medicine to treat dosage of ionizing radiation at least as carefully as we treat dosage from potent medications. In the medical professions, we do not administer unmeasured doses of powerful pharmaceuticals, and we do not take a casual view of a 5-fold, 10-fold, even 20-fold elevation in dosage of such medications.

By contrast, in both the past and the present, unmeasured doses of xrays are the rule --- not the exception (Chapter 2, Parts 2, 3a, and 3e). When sampling has been done, in which actual measurements are taken, dosage has been found to vary from one facility to another by many-fold, for the same procedure for patients of the same size. The reason for large variation is obvious from the list of numerous proven ways to reduce dosage (Box 3 at the end of this chapter). Facilities which apply all the measures can readily achieve average doses more than 5-fold lower than facilities which apply very few measures.

Certain Spinal Xrays: A Dramatic Demonstration

The potential for dose-reduction may far exceed 5-fold for some common xray exams. This has already been demonstrated for the spinal xrays employed to monitor progress in treating idiopathic adolescent scoliosis, a lateral curvature of the spine. An estimated 5% of American children, or more, have this disorder. In a most responsible way, Dr. Joel Gray and co-workers at the Mayo Clinic developed radiologic techniques for scoliosis monitoring which can reduce measured xray dose to various organs as follows (Gray 1983 in J. of Bone & Joint Surgery 65-A: 5-12):

   ● Abdominal exposure: 8-fold reduction.
   ● Thyroid exposure: 20-fold reduction (with a back to front radiograph), and 100-fold reduction (with a lateral radiograph).

● Breasts: 69-fold reduction (with a back to front radiograph), and 55-fold reduction (with a lateral radiograph).

They report, "These reductions in exposure were obtained without significant loss in the quality of the radiographs and in most instances, with an improvement in the over-all quality of the radiograph due to the more uniform exposure."

9a. Dose-Measurement: Low Cost and High Importance

Incorporated in Box 3's list, under the term "Quality Assurance," is measurement of dose-levels. Only frequent measurements can provide the feedback required to make continual dose reductions --- and also to prevent continual dose increments. The combination of frequent measurements, with an enhanced recognition that each xray photon matters, can achieve a very great deal all by themselves. Nearly everyone takes pride in doing better and better. The evidence, that a series of small improvements can amount to a big difference in result, is abundant elsewhere in medicine and pharmacology.

Fortunately, it is extremely easy to measure entrance-doses during a radiation procedure. One just presses on a small self-adhesive patch called a TLD (thermo-luminescent dosimeter), which does not interfere at all with the procedure. Moreover, the cost for a TLD, including its subsequent "reading," is just a few dollars.

We note that no major equipment purchases are required either to achieve the benefits of quality control (an estimated 2-fold reduction in average dose-level in radiography, Box 3) or to achieve better operator-techniques in fluoroscopy (an estimated 2-to-10-fold reduction in dose, Box 3). Cost is not a big obstacle to taking dose-reduction seriously. The big obstacle is the recognition that it really matters.

Mammography: A Model of Success

The importance of dose-reduction for the mammographic examination has been recognized, and such doses have been reduced by about a factor of TEN in recent years. "Where there is a will, there is a way." In certified mammography centers today, doses are routinely verified periodically, and measurements provide the feedback required, in order to achieve constant dose-reduction instead of upward creep.

9b. The Benefits of Every Procedure --- with Far Less Dose

Dose-reduction can be a truly safe measure. It is clear that average per patient doses from diagnostic and interventional radiology could be reduced by a great deal without reducing the medical BENEFITS of the procedures in any way. We can summarize from Box 3:

● Radiography: Quality-assurance (dose-reduction by an average factor of 2), beam-collimation (by a factor up to 3), rare-earth screens (by a factor of 2 to 4), rare-earth filtration (by a factor of 2 to 4), use of carbon-fibre materials (by a factor of 2), gonadal shielding (by a factor of 2 to 10 for the gonads).

● Digital Radiography: Decrease in contrast resolution, when such resolution is not needed (dose-reduction by a factor of 2 to 3), use of a pulsed system (by a factor of 2).

● Fluoroscopy: Changes in the operator's technique (dose-reduction by a factor of 2 to 10), variable aperture iris on TV camera (by a factor of 3), high and low dose-switching (by a factor of 1.5), acoustic signal related to dose-rate (by a factor of 1.3), use of a 105mm camera (by a factor of 4 to 5). Additional methods not specified in the list: Use of a circular beam-collimator when the image-receiver is circular (Chapter 2, Part 3d), adoption of "freeze-frame" or "last-image-hold" capability, and restraint in recording fluoroscopic images (Chapter 2, Part 3e).

● Part 10. An Immense Opportunity: All Benefit, No Risk

The evidence in this monograph, on an age-adjusted basis, is that most fatal cases of Cancer and Ischemic Heart Disease would not happen as they do, in the absence of xray-induced mutations.

We look forward to responses to our findings.

We have also presented findings, from outside sources, that average per patient radiation doses from diagnostic and interventional radiology could be reduced by a great deal, without reducing the medical BENEFITS of the procedures in any way. The same procedures can be done at substantially lower dose-levels (Part 9, above).

10a. Does the Public Need a Denial, "For Its Own Good" ?

One type of response to this monograph may be that the findings need to be denied immediately (without examination), lest the public refuse to accept the benefits of xray procedures.

This type of response, insulting to the public, would not be consistent with reality. In reality, the public accepts a host of dangerous medications and procedures, in exchange for their demonstrable benefits --- sometimes, for undemonstrated benefits. Very few people will forego the obvious benefits from diagnostic and interventional radiology, just because such procedures confer a risk of subsequent Cancer and IHD. The only change will probably be that people will demand that the same degree of care, now exercised with respect to dosage of potent medications, be exercised with respect to dosage of radiation from each procedure. They will want to avoid a dose-level of, say, ten rads --- if the same information could be acquired with one rad. They do not deserve "one useful part of information, and nine unnecessary parts of extra risk of Cancer and IHD." Patients will want more measurements, and fewer assumptions, about the doses delivered. But they will NOT reject the procedures themselves.

10b. Do Nothing Until the Work Is Independently Confirmed?

A second response, to the evidence in this monograph, may be that doses in diagnostic and interventional radiology should not be reduced until our work is independently confirmed.

The concept, "independent confirmation," is meaningless without equally credible, but independent, sets of data. If one is seriously interested in new prevention-measures for Cancer and Ischemic Heart Disease, then one really needs to ask: Will it ever be possible to conduct a MORE reliable evaluation --- of Fractional Causation, by medical radiation, of Cancer and IHD --- than the evaluation provided by the databases we used in this book? We doubt it, for the reasons described in Part 5b above. As for replication of our results from the SAME databases (PhysPops and age-adjusted MortRates, by Census Divisions), that could be promptly achieved.

It is worth emphasis that validity of the first part of Hypothesis-2 (medical radiation is an important cause of IHD) does not depend on the validity of the second part of Hypothesis-2 (our Unified Model of Atherogenesis --- Part 7c, above). The Unified Model will definitely need independent testing. This might consume decades. Meanwhile, why deny patients the benefits of eliminating uselessly high doses of medical radiation?

10c. The "Advocacy Issue" and the Hippocratic Oath

It is very often said that, if scientists advocate any action based on their findings, they undermine their scientific credibility. If such scientists stand to benefit financially from the actions they advocate, such suspicion occurs naturally. But even in such circumstances, if their work is presented in a way which anyone can replicate, it should be impossible for their advocacy to diminish the scientific credibility of their work.

Our findings are not encumbered either by financial interests or by any barriers to replication. We have high confidence in the scientific credibility of the results, for the reasons presented in Part 6. The findings stand on their own, whether or not we advocate any action.

I have spent a lifetime studying the causes of Ischemic Heart Disease, and then Cancer, in order to help prevent such diseases. So it would be pure hypocrisy for me to feign a lack of interest in any preventive ACTION which would be both safe and benign. And when sources, completely independent from me, set forth their findings that such action is readily feasible --- namely, significant dose-reduction in diagnostic and interventional radiology --- it would be worse than silly for me to pretend that I have no idea what action should occur. After all, as a physician, I took the Hippocratic Oath: "First, do no harm." Silence would contribute to the harm of millions of people.

10d. Why Wait? What Is the Purpose?

Although it is commonly assumed that radiation doses are "negligible" from modern medical procedures, the assumption is definitely mistaken. In reality, estimated dose-levels today from some common xray procedures are far from negligible, as illustrated in Chapter 2, Part 7e. Both the downward and upward forces upon post-1960 dose-levels are discussed in Chapter 2, Part 3. The net result is unquantifiable.

An estimated 35% to 50% of some higher-dose diagnostic procedures are currently received by patients below age 45 (details in Chapter 2, Part 3f) --- when the carcinogenic impact per dose-unit is probably stronger than it is after age 65 or so.

In diagnostic and interventional radiology, dose-reduction would be wholly safe, quite inexpensive, and guaranteed beneficial --- because induction of Cancer by ionizing radiation has been an established fact for decades. (The contribution of radiation-induced mutations, to all types of inherited afflictions, is beyond the scope of this book.) It seems to us that anyone who contemplates Part 9 of this chapter, on known methods of dose-reduction in radiology, has to ask: Why wait? What is the purpose of waiting, when only benefit, and no harm, can come from reducing uselessly high doses as rapidly as possible?

10e. A Mountain of Solid Evidence That Each Dose Matters

The fact, that xray doses are so seldom measured, reflects the false assumption that such doses do not matter. This monograph has presented a mountain of solid evidence that they do matter, enormously. And each bit of additional dose matters, because any xray photon may be the one which sets in motion the high-speed high-energy electron which causes a carcinogenic or atherogenic mutation. Such mutations rarely disappear. The higher their accumulated number in a population, the higher will be the population's mortality-rates from radiation-induced Cancer and Ischemic Heart Disease.

The xray is a proven mutagen and a proven cause of Cancer, and the evidence in this book strongly indicates that it is also a very IMPORTANT cause of Cancer and a very important atherogen. From the existing evidence, it is clear that average per patient doses from diagnostic and interventional radiology could be reduced by a great deal without reducing the medical benefits of the procedures in any way (Part 9, above): Same procedures, at lower doses. Unless effective measures are taken, to eliminate uselessly high dosage, medical radiation will continue in the next century to be a leading cause of Cancer and Ischemic Heart Disease in the United States, and will become a leading cause in the "developing" world, too.

10f. A Prudent Position from Which No One Loses, Everyone Gains

Whether diseases are common or rare, a prime reason for studying their causation is PREVENTION. Cancer and Ischemic Heart Disease, combined, accounted for 45% of all deaths in the USA during 1993 (Chapter 39, Part 4).

If we in the medical professions take the position, that we should NOT press for reducing doses from medical radiation until every question has been perfectly answered, then we can never un-do the harm inflicted during the waiting period, upon tens of millions of patients every year. By contrast, if we take the prudent position that dose-reduction should become a high priority without delay (and if humans do not start exposing themselves to some OTHER potent mutagen), the evidence in this monograph indicates that we will prevent much of the future mortality from Cancer and Ischemic Heart Disease, without causing any adverse effects on health. No one loses, everyone gains.

>>>>>>>>>>

Box 1 of Chapter 1
Final Summary for Fractional Causation, by Medical Radiation, of Cancer and Ischemic Heart Disease.

● – The range of values below represents the earliest year and the most recent year named in Column A. Values for the intervening decades are provided in the listed chapters (e.g., Ch49). The values below come from the "A" or "AA" tables in Chapters 49 – 65. "Diff-Ca" = All Cancers Except Respiratory. "AllExcGen" = All except Genital Cancers. Mortality rates in Column B are age-adjusted to the reference year 1940.

Col.A: M = Male. F = Fem.	Col.B: Nat'l Age-Adjusted Mortality Rate	Col.C: Frac. Causation by Medical Radn	Col.D: R-squared	Col.E: X-Coefficient	Col.F: Ratio of XCoef/Std.Error
Ch49, 1940-88, All-Cancer: M	Big net rise. 115.0 --> 162.7	90% --> 74%	0.95 --> 0.93	0.76 --> 0.75	11.6 --> 10.1
Ch50, 1940-88, All-Cancer: F	Net decline. 126.1 --> 111.3	58% --> 50%	0.86 --> 0.87	0.53 --> 0.34	6.6 --> 6.9
Ch51, 1940-88, Resp'y Ca: M	Enormous rise. 11.0 --> 59.7	~100% --> 74%	0.87 --> 0.78	0.12 --> 0.27	6.8 --> 5.0
Ch52, 1940-88, Resp'y Ca: F	Enormous rise. 3.3 --> 24.5	97% --> 83%	0.96 --> 0.90	0.02 --> 0.13	13.4 --> 7.8
Ch53, 1940-88, Diff-Ca: M	Approx. flat. 104.0 --> 103.0	84% --> 72%	0.93 --> 0.92	0.64 --> 0.46	10.0 --> 8.7
Ch54, 1940-88, Diff-Ca: F	Big decline. 122.8 --> 86.8	57% --> 48%	0.85 --> 0.84	0.50 --> 0.25	6.3 --> 6.1
Ch55, 1940-90, Breast-Ca: F	Flat. 23.3 --> 23.1	~100% --> 83%	0.92 --> 0.89	0.19 --> 0.12	8.7 --> 6.7
Ch56, 1940-80, AllExcGen: F	Flat. 94.0 --> 94.8	75% --> 66%	0.87 --> 0.93	0.51 --> 0.43	6.8 --> 9.6
Ch57, 1940-88, Digest-Ca: M	Big decline. 60.4 --> 38.8	97% --> 82%	0.91 --> 0.87	0.43 --> 0.20	8.3 --> 7.0
Ch58, 1940-88, Digest-Ca: F	Big decline. 50.1 --> 23.5	80% --> 68%	0.76 --> 0.86	0.29 --> 0.10	4.6 --> 6.7
Ch59, 1940-80, Urinary-Ca: M	Approx. flat. 7.4 --> 8.2	~100% --> 83%	0.92 --> 0.61	0.08 --> 0.05	9.0 --> 3.3
Ch60, 1940-80, Urinary-Ca: F	Decline. 4.0 --> 3.0	86% --> 78%	0.94 --> 0.91	0.02 --> 0.02	10.4 --> 8.5
Ch61, 1940-90, Genital-Ca: M	Some rise. 15.2 --> 16.9	79% --> 47%	0.77 --> 0.79	0.09 --> 0.05	4.9 --> 5.2
Ch63, 1940-80, Buccal-Phar: M	Approx. flat. 5.1 --> 4.6	~100% --> 81%	0.72 --> 0.73	0.04 --> 0.03	4.3 --> 4.4
Ch64, 1950-93, IHD: M	Enormous fall. 256.4 --> 131.0	79% --> 63%	0.95 --> 0.73	1.49 --> 0.50	11.2 --> 4.3
Ch65, 1950-93, IHD: F	Enormous fall. 126.5 --> 64.7	97% --> 78%	0.87 --> 0.68	0.90 --> 0.30	6.8 --> 3.9

Box 2 of Chapter 1
Comparison of Results: All Causes, NonCancers, NonCancers NonIHD, Cancers, IHD.

All the comparisons below are based on the relationship between 1940 PhysPops and 1940 MortRates, except for 3 pairs of 1950 MortRates. "Sig." means statistically significant. When XCoef/SE = 2, then P = roughly 0.05. See Chap.38.

		R-Squared	X-Coef.	XCoef/Std Err	Relationship, MortRates w. PhysPops by CensusDiv.
Ch23: All Causes Combined	Male	0.1299	Neg.	-1.02	Inverse, but not sig.
	Fem	0.2823	Neg.	-1.66	Inverse, and marginal.
Ch24: All NonCancer Combined	Male	0.2841	Neg.	-1.67	Inverse, and marginal.
	Fem	0.4362	Neg.	-2.33	Inverse, and significant.
Ch25: All NonCancer NonIHD	Male	0.7933	Neg.	-5.18	Inverse, and very sig.
	Fem	0.7037	Neg.	-4.08	Inverse, and very sig.
Ch26: Appendicitis	Male	0.0179	Neg.	-0.36	None.
	Fem	0.0010	Neg.	-0.08	None.
Ch27: CNS Vascular (Stroke)	Male	0.4000	Neg.	-2.16	Inverse, and significant.
	Fem	0.2882	Neg.	-1.68	Inverse, and marginal.
Ch28: Chronic Nephritis	Male	0.4561	Neg.	-2.42	Inverse, and significant.
	Fem	0.2687	Neg.	-1.60	Inverse, and marginal.
Ch29: Diabetes Mellitus	Male	0.6435	Pos.	3.55	Positive, and quite sig.*
	Fem	0.6005	Pos.	3.24	Positive, and quite sig.*
Ch30: Hypertensive Disease	Male	0.3564	Neg.	-1.97	Inverse, and significant.
	Fem	0.2056	Neg.	-1.35	Inverse, and very marginal.
Ch31: Influenza and Pneumonia	Male	0.8344	Neg.	-5.94	Inverse, and highly sig.
	Fem	0.8849	Neg.	-7.34	Inverse, and highly sig.
Ch32: Fatal Motor Vehicle Accid.	Male	0.0195	Neg.	-0.37	None.
	Fem	0.0003	Neg.	-0.04	None.
Ch33: Other Fatal Accidents	Male	0.0901	Neg.	-0.83	None.
	Fem	0.4440	Neg.	-2.36	Inverse, and significant.
Ch34: Rheum.Fever/Rheum.Heart	Male	0.0021	Pos.	0.12	None.
	Fem	0.0550	Pos.	0.64	None.
Ch35: Syphilis and Sequelae	Male	0.3278	Neg.	-1.85	Inverse, and marginal.
	Fem	--	--	--	--
Ch36: Tuberculosis, All Forms	Male	0.2067	Neg.	-1.35	Inverse, and very marginal.
	Fem	0.6381	Neg.	-3.51	Inverse, and quite sig.
Ch37: Ulcer: Stomach, Duoden.	Male	0.3864	Pos.	2.10	Positive, and significant.**
Ch6+7: All Cancers Combined	Male	0.9508	Pos.	11.63	Positive, and highly sig.
	Fem	0.8608	Pos.	6.58	Positive, and highly sig.
Ch8: Breast Cancer	Male	--	--	--	--
	Fem	0.9153	Pos.	8.70	Positive, and highly sig.
Ch9+10: Digestive-Syst. Cancers	Male	0.9078	Pos.	8.30	Positive, and highly sig.
	Fem	0.7550	Pos.	4.64	Positive, and very sig.
Ch11+12: Urinary-Syst. Cancers	Male	0.9208	Pos.	9.02	Positive, and highly sig.
	Fem	0.9395	Pos.	10.43	Positive, and highly sig.
Ch13+14: Genital Cancers	Male	0.7182	Pos.	4.22	Positive, and very sig.
	Fem	0.0683	Pos.	0.72	None.
Ch15: Buccal & Pharynx Cancers	Male	0.7234	Pos.	4.28	Positive, and very sig.
	Fem	--	--	--	--
Ch16+17: Respiratory-Syst. Canc	Male	0.8673	Pos.	6.76	Positive, and highly sig.
	Fem	0.9625	Pos.	13.40	Positive, and highly sig.
Ch40+41: Ischemic Heart Disease	Male	0.9475	Pos.	11.24	Positive, and highly sig.
	Fem	0.8337	Pos.	5.92	Positive, and highly sig.

* Diabetes Mellitus (DM): After the rules changed in 1949 for reporting the underlying cause of death in diabetics, DM MortRates abruptly fell in half and our R-sq. values dropped abruptly to 0.11 and 0.20 (Chap.29). The significant R-sq. values in 1940 very probably denote a correlation between PhysPop and deaths during 1940 from xray-induced Ischemic Heart Disease in people having diabetes (Chapters 29, 40, 41).

** Ulcer Deaths: The positive correlation between Ulcer Deaths in 1940 and PhysPop might be due to erroneous reporting in 1940 of deaths, truly from Stomach Cancer, as deaths from Stomach Ulcers.

Box 3 of Chapter 1
Procedures to Reduce Collective Dose Equivalent in Diagnostic Xray Examinations.

● – This box, with its title above and footnotes below, is borrowed without alteration from the 1988 UNSCEAR Report (Annex C: Exposures from Medical Uses of Radiation, Table 23 at p.282). UNSCEAR = United Nations Scientific Com'tee on the Effects of Atomic Radiation. An almost identical table appears also in the 1989 NCRP Report (Report No. 100, Table 3.21, at p.37). NCRP = National Council on Radiation Protection (USA). Details for UNSCEAR 1988, NCRP 1989, and the references cited below, are in the Reference List of this monograph.

Area	Procedure	Entrance–Dose Reduction– Factor	Reference
All Types	Elimination of medically unnecessary procedures	1.2	Cohen 1985.
	Introduction of Quality Assurance programme (general)	2*	Cohen 1985.
Radiography	Decrease in rejected films through Quality Assurance programme	1.1	Gallini 1985. Properzio 1985.
	Increase of peak kilovoltage	1.5	Wiatrowski 1983.
	Beam collimation	1 to 3	Johnson 1986. Morris 1984.
	Use of rare–earth screens	2 to 4	Kuhn 1985. Newlin 1978. Segal 1982. Wagner 1976.
	Increase of filtration	1.7	Kuhn 1985. Montanara 1986. Wiatrowski 1983.
	Rare–earth filtration	2 to 4	Tyndall 1987.
	Change from photofluorography to chest radiography	4 to 10	Jankowski 1984. Mustafa 1985. Neamiro 1983.
	Use of carbon fibre materials	2.0	Huda 1984.
	Replacement of CaWO4 screens with spot film technique	4.0	Kuhn 1985.
	Entrance exposure guidelines	1.5	Laws 1980.
	Gonadal shielding	2 to 10 **	Poretti 1985.
Pelvimetry	Use of CT topogram	5 to 10	Stanton 1983.
Fluoroscopy	Acoustic signal related to dose rate	1.3	Anderson 1985.
	Use of 105 mm camera	4 to 5	Rowley 1987.
	Radiologist technique	2 to 10	Rowley 1987.
	Variable aperture iris on TV camera	3.0	Leibovic 1983.
	High and low dose switching	1.5	Leibovic 1983.
Digital radiography	Decrease in contrast resolution	2 to 3	Rimkus 1984.
	Use of pulsed system	2	Rimkus 1984.
Computed tomography, head	Gantry angulation to exclude eye from primary beam	2 to 4 ***	Isherwood 1978.
Mammography	Intensifying screens	2 to 5	NCRP 1986. Shrivastava 1980.
	Optimal compression	1.3 – 1.5	NCRP 1986.
	Filtration	3	Hammerstein 1979.

* The role of proper training in radiation protection is extremely important. Dose reduction-factors in this regard may be large; however, they are difficult to quantify. ** Factor for gonads. *** Factor for eyes.

Figure 1-A.

All-Cancers-Combined: Dose-Response between PhysPop and MortRates.

Please refer to Parts 5a–5d of this chapter. In each graph, the line of best fit results from regressing the 1940 All-Cancer Mortality Rates (male, female) on the 1940 PhysPop values. PhysPop (physicians per 100,000 population) is a surrogate for accumulated dose from medical radiation. The nine boxy symbols denote the observed values in the Nine Census Divisions. Full details are in Chapters 6 and 7.

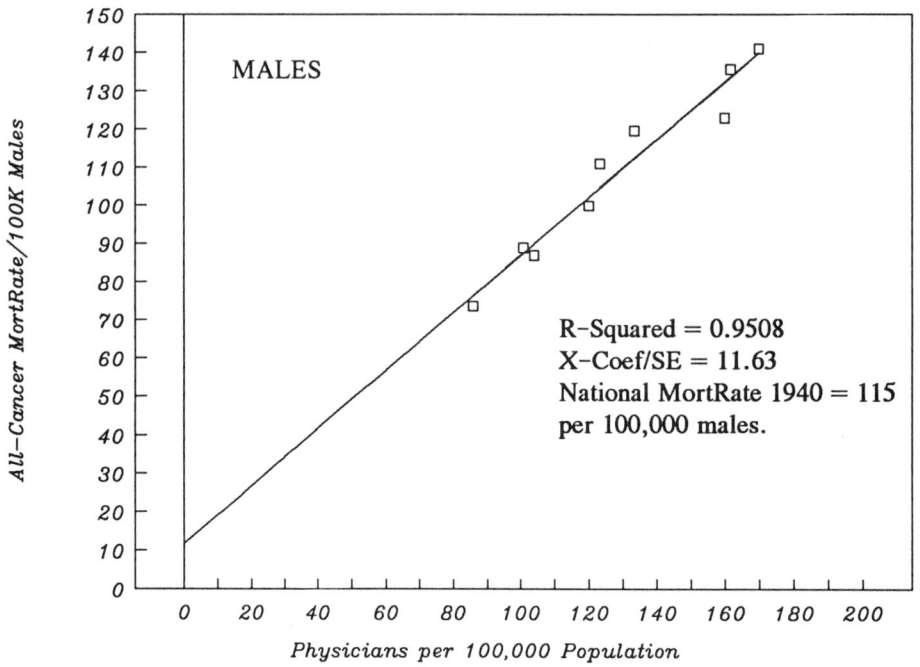

MALES

R-Squared = 0.9508
X-Coef/SE = 11.63
National MortRate 1940 = 115
per 100,000 males.

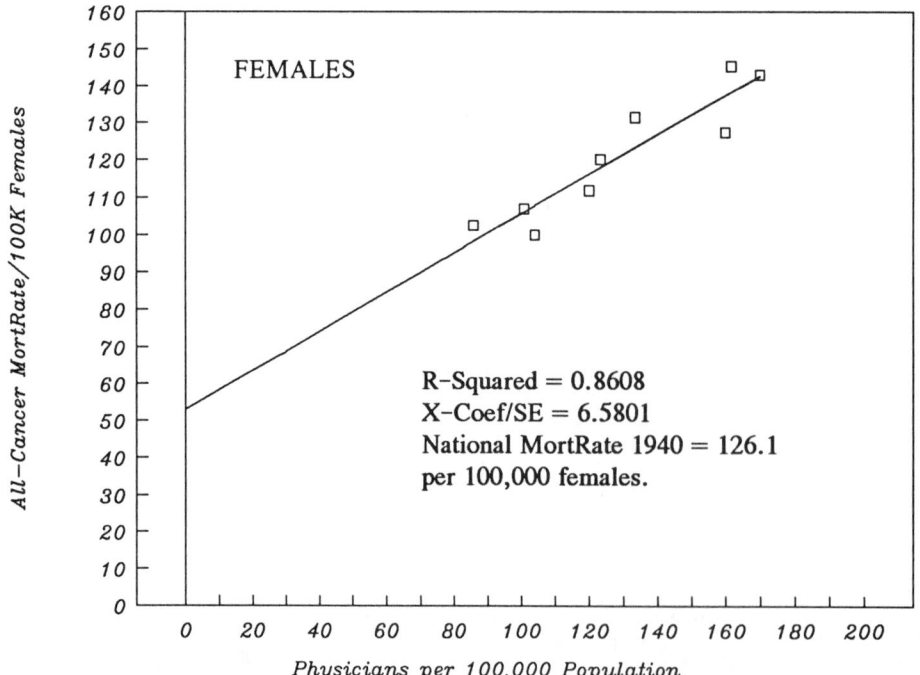

FEMALES

R-Squared = 0.8608
X-Coef/SE = 6.5801
National MortRate 1940 = 126.1
per 100,000 females.

Related text = Parts 5a – 5d.

– 24 –

Figure 1-B.
Ischemic Heart Disease: Dose-Response between PhysPop and MortRates.

Please refer to Part 5f of this chapter. In the upper graph, the line of best fit results from regressing the age-adjusted male 1950 Mortality Rates from Ischemic Heart Disease on the 1940 PhysPop values. PhysPop (physicians per 100,000 population) is a surrogate for accumulated dose from medical radiation. The nine boxy symbols denote the observed values in the Nine Census Divisions. In the lower graph (females), we show 1950 PhysPop values. When female 1950 age-adjusted IHD MortRates are paired with 1950 PhysPops, R-squared = 0.8669; with 1940 PhysPops, R-squared = 0.8337 --- a trivial difference. Full details are in Chapters 40 and 41.

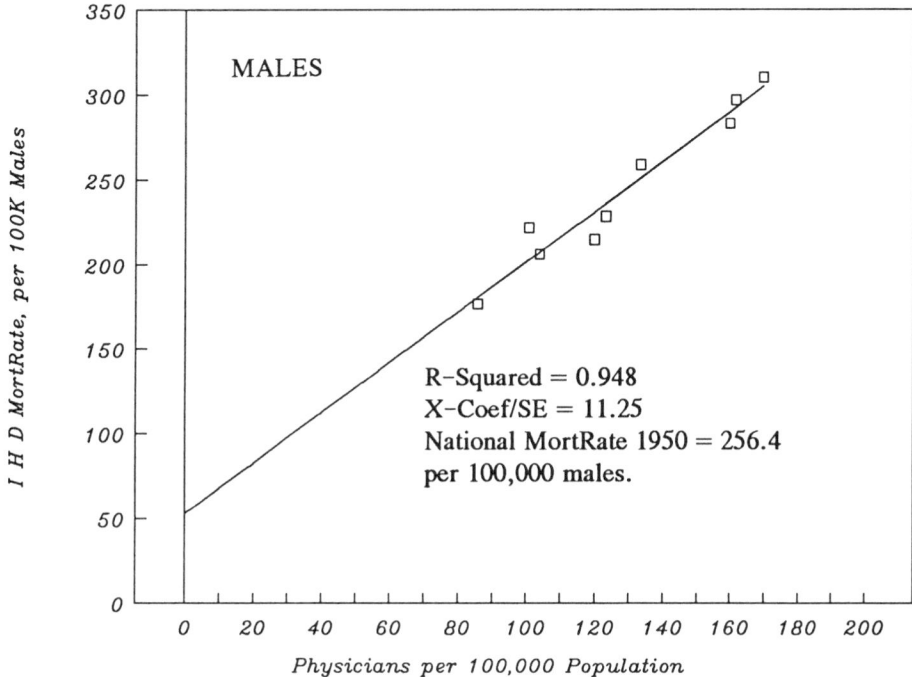

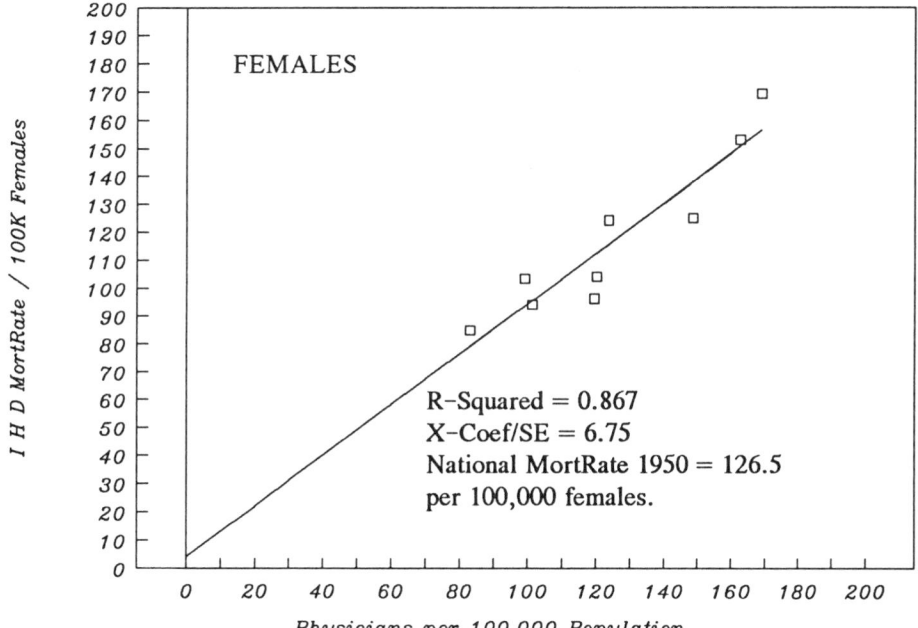

Related text = Part 5f.

Figure 1-C.
NonCancer NonIHD Deaths: Dose-Response between PhysPop and MortRates.

Please refer to Part 5f of this chapter. In each graph, the line of best fit results from regressing the 1950 age-adjusted NonCancer NonIHD MortRates (male, female) on the 1940 PhysPop values. PhysPop (physicians per 100,000 population) is a surrogate for accumulated dose from medical radiation. The nine boxy symbols denote the observed values in the Nine Census Divisions. The dose-response is inverse (negative). Full details are in in Chapter 25.

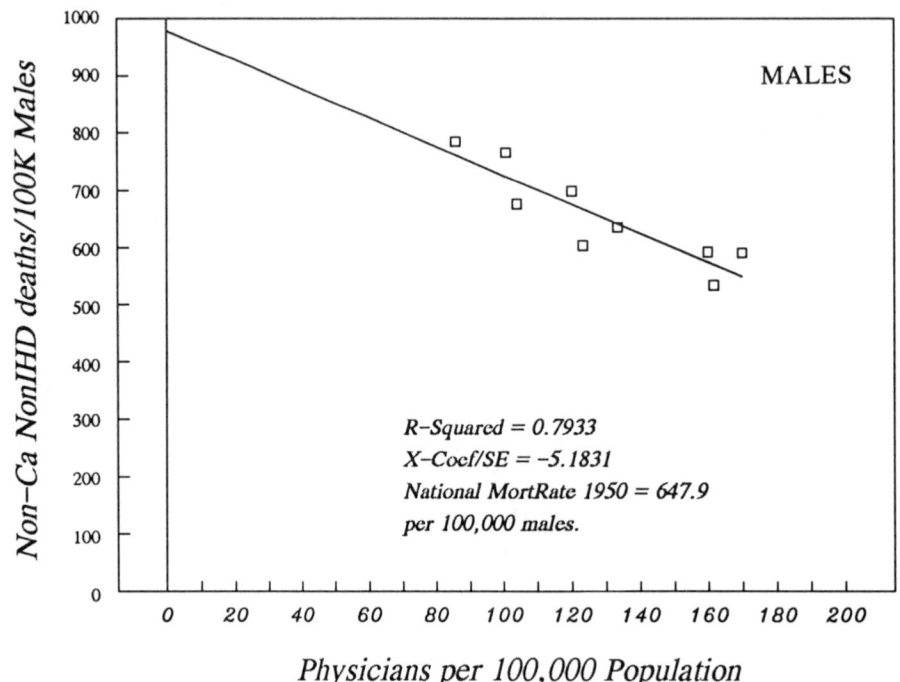

R-Squared = 0.7933
X-Coef/SE = -5.1831
National MortRate 1950 = 647.9
per 100,000 males.

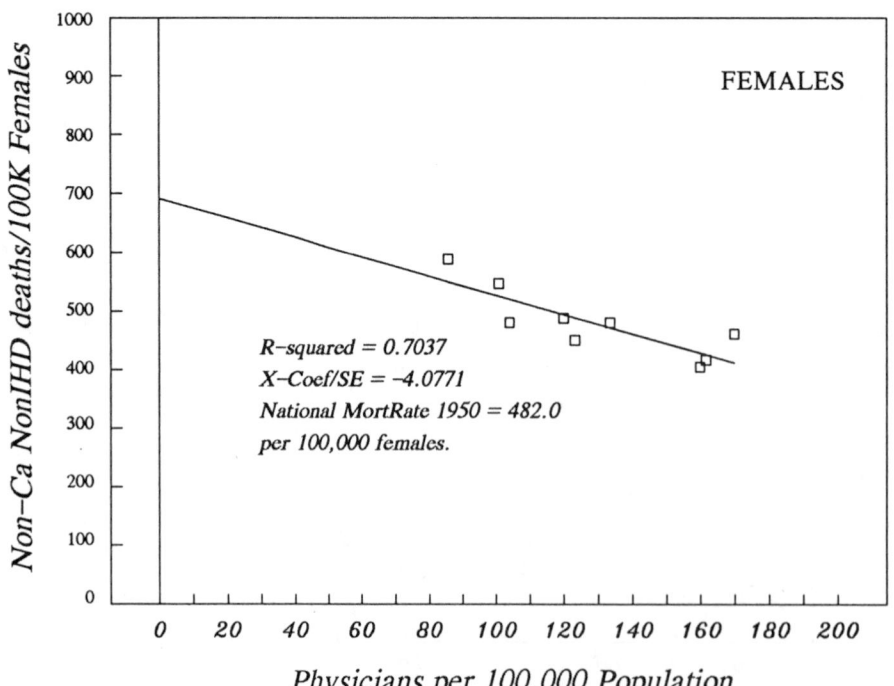

R-squared = 0.7037
X-Coef/SE = -4.0771
National MortRate 1950 = 482.0
per 100,000 females.

Related text = Part 5f.

CHAPTER 2

Pre-1960 and Post-1960 Uses of Medical Radiation, and Its Carcinogenic Action

● Part 1. Is Hypothesis-1 Long Overdue?

Hypothesis-1 proposes that exposure to medical radiation is a highly important cause (probably the principal cause) of cancer-mortality in the United States during the Twentieth Century --- even though medical radiation is only rarely mentioned in lists of "risk factors" for Cancer.

Then how did we reach the point of deciding that such an idea deserved someone's careful examination? Very slowly, as Part 9 of this chapter relates. Perhaps the conception and testing of Hypothesis-1 is long overdue.

Hypothesis-1 becomes a proposition "demanding" evaluation when two types of knowledge COMBINE: Knowledge about some history of medicine in the United States during the Twentieth Century, and knowledge about the evidence that xrays and other ionizing radiations are proven carcinogens --- indeed, are mutagens with some uniquely potent properties. Many people are versed in one of these fields, but not the other.

On both topics, this chapter provides some basic orientation, with references to ample supporting evidence. Parts 2 and 3 describe a little medical history, and Parts 4, 5, 6, 7, and 8 state some of the key knowledge about radiation carcinogenesis.

This book presents a powerful test of Hypothesis-1 and concludes that the evidence strongly supports the hypothesis. The same evidence is the basis for Hypothesis-2.

● Part 2. 1896-1960: Rapid and Widespread Embrace of Xrays in Medicine

Wilhelm Konrad Roentgen discovered the xray on November 8, 1895 (Roentgen 1895). "The ray," as it was often called, immediately caused a sensation among physicians and the general public. Commemorating the hundredth anniversary of Dr. Roentgen's discovery, Dr. Ronald G. Evens provides some vivid details in his "Roentgen Retrospective," in the Journal of the American Medical Association (Evens 1995). Referring to the USA, Evens writes (1995, p.912):

"By the time of the appearance of the first American clinical diagnostic radiograph [also called roentgenograph and skiagraph], made at Dartmouth College by Dr. Edwin Frost on February 3, 1896, physicians were becoming increasingly aware of the extraordinary potential for the new discovery. By April, 'xray mania' had seized the United States. Xray studios had opened for 'bone portraits,' and countless photographers and electricians had set up shop as 'skiagraphers.'" Thomas Edison became an enthusiast in 1896, and attempted to xray the human brain "at work" (Evens 1995, p.914).

2a. The Xray in Medicine: Diagnostic, Therapeutic and Interventional Uses

In medicine, a journal entitled Archives of Clinical Skiagraphy made its appearance in April/May 1896 (London), and the American Xray Journal began publication in 1897. In 1900, the American Roentgen Ray Society was founded. "Soon, the appearance of xray machines in general practitioners' offices across the United States would underline the notion that a new technology was available to diagnose any and every ailment. Some physicians even thought it would eliminate the need for laboratory analysis in medicine" (Evens 1995, p.915).

The xray was employed immediately not only for diagnosis of medical problems, but also for treatment. There was hope that xrays would cure Tuberculosis, Cancer and every other affliction. The ten years, up to 1906, were described as follows by Dr. George MacKee, a great figure in dermatology and an enthusiast for reasonable radiation therapies (from MacKee 1938, p.16):

"During those years the rays, to a large extent, were empirically used and they were tried out on nearly every chronic disease. The literature was misleading, as it was full of case reports of wonderful cures, the occasional paper from the pen of a good man being ignored or overlooked by the average xray operator of the period and in spite of repeated warnings from capable men, the 'radiomaniacs' held the reins."

Although of course the chaos of the first ten years subsided, enthusiasm for diagnostic, therapeutic, and interventional uses of xrays did not subside, as Parts 2c and 2d indicate.

Interventional Radiology

A term is needed, to identify uses of medical radiation which are neither strictly diagnostic nor directly therapeutic. Such a term, loosely used, is "interventional radiology." Examples include xray-use in setting broken bones, locating foreign objects, placing catheters and needles, and helping to guide many types of surgical procedures. In the past, xrays were used also to guide the deliberate collapsing of a lung, in patients who were trying to recover from Pulmonary Tuberculosis.

2b. The Skin as the Initial Dose-Meter (Dosimeter)

Appendix A of this book defines the commonly used dose-units (rad, roentgen, centi-gray, and others), and dose-ranges for what is regarded as low, moderate, and high dosage.

But when xrays were introduced into medicine, it was far from clear how to measure the xray doses given to patients, and what was biologically "too much." Everything was figured out by trial and error. Today, it is regarded as a rare event when the skin of a patient gets damaged by medical xrays. But for many years during the first half of the Twentieth Century, the skin was often the dose-meter. The reddening or burning of skin on enough patients gradually established the fact that excessive dosage could occur. Indeed, the early dose-unit in medicine was the "erythema dose" --- the dose-level which generally provokes a morbid reddening of the skin (estimated today as a dose of about 200 rads for temporary erythema, 600 rads for main erythema, and 1,500 rads for late erythema; FDA 1994, Table 2). In 1926, "erythema dose" was a term still in use in medical journals. For example (Husik 1926, p.859):

"It is now the routine treatment to radiograph all children between one and fourteen years of age booked for tonsil and adenoid operations at the throat department of the Massachusetts General Hospital and the Massachusetts Eye and Ear Infirmary. All children showing [on the diagnostic film] a broad superior mediastinum are considered as suspicious cases, and are given four xray treatments of a third of an erythema dose. The treatments are repeated at intervals of ten days." (The purpose of such "treatments" was to shrink the thymus gland, for it was widely believed that patients with smaller thymus glands had a lower chance of sudden death under anesthesia; Gofman 1995/96, Chapter 10.)

2c. Popularity of Fluoroscopy (Roentgenoscopy)

The fluoroscope is an xray machine which leaves the xray beam "on" while the physician examines the motions of a patient's organs, and/or the motions of various instruments and catheters (during surgical and other procedures). Because the beam stays "on," the fluoroscope has the potential to deliver high xray doses.

During World War One, the Army managed to reduce the size and complexity of fluoroscopes, which were used in field hospitals during bone-setting and removal of bullets and other debris. After the war, in the 1920s, fluoroscopy (also called roentgenoscopy) became an enormously popular procedure not only among radiologists (roentgenologists), but also among many kinds of physicians.

The fluoroscope produces information instantly, without the delay, expense, and training required to develop xray-exposed films.

Routine Use of Fluoroscopes in Office Practice

In 1922, Dr. Louis Bishop made the following prediction before the Medical Society of the Greater City of New York (Bishop 1922): "Fluoroscopy, I venture to assert, will become a routine measure in every physician's office before long." In 1923, Dr. Preston Hickey reported to the American Roentgen Ray Society as follows (Hickey 1923):

"It is interesting to note also the large number of internists who have placed fluoroscopes in their offices, not with the idea of specializing in xray work, but simply wishing to have conveniently at hand an xray control of their physical findings. Here again, the simplified apparatus which has developed from war-time practice is conspicuous." By 1937, Dr. Eugene Leddy of the Mayo Clinic reported (Leddy 1937, p.924):

"In fact, roentgenologic methods of diagnosis are so important that no investigation of a patient is considered complete without roentgenologic examinations, which generally include roentgenoscopy [fluoroscopy]. These studies are often carried out by a general practitioner or surgeon in his office because of lack of facilities for expert study nearby or because the physician sees no need to refer the patient to a roentgenologist."

Operation of Fluoroscopes in Pediatric Offices

By 1940 (perhaps much earlier), some pediatricians (not all) included fluoroscopy as part of every "well-baby" visit. In 1942, Dr. Franz Buschke and Herbert M. Parker wrote (Buschke 1942):

"Recently we became aware of the fact that apparently a number of pediatricians include fluoroscopy in the monthly routine examinations of infants in their care during the first and second years of life." This pediatric practice is confirmed in Pifer 1963 and in Blatz 1970. Dr. Hanson Blatz, who was New York City's chief of Radiation Control, reported (Blatz 1970): "When we questioned this practice, pediatricians would say, 'Well, the parents expect it. They think if we don't fluoroscope the patients, they are not getting a complete examination'."

After studying the radiation output of seven fluoroscopes in the offices of "reputable pediatricians selected at random," Buschke and Parker estimated (Buschke 1942, p.527): "If the average rapid fluoroscopy by an experienced and well-adapted examiner takes twenty seconds, about 8.3 roentgens [entrance dose] will be delivered at this rate or 100 roentgens during the first year of life." The roentgen is a dose-unit which is approximately equivalent to a rad (Appendix-A, Part 2).

Of course, not all examiners were well trained with fluoroscopic machines. In the seven pediatric offices visited by Buschke and Parker, "none of them knew the output of their machine" (Buschke 1942, p.525). And (p.527): "In another place under the direction of one of the best radiologists, we found that the output differed with the operator." The dose-rate differed by nearly a factor of 2.

Operation of Fluoroscopes in Hospitals

Fluoroscopy was popular not only in medical offices, but also in hospitals --- for diagnostic and surgical uses. Carl B. Braestrup, of the Physics Laboratory of the New York City Department of Hospitals, was persistent in warning about careless use of fluoroscopes. In an address to the New York Roentgen Society, he reported (Braestrup 1942, p.210):

"During the past years, we have measured the roentgen output of large numbers of fluoroscopes, using the settings at which they are normally operated ... and have found a very wide variation ... Attention is called particularly to test B-116, where the R [roentgen] per minute at the panel was 127, that is, an erythema dose would be reached in about three minutes. Such a unit could be classified as a lethal diagnostic weapon and yet there are many of these still in use." And (Braestrup 1942, p.213):

"Of the various types of radiologic equipment, the mobile unit probably has been responsible for more radiation damage than any other piece of apparatus. These accidents have in most cases occurred while the mobile unit was used for fluoroscopy by surgeons, who apparently did not realize the high output obtained at short distances." In an attempt to prevent some injuries, a limit of 100 roentgens per fluoroscopic examination was set in New York City hospitals (Braestrup 1969). "Recommended" would be a better word than "set," for even today, radiation doses during fluoroscopy are seldom measured (Part 3d).

Estimated Dose per Fluoroscopic Procedure at Mid-Century

In 1953, Dade W. Moeller (then of the Public Health Service; later, president of the Health Physics Society) published an estimate that the average entrance dose per fluoroscopic examination was about 65 roentgens at mid-century (Moeller 1953, pp.58-59). Our Appendix-K explores the implications of the Moeller estimate.

2d. Diagnostic Films: Slow Film-Speeds and Wide Beams

In addition to fluoroscopy, physicians made use of a vast number of diagnostic xray photographs ("films"). Most of the common diagnostic examinations used today were also used well before mid-century. But in terms of cancer hazard, the hazard caused per film was higher in the past, because dose was higher and because a larger area was exposed. One reason that the dose was higher in the past is that the films were "slower" and exposure required more "light" (more xray photons). A larger area was exposed because few were trained to confine the xray beam to the area of the film, and certainly not to the organs whose picture was needed. In addition to the organs which were irradiated on purpose, most of the torso and neck were often irradiated simultaneously. We surmise (but do not know) that dental xrays also exposed much more area than needed.

Pre-Birth Irradiation

Fetal irradiation was quite common. "Roentgenographic evaluation of the relative size of the fetal head and maternal pelvis has been used clinically almost since the advent of medical radiography" (Kelly 1975). The estimated frequency of xray pelvimetry in the 1947-1970 period was 1 birth out of every 13.5 births in the USA (Gofman 1995/96, pp.88-89, based on MacMahon 1962 and Kelly 1975).

2e. Radiotherapy of Benign Diseases: "Every Disease There Is"

Therapeutic irradiation for non-malignant conditions began soon after the xray's discovery. Radium, which was discovered in 1898, was sometimes used as a source of gamma rays, but unlike xrays, radium was scarce and expensive. A few examples of the ailments treated by high-dose medical radiation can illustrate the range of applications, without implying that radiation was tried on EVERY case:

Acute postpartum mastitis, ankylosing spondylitis, arthritis, asthma, excessive menstrual bleeding, herpes zoster(shingles), hyper-thyroidism, neuritis, pneumonia, pyogenic (pus-forming) infections, skin disorders of numerous variety (see below), sore shoulders (bursitis, tendonitis), stomach ulcers, swollen lymphoid tissues (e.g., "swollen adenoids"), thymus-gland enlargement (widely believed, from about 1915 to 1945, to be associated with sudden death under anesthesia, and with sudden infant death), thyroiditis, tuberculous lesions of practically every organ, and whooping cough. Documentation and references can be found in Gofman 1995/96.

In 1965, Dr. Stephen B. Dewing, a radiologist, authored a fine book in which he wrote (Dewing 1965, p.ix): "It has been said that radiation therapy has been used promiscuously, on every disease there is, and probably so."

Skin disorders deserve a paragraph of their own. By 1922, over 80 skin disorders were being treated with high-dose radiation (MacKee 1922). And this continued (MacKee 1938). Very few of these conditions were malignant. They included acne vulgaris, actinomycosis (a fungus), eczema, incessant itching, lichen planus, psoriasis, neurodermatitis, and ringworm of the scalp. Typical therapeutic doses began at about 85 roentgens per week, and could accumulate up to 1,400 roentgens per regime (Sulzberger 1952, p.639).

"Super-Soft" Xrays and a Mistaken Generalization

In 1925, Dr. Gustav Bucky introduced the use of "grenz rays" (also called "super-soft roentgen rays") for some skin disorders. Super-soft xrays lie on the continuum between xrays and ultraviolet rays, and most of them penetrate only about 2 millimeters of tissue. By contrast, "superficial roentgen rays" (which come from xray machines operated at peak kilovoltages in the 60-100 kilovolt range), penetrate more deeply. As of 1952, "most dermatoses" were treated with "superficial roentgen-ray treatments" --- not by grenz rays, according New York University's head of Dermatology (Sulzberger 1952, p.639).

Perhaps it is the concept of super-soft non-penetrating xrays which accounts for the mistaken idea, still circulating in some medical circles, that medical xrays in general are "too weak" to cause Cancer. Therefore, a reminder may be appropriate: Whenever a medical xray procedure exposes a film (or other image-maker) on the opposite side of a patient, such exposure is proof that some of the the xrays fully penetrated the patient --- not just the top 2 millimeters. Medical xrays are definitely not "too weak" to penetrate and to leave carcinogenic damage in the internal organs.

2f. Were People Too Poor to Visit Physicians?

Today, most medical care in the United States is paid for by a third party --- some variety of private and government insurance. Some readers might assume that in the 1900-1960 period, when such arrangements were absent or less common, few people could afford to visit physicians. The following estimate for the year 1950 may indicate otherwise, although the estimate is not elaborated by income-level or by "service." The estimate is that there were 150,000 practicing physicians, who performed 750,000,000 medical services per year, when the population was 150,000,000 people (Donaldson 1951, p.931). If valid, the figures mean an average of 5 "medical services" in a year for each man, woman, and child.

In addition, in 1949, allegedly 60 million people (40% of the U.S. population) visited a dentist, according to Dr. Dade W. Moeller and colleagues (Moeller 1953, p.59). These authors report that 84 million dental xray films were used in 1949: "The average exposure to the patient per film is about 5 roentgens, most of the exposure being limited to the mouth of the patient" (Moeller 1953, p.59).

2g. Emphatic Assurances of Safety

Since virtually no one keeled over as a result of diagnostic, interventional, and therapeutic xray usage, the xray was repeatedly declared harmless. "Absolutely no danger." "Harmless." "No reports of harmful effects." "So far as we know, harmless both as to immediate and remote effects." Even 2,000 roentgens, delivered to ulcer patients over 12 days, was a dose pronounced "perfectly safe" (Ricketts 1951, p.381). The context of such statements is presented in Gofman 1995/96.

The medical professions did not think about delayed consequences, like Cancer, despite some evidence from experimental animals of xray-induced Cancer. By the 1940s, a few experts were trying to discourage pediatricians from fluoroscoping well-babies every month during check-ups, lest gonadal irradiation cause INHERITED afflictions in the next generation (Buschke 1942, pp.527-532). Concern about xray-induced CANCER was hardly voiced before the late 1950s, and by then, radiation health-science was very deeply entangled with the nuclear aspects of national security.

Meanwhile, during the 1940s and 1950s, the Defense Department and the Atomic Energy Commission had staffed themselves and their numerous research arms with radiation experts transferred from medicine --- the very same people who were confident that even very high doses of xrays did no harm.

2h. A Rather Strong Warning in 1959 to the Medical Profession

Above-ground nuclear bomb-tests in Nevada during the 1950s had deposited radioactive fallout, unevenly, nearly from coast to coast. It caused a furor --- especially because milk was contaminated by strontium-90. Dr. Linus Pauling and others were warning about long-term health effects, particularly radiation-induced inherited afflictions and radiation-induced Cancers. What was their evidence?

By 1927, H.J. Muller had established, in the fruit fly, that ionizing radiation induced heritable mutations. Radiation-induced malformations and radiation-induced Cancer had been demonstrated in some experimental animals. Human evidence in the 1950s was thin --- because remarkably little epidemiologic inquiry had been undertaken, to find out if there were delayed effects from medical radiation. But evidence was far from absent. For example, human evidence of radiation-induced Cancer already included the following (and more):

- Bomb-induced Leukemia in Hiroshima-Nagasaki.
- Xray-induced Skin Cancers in radiologists.
- Xray-induced childhood Leukemia and childhood Cancer in children irradiated before birth.
- Xray-induced Thyroid Cancer following childhood radiotherapy for "enlarged thymus."
- Thorium-induced Liver Cancer in medical patients who had received thorotrast (used as a "contrast medium" to enhance diagnostic information from certain types of fluoroscopic procedures).
- Radon-and-radon-daughter-induced Lung Cancer in uranium miners.
- Radium-induced Bone Cancer in radium dial-painters and others.

The furor over radioactive fallout resulted in a 1956 report from the National Academy of Sciences entitled "The Biological Effects of Radiation: A Report to the Public," followed by a 1958 report from the United Nations. The evidence already indicated that children are probably more vulnerable than adults to radiation carcinogenesis. In 1959, Dr. Russell Morgan (Chairman of Radiology at Johns Hopkins Medical School) chaired a National Advisory Committee on Radiation for the U.S. Public Health Service. In its 20-page report, to the Surgeon General of the Public Health Service, the Committee began (PHS 1959, p.1):

"During the past several years, a number of scientific bodies, including the National Academy of Sciences of the United States (NAS 1956) and the United Nations Scientific Committee on the Effects of Atomic Radiation (UNSCEAR 1958), have reported extensively on the influence of ionizing radiation on biological systems. From these reports it is evident that serious health problems may be created by undue radiation exposure and that every practical means should be adopted to limit such exposure both to the individual and to the population at large." And (PHS 1959, pp.1-2):

"The principal sources of ionizing radiation which have been created or developed by man include xray machines, nuclear reactors and their radioisotopic byproducts, high-energy particle accelerators, a number of concentrated forms of naturally occurring radioactive materials, and the fallout constituents of nuclear weapons ...Most of the ionizing radiation received by the population today, other than that received from natural sources, has been from xray machines employed by the health professions."

While the general public may not have realized that radioactive fallout, nuclear pollution, and medical radiation all deliver ionizing radiation, the authors of the 1959 report were explicit on that fact (above). And so, we have chosen the next year, 1960, as the year in which the medical profession was warned that it should stop issuing emphatic assurances, to itself and to its clients, about the safety of medical radiation.

● Part 3. 1960 to Present: Some Changes in Usage of Medical Radiation

Before our overview begins, of post-1960 practices in medical radiation, a comment belongs here about Hypothesis-1 and the pre-1960 period. What happened in the pre-1960 period has a direct impact not only on the 1900-1960 death rates from radiation-induced cancer, but also on such death-rates from 1960 to the present year --- a fact which is documented by Part 8 of this chapter.

In 1990, over 50% of the age-adjusted cancer death-rate (USA) came from people who died of Cancer at age 65 and older. Over 93% comes from people who died at age 45 and older. Their lifetime exposure to medical radiation was very probably NOT limited to post-1960 practices. This statement will be true even well beyond the year 2000. The age-distribution of the 1990 age-adjusted cancer mortality-rate is shown in Chapter 4, Box 4.

3a. Effect of the 1956, 1958, and 1959 Warnings

After human evidence of radiation carcinogenesis began appearing, did it cause a big reduction in the population's average annual per capita exposure from medical radiation?

Parts 3b and 3c show that, during the past 40 years, some events have operated in the direction of REDUCING the population's average per capita dose from medical radiation, but during the same years, some other events have operated in the direction of ADDING to the per capita dose. If we assume that the NET effect is a reduction in the population's average per capita dose, we still lack justification for assuming it is a "big reduction." We are unaware of any reliable quantification of the population's average per capita dose from medical radiation, for any period, past or present. If such a statement seems shocking, readers need to consider these points:

● Even today, there is great uncertainty about something as basic as the NUMBER of diagnostic xrays given per year in the USA. The annual number for 1985-1990 was at least 800 diagnostic xray exams per thousand population, excluding dental xrays and nuclear medicine (UNSCEAR 1993, Table 6, p.279). That estimate "could be an underestimate by up to 60%" (UNSCEAR 1993, p.229/46).

● With regard to the average DOSE per diagnostic examination, measurement and recording were not --- and are not --- required. Today, at some facilities, dose-estimates and recording are routine, but this is not the standard practice. Dose-measurements (as distinct from expected doses, calculated by rules in a handbook) are extremely rare, even though measurement of entrance dose is not at all difficult these days.

● The ratio of measured dose over expected dose in the USA was found in a government survey to range from 0.1 to 4.0 (Wochos 1977 + Wochos 1979, p.134). In 1989, the National Council (USA) on Radiation Protection and Measurements (NCRP) warned that there may be very large disparities between true doses and expected internal-organ doses, based on commonly used "Monte Carlo methods." NCRP cites an Italian report showing that actual breast, thyroid, and testicular xray-doses from certain medical procedures "were higher by factors of 4 to 50 than Monte Carlo calculations would suggest" (NCRP 1989, p.35). The NCRP is described in our Reference List.

● Post-1960 sampling, by measurements, repeatedly shows that diagnostic doses differ by many-fold from facility to facility, and even from room to room, for the same xray procedure on patients of the same size (Wochos 1977 + Wochos 1979, p.134 + Suntharalingham 1982, among others). The reason for large variation in diagnostic doses will be clear to anyone who has examined Box 1 of Chapter 1. Facilities which implement the known ways to reduce doses, can give doses which are 10 to 50 times lower than places which do not.

● Because neither the frequency of diagnostic exams nor the average doses from them are known, we warn against believing any of the published estimates of a population's average per capita dose, (e.g. UNSCEAR 1993, p.302). But for anyone who does believe such estimates, we present the following comparison. Calculating from sales of xray film in the USA and some other data, the 1959 PHS report (PHS 1959, p.3) estimated that per capita annual whole-body dose in 1955 was 135 milli-rems from diagnostic xrays. For the 1980s, the most nearly comparable estimate in the NCRP Report Number 100 (NCRP 1989, p.44) is 115 milli-rads --- which is not a "big reduction" from 135 milli-rads.

● Moreover, diagnostic examinations contribute only part of the dosage from medical radiation. In both past and present, interventional fluoroscopy (e.g., during surgery) has made an unmeasured but large contribution to the xray dosage. And until about 1960, radiotherapy for a great variety of NON-malignant disorders also made an unmeasured but large contribution to the xray dosage (details in Gofman 1995/96).

● How little is known about dosage became clear to us recently, when we attempted to make a responsible estimate of medical radiation-dose, accumulated by the average female breasts between 1920-1960. That endeavor began with many months of combing through the magnificent collections of old medical journals in the University of California San Francisco Medical Library, and ended in Gofman 1995/96 with about 150 final pages of cautious assumptions about the frequency and typical dosage of just a few of the breast-irradiating procedures (excluding cancer therapy).

● Neither we nor anyone else is in a position to quantify the effect, of the 1956-1959 warnings, on the population's average per capita dose from medical radiation. Everyone needs to be careful about hasty assumptions.

3b. Five Forces toward Reduction of Average Per Capita Dose

1. Most therapeutic uses of medical radiation, for treatment of non-malignant conditions, have been abandoned.

2. Most of the "old" diagnostic exams (which are still useful) are now administered at lower dose and with less area receiving the dose. According to Johnson and Goetz (Johnson 1986), between 1964 and 1983, operators learned to use more care in collimating the xray beam, with the goal of reducing the area irradiated down to the size of the film. Additional reduction in area would be achieved if xray beams were collimated to the body-part needing examination, rather than to the edges of the film (Rosenstein 1979; Discussion in Gofman 1985, p.358-359).

3. Fluoroscopy is rarely if ever used now for routine check-ups of asymptomatic patients. Of course, fluoroscopy is still used (with contrast media) in common diagnostic exams like the Barium Swallow, Upper GastroIntestinal Series, Small Bowel Series, Barium Enema, Gallbladder (Cholecystogram), Cystogram-Urethrogram, Fallopian Tubes (Hysterosalpingography), Intravenous Pyelogram (I.V.P.), Retrograde Pyelogram, and all the vessel-studies (cardiac angiography, celiac angiography, cerebral angiography, pulmonary angiography, renal angiography, etc.). Additional information on such exams is available in Gofman 1985.

4. Reduced use of pelvimetry has reduced in-utero and maternal irradiation from that source.

5. Widespread population-screening for Tuberculosis became unnecessary in the USA, and this event eliminated the associated medical irradiation from repeated chest xrays (and sometimes chest fluoroscopy). Chest xrays in the past, especially from mobile units, gave doses about 100 times higher than chest xrays today.

3c. Seven Forces toward Increase of Average Per Capita Dose

On the other hand, other forces have been operating since 1960 in the opposite direction:

1. Increasing Number of Exams per Thousand Population.

Between 1964 and 1980, the estimated annual number of diagnostic xray procedures per thousand population (USA) increased from 580/1,000 to 790/1,000, according to NCRP 1980 (p.15, Table 3.7, citing Mettler 1987). This is an upward change by a factor of 1.36 --- partly due to inclusion in 1980 of estimates for chiropractic and podiatry. NCRP 1989 (p.69) also estimates that average per capita dose from diagnostic medical radiation to adult bone marrow (which provides a fair approximation of whole-body dose) increased by about 38% during the 1964-1980 period.

According to the same report (NCRP 1989, p.11, citing Wolfman 1986), the total sheets of medical xray film sold annually in the USA, per capita, rose from 1.38 (in 1963) to 3.79 (in 1980). This is an upward change by a factor of 2.75. Did the number of exams per capita rise by 1.36 fold, while sheets of medical xray film per capita rose by 2.75-fold? The correct way to reconcile the two change-factors is certainly not clear. It does seem reasonable to conclude, however, that a very considerable increase in xray exposures per capita did occur.

2. Introduction of Computed Tomography (the CT Exam).

Xray doses to patients from CT exams are typically, but not always, about 10 times higher than from "conventional" diagnostic xray examinations (UNSCEAR 1993, p.235/81). And the trend for CT doses has been upward. Why? "The number of slices imaged on each patient has risen as the time required to perform scans and reconstruct images has decreased" (UNSCEAR 1993, p.244/141). Currently under debate is expanding the use of "ultra-fast" CT scans, with "stop-motion" capability, to detect calcium deposits in coronary arteries.

3. Introduction of Digital Radiography.

Progressively more powerful and cheaper computers have resulted in great expansion of digital radiography, which accounted for 15%-30% of xray examinations by 1993 (UNSCEAR 1993 p.242/132). Among other benefits, digital radiography saves the time and money associated with films, chemicals, and archiving. Digital computed radiography has the potential to reduce xray dosage and area irradiated (UNSCEAR 1993 p.242/132; also p.238/100), and to enable image-sharing by wire.

On the other hand, "Persistent anecdotal evidence indicates that some of the dose reduction per image in computed radiography may be offset by a tendency of radiologists to obtain more images per patient than they would have done with conventional film/screen systems ... [Also, compared with conventional radiography] considerable over-exposure can go undetected in a digital system unless exposure is specifically monitored" (UNSCEAR 1993 p.243/134). The 1989 NCRP Report comments that the capability of digital systems, to provide more shades of gray than needed in various diagnostic circumstances, increases the dose by 5 to 10 fold over what it need be (NCRP 1989, p.36).

4. Expansion of Nuclear Medicine.

Nuclear medicine involves placement of radio-nuclides inside the body for diagnostic, interventional, or therapeutic purposes. The estimated number of diagnostic nuclear-medicine exams per thousand population, USA, doubled between 1972 and 1982, and the annual rate was estimated at 26 such exams per thousand population in the 1985-1990 period --- a total of 6.8 million exams per year (UNSCEAR 1993, p.306, p.275).

New uses for nuclear medicine (including pediatric uses) and new techniques in nuclear medicine continue to develop. For example, recently in trial is the placement of radioactive stents into the coronary arteries of patients, immediately after angioplasty, as an attempt to prevent re-stenosis. Also in trial is the use of nuclear medicine to diagnose Breast Cancer.

5. Increased Use of Xrays in NeoNatal Intensive Care.

The diagnostic xray examinations given to infants are generally not new. What is new is the larger number of premature and congenitally challenged infants who are now surviving long enough to receive such xrays.

6. Additional Incentives to Cut Corners.

According to Taylor (1983) and Suleiman (1992), underprocessing of xray films is a frequent cause of higher than necessary radiation doses --- higher by 50% to 300%. Since 1981, the U.S. Food and Drug Administration (FDA) has been monitoring the processing speed of over 2,000 automatic film processors in hospitals, private offices, and mammography facilities. The survey "revealed underprocessing at 33% of observed hospitals in 1987, 7% of mammography facilities in 1988, and 42% of private practices in 1989" (Suleiman 1992, p.25). "... The underprocessing component [of the data] for hospitals increased from 18% in 1984 to 33% in 1987 ... We have been told on several occasions that hospitals frequently eliminated Quality Assurance technicians to reduce costs" (Suleiman 1992, p.27).

Recent pressure on health-care providers --- to reduce referrals to specialists, and to recover some of their own costs in circuitous ways --- also may have the effect of inducing even more primary-care physicians and other non-radiologists to perform their own xray examinations (Krieger 1996). The 1989 NCRP Report comments (p.34): "In many office practices in the United States, xray examinations are performed by persons with little or no formal training in the uses of xrays or xray protection."

Even prior to the newer financial pressures on health-care providers, orthopedists, cardiologists, urologists and other specialists have often performed their own xray work --- including fluoroscopy. Chiropractic offices, too, do their own xray work in general.

7. Expanded Use of Interventional Radiology.

Xrays (including fluoroscopy) are commonly used to guide needles, wires, and catheters, and to localize renal stones in lithotripsy. Xrays are used to guide some common types of biopsies (for example, stereotactic needle biopsies). They are used in many kinds of surgical procedures, involving heart, kidney, liver, gallbladder, pancreas, and vessels (see below).

"Over the past 20 years, there has been a substantial increase in the use of xray fluoroscopy as a visualization tool for a wide range of diagnostic and therapeutic procedures," reports the Public Health Service's Center for Devices and Radiological Health (Shope 1997, p.i).

3d. The Longest Fluoroscopic Procedures

The duration of interventional fluoroscopy can still be long enough to cause serious injury of a patient's skin --- and simultaneously to cause high radiation doses to various internal organs. On September 30, 1994, the U.S. Food and Drug Administration issued a Public Health Advisory entitled, "Avoidance of Serious Xray-Induced Skin Injuries to Patients during Fluoroscopically-Guided Procedures" (FDA 1994). The Advisory provides a listing of the serious skin injuries (which increase in severity with increasing xray dose), as well as the following list of "procedures typically involving extended fluoroscopic time":

percutaneous transluminal angioplasty (coronary and other vessels),
radiofrequency cardiac catheter ablation,
vascular embolization,
stent and filter placement,
thrombolytic and fibrinolytic procedures,
percutaneous transhepatic cholangiography,
endoscopic retrograde cholangiography,
transjugular intrahepatic portosystemic shunt,
percutaneous nephrostomy,
biliary drainage,
urinary/biliary stone removal.

Procedures likely to give a patient more than 100 rads of skin-dose include radiofrequency cardiac catheter ablation, vascular embolization, transjugular intrahepatic portosystemic shunt placement, and percutaneous endovascular reconstruction (Shope 1996, p.1199).

Among procedures requiring extended fluoroscopy-time is percutaneous transluminal cardio-angioplasties --- or PTCA. Estimated average skin-dose from PTCA is about 60 rads per procedure if one stenosis is dilated, and 130 rads if two stenoses are dilated (NCRP 1989, p.31). By 1990, in the USA, the rate of PTCA each year reached an estimated 400,000 procedures (UNSCEAR 1993, p.232/69).

About 25% of dose from fluoroscopy can be pure waste, with no informational value whatsoever, because the xray beam generally falls on rectangular areas, while the image intensifier is a circle fitting inside such rectangles (NCRP 1989, p.36). In 1997, the Public Health Service was urging purchase and use of continuously adjustable, circular collimators (beam adjusters) for fluoroscopes (Shope 1997, p.14).

3e. Data Absent for an Assumption that Fluoroscopic Doses Are Falling

In 1997, the Public Health Service was warning that "Recent developments in the technology of fluoroscopic systems have resulted in ... a variety of special modes of operation and methods of recording fluoroscopic images. Some of these modes may significantly increase the entrance exposure rate to the patient " (Shope 1997, p.6). At the same time, many fluoroscopic systems now on the market offer an optional feature which could reduce radiation dose to patients: The "freeze-frame" or "last-image hold" capability. As noted, the feature is optional (Shope 1997, p.21).

Also not yet in wide use is a timing display and audible alarm on fluoroscopy machines, so that the operator could easily know the cumulative time during which the xray beam has been on, and when the usage-time during a procedure is approaching a pre-set alarm level (Shope 1997, p.20).

Recommended for years, but not yet required, is use of commercially available means to display, to the fluoroscopist, real-time DOSE-rates and cumulative DOSE to the patient's skin during a procedure (Shope 1997, p.23).

In the pre-1960 period and in the post-1960 period right up to today, fluoroscopy has been delivering by far the highest doses in non-therapeutic radiology. Yet even in 1997, there was still no system in place to quantify those doses.

3f. The Issue of Age at Exposure

Age at irradiation is another factor which interferes with efforts to compare pre-1960 and post-1960 population doses from medical radiation. Infants and young adults probably are more vulnerable to radiation carcinogenesis than older adults --- although the difference in magnitude is less than once thought (discussion in Gofman 1995/96, Chapter 3, Part 4).

It would be a big mistake to assume that medical radiation, today, is confined mainly to patients over age 65. The NCRP Report of 1989 (p.19) cites the following estimates from the FDA in 1985, for diagnostic medical xrays performed in hospitals:

Upper GastroIntestinal:	35.7% below age 45; 70% below age 65.
Cholecystography:	38.6% below age 45; 73.2% below age 65.
Barium Enema:	27.3% below age 45; 62% below age 65.
Intravenous Urography:	40.3% below age 45; 71.8% below age 65.
LumboSacral Spine:	50.8% below age 45; 79.4% below age 65.
CT Exams:	34.8% below age 45; 66.6% below age 65.
All Xrays:	47.2% below age 45; 74.2% below age 65.

NCRP 1989 (p.44) also cites a 1985 estimate that over 40% of the dose to active bone marrow, from diagnostic radiology, occurs before age 55.

In addition to problems like aching backs, curvature of the spine, and accidents, cardiovascular problems constitute a major reason for xray procedures. The variety of such problems is vast (Chapter 39, Part 4), and they are not limited to the "senior years." Today, for example, an estimated 32,000 babies per year are born with recognized heart defects (AHA 1995, p.14).

Diagnostic cardiac catheterizations were done BELOW age 45 at a rate in 1994 of about 118,000 per year; the rate was 471,000 per year in patients age 45-64, and 532,000 per year in patients over age 65 (AHA 1996, p.27). For all ages combined, the annual number increased about 3.7-fold between 1979 and 1994. Fluoroscopic xrays are used during these procedures.

Radiation doses are much higher from the PTCA (angioplasty) procedure, of course, than from diagnostic cardiac catheterizations. The PCTA procedure was done BELOW age 45 at a rate in 1994 of about 26,000 per year; the rate was about 182,000 per year in patients age 45-64, and 190,000 per year in patients over age 65 (AHA 1996, p.27). For all ages combined, the annual number increased about 4-fold between 1986 and 1994.

Such data indicate (a) that medical radiation is by no means confined to the over-65 set, and (b) that certain uses are increasing faster than the population.

3g. Profound Uncertainty about the Magnitude of Post-1960 Dose-Reduction

Some of the important differences, between the practices of pre-1960 and post-1960 radiology, have been described in Part 3. But the frequency of medical procedures, and the doses delivered (particularly during fluoroscopy), have not been measured in either era. The ubiquitous post-1960 "pie-charts" of total radiation exposure, which include average annual per capita dose from non-therapeutic uses of medical radiation, are necessarily guesstimates with respect to medical radiation.

Several post-1960 changes in radiologic practice clearly operate in the direction of reducing average annual per capita radiation dose. "We don't DO that anymore!" is a familiar refrain among today's physicians, many of whom happily embrace an assumption that today's doses are negligible from medical radiation. Such colleagues may not have realized that several post-1960 changes clearly operate in the direction of increasing average annual per capita dose from medical radiation, as shown above. The current "pie-chart" estimates for medical radiation are very probably too low by quite a bit.

Is the NET effect, of post-1960 changes, really a "big reduction" in dose? Our opinion is that a net post-1960 reduction has occurred in the average annual per capita dose from medical radiation (excluding cancer therapy), but that the magnitude of decrement is FAR from clear. Among informed people, profound uncertainty about its magnitude is likely to be permanent, given the lack of records.

● Part 4. Ionizing Radiation: A Proven Carcinogen with Some Unique Properties

Along the electromagnetic continuum of photons, from low to progressively higher energy, there are radio waves, microwaves, infra-red heat waves, visible light, ultra-violet light, xrays, and gamma rays. Xrays and gamma rays are ionizing radiations. Ionizing radiations have enough energy not only to "kick" electrons out of their normal atomic orbits, but also to endow these liberated electrons with kinetic energy which sets them into high-speed linear travel. Ultra-violet light, which lacks enough energy to penetrate to the body's internal organs, is not in the same class with medical radiation from xrays and gamma rays. Appendix-A describes alpha and beta ionizing radiations.

4a. The Unique Biological Property of Ionizing Radiation

When an xray or gamma-ray photon interacts with a molecule in living cells, the photon has enough energy not only to "kick" an electron out of its atomic orbit, but also instantly to endow the electron with such energy that it travels like a high-speed bullet through the home-cell and neighboring cells.

The damage from xrays and gamma rays does not come directly from the photon --- it comes from the high-speed high-energy electrons which are set into motion by a photon. When peak voltage across an xray tube is 90,000 electron-volts, the average energy per photon is about 30,000 electron volts. Virtually all 30,000 electron-volts get transferred to a single high-speed electron. The trail of ion pairs and excited molecules, produced by the high-speed high-energy electron along its path, is called the "primary ionization track." (Additional information in Gofman 1990, Chapter 20).

Each high-speed, high-energy electron gradually slows down, as it unloads portions of its biologically unnatural energy onto various biological molecules along its track, at irregular intervals. Such molecules include, of course, water, DNA, proteins --- whatever molecules happen to be in the path when an energy-deposit occurs. Even though each energy-deposit transfers only a portion of the electron's total energy, the single deposits very often have energies which far exceed any energy-transfer which occurs in a natural biochemical reaction. Such energy-deposits are more like grenades and small bombs.

The uniquely violent energy-transfers, caused by ionizing radiation, are simply absent in a cell's natural biochemistry. We know of no one who would dispute this statement.

4b. Repair of Chromosomal and DNA Damage: Complexity Counts

What matters, with respect to gene-based Cancers and other gene-based disorders, is MUTATION: Damage to the genetic molecules which is unrepairable, unrepaired, and enduring. By contrast, there are no mutations from damage which a cell repairs correctly.

There are reasons, in both real-world evidence and logic, to say that ionizing radiation is an especially potent mutagen. It clearly belongs to a much more potent class than the free radicals which attack genomic DNA all the time --- as shown in Appendix C.

The special potency of ionizing radiation is almost certainly due to its unique property of delivering so much extra energy, all at once, in very small regions of a cell. Dr. John F. Ward, Research Professor of Radiology at the University of California, San Diego, reports that the average energy-deposit from a high-speed high-energy electron is thought to be about 60 electron-volts, all within an area having a diameter of only 4 nanometers (Ward 1988, p.103). By comparison, the diameter of the DNA double-helix is 2 nanometers.

Double-Strand Chromosome Breakage and Mutation

As a result of such concentrated deposits of energy, a cell can experience a level of mayhem, in a segment of the DNA double helix, which far exceeds what a single free-radical can inflict upon a comparable segment. For decades, ionizing radiation has been recognized to be extremely efficient at causing double-strand chromosome breaks (e.g., Kucerova 1972, + Brewen 1973, + Sasaki 1975, +

Evans 1978, 1979, + Tonomura 1983, + Lloyd 1992). These violent double-strand ruptures, caused by ionizing radiation, are very different from the orderly double-strand breaks initiated and guided, with normal physiological energy-transfers, by enzymes for a cellular purpose. The deliberate breaks, initiated by the cell, need no repair --- whereas the messy breaks at random locations, caused by ionizing radiation, can be very difficult for cells to repair correctly. The result, of imperfect or absent repair of these double-strand breaks, is mutation.

If the pieces of a broken chromosome are re-united incorrectly, but if the break occurred at an inconsequential site, the mutation will have no biological consequences (by definition). But if the break occurred within a gene which is active in that type of cell, the incorrect reunion can cause the matching protein to be dysfunctional or non-functional. The mutated gene could be one of the many genes directly required to prevent the cell from becoming malignant. Or it could be a gene required to distribute the chromosomes correctly during cell division. Or it could be a gene required for making routine DNA repairs. If it is a repair-gene, the mutation can magnify the consequences of the cell's subsequent exposures to all mutagens (radiation and non-radiation), because of the cell's diminished ability to repair damage correctly.

The biological consequences for a cell, of acquiring a structural chromosomal mutation, depend on the site and nature of the mutation, of course. For example, removal (deletion) of just a single nucleotide can result in garbling of the nearby genetic code. A single larger deletion can result in permanent loss of partial genes or entire genes.

Imperfectly repaired chromosome-breaks cause micro-deletions, macro-deletions, terminal deletions, interstitial deletions, reciprocal translocations, dicentric chromosomes, acentric fragments, rings, inversions, insertions, and other structural re-arrangements of the chromosomes. It is a fact that many cells survive (and reproduce themselves) despite having a consequential chromosomal mutation.

Ionizing Radiation: Very Low Doubling-Dose for Chromosomal Mutations

A "doubling dose" of ionizing radiation is the dose which adds a rate (of some effect) equal to the effect's pre-existing rate. Presently, doubling-dose values for structural chromosomal mutations in human cells are reported in the range of 2 to 20 rads for radiation-induced deletions (Brewen 1973) and dicentrics (Kucerova 1972, + Evans 1979 p.523, + Lloyd 1992 Table 8) and translocations (Lucas 1999 Part 4.1 and Table 3). Some common medical procedures which deliver xray doses in the range of 2 to 20 rads, per procedure, are named in Parts 3d and 7e of this chapter.

Although some of the doubling-dose values mentioned above have large error-bands, the values suffice to indicate that very low doses of radiation readily induce structural chromosomal mutations. For example, the doubling-dose for monocentric translocations induced by gamma-rays is roughly 7.5 rads at age 24, and 15 rads at age 49 --- based on Lucas 1999, Table 3 and Figure 1, and on the observation (from Hsieh 1999) that a rad of gamma rays from cobalt-60 induces 0.00024 translocation per human lymphocyte in vitro, or 24 translocations per 100,000 cells. (Part 7a, below, cites evidence that the number of chromosomal mutations induced per rad is about 2-fold higher from xrays than from cobalt-60.) Induction-rates per 100,000 cells can be viewed in the context that, per gram of human tissue, there are roughly 675 million cells (Gofman 1990, Chapter 20, Part 2).

Laboratory techniques for detecting structural chromosomal mutations are rapidly advancing (for instance, see Lucas 1997, 1999). Observations confirm the expectation that the frequency of chromosomal mutations per 1,000 cells rises with age (for instance, see Tonumura 1983, + Tucker 1994, + Lucas 1999) --- an observation which is consistent with progressive lifelong accumulation of such lesions from exposure to ionizing radiation and nonradiation co-actors.

Genomic Instability: Inducible by Xrays and Other Types of Ionizing Radiation

Among the consequences of mutation, one of the most fearsome is genomic instability. If the original mutation involves (for instance) a gene required for repair of gene-damage or required for proper segregation of chromosomes during cell division, the cells which descend from the originally mutated cell, evolve into cells which are increasingly aberrant, genetically.

Damage, to any of the numerous genes which are part of the cell's system for maintaining genomic stability, can result in genomic instability --- a very frequent characteristic of the most aggressive cancers. Xrays and other classes of ionizing radiation are a proven cause of genomic instability (Appendix D).

The Complex and Unrepairable Injuries

The very nature of ionization tracks means that no part of the genomic DNA is protected, by shape or chemistry, from the violent energy-deposits desribed above. They can inflict their damage ANYWHERE, along any chromosome. Ionizing radiation can induce every known kind of genetic damage, common and rare, simple and complex.

The complex injuries --- including double-strand chromosome breaks --- are not always correctly repaired or repairable by a cell. The probability, that genetic injury will be complex and unrepairable, is greatly elevated by the unique capability of ionizing radiation to deliver the energy "grenades" and "bombs" described above.

4c. Evidence that Ionizing Radiation Is a Proven Human Carcinogen

Of course, many readers are not familiar with the accumulated epidemiologic evidence which shows that ionizing radiation (including the medical xray) is a proven human carcinogen. The purpose of Parts 4, 5, 6, 7 and 8 is to assure such readers that they can accept the assertion as fact. And not just for a few kinds of Cancer, but for virtually every kind of human Cancer.

In 1969, Dr. Tamplin and I warned that, "Contrary to a widespread notion that only Leukemia plus certain rare Cancers are radiation-induced in man, the evidence now points strongly to the induction of all forms of human Cancer plus Leukemia by ionizing radiation" (Gofman 1969-b, p.1). And we predicted that: "All forms of Cancer, in all probability, can be increased by ionizing radiation ..." (Gofman 1969-b, p.1).

From "Controversial Supposition" to "Accepted Wisdom"

Our warning met resistance by most of the radiation community for over a decade. By 1980, the evidence was acknowledged by the BEIR-3 Committee of the National Research Council (USA). BEIR is the acronym for Biological Effects of Ionizing Radiation. BEIR-3's Subcommittee on Somatic Effects had sixteen members, who wrote as follows (BEIR 1980, Section 5, Summary and Conclusions on "Somatic Effects: Cancer"):

● - "The Committee considers cancer induction to be the most important somatic effect of low-dose ionizing radiation ..." And:

● - "Cancers induced by radiation are indistinguishable from those occurring naturally; hence their existence can be inferred only on the basis of a statistical excess above the natural incidence." And:

● - "Cancer may be induced by radiation in nearly all the tissues of the human body."

The Chairman of the entire BEIR-3 Committee and also of the Somatic Effects Subcommittee was Edward P. Radford, M.D., then professor of epidemiology at the Graduate School of Public Health, University of Pittsburgh. Two years later, as a participant in a "roundtable" on medical irradiation for the New York Times, Dr. Radford stated (Radford 1982):

● - "The point that I feel is important is the consistency with which radiation has proved to be carcinogenic in man. It is far and away the most consistent agent that we know of to cause Cancer of any type." Subsequent human evidence continued to fortify the conclusion.

● - In 1988, UNSCEAR (the United Nations Scientific Committee on the Effects of Atomic Radiation) wrote: "It now appears that most (indeed, probably all) organs are vulnerable to radiation-induced Cancer, given the right conditions of exposure" (UNSCEAR 1988, p.460/para.394).

The power of ionizing radiation to increase virtually all forms of human Cancer is simply not in dispute anymore. For the convenience of readers, our Reference List flags --- with a dot in the margin --- some reports and papers which provide extensive bibliographies from which anyone can reconstruct the sequence in which the proof developed. Medical xrays are the source of much of the evidence.

National Cancer Institute + American Cancer Society + World Health Organization

● - In 1990, the National Cancer Institute (USA) issued a 12-page booklet entitled "Everything Doesn't Cause Cancer" (NIH publication 90-2039; NCI 1990 in our Reference List). NCI starts its booklet with a statement on page 1: "Cancer-causing agents also include xrays, sunlight, and certain viruses." At page 5, the booklet lists "radiation and radioactive materials" as proven human carcinogens. And at page 12, the booklet advises: "Don't ask for an xray if your doctor or dentist does not recommend it. If you need an xray, be sure xray shields are used if possible to protect other parts of your body."

● - In 1992, the American Cancer Society issued the following advice under the title, "Guidelines for the Wise Use of Medical Xrays" (ACS 1992): "Fluoroscopy delivers larger doses of xray than that used in standard films. If there is an alternative means of making a diagnosis, fluoroscopy should be avoided."

● - In 1996, the World Health Organization issued its 1996 report entitled "The World Health Report 1996." The section on Cancer states (WHO 1996, p.59): "An estimated 6.6 million people died of Cancer [worldwide] in 1995, and 10 million new cases were diagnosed. It is generally believed that environmental and lifestyle factors, as well as common practices such as diagnostic radiographic procedures, are largely responsible for this disease. In addition, the link between infectious diseases and Cancer is becoming increasingly clear, opening up new possibilities for prevention."

What about the IARC Monographs?

We anticipated that some readers might ask, "Why is ionizing radiation missing from the monographs issued by the International Agency for Research on Cancer?"

We put the question directly to IARC in November 1996. IARC (located in Lyon, France) has been trying to classify various carcinogens for decades, and its monographs are well known in biomedical libraries. The two-paragraph reply from IARC, dated November 25, 1996, is signed by Jerry M. Rice, Ph.D., Chief, Unit of Carcinogen Identification and Evaluation. It says, in its entirety:

"In answer to your note of 14 November, addressed to the IARC Librarian, it is true that radon is the only source of ionizing radiation that has been evaluated to date in the IARC Monographs on the Evaluation of Carcinogenic Risks to Humans." And:

"As is stated on page 1 of every volume of the Monographs, as a 'Note to the Reader,' the fact that an agent has not yet been evaluated in a Monograph does not mean that it is not carcinogenic. It is simply a historical fact that the Monographs Programme began in 1971 with chemicals, and has only in recent years begun to broaden its focus to include biological and physical agents. We expect to direct increasing attention to physical agents, including ionizing radiation, during the next several years. Thank you for your interest in IARC Monographs."

● Part 5. Is the Carcinogenic Power, per Rad of Radiation, the Same at All Dose-Levels?

In 1950, a prospective study was initiated in order to find out what would happen to the health of survivors of the 1945 atomic bombings of Hiroshima-Nagasaki. That study, sponsored jointly by the U.S. and Japanese Governments, is on-going --- for about half of the participants are still alive. Updated results are issued every 4 or 5 years. The study is managed by the Radiation Effects Research Foundation (RERF), with headquarters in Hiroshima and a contact point in Washington DC at the National Academy of Sciences.

Several features of the A-Bomb Life-Span Study make the study uniquely informative: (a) Participants of both genders and all ages at the time of the bombings, (b) Radiation exposure ranging from very low to very high doses, (c) Irradiation of all organs (not just some), and (d) Very long follow-up time. Because of its comprehensive nature, the A-Bomb Study continues to be the principal source of information concerning many aspects of radiation-induced cancer --- including the shape of the dose-response at low and moderate dose-levels.

5a. Primarily a Study of LOW Radiation Doses at Hiroshima-Nagasaki

When the study-group is described by the ages, the distribution of its 91,231 participants is as follows (from Gofman 1990, Table 4-A):
Table 4-A):
- - 18,402 persons age 0-9 years old in 1945.
- - 19,224 persons age 10-19 years old in 1945.
- - 17,691 persons age 20-34 years old in 1945.
- - 20,903 persons age 35-49 years old in 1945.
- - 15,011 persons above age 50 in 1945.

Contrary to common assumption, very few of the participants (about 3%) in the A-Bomb Life Span Study received high doses of ionizing radiation from the bombings (Pierce 1996-a, p.632-633). In general, doses at or below 10 rads (centi-grays) are called "low," and doses at or above 100 rads are called "high" (Appendix-A of this book). The rad and the centi-gray are identical dose-units. We and many others regard the simpler name as preferable. For the past decade, the centi-Sievert (cSv) has been treated as closely equivalent to the rad and centi-gray (cGy), with respect to the A-Bomb Study --- an issue discussed in Part 7 of this chapter.

Of the 91,231 participants listed above, average absorbed internal organ-doses were distributed as follows (Gofman 1990, Table 13-A, Column C):

- - 37,173 received 0.1 rad of bomb-radiation.
- - 28,855 received 1.9 rad of bomb-radiation.
- - 14,943 received 14.6 rads of bomb-radiation.
- - 4,225 received 40.6 rads of bomb-radiation.
- - 3,128 received 74.2 rads of bomb-radiation.
- - 2,907 received 197.0 rads of bomb-radiation.

Anyone claiming that the A-Bomb Study can elucidate response only to HIGH doses of ionizing radiation, just can not be familiar with the study. It is primarily a study of response to LOW doses of bomb-radiation.

5b. Shape of the Dose-Response in the A-Bomb Study, and in High-Dose Data

What does the A-Bomb Life-Span Study reveal about the shape of the dose-response for solid Cancers (in other words, excluding Leukemia)? Before the answer, definitions are needed for the relevant terms.

Terms: A positive linear dose-response means, of course, that response is directly proportional to dose, because the carcinogenic power of each incremental dose-unit (e.g., rad) is the same throughout the entire dose-range, from zero dose to very high doses. Positive linear dose-responses are depicted in Figures 1-A and 1-B of Chapter 1; discussion of their shape occurs in Chapter 5, Part 5d. By contrast, a supra-linear dose-response has curvature such that the curve lies ABOVE a straight line drawn between any two points along the curve. When such a connecting line has an upward slope, each rad at the lower dose-point is more carcinogenic on the average than each rad at the higher dose-point. The linear-quadratic dose-response (if the quadratic term is positive rather than negative) means that each rad is less carcinogenic at low total doses than at high total doses.

- 1990. In Gofman 1990, we presented a step-by-step analysis of the A-Bomb Life-Span Study data, 1950-1982, which shows that the dose-response in those data for all types of solid Cancers, combined, has a supra-linear shape at doses above about 5 rads (Gofman 1990, esp. Chapter 14). Analysts at RERF also reported supra-linearity, but they concluded that the supra-linear dose-response was not statistically superior to the linear dose-response in fitting the observations (Shimizu 1987, pp.28-30, + Shimizu 1988, pp.50-51, p.53, Table 19). In reality, the dosimetry in the A-Bomb Survivor Study has been and remains quite uncertain. Therefore, it is impossible for anyone to know whether the supra-linearity therein is based on biology or on mistaken dose-estimates.

- - 1990. The BEIR Committee (Committee on the Biological Effects of Radiation, of the National Research Council) reported its analysis of the 1950-1985 data from the A-Bomb Life-Span Study: "The dose-dependent excess of mortality from all cancer other than leukemia, shows no departure from linearity in the range below 4 sievert [approximately 400 rads], whereas the mortality data for leukemia are compatible with a linear-quadratic dose response relationship" (BEIR 1990, p.5). For now, we can ignore BEIR's questionable distinction about Leukemia, because Leukemia has accounted for a very small fraction of U.S. cancer mortality. Solid Cancers have accounted for the overwhelming share of cancer deaths. In 1940, Leukemia accounted for 3.2% of the cancer mortality rate (Grove 1968, p.700 & p.676). In 1998, the fraction of cancer deaths (USA) due to Leukemia is estimated at 3.8% (Landis 1998, p.13, Table 4).

● - 1994. The UNSCEAR Committee (United Nations Scientific Committee on the Effects of Atomic Radiation) reached the same conclusion as did the BEIR Committee, from the A-Bomb Life-Span Study: "The life span study data for solid tumours from 1950 to 1987 are consistent with linearity between 0.2 Sv [approximately 20 rads] and 4 Sv [approximately 400 rads] ..." (UNSCEAR 1994, p.89/402). The report presents a graph which depicts linearity in the data down to zero dose (UNSCEAR 1994, p.157). However, in the range between 0 dose and 20 rads, the authors say the findings lack statistical significance (UNSCEAR 1994, p.89/406).

● - 1996. RERF analysts, Donald Pierce and co-workers, report their findings on solid Cancers from a longer follow-up period (1950-1990) in the A-Bomb Life-Span Study. They state (Pierce 1996-b, p.9): "The dose response is quite linear up to about 3 Sv [up to about 300 rads] ... These data do not suggest the existence of a threshold below which there is no excess risk." The RERF analysts also report (Pierce 1996-b, pp.9-10) that the 1950-1990 data, "taken at face value," show supra-linearity and indicate that the cancer rate per cSv (rad) grows progressively more severe as dose DECREASES in the region between about 35 rads and zero dose --- but they reject the finding as statistically inferior to the linear dose-response.

The RERF analysts, working with later data than the analysts in UNSCEAR 1994, find statistically significant excess Cancer even at doses as low as about 5 cSv --- about 5 rads (Pierce 1996-b, p.10). We have not yet independently checked the 5-rad finding from the 1950-1990 raw data.

Findings from Higher Doses: A Path to Underestimating Risk at Low Doses

Although the data in the A-Bomb Study are sparse at high doses (Part 5a), other types of data have led analysts to general agreement that dose-response for radiation carcinogenesis and mutagenesis is curved in a supra-linear fashion when acute high doses are included (NCRP 1980, p.17, p.160, + BEIR 1990, pp.141-142 for carcinogenesis, + UNSCEAR 1993, p.9/42 for carcinogenesis). This means that risk per rad, from xrays received at low doses, will be underestimated whenever such analysis is based on observing medical patients who received acute high doses.

5c. Peril for the A-Bomb Database: Recent Actions

It worries us that recent actions have needlessly placed the credibility of future results from the Atomic-Bomb Survivor Study in great peril (Gofman 1988 + 1990 + 1992 + 1995/96).

Since 1986, both the number of participants in the study and their dose-estimates have been altered several times, after the results of decades of follow-up were already known. Epidemiologists worldwide recognize that retroactive changes in dose-assignments and shuffling of dose-cohorts create the opportunity for bias to enter any study. With enough retroactive changes, the "findings" can become whatever the fiddlers desire. Therefore, in order to prevent suspicion, well-established rules, which create barriers against entry of bias, are normal practice in prospective studies. Unfortunately, during the past decade, several such barriers have been demolished in the A-Bomb Study.

The impending crisis in the A-Bomb Study developed because of over-estimated doses delivered by neutrons --- especially in Hiroshima. Indeed, we discerned that there must have been errors involving neutrons in the pre-1986 dosimetry (Gofman 1981, p.246), and we applaud correction of the neutron-errors by what is called the study's "DS86" dosimetry. What worries us is the way in which use of "the new dosimetry" has unnecessarily become the occasion for removing some significant barriers against potential bias. For instance, many former participants have been discarded and thousands of new ones added from a "reserve." The former dose-estimates and dose-cohorts are no longer any part of RERF analyses. In short, current practice deprives the A-Bomb Study of its continuity, its anchor, its permanent architecture --- and thus, of its above-suspicion status.

A potential remedy for this problem would be to use "DS86" dosimetry as part of "constant-cohort, dual-dosimetry" analyses, in which the former dose-estimates (1965 vintage) and former dose-cohorts provide the continuity in future follow-up studies, and comparable results are calculated also from "the new dosimetry." This practice would simultaneously eliminate suspicions of bias AND deliver the benefits of improved dosimetry. In the computer age, the extra set of dose-estimates (DS86) is easily handled. Indeed, with the excellent cooperation of Dr. Donald Pierce at RERF, we acquired the data we needed to demonstrate "constant-cohort, dual-dosimetry" analysis of the 1950-1982 cancer-data (in Gofman 1990). We need not claim that "constant-cohort, dual-dosimetry" is the ONLY possible solution to the A-Bomb Study's impending credibility problem, but we do claim that an approach OTHER than current practice is urgently needed, in order to keep this unique and painfully acquired Japanese database forever above suspicion.

● Part 6. Absence of Any Threshold Dose: "Risk" versus Rate

Because ionizing radiation is a proven cause of human Cancer, one might assume that virtually no one would deny that the use of medical radiation has caused and is causing radiation-induced Cancers. But there are some in medicine who try to limit such an admission to "high-dose" medical radiation. Such people continue to hope for a threshold-dose, below which they speculate that REPAIR of radiation-injury may prevent any radiation-induced Cancer. They claim that no one can know for sure about very low doses. They are mistaken. IT IS POSSIBLE TO KNOW.

6a. A Five-Point Summary of the Evidence that No Threshold Exists

The nature of the evidence, that no threshold-dose exists for radiation carcinogenesis, can be summarized by five points (Gofman 1990, Chapter 18):

● Point ONE: The radiation dose from xrays, gamma rays, and beta particles is delivered by high-speed electrons, traveling through human cells and creating primary ionization tracks. Whenever there is ANY radiation dose, it means some cells and cell-nuclei are being traversed by electron-tracks. There are roughly 675 million typical cells in 1 cubic centimeter.

● Point TWO: Every track --- without any help from another track --- has a chance of inflicting chromosomal or gene-damage, if the track traverses a cell-nucleus.

● Point THREE: There are no fractional electrons. This means that the passage of one primary ionization track is the lowest conceivable dose and dose-rate which a cell-nucleus can experience from ionizing radiation.

● Point FOUR: There is solid epidemiologic evidence that extra human Cancer does occur from radiation exposures which deliver just one or a few tracks per cell-nucleus, on the average. Such evidence shows that the cell's repair-system is fallible even when it is confronted only by a minimal challenge.

● Point FIVE: The combination, of real-world evidence from epidemiology and from track-analysis, establishes that there is NO dose or dose-rate low enough to guarantee correct repair of every carcinogenic injury inflicted by ionizing radiation. Some injuries are just unrepaired, unrepairable, or misrepaired.

6b. Three Remarkably Similar Reports on the Safe-Dose Fallacy

The threshold hypothesis, with respect to radiation carcinogenesis, has been invalidated in three major reports: Gofman 1990, UNSCEAR 1993, and NRPB 1995. (UNSCEAR is the United Nations Scientific Committee on the Effects of Atomic Radiation. NRPB is Britain's National Radiological Protection Board.)

The key to each report is the insight that the appropriate way to define the lowest possible dose and dose-rate of ionizing radiation is NOT in fractions of a rad or centi-gray. The relevant definition occurs in tracks per cell-nucleus.

● - Gofman 1990, p.19-1: "Because the minimal event in dose-delivery of ionizing radiation is a single track, we can define the least possible disturbance to a single cell-nucleus: It is the traversal of the nucleus by just one primary ionization track." Traversal occurs in a tiny fraction of a second. To test the threshold hypothesis, no one needs impossible-to-obtain epidemiologic studies, at tissue-doses like 10 milli-rads or 10 micro-rads --- because minimal challenge to a cell's repair system occurs at much higher tissue-doses. From Gofman 1990, Table 20-M:

Radiation	Average number of tracks/nucleus		Tissue-dose in rads (centi-grays)
30 KeV medical xrays	1	track	0.75 rad (750 milli-rads)
	10	tracks	7.48 rad
1608 KeV gamma-rays, as at Hiroshima-Nagasaki	1	track	0.185 rad (185 milli-rads)
	10	tracks	1.85 rad

● - UNSCEAR 1993, p.680/321: "Photons deposit energy in cells in the form of tracks, comprising ionizations and excitations from energetic electrons, and the smallest insult each cell can receive is the energy deposited from one electron entering or being set in motion within a cell."

● - NRPB 1995, p.58/27: "It may be argued ... that a single radiation track (the lowest dose and dose-rate possible) traversing the nucleus of an appropriate target cell, has a finite probability, albeit low, of generating the specific damage that will result in tumour-initiating mutation."

For the convenience of readers, Appendix B provides extensive excerpts from all three reports. Here, we will present just the conclusions:

Gofman 1990, p.18-2: "Human epidemiological evidence shows that repair FAILS to prevent radiation-induced Cancer, even at doses where the repair-system has to deal with only one or a few tracks at a time, and even at dose-rates which allow ample time for repair before arrival of additional tracks ... Such evidence is proof, by any reasonable standard, that there is no dose or dose-rate which is safe ..."

UNSCEAR 1993, p.636/84: "It is highly unlikely that a dose threshold exists for the initial molecular damage to DNA, because a single track from any ionizing radiation has a finite probability of producing a sizable cluster of atomic damage directly in, or near, the DNA. Only if the resulting molecular damage, plus any associated damage from the same track, were always repaired with total efficiency could there be any possibility of a dose threshold for consequent cellular effects." And (p.680-681/323): "Biological effects are believed to arise predominantly from residual DNA changes that originate from radiation damage to chromosomal DNA. It is the repair response of the cell that determines its fate. The majority of damage is repaired, but it is the remaining unrepaired or misrepaired damage that is then considered responsible for cell killing, chromosomal aberrations, mutations, transformations, and cancerous changes."

NRPB 1995, p.60/36: "For double-strand DNA damage, there is good reason to believe that repair has an error-prone mutagenic component irrespective of damage-abundance and, by implication, will, even at very low doses, contribute to tumour risk." And (p.61/38): "It may be concluded ... that existing data from both in vitro and in vivo [radiation] studies support a linear rather than a threshold-type response for neoplasia-initiating gene mutations." And (p.68/80): "In consideration of a broad body of relevant cellular and molecular data, it is concluded that the weight of the evidence, in respect of the induction of the majority of common human tumours, falls decisively in favor of the thesis that, at low doses and low dose rates, tumorigenic risk rises as a simple function of dose without a low dose interval within which risk may be discounted."

6c. Dr. Dale L. Preston: An Additional View on the Safe-Dose Fallacy

The bottom line is that exposure to ionizing radiation creates violent random events at the cellular level. Repair of the resulting genetic damage is sometimes absent or imperfect. The failure of correct repair is not due to saturation of a cell's repair system (Gofman 1990, Chapter 18, Parts 2, 3, 4). Failure is due to the exotic nature of the lesions. In studies of radiation-induced human Cancer and of radiation-induced chromosomal mutations in human cells, generally the dose-response is linear. Linearity almost certainly means that the unrepairable fraction of genetic lesions is constant, even when there is an average of just one track per cell-nucleus.

The newest evidence on linearity has led Dr. Dale L. Preston, also, to speak out on the threshold-issue. Dr. Preston is a major analyst at RERF and was also Scientific Advisor to the BEIR-5 Committee of the National Research Council. Referring to the 1950-1990 evidence from Hiroshima-Nagasaki (Part 5b, above), Preston states:

"For solid Cancers, there is simply no way of looking at the RERF data which suggests the existence of a threshold, or even a lower risk per unit dose in the low-dose range. The lack of such evidence is not due to a relative paucity of survivors in the low-dose range, as more than 85% of the survivors have dose estimates <0.2 Sv. In fact, taken at face value, these data are quite inconsistent with the existence of a threshold, or the adequacy of a linear-quadratic dose response for solid cancers" (Preston 1997).

6d. An Important Distinction: Risk versus Rate

The proof exists, by any reasonable standard of biomedical proof, that there is no threshold-dose for radiation carcinogenesis.

Thus, there is no dose-level or dose-rate for a population which is harmless. A radiation dose which gives an individual "just a risk" of radiation-induced Cancer --- say, 1 chance in 1,000 --- is the same dose which gives a RATE of 1,000 radiation-induced Cancers among a million people irradiated at a comparable age. In such a case, all one million irradiated people are "at risk," and later, one thousand of them develop radiation-induced Cancers. Radiation-induced Cancer is "a maybe" for each individual but a certainty for the group.

● Part 7. Xrays: More Carcinogenic than Gamma Rays at Equal Doses

For about two decades, experimental evidence has been accumulating that xrays inflict more chromosomal mutations per 1,000 cells than do gamma rays, at equal tissue-doses. It follows (from more mutations) that xrays are more carcinogenic than gamma rays, at equal tissue-doses. We warned that medical xrays are about twice as carcinogenic as gamma rays, at equal rad-dose, in Gofman 1990 (p.13-4, p.20-5, p.25-15).

7a. Other Publications: Xrays Two-Fold to Four-Fold More Injurious

The BEIR-5 Committee, of the National Research Council, acknowledged a factor of two as follows (BEIR 1990, p.218):

"Most human exposures to low-LET ionizing radiation are to xrays, while the A-bomb survivors survived low-LET radiation in the form of high energy gamma rays. These are reported to be only about half as effective [injurious per rad] as ortho-voltage x-rays (ICRU 1986). While that is not a conclusion of this Committee, which did not consider the question in detail, it could be argued that since the risk estimates [for Cancer] that are presented in this report are derived chiefly (or exclusively) from the Japanese experience, they should be doubled as they may be applied to medical, industrial, or other xray exposures." Note: LET and ortho-voltage xrays are defined in Part 7b.

In 1995, Tore Straume's analysis of evidence indicated that xrays may be FOUR TIMES as harmful as Hiroshima-Nagasaki gamma rays, at equal rad-doses (Straume 1995). Dr. Straume, who was then at the Livermore National Laboratory, used experimental evidence produced at the Harwell Lab of Britain's National Radiological Protection Board: Prosser 1983, + Lloyd 1986, + Purrott 1977. The evidence consists of dose-responses for dicentric chromosomes induced by ionizing radiation in human lymphocytes (in vitro), evaluated at the first post-irradiation cell-division (Straume 1995, Figure 2). Dicentric chromosomes (having two centromeres) result from misrepaired double-strand chromosome breakage in two separate chromosomes. The frequency of post-irradiation dicentrics has been one standard measure of radiation mutagenesis for decades. Straume's analysis showed the following results, relative to the atom-bomb gamma rays and at equal rad-dose (Straume 1995, p.955):

   ● - Cobalt-60 gammas rays are about 2-fold more injurious than A-bomb gamma rays.
   ● - 250 kVp xrays (called "orthovoltage xrays," and having an average energy of 83 KeV) are about 4-fold more injurious than A-bomb gamma rays.
   ● - Tritium beta rays (average energy of 5.7 KeV) are about 5-fold more injurious than A-bomb gamma rays.
   ● - Using data from almost entirely different studies, Joe Lucas and co-workers find that xrays produce about 2-fold more dicentrics per dose-unit than cobalt-60 gamma rays (Lucas 1995, Figure 3).

   Straume comments (Straume 1995, p.955): "It is well known that biological effectiveness [damage per dose-unit] decreases as radiation energy increases, i.e., becomes less densely ionizing (Dobson 1976, + Bond 1978, + NCRP 1980, + Borek 1983, + ICRU 1986, + Brenner 1989, + NCRP 1990)."

   And, Straume p.955: "The dependence of human Cancer dose-response relationships on radiation energy has not been established and therefore may or may not be equivalent to that for the model endpoint (dicentrics) used here. It is, however, established that the energy dependence of dicentrics compares well with those of a broad range of other biological endpoints (NCRP 1990), including that for malignant cell transformation (Borek 1983), and is a convenient endpoint that has been well characterized and widely used for similar purposes (e.g., see ICRU 1986)."

   Quite explicitly, Straume warns that health consequences from xrays and tritium (radioactive hydrogen) may be larger, by a factor of 4 to 5, than the harm from an equal dose of Hiroshima-Nagasaki gamma rays (Straume 1995, p.956).

7b. An Independent Check on the Four-Fold Estimate

   We wondered: Is a four-fold disparity in mutations reasonable? Credible? We were able to make an independent check on the reasonableness, by consulting our own work in Gofman 1990.

   As noted by Straume (Part 7a), there is a large body of evidence on ionizing radiation showing that the biological damage per rad rises with the DENSITY of the energy-deposits left by the high-speed particles along their tracks. LET (Linear Energy Transfer) is a common measure of such density, for LET is defined as the average amount of energy lost per unit of track-length. For example, LET can be measured in KeV per micrometer. Xrays, gamma rays, and beta particles are low-LET radiations because the distance between energy-deposits is large, on the scale of a typical cell-nucleus, relative to such distance from alpha-particle radiation (a high-LET radiation).

   The relative intensity of low-LET radiations, in interacting with biological soft-tissues, is reflected in the number of cell-nuclei which must be traversed by electrons in order to deposit one rad of energy. The more cell-nuclei required, the less intense is the interaction.

   In order to test the threshold hypothesis in Gofman 1990, we calculated the number of traversals of cell-nuclei, by the electrons set in motion by photons of various energies, per rad of tissue-dose delivered. Column C, below, shows the values from Gofman 1990, and Column E shows how many-fold MORE traversals are required for the A-bomb gammas to deliver 1 rad of tissue-dose:

Col.A Type of Photon	Col.B Table in Gofman 1990	Col.C Number of Cell-Nuclei Traversals per Rad (cGy)	Col.D Hiro-Naga Divided by Other	Col.E Factor of Disparity
30 KeV xrays, mean energy.	Table 20-Eye	0.903 billion	3.65/0.903	4.042
83 KeV xrays, mean energy.	Table 20-O	1.54 billion	3.65/1.54	2.370
100 KeV xrays, mean energy.	Table 20-O	1.64 billion	3.65/1.64	2.226
596 KeV gammas (radium-226).	Table 20-Eye	1.98 billion	3.65/1.98	1.843
662 KeV gammas (cesium-137).	Table 20-Eye	2.13 billion	3.65/2.13	1.714
1608 KeV gammas (Hiro-Naga).	Table 20-Eye	3.65 billion	3.65/3.65	1.000

From Column E on the previous page, we note two results in particular:

Medical xrays, of 30 KeV average energy (from a peak kilovoltage of approximately 90), interact with human soft-tissue cells about 4-fold more intensely than do the Hiroshima-Nagasaki gamma rays. Ortho-voltage xrays, of 83 KeV mean energy (from a peak kilovoltage of approximately 250), interact with human cells about 2.4-fold more intensely than A-bomb gamma rays. Therefore, on the basis of both track-analysis (Part 7b) and observed chromosomal mutations (Part 7a), it is realistic to accept the warning that xrays are much more injurious, per rad, than the Hiroshima-Nagasaki gamma rays.

7c. What Do the Epidemiologic Data Reveal?

Readers may wonder why the relative carcinogenic potency per rad, of medical xrays versus gamma rays from Hiroshima-Nagasaki, has not been established directly from epidemiologic studies. The principal problem has been described in Parts 2 and 3 of this chapter: The unreliability or complete absence of records concerning which patients received which doses from medical radiation.

This is not a small obstacle. For example, a SIX-FOLD disparity exists in the rate of radiation-induced Breast Cancers, per rad of medical xrays, among various studies of female tuberculosis patients in North America who received serial fluoroscopies. The most likely explanation is assignment of retroactive dose-estimates to the wrong patients. After years of study, the BEIR Committee admitted defeat in trying to reconcile the different studies (BEIR 1990, p.255). Then WHICH value of xray potency, estimated from such studies, would make a reliable comparison with the results on radiation-induced Breast Cancer from Hiroshima and Nagasaki gamma rays? The uncertainty is very large.

This is just one illustration of why analysts consider experimental data, with well-measured doses, to be more reliable than epidemiology when they try to estimate the relative carcinogenic potency of xrays and gamma rays of various energies.

7d. The A-Bomb Survivors: Bomb-Rads Converted to Medical Rads

There is a large body of experimental evidence (cited and independently tested in Parts 7a and 7b) which indicates that 0.25 to 0.5 rad of tissue-dose, received from medical xrays, has as much mutagenic (therefore carcinogenic) impact as 1 rad received from the Hiroshima-Nagasaki bombs. If we average the two fractions, we have 0.375. Thus, a reasonable conversion-factor, from bomb-dose to xray dose, is 0.375 medical-rad per bomb-rad.

Using the conversion factor of 0.375, we can express the list of bomb-doses from Part 5a in equivalent xray doses, as follows (we use * to denote multiplication):

- − 37,173 survivors: (0.1 * 0.375) = 0.04 rad xray-equivalent (40 milli-rads).
- − 28,855 survivors: (1.9 * 0.375) = 0.71 rad xray-equivalent.
- − 14,943 survivors: (14.6 * 0.375) = 5.48 rads xray-equivalent.
- − 4,225 survivors: (40.6 * 0.375) = 15.23 rads xray-equivalent.
- − 3,128 survivors: (74.2 * 0.375) = 27.83 rads xray equivalent.
- − 2,907 survivors: (197.0 * 0.375) = 73.88 rads xray equivalent.

7e. Some Common Current Medical Procedures, with Approximate Dose-Levels

Irradiation from the Hiroshima-Nagasaki bombings exposed the entire body of the survivors. By contrast, xray exposure from a diagnostic or interventional medical procedure today may irradiate most of the head, or a quarter to three quarters of the torso, but almost never the entire body. However, during a lifespan of various diagnostic and interventional medical procedures, a person today can readily accumulate doses to specific organs which far exceed the comparable organ-doses received by most of the A-bomb survivors. Some common procedures, with approximate dose-levels:

● CT Scan: A CT scan (torso) typically delivers an entrance dose of 6 rads from xrays (NCRP 1989, p.33). Within the scanned "slices," the ratio of near-surface organ-dose over body-center organ-dose is approximately 6 to 1 (Gofman 1985, pp.248-249). This ratio would make the range of internal tissue-doses per examination about 1 to 5 rads.

● Upper GI: A male adult receives about 1.1 rad of absorbed xray dose to his stomach during a well-conducted Upper Gastro-Intestinal examination (FDA 1992, p.27). During such an exam, many addtional organs are also irradiated, including thyroid gland, esophagus, breasts, lung, active bone marrow, large intestine, liver, kidney, and pancreas.

● Interventional Fluoroscopy: Any procedure involving "extended fluoroscopy time" (Part 3d) may deliver xray doses to some internal tissues of 15, 25, even 50 medical rads or more.

● Thallium-201 Injection (a common heart-exam): Dose from thallium-201 comes from a complex mixture of gamma rays and xrays, so the following doses are not directly comparable to the others above. From administration of 2 milli-Curies to a 70-kilogram adult, approximate dose to the kidneys is 2.5 rads, thyroid 1.3 rad, liver 1.2 rad, heart wall 1.1 rad, testes 1.1 rad, ovaries 0.99 rad, stomach wall 0.84 rad, upper large intestine wall 0.54 rad, lower large intestine wall 0.46 rad, whole-body dose 0.45 rad ("Technical Product Data" from a major supplier).

● **Part 8. Variable Latency-Periods for Radiation-Induced Cancer**

If an exposure to ionizing radiation causes a genetic mutation which is carcinogenic in a cell (certainly NOT all mutations are carcinogenic), then the elapsed time between mutation and manifestation of the radiation-induced Cancer is formally called a "latency period." After its production, the carcinogenic mutation is always present, like an inventory waiting for delivery. Therefore, we like to refer to latency periods as "delivery times." They are extremely variable in duration.

8a. The Variable Delivery Times in Mixed-Age Populations

Participants of all ages in the A-Bomb Study received the bomb-irradiation in August 1945 --- and the radiation-induced Cancers were still being delivered 45 years later in a dose-dependent fashion (Part 5b). Indeed, 22% of ALL the fatal bomb-induced solid Cancers, in the A-Bomb Study, occurred during the 1986-1990 period according to the RERF analysts (Pierce 1996-b, p.1, p.5, and Table 3).

It is highly reasonable to expect deliveries of additional radiation-induced fatal Cancers during the post-1990 follow-up years --- because 56% of the initial participants were still alive at the beginning of 1991 (Pierce 1996-b, p.6, Table 4). The database currently in use by RERF analysts consists of 86,572 initial participants, whose age-distribution is similar to the age-distribution of an earlier database (91,231 initial participants) shown in Part 5a of this chapter.

8b. Duration of the Carcinogenic Impact, for People of Same Age

A very important insight has emerged from continuous study, since 1950, of the various age-groups within the A-Bomb Survivors:

When a group of people of the same age is irradiated at the same time, the excess (radiation-induced) Cancers do not occur at the same time. Each irradiated age-group "delivers" or manifests its extra Cancers gradually, over many years (see Gofman 1990, Table 17-B, for example). This is not in dispute. In other words, the duration of the delivery time varies from one irradiated individual to another. How short is the shortest delivery time? We discuss the question of "a minimum latency period" in Chapter 5, Part 4.

Once deliveries begin in an irradiated group, how long do the deliveries continue? For most age-groups, probably "forever." As the follow-up study of the A-Bomb Survivors grows ever longer, the evidence grows ever stronger from those who were relatively young in 1945, that the carcinogenic impact of exposure to ionizing radiation probably endures (though not necessarily at a constant level) for the subsequent lifespan. Because about half of the participants in the A-Bomb Study are still alive, no one can say this with certainty, however. The RERF analysts assume a lifetime impact, and they make their lifetime risk-estimates accordingly (Pierce 1996-b, pp.12-14, p.21). So did the BEIR-5

Committee (BEIR 1990). We used lifetime assumptions when making independent risk-estimates in Gofman 1981 and Gofman 1990.

8c. Persistence of Radiation-Induced Mutations

The observation, that the carcinogenic effect of exposure to radiation endures for most (probably all) of the remaining lifespan, is consistent with another observation in the A-Bomb Study: In 1992 and 1993, Lucas and Kodama reported that, among the living A-bomb survivors, a positive dose-response between bomb-dose and number of chromosomal mutations was still apparent (Lucas 1992 Figure 6, + Kodama 1993). Radiation-induced genetic mutations are the CAUSE of the radiation-induced Cancers which are gradually delivered as clinically manifest malignancies. Thus, the clinical evidence and the cell-studies are consistent: In irradiated groups, the carcinogenic impact of exposure to ionizing radiation endures for most (probably all) of the group's remaining lifespan.

The persistence of radiation-induced mutations means, of course, that a person accumulates more and more of them with each additional exposure to ionizing radiation.

● Part 9. A Very Slow Arrival at Conceiving and Testing Hypothesis-1

We have been very slow in arriving at Hypothesis-1.

9a. A Hunch in 1971 ... and a Missed Insight

Back in 1971, it occurred to us that medical irradiation had to account for some significant part of the cancer problem (Gofman + Tamplin 1971, p.266):

"Medical uses of xrays presently are a major source of population exposure and are undoubtedly responsible for a significant part of our currently experienced cancer mortality rate. Morgan's suggestions for feasible reduction in medical xray exposure, without loss of medical diagnostic information, deserve immediate action (Morgan 1971)." The reference is to one of many articles on this topic by Dr. Karl Z. Morgan, a man of immense integrity who is widely recognized as the "father of the health physics profession." Morgan's remarkable memoir is now available (Morgan and Peterson 1999).

Also in 1971, we pointed out that xray induced Cancers are routinely treated as part of the "spontaneous" or "background" cancer-rate rather than as a radiation-induced rate (Gofman + Tamplin 1971, p.244). We noted then that this treatment can lead to underestimates of radiation's role in cancer causation. Much later we realized what a huge epidemiological pitfall such treatment might represent (Gofman 1995/96, Chapter 41, Part 2):

"... exposure of a stable population to a constant level of ionizing radiation would --- at equilibrium --- cause no INCREASE per rad in the apparent 'spontaneous' rate, even if radiation were causing 100 percent of the [Cancer] problem."

9b. A Neglected Observation in Gofman 1981

In a 1976 paper, Frigerio and Stowe had claimed that they found a "consistent and continuous" INVERSE relationship between levels of natural background radiation and cancer mortality-rates in the 50 United States (Frigerio 1976, p.385). In Gofman 1981 (p.568), we grouped the states into 3 classes (high, medium, or low background radiation dose) and demonstrated the serious fault in their conclusion.

Then, keeping the same three groups of states, we looked at other data provided in the same paper by Frigerio and Stowe. We did a mini-analysis of physician-density versus cancer mortality rates, and we showed for the three groups that: "The values of physicians per 1,000 persons parallel the cancer death rates almost perfectly, and the background radiation data definitely do not" (Gofman 1981, p.569).

The higher the density of physicians, the higher the cancer death rate. We --- and everyone else, too --- failed to explore that "smoking gun" on page 569 of Gofman 1981.

9c. An Estimate in Gofman + O'Connor 1985

In our 1985 book, Gofman and O'Connor provided estimates of the personal cancer-risk associated with about 40 types of common diagnostic xray exams (Gofman 1985). The book was praised in the New England Journal of Medicine (Greenfield 1986) and attacked in the Journal of the American Medical Association (Adler 1986).

In Chapter 17 of that book, we tried to estimate how many FUTURE cases of Cancer could be prevented in the USA, if the average dose received from diagnostic xray exams were reduced to 33%. A reduction of this magnitude had been demonstrated, by Dr. Kenneth Taylor and colleagues in Ontario, Canada (Taylor 1979 + 1983), to be readily achieved in actual radiology facilities --- without any loss of diagnostic quality and without purchases of expensive equipment.

To prepare our estimate, of course we had to begin with the unreliable type of estimates (on frequencies of common diagnostic exams, and average doses therefrom) which we describe in Part 3a of this chapter. Such estimates were an unreliable foundation for our calculations. Moreover, they excluded all angiographies, CT scans, and (by definition) all interventional radiology. Nonetheless, we felt that even an underestimate would be preferable to no estimate at all.

Our estimate was that about 50,000 cases of future Cancer per year could be prevented in the USA by cutting average dose per exam, from diagnostic radiology, to 33% of the supposed average prevailing dose. Subsequent work (Gofman 1995/96 and this book) indicates that 50,000 cases was a vast underestimate. But it was the best estimate that we could provide in 1985, from using the customary but unreliable input. The result: Our own estimate in 1985 did not provoke us into considering that medical irradiation might be the PRINCIPAL cause of Cancer (USA) in the Twentieth Century.

9d. An Estimate in Gofman 1990

In Gofman 1990, we made two comments about the role of ionizing radiation (all sources) in the total cancer problem. We were still far short of conceiving Hypothesis-1.

First, we provided a list of some 13 medical uses of xrays and radium, plus some non-medical sources of exposure (use of radium-dials, fluoroscopic shoe-fitters, and tobacco whose smoke contains decay-products from uranium), and we said: "One needs to wonder seriously how much of the current cancer-rate is due to past exposure to ionizing radiation from such practices. It could be a meaningful part of the so-called 'spontaneous' rate" (Gofman 1990, p.24-20).

And in the next chapter (p.25-15), we combined (a) BEIR 1990's estimates of annual doses from radon, other natural radiation, medical xrays, and "all other" sources, with (b) our own estimates of Cancers per unit of dose, and thus we arrived at (c) the "ball-park estimate" that about 25% of cancer mortality is radiation-induced --- excluding any radiation-induced inherited predisposition.

9e. An Estimate at the AAAS Symposium of 1994

An event in February 1994 was crucial in the arrival at Hypothesis-1. At the invitation of Nancy Evans and Breast Cancer Action, I was invited to be a panelist for the symposium on Breast Cancer at the national meeting of the American Association for the Advancement of Science. Other panelists were Dr. Graham Colditz, Dr. Devra Lee Davis, Dr. Samuel Epstein, and Dr. Elihu Richter.

My assigned topic was "Ionizing Radiation and Breast Cancer." Radiation-induced Breast Cancer happens to be prominent in the literature on radiation carcinogenesis (see Gofman 1981, BEIR 1990, for instance). Indeed, it provides a large share of the evidence of radiation carcinogenesis at the lowest possible dose and dose-rate per exposure (Gofman 1990).

During preparation of the AAAS presentation, we decided to attempt a very rough estimate of what share of the current Breast-Cancer problem is attributable to radiation exposures. When we began trying to quantify it, we realized that it was much too big a task to complete in the available time. We presented just a preliminary estimate --- which was about 35 percent.

Even we (and certainly others) were startled by such a high estimate. If the estimate was in the right "ball-park," it would point at a way to make an appreciable dent in the FUTURE Breast-Cancer problem. So we decided to continue our inquiry. This meant exploration in the nearly deserted basement of the excellent medical-school library at the University of California, San Francisco.

Almost no one goes to the basement, because that is where the OLD issues of the medical journals reside. Our particular inquiry required extensive use of such journals. What happened decades ago matters, because of the long and variable latency-time for radiation-induced Cancer (Part 8, above). Illustration: The Breast Cancers, caused by radiation-induced mutations received in 1920, were gradually delivered over decades --- some cases not until 1965 (or later).

9f. An Estimate in April 1995

After a year of concentrated effort, we produced the monograph, "Preventing Breast Cancer: The Story of a Major, Proven, Preventable Cause of This Disease, First Edition" (Gofman 1995). The bottom line was that we concluded 35 percent to be a serious underestimate. Our best estimate in 1995 was that about 75 percent of Breast-Cancer cases (USA), recent and current, are due to earlier medical radiation (much of it received during the years 1920-1960).

Every step in our analysis was shown, and the unavoidable assumptions and uncertainties were made explicit.

By the end of 1995, the 75 percent estimate had received lots of peer-review. The criticisms involved the unavoidable assumptions. Most colleagues preferred assumptions which gave much lower estimates, but they were unable to show any basis for thinking that their assumptions were more likely to be right than our assumptions. A few of their competing assumptions were not reasonble, in view of existing evidence. Reasonable or not, all criticisms of which we were aware were included in the Second Edition (Gofman 1996).

9g. Arrival at Hypothesis-1 --- Almost Inescapable, Now

Since no one showed that our 75 percent estimate for Breast Cancer was either wrong or unlikely to be right, we were "stuck" with CONTINUING to believe our 75% estimate. The result: We knew it would be irrational, and even irresponsible, for us to evaluate ONLY Breast Cancer. Thus, Hypothesis-1 insisted upon its own birth and upon our respectful consideration.

Hypothesis-1: Medical radiation is a highly important cause (probably the principal cause) of cancer-mortality among U.S. males and females during the Twentieth Century.

Could we find any data capable of testing this hypothesis vigorously? At the outset, we were stumped. We could not possibly undertake all the work, required for our evaluation of xray-induced Breast Cancer (Gofman 1995/96), for every other kind of Cancer. Even if we could, we would still end up with vast gaps in the evidence on the frequency and organ-dosage from medical xrays --- as we did in the Breast-Cancer book. Such gaps would have to be filled by some assumptions, again.

Finally, we remembered the neglected "smoking gun" described in Part 9b, above. We decided to try testing Hypothesis-1 by combining two databases, which were each collected without any conceivable bias about Hypothesis-1: Physicians per 100,000 population, by Census Divisions (Chapter 3), and age-adjusted cancer mortality-rates per 100,000 population, also by Census Divisions (Chapter 4).

>>>>>>>>>>

Definition of PhysPop

"PhysPop" is our abbreviation for "Number of Physicians per 100,000 Population." When pressed for space: PP. When really pressed for space: pp.

● Part 1. Purpose and Data for Our Dose-Response Studies

The titles of this chapter and of the book itself both emphasize the term "dose-response study." Such studies address questions like, "When all other things are equal, does the cancer mortality-rate rise as exposure from ionizing radiation rises?" Yes. That fact has been established for decades (Chapter 2, Part 4c). Additional proof, that ionizing radiation is a cause of Cancer, is not needed.

Then what is our interest in additional dose-response studies?

Our interest is in exploring Hypothesis-1, that specifically MEDICAL radiation --- which is readily controllable by humans --- is a highly important cause of Twentieth Century cancer-mortality in the United States. The new set of dose-response studies, contemplated for this monograph, might be able provide a basis for estimating the MAGNITUDE of that causal role. Is the role trivial, as so often claimed, or is it highly important? We decided to attempt dose-response studies based on the input described below.

1a. The Input-Data for Our Dose-Response Studies

The minimum requirements for a dose-response study include data on responses and doses, of course.

Cancer is the relevant response for testing our Hypothesis-1. Cancer mortality-rates per 100,000 population are available for the United States, by states and by the Nine Census Divisions. These rates provide our input-data on response (details in Chapter 4).

And what about input-data on dose? A fundamental premise of our studies is that the more physicians per 100,000 population, the more radiation procedures per 100,000 population will be ordered. Such procedures are initiated by a physician, even if someone else actually performs a procedure. Thus, we arrive at the premise that average radiation dose per capita FROM MEDICAL PROCEDURES during a specific year, is approximately proportional to the number of physicians per 100,000 population during the same year.

This common-sense premise is supported by the numerous authors of the 1988 and 1993 reports of the United Nations Committee on the Effects of Atomic Radiation. In their 1993 report, in Annex C on medical radiation exposures worldwide, they state (UNSCEAR 1993, pp.223-224/Para.10): "In the UNSCEAR 1988 Report, a good correlation was shown to exist between the number of xray examinations per unit of population and the number of physicians per unit of population." And they depict a linear correlation in Figure 1 (UNSCEAR 1993, p.347). The premise is also supported by the evidence already provided specifically for the USA in our Chapter 2, Part 3c, Point 1. The substantial increases, described there, in xray procedures per 1,000 population and in per capita sales of medical xray film, occurred during the period of a rapid increase in the number of physicians per 1,000 (or per 100,000) population --- namely, following federal enactment of medical entitlements in the mid-1960s.

Our input-data on dose will be numbers of physicians per 100,000 population, by Census Divisions (USA), as explained in Part 1b below. At the outset, we did not know at what year such records began to be kept, relative to the discovery of xrays in 1895. We were able to obtain data starting in 1921.

1b. PhysPop as a Surrogate for Medical Radiation Dose, by Census Divisions

Using the premise from Part 1a, we can state:

Average radiation dose (in rads) per capita from medical procedures = (k)(PhysPop), where k is a conversion-factor from physicians per 100,000 population into average number of rads received per capita during the same year. ("Rad" is defined in Appendix A.) We approximate that k has the same value nationwide at any one time. There is no requirement for the value of k to remain the same, decade after decade.

At any one time, in each of the Nine Census Divisions, the magnitude of average per capita dose from medical radiation is proportional to (k) times (PhysPop for that particular Census Division). Thus, we can (and we do) use the PhysPop values for individual Census Divisions as a surrogate for average per capita doses received in such Census Divisions. If a PhysPop value in the First Census Division is 1.43 times bigger than the PhysPop value in the Ninth Census Division, then the resulting average dose per person is 1.43 times higher in Census Region One than in Census Region Nine --- as a good approximation. PhysPop values reveal the RELATIVE size of average per capita doses in the Nine Census Divisions.

Our studies never require the quantification of k. Thus, our studies permit the possibility that average per capita dose could decrease in every Census Division, during a period when PhysPop values could simultaneously increase. For example, if PhysPop rose 2-fold while average dose-level of radiation per procedure fell 3-fold, then the average per capita dose would decrease to 2/3 of its earlier level: Dose = (k/3)(2PhysPop) = (2/3)(k)(PhysPop).

1c. Two Special Merits of Using PhysPop Values as Dose-Surrogates

Dose-response studies, based on the relative size of doses, of course can be fully as valid as studies based on absolute dose-values. Because the absolute doses from medical radiation in the past and present are highly uncertain and forever debatable (Chapter 2, Part 3), studies based on a reasonable approximation of the relative size of doses (PhysPop values) can be the MOST reliable. Indeed, one of the major scientific strengths of this monograph is its independence from anyone's estimates of absolute doses.

A second strength of the PhysPop method deserves some discussion:

Epidemiologic research on the health effects of ionizing radiation is sometimes characterized by input-data which are vulnerable to potentially biased, after-the-fact adjustments of dosage and responses, and after-the-fact exclusion of selected groups or cases as "unqualified" for retention in a study, or after-the-fact inclusion of "reserve" samples. Even retroactive shuffling of dose-cohorts --- after they have produced a dose-response --- is now a chronic practice in one of the world's most important radiation databases, the Atomic-Bomb Survivor Database (discussion in Chapter 2, Part 5c).

Such practices, as well as the fact that so much radiation research is funded by governments which are far from neutral about the hazards of ionizing radiation, necessarily create doubt about the trustworthiness of the raw databases themselves. The FIRST obligation of objective scientists is to seek assurance that they do not work with biased data which will produce misleading results. For example, few objective analysts on the smoking-issue would rely on data from a database sponsored by, and thus controlled by, the tobacco industry.

So, in the world of radiation epidemiology, the radiation studies which are presented in this monograph have a special foundation of credibility: The inputs for both dose (PhysPops) and response (Mortality-Rates) are data collected over decades by people with no conceivable intent or ability to bias the outcome of a radiation study. The data are public and not vulnerable to successful alteration.

● Part 2. Some Reasons for Expecting Our Dose-Response Concept to Fail

We were aware, at the outset, that the merits described in Part 1c could not eliminate the several reasons to bet AGAINST detecting any dose-response in such data. But researchers who demand a guarantee before they begin, rarely begin. Pessimism is paralytic. And sometimes irrational. It can be unreasonable to assume that all imaginable obstacles, to obtaining useful information, will actually materialize.

"Whatever you want to do, if you overanalyze it --- if you start looking for all the pluses and all the minuses --- you might never start." So spoke Dr. Herb Boyer, molecular biologist, and co-founder of Genentech Inc., a pioneering enterprise in the biotechnology world. The occasion: An interview on Genentech's 20th anniversary in 1996 (Boyer 1996).

Nonetheless, Part 2 will briefly describe some of the potential obstacles, as a guide to whether or not they materialized, and as a guide to some of our decisions.

2a. Inconsistent Studies on Natural Background Radiation

What made us ever imagine that a dose-response from medical irradiation might be DETECTABLE by geographical regions, when numerous attempts to find a dose-response from geographical differences in natural background radiation have been conflicting and non-definitive?

The idea probably occurred to us because of our 1981 analysis of the Frigerio paper, described in Chapter 2, Part 9b. Moreover, as a result of our work on the 1995 breast-cancer book, we had learned how MUCH medical irradiation has been used. So we thought that the average per capita dose per year from medical radiation might exceed the annual dose from natural background sources by enough to "show up." This thinking was related to the fact that medical x-rays are 2-fold to 4-fold more harmful (biologically) than the gamma rays from natural background sources, as discussed in Chapter 2, Part 7. So we decided to take the next step, and to examine the PhysPop data.

2b. The Necessity of DIFFERENCES in PhysPop Values by Census Divisions

Of course we did not know, until we obtained and studied the data in this chapter, that sufficient DIFFERENCES would exist in the doses (PhysPops) on a geographical basis. It is impossible to do a very useful dose-response study without appreciable differences in doses! Medical irradiation could be the paramount cause of the cancer-problem, and still we would obtain no hint of such a fact from our proposed dose-response study --- if the doses were about the same in all Nine Census Divisions.

A dose-response study typically plots, on a graph, a proposed cause on the horizontal x-axis, versus a proposed consequence on the vertical y-axis. If real-world evidence shows that a series of increments in the proposed cause, goes with a series of increments in the proposed consequence, the causal presumption is reasonable unless a better explanation can be demonstrated. The causal presumption is especially reasonable when the proposed cause (ionizing radiation) is already a PROVEN cause of the effect (excess cancer-mortality).

So our very first task was to find out if there would be any appreciable DIFFERENCES in the dose-input (PhysPops in the Nine Census Divisions) for our proposed study. In Part 5 of this chapter, we will discuss the range of differences we found in PhysPops, and how the range changed over the 1921-1992 period.

2c. Annual Radiation Dose versus Accumulated Radiation Dose

The chance, that a cell acquires a new (non-inherited) carcinogenic mutation due to ionizing radiation, is proportional to the cell's ACCUMULATED radiation dose. We knew at the outset that a PhysPop value for a single year would not be proportional to the average ACCUMULATED total dose in one census region, versus all other regions --- unless the regional PhysPop values retained their proportionality with EACH OTHER over time. This aspect of our studies is explored in Parts 6 and 7 (below).

PhysPops, as informative surrogates for ACCUMULATED doses, were threatened in yet another way. Even if PhysPop rankings in the various Census Divisions happened to remain stable long enough to produce some discernible differences in the radiation consequences, we needed to worry about the impact of the population's mobility.

The Potential Problem from Migration between Census Divisions

Whenever people move from a Census Division of higher PhysPop value to a Census Division of lower PhysPop value, they carry their cancer-risk with them. Because they mix their higher accumulated dose (and their higher risk of radiation-induced Cancer) with the new population's lower accumulated dose (and lower risk of radiation-induced Cancer), such migration necessarily degrades PhysPop as a measure of the relative magnitude of accumulated dose received by people dying within those two Census Divisions. And the same potential problem applies to migration from low PhysPop to high PhysPop Census Divisions. The concern would essentially vanish if all radiation-induced Cancers were delivered within 2 or 3 years after irradiation. The potential problem occurs because latency periods (delivery times) for radiation-induced Cancers are spread over decades, as discussed in Chapter 2, Part 8. As more decades pass, more people migrate between Census Divisions. By contrast, the migration which occurs WITHIN single Census Divisions creates no problem at all for our proposed dose-response studies.

2d. Distribution of the Combined Impact from Other Carcinogens

We knew at the outset, of course, that a dose-response to medical irradiation could be obscured in our proposed studies, if the combined force of carcinogens OTHER THAN MEDICAL IRRADIATION were to have an UNEQUAL impact on the cancer mortality-rates of the Nine Census Divisions. This is a nearly universal hazard in epidemiology. For example, in the Atomic-Bomb Survivor Study, one can (and must) assure that the groups receiving different doses of bomb-radiation are comparable in age-and-sex distribution. But there is no way to force the dose-cohorts to be comparable in their lifetime exposures to all non-bomb carcinogens (known and unknown) before and after August 1945. And in our own studies, there is no way to force the nine populations in the Nine Census Divisions to be comparable in their lifetime exposures to all non-xray carcinogens (known and unknown). In such studies (and many others), one hopes that providence has distributed the extraneous non-comparabilities in such a way that their combined carcinogenic force is nearly equal ("matched"in all dose-cohorts. Otherwise, these unequal impacts can distort the true dose-response between the two variables under study (Chapter 5, Part 7, and Chapter 48).

● Part 3. PhysPop Data for the Nine Census Divisions, 1921-1992

Overlapping sources exist for data on the number of physicians per 100,000 population in the USA. They include the U.S. Government, the American Medical Association, and the American Hospital Association. (We did not happen to use any AHA publications.) We have found data back as far as 1921.

3a. The "Universal" PhysPop Table: Table 3-A (Four Pages)

Table 3-A is located, of course, after the text of this chapter. It is the Universal PhysPop Table covering the years 1921-1993, for the Nine Census Divisions of the USA. The word "universal" calls attention to the fact that the PhysPops are the same no matter what cause of death is compared with them. Thus, this single table is the origin of x-axis data for numerous chapters of this book.

The table covers general practitioners and specialists combined. The details are provided in Parts 3c and 3d. We did not find data for every calendar year between 1921 and 1992. The years for which we have found data are flagged "+" in the Universal PhysPop Table 3-A. The years for which we obtained values by interpolation, are unflagged.

The data on PhysPops are often presented state-by-state in various sources. In combining data from various states, to obtain the average PhysPop value for an entire Census Divisions, we weighted each state's PhysPop value by the contemporaneous size of the state's population (details in Part 3d).

3b. Which States Belong to Which Census Divisions?

Because we were searching for data on the Nine Census Divisions from 1895 onward, the fact that Alaska and Hawaii did not become states until after World War Two seemed like a probable complication. In view of their small populations, we decided at the outset to exclude Alaska and Hawaii from consideration. For consistency, we also excluded the District of Columbia, which is not a state and whose population has always been small, too. So, these three entities are omitted from our Universal PhysPop Table 3-A. Below, we list the states (total = 48) in each of the Nine Census Divisions (from PHS 1995, p.302, for example). Populations of each Census Division, by decades, are shown in our Table 3-B.

● EAST NORTH CENTRAL: Illinois, Indiana, Michigan, Ohio, Wisconsin. 5 states.

● EAST SOUTH CENTRAL: Alabama, Kentucky, Mississippi, Tennessee. 4 states.

● MIDDLE ATLANTIC: New Jersey, New York, Pennsylvania. 3 states.

● MOUNTAIN: Arizona, Colorado, Idaho, Montana, Nevada, New Mexico, Utah, Wyoming. 8 states.

● NEW ENGLAND: Connecticut, Maine, Massachusetts, New Hampshire, Rhode Island, Vermont. 6 states.

● PACIFIC: California, Oregon, Washington. 3 states.

● SOUTH ATLANTIC: Delaware, Florida, Georgia, Maryland, North Carolina, South Carolina, Virginia, West Virginia. 8 states.

● WEST NORTH CENTRAL: Iowa, Kansas, Minnesota, Missouri, Nebraska, North Dakota, South Dakota. 7 states.

● WEST SOUTH CENTRAL: Arkansas, Louisiana, Oklahoma, Texas. 4 states.

3c. Evolution of Four Categories of PhysPops

Over the years, the reports of PhysPop values gradually developed multiple categories. For instance, distinction is made between federal and nonfederal physicians. Federal physicians are those on active duty with the armed forces, the Public Health Service, the Veterans' Administration, the Indian Service, and other federal agencies (Pennell 1952, p.10). Some of the distinctions developed due to "manpower" forecasts, for wartime and for new federal programs such as Medicare and Medicaid, which were enacted in 1965. By 1965, PhysPop values came in four varieties, based on:

1. ● Total physicians --- active + inactive, federal + nonfederal, in the USA and its possessions (Canal Zone prior to 1980, Pacific islands, Puerto Rico, Virgin Islands). Specified in AMA 1993, Table A-16, p.32.

2. ● Total patient-care physicians, federal and nonfederal. Not available until 1965 (AMA 1993, Table A-16, p.32).

3. ● Total nonfederal physicians --- active + inactive --- in the 50 states and D.C. Specified in AMA 1993, Table A-17, p.33.

4. ● Nonfederal patient-care physicians, 50 states and D.C.

The relative sizes of these four types of PhysPops --- for the years 1965, 1970, 1975, 1980, 1985, and 1992 --- are graphed in AMA 1993, Figure 4, p.15. We have reproduced the AMA graph at the end of this chapter, as Figure 3-D. It is evident that the four types are very tightly correlated with each other.

3d. Sources Used for PhysPop Values from 1921 Onwards

Our choices for the Universal PhysPop Table 3-A were determined by what was available to us in the literature either by states or by Census Divisions. By states, the AMA tables offer only one "choice": Total nonfederal PhysPops. As of 1949, only a very small share of the total was inactive (see below). The fraction was low also in 1975, 1985, 1990 (see Part 3e).

PhysPops for 1921-1949

PhysPop data from 1921 through 1949 are from Pennell 1952 in our Reference List. This is Public Health Service Publication 263, Section 1, prepared by Maryland Y. Pennell and Marion E. Altenderfer, and entitled "The Health Manpower Source Book Section 1. Physicians." Its Table 2 (at page 14) is Physician-Population Ratios in the United States, by Region and State: 1921-1949. Pennell and Altenderfer based their table on the following sources in our Reference List: AMA 1950, Census Bureau 1951, and Linder 1947.

Pennell's Table 2 does not specify any subset of physicians. By comparing numbers in Pennell's Table 2 with numbers in Pennell's Tables 1 and 4, we can establish that the PhysPops in Table 2 are for total physicians (active + inactive, federal + nonfederal).

Separately, and only for the year 1949, Pennell provides the composition of the total 201,277 physicians: 179,041 active nonfederal + 12,536 federal + 9,700 retired.

PhysPops for 1959

PhysPop data for 1959 are from Stewart 1960 in our Reference List. This is Public Health Service Publication 263, Section 10, prepared by William H. Stewart and Maryland Y. Pennell, and entitled "The Health Manpower Source Book Section 10. Physicians' Age, Type of Practice, and Location." On page 26 is its table, Physician-Population Ratio in Each State, and Age of Physicians: Non-Federal Physicians per 100,000 Civilian Population, 1959. Stewart 1960 (p.1) bases the number and location of physicians (mid-1959) on data supplied to the Public Health Service by the American Medical Association, and rates per 100,000 population, by states, on mid-1959 population data from the Census Bureau (1959). We used 1959 population data from Grove 1968 (Table 74) in order to obtain population-weighted PhysPop values for the Nine Census Divisions.

PhysPops for 1963, 1965, 1970, 1975, 1980, 1981

PhysPop data for 1963, 1965, 1970, 1975, 1980, and 1981 all are taken from AMA 1982 in our Reference List, Table A-7, Non-Federal Physicians, Civilian Population, Physician-Population Ratios for Selected Years 1963-1981. That table provides all the data we need to calculate population-weighted PhysPop values for the Nine Census Divisions.

PhysPops for 1983

PhysPop data for 1983 are taken from AMA 1986 in our Reference List, Table A-9, Non-Federal Physicians, Civilian Population, and Physician/Population Ratios for Selected Years 1963-1985. That table provides all the data we need to calculate population-weighted PhysPop values for the Nine Census Divisions.

PhysPops for 1985, 1990, 1993

PhysPop data for 1985, 1990, and the start of 1993 are taken from AMA 1994 in our Reference List, Table A-18, Non-Federal Physicians, Civilian Population, Physician/Population Ratios for Selected Years 1970-1993. That table provides all the data we need to calculate population-weighted PhysPop values for the Nine Census Divisions.

3e. Difference between PhysPops from Table 3-A and from PHS 1995

In "Health, United States, 1995 (PHS 1995, pp.218-219), there is Table 97 which presents PhysPops for 1975, 1985, 1990, and 1994 by Census Divisions and by states. We note the word "active" in Table 97's title: Active Nonfederal Physicians and Doctors of Medicine in Patient Care per 10,000 Civilian Population ... 1975, 1985, 1990, and 1994. Of course, we multiply PhysPop values in Table 97 by ten, in order to convert them to the more customary "per 100,000" population. In our Universal PhysPop Table 3-A, the PhysPop values come from the combination of active plus inactive physicians. Not surprisingly, our PhysPops for 1975, 1985, and 1990 are higher than those presented in PHS 1995, Table 97.

Would analysts reach the same conclusions that we do, about the relationship of PhysPops with biological phenomena (such as cancer mortality-rates), if they used the ratios from Table 97, instead of the ratios from our Universal PhysPop Table 3-A?

The answer is yes. They would reach the same conclusions, because the correlations between the two sets of data are so very high. We demonstrate this by the three regression analyses which follow and which produce correlation coefficients (R) of 0.9916, 0.9855, and 0.9863. Our studies rely on the RELATIVE magnitudes rather than absolute magnitudes of the nine PhysPop values (Parts 1b and 1c), and such high correlations between Table 97 and Table 3-A mean that the relative magnitudes among the PhysPop values are extremely similar in Table 97 and Table 3-A. (Readers who are unfamiliar with linear regression analysis will find an introduction to the topic in Part 7 of this chapter, and more explanation in Chapter 5, Part 5).

Below, listed by the Nine Census Divisions, are the PhysPop values per 100,000 population from PHS 1995, Table 97 (including active doctors of osteopathy), and to the right of them, the values for the matching Census Divisions from our Universal PhysPop Table 3-A.

YEAR = 1975	Tab 97	Tab 3-A	YEAR = 1975	
Pacific	179	208	Regression Output:	
New England	191	215	Constant	-11.5257
West North Central	133	141	Std Err of Y Est	5.2312
Mid-Atlantic	195	213	R Squared	0.9832
East North Central	139	146	No. of Observations	9
Mountain	143	156	Degrees of Freedom	7
West South Central	119	128		
East South Central	105	117	X Coefficient(s)	1.1784
South Atlantic	140	156	Std Err of Coef.	0.0582
Unweighted Avg.	149.3	164.4	R =	0.9916
Ratio (Tab3A/Tab97) =	1.10			

YEAR = 1985	Tab 97	Tab 3-A	YEAR = 1985	
Pacific	225	256	Regression Output:	
New England	267	293	Constant	-13.6604
West North Central	183	186	Std Err of Y Est	8.5884
Mid-Atlantic	261	276	R Squared	0.9712
East North Central	193	195	No. of Observations	9
Mountain	178	193	Degrees of Freedom	7
West South Central	164	171		
East South Central	150	162	X Coefficient(s)	1.1391
South Atlantic	197	216	Std Err of Coef.	0.0741
Unweighted Avg.	202.0	216.4	R =	0.9855
Ratio (Tab3A/Tab97) =	1.07			

YEAR = 1990	Tab 97	Tab 3-A	YEAR = 1990	
Pacific	234	265	Regression Output:	
New England	290	320	Constant	-14.3549
West North Central	198	203	Std Err of Y Est	8.7876
Mid-Atlantic	284	298	R Squared	0.9729
East North Central	206	209	No. of Observations	9
Mountain	193	208	Degrees of Freedom	7
West South Central	178	184		
East South Central	168	182	X Coefficient(s)	1.1342
South Atlantic	217	234	Std Err of Coef.	0.0716

Table continues, next page

Unweighted Avg.   218.7   233.7   R =   0.9863
   Ratio (Tab3A/Tab97) =   1.07

● Part 4. Designation of the "High-5" and "Low-4" Census Divisions

       In the Universal PhysPop Table 3-A, the PhysPop values for 1921 are presented in order of size, from the highest value in the Pacific Division (165.11 physicians per 100,000 population) to the lowest value in the South Atlantic Division (110.32 physicians per 100,000 population).

       We can (and do) retain the 1921 sequence of the Census Divisions, even though the PhysPop values do not remain ranked in that order during all subsequent years. Use of the 1921 sequence leads to two terms used in Table 3-A (and used also in our tables of mortality rates): High-5 and Low-4.

Definition of "High-Five" and "Low-Four" Census Divisions

       ● The term "High-5" always refers to the first Five Census Divisions listed in the Universal PhysPop Table for 1921: Pacific, New England, West North Central, Mid-Atlantic, East North Central. Since PhysPop values are surrogates for average per capita dose from medical radiation (Part 1b), the term High-5 refers to the Census Divisions with the highest average doses per capita from medical irradiation in 1921. Our shortest abbreviation is Hi5.

       ● The term "Low-4" always refers to the last Four Census Divisions listed in the Universal PhysPop Table for 1921: Mountain, West South Central, East South Central, South Atlantic. Since PhysPop values are surrogates for average per capita dose from medical radiation (Part 1b), the term Low-4 refers to the Census Divisions with the lowest average doses per capita from medical irradiation in 1921. Our shortest abbreviation is Lo4.

A Point to Keep in Mind, and the Next Question

       A point to keep in mind is that High-5 and Low-4 are two Census-Division sets whose members were determined by their PHYSPOP rankings in 1921, not by their cancer mortality-rates in 1921. What happens to High-5 and Low-4 PhysPop values, as the interval after 1921 grows ever longer? We will explore that issue in Part 5, below.

● Part 5. Dose-Differences: What Does the Evidence Show?

       For PhysPop, which is the dose-surrogate in our dose-response studies, there are pages of entries in the Universal Table. Do these entries reflect sufficient DIFFERENCES in dose among the Nine Census Divisions --- and are differences maintained long enough in their rank order --- to produce detectable differences in cancer consequences?

       To facilitate getting a grasp on the issue of PhysPop differences and their duration, we calculated average values for the High-5 and Low-4 Divisions in each column. In the Universal PhysPop Table 3-A, the nine main entries for the Nine Census Division are weighted averages (Part 3a), but the High-5 and Low-4 averages (located beneath the main entries) are not population-weighted. They are provided just as approximations which can supply an overview for each particular year, and for changes over time. Table 3-A also shows the ratio of Hi5/Lo4.

5a. Revelations about PhysPop Behavior, 1921 through 1990: Figure 3-A

       Figure 3-A presents two graphs which plot annual PhysPop behavior from 1921 through 1990, in terms of High-5 and Low-4 groupings. These graphs provide a visual overview. No values from the graphs are ever used in calculations. Therefore, readers need not worry at all about some minor differences between the graphs and Table 3-A. The graphs reflect our early exploration --- before we had every final PhysPop value of Table 3-A --- of a question which would determine whether or not to proceed with the project: Was there a persistent dose-difference between the Census Divisions?

       The upper graph plots the annual High-5 and Low-4 averages, separately, 1921-1990. It is clear that they are relatively flat until almost 1970, when both of them take off like rockets to much higher values. The steep rise in PhysPop values occurred after the 1965 enactment of Medicare and other federal programs.

The lower graph plots the annual RATIOs of average High-5 PhysPop over average Low-4 PhysPop. The ratio tells us how many-fold larger High-5 PhysPop is, compared with Low-4. At a value of 1.0, of course, their magnitudes would be equal. The graph produces some very important information.

First, because the ratios never fall to 1.0, it immediately assures us that average annual High-5 doses always remained higher than the average annual Low-4 doses. So there has been an annual dose-difference, from 1921 to 1990.

Second, the graph of Hi5/Lo4 PhysPop ratios shows us that there was an extended period of relative PhysPop stability from about 1933 through 1968. From Table 3-A (which has the final PhysPop values), we know that the ratio of High-5 PhysPop over Low-4 PhysPop was 1.37 in 1933; then the ratio rose to a maximum value of 1.46 in 1940; by 1968 the ratio had returned to 1.37. In other words, between 1933 and 1968, the range for the Hi5/Lo4 ratio stayed within the limits of 1.37 and 1.46.

5b. What the Ratios Fail to Show

The Hi5/Lo4 ratios obscure the full magnitude of the differences between PhysPops. Although the maximum Hi5/Lo4 ratio is 1.46, the ratio comes from averages. Two examples illustrate the point. In 1921, when the Hi5/Lo4 ratio was only 1.18, the ratio of Pacific Division over South Atlantic was (165.11 / 110.32), or 1.50. In 1950, when the Hi5/Lo4 ratio was 1.44, the ratio of the Mid-Atlantic Division over East South Central was (168.81 / 83.25) = 2.03.

In addition, the Hi5/Lo4 ratios are crude enough to obscure shifts of PhysPop rank WITHIN both the High-5 and the Low-4 groups. Therefore, Box 1 provides a separate study of changes in PhysPop ranking for the 1921-1990 period. What emerges from Part 2 of Box 1 is that there is remarkable stability in PhysPop ranking, when the Nine Census Divisions are viewed as three "Trios": TopTrio, MidTrio, LowTrio. For example, in the 1931-1990 period, only two of the Nine Divisions (West North Central and South Atlantic) ever "migrate" from their 1940 Trio into another Trio. (Details in Box 1.)

● Part 6. "Lockstep" --- The Ideal Relation among All Sets of PhysPops

The formal definition of "lockstep" is: A method of marching in such close file that the corresponding legs of the marchers must keep step precisely.

We are going to bend the term, so that "lockstep" refers to a set of PROPORTIONS (ratios) whose values persist unchanged through time. For example, PhysPop "lockstep" would mean that the proportions observed among the nine PhysPop values in 1921, and the proportions observed in every subsequent year, are the same. PhysPop "lockstep" would mean that the RELATIVE magnitudes are constant among the nine PhysPop values, even when the absolute values rise or fall. Part 6b will provide an illustration.

Box 1 already demonstrates that PERFECT "lockstep" for PhysPop values does not occur, for perfection would tolerate no changes in Hi5/Lo4 ratios over time (see Part 7d) and no changes in rank order of the Divisions over time.

6a. The Ideal Data for Our Proposed Study

Researchers always wish for "better data." Under ideal circumstances for our inquiry, no migration of populations from one Census Division to another would have occurred after 1895, and for the entire century, the PhysPops of the Nine Census Divisions would have retained a fixed proportionality with each other: "Lockstep."

Under such circumstances, the nine average doses of medical radiation, accumulated by any particular year, would always be in that fixed proportion to each other --- regardless of their absolute values in rads. And the nine irradiated populations would gradually DELIVER the consequent radiation-induced cancers in the same Census Division where they were irradiated --- in proportion to dose (Chapter 5, Part 5d). The changing age-distribution of the population since 1895 would not distort that expectation because cancer mortality-rates, by Census Divisions, are age-adjusted to a fixed year (Chapter 4).

A Note about "Ideal" Data

In this chapter, and later, we sometimes refer to "ideal" data or circumstances. We feel impelled to emphasize, for students who may not have done any research yet themselves, that the term "ideal" does not imply any bias or passion. To imagine conditions "exactly as one would desire" (see below), unclouded by real-world perturbations, can be so crucial to elucidating a topic that it is a regular feature of science.

For example, chemists and physicists refer to "an ideal gas," "ideal conditions," and "the perfect gas law." We quote from Mahan's "University Chemistry" text (Mahan 1975, p.43-44): "The expression $PV = nRT$ is obeyed by all gases in the limit of low densities and high temperatures --- 'ideal' conditions under which the forces between molecules are of minimum importance. Consequently, [$PV = nRT$] is known as the perfect gas law, or the ideal gas equation of state."

In other words, the law is valid under ideal conditions, but does not make perfect predictions under real-world conditions. This use of the word "ideal" is in full harmony with the dictionary definition which says that "ideal" means: "Existing as an idea, model, or archetype ...; thought of as perfect or a perfect model; exactly as one would desire ...; having the nature of an idea or conception; identifying or illustrating an idea or conception" (Webster 1954, p.720).

6b. Figure 3-B: Retention of Perfect Proportionality ("Lockstep")

With respect to evaluating retention (or non-retention) of PhysPop proportionalities through time, we will use Figures 3-B and 3-C as illustrations. (The term "linear regression," in the titles of these two figures, may be unfamiliar to some readers. But the point of Part 6b can be understood without any understanding of linear regression.) Figures 3-B and 3-C each compare the set of 1921 PhysPops with the set of 1940 PhysPops, but in different ways. When readers understand the two figures, they will understand our Table 3-C, which shows with great simplicity how 21 sets of PhysPops, from 1921 through 1993, compare with EVERY OTHER set of PhysPops through most of this century.

In Figure 3-B, Column B presents the 1921 PhysPop values and Column C presents the 1940 PhysPop values, from the Universal PhysPop Table 3-A. The numerical values for South Atlantic changed from 110.32 to 100.74. The ratio (1940 / 1921) is 100.74 / 110.32, or 0.9131617.

If the nine PhysPops had the same proportions with each other in 1940 as they had with each other in 1921, we could multiply every 1921 PhysPop value by 0.9131617 to discover what the values would have been in 1940. We put these "ideal" values in Column D of Figure 3-B. Column D entries = (Column B entries times 0.9131617).

Some Consequences of Retaining PERFECT Proportionality

The D-Column values in Figure 3-B are the "ideal" values which we would have preferred to find in 1940. We would have preferred them to the REAL values in Column C, because every value in Column D still stands in the same proportion to every other value in Column D, as every value in Column B stands to every other value in Column B. We can demonstrate this "lockstepping" for any two Census Divisions. For example:

- (WNoCent 1940 Ideal / Mountain 1940 Ideal) = (128.69 / 123.62) = 1.041.

- (WNoCent 1921 / Mountain 1921) = (140.93 / 135.38) = 1.041.

And because the sets of real 1921 data and ideal 1940 data have the same internal proportionalities, it is also true that cross-ratios for every pair must be the same. Example:

- (NewEngl 1940 Ideal / NewEngl 1921 Real) = (129.89 / 142.24) = 0.913.

- (WSoCentral 1940 Ideal / WSoCentral 1921 Real) = (114.28 / 125.15) = 0.913.

And because the sets of real 1921 data and ideal 1940 data have the same internal proportionalities (we have created perfect "lockstep"), the two sets of data have a perfect linear correlation WITH EACH OTHER. Part 7 discusses "perfect linear correlation" and linear regression analysis, for readers who are unfamiliar with these terms.

● Part 7. "Lockstep" --- Reality-Checks by Regression Analysis

The technique of data regression is a branch of mathematics which can evaluate the correlation between two sets of values (for example, a set called "x" and a set called "y"). Regression analysis will be covered in considerably more detail in Chapter 5 (Parts 5, 6, 7). For now, we need touch on only a few aspects of regression analysis.

7a. Equation of Best Fit, Line of Best Fit, and the R-Squared Value

In linear regression analysis, the input data are a finite set of x-values and the corresponding y-values --- as shown in Columns B and D of Figure 3-B. The output includes three values of interest to us here: The X-Coefficient, the Constant, and the R-squared value.

Equation of Best Fit: How It Relates to Part 6b

In linear regression analysis, the equation of best fit is the equation for a straight line: $(y) = (m * x) + (c)$. Note: * is the symbol for multiplication in this book. The regression output (boxed in Figure 3-B) provides the values for "m" (the X-Coefficient) and for "c" (the Constant). Users of the equation can then specify additional values for "x" (values additional to the regression's input values) and calculate what the corresponding values for "y" would be if (repeat, if) there were a PERFECT correlation between the x-values and the y-values.

Example: If x = 80 (a value NOT in Col.B), what would the matching y-value be? We use the equation for a straight line: y = (X-Coefficient * x) + (Constant). The boxed output in Figure 3-B tells us the X-Coefficient = 0.91316 --- a number already seen in Part 6b. And the output tells us that the Constant = zero. So, when x = 80, y = 73, because: y = (0.91316 * 80) + zero.

This example is only what we already demonstrated in Part 6b --- except 80 is an ADDITIONAL value of x not used in Part 6b. The X-Coefficient in Part 6b is 0.91316 --- we just didn't give it the formal name there. And because we made x and y directly proportional in Part 6b --- when we said (y = 0.91316 * x) --- then zero is the only possible value for the Constant. Thus, it is no surprise at all that the regression output produced zero as the value of the Constant.

Line of Best Fit, and Graphing

In making x,y graphs, it is customary to measure the x-values along the horizontal axis, and to measure the y-values along the vertical axis.

The line of best fit (the straight line seen in Figure 3-B) simply depicts a long series of x,y pairs, calculated by using the equation of best fit. The point, which depicts y=73 when x=80, is part of the straight line in Figure 3-B.

The R-Squared Value: A Key Measure of Correlation

Regression output also provides a value for R-squared, which is the output of real interest in this chapter. The R-squared value is a measure of how closely the x-input and the y-input are correlated. Only a PERFECT correlation has an R-squared value of 1.00. Imperfect correlations produce R-squared values between 1.00 and zero. The value "R" --- also called the correlation coefficient (Part 3e) --- is the square root of R-squared.

Since we insured in Part 6b that our x,y pairs are perfectly proportional to each other, they are also perfectly correlated with each other. And thus it is no surprise at all that the regression output in Figure 3-B produces an R-squared value of 1.00. When R-squared = 1.00, every pair of x,y values sits right upon the line of best fit, with no scatter. In Figure 3-B, the nine boxy symbols are indeed upon the line of best fit.

7b. Figure 3-C: Degradation of Perfect Proportionality

Figure 3-C moves from the "ideal" world, depicted in Figure 3-B, into the real world. Figure 3-C shows no "ideal" values. It shows only real-world input-data: The PhysPop set of 1921 and the PhysPop set of 1940.

The two graphs in Figure 3-C look quite different from the graph in Figure 3-B. The nine boxy symbols show some SCATTER around the lines of best fit. The scatter reflects the inferior correlation compared with the "ideal" correlation (R-sq = 0.58 here, compared with 1.00 in Figure 3-B).

Thus, R-squared is an evaluation of how much the 1940 PhysPop values have strayed from the proportions which they had with each other in 1921. Quite obviously, the 1921 and 1940 sets of PhysPops are not in perfect "lockstep."

A Point about Correlations

If regression analysis is employed to study a cause-effect relationship, it is customary to designate the proposed cause as the x-axis variable. However, the correlation between two sets of numbers is whatever it is, independent of human choices to call one set "x" and the other set "y." Figure 3-C demonstrates this point by reversing the designations of the two sets of PhysPops. The R-squared values in both figures turn out the same, as they must. However, other things have changed --- such as the Constant (the value of the y-intercept) and the X-Coefficient (the slope of the best-fit line).

7c. Table 3-C: How Sets of PhysPops Correlate through Time

Table 3-C is "How Sets of PhysPops Correlate through Time." At its top are 21 sets of PhysPop values. They are the input data for approximately 200 separate regression analyses, whose R-squared values are reported in the body of Table 3-C.

Because Table 3-C (like Figure 3-A) was part of our early exploration, we had not yet obtained all the PhysPop sources which we subsequently obtained. So, not every PhysPop value in Table 3-C is an exact match for the corresponding final value in Table 3-A. The differences often come from a mixture in Table 3-C of the four different types of PhysPop values described in Part 3c. The purpose of Table 3-C was to ascertain if PhysPops were hopelessly deviant from "lockstepping" --- and since the four types of PhysPop values are so highly correlated with each other, Table 3-C is not misleading. When we undertook our subsequent dose-response studies, we used PhysPop values only from Table 3-A --- as readers can verify for themselves.

Due to Table 3-C's early origin, it does not put the Census Divisions in the same sequence as Table 3-A. Of course, the sequence has no impact whatsoever on the regression output, as long as the x and y sets of PhysPops are in the SAME sequence with respect to Census Divisions.

The Grid of R-Squared Values

Beneath the raw PhysPop data is a grid of R-squared values. For instance, where the COLUMN for 1980 intersects the ROW for 1934, the R-squared value of 0.72 comes from the regression output when the 1980 PhysPops (directly above) are regressed on the 1934 PhysPops (in a column far to the right). The R-squared value would be the same if we had regressed the 1934 PhysPops (as the y-set) on the 1980 PhysPops (as the x-set), as pointed out in Part 7b.

Readers can quickly orient themselves in Table 3-C by knowing that, when the PhysPops of 1921 are regressed upon the PhysPops of 1921, there has to be a PERFECT correlation --- and it shows up as an R-squared value of 1.00 where the COLUMN for 1921 intersects the ROW for 1921.

Because Table 3-C describes every comparison between two sets of PhysPops by an R-squared value, everyone can readily see the decrement in "lockstepping" over any chosen interval of time. The approximately 200 regression analyses are not shown.

7d. Consistency between Figure 3-A and Table 3-C

If successive PhysPop sets had retained a fixed proportionality ("lockstep") over time, the Hi5/Lo4 ratio depicted in Figure 3-A would be perfectly flat. The ratio would be the same, year after year. We can illustrate this quickly.

The Hi5/Lo4 PhysPop ratio for 1921 is 1.18 --- provided in the Universal Table 3-A. The "ideal" 1940 values from Figure 3-B (Column D) reflect perfect "lockstep" with the 1921 values. We compute the Hi5 average PhysPop as 131.792. The Lo4 average PhysPop is 112.00. The Hi5/Lo4 PhysPop ratio is (131.792 / 112.00), or 1.18 for the "ideal" entries too. Change in Hi5/Lo4 PhysPop ratios, over time, reflects DEVIATION from "lockstep."

In Figure 3-A and in the text (Part 5a), we pointed to the period of 1933 through 1968 as a period when the Hi5/Lo4 PhysPop ratio was relatively constant. This means that Table 3-C should show high R-squared values during this same period, in the vertical column for 1967. It does. The lowest R-squared value is 0.82, at the intersect of the 1967 column with the 1934 row.

● **Part 8. Two Crucial Aspects of PhysPop History**

Earlier in this chapter (Part 2b), we pointed out that our proposed dose-response studies require the existence of appreciable DIFFERENCES in PhysPops (our dose-surrogates) among the Nine Census Divisions. PhysPop history might NOT have delivered differences. It is just happenstance that such differences occurred (Parts 5a and 5b).

It is also happenstance that, during the years after the introduction of medical radiation, chaos did NOT characterize the relationships between successive sets of PhysPops. Chaos would have prevented PhysPops from representing relative ACCUMULATED dose-differences in the Nine Census Divisions. Although the R-squared value (in Table 3-C) of 0.58 between the 1921 PhysPops and the 1940 PhysPops is not "great," it's far from being a value of 0.02. By 1927-1929, the correlations with 1940 become very respectable. And for the entire stretch from 1933-1967, successive sets of PhysPops were close to retaining "lockstep" proportionality with each other (Table 3-C, Figure 3-A).

If PhysPop history had not met the requirements for dose-response studies, it might have been forever impossible for anyone to detect the particular consequences which are uncovered in this book from the introduction of radiation into medicine.

>>>>>>>>>>

Box 1 of Chap. 3
Summary of PhysPop Values by Decades, and Their Ranking by Census Divisions.

● **Part 1.** Census Divisions, in our permanent order, with corresponding PhysPop values.

From Table 3-A.	1921 PhysPop	1931 PhysPop	1940 PhysPop	1950 PhysPop	1960 PhysPop	1970 PhysPop	1980 PhysPop	1990 PhysPop
Pacific	165.11	159.97	159.72	148.60	158.74	183.83	235.84	265.09
New England	142.24	142.35	161.55	162.51	164.37	186.51	254.37	319.88
West North Central	140.93	126.50	123.14	120.06	111.25	123.77	165.86	202.78
Mid-Atlantic	137.29	140.82	169.76	168.71	162.65	192.00	237.41	297.79
East North Central	136.06	128.59	133.36	123.69	114.56	127.17	169.79	208.54
Mountain	135.38	118.89	119.89	119.38	112.93	137.27	177.76	208.20
West South Central	125.15	105.95	103.94	101.34	101.65	113.20	153.18	184.34
East South Central	119.76	96.73	85.83	83.05	88.00	100.89	139.51	182.42
South Atlantic	110.32	99.59	100.74	99.07	105.36	130.70	187.22	234.48
Average ALL	134.70	124.38	128.66	125.16	124.39	143.93	191.22	233.72
Average High-Five	144.33	139.65	149.51	144.71	142.31	162.66	212.65	258.82
Average Low-Four	122.65	105.29	102.60	100.71	101.99	120.52	164.42	202.36
Ratio (Hi5/Lo4)	1.18	1.33	1.46	1.44	1.40	1.35	1.29	1.28

● **Part 2.** Census Divisions, in shifting order, sorted by descending PhysPop values.

	1921 PhysPop	1931 PhysPop	1940 PhysPop	1950 PhysPop	1960 PhysPop	1970 PhysPop	1980 PhysPop	1990 PhysPop
Top Trio	Pac NewEng WNoCen	Pac NewEng MidAtl	MidAtl NewEng Pac	MidAtl NewEng Pac	NewEng MidAtl Pac	MidAtl NewEng Pac	NewEng MidAtl Pac	NewEng MidAtl Pac
Mid Trio	MidAtl ENoCen Mtn	ENoCen WNoCen Mtn	ENoCen WNoCen Mtn	ENoCen WNoCen Mtn	ENoCen Mtn WNoCen	Mtn SoAtl ENoCen	SoAtl Mtn ENoCen	SoAtl ENoCen Mtn
Low Trio	WSoCen ESoCen SoAtl	WSoCen SoAtl ESoCen	WSoCen SoAtl ESoCen	WSoCen SoAtl ESoCen	SoAtl WSoCen ESoCen	WNoCen WSoCen ESoCen	WNoCen WSoCen ESoCen	WNoCen WSoCen ESoCen

Above, in Part 2, where the Nine Census Divisions are sorted by descending PhysPop values, we have labeled them as three "Trios": Top, Mid, Low --- reflecting, RELATIVELY, the highest to lowest average per capita dosage from medical radiation.

During the 1931-1990 period, only two of the nine Divisions (West North Central and South Atlantic) ever "migrated" from their 1931 Trio into an adjacent Trio. Measured in terms of Trios, remarkable stability occurred for sixty years in PhysPop ranking. When the overview includes the 1921 values, then the Mid-Atlantic Division becomes a migrant too, and West North Central makes an additional move.

Figure 3–A.
Behavior of Hi5 and Lo4 PhysPops through Time.

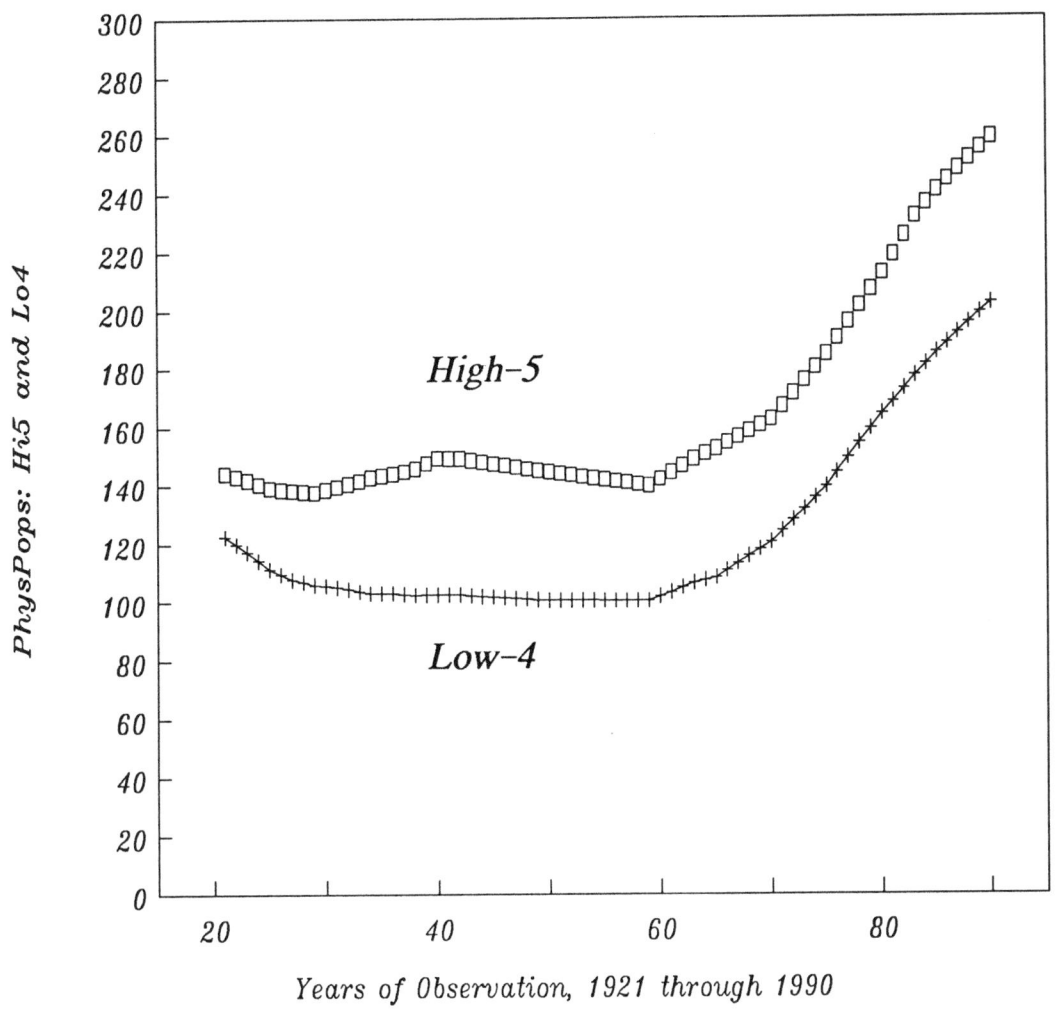

Years of Observation, 1921 through 1990

Related text = Part 5a.

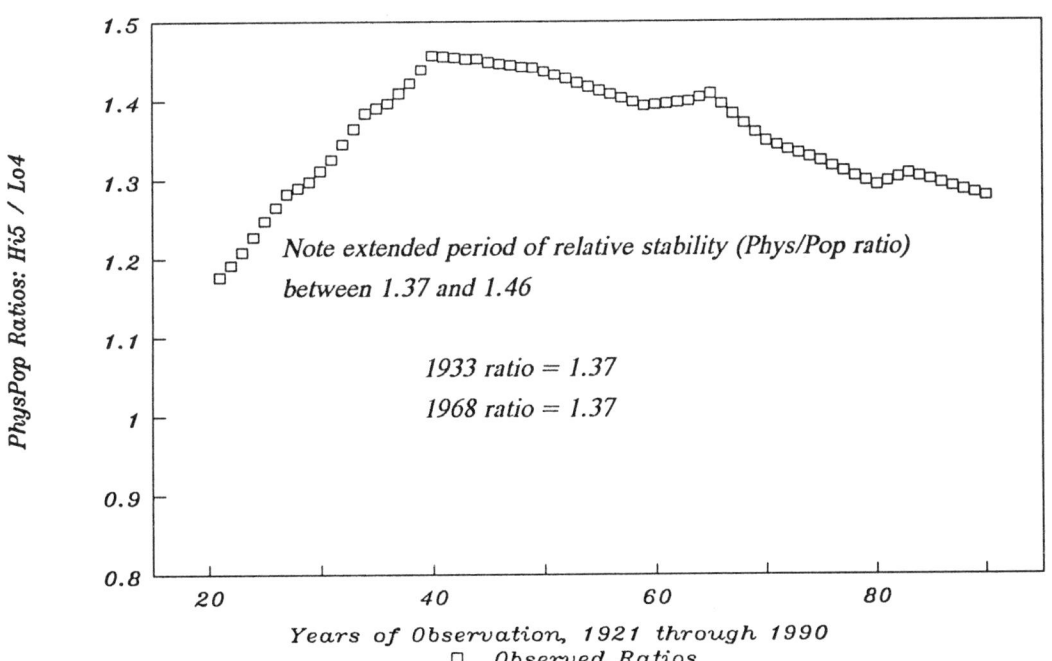

*Note extended period of relative stability (Phys/Pop ratio)
between 1.37 and 1.46*

1933 ratio = 1.37
1968 ratio = 1.37

Years of Observation, 1921 through 1990
▫ *Observed Ratios*

Figure 3-B.
Complete "Lockstepping" of PhysPop Proportions over Time.
Linear Regression of Two Perfectly Correlated and Perfectly Proportional Sets of Data.

● For the regression analysis below, the x-variable input is Col.B: Actual PhysPop values for 1921. The y-variable input is Col.D: Ideal (synthetic) PhysPop values for 1940. Output for this regression is shown to the right. Both input and output are depicted by the graph. For discussion of regression analysis and its depiction, please consult Part 7 of the text and the Index.

● The nine boxy symbols on the graph depict the nine pairs of input from Columns B and D. The line of best fit depicts the output for this perfect correlation. (R-squared = 1.00). All nine boxes sit right on the line, with no scatter. Boxes overlap when input-pairs have similar values. Because perfect proportionality exists between Columns B and D, the Constant = 0 in the best-fit equation. The line of best-fit goes right through the origin (y = 0, when x = 0).

Col.A	Col.B	Col.C	Col.D	Data Regression
	Real	Real	Ideal	Regression of Ideal 1940 PhysPops (Col.D)
Census	1921	1940	1940	upon Real 1921 PhysPops (Col.B)
Division	Phys	Phys	Phys	
	Pops	Pops	Pops	Regression Output:
Pacific	165.11	159.72	150.77	Constant 0.000000000
New England	142.24	161.55	129.89	Std Err of Y Est 0.0000014617
West North Central	140.93	123.14	128.69	R Squared 1.000000
Mid Atlantic	137.29	169.76	125.37	No. of Observations 9
East North Central	136.06	133.36	124.24	Degrees of Freedom 7
Mountain	135.38	119.89	123.62	
West South Central	125.15	103.94	114.28	X Coefficient 0.9131617114
East South Central	119.76	85.83	109.36	Std Err of Coefficient 0.0000000332
South Atlantic	110.32	100.74	100.74	

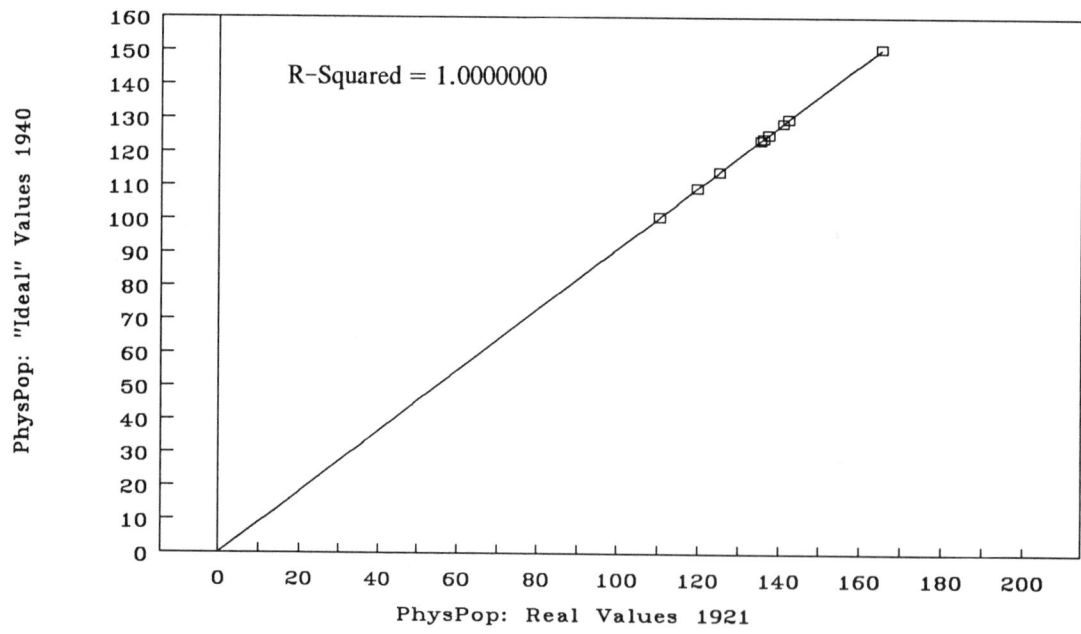

Regression of "Ideal" PhysPop 1940 upon Real PhysPop 1921

R-Squared = 1.0000000

PhysPop: "Ideal" Values 1940

PhysPop: Real Values 1921

Figure 3–C.
Imperfect Retention of PhysPop Proportions over Time.
Linear Regressions with 1921 and 1940 PhysPops from Table 3–A.

• For the first regression analysis below, the x-variable input is the set of real PhysPop values for 1921. The y-variable input is the PhysPop set of 1940. For the second regression analysis, we switch. The x-input is 1940 and the y-input is 1921. BOTH regressions produce R-squared = 0.58, because the correlation between two fixed sets of numbers is fixed. The leftside graph depicts the first regression, and the rightside graph depicts the second. Because the correlation is not perfect, the nine boxy symbols do not all sit exactly upon the line of best fit. There is some scatter around the line.

Census Division	1921 Real PhysPops "x"	1940 Real PhysPops "y"	Data Regression Regression Output:	
Pacific	165.11	159.72	Constant	−67.425
New England	142.24	161.55	Std Err of Y Est	20.641
West North Central	140.93	123.14	R Squared	0.579
Mid–Atlantic	137.29	169.76	No. of Observations	9
East North Central	136.06	133.36	Degrees of Freedom	7
Mountain	135.38	119.89		
West South Central	125.15	103.94	X Coefficient	1.4558
East South Central	119.76	85.83	Std Err of Coef.	0.4688
South Atlantic	110.32	100.74	X–Coeff. / S.E. =	3.1050

Census Division	1940 Real PhysPops "x"	1921 Real PhysPops "y"	Data Regression Regression Output:	
Pacific	159.72	165.11	Constant	83.491
New England	161.55	142.24	Std Err of Y Est	10.792
West North Central	123.14	140.93	R Squared	0.579
Mid–Atlantic	169.76	137.29	No. of Observations	9
East North Central	133.36	136.06	Degrees of Freedom	7
Mountain	119.89	135.38		
West South Central	103.94	125.15	X Coefficient	0.3980
East South Central	85.83	119.76	Std Err of Coef.	0.1282
South Atlantic	100.74	110.32	X–Coeff. / S.E. =	3.1050

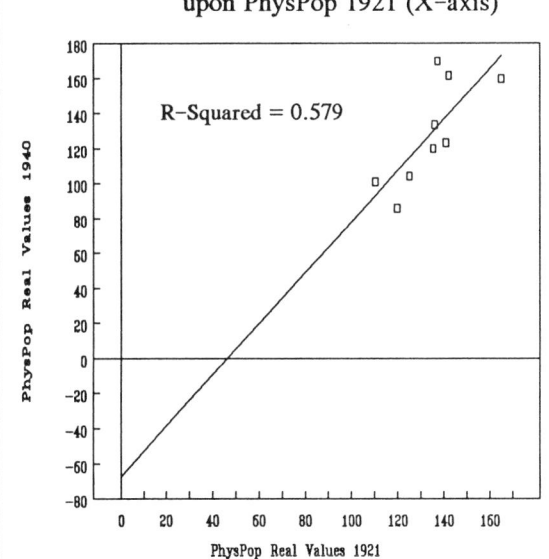

Regression of PhysPop 1940 (Y–axis) upon PhysPop 1921 (X–axis)

R–Squared = 0.579

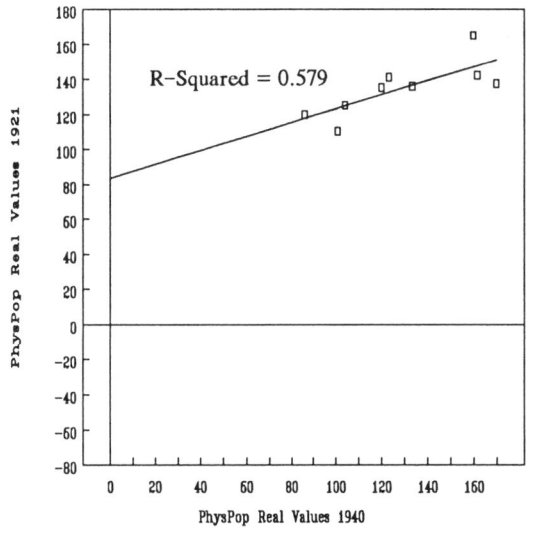

Regression of PhysPop 1921 (Y–axis) upon PhysPop 1940 (X–axis)

R–Squared = 0.579

Figure 3–D.
Comparison of Four Types of PhysPop Values.

This figure is reproduced from p.15 of AMA 1993 in our Reference List:
Physician Characteristics and Distribution in the U.S., 1993 Edition,
by Roback + Randolph + Seidman of the American Medical Association,
Department of Physician Data Services.

Related text = Part 3c.

Figure 4

Trends in Physician/Population Ratios for Selected Years 1965–1992

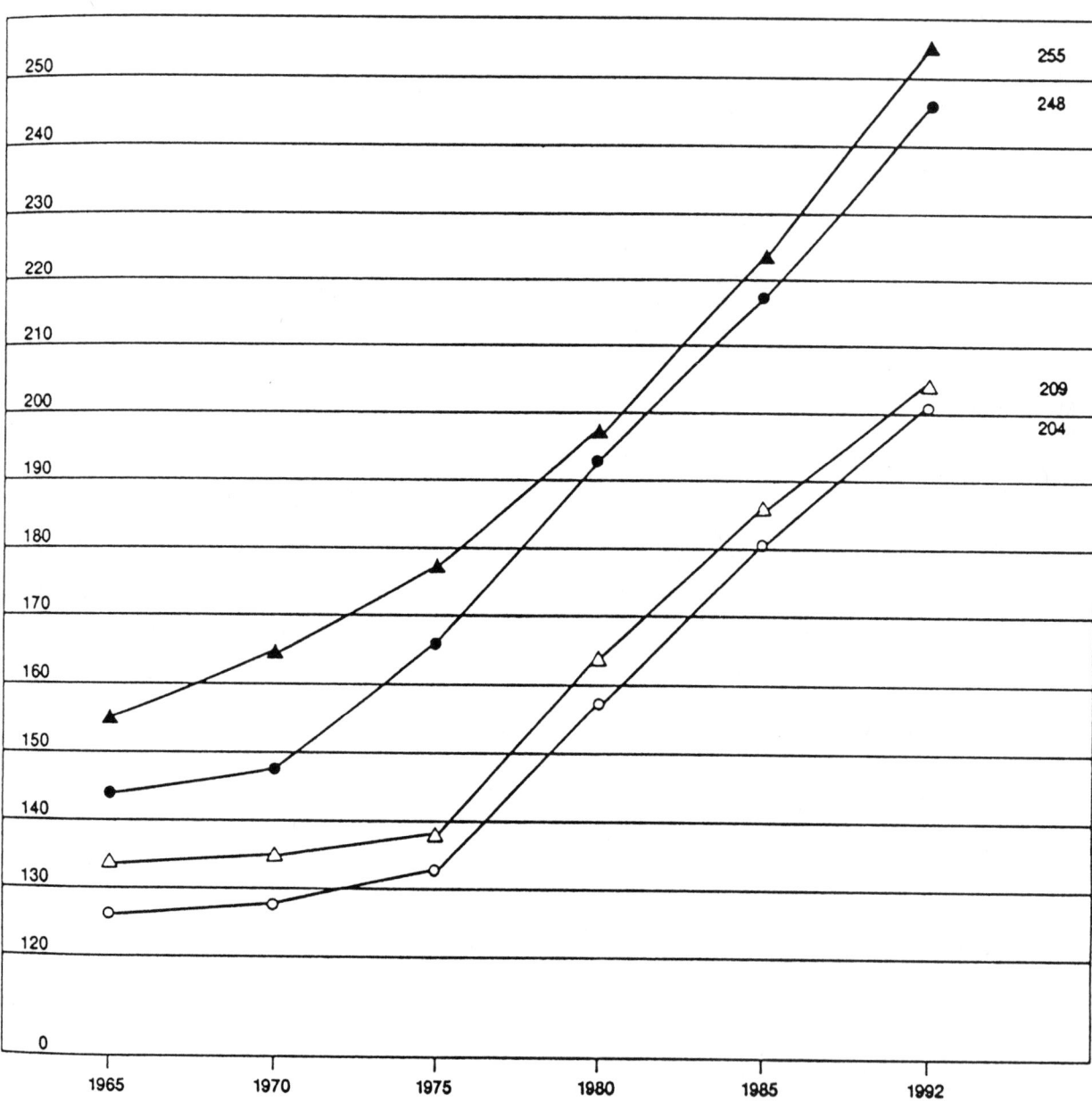

▲ Total Phys. per 100,000 Total Pop. in *U.S. & Possessions*
△ Total Patient Care Phys. per 100,000 Total Pop. in *U.S. & Possessions*
● Non-Federal Phys. per 100,000 Civilian Pop. in *U.S. only*
○ Non-Federal Patient Care Phys. per 100,000 Civ. Pop. in *U.S. only*

Sources: Table A-16 and Table A-17.

Table 3-A

Universal PhysPop Table

First page of four

Related text = Parts 3a + 3b

● PhysPop values are numbers of physicians per 100,000 population. Entries are for general practitioners and specialists combined --- 1921 through 1993 (details in text). Sources of the data are provided in the text, Part 3d. The years which are flagged with a "+" sign present prime data. Entries for the unflagged years have been interpolated.

● The particular states belonging to each Census Division are listed in the text, Part 3b. PhysPop entries for the Nine Census Divisions have been weighted by state populations, whereas the three rows of averages are non-weighted. High-5 and Low-4 are defined in the text, Part 4.

● This single table is the source of data for numerous chapters of this book. The term "universal" in the table's title emphasizes that the PhysPops are the same, regardless of which cause of death is compared with them.

Census Division	1921+	1922	1923+	1924	1925+	1926	1927+	1928	1929+	1930
Pacific	165.11	164.09	163.06	162.36	161.67	159.75	157.83	157.24	156.64	158.30
New England	142.24	139.82	137.39	137.85	138.31	137.91	137.50	137.98	138.46	140.40
West North Central	140.93	139.62	138.31	136.11	133.92	132.73	131.54	130.13	128.72	127.61
Mid-Atlantic	137.29	138.11	138.92	136.64	134.36	136.38	138.40	138.45	138.49	139.65
East North Central	136.06	133.94	131.82	129.68	127.54	126.86	126.18	126.35	126.51	127.55
Mountain	135.38	132.95	130.51	126.40	122.30	120.52	118.75	118.72	118.68	118.79
West South Central	125.15	122.16	119.16	116.00	112.83	110.54	108.25	106.92	105.60	105.77
East South Central	119.76	116.46	113.16	110.19	107.22	104.64	102.07	100.74	99.41	98.07
South-Atlantic	110.32	108.56	106.79	105.20	103.61	102.87	102.13	101.50	100.86	100.23
Average ALL	134.70	132.85	131.01	128.94	126.86	125.80	124.74	124.22	123.71	124.04
Average High-Five	144.33	143.11	141.90	140.53	139.16	138.73	138.29	138.03	137.76	138.70
Average Low-Four	122.65	120.03	117.41	114.45	111.49	109.64	107.80	106.97	106.14	105.72
Ratio (High/Low)	1.18	1.19	1.21	1.23	1.25	1.27	1.28	1.29	1.30	1.31

Census Division	1931+	1932	1933	1934+	1935	1936+	1937	1938+	1939	1940+
Pacific	159.97	160.01	160.05	160.09	159.26	158.44	158.03	157.62	158.64	159.72
New England	142.35	144.43	146.51	148.60	149.39	150.18	152.13	154.08	157.82	161.55
West North Central	126.50	126.32	126.14	125.96	126.05	126.14	125.54	124.95	124.06	123.14
Mid-Atlantic	140.82	143.75	146.69	149.62	152.33	155.05	157.87	160.69	165.19	169.76
East North Central	128.59	128.84	129.10	129.36	129.89	130.42	131.20	131.98	132.66	133.36
Mountain	118.89	118.32	117.74	117.16	118.48	119.80	119.84	119.88	119.95	119.89
West South Central	105.95	105.53	105.11	104.68	104.10	103.52	103.15	102.79	103.37	103.94
East South Central	96.73	95.15	93.58	92.00	90.97	89.94	89.07	88.21	87.03	85.83
South-Atlantic	99.59	99.20	98.80	98.41	98.78	99.16	99.21	99.26	100.06	100.74
Average ALL	124.38	124.62	124.86	125.10	125.47	125.85	126.23	126.61	127.64	128.66
Average High-Five	139.65	140.67	141.70	142.72	143.38	144.04	144.96	145.87	147.68	149.51
Average Low-Four	105.29	104.55	103.81	103.06	103.08	103.10	102.82	102.53	102.60	102.60
Ratio (High/Low)	1.33	1.35	1.37	1.38	1.39	1.40	1.41	1.42	1.44	1.46

Table 3-A Part 2

Universal PhysPop Table

● PhysPop values are numbers of physicians per 100,000 population. Entries are for general practitioners and specialists combined --- 1921 through 1993 (details in text). Sources of the data are provided in the text, Part 3d. The years which are flagged with a "+" sign present prime data. Entries for the unflagged years have been interpolated.

● The particular states belonging to each Census Division are listed in the text, Part 3b. PhysPop entries for the Nine Census Divisions have been weighted by state populations, whereas the three rows of averages are non-weighted. High-5 and Low-4 are defined in the text, Part 4.

● This single table is the source of data for numerous chapters of this book. The term "universal" in the table's title emphasizes that the PhysPops are the same, regardless of which cause of death is compared with them.

Census Division	1941	1942+	1943	1944	1945	1946	1947	1948	1949+	1950
Pacific	152.84	145.95	146.22	146.48	146.74	147.00	147.27	147.53	147.79	148.60
New England	162.77	163.99	163.77	163.55	163.33	163.11	162.88	162.66	162.44	162.51
West North Central	125.09	127.05	126.21	125.38	124.54	123.76	122.87	122.04	121.20	120.06
Mid-Atlantic	172.19	174.63	173.93	173.23	172.53	171.83	171.13	170.43	169.73	168.71
East North Central	134.12	134.89	133.48	132.06	130.65	129.24	127.82	126.41	125.00	123.69
Mountain	118.18	116.46	117.01	117.55	118.09	118.64	119.18	119.72	120.27	119.38
West South Central	104.41	104.88	104.42	103.31	103.52	103.06	102.61	102.16	101.40	101.34
East South Central	86.16	86.49	85.94	85.39	84.84	84.29	83.74	83.19	82.64	83.05
South-Atlantic	101.71	102.68	102.10	101.53	100.95	100.38	99.80	99.22	98.65	99.07
Average ALL	128.61	128.56	128.12	127.61	127.24	126.81	126.37	125.93	125.46	125.16
Average High-Five	149.40	149.30	148.72	148.14	147.56	146.99	146.39	145.81	145.23	144.71
Average Low-Four	102.62	102.63	102.37	101.95	101.85	101.59	101.33	101.07	100.74	100.71
Ratio (High/Low)	1.46	1.45	1.45	1.45	1.45	1.45	1.44	1.44	1.44	1.44

Census Division	1951	1952	1953	1954	1955	1956	1957	1958	1959+	1960
Pacific	149.40	150.21	151.01	151.82	152.62	153.43	154.23	155.04	155.84	158.74
New England	162.59	162.66	162.74	162.81	162.88	162.96	163.03	163.11	163.18	164.37
West North Central	118.09	117.77	116.62	115.48	114.34	113.19	112.05	110.90	109.76	111.25
Mid-Atlantic	167.68	166.66	165.81	164.62	163.59	162.57	161.55	160.52	159.50	162.65
East North Central	122.37	121.06	119.75	118.44	117.12	115.81	114.50	113.18	111.87	114.56
Mountain	118.51	117.64	116.76	115.88	115.00	114.12	113.25	112.37	111.49	112.93
West South Central	101.28	101.21	101.15	101.09	101.03	100.97	100.90	100.84	100.78	101.65
East South Central	83.46	83.86	84.27	84.68	85.09	85.50	85.90	86.31	86.72	88.00
South-Atlantic	99.49	99.91	100.33	100.75	101.16	101.58	102.00	102.42	102.84	105.36
Average ALL	124.76	124.55	124.27	123.95	123.65	123.35	123.05	122.74	122.44	124.39
Average High-Five	144.03	143.67	143.19	142.63	142.11	141.59	141.07	140.55	140.03	142.31
Average Low-Four	100.69	100.66	100.63	100.60	100.57	100.54	100.51	100.49	100.46	101.99
Ratio (High/Low)	1.43	1.43	1.42	1.42	1.41	1.41	1.40	1.40	1.39	1.40

Table 3-A Part 3

Universal PhysPop Table
Third page of four

● PhysPop values are numbers of physicians per 100,000 population. Entries are for general practitioners and specialists combined --- 1921 through 1993 (details in text). Sources of the data are provided in the text, Part 3d. The years which are flagged with a "+" sign present prime data. Entries for the unflagged years have been interpolated.

● The particular states belonging to each Census Division are listed in the text, Part 3b. PhysPop entries for the Nine Census Divisions have been weighted by state populations, whereas the three rows of averages are non-weighted. High-5 and Low-4 are defined in the text, Part 4.

● This single table is the source of data for numerous chapters of this book. The term "universal" in the table's title emphasizes that the PhysPops are the same, regardless of which cause of death is compared with them.

Census Division	1961	1962	1963+	1964	1965+	1966	1967	1968	1969	1970+
Pacific	161.64	164.55	167.45	167.54	167.62	170.86	174.10	177.35	180.59	183.83
New England	165.56	166.75	167.94	170.52	173.09	175.77	178.46	181.14	183.83	186.51
West North Central	112.74	114.24	115.73	118.25	120.76	121.36	121.96	122.57	123.17	123.77
Mid-Atlantic	165.80	168.94	172.09	175.22	178.34	181.07	183.80	186.54	189.27	192.00
East North Central	117.25	119.94	122.63	123.16	123.69	124.39	125.08	125.78	126.47	127.17
Mountain	114.37	115.81	117.25	117.26	117.26	121.26	125.26	129.27	133.27	137.27
West South Central	102.52	103.38	104.25	104.28	104.31	106.09	107.87	109.64	111.42	113.20
East South Central	89.28	90.57	91.85	92.98	94.11	95.47	96.82	98.18	99.53	100.89
South-Atlantic	107.88	110.39	112.91	115.41	117.91	120.47	123.03	125.58	128.14	130.70
Average ALL	126.34	128.29	130.23	131.62	133.01	135.19	137.38	139.56	141.74	143.93
Average High-Five	144.60	146.88	149.17	150.93	152.70	154.69	156.68	158.67	160.66	162.66
Average Low-Four	103.51	105.04	106.57	107.48	108.40	110.82	113.24	115.67	118.09	120.52
Ratio (High/Low)	1.40	1.40	1.40	1.40	1.41	1.40	1.38	1.37	1.36	1.35

Census Division	1971	1972	1973	1974	1975+	1976	1977	1978	1979	1980+
Pacific	188.70	193.57	198.45	203.32	208.19	213.72	219.25	224.78	230.31	235.84
New England	192.25	197.99	203.72	209.46	215.20	223.03	230.87	238.70	246.54	254.37
West North Central	127.20	130.63	134.06	137.49	140.92	145.91	150.90	155.88	160.87	165.86
Mid-Atlantic	196.24	200.47	204.71	208.94	213.18	218.03	222.87	227.72	232.56	237.41
East North Central	130.94	134.72	138.49	142.27	146.04	150.79	155.54	160.29	165.04	169.79
Mountain	141.06	144.85	148.65	152.44	156.23	160.54	164.84	169.15	173.45	177.76
West South Central	116.19	119.18	122.18	125.17	128.16	133.16	138.17	143.17	148.18	153.18
East South Central	104.18	107.47	110.75	114.04	117.33	121.77	126.20	130.64	135.07	139.51
South-Atlantic	135.78	140.86	145.94	151.02	156.10	162.32	168.55	174.77	181.00	187.22
Average ALL	148.06	152.19	156.33	160.46	164.59	169.92	175.24	180.57	185.89	191.22
Average High-Five	167.07	171.48	175.89	180.30	184.71	190.30	195.89	201.47	207.06	212.65
Average Low-Four	124.30	128.09	131.88	135.67	139.45	144.45	149.44	154.43	159.42	164.42
Ratio (High/Low)	1.34	1.34	1.33	1.33	1.32	1.32	1.31	1.30	1.30	1.29

Table 3-A Part 4

Universal PhysPop Table   Fourth page of four

● PhysPop values are numbers of physicians per 100,000 population. Entries are for general practitioners and specialists combined --- 1921 through 1993 (details in text). Sources of the data are provided in the text, Part 3d. The years which are flagged with a "+" sign present prime data. Entries for the unflagged years have been interpolated.

● The particular states belonging to each Census Division are listed in the text, Part 3b. PhysPop entries for the Nine Census Divisions have been weighted by state populations, whereas the three rows of averages are non-weighted. High-5 and Low-4 are defined in the text, Part 4.

● This single table is the source of data for numerous chapters of this book. The term "universal" in the table's title emphasizes that the PhysPops are the same, regardless of which cause of death is compared with them.

Census Division	1981+	1982	1983+	1984	1985+	1986	1987	1988	1989	1990+
Pacific	241.07	245.83	250.59	253.18	255.78	257.64	259.50	261.37	263.23	265.09
New England	261.79	270.07	278.35	285.44	292.52	298.00	303.47	308.94	314.41	319.88
West North Central	170.49	175.13	179.76	183.06	186.36	189.65	192.93	196.21	199.50	202.78
Mid-Atlantic	245.75	255.00	264.24	270.03	275.83	280.22	284.61	289.01	293.40	297.79
East North Central	174.96	180.94	186.91	190.82	194.72	197.49	200.25	203.01	205.78	208.54
Mountain	182.02	184.91	187.80	190.17	192.53	195.67	198.80	201.93	205.07	208.20
West South Central	156.72	160.32	163.92	167.48	171.04	173.70	176.36	179.02	181.68	184.34
East South Central	144.39	148.87	153.34	157.67	162.00	166.09	170.17	174.25	178.34	182.42
South-Atlantic	191.23	197.83	204.43	210.15	215.86	219.59	223.31	227.03	230.76	234.48
Average ALL	196.49	202.10	207.70	212.00	216.30	219.78	223.27	226.75	230.24	233.72
Average High-Five	218.81	225.39	231.97	236.51	241.04	244.60	248.15	251.71	255.26	258.82
Average Low-Four	168.59	172.98	177.37	181.37	185.36	188.76	192.16	195.56	198.96	202.36
Ratio (High/Low)	1.30	1.30	1.31	1.30	1.30	1.30	1.29	1.29	1.28	1.28

Census Division	1991	1992+	1993+ *
Pacific	266.57	268.05	269.50
New England	327.11	334.35	343.80
West North Central	209.48	216.17	219.00
Mid-Atlantic	307.67	317.56	323.60
East North Central	215.02	221.50	225.40
Mountain	211.23	214.26	218.30
West South Central	189.43	194.53	195.40
East South Central	188.38	194.33	196.70
South-Atlantic	239.45	244.41	247.80
Average ALL	239.37	245.02	248.83
Average High-Five	265.17	271.53	276.26
Average Low-Four	207.12	211.88	214.55
Ratio (High/Low)	1.28	1.28	1.29

* 1993 entries are for January 1, 1993, from Roback 1994.

Table 3-B

Population Sizes of the Census Divisions: 1910 through 1990

Census Division	1910 Pop	1910 Fraction	1920 Pop	1920 Fraction	1930 Pop	1930 Fraction
Pacific	4,192,304	0.0457	5,566,851	0.0529	8,194,433	0.0670
New England	6,652,675	0.0725	7,400,909	0.0703	8,268,680	0.0676
West North Central	11,637,921	0.1269	12,544,249	0.1192	13,296,915	0.1086
MidAtlantic	19,315,892	0.2105	22,261,144	0.2115	26,260,750	0.2146
East North Central	18,250,621	0.1989	21,475,543	0.2040	25,297,185	0.2067
Mountain	2,633,517	0.0287	3,336,101	0.0317	3,702,789	0.0303
West South Central	8,784,534	0.0958	10,242,224	0.0973	12,176,830	0.0995
East South Central	8,409,901	0.0917	8,893,307	0.0845	9,887,214	0.0808
South Atlantic	11,864,826	0.1293	13,552,701	0.1287	15,306,720	0.1251
	91,742,191	1.0000	105,273,029	1.0000	122,391,516	1.0000

Census Division	1940 Pop	1940 Fraction	1950 Pop	1950 Fraction	1960 Pop	1960 Fraction
Pacific	9,733,262	0.0739	14,486,527	0.0961	21,198,044	0.1182
New England	8,437,290	0.0641	9,314,453	0.0618	10,509,367	0.0586
West North Central	13,516,990	0.1027	14,061,394	0.0933	15,394,115	0.0858
MidAtlantic	27,539,487	0.2092	30,163,533	0.2002	34,168,452	0.1905
East North Central	26,626,342	0.2022	30,399,368	0.2017	36,225,024	0.2020
Mountain	4,150,003	0.0315	5,074,998	0.0337	6,855,060	0.0382
West South Central	13,064,525	0.0992	14,537,572	0.0965	16,951,255	0.0945
East South Central	10,778,225	0.0819	11,477,181	0.0762	12,050,126	0.0672
South Atlantic	17,823,151	0.1354	21,182,335	0.1406	25,971,732	0.1448
	131,669,275	1.0000	150,697,361	1.0000	179,323,175	1.0000

Census Division	1970 Pop	1970 Fraction	1980 Pop	1980 Fraction	1990 Pop	1990 Fraction
Pacific	26,087,000	0.1293	31,523,000	0.1398	37,837,000	0.1535
New England	11,781,000	0.0584	12,322,000	0.0546	12,998,000	0.0527
West North Central	16,240,000	0.0805	17,124,000	0.0759	17,777,000	0.0721
MidAtlantic	37,149,000	0.1842	36,770,000	0.1630	37,660,000	0.1527
East North Central	40,212,000	0.1993	41,636,000	0.1846	42,232,000	0.1713
Mountain	8,230,000	0.0408	11,319,000	0.0502	13,398,000	0.0543
West South Central	19,132,000	0.0948	23,669,000	0.1049	26,797,000	0.1087
East South Central	12,723,000	0.0631	14,573,000	0.0646	15,313,000	0.0621
South Atlantic	30,169,000	0.1496	36,621,000	0.1624	42,540,000	0.1725
	201,723,000	1.0000	225,557,000	1.0000	246,552,000	1.0000

Some sources provided entries to the last digit, but no one should take seriously any such implied accuracy of census-taking. Sources: For 1910, 1920, 1930: World Almanac 1991, p.553. For 1940, 1950, 1960: Grove 1968, Table 74. For 1970, 1980: Roback 1990. For 1990: Roback 1994. Entries above exclude no one by color or "race."

Table 3-C. How Sets of Physpops Correlate through Time.

Related text = Part 7c.

The 9 Census Div	1993 Phys/Pop	1992 Phys/Pop	1990 Phys/Pop	1985 Phys/Pop	1980 Phys/Pop	1975 Phys/Pop	1967 Phys/Pop	1965 Phys/Pop	1963 Phys/Pop	1949 Phys/Pop	1942 Phys/Pop	1940 Phys/Pop	1938 Phys/Pop	1936 Phys/Pop	1934 Phys/Pop	1931 Phys/Pop	1929 Phys/Pop	1927 Phys/Pop	1925 Phys/Pop	1923 Phys/Pop	1921 Phys/Pop
New England	343.8	334.3	319.9	292.5	254.2	215.1	174.5	168.6	167.1	162.4	164.0	161.6	154.1	150.2	148.6	142.3	138.5	137.5	138.3	137.4	142.2
Middle Atlantic	323.6	317.6	297.8	275.8	237.3	213.1	178.2	173.1	168.7	169.7	174.6	169.8	160.7	155.1	149.6	140.8	138.5	138.4	134.5	138.9	137.3
East North Centr	225.4	221.5	208.6	194.7	169.8	145.9	124.5	121.3	118.2	125.0	134.9	133.4	132.0	130.4	129.4	128.6	126.5	126.2	127.5	131.8	136.1
West North Centr	219.0	216.2	202.8	186.4	165.8	140.8	119.1	116.2	114.0	121.2	127.1	123.1	125.0	126.1	126.0	125.5	128.7	131.5	133.9	138.3	140.9
South Atlantic	247.8	244.4	234.5	215.9	187.0	156.0	122.7	118.4	113.0	98.7	102.7	100.7	99.3	99.2	98.4	99.6	100.9	102.1	103.6	106.8	110.3
East South Centr	196.7	194.3	182.4	162.0	139.7	117.4	93.4	90.5	89.2	83.2	86.5	85.8	88.2	89.9	92.0	96.7	99.4	102.1	107.2	113.2	119.8
West South Centr	195.4	194.5	184.3	171.0	153.3	128.0	106.2	103.4	102.5	102.2	104.9	103.9	102.8	103.5	104.7	106.0	105.6	108.2	112.8	119.2	125.2
Mountain	218.3	214.3	208.2	192.5	177.5	155.9	125.1	121.0	117.8	119.7	116.1	119.9	119.9	119.8	117.2	118.9	118.7	118.7	122.3	130.5	135.4
Pacific	269.5	268.0	265.1	255.8	236.2	208.1	167.3	161.4	159.6	147.5	146.0	159.7	157.6	158.4	160.1	160.0	156.6	157.8	161.7	163.1	165.1

Correlation of Each Phys/Pop with All Other Phys/Pops (Measured in R-Squared).

Year	1993	1992	1990	1985	1980	1975	1967	1965	1963	1949	1942	1940	1938	1936	1934	1931	1929	1927	1925	1923	1921
Phys/Pop 21	0.15	0.16	0.20	0.25	0.34	0.38	0.41	0.41	0.44	0.48	0.43	0.58	0.65	0.72	0.77	0.87	0.88	0.90	0.96	0.98	1.00
Phys/Pop 23	0.20	0.21	0.25	0.31	0.40	0.45	0.49	0.49	0.52	0.56	0.51	0.67	0.78	0.83	0.84	0.91	0.94	0.95	0.96	1.00	
Phys/Pop 25	0.28	0.29	0.33	0.40	0.49	0.53	0.56	0.56	0.59	0.61	0.57	0.71	0.77	0.83	0.88	0.95	0.97	0.98	1.00		
Phys/Pop 27	0.38	0.39	0.43	0.49	0.58	0.62	0.67	0.67	0.69	0.72	0.69	0.81	0.87	0.92	0.95	0.98	0.99	1.00			
Phys/Pop 29	0.42	0.43	0.47	0.53	0.61	0.66	0.71	0.71	0.73	0.76	0.73	0.85	0.90	0.94	0.97	0.99	1.00				
Phys/Pop 31	0.45	0.46	0.50	0.57	0.65	0.69	0.74	0.74	0.76	0.79	0.76	0.88	0.92	0.96	0.98	1.00					
Phys/Pop 34	0.56	0.57	0.60	0.66	0.72	0.77	0.82	0.83	0.85	0.89	0.87	0.95	0.96	0.99	1.00						
Phys/Pop 36	0.60	0.61	0.63	0.69	0.74	0.79	0.85	0.86	0.87	0.93	0.91	0.98	0.99	1.00							
Phys/Pop 38	0.65	0.65	0.67	0.72	0.77	0.81	0.88	0.89	0.90	0.96	0.94	0.99	1.00								
Phys/Pop 40	0.71	0.72	0.73	0.78	0.81	0.85	0.91	0.92	0.93	0.98	0.96	1.00									
Phys/Pop 42	0.76	0.76	0.74	0.77	0.76	0.79	0.88	0.89	0.89	0.98	1.00										
Phys/Pop 49	0.77	0.78	0.78	0.81	0.82	0.85	0.92	0.93	0.89	1.00											
Phys/Pop 63	0.87	0.88	0.90	0.94	0.96	0.98	1.00	1.00	1.00												
Phys/Pop 65	0.87	0.88	0.91	0.94	0.96	0.98	1.00	1.00													
Phys/Pop 67	0.87	0.89	0.91	0.95	0.96	0.99	1.00														
Phys/Pop 75	0.88	0.89	0.93	0.96	0.99	1.00															
Phys/Pop 80	0.90	0.91	0.95	0.98	1.00																
Phys/Pop 85	0.96	0.97	0.99	1.00																	
Phys/Pop 90	0.99	0.99	1.00																		
Phys/Pop 92	0.99	1.00																			
Phys/Pop 93	1.00																				
Years	1993	1992	1990	1985	1980	1975	1967	1965	1963	1949	1942	1940	1938	1936	1934	1931	1929	1927	1925	1923	1921

Table 3-C: The 21 sets of Phys/Pop values for 1921–1993 match Table 3-A, with some exceptions (please see text of Part 7c). The rows of R-squared values come from regressing one PhysPop set upon another PhysPop set. The intersection of a column with a row reveals which two sets produced the R-squared value.

Definition of Mortality Rate

From here on, we will abbreviate "age-adjusted annual mortality-rate, per 100,000 population (everyone)" as mortality rate. For brevity, MortRate. When pressed for space: MR or mr. The word "annual" is rarely expressed in connection with mortality rates, but they are rates per year. By "everyone," we mean there are no exclusions by color or "race."

Location of MortRate Tables in This Book

Unlike the Universal PhysPop Table 3-A, which is used repeatedly in chapter after chapter, the tables of mortality rates have to be specific for each cause of death studied. So readers will find such rates located at the end of the appropriate chapters. The first one occurs at the end of Chapter 6. Because age-adjusted mortality rates include ALL AGES at death, the MortRates are designated "male" or "female," rather than "rates for men" or "rates for women."

Meaning and Utility of High-5 and Low-4 MortRates

Readers will see High-5 and Low-4 MortRates in our MortRate tables. The High-5 MortRate always refers to the average MortRate in the Pacific, New England, West North Central, Mid-Atlantic, and East North Central Census Divisions combined. The Low-4 MortRate always refers to the average MortRate in the Mountain, West South Central, East South Central, and South Atlantic Census Divisions combined. UNLIKE the MortRates for individual Census Divisions, the High-5 and Low-4 MortRates are not population-weighted. They are approximations, useful in checking trends. For instance, as the Hi5/Lo4 ratio for male All-Cancer MortRates falls from 1.44 in 1940 to 0.96 in 1988 (Table 6-A), a reader knows instantly that major changes are occurring in the geographical distribution of cancer MortRates. Chapter 3, Part 4, explains how the permanent order in our list of Census Divisions was set in the first place.

● Part 1. Were Mortality-Rates Always Recorded?

We start our dose-response studies by comparing the behavior of CANCER mortality-rates and PhysPop values. Then we must do identical analyses of noncancer mortality-rates, to find out if cancer and noncancer MortRates behave DIFFERENTLY with respect to PhysPop values. (They do.)

For our proposed dose-response studies, we would like to know the age-adjusted cancer MortRates and the PhysPops in the Nine Census Divisions, all the way back at least to 1895 --- the year in which Roentgen published his discovery of the xray (Roentgen 1895). But that is not possible. Although it is pleasant to assume that Vital Statistics on births, deaths, causes of death, and so forth, have "always" been there, reality is very different.

The death registration-system evolved gradually, from only 10 states in 1900, to all 48 states in 1933 --- with Alaska added in 1959 and Hawaii in 1960. From the 1968 Grove book (p.7), we quote:

"The annual collection of mortality statistics for the national death-registration area began with the calendar year 1900. For the year 1900, figures were obtained from well-established registration areas which had adopted model laws and where it was believed that registration was at least 90% complete. [Example: Mass 1890 in our Reference List.] Each area had been requested to adopt a recommended death certificate by January 1, 1900. The death registration area for 1900 consisted of 10 states, the District of Columbia, and a number of cities located in nonregistration states ... From this time on, the Bureau [of the Census] completely abandoned the 50-year effort to obtain mortality information by census counts and relied solely on registration records." From Grove's Table B, we identify (below) the sequence in which the nationwide registry-system evolved, by Census Divisions. We exclude Alaska, Hawaii, and the District of Columbia, as stated in Chapter 3, Part 3b.

● EAST NORTH CENTRAL: Illinois 1918, Indiana 1900, Michigan 1900, Ohio 1909, Wisconsin 1908.

● EAST SOUTH CENTRAL: Alabama 1925, Kentucky 1911, Mississippi 1919, Tennessee 1917.

● MIDDLE ATLANTIC: New Jersey 1880, New York 1890, Pennsylvania 1906.

● MOUNTAIN: Arizona 1926, Colorado 1906, Idaho 1922, Montana 1910, Nevada 1929, New Mexico 1929, Utah 1910, Wyoming 1922.

● NEW ENGLAND: Connecticut 1890, Maine 1900, Massachusetts 1880, New Hampshire 1890, Rhode Island 1890, Vermont 1890.

● PACIFIC: California 1906, Oregon 1918, Washington 1908.

● SOUTH ATLANTIC: Delaware 1890 (dropped in 1900, re-admitted in 1919), Florida 1919, Georgia 1922, Maryland 1906, North Carolina 1916, South Carolina 1916, Virginia 1913, West Virginia 1925.

● WEST NORTH CENTRAL: Iowa 1923, Kansas 1914, Minnesota 1910, Missouri 1911, Nebraska 1920, North Dakota 1924, South Dakota 1906 (dropped in 1910; re-admitted in 1930).

● WEST SOUTH CENTRAL: Arkansas 1927, Louisiana 1918, Oklahoma 1928, Texas 1933.

● Part 2. Our Sources for Mortality Rates

The age-adjusted mortality rates in this book, per 100,000 population (all colors) come from three sources.

2a. Mortality Rates for 1940, 1950, 1960

The rates for 1940, 1950, and 1960 come from Grove 1968 in our Reference List. The book is "Vital Statistics Rates in the United States 1940-1960" by Robert D. Grove and Alice M. Hetzel, of the U.S. Public Health Service, National Center for Health Statistics (NatCtrHS).

In Grove 1968 is Table 67, entitled Age-Adjusted Death Rates for 32 Selected Causes by Color and Sex: United States, Each Division and State, 1940, 1950, and 1960. In Part 5 of this chapter, we show what Table 67 covers. The mammoth table begins on page 663 of Grove 1968, and continues through page 770.

Table 67 presents MortRates for 1940, 1950, 1960 which are age-adjusted to 1940 (discussion in Part 4, below). We use the rates for everyone combined (no exclusions by color or "race"). Our exclusion of Alaska, Hawaii, and the District of Columbia (Chapter 3, Part 3b) merits a comment, here:

Grove states explicitly (p.40) that the 1940 and 1950 MortRates exclude Alaska and Hawaii, and that Alaska and Hawaii are both INCLUDED in the 1960 rates. So, 1960 is the one year in this book for which the MortRate (Pacific Census Division) includes Alaska and Hawaii. Although we would have preferred to be absolutely consistent, this 1960 exception must be inconsequential, since the

1960 populations of Alaska and Hawaii (226,167 and 632,772 respectively) were too small to have much impact on the 1960 population-weighted MortRate of the entire Pacific Census Division (population = 21 million, from Table 3-B). As for the District of Columbia (which is part of the South Atlantic Census Division), Grove's handling is unclear to us. What is clear is that the population of the District (663,091 in 1940, 763,956 in 1960) was too small to have much impact on the 1940-1960 population-weighted MortRates of the entire South Atlantic Census Division (population = 18 million in 1940, 21 million in 1950, and 26 million in 1960; from Table 3-B).

By contrast with the post-1960 MortRates by CENSUS DIVISIONS, the 1960 and post-1960 consolidated NATIONAL MortRates in our tables do include Alaska, Hawaii, and the District of Columbia. As readers observe what we do with all these numbers, they will see that this inconsistency can be of no consequence in testing Hypotheses 1+2.

2b. Mortality Rates for 1970, 1980, 1990

Unfortunately, there is no publication comparable to Grove 1968 for the years after 1960. The reason is said to be insufficient funding for the work.

● 1980: The National Center for Health Statistics in Hyattsville, Maryland, has an exemplary policy of trying to provide data for research anyway. We extend our deepest appreciation for the cooperation of Dr. Harry Rosenberg and Dr. Jeff Maurer of the Center. They provided data for 1979-1981 which is comparable to Table 67 of Grove 1968. These data (over 150 pages of printout) are NatCtrHS 1980 in our Reference List. The printout covers many, but not all, of the causes of death we examined from Grove 1968. Each page begins with the following header:

"Average age-adjusted death rates and standard errors (SE) for major causes, by race and sex; United States and rank for each state, 1979-1981. Based on age-specific death rates per 100,000 population ... using as standard population the age distribution of total U.S. population as enumerated in 1940." So these data (all "races" combined) provide input which can be regarded as 1980 data.

● 1970: At the time we needed all these data, Drs. Rosenberg and Maurer were unable to provide for 1970 what they had provided for 1980. Therefore, we produced synthetic values for 1970 by taking the average of 1960 and 1980 in each Census Division. As readers observe what we do with all these numbers, they will see that the approximation for 1970 is of no consequence in testing Hypotheses 1+2.

● 1990: If we were starting this project NOW, the acquisition of 1990 health data would be much easier. But when we DID start, Drs. Rosenberg and Maurer recommended that we patch together whatever we could find for 1990 from a feature called State Maps in the publication "Monthly Vital Statistics Report," which is MVS in our Reference List. On an irregular basis, issues of MVS used to provide (until recently) age-adjusted mortality rates with the 1940 reference year, by gender and states, for variable sets of years such as 1987-1989 and 1989-1991 --- for many but not for all of the entities in Grove's Table 67. One of the missing entities: Ischemic Heart Disease.

As our work progressed, so did the online resources of the National Center for Health Statistics (which is a subdivision of the federal CDC --- Centers for Disease Control and Prevention). Ultimately, it became possible to extract the 1993 Ischemic Heart Disease MortRates (all-race males, all-race females, age-adjusted to 1940, by states) from the CDC's online "WONDER" database.

We are comfortable about treating MortRates in the 1988-1993 range as if they were 1990 MortRates, because such rates seldom change appreciably over short periods of time.

Unlike the 1940, 1950, and 1960 data from Grove 1968, the MortRates per 100,000 population for 1980 and "1990" were not "pre-packaged" by Census Divisions. They were obtained by states. We grouped states into their proper Census Divisions (Part 1), and then used the 1980 or 1990 populations of each state to obtain population-weighted MortRates for each Division.

2c. Some Incomplete Sections in Our Mortality-Rate Tables

Some of the MortRate tables in this book are incomplete, our entries for 1970 are interpolations between 1960 and 1980, and instead of all-1990 rates, sometimes we obtained 1988 or 1993 rates, as noted above. If we were starting NOW "from scratch," we are confident that we would be able to acquire more of the missing data, because recently the federal government has greatly improved online access to its current health statistics. See, for instance: < http://wonder.cdc.gov/WONDER/ >, and also the "FedStats" site, < http://www.fedstats.gov >. Online availability of older data may (or may not) improve too.

As readers see the nature of our analyses, they will see for themselves why we are correct in asserting that the remaining gaps or interpolations in our MortRate tables are only marginally relevant to examining Hypotheses 1+2. For instance, the MVS Reports did not provide the 1990 MortRates by Census Divisions for Urinary-System Cancers, for female Genital Cancers, and for male Buccal/Pharynx Cancers. These arbitrary gaps have no effect on Hypothesis-1, which concerns All Cancer Combined --- for which we do have data. Out of curiosity (not out of necessity), we opted to examine as many components of All Cancers Combined as the available data permitted at the time. The fact that some data were not available does not weaken the examination of Hypothesis-1.

In Section Three, we undertook to find out if cancer and noncancer MortRates behave differently with respect to PhysPop values. We found ample data (Chapter 24) to establish that they do. Again out of curiosity (not out of necessity), we opted to examine as many components of All Noncancers Combined as the available data permitted at the time. The fact that we are missing post-1960 MortRates for Appendicitis, Hypertensive Disease, Rheumatic Fever & Rheumatic Heart Disease, Syphilis, Tuberculosis, and Ulcers, cannot undermine what we needed to establish (and did establish) in Chapters 24 and 25, with respect to Hypotheses 1+2.

2d. Changes in ICD Numbers over Time

In our MortRate tables, we show numbers from the International Classification of Diseases (ICD) whenever possible, so that anyone can check exactly what the MortRates cover. In Part 5 of this chapter, we show in detail how disorders are grouped together.

Most of our MortRate tables cover 1940-1990. During those decades, the numbers assigned to various diseases were altered in the ICD system. ICD/7 indicates the Seventh Revision of ICD, and ICD/9 indicates the Ninth Revision. In the MortRate tables of this book, the ICD numbers for the 1940-1960 entries (ICD/7) often differ from the ICD/9 numbers stated for 1980. The continuity in MortRates from decade to decade is maintained by definition of the named entities, not by ICD numbers.

2e. Are There Worrisome Discrepancies between Sources?

It is easy to conjure up APPARENT disparities in mortality rates among various sources: The government's publications, its online databases, and non-government publications. We have a few comments about this:

● - Most often, apparent discrepancies disappear when one stops trying to compare two non-comparable tables!

● - In Part 2a of this chapter, we inform readers that MortRates in our tables, by Census Divisions, exclude Alaska, Hawaii, and the District of Columbia --- which means that our tables are not comparable to tables which include those entities.

● - In Part 4a of this chapter, we remind readers never to compare two sets of MortRates which are age-adjusted to different reference years.

● - In Chapter 23, Part 2b, we warn readers not to compare death-rates from 32-Causes-of-Death-Combined, with death-rates in other publications for ALL Causes of Death Combined.

● - In Section Five, we warn readers not to compare MortRates for Ischemic Heart Disease (or Coronary Heart Disease) with MortRates for "Diseases of the Heart" in other publications. These are not the same entities. Apples and oranges.

● - The National Center for Health Statistics deserves enormous credit for the overwhelming agreement among its own databases. Apparent discrepancies are rare, and seem to have explanations if one pursues the issue far enough.

● - The minor discrepancies which remain among various databases certainly do not have a magnitude capable of invalidating the findings in this book. The discrepancies are inconsequential.

● **Part 3. Estimated Frequency of Misdiagnosis on Death Certificates**

Some efforts have been made to estimate the rate at which Cancer is under-reported on death certificates. For example, the handlers of the Atomic-Bomb Survivor Database decided that they needed to multiply the reported rates for malignancies by a factor of 1.23, in order to compensate for "incomplete death-certificate ascertainment of cancer" (discussion in Gofman 1990, p.11-3). Their decision was based on a 1973 study comparing autopsy and death-certificate diagnoses (Steer 1973; BEIR 1980, p.196; additional references in Pierce 1996-b, p.3).

Such underascertainment in the past is not surprising to us. We are directly aware that, in some circles, there was almost a taboo against admitting that a family member died of Cancer. In addition, it can be medically difficult sometimes, without an autopsy, to distinguish between various causes of death. For instance, undoubtedly some death certificates saying cirrhosis of the liver really should say metastatic Cancer, and some certificates attributing death to stomach ulcer should really say Cancer.

3a. USA: Estimated Frequency of "Missed Cancer Diagnoses"

On the problem of under-reporting of Cancer, and misdiagnosis of a cancer's original site, Rolla B. Hill and Robert E. Anderson contributed an eye-opening article entitled "The Autopsy in Oncology" (Hill 1992). In their article, they provide Table 2, entitled "Literature Sources for Clinical Cancer Diagnosis Compared with Autopsy Findings." The table describes eleven reports, covering various years between 1917 to 1988. The first report in their list is by H.G. Wells (Wells 1923). They state (Hill 1992, p.48):

"H.G. Wells reviewed 578 autopsies of cancer patients from Chicago hospitals during 1917 to 1922 and found that in 31 percent, Cancer was first discovered at autopsy (clinical false negatives) and that 6 percent of those diagnosed clinically were not present at autopsy (clinical false positives)." And (Hill 1992, p.48):

"It is difficult to accept the fact that subsequent studies, including the most recent, have revealed only slight improvement in the overall incidence of missed cancer diagnoses. It would certainly seem that modern advances and invasive techniques have improved diagnostic ability. Yet the data in Table 2 indicate that this instinctive reaction is faulty ... Parallel findings have been reported for non-oncologic diagnoses, including evidence that modern imaging and invasive techniques must have had little overall effect on the discrepancy rate (Goldman 1983)." And (Hill 1992, p.51):

"Unfortunately, studies of the validity of death-certificate diagnoses suggest that they may be wrong up to 50 percent of the time." For this assertion, Hill and Anderson cite 12 references not already cited in their Table 2. The opportunity to learn from autopsies has diminished dramatically. "In 1950, more than one half of the patients dying in U.S. hospitals were autopsied (Roberts 1978); in 1985, the comparable figure was only 10 percent (MMWR 1988), and it seems to have dropped slightly since then" (Hill 1992, p.47).

3b. Specific Cancers: "We Do Not Know the True Prevalence of Any Type of Cancer"

Referring to Cancer, Hill and Anderson state (Hill 1992, p.49):

"Although it is discouraging ... to recognize that there has been little improvement during the past six to seven decades in overall diagnostic discrepancy rates in cancer patients, this does not mean that the situation has been static. Improvements with one type of tumor are matched by erosions elsewhere (Anderson 1989). Leukemia, for example, is currently rarely missed or overdiagnosed in patients who die and are autopsied, and the ability to make this diagnosis correctly has improved considerably over the past 50 years. In contrast, gastric carcinoma and carcinoma of the bronchus and lung have become appreciably harder to diagnose." Hill and Anderson point out (Hill 1992, p.51):

"Millions of dollars are committed each year to protocol studies, often funded by the National Cancer Institute, in which patients with specific cancers are entered randomly into clinical trials. The results of various approaches to treatment are compared, and the one that seems to hold the most promise becomes the treatment of choice ... It is an arresting fact, however, that performance of autopsy is not built into these studies, and there is no funding to cover the autopsy costs." They report the cost at about $1,000 per autopsy (p.54). Their article makes a compelling case for the recommendation (p.49): "A program must be designed to evaluate the diagnostic error, determine its cause, provide feedback to the appropriate persons, and ensure that the situation improves (Anderson 1990)." Hill and Anderson warn (1992, p.52, emphasis in the original):

"The truth is that WE DO NOT KNOW the true prevalence of any type of cancer. We do not even know how bad our present data are --- we just know that they are inaccurate and probably unreliable." Burton 1998 presents additional data supporting that view (Burton 1998).

It seems reasonable to assume that, when clinicians create a death certificate in the absence of an autopsy, they correctly distinguish Cancer from NonCancer as the cause more often than they correctly name the ORIGINAL SITE of a fatal Cancer.

● Part 4. Age-Adjustment: "Matching" of the Nine Census Divisions

In our dose-response studies, the nine sets of people who receive different average doses of medical radiation are the populations of the Nine Census Divisions. The responses we are examining are the nine separate mortality-rates, by Census Divisions, per 100,000 population.

In order to evaluate the extent to which variation in medical radiation explains variation in MortRates, the MortRates must be validly COMPARABLE among the Nine Census Divisions. If the Pacific Division were populated ONLY by children and the South Atlantic Division were populated ONLY by retired people, their mortality-rates per 100,000 population would clearly not be comparable --- because increased age is such an important correlate of death. In reality, every Census Division has a mixture of all possible ages, but not in the same proportions. The distribution of ages is never identical ("matched") across all Nine Census Divisions. Nor is age-distribution constant over the decades.

4a. The Reference Year for Age-Adjusted MortRates

Because the age-distribution differs among the populations of the Census Divisions, and because age-distributions change with time, what we need to study are AGE-ADJUSTED mortality rates --- always adjusted to a standard population of a specified reference-year. Our sources use 1940 as the reference year, and so we use 1940 consistently throughout the MortRate tables of this book.

Age-adjustment of MortRates can be done to any reference year, so it is important never to ASSUME that mortality-rates are age-adjusted to the same reference year, and never to compare mortality rates which have been age-adjusted to DIFFERENT reference years. While the National Center for Health Statistics and the American Heart Association use 1940 as the reference year for mortality-rates (confirmed in PHS 1995, p.298, and in AHA 1995, p.23), the National Cancer Institute and the American Cancer Society generally use 1970 as the reference year for their recent publications. Some of the difficulties, caused by use of two standards, are apparent in a 1992 report of the Public Health Service, entitled "Health: United States 1992" (PHS 1992, discussion at pages 327-330). We note that in a 1997 article, "Cancer Undefeated," Bailar and Gornik have opted to age-adjust their data to 1990 as the reference year (Bailar 1997, p.1570).

4b. Illustrative Calculation of an Age-Adjusted MortRate: Boxes 1 and 2

When 1940 is the reference year, the actual population of 1940 provides the Standard Million Population. Grove 1968, p.37, provides the "Enumerated Population of the United States, 1940," by age-bands. We show it in Box 1. The same data are available in PHS 1995 (p.298).

In Box 2, we show an example of how the Standard Million Population of 1940 can be applied, to arrive at an age-adjusted mortality rate. The illustration in Box 2 is generic. Column A could be used for age-specific MortRates from any disease, in any year, in any geographical unit. We chose to use, in Column A, the actual 1940 MortRates by age-bands, due to "Cancer and Other Malignant

Tumors (45-55)" in the United States, from Linder 1947, Table 14, p.250. The sum of 120.15 in Box 2 is in excellent agreement with the 1940 national rate of 120.3 per 100,000 in Linder's Table 14. The small discrepancy is undoubtedly due to trailing digits.

Policy on Trailing Digits, Significant Figures

Calculations can never yield answers which are more precise than the least precise of the input numbers. Nonetheless, during a series of calculations, it is wise to retain all the trailing digits to the right of the decimal point. For example, when a measured value of 2 is divided by 3.1, the computer obtains an answer of 0.6451612. We, and most other sources, show readers SOME of the extra digits --- so that readers can reach answers which resemble the answers yielded by calculations done with the trailing digits. When readers and their sources reach slightly different answers, the reason almost certainly lies in additional "trailing digits." Extra digits provided in this book (and in many other sources) do NOT signify that measurements are highly precise. In other words, they are not "significant figures" in the formal sense.

Grove's Note about Table 67

Mortality rates for 1940, 1950, and 1960 in this book are taken from Table 67 of Grove 1968, as stated in Part 2a above. Grove comments (at p.35) that "The age-adjusted rates shown in Table 67 are computed from specific rates of the following broad age groups: Under 25, 25-44, 45-64, and 65 and over. Consequently, the age-adjusted rates shown in this table differ from those shown in other tables [which used the age-bands shown in our Box 1]. The rates for the total population and for each color-sex group are adjusted separately, using the same standard population."

Grove's Definition of Age-Adjusted Death Rates

Grove defines age-adjusted death rates at p.35: "Age-adjusted death rates published in this volume are computed by the direct method. They are the death rates that would have been observed if the age-specific rates for the given year, computed for specified age intervals, had prevailed in a population whose age distribution was the same as that of the standard population. The entire enumerated population of the United States in 1940 is used as the standard."

4c. Comparability of Census Divisions: Matched for Age

The use of age-adjusted mortality rates is the equivalent of matching the populations of all Nine Census Divisions for age. In our dose-response studies, the MortRates from one Census Division to another are properly comparable with respect to the age-issue, because age-adjusted MortRates apply to 100,000 people with the SAME distribution of ages. If the population of one Census Division has a higher fraction of babies than another, or more elderly people, it does not matter --- because the actual number of deaths for a given age-band has been adjusted to conform to the age-distribution of the Standard Million Population of 1940, as illustrated in Part 4b above.

Changes in Population-SIZES over Time

As our studies progress from 1940 to later decades, the input remains comparable over TIME --- despite some rather dramatic changes in the relative population-sizes among the Nine Census Divisions (shown in Table 3-B of Chapter 3) --- because the comparisons are based on Physicians per 100,000 population and age-adjusted MortRates per 100,000 population.

4d. Age-Specific vs. Age-Adjusted MortRates; Boxes 3 and 4

The need for age-adjustment of MortRates does not occur if analysts are comparing rates per 100,000 people of the SAME age, of course. For example, a mortality rate per 100,000 people of age 45 could be directly compared across Census Divisions and across time. The need for age-adjustment occurs when comparisons involve populations having mixtures of ages (for instance, entire populations of Census Divisions).

Box 3 uses Breast Cancer mortality rates to illustrate the relationship among three types of MortRates: Age-adjusted, crude, and age-specific --- over the 1950-1990 period, by decades. While the age-adjusted National MortRate changes very little over time, age-specific rates are falling over time in some age-groups and rising in others, per 100,000 persons of those age-groups.

Box 4 shows how the age-specific 1990 National All-Cancer MortRates (a) provide the basis for the age-adjusted 1990 National All-Cancer MortRate, and (b) provide the basis for learning which age-groups account for various FRACTIONS of the age-adjusted 1990 National All-Cancer MortRate. We needed to know such fractions in Chapter 2, Part 3.

● **Part 5.   Causes of Death Reported in Grove 1968, with Subsets and Numbers**

In Grove 1968, Table 67 ("Age-Adjusted Death Rates," pp.663-770) covers a wide range of entities of interest for our work in this book: Malignancies, acute infections, chronic infections, metabolic disorders, cardiovascular disorders, and accidental deaths --- "32 selected causes" in all.

The list below of the 32 causes is compiled by us from Grove's Table 65, Section F, pp.595-603. The numbers in parentheses are the numbers assigned to the various entities in the International Statistical Classification of Diseases, Injuries, and Causes of Death: Seventh Revision, 1958 --- a publication of the World Health Organization (WHO). So far, nine revisions exist. Their dates of adoption and of use are presented in our Reference List, under the entry WHO 1958.

Grove notes (p.44) that differences between the Sixth and Seventh Revisions are minor, but warns that changes initiated by the Sixth Revision (1949) in "the method followed in selecting the underlying cause of death ... significantly affected the statistics for some causes, e.g. diabetes and nephritis."

In the list of 32 selected causes of death, below, the ICD numbers in parenthesis are from the Seventh Revision. Some causes listed below are subsets of larger groups, and others are combined when they are reported in Grove's Table 67.

1. Tuberculosis, all forms (001-019)
> Tuberculosis of respiratory system (001-008)
> Tuberculosis of meninges and central nervous system (010)
> Tuberculosis of intestines, peritoneum, and mesenteric glands (011)
> Tuberculosis of bones and joints (012,013)
> Tuberculosis of other organs and systems (014-018)
> Disseminated tuberculosis (019)

2. Tuberculosis of the respiratory system (001-008)

3. Syphilis and its sequelae (020-029)
> Congenital syphilis (020)
> Early syphilis (021)
> Aneurysm of Aorta (022)
> Other cardiovascular syphilis (023)
> Tabes Dorsalis (024)
> General paralysis of the insane (025)
> Other syphilis of central nervous system (026)
> All other syphilis (027-029)

4. Malignant Neoplasms, including neoplasms of lymphatic and hematopoietic tissues, (140-205). Also see entries 5 through 12, below.

5. Malignant Neoplasm of buccal cavity and pharynx (140-148)
> Of lip (140)
> Of tongue (141)
> Of other and unspecified parts of buccal cavity (142-144)
> Of pharynx (145-148)

6. Malignant Neoplasm of digestive organs and peritoneum, not specified as secondary (150-156A, 157-159)

Of esophagus (150)
Of stomach (151)
Of small intestine, including duodenum (152)
Of large intestine, except rectum (153)
   Diverse segments given as 153.1, 153.2, etc.
Of rectum (154)
Of biliary passages and of liver (stated to be primary site) (155)
   Liver   (155.0)
   Other and multiple sites of biliary passages (155.1, 155.8)
Of liver not stated whether primary or secondary (156A)
Of pancreas (157)
Of peritoneum and unspecified digestive organs (158, 159)

7. Malignant Neoplasm of respiratory system, not specified as secondary (160-164)
   Of larynx (161)
   Of bronchus and trachea, and of lung specified as primary (162)
   Of lung, unspecified as to whether primary or secondary (163)
   Of other parts of of respiratory system (160, 164)

8. Malignant Neoplasm of breast (170)

9. Malignant Neoplasm of genital organs (171-179)
   Of cervix uteri (171)
   Of other and unspecified parts of uterus (172-174)
   Of ovary, fallopian tube, and broad ligament (175)
   Of other, and unspecified female genital organs (176)
   Of prostate (177)
   Of testis, and of other and unspecified genital organs (178,179)

10. Malignant Neoplasm of urinary organs (180,181)
   Of kidney (180)
   Of bladder and other urinary organs (181)

11. Leukemia and Aleukemia (204)

12. Lymphosarcoma and other neoplasms of lymphatic and hematopoietic tissues, (200-203, 205)
   Lymphosarcoma and reticulosarcoma (200)
   Hodgkin's Disease (201)
   Other neoplasms of lymphatic and hematopoietic tissues (202, 203, 205)

13. Diabetes Mellitus (260)

14. Major cardiovascular-renal diseases (330-334, 400-468, 592-594).
  Also see entries 15 through 21, below.

15. Vascular Lesions Affecting Central Nervous System (330-334)

16. Rheumatic Fever and Chronic Rheumatic Heart Disease (400-402, 410-416)
   Rheumatic Fever (400-402)
   Chronic Rheumatic Heart Disease (410-416)

17. Arteriosclerotic Heart Disease, including coronary disease (420)
   Arteriosclerotic heart disease so described (420.0)
   Heart Disease specified as involving coronary arteries (420.1)
   Angina Pectoris without mention of coronary disease (420.2)

18. Nonrheumatic Chronic Endocarditis and other myocardial degeneration (421, 422)

19. Hypertensive Disease (440-447)

20. Hypertensive Heart Disease (440–443)
    Hypertensive heart disease with arteriolar nephrosclerosis (442)
    Other hypertensive heart disease (440, 441, 443)
        Other hypertensive disease (444–447)

21. Chronic and Unspecified Nephritis and other renal sclerosis (592–594)

22. Influenza and Pneumonia, except Pneumonia of newborn (480–493)
    Influenza (480–483)
    Pneumonia, except pneumonia of newborn (490–493)

23. Ulcer of stomach and duodenum (540, 541)

24. Appendicitis (550–553)

25. Hernia and intestinal obstruction (560, 561, 570)

26. Gastritis, duodenitis, enteritis, and colitis, except diarrhea of newborn (543, 571, 572)

27. Cirrhosis of liver (581)

28. Hyperplasia of prostate (610)

29. Motor vehicle accidents (E810–E835)

30. Other accidents (E800–E802, E840–E962)

31. Suicide (E963, E970–E979)

32. Homicide (E963, E970–E879)

>>>>>>>>>>

Box 1 of Chap. 4
The "Standard Million Population, 1940."
Enumerated Population of the United States, 1940, by Age-Bands.

Age-Band, in Years	Actual 1940 Population	Fraction of Total	Standard Million, 1940
All Combined	131,669,275	1.000000	1,000,000
Under 1 yr	2,020,174	0.015343	15,343
1-4	8,521,350	0.064718	64,718
5-14	22,430,557	0.170355	170,355
15-24	23,921,358	0.181678	181,677
25-34	21,339,026	0.162065	162,066
35-44	18,333,220	0.139237	139,237
45-54	15,512,071	0.117811	117,811
55-64	10,572,205	0.080294	80,294
65-74	6,376,189	0.048426	48,426
75-84	2,278,373	0.017304	17,304
85 and up	364,752	0.002770	2,770
Sums for check	131,669,275	1.000000	1,000,000

The tabulation above is based on Grove 1968, page 37. Comparable information, for the years 1950-1993, is provided in PHS 1995 (p.79).

Box 2 of Chap. 4
Illustrative Calculation of an Age-Adjusted MortRate.

● Col.A shows the 1940 National All-Cancer MortRates per 100,000 (100K) population in each age-band (from Linder 1947, Table 14, p.250).
● Col.B (from Box 1) shows what fraction of the total 1940 population comes from each age-band.
● Col.C = [Col.A * Col.B]. The * denotes multiplication. The sum of Col.C entries is the weighted average of the separate MortRates.
This sum, 120.15 cancer-deaths per 100,000 population, is the age-adjusted National All-Cancer MortRate for 1940. See Text, Part 4b.

Age-Band	Col.A Rate/100K	Col.B Fraction	Col.C = A * B
Under 1 yr	4.4	0.015343	0.06751
1-4	4.8	0.064718	0.31065
5-14	3.0	0.170355	0.51107
15-24	5.4	0.181678	0.98106
25-34	17.3	0.162065	2.80373
35-44	61.1	0.139237	8.50737
45-54	168.8	0.117811	19.88647
55-64	369.6	0.080294	29.67653
65-74	695.2	0.048426	33.66561
75-84	1,161.0	0.017304	20.08966
85 and up	1,319.0	0.002770	3.65391
Sums		1.000000	120.1536

• These annual National Breast-Cancer MortRates, per 100,000 resident population (all "races," no exclusions) are in "Health: United States 1995" from the Public Health Service (PHS 1995 in our Reference List), Table 41, p.138. The National rates in Row 1 are age-adjusted to the 1940 reference year, and are in very close agreement with the female National rates in Table 8-B. The "crude" National rates in Row 2 are per 100,000 population without any age-adjustment.

• While the age-adjusted National MortRate hardly changes during the 1950-1990 period, the MortRates rose in some age-bands, and declined in others --- as quantified by the ratios in the righthand column.

Row	Age	1950	1960	1970	1980	1990	1991-93	Ratio 1990 / 1950
1	All ages, age-adju	22.2	22.3	23.1	22.7	23.1	22.1	1.04
2	All ages, crude	24.7	26.1	28.4	30.6	34.0	33.2	
3	Under 25 years	--	--	--	--	--	--	--
4	Ages 25-34	3.8	3.8	3.9	3.3	2.9	2.8	0.76
5	Ages 35-44	20.8	20.2	20.4	17.9	17.8	16.1	0.86
6	Ages 45-54	46.9	51.4	52.6	48.1	45.4	43.0	0.97
7	Ages 55-64	70.4	70.8	77.6	80.5	78.6	75.0	1.12
8	Ages 65-74	94.0	90.0	93.8	101.1	111.7	107.9	1.19
9	Ages 75-84	139.8	129.9	127.4	126.4	146.3	144.1	1.05
10	Ages 85+	195.5	191.9	157.1	169.3	196.8	199.9	1.01

• Col.A entries are the observed age-specific 1990 National All-Cancer MortRates per 100,000 population of each age-band, for both sexes combined. Source: "Health: United States 1995" from the Public Health Service (PHS 1995, Table 39, p.132).

• Col.B is the weighting factor, from the "Standard Million Population, 1940" (see Chapter 4, Box 1, "Fraction of Total").

• Col.C is the product of Col.A times Col.B, and the SUM of Col.C is the 1990 age-adjusted MortRate: 135 per 100,000 population. This rate matches the entry for Both Sexes in our Table 6-B.

• Col.D divides each entry in Col.C by 135, and thus determines the fraction of the National Age-Adjusted Rate (135 per 100,000) contributed by each age-band. The sum of the fractions for age 45 and older = 0.92780, or 93%.

Age-Band	Col.A AgeSpecific Rate/100K	Col.B Weighting Factor	Col.C = A times B (A * B)	Col.D = Col.C / 135
Under 1 yr	2.3	0.015343	0.03529	0.00026
1-4	3.5	0.064718	0.22651	0.00168
5-14	3.1	0.170355	0.52810	0.00391
15-24	4.9	0.181678	0.89022	0.00659
25-34	12.6	0.162065	2.04202	0.01513
35-44	43.3	0.139237	6.02896	0.04466
45-54	158.9	0.117811	18.72015	0.13867
55-64	449.6	0.080294	36.10002	0.26741
65-74	872.3	0.048426	42.24182	0.31290
75-84	1,348.5	0.017304	23.33411	0.17285
85 and up	1,752.9	0.002770	4.85591	0.03597
Sums		1.000000	135.0031	1.0000

Related text = Part 4d.

CHAPTER 5

Dose-Response, Linear Regression, and Some Other Key Concepts in Our Analyses

● Part 1. Radiation-Induced Cancer: The "Build-Up" Years

Because ionizing radiation is a carcinogen (Chapter 2, Part 4), its introduction into medicine, in 1896, had to cause radiation-induced Cancers. The Cancers, caused by medical radiation received during 1896, did not all appear at once. Like products dispensed from an inventory, the Cancers were delivered gradually (Chapter 2, Part 8). And the Cancers caused by medical radiation received during 1897 were also delivered gradually. And the Cancers caused by medical radiation received during 1898 were gradually delivered, too. We need not name every year for a century.

1a. Figure 5-A: The "Build-Up" Years

Figure 5-A depicts the effect of gradual delivery: A period of "build-up" in the annual delivery of radiation-induced Cancer. Figure 5-A refers to cancer incidence, with no arbitrary interval between radiation exposure and diagnosis of radiation-induced cancer (discussion in Part 4, below). We emphasize that Figure 5-A is a diagram in which we arbitrarily:

   (a) show years of annual irradiation only through 1951;
   (b) make 120 cases (per 100,000 population) the total cancer-consequence from each year's irradiation;
   (c) make 40 years the maximum delivery-time for the 120 cases produced by a single year's irradiation;
   (d) make delivery of the 120 cases occur at a constant annual rate: 3 cases per 100,000 population, annually, for 40 years. This is equivalent to having 40 different latency periods before diagnosis occurs.

As a result of these choices, the annual deliveries of Cancer induced by medical irradiation show a build-up in Figure 5-A as follows:

During 1896: 3 cases delivered per 100,000 population.
During 1897: 6 cases delivered per 100,000 population.
During 1898: 9 cases delivered per 100,000 population.
During 1899: 12 cases delivered per 100,000 population.
During 1900: 15 cases delivered per 100,000 population.
   ..... The 40th year is 1935.
During 1935: 120 cases delivered per 100,000 population.

Although the numbers in Figure 5-A are merely illustrative, they make a key point: The introduction and maintenance of medical radiation necessarily caused a gradual build-up in the number of RADIATION-induced cases of Cancer per 100,000 population.

1b. Equality of Response: A Simplifying Assumption in Figure 5-A

Figure 5-A depicts an "ideal" situation in which the magnitude of the average radiation dose is the same every year, and the magnitude of the response is the same (120 radiation-induced Cancers per 100,000 population).

Equality of response over decades is a condition invoked for simplification. In reality, several lines of evidence indicate that the magnitude of carcinogenic response, per rad of radiation exposure, can be modulated (altered) by the intensity of exposure to nonradiation co-actors and by the absence of such co-actors. Therefore, the magnitude of cancer-response, per unit of radiation exposure, can vary over time according to the abundance or paucity of co-actors.

The Introduction has already discussed the widely accepted concept that more than one cause is necessary to produce a case of fatal Cancer. Ways in which carcinogenic co-actors can multiply each other's potency is a topic deferred to Chapters 49, 67, and Appendix-M. Here, we simply point out that --- when Figure 5-A depicts a response of constant size to a radiation dose of constant size, decade after decade --- we have invoked the "ideal" assumption that exposure to co-actors is also constant decade after decade.

● Part 2. Equilibrium Years: Flat Rates of Radiation-Induced Cancer

When the production-rate and the delivery-rate of Cancer are equal in the SAME CALENDAR YEAR --- despite the variable and extended latency periods --- it is because equilibrium has occurred between two opposite drives. During the equilibrium years, successive columns add one box at the top --- and subtract one box at the bottom.

Equilibrium is first reached in Figure 5-A in the year 1935. That is the first year in which 120 new Cancers/100,000 population are PRODUCED (for gradual delivery) and also 120 radiation-induced Cancers are DELIVERED (from earlier years of production plus 1935 production).

Since nothing changes in our ideal model, Figure 5-A shows that equilibrium continues through 1951. Equilibrium would continue INDEFINITELY if the average radiation dose were maintained at a constant level "forever" --- but due to the size of our page, Figure 5-A completely terminates medical irradiation after 1951.

Flat Cancer-Rates and the "Law of Equality"

The "Law of Equality" states: If an age-matched population receives the same level of irradiation and same exposure to co-actors year after year, ultimately a state of equilibrium will be reached when the annual delivery-rate of radiation-induced Cancer per 100,000 population is equal to the annual production-rate of radiation-induced Cancer per 100,000 population, and the same annual delivery-rate will endure indefinitely, if the same annual production-rate is maintained.

In other words, the "Law of Equality" leads toward FLAT rates of radiation-induced Cancer. In Figure 5-A, the rate in every year of equilibrium is 120 cases/100,000 population. The equilibrium years, in Figure 5-A, are limited to 1935 through 1951 --- simply because of the size of the page.

Does the "Law of Equality" depend on assuming that the delivery rate of a single year's production occurs in equal parts --- such as 3 cases each year as shown in Figure 5-A? No. The law is also valid for delivery in unequal parts over the specified timespan. This is demonstrated in Gofman 1995/96, where all of Chapter 4 is devoted to the "Law of Equality."

● Part 3. The "Build-Down" Phase: An Exceedingly Gradual Phenomenon

Due to the size of our page, Figure 5-A COMPLETELY terminates irradiation of the population after 1951. But deliveries of radiation-induced Cancer continue, because in 1951, deliveries from irradiation received during the 1940s and 1930s and 1920s and even earlier, were not yet complete. The total annual deliveries can decline only GRADUALLY, even when there is NO additional production.

The build-down of deliveries can be quantified by counting the vertical boxes in the post-1951 columns. Each year, just one box is coming off at the bottom of successive columns. So total delivery declines to 117 cases in 1952, 114 cases in 1953, 111 cases in 1952, and so forth.

3a. How Does the "Termination" Model Relate to Reality?

Depiction of this build-down phase should drive home an important point. If all medical radiation were abruptly and permanently terminated (which we certainly do not advocate), and if exposure to co-carcinogens were held constant, the resulting reduction in cancer mortality rates would happen gradually over about 50 years --- due to delivery of radiation-induced cases already "in the pipeline." The gradual build-down depicted in Figure 5-A is an important reminder, that uselessly high doses of xrays administered TODAY will still be causing Cancers 10, 20, 30, 40 (and more) years from now.

3b. Real-World Status of Delivery-Schedules

For radiation-induced cancer cases, delivery-intervals after irradiation are necessarily much clearer in studies of excess cases due to exposure at a SINGLE time (such as radio-iodine exposure from the Chernobyl accident, or gamma-ray exposure from the Hiroshima-Nagasaki bombs), than in studies of excess Cancer due to chronic exposure (such as occupational exposures spread over years or decades). In the latter, it is impossible to know in WHICH years the radiation produced the carcinogenic lesions. Thus, analysts rely heavily on the Atomic-Bomb Study for our knowledge of delivery-intervals and duration (Chapter 2, Part 8).

The observation, of excess Cancers (meaning radiation-induced cases) from a particular radiation event, refers to the excess number compared with the number occurring in comparable "control" groups NOT exposed to extra radiation from the particular radiation event. Of course, part of the "background" cancer-rate in the control groups is radiation-induced too --- by OTHER sources of radiation exposure.

The total cancer-rate (radiation-induced cases plus cases which would occur anyway) climbs with advancing age --- as illustrated for 1940 in Chapter 4 (Box 2), and for 1990 in Chapter 4 (Box 4). To the extent that radiation-induced cases occur at approximately the same ages as radiation-unaided cases (Gofman 1971, p.244; BEIR 1990, p.5), then the AVERAGE interval between irradiation and delivery of radiation-induced cases will be longer for cases induced during childhood than for cases induced at ages near or beyond age 55.

Even though delivery-schedules vary by age at irradiation, delivery-schedules for radiation-induced Cancers will be the same, per 100,000 population, in all Nine Census Divisions --- because the Divisions have been "matched" for age by use of age-adjusted cancer MortRates.

● Part 4. Is There Really Any "Minimum Latency Period" ?

The notion that there exists a "minimum latency period" of 5 to 20 years after radiation exposure, before any radiation-induced Cancer is manifest, is almost certainly mistaken (Gofman 1981, Gofman 1994, Gofman 1995/96). The limited evidence at hand shows that atomic-bomb-induced Leukemia showed up before five years (BEIR 1972, p.101; and UNSCEAR 1986, p.222, Fig. 24), and that Chernobyl-induced Thyroid Cancer also showed up within five years (Kazakov 1992; Baverstock 1992; WHO 1995).

Indeed, in a non-Chernobyl radio-iodine study (of 38,000 medical patients who received diagnostic doses of iodine-131), Holm et al mention a large excess of Thyroid Cancer observed during the first five years after administration of the iodine-131 (Holm 1988). However, Holm et al count NONE of these Cancers as caused by the radio-iodine. With a single sentence, the cases are simply discarded from the study --- a decision which appears to be a highly questionable prejudgment (full analysis in Gofman 1990, Chapter 22, Part 5).

After the atomic bombings in August 1945, the A-Bomb Survivors Study is silent about solid Cancers until 1950. The study's first report on solid Cancers covers the period 1950 through 1954. It shows that, by then, the 25,203 exposed survivors (all ages and all doses, combined) had a solid cancer MortRate which was already 11% higher than the rate among the 66,028 participants in the reference-group (details in Gofman 1990, Table 17-A). If the study had involved 130,000,000 to 250,000,000 participants --- as our study of the entire U.S. population does --- then a radiation-induced excess MortRate of fatal solid Cancers might have been detectable within 1 to 2 years after the bombings.

We know of no studies which are capable of ESTABLISHING that a five-year "minimum latency period" truly occurs, between exposure of a mixed-age population to extra ionizing radiation and delivery of fatal cases of radiation-induced solid Cancers.

Some Biology-Based Logic for Expecting No Minimum

We know of no biological basis for EXPECTING any minimum latency period for Cancer in a large mixed-age population. By contrast, we know of some reasons for expecting NO such minimum.

In molecular biology, evidence is accumulating that a cell becomes malignant only after its chromosomes have accumulated SEVERAL carcinogenic abnormalities (see Appendix D, for instance). Some of these genetic abnormalities may be inherited, and others may be acquired at any age after conception. If a cell, which has already accumulated a full set of carcinogenic lesions except for one, receives the final necessary lesion from a radiation-exposure, the delivery-time for that particular case of radiation-induced Cancer could be extremely short.

Almost certainly, carcinogenic genetic lesions have a range of effects, from a mild predisposition to Cancer, to a virtual guarantee of a rapidly lethal malignancy. It is very reasonable to expect that the speed of cancer development varies with the particular areas of chromosomal damage or chromosomal deletion which are present in a cell. For this reason, too, it is very reasonable to expect that some radiation-induced Cancers will be delivered almost immediately as overt, clinical cases.

Unless strong epidemiologic evidence develops someday in favor of a minimum latency period for radiation-induced Cancers, we think the most reasonable assumption is NO minimum latency period in populations of mixed ages.

● Part 5. Dose-Response: Linearity and Regression Analysis

In Figures 1-A and 1-B of Chapter 1, the boxy symbols show the nine pairs of PhysPop values and MortRates (one pair for each of the Nine Census Divisions). Those PhysPop values and MortRates have a strong linear relationship with each other --- which is clear because the boxy symbols cluster so closely around a straight line. If this were a PERFECT linear correlation, the boxy symbols would fall directly upon a single straight line, with no scatter at all.

5a. The Linear Dose-Response: Meaning and Expression

In a perfect linear relationship, one additional unit of dose adds exactly the same number of fatal cases to the MortRate, no matter whether the total dose is low or high. Suppose that each dose-unit adds 6 fatal cancers to the cancer MortRate. Then 3 additional units of dose will add 18 fatal Cancers to the MortRate. Thus, increment in MortRate (18 cases) is proportional to the increment in dose (3 units), and the constant of proportionality (which relates dose to MortRate) is 6 cases per unit of dose. 18 additional fatal cases = (6 additional cases / dose-unit) times (3 additional dose-units). The dose-units cancel out in this equation, so that additional cases = additional cases.

When dose-response is linear, each MortRate is related to its corresponding dose by the equation for a straight line: $y = mx + c$.

● The y-variable is the MortRate, expressed in "cases per 100,000 population," for example.
● The x-variable is the corresponding dose, expressed in dose-units (for example, in PhysPop values in this book).
● "m" is the coefficient of proportionality (also called the X-Coefficient), expressed as "fatal

cases per dose-unit." Thus, potency per dose-unit is the SAME at all dose-levels.

• "c" is a Constant, expressed in the same MortRate units as "y." The Constant quantifies the number of cases in the total MortRate which are NOT related to dose.

When the value of the Constant is greater than zero, then each MortRate is proportional to dose only after the Constant is SUBTRACTED from the total MortRate: $(y - c) = mx$. If the value of the constant is zero, there is nothing to subtract, and the ENTIRE MortRate is proportional to dose. In that case, $y = mx$.

5b. Figure 5-B: Perfect Proportionality (MX Model)

Figure 5-B, which is located at the end of this chapter, illustrates what we call the MX model of dose-response --- an abbreviation of the equation $y = mx$. This is the model in which the ENTIRE MortRate (y) is directly proportional to PhysPop (x). In other words, the MX model reflects the concept that medical radiation became a contributing cause to nearly all cases of fatal Cancer, with nearly no cases unaided by medical radiation.

5c. The X-Values and Y-Values for Figure 5-B (MX Model)

In Figure 5-B, the x-values are the nine real PhysPops of 1940 (from the Universal PhysPop Table 3-A). The y-values are an unreal set of MortRates. We have arbitrarily made the highest MortRate equal to 120 radiation-induced cancers per 100,000 population. Readers have seen that rate before. It is the annual delivery-rate of cancer depicted in Figure 5-A during the equilibrium years. In Figure 5-B, we pair it with the highest 1940 PhysPop value, which is 169.76 in the Mid-Atlantic Division.

To obtain eight other illustrative MortRate values, which must be perfectly proportional in the MX model to the eight other PhysPops of 1940, we do exactly what we did in Chapter 3, Part 6b. We take the ratio of the y-variable over the corresponding x-variable: $(120 / 169.76) = 0.7068803$. Then we multiply each of the eight PhysPops by 0.7068803 to obtain their matching MortRates --- thus making the pairs of x,y values perfectly proportional to each other $(y = mx)$. The value of m (the X-Coefficient) is 0.7068803.

Some Ratios Resulting from Perfect Proportionality

As a result, the proportionalities demonstrated in Chapter 3, Part 6b, apply here too. We already know that the ratios of the MortRates over the PhysPops are 0.7068803 in every Census Division, because we just made them so. In addition, any two MortRates will have the same ratio as their corresponding PhysPops. For example, we can compare the New England Division with the Mountain Division (data in Figure 5-B). The MortRate ratio is (114.1969 / 84.74820), or 1.347. The corresponding PhysPop ratio is (161.55 / 119.89), or 1.347. The same.

It follows, IN THE MX MODEL, that the ratio of PhysPop Hi5/Lo4, and the ratio of MortRate Hi5/Lo4, must be the same. (The Hi5/Lo4 ratio was introduced in Chapter 3, Part 4.) From the Universal PhysPop Table 3-A, we find that the Hi5/Lo4 PhysPop ratio for 1940 is 1.46. When we calculate the Hi5 average and the Lo4 average for the synthetic MortRates in Figure 5-B, we obtain 105.68 and 72.53, respectively. Their ratio is also 1.46.

5d. Linearity: Interpreting the Absence of Curvature

Before proceeding to linear regression analysis, we want to comment on the strong linear relationships between MortRates and PhysPop values, already depicted in Figures 1-A and 1-B of Chapter 1. In view of the data discussed in Chapter 2, Part 5b, how do we interpret the observation that these correlations are linear rather than curved?

We refer to the nature of PhysPop itself:

PhysPop is proportional to average accumulated per capita population dose from medical radiation because the more physicians there are per 100,000 population, the more radiation procedures are done per 100,000 persons. The increase in procedures occurs chiefly because MORE persons per 100,000 receive such attention --- not because the SAME persons get irradiated more often. In other words, the average per-PATIENT dose is about the same in the Census Divisions with low PhysPop values as in Divisions with high PhysPop values, but the average per-CAPITA dose is higher in high-PhysPop Census Divisions than in low PhysPop Divisions because there are more PATIENTS per 100,000 population in high-PhysPop Divisions.

At the cellular level where xray-induced mutations occur, the average per-patient dose-level is likely to be very similar in all Nine Census Divisions. Therefore, the observed absence of curvature (e.g., the absence of supra-linearity) matches expectation, in dose-responses between PhysPop and MortRates.

5e. Linear Regression Analysis: Best-Fit Equation, Best-Fit Line (MX Model)

Regression analysis is a branch of mathematics which can evaluate the correlation between sets of x,y pairs. Part 5a has already emphasized that, in a linear dose-response, each MortRate (y) is related to its corresponding PhysPop (x) by the equation for a straight line: $y = mx + b$.

In earlier decades, we had to do the calculations for regression analysis by hand. Now, we can just enter the two columns of data (the x-values and the corresponding y-values) into the proper location of a computer spreadsheet, and use a regression-analysis program to do the calculations for us. The program which we use, in the Lotus 123 spreadsheet, produces standard output from the method of least squares. The program is described in the Lotus Journal by Chuck Sullivan, a systems engineer for the Lotus Development Corporation (Sullivan 1986). Every regression analysis has input and output.

Obtaining the Equation of Best Fit, from Figure 5-B

The input-data for the regression analysis of Figure 5-B: The x-values are the nine real 1940 PhysPops, and the y-values are nine corresponding MortRates, calculated in Part 5c in order to illustrate a PERFECT linear correlation. The additional x-entries and "Best-Fit Calculated MortRates" in Figure 5-B are needed for graphing, as explained below.

The regression output: The output is located at the top-right of Figure 5-B. From it, we obtain the values of the X-Coefficient and the Constant (discussed in Part 5a) which are required in order to write the best-fit equation for this set of data. Patterned on the straight-line equation, $y = mx + c$, the equation of best fit for Figure 5-B is:

MortRate = (0.7068803 * PhysPop) + Zero. [* denotes multiplication.]

Generating the LINE of Best-Fit from the Best-Fit Equation

Using this best-fit equation, we can "plug in" any value for PhysPop, and calculate a corresponding MortRate. Each MortRate requires a separate calculation. To distinguish such MortRates from real-world observations ("observed MortRates"), it is customary to call them "calculated" or "estimated" or "best-fit" MortRates.

By using such calculations, we obtained the column of best-fit MortRates in Figure 5-B --- including MortRates when PhysPop = 90, when PhysPop = 80, when PhysPop = 70 ... right down to PhysPop = 0. The LINE OF BEST FIT, which is graphed in Figure 5-B, connects these pairs of x,y values (various PhysPops, best-fit MortRates).

5f. The X-Coefficient and the Constant (MX Model)

X-Coefficient: Because in Part 5c, we made the pairs of x,y values perfectly proportional to each other (Mort Rate = 0.7068803 times PhysPop, where PhysPop is the x-variable), the regression output had to produce 0.7068803 as the "X-Coefficient." The X-Coefficient is simply "m" in the equation, $y = mx$. Re-arranged: $m = y/x$. So "m" evaluates how many units of y (the MortRate) occur per unit of x (dose). In short, the X-Coefficient describes how steep the slope is, of the best-fit line.

Constant: The Constant is "c" in the straight-line equation, $y = mx + c$. The Constant is the value of y, when x = zero. The value of the Constant (the c-value) never changes --- which gives it the name "Constant." When the x-value changes to a new value, the X-Coefficient ("m") determines the new value of the product, "mx", which gets added to the c-value to produce the new and corresponding best-fit y-value.

In Part 5c, we made every MortRate = (0.7068803 times PhysPop), a procedure for which the equation is y = mx. So, when the resulting pairs of x,y values were fed into the linear regression analysis, there was no "room" for any c-value other than zero in the regression's straight-line equation (y = mx + c). Quite predictably, the regression output in Figure 5-B shows the value of the Constant to be zero.

In the MX model, the Constant has a value of zero, and so the ENTIRE value of the MortRate is directly proportional to every corresponding value of PhysPop (Part 5b).

The Y-Axis Intercept and the "Origin"

Because the Constant is the value of y, when x = zero, the Constant is the value of y wherever the best-fit line intercepts the vertical y-axis. Thus, in our graphs, the Constant (also called "the y-intercept") is the value of the MORTRATE, when the value of PHYSPOP is zero. The spot where both y = 0 and x = 0 is called the "origin" in such graphs.

5g. The R-Squared Value and the "Std Err of Coef" (MX Model)

The regression output at top-right of Figure 5-B provides some measures of how good (how strong) the x,y correlation is.

R-Squared Value: The R-squared value measures the "goodness of fit" between the line of best fit and the pairs of input-data. The input-pairs are depicted by the boxy symbols in our graphs. Only a PERFECT correlation produces an R-squared value of 1.00 from regression analysis, as we emphasized in Chapter 3, Parts 6 and 7. Imperfect correlations generate R-squared values less than 1.00. A rule of thumb is that R-squared values below 0.3 are not considered to be statistically significant (at about the 90% confidence level). As readers study the chapters on non-malignancies in this book, they will see some R-squared values quite a bit lower than 0.3 --- meaning no detectable correlation whatsoever between the x,y pairs.

Standard Error of the X-Coefficient: The Standard Error (SE) of the X-Coefficient is an indicator of how reliable is the SLOPE of the best-fit line. The certainty of a slope and the strength of a correlation diminish as the distance grows between the best-fit line and some of the boxy symbols, of course.

"The smaller, the better," is the rule for the size of the Standard Error (SE) of the X-Coefficient, relative to the size of the X-Coefficient itself. In Figure 5-B, the MX model produces "zero" as the SE of the X-Coefficient, because the slope of the best-fit line is not in any doubt when there is a perfect correlation (R-squared = 1.00).

90% Confidence Limits on the X-Coefficient

The 90% confidence-limits (CLs) on the X-Coefficient are calculated from the SE. The upper limit is (X-Coef) + (1.645 times SE) and the lower limit is (X-Coef) - (1.645 times SE).

For example, if the X-Coefficient from regression output is (0.203) and its Standard Error is (0.045), then (at the 90% CL) the upper limit on the X-Coefficient is (0.203) + (1.645 times 0.045) = (0.203 + 0.074) = (0.277). The lower CL is (0.203) - (1.645 times 0.045) = (0.203 - 0.074) = (0.129). In other words, if a great number of samples were measured and regressed, 90% of the X-Coefficients would fall in the range of 0.129 through 0.277. However, the central value (provided by the regression output) is the most likely value --- and therefore, the central value is often called "the best value."

Ratio of the X-Coefficient over Its Standard Error (SE)

In the example above, the ratio of the X-Coefficient over its SE is (0.203 / 0.045), or 4.51. A rule of thumb is that the value of the X-Coefficient over its SE needs to be at least 2.0 before the X-Coefficient is regarded as reasonably reliable. In our dose-response studies, we will calculate the ratio for each regression. Readers will see see some ratios as high as 5 and higher --- which means that those slopes are highly reliable.

5h.   Effect of a Single Deviant Datapoint upon the Constant

Whenever real-world data fit the MX model of dose-response rather closely, but not perfectly, the effect of a single deviant datapoint (boxy symbol) upon the Constant deserves appreciation.

For example, if we move only a high datapoint in Figure 5-B far above the best-fit line, we would need a new regression analysis using the altered input-data. The new regression analysis would produce a steeper slope and a NEGATIVE Constant, instead of a Constant of zero. The new best-fit line would intersect the vertical y-axis BELOW the origin. We would see a similar result if we had moved a low datapoint to a new location far BELOW the best-fit line. In similar fashion, different "moves" of a single datapoint could tip the slope to be LESS steep, in which case the Constant of zero (which characterizes perfect proportionality) would rise to a positive value. In real-world data, single "out-lying" datapoints can have such effects.

5i.   What Results Would We Expect from Ideal Research Circumstances?

To obtain an overview of the architecture of our dose-response studies, between PhysPop and cancer MortRates, it is useful to imagine that real-world conditions will be "ideal" for such studies.

"Ideal" conditions would resemble the conditions described for Figure 5-A. However, Figure 5-A refers only to ONE population. Our studies compare the NINE different populations in the Nine Census Divisions. "Ideally," there would be no migration among the Census Divisions, and each separate population would receive exposure to a CONSTANT annual average per capita dose of medical radiation, decade after decade, with constant levels of co-actors decade after decade.

Under such conditions, what should we expect to observe with respect to the dose-response relationship between PhysPop and cancer MortRates, after the introduction of radiation into medicine in 1896?

We would expect to observe a positive and linear dose-response, by Census Divisions, between the nine MortRates and the nine corresponding PhysPops, decade after decade. If regression analysis produced a Constant greater than zero, we would subtract the Constant from each of the nine Observed MortRates, and we would expect the nine remaining MortRate values to stay always in the same proportions with each other as the fixed proportions among the nine PhysPop values. In other words, we would expect the variation in cause to control the variation in effect.

The same expectation can be expressed somewhat differently. Under ideal research conditions, we would have nine separate populations which never mix from one Census Division to another, and each population would constantly receive its own, fixed, per capita average dose of medical radiation, decade after decade. Each of the nine, different, average per capita doses would produce its own separate stream of radiation-induced Cancers in the population of its own Census Division. Under such conditions, of course we would expect that these nine separate streams of radiation-induced Cancer (expressed as excess age-adjusted cancer MortRates per 100,000 population) would have proportions with each OTHER which mirror the proportions that the nine causal doses of medical radiation have with each other.

It remained for us to learn, just how severely REAL-world research conditions might depart from the ideal, as we undertook to examine much of a century.

● Part 6.   Dose-Response: Perfect Correlation without Perfect Proportionality

In contrast to the MX model of dose-response, the MX+C model reflects the concept that medical radiation does not contribute to EVERY case of fatal Cancer. The Constant quantifies the number of cases which occur without help from medical radiation.

6a. Figure 5-C: One Alteration in the Input Data of Figure 5-B

Figure 5-C, located at the end of this chapter, depicts the MX+C model of dose-response. It is designed to be exactly like Figure 5-B except for ONE type of alteration. Every MortRate in Figure 5-B has had 20 Cancers (per 100,000 population) added, for Figure 5-C. In other words, we have given the Constant a value of 20. When PhysPop = zero, the cancer MortRate is 20 cases (per 100,000 population). In Figure 5-C, the input-data for the x-variable (the nine PhysPops) are the same as in Figure 5-B.

How does the regression output differ in Figure 5-C from the output in Figure 5-B?

Only the Constant has changed, from zero to 20. But the slope of the best-fit line is still the same, with the X-Coefficient at 0.7068803, and with the standard error still at zero. And the correlation between the pairs of x,y variables is still perfect, with an R-squared value of 1.00.

The equation of best fit is now: MortRate = (0.7068803 times PhysPop) + 20. And with that equation, we calculated MortRates in order to graph the line of best fit. The graph shows the y-intercept at 20, of course. And the nine pairs of actual input-data (the nine boxy symbols) sit right upon the line of best fit, with no scatter, because R-squared = 1.00.

6b. Perfect Correlation without Perfect Proportionality (MX+C Model)

In Figure 5-B, we illustrated perfect proportionality between the entire MortRate and PhysPop (y = mx), as well as perfect correlation (R-Squared = 1.00).

By contrast, Figure 5-C illustrates perfect correlation between PhysPops and MortRates (R-squared = 1.00), but not perfect proportionality between the ENTIRE MortRates and their PhysPops. In order to see the proportionality between dose and response, one must first SUBTRACT the Constant from each MortRate, because the Constant represents a contribution to each MortRate which occurs "anyway" (even when dose = zero) and such a contribution is NOT proportional to dose.

6c. Can Perfect Correlation Persist, If X-Values Rise and Y-Values Fall?

The answer to the question in the subtitle is "Yes." To illustrate, we will do three linear regressions below. The first one reproduces the regression in Figure 5-C, so that we begin with "old" values (for x and y) which already have demonstrated their perfect correlation. In the second regression, each x-value of the first regression has been multiplied by 1.4, but the y-values stay as they are in the first regression. In the third regression, the x-values stay as they are in the second regression, but each y-value is multiplied by 0.8. So, the third regression shows a perfect correlation persisting even after all the x-values rose by one factor (1.4) and all the y-values fell by another factor (0.8).

Old-x	Old-y		#1. Regression Output:	
159.72	132.90		Constant	19.9974
161.55	134.20		Std Err of Y Est	0.0029
123.14	107.05		R Squared	1.0000
169.76	140.00		No. of Observations	9
133.36	114.27		Degrees of Freedom	7
119.89	104.75			
103.94	93.47		X Coefficient(s)	0.7069
85.83	80.67		Std Err of Coef.	0.0000
100.74	91.21	Except for rounding, input and output are the same as Figure 5-C.		
new-x	old-y		#2. Regression Output:	
223.61	132.90		Constant	19.9942
226.17	134.20		Std Err of Y Est	0.0031
172.40	107.05		R Squared	1.0000
237.66	140.00		No. of Observations	9
186.70	114.27		Degrees of Freedom	7
167.85	104.75			
145.52	93.47		X Coefficient(s)	0.5049
120.16	80.67		Std Err of Coef.	0.0000
141.04	91.21	Note: X-values are 1.4 times x-values in #1.		
		Note: This X-Coef = (0.7069 from #1) divided by 1.4 = 0.5049		

new-x	new-y
223.61	106.32
226.17	107.36
172.40	85.64
237.66	112.00
186.70	91.42
167.85	83.80
145.52	74.78
120.16	64.54
141.04	72.97

#3.   Regression Output:

Constant	15.9953
Std Err of Y Est	0.0025
R Squared	1.0000
No. of Observations	9
Degrees of Freedom	7
X Coefficient(s)	0.4040
Std Err of Coef.	0.0000

Note: Y-values are 0.8 times y-values in #2.
Note: This X-Coef = (0.5049 from #2) * 0.8 = 0.4039
Note: This Constant = (19.9942 from #2) * 0.8 = 15.9954

● **Part 7. Dose-Response: Effects of Imperfect Matching across Dose-Groups**

For multi-cause diseases such as Cancer and Ischemic Heart Disease, we can define co-actors as necessary co-causes in producing single cases of those diseases (Introduction, Parts 4 and 5). When analysts want to study the dose-response between ONE co-actor (for instance, medical radiation) and the mortality rate from the disease, they hope to compare study-groups which differ in dosage of the ONE co-actor but which are alike ("matched") with respect to the other co-actors (for instance, smoking). In this book, the study-groups (or dose-groups) are the populations of the Nine Census Divisions.

7a. The Real World: Imperfect Matching across Dose-Groups

In the real world of cancer-studies, perfect matching across dose-groups is never possible. Practical obstacles are immense. In addition, all causes of Cancer are probably not even recognized yet, and it would be impossible to match dose-groups for unrecognized co-actors. For both reasons, imperfect matching always occurs.

Imperfect matching for co-actors can interfere with detection of a positive correlation which is truly present, or can produce an apparent correlation which is spurious. The power of "confounding variables" is a major concern for all analysts. In this book, we need not worry about finding a spurious positive correlation (between medical radiation and cancer MortRates), because a causal relationship between ionizing radiation and fatal Cancer has been well established by a multitude of earlier studies (Chapter 2, Part 4c). But we need to appreciate the power of imperfect matching to OBSCURE the correlation in a set of data.

7b. Figure 5-D: Inconsistency with the "Correlation Axiom"

Comparison of Figures 5-B and Figure 5-D, at the end of this chapter, illustrates how imperfect matching for co-actors can change a perfect correlation (R-Squared = 1.00) into an imperfect correlation with an R-Squared value of 0.7112.

Figure 5-D uses the real 1940 PhysPops as the x-values, as did Figure 5-B. However, Figure 5-D depicts the consequence of Census Divisions which are imperfectly matched for co-actors. The UNEQUAL average exposure to nonradiation co-actors, in the Nine Census Divisions, can degrade the PhysPop-MortRate correlation in two ways. One: Xray potency per rad is modulated differently in the various Census Divisions (Chapter 6, Part 6; and Chapter 49, Part 2). Two: The number of cases in which xrays are not a co-actor may differ across the Census Divisions. As a result of one or both phenomena, the MortRates from Figure 5-B increase by irregular numbers (purely illustrative) as follows:

	MortRate Fig.5-B	Increments in MortRate due to Imperfect Matching of Co-Actors		"y" in Figure 5-D
Pacific Division:	112.9029	+ 25	=	137.9029
New England:	114.1965	+ 11	=	125.1965
West North Central:	87.0452	+ 20	=	107.0452
Mid-Atlantic:	120.0000	+ 17	=	137.0000
East North Central:	94.2696	+ 35	=	129.2696
Mountain:	84.7479	+ 11	=	95.7479
West South Central:	73.4731	+ 21	=	94.4731
East South Central:	60.6715	+ 45	=	105.6715
South Atlantic:	71.2111	+ 31	=	102.2111

As a result of imperfect matching, the R-squared value of 1.00 in Figure 5-B falls to 0.7112 in Figure 5-D. The true biological correlation is OBSCURED (but not changed) by imperfect matching of co-actors across the dose-groups. Imperfect matching is not consistent with what we can abbreviate as the "Correlation Axiom," below.

The Correlation Axiom

Correlation Axiom: Increment in cancer MortRate is perfectly proportional to increment in radiation dose (PhysPop), provided that co-actors are perfectly matched across the dose-groups. The Correlation Axiom describes (a) the linear dose-response, and (b) the matching of dose-groups --- which is a fundamental principle of dose-response research, even though it is never fully achievable (Part 7a; also Chapter 3, Part 2d).

7c. Figure 5-E: A Truly Positive Correlation Which Looks Negative

Imperfect matching for co-actors can interfere --- much more severely than illustrated in Figure 5-D --- with detection of a positive correlation which is truly present. With Figure 5-E, we will demonstrate how imperfect matching can even make a truly positive correlation appear negative.

We are preparing to study the dose-response between PhysPop (surrogate for medical radiation) and cancer MortRates. Suppose that a carcinogenic co-actor, such as smoking, occurs with the most intensity where PhysPop values are the lowest, and with the least intensity where PhysPop values are the highest. In other words, suppose there is an INVERSE relationship between PhysPop and smoking. In such a situation, smoking will increase the cancer MortRates more in Census Divisions with low PhysPop values than in Census Divisions with high PhysPop values. Below, starting with the values from Figure 5-B, we arrange the Census Divisions in descending order of their 1940 PhysPop values, and then we add to the 1940 cancer MortRates from Figure 5-B in a way inverse to the trend of PhysPop values:

	1940 PhysPop	MortRate Fig.5-B	Increments in MortRate due to Imperfect Matching of Co-Actors		"y" Fig.5-E	Regression Output:	
Mid-Atl	169.76	120.0	+ 20	=	140.0	Constant	176.7119
New Eng	161.55	114.2	+ 30	=	144.2	Std Err of Y Est	4.8615
Pacific	159.72	112.9	+ 40	=	152.9	R Squared	0.6322
ENoCen	133.36	94.3	+ 50	=	144.3	No. of Observations	9
WNoCen	123.14	87.0	+ 60	=	147.0	Degrees of Freedom	7
Mtn	119.89	84.7	+ 70	=	154.7		
WSoCen	103.94	73.5	+ 80	=	153.5	X Coefficient(s)	-0.2003
SoAtlan	100.74	71.2	+ 90	=	161.2	Std Err of Coef.	0.0577
ESoCen	85.83	60.7	+ 100	=	160.7	X-Coef / S.E. =	-3.4686

The regression-output in Figure 5-E shows that the sign on the X-Coefficient has become 0.20 with a NEGATIVE sign, which means that when PhysPop increases by one unit, cancer MortRate FALLS by 0.2 unit. In other words, the true POSITIVE correlation between PhysPop and cancer MortRate has been so well concealed by the non-matched co-factor (smoking), that the OBSERVED correlation between PhysPop and cancer MortRates will be INVERSE in such a situation. But imperfect matching of co-actors is just an error, an inconsistency with the Correlation Axiom. Such errors have no power to repeal the laws of physics and human biology --- the laws which established the Correlation Axiom for ionizing radiation (PhysPop) in the first place.

● Part 8. Real-World "Entropic Circumstances" Which Reduce Observed Correlations

The ideal MX model and the ideal MX+C model both reflect perfect correlation between dose and response. They are very orderly models. But in the real world, order is opposed by the tendency toward disorder. Most systems move spontaneously from states of order toward states of disorder. In chemistry, the molecular chaos of a substance or a system is measured by a property called "entropy."

What Do We Mean in This Book by "Entropic Circumstances"?

In this book, we need a name for the group of real-world events which perturb the orderly, ideal models of this chapter. Our name is "entropic circumstances." Entropic circumstances operate generally AGAINST order --- they DO NOT CREATE order. (Weiss 1998 describes some recent insights about entropy.)

8a.   Some Specific Entropic Circumstances of Concern

For our dose-response studies, we know that two entropic circumstances of great concern have to be migration of populations from one Census Division to another (discussion in Chapter 3, Part 2c), and PhysPop deviations from "lockstep" over time (discussion in Chapter 3, Parts 2c and 8).

Both migration and deviations from PhysPop "lockstep" degrade PhysPops as surrogates for ACCUMULATED radiation dose-differences from medical applications. Neither migration nor deviations from PhysPop "lockstep" would be serious problems in our dose-response studies if complete delivery of radiation-induced cancers occurred within 2 or 3 years. They become problems because of the very gradual delivery-times for radiation-induced cancers --- with such delivery-times stretching over at least 40 years (or longer) for mixed-age populations. By comparison, other entropic circumstances may be less important --- and we emphasize "may."

8b.   Finding the Maximum Real-World Correlations (PPs with MRs)

Because entropic circumstances operate AGAINST orderly phenomena (such as correlations), entropic circumstances reduce R-Squared values. Therefore, if we seek the best approximation of the real dose-response relationship, between PhysPops and cancer MortRates, we will seek and accept the HIGHEST values of R-squared which survive erosion by entropic circumstances.

1940 is our first year of MortRate data with all 48 states represented. And 1921 is the year of our earliest PhysPop data. In our search for the strongest correlation, we regressed the 1940 MortRates serially on every set of prime (not interpolated) PhysPop data between 1921 and 1940 --- including the 1940 PhysPops. Although cancer MORTALITY during 1940 can hardly be influenced by medical radiation received during 1940, the 1940 PhysPops are nearly in "lockstep" with the PhysPops of many preceding years (Chapter 3, Table 3-C) --- and thus, 1940 PhysPops reflect the approximate differences in accumulated dose of medical radiation from many PRIOR years.

● Part 9.   Estimating the Impact of Medical Radiation on Cancer MortRates

We undertook this project in order to explore Hypothesis-1, that medical irradiation is the principal cause of cancer mortality in the USA during the Twentieth Century. We remind readers that we are not trying to ESTABLISH the existence of a positive correlation between ionizing radiation and cancer mortality. That was proven many years ago. Instead, we are making use of that knowledge to test Hypothesis-1.

We begin, in Section Two of this book, by looking at what we can learn about Hypothesis-1 from regressing 1940 cancer MortRates on earlier PhysPops. In Section Five of this book, we examine the whole 1940-1990 period. We arrive at estimated Fractional Causation of cancer mortality by medical radiation. Such results clearly support Hypothesis-1.

>>>>>>>>>>

Figure 5–A. Annual Delivery–Rates of Radiation–Induced Cancer

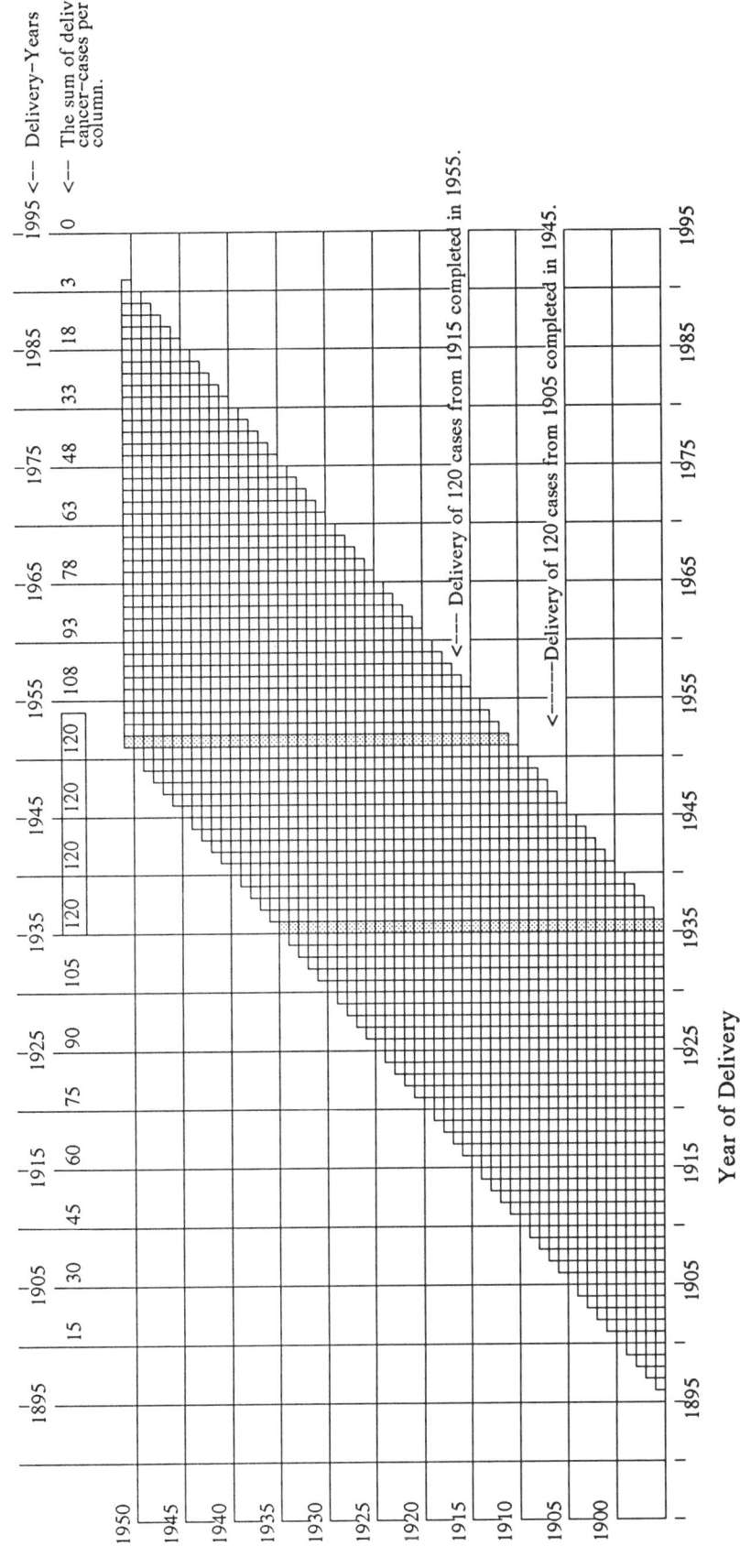

● – Each box in the grid represents 3 cases of radiation–induced cancer per 100,000 population (mixed ages).

● – Each horizontal row of 40 boxes represents gradual delivery of 120 cancers per 100,000 population. In this illustration, 120 is the number of cases produced by the radiation received during a single calendar–year. These 120 cases are delivered gradually at the rate of 3 cases per year for 40 years.

● – Each vertical column represents the number of radiation–induced cancers delivered during a single calendar–year, per 100,000 population, from all earlier years of irradiation. All boxes in a column were produced by radiation received in different calendar–years.

● – Both SHADED columns have 40 vertical boxes (representing 120 cancers) as do the columns BETWEEN the two shaded columns. Such columns demonstrate the "Law of Equality": The annual radiation– induced delivery of 120 = the annual radiation–induced production of 120.

1995 <--- Delivery–Years

0 <--- The sum of delivered cancer–cases per column.

<------- Delivery of 120 cases from 1915 completed in 1955.

<------- Delivery of 120 cases from 1905 completed in 1945.

Year of Delivery

Year of Irradiation

Figure 5–A

Related text = Parts 1, 2, + 3.

Figure 5-B. The MX Model of Dose-Response

Census Divisions	1940 "x" PhysPops	1940 "y" MortRates	Best-Fit Calc. MortRates		
Pacific	159.72	112.9029	112.9029	Regression Output:	
New England	161.55	114.1965	114.1965	Constant	0.000000
West No. Central	123.14	87.0452	87.0452	Std Err of Y Est	0.000000
Mid–Atlantic	169.76	120.0000	120.0000	R Squared	1.000000
East No. Central	133.36	94.2696	94.2696	No. of Observations	9
Mountain	119.89	84.7479	84.7479	Degrees of Freedom	7
West So. Central	103.94	73.4731	73.4731		
East So. Central	85.83	60.6715	60.6715	X Coefficient(s)	0.706880
South Atlantic	100.74	71.2111	71.2111	Std Err of Coef.	0.000000
Additional PhysPops	90.00		63.6192		
--- not "observed" ---	80.00		56.5504		
down to zero PhysPop	70.00		49.4816		
(zero medical radiation).	60.00		42.4128		
For each, we calculate	50.00		35.3440		
a best–fit MortRate.	40.00		28.2752		
These additional x,y pairs	30.00		21.2064		
are also part of the	20.00		14.1376		
best–fit line .	10.00		7.0688		
	0		0.0000		

Related text = Part 5.

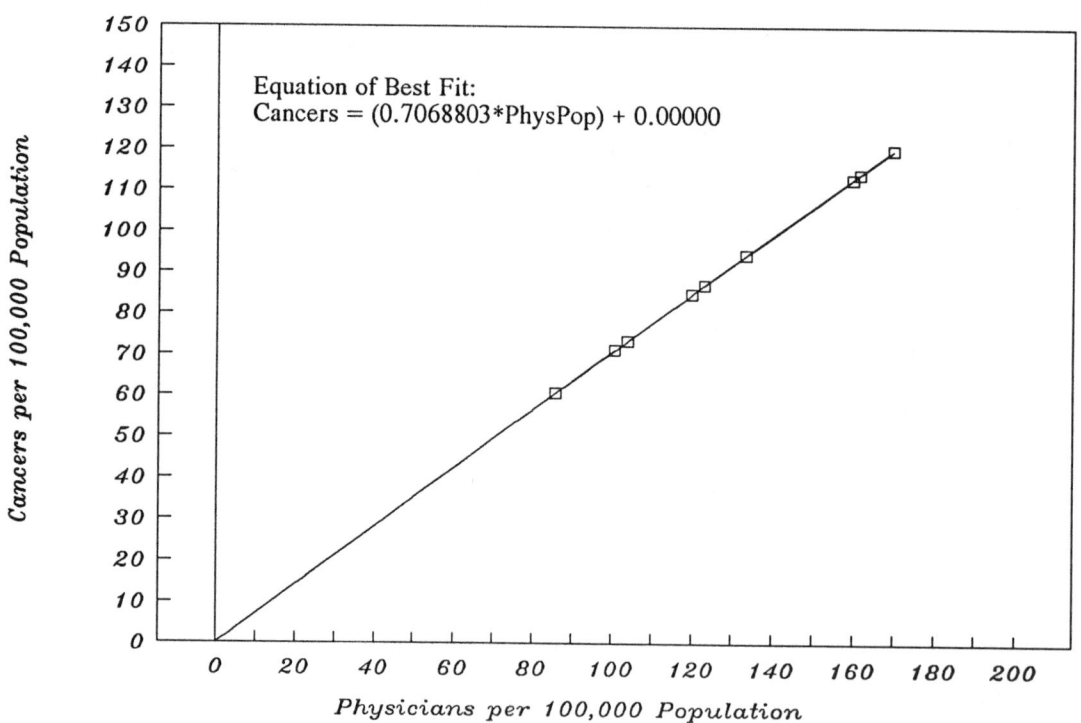

Equation of Best Fit:
Cancers = (0.7068803*PhysPop) + 0.00000

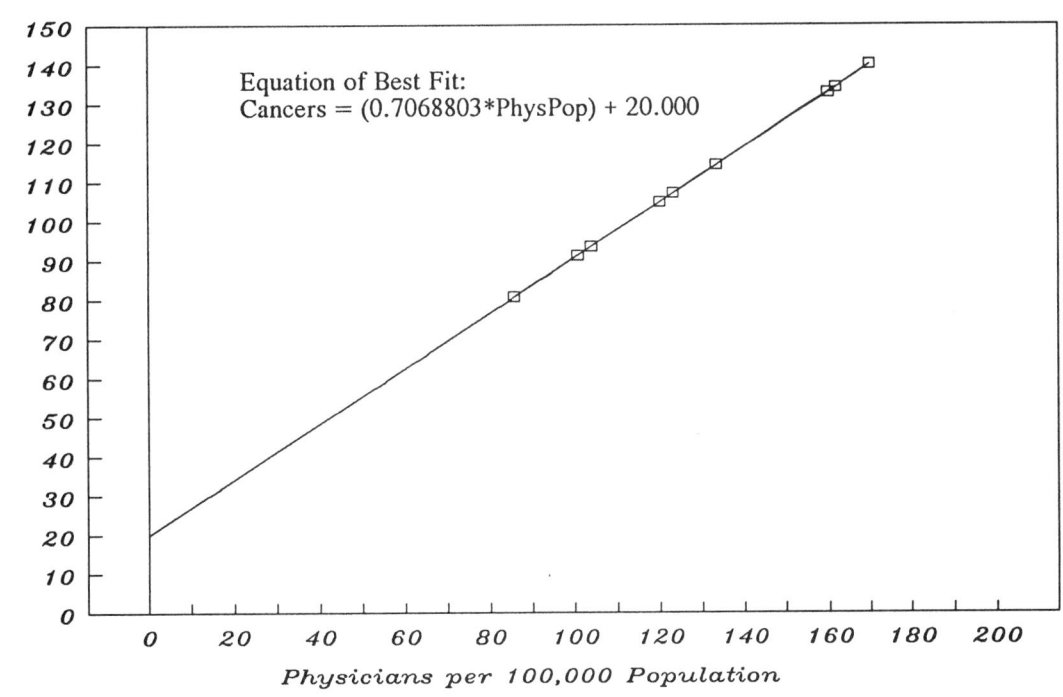

Figure 5-C.   The MX+C Model of Dose-Response.

Census Divisions	1940 "x" PhysPops	1940 "y" MortRates	Best-Fit Calc. MortRates		
Pacific	159.72	132.9029	132.9029	Regression Output:	
New England	161.55	134.1965	134.1965	Constant	20.000
West No. Central	123.14	107.0452	107.0452	Std Err of Y Est	0.0000
Mid-Atlantic	169.76	140.0000	140.0000	R Squared	1.000000
East No. Central	133.36	114.2696	114.2696	No. of Observations	9
Mountain	119.89	104.7479	104.7479	Degrees of Freedom	7
West So. Central	103.94	93.4731	93.4731		
East So. Central	85.83	80.6715	80.6715	X Coefficient(s)	0.706880
South Atlantic	100.74	91.2111	91.2111	Std Err of Coef.	0.000000
Additional PhysPops	90.00		83.6192		
--- not "observed" ---	80.00		76.5504		
down to zero PhysPop	70.00		69.4816		
(zero medical radiation).	60.00		62.4128		
For each, we calculate	50.00		55.3440		
a best-fit MortRate.	40.00		48.2752		
These additional x,y pairs	30.00		41.2064		
are also part of the	20.00		34.1376		
best-fit line .	10.00		27.0688		
	0		20.0000		

Related text = Part 6a.

Equation of Best Fit:
Cancers = (0.7068803*PhysPop) + 20.000

Cancers per 100,000 Population

Physicians per 100,000 Population

Figure 5-D. Effect of Imperfect Matching of Dose-Groups

- Regression input for the x-variable (PhysPop) is the same as in Figure 5-B.

- Regression input for the y-variable (MortRate) comes from the text of Chapter 5, Part 7b. The MortRates differ from Figure 5-B in a manner which reflects Census Divisions which are imperfectly matched for radiation's carcinogenic co-actors.

- Each Best-Fit MortRate (to make the graph) is calculated with the equation of best fit provided by the regression output: MortRate = (0.4929 * PhysPop) + 51.5299.

	1940 PhysPop "x"	Part 7b MortRate "y"	Best-Fit Calc. MortRates		
Pacific	159.72	137.9	130.3	Regression Output:	
New England	161.55	125.2	131.2	Constant	51.5299
West No. Central	123.14	107.0	112.2	Std Err of Y Est	9.9955
Mid-Atlantic	169.76	137.0	135.2	R Squared	0.7112
East No. Central	133.36	129.3	117.3	No. of Observations	9
Mountain	119.89	95.7	110.6	Degrees of Freedom	7
West So. Central	103.94	94.5	102.8		
East So. Central	85.83	105.7	93.8	X Coefficient(s)	0.4929
South Atlantic	100.74	102.2	101.2	Std Err of Coef.	0.1187
Additional PhysPops	70.00		86.0	XCoef/SE	4.1523
--- not "observed" ---	60.00		81.1		
down to zero PhysPop	50.00		76.2		
(zero medical radiation).	40.00		71.2		
For each, we calculate	30.00		66.3		
a best-fit MortRate.	20.00		61.4		
These additional x,y pairs	10.00		56.5		
are also part of the	0		51.5		
best-fit line.					

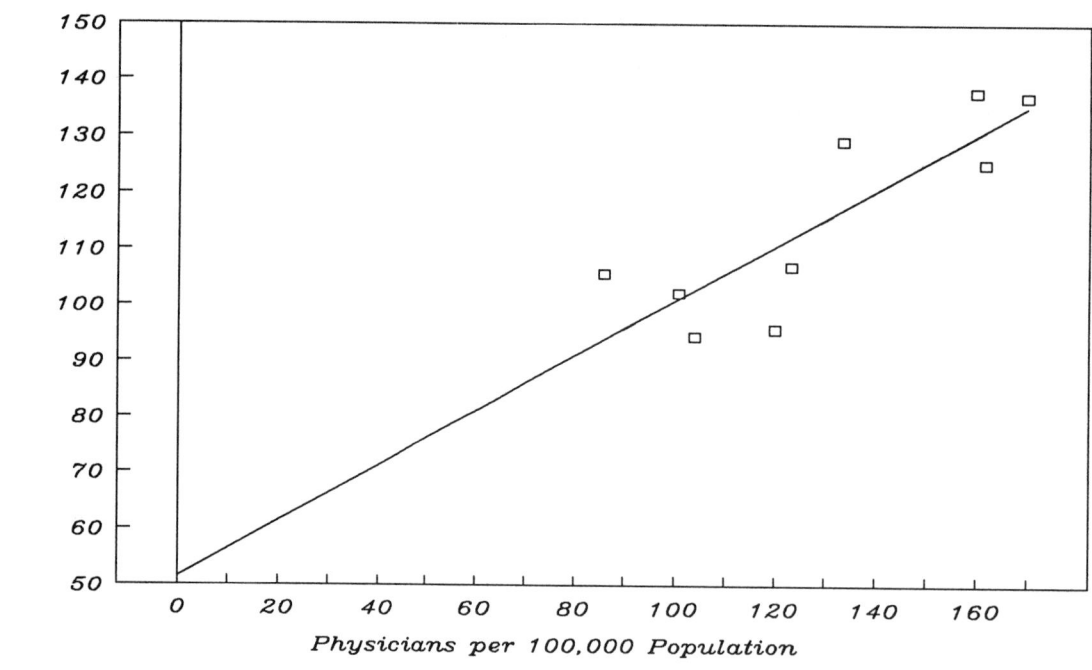

Physicians per 100,000 Population

Cancers per 100,000 Population

● Regression input for the x–variable (PhysPop) is the same as in Figure 5–B. The sequence here is in order of descending values. (Sequence does not affect regression output.)

● Regression input for the y–variable (MortRate) comes from the text of Chapter 5, Part 7c. The MortRates differ from Figure 5–B in a manner which reflects an inverse relationship between PhysPop and intensity of a co–actor across the Census Divisions.

	1940 "x" PhysPops	Part 7C "y" MortRates	Best Fit Calc. MortRates	Regression Output:	
Mid–Atlantic	169.76	140.0	142.7	Regression Output:	
New England	161.55	144.2	144.4	Constant	176.7119
Pacific	159.72	152.9	144.7	Std Err of Y Est	4.8615
East No. Central	133.36	144.3	150.0	R Squared	0.6322
West No. Central	123.14	147.0	152.0	No. of Observations	9
Mountain	119.89	154.7	152.7	Degrees of Freedom	7
West So. Central	103.94	153.5	155.9		
South Atlantic	100.74	161.2	156.5	X Coefficient(s)	–0.2003
East So. Central	85.83	160.7	159.5	Std Err of Coef.	0.0577
Additional PhysPops	70.00		162.7	X–Coef / S.E. =	–3.4686
— not "observed" —	60.00		164.7		
down to zero PhysPop	50.00		166.7		
(zero medical radiation).	40.00		168.7		
For each, we calculate	30.00		170.7		
a best–fit MortRate.	20.00		172.7		
These additional x,y pairs	10.00		174.7		
are also part of the	0		176.7		
best–fit line.					

Related text = Part 7c.

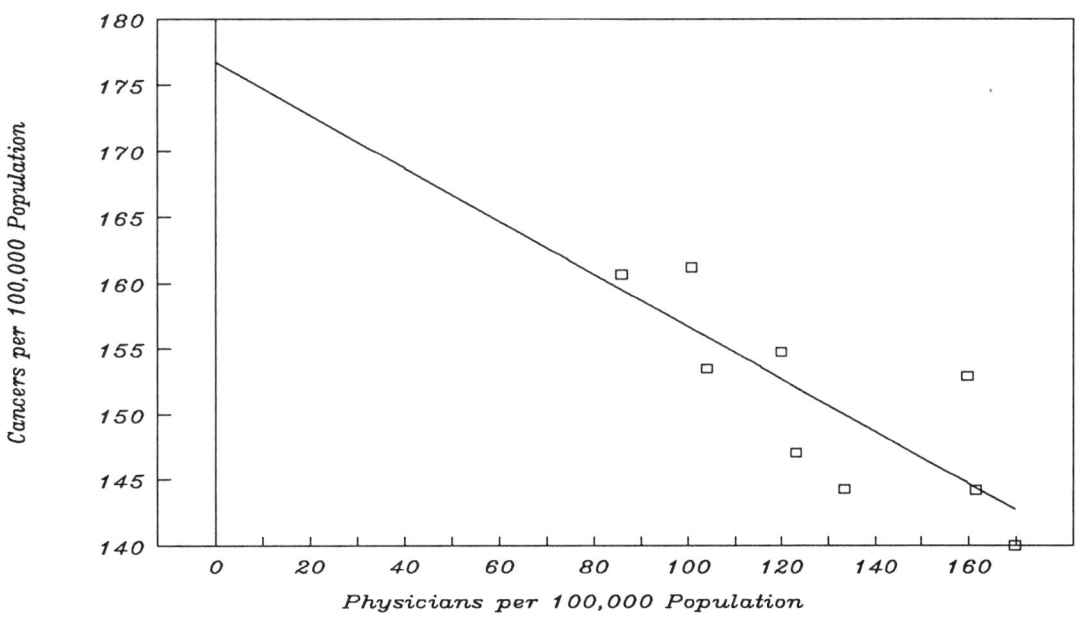

- 105 -

All-Cancers-Combined, Males: Relation with Medical Radiation

Reminder: Boxes, Figures, and Tables are located at the end of each chapter.

The term "All-Cancers" includes all malignancies, no matter how uncommon. After Chapters 6 and 7 on All-Cancers, we limit the cancer-chapters to malignancies (or to groups of malignancies) where the number of annual deaths per 100,000 population has been large enough to make the numbers relatively reliable. Even so, in some of our cancer-chapters, the "small numbers problem" is worrisome. The smaller the numbers per 100,000, the greater are the impacts of random fluctuations and of various types of reporting errors.

"All-Cancers, Males" and "All-Cancers, Females" include, of course, those malignancies which are subsequently examined in separate chapters, as well as all the malignancies (such as leukemia) which are NOT examined in separate chapters.

Hypothesis-1: All-Cancers (Combined) vs. Specific Cancers

Hypothesis-1 is that medical radiation is the principal cause of cancer-mortality in the United States during the Twentieth Century.

It deserves emphasis that Hypothesis-1 concerns cancer in the aggregate --- All-Cancers (combined). We explore subsets in this book in order to learn their roles in the overall result, but Hypothesis-1 does not demand that the impact of medical radiation be the same for every type of cancer, or for the two sexes. Indeed, because Hypothesis-1 leaves plenty of room for contributions by nonradiation carcinogens (Part 6 of this chapter), we expect to observe some biology-based differences (not just statistical noise) among the cancer subsets which we explore.

Chapter 6 as the General Model for Other Chapters

The "materials and methods" of our studies have been set forth in Chapters 3, 4, and 5. Chapter 6 demonstrates the first results. Chapter 6 also provides the general model for studies in the rest of this monograph. Chapter 6 explains the boxes, figures, and table which will be standard items in many subsequent chapters --- where these standard items will need no text. Chapter 6 includes various comments which apply also to later chapters, but which will seldom be repeated.

For everyone's convenience, Chapter 22, Box 1, tabulates the results from Chapters 6 through 21 --- for easy comparison with each other.

● Part 1. All-Cancer Mortality Rates, Males

At the end of this chapter are Tables 6-A and 6-B. Table 6-A provides the mortality rates by the Nine Census Divisions, 1940-1988, and Table 6-B provides the NATIONAL mortality-rates,

1940-1988. Although this chapter requires the rates only for 1940, the post-1940 rates will be used in Section Five of the book. The ICD numbers change with time (Chapter 4, Part 2d).

● **Part 2. How the Dose-Response Develops, 1921-1940**

In Part 2, we regress the 1940 MortRates (from Table 6-A) upon the non-interpolated sets of PhysPop values 1921-1940 (from the Universal PhysPop Table 3-A). The summary-results of all the regression analyses are presented in Box 1. The correlations in Box 1, between PhysPop and the 1940 All-Cancer MortRates, steadily improve --- and Chapter 22 (Part 2) discusses why.

In our dose-response studies, PhysPops for the Nine Census Divisions represent the relative radiation doses accumulated from medical procedures, so they are the x-values in our linear regression analyses. The corresponding MortRates for the Nine Census Divisions are the responses to be studied, so they are the matching y-values. Both the x and the y variables have the denominator "per 100,000 population." The 1940 MortRates are the y-input for all ten regression analyses in this chapter.

The strongest dose-response relationship (Part 2j) has an R-squared value of 0.951 and a ratio of 11.6 for the X-Coefficient over its Standard Error. The strength of the correlation is rather dazzling. What follows are the linear regression analyses from which Box 1's summary arises.

Readers need to avoid a pitfall, as they inspect these regressions. The pitfall would be to imagine that the regressions examine correlations between various PRE-1940 cancer MortRates and PhysPop. No. There is only one set of MortRates --- the 1940 set --- because complete nationwide cancer-MortRate data do not exist for 1930 or 1920 or earlier (Chapter 4, Part 1). Therefore, the regressions examine how the 1921 to 1940 PhysPops "line up with" (correlate with) a single "end-point": The 1940 cancer MortRates. We can not predict WHICH set of PhysPops will display the highest observed correlation with the 1940 MortRates. It is worth remembering that the 1940 cancer MortRates are influenced by medical radiation received BEFORE 1921, as well as after 1921 --- because latency periods can last 40 years or longer in irradiated populations of mixed ages (Chapter 2, Part 8a).

● - Part 2a.	x 1921 PhysPop	y 1940 MortRate	All-Cancers, Males Regression Output:	
Pacific	165.11	122.9	Constant	-27.0754
New England	142.24	135.5	Std Err of Y Est	18.0748
West North Central	140.93	110.9	R Squared	0.4630
Mid-Atlantic	137.29	140.9	No. of Observations	9
East North Central	136.06	119.6	Degrees of Freedom	7
Mountain	135.38	99.8		
West South Central	125.15	86.9	X Coefficient(s)	1.0086
East South Central	119.76	73.6	Std Err of Coef.	0.4105
South Atlantic	110.32	88.9	Coefficient / S.E.	2.4568

● - Part 2b.	1923 PhysPop	1940 MortRate	All-Cancers, Males Regression Output:	
Pacific	163.06	122.9	Constant	-24.8337
New England	137.39	135.5	Std Err of Y Est	16.6440
West North Central	138.31	110.9	R Squared	0.5447
Mid-Atlantic	138.92	140.9	No. of Observations	9
East North Central	131.82	119.6	Degrees of Freedom	7
Mountain	130.51	99.8		
West South Central	119.16	86.9	X Coefficient(s)	1.0198
East South Central	113.16	73.6	Std Err of Coef.	0.3524
South Atlantic	106.79	88.9	Coefficient / S.E.	2.8937

● - Part 2c.	1925 PhysPop	1940 MortRate	All-Cancers, Males Regression Output:	
Pacific	161.67	122.9	Constant	-16.5482
New England	138.31	135.5	Std Err of Y Est	15.7102
West North Central	133.92	110.9	R Squared	0.5943
Mid-Atlantic	134.36	140.9	No. of Observations	9
East North Central	127.54	119.6	Degrees of Freedom	7
Mountain	122.30	99.8		

West South Central	112.83	86.9	X Coefficient(s)	0.9879
East South Central	107.22	73.6	Std Err of Coef.	0.3085
South Atlantic	103.61	88.9	Coefficient / S.E.	3.2024

● – Part 2d.	1927	1940	All-Cancers, Males	
	PhysPop	MortRate	Regression Output:	
Pacific	157.83	122.9	Constant	-20.9399
New England	137.50	135.5	Std Err of Y Est	13.1094
West North Central	131.54	110.9	R Squared	0.7175
Mid-Atlantic	138.40	140.9	No. of Observations	9
East North Central	126.18	119.6	Degrees of Freedom	7
Mountain	118.75	99.8		
West South Central	108.25	86.9	X Coefficient(s)	1.0399
East South Central	102.07	73.6	Std Err of Coef.	0.2466
South Atlantic	102.13	88.9	Coefficient / S.E.	4.2168

● – Part 2e.	1929	1940	All-Cancers, Males	
	PhysPop	MortRate	Regression Output:	
Pacific	156.64	122.9	Constant	-19.27093
New England	138.46	135.5	Std Err of Y Est	12.0934
West North Central	128.72	110.9	R Squared	0.7596
Mid-Atlantic	138.49	140.9	No. of Observations	9
East North Central	126.51	119.6	Degrees of Freedom	7
Mountain	118.68	99.8		
West South Central	105.60	86.9	X Coefficient(s)	1.0351
East South Central	99.41	73.6	Std Err of Coef.	0.2201
South Atlantic	100.86	88.9	Coefficient / S.E.	4.7032

● – Part 2f.	1931	1940	All-Cancers, Males	
	PhysPop	MortRate	Regression Output:	
Pacific	159.97	122.9	Constant	-10.4041
New England	142.35	135.5	Std Err of Y Est	11.4992
West North Central	126.50	110.9	R Squared	0.7827
Mid-Atlantic	140.82	140.9	No. of Observations	9
East North Central	128.59	119.6	Degrees of Freedom	7
Mountain	118.89	99.8		
West South Central	105.95	86.9	X Coefficient(s)	0.9582
East South Central	96.73	73.6	Std Err of Coef.	0.1909
South Atlantic	99.59	88.9	Coefficient / S.E.	5.0207

● – Part 2g.	1934	1940	All-Cancers, Males	
	PhysPop	MortRate	Regression Output:	
Pacific	160.09	122.9	Constant	-2.6000
New England	148.60	135.5	Std Err of Y Est	8.8299
West North Central	125.96	110.9	R Squared	0.8718
Mid-Atlantic	149.62	140.9	No. of Observations	9
East North Central	129.36	119.6	Degrees of Freedom	7
Mountain	117.16	99.8		
West South Central	104.68	86.9	X Coefficient(s)	0.8903
East South Central	92.00	73.6	Std Err of Coef.	0.1290
South Atlantic	98.41	88.9	Coefficient / S.E.	6.9009

● – Part 2h.	1936	1940	All-Cancers, Males	
	PhysPop	MortRate	Regression Output:	
Pacific	158.44	122.9	Constant	-1.4212
New England	150.18	135.5	Std Err of Y Est	7.3226
West North Central	126.14	110.9	R Squared	0.9119
Mid-Atlantic	155.05	140.9	No. of Observations	9
East North Central	130.42	119.6	Degrees of Freedom	7
Mountain	119.80	99.8		
West South Central	103.52	86.9	X Coefficient(s)	0.8756
East South Central	89.94	73.6	Std Err of Coef.	0.1029
South Atlantic	99.16	88.9	Coefficient / S.E.	8.5104

● – Part 2i.	1938	1940	All-Cancers, Males	
	PhysPop	MortRate	Regression Output:	
Pacific	157.62	122.9	Constant	3.0512
New England	154.08	135.5	Std Err of Y Est	6.0043
West North Central	124.95	110.9	R Squared	0.9407

Mid-Atlantic	160.69	140.9	No. of Observations	9
East North Central	131.98	119.6	Degrees of Freedom	7
Mountain	119.88	99.8		
West South Central	102.79	86.9	X Coefficient(s)	0.8351
East South Central	88.21	73.6	Std Err of Coef.	0.0792
South Atlantic	99.26	88.9	Coefficient / S.E.	10.5419

● – Part 2j.	1940	1940	All-Cancers, Males	
	PhysPop	MortRate	Regression Output:	
Pacific	159.72	122.9	Constant	11.5484
New England	161.55	135.5	Std Err of Y Est	5.4727
West North Central	123.14	110.9	R Squared	0.9508
Mid-Atlantic	169.76	140.9	No. of Observations	9
East North Central	133.36	119.6	Degrees of Freedom	7
Mountain	119.89	99.8		
West South Central	103.94	86.9	X Coefficient(s)	0.7557
East South Central	85.83	73.6	Std Err of Coef.	0.0650
South Atlantic	100.74	88.9	Coefficient / S.E.	11.6275

● Part 3. Maximum Relationship (Box 1), Best-Fit Equation (Box 2), and Graph

The regression analysis of Part 2j produces the strongest correlation --- which is the appropriate one to use (Chapter 5, Part 8b; and Chapter 22, Part 5). From the output, we can write the best-fit equation, as we did for the MX and MX+C models in Chapter 5. (We use the symbol * to denote multiplication.)

● – All-Cancer MortRate, Males = (X-Coefficient * PhysPop) + Constant.
● – All-Cancer MortRate, Males = (0.7557 * PhysPop) + 11.55.

Using the equation of best fit, we can calculate a best-fit MortRate for any value of PhysPop. In Box 2, we show best-fit MortRates which have been calculated for the nine actual PhysPop values of Part 2j, and also for lower PhysPop values, down to zero PhysPop (Chapter 5, Part 5e).

Figure 6-A, shows the line of best fit --- which connects these pairs of x,y values (various PhysPops, best-fit MortRates). The graph also shows nine boxy symbols (the nine actual observations from Part 2j). Per 100K means per 100,000 population.

Relationship between the Census-Division List and the Graph's Boxy Symbols

Emphasis belongs on the fact that the permanent sequence of the Census-Division list has no effect upon the regression analysis and no effect upon the graph. Graph-related example: Although the Pacific Census Division is at the top of the PhysPop list, the boxy symbol which is farthest to the right on the graph does NOT represent the Pacific Census Division. That boxy symbol represents Mid-Atlantic, because by 1940, Mid-Atlantic (not Pacific) is the Division with the highest PhysPop value.

Identification of the boxy symbols is completely unnecessary for visual recognition of their scatter and sequence around the best-fit line --- and those are the features which largely determine the quality of a dose-response. However, if some readers wish to know "who" each boxy symbol represents, a good way to begin is to identify the two Census Divisions with the highest and lowest PhysPop values in Box 2 --- or the two Divisions with the highest and lowest Observed MortRates. All the boxy symbols in Figure 6-A are identified on its replica, Figure 22-C. In that same chapter, Figure 22-A depicts the dose-response between the earlier (1921) PhysPops and the 1940 MortRates. The lower quality of the dose-response in Figure 22-A, compared with Figure 22-C, is visually obvious.

Ranges of Values for the Y-Axis in Our Graphs

On the y-axis in Figure 6-A, values range from 0 to 150 annual deaths per 100,000 males. In later chapters where we study only a single group of Cancers, the height of the y-axis will be the same, but its range of values will be very much smaller. For example, in Chapter 12, the highest value on the y-axis will be 10 per 100,000. In graphs such as these, the visual steepness of best-fit lines is tied to the scales for the y-axis and x-axis. We can keep the x-scale (PhysPops) the same throughout

Section Two of the book, but we must adjust the y-scale according to the magnitude of the MortRates.

● **Part 4. Best Estimate (Box 3): 90% of Male Cancers in 1940 due to Medical Radiation**

The data have revealed a linear dose-response relationship in 1940 of immense strength between medical radiation and male cancer mortality. And now to test Hypothesis-1, we must ask:

"What would be the estimated male cancer mortality-rate in 1940 if there were NO dosage of medical radiation?"

No medical irradiation would occur if there were NO PHYSICIANS per 100,000 population. So we want to know the value of the y-variable (cancer MortRate) when the value of the x-variable (PhysPop) is equal to zero. This value is, of course, called the Constant in the regression output of Part 2j. On the graph, the Constant is the value of the MortRate where the line of best-fit intersects the y-axis. This "intercept" occurs where the value of PHYSPOP equals zero. No medical radiation at all.

Since every Census Division has physicians, there can be no real-world datapoint in our study of the male cancer MortRate when PhysPop = zero. But the calculated or "estimated" MortRate, if PhysPop were zero, certainly does not come out of thin air. It is extrapolated from nine real-world observations which reflect a very strong linear relationship. It merits emphasis that the raw data which reveal this relationship are neutral --- by which we mean they were collected long ago by people having no conceivable bias with respect to the studies in this monograph.

4a. Percentage Caused by Medical Radiation: "Fractional Causation"

Fractional Causation has been defined in this book's Introduction, Part 5. Fractional Causation is the fraction of the cancer mortality rate which would be ABSENT (prevented) in the ABSENCE of a specified carcinogen --- which is medical radiation, in the studies of this monograph. Therefore, Fractional Causation is the fraction of the cancer MortRate attributable to medical radiation --- or caused by medical radiation, in ordinary parlance. Here, Part 4a explains the procedure for obtaining the estimate of Fractional Causation, by medical radiation, of the 1940 National All-Cancer MortRate (males). The same procedure is also presented at the top of Box 3, in the format to be used in subsequent chapters.

The estimated cancer MortRate, if PhysPop were zero, is the Constant. The increments in MortRate above the Constant occur in proportion to accumulated dose of medical radiation, and such increments occur BECAUSE of medical radiation. That is the meaning of dose-response, of course. Therefore:

● The total National All-Cancer MortRate in 1940, minus the MortRate to which medical radiation did NOT contribute (the MortRate indicated by the Constant), is the MortRate induced by medical radiation. Radiation-induced cases of Cancer are defined as cases which would be absent in the absence of radiation exposure (Introduction, Part 5).

● The MortRate induced by medical radiation, divided by the entire National MortRate, is the fraction of the total caused by medical radiation (Box 3). We express that fraction as a percentage.

● When we subtract the Constant of 11.55 (Part 2j) from the National MortRate of 115.0 (Table 6-B), we have the rate of 103.45 per 100,000 from medical radiation. The fraction of the total is thus (103.45 / 115.0), or 0.8996. In other words, the "best estimate" which falls out of the data is that 90% of All-Cancer deaths in males, at approximately mid-century, are attributable to medical radiation (Box 3).

Comments: Use of 1940 PhysPops, and Treatment of Negative Constants

Use of the 1940 PhysPops: The 1940 PhysPop values are very highly correlated with the PhysPop values of 1929, 1931, 1934, 1936, and 1938, as demonstrated in Table 3-C. Although we do NOT believe that additional radiation received during 1940 contributes to the 1940 cancer mortality-rates, if the 1940 PhysPops produce the best correlation with the 1940 MortRates, we use that combination to estimate Fractional Causation (Chapter 5, Part 8b; Chapter 22, Part 5).

Negative Constants: In some chapters, the best-fit equations produce negative Constants. This is bound to happen occasionally if the true Constant is zero or near zero. In such cases, we will assume the Constant's value to be zero, since in the real world, mortality rates cannot go below zero. Chapter 22, Part 3, examines negative Constants and their probable origin.

Comment: Speculation versus the Evidence at Hand

In Chapter 1 (Part 6), we have already explained why we have a high level of confidence in our findings. Here, we discuss whether evidence should be discarded in favor of groundless speculation.

For example, we and others can speculate that, prior to 1940, exposure to some nonxray carcinogen (which we will name NX) was quite unequal in the Nine Census Divisions, AND that this failure in matching was such that, as PhysPop values rose, exposure to NX also rose (producing a positive correlation between PhysPop and NX). If this occurred, then the observed dose-response in Figure 6-A would include some fatal cases produced by co-action between medical radiation and NX, but it would also include some fatal cases produced by co-action between NX and nonxray carcinogens, without any participation by medical radiation. Thus, our estimated Fractional Causation of 90%, by medical radiation, could be too high. How much too high would depend on what fraction of persons were exposed to NX and the potency of NX in combination with nonxray co-actors. We note that not many persons would be exposed to NX who were not ALSO exposed to xrays at some time --- because the rate of xray examinations was so high in the USA (Appendix-K, Part 2).

What is the customary way to consider the "what if" types of speculation? One checks them out, if possible. But we know of no way in which anyone CAN undertake a reality-check on the speculation that some nameless carcinogen may have had a persistent positive correlation with PhysPop in the decades leading up to 1940.

Moreover, in the absence of any BASIS for suspicion that such a situation existed prior to 1940, confidence belongs with the 90% estimate which is grounded in the real-world evidence provided in Chapters 2 and 6 of this monograph.

4b. Determining the Range of the X-Coefficient (Box 3)

Our central estimate, of Fractional Causation by medical radiation, is tightly tied to the value of the Constant, so we want to know the range of the Constant's likely value.

One way to estimate the range of the Constant is to work with the Standard Error of the X-Coefficient (the slope of the best-fit line). Using the Standard Error (SE) from Part 2j, we can learn the range of values within which 90% of the measured X-Coefficients would fall, if a great number of samples were measured (Chapter 5, Part 5g). Then, we can use the two extremes of this range, in the equation of best fit, in order to calculate the matching Constants. The calculations are tabulated in Box 3, below the dotted line, and also stated below:

In Part 2j, the X-Coefficient is 0.7557 with a Standard Error of 0.0650. The confidence limits on the X-Coefficient lie 1.645 Standard Errors away from the central value of 0.7557. Therefore, the lower 90% confidence limit on the X-Coefficient is (0.7557) - (1.645 times 0.0650) = 0.6488 --- which means a flatter slope than 0.7557. We do a comparable calculation to obtain the upper 90% confidence limit, except we ADD to the central value of 0.7557 instead of subtracting from it. The upper 90% confidence limit on the X-Coefficient is (0.7557) + (1.645 times 0.0650) = 0.8626 --- which means a steeper slope than 0.7557.

4c. Fractional Causation at the High 90% Confidence Limit (Box 3)

Now we can write the equation of best fit, using the National MortRate from Table 6-B, the upper 90% confidence-limit on the X-Coefficient from Part 4b, and the National PhysPop from Box 4 (the sum of Column D).

Nat'l All-Cancer MortRate = (X-Coef. * Nat'l PhysPop) + Constant.
Then we re-arrange:
Constant = (Nat'l All-Cancer MortRate) - (X-Coef * Nat'l PhysPop)

Constant = 115.0 - (0.8626 * 132.04)
Constant = 115.0 - 113.9
Constant =   1.10

So, at the upper 90% Confidence Limit on the X-Coefficient, we subtract 1.1 (the new Constant) from 115.0 (the National MortRate), to arrive at 113.9 per 100,000 males as the All-Cancer MortRate ascribed to PhysPop (medical radiation). At this limit, Fractional Causation is (113.9 / 115.0), or 99.0% by medical radiation (Box 3).

If the Constant here had turned out negative, we would have treated it as if its value were zero, since cancer mortality-rates cannot be less than zero in the real world. With "zero" to subtract from the National All-Cancer MortRate, the Fractional Causation would have been $\sim$ 100 % by medical radiation.

4d. Fractional Causation at the Low 90% Confidence Limit (Box 3)

We repeat Part 4c, except that the X-Coefficient changes from 0.8626 to 0.6488 (Part 4b).

Constant = (Nat'l All-Cancer MortRate) - (X-Coef * Nat'l PhysPop)
Constant = (115.0) - (0.6488 * 132.04)
Constant = 115.0 - 85.7
Constant = 29.3

So, at the lower 90% Confidence Limit on the X-Coefficient, we subtract 29.3 from 115.0, to arrive at 85.7 per 100,000 males as the All-Cancer MortRate ascribed to PhysPop (medical radiation). At this limit, Fractional Causation is (85.7 / 115.0) = 0.745, or 74.5 % by medical radiation (Box 3).

All the steps to obtain the best estimate of Fractional Causation, and the Confidence-Limits, are abbreviated in Box 3. In subsequent chapters, Box 3 by itself will suffice.

● Part 5. Looking for Consistencies (Box 4): Error-Checks on Input and Output

We can use the National 1940 MortRate for All-Cancers (male) from Table 6-B, in order to verify that we have not made any serious errors in our work so far. Absent errors, our own work must produce a National MortRate which is reasonably close to the value in Table 6-B.

In calculating the National All-Cancer MortRate (male) in 1940 for the USA as a whole, we must weight the MortRate in each Census Division by multiplying it by its share of the total population. We also needed, for Parts 4c and 4d of this chapter, the weighted-average National PhysPop in 1940. We do both calculatons in Box 4.

The National MortRate of 112.65 (calculated in Box 4, Column F) is in good agreement with the National MortRate of 115.0 in Table 6-B. We can also check the reasonableness of our 1940 National PhysPop value (calculated in Box 4) and the X-Coefficient and Constant in our dose-response study of 1940 PhysPops and 1940 MortRates (Part 2j). If these three values are "good," then they too should produce a reasonable national MortRate for 1940, when we insert them into the appropriate best-fit equation (1940 PhysPops with 1940 MortRates):

1940 Nat'l All-Cancer MortRate = (X-Coefficient * Nat'l PhysPop) + Constant.
1940 Nat'l All-Cancer MortRate = (0.7557 * 132.04) + 11.55
1940 Nat'l All-Cancer MortRate = 111.33. This value, too, is in reasonable agreement with 115.0 from Table 6-B, and we are assured that there cannot be any serious errors in the work so far.

All the checks in Part 5 are abbreviated in Box 4. In subsequent chapters, Box 4 by itself will suffice.

● Part 6. Fractional Causation by Medical Radiation and by NonXray Causes: Co-Action

The estimate, that 90% of the males' National All-Cancer MortRate in 1940 was caused by medical radiation, may result in readers thinking, "That leaves only 10% for other causes!" Not true.

Because of co-action, the finding that Cancer has a high Fractional Causation by medical radiation does not limit any other cause of Cancer to a low Fractional Causation. This important point, already explained in the book's Introduction (Part 5), merits the additional discussion of co-action below.

6a. Co-Action among Causes: Views from the BEIR Reports of 1990 and 1999

With respect to the likelihood that co-action occurs between causes of Cancer, the Introduction (Part 4) has already presented "the general wisdom." Here, we add the views from the two most recent BEIR Reports. Of course, not all readers of this monograph will be familiar with the BEIR Reports. They are a series of six monographs issued during the 1972 - 1999 period by six (different) Committees on the Biological Effects of Ionizing Radiation --- all funded by the federal government and organized under the National Research Council.

BEIR 1990 states (p.152): "As discussed in the preceding section, the carcinogenic process includes the successive stages of initiation and promotion. The latter phase, promotion, appears to be particularly susceptible to modulation, with cigarette smoking being a conspicuous example of a modulating factor. Susceptibility to the carcinogenic effects of radiation can thus be affected by a number of factors, such as genetic constitution, sex, age at initiation, physiological state, smoking habits, drugs, and various other physical and chemical agents (UNSCEAR 1982)." The alteration of carcinogenic potency per rad of exposure, by nonradiation agents, is not speculation; there is experimental evidence (for instance, Segaloff 1971; BEIR 1990, pp.145-147). Our Chapter 49 (Part 2) discusses HOW co-actors can modify each other's potency.

BEIR 1999 (p.5) re-affirms the same view as BEIR 1990, even though the BEIR Committees which issued the two reports differ almost completely in their memberships: "Radiation carcinogenesis, in common with any other form of cancer induction, is likely to be a complex multistep process that can be influenced by other agents and genetic factors at each step."

6b. The Meaning of Co-Action: More Than One Cause per Case

The statements from BEIR reflect co-action among multiple causes. For example (from BEIR 1990): "Susceptibility to the carcinogenic effects of radiation can thus be affected by ... various other physical and chemical agents."   This can be true only if the radiation and nonradiation agents are co-actors in the same case.

If a factor contributes to an outcome (say, a death from Stomach Cancer at age 55), the meaning of "contributes" is that the factor is a necessary cause of the outcome. If the additional factor is NOT necessary --- if the outcome would happen as it does anyway --- then the factor contributes nothing to the outcome. A contributor is a cause. And multiple causes per case are co-actors.

Box 1 of Chap. 6
Summary: Regression Outputs for All-Cancers, Males.

Below are the summary-results from regressing the 1940 cancer MortRates upon the ten sets of PhysPops (1921-1940), as presented in Parts 2a-2j of this chapter. We are searching for the maximum correlation. Even the maximum will tend to understate the true correlation (Chapter 5, Part 8b).

Part	PhysPop	R-squared	Constant	X-Coef	Std Err	X-Coef/SE
2a	1921	0.4630	-27.08	1.0086	0.4105	2.4568
2b	1923	0.5447	-24.83	1.0198	0.3524	2.8937
2c	1925	0.5943	-16.55	0.9879	0.3085	3.2024
2d	1927	0.7175	-20.94	1.0399	0.2466	4.2168
2e	1929	0.7596	-19.27	1.0351	0.2201	4.7032
2f	1931	0.7827	-10.40	0.9582	0.1909	5.0207
2g	1934	0.8718	-2.60	0.8903	0.1290	6.9009
2h	1936	0.9119	-1.42	0.8756	0.1029	8.5104
2i	1938	0.9407	3.05	0.8351	0.0792	10.5419
2j --->	1940 Max	0.9508	11.55	0.7557	0.0650	11.6275

Box 2 of Chap. 6
Input-Data for Figure 6-A. All-Cancers. Males.

Part 2j, Best-Fit Equation: Calc. MortRate = (0.7557 * PhysPop) + (11.55)

Census Divisions	1940 Observed PhysPops	1940 Observed MortRates	Best-Fit Calc. MortRates
Pacific	159.72	122.9	132.250
New England	161.55	135.5	133.633
West No. Central	123.14	110.9	104.607
Mid-Atlantic	169.76	140.9	139.838
East No. Central	133.36	119.6	112.330
Mountain	119.89	99.8	102.151
West So. Central	103.94	86.9	90.097
East So. Central	85.83	73.6	76.412
South Atlantic	100.74	88.9	87.679
Additional PhysPops	70.00		64.449
--- not "observed" ---	60.00		56.892
down to zero PhysPop	50.00		49.335
(zero medical radiation).	40.00		41.778
For each, we calculate	30.00		34.221
a best-fit MortRate.	20.00		26.664
These additional x,y pairs	10.00		19.107
are also part of the	0		11.550
best-fit line (Chap 5, Part 5e).			

Box 3 of Chap. 6
Presumptive Fraction of Cancer MortRate Attributable to Medical Radiation.

Please see text in Chapter 6, Parts 4 and 6.

All-Cancers. MALES.   * denotes multiplication.

- MALE National MortRate (MR) 1940, from Table 6-B   115.0   National MortRate
- Constant, from regression, Part 2j   11.5484 Constant
- Fractional Causation, Best Est. = (Natl MR – Constant) / Natl MR   90.0%   Frac. Causation

....

  90% Confidence-Limits (C.L.) on Fractional Causation. See text in Chapter 6, Part 4b, please.

X-Coefficient, from Part 2j   0.7557 X-Coef., Best Est.
Standard Error (SE) of X-Coefficient, from Part 2j   0.0650 Standard Error

Upper 90% C.L. on X-Coef. = (Coef) + (1.645 * SE) =   0.8626 New X-Coefficient
New Constant = (Natl MR) – (New X-Coef * 1940 Natl PhysPop) =   1.0990 New Constant
Frac. Causation, High-Limit = (Natl MR – New Constant) / Natl MR =   99.0%   New Frac. Caus'n.

Lower 90% C.L. on X-Coef. = (Coef) – (1.645 * SE) =   0.6488 New X-Coefficient
New Constant = (Natl MR) – (New X-Coef * 1940 Natl PhysPop) =   29.3351 New Constant
Frac. Causation, Low-Limit = (Natl MR – New Constant) / Natl MR =   74.5%   New Frac. Caus'n.

Box 4 of Chap. 6
Error-Check on Our Own Work: All-Cancers, Males.

Please see text in Chapter 6, Part 5.

Below, Columns A, C, and E come directly from the regression input in Part 2j. Column B, the fraction of the whole 1940 population in each Census Division, comes from Table 3-B in Chapter 3. Each Column-D entry is the product of (B-entry times C-entry). Each Column-F entry is the product of (B-entry times E-entry). PhysPops and MortRates are each "per 100,000."

The Weighted-Avg. Nat'l PhysPop, 1940, is the sum of Column-D entries =   132.04

The Weighted-Avg. Nat'l Male MortRate, 1940, is sum of Col.F entries =   112.65
The Nat'l Male MortRate is also (X-Coef * Nat'l PhysPop) + Constant =   111.33
Comparison: The Nat'l Male MortRate, 1940, in Table 6-B =   115.00

(A) Census Division	(B) Pop'n Fraction	(C) PhysPop 1940	(D) 1940 Weighted PhysPop	(E) MortRate 1940	(F) Weighted MortRate
Pacific	0.0739	159.72	11.80	122.9	9.08
New England	0.0641	161.55	10.36	135.5	8.69
West No. Central	0.1027	123.14	12.65	110.9	11.39
Mid-Atlantic	0.2092	169.76	35.51	140.9	29.48
East No. Central	0.2022	133.36	26.97	119.6	24.18
Mountain	0.0315	119.89	3.78	99.8	3.14
West So. Central	0.0992	103.94	10.31	86.9	8.62
East So. Central	0.0819	85.83	7.03	73.6	6.03
South Atlantic	0.1354	100.74	13.64	88.9	12.04
Sums	1.0000		132.04		112.65

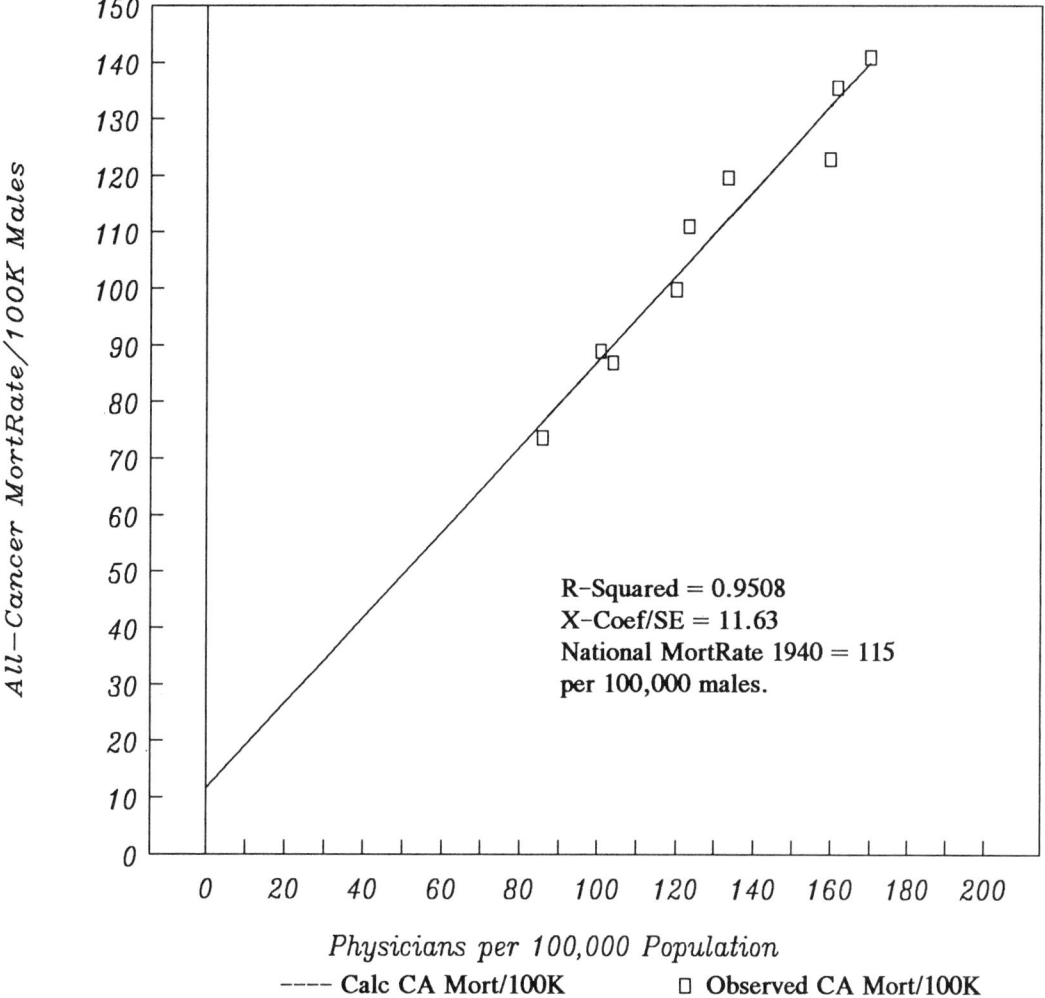

1940 All–Cancer Mortality-Rates versus
1940 PhysPop Values for the 9 Census Divisions, USA.
Dose–Response Relationship
PhysPop is a surrogate for accumulated dose from medical irradiation.

R-Squared = 0.9508
X-Coef/SE = 11.63
National MortRate 1940 = 115
per 100,000 males.

Related text = Parts 3 + 4.

Physicians per 100,000 Population

---- Calc CA Mort/100K   □ Observed CA Mort/100K

On the X–axis, PhysPop values = Physicians per 100,000 Population in the Nine Census Divisions of the USA Population, Year 1940. This variable is a surrogate for accumulated radiation dose --- the more physicians per 100,000 people, the more radiation procedures are done per 100,000 people.

On the Y–axis, All–Cancer Mortality-Rate per 100,000 males = the reported rates in USA Vital Statistics for the Nine Census Divisions, Year 1940.

Shown above is the strongest relationship between these two variables (Part 2j). The nine datapoints (boxy symbols) were collected long ago for other purposes, and are free from potential bias with respect to this dose-response study. Fractional causation is (Natl MortRate minus the Y–intercept) / (Natl MortRate).

Fractional Causation of All–Cancer Mortality–Rate in Males

by Medical Radiation = 90 % from Best Estimate (Box 3).

74.5% at Lower 90% Conf. Limit (Box 3).   99% at Upper 90% Conf. Limit (Box 3).

Table 6-A.
All-Cancer Mortality Rates by Census Divisions: Males.

Rates are annual deaths per 100,000 male population, USA, age-adjusted to the 1940 reference year. There are no exclusions by color or "race." Sources are stated in Table 6-B, and described in Chap. 4, Part 2. The Nine Census-Division MortRates are population-weighted (Chap. 4, Part 2b). The averages below them are not.

Census Division	1940	1950	1960	1970	1980	1988
Pacific	122.9	127.2	140.7	147.2	153.7	148.5
New England	135.5	152.4	164.6	167.5	170.3	167.1
West North Central	110.9	125.3	135.6	143.8	152.0	155.9
Mid-Atlantic	140.9	156.0	164.0	167.9	171.8	168.4
East North Central	119.6	138.3	150.7	160.1	169.5	171.2
Mountain	99.8	108.1	118.7	126.7	134.7	139.1
West South Central	86.9	112.7	133.8	148.3	162.9	172.9
East South Central	73.6	104.7	125.1	149.6	174.1	188.2
South Atlantic	88.9	116.3	137.1	154.2	171.4	175.8
Average, ALL	108.8	126.8	141.1	151.7	162.3	165.2
Average, High-5	126.0	139.8	151.1	157.3	163.5	162.2
Average, Low-4	87.3	110.5	128.7	144.7	160.8	169.0
Ratio, Hi5/Lo4	1.44	1.27	1.17	1.09	1.02	0.96

The declining Hi5/Lo4 ratio is explained in Chapters 48 and 49.

Table 6-B.
All-Cancer Mortality Rates, USA National.

Rates are age-adjusted to the 1940 reference year. Both sexes: Deaths per 100,000 population (males + females). Males: Deaths per 100,000 male population. Females: Deaths per 100,000 female population. No exclusions by color or "race."

	Both Sexes	Male	Female
1940	120.3	115.0	126.1
1950	127.7	132.8	123.2
1960	129.1	145.7	114.9
1970	129.8	155.1	111.7
1979-81	131.9	164.5	108.5
1987-89	135.0	162.7	111.3

● - 1940, 1950, 1960: All rates come from Grove 1968, Table 67, p.676, "Malignant neoplasms, including neoplasms of lymphatic and hematopoietic tissues (140-205)" ICD/7.
● - 1970: All rates are interpolations (Chap. 4, Parts 2b, 2c); except that the 1970 National "Both Sexes" rate comes from PHS 1995, Table 30, p.110.
● - 1980: All rates (ICD/9, 140-208) come from the reference NatCtrHS 1980.
● - 1988 rates by Divisions and National come from Monthly Vital Statistics Vol.41, No.6, November 12, 1992. Exception: National "Both Sexes" is for 1990, and comes from PHS 1995, Table 39, p.132.

All-Cancers-Combined, Females: Relation with Medical Radiation

● Part 1. Introduction

Chapter 7 follows the model of Chapter 6, bur eliminates the text.

For everyone's convenience, Chapter 22, Box 1, tabulates the results from Chapters 6 through 21 --- for easy comparison with each other.

● Part 2. How the Dose-Response Develops, 1921-1940

In Part 2, we regress the 1940 MortRates (from Table 7-A) upon the non-interpolated sets of PhysPop values 1921-1940 (from the Universal PhysPop Table 3-A). The summary-results of all the regression analyses are presented in Box 1.

● - Part 2a.	1921 PhysPop	1940 MortRate	All-Cancers, Females Regression Output:	
Pacific	165.11	127.4	Constant	33.3847
New England	142.24	145.3	Std Err of Y Est	14.5247
West North Central	140.93	120.1	R Squared	0.3566
Mid-Atlantic	137.29	142.9	No. of Observations	9
East North Central	136.06	131.4	Degrees of Freedom	7
Mountain	135.38	111.8		
West South Central	125.15	99.8	X Coefficient(s)	0.6497
East South Central	119.76	102.5	Std Err of Coef.	0.3299
South Atlantic	110.32	106.9	Coefficient / S.E.	1.9695

● - Part 2b.	1923 PhysPop	1940 MortRate	All-Cancers, Females Regression Output:	
Pacific	163.06	127.4	Constant	34.9732
New England	137.39	145.3	Std Err of Y Est	13.8135
West North Central	138.31	120.1	R Squared	0.4180
Mid-Atlantic	138.92	142.9	No. of Observations	9
East North Central	131.82	131.4	Degrees of Freedom	7
Mountain	130.51	111.8		
West South Central	119.16	99.8	X Coefficient(s)	0.6559
East South Central	113.16	102.5	Std Err of Coef.	0.2925
South Atlantic	106.79	106.9	Coefficient / S.E.	2.2423

● - Part 2c.	1925 PhysPop	1940 MortRate	All-Cancers, Females Regression Output:	
Pacific	161.67	127.4	Constant	37.9005
New England	138.31	145.3	Std Err of Y Est	13.0106
West North Central	133.92	120.1	R Squared	0.4837
Mid-Atlantic	134.36	142.9	No. of Observations	9
East North Central	127.54	131.4	Degrees of Freedom	7
Mountain	122.30	111.8		
West South Central	112.83	99.8	X Coefficient(s)	0.6542
East South Central	107.22	102.5	Std Err of Coef.	0.2555
South Atlantic	103.61	106.9	Coefficient / S.E.	2.5609

● - Part 2d.	1927 PhysPop	1940 MortRate	All-Cancers, Females Regression Output:	
Pacific	157.83	127.4	Constant	33.8179
New England	137.50	145.3	Std Err of Y Est	11.4512
West North Central	131.54	120.1	R Squared	0.6001
Mid-Atlantic	138.40	142.9	No. of Observations	9
East North Central	126.18	131.4	Degrees of Freedom	7
Mountain	118.75	111.8		
West South Central	108.25	99.8	X Coefficient(s)	0.6981

East South Central	102.07	102.5	Std Err of Coef.	0.2154
South Atlantic	102.13	106.9	Coefficient / S.E.	3.2407

● - Part 2e.

	1929	1940	All-Cancers, Females	
	PhysPop	MortRate	Regression Output:	
Pacific	156.64	127.4	Constant	34.0646
New England	138.46	145.3	Std Err of Y Est	10.7394
West North Central	128.72	120.1	R Squared	0.6482
Mid-Atlantic	138.49	142.9	No. of Observations	9
East North Central	126.51	131.4	Degrees of Freedom	7
Mountain	118.68	111.8		
West South Central	105.60	99.8	X Coefficient(s)	0.7019
East South Central	99.41	102.5	Std Err of Coef.	0.1954
South Atlantic	100.86	106.9	Coefficient / S.E.	3.5916

● - Part 2f.

	1931	1940	All-Cancers, Females	
	PhysPop	MortRate	Regression Output:	
Pacific	159.97	127.4	Constant	39.7540
New England	142.35	145.3	Std Err of Y Est	10.3504
West North Central	126.50	120.1	R Squared	0.6732
Mid-Atlantic	140.82	142.9	No. of Observations	9
East North Central	128.59	131.4	Degrees of Freedom	7
Mountain	118.89	111.8		
West South Central	105.95	99.8	X Coefficient(s)	0.6524
East South Central	96.73	102.5	Std Err of Coef.	0.1718
South Atlantic	99.59	106.9	Coefficient / S.E.	3.7978

● - Part 2g.

	1934	1940	All-Cancers, Females	
	PhysPop	MortRate	Regression Output:	
Pacific	160.09	127.4	Constant	44.2545
New England	148.60	145.3	Std Err of Y Est	8.7564
West North Central	125.96	120.1	R Squared	0.7661
Mid-Atlantic	149.62	142.9	No. of Observations	9
East North Central	129.36	131.4	Degrees of Freedom	7
Mountain	117.16	111.8		
West South Central	104.68	99.8	X Coefficient(s)	0.6127
East South Central	92.00	102.5	Std Err of Coef.	0.1279
South Atlantic	98.41	106.9	Coefficient / S.E.	4.7888

● - Part 2h.

	1936	1940	All-Cancers, Females	
	PhysPop	MortRate	Regression Output:	
Pacific	158.44	127.4	Constant	44.9599
New England	150.18	145.3	Std Err of Y Est	8.0257
West North Central	126.14	120.1	R Squared	0.8035
Mid-Atlantic	155.05	142.9	No. of Observations	9
East North Central	130.42	131.4	Degrees of Freedom	7
Mountain	119.80	111.8		
West South Central	103.52	99.8	X Coefficient(s)	0.6034
East South Central	89.94	102.5	Std Err of Coef.	0.1128
South Atlantic	99.16	106.9	Coefficient / S.E.	5.3509

● - Part 2i.

	1938	1940	All-Cancers, Females	
	PhysPop	MortRate	Regression Output:	
Pacific	157.62	127.4	Constant	47.4535
New England	154.08	145.3	Std Err of Y Est	7.1875
West North Central	124.95	120.1	R Squared	0.8424
Mid-Atlantic	160.69	142.9	No. of Observations	9
East North Central	131.98	131.4	Degrees of Freedom	7
Mountain	119.88	111.8		
West South Central	102.79	99.8	X Coefficient(s)	0.5801
East South Central	88.21	102.5	Std Err of Coef.	0.0948
South Atlantic	99.26	106.9	Coefficient / S.E.	6.1177

● - Part 2j.

	1940	1940	All-Cancers, Females	
	PhysPop	MortRate	Regression Output:	
Pacific	159.72	127.4	Constant	52.9840
New England	161.55	145.3	Std Err of Y Est	6.7550

West North Central	123.14	120.1	R Squared	0.8608
Mid–Atlantic	169.76	142.9	No. of Observations	9
East North Central	133.36	131.4	Degrees of Freedom	7
Mountain	119.89	111.8		
West South Central	103.94	99.8	X Coefficient(s)	0.5279
East South Central	85.83	102.5	Std Err of Coef.	0.0802
South Atlantic	100.74	106.9	Coefficient / S.E.	6.5801

Box 1 of Chap. 7
Summary: Regression Outputs for All–Cancers, Females.

Below are the summary-results from regressing the 1940 cancer MortRates upon the ten sets of PhysPops (1921-1940), as presented in Parts 2a-2j of this chapter. We are searching for the maximum correlation. Even the maximum will tend to understate the true correlation (Chapter 5, Part 8b).

Part	PhysPop	R-squared	Constant	X-Coef	Std Err	X-Coef/SE
2a	1921	0.3566	33.38	0.6497	0.3299	1.9695
2b	1923	0.4180	34.97	0.6559	0.2925	2.2423
2c	1925	0.4837	37.90	0.6542	0.2555	2.5609
2d	1927	0.6001	33.82	0.6981	0.2154	3.2407
2e	1929	0.6482	34.06	0.7019	0.1954	3.5916
2f	1931	0.6732	39.75	0.6524	0.1718	3.7978
2g	1934	0.7661	44.25	0.6127	0.1279	4.7888
2h	1936	0.8035	44.96	0.6034	0.1128	5.3509
2i	1938	0.8424	47.45	0.5801	0.0948	6.1177
2j --->	1940 Max	0.8608	52.98	0.5279	0.0802	6.5801

Box 2 of Chap. 7
Input-Data for Figure 7-A. All-Cancers. Females.

Part 2j, Best-Fit Equation: Calc. MortRate = (0.5279 * PhysPop) + (52.98)

Census Divisions	1940 Observed PhysPops	1940 Observed MortRates	Best-Fit Calc. MortRates
Pacific	159.72	127.4	137.296
New England	161.55	145.3	138.262
West No. Central	123.14	120.1	117.986
Mid-Atlantic	169.76	142.9	142.596
East No. Central	133.36	131.4	123.381
Mountain	119.89	111.8	116.270
West So. Central	103.94	99.8	107.850
East So. Central	85.83	102.5	98.290
South Atlantic	100.74	106.9	106.161
Additional PhysPops	70.00		89.933
--- not "observed" ---	60.00		84.654
down to zero PhysPop	50.00		79.375
(zero medical radiation).	40.00		74.096
For each, we calculate	30.00		68.817
a best-fit MortRate.	20.00		63.538
These additional x,y pairs	10.00		58.259
are also part of the	0		52.980
best-fit line (Chap 5, Part 5e).			

Box 3 of Chap. 7
Presumptive Fraction of Cancer MortRate Attributable to Medical Radiation.

Please see text in Chapter 6, Parts 4 and 6.

All-Cancers. FEMALES.   * denotes multiplication.

- FEMALE National MortRate (MR) 1940, from Table 7-B   126.1 National MortRate
- Constant, from regression, Part 2j   52.9840 Constant
- Fractional Causation, Best Est. = (Natl MR − Constant) / Natl MR   58.0% Frac. Causation

....

90% Confidence-Limits (C.L.) on Fractional Causation. See text in Chapter 6, Part 4b, please.

X-Coefficient, from Part 2j   0.5279 X-Coef., Best Est.
Standard Error (SE) of X-Coefficient, from Part 2j   0.0802 Standard Error

Upper 90% C.L. on X-Coef. = (Coef) + (1.645 * SE) =   0.6598 New X-Coefficient
New Constant = (Natl MR) − (New X-Coef * 1940 Natl PhysPop) =   38.9762 New Constant
Frac. Causation, High-Limit = (Natl MR − New Constant) / Natl MR =   69.1% New Frac. Caus'n.

Lower 90% C.L. on X-Coef. = (Coef) − (1.645 * SE) =   0.3960 New X-Coefficient
New Constant = (Natl MR) − (New X-Coef * 1940 Natl PhysPop) =   73.8160 New Constant
Frac. Causation, Low-Limit = (Natl MR − New Constant) / Natl MR =   41.5% New Frac. Caus'n.

Box 4 of Chap. 7
Error-Check on Our Own Work: All-Cancers, Females.

Please see text in Chapter 6, Part 5.

Below, Columns A, C, and E come directly from the regression input in Part 2j. Column B, the fraction of the whole 1940 population in each Census Division, comes from Table 3-B in Chapter 3. Each Column-D entry is the product of (B-entry times C-entry). Each Column-F entry is the product of (B-entry times E-entry). PhysPops and MortRates are each "per 100,000."

The Weighted-Avg. Nat'l PhysPop, 1940, is the sum of Column-D entries = 132.04

The Weighted-Avg. Nat'l Female MortRate, 1940, is sum of Col.F entries = 123.82
The Nat'l Female MortRate is also (X-Coef * Nat'l PhysPop) + Constant = 122.68
Comparison: The Nat'l Female MortRate, 1940, in Table 7-B = 126.10

(A) Census Division	(B) Pop'n Fraction	(C) PhysPop 1940	(D) 1940 Weighted PhysPop	(E) MortRate 1940	(F) Weighted MortRate
Pacific	0.0739	159.72	11.80	127.4	9.41
New England	0.0641	161.55	10.36	145.3	9.31
West No. Central	0.1027	123.14	12.65	120.1	12.33
Mid-Atlantic	0.2092	169.76	35.51	142.9	29.89
East No. Central	0.2022	133.36	26.97	131.4	26.57
Mountain	0.0315	119.89	3.78	111.8	3.52
West So. Central	0.0992	103.94	10.31	99.8	9.90
East So. Central	0.0819	85.83	7.03	102.5	8.39
South Atlantic	0.1354	100.74	13.64	106.9	14.47
Sums	1.0000		132.04		123.82

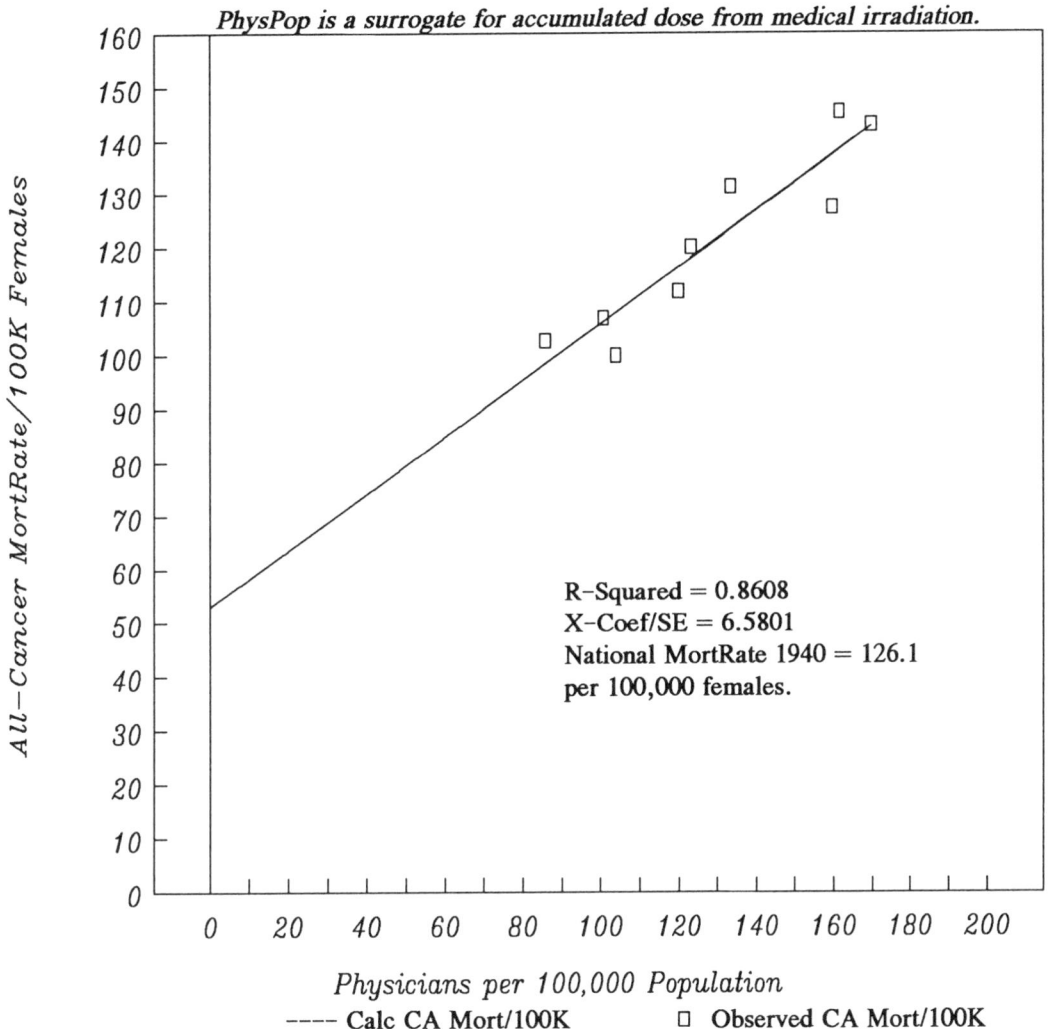

1940 All–Cancer Mortality-Rates versus 1940 PhysPop Values for the 9 Census Divisions, USA.
Dose–Response Relationship

PhysPop is a surrogate for accumulated dose from medical irradiation.

R-Squared = 0.8608
X-Coef/SE = 6.5801
National MortRate 1940 = 126.1
per 100,000 females.

Physicians per 100,000 Population

---- Calc CA Mort/100K   □ Observed CA Mort/100K

On the X–axis, PhysPop values = Physicians per 100,000 Population in the Nine Census Divisions of the USA Population, Year 1940. This variable is a surrogate for accumulated radiation dose --- the more physicians per 100,000 people, the more radiation procedures are done per 100,000 people.

On the Y–axis, All–Cancer Mortality-Rate per 100,000 females = the reported rates in USA Vital Statistics for the Nine Census Divisions, Year 1940.

Shown above is the strongest relationship between these two variables (Part 2j). The nine datapoints (boxy symbols) were collected long ago for other purposes, and are free from potential bias with respect to this dose–response study. Fractional causation is (Natl MortRate minus the Y–intercept) / (Natl MortRate).

Fractional Causation of All–Cancer Mortality-Rate in Females by Medical Radiation = 58 % from Best Estimate (Box 3).

41.5% at Lower 90% Conf. Limit (Box 3). 69 % at Upper 90 % Conf. Limit (Box 3).

Table 7-A.
All-Cancer Mortality Rates by Census Divisions: Females.

Rates are annual deaths per 100,000 female population, USA, age-adjusted to the 1940 reference year. There are no exclusions by color or "race." Sources are stated in Table 7-B, and described in Chap. 4, Part 2. The Nine Census-Division MortRates are population-weighted (Chap. 4, Part 2b). The averages below them are not.

Census Division	1940	1950	1960	1970	1980	1988
Pacific	127.4	117.7	110.1	110.2	110.4	111.5
New England	145.3	132.1	122.4	119.4	116.4	116.4
West North Central	120.1	117.1	109.3	105.1	101.0	106.8
Mid-Atlantic	142.9	137.0	127.4	122.4	117.5	118.6
East North Central	131.4	127.5	119.8	115.9	112.0	116.5
Mountain	111.8	106.0	101.0	97.9	94.9	100.4
West South Central	99.8	109.3	102.9	101.5	100.1	109.8
East South Central	102.5	110.3	104.8	104.0	103.2	112.7
South Atlantic	106.9	113.3	107.4	106.2	105.0	111.6
Average, ALL	120.9	118.9	111.7	109.2	106.7	111.6
Average, High-5	133.4	126.3	117.8	114.6	111.5	114.0
Average, Low-4	105.3	109.7	104.0	102.4	100.8	108.6
Ratio, Hi5/Lo4	1.27	1.15	1.13	1.12	1.11	1.05

Table 7-B.
All-Cancer Mortality Rates, USA National.

Rates are age-adjusted to the 1940 reference year. Both sexes: Deaths per 100,000 population (males + females). Males: Deaths per 100,000 male population. Females: Deaths per 100,000 female population. No exclusions by color or "race."

	Both Sexes	Male	Female
1940	120.3	115.0	126.1
1950	127.7	132.8	123.2
1960	129.1	145.7	114.9
1970	129.8	155.1	111.7
1979-81	131.9	164.5	108.5
1987-89	135.0	162.7	111.3

● - 1940, 1950, 1960: All rates come from Grove 1968, Table 67, p.676, "Malignant neoplasms, including neoplasms of lymphatic and hematopoietic tissues (140-205)" ICD/7.

● - 1970: All rates are interpolations (Chap. 4, Parts 2b, 2c), except that the 1970 National "Both Sexes" rate comes from PHS 1995, Table 30, p.110.

● - 1980: All rates (ICD/9, 140-208) come from the reference NatCtrHS 1980.

● - 1988 rates by Divisions and National come from Monthly Vital Statistics Vol.41, No.6, November 12, 1992. Exception: National "Both Sexes" is for 1990, and comes from PHS 1995, Table 39, p.132.

● Part 1. Introduction

We put Breast Cancer early in Section Two of this book, because of the interest in seeing how the 1940 Fractional Causation obtained in this chapter compares with the 1995 Fractional Causation (75%) obtained for this particular cancer in Gofman 1995/1996. The two studies use completely different data and completely different methods.

● Part 2. How the Dose-Response Develops, 1921-1940

In Part 2, we regress the 1940 MortRates (from Table 8-A) upon the non-interpolated sets of PhysPop values 1921-1940 (from the Universal PhysPop Table 3-A). The summary-results of all the regression analyses are presented in Box 1 nearby.

● – Part 2a.	1921 PhysPop	1940 MortRate	Breast Cancer, Females Regression Output:	
Pacific	165.11	26.7	Constant	-10.9421
New England	142.24	28.8	Std Err of Y Est	4.0114
West North Central	140.93	22.6	R Squared	0.5061
Mid-Atlantic	137.29	27.8	No. of Observations	9
East North Central	136.06	24.3	Degrees of Freedom	7
Mountain	135.38	18.6		
West South Central	125.15	15.1	X Coefficient(s)	0.2440
East South Central	119.76	15.1	Std Err of Coef.	0.0911
South Atlantic	110.32	18.3	Coefficient / S.E.	2.6780

● – Part 2b.	1923 PhysPop	1940 MortRate	Breast Cancer, Females Regression Output:	
Pacific	163.06	26.7	Constant	-9.9385
New England	137.39	28.8	Std Err of Y Est	3.7059
West North Central	138.31	22.6	R Squared	0.5784
Mid-Atlantic	138.92	27.8	No. of Observations	9
East North Central	131.82	24.3	Degrees of Freedom	7
Mountain	130.51	18.6		
West South Central	119.16	15.1	X Coefficient(s)	0.2432
East South Central	113.16	15.1	Std Err of Coef.	0.0785
South Atlantic	106.79	18.3	Coefficient / S.E.	3.0991

● – Part 2c.	1925 PhysPop	1940 MortRate	Breast Cancer, Females Regression Output:	
Pacific	161.67	26.7	Constant	-8.6344
New England	138.31	28.8	Std Err of Y Est	3.3288
West North Central	133.92	22.6	R Squared	0.6598
Mid-Atlantic	134.36	27.8	No. of Observations	9
East North Central	127.54	24.3	Degrees of Freedom	7
Mountain	122.30	18.6		
West South Central	112.83	15.1	X Coefficient(s)	0.2409
East South Central	107.22	15.1	Std Err of Coef.	0.0654
South Atlantic	103.61	18.3	Coefficient / S.E.	3.6849

● – Part 2d.	1927 PhysPop	1940 MortRate	Breast Cancer, Females Regression Output:	
Pacific	157.83	26.7	Constant	-9.1171
New England	137.50	28.8	Std Err of Y Est	2.7534
West North Central	131.54	22.6	R Squared	0.7673
Mid-Atlantic	138.40	27.8	No. of Observations	9
East North Central	126.18	24.3	Degrees of Freedom	7
Mountain	118.75	18.6		
West South Central	108.25	15.1	X Coefficient(s)	0.2488

East South Central	102.07	15.1	Std Err of Coef.		0.0518
South Atlantic	102.13	18.3	Coefficient / S.E.		4.8040

● – Part 2e.

	1929	1940	Breast Cancer, Females	
	PhysPop	MortRate	Regression Output:	
Pacific	156.64	26.7	Constant	-8.5821
New England	138.46	28.8	Std Err of Y Est	2.5197
West North Central	128.72	22.6	R Squared	0.8051
Mid-Atlantic	138.49	27.8	No. of Observations	9
East North Central	126.51	24.3	Degrees of Freedom	7
Mountain	118.68	18.6		
West South Central	105.60	15.1	X Coefficient(s)	0.2466
East South Central	99.41	15.1	Std Err of Coef.	0.0459
South Atlantic	100.86	18.3	Coefficient / S.E.	5.3774

● – Part 2f.

	1931	1940	Breast Cancer, Females	
	PhysPop	MortRate	Regression Output:	
Pacific	159.97	26.7	Constant	-6.3107
New England	142.35	28.8	Std Err of Y Est	2.4198
West North Central	126.50	22.6	R Squared	0.8203
Mid-Atlantic	140.82	27.8	No. of Observations	9
East North Central	128.59	24.3	Degrees of Freedom	7
Mountain	118.89	18.6		
West South Central	105.95	15.1	X Coefficient(s)	0.2270
East South Central	96.73	15.1	Std Err of Coef.	0.0402
South Atlantic	99.59	18.3	Coefficient / S.E.	5.6519

● – Part 2g.

	1934	1940	Breast Cancer, Females	
	PhysPop	MortRate	Regression Output:	
Pacific	160.09	26.7	Constant	-4.0298
New England	148.60	28.8	Std Err of Y Est	1.9436
West North Central	125.96	22.6	R Squared	0.8840
Mid-Atlantic	149.62	27.8	No. of Observations	9
East North Central	129.36	24.3	Degrees of Freedom	7
Mountain	117.16	18.6		
West South Central	104.68	15.1	X Coefficient(s)	0.2075
East South Central	92.00	15.1	Std Err of Coef.	0.0284
South Atlantic	98.41	18.3	Coefficient / S.E.	7.3052

● – Part 2h.

	1936	1940	Breast Cancer, Females	
	PhysPop	MortRate	Regression Output:	
Pacific	158.44	26.7	Constant	-3.4183
New England	150.18	28.8	Std Err of Y Est	1.8002
West North Central	126.14	22.6	R Squared	0.9005
Mid-Atlantic	155.05	27.8	No. of Observations	9
East North Central	130.42	24.3	Degrees of Freedom	7
Mountain	119.80	18.6		
West South Central	103.52	15.1	X Coefficient(s)	0.2014
East South Central	89.94	15.1	Std Err of Coef.	0.0253
South Atlantic	99.16	18.3	Coefficient / S.E.	7.9604

● – Part 2i.

	1938	1940	Breast Cancer, Females	
	PhysPop	MortRate	Regression Output:	
Pacific	157.62	26.7	Constant	-2.2092
New England	154.08	28.8	Std Err of Y Est	1.6613
West North Central	124.95	22.6	R Squared	0.9153
Mid-Atlantic	160.69	27.8	No. of Observations	9
East North Central	131.98	24.3	Degrees of Freedom	7
Mountain	119.88	18.6		
West South Central	102.79	15.1	X Coefficient(s)	0.1906
East South Central	88.21	15.1	Std Err of Coef.	0.0219
South Atlantic	99.26	18.3	Coefficient / S.E.	8.6965

● – Part 2j.

	1940	1940	Breast Cancer, Females	
	PhysPop	MortRate	Regression Output:	
Pacific	159.72	26.7	Constant	-0.1205
New England	161.55	28.8	Std Err of Y Est	1.6870
West North Central	123.14	22.6	R Squared	0.9126

Mid-Atlantic	169.76	27.8	No. of Observations		9
East North Central	133.36	24.3	Degrees of Freedom		7
Mountain	119.89	18.6			
West South Central	103.94	15.1	X Coefficient(s)		0.1713
East South Central	85.83	15.1	Std Err of Coef.		0.0200
South Atlantic	100.74	18.3	Coefficient / S.E.		8.5512

Box 1 of Chap. 8
Summary: Regression Outputs for Breast Cancer, Females.

Below are the summary-results from regressing the 1940 cancer MortRates upon the ten sets of PhysPops (1921-1940), as presented in Parts 2a-2j of this chapter. We are searching for the maximum correlation. Even the maximum will tend to understate the true correlation (Chapter 5, Part 8b).

Part	PhysPop	R-squared	Constant	X-Coef	Std Err	X-Coef/SE
2a	1921	0.5061	-10.94	0.2440	0.0911	2.6780
2b	1923	0.5784	-9.94	0.2432	0.0785	3.0991
2c	1925	0.6598	-8.63	0.2409	0.0654	3.6849
2d	1927	0.7673	-9.12	0.2488	0.0518	4.8040
2e	1929	0.8051	-8.58	0.2466	0.0459	5.3774
2f	1931	0.8203	-6.31	0.2270	0.0402	5.6519
2g	1934	0.8840	-4.03	0.2075	0.0284	7.3052
2h	1936	0.9005	-3.42	0.2014	0.0253	7.9604
2i --->	1938 Max	0.9153	-2.21	0.1906	0.0219	8.6965
2j	1940	0.9126	-0.12	0.1713	0.0200	8.5512

Box 2 of Chap. 8
Input-Data for Figure 8-A. Breast Cancer. Females.

Part 2i, Best-Fit Equation: Calc. MortRate = (0.1906 * PhysPop) + (-2.21)

Census Divisions	1938 Observed PhysPops	1940 Observed MortRates	Best-Fit Calc. MortRates
Pacific	157.62	26.7	27.832
New England	154.08	28.8	27.158
West No. Central	124.95	22.6	21.605
Mid-Atlantic	160.69	27.8	28.418
East No. Central	131.98	24.3	22.945
Mountain	119.88	18.6	20.639
West So. Central	102.79	15.1	17.382
East So. Central	88.21	15.1	14.603
South Atlantic	99.26	18.3	16.709
Additional PhysPops	70.00		11.132
--- not "observed" ---	60.00		9.226
down to zero PhysPop	50.00		7.320
(zero medical radiation).	40.00		5.414
For each, we calculate	30.00		3.508
a best-fit MortRate.	20.00		1.602
These additional x,y pairs	10.00		-0.304
are also part of the	0		-2.210
best-fit line (Chap 5, Part 5e).			

Box 3 of Chap. 8
Presumptive Fraction of Cancer MortRate Attributable to Medical Radiation.

Please see text in Chapter 6, Parts 4 and 6.

Breast Cancer. FEMALES.

- FEMALE National MortRate (MR) 1940, from Table 8-B   23.3 National MortRate
- Constant, from regression, Part 2i   -2.2092 Constant
- Fractional Causation, Best Est. = (Natl MR - Constant) / Natl MR   109.5% Frac. Causation
  # The Upper-Limit is 100%. Negative Constants produce values > 100%.   See Chapter 22, Part 3.

....

90% Confidence-Limits (C.L.) on Fractional Causation. See text in Chapter 6, Part 4b, please.

X-Coefficient, from Part 2i   0.1906 X-Coef., Best Est.
Standard Error (SE) of X-Coefficient, from Part 2i   0.0219 Standard Error

Upper 90% C.L. on X-Coef. = (Coef) + (1.645 * SE) =   0.2266 New X-Coefficient
New Constant = (Natl MR) - (New X-Coef * 1938 Natl PhysPop) = -5.9994 New Constant
Frac. Causation, High-Limit = (Natl MR - New Constant) / Natl MR =   1 26%. New Frac. Caus'n.
 # The Upper-Limit is 100%. Negative Constants produce values > 100%.   See Chapter 22, Part 3.

Lower 90% C.L. on X-Coef. = (Coef) - (1.645 * SE) =   0.1546 New X-Coefficient
New Constant = (Natl MR) - (New X-Coef * 1938 Natl PhysPop) = 3.3102 New Constant
Frac. Causation, Low-Limit = (Natl MR - New Constant) / Natl MR =   85.8% New Frac. Caus'n.

Box 4 of Chap. 8
Error-Check on Our Own Work: Breast Cancer, Females.

Please see text in Chapter 6, Part 5.

Below, Columns A, C, and E come directly from the regression input in Part 2i. Column
B, the fraction of the whole 1940 population in each Census Division, comes from Table
3-B in Chapter 3. Each Column-D entry is the product of (B-entry times C-entry).
Each Column-F entry is the product of (B-entry times E-entry). PhysPops and
MortRates are each "per 100,000."

The Weighted-Avg. Nat'l PhysPop, 1938, is the sum of Column-D entries = 129.30
  The 1938 PhysPop approximation is weighted by the 1940 population-fractions.
The Weighted-Avg. Nat'l Female MortRate, 1940, is sum of Col.F entries = 22.67
The Nat'l Female MortRate is also (X-Coef * Nat'l PhysPop) + Constant = 22.44
Comparison: The Nat'l Female MortRate, 1940, in Table 8-B = 23.30

(A) Census Division	(B) Pop'n Fraction	(C) PhysPop 1938	(D) 1938 Weighted PhysPop	(E) MortRate 1940	(F) Weighted MortRate
Pacific	0.0739	157.62	11.65	26.7	1.97
New England	0.0641	154.08	9.88	28.8	1.85
West No. Central	0.1027	124.95	12.83	22.6	2.32
Mid-Atlantic	0.2092	160.69	33.62	27.8	5.82
East No. Central	0.2022	131.98	26.69	24.3	4.91
Mountain	0.0315	119.88	3.78	18.6	0.59
West So. Central	0.0992	102.79	10.20	15.1	1.50
East So. Central	0.0819	88.21	7.22	15.1	1.24
South Atlantic	0.1354	99.26	13.44	18.3	2.48
Sums	1.0000		129.30		22.67

1940 Breast–Cancer Mortality–Rates versus
1938 PhysPop Values for the 9 Census Divisions, USA.
Dose–Response Relationship

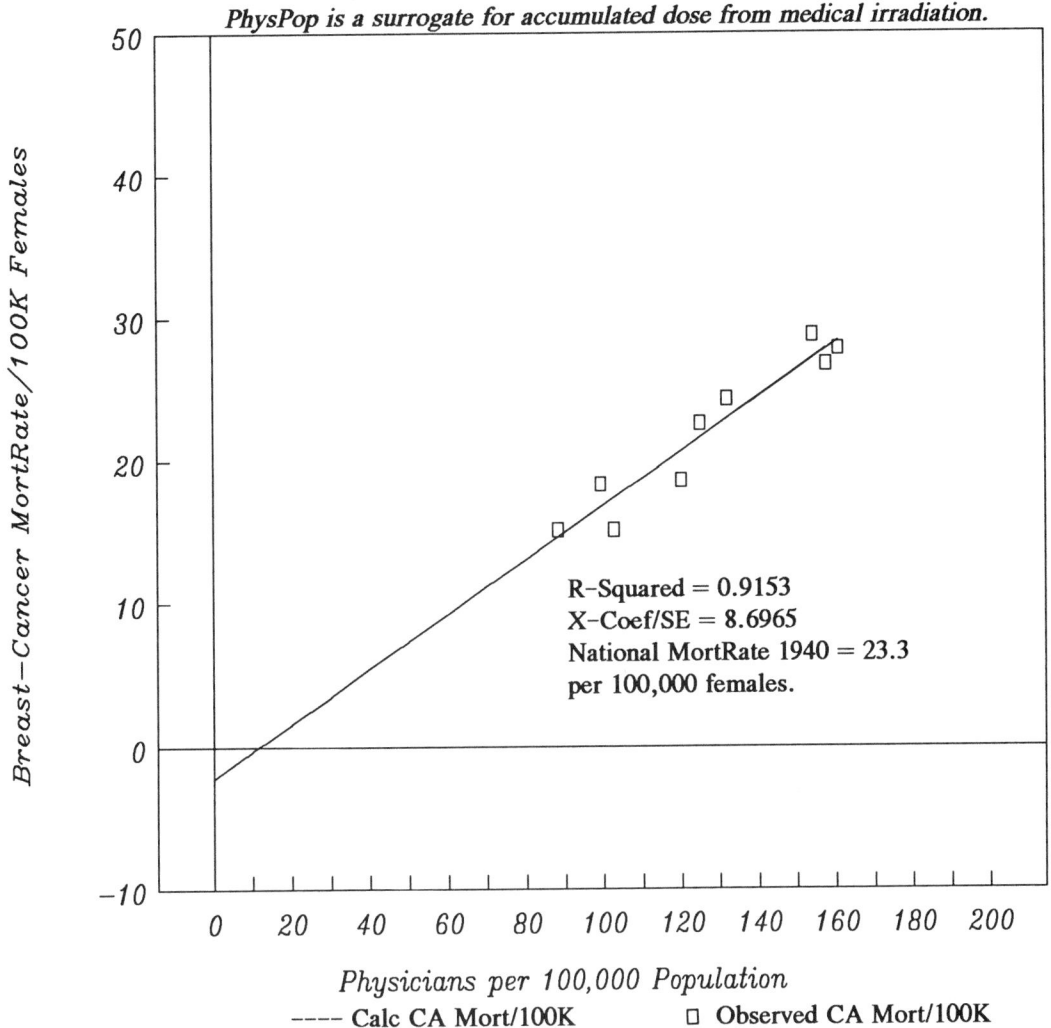

PhysPop is a surrogate for accumulated dose from medical irradiation.

R–Squared = 0.9153
X–Coef/SE = 8.6965
National MortRate 1940 = 23.3
per 100,000 females.

Physicians per 100,000 Population
---- Calc CA Mort/100K   □ Observed CA Mort/100K

On the X–axis, PhysPop values = Physicians per 100,000 Population
in the Nine Census Divisions of the USA Population, Year 1940. This
variable is a surrogate for accumulated radiation dose --- the more
physicians per 100,000 people, the more radiation procedures are done per
100,000 people.

On the Y–axis, Breast–Cancer Mortality–Rate per 100,000 females = the
reported rates in USA Vital Statistics for the Nine Census Divisions, Year 1940.

Shown above is the strongest relationship between these two
variables (Part 2i). The nine datapoints (boxy symbols) were collected long
ago for other purposes, and are free from potential bias with respect to this
dose–response study. Fractional causation is (Natl MortRate minus the
Y–intercept) / (Natl MortRate).

Fractional Causation of Breast–Cancer Mortality–Rate in Females
by Medical Radiation = ~100 % from Best Estimate (Box 3).
85.8 % at Lower 90% Conf. Limit (Box 3). ~100 % at Upper 90 % Conf. Limit (Box 3).

Table 8-A.
Breast-Cancer Mortality Rates by Census Divisions: Females.

Rates are annual deaths per 100,000 female population, USA, age–adjusted to the 1940 reference year. There are no exclusions by color or "race." Sources are stated in Table 8-B, and described in Chap. 4, Part 2. The Nine Census-Division MortRates are population–weighted (Chap. 4, Part 2b). The averages below them are not.

Census Division	1940	1950	1960	1970	1980	1990
Pacific	26.7	23.8	23.3	22.3	21.2	22.7
New England	28.8	25.8	25.9	25.3	24.7	24.3
West North Central	22.6	22.6	22.8	22.2	21.7	22.6
Mid–Atlantic	27.8	26.5	26.8	26.4	25.9	25.8
East North Central	24.3	23.5	24.3	24.2	24.0	24.1
Mountain	18.6	18.8	20.3	20.3	20.3	21.0
West South Central	15.1	16.6	17.8	18.4	18.9	20.8
East South Central	15.1	16.6	17.6	18.6	19.6	21.4
South Atlantic	18.3	18.4	19.4	20.2	21.0	22.6
Average, ALL	21.9	21.4	22.0	22.0	21.9	22.8
Average, High-5	26.0	24.4	24.6	24.1	23.5	23.9
Average, Low-4	16.8	17.6	18.8	19.4	20.0	21.5
Ratio, Hi5/Lo4	1.55	1.39	1.31	1.24	1.18	1.11

Table 8-B.
Breast-Cancer Mortality Rates, USA National.

Rates are age–adjusted to the 1940 reference year. Both sexes: Deaths per 100,000 population (males + females). Males: Deaths per 100,000 male population. Females: Deaths per 100,000 female population. No exclusions by color or "race."

	Both Sexes	Male	Female
1940	11.7	0.2	23.3
1950	11.6	0.2	22.5
1960	12.1	0.2	22.9
1970	--	--	23.1
1979-81	12.4	0.2	22.6
1989-91	--	--	23.1

● - 1940, 1950, 1960: All rates come from Grove 1968, Table 67, p.690, "Malignant neoplasms of the breast (170)" ICD/7.
● - 1970: All rates by Divisions are interpolations (Chap. 4, Parts 2b, 2c), except that the 1970 National rate for Females comes from PHS 1995, Table 41, p.138.
● - 1980: All rates (ICD/9, 174-175) come from the reference NatCtrHS 1980.
● - 1990 rates by Divisions and National come from Monthly Vital Statistics Vol.43, No.8, January 31, 1995 (MVS in our Reference List).

● - In Chapter 4, Box 3, age-SPECIFIC Breast-Cancer MortRates are shown (as an illustration of age-specific rates), by decades for the 1950-1990 period.

CHAPTER 9

Digestive System Cancers, Males: Relation with Medical Radiation

● Part 1. Introduction

Digestive-System Cancers include cancers of the esophagus, stomach, small and large intestine, rectum, biliary passages and liver, pancreas, and peritoneum (see Chapter 4, Part 5, Number 6).

● Part 2. How the Dose-Response Develops, 1921-1940

● — Part 2a.

	1921 PhysPop	1940 MortRate	Digestive Cancers, Males Regression Output:	
Pacific	165.11	63.4	Constant	-21.7593
New England	142.24	71.7	Std Err of Y Est	10.4249
West North Central	140.93	59.9	R Squared	0.4639
Mid-Atlantic	137.29	74.7	No. of Observations	9
East North Central	136.06	64.9	Degrees of Freedom	7
Mountain	135.38	52.1		
West South Central	125.15	42.3	X Coefficient(s)	0.5827
East South Central	119.76	38.2	Std Err of Coef.	0.2368
South Atlantic	110.32	43.4	Coefficient / S.E.	2.4611

● — Part 2b.

	1923 PhysPop	1940 MortRate	Digestive Cancers, Males Regression Output:	
Pacific	163.06	63.4	Constant	-20.1073
New England	137.39	71.7	Std Err of Y Est	9.6496
West North Central	138.31	59.9	R Squared	0.5407
Mid-Atlantic	138.92	74.7	No. of Observations	9
East North Central	131.82	64.9	Degrees of Freedom	7
Mountain	130.51	52.1		
West South Central	119.16	42.3	X Coefficient(s)	0.5865
East South Central	113.16	38.2	Std Err of Coef.	0.2043
South Atlantic	106.79	43.4	Coefficient / S.E.	2.8705

● — Part 2c.

	1925 PhysPop	1940 MortRate	Digestive Cancers, Males Regression Output:	
Pacific	161.67	63.4	Constant	-14.8748
New England	138.31	71.7	Std Err of Y Est	9.2016
West North Central	133.92	59.9	R Squared	0.5823
Mid-Atlantic	134.36	74.7	No. of Observations	9
East North Central	127.54	64.9	Degrees of Freedom	7
Mountain	122.30	52.1		
West South Central	112.83	42.3	X Coefficient(s)	0.5645
East South Central	107.22	38.2	Std Err of Coef.	0.1807
South Atlantic	103.61	43.4	Coefficient / S.E.	3.1240

● — Part 2d.

	1927 PhysPop	1940 MortRate	Digestive Cancers, Males Regression Output:	
Pacific	157.83	63.4	Constant	-17.3245
New England	137.50	71.7	Std Err of Y Est	7.7735
West North Central	131.54	59.9	R Squared	0.7019
Mid-Atlantic	138.40	74.7	No. of Observations	9
East North Central	126.18	64.9	Degrees of Freedom	7
Mountain	118.75	52.1		
West South Central	108.25	42.3	X Coefficient(s)	0.5937 Page 2
East South Central	102.07	38.2	Std Err of Coef.	0.1462
South Atlantic	102.13	43.4	Coefficient / S.E.	4.0599

● — Part 2e.

	1929 PhysPop	1940 MortRate	Digestive Cancers, Males Regression Output:	
Pacific	156.64	63.4	Constant	-16.3714
New England	138.46	71.7	Std Err of Y Est	7.2168

	PhysPop	MortRate		
West North Central	128.72	59.9	R Squared	0.7431
Mid-Atlantic	138.49	74.7	No. of Observations	9
East North Central	126.51	64.9	Degrees of Freedom	7
Mountain	118.68	52.1		
West South Central	105.60	42.3	X Coefficient(s)	0.5909
East South Central	99.41	38.2	Std Err of Coef.	0.1313
South Atlantic	100.86	43.4	Coefficient / S.E.	4.4995

● - Part 2f.	1931	1940	Digestive Cancers, Males	
	PhysPop	MortRate	Regression Output:	
Pacific	159.97	63.4	Constant	-10.9999
New England	142.35	71.7	Std Err of Y Est	6.9944
West North Central	126.50	59.9	R Squared	0.7587
Mid-Atlantic	140.82	74.7	No. of Observations	9
East North Central	128.59	64.9	Degrees of Freedom	7
Mountain	118.89	52.1		
West South Central	105.95	42.3	X Coefficient(s)	0.5446
East South Central	96.73	38.2	Std Err of Coef.	0.1161
South Atlantic	99.59	43.4	Coefficient / S.E.	4.6911

● - Part 2g.	1934	1940	Digestive Cancers, Males	
	PhysPop	MortRate	Regression Output:	
Pacific	160.09	63.4	Constant	-6.3946
New England	148.60	71.7	Std Err of Y Est	5.6845
West North Central	125.96	59.9	R Squared	0.8406
Mid-Atlantic	149.62	74.7	No. of Observations	9
East North Central	129.36	64.9	Degrees of Freedom	7
Mountain	117.16	52.1		
West South Central	104.68	42.3	X Coefficient(s)	0.5046
East South Central	92.00	38.2	Std Err of Coef.	0.0831
South Atlantic	98.41	43.4	Coefficient / S.E.	6.0756

● - Part 2h.	1936	1940	Digestive Cancers, Males	
	PhysPop	MortRate	Regression Output:	
Pacific	158.44	63.4	Constant	-5.7821
New England	150.18	71.7	Std Err of Y Est	4.9168
West North Central	126.14	59.9	R Squared	0.8807
Mid-Atlantic	155.05	74.7	No. of Observations	9
East North Central	130.42	64.9	Degrees of Freedom	7
Mountain	119.80	52.1		
West South Central	103.52	42.3	X Coefficient(s)	0.4968
East South Central	89.94	38.2	Std Err of Coef.	0.0691
South Atlantic	99.16	43.4	Coefficient / S.E.	7.1902

● - Part 2i.	1938	1940	Digestive Cancers, Males	
	PhysPop	MortRate	Regression Output:	
Pacific	157.62	63.4	Constant	-3.2139
New England	154.08	71.7	Std Err of Y Est	4.3256
West North Central	124.95	59.9	R Squared	0.9077
Mid-Atlantic	160.69	74.7	No. of Observations	9
East North Central	131.98	64.9	Degrees of Freedom	7
Mountain	119.88	52.1		
West South Central	102.79	42.3	X Coefficient(s)	0.4735
East South Central	88.21	38.2	Std Err of Coef.	0.0571
South Atlantic	99.26	43.4	Coefficient / S.E.	8.2970

● - Part 2j.	1940	1940	Digestive Cancers, Males	
	PhysPop	MortRate	Regression Output:	
Pacific	159.72	63.4	Constant	1.8931
New England	161.55	71.7	Std Err of Y Est	4.3237
West North Central	123.14	59.9	R Squared	0.9078
Mid-Atlantic	169.76	74.7	No. of Observations	9
East North Central	133.36	64.9	Degrees of Freedom	7
Mountain	119.89	52.1		
West South Central	103.94	42.3	X Coefficient(s)	0.4262
East South Central	85.83	38.2	Std Err of Coef.	0.0513
South Atlantic	100.74	43.4	Coefficient / S.E.	8.3009

Box 1 of Chap. 9
Summary: Regression Outputs, for Digestive-System Cancers, Males.

Below are the summary-results from regressing the 1940 cancer MortRates upon the ten sets of PhysPops (1921-1940), as presented in Parts 2a-2j of this chapter.

Part	PhysPop	R-squared	Constant	X-Coef	Std Err	X-Coef/SE
2a	1921	0.4639	-21.76	0.5827	0.2368	2.4611
2b	1923	0.5407	-20.11	0.5865	0.2043	2.8705
2c	1925	0.5823	-14.87	0.5645	0.1807	3.1240
2d	1927	0.7019	-17.32	0.5937	0.1462	4.0599
2e	1929	0.7431	-16.37	0.5909	0.1313	4.4995
2f	1931	0.7587	-11.00	0.5446	0.1161	4.6911
2g	1934	0.8406	-6.39	0.5046	0.0831	6.0756
2h	1936	0.8807	-5.78	0.4968	0.0691	7.1902
2i	1938	0.9077	-3.21	0.4735	0.0571	8.2970
2j --->	1940 Max	0.9078	1.89	0.4262	0.0513	8.3009

Box 2 of Chap. 9
Input-Data for Figure 9-A. Digestive-System Cancers. Males.

Part 2j, Best-Fit Equation: Calc. MortRate = (0.4262 * PhysPop) + (1.89)

Census Divisions	1940 Observed PhysPops	1940 Observed MortRates	Best-Fit Calc. MortRates
Pacific	159.72	63.4	69.963
New England	161.55	71.7	70.743
West No. Central	123.14	59.9	54.372
Mid-Atlantic	169.76	74.7	74.242
East No. Central	133.36	64.9	58.728
Mountain	119.89	52.1	52.987
West So. Central	103.94	42.3	46.189
East So. Central	85.83	38.2	38.471
South Atlantic	100.74	43.4	44.825
Additional PhysPops	70.00		31.724
--- not "observed" ---	60.00		27.462
down to zero PhysPop	50.00		23.200
(zero medical radiation).	40.00		18.938
For each, we calculate	30.00		14.676
a best-fit MortRate.	20.00		10.414
These additional x,y pairs	10.00		6.152
are also part of the	0		1.890
best-fit line (Chap 5, Part 5e).			

Box 3 of Chap. 9

Presumptive Fraction of Cancer MortRate Attributable to Medical Radiation.

Please see text in Chapter 6, Parts 4 and 6.

Digestive-System Cancers. MALES.

- MALE National MortRate (MR) 1940, from Table 9-B   60.4 National MortRate
- Constant, from regression, Part 2j   1.8931 Constant
- Fractional Causation, Best Est. = (Natl MR − Constant) / Natl MR   96.9% Frac. Causation

....

90% Confidence-Limits (C.L.) on Fractional Causation. See text in Chapter 6, Part 4b, please.

X-Coefficient, from Part 2j   0.4262 X-Coef., Best Est.
Standard Error (SE) of X-Coefficient, from Part 2j   0.0513 Standard Error

Upper 90% C.L. on X-Coef. = (Coef) + (1.645 * SE) =   0.5106 New X-Coefficient
New Constant = (Natl MR) − (New X-Coef * 1940 Natl PhysPop)=   −7.0196 New Constant
Frac. Causation, High-Limit = (Natl MR − New Constant) / Natl MR =   111.6% New Frac. Caus'n.
   # The Upper-Limit is 100%. Negative Constants produce values > 100%. See Chapter 22, Part 3.

Lower 90% C.L. on X-Coef. = (Coef) − (1.645 * SE) =   0.3418 New X-Coefficient
New Constant = (Natl MR) − (New X-Coef * 1940 Natl PhysPop) =   15.2672 New Constant
Frac. Causation, Low-Limit = (Natl MR − New Constant) / Natl MR =   74.7% New Frac. Caus'n.

Box 4 of Chap. 9

Error-Check on Our Own Work: Digestive-System Cancers, Males.

Please see text in Chapter 6, Part 5.

Below, Columns A, C, and E come directly from the regression input in Part 2j. Column B, the fraction of the whole 1940 population in each Census Division, comes from Table 3-B in Chapter 3. Each Column-D entry is the product of (B-entry times C-entry). Each Column-F entry is the product of (B-entry times E-entry). PhysPops and MortRates are each "per 100,000."

The Weighted-Avg. Nat'l PhysPop, 1940, is the sum of Column-D entries =   132.04

The Weighted-Avg. Nat'l Male MortRate, 1940, is sum of Col.F entries =   59.03
The Nat'l Male MortRate is also (X-Coef * Nat'l PhysPop) + Constant =   58.17
Comparison: The Nat'l Male MortRate, 1940, in Table 9-B =   60.40

(A) Census Division	(B) Pop'n Fraction	(C) PhysPop 1940	(D) 1940 Weighted PhysPop	(E) MortRate 1940	(F) Weighted MortRate
Pacific	0.0739	159.72	11.80	63.4	4.69
New England	0.0641	161.55	10.36	71.7	4.60
West No. Central	0.1027	123.14	12.65	59.9	6.15
Mid-Atlantic	0.2092	169.76	35.51	74.7	15.63
East No. Central	0.2022	133.36	26.97	64.9	13.12
Mountain	0.0315	119.89	3.78	52.1	1.64
West So. Central	0.0992	103.94	10.31	42.3	4.20
East So. Central	0.0819	85.83	7.03	38.2	3.13
South Atlantic	0.1354	100.74	13.64	43.4	5.88
Sums	1.0000		132.04		59.03

1940 Digestive–System Cancer Mortality–Rates versus 1940 PhysPop Values for the 9 Census Divisions, USA.
Dose–Response Relationship

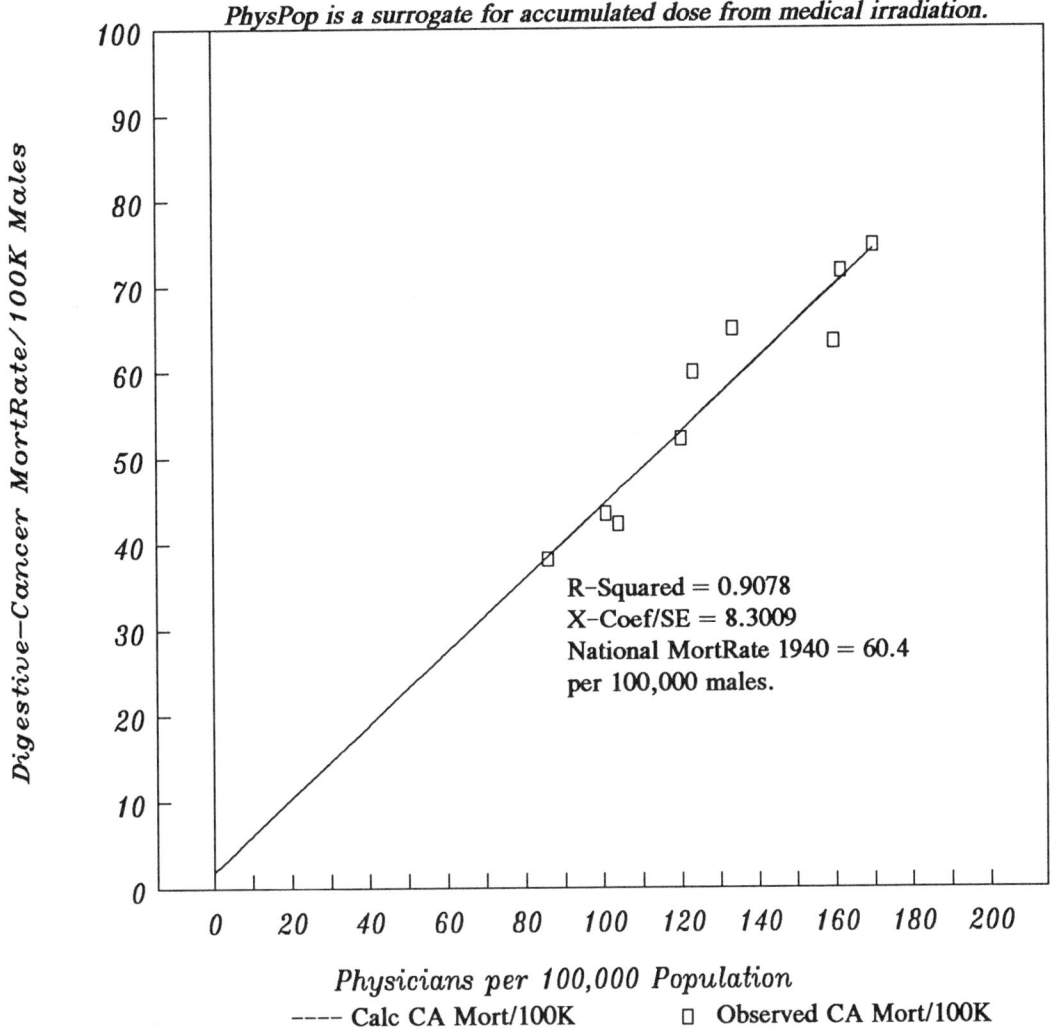

PhysPop is a surrogate for accumulated dose from medical irradiation.

R–Squared = 0.9078
X–Coef/SE = 8.3009
National MortRate 1940 = 60.4
per 100,000 males.

Physicians per 100,000 Population

---- Calc CA Mort/100K   □ Observed CA Mort/100K

On the X–axis, PhysPop values = Physicians per 100,000 Population in the Nine Census Divisions of the USA Population, Year 1940. This variable is a surrogate for accumulated radiation dose --- the more physicians per 100,000 people, the more radiation procedures are done per 100,000 people.

On the Y–axis, Digestive–Cancer Mortality–Rate per 100,000 males = the reported rates in USA Vital Statistics for the Nine Census Divisions, Year 1940.

Shown above is the strongest relationship between these two variables (Part 2i). The nine datapoints (boxy symbols) were collected long ago for other purposes, and are free from potential bias with respect to this dose–response study. Fractional causation is (Natl MortRate minus the Y–intercept) / (Natl MortRate).

Fractional Causation of Digestive–Cancer Mortality (Males) by Medical Radiation
97 % from Best Estimate (Box 3).
74.7 % at Lower 90% Conf. Limit (Box 3). ~100 % at Upper 90 % Conf. Limit (Box 3).

Table 9-A.
Digestive-System Cancer Mortality Rates by Census Divisions: Males.

Rates are annual deaths per 100,000 male population, USA, age-adjusted to the 1940 reference year. There are no exclusions by color or "race." Sources are stated in Table 9-B, and described in Chap. 4, Part 2. The Nine Census-Division MortRates are population-weighted (Chap. 4, Part 2b). The averages below them are not.

Census Division	1940	1950	1960	1970	1980	1988
Pacific	63.4	50.8	46.4	42.9	39.3	36.3
New England	71.7	66.3	58.9	52.5	46.0	42.1
West North Central	59.9	51.9	46.3	42.4	38.5	35.8
Mid-Atlantic	74.7	67.7	60.1	54.2	48.3	43.3
East North Central	64.9	60.0	53.0	48.4	43.7	40.2
Mountain	52.1	43.7	38.9	36.3	33.7	33.0
West South Central	42.3	42.9	40.6	38.8	37.0	36.5
East South Central	38.2	41.3	39.4	38.9	38.4	38.0
South Atlantic	43.4	45.2	43.1	41.6	40.1	38.5
Average, ALL	56.7	52.2	47.4	44.0	40.6	38.2
Average, High-5	66.9	59.3	52.9	48.1	43.2	39.5
Average, Low-4	44.0	43.3	40.5	38.9	37.3	36.5
Ratio, Hi5/Lo4	1.52	1.37	1.31	1.24	1.16	1.08

Table 9-B.
Digestive-System Cancer Mortality Rates, USA National.

Rates are age-adjusted to the 1940 reference year. Both sexes: Deaths per 100,000 population (males + females). Males: Deaths per 100,000 male population. Females: Deaths per 100,000 female population. No exclusions by color or "race."

	Both Sexes	Male	Female
1940	55.3	60.4	50.1
1950	48.7	55.4	42.4
1960	42.4	49.7	35.8
1970	37.7	45.7	31.0
1979-81	32.9	41.7	26.2
1987-1989	---	38.8	23.5

● - 1940, 1950, 1960: All rates come from Grove 1968, Table 67, p.684, "Malignant neoplasm of digestive organs and peritoneum, not specified as secondary (150-156A, 157-159)" ICD/7.
● - 1970: All rates by Divisions are interpolations (Chap. 4, Parts 2b, 2c).
● - 1980: All rates (ICD/9, 150-159) come from the reference NatCtrHS 1980.
● - 1988 rates by Divisions and National come from Monthly Vital Statistics Vol.41, No.9, February 16, 1993.

● Part 1. Introduction

　　　　Digestive-System Cancers include cancers of the esophagus, stomach, small and large intestine, rectum, biliary passages and liver, pancreas, and peritoneum (See Chapter 4, Part 5, Number 6).

● Part 2. How the Dose-Response Develops, 1921-1940

● − Part 2a.

	1921	1940	Digestive Cancers, Females	
	PhysPop	MortRate	Regression Output:	
Pacific	165.11	46.8	Constant	0.3717
New England	142.24	61.3	Std Err of Y Est	8.8662
West North Central	140.93	49.7	R Squared	0.3007
Mid-Atlantic	137.29	60.2	No. of Observations	9
East North Central	136.06	53.1	Degrees of Freedom	7
Mountain	135.38	47.7		
West South Central	125.15	34.5	X Coefficient(s)	0.3494
East South Central	119.76	36.3	Std Err of Coef.	0.2014
South Atlantic	110.32	37.3	Coefficient / S.E.	1.7350

● − Part 2b.

	1923	1940	Digestive Cancers, Females	
	PhysPop	MortRate	Regression Output:	
Pacific	163.06	46.8	Constant	1.1696
New England	137.39	61.3	Std Err of Y Est	8.5255
West North Central	138.31	49.7	R Squared	0.3534
Mid-Atlantic	138.92	60.2	No. of Observations	9
East North Central	131.82	53.1	Degrees of Freedom	7
Mountain	130.51	47.7		
West South Central	119.16	34.5	X Coefficient(s)	0.3531
East South Central	113.16	36.3	Std Err of Coef.	0.1805
South Atlantic	106.79	37.3	Coefficient / S.E.	1.9561

● − Part 2c.

	1925	1940	Digestive Cancers, Females	
	PhysPop	MortRate	Regression Output:	
Pacific	161.67	46.8	Constant	4.0462
New England	138.31	61.3	Std Err of Y Est	8.3113
West North Central	133.92	49.7	R Squared	0.3855
Mid-Atlantic	134.36	60.2	No. of Observations	9
East North Central	127.54	53.1	Degrees of Freedom	7
Mountain	122.30	47.7		
West South Central	112.83	34.5	X Coefficient(s)	0.3420
East South Central	107.22	36.3	Std Err of Coef.	0.1632
South Atlantic	103.61	37.3	Coefficient / S.E.	2.0956

● − Part 2d.

	1927	1940	Digestive Cancers, Females	
	PhysPop	MortRate	Regression Output:	
Pacific	157.83	46.8	Constant	1.2118
New England	137.50	61.3	Std Err of Y Est	7.5490
West North Central	131.54	49.7	R Squared	0.4931
Mid-Atlantic	138.40	60.2	No. of Observations	9
East North Central	126.18	53.1	Degrees of Freedom	7
Mountain	118.75	47.7		
West South Central	108.25	34.5	X Coefficient(s)	0.3705
East South Central	102.07	36.3	Std Err of Coef.	0.1420
South Atlantic	102.13	37.3	Coefficient / S.E.	2.6093

● − Part 2e.

	1929	1940	Digestive Cancers, Females	
	PhysPop	MortRate	Regression Output:	
Pacific	156.64	46.8	Constant	0.9492

New England	138.46	61.3	Std Err of Y Est		7.1771
West North Central	128.72	49.7	R Squared		0.5418
Mid-Atlantic	138.49	60.2	No. of Observations		9
East North Central	126.51	53.1	Degrees of Freedom		7
Mountain	118.68	47.7			
West South Central	105.60	34.5	X Coefficient(s)		0.3758
East South Central	99.41	36.3	Std Err of Coef.		0.1306
South Atlantic	100.86	37.3	Coefficient / S.E.		2.8769

● – Part 2f.

	1931	1940	Digestive Cancers, Females	
	PhysPop	MortRate	Regression Output:	
Pacific	159.97	46.8	Constant	4.2282
New England	142.35	61.3	Std Err of Y Est	7.0596
West North Central	126.50	49.7	R Squared	0.5567
Mid-Atlantic	140.82	60.2	No. of Observations	9
East North Central	128.59	53.1	Degrees of Freedom	7
Mountain	118.89	47.7		
West South Central	105.95	34.5	X Coefficient(s)	0.3474
East South Central	96.73	36.3	Std Err of Coef.	0.1172
South Atlantic	99.59	37.3	Coefficient / S.E.	2.9647

● – Part 2g.

	1934	1940	Digestive Cancers, Females	
	PhysPop	MortRate	Regression Output:	
Pacific	160.09	46.8	Constant	6.2062
New England	148.60	61.3	Std Err of Y Est	6.3037
West North Central	125.96	49.7	R Squared	0.6465
Mid-Atlantic	149.62	60.2	No. of Observations	9
East North Central	129.36	53.1	Degrees of Freedom	7
Mountain	117.16	47.7		
West South Central	104.68	34.5	X Coefficient(s)	0.3296
East South Central	92.00	36.3	Std Err of Coef.	0.0921
South Atlantic	98.41	37.3	Coefficient / S.E.	3.5781

● – Part 2h.

	1936	1940	Digestive Cancers, Females	
	PhysPop	MortRate	Regression Output:	
Pacific	158.44	46.8	Constant	5.8518
New England	150.18	61.3	Std Err of Y Est	5.7814
West North Central	126.14	49.7	R Squared	0.7027
Mid-Atlantic	155.05	60.2	No. of Observations	9
East North Central	130.42	53.1	Degrees of Freedom	7
Mountain	119.80	47.7		
West South Central	103.52	34.5	X Coefficient(s)	0.3304
East South Central	89.94	36.3	Std Err of Coef.	0.0812
South Atlantic	99.16	37.3	Coefficient / S.E.	4.0672

● – Part 2i.

	1938	1940	Digestive Cancers, Females	
	PhysPop	MortRate	Regression Output:	
Pacific	157.62	46.8	Constant	7.0152
New England	154.08	61.3	Std Err of Y Est	5.3636
West North Central	124.95	49.7	R Squared	0.7441
Mid-Atlantic	160.69	60.2	No. of Observations	9
East North Central	131.98	53.1	Degrees of Freedom	7
Mountain	119.88	47.7		
West South Central	102.79	34.5	X Coefficient(s)	0.3192
East South Central	88.21	36.3	Std Err of Coef.	0.0708
South Atlantic	99.26	37.3	Coefficient / S.E.	4.5115

● - Part 2j.

	1940 PhysPop	1940 MortRate	Digestive Cancers, Females Regression Output:	
Pacific	159.72	46.8	Constant	10.1907
New England	161.55	61.3	Std Err of Y Est	5.2483
West North Central	123.14	49.7	R Squared	0.7550
Mid-Atlantic	169.76	60.2	No. of Observations	9
East North Central	133.36	53.1	Degrees of Freedom	7
Mountain	119.89	47.7		
West South Central	103.94	34.5	X Coefficient(s)	0.2895
East South Central	85.83	36.3	Std Err of Coef.	0.0623
South Atlantic	100.74	37.3	Coefficient / S.E.	4.6442

Box 1 of Chap. 10
Summary: Regression Outputs, Digestive-System Cancers, Females.

Below are the summary-results from all the calculations of Part 2, for the 1940 MortRates regressed on PhysPop.

Part	PhysPop	R-squared	Constant	X-Coef	Std Err	X-Coef/SE
2a	1921	0.3007	0.37	0.3494	0.2014	1.7350
2b	1923	0.3534	1.17	0.3531	0.1805	1.9561
2c	1925	0.3855	4.05	0.3420	0.1632	2.0956
2d	1927	0.4931	1.21	0.3705	0.1420	2.6093
2e	1929	0.5418	0.95	0.3758	0.1306	2.8769
2f	1931	0.5567	4.23	0.3474	0.1172	2.9647
2g	1934	0.6465	6.21	0.3296	0.0921	3.5781
2h	1936	0.7027	5.85	0.3304	0.0812	4.0672
2i	1938	0.7441	7.02	0.3192	0.0708	4.5115
2j --->	1940 Max	0.7550	10.19	0.2895	0.0623	4.6442

Box 2 of Chap. 10
Input-Data for Figure 10-A. Digestive-System Cancers. Females.

Part 2j, Best-Fit Equation: Calc. MortRate = (0.2895 * PhysPop) + (10.19)

Census Divisions	1940 Observed PhysPops	1940 Observed MortRates	Best-Fit Calc. MortRates
Pacific	159.72	46.8	56.429
New England	161.55	61.3	56.959
West No. Central	123.14	49.7	45.839
Mid-Atlantic	169.76	60.2	59.336
East No. Central	133.36	53.1	48.798
Mountain	119.89	47.7	44.898
West So. Central	103.94	34.5	40.281
East So. Central	85.83	36.3	35.038
South Atlantic	100.74	37.3	39.354
Additional PhysPops	70.00		30.455
--- not "observed" ---	60.00		27.560
down to zero PhysPop	50.00		24.665
(zero medical radiation).	40.00		21.770
For each, we calculate	30.00		18.875
a best-fit MortRate.	20.00		15.980
These additional x,y pairs	10.00		13.085
are also part of the	0		10.190
best-fit line (Chap 5, Part 5e).			

Box 3 of Chap. 10
Presumptive Fraction of Cancer MortRate Attributable to Medical Radiation.

Please see text in Chapter 6, Parts 4 and 6.

Digestive-System Cancers. FEMALES.

- FEMALE National MortRate (MR) 1940, from Table 10-B   50.1 National MortRate
- Constant, from regression, Part 2j   10.1907 Constant
- Fractional Causation, Best Est. = (Natl MR − Constant) / Natl MR   79.7% Frac. Causation

....

90% Confidence-Limits (C.L.) on Fractional Causation. See text in Chapter 6, Part 4b, please.

X-Coefficient, from Part 2j   0.2895 X-Coef., Best Est.
Standard Error (SE) of X-Coefficient, from Part 2j   0.0623 Standard Error

Upper 90% C.L. on X-Coef. = (Coef) + (1.645 * SE) =   0.3920 New X-Coefficient
New Constant = (Natl MR) − (New X-Coef * 1940 Natl PhysPop) = −1.6575 New Constant
Frac. Causation, High-Limit = (Natl MR − New Constant) / Natl MR = #103.31% New Frac. Caus'n.
The Upper-Limit is 100%. Negative Constants produce values > 100%.   See Chapter 22, Part 3.

Lower 90% C.L. on X-Coef. = (Coef) − (1.645 * SE) =   0.1870 New X-Coefficient
New Constant = (Natl MR) − (New X-Coef * 1940 Natl PhysPop) = 25.4085 New Constant
Frac. Causation, Low-Limit = (Natl MR − New Constant) / Natl MR =   49.3% New Frac. Caus'n.

Box 4 of Chap. 10
Error-Check on Our Own Work: Digestive-System Cancers, Females.

Please see text in Chapter 6, Part 5.

Below, Columns A, C, and E come directly from the regression input in Part 2j. Column B, the fraction of the whole 1940 population in each Census Division, comes from Table 3-B in Chapter 3. Each Column-D entry is the product of (B-entry times C-entry). Each Column-F entry is the product of (B-entry times E-entry). PhysPops and MortRates are each "per 100,000."

The Weighted-Avg. Nat'l PhysPop, 1940, is the sum of Column-D entries = 132.04

The Weighted-Avg. Nat'l Female MortRate, 1940, is sum of Col.F entries = 48.77
The Nat'l Female MortRate is also (X-Coef * Nat'l PhysPop) + Constant = 48.42
Comparison: The Nat'l Female MortRate, 1940, in Table 10-B = 50.10

(A) Census Division	(B) Pop'n Fraction	(C) PhysPop 1940	(D) 1940 Weighted PhysPop	(E) MortRate 1940	(F) Weighted MortRate
Pacific	0.0739	159.72	11.80	46.8	3.46
New England	0.0641	161.55	10.36	61.3	3.93
West No. Central	0.1027	123.14	12.65	49.7	5.10
Mid-Atlantic	0.2092	169.76	35.51	60.2	12.59
East No. Central	0.2022	133.36	26.97	53.1	10.74
Mountain	0.0315	119.89	3.78	47.7	1.50
West So. Central	0.0992	103.94	10.31	34.5	3.42
East So. Central	0.0819	85.83	7.03	36.3	2.97
South Atlantic	0.1354	100.74	13.64	37.3	5.05
Sums	1.0000		132.04		48.77

1940 Digestive–System Cancer Mortality–Rates versus 1940 PhysPop Values for the 9 Census Divisions, USA.

Dose–Response Relationship

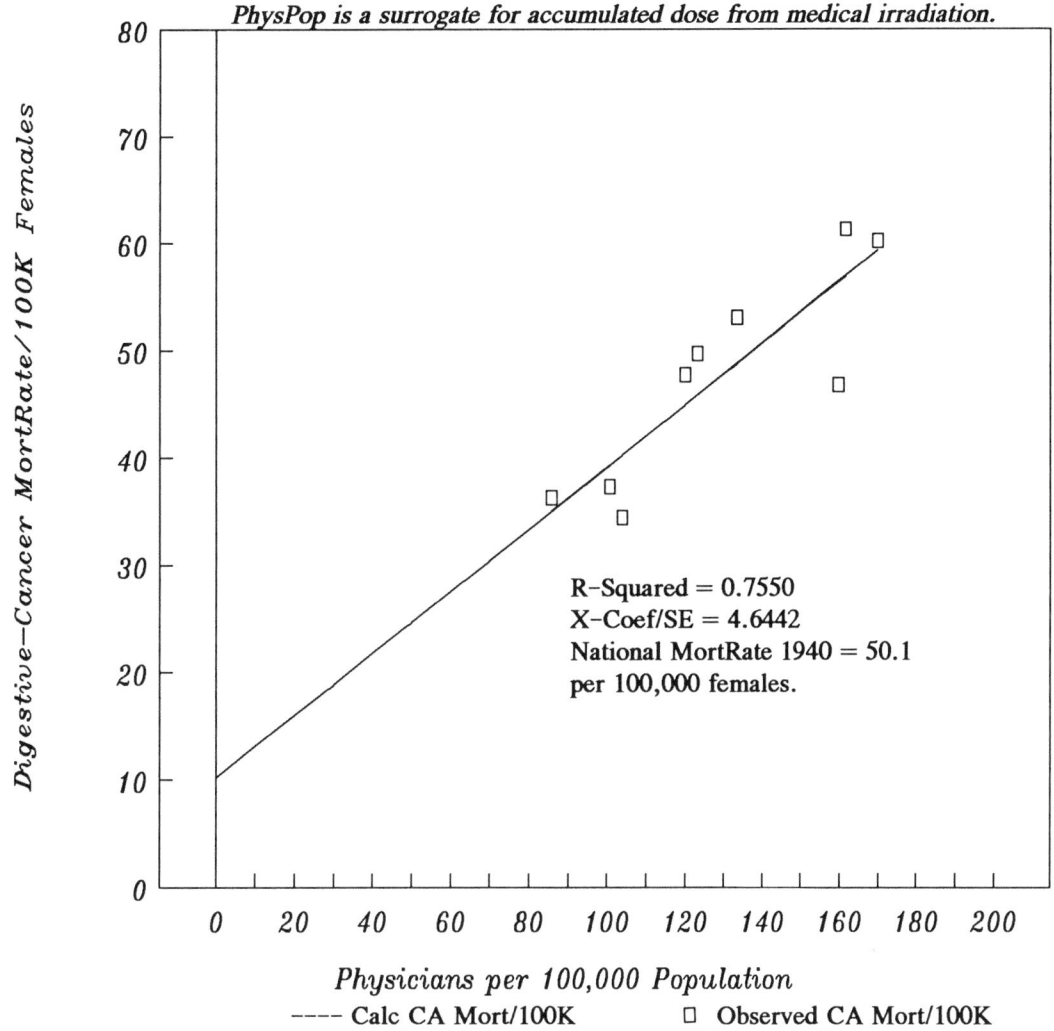

PhysPop is a surrogate for accumulated dose from medical irradiation.

R–Squared = 0.7550
X–Coef/SE = 4.6442
National MortRate 1940 = 50.1
per 100,000 females.

Physicians per 100,000 Population

---- Calc CA Mort/100K   □ Observed CA Mort/100K

On the X-axis, PhysPop values = Physicians per 100,000 Population in the Nine Census Divisions of the USA Population, Year 1940. This variable is a surrogate for accumulated radiation dose --- the more physicians per 100,000 people, the more radiation procedures are done per 100,000 people.

On the Y-axis, Digestive–Cancer Mortality–Rate per 100,000 females = the reported rates in USA Vital Statistics for the Nine Census Divisions, Year 1940.

Shown above is the strongest relationship between these two variables (Part 2j). The nine datapoints (boxy symbols) were collected long ago for other purposes, and are free from potential bias with respect to this dose-response study. Fractional causation is (Natl MortRate minus the Y-intercept) / (Natl MortRate).

Fractional Causation of Digest.–Cancer Mortality (Females) by Medical Radiation
80 % from Best Estimate (Box 3).
49 % at Lower 90% Conf. Limit (Box 3). ~100 % at Upper 90 % Conf. Limit (Box 3).

Table 10-A.

Digestive-System Cancer Mortality Rates by Census Divisions: Females.

Rates are annual deaths per 100,000 female population, USA, age-adjusted to the 1940 reference year. There are no exclusions by color or "race." Sources are stated in Table 10-B, and described in Chap. 4, Part 2. The Nine Census-Division MortRates are population-weighted (Chap. 4, Part 2b). The averages below them are not.

Census Division	1940	1950	1960	1970	1980	1988
Pacific	46.8	37.3	32.5	28.9	25.4	22.8
New England	61.3	48.9	40.7	34.7	28.7	24.7
West North Central	49.7	40.4	34.1	29.3	24.6	21.8
Mid-Atlantic	60.2	51.1	42.9	36.5	30.1	26.0
East North Central	53.1	44.7	38.5	32.8	27.1	24.2
Mountain	47.7	34.8	30.5	26.3	22.2	21.1
West South Central	34.5	33.3	29.6	26.6	23.6	21.5
East South Central	36.3	34.5	29.9	27.2	24.4	23.3
South Atlantic	37.3	34.9	30.6	27.5	24.4	22.8
Average, ALL	47.4	40.0	34.4	30.0	25.6	23.1
Average, High-5	54.2	44.5	37.7	32.4	27.2	23.9
Average, Low-4	39.0	34.4	30.2	26.9	23.7	22.2
Ratio, Hi5/Lo4	1.39	1.29	1.25	1.21	1.15	1.08

Table 10-B.

Digestive-System Cancer Mortality Rates, USA National.

Rates are age-adjusted to the 1940 reference year. Both sexes: Deaths per 100,000 population (males + females). Males: Deaths per 100,000 male population. Females: Deaths per 100,000 female population. No exclusions by color or "race."

	Both Sexes	Male	Female
1940	55.3	60.4	50.1
1950	48.7	55.4	42.4
1960	42.4	49.7	35.8
1970	37.7	45.7	31.0
1979-1981	32.9	41.7	26.2
1987-1989	--	38.8	23.5

● - 1940, 1950, 1960: All rates come from Grove 1968, Table 67, p.684, "Malignant neoplasm of digestive organs and peritoneum, not specified as secondary (150-156A, 157-159)" ICD/7.

● - 1970: All rates are interpolations (Chap. 4, Parts 2b, 2c).

● - 1980: All rates (ICD/9, 150-159) come from the reference NatCtrHS 1980.

● - 1988 rates by Divisions and National come from Monthly Vital Statistics Vol.41, No.9, February 16, 1993.

● Part 1. Introduction

Urinary-System Cancers include cancers of the kidney, bladder, "and other urinary organs" (Chapter 4, Part 5, Number 10).

This study produces negative Constants for the central estimate and for both of the confidence-limits on the X-Coefficient --- as shown in Box 3. In this situation, we hesitate to use any value for Fractional Causation in Figure 11-A. Instead, we will say that the true Fractional Causation is far more likely to be near 100% than to be a low percentage. The dose-response in Part 2j is highly significant.

● Part 2. How the Dose-Response Develops, 1921-1940

● - Part 2a.

	1921 PhysPop	1940 MortRate	Urinary-System Ca, Males Regression Output:	
Pacific	165.11	8.1	Constant	-5.9634
New England	142.24	9.1	Std Err of Y Est	1.9211
West North Central	140.93	6.7	R Squared	0.4030
Mid-Atlantic	137.29	10.2	No. of Observations	9
East North Central	136.06	8.1	Degrees of Freedom	7
Mountain	135.38	6.5		
West South Central	125.15	4.3	X Coefficient(s)	0.0948
East South Central	119.76	3.0	Std Err of Coef.	0.0436
South Atlantic	110.32	5.3	Coefficient / S.E.	2.1736

● - Part 2b.

	1923 PhysPop	1940 MortRate	Urinary-System Ca, Males Regression Output:	
Pacific	163.06	8.1	Constant	-5.9647
New England	137.39	9.1	Std Err of Y Est	1.7752
West North Central	138.31	6.7	R Squared	0.4902
Mid-Atlantic	138.92	10.2	No. of Observations	9
East North Central	131.82	8.1	Degrees of Freedom	7
Mountain	130.51	6.5		
West South Central	119.16	4.3	X Coefficient(s)	0.0975
East South Central	113.16	3.0	Std Err of Coef.	0.0376
South Atlantic	106.79	5.3	Coefficient / S.E.	2.5942

● - Part 2c.

	1925 PhysPop	1940 MortRate	Urinary-System Ca, Males Regression Output:	
Pacific	161.67	8.1	Constant	-5.0874
New England	138.31	9.1	Std Err of Y Est	1.7094
West North Central	133.92	6.7	R Squared	0.5273
Mid-Atlantic	134.36	10.2	No. of Observations	9
East North Central	127.54	8.1	Degrees of Freedom	7
Mountain	122.30	6.5		
West South Central	112.83	4.3	X Coefficient(s)	0.0938
East South Central	107.22	3.0	Std Err of Coef.	0.0336
South Atlantic	103.61	5.3	Coefficient / S.E.	2.7943

● - Part 2d.

	1927 PhysPop	1940 MortRate	Urinary-System Ca, Males Regression Output:	
Pacific	157.83	8.1	Constant	-5.6854
New England	137.50	9.1	Std Err of Y Est	1.4594
West North Central	131.54	6.7	R Squared	0.6554
Mid-Atlantic	138.40	10.2	No. of Observations	9
East North Central	126.18	8.1	Degrees of Freedom	7
Mountain	118.75	6.5		

West South Central	108.25	4.3	X Coefficient(s)		0.1002
East South Central	102.07	3.0	Std Err of Coef.		0.0275
South Atlantic	102.13	5.3	Coefficient / S.E.		3.6490

● – Part 2e.	1929	1940	Urinary–System Ca, Males	
	PhysPop	MortRate	Regression Output:	
Pacific	156.64	8.1	Constant	−5.6251
New England	138.46	9.1	Std Err of Y Est	1.3498
West North Central	128.72	6.7	R Squared	0.7052
Mid–Atlantic	138.49	10.2	No. of Observations	9
East North Central	126.51	8.1	Degrees of Freedom	7
Mountain	118.68	6.5		
West South Central	105.60	4.3	X Coefficient(s)	0.1005
East South Central	99.41	3.0	Std Err of Coef.	0.0246
South Atlantic	100.86	5.3	Coefficient / S.E.	4.0924

● – Part 2f.	1931	1940	Urinary–System Ca, Males	
	PhysPop	MortRate	Regression Output:	
Pacific	159.97	8.1	Constant	−4.7933
New England	142.35	9.1	Std Err of Y Est	1.2911
West North Central	126.50	6.7	R Squared	0.7303
Mid–Atlantic	140.82	10.2	No. of Observations	9
East North Central	128.59	8.1	Degrees of Freedom	7
Mountain	118.89	6.5		
West South Central	105.95	4.3	X Coefficient(s)	0.0933
East South Central	96.73	3.0	Std Err of Coef.	0.0214
South Atlantic	99.59	5.3	Coefficient / S.E.	4.3539

● – Part 2g.	1934	1940	Urinary–System Ca, Males	
	PhysPop	MortRate	Regression Output:	
Pacific	160.09	8.1	Constant	−4.0741
New England	148.60	9.1	Std Err of Y Est	1.0558
West North Central	125.96	6.7	R Squared	0.8197
Mid–Atlantic	149.62	10.2	No. of Observations	9
East North Central	129.36	8.1	Degrees of Freedom	7
Mountain	117.16	6.5		
West South Central	104.68	4.3	X Coefficient(s)	0.0870
East South Central	92.00	3.0	Std Err of Coef.	0.0154
South Atlantic	98.41	5.3	Coefficient / S.E.	5.6404

● – Part 2h.	1936	1940	Urinary–System Ca, Males	
	PhysPop	MortRate	Regression Output:	
Pacific	158.44	8.1	Constant	−4.0632
New England	150.18	9.1	Std Err of Y Est	0.8826
West North Central	126.14	6.7	R Squared	0.8740
Mid–Atlantic	155.05	10.2	No. of Observations	9
East North Central	130.42	8.1	Degrees of Freedom	7
Mountain	119.80	6.5		
West South Central	103.52	4.3	X Coefficient(s)	0.0864
East South Central	89.94	3.0	Std Err of Coef.	0.0124
South Atlantic	99.16	5.3	Coefficient / S.E.	6.9672

● – Part 2i.	1938	1940	Urinary–System Ca, Males	
	PhysPop	MortRate	Regression Output:	
Pacific	157.62	8.1	Constant	−3.6578
New England	154.08	9.1	Std Err of Y Est	0.7547
West North Central	124.95	6.7	R Squared	0.9079
Mid–Atlantic	160.69	10.2	No. of Observations	9
East North Central	131.98	8.1	Degrees of Freedom	7
Mountain	119.88	6.5		
West South Central	102.79	4.3	X Coefficient(s)	0.0827
East South Central	88.21	3.0	Std Err of Coef.	0.0100
South Atlantic	99.26	5.3	Coefficient / S.E.	8.3046

● – Part 2j.	1940	1940	Urinary–System Ca, Males	
	PhysPop	MortRate	Regression Output:	
Pacific	159.72	8.1	Constant	−2.8335

New England	161.55	9.1	Std Err of Y Est	0.6997	
West North Central	123.14	6.7	R Squared	0.9208	
Mid-Atlantic	169.76	10.2	No. of Observations	9	
East North Central	133.36	8.1	Degrees of Freedom	7	
Mountain	119.89	6.5			
West South Central	103.94	4.3	X Coefficient(s)	0.0750	
East South Central	85.83	3.0	Std Err of Coef.	0.0083	
South Atlantic	100.74	5.3	Coefficient / S.E.	9.0208	

Box 1 of Chap. 11
Summary: Regression Outputs, Urinary-System Cancers, Males.

Below are the summary-results from all the calculations of Part 2, for the 1940 MortRates regressed on PhysPop.

Part	PhysPop	R-squared	Constant	X-Coef	Std Err	X-Coef/SE
2a	1921	0.4030	-5.96	0.0948	0.0436	2.1736
2b	1923	0.4902	-5.96	0.0975	0.0376	2.5942
2c	1925	0.5273	-5.09	0.0938	0.0336	2.7943
2d	1927	0.6554	-5.69	0.1002	0.0275	3.6490
2e	1929	0.7052	-5.63	0.1005	0.0246	4.0924
2f	1931	0.7303	-4.79	0.0933	0.0214	4.3539
2g	1934	0.8197	-4.07	0.0870	0.0154	5.6404
2h	1936	0.8740	-4.06	0.0864	0.0124	6.9672
2i	1938	0.9079	-3.66	0.0827	0.0100	8.3046
2j --->	1940 Max	0.9208	-2.83	0.0750	0.0083	9.0208

Box 2 of Chap. 11
Input-Data for Figure 11-A. Urinary-System Cancers. Males.

Part 2j, Best-Fit Equation: Calc. MortRate = (0.0750 * PhysPop) + (-2.83)

Census Divisions	1940 Observed PhysPops	1940 Observed MortRates	Best-Fit Calc. MortRates
Pacific	159.72	8.1	9.149
New England	161.55	9.1	9.286
West No. Central	123.14	6.7	6.406
Mid-Atlantic	169.76	10.2	9.902
East No. Central	133.36	8.1	7.172
Mountain	119.89	6.5	6.162
West So. Central	103.94	4.3	4.966
East So. Central	85.83	3.0	3.607
South Atlantic	100.74	5.3	4.725
Additional PhysPops	70.00		2.420
--- not "observed" ---	60.00		1.670
down to zero PhysPop	50.00		0.920
(zero medical radiation).	40.00		0.170
For each, we calculate	30.00		-0.580
a best-fit MortRate.	20.00		-1.330
These additional x,y pairs	10.00		-2.080
are also part of the	0		-2.830
best-fit line (Chap 5, Part 5e).			

Please see text in Chapter 6, Parts 4 and 6.

Urinary-System Cancers. MALES.

- MALE National MortRate (MR) 1940, from Table 11-B   7.4   National MortRate
- Constant, from regression, Part 2j   -2.8335   Constant
- Fractional Causation, Best Est. = (Natl MR − Constant) / Natl MR   138.3 %   Frac. Causation

\# The Upper-Limit is 100%. Negative Constants produce values > 100%.   See Chapter 22, Part 3.

....

90% Confidence-Limits (C.L.) on Fractional Causation. See text in Chapter 6, Part 4b, please.

X-Coefficient, from Part 2j   0.0750   X-Coef., Best Est.

Standard Error (SE) of X-Coefficient, from Part 2j   0.0083   Standard Error

Upper 90% C.L. on X-Coef. = (Coef) + (1.645 * SE) =   0.0887   New X-Coefficient

New Constant = (Natl MR) − (New X-Coef * 1940 Natl PhysPop) =   -4.3058   New Constant

Frac. Causation, High-Limit = (Natl MR − New Constant) / Natl MR =   158.2 %   New Frac. Caus'n.

\# The Upper-Limit is 100%. Negative Constants produce values > 100%.   See Chapter 22, Part 3.

Lower 90% C.L. on X-Coef. = (Coef) − (1.645 * SE) =   0.0613   New X-Coefficient

New Constant = (Natl MR) − (New X-Coef * 1940 Natl PhysPop) =   -0.7002   New Constant

Frac. Causation, Low-Limit = (Natl MR − New Constant) / Natl MR =   109.5 %   New Frac. Caus'n.

\# The Upper-Limit is 100%. Negative Constants produce values > 100%.   See Chapter 22, Part 3.

Box 4 of Chap. 11
Error-Check on Our Own Work: Urinary-System Cancers, Males.

Please see text in Chapter 6, Part 5.

Below, Columns A, C, and E come directly from the regression input in Part 2j. Column B, the fraction of the whole 1940 population in each Census Division, comes from Table 3-B in Chapter 3. Each Column-D entry is the product of (B-entry times C-entry). Each Column-F entry is the product of (B-entry times E-entry). PhysPops and MortRates are each "per 100,000."

The Weighted-Avg. Nat'l PhysPop, 1940, is the sum of Column-D entries =   132.04

The Weighted-Avg. Nat'l Male MortRate, 1940, is sum of Col.F entries =   7.24

The Nat'l Male MortRate is also (X-Coef * Nat'l PhysPop) + Constant =   7.07

Comparison: The Nat'l Male MortRate, 1940, in Table 11-B =   7.40

(A) Census Division	(B) Pop'n Fraction	(C) PhysPop 1940	(D) 1940 Weighted PhysPop	(E) MortRate 1940	(F) Weighted MortRate
Pacific	0.0739	159.72	11.80	8.1	0.60
New England	0.0641	161.55	10.36	9.1	0.58
West No. Central	0.1027	123.14	12.65	6.7	0.69
Mid-Atlantic	0.2092	169.76	35.51	10.2	2.13
East No. Central	0.2022	133.36	26.97	8.1	1.64
Mountain	0.0315	119.89	3.78	6.5	0.20
West So. Central	0.0992	103.94	10.31	4.3	0.43
East So. Central	0.0819	85.83	7.03	3.0	0.25
South Atlantic	0.1354	100.74	13.64	5.3	0.72
Sums	1.0000		132.04		7.24

1940 Urinary–System Cancer Mortality–Rates versus
1940 PhysPop Values for the 9 Census Divisions, US
Dose–Response Relationship

PhysPop is a surrogate for accumulated dose from medical irradiation.

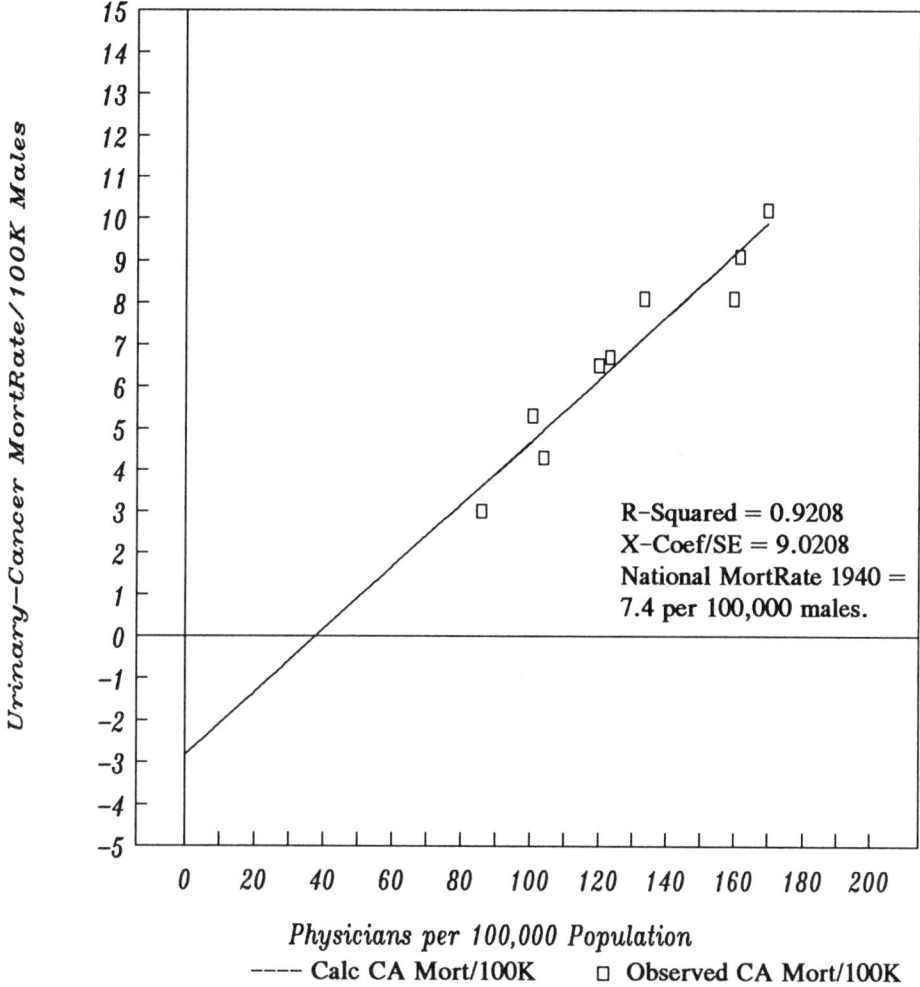

R–Squared = 0.9208
X–Coef/SE = 9.0208
National MortRate 1940 =
7.4 per 100,000 males.

Physicians per 100,000 Population
---- Calc CA Mort/100K   □ Observed CA Mort/100K

On the X-axis, PhysPop values = Physicians per 100,000 Population in the Nine Census Divisions of the USA Population, Year 1940. This variable is a surrogate for accumulated radiation dose --- the more physicians per 100,000 people, the more radiation procedures are done per 100,000 people.

On the Y-axis, Urinary-Cancer Mortality-Rate per 100,000 males = the reported rates in USA Vital Statistics for the Nine Census Divisions, Year 1940.

Shown above is the strongest relationship between these two variables (Part 2j). The nine datapoints (boxy symbols) were collected long ago for other purposes, and are free from potential bias with respect to this dose-response study.

Fractional Causation of Urinary–System Cancer Mortality–Rate (Male) by Medical Radiation: ~100 % is far more likely than a low percent. See Text, Part 1.

Table 11-A.

Urinary-System Cancer Mortality Rates by Census Divisions: Males.

Rates are annual deaths per 100,000 male population, USA, age-adjusted to the 1940 reference year. There are no exclusions by color or "race." Sources are stated in Table 11-B, and described in Chap. 4, Part 2. The Nine Census-Division MortRates are population-weighted (Chap. 4, Part 2b). The averages below them are not.

Census Division	1940	1950	1960	1970	1980	1990
Pacific	8.1	8.4	8.2	8.0	7.7	--
New England	9.1	10.5	10.7	10.1	9.5	--
West North Central	6.7	7.2	8.3	8.1	7.9	--
Mid-Atlantic	10.2	10.5	10.2	9.7	9.2	--
East North Central	8.1	8.6	9.4	9.1	8.7	--
Mountain	6.5	6.1	7.8	7.4	7.0	--
West South Central	4.3	5.8	6.6	6.8	7.0	--
East South Central	3.0	5.0	5.2	6.3	7.3	--
South Atlantic	5.3	6.1	6.9	7.4	7.8	--
Average, ALL	6.8	7.6	8.1	8.1	8.0	--
Average, High-5	8.4	9.0	9.4	9.0	8.6	--
Average, Low-4	4.8	5.8	6.6	7.0	7.3	--
Ratio, Hi5/Lo4	1.77	1.57	1.41	1.29	1.18	--

Table 11-B.

Urinary-System Cancer Mortality Rates, USA National.

Rates are age-adjusted to the 1940 reference year. Both sexes: Deaths per 100,000 population (males + females). Males: Deaths per 100,000 male population. Females: Deaths per 100,000 female population. No exclusions by color or "race."

	Both Sexes	Male	Female
1940	5.7	7.4	4.0
1950	6.0	8.1	3.9
1960	5.9	8.5	3.6
1970	5.6	8.35	3.3
1979-81	5.2	8.2	3.0
1990	--	--	--

● - 1940, 1950, 1960: All rates come from Grove 1968, Table 67, p.697, "Malignant neoplasm of urinary organs (180-181)" ICD/7.
● - 1970: All rates are interpolations (Chap. 4, Parts 2b, 2c).
● - 1980: All rates (ICD/9, 188-189) come from the reference NatCtrHS 1980.
● - 1990: No data obtained. Please see Chap.4, Part 2c.

● Part 1. Introduction

Urinary-System Cancers include cancers of the kidney, bladder, "and other urinary organs" (Chapter 4, Part 5, Number 10).

● Part 2. How the Dose-Response Develops, 1921-1940

● - Part 2a.

	1921 PhysPop	1940 MortRate	Urinary-System Ca, Females Regression Output:	
Pacific	165.11	4.1	Constant	-0.4804
New England	142.24	4.7	Std Err of Y Est	0.6216
West North Central	140.93	3.7	R Squared	0.4148
Mid-Atlantic	137.29	4.9	No. of Observations	9
East North Central	136.06	4.1	Degrees of Freedom	7
Mountain	135.38	3.5		
West South Central	125.15	3.1	X Coefficient(s)	0.0314
East South Central	119.76	2.7	Std Err of Coef.	0.0141
South Atlantic	110.32	3.0	Coefficient / S.E.	2.2273

● - Part 2b.

	1923 PhysPop	1940 MortRate	Urinary-System Ca, Females Regression Output:	
Pacific	163.06	4.1	Constant	-0.4168
New England	137.39	4.7	Std Err of Y Est	0.5807
West North Central	138.31	3.7	R Squared	0.4894
Mid-Atlantic	138.92	4.9	No. of Observations	9
East North Central	131.82	4.1	Degrees of Freedom	7
Mountain	130.51	3.5		
West South Central	119.16	3.1	X Coefficient(s)	0.0318
East South Central	113.16	2.7	Std Err of Coef.	0.0123
South Atlantic	106.79	3.0	Coefficient / S.E.	2.5901

● - Part 2c.

	1925 PhysPop	1940 MortRate	Urinary-System Ca, Females Regression Output:	
Pacific	161.67	4.1	Constant	-0.1529
New England	138.31	4.7	Std Err of Y Est	0.5556
West North Central	133.92	3.7	R Squared	0.5326
Mid-Atlantic	134.36	4.9	No. of Observations	9
East North Central	127.54	4.1	Degrees of Freedom	7
Mountain	122.30	3.5		
West South Central	112.83	3.1	X Coefficient(s)	0.0308
East South Central	107.22	2.7	Std Err of Coef.	0.0109
South Atlantic	103.61	3.0	Coefficient / S.E.	2.8242

● - Part 2d.

	1927 PhysPop	1940 MortRate	Urinary-System Ca, Females Regression Output:	
Pacific	157.83	4.1	Constant	-0.3300
New England	137.50	4.7	Std Err of Y Est	0.4767
West North Central	131.54	3.7	R Squared	0.6558
Mid-Atlantic	138.40	4.9	No. of Observations	9
East North Central	126.18	4.1	Degrees of Freedom	7
Mountain	118.75	3.5		
West South Central	108.25	3.1	X Coefficient(s)	0.0328
East South Central	102.07	2.7	Std Err of Coef.	0.0090
South Atlantic	102.13	3.0	Coefficient / S.E.	3.6521

● - Part 2e.

	1929 PhysPop	1940 MortRate	Urinary-System Ca, Females Regression Output:	
Pacific	156.64	4.1	Constant	-0.2984

New England	138.46	4.7	Std Err of Y Est	0.4439
West North Central	128.72	3.7	R Squared	0.7015
Mid-Atlantic	138.49	4.9	No. of Observations	9
East North Central	126.51	4.1	Degrees of Freedom	7
Mountain	118.68	3.5		
West South Central	105.60	3.1	X Coefficient(s)	0.0328
East South Central	99.41	2.7	Std Err of Coef.	0.0081
South Atlantic	100.86	3.0	Coefficient / S.E.	4.0562

● – Part 2f.	1931	1940	Urinary-System Ca, Females	
	PhysPop	MortRate	Regression Output:	
Pacific	159.97	4.1	Constant	−0.0396
New England	142.35	4.7	Std Err of Y Est	0.4213
West North Central	126.50	3.7	R Squared	0.7312
Mid-Atlantic	140.82	4.9	No. of Observations	9
East North Central	128.59	4.1	Degrees of Freedom	7
Mountain	118.89	3.5		
West South Central	105.95	3.1	X Coefficient(s)	0.0305
East South Central	96.73	2.7	Std Err of Coef.	0.0070
South Atlantic	99.59	3.0	Coefficient / S.E.	4.3637

● – Part 2g.	1934	1940	Urinary-System Ca, Females	
	PhysPop	MortRate	Regression Output:	
Pacific	160.09	4.1	Constant	0.1704
New England	148.60	4.7	Std Err of Y Est	0.3327
West North Central	125.96	3.7	R Squared	0.8323
Mid-Atlantic	149.62	4.9	No. of Observations	9
East North Central	129.36	4.1	Degrees of Freedom	7
Mountain	117.16	3.5		
West South Central	104.68	3.1	X Coefficient(s)	0.0287
East South Central	92.00	2.7	Std Err of Coef.	0.0049
South Atlantic	98.41	3.0	Coefficient / S.E.	5.8947

● – Part 2h.	1936	1940	Urinary-System Ca, Females	
	PhysPop	MortRate	Regression Output:	
Pacific	158.44	4.1	Constant	0.1916
New England	150.18	4.7	Std Err of Y Est	0.2829
West North Central	126.14	3.7	R Squared	0.8788
Mid-Atlantic	155.05	4.9	No. of Observations	9
East North Central	130.42	4.1	Degrees of Freedom	7
Mountain	119.80	3.5		
West South Central	103.52	3.1	X Coefficient(s)	0.0283
East South Central	89.94	2.7	Std Err of Coef.	0.0040
South Atlantic	99.16	3.0	Coefficient / S.E.	7.1239

● – Part 2i.	1938	1940	Urinary-System Ca, Females	
	PhysPop	MortRate	Regression Output:	
Pacific	157.62	4.1	Constant	0.3153
New England	154.08	4.7	Std Err of Y Est	0.2330
West North Central	124.95	3.7	R Squared	0.9177
Mid-Atlantic	160.69	4.9	No. of Observations	9
East North Central	131.98	4.1	Degrees of Freedom	7
Mountain	119.88	3.5		
West South Central	102.79	3.1	X Coefficient(s)	0.0272
East South Central	88.21	2.7	Std Err of Coef.	0.0031
South Atlantic	99.26	3.0	Coefficient / S.E.	8.8378

● – Part 2j.	1940	1940	Urinary-System Ca, Females	
	PhysPop	MortRate	Regression Output:	
Pacific	159.72	4.1	Constant	0.5714
New England	161.55	4.7	Std Err of Y Est	0.1998
West North Central	123.14	3.7	R Squared	0.9395
Mid-Atlantic	169.76	4.9	No. of Observations	9
East North Central	133.36	4.1	Degrees of Freedom	7
Mountain	119.89	3.5		
West South Central	103.94	3.1	X Coefficient(s)	0.0247
East South Central	85.83	2.7	Std Err of Coef.	0.0024
South Atlantic	100.74	3.0	Coefficient / S.E.	10.4305

Box 1 of Chap. 12
Summary: Regression Outputs, Urinary-System Cancers, Females.

Below are the summary-results from all the calculations of Part 2, for the 1940 MortRates regressed on PhysPop.

Part	PhysPop	R-squared	Constant	X-Coef	Std Err	X-Coef/SE
2a	1921	0.4148	-0.48	0.0314	0.0141	2.2273
2b	1923	0.4894	-0.42	0.0318	0.0123	2.5901
2c	1925	0.5326	-0.15	0.0308	0.0109	2.8242
2d	1927	0.6558	-0.33	0.0328	0.0090	3.6521
2e	1929	0.7015	-0.30	0.0328	0.0081	4.0562
2f	1931	0.7312	-0.04	0.0305	0.0070	4.3637
2g	1934	0.8323	0.17	0.0287	0.0049	5.8947
2h	1936	0.8788	0.19	0.0283	0.0040	7.1239
2i	1938	0.9177	0.32	0.0272	0.0031	8.8378
2j --->	1940 Max	0.9395	0.57	0.0247	0.0024	10.4305

Box 2 of Chap. 12
Input-Data for Figure 12-A. Urinary-System Cancers. Females.

Part 2j, Best-Fit Equation: Calc. MortRate = (0.0247 * PhysPop) + (0.57)

Census Divisions	1940 Observed PhysPops	1940 Observed MortRates	Best-Fit Calc. MortRates
Pacific	159.72	4.1	4.515
New England	161.55	4.7	4.560
West No. Central	123.14	3.7	3.612
Mid-Atlantic	169.76	4.9	4.763
East No. Central	133.36	4.1	3.864
Mountain	119.89	3.5	3.531
West So. Central	103.94	3.1	3.137
East So. Central	85.83	2.7	2.690
South Atlantic	100.74	3.0	3.058
Additional PhysPops	70.00		2.299
--- not "observed" ---	60.00		2.052
down to zero PhysPop	50.00		1.805
(zero medical radiation).	40.00		1.558
For each, we calculate	30.00		1.311
a best-fit MortRate.	20.00		1.064
These additional x,y pairs	10.00		0.817
are also part of the	0		0.570
best-fit line (Chap 5, Part 5e).			

Box 3 of Chap. 12

Presumptive Fraction of Cancer MortRate Attributable to Medical Radiation.

Please see text in Chapter 6, Parts 4 and 6.

Urinary-System Cancers. FEMALES.

- FEMALE National MortRate (MR) 1940, from Table 12-B   4.0 National MortRate
- Constant, from regression, Part 2j   0.5714 Constant
- Fractional Causation, Best Est. = (Natl MR – Constant) / Natl MR   85.7% Frac. Causation

....

90% Confidence-Limits (C.L.) on Fractional Causation. See text in Chapter 6, Part 4b, please.

X-Coefficient, from Part 2j   0.0247 X-Coef., Best Est.
Standard Error (SE) of X-Coefficient, from Part 2j   0.0024 Standard Error

Upper 90% C.L. on X-Coef. = (Coef) + (1.645 * SE) =   0.0286 New X-Coefficient
New Constant = (Natl MR) – (New X-Coef * 1940 Natl PhysPop) =   0.2173 New Constant
Frac. Causation, High-Limit = (Natl MR – New Constant) / Natl MR =   94.6% New Frac. Caus'n.

Lower 90% C.L. on X-Coef. = (Coef) – (1.645 * SE) =   0.0208 New X-Coefficient
New Constant = (Natl MR) – (New X-Coef * 1940 Natl PhysPop) =   1.2599 New Constant
Frac. Causation, Low-Limit = (Natl MR – New Constant) / Natl MR =   68.5% New Frac. Caus'n.

Box 4 of Chap. 12

Error-Check on Our Own Work: Urinary-System Cancers, Females.

Below, Columns A, C, and E come directly from the regression input in Part 2j. Column B, the fraction of the whole 1940 population in each Census Division, comes from Table 3-B in Chapter 3. Each Column-D entry is the product of (B-entry times C-entry). Each Column-F entry is the product of (B-entry times E-entry). PhysPops and MortRates are each "per 100,000."

The Weighted-Avg. Nat'l PhysPop, 1940, is the sum of Column-D entries =   132.04

The Weighted-Avg. Nat'l Female MortRate, 1940, is sum of Col.F entries =   3.88
The Nat'l Female MortRate is also (X-Coef * Nat'l PhysPop) + Constant =   3.83
Comparison: The Nat'l Female MortRate, 1940, in Table 12-B =   4.00

(A) Census Division	(B) Pop'n Fraction	(C) PhysPop 1940	(D) Weighted PhysPop	(E) MortRate 1940	(F) Weighted MortRate
Pacific	0.0739	159.72	11.80	4.1	0.30
New England	0.0641	161.55	10.36	4.7	0.30
West No. Central	0.1027	123.14	12.65	3.7	0.38
Mid-Atlantic	0.2092	169.76	35.51	4.9	1.03
East No. Central	0.2022	133.36	26.97	4.1	0.83
Mountain	0.0315	119.89	3.78	3.5	0.11
West So. Central	0.0992	103.94	10.31	3.1	0.31
East So. Central	0.0819	85.83	7.03	2.7	0.22
South Atlantic	0.1354	100.74	13.64	3.0	0.41
Sums	1.0000		132.04		3.88

1940 Urinary–System Cancer Mortality–Rates versus 1940 PhysPop Values for the 9 Census Divisions, USA.

Dose–Response Relationship

PhysPop is a surrogate for accumulated dose from medical irradiation.

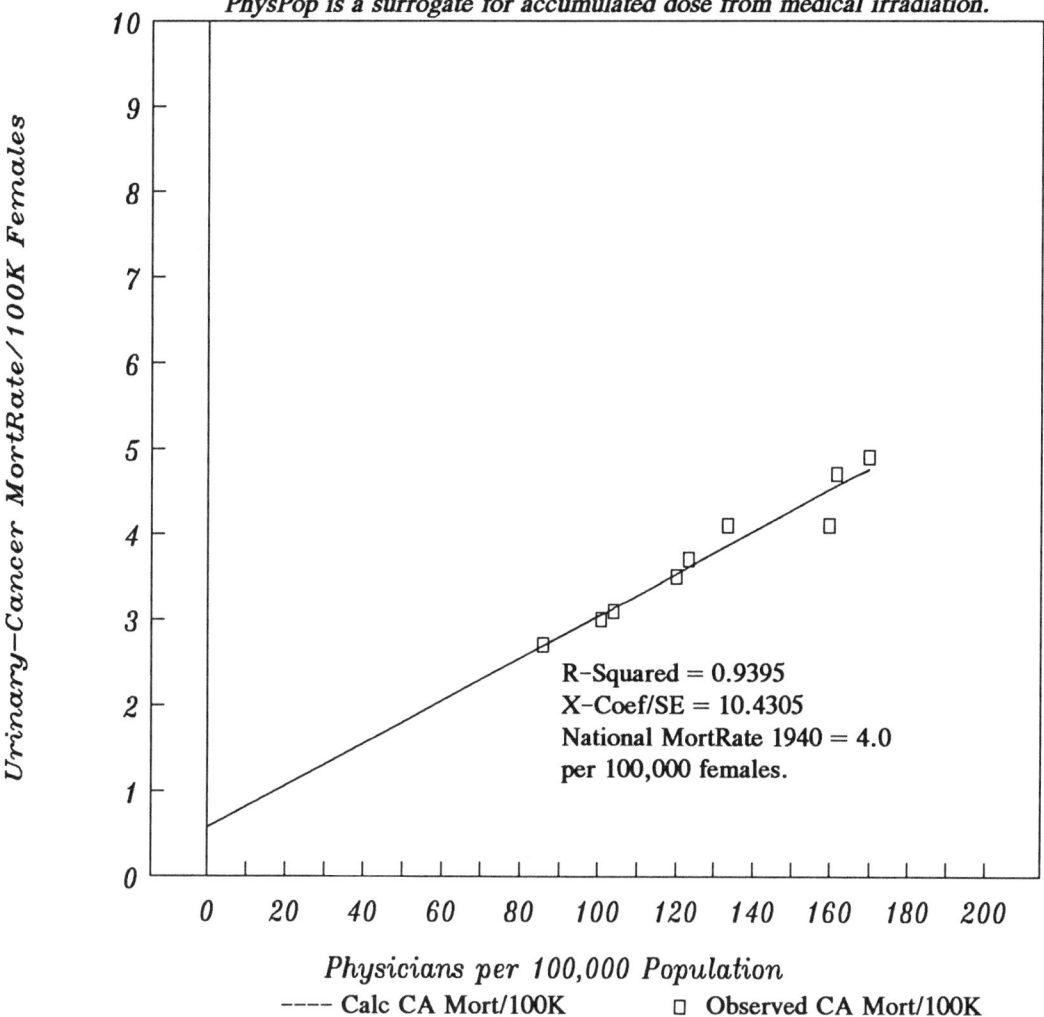

R-Squared = 0.9395
X-Coef/SE = 10.4305
National MortRate 1940 = 4.0
per 100,000 females.

Physicians per 100,000 Population

---- Calc CA Mort/100K   □ Observed CA Mort/100K

On the X–axis, PhysPop values = Physicians per 100,000 Population in the Nine Census Divisions of the USA Population, Year 1940. This variable is a surrogate for accumulated radiation dose --- the more physicians per 100,000 people, the more radiation procedures are done per 100,000 people.

On the Y–axis, Urinary–Cancer Mortality–Rate per 100,000 females = the reported rates in USA Vital Statistics for the Nine Census Divisions, Year 1940.

Shown above is the strongest relationship between these two variables (Part 2j). The nine datapoints (boxy symbols) were collected long ago for other purposes, and are free from potential bias with respect to this dose–response study. Fractional Causation is (Natl MortRate minus the Y–intercept) / (Natl MortRate).

Fractional Causation of Urinary–Cancer Mortality (Females) by Medical Radiation
85.7 % from Best Estimate (Box 3).

68 % at lower 90 % confidence limit (Box 3). 94.6 % at upper 90% confidence limit (Box 3).

Table 12-A.

Urinary-System Cancer Mortality Rates by Census Divisions: Females.

Rates are annual deaths per 100,000 female population, USA, age-adjusted to the 1940 reference year. There are no exclusions by color or "race." Sources are stated in Table 12-B, and described in Chap. 4, Part 2. The Nine Census-Division MortRates are population-weighted (Chap. 4, Part 2b). The averages below them are not.

Census Division	1940	1950	1960	1970	1980	1990
Pacific	4.1	3.9	3.3	3.1	2.8	--
New England	4.7	3.9	3.9	3.7	3.4	--
West North Central	3.7	3.6	3.3	3.2	3.0	--
Mid-Atlantic	4.9	4.5	4.0	3.6	3.2	--
East North Central	4.1	4.2	3.9	3.5	3.0	--
Mountain	3.5	3.5	3.4	3.0	2.5	--
West South Central	3.1	3.4	3.2	3.0	2.8	--
East South Central	2.7	3.6	3.0	2.9	2.8	--
South Atlantic	3.0	3.6	3.3	3.1	2.9	--
Average, ALL	3.8	3.8	3.5	3.2	2.9	--
Average, High-5	4.3	4.0	3.7	3.4	3.1	--
Average, Low-4	3.1	3.5	3.2	3.0	2.8	--
Ratio, Hi5/Lo4	1.40	1.14	1.14	1.13	1.12	--

Table 12-B.

Urinary-System Cancer Mortality Rates, USA National.

Rates are age-adjusted to the 1940 reference year. Both sexes: Deaths per 100,000 population (males + females). Males: Deaths per 100,000 male population. Females: Deaths per 100,000 female population. No exclusions by color or "race."

	Both Sexes	Male	Female
1940	5.7	7.4	4.0
1950	6.0	8.1	3.9
1960	5.9	8.5	3.6
1970	5.6	8.4	3.3
1979-81	5.2	8.2	3.0
1990	--	--	--

● - 1940, 1950, 1960: All rates come from Grove 1968, Table 67, p.697, "Malignant neoplasm of urinary organs (180-181)" ICD/7.
● - 1970: All rates are interpolations (Chap. 4, Parts 2b, 2c).
● - 1980: All rates (ICD/9, 188-189) come from the reference NatCtrHS 1980.
● - 1990: No data obtained. Please see Chap.4, Part 2c.

CHAPTER 13

Genital Cancers, Males: Relation with Medical Radiation

● Part 1.   Introduction

        Male Genital Cancers include cancers of the prostate and testis (see Chapter 4, Part 5, Number 9).

● Part 2.   How the Dose-Response Develops, 1921-1940

● - Part 2a.	1921 PhysPop	1940 MortRate	Genital Cancers, Males Regression Output:	
Pacific	165.11	17.2	Constant	-3.0904
New England	142.24	18.2	Std Err of Y Est	1.7784
West North Central	140.93	16.5	R Squared	0.6097
Mid-Atlantic	137.29	15.8	No. of Observations	9
East North Central	136.06	15.8	Degrees of Freedom	7
Mountain	135.38	15.8		
West South Central	125.15	11.6	X Coefficient(s)	0.1336
East South Central	119.76	10.4	Std Err of Coef.	0.0404
South Atlantic	110.32	12.8	Coefficient / S.E.	3.3066

● - Part 2b.	1923 PhysPop	1940 MortRate	Genital Cancers, Males Regression Output:	
Pacific	163.06	17.2	Constant	-1.7908
New England	137.39	18.2	Std Err of Y Est	1.7122
West North Central	138.31	16.5	R Squared	0.6382
Mid-Atlantic	138.92	15.8	No. of Observations	9
East North Central	131.82	15.8	Degrees of Freedom	7
Mountain	130.51	15.8		
West South Central	119.16	11.6	X Coefficient(s)	0.1274
East South Central	113.16	10.4	Std Err of Coef.	0.0363
South Atlantic	106.79	12.8	Coefficient / S.E.	3.5139

● - Part 2c.	1925 PhysPop	1940 MortRate	Genital Cancers, Males Regression Output:	
Pacific	161.67	17.2	Constant	-0.5537
New England	138.31	18.2	Std Err of Y Est	1.6140
West North Central	133.92	16.5	R Squared	0.6785
Mid-Atlantic	134.36	15.8	No. of Observations	9
East North Central	127.54	15.8	Degrees of Freedom	7
Mountain	122.30	15.8		
West South Central	112.83	11.6	X Coefficient(s)	0.1218
East South Central	107.22	10.4	Std Err of Coef.	0.0317
South Atlantic	103.61	12.8	Coefficient / S.E.	3.8437

● - Part 2d.	1927 PhysPop	1940 MortRate	Genital Cancers, Males Regression Output:	
Pacific	157.83	17.2	Constant	-0.1623
New England	137.50	18.2	Std Err of Y Est	1.4890
West North Central	131.54	16.5	R Squared	0.7264
Mid-Atlantic	138.40	15.8	No. of Observations	9
East North Central	126.18	15.8	Degrees of Freedom	7
Mountain	118.75	15.8		
West South Central	108.25	11.6	X Coefficient(s)	0.1207
East South Central	102.07	10.4	Std Err of Coef.	0.0280
South Atlantic	102.13	12.8	Coefficient / S.E.	4.3110

● - Part 2e.	1929 PhysPop	1940 MortRate	Genital Cancers, Males Regression Output:	
Pacific	156.64	17.2	Constant	0.1163
New England	138.46	18.2	Std Err of Y Est	1.3938

West North Central	128.72	16.5	R Squared	0.7603
Mid-Atlantic	138.49	15.8	No. of Observations	9
East North Central	126.51	15.8	Degrees of Freedom	7
Mountain	118.68	15.8		
West South Central	105.60	11.6	X Coefficient(s)	0.1195
East South Central	99.41	10.4	Std Err of Coef.	0.0254
South Atlantic	100.86	12.8	Coefficient / S.E.	4.7115

● – Part 2f.

	1931	1940	Genital Cancers, Males	
	PhysPop	MortRate	Regression Output:	
Pacific	159.97	17.2	Constant	1.3504
New England	142.35	18.2	Std Err of Y Est	1.3958
West North Central	126.50	16.5	R Squared	0.7596
Mid-Atlantic	140.82	15.8	No. of Observations	9
East North Central	128.59	15.8	Degrees of Freedom	7
Mountain	118.89	15.8		
West South Central	105.95	11.6	X Coefficient(s)	0.1089
East South Central	96.73	10.4	Std Err of Coef.	0.0232
South Atlantic	99.59	12.8	Coefficient / S.E.	4.7024

● – Part 2g.

	1934	1940	Genital Cancers, Males	
	PhysPop	MortRate	Regression Output:	
Pacific	160.09	17.2	Constant	2.8680
New England	148.60	18.2	Std Err of Y Est	1.3830
West North Central	125.96	16.5	R Squared	0.7640
Mid-Atlantic	149.62	15.8	No. of Observations	9
East North Central	129.36	15.8	Degrees of Freedom	7
Mountain	117.16	15.8		
West South Central	104.68	11.6	X Coefficient(s)	0.0962
East South Central	92.00	10.4	Std Err of Coef.	0.0202
South Atlantic	98.41	12.8	Coefficient / S.E.	4.7598

● – Part 2h.

	1936	1940	Genital Cancers, Males	
	PhysPop	MortRate	Regression Output:	
Pacific	158.44	17.2	Constant	3.1726
New England	150.18	18.2	Std Err of Y Est	1.3490
West North Central	126.14	16.5	R Squared	0.7754
Mid-Atlantic	155.05	15.8	No. of Observations	9
East North Central	130.42	15.8	Degrees of Freedom	7
Mountain	119.80	15.8		
West South Central	103.52	11.6	X Coefficient(s)	0.0932
East South Central	89.94	10.4	Std Err of Coef.	0.0190
South Atlantic	99.16	12.8	Coefficient / S.E.	4.9160

● – Part 2i.

	1938	1940	Genital Cancers, Males	
	PhysPop	MortRate	Regression Output:	
Pacific	157.62	17.2	Constant	3.9334
New England	154.08	18.2	Std Err of Y Est	1.3946
West North Central	124.95	16.5	R Squared	0.7600
Mid-Atlantic	160.69	15.8	No. of Observations	9
East North Central	131.98	15.8	Degrees of Freedom	7
Mountain	119.88	15.8		
West South Central	102.79	11.6	X Coefficient(s)	0.0866
East South Central	88.21	10.4	Std Err of Coef.	0.0184
South Atlantic	99.26	12.8	Coefficient / S.E.	4.7078

● – Part 2j.

	1940	1940	Genital Cancers, Males	
	PhysPop	MortRate	Regression Output:	
Pacific	159.72	17.2	Constant	5.1480
New England	161.55	18.2	Std Err of Y Est	1.5112
West North Central	123.14	16.5	R Squared	0.7182
Mid-Atlantic	169.76	15.8	No. of Observations	9
East North Central	133.36	15.8	Degrees of Freedom	7
Mountain	119.89	15.8		
West South Central	103.94	11.6	X Coefficient(s)	0.0758
East South Central	85.83	10.4	Std Err of Coef.	0.0179
South Atlantic	100.74	12.8	Coefficient / S.E.	4.2234

Box 1 of Chap. 13
Summary: Regression Outputs, Genital Cancers, Males.

Below are the summary-results from all the calculations of Part 2, for the 1940 MortRates regressed on PhysPop.

Part	PhysPop	R-squared	Constant	X-Coef	Std Err	X-Coef/SE
2a	1921	0.6097	-3.09	0.1336	0.0404	3.3066
2b	1923	0.6382	-1.79	0.1274	0.0363	3.5139
2c	1925	0.6785	-0.55	0.1218	0.0317	3.8437
2d	1927	0.7264	-0.16	0.1207	0.0280	4.3110
2e	1929	0.7603	0.12	0.1195	0.0254	4.7115
2f	1931	0.7596	1.35	0.1089	0.0232	4.7024
2g	1934	0.7640	2.87	0.0962	0.0202	4.7598
2h ---->	1936 Max	0.7754	3.17	0.0932	0.0190	4.9160
2i	1938	0.7600	3.93	0.0866	0.0184	4.7078
2j	1940	0.7182	5.15	0.0758	0.0179	4.2234

Box 2 of Chap. 13
Input-Data for Figure 13-A. Genital Cancers. Males.

Part 2h, Best-Fit Equation: Calc. MortRate = (0.0932 * PhysPop) + (3.17)

Census Divisions	1940 Observed PhysPops	1940 Observed MortRates	Best-Fit Calc. MortRates
Pacific	159.72	17.2	18.056
New England	161.55	18.2	18.226
West No. Central	123.14	16.5	14.647
Mid-Atlantic	169.76	15.8	18.992
East No. Central	133.36	15.8	15.599
Mountain	119.89	15.8	14.344
West So. Central	103.94	11.6	12.857
East So. Central	85.83	10.4	11.169
South Atlantic	100.74	12.8	12.559
Additional PhysPops	70.00		9.694
--- not "observed" ---	60.00		8.762
down to zero PhysPop	50.00		7.830
(zero medical radiation).	40.00		6.898
For each, we calculate	30.00		5.966
a best-fit MortRate.	20.00		5.034
These additional x,y pairs	10.00		4.102
are also part of the	0		3.170
best-fit line (Chap 5, Part 5e).			

```
┌─────────────────────────────────────────────────────────────────────────────┐
│                                                                             │
│                            Box 3 of Chap. 13                                │
│        Presumptive Fraction of Cancer MortRate Attributable to Medical Radiation. │
├─────────────────────────────────────────────────────────────────────────────┤
```

Please see text in Chapter 6, Parts 4 and 6.

Genital Cancers. MALES.

- MALE National MortRate (MR) 1940, from Table 13-B   15.2   National MortRate
- Constant, from regression, Part 2h   3.1726   Constant
- Fractional Causation, Best Est. = (Natl MR − Constant) / Natl MR   79.1%   Frac. Causation

....

    90% Confidence-Limits (C.L.) on Fractional Causation. See text in Chapter 6, Part 4b, please.

X-Coefficient, from Part 2h   0.0932   X-Coef., Best Est.
Standard Error (SE) of X-Coefficient, from Part 2h   0.0190   Standard Error

Upper 90% C.L. on X-Coef. = (Coef) + (1.645 * SE) =   0.1245   New X-Coefficient
New Constant = (Natl MR) − (New X-Coef * 1936 Natl PhysPop) =   −0.7215   New Constant
Frac. Causation, High-Limit = (Natl MR − New Constant) / Natl MR =   104.7% #   New Frac. Caus'n.
  # The Upper-Limit is 100%. Negative Constants produce values > 100%. See Chapter 22, Part 3.

Lower 90% C.L. on X-Coef. = (Coef) − (1.645 * SE) =   0.0619   New X-Coefficient
New Constant = (Natl MR) − (New X-Coef * 1936 Natl PhysPop) =   7.2754   New Constant
Frac. Causation, Low-Limit = (Natl MR − New Constant) / Natl MR =   52.1%   New Frac. Caus'n.

```
┌─────────────────────────────────────────────────────────────────────────────┐
│                            Box 4 of Chap 13.                                 │
│          Error-Check on Our Own Work:  Genital Cancers, Males.               │
├─────────────────────────────────────────────────────────────────────────────┤
```

Please see text in Chapter 6, Part 5.

Below, Columns A, C, and E come directly from the regression input in Part 2h. Column B, the fraction of the whole 1940 population in each Census Division, comes from Table 3-B in Chapter 3. Each Column-D entry is the product of (B-entry times C-entry). Each Column-F entry is the product of (B-entry times E-entry). PhysPops and MortRates are each "per 100,000."

The Weighted-Avg. Nat'l PhysPop, 1936, is the sum of Column-D entries =   127.93
    The 1936 PhysPop approximation is weighted by the 1940 population-fractions.
The Weighted-Avg. Nat'l Male MortRate, 1940, is sum of Col.F entries =   14.87
The Nat'l Male MortRate is also (X-Coef * Nat'l PhysPop) + Constant =   15.10
Comparison: The Nat'l Male MortRate, 1940, in Table 13-B =   15.20

(A) Census Division	(B) Pop'n Fraction	(C) PhysPop 1936	(D) Weighted PhysPop	(E) MortRate 1940	(F) Weighted MortRate
Pacific	0.0739	158.44	11.71	17.2	1.27
New England	0.0641	150.18	9.63	18.2	1.17
West No. Central	0.1027	126.14	12.95	16.5	1.69
Mid-Atlantic	0.2092	155.05	32.44	15.8	3.31
East No. Central	0.2022	130.42	26.37	15.8	3.19
Mountain	0.0315	119.80	3.77	15.8	0.50
West So. Central	0.0992	103.52	10.27	11.6	1.15
East So. Central	0.0819	89.94	7.37	10.4	0.85
South Atlantic	0.1354	99.16	13.43	12.8	1.73
Sums	1.0000		127.93		14.87

1940 Genital–System Cancer Mortality–Rates versus 1936 PhysPop Values for the 9 Census Divisions, USA.
Dose–Response Relationship

PhysPop is a surrogate for accumulated dose from medical irradiation.

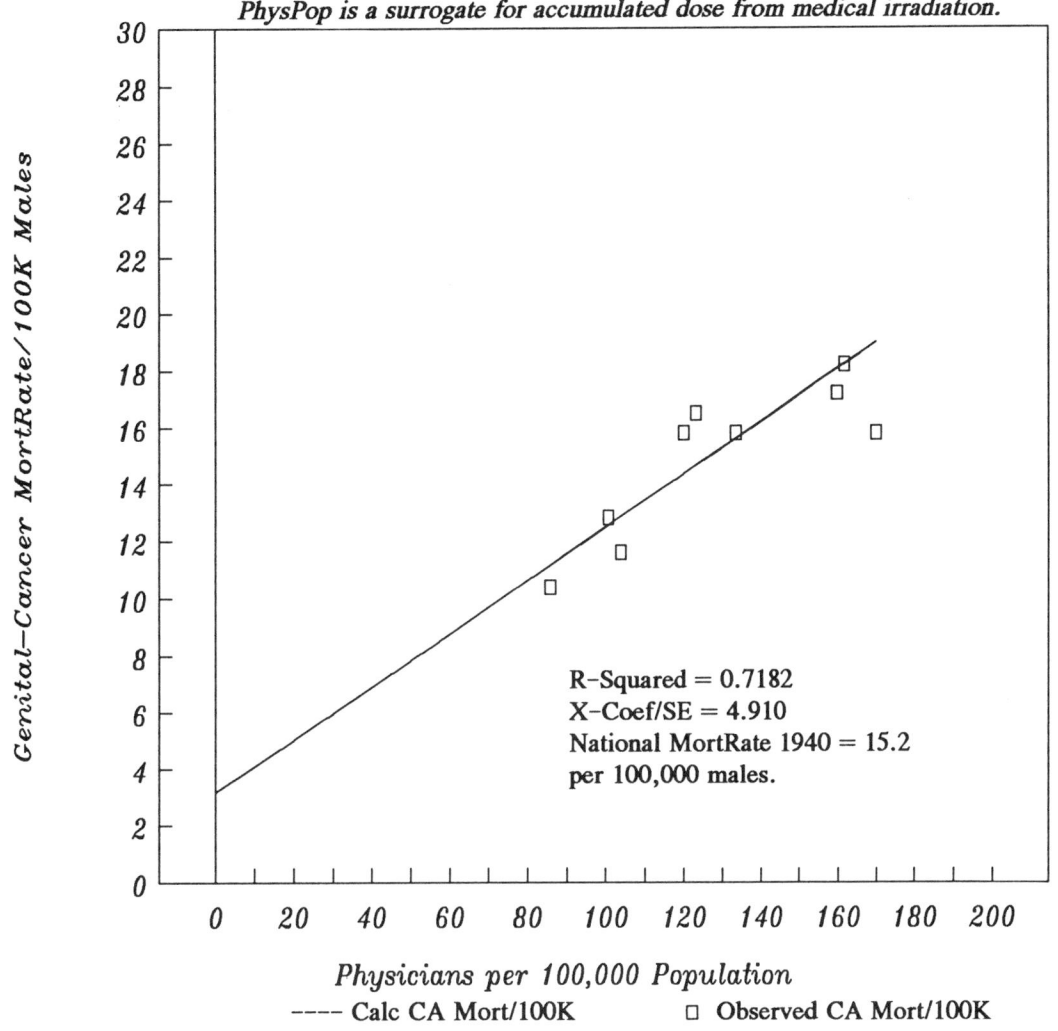

R-Squared = 0.7182
X-Coef/SE = 4.910
National MortRate 1940 = 15.2
per 100,000 males.

Physicians per 100,000 Population
---- Calc CA Mort/100K   □ Observed CA Mort/100K

On the X-axis, PhysPop values = Physicians per 100,000 Population in the Nine Census Divisions of the USA Population, Year 1940. This variable is a surrogate for accumulated radiation dose --- the more physicians per 100,000 people, the more radiation procedures are done per 100,000 people.

On the Y-axis, Genital-Cancer Mortality-Rate per 100,000 males = the reported rates in USA Vital Statistics for the Nine Census Divisions, Year 1940.

Shown above is the strongest relationship between these two variables (Part 2h). The nine datapoints (boxy symbols) were collected long ago for other purposes, and are free from potential bias with respect to this dose-response study. Fractional Causation is (Natl MortRate minus the Y-intercept) / (Natl MortRate).

Fractional Causation of Genital–Cancer Mortality (Males) by Medical Radiation

79 % from Best Estimate (Box 3).

52 % at lower 90 % confidence limit (Box 3). ∼100 % at upper 90% confidence limit (Box 3).

Table 13-A.
Genital Cancer Mortality Rates by Census Divisions: Males.

Rates are annual deaths per 100,000 male population, USA, age-adjusted to the 1940 reference year. There are no exclusions by color or "race." Sources are stated in Table 13-B, and described in Chap. 4, Part 2. The Nine Census-Division MortRates are population-weighted (Chap. 4, Part 2b). The averages below them are not.

Census Division	1940	1950	1960	1970	1980	1990
Pacific	17.2	14.0	13.4	13.9	14.4	15.9
New England	18.2	16.6	15.7	15.3	14.8	16.6
West North Central	16.5	16.6	15.4	14.8	14.2	16.3
Mid-Atlantic	15.8	14.2	13.6	14.2	14.8	16.8
East North Central	15.8	15.1	15.1	15.2	15.3	17.2
Mountain	15.8	13.6	15.2	14.9	14.5	16.6
West South Central	11.6	13.3	14.6	14.5	14.3	16.7
East South Central	10.4	14.7	15.9	15.7	15.4	17.5
South Atlantic	12.8	14.7	14.6	15.5	16.4	18.6
Average, ALL	14.9	14.8	14.8	14.9	14.9	16.9
Average, High-5	16.7	15.3	14.6	14.7	14.7	16.6
Average, Low-4	12.7	14.1	15.1	15.1	15.2	17.4
Ratio, Hi5/Lo4	1.32	1.09	0.97	0.97	0.97	0.95

Table 13-B.
Genital Cancer Mortality Rates, USA National.

Rates are age-adjusted to the 1940 reference year. Both sexes: Deaths per 100,000 population (males + females). Males: Deaths per 100,000 male population. Females: Deaths per 100,000 female population. No exclusions by color or "race."

	Both Sexes	Male	Female
1940	23.5	15.2	32.1
1950	21.0	14.9	27.2
1960	18.5	14.6	22.4
1970	--	14.8	18.0
1979-81	--	15.0	13.7
1989-1991	--	16.9	--

● - 1940, 1950, 1960: All rates come from Grove 1968, Table 67, p.693, "Malignant neoplasms of genital organs (171-179)" ICD/7.
  ● - 1970: All rates are interpolations (Chap. 4, Parts 2b, 2c).
  ● - 1980: All rates (ICD/9, 179-187) come from the reference NatCtrHS 1980.
  ● - 1990 male rates by Divisions and National come from Monthly Vital Statistics Vol.43, No.8, January 31, 1995. Females: Not available.

CHAPTER 14

Genital Cancers, Females: Relation with Medical Radiation, and Discussion

Part 1.   How a Dose-Response Fails to Develop, 1921-1940
Part 2.   Division of This Cancer MortRate into Uterus vs. Other
Part 3.   Findings in the Most Recent Report on A-Bomb Survivors
Part 4.   Ovarian Cancer and Talcum Powder
Part 5.   Cervical Cancer and Human Papilloma Virus ... and Co-Actors
Part 6.   A Likely Explanation for the Absent Dose-Response

Box 1.   Summary-Results from All Ten Regression Analyses.
Box 2.   Input-Data for Graph of Figure 14-A.
Figure 14-A.   Graph of the Strongest Dose-Response.
Tables 14-A, 14-B.   Genital Cancer MortRates, 1940-1980.

● **Part 1.   How a Dose-Response Fails to Develop, 1921-1940**

Female genital cancers include cancers of the cervix uteri, corpus uteri, ovaries, fallopian tubes, broad ligament, and other female genital organs (see Chapter 4, Part 5, Number 9).

Inspection, of the 1940 MortRates in Table 14-A, shows that the 1940 Hi5/Lo4 MortRate ratio is near unity (1.04). We have not seen such a low ratio in any of the previous chapters. Because the response (cancer MortRate) is so nearly alike in the Nine Census Divisions, it is highly unlikely that any significant dose-response exists in these data. And indeed, regression analysis confirms the absence of any dose-response. (We do not show the standard ten regressions in this chapter. Of course, the y-values are always the 1940 MortRates from Table 14-A, and the x-values are always the ten familiar sets of PhysPops, from Table 3-A.)

The summary-results of the regression analyses are presented in Box 1. In the maximum relationship (2j), the two measures of significance are both exceedingly low. The highest R-squared value is 0.0683. The highest ratio of the X-Coefficient over its Standard Error is 0.7163.

Box 2 prepares the input for Figure 14-A. As expected, Figure 14-A shows nine boxy symbols which predict a line of best fit which is nearly flat. In other words, an increment in dose (PhysPop) hardly produces any increment in response (MortRate).

Before moving to other considerations, we further explored the absence of a dose-response for female Genital Cancers by regressing the non-white 1940 MortRates upon PhysPop. Like the all-race MortRates for 1940, the non-white MortRates come from Grove 1968, Table 67. Although the Hi5/Lo4 ratio for the non-white study-group is 1.35 instead of 1.04, the correlation of the non-white 1940 MortRates with PhysPop is just as poor as it is for the all-race MortRates:

Census Div.	1940 PhysPop	1940 MortRate	Genital Ca, non-white Females: Regression Output:	
Pacific	159.72	33.9	Constant	36.7668
New England	161.55	60.4	Std Err of Y Est	13.3506
WestNoCentral	123.14	65.0	R Squared	0.0484
Mid-Atlantic	169.76	55.8	No. of Observations	9
EastNoCentral	133.36	61.3	Degrees of Freedom	7
Mountain	119.89	27.5		
WestSoCentral	103.94	45.6	X Coefficient(s)	0.0946
EastSoCentral	85.83	46.0	Std Err of Coef.	0.1586
So Atlantic	100.74	44.9	XCoef/S.E. =	0.5964

Among all the cancers studied in this monograph, female Genital Cancers turn out to be the ONLY group of cancers whose 1940 MortRates have no significant relationship with PhysPop (Chapter 22, Box 1). What are the possible explanations?

"Small numbers" are clearly not the explanation for the absence of any significant dose-response. The 1940 female MortRates for Genital Cancers are many times higher than the female MortRates for Urinary Cancer in Chapter 12 --- where a strong dose-response develops by 1940. We see no reason to think that the absent dose-response reflects misdiagnosis of female Genital Cancers. And there is no reason to believe that the female pelvis escaped exposure to medical irradiation, inasmuch as it received exposure from xrays of the hip, lumbar spine, lumbo-sacral spine, kidney-ureter-bladder, lower gastro-intestinal tract, pelvimetry, etc. We note that significant dose-responses occur for both digestive-system and urinary-system cancers in females (Chapters 10 and 12).

For female genital cancers, the absence of a dose-response is not a marginal matter. The results do not fall BARELY below some arbitrary level of statistical significance. The difference from every other set of cancers is like "night versus day."

● Part 2. The Subsets of Female Genital Cancers

We wish to ascertain which was the most important subset of female Genital Cancers in 1940. Grove 1968 (in Table 65, p.589), provides National 1940 MortRates for three groups of female Genital Cancers per 100,000 population (male + female), which is why the total rate of 16.0 (shown below) is approximately HALF of the National values in our Table 14-B. In 1960, Grove's Table 65 (p.597) provides separate entries for four groups of female Genital Cancers, so we will examine the 1960 entries, too. Although the 1960 entries are not age-adjusted to the 1940 reference year, we can use them "as is" to calculate percentages:

100.0%	16.0	= 1940 MortRate for all female Genital Cancers.
79.4%	12.7	= 1940 MortRate for cancer of the uterus (corpus + cervix).
17.5%	2.8	= 1940 MortRate for cancer of ovary, fallopian tube, and parametrium.
3.1%	0.5	= 1940 MortRate for cancer of vagina, vulva, and unspecified sites.

100.0%	13.0	= 1960 MortRate for all female Genital Cancers.
36.2%	4.7	= 1960 MortRate, malignant neoplasm of cervix uteri.
25.4%	3.3	= 1960 MortRate, malignant neoplasm of other parts of the uterus.
34.6%	4.5	= 1960 MortRate, malig. neoplasm of ovary, fallopian tube, broad ligament.
3.8%	0.5	= 1960 MortRate, malig. neoplasm of unspecifed female genital organs.

In both 1940 and 1960, the two major parts of the uterus (cervix and corpus) dominate the cancer death-rates in the grouping called female Genital Cancers. And in the uterus, the cervix accounts for more cancer mortality in 1960 than does the corpus. It is noteworthy that, by 1988, age-adjusted MortRates from Ovarian Cancer and from Uterine Cancer (including Cervical Cancer) had become approximately equal (our Figure 67-A). During the 1940-1988 period, MortRates from Ovarian Cancers increased somewhat, while MortRates from Uterine Cancers fell drastically. A contribution to the latter's fall may have come from a high rate of womb-removal in the USA.

● Part 3. Findings in the Most Recent Report on A-Bomb Survivors

Naturally, we wondered what the findings are for induction of female Genital Cancers by bomb-radiation in the A-Bomb Survivor Study.

The most recent follow-up (1950-1990) of the A-Bomb Survivors is presented in Pierce 1996-b. There (at page 15), we find Table X, "Numbers of Cancer Deaths and One-Sided P-Values for a Dose-Effect." The P-value for cancer of the uterus (cervix + corpus) is presented as 0.092. The P-value for cancer of the ovary is presented as 0.010. With one result (uterus) statistically suggestive and the other (ovary) statistically significant, the message is VERY different from the findings in our Box 1 --- where statistical significance is not even approached at all.

This difference is a clue of some sort. One would expect the statistically stronger results to occur in the larger study, all other things being equal. But we find the reverse, even though our xray study has millions of female participants and the A-Bomb Study has only about 50,000 females --- most of whom were hardly irradiated at all by the bomb (Chapter 2, Part 5a).

In view of the results in the A-Bomb Study, as well as the 1940 results in our studies for all cancers EXCEPT female Genital Cancers, we would expect to find a significant and positive dose-response relationship between medical radiation (PhysPop) and female Genital Cancers. Therefore, one ought to ask: What might make it appear that this relationship does NOT exist, if it DOES exist?

If the nine dose-groups (the populations of the Nine Census Divisions) happen to be badly matched for nonxray causes of female Genital Cancers, the bad matching could explain the complete absence of a dose-response in our Box 1 of this chapter. The effect of poor matching can vary in degree, of course. In Chapter 5, our Figure 5-D illustrates degradation of statistical strength from perfection (R-squared = 1.00) to a strength which is still highly significant (R-squared = 0.7112). Our Figure 5-E illustrates that an INVERSE relationship, between PhysPop and some other powerful cause of the same cancers, can even make a truly positive dose-response between medical radiation and those cancers appear to be negative. Between these extreme illustrations lies a range of concealment where poor matching just makes a signficant positive dose-response appear to be non-existent (so that the line of best-fit is approximately flat, with about the same response at all dose-levels).

Parts 4 and 5 discuss, for illustrative purposes, two of the several nonxray causes of female Genital Cancers which might be badly matched across our nine dose-groups. We emphasize "might," because we doubt very much that any useful data exist on the geographical distribution of these two causes before (or after) 1940.

● Part 4. Ovarian Cancer and Talcum Powder

For decades, there has been suspicion that the substance, talc, might be gaining access to the ovary by the vagina-uterus-oviduct route, and that talc might be an ovarian carcinogen.

4a. The Background as Presented by Cramer et al, 1982

Daniel W. Cramer and associates begin their 1982 paper by stating (Cramer 1982, p.372): "The possibility that ovarian cancer may be caused by exposure to certain hydrous magnesium silicates such as talc and asbestos has been raised by several researchers (Graham 1967 + Henderson 1971 + Longo 1979). The lack of epidemiologic studies regarding this hypothesis prompted us to investigate talc exposures in a case-control study of ovarian cancer." In their discussion, they state (Cramer 1982, p.375):

"The argument linking talc and ovarian cancer includes four elements: The chemical relationship between talc and asbestos, asbestos as a cause of pleural and peritoneal mesotheliomas, the possible relationship between epithelial ovarian cancers and mesotheliomas, and the ability of talc to enter the pelvic cavity. The mineral talc is a specific hydrous magnesium silicate chemically related to several asbestos group minerals and occurring in nature with them. Generic 'talc' is seldom pure and may be contaminated with asbestos, particularly in powders formulated prior to 1976 (Cralley 1968 + Rohl 1976)." And they add (Cramer 1982, p.375):

"Although greeted with skepticism, the finding of talc particles embedded in normal and abnormal ovaries suggests that talc is a substance that can enter the pelvic cavity via the vagina (Henderson 1971)."

4b. Results of the Talc Study by Cramer et al. 1982

The Cramer study compared 215 white females with epithelial ovarian cancers with 215 control women from the general population matched by age (average age = 53), race, residence, educational level, and religion. They were not matched for parity, however (Table 1). Relative risks had to be adjusted for this and some other "potential confounders" (Cramer 1982, p.373). Results:

"Ninety-two of the cases (42.8%) regularly used talc either as a dusting powder on the perineum or on sanitary napkins compared with 61 (28.4%) controls. Adjusted for parity and menopausal status, this difference yielded a relative risk of 1.92 (P < 0.003) for ovarian cancer associated with these practices. Women who had regularly engaged in both practices had an adjusted relative risk of 3.28 (P < 0.001) compared to women with neither exposure. This provides some support for an association between talc and ovarian cancer ... The authors also investigated opportunities for potential talc exposure from rubber products such as condoms or diaphragms or from pelvic surgery. No significant differences were noted between cases and controls in these exposures ..." (Cramer 1982, p.372). In the end, they conclude (Cramer 1982, p.376):

"If talc is involved in the etiology of ovarian cancer, it is not clear whether this derives from the asbestos content of talc or from the uniqueness of the ovary which might make it susceptible to carcinogenesis from both talc and other particulates." And (p.376): "It is hoped that this report will stimulate further study of talc exposure in relation to ovarian cancer."

Exposure to genital powders is very common among American women (Part 4c, below).

4c. Results of the Powder and Spray Study by Cook et al. 1997

Linda S. Cook and colleagues begin their 1997 paper as follows (Cook 1997, p.459):

"Studies documenting the migration of carbon particles and radioactive particulate agents from the vagina to the ovaries (2 references), as well as those that have identified talc-like particles more frequently in ovarian tumors than in normal human ovarian tissue (1 reference), have raised concern that genital powder exposure may increase a woman's risk of developing ovarian cancer. While the results of several epidemiological studies have suggested elevated risks for ovarian cancer among women with genital powder exposures (8 references), results have been inconsistent for particular methods of powder application (1 reference). In this population-based case-control study, information on the method, duration, and frequency of powder application was collected to evaluate the impact of genital powder exposures on the risk of epithelial ovarian cancer." Their study consisted of 313 ovarian cancer cases and 422 controls. One of the featured results (Cook 1997, p.459):

"After adjustment for age and other methods of genital powder application (none vs. any), an elevated relative risk of ovarian cancer was noted only for women with a history of perineal dusting (RR = 1.6, 95% CI 1.1-2.3) or use of genital deodorant spray (RR = 1.9, 95% CI 1.1-3.1). These results offer support for the hypothesis, raised by prior epidemiologic studies, that powder exposure from perineal dusting contributes to the development of ovarian cancer, and they suggest that use of genital deodorant sprays may do so as well."

At the end of their paper, Cook and co-workers urge additional studies, and point out (Cook 1997, p.465):

"The prevalence of genital powder exposure reported among control women in this and other studies conducted in the United States ranges from 28 percent to 51 percent (5 references). Given such a common practice, even the modest elevation of ovarian cancer risk associated with genital powder application suggested by most of the epidemiologic studies could have a notable impact on the incidence of ovarian cancer in the United States."

● Part 5. Cervical Cancer and Human Papilloma Virus ... and Co-Actors

There seems to be little doubt that infection of the female genital tract, with human papilloma viruses (HPV), plays a very important role in the causation of squamous-cell carcinoma of the cervix ("cervical cancer"). Dr. Keerti Shah comments on some of the recent findings in an editorial in the New England Journal of Medicine (Nov. 6, 1997). Indeed, Shah states (Shah 1997, p.1387): "In all parts of the world, infections with genital HPVs appear to account for nearly 100 percent of cervical cancers (Bosch 1995 and unpublished data)." And (p.1387): "In most cancers, the HPV genome is integrated into the cellular DNA."

Among about 30 strains of HPV which can infect the cervix, only a few strains --- most especially HPV-16, 18, 31, and 45 --- seem to be carcinogenic. A 1998 study indicates that infection even by these "high-risk" HPV strains often clears up (Ho 1998, p.424, Table 1.)

Several lines of evidence indicate that the HPV virus needs help from carcinogenic co-actors, in order to produce a case of fatal cervical cancer (ZurHausen 1998). For example, work by Apple (1994, 1995) suggests that a woman's particular mixture, of inherited genes for HLA proteins, has an influence on her risk of developing cervical cancer after cervical infection with HPV-16. Other recent work (Storey 1998) suggests that women who inherit a particular variant of the p53 gene are most at risk for the CONSEQUENCES of infection by the "high-risk" HPV strains.

Prokopczyk, another investigator into the etiology of cervical cancer, explicitly asserts that HPV infection by itself is not enough to cause cervical cancer in women. He suggests: "There must be another factor initially damaging the cervical DNA" (Prokopczyk 1996, p.282). Some experimental work with mice (Arbeit 1996) also seems to suggest that HPV alone does not suffice. According to Prokopczyk and colleagues (Prokopczyk 1997, p.869):

"HPV-modified DNA has been detected in up to 93% of cervical tumor specimens (IARC 1995 + Bosch 1995). However, because HPV infections are widespread in the general population and HPV-immortalized cell lines are generally not tumorigenic, HPV infection likely interacts with one or more co-factors before cancer develops." They mention deficiency in micronutrients, lower socioeconomic status, use of oral contraceptives, and cigarette smoking as co-factors which have been explored.

With respect to smoking, they state (Prokopcyzk 1997, p.869): "Winkelstein (1990) reviewed 18 studies of cigarette smoking and cervical cancer: 15 of these studies supported an increased risk (up to 4.3-fold higher) of cervical cancer among smokers, and several of these studies demonstrated a dose-response relationship (2 references). Environmental exposure to cigarette smoke has also been suggested to increase the risk of cervical cancer (Slattery 1989) ..." Later (at p.872), Prokopcyzk et al report that "Smoking-related DNA damage has been demonstrated by several studies (5 references) through P-32 postlabeling techniques; however, structures of these putative adducts remain unknown."

These workers undertook a small pilot study in which they found that that cervical mucus, from women who smoke, contains a significantly higher concentration of a carcinogenic tobacco-specific nitrosamine (NNK) than cervical mucus from nonsmokers (Prokopcyzk 1997, p.871, Table 1).

● Part 6. A Likely Explanation for the Absent Dose-Response

Cancer is a disease having multiple causes. Indeed, co-action among two or more causes may be required to produce most of the fatal cases. In every chapter of Section 2 except this chapter, we find that medical radiation was a NECESSARY cause in a very high fraction of all cancers which were fatal in 1940.

We doubt very much that medical radiation plays no role at all in female Genital Cancers. We think the probable explanation, for the absence of any dose-response in these data, is bad matching (across the Nine Census Divisions) of some co-actors which are potent in causing female Genital Cancers but are not potent in causing the other cancers. There is no doubt that badly matched dose-groups can mask a true dose-response beyond detection. This is such a common pitfall, in human epidemiological research, that it is reasonable to suspect it to be the explanation here.

>>>>>>>>>>

Box 1 of Chap. 14

Box 1 of Chap. 14
Summary: Regression Outputs, Genital Cancers, Females.

Below are the summary-results from regressing the 1940 cancer MortRates upon the ten sets of PhysPops (1921-1940).

Part	PhysPop	R-squared	Constant	X-Coef	Std Err	X-Coef/SE
2a	1921	0.0006	31.06	0.0034	0.0529	0.0643
2b	1923	0.0020	30.77	0.0058	0.0493	0.1171
2c	1925	0.0163	29.56	0.0155	0.0454	0.3410
2d	1927	0.0262	29.18	0.0188	0.0433	0.4340
2e	1929	0.0318	29.05	0.0200	0.0417	0.4798
2f	1931	0.0427	28.89	0.0211	0.0378	0.5587
2g	1934	0.0534	28.92	0.0208	0.0331	0.6285
2h	1936	0.0458	29.19	0.0185	0.0320	0.5793
2i	1938	0.0539	29.13	0.0189	0.0299	0.6316
2j --->	1940 Max	0.0683	29.06	0.0191	0.0267	0.7163

Box 2 of Chap. 14
Input-Data for Figure 14-A. Genital Cancers. Females.

Part 2j, Best-Fit Equation: Calc. MortRate = (0.0191 * 1940 PhysPop) + (29.06)

Census Divisions	1940 Observed PhysPops	1940 Observed MortRates	Best-Fit Calc. MortRates
Pacific	159.72	33.1	32.111
New England	161.55	32.8	32.146
West No. Central	123.14	28.4	31.412
Mid-Atlantic	169.76	32.7	32.302
East No. Central	133.36	33.2	31.607
Mountain	119.89	27.8	31.350
West So. Central	103.94	30.0	31.045
East So. Central	85.83	33.2	30.699
South Atlantic	100.74	32.5	30.984
Additional PhysPops	70.00		30.397
--- not "observed" ---	60.00		30.206
down to zero PhysPop	50.00		30.015
(zero medical radiation).	40.00		29.824
For each, we calculate	30.00		29.633
a best-fit MortRate.	20.00		29.442
These additional x,y pairs	10.00		29.251
are also part of the	0		29.060
best-fit line (Chap 5, Part 5e).			

1940 Genital–System Cancer Mortality–Rates versus 1940 PhysPop Values for the 9 Census Divisions, USA.

Dose–Response Relationship

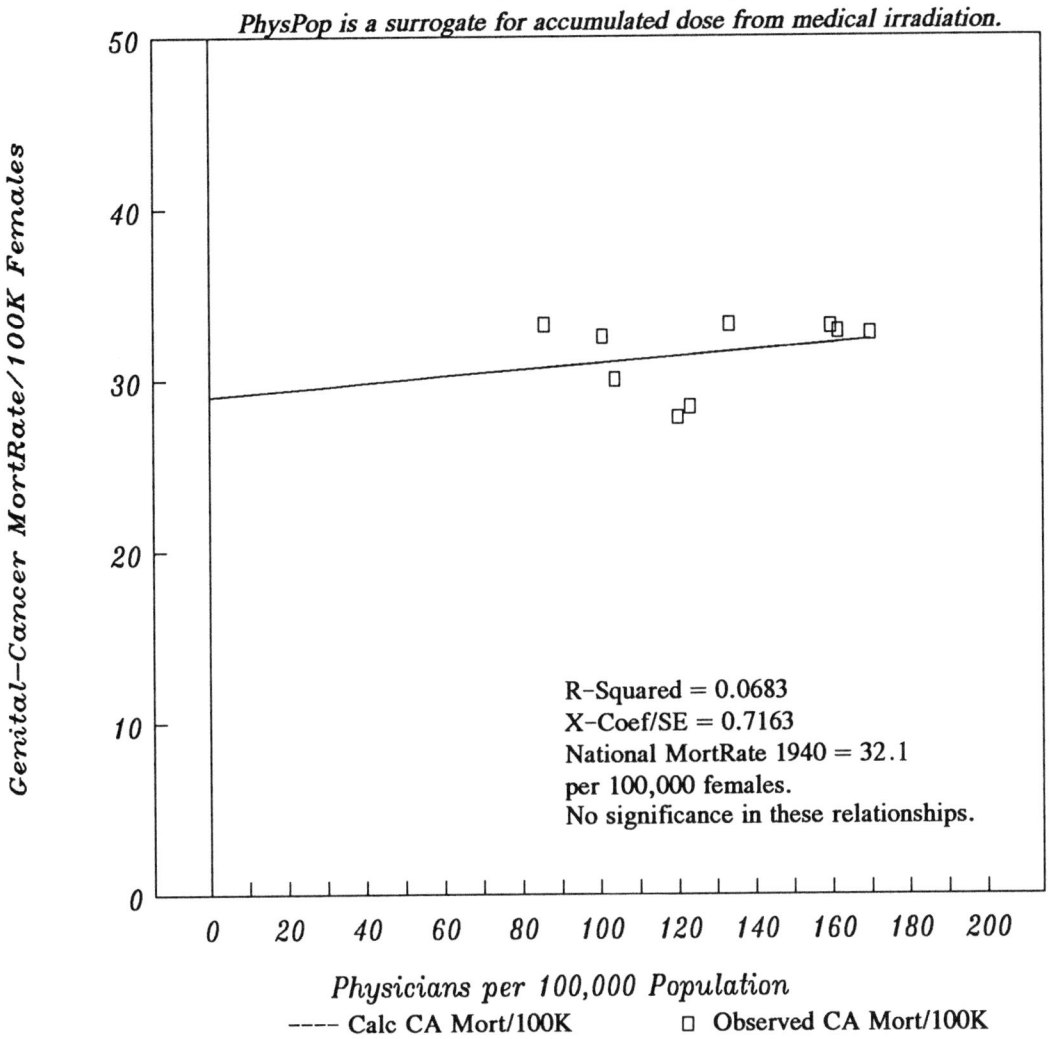

PhysPop is a surrogate for accumulated dose from medical irradiation.

R–Squared = 0.0683
X–Coef/SE = 0.7163
National MortRate 1940 = 32.1
per 100,000 females.
No significance in these relationships.

Physicians per 100,000 Population
---- Calc CA Mort/100K   □ Observed CA Mort/100K

On the X–axis, PhysPop values = Physicians per 100,000 Population in the Nine Census Divisions of the USA Population, Year 1940. This variable is a surrogate for accumulated radiation dose --- the more physicians per 100,000 people, the more radiation procedures are done per 100,000 people.

On the Y–axis, Genital–Cancer Mortality–Rate per 100,000 females = the reported rates in USA Vital Statistics for the Nine Census Divisions, Year 1940, for all "races" combined, no exclusions.

Shown above is the strongest relationship between these two variables (Part 2j). The nine datapoints (boxy symbols) were collected long ago for other purposes, and are free from potential bias with respect to this dose–response study. There is no dose–response relationship detected with these data, so the presumptive Fractional Causation in 1940 by medical radiation is zero.

Table 14-A.

Genital Cancer Mortality Rates by Census Divisions: Females.

Rates are annual deaths per 100,000 female population, USA, age-adjusted to the 1940 reference year. Sources are provided in Chapter 4, Part 2. There are no exclusions by color or "race." The tabulation includes averages (not population-weighted) and the Hi5/Lo4 ratios --- explained at the outset of Chapter 4.

Census Division	1940	1950	1960	1970	1980	1990
Pacific	33.1	25.5	20.0	16.7	13.3	--
New England	32.8	25.1	21.7	17.6	13.4	--
West North Central	28.4	23.4	20.3	16.8	13.3	--
Mid-Atlantic	32.7	27.2	22.2	18.3	14.3	--
East North Central	33.2	28.3	24.2	19.4	14.5	--
Mountain	27.8	23.6	18.4	15.0	11.7	--
West South Central	30.0	27.4	22.1	17.3	12.5	--
East South Central	33.2	29.7	24.7	19.5	14.3	--
South Atlantic	32.5	29.9	24.0	18.8	13.5	--
Average, ALL	31.5	26.7	22.0	17.7	13.4	--
Average, High-5	32.0	25.9	21.7	17.7	13.8	--
Average, Low-4	30.9	27.7	22.3	17.7	13.0	--
Ratio, Hi5/Lo4	1.04	0.94	0.97	1.00	1.06	--

Table 14-B.

Genital Cancer Mortality Rates, USA National.

Rates are age-adjusted to the 1940 reference year. Both sexes: Deaths per 100,000 population (males + females). Males: Deaths per 100,000 male population. Females: Deaths per 100,000 female population. No exclusions by color or "race."

	Both Sexes	Male	Female
1940	23.5	15.2	32.1
1950	21.0	14.9	27.2
1960	18.5	14.6	22.4
1970	--	14.8	18.0
1979-81	13.5	15.0	13.7
1990	--	16.9	--

● - 1940, 1950, 1960: All rates come from Grove 1968, Table 67, p.693, "Malignant neoplasms of genital organs (171-179)" ICD/7.
 ● - 1970: All rates are interpolations (Chap. 4, Parts 2b, 2c).
 ● - 1980: All rates (ICD/9, 179-187) come from the reference NatCtrHS 1980.
 ● - 1990 male rates by Divisions and National come from Monthly Vital Statistics Vol.43, No.8, January 31, 1995. Females: Not available.

CHAPTER 15

Buccal-Cavity & Pharynx Cancers, Males: Relation with Medical Radiation

● Part 1. Introduction

Buccal-Cavity and Pharynx Cancers include cancers of the lip, tongue, unspecified parts of the buccal cavity, and the pharynx (see Chapter 4, Part 5, Number 5). The MortRates in Table 15-A clearly present a "small numbers problem." The problem is so severe for the females --- as indicated by their lower national MortRate in Table 15-B --- that we did not analyse the female data.

● Part 2. How the Dose-Response Develops, 1921-1940

● - Part 2a.

	1921 PhysPop	1940 MortRate	Buccal + Pharynx Ca, Males Regression Output:	
Pacific	165.11	5.3	Constant	-0.3152
New England	142.24	6.4	Std Err of Y Est	1.2889
West North Central	140.93	4.6	R Squared	0.1884
Mid-Atlantic	137.29	6.9	No. of Observations	9
East North Central	136.06	4.8	Degrees of Freedom	7
Mountain	135.38	2.8		
West South Central	125.15	4.0	X Coefficient(s)	0.0373
East South Central	119.76	3.3	Std Err of Coef.	0.0293
South Atlantic	110.32	4.3	Coefficient / S.E.	1.2746

● - Part 2b.

	1923 PhysPop	1940 MortRate	Buccal + Pharynx Ca, Males Regression Output:	
Pacific	163.06	5.3	Constant	-0.6198
New England	137.39	6.4	Std Err of Y Est	1.2327
West North Central	138.31	4.6	R Squared	0.2577
Mid-Atlantic	138.92	6.9	No. of Observations	9
East North Central	131.82	4.8	Degrees of Freedom	7
Mountain	130.51	2.8		
West South Central	119.16	4.0	X Coefficient(s)	0.0407
East South Central	113.16	3.3	Std Err of Coef.	0.0261
South Atlantic	106.79	4.3	Coefficient / S.E.	1.5589

● - Part 2c.

	1925 PhysPop	1940 MortRate	Buccal + Pharynx Ca, Males Regression Output:	
Pacific	161.67	5.3	Constant	-0.7436
New England	138.31	6.4	Std Err of Y Est	1.1670
West North Central	133.92	4.6	R Squared	0.3346
Mid-Atlantic	134.36	6.9	No. of Observations	9
East North Central	127.54	4.8	Degrees of Freedom	7
Mountain	122.30	2.8		
West South Central	112.83	4.0	X Coefficient(s)	0.0430
East South Central	107.22	3.3	Std Err of Coef.	0.0229
South Atlantic	103.61	4.3	Coefficient / S.E.	1.8763

● - Part 2d.

	1927 PhysPop	1940 MortRate	Buccal + Pharynx Ca, Males Regression Output:	
Pacific	157.83	5.3	Constant	-1.2380
New England	137.50	6.4	Std Err of Y Est	1.0624
West North Central	131.54	4.6	R Squared	0.4486
Mid-Atlantic	138.40	6.9	No. of Observations	9
East North Central	126.18	4.8	Degrees of Freedom	7
Mountain	118.75	2.8		
West South Central	108.25	4.0	X Coefficient(s)	0.0477
East South Central	102.07	3.3	Std Err of Coef.	0.0200
South Atlantic	102.13	4.3	Coefficient / S.E.	2.3862

● – Part 2e.

	1929 PhysPop	1940 MortRate	Buccal + Pharynx Ca, Males Regression Output:	
Pacific	156.64	5.3	Constant	−1.1282
New England	138.46	6.4	Std Err of Y Est	1.0421
West North Central	128.72	4.6	R Squared	0.4695
Mid–Atlantic	138.49	6.9	No. of Observations	9
East North Central	126.51	4.8	Degrees of Freedom	7
Mountain	118.68	2.8		
West South Central	105.60	4.0	X Coefficient(s)	0.0472
East South Central	99.41	3.3	Std Err of Coef.	0.0190
South Atlantic	100.86	4.3	Coefficient / S.E.	2.4891

● – Part 2f.

	1931 PhysPop	1940 MortRate	Buccal + Pharynx Ca, Males Regression Output:	
Pacific	159.97	5.3	Constant	−0.7783
New England	142.35	6.4	Std Err of Y Est	1.0182
West North Central	126.50	4.6	R Squared	0.4935
Mid–Atlantic	140.82	6.9	No. of Observations	9
East North Central	128.59	4.8	Degrees of Freedom	7
Mountain	118.89	2.8		
West South Central	105.95	4.0	X Coefficient(s)	0.0441
East South Central	96.73	3.3	Std Err of Coef.	0.0169
South Atlantic	99.59	4.3	Coefficient / S.E.	2.6115

● – Part 2g.

	1934 PhysPop	1940 MortRate	Buccal + Pharynx Ca, Males Regression Output:	
Pacific	160.09	5.3	Constant	−0.6732
New England	148.60	6.4	Std Err of Y Est	0.8985
West North Central	125.96	4.6	R Squared	0.6056
Mid–Atlantic	149.62	6.9	No. of Observations	9
East North Central	129.36	4.8	Degrees of Freedom	7
Mountain	117.16	2.8		
West South Central	104.68	4.0	X Coefficient(s)	0.0430
East South Central	92.00	3.3	Std Err of Coef.	0.0131
South Atlantic	98.41	4.3	Coefficient / S.E.	3.2784

● – Part 2h.

	1936 PhysPop	1940 MortRate	Buccal + Pharynx Ca, Males Regression Output:	
Pacific	158.44	5.3	Constant	−0.5979
New England	150.18	6.4	Std Err of Y Est	0.8714
West North Central	126.14	4.6	R Squared	0.6290
Mid–Atlantic	155.05	6.9	No. of Observations	9
East North Central	130.42	4.8	Degrees of Freedom	7
Mountain	119.80	2.8		
West South Central	103.52	4.0	X Coefficient(s)	0.0422
East South Central	89.94	3.3	Std Err of Coef.	0.0122
South Atlantic	99.16	4.3	Coefficient / S.E.	3.4453

● – Part 2i.

	1938 PhysPop	1940 MortRate	Buccal + Pharynx Ca, Males Regression Output:	
Pacific	157.62	5.3	Constant	−0.4713
New England	154.08	6.4	Std Err of Y Est	0.8197
West North Central	124.95	4.6	R Squared	0.6718
Mid–Atlantic	160.69	6.9	No. of Observations	9
East North Central	131.98	4.8	Degrees of Freedom	7
Mountain	119.88	2.8		
West South Central	102.79	4.0	X Coefficient(s)	0.0409
East South Central	88.21	3.3	Std Err of Coef.	0.0108
South Atlantic	99.26	4.3	Coefficient / S.E.	3.7853

● – Part 2j.

	1940 PhysPop	1940 MortRate	Buccal + Pharynx Ca, Males Regression Output:	
Pacific	159.72	5.3	Constant	−0.2081
New England	161.55	6.4	Std Err of Y Est	0.7525
West North Central	123.14	4.6	R Squared	0.7234
Mid–Atlantic	169.76	6.9	No. of Observations	9
East North Central	133.36	4.8	Degrees of Freedom	7

Mountain	119.89	2.8		
West South Central	103.94	4.0	X Coefficient(s)	0.0382
East South Central	85.83	3.3	Std Err of Coef.	0.0089
South Atlantic	100.74	4.3	Coefficient / S.E.	4.2782

Box 1 of Chap. 15
Summary: Regression Outputs, Buccal Cav. & Pharynx Cancers, Males.

Below are the summary-results from regressing the 1940 cancer MortRates upon the ten sets of PhysPops (1921-1940), as presented in Parts 2a-2j of this chapter.

Part	PhysPop	R-squared	Constant	X-Coef	Std Err	X-Coef/SE
2a	1921	0.1884	-0.32	0.0373	0.0293	1.2746
2b	1923	0.2577	-0.62	0.0407	0.0261	1.5589
2c	1925	0.3346	-0.74	0.0430	0.0229	1.8763
2d	1927	0.4486	-1.24	0.0477	0.0200	2.3862
2e	1929	0.4695	-1.13	0.0472	0.0190	2.4891
2f	1931	0.4935	-0.78	0.0441	0.0169	2.6115
2g	1934	0.6056	-0.67	0.0430	0.0131	3.2784
2h	1936	0.6290	-0.60	0.0422	0.0122	3.4453
2i	1938	0.6718	-0.47	0.0409	0.0108	3.7853
2j --->	1940 Max	0.7234	-0.21	0.0382	0.0089	4.2782

Box 2 of Chap. 15
Input-Data for Figure 15-A. Buccal Cav. & Pharynx Cancers. Males.

Part 2j, Best-Fit Equation: Calc. MortRate = (0.0382 * PhysPop) + (-0.21)

Census Divisions	1940 Observed PhysPops	1940 Observed MortRates	Best-Fit Calc. MortRates
Pacific	159.72	5.3	5.891
New England	161.55	6.4	5.961
West No. Central	123.14	4.6	4.494
Mid-Atlantic	169.76	6.9	6.275
East No. Central	133.36	4.8	4.884
Mountain	119.89	2.8	4.370
West So. Central	103.94	4.0	3.761
East So. Central	85.83	3.3	3.069
South Atlantic	100.74	4.3	3.638
Additional PhysPops	70.00		2.464
--- not "observed" ---	60.00		2.082
down to zero PhysPop	50.00		1.700
(zero medical radiation).	40.00		1.318
For each, we calculate	30.00		0.936
a best-fit MortRate.	20.00		0.554
These additional x,y pairs	10.00		0.172
are also part of the	0		-0.210
best-fit line (Chap 5, Part 5e).			

```
┌────────────────────────────────────────────────────────────────────────────┐
│                             Box 3 of Chap. 15                                │
│         Presumptive Fraction of Cancer MortRate Attributable to Medical Radiation. │
├════════════════════════════════════════════════════════════════════════════┤
```

Please see text in Chapter 6, Parts 4 and 6.

Buccal-Cavity + Pharynx Cancers. MALES.

- MALE National MortRate (MR) 1940, from Table 15-B   5.1   National MortRate
- Constant, from regression, Part 2j   -0.2081   Constant
- Fractional Causation, Best Est. = (Natl MR - Constant) / Natl MR   104.1% #   Frac. Causation

\# The Upper-Limit is 100%. Negative Constants produce values > 100%. See Chapter 22, Part 3.

....

90% Confidence-Limits (C.L.) on Fractional Causation. See text in Chapter 6, Part 4b, please.

X-Coefficient, from Part 2j   0.0382   X-Coef., Best Est.
Standard Error (SE) of X-Coefficient, from Part 2j   0.0089   Standard Error

Upper 90% C.L. on X-Coef. = (Coef) + (1.645 * SE) =   0.0528   New X-Coefficient
New Constant = (Natl MR) - (New X-Coef * 1940 Natl PhysPop) =   -1.8771   New Constant
Frac. Causation, High-Limit = (Natl MR - New Constant) / Natl MR =   136.8% #   New Frac. Caus'n.

\# The Upper-Limit is 100%. Negative Constants produce values > 100%. See Chapter 22, Part 3.

Lower 90% C.L. on X-Coef. = (Coef) - (1.645 * SE) =   0.0236   New X-Coefficient
New Constant = (Natl MR) - (New X-Coef * 1940 Natl PhysPop) =   1.9892   New Constant
Frac. Causation, Low-Limit = (Natl MR - New Constant) / Natl MR =   61.0%   New Frac. Caus'n.

```
┌────────────────────────────────────────────────────────────────────────────┐
│                             Box 4 of Chap. 15                                │
│        Error-Check on Our Own Work:  Buccal-Cav. & Pharynx Cancers, Males.   │
├════════════════════════════════════════════════════════════════════════════┤
```

Please see text in Chapter 6, Part 5.

Below, Columns A, C, and E come directly from the regression input in Part 2j. Column B, the fraction of the whole 1940 population in each Census Division, comes from Table 3-B in Chapter 3. Each Column-D entry is the product of (B-entry times C-entry). Each Column-F entry is the product of (B-entry times E-entry). PhysPops and MortRates are each "per 100,000."

The Weighted-Avg. Nat'l PhysPop, 1940, is the sum of Column-D entries =   132.04

The Weighted-Avg. Nat'l Male MortRate, 1940, is sum of Col.F entries =   5.03
The Nat'l Male MortRate is also (X-Coef * Nat'l PhysPop) + Constant =   4.84
Comparison: The Nat'l Male MortRate, 1940, in Table 15-B =   5.10

(A) Census Division	(B) Pop'n Fraction	(C) PhysPop 1940	(D) Weighted PhysPop	(E) MortRate 1940	(F) Weighted MortRate
Pacific	0.0739	159.72	11.80	5.3	0.39
New England	0.0641	161.55	10.36	6.4	0.41
West No. Central	0.1027	123.14	12.65	4.6	0.47
Mid-Atlantic	0.2092	169.76	35.51	6.9	1.44
East No. Central	0.2022	133.36	26.97	4.8	0.97
Mountain	0.0315	119.89	3.78	2.8	0.09
West So. Central	0.0992	103.94	10.31	4.0	0.40
East So. Central	0.0819	85.83	7.03	3.3	0.27
South Atlantic	0.1354	100.74	13.64	4.3	0.58
Sums	1.0000		132.04		5.03

1940 Buccal/Pharynx Mortality–Rates versus
1940 PhysPop Values for the 9 Census Divisions, USA.
Dose–Response Relationship

PhysPop is a surrogate for accumulated dose from medical irradiation.

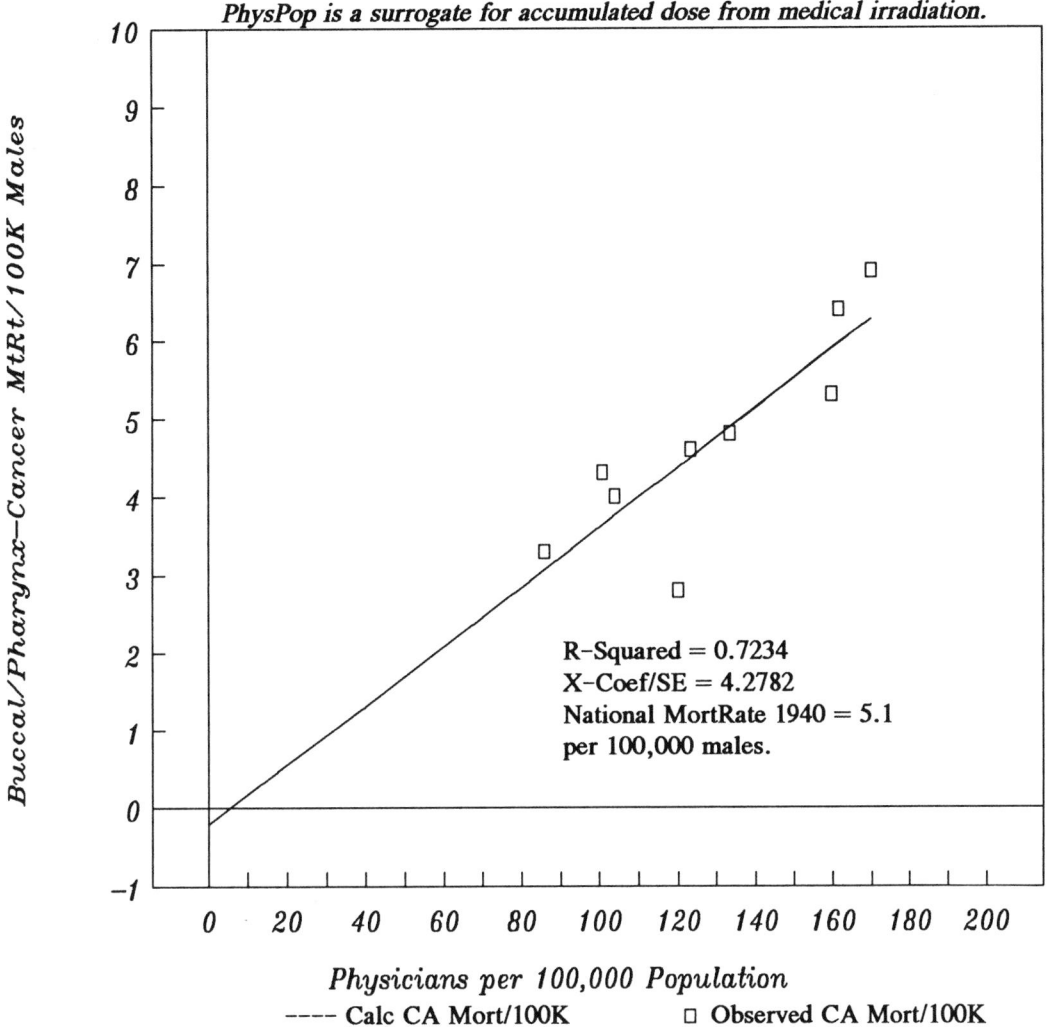

R-Squared = 0.7234
X-Coef/SE = 4.2782
National MortRate 1940 = 5.1
per 100,000 males.

Physicians per 100,000 Population

---- Calc CA Mort/100K   □ Observed CA Mort/100K

On the X–axis, PhysPop values = Physicians per 100,000 Population in
the Nine Census Divisions of the USA Population, Year 1940. This variable is
a surrogate for accumulated radiation dose --- the more physicians per 100,000
people, the more radiation procedures are done per 100,000 people.

On the Y–axis, Buccal/Pharnyx Mortality–Rate per 100,000 males = the
reported rates in USA Vital Statistics for the Nine Census Divisions, Year 1940.

Shown above is the strongest relationship between these two
variables (Part 2j). The nine datapoints (boxy symbols) were collected long
ago for other purposes, and are free from potential bias with respect to this
dose–response study. Fractional causation is (Natl MortRate minus the
Y–intercept) / (Natl MortRate).

Fractional Causation of Buccal/Pharynx Cancer Mortality–Rate in Males
by Medical Radiation = ~100 % from Best Estimate (Box 3).

61 % at Lower 90% Conf. Limit (Box 3). ~100 % at Upper 90 % Conf. Limit (Box 3).

Table 15-A.
Buccal-Cav. & Pharynx Cancer MortRates by Census Divisions: Males.

Rates are annual deaths per 100,000 male population, USA, age-adjusted to the 1940 reference year. There are no exclusions by color or "race." Sources are stated in Table 15-B, and described in Chap. 4, Part 2. The Nine Census-Division MortRates are population-weighted (Chap. 4, Part 2b). The averages below them are not.

Census Division	1940	1950	1960	1970	1980	1990
Pacific	5.3	4.7	4.6	4.4	4.2	--
New England	6.4	6.5	6.6	6.2	5.7	--
West North Central	4.6	4.5	4.0	3.8	3.5	--
Mid-Atlantic	6.9	5.9	5.4	5.3	5.1	--
East North Central	4.8	4.9	4.9	4.8	4.6	--
Mountain	2.8	3.2	2.6	2.8	2.9	--
West South Central	4.0	4.3	3.9	4.1	4.2	--
East South Central	3.3	4.2	4.2	4.3	4.4	--
South Atlantic	4.3	4.7	4.5	4.8	5.0	--
Average, ALL	4.7	4.8	4.5	4.5	4.4	--
Average, High-5	5.6	5.3	5.1	4.9	4.6	--
Average, Low-4	3.6	4.1	3.8	4.0	4.1	--
Ratio, Hi5/Lo4	1.56	1.29	1.34	1.23	1.12	--

Table 15-B.
Buccal-Cavity & Pharynx Cancer Mortality Rates, USA National.

Rates are age-adjusted to the 1940 reference year. Both sexes: Deaths per 100,000 population (males + females). Males: Deaths per 100,000 male population. Females: Deaths per 100,000 female population. No exclusions by color or "race."

	Both Sexes	Male	Female
1940	3.1	5.1	1.1
1950	3.0	5.0	1.2
1960	3.0	4.7	1.3
1970	--	4.65	1.4
1979-81	2.9	4.6	1.5
1989-91	--	--	--

● - 1940, 1950, 1960: All rates come from Grove 1968, Table 67, p.679, "Malignant neoplasm of buccal cavity and pharynx (140-148)," ICD/7.
● - 1970: All rates are interpolations (Chap. 4, Parts 2b, 2c).
● - 1980: All rates (ICD/9, 140-149) come from the reference NatCtrHS 1980.
● - 1990: No data obtained. Please see Chap. 4, Part 2c.

● Part 1. Introduction

Respiratory-System Cancers include cancers of the larynx, bronchus and trachea, of lung specified as primary, of lung unspecified as to whether primary or secondary, and of other parts of the respiratory-system (see Chapter 4, Part 5, Number 7).

This study produces negative Constants for the central estimate and for both of the confidence-limits on the X-Coefficient --- as shown in Box 3. In this situation, we hesitate to use any value for Fractional Causation in Figure 16-A. Instead, we will say that the true Fractional Causation is far more likely to be near 100% than to be a low percentage. The dose-response in Part 2j is highly significant.

● Part 2. How the Dose-Response Develops, 1921-1940

● - Part 2a.	1921 PhysPop	1940 MortRate	Respiratory-System Ca, Males Regression Output:	
Pacific	165.11	12.0	Constant	-6.1174
New England	142.24	13.5	Std Err of Y Est	3.4686
West North Central	140.93	7.7	R Squared	0.2466
Mid-Atlantic	137.29	17.1	No. of Observations	9
East North Central	136.06	10.6	Degrees of Freedom	7
Mountain	135.38	7.8		
West South Central	125.15	7.6	X Coefficient(s)	0.1192
East South Central	119.76	4.9	Std Err of Coef.	0.0788
South Atlantic	110.32	8.3	Coefficient / S.E.	1.5136

● - Part 2b.	1923 PhysPop	1940 MortRate	Respiratory-System Ca, Males Regression Output:	
Pacific	163.06	12.0	Constant	-6.9754
New England	137.39	13.5	Std Err of Y Est	3.2642
West North Central	138.31	7.7	R Squared	0.3328
Mid-Atlantic	138.92	17.1	No. of Observations	9
East North Central	131.82	10.6	Degrees of Freedom	7
Mountain	130.51	7.8		
West South Central	119.16	7.6	X Coefficient(s)	0.1291
East South Central	113.16	4.9	Std Err of Coef.	0.0691
South Atlantic	106.79	8.3	Coefficient / S.E.	1.8685

● - Part 2c.	1925 PhysPop	1940 MortRate	Respiratory-System Ca, Males Regression Output:	
Pacific	161.67	12.0	Constant	-6.2251
New England	138.31	13.5	Std Err of Y Est	3.1543
West North Central	133.92	7.7	R Squared	0.3769
Mid-Atlantic	134.36	17.1	No. of Observations	9
East North Central	127.54	10.6	Degrees of Freedom	7
Mountain	122.30	7.8		
West South Central	112.83	7.6	X Coefficient(s)	0.1275
East South Central	107.22	4.9	Std Err of Coef.	0.0619
South Atlantic	103.61	8.3	Coefficient / S.E.	2.0578

● - Part 2d.	1927 PhysPop	1940 MortRate	Respiratory-System Ca, Males Regression Output:	
Pacific	157.83	12.0	Constant	-7.7067
New England	137.50	13.5	Std Err of Y Est	2.8082
West North Central	131.54	7.7	R Squared	0.5062
Mid-Atlantic	138.40	17.1	No. of Observations	9
East North Central	126.18	10.6	Degrees of Freedom	7

Mountain	118.75	7.8		
West South Central	108.25	7.6	X Coefficient(s)	0.1415
East South Central	102.07	4.9	Std Err of Coef.	0.0528
South Atlantic	102.13	8.3	Coefficient / S.E.	2.6786

....

| ● – Part 2e. | 1929 | 1940 | Respiratory–System Ca, Males | |
	PhysPop	MortRate	Regression Output:	
Pacific	156.64	12.0	Constant	–7.6693
New England	138.46	13.5	Std Err of Y Est	2.6878
West North Central	128.72	7.7	R Squared	0.5476
Mid–Atlantic	138.49	17.1	No. of Observations	9
East North Central	126.51	10.6	Degrees of Freedom	7
Mountain	118.68	7.8		
West South Central	105.60	7.6	X Coefficient(s)	0.1424
East South Central	99.41	4.9	Std Err of Coef.	0.0489
South Atlantic	100.86	8.3	Coefficient / S.E.	2.9109

....

| ● – Part 2f. | 1931 | 1940 | Respiratory–System Ca, Males | |
	PhysPop	MortRate	Regression Output:	
Pacific	159.97	12.0	Constant	–6.7703
New England	142.35	13.5	Std Err of Y Est	2.5697
West North Central	126.50	7.7	R Squared	0.5865
Mid–Atlantic	140.82	17.1	No. of Observations	9
East North Central	128.59	10.6	Degrees of Freedom	7
Mountain	118.89	7.8		
West South Central	105.95	7.6	X Coefficient(s)	0.1344
East South Central	96.73	4.9	Std Err of Coef.	0.0426
South Atlantic	99.59	8.3	Coefficient / S.E.	3.1510

....

| ● – Part 2g. | 1934 | 1940 | Respiratory–System Ca, Males | |
	PhysPop	MortRate	Regression Output:	
Pacific	160.09	12.0	Constant	–6.2805
New England	148.60	13.5	Std Err of Y Est	2.1708
West North Central	125.96	7.7	R Squared	0.7049
Mid–Atlantic	149.62	17.1	No. of Observations	9
East North Central	129.36	10.6	Degrees of Freedom	7
Mountain	117.16	7.8		
West South Central	104.68	7.6	X Coefficient(s)	0.1297
East South Central	92.00	4.9	Std Err of Coef.	0.0317
South Atlantic	98.41	8.3	Coefficient / S.E.	4.0891

....

| ● – Part 2h. | 1936 | 1940 | Respiratory–System Ca, Males | |
	PhysPop	MortRate	Regression Output:	
Pacific	158.44	12.0	Constant	–6.3345
New England	150.18	13.5	Std Err of Y Est	1.9653
West North Central	126.14	7.7	R Squared	0.7581
Mid–Atlantic	155.05	17.1	No. of Observations	9
East North Central	130.42	10.6	Degrees of Freedom	7
Mountain	119.80	7.8		
West South Central	103.52	7.6	X Coefficient(s)	0.1294
East South Central	89.94	4.9	Std Err of Coef.	0.0276
South Atlantic	99.16	8.3	Coefficient / S.E.	4.6842

....

| ● – Part 2i. | 1938 | 1940 | Respiratory–System Ca, Males | |
	PhysPop	MortRate	Regression Output:	
Pacific	157.62	12.0	Constant	–5.9557
New England	154.08	13.5	Std Err of Y Est	1.7390
West North Central	124.95	7.7	R Squared	0.8106
Mid–Atlantic	160.69	17.1	No. of Observations	9
East North Central	131.98	10.6	Degrees of Freedom	7
Mountain	119.88	7.8		
West South Central	102.79	7.6	X Coefficient(s)	0.1256
East South Central	88.21	4.9	Std Err of Coef.	0.0229
South Atlantic	99.26	8.3	Coefficient / S.E.	5.4740

....

● – Part 2j.	1940 PhysPop	1940 MortRate	Respiratory-System Ca, Males Regression Output:	
Pacific	159.72	12.0	Constant	-5.1002
New England	161.55	13.5	Std Err of Y Est	1.4558
West North Central	123.14	7.7	R Squared	0.8673
Mid–Atlantic	169.76	17.1	No. of Observations	9
East North Central	133.36	10.6	Degrees of Freedom	7
Mountain	119.89	7.8		
West South Central	103.94	7.6	X Coefficient(s)	0.1169
East South Central	85.83	4.9	Std Err of Coef.	0.0173
South Atlantic	100.74	8.3	Coefficient / S.E.	6.7636

Box 1 of Chap. 16
Summary: Regression Outputs, Respiratory-System Cancers, Males.

Below are the summary-results from regressing the 1940 cancer MortRates upon the ten sets of PhysPops (1921-1940), as presented in Parts 2a-2j of this chapter.

Part	PhysPop	R-squared	Constant	X-Coef	Std Err	X-Coef/SE
2a	1921	0.2466	-6.12	0.1192	0.0788	1.5136
2b	1923	0.3328	-6.98	0.1291	0.0691	1.8685
2c	1925	0.3769	-6.23	0.1275	0.0619	2.0578
2d	1927	0.5062	-7.71	0.1415	0.0528	2.6786
2e	1929	0.5476	-7.67	0.1424	0.0489	2.9109
2f	1931	0.5865	-6.77	0.1344	0.0426	3.1510
2g	1934	0.7049	-6.28	0.1297	0.0317	4.0891
2h	1936	0.7581	-6.33	0.1294	0.0276	4.6842
2i	1938	0.8106	-5.96	0.1256	0.0229	5.4740
2j --->	1940 Max	0.8673	-5.10	0.1169	0.0173	6.7636

Box 2 of Chap. 16
Input-Data for Figure 16-A. Respiratory-System Cancers. Males.

Part 2j, Best-Fit Equation: Calc. MortRate = (0.1169 * PhysPop) + (-5.10)

Census Divisions	1940 Observed PhysPops	1940 Observed MortRates	Best-Fit Calc. MortRates
Pacific	159.72	12.0	13.571
New England	161.55	13.5	13.785
West No. Central	123.14	7.7	9.295
Mid–Atlantic	169.76	17.1	14.745
East No. Central	133.36	10.6	10.490
Mountain	119.89	7.8	8.915
West So. Central	103.94	7.6	7.051
East So. Central	85.83	4.9	4.934
South Atlantic	100.74	8.3	6.677
Additional PhysPops	70.00		3.083
--- not "observed" ---	60.00		1.914
down to zero PhysPop	50.00		0.745
(zero medical radiation).	40.00		-0.424
For each, we calculate	30.00		-1.593
a best-fit MortRate.	20.00		-2.762
These additional x,y pairs	10.00		-3.931
are also part of the	0		-5.100
best-fit line (Chap 5, Part 5e).			

Box 3 of Chap. 16
Presumptive Fraction of Cancer MortRate Attributable to Medical Radiation.

Please see text in Chapter 6, Parts 4 and 6.

Respiratory-System Cancers. MALES.

- MALE National MortRate (MR) 1940, from Table 16-B   11.0   National MortRate
- Constant, from regression, Part 2j   −5.1002   Constant
- Fractional Causation, Best Est. = (Natl MR − Constant) / Natl MR   146.4% # Frac. Causation
  # The Upper-Limit is 100%. Negative Constants produce values > 100%. See Chapter 22, Part 3.

··

90% Confidence-Limits (C.L.) on Fractional Causation. See text in Chapter 6, Part 4b, please.

X-Coefficient, from Part 2j   0.1169   X-Coef., Best Est.
Standard Error (SE) of X-Coefficient, from Part 2j   0.0173   Standard Error

Upper 90% C.L. on X-Coef. = (Coef) + (1.645 * SE) =   0.1454   New X-Coefficient
New Constant = (Natl MR) − (New X-Coef * 1940 Natl PhysPop) =   −8.1931   New Constant
Frac. Causation, High-Limit = (Natl MR − New Constant) / Natl MR =   174.5% # New Frac. Caus'n.
  # The Upper-Limit is 100%. Negative Constants produce values > 100%. See Chapter 22, Part 3.

Lower 90% C.L. on X-Coef. = (Coef) − (1.645 * SE) =   0.0884   New X-Coefficient
New Constant = (Natl MR) − (New X-Coef * 1940 Natl PhysPop) =   −0.6778   New Constant
Frac. Causation, Low-Limit = (Natl MR − New Constant) / Natl MR =   106.2% # New Frac. Caus'n.
  # The Upper-Limit is 100%. Negative Constants produce values > 100%. See Chapter 22, Part 3.

Box 4 of Chap. 16
Error-Check on Our Own Work: Respiratory-System Cancer, Males.

Please see text in Chapter 6, Part 5.

Below, Columns A, C, and E come directly from the regression input in Part 2j. Column B, the fraction of the whole 1940 population in each Census Division, comes from Table 3-B in Chapter 3. Each Column-D entry is the product of (B-entry times C-entry). Each Column-F entry is the product of (B-entry times E-entry). PhysPops and MortRates are each "per 100,000."

The Weighted-Avg. Nat'l PhysPop, 1940, is the sum of Column-D entries =   132.04

The Weighted-Avg. Nat'l Male MortRate, 1940, is sum of Col.F entries =   10.79
The Nat'l Male MortRate is also (X-Coef * Nat'l PhysPop) + Constant =   10.34
Comparison: The Nat'l Male MortRate, 1940, in Table 16-B =   11.00

(A) Census Division	(B) Pop'n Fraction	(C) PhysPop 1940	(D) Weighted PhysPop	(E) MortRate 1940	(F) Weighted MortRate
Pacific	0.0739	159.72	11.80	12.0	0.89
New England	0.0641	161.55	10.36	13.5	0.87
West No. Central	0.1027	123.14	12.65	7.7	0.79
Mid-Atlantic	0.2092	169.76	35.51	17.1	3.58
East No. Central	0.2022	133.36	26.97	10.6	2.14
Mountain	0.0315	119.89	3.78	7.8	0.25
West So. Central	0.0992	103.94	10.31	7.6	0.75
East So. Central	0.0819	85.83	7.03	4.9	0.40
South Atlantic	0.1354	100.74	13.64	8.3	1.12
Sums	1.0000		132.04		10.79

1940 Respiratory Cancer Mortality–Rates versus
1940 PhysPop Values for the 9 Census Divisions, USA.
Dose–Response Relationship

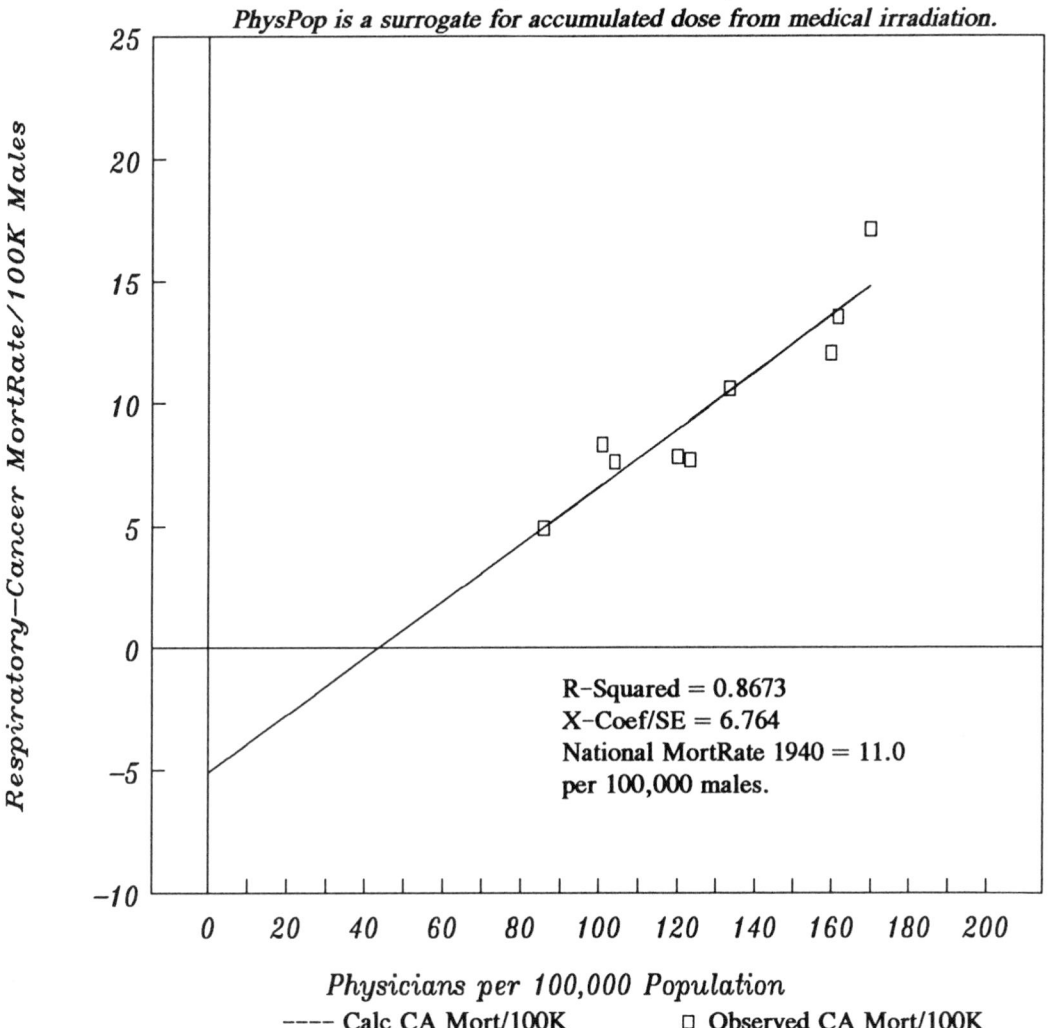

PhysPop is a surrogate for accumulated dose from medical irradiation.

R–Squared = 0.8673
X–Coef/SE = 6.764
National MortRate 1940 = 11.0
per 100,000 males.

Physicians per 100,000 Population
---- Calc CA Mort/100K   □ Observed CA Mort/100K

On the X–axis, PhysPop values = Physicians per 100,000 Population in the Nine Census Divisions of the USA Population, Year 1940. This variable is a surrogate for accumulated radiation dose --- the more physicians per 100,000 people, the more radiation procedures are done per 100,000 people.

On the Y–axis, Respiratory Cancer Mortality–Rate per 100,000 males = the reported rates in USA Vital Statistics for the Nine Census Divisions, Year 1940.

Shown above is the strongest relationship between these two variables (Part 2j). The nine datapoints (boxy symbols) were collected long ago for other purposes, and are free from potential bias with respect to this dose–response study.

Fractional Causation of Respiratory Cancer Mortality–Rate (Male) by Medical Radiation: ∼100 % is far more likely than a low percent. See Text, Part 1.

Table 16-A.
Respiratory-System Cancer MortRates by Census Divisions: Males.

Rates are annual deaths per 100,000 male population, USA, age-adjusted to the 1940 reference year. There are no exclusions by color or "race." Sources are stated in Table 16-B, and described in Chap. 4, Part 2. The Nine Census-Division MortRates are population-weighted (Chap. 4, Part 2b). The averages below them are not.

Census Division	1940	1950	1960	1970	1980	1988
Pacific	12.0	21.1	34.9	44.2	53.5	50.7
New England	13.5	23.6	38.1	47.7	57.3	56.3
West North Central	7.7	16.5	28.4	41.1	53.7	56.2
Mid-Atlantic	17.1	28.4	40.6	49.5	58.4	57.5
East North Central	10.6	21.8	35.7	48.6	61.4	62.3
Mountain	7.8	16.7	25.5	34.6	43.6	44.2
West South Central	7.6	19.0	34.9	48.9	62.8	67.9
East South Central	4.9	14.7	29.0	49.9	70.8	79.1
South Atlantic	8.3	19.8	35.7	50.5	65.2	68.5
Average, ALL	9.9	20.2	33.6	46.1	58.5	60.3
Average, High-5	12.2	22.3	35.5	46.2	56.9	56.6
Average, Low-4	7.2	17.6	31.3	45.9	60.6	64.9
Ratio, Hi5/Lo4	1.70	1.27	1.14	1.01	0.94	0.87

Table 16-B.
Respiratory-System Cancer Mortality Rates, USA National.

Rates are age-adjusted to the 1940 reference year. Both sexes: Deaths per 100,000 population (males + females). Males: Deaths per 100,000 male population. Females: Deaths per 100,000 female population. No exclusions by color or "race."

	Both Sexes	Male	Female
1940	7.2	11.0	3.3
1950	13.0	21.6	4.6
1960	19.5	35.2	5.3
1970	28.4	47.3	11.7
1979-81	36.1	59.4	18.0
1987-89	--	59.7	24.5

● - 1940, 1950, 1960: All rates come from Grove 1968, Table 67, p.686, "Malignant neoplasm of respiratory system, not specified as secondary (160-164)," ICD/7.
  ● - 1970: All rates by Divisions are interpolations (Chap. 4, Parts 2b, 2c), except that the 1970 National "Both Sexes" rate comes from PHS 1995, Table 30, p.110.
  ● - 1980: All rates (ICD/9, 160-165) come from the reference NatCtrHS 1980.
  ● - 1990: All rates for 1987-1989 come from Monthly Vital Statistics Vol.41, No.7, December 1992. The 1988 rates are an acceptable approximation for 1990 (Chap.4, Part 2b.)

● Part 1. Introduction

Respiratory-System Cancers include cancers of the larynx, bronchus and trachea, of lung specified as primary, of lung unspecified as to whether primary or secondary, and of other parts of the respiratory-system (see Chapter 4, Part 5, Number 7). Although the 1940 female MortRates present a severe "small numbers problem," we analyze these data here because the "small numbers" will not persist --- as shown in Table 17-A.

● Part 2. How the Dose-Response Develops, 1921-1940

● - Part 2a.	1921 PhysPop	1940 MortRate	Respiratory-System Ca, Females Regression Output:	
Pacific	165.11	3.8	Constant	-1.4119
New England	142.24	4.1	Std Err of Y Est	0.5265
West North Central	140.93	3.1	R Squared	0.5358
Mid-Atlantic	137.29	4.2	No. of Observations	9
East North Central	136.06	3.2	Degrees of Freedom	7
Mountain	135.38	2.9		
West South Central	125.15	2.4	X Coefficient(s)	0.0340
East South Central	119.76	2.4	Std Err of Coef.	0.0120
South Atlantic	110.32	2.4	Coefficient / S.E.	2.8427

● - Part 2b.	1923 PhysPop	1940 MortRate	Respiratory-System Ca, Females Regression Output:	
Pacific	163.06	3.8	Constant	-1.2648
New England	137.39	4.1	Std Err of Y Est	0.4823
West North Central	138.31	3.1	R Squared	0.6104
Mid-Atlantic	138.92	4.2	No. of Observations	9
East North Central	131.82	3.2	Degrees of Freedom	7
Mountain	130.51	2.9		
West South Central	119.16	2.4	X Coefficient(s)	0.0338
East South Central	113.16	2.4	Std Err of Coef.	0.0102
South Atlantic	106.79	2.4	Coefficient / S.E.	3.3120

● - Part 2c.	1925 PhysPop	1940 MortRate	Respiratory-System Ca, Females Regression Output:	
Pacific	161.67	3.8	Constant	-0.9861
New England	138.31	4.1	Std Err of Y Est	0.4473
West North Central	133.92	3.1	R Squared	0.6649
Mid-Atlantic	134.36	4.2	No. of Observations	9
East North Central	127.54	3.2	Degrees of Freedom	7
Mountain	122.30	2.9		
West South Central	112.83	2.4	X Coefficient(s)	0.0327
East South Central	107.22	2.4	Std Err of Coef.	0.0088
South Atlantic	103.61	2.4	Coefficient / S.E.	3.7266

● - Part 2d.	1927 PhysPop	1940 MortRate	Respiratory-System Ca, Females Regression Output:	
Pacific	157.83	3.8	Constant	-1.0503
New England	137.50	4.1	Std Err of Y Est	0.3685
West North Central	131.54	3.1	R Squared	0.7726
Mid-Atlantic	138.40	4.2	No. of Observations	9
East North Central	126.18	3.2	Degrees of Freedom	7
Mountain	118.75	2.9		
West South Central	108.25	2.4	X Coefficient(s)	0.0338
East South Central	102.07	2.4	Std Err of Coef.	0.0069
South Atlantic	102.13	2.4	Coefficient / S.E.	4.8768

● – Part 2e.	1929	1940	Respiratory-System Ca, Females	
	PhysPop	MortRate	Regression Output:	
Pacific	156.64	3.8	Constant	−0.9727
New England	138.46	4.1	Std Err of Y Est	0.3379
West North Central	128.72	3.1	R Squared	0.8088
Mid-Atlantic	138.49	4.2	No. of Observations	9
East North Central	126.51	3.2	Degrees of Freedom	7
Mountain	118.68	2.9		
West South Central	105.60	2.4	X Coefficient(s)	0.0335
East South Central	99.41	2.4	Std Err of Coef.	0.0061
South Atlantic	100.86	2.4	Coefficient / S.E.	5.4408

....

● – Part 2f.	1931	1940	Respiratory-System Ca, Females	
	PhysPop	MortRate	Regression Output:	
Pacific	159.97	3.8	Constant	−0.6736
New England	142.35	4.1	Std Err of Y Est	0.3206
West North Central	126.50	3.1	R Squared	0.8279
Mid-Atlantic	140.82	4.2	No. of Observations	9
East North Central	128.59	3.2	Degrees of Freedom	7
Mountain	118.89	2.9		
West South Central	105.95	2.4	X Coefficient(s)	0.0309
East South Central	96.73	2.4	Std Err of Coef.	0.0053
South Atlantic	99.59	2.4	Coefficient / S.E.	5.8033

....

● – Part 2g.	1934	1940	Respiratory-System Ca, Females	
	PhysPop	MortRate	Regression Output:	
Pacific	160.09	3.8	Constant	−0.3885
New England	148.60	4.1	Std Err of Y Est	0.2381
West North Central	125.96	3.1	R Squared	0.9051
Mid-Atlantic	149.62	4.2	No. of Observations	9
East North Central	129.36	3.2	Degrees of Freedom	7
Mountain	117.16	2.9		
West South Central	104.68	2.4	X Coefficient(s)	0.0284
East South Central	92.00	2.4	Std Err of Coef.	0.0035
South Atlantic	98.41	2.4	Coefficient / S.E.	8.1685

....

● – Part 2h.	1936	1940	Respiratory-System Ca, Females	
	PhysPop	MortRate	Regression Output:	
Pacific	158.44	3.8	Constant	−0.3215
New England	150.18	4.1	Std Err of Y Est	0.2032
West North Central	126.14	3.1	R Squared	0.9309
Mid-Atlantic	155.05	4.2	No. of Observations	9
East North Central	130.42	3.2	Degrees of Freedom	7
Mountain	119.80	2.9		
West South Central	103.52	2.4	X Coefficient(s)	0.0277
East South Central	89.94	2.4	Std Err of Coef.	0.0029
South Atlantic	99.16	2.4	Coefficient / S.E.	9.7080

....

● – Part 2i.	1938	1940	Respiratory-System Ca, Females	
	PhysPop	MortRate	Regression Output:	
Pacific	157.62	3.8	Constant	−0.1666
New England	154.08	4.1	Std Err of Y Est	0.1681
West North Central	124.95	3.1	R Squared	0.9527
Mid-Atlantic	160.69	4.2	No. of Observations	9
East North Central	131.98	3.2	Degrees of Freedom	7
Mountain	119.88	2.9		
West South Central	102.79	2.4	X Coefficient(s)	0.0263
East South Central	88.21	2.4	Std Err of Coef.	0.0022
South Atlantic	99.26	2.4	Coefficient / S.E.	11.8728

....

● - Part 2j.

	1940 PhysPop	1940 MortRate
Pacific	159.72	3.8
New England	161.55	4.1
West North Central	123.14	3.1
Mid-Atlantic	169.76	4.2
East North Central	133.36	3.2
Mountain	119.89	2.9
West South Central	103.94	2.4
East South Central	85.83	2.4
South Atlantic	100.74	2.4

Respiratory-System Ca, Females
Regression Output:

Constant	0.1019
Std Err of Y Est	0.1496
R Squared	0.9625
No. of Observations	9
Degrees of Freedom	7
X Coefficient(s)	0.0238
Std Err of Coef.	0.0018
Coefficient / S.E.	13.4046

Box 1 of Chap. 17
Summary: Regression Outputs, Respiratory-System Cancers, Females.

Below are the summary-results from regressing the 1940 cancer MortRates upon the ten sets of PhysPops (1921-1940), as presented in Parts 2a-2j of this chapter.

Part	PhysPop	R-squared	Constant	X-Coef	Std Err	X-Coef/SE
2a	1921	0.5358	-1.41	0.0340	0.0120	2.8427
2b	1923	0.6104	-1.26	0.0338	0.0102	3.3120
2c	1925	0.6649	-0.99	0.0327	0.0088	3.7266
2d	1927	0.7726	-1.05	0.0338	0.0069	4.8768
2e	1929	0.8088	-0.97	0.0335	0.0061	5.4408
2f	1931	0.8279	-0.67	0.0309	0.0053	5.8033
2g	1934	0.9051	-0.39	0.0284	0.0035	8.1685
2h	1936	0.9309	-0.32	0.0277	0.0029	9.7080
2i	1938	0.9527	-0.17	0.0263	0.0022	11.8728
2j --->	1940 Max	0.9625	0.10	0.0238	0.0018	13.4046

Box 2 of Chap. 17
Input-Data for Figure 17-A. Respiratory-System Cancers. Females.

Part 2j, Best-Fit Equation: Calc. MortRate = (0.0238 * PhysPop) + (0.10)

Census Divisions	1940 Observed PhysPops	1940 Observed MortRates	Best-Fit Calc. MortRates
Pacific	159.72	3.8	3.901
New England	161.55	4.1	3.945
West No. Central	123.14	3.1	3.031
Mid-Atlantic	169.76	4.2	4.140
East No. Central	133.36	3.2	3.274
Mountain	119.89	2.9	2.953
West So. Central	103.94	2.4	2.574
East So. Central	85.83	2.4	2.143
South Atlantic	100.74	2.4	2.498
Additional PhysPops	70.00		1.766
--- not "observed" ---	60.00		1.528
down to zero PhysPop	50.00		1.290
(zero medical radiation).	40.00		1.052
For each, we calculate	30.00		0.814
a best-fit MortRate.	20.00		0.576
These additional x,y pairs	10.00		0.338
are also part of the	0		0.100
best-fit line (Chap 5, Part 5e).			

Box 3 of Chap. 17
Presumptive Fraction of Cancer MortRate Attributable to Medical Radiation.

Please see text in Chapter 6, Parts 4 and 6.

Respiratory-System Cancers. FEMALES.

- FEMALE National MortRate (MR) 1940, from Table 17-B   3.3   National MortRate
- Constant, from regression, Part 2j   0.1019   Constant
- Fractional Causation, Best Est. = (Natl MR – Constant) / Natl MR   96.9%   Frac. Causation

....

90% Confidence-Limits (C.L.) on Fractional Causation. See text in Chapter 6, Part 4b, please.

X-Coefficient, from Part 2j	0.0238	X-Coef., Best Est.
Standard Error (SE) of X-Coefficient, from Part 2j	0.0018	Standard Error

Upper 90% C.L. on X-Coef. = (Coef) + (1.645 * SE) =   0.0268   New X-Coefficient
New Constant = (Natl MR) – (New X-Coef * 1940 Natl PhysPop) =   –0.2335   New Constant
Frac. Causation, High-Limit = (Natl MR – New Constant) / Natl MR =   107.1% #   New Frac. Caus'n.
   # The Upper-Limit is 100%. Negative Constants produce values > 100%. See Chapter 22, Part 3.

Lower 90% C.L. on X-Coef. = (Coef) – (1.645 * SE) =   0.0208   New X-Coefficient
New Constant = (Natl MR) – (New X-Coef * 1940 Natl PhysPop) =   0.5484   New Constant
Frac. Causation, Low-Limit = (Natl MR – New Constant) / Natl MR =   83.4%   New Frac. Caus'n.

Box 4 of Chap. 17
Error-Check on Our Own Work: Respiratory-System Cancer, Females.

Please see text in Chapter 6, Part 5.

Below, Columns A, C, and E come directly from the regression input in Part 2j. Column B, the fraction of the whole 1940 population in each Census Division, comes from Table 3-B in Chapter 3. Each Column-D entry is the product of (B-entry times C-entry). Each Column-F entry is the product of (B-entry times E-entry). PhysPops and MortRates are each "per 100,000."

The Weighted-Avg. Nat'l PhysPop, 1940, is the sum of Column-D entries =   132.04

The Weighted-Avg. Nat'l Female MortRate, 1940, is sum of Col.F entries =   3.24
The Nat'l Female MortRate is also (X-Coef * Nat'l PhysPop) + Constant =   3.24
Comparison: The Nat'l Female MortRate, 1940, in Table 17-B =   3.30

(A) Census Division	(B) Pop'n Fraction	(C) PhysPop 1940	(D) Weighted PhysPop	(E) MortRate 1940	(F) Weighted MortRate
Pacific	0.0739	159.72	11.80	3.8	0.28
New England	0.0641	161.55	10.36	4.1	0.26
West No. Central	0.1027	123.14	12.65	3.1	0.32
Mid-Atlantic	0.2092	169.76	35.51	4.2	0.88
East No. Central	0.2022	133.36	26.97	3.2	0.65
Mountain	0.0315	119.89	3.78	2.9	0.09
West So. Central	0.0992	103.94	10.31	2.4	0.24
East So. Central	0.0819	85.83	7.03	2.4	0.20
South Atlantic	0.1354	100.74	13.64	2.4	0.32
Sums	1.0000		132.04		3.24

1940 Respiratory Cancer Mortality–Rates versus
1940 PhysPop Values for the 9 Census Divisions, USA.
Dose–Response Relationship

PhysPop is a surrogate for accumulated dose from medical irradiation

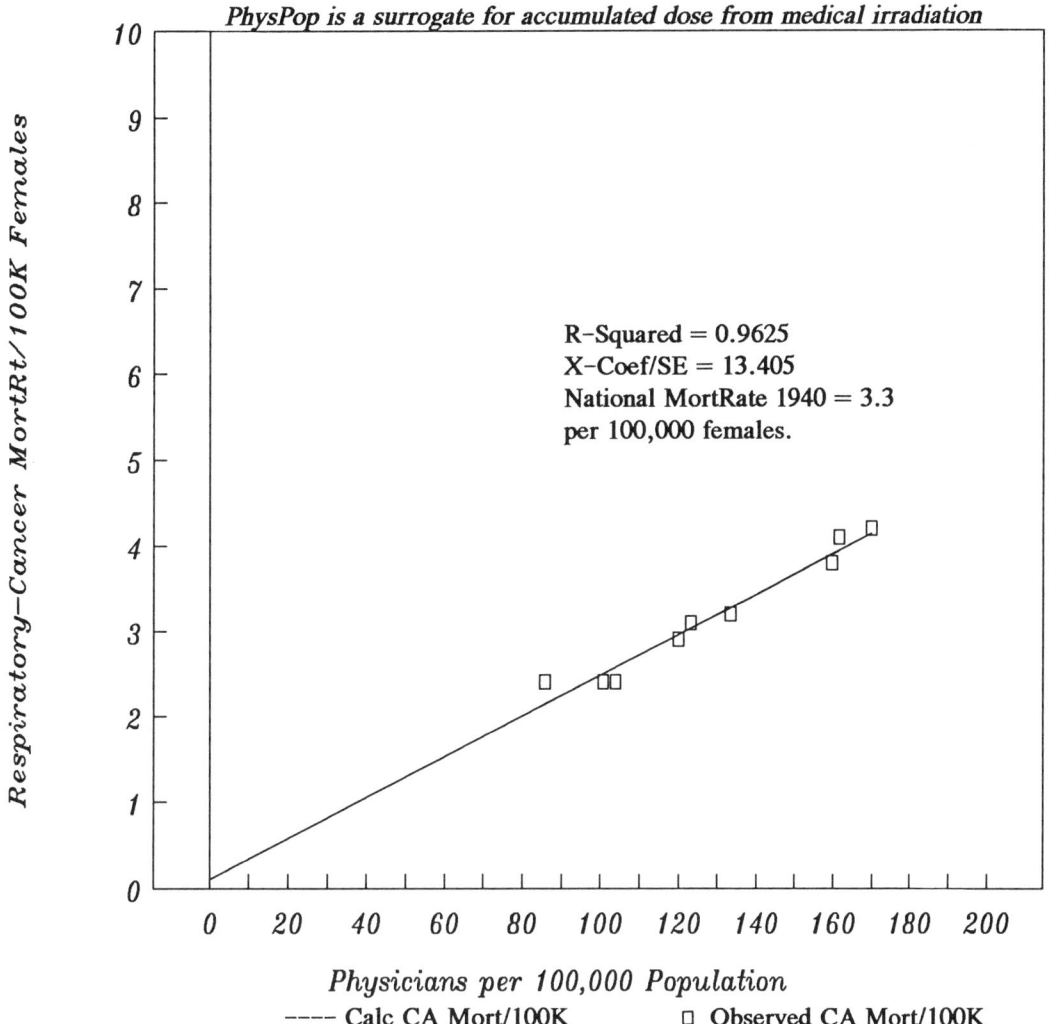

R-Squared = 0.9625
X-Coef/SE = 13.405
National MortRate 1940 = 3.3
per 100,000 females.

Physicians per 100,000 Population

---- Calc CA Mort/100K   □ Observed CA Mort/100K

On the X–axis, PhysPop values = Physicians per 100,000 Population in the Nine Census Divisions of the USA Population, Year 1940. This variable is a surrogate for accumulated radiation dose --- the more physicians per 100,000 people, the more radiation procedures are done per 100,000 people.

On the Y–axis, Respiratory Cancer Mortality–Rate per 100,000 females = the reported rates in USA Vital Statistics for the Nine Census Divisions, Year 1940.

Shown above is the strongest relationship between these two variables (Part 2j). The nine datapoints (boxy symbols) were collected long ago for other purposes, and are free from potential bias with respect to this dose–response study. Fractional causation is (Natl MortRate minus the Y–intercept) / (Natl MortRate).

Fractional Causation of Respiratory Cancer Mort–Rate (Female) by Medical Rad'n
97 % from Best Estimate (Box 3).

83 % at lower 90 % confidence limit (Box 3).   ∼100 % at upper 90 % confidence limit (Box 3).

Table 17-A.
Respiratory-System Cancer MortRates by Census Divisions: Females.

Rates are annual deaths per 100,000 female population, USA, age-adjusted to the 1940 reference year. There are no exclusions by color or "race." Sources are stated in Table 17-B, and described in Chap. 4, Part 2. The Nine Census-Division MortRates are population-weighted (Chap. 4, Part 2b). The averages below them are not.

Census Division	1940	1950	1960	1970	1980	1988
Pacific	3.8	4.4	5.9	13.6	21.2	27.8
New England	4.1	4.1	5.6	12.1	18.5	26.9
West North Central	3.1	4.8	4.4	9.7	15.0	23.1
Mid-Atlantic	4.2	5.0	6.0	12.3	18.5	25.8
East North Central	3.2	4.5	5.1	11.6	18.1	26.4
Mountain	2.9	4.2	4.1	9.5	14.9	22.2
West South Central	2.4	4.3	5.2	11.3	17.3	26.6
East South Central	2.4	4.7	4.7	10.9	17.0	26.6
South Atlantic	2.4	4.7	5.0	11.5	17.9	26.6
Average, ALL	3.2	4.5	5.1	11.4	17.6	25.8
Average, High-5	3.7	4.6	5.4	11.8	18.3	26.0
Average, Low-4	2.5	4.5	4.8	10.8	16.8	25.5
Ratio, Hi5/Lo4	1.46	1.02	1.14	1.10	1.09	1.02

● - 1940: Although the MortRates for WestSoCentral, EastSoCentral, and SouthAtlantic are identical, they are truly the entries for these Census Divisions in Grove 1968, Table 67, page 687.

● - 1950: These entries are such that the Hi5/Lo4 Ratio suddenly drops from 1.46 in 1940 to 1.02 in 1950. This seems unlikely to be correct, and may result from random fluctuations in small numbers, or from reporting-errors. On the other hand, the values may be accurate. In any case, the official values have been copied correctly by us from Grove 1968.

● - 1988: Although the MortRates for WestSoCentral, EastSoCentral, and SouthAtlantic are identical (again), we have double-checked the state-values from the government, as well as our own calculations which combined these various state-values into Census Divisions. We find no errors.

Table 17-B.
Respiratory-System Cancer Mortality Rates, USA National.

Rates are age-adjusted to the 1940 reference year. Both sexes: Deaths per 100,000 population (males + females). Males: Deaths per 100,000 male population. Females: Deaths per 100,000 female population. No exclusions by color or "race."
color or "race."

	Both Sexes	Male	Female
1940	7.2	11.0	3.3
1950	13.0	21.6	4.6
1960	19.5	35.2	5.3
1970	28.4	47.3	11.7
1979-81	36.1	59.4	18.0
1987-89	--	59.7	24.5

● - 1940, 1950, 1960: All rates come from Grove 1968, Table 67, p.686, "Malignant neoplasm of respiratory system, not specified as secondary (160-164)," ICD/7.

● - 1970: All rates by Divisions are interpolations (Chap. 4, Parts 2b, 2c), except that the 1970 National "Both Sexes" rate comes from PHS 1995, Table 30, p.110.

● - 1980: All rates (ICD/9, 160-165) come from the reference NatCtrHS 1980.

● - 1990: All rates for 1987 -1989 come from Monthly Vital Statistics Vol.41, No.7, December 1992. The 1988 rates are an acceptable approximation for 1990 (Chap.4, Part 2b.)

● Part 1.   Introduction

        Difference-Cancers are All-Cancers-Minus-Respiratory-System Cancers.   The dramatic
increase in Respiratory-System Cancers since 1940 has put such cancers into a class by themselves.
By subtracting Respiratory-System Cancers from All-Cancers, we can observe how all the REST of
the cancers behave.   We are not alone in creating a cancer category for "All-Minus-Respiratory."   The
National Cancer Institute regularly presents an entry for "All Except Lung" in its reports from the
SEER Program (Surveillance, Epidemiology, and End Results Program --- for example, see SEER
1997, p.45).

● Part 2.   How the Dose-Response Develops, 1921-1940

● - Part 2a.	1921	1940	Difference-Cancers, Males	
	PhysPop	MortRate	Regression Output:	
Pacific	165.11	110.9	Constant	-20.9581
New England	142.24	122.0	Std Err of Y Est	14.9455
West North Central	140.93	103.2	R Squared	0.4951
Mid-Atlantic	137.29	123.8	No. of Observations	9
East North Central	136.06	109.0	Degrees of Freedom	7
Mountain	135.38	92.0		
West South Central	125.15	79.3	X Coefficient(s)	0.8894
East South Central	119.76	68.7	Std Err of Coef.	0.3395
South Atlantic	110.32	80.6	Coefficient / S.E.	2.6199

● - Part 2b.	1923	1940	Difference-Cancers, Males	
	PhysPop	MortRate	Regression Output:	
Pacific	163.06	110.9	Constant	-17.8582
New England	137.39	122.0	Std Err of Y Est	13.7707
West North Central	138.31	103.2	R Squared	0.5714
Mid-Atlantic	138.92	123.8	No. of Observations	9
East North Central	131.82	109.0	Degrees of Freedom	7
Mountain	130.51	92.0		
West South Central	119.16	79.3	X Coefficient(s)	0.8907
East South Central	113.16	68.7	Std Err of Coef.	0.2916
South Atlantic	106.79	80.6	Coefficient / S.E.	3.0546

● - Part 2c.	1925	1940	Difference-Cancers, Males	
	PhysPop	MortRate	Regression Output:	
Pacific	161.67	110.9	Constant	-10.3231
New England	138.31	122.0	Std Err of Y Est	12.9650
West North Central	133.92	103.2	R Squared	0.6200
Mid-Atlantic	134.36	123.8	No. of Observations	9
East North Central	127.54	109.0	Degrees of Freedom	7
Mountain	122.30	92.0		
West South Central	112.83	79.3	X Coefficient(s)	0.8604
East South Central	107.22	68.7	Std Err of Coef.	0.2546
South Atlantic	103.61	80.6	Coefficient / S.E.	3.3798

● - Part 2d.	1927	1940	Difference-Cancers, Males	
	PhysPop	MortRate	Regression Output:	
Pacific	157.83	110.9	Constant	-13.2331
New England	137.50	122.0	Std Err of Y Est	10.7969
West North Central	131.54	103.2	R Squared	0.7365
Mid-Atlantic	138.40	123.8	No. of Observations	9
East North Central	126.18	109.0	Degrees of Freedom	7
Mountain	118.75	92.0		
West South Central	108.25	79.3	X Coefficient(s)	0.8984

East South Central	102.07	68.7	Std Err of Coef.	0.2031
South Atlantic	102.13	80.6	Coefficient / S.E.	4.4232

● – Part 2e.	1929	1940	Difference–Cancers, Males	
	PhysPop	MortRate	Regression Output:	
Pacific	156.64	110.9	Constant	−11.6016
New England	138.46	122.0	Std Err of Y Est	9.9319
West North Central	128.72	103.2	R Squared	0.7770
Mid–Atlantic	138.49	123.8	No. of Observations	9
East North Central	126.51	109.0	Degrees of Freedom	7
Mountain	118.68	92.0		
West South Central	105.60	79.3	X Coefficient(s)	0.8927
East South Central	99.41	68.7	Std Err of Coef.	0.1807
South Atlantic	100.86	80.6	Coefficient / S.E.	4.9390

● – Part 2f.	1931	1940	Difference–Cancers, Males	
	PhysPop	MortRate	Regression Output:	
Pacific	159.97	110.9	Constant	−3.6338
New England	142.35	122.0	Std Err of Y Est	9.5091
West North Central	126.50	103.2	R Squared	0.7956
Mid–Atlantic	140.82	123.8	No. of Observations	9
East North Central	128.59	109.0	Degrees of Freedom	7
Mountain	118.89	92.0		
West South Central	105.95	79.3	X Coefficient(s)	0.8238
East South Central	96.73	68.7	Std Err of Coef.	0.1578
South Atlantic	99.59	80.6	Coefficient / S.E.	5.2199

● – Part 2g.	1934	1940	Difference–Cancers, Males	
	PhysPop	MortRate	Regression Output:	
Pacific	160.09	110.9	Constant	3.6805
New England	148.60	122.0	Std Err of Y Est	7.4329
West North Central	125.96	103.2	R Squared	0.8751
Mid–Atlantic	149.62	123.8	No. of Observations	9
East North Central	129.36	109.0	Degrees of Freedom	7
Mountain	117.16	92.0		
West South Central	104.68	79.3	X Coefficient(s)	0.7606
East South Central	92.00	68.7	Std Err of Coef.	0.1086
South Atlantic	98.41	80.6	Coefficient / S.E.	7.0037

● – Part 2h.	1936	1940	Difference–Cancers, Males	
	PhysPop	MortRate	Regression Output:	
Pacific	158.44	110.9	Constant	4.9134
New England	150.18	122.0	Std Err of Y Est	6.2783
West North Central	126.14	103.2	R Squared	0.9109
Mid–Atlantic	155.05	123.8	No. of Observations	9
East North Central	130.42	109.0	Degrees of Freedom	7
Mountain	119.80	92.0		
West South Central	103.52	79.3	X Coefficient(s)	0.7463
East South Central	89.94	68.7	Std Err of Coef.	0.0882
South Atlantic	99.16	80.6	Coefficient / S.E.	8.4596

● – Part 2i.	1938	1940	Difference–Cancers, Males	
	PhysPop	MortRate	Regression Output:	
Pacific	157.62	110.9	Constant	9.0069
New England	154.08	122.0	Std Err of Y Est	5.4086
West North Central	124.95	103.2	R Squared	0.9339
Mid–Atlantic	160.69	123.8	No. of Observations	9
East North Central	131.98	109.0	Degrees of Freedom	7
Mountain	119.88	92.0		
West South Central	102.79	79.3	X Coefficient(s)	0.7095
East South Central	88.21	68.7	Std Err of Coef.	0.0714
South Atlantic	99.26	80.6	Coefficient / S.E.	9.9430

● – Part 2j.

	1940 PhysPop	1940 MortRate
Pacific	159.72	110.9
New England	161.55	122.0
West North Central	123.14	103.2
Mid–Atlantic	169.76	123.8
East North Central	133.36	109.0
Mountain	119.89	92.0
West South Central	103.94	79.3
East South Central	85.83	68.7
South Atlantic	100.74	80.6

Difference–Cancers, Males
Regression Output:

Constant	16.6486
Std Err of Y Est	5.3951
R Squared	0.9342
No. of Observations	9
Degrees of Freedom	7
X Coefficient(s)	0.6388
Std Err of Coef.	0.0641
Coefficient / S.E.	9.9695

Box 1 of Chap. 18
Summary: Regression Outputs, "Difference" Cancers, Males.

Below are the summary-results from regressing the 1940 cancer MortRates upon the ten sets of PhysPops (1921-1940), as presented in Parts 2a-2j of this chapter.

Part	PhysPop	R-squared	Constant	X-Coef	Std Err	X-Coef/SE
2a	1921	0.4951	−20.96	0.8894	0.3395	2.6199
2b	1923	0.5714	−17.86	0.8907	0.2916	3.0546
2c	1925	0.6200	−10.32	0.8604	0.2546	3.3798
2d	1927	0.7365	−13.23	0.8984	0.2031	4.4232
2e	1929	0.7770	−11.60	0.8927	0.1807	4.9390
2f	1931	0.7956	−3.63	0.8238	0.1578	5.2199
2g	1934	0.8751	3.68	0.7606	0.1086	7.0037
2h	1936	0.9109	4.91	0.7463	0.0882	8.4596
2i	1938	0.9339	9.01	0.7095	0.0714	9.9430
2j --->	1940 Max	0.9342	16.65	0.6388	0.0641	9.9695

Box 2 of Chap. 18
Input-Data for Figure 18-A. "Difference" Cancers. Males.

Part 2j, Best-Fit Equation: Calc. MortRate = (0.6388 * PhysPop) + (16.65)

Census Divisions	1940 Observed PhysPops	1940 Observed MortRates	Best-Fit Calc. MortRates
Pacific	159.72	110.9	118.679
New England	161.55	122.0	119.848
West No. Central	123.14	103.2	95.312
Mid-Atlantic	169.76	123.8	125.093
East No. Central	133.36	109.0	101.840
Mountain	119.89	92.0	93.236
West So. Central	103.94	79.3	83.047
East So. Central	85.83	68.7	71.478
South Atlantic	100.74	80.6	81.003
Additional PhysPops	70.00		61.366
--- not "observed" ---	60.00		54.978
down to zero PhysPop	50.00		48.590
(zero medical radiation).	40.00		42.202
For each, we calculate	30.00		35.814
a best-fit MortRate.	20.00		29.426
These additional x,y pairs	10.00		23.038
are also part of the	0		16.650
best-fit line (Chap 5, Part 5e).			

Box 3 of Chap. 18
Presumptive Fraction of Cancer MortRate Attributable to Medical Radiation.

Please see text in Chapter 6, Parts 4 and 6.

Difference-Cancers. MALES.

- MALE National MortRate (MR) 1940, from Table 18-B   104.0   National MortRate
- Constant, from regression, Part 2j   16.6486   Constant
- Fractional Causation, Best Est. = (Natl MR − Constant) / Natl MR   84.0%   Frac. Causation

....

   90% Confidence-Limits (C.L.) on Fractional Causation. See text in Chapter 6, Part 4b, please.

X-Coefficient, from Part 2j   0.6388   X-Coef., Best Est.
Standard Error (SE) of X-Coefficient, from Part 2j   0.0641   Standard Error

Upper 90% C.L. on X-Coef. = (Coef) + (1.645 * SE) =   0.7442   New X-Coefficient
New Constant = (Natl MR) − (New X-Coef * 1940 Natl PhysPop) =   5.7300   New Constant
Frac. Causation, High-Limit = (Natl MR − New Constant) / Natl MR =   94.5%   New Frac. Caus'n.

Lower 90% C.L. on X-Coef. = (Coef) − (1.645 * SE) =   0.5334   New X-Coefficient
New Constant = (Natl MR) − (New X-Coef * 1940 Natl PhysPop) =   33.5757   New Constant
Frac. Causation, Low-Limit = (Natl MR − New Constant) / Natl MR =   67.7%   New Frac. Caus'n.

Box 4 of Chap. 18
Error-Check on Our Own Work: "Difference" Cancers, Males.

Please see text in Chapter 6, Part 5.

Below, Columns A, C, and E come directly from the regression input in Part 2j. Column B, the fraction of the whole 1940 population in each Census Division, comes from Table 3-B in Chapter 3. Each Column-D entry is the product of (B-entry times C-entry). Each Column-F entry is the product of (B-entry times E-entry). PhysPops and MortRates are each "per 100,000."

The Weighted-Avg. Nat'l PhysPop, 1940, is the sum of Column-D entries =   132.04

The Weighted-Avg. Nat'l Male MortRate, 1940, is sum of Col.F entries =   101.86
The Nat'l Male MortRate is also (X-Coef * Nat'l PhysPop) + Constant =   101.00
Comparison: The Nat'l Male MortRate, 1940, in Table 18-B =   104.00

(A) Census Division	(B) Pop'n Fraction	(C) PhysPop 1940	(D) Weighted PhysPop	(E) MortRate 1940	(F) Weighted MortRate
Pacific	0.0739	159.72	11.80	110.9	8.20
New England	0.0641	161.55	10.36	122.0	7.82
West No. Central	0.1027	123.14	12.65	103.2	10.60
Mid-Atlantic	0.2092	169.76	35.51	123.8	25.90
East No. Central	0.2022	133.36	26.97	109.0	22.04
Mountain	0.0315	119.89	3.78	92.0	2.90
West So. Central	0.0992	103.94	10.31	79.3	7.87
East So. Central	0.0819	85.83	7.03	68.7	5.63
South Atlantic	0.1354	100.74	13.64	80.6	10.91
Sums	1.0000		132.04		101.86

1940 "Difference" Cancer Mortality-Rates versus 1940 PhysPop Values for the 9 Census Divisions, USA.
Dose–Response Relationship

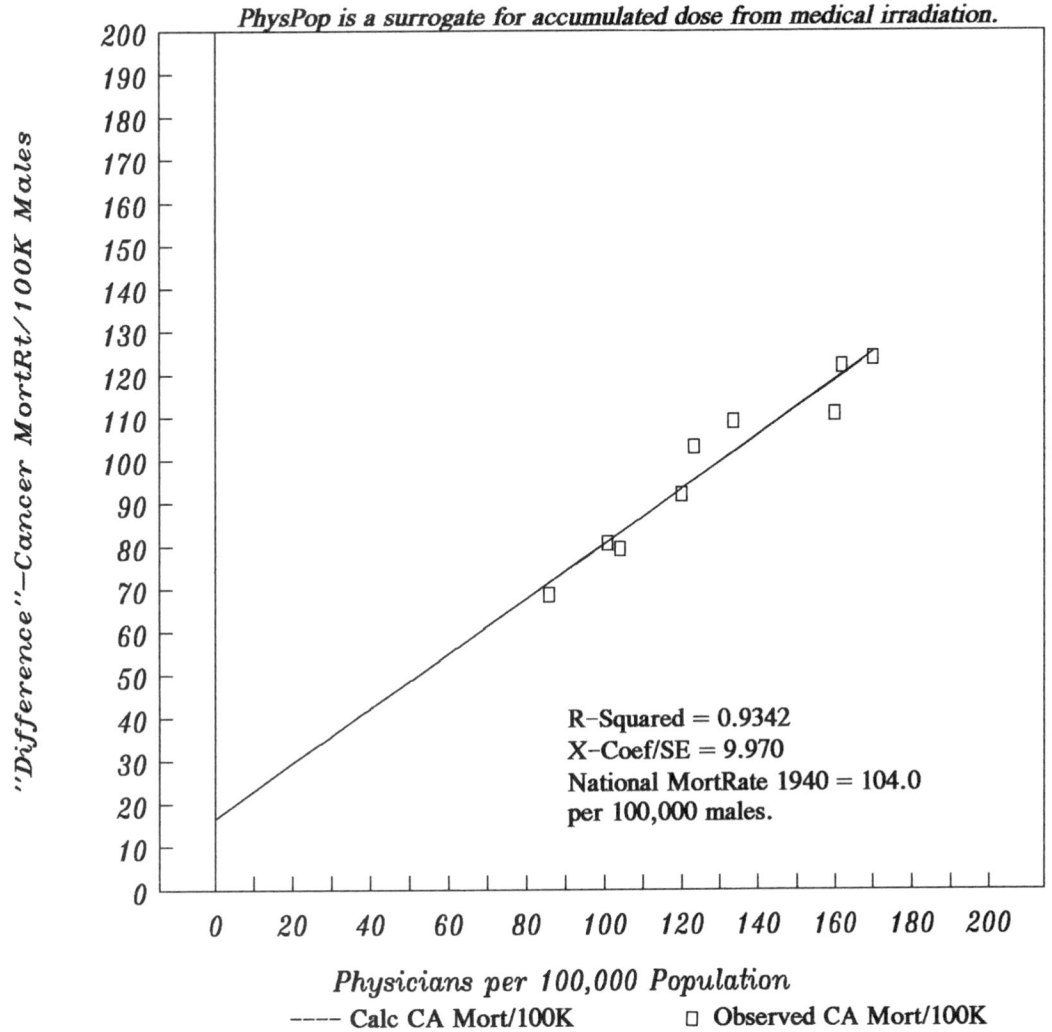

PhysPop is a surrogate for accumulated dose from medical irradiation.

R-Squared = 0.9342
X-Coef/SE = 9.970
National MortRate 1940 = 104.0
per 100,000 males.

Physicians per 100,000 Population

---- Calc CA Mort/100K   □ Observed CA Mort/100K

On the X-axis, PhysPop values = Physicians per 100,000 Population in the Nine Census Divisions of the USA Population, Year 1940. This variable is a surrogate for accumulated radiation dose --- the more physicians per 100,000 people, the more radiation procedures are done per 100,000 people.

On the Y-axis, "Difference" Cancer Mortality-Rate per 100,000 males = the reported rates in USA Vital Statistics for the Nine Census Divisions, Year 1940.

Shown above is the strongest relationship between these two variables (Part 2j). The nine datapoints (boxy symbols) were collected long ago for other purposes, and are free from potential bias with respect to this dose-response study. Fractional causation is (Natl MortRate minus the Y-intercept) / (Natl MortRate).

Fractional Causation of "Difference" Cancer Mort-Rate (Male) by Medical Rad'n
84 % from Best Estimate (Box 3).

68 % at lower 90 % confidence limit (Box 3). ~94 % at upper 90 % confidence limit (Box 3).

Table 18-A.
"Difference" Cancer MortRates by Census Divisions: Males.

"Difference" Cancers are (All-Cancers minus Respiratory-System Cancers). The entries below are the corresponding entries in Table 6-A (All-Cancers, Male) minus the corresponding entries in Table 16-A (Respiratory-System Cancers, Male). Rates are annual deaths per 100,000 male population, USA, age-adjusted to the 1940 reference year. There are no exclusions by color or "race."

Census Division	1940	1950	1960	1970	1980	1988
Pacific	110.9	106.1	105.8	103.0	100.2	97.8
New England	122.0	128.8	126.5	119.8	113.0	110.8
West North Central	103.2	108.8	107.2	102.8	98.3	99.7
Mid-Atlantic	123.8	127.6	123.4	118.4	113.4	110.9
East North Central	109.0	116.5	115.0	111.6	108.1	108.9
Mountain	92.0	91.4	93.2	92.2	91.1	94.9
West South Central	79.3	93.7	98.9	99.5	100.1	105.0
East South Central	68.7	90.0	96.1	99.7	103.3	109.1
South Atlantic	80.6	96.5	101.4	103.7	106.2	107.3
Average, ALL	98.8	106.6	107.5	105.6	103.7	104.9
Average, High-5	113.8	117.6	115.6	111.1	106.6	105.6
Average, Low-4	80.2	92.9	97.4	98.8	100.2	104.1
Ratio, Hi5/Lo4	1.42	1.27	1.19	1.12	1.06	1.01

Table 18-B.
"Difference" Cancer Mortality Rates, USA National.

Annual MortRates in Table 18-B are obtained by subtracting Table 16-B from Table 6-B.

Rates are age-adjusted to the 1940 reference year. Both sexes: Deaths per 100,000 population (males + females). Males: Deaths per 100,000 male population. Females: Deaths per 100,000 female population. No exclusions by color or "race."

	Both Sexes	Male	Female
1940	113.1	104.0	122.8
1950	114.7	111.2	118.6
1960	109.6	110.5	109.6
1970	101.4	107.8	100.0
1979-81	95.8	105.1	90.5
1987-89	--	103.0	86.8

● - Sources are stated in Table 16-B and Table 6-B.

CHAPTER 19

"Difference" Cancers, Females: Relation with Medical Radiation

● Part 1. Introduction

　　　Difference-Cancers are All-Cancers-Minus-Respiratory-System Cancers.　Please see Chapter 18, Part 1.

● Part 2. How the Dose-Response Develops, 1921-1940

● - Part 2a.

	1921 PhysPop	1940 MortRate	Difference-Cancers, Females Regression Output:	
Pacific	165.11	123.6	Constant	34.7966
New England	142.24	141.2	Std Err of Y Est	14.0275
West North Central	140.93	117.0	R Squared	0.3479
Mid-Atlantic	137.29	138.7	No. of Observations	9
East North Central	136.06	128.2	Degrees of Freedom	7
Mountain	135.38	108.9		
West South Central	125.15	97.4	X Coefficient(s)	0.6157
East South Central	119.76	100.1	Std Err of Coef.	0.3186
South Atlantic	110.32	104.5	Coefficient / S.E.	1.9326

● - Part 2b.

	1923 PhysPop	1940 MortRate	Difference-Cancers, Females Regression Output:	
Pacific	163.06	123.6	Constant	36.2380
New England	137.39	141.2	Std Err of Y Est	13.3595
West North Central	138.31	117.0	R Squared	0.4085
Mid-Atlantic	138.92	138.7	No. of Observations	9
East North Central	131.82	128.2	Degrees of Freedom	7
Mountain	130.51	108.9		
West South Central	119.16	97.4	X Coefficient(s)	0.6220
East South Central	113.16	100.1	Std Err of Coef.	0.2829
South Atlantic	106.79	104.5	Coefficient / S.E.	2.1989

● - Part 2c.

	1925 PhysPop	1940 MortRate	Difference-Cancers, Females Regression Output:	
Pacific	161.67	123.6	Constant	38.8866
New England	138.31	141.2	Std Err of Y Est	12.5952
West North Central	133.92	117.0	R Squared	0.4743
Mid-Atlantic	134.36	138.7	No. of Observations	9
East North Central	127.54	128.2	Degrees of Freedom	7
Mountain	122.30	108.9		
West South Central	112.83	97.4	X Coefficient(s)	0.6215
East South Central	107.22	100.1	Std Err of Coef.	0.2473
South Atlantic	103.61	104.5	Coefficient / S.E.	2.5130

● - Part 2d.

	1927 PhysPop	1940 MortRate	Difference-Cancers, Females Regression Output:	
Pacific	157.83	123.6	Constant	34.8682
New England	137.50	141.2	Std Err of Y Est	11.1182
West North Central	131.54	117.0	R Squared	0.5904
Mid-Atlantic	138.40	138.7	No. of Observations	9
East North Central	126.18	128.2	Degrees of Freedom	7
Mountain	118.75	108.9		
West South Central	108.25	97.4	X Coefficient(s)	0.6643
East South Central	102.07	100.1	Std Err of Coef.	0.2092
South Atlantic	102.13	104.5	Coefficient / S.E.	3.1762

● - Part 2e.

	1929 PhysPop	1940 MortRate	Difference-Cancers, Females Regression Output:	
Pacific	156.64	123.6	Constant	35.0373

New England	138.46	141.2	Std Err of Y Est		10.4406
West North Central	128.72	117.0	R Squared		0.6388
Mid–Atlantic	138.49	138.7	No. of Observations		9
East North Central	126.51	128.2	Degrees of Freedom		7
Mountain	118.68	108.9			
West South Central	105.60	97.4	X Coefficient(s)		0.6685
East South Central	99.41	100.1	Std Err of Coef.		0.1900
South Atlantic	100.86	104.5	Coefficient / S.E.		3.5183

● – Part 2f.	1931	1940	Difference–Cancers, Females		
	PhysPop	MortRate	Regression Output:		
Pacific	159.97	123.6	Constant		40.4276
New England	142.35	141.2	Std Err of Y Est		10.0708
West North Central	126.50	117.0	R Squared		0.6639
Mid–Atlantic	140.82	138.7	No. of Observations		9
East North Central	128.59	128.2	Degrees of Freedom		7
Mountain	118.89	108.9			
West South Central	105.95	97.4	X Coefficient(s)		0.6215
East South Central	96.73	100.1	Std Err of Coef.		0.1671
South Atlantic	99.59	104.5	Coefficient / S.E.		3.7185

● – Part 2g.	1934	1940	Difference–Cancers, Females		
	PhysPop	MortRate	Regression Output:		
Pacific	160.09	123.6	Constant		44.6430
New England	148.60	141.2	Std Err of Y Est		8.5632
West North Central	125.96	117.0	R Squared		0.7570
Mid–Atlantic	149.62	138.7	No. of Observations		9
East North Central	129.36	128.2	Degrees of Freedom		7
Mountain	117.16	108.9			
West South Central	104.68	97.4	X Coefficient(s)		0.5843
East South Central	92.00	100.1	Std Err of Coef.		0.1251
South Atlantic	98.41	104.5	Coefficient / S.E.		4.6697

● – Part 2h.	1936	1940	Difference–Cancers, Females		
	PhysPop	MortRate	Regression Output:		
Pacific	158.44	123.6	Constant		45.2814
New England	150.18	141.2	Std Err of Y Est		7.8708
West North Central	126.14	117.0	R Squared		0.7947
Mid–Atlantic	155.05	138.7	No. of Observations		9
East North Central	130.42	128.2	Degrees of Freedom		7
Mountain	119.80	108.9			
West South Central	103.52	97.4	X Coefficient(s)		0.5757
East South Central	89.94	100.1	Std Err of Coef.		0.1106
South Atlantic	99.16	104.5	Coefficient / S.E.		5.2055

● – Part 2i.	1938	1940	Difference–Cancers, Females		
	PhysPop	MortRate	Regression Output:		
Pacific	157.62	123.6	Constant		47.6201
New England	154.08	141.2	Std Err of Y Est		7.0749
West North Central	124.95	117.0	R Squared		0.8341
Mid–Atlantic	160.69	138.7	No. of Observations		9
East North Central	131.98	128.2	Degrees of Freedom		7
Mountain	119.88	108.9			
West South Central	102.79	97.4	X Coefficient(s)		0.5538
East South Central	88.21	100.1	Std Err of Coef.		0.0933
South Atlantic	99.26	104.5	Coefficient / S.E.		5.9331

● – Part 2j.	1940	1940	Difference–Cancers, Females		
	PhysPop	MortRate	Regression Output:		
Pacific	159.72	123.6	Constant		52.8821
New England	161.55	141.2	Std Err of Y Est		6.6648
West North Central	123.14	117.0	R Squared		0.8528
Mid–Atlantic	169.76	138.7	No. of Observations		9
East North Central	133.36	128.2	Degrees of Freedom		7
Mountain	119.89	108.9			
West South Central	103.94	97.4	X Coefficient(s)		0.5041
East South Central	85.83	100.1	Std Err of Coef.		0.0792
South Atlantic	100.74	104.5	Coefficient / S.E.		6.3682

Box 1 of Chap. 19
Summary: Regression Outputs, "Difference" Cancers, Females.

Below are the summary-results from regressing the 1940 cancer MortRates upon the ten sets of PhysPops (1921-1940), as presented in Parts 2a-2j of this chapter.

Part	PhysPop	R-squared	Constant	X-Coef	Std Err	X-Coef/SE
2a	1921	0.3479	34.80	0.6157	0.3186	1.9326
2b	1923	0.4085	36.24	0.6220	0.2829	2.1989
2c	1925	0.4743	38.89	0.6215	0.2473	2.5130
2d	1927	0.5904	34.87	0.6643	0.2092	3.1762
2e	1929	0.6388	35.04	0.6685	0.1900	3.5183
2f	1931	0.6639	40.43	0.6215	0.1671	3.7185
2g	1934	0.7570	44.64	0.5843	0.1251	4.6697
2h	1936	0.7947	45.28	0.5757	0.1106	5.2055
2i	1938	0.8341	47.62	0.5538	0.0933	5.9331
2j --->	1940 Max	0.8528	52.88	0.5041	0.0792	6.3682

Box 2 of Chap. 19
Input-Data for Figure 19-A. "Difference" Cancers. Females.

Part 2j, Best-Fit Equation: Calc. MortRate = (0.5041 * PhysPop) + (52.88)

Census Divisions	1940 Observed PhysPops	1940 Observed MortRates	Best-Fit Calc. MortRates
Pacific	159.72	123.6	133.395
New England	161.55	141.2	134.317
West No. Central	123.14	117.0	114.955
Mid-Atlantic	169.76	138.7	138.456
East No. Central	133.36	128.2	120.107
Mountain	119.89	108.9	113.317
West So. Central	103.94	97.4	105.276
East So. Central	85.83	100.1	96.147
South Atlantic	100.74	104.5	103.663
Additional PhysPops	70.00		88.167
--- not "observed" ---	60.00		83.126
down to zero PhysPop	50.00		78.085
(zero medical radiation).	40.00		73.044
For each, we calculate	30.00		68.003
a best-fit MortRate.	20.00		62.962
These additional x,y pairs	10.00		57.921
are also part of the	0		52.880
best-fit line (Chap 5, Part 5e).			

Box 3 of Chap. 19

Presumptive Fraction of Cancer MortRate Attributable to Medical Radiation.

Please see text in Chapter 6, Parts 4 and 6.

Difference-Cancers. FEMALES.

- FEMALE National MortRate (MR) 1940, from Table 19-B   122.8   National MortRate
- Constant, from regression, Part 2j   52.8821   Constant
- Fractional Causation, Best Est. = (Natl MR − Constant) / Natl MR   56.9%   Frac. Causation

....

90% Confidence-Limits (C.L.) on Fractional Causation. See text in Chapter 6, Part 4b, please.

X-Coefficient, from Part 2j	0.5041	X-Coef., Best Est.
Standard Error (SE) of X-Coefficient, from Part 2j	0.0792	Standard Error

Upper 90% C.L. on X-Coef. = (Coef) + (1.645 * SE) =   0.6344   New X-Coefficient
New Constant = (Natl MR) − (New X-Coef * 1940 Natl PhysPop) =   39.0359   New Constant
Frac. Causation, High-Limit = (Natl MR − New Constant) / Natl MR =   68.2%   New Frac. Caus'n.

Lower 90% C.L. on X-Coef. = (Coef) − (1.645 * SE) =   0.3738   New X-Coefficient
New Constant = (Natl MR) − (New X-Coef * 1940 Natl PhysPop) =   73.4413   New Constant
Frac. Causation, Low-Limit = (Natl MR − New Constant) / Natl MR =   40.2%   New Frac. Caus'n.

Box 4 of Chap. 19

Error-Check on Our Own Work: "Difference" Cancers, Females.

Below, Columns A, C, and E come directly from the regression input in Part 2j. Column B, the fraction of the whole 1940 population in each Census Division, comes from Table 3-B in Chapter 3. Each Column-D entry is the product of (B-entry times C-entry). Each Column-F entry is the product of (B-entry times E-entry). PhysPops and MortRates are each "per 100,000."

The Weighted-Avg. Nat'l PhysPop, 1940, is the sum of Column-D entries =   132.04

The Weighted-Avg. Nat'l Female MortRate, 1940, is sum of Col.F entries =   120.58
The Nat'l Female MortRate is also (X-Coef * Nat'l PhysPop) + Constant =   119.44
Comparison: The Nat'l Female MortRate, 1940, in Table 19-B =   122.80

(A) Census Division	(B) Pop'n Fraction	(C) PhysPop 1940	(D) Weighted PhysPop	(E) MortRate 1940	(F) Weighted MortRate
Pacific	0.0739	159.72	11.80	123.6	9.13
New England	0.0641	161.55	10.36	141.2	9.05
West No. Central	0.1027	123.14	12.65	117.0	12.02
Mid-Atlantic	0.2092	169.76	35.51	138.7	29.02
East No. Central	0.2022	133.36	26.97	128.2	25.92
Mountain	0.0315	119.89	3.78	108.9	3.43
West So. Central	0.0992	103.94	10.31	97.4	9.66
East So. Central	0.0819	85.83	7.03	100.1	8.20
South Atlantic	0.1354	100.74	13.64	104.5	14.15
Sums	1.0000		132.04		120.58

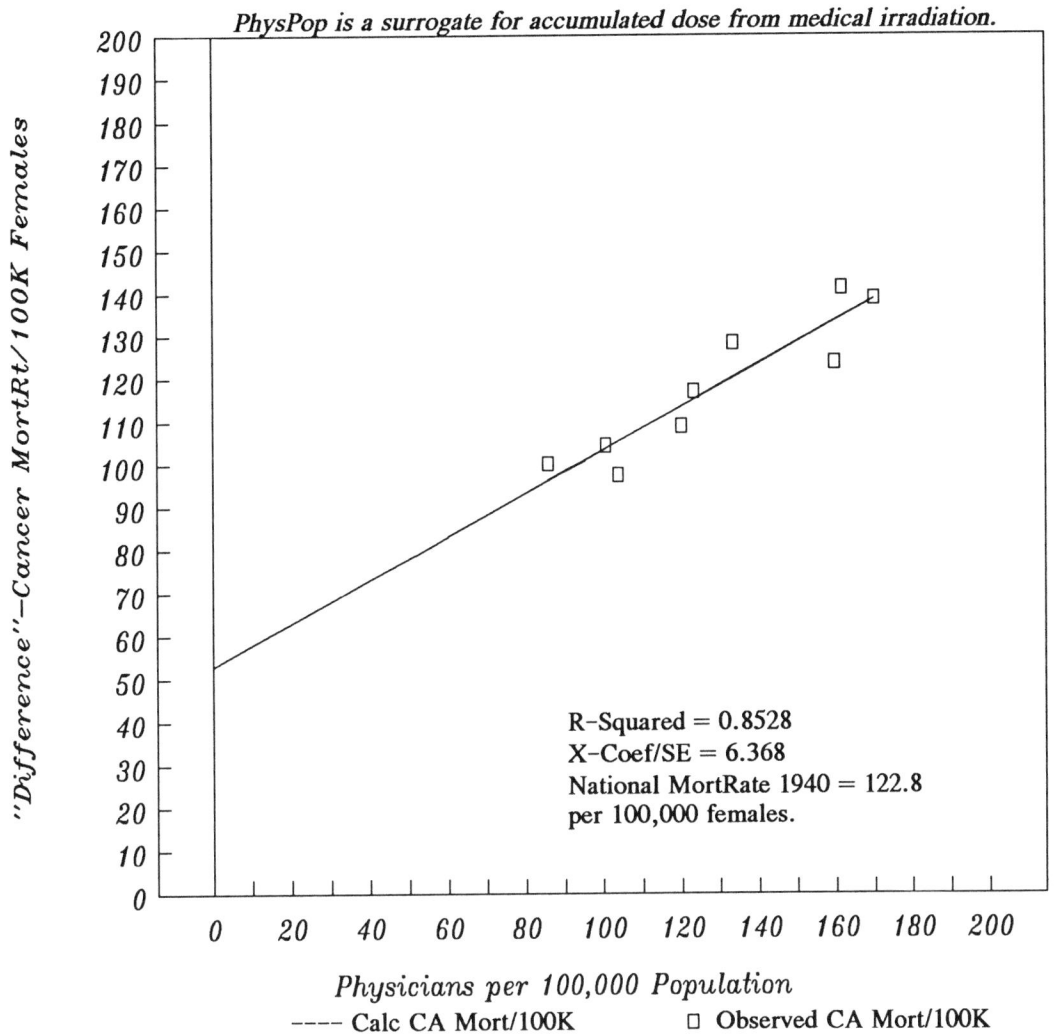

1940 "Difference" Cancer Mortality-Rates versus
1940 PhysPop Values for the 9 Census Divisions, USA.
Dose–Response Relationship
PhysPop is a surrogate for accumulated dose from medical irradiation.

R-Squared = 0.8528
X-Coef/SE = 6.368
National MortRate 1940 = 122.8
per 100,000 females.

Physicians per 100,000 Population
---- Calc CA Mort/100K   □ Observed CA Mort/100K

On the X–axis, PhysPop values = Physicians per 100,000 Population in the Nine Census Divisions of the USA Population, Year 1940. This variable is a surrogate for accumulated radiation dose --- the more physicians per 100,000 people, the more radiation procedures are done per 100,000 people.

On the Y–axis, "Difference" Cancer Mortality-Rate per 100,000 females = the reported rates in USA Vital Statistics for the Nine Census Divisions, Year 1940.

Shown above is the relationship between these two variables (Part 2j). The nine datapoints (boxy symbols) were collected long ago for other purposes, and are free from potential bias with respect to this dose–response study. Fractional causation is (Natl MortRate minus the Y–intercept) / (Natl MortRate).

Fractional Causation of "Difference" Cancer Mort–Rate (Female) by Medical Rad'n
57 % from Best Estimate (Box 3).

40 % at lower 90 % confidence limit (Box 3).   ~68 % at upper 90 % confidence limit (Box 3).

Table 19-A.
"Difference" Cancer MortRates by Census Divisions: Females.

"Difference" Cancers are (All-Cancers minus Respiratory-System Cancers). The entries below are the corresponding entries in Table 7-A (All-Cancers, Female) minus the corresponding entries in Table 17-A (Respiratory-System Cancers, Female). Rates are annual deaths per 100,000 female population, USA, age-adjusted to the 1940 reference year. There are no exclusions by color or "race."

Census Division	1940	1950	1960	1970	1980	1988
Pacific	123.6	113.3	104.2	96.7	89.2	83.7
New England	141.2	128.0	116.8	107.4	97.9	89.5
West North Central	117.0	112.3	104.9	95.4	86.0	83.7
Mid-Atlantic	138.7	132.0	121.4	110.2	99.0	92.8
East North Central	128.2	123.0	114.7	104.3	93.9	90.1
Mountain	108.9	101.8	96.9	88.4	80.0	78.2
West South Central	97.4	105.0	97.7	90.3	82.8	83.2
East South Central	100.1	105.6	100.1	93.2	86.2	86.1
South Atlantic	104.5	108.6	102.4	94.8	87.1	85.0
Average, ALL	117.7	114.4	106.6	97.8	89.1	85.8
Average, High-5	113.8	117.6	115.6	111.1	106.6	105.6
Average, Low-4	80.2	92.9	97.4	98.8	100.2	104.1
Ratio, Hi5/Lo4	1.42	1.27	1.19	1.12	1.06	1.01

Table 19-B.
"Difference" Cancer Mortality Rates, USA National.

Annual MortRates in Table 19-B are obtained by subtracting Table 17-B from Table 7-B.

Rates are age-adjusted to the 1940 reference year. Both sexes: Deaths per 100,000 population (males + females). Males: Deaths per 100,000 male population. Females: Deaths per 100,000 female population. No exclusions by color or "race."

	Both Sexes	Male	Female
1940	113.1	104.0	122.8
1950	114.7	111.2	118.6
1960	109.6	110.5	109.6
1970	101.4	107.8	100.1
1979-81	95.8	105.1	90.5
1987-89	--	103.0	86.8

● – Sources are stated in Table 17-B and Table 7-B.

CHAPTER 20

"All-Cancer-Except-Genital-Cancer": Relation with Medical Radiation

● Part 1.   "All-Cancer-Except-Genital-Cancer": A Result of Chapter 14

In Chapter 14, we found that the MortRate for female Genital Cancers showed no significant relationship, either positive or negative, with medical radiation (discussion in Chapter 14).   In this chapter, we obtain the estimated Fractional Causation by medical radiation for female All-Cancer deaths in 1940 except for Genital-Cancer deaths --- a group which can be called "All-Cancer-Except-Genital" or "All-Minus-Genital."   The "All-Minus-Genital" group accounts for 74.5% of all cancer-deaths in 1940 among females (Table 20-A, Row 26, at the end of this chapter).

Although Chapter 13 uncovered a strong relationship between medical radiation and MALE Genital-Cancer MortRates in 1940, we do the analysis here for All-Minus-Genital for males too. All-Minus-Genital accounts for 86.8% of All-Cancer deaths in 1940 among males (Table 20-A, Row 13).

● Part 2. Regression Analysis, with Estimated Fractional Causation

Below, we show the linear regression analyses for the 1940 MortRates regressed upon the 1940 PhysPops --- males first (Part 2a), then females (Part 2b).   We omit the "build-up" years which use PhysPops prior to 1940.   The Universal PhysPop Table 3-A, and Table 20-A, provide the input-data for these regressions.   Box 1 does not exist in this chapter.

Part 2a. MALES.	1940 PhysPop	1940 MortRate	MALES: All-Minus-Genital Regression Output:	
Pacific	159.72	105.7	Constant	6.4004
New England	161.55	117.3	Std Err of Y Est	5.1290
West North Central	123.14	94.4	R Squared	0.9468
Mid-Atlantic	169.76	125.1	No. of Observations	9
East North Central	133.36	103.8	Degrees of Freedom	7
Mountain	119.89	84.0		
West South Central	103.94	75.3	X Coefficient(s)	0.6799
East South Central	85.83	63.2	Std Err of Coef.	0.0609
South Atlantic	100.74	76.1	Coefficient / S.E.	11.1621

Part 2b. FEMALES.	1940 PhysPop	1940 MortRate	FEMALES: All-Minus-Genital Regression Output:	
Pacific	159.72	94.3	Constant	23.9237
New England	161.55	112.5	Std Err of Y Est	6.3277
West North Central	123.14	91.7	R Squared	0.8675
Mid-Atlantic	169.76	110.2	No. of Observations	9
East North Central	133.36	98.2	Degrees of Freedom	7
Mountain	119.89	84.0		
West South Central	103.94	69.8	X Coefficient(s)	0.5087
East South Central	85.83	69.3	Std Err of Coef.	0.0751
South Atlantic	100.74	74.4	Coefficient / S.E.	6.7698

>>>>>>>>>>

Box 4 will precede Box 3.

Box 2 of Chap. 20
Input-Data for Figure 20-A (males), Figure 20-B (females).

Part 2a, Best-Fit Equation: Calc. MortRate = (0.6799 * PhysPop) + (6.4004)
Part 2b, Best-Fit Equation: Calc. MortRate = (0.5087 * PhysPop) + (23.9237)

Census Divisions	1940 Observed PhysPops	1940 Observed MortRates MALES	Best-Fit Calc. MortRates MALES	1940 Observed MortRates FEMALE	Best-Fit Calc. MortRates FEMALES
Pacific	159.72	105.7	114.994	94.3	105.173
New England	161.55	117.3	116.238	112.5	106.104
West No. Central	123.14	94.4	90.123	91.7	86.565
Mid-Atlantic	169.76	125.1	121.820	110.2	110.281
East No. Central	133.36	103.8	97.072	98.2	91.764
Mountain	119.89	84.0	87.914	84.0	84.912
West So. Central	103.94	75.3	77.069	69.8	76.798
East So. Central	85.83	63.2	64.756	69.3	67.585
South Atlantic	100.74	76.1	74.894	74.4	75.170
Additional PhysPops	70.00		53.993		59.533
--- not "observed" --	60.00		47.194		54.446
down to zero PhysPop	50.00		40.395		49.359
(zero med. radiation).	40.00		33.596		44.272
For each, we calculate	30.00		26.797		39.185
a best-fit MortRate.	20.00		19.998		34.098
These additional x,y	10.00		13.199		29.011
pairs are also part of	0		6.400		23.924
the best-fit line (Chap 5, Part 5e).					

Box 4 of Chap. 20
Error-Check on Our Own Work: All-Minus-Genital.

The Weighted-Avg. Nat'l PhysPop, 1940, is the sum of Column-D entries =	132.04
The Weighted-Avg. Nat'l MALE MortRate of 1940 is the sum of Column-F entries =	97.78
The Nat'l Male MortRate is also (X-Coef * 1940 Nat'l PhysPop) + Constant =	96.17
Comparison: The National MALE MortRate of 1940, in Table 20-A, Row 12 =	99.80
The Weighted-Avg. Nat'l FEM. MortRate of 1940 is the sum of Column-H entries =	91.83
The Nat'l Female MortRate is also (X-Coef * 1940 Nat'l PhysPop) + Constant =	91.09
Comparison: The National FEM. MortRate of 1940, in Table 20-A, Row 25 =	94.00

(A) Census Division	(B) Pop'n Fraction	(C) PhysPop 1940	(D) Weighted PhysPop	(E) MALE MortRate 1940	(F) MALE Weighted MortRate	(G) FEM. MortRate 1940	(H) FEM. Weighted MortRate
Pacific	0.0739	159.72	11.80	105.7	7.81	94.3	6.97
New England	0.0641	161.55	10.36	117.3	7.52	112.5	7.21
West No. Central	0.1027	123.14	12.65	94.4	9.69	91.7	9.42
Mid-Atlantic	0.2092	169.76	35.51	125.1	26.17	110.2	23.05
East No. Central	0.2022	133.36	26.97	103.8	20.99	98.2	19.86
Mountain	0.0315	119.89	3.78	84.0	2.65	84.0	2.65
West So. Central	0.0992	103.94	10.31	75.3	7.47	69.8	6.92
East So. Central	0.0819	85.83	7.03	63.2	5.18	69.3	5.68
South Atlantic	0.1354	100.74	13.64	76.1	10.30	74.4	10.07
Sums	1.0000		132.04		97.78		91.83

Please see text in Chapter 6, Parts 4 and 6.

All–Minus–Genital Cancers. MALES.

● MALE National MortRate (MR) 1940, from Table 20-A, Row 12	99.8	National MortRate
● Constant, from regression, Part 2a	6.4004	Constant
● Fractional Causation, Best Est. = (Natl MR – Constant) / Natl MR	93.6%	Frac. Causation

90% Confidence-Limits (C.L.) on Fractional Causation. See text in Chapter 6, Part 5.

X-Coefficient, from Part 2a	0.6799	X-Coef., Best Est.
Standard Error (SE) of X-Coefficient, from Part 2a	0.0609	Standard Error
Upper 90% C.L. on X-Coef. = (Coef) + (1.645 * SE) =	0.7801	New X-Coefficient
New Constant = (Natl MR) – (New X-Coef * 1940 Natl PhysPop) =	-3.2018	New Constant
Frac. Caus'n, High-Limit = (Natl MR – New Constant) / Natl MR =	103.2% #	New Frac. Caus'n.

The Upper-Limit is 100%. Negative Constants produce values > 100%. See Chapter 22, Part 3.

Lower 90% C.L. on X-Coef. = (Coef) – (1.645 * SE) =	0.5797	New X-Coefficient
New Constant = (Natl MR) – (New X-Coef * 1940 Natl PhysPop) =	23.2538	New Constant
Frac. Caus'n, Low-Limit = (Natl MR – New Constant) / Natl MR =	76.7%	New Frac. Caus'n.

All–Minus–Genital Cancers. FEMALES.

● FEMALE National MortRate 1940, from Table 20-A, Row 25	94.0	National MortRate
● Constant, from regression, Part 2b	23.9237	Constant
● Fractional Causation, Best Est. = (Natl MR – Constant) / Natl MR	74.5%	Frac. Causation

90% Confidence-Limits (C.L.) on Fractional Causation. See text in Chapter 6, Part 5.

X-Coefficient, from Part 2b	0.5087	X-Coef., Best Est.
Standard Error (SE) of X-Coefficient, from Part 2b	0.0751	Standard Error
Upper 90% C.L. on X-Coef. = (Coef) + (1.645 * SE) =	0.6322	New X-Coefficient
New Constant = (Natl MR) – (New X-Coef * 1940 Natl PhysPop) =	10.5191	New Constant
Frac. Caus'n, High-Limit = (Natl MR – New Constant) / Natl MR =	88.8%	New Frac. Caus'n.
Lower 90% C.L. on X-Coef. = (Coef) – (1.645 * SE) =	0.3852	New X-Coefficient
New Constant = (Natl MR) – (New X-Coef * 1940 Natl PhysPop) =	43.1434	New Constant
Frac. Caus'n, Low-Limit = (Natl MR – New Constant) / Natl MR =	54.1%	New Frac. Caus'n.

1940 (All Minus Gen) Cancer Mortality-Rates versus 1940 PhysPop Values for the 9 Census Divisions, USA.

Dose–Response Relationship

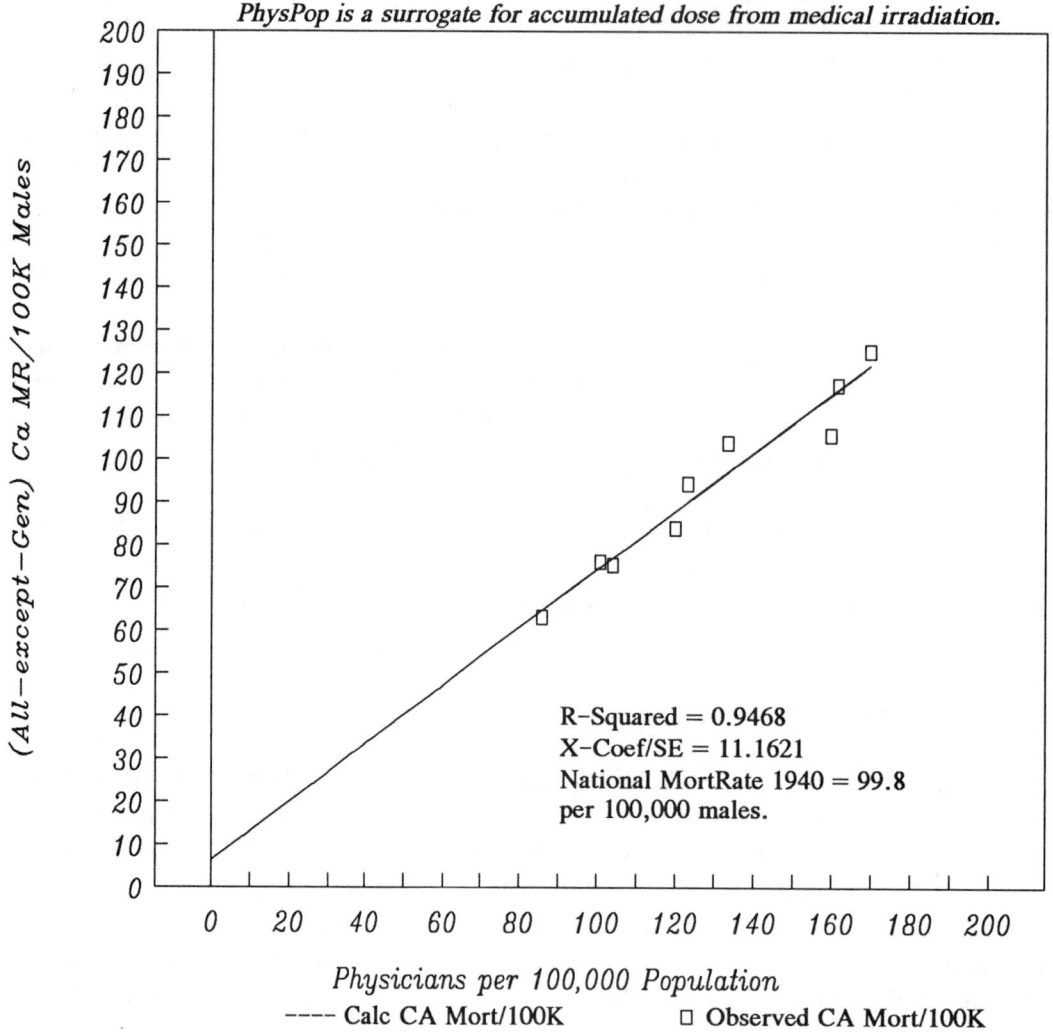

PhysPop is a surrogate for accumulated dose from medical irradiation.

R-Squared = 0.9468
X-Coef/SE = 11.1621
National MortRate 1940 = 99.8
per 100,000 males.

Y-axis: (All–except–Gen) Ca MR/100K Males

X-axis: *Physicians per 100,000 Population*

---- Calc CA Mort/100K   □ Observed CA Mort/100K

On the X-axis, PhysPop values = Physicians per 100,000 Population in the Nine Census Divisions of the USA Population, Year 1940. This variable is a surrogate for accumulated radiation dose --- the more physicians per 100,000 people, the more radiation procedures are done per 100,000 people.

On the Y-axis, Cancer Mortality-Rate per 100,000 males = the reported rates in USA Vital Statistics for the Nine Census Divisions, Year 1940.

Shown above is the relationship between these two variables (Part 2a). The nine datapoints (boxy symbols) were collected long ago for other purposes, and are free from potential bias with respect to this dose-response study. Fractional causation is (Natl MortRate minus the Y-intercept) / (Natl MortRate).

Fractional Causation of "All–Except–Genital Ca" (Male) by Medical Rad'n
93.6 % from Best Estimate (Box 3).
~77% at lower 90 % confidence limit (Box 3).   ~100 % at upper 90 % confidence limit (Box 3).

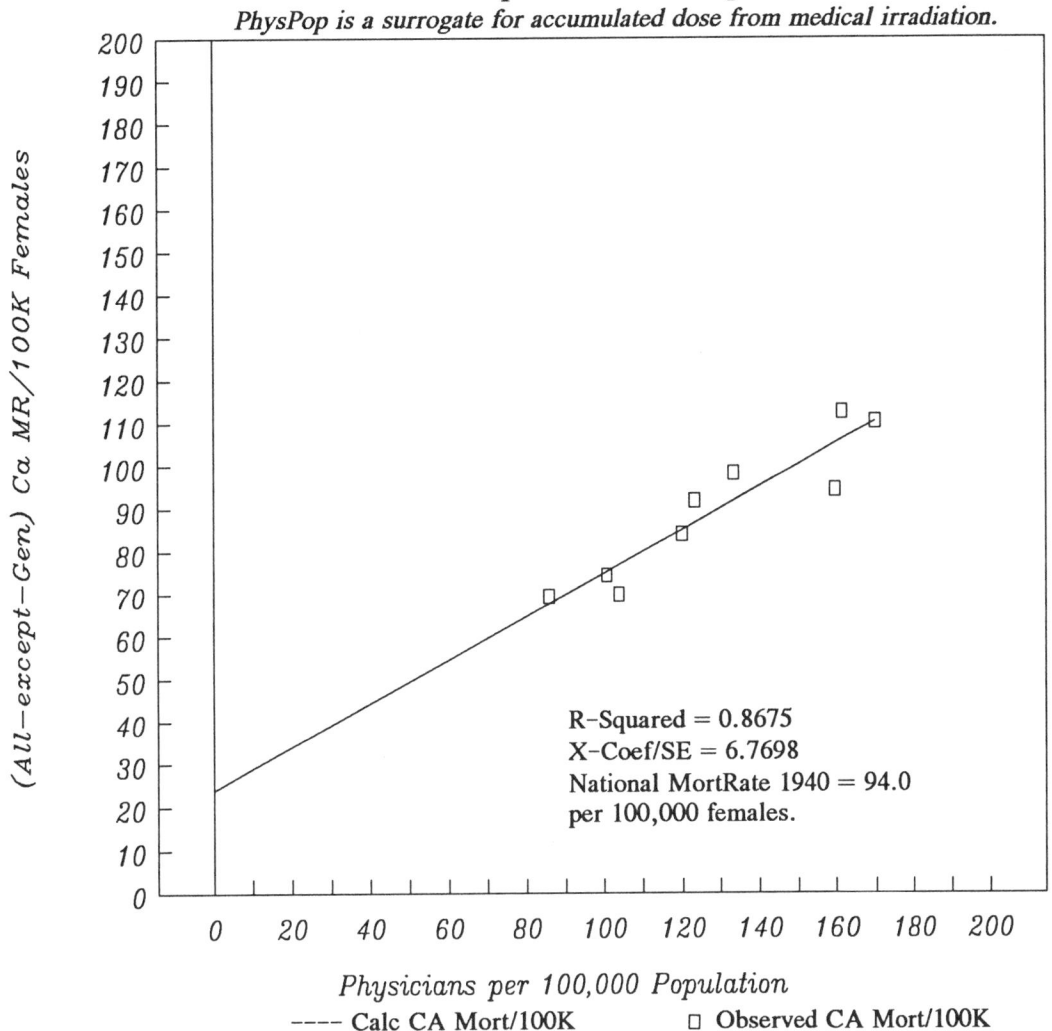

1940 (All Minus Gen) Cancer Mortality-Rates versus 1940 PhysPop Values for the 9 Census Divisions, USA.
Dose–Response Relationship
PhysPop is a surrogate for accumulated dose from medical irradiation.

R–Squared = 0.8675
X–Coef/SE = 6.7698
National MortRate 1940 = 94.0
per 100,000 females.

Physicians per 100,000 Population
---- Calc CA Mort/100K   □ Observed CA Mort/100K

On the X–axis, PhysPop values = Physicians per 100,000 Population in the Nine Census Divisions of the USA Population, Year 1940. This variable is a surrogate for accumulated radiation dose --- the more physicians per 100,000 people, the more radiation procedures are done per 100,000 people.

On the Y–axis, Cancer Mortality-Rate per 100,000 females = the reported rates in USA Vital Statistics for the Nine Census Divisions, Year 1940.

Shown above is the relationship between these two variables (Part 2b). The nine datapoints (boxy symbols) were collected long ago for other purposes, and are free from potential bias with respect to this dose-response study. Fractional causation is (Natl MortRate minus the Y-intercept) / (Natl MortRate).

Fractional Causation of "(All–Except–Genital) Ca" (Female) by Medical Rad'n
74.5 % from Best Estimate (Box 3).
~54% at lower 90 % confidence limit (Box 3). ~89 % at upper 90 % confidence limit (Box 3).

Table 20-A.
All-Cancer-Except-Genital MortRates: Males, Females.

"All-Cancer-Except-Genital" male mortality rates (MRs) below are the corresponding entries in Table 6-A+B (All-Cancers, Male) minus the corresponding entries in Table 13-A+B (Genital Cancers, Male). Rates are annual deaths per 100,000 male population, USA, age-adjusted to the 1940 reference year. There are no exclusions by color or "race." Corresponding comments apply to female values below (calculated from Chapters 7 and 14).

------------------------ MALES --

Census Division Row	Males	1940	1950	1960	1970	1980	1990
1	Pacific	105.7	113.2	127.3	133.3	139.3	132.6
2	New England	117.3	135.8	148.9	152.3	155.5	150.5
3	West North Central	94.4	108.7	120.2	129.0	137.8	139.6
4	Mid-Atlantic	125.1	141.8	150.4	153.7	157.0	151.6
5	East North Central	103.8	123.2	135.6	144.9	154.2	154.0
6	Mountain	84.0	94.5	103.5	111.9	120.2	122.5
7	West South Central	75.3	99.4	119.2	133.9	148.6	156.2
8	East South Central	63.2	90.0	109.2	134.0	158.7	170.7
9	South Atlantic	76.1	101.6	122.5	138.7	155.0	157.2
10	Natl, All-Cancer MR	115.0	132.8	145.7	155.1	164.5	162.7
11	Natl, Genital-Cancer MR	15.2	14.9	14.6	14.8	15.0	16.9
12	Natl, All-Minus-Genital MR	99.8	117.9	131.1	140.3	149.5	145.8
13	Percent, Row 12/Row10	86.8%	88.8%	90.0%	90.5%	90.9%	89.6%

------------------------ FEMALES --

Census Division Row	Females	1940	1950	1960	1970	1980	1990
14	Pacific	94.3	92.2	90.1	93.6	97.1	--
15	New England	112.5	107.0	100.7	101.9	103.0	--
16	West North Central	91.7	93.7	89.0	88.3	87.7	--
17	Mid-Atlantic	110.2	109.8	105.2	104.2	103.2	--
18	East North Central	98.2	99.2	95.6	96.6	97.5	--
19	Mountain	84.0	82.4	82.6	82.9	83.2	--
20	West South Central	69.8	81.9	80.8	84.2	87.6	--
21	East South Central	69.3	80.6	80.1	84.5	88.9	--
22	South Atlantic	74.4	83.4	83.4	87.5	91.5	--
23	Natl, All-Cancer MR	126.1	123.2	114.9	111.7	108.5	111.3
24	Natl, Genital-Cancer MR	32.1	27.2	22.4	18.0	13.7	--
25	Natl, All-Minus-Genital MR	94.0	96.0	92.5	93.7	94.8	--
26	Percent, Row 25/Row23	74.5%	77.9%	80.5%	83.9%	87.4%	--

CHAPTER 21

"All-Cancer-Except-(Genital+Respiratory)": Relation with Medical Radiation

● Part 1.   "All-Cancer-Except-(Genital+Respiratory)"

The group, "All-Cancer-Except-(Genital+Respiratory)," is the same as "Difference-Cancers-Minus-Genital-Cancers." With respect to our PhysPop-MortRate analysis, this chapter explores what happens to the dose-response, the Constant, and the Fractional Causation of cancer by medical radiation, if BOTH Genital and Respiratory-System cancers are subtracted from the All-Cancer MortRates.

After both subtractions are made, we are still dealing with 77.2% of all the male cancer-deaths in 1940 (Table 21-A, Row 14, at the end of the chapter). And we are still dealing with 71.9% of all the female cancer-deaths in 1940 (Table 21-A, Row 28).

● Part 2. Regression Analysis, with Estimated Fractional Causation

Below, we show the linear regression analyses for the 1940 MortRates regressed upon the 1940 PhysPops --- males first (Part 2a), then females (Part 2b). We omit the "build-up" years which use PhysPops prior to 1940. The Universal PhysPop Table 3-A, and Table 21-A, provide the input data for these regressions.

Part 2a. MALES.	1940 PhysPop	1940 MortRate	MALES: All-Minus-(Gen+Respy) Regression Output:	
Pacific	159.72	93.7	Constant	11.5006
New England	161.55	103.8	Std Err of Y Est	4.7251
West North Central	123.14	86.7	R Squared	0.9350
Mid-Atlantic	169.76	108	No. of Observations	9
East North Central	133.36	93.2	Degrees of Freedom	7
Mountain	119.89	76.2		
West South Central	103.94	67.7	X Coefficient(s)	0.5630
East South Central	85.83	58.3	Std Err of Coef.	0.0561
South Atlantic	100.74	67.8	Coefficient / S.E.	10.0324

Part 2b. FEMALES.	1940 PhysPop	1940 MortRate	FEMALES: All-Minus-(Gen+Respy) Regression Output:	
Pacific	159.72	90.5	Constant	23.8218
New England	161.55	108.4	Std Err of Y Est	6.2443
West North Central	123.14	88.6	R Squared	0.8593
Mid-Atlantic	169.76	106.0	No. of Observations	9
East North Central	133.36	95.0	Degrees of Freedom	7
Mountain	119.89	81.1		
West South Central	103.94	67.4	X Coefficient(s)	0.4849
East South Central	85.83	66.9	Std Err of Coef.	0.0742
South Atlantic	100.74	72.0	Coefficient / S.E.	6.5390

● - Box 1 does not exist for this chapter.
● - Box 2 does not exist for this chapter. There are no graphs.
● - Box 3 shows the Fractional Causation by medical radiation.
● - Box 4 shows the two Error-Checks on our work.
● - Table 21-A provides the MortRates for Parts 2a and 2b.

>>>>>>>>>>

Please see text in Chapter 6, Parts 4 and 6.

All-Cancer-Except-(Genital+Respiratory). MALES.

- MALE National MortRate (MR) 1940, from Table 21-A, Row 13   88.8   National MortRate
- Constant, from regression, Part 2a   11.5006   Constant
- Fractional Causation, Best Est. = (Natl MR - Constant) / Natl MR   87.0%   Frac. Causation

....

    90% Confidence-Limits (C.L.) on Fractional Causation. See text in Chapter 6, Part 5.

X-Coefficient, from Part 2a	0.5630	X-Coef., Best Est.
Standard Error (SE) of X-Coefficient, from Part 2a	0.0561	Standard Error
Upper 90% C.L. on X-Coef. = (Coef) + (1.645 * SE) =	0.6553	New X-Coefficient
New Constant = (Natl MR) - (New X-Coef * 1940 Natl PhysPop) =	2.2762	New Constant
Frac. Caus'n, High-Limit = (Natl MR - New Constant) / Natl MR =	97.4%	New Frac. Caus'n.
Lower 90% C.L. on X-Coef. = (Coef) - (1.645 * SE) =	0.4707	New X-Coefficient
New Constant = (Natl MR) - (New X-Coef * 1940 Natl PhysPop) =	26.6467	New Constant
Frac. Caus'n, Low-Limit = (Natl MR - New Constant) / Natl MR =	70.0%	New Frac. Caus'n.

All-Cancer-Except-(Genital+Respiratory). FEMALES.

- FEMALE National MortRate 1940, from Table 21-A, Row 27   90.7   National MortRate
- Constant, from regression, Part 2b   23.8218   Constant
- Fractional Causation, Best Est. = (Natl MR - Constant) / Natl MR   73.7%   Frac. Causation

....

    90% Confidence-Limits (C.L.) on Fractional Causation. See text in Chapter 6, Part 5.

X-Coefficient, from Part 2b	0.4849	X-Coef., Best Est.
Standard Error (SE) of X-Coefficient, from Part 2b	0.0742	Standard Error
Upper 90% C.L. on X-Coef. = (Coef) + (1.645 * SE) =	0.6070	New X-Coefficient
New Constant = (Natl MR) - (New X-Coef * 1940 Natl PhysPop) =	10.5571	New Constant
Frac. Caus'n, High-Limit = (Natl MR - New Constant) / Natl MR =	88.4%	New Frac. Caus'n.
Lower 90% C.L. on X-Coef. = (Coef) - (1.645 * SE) =	0.3628	New X-Coefficient
New Constant = (Natl MR) - (New X-Coef * 1940 Natl PhysPop) =	42.7905	New Constant
Frac. Caus'n, Low-Limit = (Natl MR - New Constant) / Natl MR =	52.8%	New Frac. Caus'n.

Please see text in Chapter 6, Part 5.

Below, Columns A, C, E, and G come directly from the regression input in Parts 2a (males) and 2b (females). Column B, the fraction of the whole 1940 population in each Census Division, comes from Table 3-B in Chapter 3. Each Column-D entry is the product of (B-entry times C-entry). Each Column-F entry is the product of (B-entry times E-entry). Each Column-H entry is the product of (B-entry times G-entry). PhysPops and MortRates are each "per 100,000."

The Weighted-Avg. Nat'l PhysPop, 1940, is the sum of Column-D entries =   132.04

The Weighted-Avg. Nat'l MALE MortRate of 1940 is the sum of Column-F entries =   86.99
The Nat'l Male MortRate is also (X-Coef * 1940 Nat'l PhysPop) + Constant =   85.84
Comparison: The National MALE MortRate of 1940, in Table 21-A, Row 13 =   88.80

The Weighted-Avg. Nat'l FEM. MortRate of 1940 is the sum of Column-H entries =   88.59
The Nat'l Female MortRate is also (X-Coef * 1940 Nat'l PhysPop) + Constant =   87.85
Comparison: The National FEM. MortRate of 1940, in Table 21-A, Row 27 =   90.70

(A) Census Division	(B) Pop'n Fraction	(C) PhysPop 1940	(D) Weighted PhysPop	(E) MALE MortRate 1940	(F) MALE Weighted MortRate	(G) FEM. MortRate 1940	(H) FEM. Weighted MortRate
Pacific	0.0739	159.72	11.80	93.7	6.92	90.5	6.69
New England	0.0641	161.55	10.36	103.8	6.65	108.4	6.95
West No. Central	0.1027	123.14	12.65	86.7	8.90	88.6	9.10
Mid-Atlantic	0.2092	169.76	35.51	108.0	22.59	106.0	22.18
East No. Central	0.2022	133.36	26.97	93.2	18.85	95.0	19.21
Mountain	0.0315	119.89	3.78	76.2	2.40	81.1	2.55
West So. Central	0.0992	103.94	10.31	67.7	6.72	67.4	6.69
East So. Central	0.0819	85.83	7.03	58.3	4.77	66.9	5.48
South Atlantic	0.1354	100.74	13.64	67.8	9.18	72.0	9.75
Sums	1.0000		132.04		86.99		88.59

This chapter contains no graph.

Table 21-A.
All-Cancer-Except-(Genital+Respiratory) MortRates: Males, Females.

"All-Cancer-Except-(Genital+Respiratory) male mortality rates (MRs) below are the same as "Difference-Cancers-Minus-Genital" male MortRates. So the male entries below are the rates from Table 18-A+B (Diff-Cancers, Male) minus the corresponding rates in Table 13-A+B (Genital Cancers, Male). Rates are annual deaths per 100,000 male population, USA, age-adjusted to the 1940 reference year. There are no exclusions by color or "race." The female rates below derive from Tables 19-A+B and 14-A+B.

---------------------------- MALES ---

| Census Division   Males | 1940 | 1950 | 1960 | 1970 | 1980 | 1990 |
Row						
1	Pacific	93.7	92.1	92.4	89.1	85.8	81.9
2	New England	103.8	112.2	110.8	104.6	98.2	94.2
3	West North Central	86.7	92.2	91.8	88.0	84.1	83.4
4	Mid-Atlantic	108.0	113.4	109.8	104.2	98.6	94.1
5	East North Central	93.2	101.4	99.9	96.4	92.8	91.7
6	Mountain	76.2	77.8	78.0	77.3	76.6	78.3
7	West South Central	67.7	80.4	84.3	85.0	85.8	88.3
8	East South Central	58.3	75.3	80.2	84.1	87.9	91.6
9	South Atlantic	67.8	81.8	86.8	88.2	89.8	88.7
10	Natl, All-Cancer MR	115.0	132.8	145.7	155.1	164.5	162.7
11	Natl, Resp'y-Canc MR	11.0	21.6	35.2	47.3	59.4	59.7
12	Natl, Genital-Canc MR	15.2	14.9	14.6	14.8	15.0	16.9
13	Natl, All-But-(Gen+Rsp)	88.8	96.3	95.9	93.0	90.1	86.1
14	Percent, Row 13/Row 10	77.2%	72.5%	65.8%	60.0%	54.8%	52.9%

---------------------------- FEMALES ---

| Census Division   Females | 1940 | 1950 | 1960 | 1970 | 1980 | 1990 |
Row						
15	Pacific	90.5	87.8	84.2	80.0	75.9	--
16	New England	108.4	102.9	95.1	89.8	84.5	--
17	West North Central	88.6	88.9	84.6	78.6	72.7	--
18	Mid-Atlantic	106.0	104.8	99.2	91.9	84.7	--
19	East North Central	95.0	94.7	90.5	85.0	79.4	--
20	Mountain	81.1	78.2	78.5	73.4	68.3	--
21	West South Central	67.4	77.6	75.6	73.0	70.3	--
22	East South Central	66.9	75.9	75.4	73.7	71.9	--
23	South Atlantic	72.0	78.7	78.4	76.0	73.6	--
24	Natl, All-Cancer MR	126.1	123.2	114.9	111.7	108.5	111.3
25	Natl, Resp'y-Canc MR	3.3	4.6	5.3	11.7	18.0	24.5
26	Natl, Genital-Canc MR	32.1	27.2	22.4	18.0	13.7	--
27	Natl, All-But-(Gen+Rsp)	90.7	91.4	87.2	82.0	76.8	--
28	Percent, Row 27/Row 24	71.9%	74.2%	75.9%	73.4%	70.8%	--

Part 1. Strong Support for Hypothesis-1 at Mid-Century
Part 2. Biological Basis for the Steady Improvement in Correlations
Part 3. Are the Negative Constants a Worry?
Part 4. An Extremely Large, "Blind," Prospective Dose-Response Study
Part 5. Fractional Causation: Why We Used 1940 PhysPops with 1940 MortRates
Part 6. Ockham's Razor: The Law of Minimum Hypotheses
Part 7. Comment on the Results So Far, and on a "Bonus"

Box 1. Comparison: Fractional-Causation Estimates from Chapters 6-21.
Box 2. Summary from Chapters 6, 7 and 8: Regression-Results, 1921 Onward
Box 3. Companion for Box 2: Results when Negative Constants Are Banished.
Box 4. Comparison: Predicted National 1940 Cancer MortRates vs. Observed Rates.
    In this chapter, Box 4 is located after the Figures.
Figure 22-A+B. Dose-Response between 1921 PhysPops and 1940 MortRates.
Figure 22-C. Dose-Response between 1940 PhysPops and 1940 MortRates.

● Part 1. Strong Support for Hypothesis-1 at Mid-Century

Chapters 6 through 21 have uncovered strong, positive dose-response relationships between PhysPop (medical radiation) and cancer MortRates --- with the exception of a single subset: Female Genital Cancers. The findings are summarized in Box 1.

The estimates of Fractional Causation in Box 1 certainly support the hypothesis that medical radiation was a highly important cause (probably the principal cause) of cancer-mortality in the USA in 1940. We discuss the period before 1940 later in this chapter (Part 5b). We consider the period after 1940 in Section Five of this book.

1a. Important Reminders about the Meaning of Fractional Causation

In Box 1, the G-Column presents the estimates of Fractional Causation by medical radiation of the corresponding 1940 National All-Cancer MortRate. Each estimate of Fractional Causation is an estimate of the percentage of cancer deaths which would NOT have occurred, if medical radiation had been absent.

It is worth repeating at the outset of this summary that a radiation-induced cancer MortRate does not mean that radiation is the ONLY agent contributing to such cases (Introduction, Part 5). It follows that, for cancer and other diseases having multiple causes, high Fractional Causation by medical radiation does not necessarily mean that other carcinogens have low Fractional Causations (Introduction, Part 5).

We emphasize also that, when an entry of ∿ 100% occurs in Column G, such a finding is fully consistent with the fact that cancers of these organs occurred before introduction of radiation into medicine. Other causes of such cancers (including radiation exposure from nature itself) have been operative both before and after the introduction of medical radiation. A finding, of ∿ 100% Fractional Causation by medical radiation in 1940, means that by 1940, a very low fraction of such deaths would have occurred without medical radiation as a co-actor.

1b. Estimates Supported by High R-Squared Values and Ratios

The strong, positive correlations in Chapters 6 through 21 indicate that the variation in accumulated radiation dose (PhysPop) is causing most of the the variation in the 1940 cancer MortRates among the Nine Census Divisions. But the purpose of this work is certainly not to re-invent the wheel.

No further evidence is needed to establish the fact that ionizing radiation is a cause of nearly all types of human cancer. That has been firmly established during several decades from other evidence (Chapter 2, Part 4).

The purpose of this work is to see if we have found objective databases from which it is possible to estimate HOW IMPORTANT medical radiation has been in causing the cancer mortality of the USA. And we submit that the high R-squared values (Box 1, Column E) and high X-Coef/SE ratios (Box 1, Column F) support considerable confidence that the resulting best-estimates of Fractional Causation in Box 1 are MEANINGFUL.

1c. Hypothesis-1: Independent of Cancer-Trends over Time

Hypothesis-1 proposes that "Medical radiation is a highly important cause (probably the principal cause) of cancer-mortality in the United States during the Twentieth Century."

It is important to recognize that Hypothesis-1 addresses the fraction of the cancer-deaths which DO occur, not whether the absolute number of cases (age-adjusted) per 100,000 is rising or falling between 1900 and 1999. Still, as we complete our analyses up to 1940, many readers will want to know that the available data (incomplete) on cancer MortRates before 1940 indicate that age-adjusted All-Cancer MortRates rose dramatically between 1900 and 1940 (details in Chapter 67).

The same incomplete data indicate that, between 1930 and 1940, MortRates for some cancers were falling --- especially cancers of the stomach, liver, and uterus (cervix+corpus) (ACS-CA 1992, pp.28-29). The big increase in the age-adjusted All-Cancer MortRate between 1900 and 1940 occurred DESPITE the net decrease for some specific cancers in pre-1940 MortRates.

The fact that age-adjusted MortRates simultaneously rise for some cancers, fall for others, and remain flat for others, is very strong evidence that causes OTHER than medical radiation contribute with medical radiation to produce a cancer's MortRate. We have emphasized earlier (Introduction, Part 5) that, for diseases which have multiple causes per case, the fraction of deaths due to ONE of the causes can be estimated by evaluating what the MortRate would be if that contributing cause were absent (e.g., if PhysPop = zero). And that is how we have estimated the Fractional Causation due to medical radiation in Chapters 6 through 21.

● Part 2. Biological Basis for the Steady Improvement in Correlations

Wilhelm Roentgen discovered xrays at the end of 1895, and the use of xrays in medicine was promptly initiated in a large way (Chapter 2, Part 2). Thus, in 1896, a new carcinogen (medical radiation) was introduced into the U.S. population --- a population which had a pre-existing cancer MortRate due to ancestral and direct exposures to natural background radiation, viruses, and carcinogenic chemicals (probably including some chemicals of viral, bacterial and fungal origin).

2a. The Mounting Response to Medical Radiation: Figure 5-A Revisited

During every year from 1896 onward, some fraction of the population received new exposures to medical radiation, and each annual set of exposures had its OWN trail of cancer consequences, spread over at least 40 years (Chapter 2, Part 8). Such trails are indicated by the horizontal rows in Figure 5-A of Chapter 5.

In Figure 5-A, the vertical columns tell their own story. For instance, the vertical stack of 20 "cancer boxes" for the year 1915 depicts why the rate of radiation-induced cancer in 1915 is influenced by ALL the doses of medical radiation delivered in 1896 through 1915: Each year of irradiation contributes a separate "cancer box" to the column which represents the rate of radiation-induced cancer delivered during 1915.

Worth attention, too, is Figure 5-A's column for 1935. By using any slip of paper as a measure, readers can confirm that there are many more "cancer boxes" (40 boxes) in the 1935 column than in the 1915 column (20 boxes) --- as a result of case-delivery during 1935 from an increasing number of irradiation-years. In other words, the annual rate of radiation-induced cancer is higher by

1935 than it was in 1915, even though the annual average radiation dose has been steady (in the model for Figure 5-A).

Some Distinctions between Figure 5-A and Our Real-World Studies

Of course, Figure 5-A is a simplified model which differs in many details from our real-world studies. For example:

(1) Figure 5-A approximates the consequences of introducing annual medical radiation into ONE population of mixed ages, whereas in our real-world dose-response studies, annual medical radiation has been introduced into NINE such populations, having nine DIFFERENT average dose-levels.

(2) Figure 5-A is for cancer incidence (including nonfatal cases), whereas our real-world data are for cancer MortRates.

(3) Figure 5-A has illustrative rates of radiation-induced cancer for every year, 1896-1991, whereas our real-world cancer MortRates are for 1940 only.

(4) Because of space-limits, Figure 5-A shows 1951 as the last year in which any medical irradiation occurred, whereas in reality, no cessation in use of medical radiation has ever occurred.

2b. Box 2: How Correlations Improve As the PhysPop Year Advances

Our studies reveal that the relationship, between PhysPop and 1940 cancer MortRates, tightens as PhysPop-years advance from 1921 toward 1940. To provide ourselves and readers with a convenient way to review this finding, Box 2 reproduces the summary of results from All-Cancers-Combined, and from Breast Cancer separately. The inclusion of Breast Cancer is due to the high level of interest in that specific cancer. The inclusion of a row for "whites only" has a purpose too.

The "Whites Only" Rows in Box 2

Some readers may wonder whether the correlations we have uncovered in Chapters 6 through 21 are somehow based on the geographic distribution of white and black "races." We have explored that possibility. All the work presented in Chapters 6 through 19 was done also for "whites only." The correlations are very similar, as indicated in Box 2 for All-Cancers (and Breast Cancer), for the 1940-1940 analyses. Since "whites only" account for the overwhelming share of cancer-deaths in 1940, and our "whites only" analyses so closely mirror our "all-race" analyses, we have assurance that the correlations we uncovered are NOT somehow due to the geographic distribution of "blacks." Even if the correlations had differed appreciably, Hypothesis-1 refers to cancer mortality for the United States as a WHOLE, and requires use of the "all-race" data.

The Initial Correlation: 1921 PhysPops with 1940 Cancer MortRates

Box 2 shows that even the 1921 PhysPops have a statistically significant correlation with the 1940 MortRates. (The X-Coef/SE ratio is 2.0 or higher.) How can the 1921 PhysPops correlate as well as they do, with the 1940 cancer MortRates, when Figure 5-A shows (a) that radiation given during 1921 contributes only ONE of the 40 "cancer boxes" in the column which depicts delivery of radiation-induced cancer during 1940, and (b) that the overwhelming share of radiation-induced cases delivered during 1940 is coming from radiation received in years before and after 1921?

The answer is this. The correlation is biologically reasonable BECAUSE the 1921 PhysPops are almost certainly correlated with earlier PhysPops (which we do not have) and are definitely correlated with later PhysPops. In Chapter 3, our Table 3-C shows the correlations between the 1921 PhysPops and the later 1923, 1925, 1927, 1929, 1931, 1934, 1936, 1938, and 1940 PhysPops.

2c. Explanation of the Tightening PhysPop-MortRate Correlations

While the 1921 PhysPops already correlate rather well with the 1940 cancer MortRates, the post-1921 PhysPops correlate even better with the 1940 cancer MortRates (Box 2). Why better?

Biology and demography combine to provide the explanation.

● It is a biological fact that medical radiation received not only before 1921, but also AFTER 1921, has an impact on the 1940 cancer MortRates (Chapter 2, Part 8).

● It is a demographic fact that PhysPop proportions (dose proportions) changed among the Nine Census Divisions between 1921 and 1940. If the 1921 PhysPop values had persisted WITHOUT change in proportion until 1940, those unchanged PhysPop proportions would have "driven" the nine cancer MortRates of 1940 into proportions somewhat DIFFERENT from the proportions actually observed in 1940 among the Nine Census Divisions.

But in the real world, between 1921 and 1940, the "spread" among the PhysPop values grew (Table 3-A). In 1921, (Pacific PhysPop / SouthAtlantic PhysPop) produced the biggest ratio: (165.11 / 110.32) = 1.50. In 1940, (MidAtlantic PhysPop / EastSouthCentral PhysPop) produced an appreciably bigger ratio: (169.76 / 85.83) = 2.00. The Hi5/Lo4 ratio changed from 1.18 to 1.46 during those years. Variation in a cause produces variation in its effect, and it follows that the greater post-1921 spread in PhysPop would cause (biologically) a greater spread in the 1940 cancer MortRates than the 1921 PhysPops would cause.

Because the Observed 1940 cancer MortRates in the Nine Census Divisions are affected by post-1921 changes in the relative strength of the biological CAUSAL agent (PhysPop), it is not surprising that the post-1921 measurements of that agent correlate better with those MortRates than does the 1921 measurement. We would expect post-1921 PhysPops to explain the 1940 outcome better --- and they do.

2d. Visual Evidence: Radiation Driving x,y Datapoints into Line (Figures 22-A + C)

Box 2 shows that the R-squared values and the reliability of the slope (as measured by the X-Coef/SE) improve progressively as PhysPop approaches 1940. One can SEE the improvement in correlation, between 1921 and 1940, by comparing Figures 22-A and 22-C. The MortRates (y-values) for 1940 are identical in both graphs, of course. Only PhysPops (x-values) change --- and such changes cause the boxy symbols to move laterally but not vertically.

Figure 22-C depicts a much tighter dose-response than Figure 22-A, between PhysPops and the MortRates. All of the nine real-world datapoints in Figure 22-C lie close to the line of best fit. The cumulative consequences of 44 years of medical radiation have been gradually causing the x,y datapoints to line up in this way. The fact, that a cause drives x,y datapoints toward a line of best fit, is the essence of any prospective study which uncovers a linear dose-response.

2e. The Power of This New Carcinogen

After Roentgen's discovery of the xray in 1895, PhysPop became approximately proportional to the biological agent called medical radiation. The reality summarized in Box 1 is that this new carcinogen, medical radiation, had the power to make variation in the 1940 cancer MortRates, among the Nine Census Divisions, correlate almost perfectly and positively with variation in PhysPop. The goodness of the correlation says that the 1940 cancer death-rates were virtually set in concrete by PhysPop.

In striking contrast with the positive correlations in Box 1, Chapter 25 will reveal a significant but negative correlation between PhysPop and the 1940 MortRates from all NonCancer NonIHD causes of death combined.

● Part 3   Are the Negative Constants a Worry?

In our graphs, which are based on equations of best fit, the Constant (y-axis intercept) represents the value of the cancer MortRate when PhysPop equals zero. Biologically, there is no such thing as a cancer MortRate BELOW zero. Therefore, should we worry about the string of negative constants in Box 2 for All-Cancer, Males, and for Breast Cancer?

Those who work with numbers realize that a few "outliers" --- datapoints which are "way out of line" in a series of observations --- are capable of tilting a best-fit slope. In epidemiology, a few outliers are no justification for disbelieving the bigger picture.

Because records were not kept, no one can ever plot datapoints for PhysPop-MortRate pairs in 1900, by the Nine Census Divisions. We can never know the distribution out of which developed the distribution in Figure 22-A: 1921 PhysPops paired with 1940 cancer MortRates (male). It is likely that the outliers in Figure 22-A, which produce negative Constants in Box 2 for All-Cancers Male, would be traceable to a few pre-xray datapoints near the turn of the century. For both All-Cancers Male and for Breast Cancer, the negative Constants in Box 2 move inexorably toward positive values, with later PhysPop years.

3a. Demonstration That the Negative Constants Do Not Mislead about Correlation

We have explored what happens if we BANISH negative Constants from our analyses. This can be done in regression analysis by equations which provide the best-fit output after one forces the Constant to be ZERO. Setting the Constant equal to zero is equivalent to asking: How well do the observations fit the MX linear model instead of the MX + C linear model? (Chapter 5, Parts 5 and 6.)

Box 3 provides the answers in a form very easily compared with Box 2. Readers can see for themselves:

● - All-Cancers, Male: The R-squared values in every row are very nearly the same, whether the Constant is negative or zero. So the negative Constants have virtually no impact on the strength of the correlations. A comparison of Figure 22-A with Figure 22-B shows the very similar relationship, between the two different lines of best fit and the single set of boxy symbols (the real-world observed pairs of 1921 PhysPops with 1940 MortRates).

● - All-Cancers, Female: There were no negative Constants to consider. We show the effect, of forcing the positive Constants to equal zero, just to satisfy curiosity.

● - Breast Cancer: Forcing the line of best fit to go through zero makes the fit a little worse for a while --- as signaled by the lower R-squared values in Box 3 than in Box 2. By 1934, there is very little difference in R-squared values between the two types of regression analysis. So the negative Constants have virtually no impact on the strength of the correlations.

3b. A Dramatic Visual Contrast: Outliers Move into Line

Figure 22-A shows the 1940 cancer MortRates, male, regressed on 1921 PhysPops. It is obvious that there are two datapoints which are very much out of line. Mid-Atlantic lies far ABOVE the line of best fit, and East South Central lies well BELOW it. So the line of best-fit is steep enough to produce a negative Constant, by intersecting the y-axis (MortRate) below zero.

The contrast between Figure 22-A and Figure 22-C is easy to see. Of course, the MortRates (y-values) for 1940 are identical in both graphs. Because PhysPops (the x-values) DIFFER in the two graphs, the boxy symbols move laterally but not vertically.

The result: In Figure 22-C, the worst outliers are gone. The real-world observations (the boxy symbols) now lie close to the best-fit line, and the best-fit line has a new slope which makes the Constant POSITIVE. Box 2 confirms that the slope is less steep in Figure 22-C than in 22-A: The best-fit equation for Figure 22-C has an X-Coefficient of 0.7557, whereas the best-fit equation for Figure 22-A has an X-Coefficient of 1.0086.

● Part 4. An Extremely Large, "Blind," Prospective Dose-Response Study

In the world of medicine and pharmacology, the "gold standard" for establishing certain types of cause-and-effect is the "blind" prospective dose-response study. Although a dose-response can never prove causation in the STRICTEST definition of proof, it can provide circumstantial evidence "beyond a reasonable doubt" --- and all other things being equal, the larger is the study, the more reliable are the results.

As noted in Part 1b, we were not seeking additional proof that ionizing radiation is a cause of human cancer when we undertook the studies in this book. Additions are not needed. Instead, we undertook the work in order to evaluate Hypothesis-1. Nonetheless, it is well worth noting that in the process, we HAVE provided powerful additional proof.

Our combination of PhysPop with cancer MortRates, by Census Divisions, represents one of the largest "blind" prospective dose-response studies imaginable. Yet the prospective nature of our study would not be evident to readers if they focus only on the results of 1940 MortRates paired with 1940 PhysPops. And so we call attention to the dose-responses in Box 2, between the 1940 cancer MortRates and PhysPops of years EARLIER than 1940. The dose-responses become statistically significant when the ratio, X-Coef/SE, reaches about 2.0. Almost all results in Box 2 are considerably stronger than a ratio of 2.0.

The 1940 cancer MortRates in the Nine Census Divisions grew out of populations for whom the x-variable (PhysPop) was measured up to 19 years BEFORE measurement of the outcome (1940 cancer MortRates). Even in 1921, variation in PhysPop explains much of the variation in 1940 cancer MortRates.

Separately, Box 4 considers the 1940 NATIONAL cancer MortRates, and demonstrates that:

- The 1921 PhysPops predict the Observed National MortRates for 1940 quite well.
- The 1931 PhysPops predict the same rates even better.
- The 1938 PhysPops predict them better yet. Why improvement occurs is discussed in Parts 2c and 2d, above.

● Part 5. Fractional Causation: Why We Used 1940 PhysPops with 1940 MortRates

The fact that we used 1940 PhysPops with 1940 MortRates, in order to calculate Fractional Causation, deserves some comment here.

There is very probably no MINIMUM incubation-time (latency period) between time of irradiation and delivery of cancer (discussion in Chapter 5, Part 4). Nonetheless, there is almost always at least a year between DIAGNOSIS of a cancer, and DEATH from that cancer. Then why did we "mate" 1940 PhysPops with 1940 MortRates, when a 1940 change in PhysPop-proportions (compared with PhysPop-proportions in 1938) could have no biological impact on the 1940 cancer MortRates?

5a. Consequences of the Competing Alternatives

We were searching for the MAXIMUM detectable correlations remaining in the data, after operation of migration, changes in PhysPop proportions, and other entropic circumstances which conceal the true strength of a relationship (Chapter 5, Part 8). Regression analyses revealed that the very best correlations between PhysPop and All-Cancer MortRates, both for males and for females, occur when the 1940 PhysPops are the input for the x-axis. The improvement in correlation, produced by the 1940 PhysPops compared with the 1938 PhysPops, is in fact TRIVIAL --- as shown by the R-squared values and X-Coef/SE ratios in Box 2.

In order to avoid pairing 1940 cancer MortRates with PhysPops of the same calendar-year, we could have paired 1940 PhysPops with 1942 cancer MortRates --- but we don't have 1942 cancer MortRates by gender and Census Divisions.

The other alternative, in order to avoid same-year pairs, would have been to use the results from pairing 1940 cancer MortRates with the 1938 PhysPops, or the 1936 PhysPops. If we had chosen a pre-1940 set of PhysPops, the estimated Fractional Causation by medical radiation would have been HIGHER for both males and females, because for both genders, the Constants were LOWER in 1938 and in 1936 than in 1940 (Box 2). So our decision to use the 1940 PhysPops was in the direction of LOWER estimates of Fractional Causation. Our choice was also in the direction of somewhat tighter confidence-limits, because in 1940, the ratios of X-Coef/SE were somewhat higher than they were in 1938, for both genders (Box 2). Those who may prefer use of the 1936 or 1938 PhysPops of course can use them to obtain higher estimates of Fractional Causation. When the maximum correlation did occur with 1936 PhysPops (male Genital Cancers) or with 1938 PhysPops (female Breast Cancer), we have already used those pre-1940 PhysPops.

5b. What about Fractional Causation of Pre-1940 Cancer MortRates?

Hypothesis-1 embraces the entire Twentieth Century. Yet complete cancer MortRates for each state and gender are not available before 1940 (Chapter 4, Part 1). Then what can we say about Fractional Causation of 1910, 1920, 1930 cancer mortality by medical radiation? Only this:

   ● - In 1896, Fractional Causation by medical radiation was zero.

   ● - In 1940, the best estimates of Fractional Causation by medical radiation (Box 1) are about 90% for males, and 58% for females (or 75% for females, if Genital Cancers are excluded).

   ● - It follows that between 1896 and 1940, Fractional Causation of cancer MortRates by medical radiation had to rise, from zero percent, toward 90% (male estimate) and 58% (female estimate). It seems reasonable to suggest that by 1920 (the midpoint between 1900 and 1940), perhaps Fractional Causation was one-third of its 1940 value. This would mean that about 30% of the 1920 male All-Cancer MortRate was radiation-induced by physicians, and about 20% of the 1920 female All-Cancer MortRate was radiation-induced by physicians.

● Part 6. Ockham's Razor: The Law of Minimum Hypotheses

Every hypothesis in science is viewed in the light of a famous principle, which deserves explicit attention here.

6a. The Law of Minimum Hypotheses: "Ockham's Razor"

The Law of Minimum Hypotheses, in logic and science, has various formulations. One example: To explain a phenomenon, invoke only as many explanations as required. Or: Avoid fabricating many explanations if one suffices.

The Law of Minimum Hypotheses is also known as "Ockham's Razor," because it was stated (in Latin) by Wilhelm of Ockham in the Fourteenth Century: "Entities [explanations] should not be multiplied beyond what is needed."

6b. The Hypothesis under Examination: Size of Effect, Not Effect Itself

The hypothesis under examination here is that "Medical radiation is a highly important cause (probably the principal cause) of cancer-mortality in the United States during the Twentieth Century" (Hypothesis-1). The issue is the SIZE of medical radiation's impact on the total cancer MortRate. The OCCURRENCE of an impact is beyond doubt (Chapter 2).

Our findings, about the size of medical radiation's impact on the 1940 cancer MortRates, are summarized in Box 1. These findings are based on irrefutable, positive correlations between PhysPop and the 1940 cancer MortRates. We know of no basis for either speculating or assuming that the estimates in Box 1 are too high (Chapter 6, Part 4a).

Therefore, we remind readers of Ockham's Razor. There exists an IDENTIFIED and proven carcinogen which is proportional to PhysPop: Medical radiation. In Chapter 2, we summarized some of the facts about the manner in which xrays really have been used in medicine, and about the special biological properties of ionizing radiation. If one contemplates such facts, the findings in Box 1 seem of reasonable magnitude --- and not in need of additional explanations.

● Part 7. Comment on the Results So Far, and on a "Bonus"

The giant prospective study, presented in Chapters 6 through 21, evaluates the impact by 1940 of an event in history which will never recur: Introduction of ionizing radiation into United States medicine.

By 1940, population was about 132 million (Table 3-B). Unlike studies where almost an entire population is used to approximate a "control group," our study divides the entire population into nine groups --- ALL of which are exposed, because there is no Census Division where medical radiation is absent. Probably our study is one of the largest prospective dose-response studies ever conducted in this country.

Additionally, the databases we have used, by their very nature, exclude intentional bias. It deserves emphasis that the FIRST obligation of objective investigators is to assure that they are working with trustworthy data --- because even Einstein himself would produce false answers, if he were working with a tainted database.

The studies in Chapters 6 through 21 constitute some of the most powerful evidence ever assembled confirming that ionizing radiation is a potent cause of virtually all types of human cancer. We regard this confirmation as a "bonus" from the work, since our studies were undertaken for a different purpose: To test Hypothesis-1.

We would have liked to have had earlier cancer MortRates in every state, but the available data have clearly sufficed to address Hypothesis-1. We conclude that the findings strongly indicate that, during the first half of the Twentieth Century, medical radiation became a highly important cause (probably the principal cause) of cancer mortality in the United States.

>>>>>>>>>>

Box 1 of Chap. 22
Comparison of Results from Chapters 6 through 21.

Cancer	Col.A Natl MR 1940	Col.B Share of ALL	Col.C PP-Year Constant	Col.D for Max	Col.E R-sq.	Col.F Ratio X-Coef. / SE	Col.G Best-Est. FracCausn	Col.H 90% C.L. on Frac. Causation
All-Cancers								
Males: Ch6	115.0	All	11.6	1940	0.95	11.63	90%	74% - 99%
Females: Ch7	126.1	All	53.0	1940	0.86	6.58	58%	41% - 69%
Breast Cancer								
Females: Ch8	23.3	0.18	-2.2	1938	0.92	8.70	∼100%	86% - ∼ 100%
Digestive-System								
Males: Ch9	60.4	0.53	1.9	1940	0.91	8.30	97%	75% - ∼ 100%
Females: Ch10	50.1	0.40	10.2	1940	0.76	4.64	80%	49% - ∼ 100%
Urinary-System								
Males: Ch11	7.4	0.06	-2.8	1940	0.92	9.02	∼100%	See Chap. 11 text.
Females: Ch12	4.0	0.03	0.6	1940	0.94	10.43	86%	68% - 95%
Genital			15.2					
Males: Ch13	15.2	0.13	3.2	1936	0.78	4.92	79%	52% - ∼ 100%
Females: Ch14	32.1	0.25	29.1	1940	0.07	0.72	∼0%	See Chap. 14 text.
Buccal-Pharynx								
Males: Ch15	5.1	0.04	-0.2	1940	0.72	4.28	∼100%	61% - ∼ 100%
Respiratory-System								
Males: Ch16	11.0	0.10	-5.1	1940	0.87	6.76	∼100%	See Chap. 16 text.
Females: Ch17	3.3	0.03	0.1	1940	0.96	13.40	97%	83% - ∼ 100%
"Difference" Ca.								
Males: Ch18	104.0	0.90	16.7	1940	0.93	9.97	84%	68% - 94%
Females: Ch19	122.8	0.97	52.9	1940	0.85	6.37	57%	40% - 68%
All-Except-Genital								
Males: Ch20	99.8	0.87	6.4	1940	0.95	11.16	94%	77% - ∼ 100%
Females: Ch20	94.0	0.75	23.9	1940	0.87	6.77	75%	54% - 89%
All-Except-(Gen+Respy)								
Males: Ch21	88.8	0.77	11.5	1940	0.94	10.03	87%	70% - 97%
Females: Ch21	90.7	0.72	23.8	1940	0.86	6.54	74%	53% - 88%

- Col.A: National age-adjusted MortRates are deaths per 100,000 population.
- Col.B: Each entry is the ratio of 2 values from Col.A. Example: In 1940, Breast Cancer accounts for (23.3 / 126.1), or 0.18 of the female All-Cancer MortRate.
- Col.G: These percentages are best estimates (most likely values) of Fractional Causation by medical radiation of the corresponding 1940 National MortRate. Please see text (Part 1a of Chapter 22).
- When an entry of ∼100% occurs in Column G, such a finding is fully consistent with the fact that cancers of these organs occurred before introduction of radiation into medicine. Other causes of such cancers (including radiation exposure from nature itself) have been operative both before and after the introduction of medical radiation. A finding, of ∼ 100% Fractional Causation by medical radiation in 1940, means that by 1940, a very low fraction of such deaths would have occurred without medical radiation as a co-actor.

● Below are the summary-results for the 1940 MortRates regressed on PhysPops, by Census Divisions. For the 1940-1940 pairs, we also show the output when the analysis was done with data for "Whites Only" (Text, Part 2b).

ALL-CANCERS, MALE. From Chapter 6, Box. 1...

Part	PhysPop	R-squared	Constant	X-Coef	Std Err	X-Coef/SE
2a	1921	0.4630	−27.08	1.0086	0.4105	2.4568
2b	1923	0.5447	−24.83	1.0198	0.3524	2.8937
2c	1925	0.5943	−16.55	0.9879	0.3085	3.2024
2d	1927	0.7175	−20.94	1.0399	0.2466	4.2168
2e	1929	0.7596	−19.27	1.0351	0.2201	4.7032
2f	1931	0.7827	−10.40	0.9582	0.1909	5.0207
2g	1934	0.8718	−2.60	0.8903	0.1290	6.9009
2h	1936	0.9119	−1.42	0.8756	0.1029	8.5104
2i	1938	0.9407	3.05	0.8351	0.0792	10.5419
2j-->	1940 Max	0.9508	11.55	0.7557	0.0650	11.6275
Whites:	1940	0.9473	23.83	0.6740	0.0601	11.2146

ALL-CANCERS, FEMALE. From Chapter 7, Box 1.......................................

Part	PhysPop	R-squared	Constant	X-Coef	Std Err	X-Coef/SE
2a	1921	0.3566	33.38	0.6497	0.3299	1.9695
2b	1923	0.4180	34.97	0.6559	0.2925	2.2423
2c	1925	0.4837	37.90	0.6542	0.2555	2.5609
2d	1927	0.6001	33.82	0.6981	0.2154	3.2407
2e	1929	0.6482	34.06	0.7019	0.1954	3.5916
2f	1931	0.6732	39.75	0.6524	0.1718	3.7978
2g	1934	0.7661	44.25	0.6127	0.1279	4.7888
2h	1936	0.8035	44.96	0.6034	0.1128	5.3509
2i	1938	0.8424	47.45	0.5801	0.0948	6.1177
2j-->	1940 Max	0.8608	52.98	0.5279	0.0802	6.5801
Whites:	1940	0.8638	51.35	0.5352	0.0803	6.6650

BREAST CANCER, FEMALE. From Chapter 8, Box 1.....................................

Part	PhysPop	R-squared	Constant	X-Coef	Std Err	X-Coef/SE
2a	1921	0.5061	−10.94	0.2440	0.0911	2.6780
2b	1923	0.5784	−9.94	0.2432	0.0785	3.0991
2c	1925	0.6598	−8.63	0.2409	0.0654	3.6849
2d	1927	0.7673	−9.12	0.2488	0.0518	4.8040
2e	1929	0.8051	−8.58	0.2466	0.0459	5.3774
2f	1931	0.8203	−6.31	0.2270	0.0402	5.6519
2g	1934	0.8840	−4.03	0.2075	0.0284	7.3052
2h	1936	0.9005	−3.42	0.2014	0.0253	7.9604
2i-->	1938 Max	0.9153	−2.21	0.1906	0.0219	8.6965
2j	1940	0.9126	−0.12	0.1713	0.0200	8.5512
Whites:	1940	0.9184	0.3566	0.1683	0.0190	8.8740

Box 3 of Chap. 22
Companion for Box 2: Results When Negative Constants Are Banished.

Related text = Part 3a.

● Below are the summary-results for the 1940 MortRates regressed on PhysPops, by Census Divisions, when the Constant is forced to equal Zero. Regressions are not shown. They use exactly the same input presented in Chapters 6, 7, 8. Although entries for (X-Coef/SE) below should not be compared with corresponding entries in Box 2, comparisons within each box are valid.

ALL-CANCERS, MALE...

	PhysPop	R-Squared	Constant	X-Coef.	Std Err	X-Coef/SE
2a	1921	0.4449	0	0.8100	0.0423	19.1511
2b	1923	0.5261	0	0.8330	0.0401	20.7611
2c	1925	0.5840	0	0.8597	0.0388	22.1842
2d	1927	0.6992	0	0.8754	0.0335	26.1476
2e	1929	0.7428	0	0.8827	0.0312	28.2985
2f	1931	0.7769	0	0.8767	0.0288	30.4032
2g	1934	0.8714	0	0.8702	0.0217	40.1186
2h	1936	0.9118	0	0.8647	0.0178	48.4724
2i	1938	0.9400	0	0.8583	0.0146	58.7986
2j	1940	0.9380	0	0.8414	0.0145	57.8434

ALL-CANCERS, FEMALE. ....

	PhysPop	R-Squared	Constant	X-Coef.	Std Err	X-Coef/SE
2a	1921	0.3053	0	0.8947	0.0347	25.7607
2b	1923	0.3498	0	0.9190	0.0345	26.6365
2c	1925	0.3845	0	0.9477	0.0346	27.3856
2d	1927	0.5114	0	0.9639	0.0313	30.7726
2e	1929	0.5507	0	0.9714	0.0303	32.0982
2f	1931	0.5157	0	0.9639	0.0312	30.9080
2g	1934	0.5189	0	0.9551	0.0308	31.0132
2h	1936	0.5316	0	0.9484	0.0302	31.4336
2i	1938	0.5043	0	0.9406	0.0308	30.5475
2j	1940	0.3607	0	0.9210	0.0343	26.8662

BREAST CANCER, FEMALE..

	PhysPop	R-Squared	Constant	X-Coef.	Std Err	X-Coef/SE
2a	1921	0.4506	0	0.1637	0.0097	16.8158
2b	1923	0.5229	0	0.1684	0.0093	18.0788
2c	1925	0.6080	0	0.1740	0.0087	19.9874
2d	1927	0.7024	0	0.1772	0.0077	22.9946
2e	1929	0.7427	0	0.1787	0.0072	24.7569
2f	1931	0.7804	0	0.1775	0.0066	26.8191
2g	1934	0.8635	0	0.1763	0.0052	34.0861
2h	1936	0.8847	0	0.1751	0.0047	37.1150
2i	1938	0.9079	0	0.1738	0.0042	41.5408
2j	1940	0.9126	0	0.1704	0.0040	42.6609

Box 4 is located after the Figures.

A.

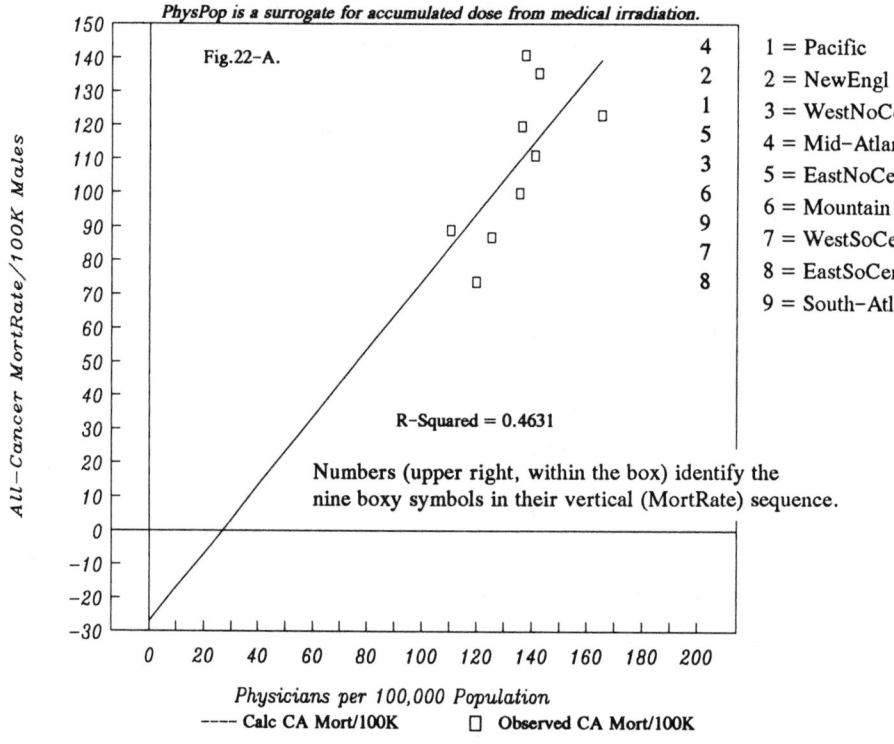

1940 All–Cancer Mortality–Rates versus
1921 PhysPop Values for the 9 Census Divisions, USA.
Dose–Response Relationship

PhysPop is a surrogate for accumulated dose from medical irradiation.

Fig.22-A.

R–Squared = 0.4631

Numbers (upper right, within the box) identify the
nine boxy symbols in their vertical (MortRate) sequence.

1 = Pacific
2 = NewEngl
3 = WestNoCen
4 = Mid–Atlan
5 = EastNoCen
6 = Mountain
7 = WestSoCen
8 = EastSoCen
9 = South–Atl

Physicians per 100,000 Population
---- Calc CA Mort/100K   ☐ Observed CA Mort/100K

B.

Same as A,
except line
of best –fit
has been
forced to
go through
the origin.

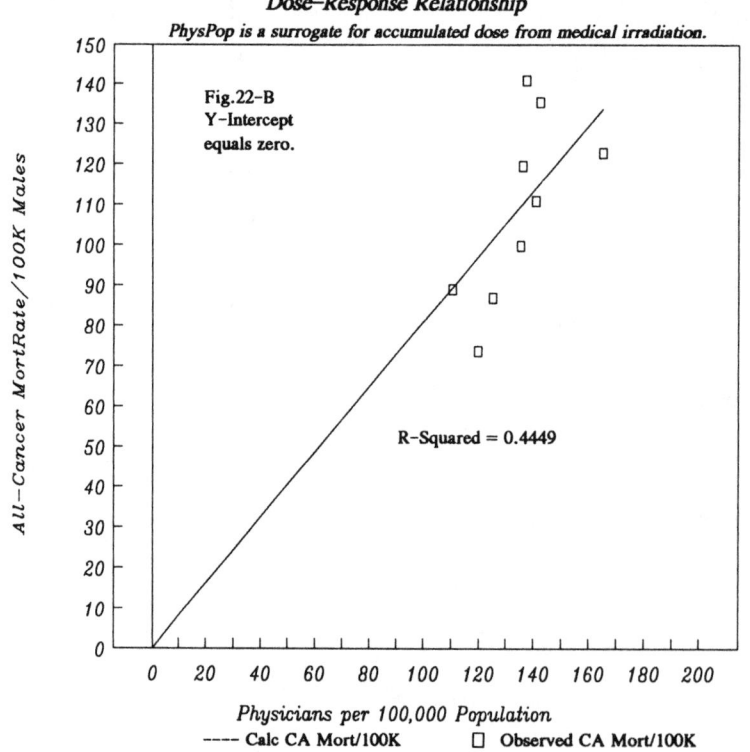

1940 All–Cancer Mortality–Rates versus
1921 PhysPop Values for the 9 Census Divisions, USA.
Dose–Response Relationship

PhysPop is a surrogate for accumulated dose from medical irradiation.

Fig.22-B
Y–Intercept
equals zero.

R–Squared = 0.4449

Physicians per 100,000 Population
---- Calc CA Mort/100K   ☐ Observed CA Mort/100K

Related text = Part 3a.

1940 All–Cancer Mortality–Rates versus
1940 PhysPop Values for the 9 Census Divisions, USA.
Dose–Response Relationship

PhysPop is a surrogate for accumulated dose from medical irradiation.

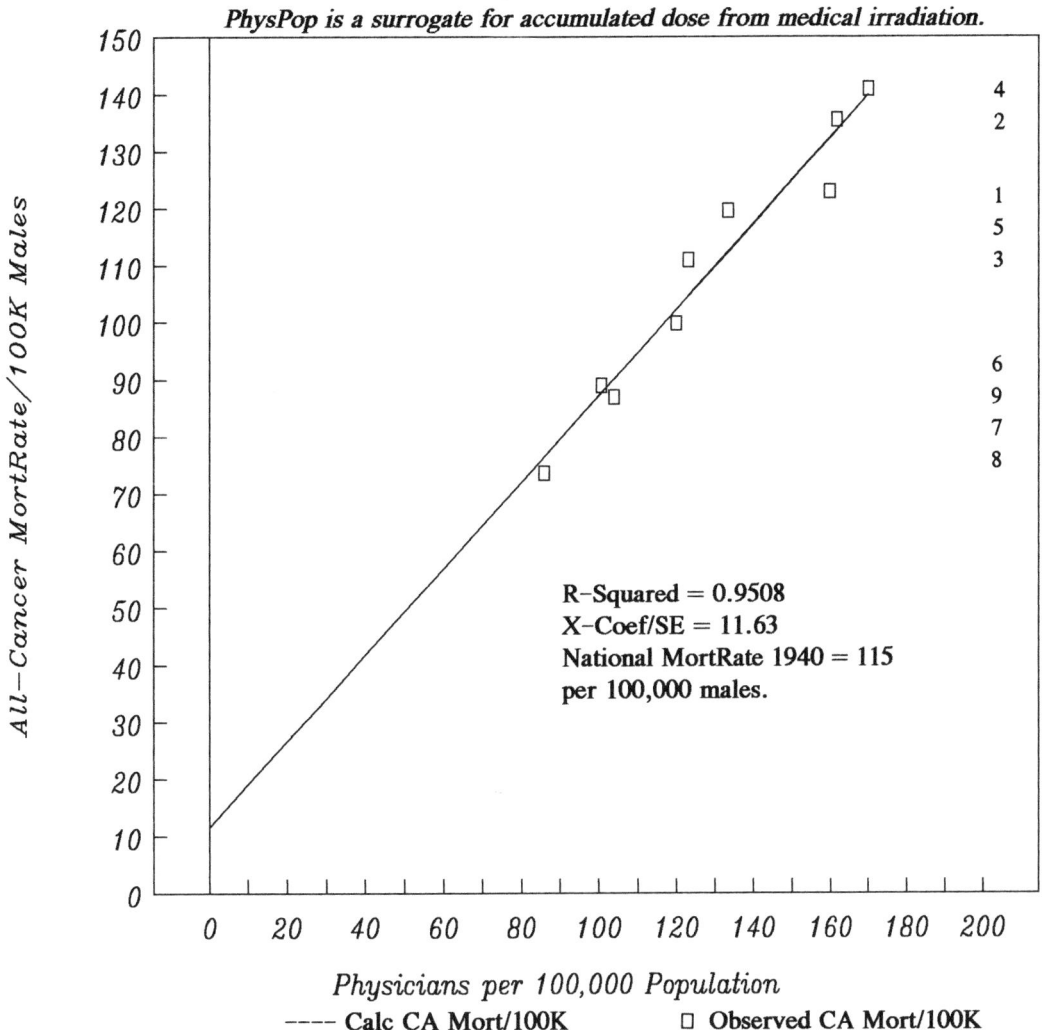

1 = Pacific	
2 = NewEngl	
3 = WestNoCen	
4 = Mid–Atlan	
5 = EastNoCen	
6 = Mountain	
7 = WestSoCen	
8 = EastSoCen	
9 = South–Atl	

R–Squared = 0.9508
X–Coef/SE = 11.63
National MortRate 1940 = 115
per 100,000 males.

Physicians per 100,000 Population

---- Calc CA Mort/100K   □ Observed CA Mort/100K

Related text = Parts 2d + 3b.

On the X–axis, PhysPop values = Physicians per 100,000 Population
in the Nine Census Divisions of the USA Population, Year 1940. This
variable is a surrogate for accumulated radiation dose --- the more
physicians per 100,000 people, the more radiation procedures are done per
100,000 people.

On the Y–axis, All–Cancer Mortality–Rate per 100,000 males = the
reported rates in USA Vital Statistics for the Nine Census Divisions, Year 1940.

Figure 22–A, nearby, shows that the essence of the relationship
between PhysPop and All–Cancer MortRates was already present by 1921.
Indeed, the 1921 PhysPops predict the male's 1940 National All–Cancer
MortRate quite well (text, Part 4 and Box 4).

Above, Figure 22–C shows that, by 1940, the tightness of the
correlation had improved to near perfection (text, Parts 2d and 2e).
Because the 1940 MortRates are the same in Figures 22–A and 22–C, only
lateral differences occur in the positions of the nine boxy symbols.

In the upper part of this box, we calculate the national PhysPop values for 1921, 1931, and 1938. We already know that for 1940, the national PhysPop value is 132.04 (Chapter 6, Box 4). Pop'n Fractions: Table 3-B.

(A) Census Division	(B) 1920 Pop'n Fraction	(C) PhysPop 1921	(D) 1921 Weighted PhysPop	(E) 1930 Pop'n Fraction	(F) PhysPop 1931	(G) 1931 Weighted PhysPop	(H) 1940 Pop'n Fraction	(I) PhysPop 1938	(J) 1938 Weighted PhysPop
Pacific	0.0529	165.11	8.73	0.0670	159.97	10.72	0.0739	157.62	11.65
New England	0.0703	142.24	10.00	0.0676	142.35	9.62	0.0641	154.08	9.88
West No. Centra	0.1192	140.93	16.80	0.1086	126.50	13.74	0.1027	124.95	12.83
Mid-Atlantic	0.2115	137.29	29.04	0.2146	140.82	30.22	0.2092	160.69	33.62
East No. Central	0.2040	136.06	27.76	0.2067	128.59	26.58	0.2022	131.98	26.69
Mountain	0.0317	135.38	4.29	0.0303	118.89	3.60	0.0315	119.88	3.78
West So. Central	0.0973	125.15	12.18	0.0995	105.95	10.54	0.0992	102.79	10.20
East So. Central	0.0845	119.76	10.12	0.0808	96.73	7.82	0.0819	88.21	7.22
South Atlantic	0.1287	110.32	14.20	0.1251	99.59	12.46	0.1354	99.26	13.44
Sums	1.0001		133.11	1.0002		125.30	1.0001		129.30

All the predictions below use the equation of best fit:
Cancer MortRate 1940 = (Xcoef * Natl PhysPop) + Constant.
The values for Xcoef and Constant come from Part 2 of Chapters 6, 7, 8.
For the zero-intercept calculations, Xcoefs and Constants come from Chapter 22, Box 3.

		PREDICTED	OBSERVED
● - MALES, ALL-CANCERS.			
1921 Best-Fit Eq. MALE MR 1940 =	(1.0086*133.11)+(-27.0754) =	107.2	.... Observed = 115
MALES All-Canc. w. zero intercept=	(0.8100*133.11) =	107.8	.... Observed = 115
1931 Best-Fit Eq. MALE MR 1940 =	(0.9582*125.3)+(-10.4041) =	109.7	.... Observed = 115
MALES All-Canc. w. zero intercept=	(0.8767*125.3) =	109.9	.... Observed = 115
1938 Best-Fit Eq. MALE MR 1940 =	(0.8351*129.3)+3.0512 =	111.0	.... Observed = 115
MALES All-Canc. w. zero intercept=	(0.8583*129.3) =	111.0	.... Observed = 115
1940 Best-Fit Eq. MALE MR 1940 =	(0.7557*132.04)+11.55 =	111.3	.... Observed = 115

		PREDICTED	OBSERVED
● - FEMALES, ALL-CANCERS.			
1921 Best-Fit Eq. FEM. MR 1940 =	(0.6497*133.11)+(33.3847) =	119.9	.... Observed = 126.1
FEMALES All-Ca. w. zero intercept=	(0.8947*133.11) =	119.1	.... Observed = 126.1
1931 Best-Fit Eq. FEM. MR 1940 =	(0.6524*125.3)+(39.754) =	121.5	.... Observed = 126.1
FEMALES All-Ca. w. zero intercept=	(0.9639*125.3) =	120.8	.... Observed = 126.1
1938 Best-Fit Eq. FEM. MR 1940 =	(0.5801*129.3)+47.4535 =	122.5	.... Observed = 126.1
FEMALES All-Ca. w. zero intercept=	(0.9406*129.3) =	121.6	.... Observed = 126.1
1940 Best-Fit Eq. FEMALE MR 1940 =	(0.5279*132.04)+52.984 =	122.7	.... Observed = 126.1

		PREDICTED	OBSERVED
● - FEMALES, BREAST CANCER.			
1921 Best-Fit Eq. FEM. MR 1940 =	(0.2440*133.11)+(-10.9421) =	21.5	.... Observed = 23.3
FEMALES Breast Ca w. zero intercept	(0.1637*133.11) =	21.8	.... Observed = 23.3
1931 Best-Fit Eq. FEM. MR 1940 =	(0.2270*125.3)+(-6.3107) =	22.1	.... Observed = 23.3
FEMALES Breast Ca w. zero intercept	(0.1775*125.3) =	22.2	.... Observed = 23.3
1938 Best-Fit Eq. FEM. MR 1940 =	(0.1906*129.3)+(-2.2092) =	22.4	.... Observed = 23.3
FEMALES Breast Ca w. zero intercept	(0.1738*129.3) =	22.5	.... Observed = 23.3
1940 Best-Fit Eq. FEMALE MR 1940 =	(0.1713*132.04)+(-0.1205) =	22.5	.... Observed = 23.3
FEMALES Breast Ca w. zero intercept	(0.1704*132.04) =	22.5	.... Observed = 23.3

Related text = Part 4.

Part 1. The Purpose of Section Three in This Book
Part 2. Parallel Analyses for Malignancies and Non-Malignancies
Part 3. Regression-Outputs: The Remarkable X-Coefficients in Box 1

Box 1. Summary: Regression Outputs, All Causes of Death Combined.
Figures 23-A+B. Graphs (Male, Female): 1940 MortRates with 1940 PhysPops.
Table 23-A. All Causes of Death Combined: Rates by Census Divisions and National.

● **Part 1. The Purpose of Section Three of This Book**

Now we begin Section Three of our inquiry. Our purpose is to learn whether or not the extremely strong positive correlation, observed between 1940 PhyPops and 1940 cancer mortality-rates, also occurred (a) for Non-Malignancies as a group, and (b) for specific types of Non-Malignancies. This inquiry was undertaken as an independent check on the concept set forth in Chapter 3: That we could explore the relationship, between the average per capita accumulated dose of medical radiaton and cancer MortRates, by studying the relationship between PHYSPOP and cancer MortRates.

1a. The Logic: Expectation of a Contrast

Our reasoning begins with the current "general wisdom": The only proven cause of DEATH, inducible by ionizing radiation in irradiated people, is fatal cancer. The established exception is prompt noncancer death in a person who has received an extremely high dose of ionizing radiation all at once (acute exposure) to most or all of the body. The prompt deaths in 1945, among many persons who briefly survived the Hiroshima-Nagasaki bombings, and the prompt deaths in 1986 among some of the firemen who tried to extinguish the fire at the Chernobyl nuclear reactor, are examples. High, acute, whole-body doses are not part of medical practice, without bone-marrow transplantation. One additional exception would be death due to unrecognized and therefore untreated radiation-induced myxedema (severe hypo-thyroidism), resulting in coma and death.

The "general wisdom" above has the following implication, for our study of the role of medical radiation in cancer MortRates:

We would expect the PhysPop-Cancer relationship to DIFFER from the PhysPop-NonCancer relationship, if PhysPop is approximately proportional to accumulated dose from medical radiation. If PhysPop FAILS to produce strong positive correlations with noncancer causes of death, while PhysPop DOES produce strong positive correlations with cancer MortRates, the contrast will be a powerful, independent, confirmatory piece of evidence that the dose-response, between PhysPop and cancer MortRates, is a dose-response between medical radiation and cancer MortRates.

1b. Overview of the Results

Results? During the first four decades after ionizing radiation was introduced into medicine, nearly all Non-Malignancies behaved VERY DIFFERENTLY from Malignancies, with respect to PhysPop --- as already indicated in Chapter 4, Part 1. The numerous chapters in Section Three provide the data which support our earlier statement. For everyone's convenience, the results are tabulated for comparison in Chapter 38, Box 1.

1c. An Important Exception: Ischemic Heart Disease

The big exception to our finding about Non-Malignancies is Ischemic Heart Disease (IHD). We discovered that it behaves VERY SIMILARLY to the Malignancies, with respect to PhysPop. This startling result is not a marginal finding --- it is statistically very strong. It led us to propose

Hypothesis-2: "Medical radiation, received even at very low and moderate doses , is an important cause of Ischemic Heart Disease..."

The data for Ischemic Heart Disease are provided, analyzed, and discussed in their own sections of this book: Section Four.

● Part 2. Parallel Analyses for Malignancies and Non-Malignancies

In Chapters 23 through 37, we have done exactly the same linear regressions of MortRates upon PhysPops as we did for cancers in Chapters 6 through 21. The very same PhysPop values are used from the Universal PhysPop Table 3-A. And the MortRates in Chapters 23 through 37 come from the same sources described in Chapter 4. In short, the results in Section Three can be validly compared with the results in Section Two.

The two sets of chapters look different, however. For Non-Malignancies, we do not show each regression analysis separately. By now, readers must be saturated with the format. We present the regression OUTPUT in Box 1 of each chapter. The INPUT-data are provided by the MortRate table at the end of each chapter, and by the Universal PhysPop Table 3-A, which means that we present all the data which readers need if they wish to verify the work independently. In addition, Box 1 extends the analyses for Non-Malignancies beyond 1940 --- which we will do in Section Five for the Malignancies and Ischemic Heart Disease.

2a. The Tables of Mortality Rates

We begin here in Chapter 23 with All Causes of Death Combined (please see Table 23-A, at the end of this chapter). In Chapter 24, we subtract All-Cancers, to obtain a good approximation for All Non-Malignancies.

In the MortRate tables of Section Three, as in Section Two, the MortRate entries for the Nine Census Divisions are population-weighted, whereas the averages below them are not. "High-5" continues to refer to the first five Census Divisions in the list (Pacific, New England, West North Central, Mid-Atlantic, East North Central). "Low-4" refers to the last four (Mountain, West South Central, East South Central, South Atlantic).

2b. Some Approximations in Table 23-A

Although it is convenient to label Table 23-A as "All Causes of Death Combined," in reality, it seems that not quite every cause of death is included. The MortRates available in Table 67 of Grove 1968, for the years 1940, 1950, and 1960 by Census Divisions, cover "32 Selected Causes of Death." These causes are listed with their ICD numbers (7th Revision) in our Chapter 4, Part 5. Nearly every cause of death is included: Major illnesses (cancer included), accidents, suicide, homicide, miscellaneous, and cause unknown.

By contrast with Grove's Table 67 (which consumes over 100 pages), Grove's Table 54 reports consolidated national "Age-Adjusted Death Rates" for those same years in just a single page (Grove p.317). Because Table 54 provides no information by Census Divisions or by states, it is useless for our studies.

Nonetheless, we mention Table 54 here because its consolidated national rates (total death-rates) for 1950 and 1960 show up in recent government publications such as "Health United States, 1995" (PHS 1995, Table 36, p.122) --- and neither the PHS entries nor their sources in Grove's Table 54 are identical with the consolidated national rates in Grove's Table 67, which we use in our Table 23-A. The disparity between Grove's Table 54 and Grove's Table 67, with respect to the TOTAL death-rates in 1940, 1950, and 1960, causes us to infer that the entries for "All Causes" in Grove's Table 67 are for the combination of All 32 Selected Causes, and not for absolutely every cause.

Fortunately, a "first approximation" of the TOTAL death rates is all that we need to execute the inquiry which we described above in Part 1a. For this purpose, our Table 23-A uses the "All Causes" entries from Grove's Table 67 for the years 1940, 1950, 1960. For the year 1980, our Table 23-A uses the "All Causes" entries from reference NatCtrHS 1980. In that document, the 1980 All-Cause rates --- though labeled "All Causes" --- may be a first approximation which does not provide exact continuity for Grove's "All Causes." However, the nature of our Section Three does not demand a perfect match.

● **Part 3. Regression-Outputs: The Remarkable X-Coefficients in Box 1**

Box 1 presents the output from all the linear regressions. The first ten lines of each group (male, female) are the outputs from regressing the 1940 MortRates of Table 23-A upon the PhysPops of 1921-1940, in parallel with our analyses for cancer. The remaining lines in each group regress the 1950 MortRates upon the 1950 PhysPops, the 1960 MortRates upon the 1960 PhysPops, etc.

3a. Inverse Relationships: X-Coefficients with Negative Signs

In great contrast to the cancer chapters, there are no statistically significant values for R-squared or for the X-Coefficient/SE ratio, when the 1940 MortRates for All-Causes-Combined are regressed upon 1940 PhysPops.

However, in the FULL female set of regression output (Box 1), there are some statistically significant relationships, as indicated by X-Coef/SE ratios above ~ 2.0.

And what else is remarkable? These significant relationships are INVERSE relationships between PhysPop and MortRates, as revealed by the negative sign on the X-Coefficient. The higher is the physician-density, the lower is the MortRate. Moreover, we should not ignore the statistically "non-significant" output when we are pondering positive versus negative correlations. In a "sign" test, the totality of information is to be considered. All ten correlations for the males, and all ten correlations for the females, produce X-Coefficients with a negative sign. The constancy of direction is itself a test of significance --- in this case, for an INVERSE relationship between All Causes of Death Combined, and PhysPop. And this occurs even though All-Causes-Combined includes cancer, which produces a consistently POSITIVE X-Coefficient.

The inverse correlations suggested in Box 1 are consistent with the finding in Table 23-A that the Hi5/Lo4 MortRate ratios are consistently below 1.0, whereas the Hi5/Lo4 PhysPop ratios in Table 3-A are consistently above 1.0.

The results in Box 1 are extremely different from the results for cancer alone. We remind readers that the cancer regressions (both statistically significant ones, and non-significant ones) produce only positive X-Coefficients in Chapters 6 through 21.

3b. The Graphs of 1940 MortRates Regressed on 1940 PhysPops

For our graphs, we pick the 1940-1940 combination of variables, so that these graphs can be compared with the graphs for cancer. The graphs are prepared in the way demonstrated repeatedly in Box 2 of the cancer chapters. In all the graphs, PhysPop is a surrogate for average accumulated dose from medical radiation.

Figures 23-A+B (males, females) depict the best-fit lines when the 1940 MortRates are regressed upon the 1940 PhysPops. Readers can appreciate the scatter of the boxy symbols, which represent the nine real-world datapoints, and can note the direction of the best-fit line (up, flat, or down).

When the sign of the X-Coefficient in Box 1 is negative for the 1940-1940 combination, the direction of the best-fit line is downward. When the sign is positive (in some other chapters), the direction is upward. We remind readers (from Chapter 6, Part 3) that the visual steepness of the best-fit line --- but not its direction --- is tied to the scales for the y-axis and x-axis. A flat line of best-fit indicates that the y-variable (MortRate) does not respond to increases in the x-variable (PhysPop).

Box 1 of Chap.23

Summary: Regression Outputs, All Causes of Death Combined.

Below are the summary-results from all the regressions of MortRates upon PhysPops. MortRates are from Table 23-A, and PhysPops are from Table 3-A.

MALES

Year MortRate	Year PhysPop	R-squared	Constant	X-Coef	Std Err	X-Coef/SE
1940	1921	0.2634	1609.61	-3.0177	1.9073	-1.5822
1940	1923	0.2459	1559.28	-2.7183	1.7991	-1.5110
1940	1925	0.2338	1514.92	-2.4576	1.6818	-1.4613
1940	1927	0.2271	1492.64	-2.3208	1.6181	-1.4343
1940	1929	0.2112	1470.96	-2.1649	1.5815	-1.3689
1940	1931	0.1900	1436.09	-1.8729	1.4615	-1.2815
1940	1934	0.1839	1406.03	-1.6218	1.2915	-1.2558
1940	1936	0.1736	1393.86	-1.5154	1.2498	-1.2125
1940	1938	0.1596	1375.91	-1.3646	1.1833	-1.1532
1940	1940	0.1299	1345.73	-1.1082	1.0839	-1.0224
1950	1950	0.0997	1085.78	-0.4912	0.5580	-0.8803
1960	1960	0.0650	1032.23	-0.3759	0.5388	-0.6977
1970	1970	0.1437	947.88	-0.5112	0.4717	-1.0838
1980	1980	0.2343	882.58	-0.6247	0.4269	-1.4636
1990	1990	--	--	--	--	--

FEMALES

Year MortRate	Year PhysPop	R-squared	Constant	X-Coef	Std Err	X-Coef/SE
1940	1921	0.7010	1514.19	-4.3050	1.0626	-4.0514
1940	1923	0.6674	1447.39	-3.9161	1.0448	-3.7482
1940	1925	0.6377	1384.66	-3.5497	1.0112	-3.5103
1940	1927	0.5753	1337.23	-3.2299	1.0490	-3.0790
1940	1929	0.5443	1310.32	-3.0393	1.0511	-2.8915
1940	1931	0.5122	1268.79	-2.6890	0.9918	-2.7113
1940	1934	0.4390	1208.49	-2.1916	0.9364	-2.3405
1940	1936	0.4000	1187.51	-2.0117	0.9313	-2.1602
1940	1938	0.3468	1157.01	-1.7588	0.9123	-1.9278
1940	1940	0.2823	1118.10	-1.4283	0.8609	-1.6592
1950	1950	0.0654	768.18	-0.4147	0.5927	-0.6997
1960	1960	0.0072	651.92	-0.1041	0.4624	-0.2251
1970	1970	0.0118	542.80	-0.0912	0.3152	-0.2894
1980	1980	0.0346	441.82	-0.1122	0.2240	-0.5009
1990	1990	--	--	--	--	--

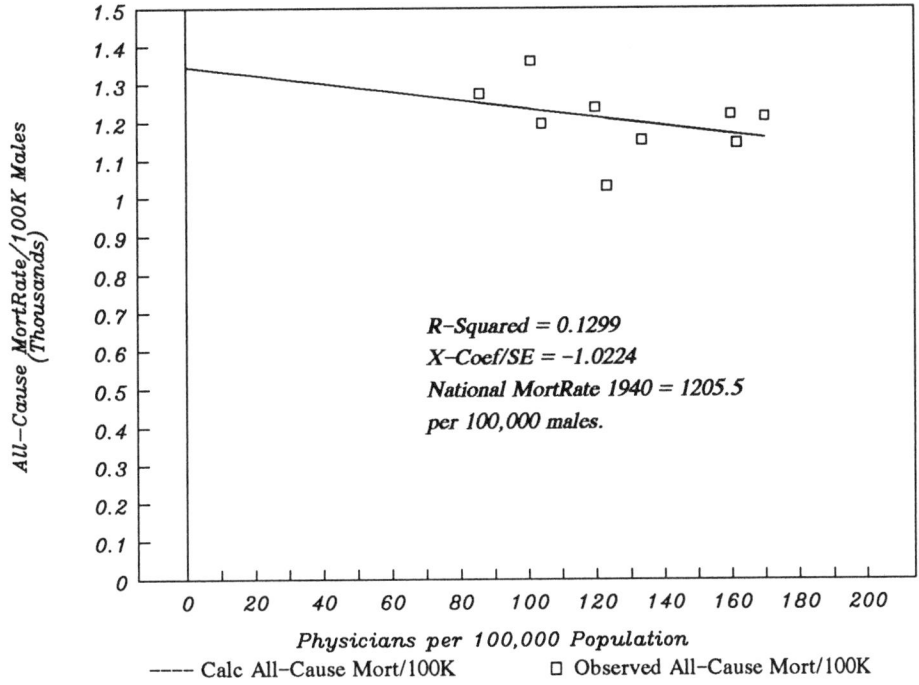

1940 MortRate, Males, All-Causes Combined, versus
1940 PhysPop Values for the 9 Census Divisions, USA.
No provable relationship.
PhysPop is a surrogate for accumulated dose from medical irradiation.

R-Squared = 0.1299
X-Coef/SE = -1.0224
National MortRate 1940 = 1205.5
per 100,000 males.

All-Cause MortRate/100K Males
(Thousands)

Physicians per 100,000 Population
---- Calc All-Cause Mort/100K   □ Observed All-Cause Mort/100K

All Causes of Death, Combined   Figure 23-B

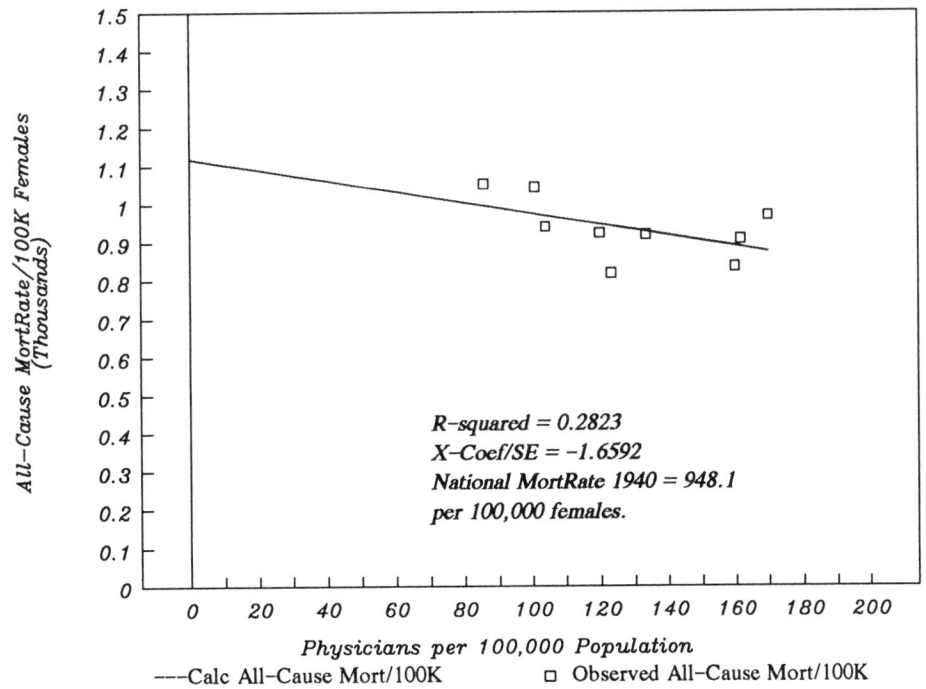

1940 MortRate, Females, All-Causes Combined, versus
1940 PhysPop Values for the 9 Census Divisions, USA.
No Provable Relationship
PhysPop is a surrogate for accumulated dose from medical irradiation

R-squared = 0.2823
X-Coef/SE = -1.6592
National MortRate 1940 = 948.1
per 100,000 females.

All-Cause MortRate/100K Females
(Thousands)

Physicians per 100,000 Population
----Calc All-Cause Mort/100K   □ Observed All-Cause Mort/100K

Table 23-A

All Causes of Death Combined: Rates by Census Divisions and National.

The term "All Causes of Death Combined" is an approximation (text, Part 2b). Annual rates per 100,000 are age-adjusted to the 1940 reference year. There are no exclusions by color or "race." Entries for the Nine Census-Divisions are population-weighted; the averages below them are not. "National Rates" for both sexes: Deaths per 100,000 population (males + females). Males: Deaths per 100,000 male population. Females: Deaths per 100,000 female population. Sources and omissions are discussed in Chapter 4, Part 2.

MALES ....

Census Division	1940	1950	1960	1970	1980	1990
Pacific	1219.8	1002.7	934.2	820.7	707.1	--
New England	1144.3	983.6	991.1	856.6	722.0	--
West North Central	1031.3	957.1	923.7	815.1	706.5	--
Mid-Atlantic	1213.0	1056.9	1007.2	893.5	779.8	--
East North Central	1151.4	1032.4	986.9	882.3	777.7	--
Mountain	1238.2	1021.0	957.6	833.9	710.2	--
West South Central	1196.2	994.7	979.6	891.0	802.4	--
East South Central	1273.8	1066.4	1035.8	940.6	845.4	--
South Atlantic	1360.3	1103.9	1053.2	935.1	817.0	--
Average, ALL	1203.1	1024.3	985.5	874.3	763.1	--
Average, High-5	1152.0	1006.5	968.6	853.6	738.6	--
Average, Low-4	1267.1	1046.5	1006.5	900.1	793.8	--
Ratio, Hi5/Lo4	0.91	0.96	0.96	0.95	0.93	--

FEMALES ....

Census Division	1940	1950	1960	1970	1980	1990
Pacific	835.3	647.4	593.7	501.2	408.6	--
New England	907.1	702.6	648.7	524.6	400.5	--
West North Central	818.8	672.2	601.3	490.5	379.7	--
Mid-Atlantic	968.9	767.6	678.1	561.3	444.5	--
East North Central	919.5	733.0	652.5	544.6	436.6	--
Mountain	923.9	691.4	600.1	497.2	394.3	--
West South Central	939.9	684.6	619.8	526.3	432.8	--
East South Central	1052.5	783.6	684.6	565.4	446.2	--
South Atlantic	1043.1	764.0	671.9	556.0	440.1	--
Average, ALL	934.3	716.3	639.0	529.7	420.4	--
Average, High-5	889.9	704.6	634.9	524.4	414.0	--
Average, Low-4	989.9	730.9	644.1	536.2	428.4	--
Ratio, Hi5/Lo4	0.90	0.96	0.99	0.98	0.97	--

NATIONAL RATES ....

	1940	1950	1960	1970	1980	1990
Both Sexes	1077.2	881.5	811.5	694.9	578.3	520
Males	1205.5	1037.1	990.4	879.3	768.1	--
Females	948.1	731.7	648.2	537.3	426.4	--

● - 1940, 1950, 1960: All rates come from Grove 1968, Table 67, Page 663, "All Causes." No ICD numbers were given. Please see Chap. 23, Part 2b; also Chap. 4, Part 5.
● - 1970 rates are interpolations (Chap. 4, Parts 2b, 2c).
● - 1980: All rates come from the reference NatCtrHS 1980. No IDC numbers were given. Please see Chap. 23, Part 2b.
● - 1990: "Both sexes" national rate is from PHS 1995, Table 36, p.122. No other data were obtained. Please see Chap. 4, Part 2c.

CHAPTER 24

All Causes of Death Except Cancer: Relation with Medical Radiation

MortRates for All Causes of Death Except Cancer

To obtain the MortRates for "All Causes of Death except Cancer," we begin with the MortRates per 100,000 for All-Causes-Combined in Table 23-A, and we subtract the matching rates per 100,000 for All-CANCERS-Combined from Table 6-A (males) or Table 7-A (females). The resulting rates are provided in Table 24-A at the end of this chapter.

Example for Males, 1940, Pacific Census Division: The entry of 1096.9 = (1219.8 from Table 23-A) minus (122.9 from Table 6-A). The National MortRates in Table 24-A are obtained in similar fashion.

Box 1: The Summary-Results of All the Regressions

Following the pattern of Chapter 23, we regressed all the MortRates of Table 24-A upon the PhysPops indicated in Box 1, where the results are presented. Figures 24-A and 24-B depict the best-fit line when the 1940 MortRates (male, female) are regressed upon the 1940 PhysPops.

Subtraction of the All-Cancer MortRates, from All Causes of Death Combined, results in considerably stronger correlations (higher R-squared values in Box 1) than in the previous chapter for All Causes Combined. In other words, removal of the cancer rates leaves noncancer mortality rates in 1940 having a stronger relationship with PhysPop than in Chapter 23 --- and the relationship is INVERSE (negative).

By contrast, let us consider Box 1 in Chapters 6 and 7 for All-Cancer mortality. There, a POSITIVE correlation is increasing to highly significant levels, as one approaches the 1940 PhysPops. In relation to PhysPop, the DIFFERENCE in behavior between cancer mortality-rates and noncancer mortality-rates is extremely clear for both males and females.

Box 1 is on the next page.

>>>>>>>>>>

Below are the summary-results from all the regressions of MortRates upon PhysPops. MortRates are from Table 24-A, and PhysPops are from Table 3-A.

MALES ....

Year MortRate	Year PhysPop	R-squared	Constant	X-Coef	Std Err	X-Coef/SE
1940	1921	0.3624	1636.68	-4.0263	2.0184	-1.9947
1940	1923	0.3594	1584.12	-3.7382	1.8861	-1.9819
1940	1925	0.3551	1531.47	-3.4455	1.7550	-1.9632
1940	1927	0.3681	1513.58	-3.3607	1.6643	-2.0193
1940	1929	0.3566	1490.23	-3.2000	1.6246	-1.9696
1940	1931	0.3356	1446.49	-2.8311	1.5057	-1.8803
1940	1934	0.3409	1408.63	-2.5122	1.3202	-1.9029
1940	1936	0.3340	1395.28	-2.3911	1.2763	-1.8735
1940	1938	0.3206	1372.86	-2.1997	1.2103	-1.8175
1940	1940	0.2841	1334.18	-1.8639	1.1184	-1.6667
1950	1950	0.3487	1029.08	-1.0511	0.5429	-1.9360
1960	1960	0.2745	943.07	-0.7938	0.4877	-1.6275
1970	1970	0.2950	820.91	-0.6831	0.3991	-1.7115
1980	1980	0.3591	725.53	-0.6520	0.3293	-1.9803
1990	1990	--	--	--	--	--

FEMALES ....

Year MortRate	Year PhysPop	R-squared	Constant	X-Coef	Std Err	X-Coef/SE
1940	1921	0.7650	1480.81	-4.9547	1.0378	-4.7742
1940	1923	0.7495	1412.42	-4.5719	0.9990	-4.5763
1940	1925	0.7369	1346.76	-4.2039	0.9494	-4.4279
1940	1927	0.7009	1303.41	-3.9280	0.9698	-4.0505
1940	1929	0.6795	1276.26	-3.7413	0.9712	-3.8520
1940	1931	0.6516	1229.04	-3.3415	0.9235	-3.6184
1940	1934	0.5922	1164.24	-2.8042	0.8796	-3.1882
1940	1936	0.5569	1142.55	-2.6152	0.8817	-2.9659
1940	1938	0.5053	1109.56	-2.3389	0.8747	-2.6738
1940	1940	0.4362	1065.12	-1.9562	0.8406	-2.3271
1950	1950	0.2008	686.511	-0.7124	0.5371	-1.3264
1960	1960	0.0876	568.964	-0.3350	0.4088	-0.8196
1970	1970	0.1168	458.902	-0.2669	0.2774	-0.9622
1980	1980	0.1916	362.017	-0.2530	0.1964	-1.2883
1990	1990	--	--	--	--	--

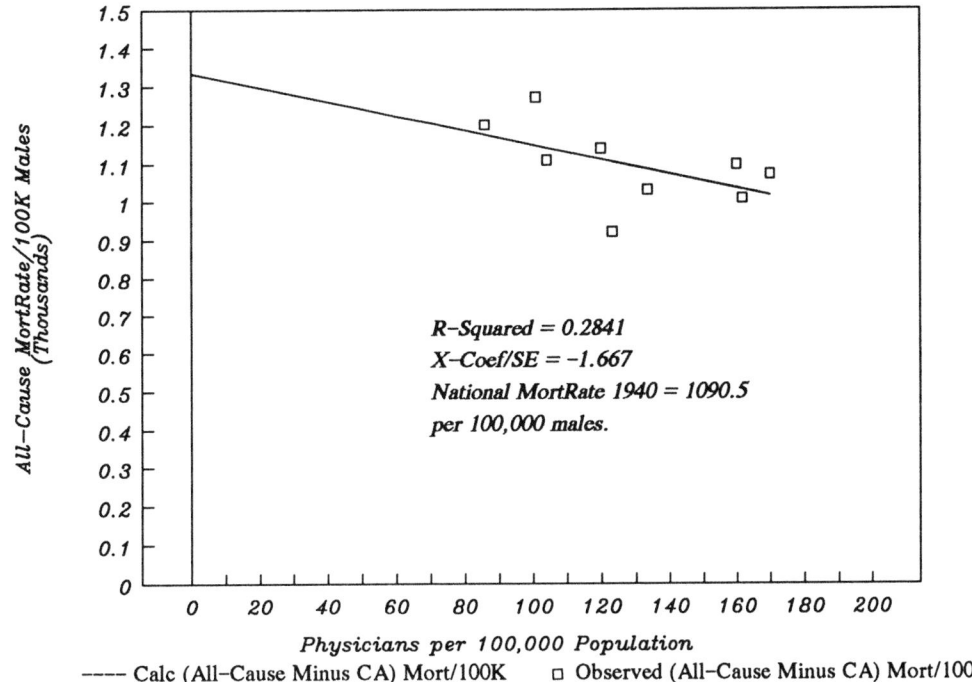

*1940 MortRate, Males, All-Causes Minus Cancer, versus
1940 PhysPop Values for the 9 Census Divisions, USA.
Relationship of Borderline Significance.*
PhysPop is a surrogate for accumulated dose from medical irradiation.

R-Squared = 0.2841
X-Coef/SE = -1.667
National MortRate 1940 = 1090.5
per 100,000 males.

---- Calc (All-Cause Minus CA) Mort/100K   □ Observed (All-Cause Minus CA) Mort/100K

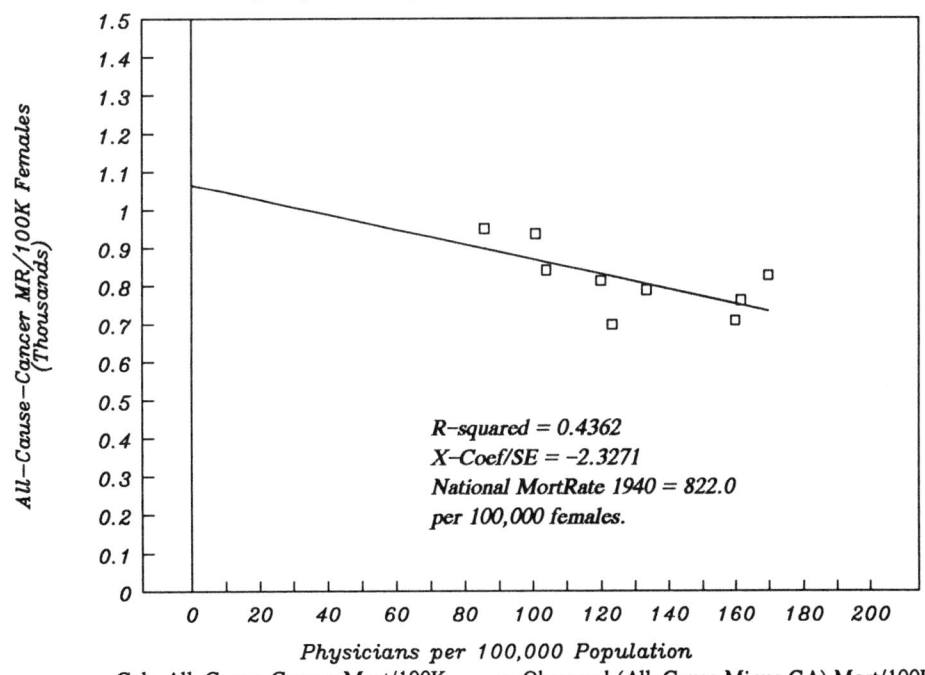

*1940 MortRate, Females, All-Causes Minus Cancer, versus
1940 PhysPop Values for the 9 Census Divisions, USA.
Significant INVERSE Relationship*
PhysPop is a surrogate for accumulated dose from medical irradiation

R-squared = 0.4362
X-Coef/SE = -2.3271
National MortRate 1940 = 822.0
per 100,000 females.

---Calc All-Cause-Cancer Mort/100K   □ Observed (All-Cause Minus CA) Mort/100K

Table 24-A

These rates are the "All Causes of Death" rates in Table 23-A minus the "All-Cancer" rates in Table 6-A (males) or Table 7-A (females). Rates per 100,000 are age-adjusted to the 1940 reference year. There are no exclusions by color or "race."

MALES ....

Census Division	1940	1950	1960	1970	1980	1990
Pacific	1096.9	875.5	793.5	673.5	553.4	--
New England	1008.8	831.2	826.5	689.1	551.7	--
West North Central	920.4	831.8	788.1	671.3	554.5	--
Mid-Atlantic	1072.1	900.9	843.2	725.6	608.0	--
East North Central	1031.8	894.1	836.2	722.2	608.2	--
Mountain	1138.4	912.9	838.9	707.2	575.5	--
West South Central	1109.3	882.0	845.8	742.7	639.5	--
East South Central	1200.2	961.7	910.7	791.0	671.3	--
South Atlantic	1271.4	987.6	916.1	780.9	645.6	--
Average, ALL	1094.4	897.5	844.3	722.6	600.9	--
Average, High-5	1026.0	866.7	817.5	696.3	575.2	--
Average, Low-4	1179.8	936.1	877.9	755.4	633.0	--
Ratio, Hi5/Lo4	0.87	0.93	0.93	0.92	0.91	--

FEMALES ....

Census Division	1940	1950	1960	1970	1980	1990
Pacific	707.9	529.7	483.6	391.0	298.2	--
New England	761.8	570.5	526.3	405.2	284.1	--
West North Central	698.7	555.1	492.0	385.4	278.7	--
Mid-Atlantic	826.0	630.6	550.7	438.9	327.0	--
East North Central	788.1	605.5	532.7	428.7	324.6	--
Mountain	812.1	585.4	499.1	399.3	299.4	--
West South Central	840.1	575.3	516.9	424.8	332.7	--
East South Central	950.0	673.3	579.8	461.4	343.0	--
South Atlantic	936.2	650.7	564.5	449.8	335.1	--
Average, ALL	813.4	597.3	527.3	420.5	313.6	--
Average, High-5	756.5	578.3	517.1	409.8	302.5	--
Average, Low-4	884.6	621.2	540.1	433.8	327.6	--
Ratio, Hi5/Lo4	0.86	0.93	0.96	0.94	0.92	--

NATIONAL RATES ....

Rates are age-adjusted to the 1940 reference year. Both sexes: Deaths per 100,000 population (males + females). Males: Deaths per 100,000 male population. Females: Deaths per 100,000 female population. No exclusions by color or "race."

	1940	1950	1960	1970	1980	1990
Both Sexes	956.9	753.8	682.4	564.3	446.4	--
Males	1090.5	904.3	844.7	724.2	603.6	--
Females	822.0	608.5	533.3	425.6	317.9	--

CHAPTER 25

All Causes of Death Except (Cancer-and-IHD): Relation with Medical Radiation

• Part 1. MortRate-PhysPop Correlations for NonCancer NonIHD Causes of Death

Chapter 24 revealed that MortRates from the disease, Cancer --- known to be inducible by ionizing radiation --- had a very different relationship in 1940 with PhysPop than all the remaining causes of death. We could have stopped there. To have a shorter book was a great temptation. But so was another temptation. It is almost as compelling, to look for information carried in abstract MortRates, as it has been for us to look for information carried in various molecules and chromosomes in our experimental laboratories. So we undertook to find out what might be the relationship between PhysPop and DISTINCT NonCancer causes of death. The findings are summarized in Chapter 38, Box 1.

1a. Large Surprise from a Closer Look at the NonCancer MortRates

To our astonishment, our closer look at the NonCancer MortRates revealed that Ischemic Heart Disease (IHD) --- also called Coronary Heart Disease and Coronary Atherosclerosis and Coronary Artery Disease --- is a cause of death whose relationship with PhysPop closely resembles the relationship of CANCER with PhysPop.

The results for Ischemic Heart Disease are the last entries in Chapter 38, Box 1. The regression analyses themselves, of IHD MortRates regressed on PhysPop, are fully shown in Chapters 40 and 41 of this book.

The finding, of a strong positive correlation between IHD MortRates and PhysPop, compelled us to return to the MortRates in Table 24-A, and to subtract from them the MortRates for Ischemic Heart Disease from Chapters 40 and 41. The subtractions create Table 25-A: MortRates for All Causes of Death Except (Cancer+IHD) --- more simply called the NonCancer NonIHD MortRates.

Because 1950 is the first year for which mortality rates by states or by Census Divisions are available for Ischemic Heart Disease, Table 25-A necessarily excludes 1940. Also it excludes 1990, because Table 23-A (All Causes) lacks the required entries.

1b. Summary: The NonCancer NonIHD MortRates Regressed on PhysPop

Following the patterns of Chapters 23 and 24, we regressed the NonCancer NonIHD MortRates of Table 25-A upon the PhysPops indicated in Box 1, where the results are presented.

Figures 25-A and 25-B depict the best-fit line when the 1950 NonCancer NonIHD MortRates (male, female) are regressed upon the 1940 PhysPops. The slopes are clearly negative, and the nine boxy symbols all lie close to the best-fit line. The strength of the negative correlation is quite high. For males, the R-squared value is 0.7933, and the ratio of X-Coefficent/StandardError is -5.1831. For females, the R-squared value is 0.7037 and the ratio of the X-Coefficient/StandardError is -4.0771.

The very sharp contrast of these findings for NonCancer NonIHD MortRates, versus the findings for All-Cancers and for Ischemic Heart Disease, are shown on the first page of Chapter 38.

• Part 2. The BENEFIT of High PhysPops for NonCancer NonIHD Afflictions

The signficant NEGATIVE correlation, by Census Divisions, of the NonCancer NonIHD MortRates with PhysPop, supports what many physicians may regard as self-evident: The more physicians there are per 100,000 population, the better a population fares with many health problems.

The fact, that more physicians per 100,000 population means more radiation procedures per 100,000 population (Chapter 3, Part 1a), probably helps to explain the negative correlations in Box 1 of this chapter with respect to NonCancer NonIHD death-rates. We certainly concur that diagnostic and interventional medical radiation can supply life-extending information in the appropriate circumstances. It is likely that medical radiation should claim part of the CREDIT for the beneficial negative correlations uncovered in this chapter.

Because PhysPop is approximately proportional to average per capita accumulated dose of medical radiation, there is no conflict between observing a strong NEGATIVE dose-response between PhysPop and NonCancer NonIHD causes of death --- causes which are thought NOT inducible by ionizing radiation --- while simultaneously observing strong POSITIVE dose-responses between PhysPop and Cancer (known to be radiation-inducible) and between PhysPop and Ischemic Heart Disease (proposed in this book to be radiation-inducible).

Indeed, the FACT that Cancer and Ischemic Heart Disease behave differently from the combined NonCancer NonIHD causes of death, with respect to PhysPop, is an observation which "demands" an explanation. In addition to Hypothesis-1, we propose Hypothesis-2: Radiation-induced mutations of genes and chromosomes have a causal role in Ischemic Heart Disease (as they do in Cancer). In Chapter 45, we suggest a model of how mutations may work, causally, in IHD mortality.

Below are the summary-results from all the regressions of MortRates upon PhysPops.
MortRates are from Table 25-A, and PhysPops are from Table 3-A.

MALES ....

Year MortRate	Year PhysPop	R-squared	Constant	Coeff.	Std Err	Coeff/S.E
1950	1921	0.5985	1219.035	-4.1998	1.3001	-3.2305
1950	1923	0.6286	1179.090	-4.0129	1.1658	-3.4423
1950	1925	0.6806	1144.522	-3.8717	1.0025	-3.8619
1950	1927	0.7393	1135.594	-3.8660	0.8677	-4.4558
1950	1929	0.7525	1120.106	-3.7731	0.8179	-4.6130
1950	1931	0.7668	1085.387	-3.4736	0.7241	-4.7970
1950	1934	0.8145	1047.616	-3.1517	0.5685	-5.5439
1950	1936	0.8106	1033.875	-3.0237	0.5524	-5.4736
1950	1938	0.8079	1012.178	-2.8342	0.5224	-5.4254
1950	1940	0.7933	978.617	-2.5282	0.4878	-5.1830
1950	1950	0.7832	971.485	-2.5419	0.5055	-5.0281
1960	1960	0.6340	766.815	-1.7513	0.5030	-3.4821
1970	1970	0.5343	633.823	-1.1346	0.4004	-2.8337
1980	1980	0.4965	536.879	-0.7576	0.2883	-2.6275
1990	1990	--	--	--	--	--

FEMALES ....

Year MortRate	Year PhysPop	R-squared	Constant	Coeff.	Std Err	Coeff/S.E
1950	1921	0.7430	915.173	-3.2292	0.7179	-4.4984
1950	1923	0.7534	877.381	-3.0315	0.6556	-4.6242
1950	1925	0.7850	844.257	-2.8696	0.5676	-5.0555
1950	1927	0.7922	824.712	-2.7618	0.5346	-5.1658
1950	1929	0.7929	810.869	-2.6729	0.5163	-5.1768
1950	1931	0.8050	785.715	-2.4563	0.4569	-5.3757
1950	1934	0.7971	749.385	-2.1517	0.4103	-5.2443
1950	1936	0.7716	736.427	-2.0359	0.4186	-4.8632
1950	1938	0.7390	717.055	-1.8707	0.4202	-4.4519
1950	1940	0.7037	691.625	-1.6432	0.4030	-4.0771
1950	1950	0.6650	682.530	-1.6165	0.4336	-3.7280
1960	1960	0.5617	501.286	-0.9511	0.3176	-2.9950
1970	1970	0.4823	385.722	-0.5845	0.2289	-2.5535
1980	1980	0.4056	288.360	-0.3566	0.1632	-2.1857
1990	1990	--	--	--	--	--

1950 MortRate, Males, All-Cause-(CA+IHD), versus
1940 PhysPop Values for the 9 Census Divisions, USA.
Highly Significant INVERSE Relationship.
PhysPop is a surrogate for accumulated dose from medical irradiation.

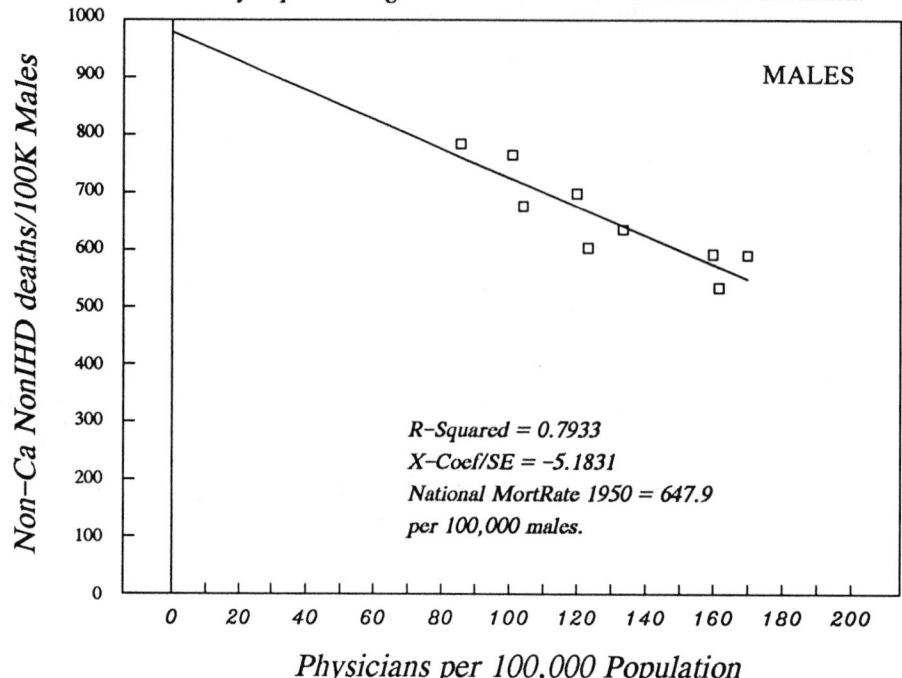

MALES

R-Squared = 0.7933
X-Coef/SE = −5.1831
National MortRate 1950 = 647.9
per 100,000 males.

Non-Ca NonIHD deaths/100K Males

Physicians per 100,000 Population

1950 MortRate, Females, All-Causes − (CA+IHD), versus
1940 PhysPop Values for the 9 Census Divisions, USA.
Highly Significant INVERSE Relationship
PhysPop is a surrogate for accumulated dose from medical irradiation

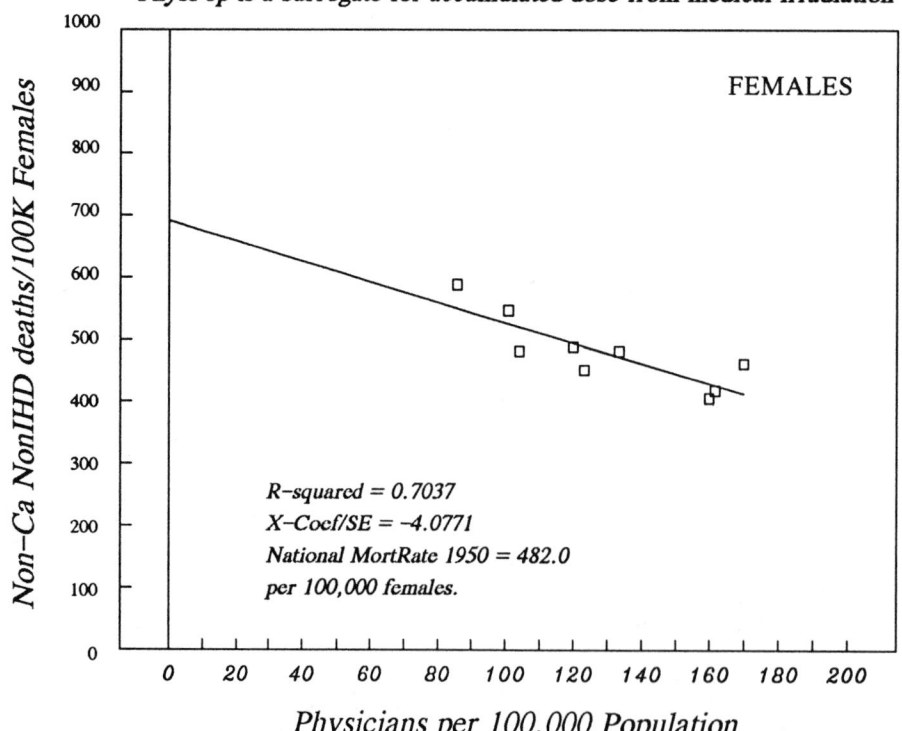

FEMALES

R-squared = 0.7037
X-Coef/SE = −4.0771
National MortRate 1950 = 482.0
per 100,000 females.

Non-Ca NonIHD deaths/100K Females

Physicians per 100,000 Population

Table 25-A

All Causes of Death except (Cancer + IHD Deaths): Rates by Census Divisions and National.

These annual MortRates are the "All Causes of Death Minus Cancer" MortRates in Table 24-A, minus the annual Ischemic Heart Disease MortRates in Table 40-A (males) or Table 41-A (females). The net result is "All Causes of Death Except (Cancer + IHD Deaths) --- which we generally call "All NonCancer NonIHD MortRates." Rates are per 100,000 population, age-adjusted to the 1940 reference year. There are no exclusions by color or "race."

	1940	1950	1960	1970	1980	1990
Pacific	--	592.3	509.3	442.5	375.7	--
New England	--	534.1	479.4	403.8	328.2	--
West North Central	--	603.4	504.0	426.2	348.4	--
Mid-Atlantic	--	590.6	488.2	424.8	361.4	--
East North Central	--	635.2	515.4	448.1	380.8	--
Mountain	--	698.1	582.1	492.0	401.9	--
West South Central	--	675.9	576.4	510.8	445.0	--
East South Central	--	784.9	656.2	554.2	452.1	--
South Atlantic	--	765.6	629.7	532.3	434.7	--
Average, ALL	--	653.3	549.0	470.5	392.0	--
Average, High-5	--	591.1	499.3	429.1	358.9	--
Average, Low-4	--	731.1	611.1	522.3	433.4	--
Ratio, Hi5/Lo4	--	0.81	0.82	0.82	0.83	--

FEMALES ....

	1940	1950	1960	1970	1980	1990
Pacific	--	404.7	350.2	283.6	216.8	--
New England	--	417.3	350.0	267.3	184.5	--
West North Central	--	451.0	356.2	274.5	192.6	--
Mid-Atlantic	--	461.2	361.0	284.0	206.9	--
East North Central	--	481.3	370.5	294.2	217.8	--
Mountain	--	489.2	380.2	302.8	225.2	--
West South Central	--	481.3	393.0	319.3	245.6	--
East South Central	--	588.6	453.6	350.7	247.8	--
South Atlantic	--	547.3	432.1	338.2	244.3	--
Average ALL	--	480.2	383.0	301.6	220.2	--
Average High-Five	--	443.1	357.6	280.7	203.7	--
Average Low-Four	--	526.6	414.7	327.7	240.7	--
Ratio (High/Low)	--	0.84	0.86	0.86	0.85	--

NATIONAL RATES ....

Rates are age-adjusted to the 1940 reference year. Both sexes: Deaths per 100,000 population (males + females). Males: Deaths per 100,000 male population. Females: Deaths per 100,000 female population. No exclusions by color or "race."

	1940	1950	1960	1970	1980	1990
Both Sexes	--	563.8	456.9	377.6	298.3	--
Males	--	647.9	538.2	464.5	390.8	--
Females	--	482.0	380.8	300.8	220.7	--

CHAPTER 26

Appendicitis, Deaths: Relation with Medical Radiation

Box 1 of Chap. 26
Summary: Regression Outputs, Appendicitis Deaths

Below are the summary-results from all the regressions of MortRates upon PhysPops.
MortRates are from Table 26-A, and PhysPops are from Table 3-A.

MALES ....

Year MortRate	Year PhysPop	R-squared	Constant		Coeff.	Std Err	Coeff/S.E
1940	1921	0.0221	9.7705		0.0211	0.0530	0.3981
1940	1923	0.0072	11.1403		0.0112	0.0498	0.2256
1940	1925	0.0000	12.7086		-0.0008	0.0463	-0.0166
1940	1927	0.0035	13.4821		-0.0070	0.0443	-0.1576
1940	1929	0.0030	13.3746		-0.0062	0.0429	-0.1440
1940	1931	0.0037	13.3977		-0.0063	0.0391	-0.1618
1940	1934	0.0097	13.7368		-0.0090	0.0343	-0.2624
1940	1936	0.0075	13.5651		-0.0076	0.0330	-0.2296
1940	1938	0.0103	13.6695		-0.0084	0.0310	-0.2700
1940	1940	0.0179	13.8878		-0.0099	0.0278	-0.3574
1950	1950	0.3168	3.3804		-0.0075	0.0042	-1.8018
1960	1960	0.0179	1.0540		0.0006	0.0015	0.3577
1970	1970	--	--	--	--	--	--
1980	1980	--	--	--	--	--	--
1990	1990	--	--	--	--	--	--

FEMALES ....

Year MortRate	Year PhysPop	R-squared	Constant		Coeff.	Std Err	Coeff/S.E
1940	1921	0.0057	6.9336		0.0058	0.0288	0.2005
1940	1923	0.0024	7.2571		0.0035	0.0269	0.1289
1940	1925	0.0016	8.0491		-0.0027	0.0249	-0.1068
1940	1927	0.0025	8.1055		-0.0032	0.0239	-0.1324
1940	1929	0.0010	7.9501		-0.0019	0.0231	-0.0835
1940	1931	0.0015	7.9810		-0.0022	0.0211	-0.1029
1940	1934	0.0023	8.0060		-0.0024	0.0186	-0.1271
1940	1936	0.0001	7.7709		-0.0005	0.0179	-0.0266
1940	1938	0.0001	7.7632		-0.0004	0.0168	-0.0245
1940	1940	0.0010	7.8731		-0.0013	0.0151	-0.0835
1950	1950	0.1253	1.649		-0.0026	0.0026	-1.0015
1960	1960	0.3477	0.907		-0.0026	0.0013	-1.9316
1970	1970	--	--	--	--	--	--
1980	1980	--	--	--	--	--	--
1990	1990	--	--	--	--	--	--

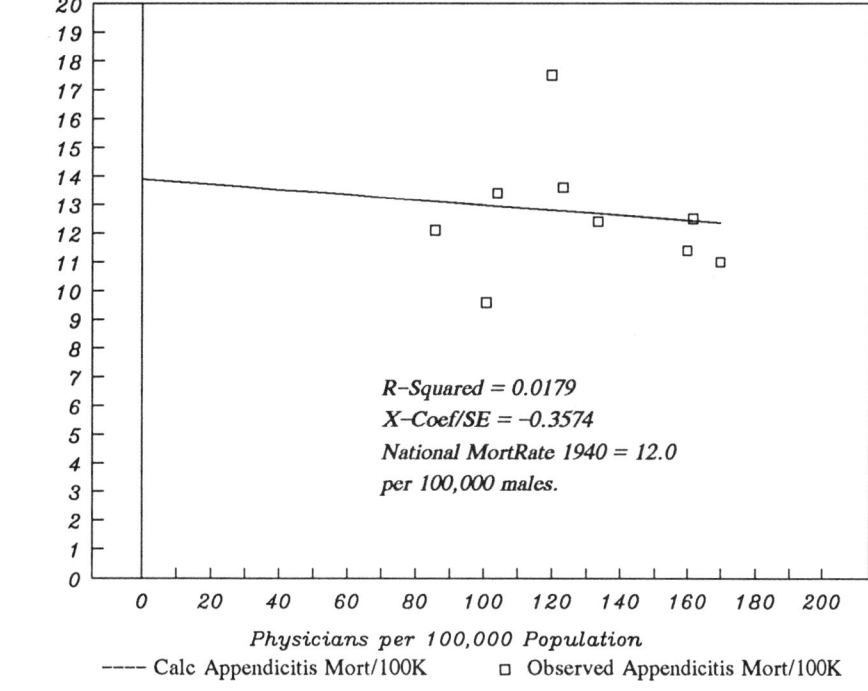

1940 MortRate, Males, Appendicitis, Deaths, versus
1940 PhysPop Values for the 9 Census Divisions, USA.
No Significant Relationship.
PhysPop is a surrogate for accumulated dose from medical irradiation.

R-Squared = 0.0179
X-Coef/SE = -0.3574
National MortRate 1940 = 12.0
per 100,000 males.

---- Calc Appendicitis Mort/100K   □ Observed Appendicitis Mort/100K

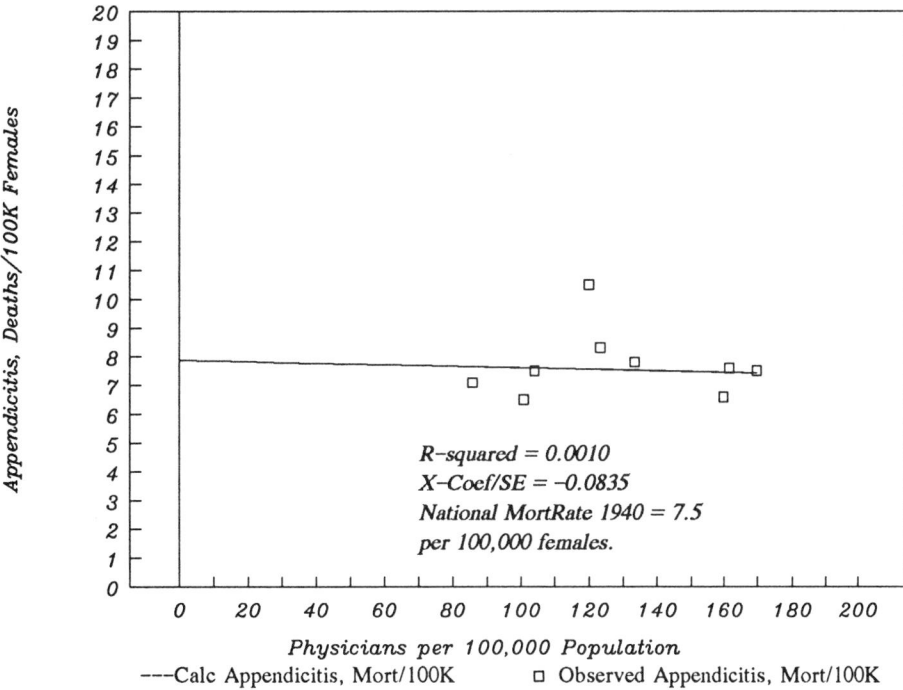

1940 MortRate, Females, Appendicitis, Deaths, versus
1940 PhysPop Values for the 9 Census Divisions, USA.
No Significant Relationship
PhysPop is a surrogate for accumulated dose from medical irradiation

R-squared = 0.0010
X-Coef/SE = -0.0835
National MortRate 1940 = 7.5
per 100,000 females.

---Calc Appendicitis, Mort/100K   □ Observed Appendicitis, Mort/100K

Table 26-A
Death from Appendicitis: Rates by Census Divisions and National

Annual death-rates per 100,000 are age-adjusted to the 1940 reference year. There are no exclusions by color or "race." Entries for the Nine Census Divisions are population-weighted; the averages below them are not. "National Rates" for both sexes: Deaths per 100,000 population (males + females). Males: Deaths per 100,000 male population. Females: Deaths per 100,000 female population.

MALES ....

Census Division	1940	1950	1960	1970	1980	1990
Pacific	11.4	1.8	1.0	--	--	--
New England	12.5	1.9	1.1	--	--	--
West North Central	13.6	2.5	1.1	--	--	--
Mid-Atlantic	11.0	2.4	1.3	--	--	--
East North Central	12.4	2.6	1.3	--	--	--
Mountain	17.5	3.1	1.0	--	--	--
West South Central	13.4	2.6	1.0	--	--	--
East South Central	12.1	2.5	1.1	--	--	--
South Atlantic	9.6	2.6	1.2	--	--	--
Average ALL	12.6	2.4	1.1	--	--	--
Average High-Five	12.2	2.2	1.2	--	--	--
Average Low-Four	13.2	2.7	1.1	--	--	--
Ratio (High/Low)	0.93	0.83	1.08	--	--	--

FEMALES ....

	1940	1950	1960	1970	1980	1990
Pacific	6.6	0.9	0.5	--	--	--
New England	7.6	1.2	0.3	--	--	--
West North Central	8.3	1.2	0.6	--	--	--
Mid-Atlantic	7.5	1.5	0.7	--	--	--
East North Central	7.8	1.4	0.6	--	--	--
Mountain	10.5	1.5	0.6	--	--	--
West South Central	7.5	1.2	0.7	--	--	--
East South Central	7.1	1.6	0.7	--	--	--
South Atlantic	6.5	1.4	0.6	--	--	--
Average ALL	7.7	1.3	0.6	--	--	--
Average High-Five	7.6	1.2	0.5	--	--	--
Average Low-Four	7.9	1.4	0.7	--	--	--
Ratio (High/Low)	0.96	0.87	0.83	--	--	--

NATIONAL RATES ....

	1940	1950	1960	1970	1980	1990
Both Sexes	9.8	1.9	0.9	--	--	--
Males	12.0	2.5	1.4	--	--	--
Females	7.5	1.4	0.5	--	--	--

● - 1940, 1950, 1960: All rates come from Grove 1968, Table 67, Pages 741-744, "Appendicitis (550-553)" ICD/7.

● - Omitted data are only marginally related to Hypotheses 1+2, as noted in Chapter 4, Part 2c.

CHAPTER 27

CNS Vascular Lesions, Deaths: Relation with Medical Radiation

---

Box 1 of Chap. 27
Summary: Regression Outputs, Deaths due to Central Nervous System Vascular Lesions

Below are the summary-results from all the regressions of MortRates upon PhysPops. MortRates are from Table 26-A, and PhysPops are from Table 3-A.

MALES ....

Year MortRate	Year PhysPop	R-squared	Constant	Coeff.	Std Err	Coeff/S.E
1940	1921	0.5421	193.710	-0.7601	0.2640	-2.8788
1940	1923	0.5419	184.147	-0.7084	0.2462	-2.8774
1940	1925	0.4434	166.720	-0.5942	0.2517	-2.3613
1940	1927	0.4431	162.328	-0.5691	0.2412	-2.3601
1940	1929	0.4451	159.599	-0.5518	0.2329	-2.3697
1940	1931	0.4448	153.904	-0.5031	0.2125	-2.3680
1940	1934	0.4357	146.170	-0.4384	0.1885	-2.3249
1940	1936	0.4470	145.069	-0.4270	0.1795	-2.3788
1940	1938	0.4323	141.248	-0.3942	0.1708	-2.3088
1940	1940	0.4000	135.252	-0.3414	0.1580	-2.1601
1950	1950	0.4472	132.943	-0.3228	0.1357	-2.3795
1960	1960	0.5130	135.143	-0.3660	0.1348	-2.7156
1970	1970	0.5226	100.265	-0.2316	0.0837	-2.7682
1980	1980	0.4458	66.878	-0.1183	0.0499	-2.3729
1990	1990	0.3417	42.580	-0.0532	0.0279	-1.9061

FEMALES ....

Year MortRate	Year PhysPop	R-squared	Constant	Coeff.	Std Err	Coeff/S.E
1940	1921	0.5588	169.367	-0.5889	0.1978	-2.9776
1940	1923	0.5502	161.414	-0.5447	0.1862	-2.9259
1940	1925	0.4371	147.169	-0.4503	0.1931	-2.3315
1940	1927	0.4104	142.185	-0.4180	0.1894	-2.2073
1940	1929	0.3975	139.278	-0.3980	0.1852	-2.1491
1940	1931	0.3882	134.657	-0.3587	0.1702	-2.1075
1940	1934	0.3558	127.864	-0.3023	0.1537	-1.9665
1940	1936	0.3540	126.536	-0.2900	0.1481	-1.9584
1940	1938	0.3267	123.158	-0.2615	0.1419	-1.8429
1940	1940	0.2882	118.496	-0.2211	0.1314	-1.6835
1950	1950	0.3109	112.772	-0.1927	0.1084	-1.7770
1960	1960	0.4617	113.406	-0.2418	0.0987	-2.4504
1970	1970	0.4847	82.261	-0.1534	0.0598	-2.5659
1980	1980	0.3875	52.030	-0.0784	0.0373	-2.1044
1990	1990	0.3719	34.472	-0.0386	0.0189	-2.0358

1940 MortRate, Males, CNS Vascular Deaths, versus
1940 PhysPop Values for the 9 Census Divisions, USA.
Significant INVERSE Relationship.
PhysPop is a surrogate for accumulated dose from medical irradiation.

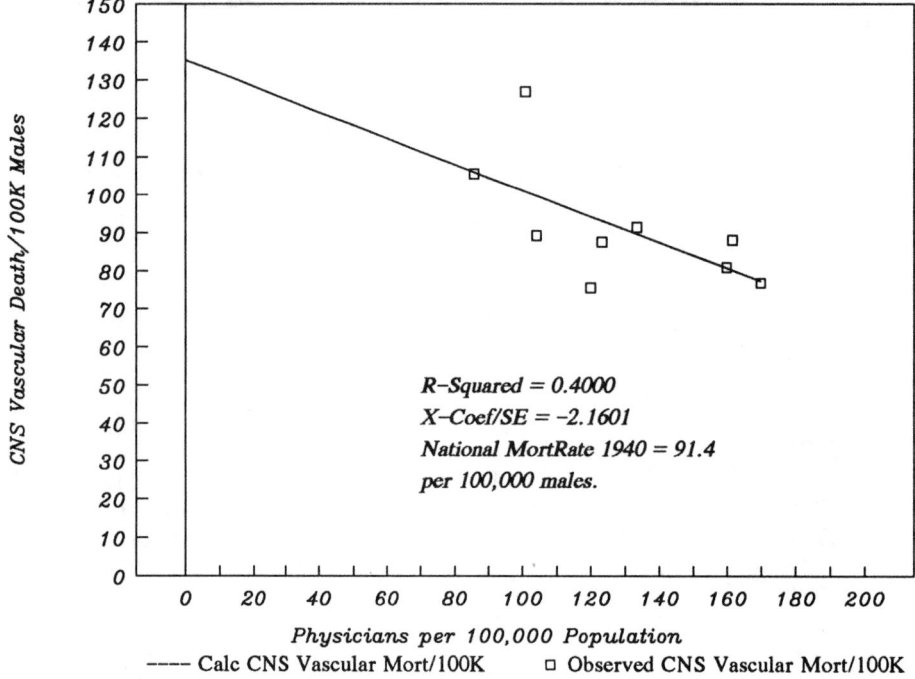

R–Squared = 0.4000
X–Coef/SE = –2.1601
National MortRate 1940 = 91.4
per 100,000 males.

---- Calc CNS Vascular Mort/100K   □ Observed CNS Vascular Mort/100K

1940 MortRate, Females, CNS Vascular Deaths, versus
1940 PhysPop Values for the 9 Census Divisions, USA.
Marginally Significant. All sign tests show Inverse Relationship
PhysPop is a surrogate for accumulated dose from medical irradiation

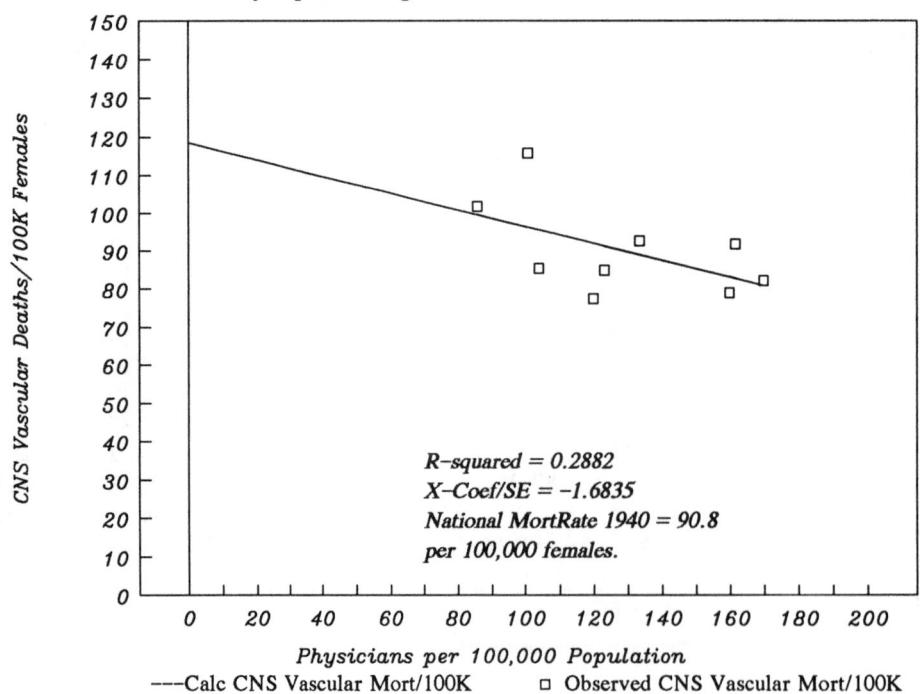

R–squared = 0.2882
X–Coef/SE = –1.6835
National MortRate 1940 = 90.8
per 100,000 females.

----Calc CNS Vascular Mort/100K   □ Observed CNS Vascular Mort/100K

Table 27-A
Vascular Lesions Affecting the Central Nervous System

Annual death-rates per 100,000 are age-adjusted to the 1940 reference year. There are no exclusions by color or "race." Entries for the Nine Census Divisions are population-weighted; the averages below them are not. "National Rates" for both sexes: Deaths per 100,000 population (males + females). Males: Deaths per 100,000 male population. Females: Deaths per 100,000 female population.

MALES ....

Census Division	1940	1950	1960	1970	1980	1990
Pacific	80.9	83.6	81.5	61.3	41.1	30.6
New England	88.0	85.3	81.0	59.1	37.2	24.9
West North Central	87.6	97.9	89.7	65.8	41.8	28.9
Mid-Atlantic	76.8	79.8	73.2	56.1	39.0	26.7
East North Central	91.5	95.5	90.2	68.0	45.8	30.7
Mountain	75.5	74.9	72.9	54.1	35.2	24.4
West South Central	89.3	88.3	93.6	71.9	50.1	33.1
East South Central	105.5	108.9	117.8	87.6	57.3	38.4
South Atlantic	126.9	118.7	106.6	78.7	50.8	33.7
Average ALL	91.3	92.5	89.6	66.9	44.3	30.2
Average High-Five	85.0	88.4	83.1	62.1	41.0	28.4
Average Low-Four	99.3	97.7	97.7	73.0	48.4	32.4
Ratio (High/Low)	0.86	0.91	0.85	0.85	0.85	0.88

FEMALES ....

	1940	1950	1960	1970	1980	1990
Pacific	78.9	81.3	78.9	58.0	37.0	27.1
New England	91.7	85.0	77.7	54.3	30.9	21.1
West North Central	84.8	92.1	84.6	59.3	33.9	23.4
Mid-Atlantic	82.1	83.1	71.7	52.6	33.4	22.7
East North Central	92.6	93.0	84.7	61.2	37.6	25.8
Mountain	77.5	75.4	70.5	50.9	31.2	22.6
West South Central	85.3	80.5	84.3	63.2	42.0	28.3
East South Central	101.7	103.8	103.0	74.3	45.6	30.4
South Atlantic	115.8	103.7	94.5	68.1	41.7	27.7
Average ALL	90.0	88.7	83.3	60.2	37.0	25.5
Average High-Five	86.0	86.9	79.5	57.0	34.6	24.0
Average Low-Four	95.1	90.9	88.1	64.1	40.1	27.3
Ratio (High/Low)	0.90	0.96	0.90	0.89	0.86	0.88

NATIONAL RATES ....

	1940	1950	1960	1970	1980	1990
Both Sexes	91.1	91.6	85.7	71.7	57.6	28.0
Males	91.4	93.4	88.9	76.7	64.4	30.3
Females	90.8	90.0	82.8	67.8	52.8	25.6

● - 1940, 1950, 1960: All rates come from Grove 1968, Table 67, Pages 713-716, "Vascular Lesions Affecting Central Nervous System (330-334)" ICD/7.
  ● - 1970 rates are interpolations (Chap. 4, Parts 2b, 2c).
  ● - 1980: All rates (ICD/9, 430-438) come from the reference NatCtrHS 1980.
  ● - 1990: Rates are for 1987-1989, from Monthly Vital Statistics Report, Vol.43, No.2, July 21, 1994. "Both sexes" combined is approximated as the average of male and female values. Details in Chap. 4, Part 2b.

Box 1 of Chap. 28
Summary: Regression Outputs, Deaths due to Chronic and Unspecified Nephritis

---

Below are the summary-results from all the regressions of MortRates upon PhysPops.
MortRates are from Table 28-A, and PhysPops are from Table 3-A. The MortRates for
1950 and beyond were not reported on the same basis as the rates for 1940 (please see
Chapter 4, Part 5).

MALES ....

Year MortRate	Year PhysPop	R-squared	Constant	Coeff.	Std Err	Coeff/S.E
1940	1921	0.6768	248.885	-1.2204	0.3188	-3.8284
1940	1923	0.6414	229.620	-1.1077	0.3130	-3.5388
1940	1925	0.5771	208.098	-0.9743	0.3152	-3.0905
1940	1927	0.5597	199.162	-0.9192	0.3082	-2.9829
1940	1929	0.5692	195.433	-0.8967	0.2949	-3.0409
1940	1931	0.5687	186.178	-0.8175	0.2691	-3.0380
1940	1934	0.5387	172.126	-0.7005	0.2450	-2.8593
1940	1936	0.5339	168.897	-0.6706	0.2368	-2.8319
1940	1938	0.5084	162.289	-0.6144	0.2284	-2.6905
1940	1940	0.4561	151.903	-0.5239	0.2162	-2.4230
1950	1950	0.6537	33.894	-0.1400	0.0385	-3.6351
1950	1960	0.6543	12.790	-0.0464	0.0127	-3.6402
1950	1970	--	--	--	--	--
1950	1980	--	--	--	--	--
1950	1990	--	--	--	--	--

FEMALES ....

Year MortRate	Year PhysPop	R-squared	Constant	Coeff.	Std Err	Coeff/S.E
1940	1921	0.6772	189.917	-0.8788	0.2293	-3.8319
1940	1923	0.6063	173.109	-0.7752	0.2361	-3.2831
1940	1925	0.5344	157.170	-0.6749	0.2381	-2.8347
1940	1927	0.4683	147.050	-0.6053	0.2438	-2.4831
1940	1929	0.4554	142.976	-0.5774	0.2387	-2.4192
1940	1931	0.4429	136.140	-0.5193	0.2202	-2.3589
1940	1934	0.3849	124.867	-0.4262	0.2036	-2.0931
1940	1936	0.3595	121.400	-0.3962	0.1998	-1.9823
1940	1938	0.3211	116.051	-0.3515	0.1932	-1.8197
1940	1940	0.2687	108.786	-0.2895	0.1805	-1.6038
1950	1950	0.7010	28.068	-0.1156	0.0285	-4.0513
1950	1960	0.5441	8.911	-0.0314	0.0109	-2.8903
1950	1970	--	--	--	--	--
1950	1980	--	--	--	--	--
1950	1990	--	--	--	--	--

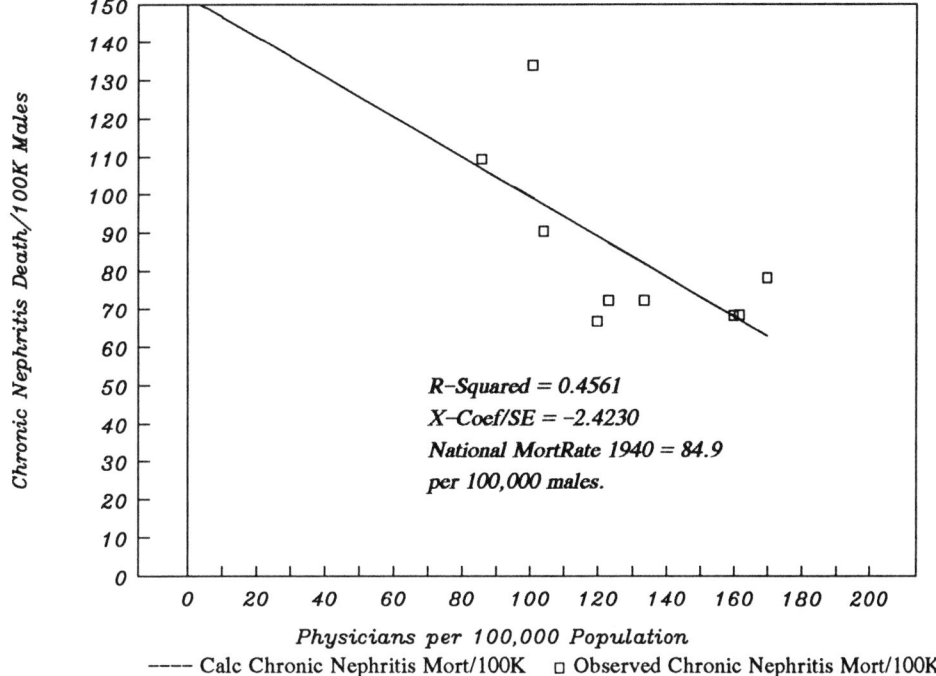

1940 MortRate, Males, Chronic Nephritis Deaths, versus
1940 PhysPop Values for the 9 Census Divisions, USA.
Significant INVERSE Relationship.
PhysPop is a surrogate for accumulated dose from medical irradiation.

R-Squared = 0.4561
X-Coef/SE = -2.4230
National MortRate 1940 = 84.9
per 100,000 males.

Physicians per 100,000 Population
---- Calc Chronic Nephritis Mort/100K   □ Observed Chronic Nephritis Mort/100K

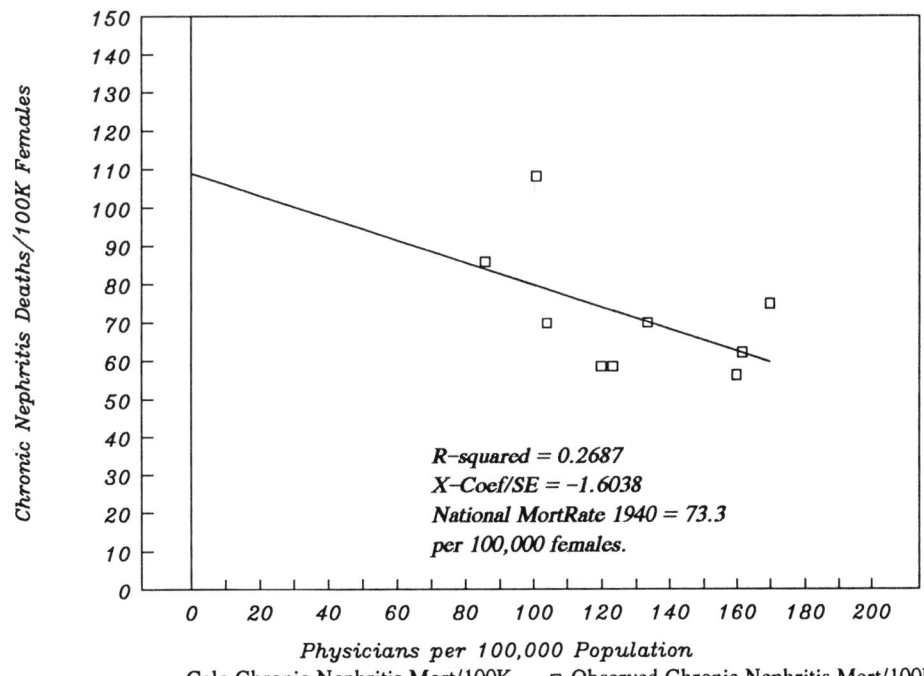

1940 MortRate, Females, Chronic Nephritis Deaths, versus
1940 PhysPop Values for the 9 Census Divisions, USA.
Marginally Significant. All sign tests show Inverse Relationship
PhysPop is a surrogate for accumulated dose from medical irradiation

R-squared = 0.2687
X-Coef/SE = -1.6038
National MortRate 1940 = 73.3
per 100,000 females.

Physicians per 100,000 Population
---Calc Chronic Nephritis Mort/100K   □ Observed Chronic Nephritis Mort/100K

NOTE: The MortRates for 1950 and beyond were not reported on the same basis as the rates for 1940 (please see Chapter 4, Part 5).

Annual death-rates per 100,000 are age-adjusted to the 1940 reference year. There are no exclusions by color or "race." Entries for the Nine Census Divisions are population-weighted; the averages below them are not. "National Rates" for both sexes: Deaths per 100,000 population (males + females). Males: Deaths per 100,000 male population. Females: Deaths per 100,000 female population.

MALES ....

Census Division	1940	1950	1960	1970	1980	1990
Pacific	68.4	9.6	4.7	--	--	--
New England	68.5	12.2	5.6	--	--	--
West North Central	72.4	15.4	6.7	--	--	--
Mid-Atlantic	78.2	14.7	6.2	--	--	--
East North Central	72.4	15.4	6.3	--	--	--
Mountain	66.8	12.7	6.7	--	--	--
West South Central	90.4	19.7	7.9	--	--	--
East South Central	109.4	25.1	9.8	--	--	--
South Atlantic	134.0	22.6	9.3	--	--	--
Average ALL	84.5	16.4	7.0	--	--	--
Average High-Five	72.0	13.5	5.9	--	--	--
Average Low-Four	100.2	20.0	8.4	--	--	--
Ratio (High/Low)	0.72	0.67	0.70	--	--	--

FEMALES ....

	1940	1950	1960	1970	1980	1990
Pacific	56.3	7.8	3.4	--	--	--
New England	62.1	10.3	4.4	--	--	--
West North Central	58.5	12.3	4.5	--	--	--
Mid-Atlantic	74.9	11.9	4.3	--	--	--
East North Central	70.0	13.1	4.6	--	--	--
Mountain	58.5	11.8	4.1	--	--	--
West South Central	69.8	16.3	6.2	--	--	--
East South Central	85.8	20.8	7.1	--	--	--
South Atlantic	108.0	18.1	6.4	--	--	--
Average ALL	71.5	13.6	5.0	--	--	--
Average High-Five	64.4	11.1	4.2	--	--	--
Average Low-Four	80.5	16.8	6.0	--	--	--
Ratio (High/Low)	0.80	0.66	0.71	--	--	--

NATIONAL RATES ....

	1940	1950	1960	1970	1980	1990
Both Sexes	79.1	14.8	5.9	--	--	--
Males	84.9	16.2	6.9	--	--	--
Females	73.3	13.4	4.9	--	--	--

● – 1940, 1950, 1960: All rates come from Grove 1968, Table 67, Pages 730-734, "Chronic and unspecified nephritis and other renal sclerosis (592-594)" ICD/7.

● – Omitted data are only marginally related to examination of Hypotheses 1+2, as noted in Chapter 4, Part 2c.

CHAPTER 29

Diabetes Mellitus, Deaths: Relation with Medical Radiation

Box 1 of Chap. 29
Summary: Regression Outputs, Deaths due to Diabetes Mellitus.

Below are the summary–results from all the regressions of MortRates upon PhysPops. MortRates come from Table 29–A, and PhysPops come from Table 3–A.

We think that the positive correlation, which grows between PhysPop (1921 to 1940) and 1940 MortRates for Diabetes Mellitus, is very probably a correlation between PhysPop and xray-induced deaths during 1940 from Ischemic Heart Disease (IHD) in people having diabetes. Chapters 40, 41 show the PhysPop-IHD correlations in 1950. In 1940, IHD was not reported as a distinct cause of death. Later, it was learned that many diabetics have elevated levels of atherogenic lipoproteins, and that their MortRate from IHD is high. Today, almost 75% of diabetics die from IHD (Bierman 1992, p.647). In 1949, the rules were altered for reporting the underlying cause of death in diabetics, and by 1950, IHD was reported as a distinct cause of death. In 1950, reported MortRates from Diabetes Mellitus fell to half of the 1940 values (Table 29–A), and the correlations with PhysPop (below) abruptly dropped to 0.11 and 0.20.

MALES ....

Year MortRate	Year PhysPop	R-squared	Constant	Coeff.	Std Err	Coeff/S.E
1940	1921	0.1046	6.742	0.0875	0.0968	0.9045
1940	1923	0.1653	5.095	0.1026	0.0871	1.1774
1940	1925	0.2190	4.645	0.1095	0.0781	1.4008
1940	1927	0.3362	2.323	0.1300	0.0690	1.8827
1940	1929	0.3694	2.230	0.1318	0.0651	2.0250
1940	1931	0.3853	3.265	0.1228	0.0586	2.0949
1940	1934	0.4969	3.181	0.1227	0.0467	2.6296
1940	1936	0.5458	2.967	0.1237	0.0426	2.9004
1940	1938	0.5999	3.119	0.1218	0.0376	3.2398
1940	1940	0.6435	3.930	0.1135	0.0319	3.5545
1950	1950	0.1127	7.966	0.0226	0.0239	0.9429
1960	1960	0.0257	10.427	0.0099	0.0230	0.4298
1970	1970	0.0131	10.046	0.0050	0.0163	0.3049
1980	1980	0.0076	9.359	0.0027	0.0115	0.2310
1990	1990	0.0396	13.971	-0.0070	0.0130	-0.5369

FEMALES ....

Year MortRate	Year PhysPop	R-squared	Constant	Coeff.	Std Err	Coeff/S.E
1940	1921	0.0648	7.831	0.1563	0.2243	0.6967
1940	1923	0.1205	2.859	0.1986	0.2028	0.9794
1940	1925	0.1458	3.184	0.2025	0.1853	1.0929
1940	1927	0.2556	-3.173	0.2569	0.1657	1.5502
1940	1929	0.2901	-3.885	0.2648	0.1566	1.6915
1940	1931	0.3088	-2.115	0.2492	0.1409	1.7684
1940	1934	0.4224	-3.217	0.2566	0.1134	2.2626
1940	1936	0.4875	-4.478	0.2650	0.1027	2.5802
1940	1938	0.5502	-4.597	0.2644	0.0903	2.9264
1940	1940	0.6005	-3.113	0.2486	0.0766	3.2441
1950	1950	0.2039	7.071	0.0681	0.0509	1.3390
1960	1960	0.0127	12.863	0.0115	0.0382	0.3001
1970	1970	0.0113	12.774	-0.0065	0.0231	-0.2824
1980	1980	0.1942	12.677	-0.0173	0.0133	-1.2987
1990	1990	0.3638	16.307	-0.0225	0.0113	-2.0008

1940 MortRate, Males, Diabetes Mellitus Deaths, versus
1940 PhysPop Values for the 9 Census Divisions, USA.
Significant DIRECT Relationship.
PhysPop is a surrogate for accumulated dose from medical irradiation.

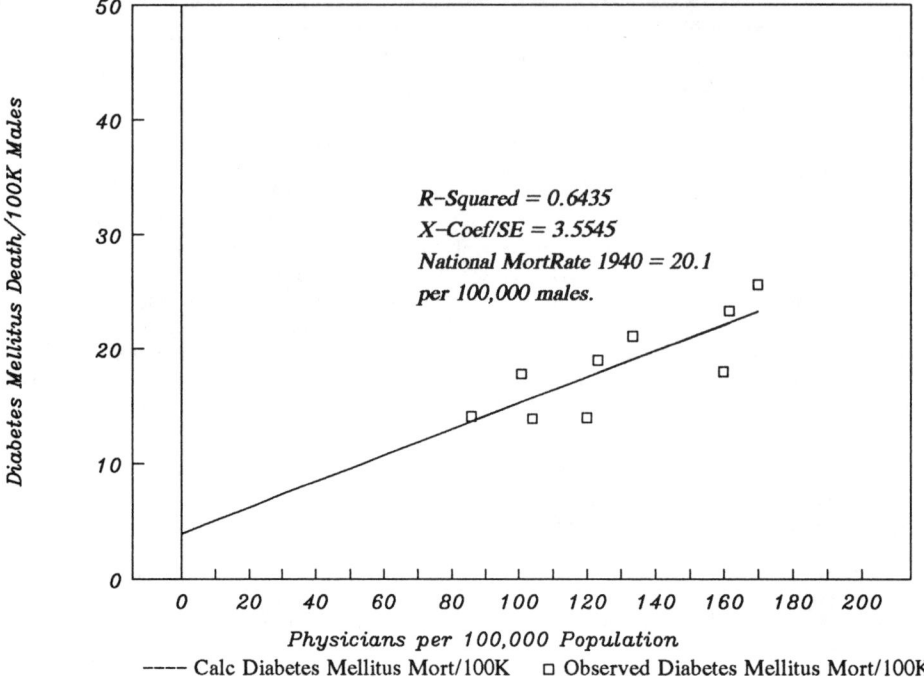

---- Calc Diabetes Mellitus Mort/100K   □ Observed Diabetes Mellitus Mort/100K

1940 MortRate, Females, Diabetes Mellitus Deaths, versus
1940 PhysPop Values for the 9 Census Divisions, USA.
Significant DIRECT Relationship.
PhysPop is a surrogate for accumulated dose from medical irradiation

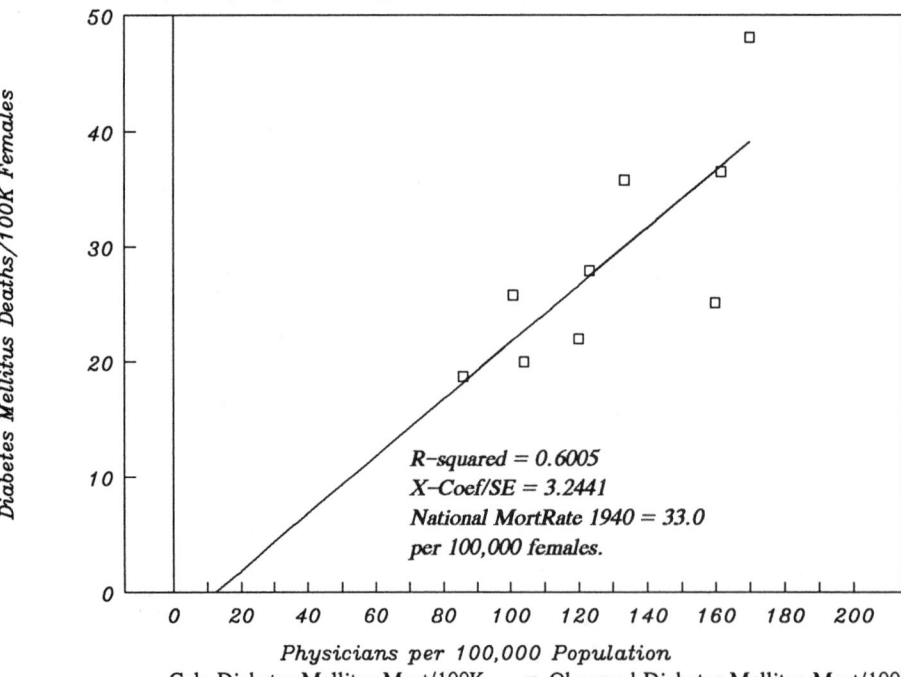

---Calc Diabetes Mellitus Mort/100K   □ Observed Diabetes Mellitus Mort/100K

Table 29-A
Diabetes Mellitus: Death Rates by Census Divisions and National

Abrupt drop, 1940 vs. 1950: The rules were altered in 1949 for reporting the underlying cause of death in diabetics (Chaper 4, Part 5).

Annual death-rates per 100,000 are age-adjusted to the 1940 reference year. There are no exclusions by color or "race." Entries for the Nine Census Divisions are population-weighted; the averages below them are not. "National Rates" for both sexes: Deaths per 100,000 population (males + females). Males: Deaths per 100,000 male population. Females: Deaths per 100,000 female population.

MALES ....

Census Division	1940	1950	1960	1970	1980	1990
Pacific	18.0	7.8	8.6	8.3	7.9	9.0
New England	23.3	12.6	13.2	11.7	10.2	12.6
West North Central	19.0	11.6	11.6	10.2	8.7	11.0
Mid-Atlantic	25.6	12.5	13.7	12.8	11.8	12.9
East North Central	21.1	13.7	13.7	12.6	11.4	14.2
Mountain	14.0	8.8	9.6	9.3	8.9	11.9
West South Central	13.9	9.9	11.8	11.0	10.1	15.0
East South Central	14.1	9.3	10.5	10.1	9.6	11.7
South Atlantic	17.8	10.9	12.2	11.2	10.2	12.7
Average ALL	18.5	10.8	11.7	10.8	9.9	12.3
Average High-Five	21.4	11.6	12.2	11.1	10.0	11.9
Average Low-Four	15.0	9.7	11.0	10.4	9.7	12.8
Ratio (High/Low)	1.43	1.20	1.10	1.07	1.03	0.93

FEMALES ....

	1940	1950	1960	1970	1980	1990
Pacific	25.1	9.1	9.8	8.4	7.0	11.0
New England	36.5	19.3	15.3	11.7	8.0	9.4
West North Central	27.9	16.3	13.3	10.4	7.5	9.9
Mid-Atlantic	48.1	21.6	18.2	14.5	10.8	11.0
East North Central	35.8	21.3	18.8	15.0	11.1	12.4
Mountain	22.0	12.1	11.1	9.9	8.6	10.5
West South Central	20.0	13.3	15.1	12.9	10.6	14.4
East South Central	18.7	11.7	12.8	11.9	10.9	12.2
South Atlantic	25.8	15.7	14.2	12.1	9.9	11.5
Average ALL	28.9	15.6	14.3	11.8	9.4	11.4
Average High-Five	34.7	17.5	15.1	12.0	8.9	10.7
Average Low-Four	21.6	13.2	13.3	11.7	10.0	12.2
Ratio (High/Low)	1.60	1.33	1.13	1.03	0.89	0.88

NATIONAL RATES ....

	1940	1950	1960	1970	1980	1990
Both Sexes	26.6	14.5	13.9	11.9	9.9	11.8
Males	20.1	11.5	12.2	11.2	10.1	12.4
Females	33.0	17.3	15.4	12.6	9.7	11.1

● - 1940, 1950, 1960; All rates come from Grove 1968, Table 67, Pages 706-709, "Diabetes Mellitus (260)" ICD/7.
● - 1970 rates are interpolations (Chap. 4, Parts 2b, 2c).
● - 1980: All rates (ICD/9, Diabetes Mellitus (250)) come from reference NatCtrHS 1980.
● - 1990: Rates are for 1989-91, from Monthly Vital Statistics Report, Vol.43, No.4, October 4, 1994. "Both sexes" combined is approximated as the average of male and female values. Details in Chap. 4, Part 2b.

Box 1 of Chap. 30
Summary: Regression Outputs, Deaths due to Hypertensive Disease

Below are the summary-results from all the regressions of MortRates upon PhysPops.
MortRates are from Table 30-A, and PhysPops are from Table 3-A. No 1940
MortRates were available.

MALES ....

Year MortRate	Year PhysPop	R-squared	Constant	Coeff.	Std Err	Coeff/S.E
1950	1921	0.5421	164.2534	-0.8039	0.2793	-2.8786
1950	1923	0.5152	151.6962	-0.7307	0.2679	-2.7275
1950	1925	0.4405	135.4498	-0.6265	0.2669	-2.3475
1950	1927	0.4249	129.5071	-0.5896	0.2592	-2.2744
1950	1929	0.4341	127.2765	-0.5764	0.2488	-2.3172
1950	1931	0.4362	121.5148	-0.5270	0.2264	-2.3273
1950	1934	0.4133	112.4607	-0.4516	0.2034	-2.2207
1950	1936	0.4202	111.0723	-0.4379	0.1944	-2.2521
1950	1938	0.3982	106.6419	-0.4003	0.1860	-2.1523
1950	1940	0.3564	99.8226	-0.3409	0.1731	-1.9690
1950	1950	0.3458	98.4891	-0.3398	0.1766	-1.9236
1960	1960	0.2406	55.0867	-0.1728	0.1161	-1.4891
1970	1970	--	--	--	--	--
1980	1980	--	--	--	--	--
1990	1990	--	--	--	--	--

FEMALES ....

Year MortRate	Year PhysPop	R-squared	Constant	Coeff.	Std Err	Coeff/S.E
1950	1921	0.5351	139.6800	-0.6175	0.2175	-2.8384
1950	1923	0.4809	128.0052	-0.5457	0.2143	-2.5464
1950	1925	0.4076	115.6127	-0.4659	0.2123	-2.1944
1950	1927	0.3539	108.3911	-0.4159	0.2124	-1.9581
1950	1929	0.3484	105.8938	-0.3992	0.2064	-1.9345
1950	1931	0.3421	101.3833	-0.3608	0.1891	-1.9078
1950	1934	0.2930	93.2838	-0.2940	0.1726	-1.7034
1950	1936	0.2817	91.3934	-0.2772	0.1673	-1.6570
1950	1938	0.2487	87.4682	-0.2445	0.1606	-1.5222
1950	1940	0.2056	82.2607	-0.2001	0.1487	-1.3461
1950	1950	0.1905	80.9067	-0.1949	0.1519	-1.2833
1960	1960	0.1649	50.3462	-0.1217	0.1035	-1.1758
1970	1970	--	--	--	--	--
1980	1980	--	--	--	--	--
1990	1990	--	--	--	--	--

1950 MortRate, Males, Hypertensive Disease Deaths, versus
1940 PhysPop Values for the 9 Census Divisions, USA.
Marginally Significant INVERSE Relationship.
PhysPop is a surrogate for accumulated dose from medical irradiation.

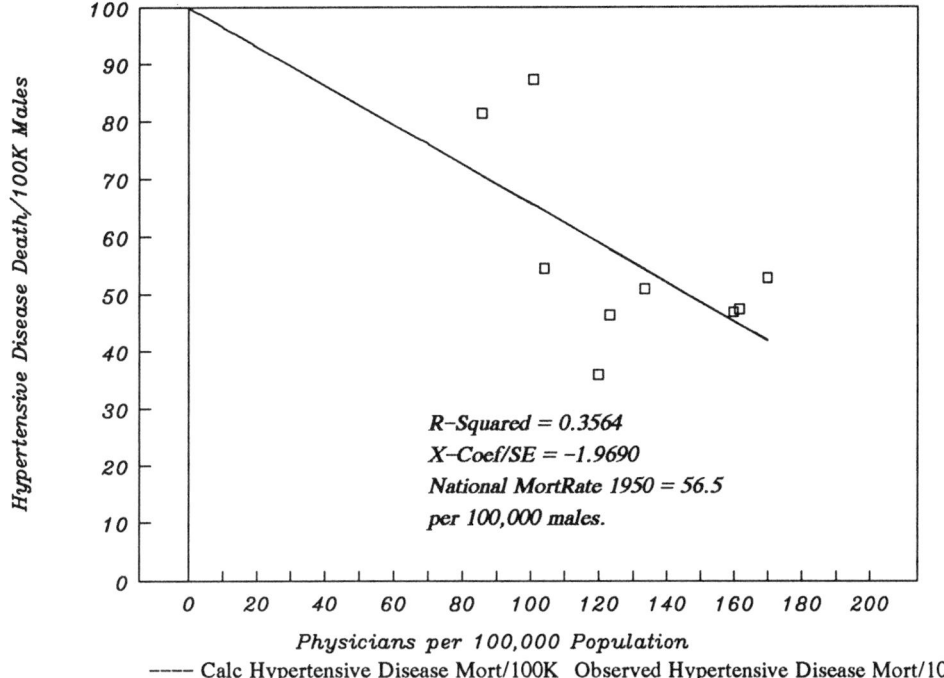

R-Squared = 0.3564
X-Coef/SE = -1.9690
National MortRate 1950 = 56.5
per 100,000 males.

---- Calc Hypertensive Disease Mort/100K   Observed Hypertensive Disease Mort/100K   □

1950 MortRate, Females, Hypertensive Disease Deaths, versus
1940 PhysPop Values for the 9 Census Divisions, USA.
Not provably significant, but sign tests are all in same direction.
PhysPop is a surrogate for accumulated dose from medical irradiation

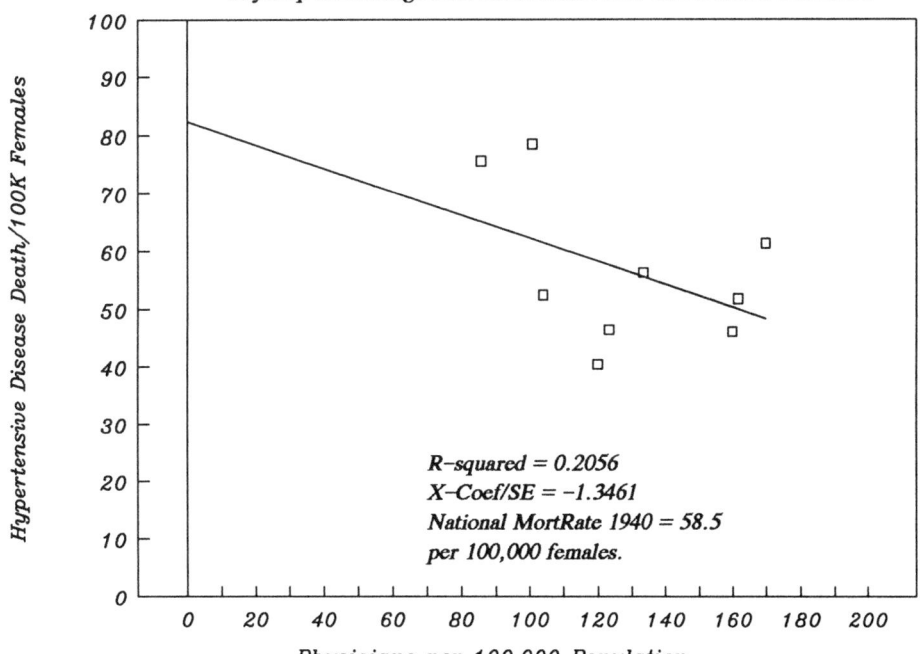

R-squared = 0.2056
X-Coef/SE = -1.3461
National MortRate 1940 = 58.5
per 100,000 females.

---Calc Hypertensive Disease Mort/100K   Observed Hypertensive Disease Mort/100K   □

Table 30-A
Hypertensive Disease: Death Rates by Census Divisions and National.

NOTE: Grove 1968 did not supply MortRates for 1940.

Annual death-rates per 100,000 are age-adjusted to the 1940 reference year. There are no exclusions by color or "race." Entries for the Nine Census Divisions are population-weighted; the averages below them are not. "National Rates" for both sexes: Deaths per 100,000 population (males + females). Males: Deaths per 100,000 male population. Females: Deaths per 100,000 female population.

MALES ....

Census Division	1940	1950	1960	1970	1980	1990
Pacific	--	46.9	25.8	--	--	--
New England	--	47.4	26.7	--	--	--
West North Central	--	46.4	26.3	--	--	--
Mid-Atlantic	--	52.8	34.3	--	--	--
East North Central	--	51.0	33.8	--	--	--
Mountain	--	36.0	21.7	--	--	--
West South Central	--	54.5	33.5	--	--	--
East South Central	--	81.4	49.2	--	--	--
South Atlantic	--	87.3	51.0	--	--	--
Average ALL	--	56.0	33.6	--	--	--
Average High-Five	--	48.9	29.4	--	--	--
Average Low-Four	--	64.8	38.9	--	--	--
Ratio (High/Low)	--	0.75	0.76	--	--	--

FEMALES ....

	1940	1950	1960	1970	1980	1990
Pacific	--	46.1	27.4	--	--	--
New England	--	51.7	30.7	--	--	--
West North Central	--	46.4	28.2	--	--	--
Mid-Atlantic	--	61.3	38.8	--	--	--
East North Central	--	56.2	35.5	--	--	--
Mountain	--	40.4	24.5	--	--	--
West South Central	--	52.4	34.7	--	--	--
East South Central	--	75.6	48.9	--	--	--
South Atlantic	--	78.5	48.2	--	--	--
Average ALL	--	56.5	35.2	--	--	--
Average High-Five	--	52.3	32.1	--	--	--
Average Low-Four	--	61.7	39.1	--	--	--
Ratio (High/Low)	--	0.85	0.82	--	--	--

NATIONAL RATES ....

	1940	1950	1960	1970	1980	1990
Both Sexes	--	57.6	35.7	--	--	--
Males	--	56.5	34.6	--	--	--
Females	--	58.5	36.5	--	--	--

● - 1950, 1960: All rates come from Grove 1968, Table 67, Pages 725-728, "Hypertensive Disease (440-447)" ICD/7.
   ● - Omitted data are only marginally related to examination of Hypotheses 1+2, as noted in Chapter 4, Part 2c.

Box 1 of Chap. 31
Summary: Regression Outputs, Influenza and Pneumonia Death Rates.

Below are the summary-results from all the regressions of MortRates upon PhysPops.
MortRates are from Table 31-A, and PhysPops are from Table 3-A.

MALES ....

Year MortRate	Year PhysPop	R-squared	Constant	Coeff.	Std Err	Coeff/S.E
1940	1921	0.6012	210.1915	-0.9650	0.2971	-3.2484
1940	1923	0.6587	203.5890	-0.9417	0.2562	-3.6760
1940	1925	0.7054	194.8508	-0.9037	0.2207	-4.0936
1940	1927	0.7906	194.5443	-0.9166	0.1783	-5.1416
1940	1929	0.8198	191.9033	-0.9029	0.1600	-5.6424
1940	1931	0.8224	182.7948	-0.8248	0.1448	-5.6941
1940	1934	0.8575	172.9551	-0.7414	0.1143	-6.4890
1940	1936	0.8691	170.5428	-0.7178	0.1053	-6.8163
1940	1938	0.8672	165.4467	-0.6732	0.0996	-6.7622
1940	1940	0.8344	156.6889	-0.5944	0.1001	-5.9381
1950	1950	0.6466	59.1645	-0.2012	0.0562	-3.5791
1960	1960	0.2526	46.7471	-0.0620	0.0403	-1.5380
1970	1970	0.0560	29.8264	-0.0139	0.0216	-0.6442
1980	1980	0.4244	13.6382	0.0155	0.0068	2.2716

FEMALES ....

Year MortRate	Year PhysPop	R-squared	Constant	Coeff.	Std Err	Coeff/S.E
1940	1921	0.6377	208.1584	-1.0557	0.3007	-3.5104
1940	1923	0.7031	201.3472	-1.0334	0.2538	-4.0711
1940	1925	0.7482	191.3687	-0.9886	0.2168	-4.5602
1940	1927	0.8350	190.7638	-1.0006	0.1681	-5.9519
1940	1929	0.8711	188.2565	-0.9886	0.1437	-6.8779
1940	1931	0.8787	178.5869	-0.9056	0.1272	-7.1204
1940	1934	0.9067	167.2585	-0.8098	0.0982	-8.2465
1940	1936	0.9210	164.7357	-0.7849	0.0869	-9.0363
1940	1938	0.9172	159.0658	-0.7354	0.0835	-8.8054
1940	1940	0.8849	149.6160	-0.6503	0.0886	-7.3358
1950	1950	0.8227	56.4194	-0.2434	0.0427	-5.6987
1960	1960	0.4112	34.9403	-0.0697	0.0315	-2.2108
1970	1970	0.2203	20.8222	-0.0214	0.0152	-1.4065
1980	1980	0.4372	7.5330	0.0088	0.0038	2.3320

1940 MortRate, Males, Influenza & Pneumonia Deaths, versus
1940 PhysPop Values for the 9 Census Divisions, USA.
HIGHLY Significant INVERSE Relationship.

PhysPop is a surrogate for accumulated dose from medical irradiation.

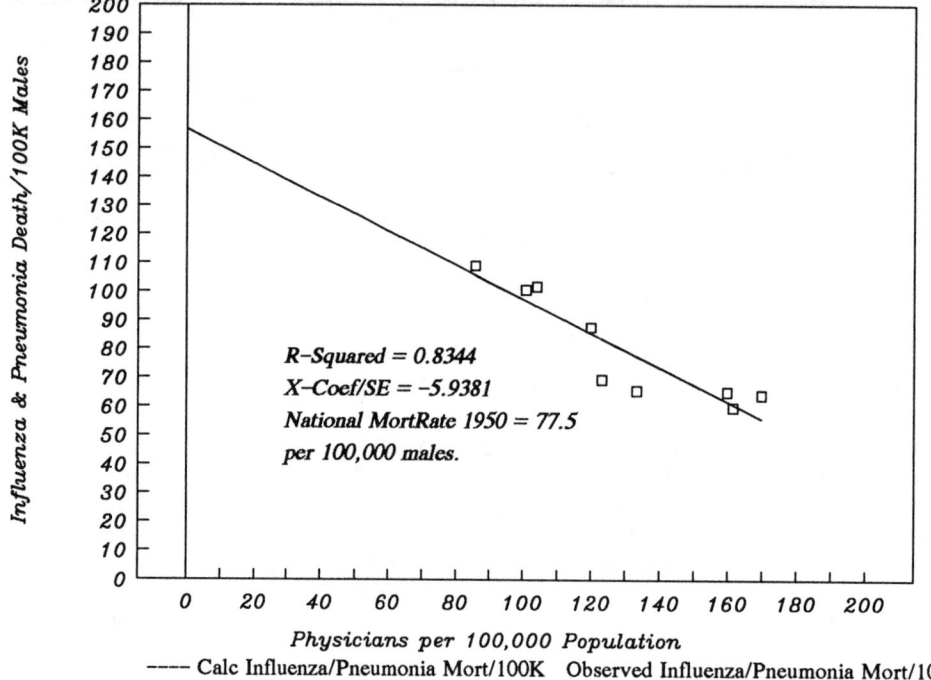

R–Squared = 0.8344
X–Coef/SE = –5.9381
National MortRate 1950 = 77.5
per 100,000 males.

Physicians per 100,000 Population

---- Calc Influenza/Pneumonia Mort/100K   Observed Influenza/Pneumonia Mort/100K   □

1940 MortRate, Females, Influenza and Pneumonia Deaths, versus
1940 PhysPop Values for the 9 Census Divisions, USA.
Very Highly significant. INVERSE DIRECTION

PhysPop is a surrogate for accumulated dose from medical irradiation

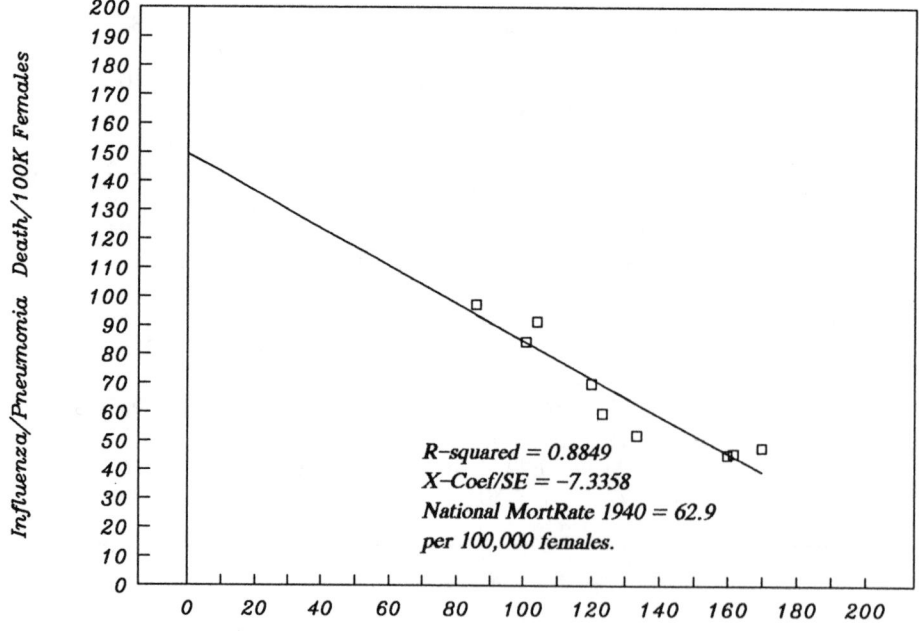

R–squared = 0.8849
X–Coef/SE = –7.3358
National MortRate 1940 = 62.9
per 100,000 females.

Physicians per 100,000 Population

---Calc Influenza/Pneumonia Mort/100K   Observed Influenza/Pneumonia Mort/100K   □

Table 31-A
___
Influenza and Pneumonia: Death Rates by Census Divisions and National.
===

Annual death-rates per 100,000 are age-adjusted to the 1940 reference year. There are no exclusions by color or "race." Entries for the Nine Census Divisions are population-weighted; the averages below them are not. "National Rates" for both sexes: Deaths per 100,000 population (males + females). Males: Deaths per 100,000 male population. Females: Deaths per 100,000 female population.

MALES ....

Census Division	1940	1950	1960	1970	1980	1990
Pacific	65.0	27.7	35.4	25.8	16.1	--
New England	59.7	22.7	40.1	29.2	18.3	--
West North Central	69.2	30.1	36.4	26.0	15.6	--
Mid-Atlantic	63.9	30.7	35.6	26.7	17.7	--
East North Central	65.5	30.2	34.9	25.3	15.7	--
Mountain	87.6	43.1	40.8	28.6	16.3	--
West South Central	101.6	38.1	40.9	28.3	15.7	--
East South Central	108.9	43.9	41.9	29.3	16.7	--
South Atlantic	100.5	39.3	45.3	31.3	17.4	--
Average ALL	80.2	34.0	39.0	27.8	16.6	--
Average High-Five	64.7	28.3	36.5	26.6	16.7	--
Average Low-Four	99.7	41.1	42.2	29.4	16.5	--
Ratio (High/Low)	0.65	0.69	0.86	0.90	1.01	--

FEMALES ....

	1940	1950	1960	1970	1980	1990
Pacific	45.2	17.5	22.9	16.3	9.6	--
New England	45.8	16.2	26.3	18.0	9.7	--
West North Central	59.7	24.5	24.8	17.0	9.1	--
Mid-Atlantic	47.7	20.1	23.4	16.7	9.9	--
East North Central	52.1	20.9	22.4	15.4	8.4	--
Mountain	70.0	31.2	27.3	18.6	9.8	--
West South Central	91.4	31.6	28.0	18.4	8.7	--
East South Central	97.3	38.7	31.6	20.4	9.1	--
South Atlantic	84.4	32.9	29.7	19.2	8.7	--
Average ALL	66.0	26.0	26.3	17.7	9.2	--
Average High-Five	50.1	19.8	24.0	16.7	9.3	--
Average Low-Four	85.8	33.6	29.2	19.1	9.1	--
Ratio (High/Low)	0.58	0.59	0.82	0.87	1.03	--

NATIONAL RATES ....

	1940	1950	1960	1970	1980	1990
Both Sexes	70.2	28.9	31.6	21.9	12.2	--
Males	77.5	33.2	38.3	27.4	16.5	--
Females	62.9	24.8	25.6	17.4	9.1	--

● - 1940, 1950, 1960: All rates come from Grove 1968, Table 67, Pages 734-737, "Influenza and Pneumonia, except Pneumonia of Newborn (480-493)" ICD/7.
  ● - 1970 rates are interpolations (Chap. 4, Parts 2b, 2c).
  ● - 1980: All rates (ICD/9, 480-487) come from reference NatCtrHS 1980.
  ● - 1990: Omitted. These data are only marginally related to examination of Hypotheses 1+2, as noted in Chapter 4, Part 2c.

CHAPTER 32

Fatal Motor Vehicle Accidents: Relation with Medical Radiation

---

Box 1 of Chap. 32
Summary: Regression Outputs, Motor Vehicle Accident Death Rates.

Below are the summary-results from all the regressions of MortRates upon PhysPops. MortRates are from Table 32-A, and PhysPops are from Table 3-A.

MALES ....

Year MortRate	Year PhysPop	R-squared	Constant	Coeff.	Std Err	Coeff/S.E
1940	1921	0.0304	25.0324	0.1212	0.2585	0.4687
1940	1923	0.0230	28.5006	0.0981	0.2420	0.4055
1940	1925	0.0113	33.2529	0.0639	0.2257	0.2830
1940	1927	0.0019	38.2431	0.0250	0.2173	0.1148
1940	1929	0.0020	38.3133	0.0246	0.2102	0.1170
1940	1931	0.0019	38.5694	0.0224	0.1917	0.1169
1940	1934	0.0020	43.8443	-0.0199	0.1687	-0.1179
1940	1936	0.0041	44.8302	-0.0276	0.1621	-0.1703
1940	1938	0.0104	46.5629	-0.0411	0.1517	-0.2711
1940	1940	0.0195	47.8900	-0.0508	0.1359	-0.3736
1950	1950	0.4761	66.1657	-0.2347	0.0931	-2.5224
1960	1960	0.5609	63.9413	-0.2422	0.0810	-2.9901
1970	1970	0.6234	58.9485	-0.1877	0.0551	-3.4039
1980	1980	0.7527	59.6950	-0.1551	0.0336	-4.6160
1990	1990	0.7311	57.2462	-0.1246	0.0286	-4.3626

FEMALES..

Year MortRate	Year PhysPop	R-squared	Constant	Coeff.	Std Err	Coeff/S.E
1940	1921	0.1128	-1.3760	0.1046	0.1109	0.9432
1940	1923	0.0914	1.2134	0.0878	0.1046	0.8389
1940	1925	0.0503	5.0483	0.0604	0.0992	0.6090
1940	1927	0.0257	7.5560	0.0413	0.0962	0.4294
1940	1929	0.0272	7.6222	0.0411	0.0930	0.4422
1940	1931	0.0265	8.1016	0.0371	0.0849	0.4367
1940	1934	0.0059	10.7888	0.0154	0.0755	0.2035
1940	1936	0.0046	11.0689	0.0130	0.0727	0.1796
1940	1938	0.0009	12.0180	0.0055	0.0683	0.0801
1940	1940	0.0003	13.0546	-0.0027	0.0615	-0.0434
1950	1950	0.1790	17.5200	-0.0506	0.0410	-1.2352
1960	1960	0.3522	20.3931	-0.0720	0.0369	-1.9507
1970	1970	0.4324	19.9743	-0.0603	0.0261	-2.3090
1980	1980	0.6130	21.5524	-0.0544	0.0163	-3.3299
1990	1990	0.7348	21.9803	-0.0469	0.0107	-4.4040

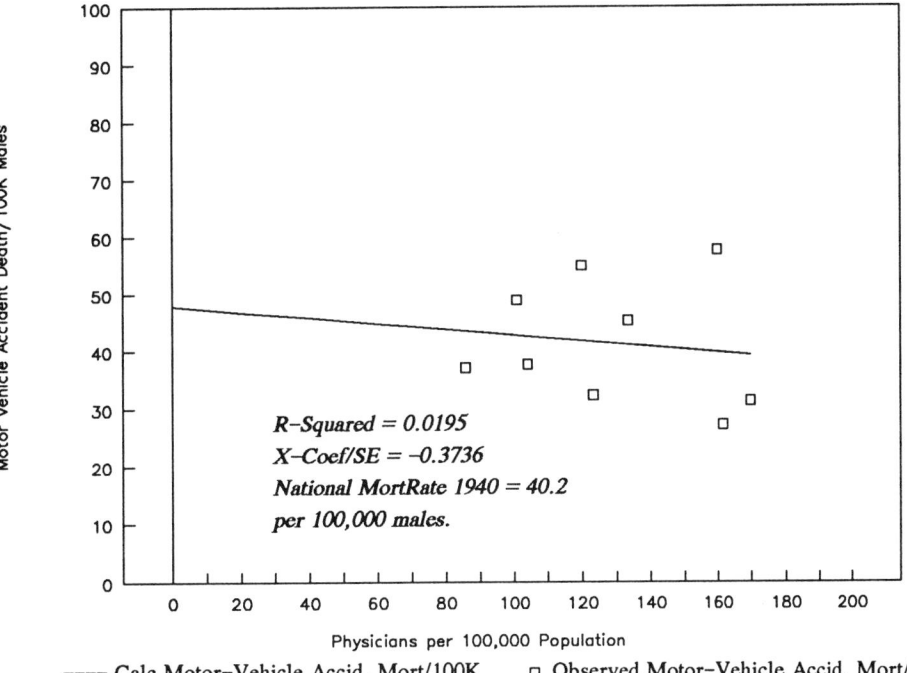

1940 MortRate, Males, Motor-Vehicle Accident Deaths, versus
1940 PhysPop Values for the 9 Census Divisions, USA.
Not Significant
PhysPop is a surrogate for accumulated dose from medical irradiation.

R-Squared = 0.0195
X-Coef/SE = -0.3736
National MortRate 1940 = 40.2
per 100,000 males.

---- Calc Motor-Vehicle Accid. Mort/100K   □ Observed Motor-Vehicle Accid. Mort/100K

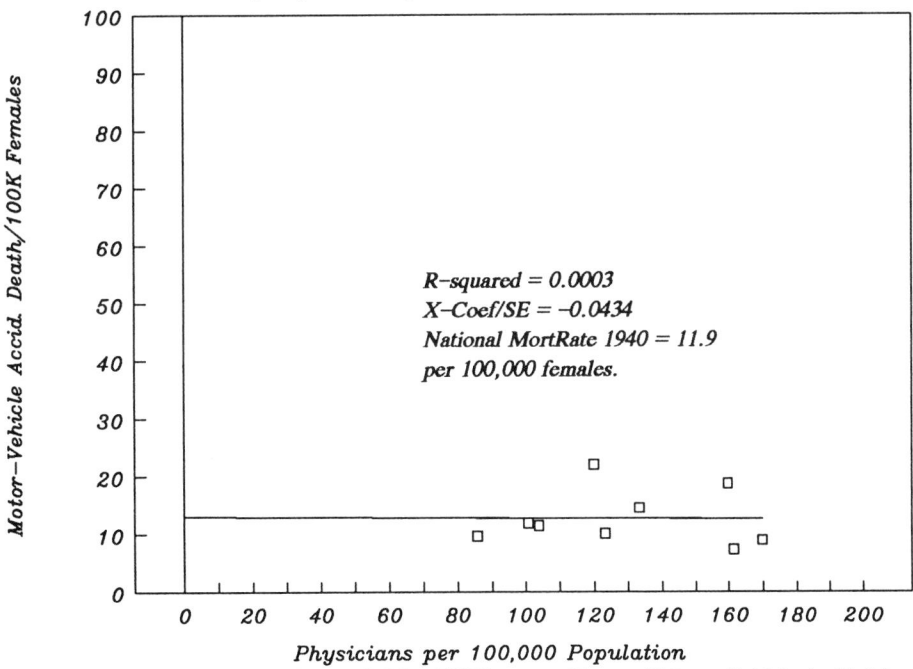

1940 MortRate, Females, Motor-Vehicle Accident Deaths, versus
1940 PhysPop Values for the 9 Census Divisions, USA.
Not Significant.
PhysPop is a surrogate for accumulated dose from medical irradiation

R-squared = 0.0003
X-Coef/SE = -0.0434
National MortRate 1940 = 11.9
per 100,000 females.

---Calc Motor-Vehicle Accid. Mort/100K   □ Observed Motor-Vehicle Accid. Mort/100K

Table 32-A
Fatal Motor-Vehicle Accidents: Rates by Census Divisions and National.

Annual death-rates per 100,000 are age-adjusted to the 1940 reference year. There are no exclusions by color or "race." Entries for the Nine Census Divisions are population-weighted; the averages below them are not. "National Rates" for both sexes: Deaths per 100,000 population (males + females). Males: Deaths per 100,000 male population. Females: Deaths per 100,000 female population.

MALES ....

Census Division	1940	1950	1960	1970	1980	1990
Pacific	57.6	42.1	35.2	31.9	28.5	25.1
New England	27.2	20.3	19.3	18.6	18.0	17.3
West North Central	32.3	34.7	34.9	31.9	28.9	25.9
Mid-Atlantic	31.3	22.0	19.9	19.5	19.1	18.7
East North Central	45.3	36.4	30.0	32.6	35.2	37.8
Mountain	54.9	51.4	47.9	42.2	36.5	30.7
West South Central	37.6	42.0	38.8	36.0	33.1	30.2
East South Central	37.1	41.0	41.8	40.5	39.1	37.8
South Atlantic	48.9	41.2	36.5	34.3	32.0	29.7
Average ALL	41.4	36.8	33.8	31.9	30.0	28.1
Average High-Five	38.7	31.1	27.9	26.9	25.9	25.0
Average Low-Four	44.6	43.9	41.3	38.2	35.2	32.1
Ratio (High/Low)	0.87	0.71	0.68	0.70	0.74	0.78

FEMALES ....

	1940	1950	1960	1970	1980	1990
Pacific	18.7	13.9	13.2	12.2	11.1	10.1
New England	7.3	5.9	6.0	6.3	6.5	6.8
West North Central	10.1	10.9	12.3	11.9	11.4	11.0
Mid-Atlantic	8.9	6.8	6.5	6.9	7.2	7.6
East North Central	14.5	11.8	10.6	10.4	10.3	10.1
Mountain	22.0	17.6	17.5	16.3	15.0	13.8
West South Central	11.4	12.6	13.4	13.0	12.6	12.2
East South Central	9.6	10.3	12.6	13.5	14.3	15.2
South Atlantic	11.9	10.9	10.8	11.3	11.8	12.3
Average ALL	12.7	11.2	11.4	11.3	11.2	11.0
Average High-Five	11.9	9.9	9.7	9.5	9.3	9.1
Average Low-Four	13.7	12.9	13.6	13.5	13.4	13.4
Ratio (High/Low)	0.87	0.77	0.72	0.70	0.69	0.68

NATIONAL RATES ....

	1940	1950	1960	1970	1980	1990
Both Sexes	26.1	22.6	20.9	20.0	19.1	18.2
Males	40.2	35.0	31.7	29.7	27.7	25.7
Females	11.9	10.6	10.6	10.6	10.7	10.7

● – 1940, 1950, 1960: All rates come from Grove 1968, Table 67, Pages 757-760, Death-rates from "Motor Vehicle Accidents (E810-E835)" ICD/7.
● – 1970 rates are interpolations; please see 1980, below.
● – 1980: For this particular table, the reference NatCtrHS 1980 provided no entries. The entire range of values between 1960 and 1990 is so small, that we think interpolations for both 1970 and 1980 can serve here as acceptable approximations.
● – 1990: Rates are for 1989-1991, from Monthly Vital Statistics Report, Vol.43, No.9, March 1, 1995.

CHAPTER 33

Other Fatal Accidents: Relation with Medical Radiation

Box 1 of Chap. 33
Summary: Regression Outputs, Other Accident Death Rates

Below are the summary-results from all the regressions of MortRates upon PhysPops.
MortRates are from Table 33-A, and PhysPops are from Table 3-A.

MALES ....

Year MortRate	Year PhysPop	R-squared	Constant	Coeff.	Std Err	Coeff/S.E
1940	1921	0.0107	71.7318	-0.0628	0.2283	-0.2752
1940	1923	0.0154	72.4819	-0.0703	0.2124	-0.3312
1940	1925	0.0375	76.1584	-0.1016	0.1947	-0.5219
1940	1927	0.0557	78.0775	-0.1187	0.1848	-0.6426
1940	1929	0.0499	76.7179	-0.1087	0.1793	-0.6065
1940	1931	0.0517	75.8137	-0.1009	0.1634	-0.6175
1940	1934	0.0791	77.0176	-0.1099	0.1417	-0.7756
1940	1936	0.0714	75.9027	-0.1004	0.1369	-0.7336
1940	1938	0.0797	75.8781	-0.0996	0.1279	-0.7787
1940	1940	0.0901	75.5351	-0.0954	0.1145	-0.8328
1950	1950	0.2279	67.2868	-0.1381	0.0961	-1.4375
1960	1960	0.4278	63.7628	-0.1756	0.0768	-2.2875
1970	1970	--	--	--	--	--
1980	1980	--	--	--	--	--
1990	1990	--	--	--	--	--

FEMALES...

Year MortRate	Year PhysPop	R-squared	Constant	Coeff.	Std Err	Coeff/S.E
1940	1921	0.5688	47.7376	-0.1219	0.0401	-3.0388
1940	1923	0.5679	46.1956	-0.1135	0.0374	-3.0332
1940	1925	0.5976	45.0231	-0.1080	0.0335	-3.2246
1940	1927	0.5633	43.8521	-0.1004	0.0334	-3.0049
1940	1929	0.5489	43.1886	-0.0959	0.0329	-2.9184
1940	1931	0.5778	42.4863	-0.0898	0.0290	-3.0953
1940	1934	0.5434	40.9082	-0.0766	0.0265	-2.8863
1940	1936	0.4941	40.1653	-0.0703	0.0269	-2.6145
1940	1938	0.4577	39.3615	-0.0635	0.0261	-2.4305
1940	1940	0.4440	38.5655	-0.0563	0.0238	-2.3642
1950	1950	0.2651	28.1325	-0.0401	0.0253	-1.5889
1960	1960	0.2646	23.0083	-0.0372	0.0235	-1.5870
1970	1970	--	--	--	--	--
1980	1980	--	--	--	--	--
1990	1990	--	--	--	--	--

1940 MortRate, Males, "Other Accidents" Deaths, versus
1940 PhysPop Values for the 9 Census Divisions, USA.
Not Significant

PhysPop is a surrogate for accumulated dose from medical irradiation.

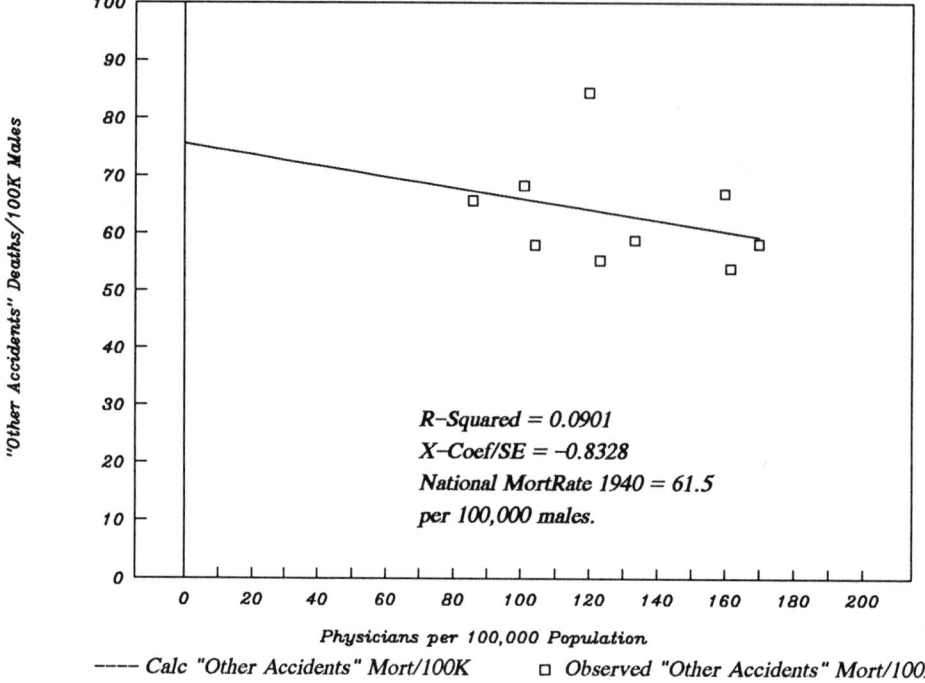

R-Squared = 0.0901
X-Coef/SE = -0.8328
National MortRate 1940 = 61.5
per 100,000 males.

---- Calc "Other Accidents" Mort/100K   □ Observed "Other Accidents" Mort/100K

1940 MortRate, Females, "Other Accidents" Deaths, versus
1940 PhysPop Values for the 9 Census Divisions, USA.
Significant. INVERSE Direction.

PhysPop is a surrogate for accumulated dose from medical irradiation

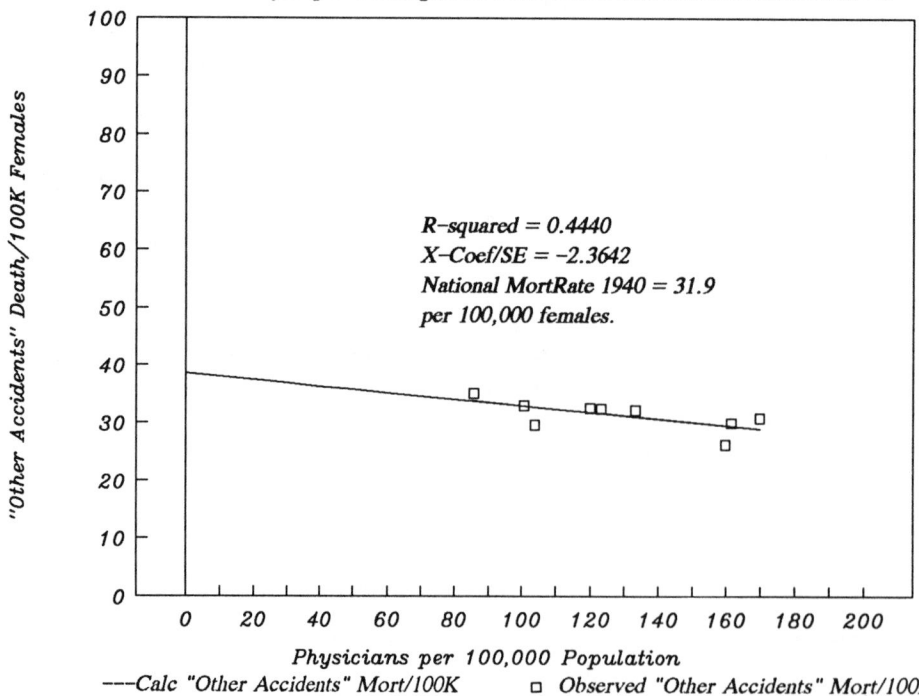

R-squared = 0.4440
X-Coef/SE = -2.3642
National MortRate 1940 = 31.9
per 100,000 females.

---Calc "Other Accidents" Mort/100K   □ Observed "Other Accidents" Mort/100K

Annual death-rates per 100,000 are age-adjusted to the 1940 reference year. There are no exclusions by color or "race." Entries for the Nine Census Divisions are population-weighted; the averages below them are not. "National Rates" for both sexes: Deaths per 100,000 population (males + females). Males: Deaths per 100,000 male population. Females: Deaths per 100,000 female population.

MALES ....

Census Division	1940	1950	1960	1970	1980	1990
Pacific	66.9	49.7	39.8	--	--	--
New England	53.9	41.7	34.7	--	--	--
West North Central	55.3	47.5	39.8	--	--	--
Mid-Atlantic	58.2	41.6	32.6	--	--	--
East North Central	58.8	43.5	33.3	--	--	--
Mountain	84.4	69.2	55.4	--	--	--
West South Central	58.0	54.5	46.2	--	--	--
East South Central	65.6	51.1	49.0	--	--	--
South Atlantic	68.3	51.2	46.5	--	--	--
Average ALL	63.3	50.0	41.9	--	--	--
Average High-Five	58.6	44.8	36.0	--	--	--
Average Low-Four	69.1	56.5	49.3	--	--	--
Ratio (High/Low)	0.85	0.79	0.73	--	--	--

FEMALES ....

	1940	1950	1960	1970	1980	1990
Pacific	26.3	17.8	17.2	--	--	--
New England	30.0	23.5	18.8	--	--	--
West North Central	32.4	25.4	16.4	--	--	--
Mid-Atlantic	30.8	22.0	15.9	--	--	--
East North Central	32.2	22.4	15.6	--	--	--
Mountain	32.5	24.8	20.2	--	--	--
West South Central	29.6	23.9	20.0	--	--	--
East South Central	35.1	24.9	21.0	--	--	--
South Atlantic	33.0	23.3	20.3	--	--	--
Average ALL	31.3	23.1	18.4	--	--	--
Average High-Five	30.3	22.2	16.8	--	--	--
Average Low-Four	32.6	24.2	20.4	--	--	--
Ratio (High/Low)	0.93	0.92	0.82	--	--	--

NATIONAL RATES ....

	1940	1950	1960	1970	1980	1990
Both Sexes	46.9	35.4	28.6	--	--	--
Males	61.5	47.9	39.7	--	--	--
Females	31.9	23.1	17.8	--	--	--

● - 1940, 1950, 1960: All rates come from Grove 1968, Table 67, Pages 760-764, Death-rates from "Other Accidents (E800-E802, E840-E962)" ICD/7.

● - Omitted data are only marginally related to examination of Hypotheses 1+2, as noted in Chapter 4, Part 2c.

CHAPTER 34

Rheumatic Fever & Rheum. Heart Disease: Relation with Medical Radiation

---

Box 1 of Chap. 34
Summary: Regression Outputs, Rheumatic Fever & Rheum. Heart Disease

Below are the summary-results from all the regressions of MortRates upon PhysPops. MortRates are from Table 34-A, and PhysPops are from Table 3-A.

MALES ....

Year MortRate	Year PhysPop	R-squared	Constant	Coeff.	Std Err	Coeff/S.E
1940	1921	0.0456	27.4733	-0.0401	0.0693	-0.5782
1940	1923	0.0254	25.7307	-0.0279	0.0653	-0.4272
1940	1925	0.0175	24.8033	-0.0215	0.0608	-0.3535
1940	1927	0.0041	23.3121	-0.0099	0.0586	-0.1688
1940	1929	0.0022	22.9462	-0.0070	0.0568	-0.1237
1940	1931	0.0046	23.2366	-0.0093	0.0517	-0.1802
1940	1934	0.0007	22.4888	-0.0033	0.0456	-0.0720
1940	1936	0.0000	22.0125	0.0005	0.0439	0.0118
1940	1938	0.0011	21.6206	0.0036	0.0412	0.0877
1940	1940	0.0021	21.5047	0.0045	0.0370	0.1202
1950	1950	0.1891	9.8946	0.0335	0.0262	1.2775
1960	1960	0.3873	3.6475	0.0467	0.0222	2.1034
1970	1970	--	--	--	--	--
1980	1980	--	--	--	--	--
1990	1990	--	--	--	--	--

FEMALES...

Year MortRate	Year PhysPop	R-squared	Constant	Coeff.	Std Err	Coeff/S.E
1940	1921	0.0092	23.9603	-0.0191	0.0748	-0.2554
1940	1923	0.0025	22.5958	-0.0092	0.0699	-0.1317
1940	1925	0.0018	22.3047	-0.0072	0.0649	-0.1113
1940	1927	0.0015	20.5846	0.0064	0.0622	0.1037
1940	1929	0.0053	19.9609	0.0115	0.0600	0.1923
1940	1931	0.0039	20.2540	0.0091	0.0548	0.1666
1940	1934	0.0155	19.3990	0.0159	0.0479	0.3318
1940	1936	0.0314	18.6489	0.0218	0.0457	0.4761
1940	1938	0.0468	18.2291	0.0250	0.0426	0.5860
1940	1940	0.0550	18.2527	0.0244	0.0382	0.6386
1950	1950	0.3732	5.1440	0.0659	0.0323	2.0417
1960	1960	0.4874	0.4857	0.0713	0.0276	2.5799
1970	1970	--	--	--	--	--
1980	1980	--	--	--	--	--
1990	1990	--	--	--	--	--

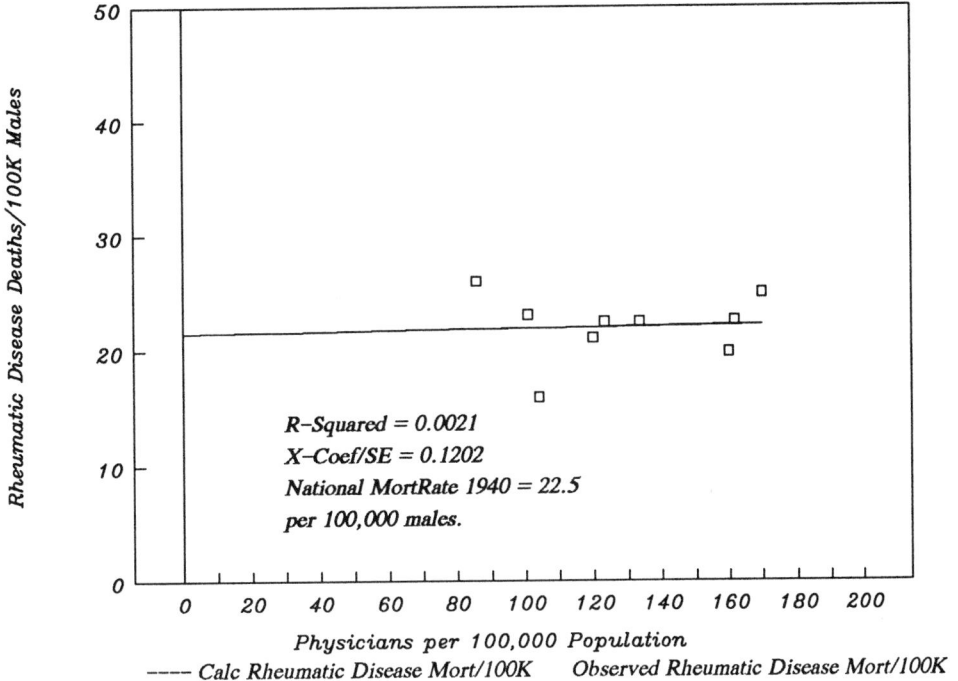

1940 MortRate, Males, "Rheumatic Disease" Deaths, versus
1940 PhysPop Values for the 9 Census Divisions, USA.
Not Significant
PhysPop is a surrogate for accumulated dose from medical irradiation.

R–Squared = 0.0021
X–Coef/SE = 0.1202
National MortRate 1940 = 22.5
per 100,000 males.

---- *Calc Rheumatic Disease Mort/100K*   *Observed Rheumatic Disease Mort/100K* □

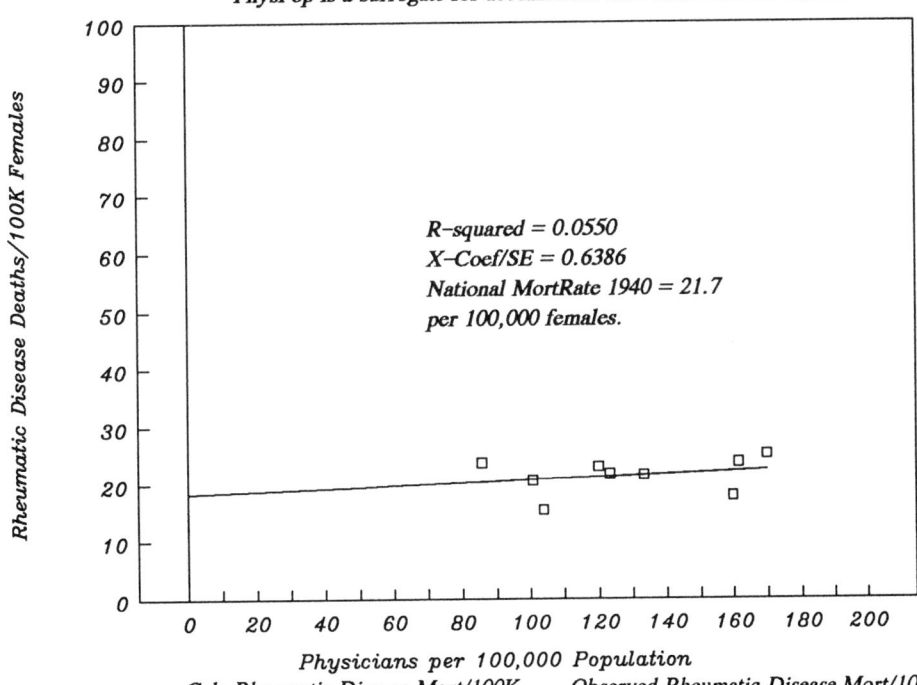

1940 MortRate, Females, "Rheumatic Disease" Deaths, versus
1940 PhysPop Values for the 9 Census Divisions, USA.
Not Significant.
PhysPop is a surrogate for accumulated dose from medical irradiation

R–squared = 0.0550
X–Coef/SE = 0.6386
National MortRate 1940 = 21.7
per 100,000 females.

----*Calc Rheumatic Disease Mort/100K*   *Observed Rheumatic Disease Mort/100K* □

Table 34-A
Rheumatic Fever and Chronic Rheumatic Heart Disease

Death Rates by Census Divisions and National.

Annual death-rates per 100,000 are age-adjusted to the 1940 reference year. There are no exclusions by color or "race." Entries for the Nine Census Divisions are population-weighted; the averages below them are not. "National Rates" for both sexes: Deaths per 100,000 population (males + females). Males: Deaths per 100,000 male population. Females: Deaths per 100,000 female population.

MALES ....

Census Division	1940	1950	1960	1970	1980	1990
Pacific	19.9	14.5	10.2	--	--	--
New England	22.6	14.0	10.7	--	--	--
West North Central	22.5	13.8	9.6	--	--	--
Mid-Atlantic	25.0	16.0	11.5	--	--	--
East North Central	22.5	15.6	9.9	--	--	--
Mountain	21.1	17.1	12.4	--	--	--
West South Central	16.0	9.2	5.6	--	--	--
East South Central	26.0	13.9	6.7	--	--	--
South Atlantic	23.1	12.7	8.5	--	--	--
Average ALL	22.1	14.1	9.5	--	--	--
Average High-Five	22.5	14.8	10.4	--	--	--
Average Low-Four	21.6	13.2	8.3	--	--	--
Ratio (High/Low)	1.04	1.12	1.25	--	--	--

FEMALES ....

	1940	1950	1960	1970	1980	1990
Pacific	17.8	13.3	10.4	--	--	--
New England	23.7	15.6	11.5	--	--	--
West North Central	21.7	11.9	8.2	--	--	--
Mid-Atlantic	25.1	16.7	12.9	--	--	--
East North Central	21.5	14.1	9.9	--	--	--
Mountain	23.0	18.1	13.2	--	--	--
West South Central	15.4	7.7	4.7	--	--	--
East South Central	23.7	12.1	6.0	--	--	--
South Atlantic	20.6	11.0	7.4	--	--	--
Average ALL	21.4	13.4	9.4	--	--	--
Average High-Five	22.0	14.3	10.6	--	--	--
Average Low-Four	20.7	12.2	7.8	--	--	--
Ratio (High/Low)	1.06	1.17	1.35	--	--	--

NATIONAL RATES ....

	1940	1950	1960	1970	1980	1990
Both Sexes	22.1	14.0	9.6	--	--	--
Males	22.5	14.4	9.6	--	--	--
Females	21.7	13.5	9.6	--	--	--

● - 1940, 1950, 1960: All rates come from Grove 1968, Table 67, Pages 716-720, Death-rates from "Rheumatic Fever and Chronic Rheumatic Heart Disease (400-402, 410-416)" ICD/7.

● - Omitted data are only marginally related to examination of Hypotheses 1+2, as noted in Chapter 4, Part 2c.

CHAPTER 35

Syphilis and Sequelae, Deaths: Relation with Medical Radiation

---

Box 1 of Chap. 35
Summary: Regression Outputs, Syphilis and Its Sequelae, Death Rates.

Below are the summary-results from all the regressions of MortRates upon PhysPops.
MortRates are from Table 35-A, and PhysPops are from Table 3-A.

MALES ....

Year MortRate	Year PhysPop	R-squared	Constant	Coeff.	Std Err	Coeff/S.E
1940	1921	0.3170	47.2321	-0.1986	0.1102	-1.8025
1940	1923	0.2964	43.9384	-0.1791	0.1043	-1.7174
1940	1925	0.3040	41.8094	-0.1681	0.0962	-1.7484
1940	1927	0.3220	41.1592	-0.1658	0.0909	-1.8232
1940	1929	0.3415	40.9113	-0.1652	0.0867	-1.9052
1940	1931	0.3336	38.9976	-0.1489	0.0795	-1.8720
1940	1934	0.3465	37.1900	-0.1336	0.0693	-1.9266
1940	1936	0.3581	36.9131	-0.1306	0.0661	-1.9759
1940	1938	0.3578	35.9960	-0.1226	0.0621	-1.9747
1940	1940	0.3278	34.0663	-0.1056	0.0572	-1.8477
1950	1950	0.4949	12.1845	-0.0412	0.0157	-2.6189
1960	1960	0.0429	2.7310	-0.0042	0.0075	-0.5600
1970	1970	--	--	--	--	--
1980	1980	--	--	--	--	--
1990	1990	--	--	--	--	--

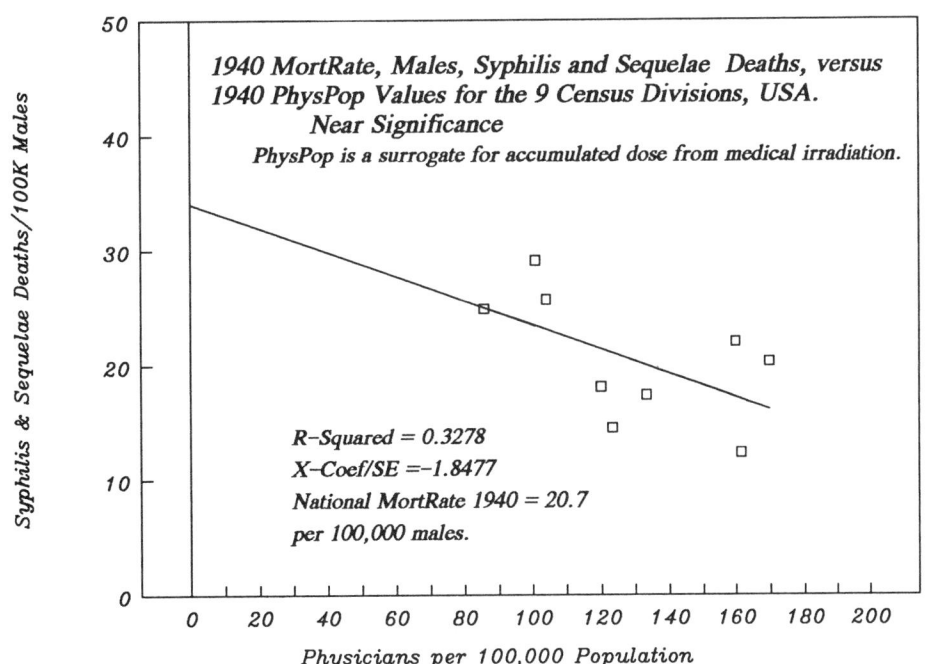

1940 MortRate, Males, Syphilis and Sequelae Deaths, versus
1940 PhysPop Values for the 9 Census Divisions, USA.
Near Significance
PhysPop is a surrogate for accumulated dose from medical irradiation.

R-Squared = 0.3278
X-Coef/SE = -1.8477
National MortRate 1940 = 20.7
per 100,000 males.

Figure 35-A

---- *Calc Syphilis and Sequelae Mort/100K*   *Observed Syphilis and Sequelae Mort/100K* □

Annual death-rates per 100,000 are age-adjusted to the 1940 reference year. There are no exclusions by color or "race." Entries for the Nine Census Divisions are population-weighted; the averages below them are not. "National Rates" for both sexes: Deaths per 100,000 population (males + females). Males: Deaths per 100,000 male population. Females: Deaths per 100,000 female population.

MALES ....

Census Division	1940	1950	1960	1970	1980	1990
Pacific	22.0	7.1	2.3	--	--	--
New England	12.3	4.7	2.0	--	--	--
West North Central	14.5	6.0	1.9	--	--	--
Mid-Atlantic	20.3	5.4	2.0	--	--	--
East North Central	17.4	6.8	1.6	--	--	--
Mountain	18.1	6.8	2.0	--	--	--
West South Central	25.7	10.0	3.6	--	--	--
East South Central	24.9	7.1	1.9	--	--	--
South Atlantic	29.1	9.4	2.6	--	--	--
Average ALL	20.5	7.0	2.2	--	--	--
Average High-Five	17.3	6.0	2.0	--	--	--
Average Low-Four	24.5	8.3	2.5	--	--	--
Ratio (High/Low)	0.71	0.72	0.78	--	--	--

FEMALES ....

	1940	1950	1960	1970	1980	1990
Pacific					--	--
New England					--	--
West North Central					--	--
Mid-Atlantic					--	--
East North Central		The National Rates below reflect			--	--
Mountain		a severe "small numbers problem"			--	--
West South Central		for the female rates, so we did not			--	--
East South Central		analyze the female data at all.			--	--
South Atlantic					--	--
Average ALL					--	--
Average High-Five					--	--
Average Low-Four					--	--
Ratio (High/Low)					--	--

NATIONAL RATES ....

	1940	1950	1960	1970	1980	1990
Both Sexes	14.4	4.6	1.4	--	--	--
Males	20.7	7.0	2.2	--	--	--
Females	7.9	2.4	0.7	--	--	--

● - 1940, 1950, 1960: All rates come from Grove 1968, Table 67, Pages 673-676, Death-rates from "Syphilis and Its Sequelae (020-029)" ICD/7.

● - Omitted data are only marginally related to examination of Hypotheses 1+2, as noted in Chapter 4, Part 2c.

CHAPTER 36

Tuberculosis, Deaths: Relation with Medical Radiation

Box 1 of Chap. 36
Summary: Regression Outputs, Tuberculosis, All Forms, Death Rates.

Below are the summary-results from all the regressions of MortRates upon PhysPops.
MortRates are from Table 36-A, and PhysPops are from Table 3-A.

MALES ....

Year MortRate	Year PhysPop	R-squared	Constant	Coeff.	Std Err	Coeff/S.E
1940	1921	0.1387	101.3459	−0.3326	0.3132	−1.0618
1940	1923	0.1563	99.6695	−0.3292	0.2890	−1.1388
1940	1925	0.1987	100.2053	−0.3442	0.2612	−1.3176
1940	1927	0.2341	101.1808	−0.3578	0.2447	−1.4625
1940	1929	0.2275	98.7660	−0.3413	0.2377	−1.4359
1940	1931	0.2100	93.7415	−0.2991	0.2192	−1.3642
1940	1934	0.2320	91.1603	−0.2767	0.1903	−1.4542
1940	1936	0.2265	89.6338	−0.2629	0.1837	−1.4317
1940	1938	0.2247	87.6750	−0.2459	0.1726	−1.4243
1940	1940	0.2067	83.8602	−0.2123	0.1572	−1.3507
1950	1950	0.1737	40.0950	−0.0892	0.0735	−1.2132
1960	1960	0.1415	12.0470	−0.0292	0.0272	−1.0741
1970	1970	--	--	--	--	--
1980	1980	--	--	--	--	--
1990	1990	--	--	--	--	--

FEMALES..

Year MortRate	Year PhysPop	R-squared	Constant	Coeff.	Std Err	Coeff/S.E
1940	1921	0.4426	120.5024	−0.6038	0.2561	−2.3577
1940	1923	0.4909	116.8396	−0.5928	0.2282	−2.5980
1940	1925	0.5301	111.6456	−0.5712	0.2033	−2.8101
1940	1927	0.6039	112.0405	−0.5841	0.1788	−3.2669
1940	1929	0.6249	110.2892	−0.5748	0.1683	−3.4153
1940	1931	0.6170	103.9682	−0.5209	0.1551	−3.3582
1940	1934	0.6527	98.1795	−0.4716	0.1300	−3.6268
1940	1936	0.6724	97.1170	−0.4604	0.1215	−3.7907
1940	1938	0.6698	93.8004	−0.4314	0.1145	−3.7685
1940	1940	0.6381	87.9447	−0.3790	0.1079	−3.5128
1950	1950	0.5942	35.9541	−0.1648	0.0515	−3.2015
1960	1960	0.3334	6.0716	−0.0243	0.0130	−1.8711
1970	1970	--	--	--	--	--
1980	1980	--	--	--	--	--
1990	1990	--	--	--	--	--

1940 MortRate, Males, Tuberculosis Deaths, versus
1940 PhysPop Values for the 9 Census Divisions, USA.
Not Significant
PhysPop is a surrogate for accumulated dose from medical irradiation.

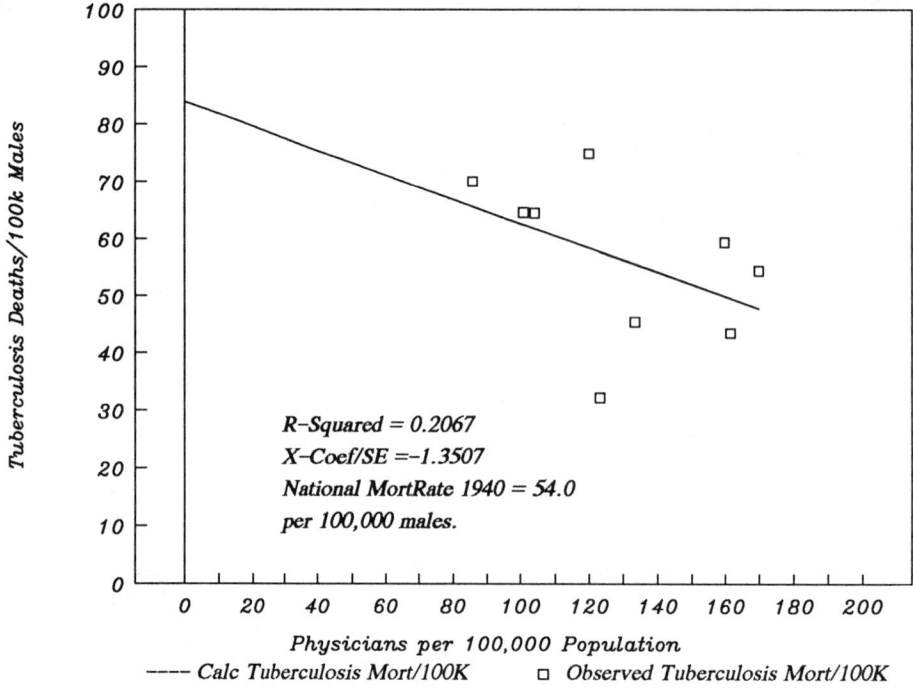

R–Squared = 0.2067
X–Coef/SE =–1.3507
National MortRate 1940 = 54.0
per 100,000 males.

Physicians per 100,000 Population
---- *Calc Tuberculosis Mort/100K*   □ *Observed Tuberculosis Mort/100K*

1940 MortRate, Females, Tuberculosis, All forms Deaths, versus
1940 PhysPop Values for the 9 Census Divisions, USA.
Highly Significant. Direction is INVERSE.
PhysPop is a surrogate for accumulated dose from medical irradiation

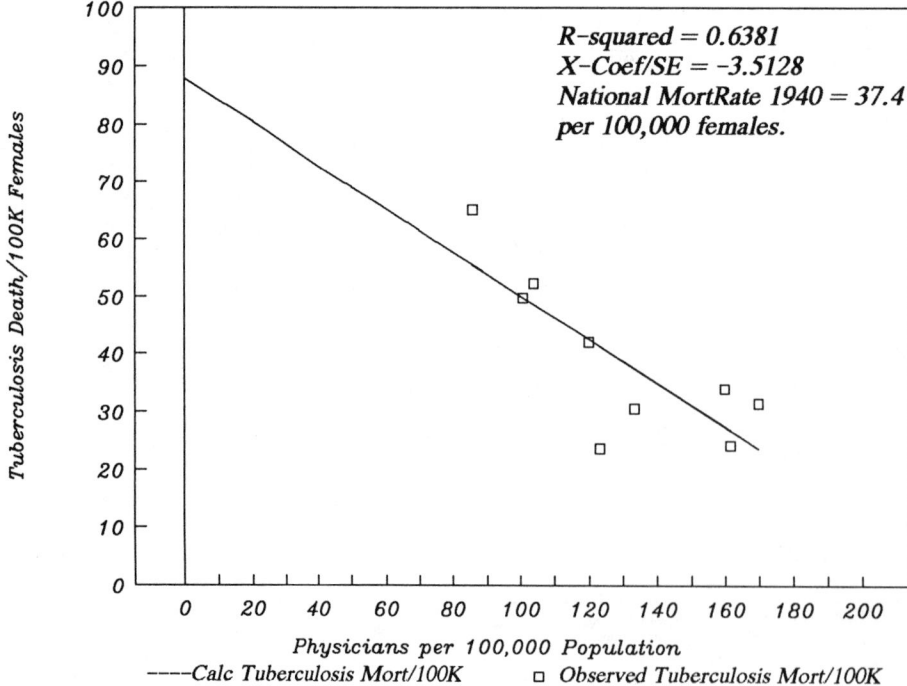

R–squared = 0.6381
X–Coef/SE = –3.5128
National MortRate 1940 = 37.4
per 100,000 females.

Physicians per 100,000 Population
----*Calc Tuberculosis Mort/100K*   □ *Observed Tuberculosis Mort/100K*

Annual death-rates per 100,000 are age-adjusted to the 1940 reference year. There are no exclusions by color or "race." Entries for the Nine Census Divisions are population-weighted; the averages below them are not. "National Rates" for both sexes: Deaths per 100,000 population (males + females). Males: Deaths per 100,000 male population. Females: Deaths per 100,000 female population.

MALES ....

Census Division	1940	1950	1960	1970	1980	1990
Pacific	59.4	26.1	6.5	--	--	--
New England	43.5	24.6	7.1	--	--	--
West North Central	32.2	16.8	4.9	--	--	--
Mid-Atlantic	54.4	32.4	9.9	--	--	--
East North Central	45.4	26.3	7.1	--	--	--
Mountain	74.9	29.9	8.8	--	--	--
West South Central	64.5	33.1	10.3	--	--	--
East South Central	70.0	38.5	12.4	--	--	--
South Atlantic	64.6	32.7	8.7	--	--	--
Average ALL	56.5	28.9	8.4	--	--	--
Average High-Five	47.0	25.2	7.1	--	--	--
Average Low-Four	68.5	33.6	10.1	--	--	--
Ratio (High/Low)	0.69	0.75	0.71	--	--	--

FEMALES ....

	1940	1950	1960	1970	1980	1990
Pacific	33.9	10.7	2.1	--	--	--
New England	24.1	9.7	2.0	--	--	--
West North Central	23.6	8.4	1.8	--	--	--
Mid-Atlantic	31.4	12.7	3.2	--	--	--
East North Central	30.5	11.8	2.4	--	--	--
Mountain	42.0	17.4	3.1	--	--	--
West South Central	52.3	21.2	3.8	--	--	--
East South Central	65.1	27.4	5.8	--	--	--
South Atlantic	49.7	18.7	3.2	--	--	--
Average ALL	39.2	15.3	3.0	--	--	--
Average High-Five	28.7	10.7	2.3	--	--	--
Average Low-Four	52.3	21.2	4.0	--	--	--
Ratio (High/Low)	0.55	0.50	0.58	--	--	--

NATIONAL RATES ....

	1940	1950	1960	1970	1980	1990
Both Sexes	45.8	21.6	5.5	--	--	--
Males	54.0	29.0	8.3	--	--	--
Females	37.4	14.5	2.9	--	--	--

● - 1940, 1950, 1960: All rates come from Grove 1968, Table 67, Pages 666-669, Death-rates from "Tuberculosis, All Forms (001-019)" ICD/7.

● - Omitted data are only marginally related to examination of Hypotheses 1+2, as noted in Chapter 4, Part 2c.

CHAPTER 37

Ulcer of Stomach and Duodenum, Deaths: Relation with Medical Radiation

---

Box 1 of Chap. 37
Summary: Regression Outputs, Ulcer of Stomach and Duodenum, Death Rates.

Below are the summary-results from all the regressions of MortRates upon PhysPops.
MortRates are from Table 37-A, and PhysPops are from Table 3-A.

MALES ....

Year MortRate	Year PhysPop	R-squared	Constant	Coeff.	Std Err	Coeff/S.E
1940	1921	0.4257	0.9566	0.0765	0.0336	2.2778
1940	1923	0.4434	1.7250	0.0728	0.0308	2.3615
1940	1925	0.4114	3.0135	0.0651	0.0294	2.2120
1940	1927	0.4088	3.5173	0.0621	0.0282	2.1999
1940	1929	0.4326	3.6185	0.0618	0.0268	2.3101
1940	1931	0.4420	4.1778	0.0570	0.0242	2.3547
1940	1934	0.4115	5.2106	0.0484	0.0219	2.2122
1940	1936	0.4210	5.3403	0.0471	0.0209	2.2559
1940	1938	0.4036	5.7852	0.0433	0.0199	2.1767
1940	1940	0.3864	6.3609	0.0381	0.0182	2.0994
1950	1950	0.5962	3.5228	0.0374	0.0116	3.2147
1960	1960	0.6064	4.4028	0.0328	0.0100	3.2842
1970	1970	--	--	--	--	--
1980	1980	--	--	--	--	--
1990	1990	--	--	--	--	--

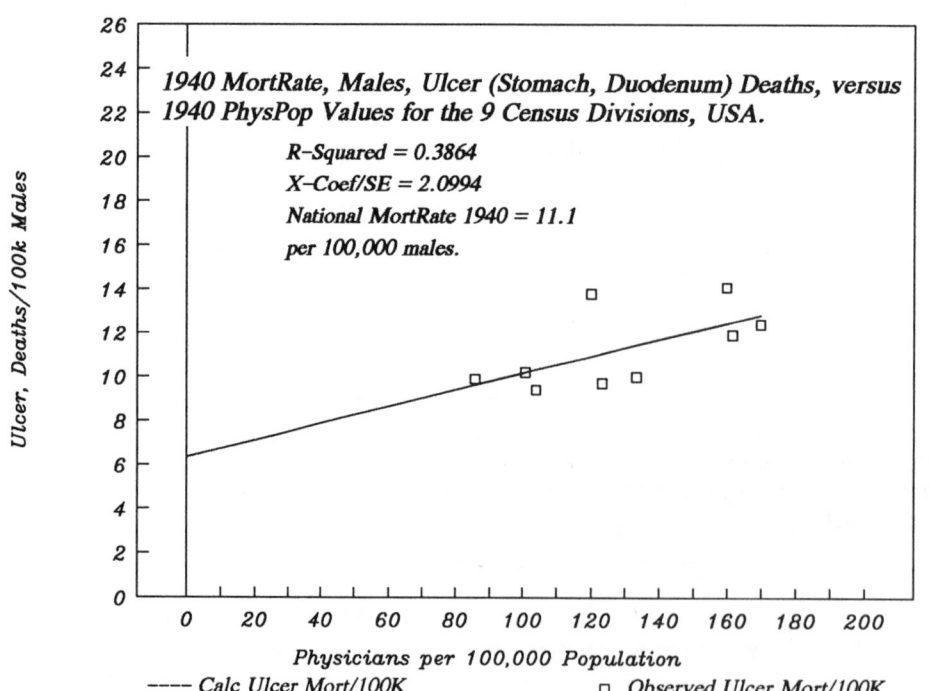

1940 MortRate, Males, Ulcer (Stomach, Duodenum) Deaths, versus 1940 PhysPop Values for the 9 Census Divisions, USA.

R-Squared = 0.3864
X-Coef/SE = 2.0994
National MortRate 1940 = 11.1
per 100,000 males.

Figure 37-A

Ulcer, Deaths/100k Males

Physicians per 100,000 Population

---- Calc Ulcer Mort/100K   □ Observed Ulcer Mort/100K

Annual death-rates per 100,000 are age-adjusted to the 1940 reference year. There are no exclusions by color or "race." Entries for the Nine Census Divisions are population-weighted; the averages below them are not. "National Rates" for both sexes: Deaths per 100,000 population (males + females). Males: Deaths per 100,000 male population. Females: Deaths per 100,000 female population.

MALES ....

Census Division	1940	1950	1960	1970	1980	1990
Pacific	14.1	8.5	8.8	--	--	--
New England	11.9	9.5	10.3	--	--	--
West North Central	9.7	7.7	7.9	--	--	--
Mid-Atlantic	12.4	9.8	9.4	--	--	--
East North Central	10.0	8.5	9.3	--	--	--
Mountain	13.8	9.9	9.2	--	--	--
West South Central	9.4	5.8	6.8	--	--	--
East South Central	9.9	6.7	6.7	--	--	--
South Atlantic	10.2	7.4	7.9	--	--	--
Average ALL	11.3	8.2	8.5	--	--	--
Average High-Five	11.6	8.8	9.1	--	--	--
Average Low-Four	10.8	7.5	7.7	--	--	--
Ratio (High/Low)	1.07	1.18	1.19	--	--	--

FEMALES ....

	1940	1950	1960	1970	1980	1990
Pacific						
New England						
West North Central						
Mid-Atlantic						
East North Central						
Mountain						
West South Central						
East South Central						
South Atlantic						
Average ALL						
Average High-Five						
Average Low-Four						
Ratio (High/Low)						

The National Rates below reflect a severe "small numbers problem" for the female rates, so we did not analyze the female data at all.

NATIONAL RATES ....

	1940	1950	1960	1970	1980	1990
Both Sexes	6.8	5.0	5.4	--	--	--
Males	11.1	8.3	8.6	--	--	--
Females	2.3	1.8	2.5	--	--	--

● - 1940, 1950, 1960: All rates come from Grove 1968, Table 67, Pages 737-740, Death-rates from "Ulcer of Stomach and Duodenum (540, 541)" ICD/7.

● - Omitted data are only marginally related to examination of Hypotheses 1+2, as noted in Chapter 4, Part 2c.

CHAPTER 38

Summary on NonCancer NonIHD Results: Facts "Demanding" an Explanation

Section 3 of this book was initiated to find out if Cancer and nonCancer MortRates differ in their response to PhysPop (Chapter 23, Part 1).

Box 1 of this chapter makes it easy to see the "bottom line" of the work in Chapters 23 through 37, and to compare it with the results for Cancer from Chapters 6 through 19, and with the results for Ischemic Heart Disease (IHD) from Chapters 40 and 41.

The results in Box 1 always arise from 1940 PhysPops as the x-variable, even in the three cases when some other PhysPop-year produced a higher correlation (see Breast Cancer, male Genital Cancers, and Female IHD). For Box 1, the 1940 MortRates are always the y-variable, with three exceptions. For All NonCancer NonIHD Causes Combined (Chapter 25), for Hypertensive Disease (Chapter 30), and for Ischemic Heart Disease (Chapters 40 and 41), the MortRates are 1950 --- because MortRates in 1940 are not available. From Box 1:

		R-squared	Coef/ SE	Relationship: MortRates by CensusDiv vs PhysPops by CensusDiv
All NonCancer NonIHD	Male	0.7933	−5.18	Inverse, and very sig.
	Fem	0.7037	− 4.08	Inverse, and very sig.
All Cancers Combined	Male	0.9508	+11.63	Positive, and highly sig.
	Fem	0.8608	+ 6.58	Positive, and highly sig.
Ischemic Heart Disease	Male	0.9475	+11.24	Positive, and highly sig.
	Fem	0.8337	+ 5.92	Positive, and highly sig.

The contrast of Cancer and Ischemic Heart Disease, with All NonCancer NonIHD Causes of Death Combined, is unmistakable. Not only are the correlations spectacular between Cancer MortRates and PhysPop, and between IHD MortRates and PhysPop, but these correlations are POSITIVE in direction while the very significant correlations for All NonCancer NonIHD Causes of Death Combined are NEGATIVE with PhysPop.

These observations are facts --- not interpretations --- and they "demand" an explanation.

The explanation proposed by this book is that physicians, by what they do, are CAUSING both the positive and inverse relationships shown above. Of course, single correlations alone can never PROVE causation, because correlation is a type of circumstantial evidence. But pieces of circumstantial evidence combined with logic can produce arguments "beyond reasonable doubt" for causation, both in science and in everyday experience.

How would the activities of physicians CAUSE the negative and positive correlations listed above?

● - NONCANCER NON-IHD DEATHS: The fact, that age-adjusted nonCancer nonIHD death-rates FALL where physician-density RISES, strongly supports the widespread expectation that "extra" medical attention per 100,000 population should reduce some types of age-adjusted death-rates. In general, the more physicians there are per 100,000 population, the more radiation procedures occur per 100,000 population. The fact, that this aspect of "extra" medical attention does not cause a POSITIVE correlation between PhysPop and nonCancer nonIHD death-rates, is consistent with the "general wisdom" that such deaths are not inducible in patients by ionizing radiation (Chapter 23, Part 1a).

● - CANCER DEATHS: The fact, that age-adjusted Cancer death-rates at mid-century RISE where physician-density RISES, can be readily explained by a few additional facts: (a) Ionizing radiation is a proven cause of human Cancer in irradiated populations, (b) Physicians irradiate populations, and (c) Physicians at mid-century rarely had effective treatment for Cancers.

    - IHD DEATHS: The fact, that age-adjusted IHD MortRates at mid-century RISE where physician-density RISES, is a startling finding --- especially in view of the contrasting behavior of the nonCancer nonIHD MortRates.   The very strong positive correlation for IHD MortRates practically "demands" that the concept of radiation-CAUSATION be taken seriously in medicine and in public health, until and unless a better explanation for this positive correlation is provided.

>>>>>>>>>>

All the comparisons below are based on the relationship between 1940 PhysPops and 1940 MortRates, except for 3 pairs of 1950 MortRates. "Sig." means statistically significant. When XCoef/SE = 2, then P = roughly 0.05.

		R-Squared	X-Coef.	XCoef/ Std Err	Relationship, MortRates w. PhysPops by CensusDiv.
Ch23: All Causes Combined	Male	0.1299	Neg.	−1.02	Inverse, but not sig.
	Fem	0.2823	Neg.	−1.66	Inverse, and marginal.
Ch24: All NonCancer Combined	Male	0.2841	Neg.	−1.67	Inverse, and marginal.
	Fem	0.4362	Neg.	−2.33	Inverse, and significant.
Ch25: All NonCancer NonIHD	Male	0.7933	Neg.	−5.18	Inverse, and very sig.
	Fem	0.7037	Neg.	−4.08	Inverse, and very sig.
Ch26: Appendicitis	Male	0.0179	Neg.	−0.36	None.
	Fem	0.0010	Neg.	−0.08	None.
Ch27: CNS Vascular (Stroke)	Male	0.4000	Neg.	−2.16	Inverse, and significant.
	Fem	0.2882	Neg.	−1.68	Inverse, and marginal.
Ch28: Chronic Nephritis	Male	0.4561	Neg.	−2.42	Inverse, and significant.
	Fem	0.2687	Neg.	−1.60	Inverse, and marginal.
Ch29: Diabetes Mellitus	Male	0.6435	Pos.	3.55	Positive, and quite sig.*
	Fem	0.6005	Pos.	3.24	Positive, and quite sig.*
Ch30: Hypertensive Disease	Male	0.3564	Neg.	−1.97	Inverse, and significant.
	Fem	0.2056	Neg.	−1.35	Inverse, and very marginal.
Ch31: Influenza and Pneumonia	Male	0.8344	Neg.	−5.94	Inverse, and highly sig.
	Fem	0.8849	Neg.	−7.34	Inverse, and highly sig.
Ch32: Fatal Motor Vehicle Accid.	Male	0.0195	Neg.	−0.37	None.
	Fem	0.0003	Neg.	−0.04	None.
Ch33: Other Fatal Accidents	Male	0.0901	Neg.	−0.83	None.
	Fem	0.4440	Neg.	−2.36	Inverse, and significant.
Ch34: Rheum.Fever/Rheum.Heart	Male	0.0021	Pos.	0.12	None.
	Fem	0.0550	Pos.	0.64	None.
Ch35: Syphilis and Sequelae	Male	0.3278	Neg.	−1.85	Inverse, and marginal.
	Fem	--	--	--	--
Ch36: Tuberculosis, All Forms	Male	0.2067	Neg.	−1.35	Inverse, and very marginal.
	Fem	0.6381	Neg.	−3.51	Inverse, and quite sig.
Ch37: Ulcer: Stomach, Duoden.	Male	0.3864	Pos.	2.10	Positive, and significant.**
Ch6+7: All Cancers Combined	Male	0.9508	Pos.	11.63	Positive, and highly sig.
	Fem	0.8608	Pos.	6.58	Positive, and highly sig.
Ch8: Breast Cancer	Male	--	--	--	--
	Fem	0.9153	Pos.	8.70	Positive, and highly sig.
Ch9+10: Digestive-Syst. Cancers	Male	0.9078	Pos.	8.30	Positive, and highly sig.
	Fem	0.7550	Pos.	4.64	Positive, and very sig.
Ch11+12: Urinary-Syst. Cancers	Male	0.9208	Pos.	9.02	Positive, and highly sig.
	Fem	0.9395	Pos.	10.43	Positive, and highly sig.
Ch13+14: Genital Cancers	Male	0.7182	Pos.	4.22	Positive, and very sig.
	Fem	0.0683	Pos.	0.72	None.
Ch15: Buccal & Pharynx Cancers	Male	0.7234	Pos.	4.28	Positive, and very sig.
	Fem	--	--	--	--
Ch16+17: Respiratory-Syst. Canc	Male	0.8673	Pos.	6.76	Positive, and highly sig.
	Fem	0.9625	Pos.	13.40	Positive, and highly sig.
Ch40+41: Ischemic Heart Disease	Male	0.9475	Pos.	11.24	Positive, and highly sig.
	Fem	0.8337	Pos.	5.92	Positive, and highly sig.

   * Diabetes Mellitus (DM): After the rules changed in 1949 for reporting the underlying cause of death in diabetics, DM MortRates abruptly fell in half and our R-sq. values dropped abruptly to 0.11 and 0.20 (Chap.29). The significant R-sq. values in 1940 very probably denote a correlation between PhysPop and deaths during 1940 from xray-induced Ischemic Heart Disease in people having diabetes (Chapters 29, 40, 41).

   ** Ulcer Deaths: The positive correlation between Ulcer Deaths in 1940 and PhysPop might be due to erroneous reporting in 1940 of deaths, truly from Stomach Cancer, as deaths from Stomach Ulcers.

CHAPTER 39

Ischemic Heart Disease: Medical Radiation as a Cause

Part 1.   Statement of Hypothesis-2
Part 2.   Various Names for Ischemic Heart Disease; ICD Numbers
Part 3.   Some Additional Terms
Part 4.   Current Place of IHD in U.S. Mortality

● Part 1.   Statement of Hypothesis-2

Hypothesis-2 is this: Medical radiation, received even at very low and moderate doses, is an important cause of Ischemic Heart Disease (IHD); the probable mechanism is radiation-induction of mutations in the coronary arteries, resulting in dysfunctional clones (mini-tumors) of smooth muscle cells.

Although this book was undertaken only to evaluate our Hypothesis-1 concerning the etiology of Cancer, Hypothesis-2 practically "fell out of the data" --- much to our surprise, as we said in Chapter 25. Evaluation of Hypothesis-2 is at least as important as evaluation of Hypothesis-1, in our opinion, in terms of the implications for explaining and reducing rates of a major disease.

We can prevent some confusion about Hypothesis-2 --- as noted in the Introduction --- by stating that (a) it was discovered decades ago that medical radiation at very high doses can damage the heart and its vessels, and that (b) the kinds of damage reported from very-high-dose radiation seldom resemble the lesions of Ischemic Heart Disease (details in Appendix J).

Ischemic Heart Disease is a complicated, multi-stage phenomenon. Passions run very high among medical investigators about its major "players" (the proven and suspected causes), and about the likely sequence of their actions. Chapter 44 discusses several of those major players. It appears that Hypothesis-2 is not in conflict with observations made by other investigators (Chapter 46).

● Part 2.   Various Names for Ischemic Heart Disease; ICD Numbers

Ischemia means a local deficiency of blood. Ischemic Heart Disease means that the heart muscle is not receiving enough blood to do its job properly.

There is widespread agreement that ischemia of the heart has its origin in what is going on in the CORONARY arteries --- namely, interference with the blood-flow in those arteries. Thus, IHD is often called coronary heart disease (CHD) and coronary artery disease (CAD). Other names incorporate suggestions of WHAT has gone wrong with the coronary arteries. Coronary arteriosclerosis: "Hardening" of the arteries. Coronary atherosclerosis: A build-up in the arterial wall of material (atherosclerotic "plaque") which finally intrudes into the lumen and obstructs the blood-flow, partially or completely.

In the Seventh International Classification of Diseases, adopted in 1955, what we now call Ischemic Heart Disease was called Arteriosclerotic Heart Disease (our Chapter 4, Part 5, Entry 17). The Ninth International Classification of Diseases, adopted in 1975 and still current, uses the name Ischemic Heart Disease and gives it numbers 410-414. There is no longer an ICD listing for Arteriosclerotic Heart Disease.

● Part 3.   Some Additional Terms

Coronary thrombus. In acute attacks of coronary ischemia, the immediate cause is usually, but not always, a thrombus (a wall-attached "blood clot") in a coronary artery, and the thrombus is blocking flow of the blood. A thrombus within the lumen is sometimes called an intraluminal thrombus, to distinguish it from a clot occurring within the atherosclerotic lesion itself. The relationship between clot-formation and atherosclerotic plaques of various types is under continuing study, and we will have more to say about this in Chapter 44, Part 7.

Embolus. An embolus is a blood-clot or other plug carried by the blood stream to a new location and forced into a smaller vessel, where it obstructs the circulation. An air bubble in the blood-stream also can be an embolus.

Angina pectoris. The pain sometimes perceived by patients, when the blood-supply to the heart muscle is transiently inadequate, is named angina pectoris. Angina pectoris and severe coronary atherosclerosis are often found together. Some episodes of angina pectoris are suspected to be due, in part, to a reversible spasm of a coronary artery.

Arrhythmia. Variations from the normal rhythm of the heart's BEAT are arrhythmias. There are several causes of arrhythmia. Ventricular fibrillation ("V-fib") is a condition of the heart's ventricles in which the individual muscle fibres take up their own independent action, producing an uncoordinated contraction and very little (if any) pumping of the blood.

Myocardial infarction (MI) is commonly called "heart attack." When part of the heart muscle (myocardium) dies due to blockage of one of the coronary arteries, this is myocardial infarction. An infarct is an area of necrosis in a tissue due to local obstruction of blood circulation to that area. Not all myocardial infarctions are fatal.

Cardiac arrest. An abrupt loss of heart function, whatever the cause, is cardiac arrest. Myocardial infarction may or may not cause cardiac arrest, and cardiac arrest may have other causes, so MI and cardiac arrest are not synonyms.

Acute IHD "Events" or Syndromes. A new myocardial infarction, unstable angina, and ischemic sudden death, each qualify as "an acute IHD event."

Other Types of Heart Disease

Hypothesis-2 concerns specifically Ischemic Heart Disease --- not hypertensive heart disease, and not rheumatic heart disease. These two other entities were considered separately, in Chapters 30 and 34. In Part 4, below, some additional and important distinctions are made.

● Part 4. Current Place of IHD in U.S. Mortality

In 1993, in the United States, the absolute number of deaths from all causes combined was 2,268,553 deaths (ACS-CA, Jan. 1997, p.18).

Also in 1993, the absolute number of deaths from all kinds of Cancer combined was 529,904 cancer deaths --- or 23.4 percent of the U.S. total (ACS-CA, Jan. 1997, p.18).

Also in 1993, the absolute number of deaths from Ischemic Heart Disease was 489,970 IHD deaths --- or 21.6 percent of the U.S. total (AHA 1995, p.9).

IHD accounted for about 51.4 percent of the 954,138 deaths in 1993 which were classed by the American Heart Association as deaths due to "cardiovascular diseases" (AHA 1995, p.2). According to the AHA (1995, p.9), there are nearly 1.5 million new or recurrent "heart attacks" per year in the USA (with approximately one-third of them being fatal).

4a. CardioVascular Diseases (CVD) versus "Diseases of the Heart"

There is plenty of room for confusion in comparing MortRates from one source versus another. The American Heart Association is very helpful by being explicit (AHA 1995, p.1):

"In compiling statistics, the AHA looks at specific cardiovascular disease categories, based on ICDA codes [International Classification of Diseases, Adapted] ... Primarily, we look at 'total cardiovascular' (total circulatory, codes 390-459). Depending on availability, data for congenital anomalies of the circulatory system (745-747) are also included. Within total CVD, the AHA follows what are considered to be the major cardiovascular diseases: Ischemic (Coronary) Heart Disease (410–414); Hypertensive Disease (401–404); Rheumatic Fever/Rheumatic Heart Disease (390–398) and Cerebrovascular Disease (Stroke) (430–438) ..." And (AHA 1995, p.1):

"... 'Diseases of the Heart' is a term commonly used by NCHS [National Center for Health Statistics] in its mortality publications and in its compilation of the 'Leading Causes of Death' ... Statistically, this category represents about 78 percent of total cardiovascular mortality." The category ("diseases of the heart") is described as follows by AHA 1995 (p.1):

ICDA		INCLUDED in "Diseases of the Heart"
390–398		Rheumatic Fever/Rheumatic Heart Disease
402		Hypertensive Heart Disease
404		Hypertensive Heart and Renal Disease
410–414		Ischemic Heart Disease
415–417		Disease of Pulmonary Circulation
420–429		Other Forms of Heart Disease

ICDA		EXCLUDED from "Diseases of the Heart"
401, 403		Hypertension with or without Renal Disease
430–438		Cerebrovascular Disease (Stroke)
440		Atherosclerosis
	Note:	Atherosclerosis is the underlying cause of IHD and of many strokes. 440 refers to atherosclerosis occurring in arterial vessels OTHER than the coronary and cerebral arteries. Note that ICD 440 neither overlaps with 410–414 for IHD nor with 430–438 for stroke.
441–448		Other Diseases of Arteries, Arterioles and
451–459		Capillaries and Veins and Lymphatics

4b. Estimated Annual Deaths from Some Other Segments of CVD

AHA 1995 (pp.14–15) provides the following estimates for annual, recent (1990–1993) deaths from some segments of Cardiovascular Disease which are NOT Ischemic Heart Disease, Hypertensive Disease, or Cerebrovascular Disease (Stroke). We have arranged them in their ICD/9 sequence. In some cases, we note the prevalence (the total number of cases which exist in a population at a specific time), as an indication of how many people may be monitored with repeated xray procedures.

390–398		Rheumatic Fever/Rheumatic Heart Disease. Approx. 1.36 million Americans today have rheumatic heart disease.	5,590
421.0		Bacterial Endocarditis	2,011
424		Valvular Heart Disease, including:	15,070
	424.0	Mitral Valve Disorders Mitral valve prolapse occurs in 4% of young men and 6–10% of young women. "It is reported most often in young women ages 14 to 30, where prevalence may exceed 10 percent."	2,044
	424.1	Aortic Valve Disorders	7,966
	424.2,3	Tricuspid and Pulmonary Valves	14
425		Cardiomyopathy	24,573
426, 427		Arrhythmias, Including:	40,843
	427.0,1,2	Tachycardia Estimated prevalence: 2.2 million.	544

	427.3	Atrial Fibrillation	4,056
		Estimated prevalence: 2 million.	
	427.4	Ventricular Fibrillation: "While ventricular fibrillation is listed as the cause of relatively few deaths, the overwhelming number of sudden deaths (which are estimated at about 250,000 per year) are thought to be from ventricular fibrillation."	1,461

428.0		Congestive Heart Failure	36,387
		About 4.7 million Americans are living with CHF.	

440-448		Arteries, Diseases of, Including Peripheral Vascular Disease and:	43,520
	440	Atherosclerosis (see "440 Note" in Part 4a)	17,090
	441	Aortic Aneurysm	16,220
	442-448	Other Diseases of the Arteries	9,117

745-747		Congenital Heart Defects	5,508
		About 32,000 babies are born each year with heart defects. "About 960,000 Americans with heart defects are alive today."	

4c. The 15 Leading Causes of Death during 1993, USA

The following list of deaths in 1993 is presented by the American Cancer Society in its journal CA (ACS-CA, Jan. 1997, p.18), where the source is cited as "Vital Statistics of the United States, 1996." An asterisk indicates that readers should consult the definitions above.

				Percent
		All Causes	2,268,553	
1	*	Heart Diseases (incl. IHD; see Part 4a)	743,460	32.8
2		Cancer	529,904	23.4
3	*	Cerebrovascular Diseases (see Part 4a)	150,108	6.6
4		Chronic Obstructive Pulmonary Diseases	101,077	4.5
5		Accidents	90,523	4.0
6		Pneumonia & Influenza	82,820	3.7
7		Diabetes Mellitus	53,894	2.4
8		HIV Infection	37,267	1.6
9		Suicide	31,102	1.4
10		Homicide	26,009	1.1
11	*	Diseases of Arteries (excl. IHD; see Part 4b)	26,005	1.1
12		Cirrhosis of Liver	25,209	1.1
13		Nephritis	23,317	1.0
14		Septicemia	20,634	0.9
15	*	Atherosclerosis (excl. IHD; see Part 4a)	17,272	0.8
		Others & ill-defined causes	309,952	13.7

>>>>>>>>>>>

CHAPTER 40

IHD in Males: The Dose-Response between Medical Radiation and IHD

Part 1. Ischemic Heart Disease, Mortality Rates, Males
Part 2. How the Dose-Response Behaves, for Medical Radiation and IHD
Part 3. Maximum Relationship: Best-Fit Equation and Graph
Part 4. Best Estimate: 79 % of Male IHD MortRate due to Medical Radiation, 1950
Part 5. Not True That "79% Leaves Only 21% for Other Causes!"

Box 1. Summary-Results from All Calculations of Part 2.
Box 2. Input-Data for Graph of Figure 40-A.
Box 3. Percent of IHD MortRate Attributable to Medical Radiation.
Box 4. Error-Check on Our Own Work.

Figure 40-A. Graph of the Strongest Dose-Response.
Figure 40-B. The 1940-1992 Trends in MortRates for Cardiovascular Diseases.
Tables 40-A, 40-B. The IHD MortRates, 1950-1993.

● Part 1. Ischemic Heart Disease, Mortality Rates, Males

● 1950, 1960: The year 1950 is the earliest one for which Grove 1968 provides data for
Ischemic Heart Disease. Grove 1968 provides age-adjusted data (1940 Standard Population) for 1950
and 1960, but none for 1940. In Grove 1968, the entity was as we show it in our Chapter 4, Part 5,
Entry 17 (with ICD/7 numbers):

Arteriosclerotic Heart Disease, including coronary disease (420).
    Arteriosclerotic heart disease so described (420.0).
    Heart Disease specified as involving coronary arteries (420.1).
    Angina Pectoris without mention of coronary disease (420.2).

● 1980: The 1980 age-adjusted MortRates (1940 Standard Population) come from the 1979-81
printout supplied to us by the National Center for Health Statistics (NatCtrHS 1980). By then, the
Ninth Revision of ICD numbers was operative, and the data in the printout are described as
follows:

Ischemic Heart Disease (410-414).
    Acute Myocardial Infarction (410).
    Other Acute and Subacute Forms of Ischemic Heart Disease (411).
    Angina Pectoris (413).
    Old Myocardial Infarction and Other Forms of Chronic Ischemic Heart Disease (412, 414).

● 1970: For 1970, no IHD MortRate data became available to us by states or Census
Divisions, for males and females separately. To obtain values for 1970, we interpolated between the
1960 and 1980 values.

● 1993: The 1993 age-adjusted MortRates (1940 Standard Population) come from the
1992-1994 printout supplied to us in 1997 by the National Center for Health Statistics (NatCtrHS 1993)
for ICD numbers 410 to 414.9. Rates were supplied for each gender by states. To obtain the
population-weighted male MortRate for each Census Division, we weighted the male rate for each
STATE according to the fraction contributed by its state's population to the population of the entire
Census Division in 1990. Female MortRates for each Census Division were obtained the same way.

● Part 2. How the Dose-Response Behaves, for Medical Radiation and IHD

In our dose-response studies, PhysPops for the Nine Census Divisions represent the relative,
average accumulated doses from medical radiation. This surrogacy has been examined in Chapter 3.

As the dose, PhysPops are the x-values in our linear regression analyses. The corresponding MortRates for the Nine Census Divisions are the responses to be studied, so they are the matching y-values. Both the x and the y variables have the denominator "per 100,000 population." Chapter 40 follows the model of Chapter 6.

In Parts 2a through 2k, we regress the 1950 MortRates for Ischemic Heart Disease (from Table 40-A) upon the non-interpolated sets of PhysPop values 1921-1940 (from the Universal PhysPop Table 3-A). This work is comparable to our work with All Cancers Combined, Chapters 6 and 7, except that we were able to use 1940 MortRates for Cancers. Here, we must use 1950 MortRates for IHD, as noted in Part 1 above. In Part 2k, we regress 1950 MortRates for IHD on 1950 PhysPops. Readers are reminded that, in the regressions below, only the PhysPops reach back into earlier decades. We are exploring which PhysPop set has the strongest correlation with a SINGLE set of MortRates (1950).

A Relationship of Immense Strength

The summary-results of all the regression analyses are presented in Box 1. The strongest dose-response relationship in Box 1 has an R-squared value of 0.948 and a ratio of 11.2 for the X-Coefficient over its Standard Error. The strength of the correlation is immense.

Figure 40-A shows the strong POSITIVE relationship between PhysPop and male IHD --- which closely resembles the strong POSITIVE relationship between PhysPop and male CANCER shown in Figure 6-A. By contrast, Figure 25-A+B shows that the nonIHD noncancer causes of death have an INVERSE correlation with PhysPop.

● - Part 2a.	x 1921 PhysPop	y 1950 MortRate	IHD, Males Regression Output:	
Pacific	165.11	283.2	Constant	-3.8662
New England	142.24	297.1	Std Err of Y Est	37.6682
West North Central	140.93	228.4	R Squared	0.3983
Mid-Atlantic	137.29	310.3	No. of Observations	9
East North Central	136.06	258.9	Degrees of Freedom	7
Mountain	135.38	214.8		
West South Central	125.15	206.1	X Coefficient(s)	1.8415
East South Central	119.76	176.8	Std Err of Coef.	0.8556
South Atlantic	110.32	222.0	Coefficient / S.E.	2.1524

● - Part 2b.	1923 PhysPop	1950 MortRate	IHD, Males Regression Output:	
Pacific	163.06	283.2	Constant	-3.3503
New England	137.39	297.1	Std Err of Y Est	34.9379
West North Central	138.31	228.4	R Squared	0.4823
Mid-Atlantic	138.92	310.3	No. of Observations	9
East North Central	131.82	258.9	Degrees of Freedom	7
Mountain	130.51	214.8		
West South Central	119.16	206.1	X Coefficient(s)	1.8893
East South Central	113.16	176.8	Std Err of Coef.	0.7398
South Atlantic	106.79	222.0	Coefficient / S.E.	2.5539

● - Part 2c.	1925 PhysPop	1950 MortRate	IHD, Males Regression Output:	
Pacific	161.67	283.2	Constant	6.1273
New England	138.31	297.1	Std Err of Y Est	32.4561
West North Central	133.92	228.4	R Squared	0.5533
Mid-Atlantic	134.36	310.3	No. of Observations	9
East North Central	127.54	258.9	Degrees of Freedom	7
Mountain	122.30	214.8		
West South Central	112.83	206.1	X Coefficient(s)	1.8764
East South Central	107.22	176.8	Std Err of Coef.	0.6373
South Atlantic	103.61	222.0	Coefficient / S.E.	2.9443

● - Part 2d.	1927 PhysPop	1950 MortRate	IHD, Males Regression Output:	
Pacific	157.83	283.2	Constant	-3.7561
New England	137.50	297.1	Std Err of Y Est	27.6261

West North Central	131.54	228.4	R Squared	0.6763	
Mid-Atlantic	138.40	310.3	No. of Observations	9	
East North Central	126.18	258.9	Degrees of Freedom	7	
Mountain	118.75	214.8			
West South Central	108.25	206.1	X Coefficient(s)	1.9876	
East South Central	102.07	176.8	Std Err of Coef.	0.5197	
South Atlantic	102.13	222.0	Coefficient / S.E.	3.8246	

● - Part 2e.	1929	1950	IHD, Males		
	PhysPop	MortRate	Regression Output:		
Pacific	156.64	283.2	Constant	-1.1046	
New England	138.46	297.1	Std Err of Y Est	25.7335	
West North Central	128.72	228.4	R Squared	0.7192	
Mid-Atlantic	138.49	310.3	No. of Observations	9	
East North Central	126.51	258.9	Degrees of Freedom	7	
Mountain	118.68	214.8			
West South Central	105.60	206.1	X Coefficient(s)	1.9828	
East South Central	99.41	176.8	Std Err of Coef.	0.4683	
South Atlantic	100.86	222.0	Coefficient / S.E.	4.2338	

● - Part 2f.	1931	1950	IHD, Males		
	PhysPop	MortRate	Regression Output:		
Pacific	159.97	283.2	Constant	13.5791	
New England	142.35	297.1	Std Err of Y Est	23.9871	
West North Central	126.50	228.4	R Squared	0.7560	
Mid-Atlantic	140.82	310.3	No. of Observations	9	
East North Central	128.59	258.9	Degrees of Freedom	7	
Mountain	118.89	214.8			
West South Central	105.95	206.1	X Coefficient(s)	1.8540	
East South Central	96.73	176.8	Std Err of Coef.	0.3981	
South Atlantic	99.59	222.0	Coefficient / S.E.	4.6569	

● - Part 2g.	1934	1950	IHD, Males		
	PhysPop	MortRate	Regression Output:		
Pacific	160.09	283.2	Constant	27.5355	
New England	148.60	297.1	Std Err of Y Est	18.7380	
West North Central	125.96	228.4	R Squared	0.8511	
Mid-Atlantic	149.62	310.3	No. of Observations	9	
East North Central	129.36	258.9	Degrees of Freedom	7	
Mountain	117.16	214.8			
West South Central	104.68	206.1	X Coefficient(s)	1.7318	
East South Central	92.00	176.8	Std Err of Coef.	0.2738	
South Atlantic	98.41	222.0	Coefficient / S.E.	6.3254	

● - Part 2h.	1936	1950	IHD, Males		
	PhysPop	MortRate	Regression Output:		
Pacific	158.44	283.2	Constant	30.6794	
New England	150.18	297.1	Std Err of Y Est	16.6019	
West North Central	126.14	228.4	R Squared	0.8831	
Mid-Atlantic	155.05	310.3	No. of Observations	9	
East North Central	130.42	258.9	Degrees of Freedom	7	
Mountain	119.80	214.8			
West South Central	103.52	206.1	X Coefficient(s)	1.6965	
East South Central	89.94	176.8	Std Err of Coef.	0.2333	
South Atlantic	99.16	222.0	Coefficient / S.E.	7.2723	

● - Part 2i.	1938	1950	IHD, Males		
	PhysPop	MortRate	Regression Output:		
Pacific	157.62	283.2	Constant	38.7532	
New England	154.08	297.1	Std Err of Y Est	14.0452	
West North Central	124.95	228.4	R Squared	0.9163	
Mid-Atlantic	160.69	310.3	No. of Observations	9	
East North Central	131.98	258.9	Degrees of Freedom	7	
Mountain	119.88	214.8			
West South Central	102.79	206.1	X Coefficient(s)	1.6225	
East South Central	88.21	176.8	Std Err of Coef.	0.1853	
South Atlantic	99.26	222.0	Coefficient / S.E.	8.7563	

● – Part 2j.	1940	1950	IHD, Males	
	PhysPop	MortRate	Regression Output:	
Pacific	159.72	283.2	Constant	53.0895
New England	161.55	297.1	Std Err of Y Est	11.1218
West North Central	123.14	228.4	R Squared	0.9475
Mid–Atlantic	169.76	310.3	No. of Observations	9
East North Central	133.36	258.9	Degrees of Freedom	7
Mountain	119.89	214.8		
West South Central	103.94	206.1	X Coefficient(s)	1.4852
East South Central	85.83	176.8	Std Err of Coef.	0.1321
South Atlantic	100.74	222.0	Coefficient / S.E.	11.2446

● – Part 2k.	1950	1950	IHD, Males	
	PhysPop	MortRate	Regression Output:	
Pacific	148.60	283.2	Constant	57.5900
New England	162.51	297.1	Std Err of Y Est	12.6311
West North Central	120.06	228.4	R Squared	0.9323
Mid–Atlantic	168.71	310.3	No. of Observations	9
East North Central	123.69	258.9	Degrees of Freedom	7
Mountain	119.38	214.8		
West South Central	101.34	206.1	X Coefficient(s)	1.4908
East South Central	83.05	176.8	Std Err of Coef.	0.1518
South Atlantic	99.07	222.0	Coefficient / S.E.	9.8212

● Part 3. Maximum Relationship: Best-Fit Equation and Graph

The regression analysis of Part 2j produces the strongest correlation. From the output, we can write the best-fit equation, as we have throughout Sections Two and Three of the book. (Reminder: We use the symbol * to denote multiplication.)

● – Ischemic Heart Disease MortRate, Males = (X-Coefficient * PhysPop) + Constant.
● – Ischemic Heart Disease MortRate, Males = (1.4852 * PhysPop) + 53.09.

Using the equation of best fit, we can calculate a best-fit MortRate for any value of PhysPop. In Box 2, we show best-fit MortRates which have been calculated for the nine actual PhysPop values of Part 2j, and also for lower PhysPop values, down to zero PhysPop (Chapter 5, Part 5e).

Figure 40-A shows the line of best fit --- which connects these pairs of x,y values (various PhysPops, best-fit MortRates). The graph also shows nine boxy symbols (the nine actual observations from Part 2j). Per 100K means per 100,000. Chapter 6, Part 3, discusses how to know which boxy symbol "belongs with" which Census Division.

● Part 4. Best Estimate: 79 % of Male IHD MortRate due to Medical Radiation, 1950

The data have revealed a linear relationship in 1950 of immense strength between medical radiation and male IHD mortality. The reasonable presumption, in the absence of a better explanation, is that the relationship is TRULY CAUSAL --- in other words, a dose-response. So then one must ask:

"What would be the estimated male IHD mortality-rate in 1950 if there were NO dosage of medical radiation?"

No medical irradiation would occur if there were NO PHYSICIANS per 100,000 population. So we want to know the value of the y-variable (IHD MortRate) when the value of the x-variable (PhysPop) is equal to zero. This value is, of course, called the Constant in the regression output of Part 2j. On the graph, the Constant is the value of the MortRate where the line of best-fit intersects the y-axis. This "intercept" occurs where the value of PHYSPOP equals zero. No medical radiation at all.

Since every Census Division has physicians, there can be no real-world datapoint in our study for the male IHD MortRate when PhysPop = zero. But the calculated or "estimated" MortRate, if PhysPop were zero, certainly does not come out of thin air. It is extrapolated from nine real-world observations which reflect a very strong linear relationship. It merits emphasis that the raw data which reveal this relationship are neutral --- by which we mean they were collected long ago by people having no conceivable bias with respect to the studies in this monograph.

4a. Percentage Attributable to Medical Radiation: "Fractional Causation"

Using our strongest dose-response result, we propose that the Ischemic Heart Disease mortality (male) attributable to medical radiation at approximately mid-century is the total National IHD MortRate in 1950 for males, minus the MortRate indicated by the Constant. The fraction attributable to medical radiation is (1950 National MortRate minus the Constant) / (1950 National MortRate).

Calculation of the Fractional Causation by medical radiation is summarized in Box 3 --- a box very familiar from Chapters 6-19 of this book.

When we subtract the Constant of 53.09 (Part 2j) from the National MortRate of 256.4 (Table 40-B), we have the rate of 203.31 per 100,000 from medical irradiation. The fraction of the total is thus (203.31 / 256.4), or 0.793. In other words, the "best estimate" which falls out of the data is that 79 % of Ischemic Heart Disease MortRate, in males, at approximately mid-century, is attributable to medical radiation. Box 3 calculates 90 % confidence limits on the central estimate, in the same manner explained by Chapter 6, Parts 4b-4d.

4b. Looking for Consistencies: Error-Checks on Input and Output

In Box 4, we use input and output generated in this chapter, to calculate the male National IHD MortRate by two separate methods. If each calculation yields a rate which is close to the National Rate provided from Vital Statistics in our Table 40-B, then we can assure ourselves and readers that we have made no serious errors here. Box 4 was introduced and explained in greater detail in Chapter 6, Part 5.

● Part 5. Not True That "79% Leaves Only 21% for OTHER Causes!"

The estimate, that 79 % of the male National IHD MortRate is caused by medical radiation, may result in readers thinking, "You leave too little room for other causes!" Not so, if co-action is taken into account.

Both Ischemic Heart Disease and Cancer are well established as multi-cause diseases. There is convincing evidence that several different causes increase the death-rate from Ischemic Heart Disease, and likewise, that several different causes increase the death-rate from Cancer. It is highly likely that each single case of IHD (or Cancer) has more than one contributing cause. And if an agent contributes to the outcome, it must be NECESSARY to the outcome --- for if the outcome would have been the same without its presence, then it contributes nothing to the outcome.

The concept of NECESSARY co-actors is an old one. For instance, in the famous 1964 "Surgeon General's Report" on cigarette smoking as a cause of Lung Cancer, the authors wrote (SurgeonGen 1964, p.31): "It is recognized that often the co-existence of several factors is required for the occurrence of a disease, and that one of the factors may play a dominant role; that is, without it, the other factors (such as genetic susceptibility) seldom lead to the occurrence of the disease."

Any contributing agent can be appropriately called "the cause" of a case, if the case would not have occurred in the absence of help from that specific agent. Thus, our finding that 79% of male IHD deaths (USA) in 1950 were caused by medical radiation does not restrict other causes to a small role. The Introduction (Part 5) of this monograph illustrates the point quantitatively.

When cases of Ischemic Heart Disease have more than one cause per case, then reducing exposure to one of the contributing co-actors reduces the impact of all its partners. The evidence in this book, that exposure to medical radiation is an important atherogen, points to a major opportunity for reducing future mortality-rates from Ischemic Heart Disease.

>>>>>>>>>>

We are searching for the maximum correlation between PhysPops of 1921-1950 and the male Ischemic Heart Disease MortRates of 1950. Even the maximum correlation will tend to understate the true correlation (Chapter 5, Part 8b). Below are the summary-results from all the calculations of Part 2, for the 1950 MortRates regressed on PhysPops of 1921-1950.

Part	PhysPop	R-squared	Constant	X-Coef	Std Err	X-Coef/SE
2a	1921	0.3983	-3.87	1.8415	0.8556	2.1524
2b	1923	0.4823	-3.35	1.8893	0.7398	2.5539
2c	1925	0.5533	6.13	1.8764	0.6373	2.9443
2d	1927	0.6763	-3.76	1.9876	0.5197	3.8246
2e	1929	0.7192	-1.10	1.9828	0.4683	4.2338
2f	1931	0.7560	13.58	1.8540	0.3981	4.6569
2g	1934	0.8511	27.54	1.7318	0.2738	6.3254
2h	1936	0.8831	30.68	1.6965	0.2333	7.2723
2i	1938	0.9163	38.75	1.6225	0.1853	8.7563
2j --->	1940 Max	0.9475	53.09	1.4852	0.1321	11.2446
2k	1950	0.9323	57.59	1.4908	0.1518	9.8212

Related text = Part 2.

Box 2 of Chap. 40

Input-Data for Figure 40-A. Ischemic Heart Disease. Males.

Part 2j, Best-Fit Equation: Calc. MortRate = (1.4852 * PhysPop) + (53.09)

Census Divisions	1940 Observed PhysPops	1950 Observed MortRates	Best-Fit Calc. MortRates
Pacific	159.72	283.2	290.306
New England	161.55	297.1	293.024
West No. Central	123.14	228.4	235.978
Mid-Atlantic	169.76	310.3	305.218
East No. Central	133.36	258.9	251.156
Mountain	119.89	214.8	231.151
West So. Central	103.94	206.1	207.462
East So. Central	85.83	176.8	180.565
South Atlantic	100.74	222.0	202.709
Additional PhysPops	70.00		157.054
--- not "observed" ---	60.00		142.202
down to zero PhysPop·	50.00		127.350
(zero medical radiation).	40.00		112.498
For each, we calculate	30.00		97.646
a best-fit MortRate.	20.00		82.794
These additional x,y pairs	10.00		67.942
are also part of the	0		53.090
best-fit line (Chap 5, Part 5e).			

Related text = Part 3.

Box 3 of Chap. 40
Percent of IHD MortRate Attributable to Medical Radiation.

Please see text in Chapter 40, Parts 4 and 5.

IHD. MALES.   * denotes multiplication.

- MALE National MortRate (MR) 1950, from Table 40-B   256.4 National MortRate
- Constant, from regression, Part 2j   53.0895 Constant
- Fractional Causation, Best Est. = (Natl MR − Constant) / Natl MR   79.3 % Frac. Causation

....

90% Confidence-Limits (C.L.) on Fractional Causation. See text in Chapter 6, Parts 4b-d, please.

X-Coefficient, from Part 2j   1.4852 X-Coef., Best Est.
Standard Error (SE) of X-Coefficient, from Part 2j   0.1321 Standard Error

Upper 90% C.L. on X-Coef. = (Coef) + (1.645 * SE) =   1.7025 New X-Coefficient
New Constant = (Natl MR) − (New X-Coef * 1940 Natl PhysPop) = 31.6013 New Constant
Frac. Caus'n, High-Limit = (Natl MR − New Constant) / Natl MR =   87.7 % New Frac. Caus'n.

Lower 90% C.L. on X-Coef. = (Coef) − (1.645 * SE) =   1.2679 New X-Coefficient
New Constant = (Natl MR) − (New X-Coef * 1940 Natl PhysPop) = 88.9871 New Constant
Frac. Caus'n, Low-Limit = (Natl MR − New Constant) / Natl MR =   65.3 % New Frac. Caus'n.

Box 4 of Chap. 40
Error-Check on Our Own Work: Ischemic Heart Disease, Males.

Please see text in Chapter 6, Part 5.

Below, Columns A, C, and F come directly from the regression input in Part 2j. Columns B and E, fractions of the whole 1940 and 1950 population in each Census Division, come from Table 3-B in Chapter 3. Each Column-D entry = (B * C). Each Column-G entry = (E) * (F). MortRates and PhysPops are each "per 100,000."

The Weighted-Avg. Nat'l PhysPop, 1940, is the sum of Column-D entries =   132.04

Weighted-Avg. Nat'l Male MortRate, 1950, is sum of Col.G entries =   253.04
Nat'l Male MortRate is also (X-Coef * 1940 Nat'l PhysPop) + Constant =   249.20
Comparison: The Nat'l Male MortRate, 1950, in Table 40-B =   256.40

(A) Census Division	(B) 1940 Pop'n Fraction	(C) PhysPop 1940	(D) 1940 Weighted PhysPop	(E) 1950 Pop,n Frac'n	(F) MortRate 1950	(G) Weighted MortRate
Pacific	0.0739	159.72	11.80	0.0961	283.2	27.22
New England	0.0641	161.55	10.36	0.0618	297.1	18.36
West No. Central	0.1027	123.14	12.65	0.0933	228.4	21.31
Mid-Atlantic	0.2092	169.76	35.51	0.2002	310.3	62.12
East No. Central	0.2022	133.36	26.97	0.2017	258.9	52.22
Mountain	0.0315	119.89	3.78	0.0337	214.8	7.24
West So. Central	0.0992	103.94	10.31	0.0965	206.1	19.89
East So. Central	0.0819	85.83	7.03	0.0762	176.8	13.47
South Atlantic	0.1354	100.74	13.64	0.1406	222.0	31.21
Sums	1.0000		132.04	1.0001		253.04

1950 Ischemic Heart Disease Mortality–Rates versus 1940 PhysPop Values for the 9 Census Divisions, USA.
Dose–Response Relationship
PhysPop is a surrogate for accumulated dose from medical irradiation.

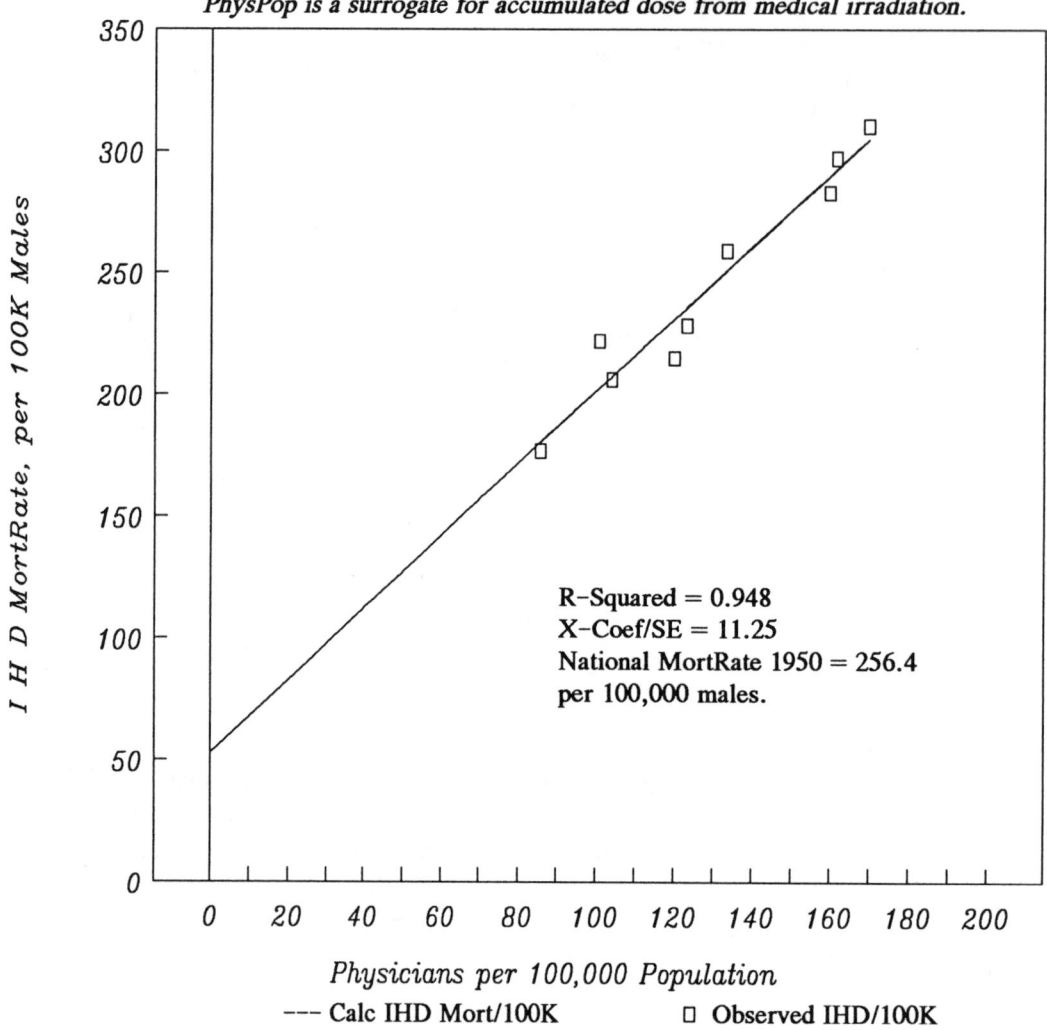

R–Squared = 0.948
X–Coef/SE = 11.25
National MortRate 1950 = 256.4
per 100,000 males.

Physicians per 100,000 Population

--- Calc IHD Mort/100K   □ Observed IHD/100K

On the X–axis, PhysPop values = Physicians per 100,000 Population in the Nine Census Divisions of the USA Population, Year 1940. This variable is a surrogate for accumulated radiation dose --- the more physicians per 100,000 people, the more radiation procedures are done per 100,000 people.

On the Y–axis, Ischemic Heart Disease Mortality–Rate per 100,000 males = the reported rates in USA Vital Statistics for the Nine Census Divisions, Year 1950.

Shown above is the strongest relationship between these two variables (Part 2j). The nine datapoints (boxy symbols) were collected long ago for other purposes, and are free from potential bias with respect to this dose–response study. Fractional causation is (Natl MortRate minus the Y–intercept) / (Natl MortRate).

Fractional Causation of Ischemic Heart Disease Mortality–Rate (Males)
by Medical Radiation = 79 % from Best Estimate (Box 3).
65% at Lower 90% Conf. Limit (Box 3).   88% at Upper 90% Conf. Limit (Box 3).

Figure 40–B

Heart and Stroke Facts: 1996 Statistical Supplement

Age-Adjusted Death Rates for Major Cardiovascular Diseases

United States: 1940–92

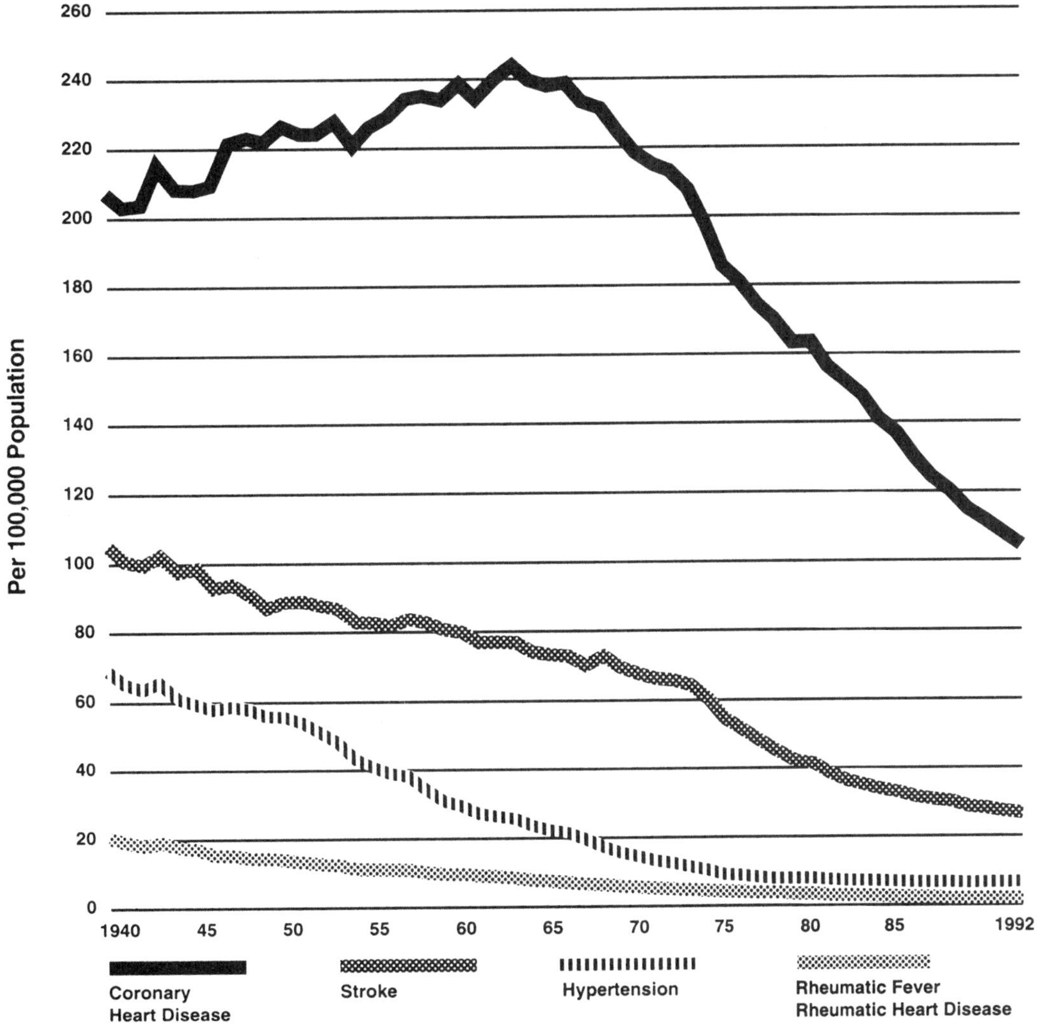

Per 100,000 Population

Coronary Heart Disease	Stroke	Hypertension	Rheumatic Fever Rheumatic Heart Disease	

Age-adjusted to 1940 U.S. population and to the 6th Revision ICDA. Source: National Center for Health Statistics and the American Heart Association.

Table 40–A.
Ischemic Heart Disease: Male Mortality Rates by Census Divisions

Rates are annual deaths per 100,000 male population, USA, age-adjusted to the 1940 reference year.
No exclusions by color or "race." The MortRate for each Census Division is population-weighted,
whereas the averages in Table 40-A are not. Chapter 4 defines "High-5" and "Low-4."
Sources: See Table 40-B.

Census Division	1940	1950	1960	1970	1980	1993
Pacific	--	283.2	284.2	231.0	177.7	112.4
New England	--	297.1	347.1	285.3	223.5	117.8
West North Central	--	228.4	284.1	245.1	206.1	129.9
Mid-Atlantic	--	310.3	355.0	300.8	246.6	147.9
East North Central	--	258.9	320.8	274.1	227.4	140.5
Mountain	--	214.8	256.8	215.2	173.6	101.2
West South Central	--	206.1	269.4	232.0	194.5	137.6
East South Central	--	176.8	254.4	236.8	219.2	145.8
South Atlantic	--	222.0	286.4	248.7	210.9	128.7
Average, ALL	--	244.2	295.4	252.1	208.8	129.1
Average, High-5	--	275.6	318.2	267.3	216.3	129.7
Average, Low-4	--	204.9	266.8	233.2	199.6	128.3
Ratio, Hi5/Lo4	--	1.34	1.19	1.15	1.08	1.01

Table 40–B.
Ischemic Heart Disease: National Mortality Rates, USA.

Rates are age-adjusted to the 1940 reference year. Both sexes: Deaths per 100,000 population (males
+ females). Males: Deaths per 100,000 male population. Females: Deaths per 100,000 female
population. No exclusions by color or "race."

	Both Sexes	Male	Female
1940	--	--	--
1950	190.0	256.4	126.5
1960	225.5 #	306.5	152.5
1970	186.8	259.7	124.9
1979–81	148.1	212.8	97.2
1992–94	94.9	131.0	64.7

\#   The peak rate occurred in 1963 (AHA 1995).

  ● – 1950, 1960: All rates are from Grove 1968, Table 67, Pages 720–722, "Arteriosclerotic Heart
Disease, including coronary disease (420)" ICD/7.
  ● – 1970: Rates are interpolations between 1960 and 1980 (Chap. 4, Parts 2b, 2c).
  ● – 1980: All rates (ICD/9, 410-414) come from the reference NatCtrHS 1980.
  ● – 1990: The 1993 rates (ICD/9, 410-414.9) come from the reference NatCtrHS 1993;
please see Chap.4, Part 2b. Exception: The 1993 Rate for both sexes (combined) comes from AHA
1996, p.8.

● Part 1. Introduction: IHD in Females

This chapter follows the model of Chapter 40, but eliminates the text.

● Part 2. How the Dose-Response Behaves, for Medical Radiation and IHD in Females

● - Part 2a.	x 1921 PhysPop	y 1950 MortRate	IHD, Females Regression Output:	
Pacific	165.11	125.0	Constant	5.8843
New England	142.24	153.2	Std Err of Y Est	27.2707
West North Central	140.93	104.1	R Squared	0.2026
Mid-Atlantic	137.29	169.4	No. of Observations	9
East North Central	136.06	124.2	Degrees of Freedom	7
Mountain	135.38	96.2		
West South Central	125.15	94.0	X Coefficient(s)	0.8259
East South Central	119.76	84.7	Std Err of Coef.	0.6194
South Atlantic	110.32	103.4	Coefficient / S.E.	1.3334

● - Part 2b.	1923 PhysPop	1950 MortRate	IHD, Females Regression Output:	
Pacific	163.06	125.0	Constant	-0.4910
New England	137.39	153.2	Std Err of Y Est	25.9957
West North Central	138.31	104.1	R Squared	0.2754
Mid-Atlantic	138.92	169.4	No. of Observations	9
East North Central	131.82	124.2	Degrees of Freedom	7
Mountain	130.51	96.2		
West South Central	119.16	94.0	X Coefficient(s)	0.8978
East South Central	113.16	84.7	Std Err of Coef.	0.5504
South Atlantic	106.79	103.4	Coefficient / S.E.	1.6310

● - Part 2c.	1925 PhysPop	1950 MortRate	IHD, Females Regression Output:	
Pacific	161.67	125.0	Constant	1.2851
New England	138.31	153.2	Std Err of Y Est	24.9726
West North Central	133.92	104.1	R Squared	0.3313
Mid-Atlantic	134.36	169.4	No. of Observations	9
East North Central	127.54	124.2	Degrees of Freedom	7
Mountain	122.30	96.2		
West South Central	112.83	94.0	X Coefficient(s)	0.9132
East South Central	107.22	84.7	Std Err of Coef.	0.4904
South Atlantic	103.61	103.4	Coefficient / S.E.	1.8623

● - Part 2d.	1927 PhysPop	1950 MortRate	IHD, Females Regression Output:	
Pacific	157.83	125.0	Constant	-11.2627
New England	137.50	153.2	Std Err of Y Est	22.4700
West North Central	131.54	104.1	R Squared	0.4586
Mid-Atlantic	138.40	169.4	No. of Observations	9
East North Central	126.18	124.2	Degrees of Freedom	7
Mountain	118.75	96.2		
West South Central	108.25	94.0	X Coefficient(s)	1.0293
East South Central	102.07	84.7	Std Err of Coef.	0.4227
South Atlantic	102.13	103.4	Coefficient / S.E.	2.4351

● - Part 2e.	1929 PhysPop	1950 MortRate	IHD, Females Regression Output:	
Pacific	156.64	125.0	Constant	-11.6926
New England	138.46	153.2	Std Err of Y Est	21.5596
West North Central	128.72	104.1	R Squared	0.5016

Mid-Atlantic	138.49	169.4	No. of Observations		9
East North Central	126.51	124.2	Degrees of Freedom		7
Mountain	118.68	96.2			
West South Central	105.60	94.0	X Coefficient(s)		1.0414
East South Central	99.41	84.7	Std Err of Coef.		0.3924
South Atlantic	100.86	103.4	Coefficient / S.E.		2.6542

● – Part 2f.	1931	1950	IHD, Females	
	PhysPop	MortRate	Regression Output:	
Pacific	159.97	125.0	Constant	-5.0973
New England	142.35	153.2	Std Err of Y Est	20.7787
West North Central	126.50	104.1	R Squared	0.5370
Mid-Atlantic	140.82	169.4	No. of Observations	9
East North Central	128.59	124.2	Degrees of Freedom	7
Mountain	118.89	96.2		
West South Central	105.95	94.0	X Coefficient(s)	0.9827
East South Central	96.73	84.7	Std Err of Coef.	0.3449
South Atlantic	99.59	103.4	Coefficient / S.E.	2.8496

● – Part 2g.	1934	1950	IHD, Females	
	PhysPop	MortRate	Regression Output:	
Pacific	160.09	125.0	Constant	-3.0909
New England	148.60	153.2	Std Err of Y Est	17.7357
West North Central	125.96	104.1	R Squared	0.6627
Mid-Atlantic	149.62	169.4	No. of Observations	9
East North Central	129.36	124.2	Degrees of Freedom	7
Mountain	117.16	96.2		
West South Central	104.68	94.0	X Coefficient(s)	0.9610
East South Central	92.00	84.7	Std Err of Coef.	0.2591
South Atlantic	98.41	103.4	Coefficient / S.E.	3.7086

● – Part 2h.	1936	1950	IHD, Females	
	PhysPop	MortRate	Regression Output:	
Pacific	158.44	125.0	Constant	-3.7018
New England	150.18	153.2	Std Err of Y Est	16.2958
West North Central	126.14	104.1	R Squared	0.7153
Mid-Atlantic	155.05	169.4	No. of Observations	9
East North Central	130.42	124.2	Degrees of Freedom	7
Mountain	119.80	96.2		
West South Central	103.52	94.0	X Coefficient(s)	0.9602
East South Central	89.94	84.7	Std Err of Coef.	0.2290
South Atlantic	99.16	103.4	Coefficient / S.E.	4.1932

● – Part 2i.	1938	1950	IHD, Females	
	PhysPop	MortRate	Regression Output:	
Pacific	157.62	125.0	Constant	-1.6209
New England	154.08	153.2	Std Err of Y Est	14.5088
West North Central	124.95	104.1	R Squared	0.7743
Mid-Atlantic	160.69	169.4	No. of Observations	9
East North Central	131.98	124.2	Degrees of Freedom	7
Mountain	119.88	96.2		
West South Central	102.79	94.0	X Coefficient(s)	0.9380
East South Central	88.21	84.7	Std Err of Coef.	0.1914
South Atlantic	99.26	103.4	Coefficient / S.E.	4.9002

● – Part 2j.	1940	1950	IHD, Females	
	PhysPop	MortRate	Regression Output:	
Pacific	159.72	125.0	Constant	4.4119
New England	161.55	153.2	Std Err of Y Est	12.4544
West North Central	123.14	104.1	R Squared	0.8337
Mid-Atlantic	169.76	169.4	No. of Observations	9
East North Central	133.36	124.2	Degrees of Freedom	7
Mountain	119.89	96.2		
West South Central	103.94	94.0	X Coefficient(s)	0.8761
East South Central	85.83	84.7	Std Err of Coef.	0.1479
South Atlantic	100.74	103.4	Coefficient / S.E.	5.9234

● – Part 2k.	1950 PhysPop	1950 MortRate
Pacific	148.60	125.0
New England	162.51	153.2
West North Central	120.06	104.1
Mid–Atlantic	168.71	169.4
East North Central	123.69	124.2
Mountain	119.38	96.2
West South Central	101.34	94.0
East South Central	83.05	84.7
South Atlantic	99.07	103.4

IHD, Females

Regression Output:

Constant	3.9806
Std Err of Y Est	11.1400
R Squared	0.8669
No. of Observations	9
Degrees of Freedom	7
X Coefficient(s)	0.9041
Std Err of Coef.	0.1339
Coefficient / S.E.	6.7531

Box 1 of Chap. 41
Summary: Regression Outputs, Female IHD MortRates Regressed on PhysPop.

We are searching for the maximum correlation between PhysPops of 1921–1950 and the female Ischemic Heart Disease MortRates of 1950. Even the maximum correlation will tend to understate the true correlation (Chapter 5, Part 8b).

Part	PhysPop	R-squared	Constant	X-Coef	Std Err	X-Coef/SE
2a	1921	0.2026	5.88	0.8259	0.6194	1.3334
2b	1923	0.2754	−0.49	0.8978	0.5504	1.6310
2c	1925	0.3313	1.29	0.9132	0.4904	1.8623
2d	1927	0.4586	−11.26	1.0293	0.4227	2.4351
2e	1929	0.5016	−11.69	1.0414	0.3924	2.6542
2f	1931	0.5370	−5.10	0.9827	0.3449	2.8496
2g	1934	0.6627	−3.09	0.9610	0.2591	3.7086
2h	1936	0.7153	−3.07	0.9602	0.2290	4.1932
2i	1938	0.7743	−1.62	0.9380	0.1914	4.9002
2j	1940	0.8337	4.41	0.8761	0.1479	5.9234
2k --->	1950 Max	0.8669	3.98	0.9041	0.1339	6.7531

Box 2 of Chap. 41
Input–Data for Figure 41-A. Ischemic Heart Disease. Females.

Part 2k, Best-Fit Equation: Calc. MortRate = (0.9041 * PhysPop) + (3.981)

Census Divisions	1950 Observed PhysPops	1950 Observed MortRates	Best–Fit Calc. MortRates
Pacific	148.60	125.0	138.330
New England	162.51	153.2	150.906
West No. Central	120.06	104.1	112.527
Mid–Atlantic	168.71	169.4	156.512
East No. Central	123.69	124.2	115.809
Mountain	119.38	96.2	111.912
West So. Central	101.34	94.0	95.602
East So. Central	83.05	84.7	79.067
South Atlantic	99.07	103.4	93.550
Additional PhysPops	70.00		67.268
--- not "observed" ---	60.00		58.227
down to zero PhysPop	50.00		49.186
(zero medical radiation).	40.00		40.145
For each, we calculate	30.00		31.104
a best-fit MortRate.	20.00		22.063
These additional x,y pairs	10.00		13.022
are also part of the	0		3.981
best-fit line (Chap 5, Part 5e).			

Box 3 of Chap. 41

Percent of IHD MortRate Attributable to Medical Radiation.

Please see text in Chapter 40, Parts 4 and 5.

IHD. FEMALES.   * denotes multiplication.

- FEMALE National MortRate (MR) 1950, from Table 41-B   126.5   National MortRate
- Constant, from regression, Part 2k   3.9806   Constant
- Fractional Causation, Best Est. = (Natl MR − Constant) / Natl MR   96.9%   Frac. Causation

....

90% Confidence-Limits (C.L.) on Fractional Causation. See text in Chapter 6, Parts 4b-d, please.

X-Coefficient, from Part 2k   0.9041   X-Coef., Best Est.
Standard Error (SE) of X-Coefficient, from Part 2j   0.1339   Standard Error

Upper 90% C.L. on X-Coef. = (Coef) + (1.645 * SE) =   1.1244   New X-Coefficient
New Constant = (Natl MR) − (New X-Coef * 1950 Natl PhysPop) = −17.7673 New Constant
Frac. Caus'n, High-Limit = (Natl MR − New Constant) / Natl MR = 114.0% New Frac. Caus'n.
 # The Upper-Limit is 100%. Negative Constants produce values > 100%. See Chapter 22, Part 3.

Lower 90% C.L. on X-Coef. = (Coef) − (1.645 * SE) =   0.6838   New X-Coefficient
New Constant = (Natl MR) − (New X-Coef * 1950 Natl PhysPop) =   38.7572 New Constant
Frac. Caus'n, Low-Limit = (Natl MR − New Constant) / Natl MR =   69.4% New Frac. Caus'n.

Box 4 of Chap. 41

Error-Check on Our Own Work: Ischemic Heart Disease, Females.

Please see text in Chapter 6, Part 5.

Below, Columns A, C, and E come directly from the regression input in Part 2k. Column B, the fraction of the whole 1950 population in each Census Division, comes from Table 3-B in Chapter 3. Each Column-D entry = (B * C). Each Column-F entry = (B * E). MortRates are each "per 100,000."

The Weighted-Avg. Nat'l PhysPop, 1950, is the sum of Column-D entries =   128.31

Weighted-Avg. Nat'l Female MortRate, 1950, is sum of Col.F entries =   123.46
Nat'l Female MortRate is also (X-Coef * 1950 Nat'l PhysPop) + Constant =   119.99
Comparison: The Nat'l Female MortRate, 1950, in Table 41-B =   126.50

(A) Census Division	(B) 1950 Pop'n Fraction	(C) PhysPop 1950	(D) 1950 Weighted PhysPop	(E) MortRate 1950	(F) Weighted MortRate
Pacific	0.0961	148.60	14.28	125.0	12.01
New England	0.0618	162.51	10.04	153.2	9.47
West No. Central	0.0933	120.06	11.20	104.1	9.71
Mid-Atlantic	0.2002	168.71	33.78	169.4	33.91
East No. Central	0.2017	123.69	24.95	124.2	25.05
Mountain	0.0337	119.38	4.02	96.2	3.24
West So. Central	0.0965	101.34	9.78	94.0	9.07
East So. Central	0.0762	83.05	6.33	84.7	6.45
South Atlantic	0.1406	99.07	13.93	103.4	14.54
Sums	1.0001		128.31		123.46

1950 Ischemic Heart Disease Mortality-Rates versus 1950 PhysPop Values for the 9 Census Divisions, USA.
Dose-Response Relationship
PhysPop is a surrogate for accumulated dose from medical irradiation.

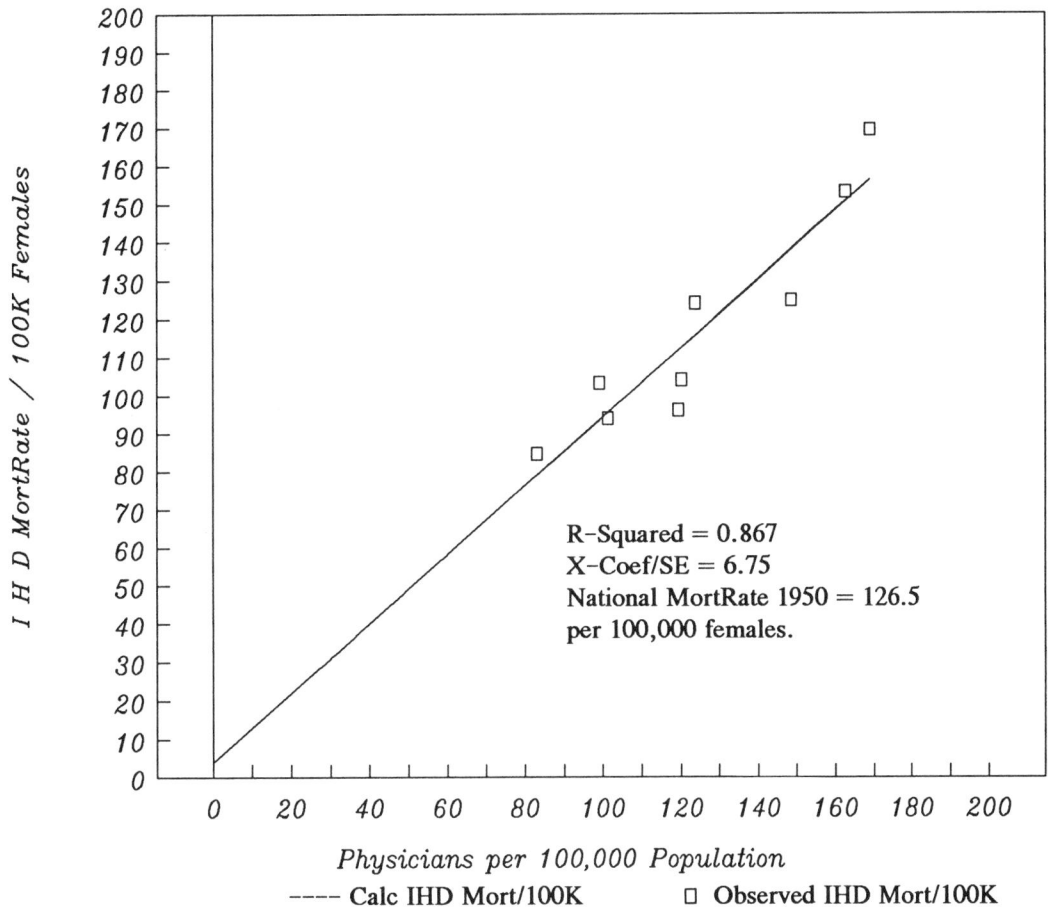

R-Squared = 0.867
X-Coef/SE = 6.75
National MortRate 1950 = 126.5
per 100,000 females.

Physicians per 100,000 Population
---- Calc IHD Mort/100K   □ Observed IHD Mort/100K

On the X-axis, PhysPop values = Physicians per 100,000 Population in the Nine Census Divisions of the USA Population, Year 1950. This variable is a surrogate for accumulated radiation dose --- the more physicians per 100,000 people, the more radiation procedures are done per 100,000 people.

On the Y-axis, Ischemic Heart Disease Mortality-Rate per 100,000 females = the reported rates in USA Vital Statistics for the Nine Census Divisions, Year 1950.

Shown above is the strongest relationship between these two variables (Part 2k). The nine datapoints (boxy symbols) were collected long ago for other purposes, and are free from potential bias with respect to this dose-response study. Fractional causation is (Natl MortRate minus the Y-intercept) / (Natl MortRate).

Fractional Causation of Ischemic Heart Disease Mortality-Rate (Females)
by Medical Radiation = 97 % from Best Estimate (Box 3).

69 % at lower 90 % Conf. Limit (Box 3). ~100 % at Upper 90 % Conf. Limit (Box 3).

Table 41-A.
Ischemic Heart Disease: Female Mortality Rates by Census Divisions

Rates are annual deaths per 100,000 female population, USA, age-adjusted to the 1940 reference year. No exclusions by color or "race." The MortRate for each Census Division is population-weighted, whereas the averages in Table 41-A are not. Chapter 4 defines "High-5" and "Low-4." Sources: See Table 41-B.

Census Division	1940	1950	1960	1970	1980	1993
Pacific	--	125.0	133.4	107.4	81.4	57.7
New England	--	153.2	176.3	138.0	99.6	55.7
West North Central	--	104.1	135.8	111.0	86.1	58.3
Mid-Atlantic	--	169.4	189.7	154.9	120.1	78.8
East North Central	--	124.2	162.2	134.5	106.8	70.2
Mountain	--	96.2	118.9	96.6	74.2	46.3
West South Central	--	94.0	123.9	105.5	87.1	66.5
East South Central	--	84.7	126.2	110.7	95.2	67.7
South Atlantic	--	103.4	132.4	111.6	90.8	61.6
Average, ALL	--	117.1	144.3	118.9	93.5	62.5
Average, High-5	--	135.2	159.5	129.1	98.8	64.1
Average, Low-4	--	94.6	125.4	106.1	86.8	60.5
Ratio, Hi5/Lo4	--	1.43	1.27	1.22	1.14	1.06

Table 41-B.
Ischemic Heart Disease: National Mortality Rates, USA.

Rates are age-adjusted to the 1940 reference year. Both sexes: Deaths per 100,000 population (males + females). Males: Deaths per 100,000 male population. Females: Deaths per 100,000 female population. No exclusions by color or "race."

	Both Sexes	Male	Female
1940	--	--	--
1950	190.0	256.4	126.5
1960	225.5 #	306.5	152.5
1970	186.8	259.7	124.9
1979-81	148.1	212.8	97.2
1992-94	94.9	131.0	64.7

\# The peak rate occurred in 1963 (AHA 1995).

● – 1950, 1960: All rates are from Grove 1968, Table 67, Pages 720-722, "Arteriosclerotic Heart Disease, including coronary disease (420)" ICD/7.
  ● – 1970: Rates are interpolations between 1960 and 1980 (Chap. 4, Parts 2b, 2c).
  ● – 1980: All rates (ICD/9, 410-414) come from the reference NatCtrHS 1980.
  ● – 1990: The 1993 rates (ICD/9, 410-414.9) come from the reference NatCtrHS 1993; please see Chap.4, Part 2b. Exception: The 1993 Rate for both sexes (combined) comes from AHA 1996, p.8.

CHAPTER 42

Similarities in the IHD and Cancer Findings: Tumors in Both Diseases?

Part 1. List of Epidemiological Similarities Found for IHD and Cancer
Part 2. Which Aspect of Physician-Density Deserves the Blame?
Part 3. The Suspicion of Multiple Mini-Tumors in Ischemic Heart Disease

● Part 1. List of Epidemiological Similarities Found for IHD and Cancer

The nature of epidemiology is such that it supplies circumstantial evidence about causation ("who done it"). Rarely is a single piece of epidemiologic evidence capable of PROVING causation or validating an hypothesis. Like circumstantial cases in criminal law, a case based on epidemiology grows stronger with each additional piece of supporting evidence. "What is the chance that all these observations would occur together, if the suspect is INNOCENT? What other explanation fits the COMBINED observations?"

Below, as a convenience, we list the similar epidemiologic observations uncovered in this book with respect to Ischemic Heart Disease and Cancer.

1a. Positive, Unmistakable Dose-Response with PhysPop at Mid-Century

At approximately mid-century, both IHD and cancer MortRates for each sex separately, by Census Divisions, have a positive and irrefutable dose-response relationship with PhysPops (numbers of physicians per 100,000 population). The maximum relationship occurs for IHD in 1950 (the first year for which we have such data), and occurs for Cancer in 1940 (the first year for which we have such data).

The MortRates used in this book include everyone (no exclusions by color or "race"). We also regressed MortRates for "whites-only" on PhysPops, and the results were barely different from the results presented in this book.

1b. Linearity and Strength of Dose-Response

For both diseases, the dose-response is linear and highly significant (Chapters 6, 7, 40, and 41). Such a dose-response for IHD is what elicited Hypothesis-2 in the first place:

	R-squared Males	Coef/SE Males	R-squared Females	Coef/SE Females
IHD, 1950:	0.95	11.24	0.87	6.75
Cancer, 1940:	0.95	11.63	0.86	6.58

It deserves emphasis that the IHD and cancer dose-responses with PhysPop are NOT HYPOTHETICAL. They are real-world facts, as are the other observations listed below. And all of them arise from neutral, objective databases --- in contrast to some databases in which radiation dosage has been retroactively revised and in which dose-cohorts have been shuffled, pruned, and augmented AFTER follow-up results are known.

1c. High Fractional Causation by PhysPop of the Entire MortRate

For both diseases, the central estimate is high and far from negligible, for the fraction of the entire MortRate due to PhysPop and its co-factors. MortRates are "per 100,000 population."

	MortRate Males	Fraction Males		MortRate Females	Fraction Females
IHD, 1950:	256.4	79%		126.5	97%
Cancer, 1940:	115.0	90%		126.1	58%

1d. Prediction of National MortRates by PhysPop Values of 10-20 Years Earlier

A positive and significant correlation, of the MortRates of 1950 (IHD) and 1940 (Cancer) with PhysPops, begins with PhysPops of much earlier years (Boxes 1 in Chapters 6, 7, 40, and 41).

Indeed, for Cancer, the 1921 PhysPops predict the National 1940 All-Cancer MortRates (male, female) quite closely --- and the 1931 PhysPops predict the 1940 National MortRates even better (Chapter 22, Box 4).

For IHD, since our earliest MortRates are for 1950, we look first at the 1931 PhysPops. How closely do the 1931 PhysPops predict the National 1950 IHD MortRates? Quite closely, as shown below.

● 1931, IHD Males: Best-fit Equation comes from Chapter 40 (Part 2f) and the National PhysPop value comes from Chapter 22 (top of Box 4):

Predicted Male National MortRate 1950 = (Xcoef * Natl PhysPop) + Constant
Predicted Male National MortRate 1950 = (1.8540 * 125.3) + 13.58
Predicted Male National MortRate 1950 =   245.9
Observed Male National MortRate 1950 =   256.4   from Table 40-B.

● 1931, IHD Females: Best-fit equation comes from Chapter 41 (Part 2f), and National PhysPop from Chapter 22, Box 4:

Predicted Female National MortRate 1950 = (Xcoef * Natl PhysPop) + Constant
Predicted Female National MortRate 1950 = (0.9827 * 125.3) + (-5.1)
Predicted Female National MortRate 1950 =   118.0
Observed Female National MortRate 1950 =   126.5   from Table 41-B.

1e. Distinction of IHD and Cancer from All Other Causes of Death

Ischemic Heart Disease and Cancer behave alike, when they "select themselves out" from other causes of death, with respect to their mid-century dose-responses with PhysPop. At mid-century, IHD and Cancer each have a highly significant and POSITIVE correlation with PhysPop. By contrast, the dose-response at mid-century between PhysPop and NonCancer NonIHD MortRates is significant and NEGATIVE (Chapter 25, Box 1; subsets are summarized in Chapter 38, Box 1). In terms of a relationship with PhysPop, Ischemic Heart Disease and Cancer clearly do not belong with the other causes of death. They belong with EACH OTHER.

This is a remarkable finding. In Census Divisions where there were more physicians per 100,000 population, the populations at mid-century fared WORSE with respect to IHD and Cancer than did the populations in Census Divisions with fewer physicians per 100,000. Yet simultaneously, populations in high-PhysPop Divisions fared BETTER than populations in low-PhysPop Divisions, with respect to the combination of all OTHER causes of death.

● **Part 2. Which Aspect of Physician-Density Deserves the Blame?**

Strong dose-responses are widely acknowledged to be strong presumptive evidence of causation, unless shown otherwise. For both Ischemic Heart Disease and Cancer, the strong dose-responses between MortRates and PhysPop, by Census Divisions, point to variation in PHYSPOP as the cause of variation in DEATH RATES (Chapter 5, Part 5a).

So we must ask: WHICH aspect of physician-density can be the cause of the observed variation in mortality?

Fortunately, we do not have to guess randomly at the answer. For Cancer, the evidence points clearly to RADIATION from medical procedures as the culprit (Chapter 22, Part 6). Both common-sense and evidence support the premise that the more physicians per 100,000 population, the more radiation procedures per 100,000 population will be ordered (Chapter 3, Part 1a). In addition, there is separate and solid evidence that:

● - Ionizing radiation is a uniquely powerful mutagen, capable of inducing every known kind of mutation, especially the complex types which --- quite unlike routine DNA damage from endogenous free radicals --- often elude successful repair (Chapter 2 and Appendices C and D).

● - There is no threshold dose with respect to the induction of unrepairable genetic damage by ionizing radiation (Chapter 2, Parts 4 and 6, and Appendix-B).

● - Xrays are an even more potent mutagen than gamma rays, per dose-unit (Chapter 2, Part 7).

● - Ionizing radiation is a proven cause of genomic instability, a feature of the most aggressive cancers (Chapter 2, Part 4b, and Appendix-D).

● - Most kinds of human cancer are inducible by ionizing radiation (Chapter 2, Part 4c).

But what about Ischemic Heart Disease?

● Part 3. The Suspicion of Multiple Mini-Tumors in Ischemic Heart Disease

In the epidemiologic features listed above in Part 1, Ischemic Heart Disease and Cancer behave like each other --- and NOT like most other causes of death. Those similarities between the two diseases are so striking that --- even if we knew nothing else about either disease --- we would suggest that the disease called Ischemic Heart Disease is closely related in etiology with the set of diseases called All-Cancers-Combined.

3a. Radiation-Induced Mini-Tumors in the Coronary Arteries

Because solid Cancers are characterized by tumors and generally by multiple genetic mutations (inherited and/or acquired), we propose the second part of Hypothesis-2:

Radiation-induction of mutations in the coronary arteries, resulting in dysfunctional clones (mini-tumors) of smooth muscle cells, is the probable mechanism by which medical radiation contributes causally to Ischemic Heart Disease.

Such a concept ought to be testable by pathologists and molecular biologists. And indeed, long before our study here, a few investigators have done work which leads them to say that multiple mini-tumors DO exist in atherosclerotic lesions of the coronary arteries (Chapter 44, Part 8).

3b. Evidence that Ionizing Radiation Induces Non-Malignant Tumors

A central feature of both malignant and non-malignant tumors is inappropriate proliferation by cells, where proliferation serves no beneficial purpose. If the net balance between cell-division and cell-death is very largely under genetic control, one might expect a potent mutagen, like ionizing radiation, to cause non-malignant tumors as well as malignant ones.

The evidence, that ionizing radiation can induce NON-malignant tumors (as well as malignancies) in humans, is compelling for thyroid nodules and adenomas, and parathyroid adenomas (Gofman 1981, pp.189-197, + BEIR 1990, pp.289-292 and pp.321-323, + Shore 1993, + Wong 1993). For non-malignant tumors of the stomach, a recent analysis of the Hiroshima-Nagasaki LifeSpan Study reveals a positive dose-response between bomb-dosage and such tumors (Ron 1995). For myoma uteri (a non-malignant tumor of the womb), a statistically significant excess in bomb-exposed females has been reported from the Hiroshima-Nagasaki Adult Health Study (Wong 1993). A study of medical xrays finds excess non-malignant skin tumors (as well as malignant ones) in irradiated medical patients (Ron 1991).

It would be hard to assess how many grants have ever been issued to LOOK for radiation-induced non-malignant tumors in various other organs. A lack of evidence may mean simply that there has been little support for such inquiries. With respect to the coronary arteries, we doubt very much that grants have been issued to look for radiation-induced mini-tumors in such arteries.

3c. Hypothesis-2: Its Two Parts Are Independent of Each Other

If tumors are a feature of BOTH Cancer and Ischemic Heart Disease, it would explain why IHD MortRates respond to PhysPop in the same way that cancer MortRates respond to PhysPop, and why the response of both diseases to PhysPop is so different from NonCancer NonIHD MortRates. Such similarity between IHD and Cancer is a "smoking gun" which deserves attention, as a starting point for further inquiry into the second part of Hypothesis-2.

There is a popular adage about similarities. The adage, which is just a variant on Ockham's Razor (Chapter 22, Part 6a), urges human beings not to scorn an obvious explanation: "If it walks like a duck, and if it quacks like a duck, and if it looks like a duck, it probably is a duck."

Ionizing radiation can induce genetic mutations and tumors. These two well-established facts may suffice to explain the findings in this book, that variation in medical radiation controls variation in the MortRates from Ischemic Heart Disease, by Census Divisions. Chapters 43 through 46 discuss how the tumor hypothesis fits into the existing knowledge about the causes of Ischemic Heart Disease. Time will tell if the second part of Hypothesis-2 is valid, but only if appropriate studies are done.

Meanwhile, the epidemiologic observations provided in this book stand independently, on their own, as evidence in favor of the first part of Hypothesis-2: Medical radiation, received even at very low and moderate doses, is an important cause of Ischemic Heart Disease.

>>>>>>>>>>

CHAPTER 43

Nature of the Atherosclerotic Lesions Underlying IHD

Part 1. The Walls of the Coronary Arteries
Part 2. Coronary Atherosclerosis: Still a Controversial Topic
Part 3. Characteristics of the "Typical" Lesions
Part 4. A Recent Warning about Defects in Clinical Evaluations

Hypothesis-2 is fully stated at the outset of Chapter 39. In order to discuss it (in Chapters 44, 45, and 46), we need to become more specific about "the coronary arteries" and about the lesions of "coronary atherosclerosis" --- which we described in Chapter 39 only as a build-up in the arterial wall of material (atherosclerotic "plaque") which finally intrudes into the lumen and obstructs the blood-flow, partially or completely.

● **Part 1. The Walls of the Coronary Arteries**

The right and left coronary arteries each depart, from the ascending aorta, very close to the aorta's exit from the the heart's left side. The right and left coronary arteries "return" to the exterior surface of the heart, where they each divide into several branches which have their own names and which are known collectively as "the coronary arteries" or just "the coronaries." The coronary arteries supply the heart with the freshly oxygenated blood which it (like other organs) requires in order to survive and to operate.

The artery is an elastic tube which consists of three major structural regions, or layers, named the intima, the media, and the adventitia. These three regions show major dissimilarities in composition. The dangerous events occur mainly, but not totally, in the intimal layer of the arterial wall.

The Intimal Layer and Internal Elastic Membrane

The inner aspect of the intimal wall (the aspect next to the flowing blood) is lined by a single layer of cells known as endothelial cells. Much has been said and written, about the role of injury to these endothelial cells in development of atherosclerosis, but little is known with certainty.

The intima, when it is healthy, is the thinnest layer of the arterial wall. The intima consists of a little cellular material and a sparse matrix of collagen fibers. There are relatively few elastic fibers in the intima.

Separating the intima and the media is a continuous structure made of elastic tissue, and referred to as the internal elastic membrane, or "the internal elastica." Approximately 1% of the area of the internal elastica consists of channels or "fenestra" (Young 1960), linking the intima with the next layer, the media.

The Media Layer

The media layer is that layer which contributes to the elastic functions of the arterial wall. Normally, the media is much thicker than the intima. There are two major elements in the media layer: (1) a thick network of elastic tissue, which is interspersed with (2) smooth muscle cells (akin to those in the intestinal walls, the uterine muscle walls, and still other organs).

The Adventitia

The adventitia is the outer coat of the arterial wall. It consists primarily of fibrous tissue of the collagen variety, rather than elastin. There are very small vessels in the adventitia, whose function appears to be the provision of nourishment and waste disposal for the arterial wall itself.

Relative Thicknesses of the Layers

In arteries with an outer diameter of about 2,000 micrometers, measurements have shown the following ranges, in micrometers (Gofman 1963, p.203): Intima, 2.5-5.8. Internal Elastica, 3.7-6.1. Media, 182-212. Adventitia, 71-91. Arteries require a blood-supply of their own, of course, and it is delivered by the "vasa vasorum" within the artery wall.

The Presence of "Collaterals"

When the lumen of an artery becomes obstructed, the body commonly attempts to generate a network of capillaries and arterioles which by-pass the obstruction. When this process (angiogenesis) is successful, the resulting new vessels are called "collaterals." The stimulation of coronary angiogenesis is an active field of research.

● Part 2. Coronary Atherosclerosis: Still a Controversial Topic

● "Thrombus formation is the proximate cause of myocardial infarction, but atherosclerosis, the chief underlying cause, is a chronic disease that progresses over decades of life" (Ridker 1997, p.973, citing Fuster 1992).

2a. Raging Debates on Initiation, Progression, Reversal of the Process

Atherosclerotic lesions or atheromata, commonly called "plaques," develop in the intima and then ultimately protrude into the lumen to various degrees. The plaques are localized, with regions free from the pathology adjacent to regions which are deeply involved. Why discontinuity occurs, needs explanation.

There is anything BUT agreement on exactly what starts the atherosclerotic process, what determines its rate of progression, and what reversibility there is for different types of the pathological material laid down in the artery wall. The differences among investigators are large, and many diverse proposals are made, and re-made, concerning the events which describe the evolution of various lesions. Moreover, confirmation of one hypothesis would not automatically rule out validity for several other hypotheses. There may be more than one route to a single result, and within any route, co-action by multiple causes may be required.

While some books and articles describe the atherosclerotic process as though it were thoroughly understood, it is not. In 1997, Attilio Maseri editorialized about atherosclerosis and ischemic events in the New England Journal of Medicine. He commented (Maseri 1997, p.1015): "Ischemic heart disease is appearing to be an ever more complex syndrome ..."

2b. An "Infinite Variety" of Lesions Observed

What does seem quite certain is that some pathologic changes occur acutely, while other changes occur in a sequence of events over months, years, and decades. Legions of investigators frequently have to study a lengthy process from evidence which represents just one particular stage --- but different stages for different investigators. They necessarily see different things and report different findings. They are often studying different "snapshots" (frames) extracted from a three-hour "film epic." Elspeth B. Smith, who has made some excellent contributions on the problem of what happens in arterial walls, has commented perceptively (Smith 1977, p.673):

"Clearly, there is a very complicated system within the intima ... This system must be infinitely variable, which would be compatible with the infinite variety of lesions found in human arteries."

2c. Efforts to Define and Classify Atherosclerotic Lesions

In any attempt to communicate, shared definitions help a lot. Chemists would not make much progress if they were at liberty to define the element "carbon" in any way they wish. So, to facilitate fruitful communication among the legions of investigators in Ischemic Heart Disease, the American Heart Association formed a committee. In 1994, the committee took over 8 pages of fine-print and a list of 213 references to present and discuss "A Definition of Initial, Fatty Streak, and Intermediate

Lesions of Atherosclerosis" (Stary 1994 in our Reference List). We quote from the first page (Stary 1994, p.840):

"In this report, we characterize lesions that precede and may initiate the development of advanced atherosclerotic lesions. Advanced lesions are defined as those in which an accumulation of lipid in the intima is associated with intimal disorganization and thickening, deformity of the arterial wall, and often with complications such as fissure, hematoma, and thrombosis. Advanced lesions may produce symptoms, but the lesions that precede them are clinically silent." And (Stary 1994, p.840):

"The precursors of advanced lesions are divided into three morphologically characteristic types. Both type I and II lesions represent small lipid deposits in the arterial intima, and type II includes those lesions generally referred to as fatty streaks. Type III represents the stage that links type II to advanced lesions."

The advanced, trouble-making lesions, also classified by the American Heart Association, are types IV, Va, Vb, Vc, and VI. (Stary 1995 in our Reference List).

● Part 3. Characteristics of the "Typical" Lesions

Despite the variety of atherosclerotic lesions, the literature is filled with some very good attempts to describe "typical" or "usual" lesions. Below, we present a few, in approximately chronological order. With some workers stressing one aspect more than another, the combined descriptions cover the topic well. It may be worth noting that the word "athero" itself is Greek for gruel or porridge. According to legend, pathologists who saw a resemblance between such food and many arterial lesions, produced the name "atherosclerosis."

3a. 1976, Earl P. Benditt's Description of a Commonly Seen Plaque

Earl P. Benditt is a professor of pathology (now emeritus) at the University of Washington School of Medicine in Seattle. He has provided a very good description of the chief morphological changes which occur with atherosclerosis development. We quote (Benditt 1976, p.96):

"The form of the atherosclerotic lesions most commonly seen in vessels of humans at autopsy is a smooth-surfaced mass raised above the level of surrounding nonatherosclerotic vascular intima. Such raised lesions which vary in color from pearl gray to faint yellow-gray, are, on histological examination, composed of cells embedded in dense extracellular connective tissue. Selective stains identify collagen as the main connective tissue fibrillar constituent and elastin usually as a minor or variable constituent. Glycosaminoglycans [GAG] are present in varying amounts in the extracellular matrix. These histological findings are confirmed by electron microscopy. The electron microscope reveals, furthermore, that the cells embedded in and responsible for the production of the extracellular matric substances have properties that identify them as smooth muscle (Haust 1960, + Geer 1961, + Doud 1964)." And (Benditt 1976, p.96):

"Lipid stains and electron microscope examination show that the smooth muscle cells usually contain very little fat, except in the deeper layers of the plaques. In some plaques, lipid is present, and substantial amounts of cholesterol can be found both in the deeper layers of the plaque and underlying the plaques. Smooth muscle cells in the deeper layers of plaques, adjacent to the atheromatous debris, frequently exhibit pathologic fatty changes, indicative of cell injury, and cells in the deepest layers frequently appear to be dying and disintegrating." And (Benditt 1976, p.96):

"All of the plaque mass lies in the intima beneath the endothelium and entirely on the luminal side of the internal elastica of the artery. These masses are usually sharply raised (but not sharply demarcated) from the adjacent intima. However, some features of the cellular mass and associated extracellular matrix set it off from the underlying media: Cells of plaques appear to be smaller than cells of normal media. The main extracellular material produced by plaque cells is collagen, whereas elastin is a major item in the aortic media. The arrangement of cells of the plaque lacks the order found in the arterial media, and the number of intercellular junctions appears to be reduced ..." And lastly (Benditt 1976, p.96):

"The impressive feature of the human lesion is the presence of an excessive and apparently useless mass of cells that resemble in many respects, but differ in the subtle ways indicated, the cells and structures that comprise the media of arteries."

We should emphasize that the "apparently useless mass of cells" and the "atheromatous debris" are by no means INNOCUOUS. This useless debris constitutes important parts of the atherosclerotic process which almost always underlies the clinical occurrence of Ischemic Heart Disease.

3b. 1977, Elspeth B. Smith on the Range of Characteristics

Dr. Elspeth B. Smith was (in 1977) in the Department of Chemical Pathology, University of Aberdeen in Scotland. Smith comments (Smith 1977, p.669):

"The plaque with its central pool of atheroma lipid is obviously heterogeneous, and its chemistry changes from region to region ..." And (1977, p.672-673): "The most typical gelatinous lesion has loosely packed, thick, linear collagen bundles lying between rather sparse smooth muscle cells. However, one encounters a complete spectrum of lesion morphology and biochemistry, extending from extremely loose lesions that seem to contain pools of plasma insudate (Haust 1971) and have an LDL [low-density lipoprotein] content five to six times greater than adjacent normal intima, to mounds of densely packed smooth muscle cells with a lower than normal LDL content which frequently have not accumulated lipid (Smith 1976). Are these originally the same lesions that have developed differently, or do the gelatinous lesions have a different origin from the pure smooth-muscle-cell proliferation ... ?"

Here is probably an example of the "snapshot" phenomenon (Part 2b). Whenever pathologists study "snapshots" of a disorder which is progressing through days, weeks, months, and many years, the "snapshots" may be very difficult to convert into the real sequence of what preceded what (or what occurred independently), and into an explanation of how nature accomplished it.

3c. 1977 & 1993, Russell Ross on Features of Atherosclerotic Lesions

The late Dr. Russell Ross was in the Department of Pathology, University of Washington School of Medicine, in Seattle. Ross and co-workers state (Ross 1977, p.676):

"Three principal events are associated with the formation of the lesions of atherosclerosis. These are: a) intimal proliferation of smooth muscle cells, b) formation by these cells of large amounts of connective tissue matrix including collagen, elastic fibre proteins, and proteoglycans, and c) deposition of intracellular and extracellular lipid that eventually results in the formation of a pool of lipid and cell debris in the deeper portion of the more extensive or complicated lesions." For comparison, we also provide a description by Ross in 1993 (Ross 1993, p.801):

"The earliest recognizable lesion of atherosclerosis is the so-called 'fatty streak,' an aggregation of lipid-rich macrophages and T lymphocytes within the innermost layer of the artery wall, the intima ... Animal observations have shown that fatty streaks precede the development of intermediate lesions, which are composed of layers of macrophages and smooth muscle cells and, in turn, develop into the more advanced, complex, occlusive lesions called fibrous plaques. The fibrous plaques increase in size and, by projecting into the arterial lumen, may impede the flow of blood. They are covered by a dense cap of connective tissue with embedded smooth muscle cells that usually overlays a core of lipid and necrotic debris. The fibrous plaques contain monocyte-derived macrophages, smooth muscle cells and T lymphocytes, many of which are activated ..." Later in the same paper, Ross takes note of another fact (Ross 1993, p.804):

"During the most advanced stages of atherogenesis as the lesions become thicker, fibrous plaques become vascularized and contain large numbers of capillary and venule-like channels."

3d. 1988, Munro and Cotran on a "Typical Cellular Plaque"

J. Michael Munro and Ramzi S. Cotran were (in 1988) in the Departments of Pathology at Brigham and Women's Hospital and Harvard Medical School, when they presented an "overreview" on the pathogenesis of atherosclerosis (Munro 1988). In their paper, they state (Munro 1988, p.250):

"Plaques exhibit histologic variability, but a typical cellular plaque consists of: a fibrous cap, composed mostly of smooth muscle cells with a few leukocytes, and relatively dense connective tissue containing elastin, collagen fibrils, proteoglycans, and basement membrane (Kramsch 1971, + Murata 1986, + Yla-Herttuala 1986); a cellular area beneath and to the side of the cap consisting of a mixture of macrophages, smooth muscle cells, and T lymphocytes (Jonasson 1986); and a deeper 'necrotic core' which contains cellular debris, extracellular lipid droplets, cholesterol crystals, and calcium deposits. This necrotic core often contains numerous large foam cells of both the macrophage and smooth muscle type. [Foam cells are cells whose cytoplasm looks "foamy" due to the presence therein of lipids.] Finally, one can sometimes see, particularly around the periphery of the lesions, proliferating small blood vessels, which are evidence of neo-vascularization from the direction of the adventitia. The relative content of fibrous tissue and lipid within a plaque is variable; coronary artery lesions are often largely fibrous."

3e. 1995, Peter Libby on the "Typical" Atherosclerotic Plaque

Peter Libby was (in 1995) in the Department of Medicine's Vascular Medicine and Atherosclerosis Unit, at Brigham and Women's Hospital in Boston, when he wrote (Libby 1995, p.2845):

"... pathological studies have shown repeatedly in humans and in experimental animals that over much of its history, growth of an atherosclerotic plaque occurs by outward, abluminal [away from the lumen] expansion (Clarkson 1994, + Glagov 1987, + Armstrong 1989). Hence, most obstructive plaques may pass through a phase that may last many years or even decades of so-called 'remodeling' without encroaching on the arterial lumen. Only after the plaque burden approaches half of the luminal area does the plaque usually protrude into the lumen, becoming visible by angiography and capable of impairing flow." Blankenhorn and Hodis also emphasize this point (Blankenhorn 1994, p.178):

"The arterial wall undergoes compensatory remodeling as lesions form (Glagov 1987) in response to changing shear stress on the endothelium (Stary 1992). As a result, extensive atherosclerosis can be present before lesions intrude into the vessel lumen (Glagov 1987)." We return now to Libby (Libby 1995, p.2845):

"Atherosclerotic plaques typically consist of a lipid-rich core in the central portion of the eccentrically thickened intima. The lipid core is bounded on its luminal aspect by a fibrous cap, at its edges by the 'shoulder' region, and on its abluminal aspect by the base of the plaque. The central, lipid-rich core of the typical lesion contains many lipid-laden macrophage foam cells derived from blood monocytes. Once resident within the arterial wall, these cells imbibe lipid, which accounts for their foamy cytoplasm."

Of course, it is the natural function of macrophages to attempt to engulf any "foreign" material --- any material which is "out of place" in the body.

● Part 4. A Recent Warning about Defects in Clinical Evaluations

David H. Blankenhorn was, at the time he gave the George Lyman Duff Memorial Lecture for the American Heart Association in November 1992, at the Atherosclerosis Research Institute of the University of Southern California School of Medicine. Soon thereafter, he became very ill with recurring prostate cancer. Before his death in May 1993, Dr. Blankenhorn requested Howard N. Hodis to complete the work entitled "Arterial Imaging and Atherosclerosis Reversal," from which we quote (Blankenhorn 1994).

Among the many important points in their paper, Blankenhorn and Hodis emphasize that the ability to DETECT whether atherosclerotic disease is progressing or regressing, as the result of various therapies, is critical to improvement of treatments (p.178). "Lesion staging is necessary for studies of progression and regression ... scales used to grade atherosclerosis from images are rudimentary. Angiography [an xray procedure] measures the internal vessel lumen, and we infer lesion severity from the extent of lumen lost."

They explain that the most widely used measure is "%S" --- meaning percent stenosis (narrowing) at the narrowest point in a vessel, compared with the diameter of a "normal" segment. And (p.178): "Clinicians typically stage coronary artery lesions on a scale of 3: complete occlusion, >= 50%S, and <50%S." A "high-grade" stenosis is regarded as >= 70%S.

The warning given by Blankenhorn and Hodis is that, not only is angiography unable to detect progression or regression for all the plaques which do not intrude into the lumen, but that too much emphasis is placed on the most stenotic lesions: "Accumulating evidence indicates that acute clinical events [myocardial infarction, unstable angina, sudden ischemic death] result from instability of small, lipid-rich plaques rather than large, fibrotic, calcified, stenosing plaques (Fuster 1992). Although large plaques tend to progress to total occlusion more frequently than small plaques, occlusion by large plaques infrequently results in acute clinical events because of the formation of collateral vessels."

They argue that more progress will be made in reducing IHD mortality when angiography is supplemented by some additional techniques which can also measure the pre-intrusive atherosclerotic lesions in study-populations and in the population at large.

>>>>>>>>>>

Our intention in this chapter is to describe some concepts which have influenced how we think about Hypothesis-2 --- especially about its second part, which will be the focus of Chapters 45 and 46.

Although it is an over-simplification, we will say that historically, scientists working on the problem of coronary atherosclerosis have tended to belong to two "schools." The "Endothelial Injury" school regards injury by various agents, to the endothelial layer of the coronaries, as the first step in the whole process. The "Lipid" school regards entrance of certain types of lipoproteins, into the intima from the circulating blood, as the first step in the whole process. Both schools are consistent with an inflammation model of atherogenesis, and so we begin with a few words about inflammation.

● **Part 1. The Inflammatory Response and the Role of Smooth Muscle Cells**

Inflammation is a basic response of the body to local injury. The exact reaction varies according to the injurious agent and other specifics. The general function of the inflammatory response is to bring plasma proteins and white cells to the injured site, so that foreign bodies can be destroyed or isolated, and the site prepared for repair. In other words, overlap occurs between "the inflammatory response" and various "immune-system responses." Repair --- generally accompanied by local proliferation of collagen-secreting fibroblasts --- is the final stage of the inflammatory process. Under some circumstances, total repair is not possible, and the final stage of the inflammatory response is containment of the problem (for instance, by formation of an abcess or of a granuloma). Calcium-deposition and growth of a new blood supply at the site are also frequent features of the inflammatory response.

1a. Smooth Muscle Cells in the Arterial Inflammatory Response

The similarity between the inflammatory process and atherogenesis was noticed as early as 1823 by P. Rayer (Rayer 1823), according to Munro and Cotran (Munro 1988, p.249). Munro and Cotran point out (Munro 1988, p.254):

"Inflammation outside the arteries is often characterized by proliferation of fibroblasts, which then deposit collagen and other substances. In the arterial intima, it is the smooth muscle cell which serves this role. Smooth muscle cell proliferation is a critical event in the development of an atheromatous plaque. Smooth muscle cells are present in fatty streaks, although they are overshadowed by macrophage-derived foam cells; in the fibrous cap of the well-developed fibro-fatty plaque, they appear to be the predominant cell type. It is evident that the smooth muscle cells synthesize the bulk of the connective tissue matrix, including collagen, elastic tissue, and the proteoglycans (Kramsch 1971, + Murata 1986, + Yla-Herttuala 1986). In addition, they can accumulate lipid and become foam cells."

It should be noted here that smooth muscle cells are multifunctional cells, capable of modulating their function, as needed, from the function almost exclusively of contraction, to the function almost exclusively of synthesis of elastic fibres, mucopolysaccharides, and myosin (Campbell 1988).

In their very interesting paper, Munro and Cotran review the several "biochemical signals which underlie smooth muscle proliferation" (p.255), and conclude (p.256) that "... there is no dearth of mechanisms to account for smooth muscle proliferation in the vessel wall in atherosclerosis."

1b. A Marker for Inflammation Found to Predict Myocardial Infarction

In 1997, Paul M. Ridker and co-workers published the results of a study of 1,100 apparently healthy men from whom a single base-line measurement of plasma C-reactive protein had been made years earlier. "C-reactive protein is an acute-phase reactant that is a marker for underlying systemic inflammation" (Ridker 1997, p.973).

These workers found a positive and significant dose-response: "The relative risk of first myocardial infarction increased significantly with each increasing quartile of base-line concentrations of C-reactive protein (P for trend across quartiles, <0.001), in such a way that the men in the highest quartile had a risk of future myocardial infarction almost three times that among those in the lowest quartile (relative risk, 2.9; 95 percent confidence interval, 1.8 to 4.6; P<0.001)." And (Ridker 1997, p.978): "C-reactive protein is not simply a short-term marker of risk, as has previously been demonstrated in patients with unstable angina (Liuzzo 1994), but is also a long-term marker of risk, even for events occurring six or more years later." Ridker also reports positive findings from a smaller female study-sample (Ridker 1998).

Within an excellent update on systemic responses to inflammation, Gabay and Kushner comment (Gabay 1999, p.452) that such findings "may reflect the presence of low-grade inflammation in coronary arteries or elsewhere, or alternatively, they may reflect pro-inflammatory or pro-thrombotic effects of C-reactive protein itself [two references]." Elsewhere (p.451), they offer an important reminder that inflammatory responses can have both beneficial and detrimental aspects.

● Part 2. The "Endothelial Injury" and the "Infection" Hypotheses

Among investigations within the "Endothelial Injury" school of atherogenesis, attention has naturally focused quite often on identifying what may injure the endothelium in the first place. We will begin with a list of candidates from Ross 1977, and then focus briefly on one of them: Infectious agents. Infection as a cause of atherosclerosis was suggested by William Osler (Osler 1908) and others, and has been under consideration for almost a century.

2a. Initial Injurant: List of Suspects from Russell Ross et al

In the 1970s, Russell Ross and his colleagues in the Department of Pathology at the University of Washington School of Medicine, produced a series of often-cited papers on the "Endothelial Injury" hypothesis (Ross 1973 + 1976 + 1977) --- also called "Response to Injury" hypothesis. The Ross 1977 paper is well summarized by its abstract (Ross 1977, p.675):

"We postulate that the lesions of atherosclerosis arise as a result of some form of 'injury' to arterial endothelium. This injury somehow results in alteration in endothelial cell-cell attachment or endothelial cell-connective tissue attachment, so that forces such as those derived from the shear in the flow of blood result in focal desquamation [tearing away] of endothelium. This is followed by adherence, aggregation, and release of platelets at the sites of local injury." And: "During the process of release, a mitogenic factor [an agent which stimulates cell-division] is secreted from the platelets which, together with plasma constituents, gains entry into the artery wall, resulting in focal intimal proliferation of smooth muscle cells. This intimal proliferation is accompanied by the synthesis of new connective tissue matrix proteins and often by the deposition of intracellular and extracellular lipids."

In the text, Ross lists all of the following as "potential sources of injury" to the intima's endothelial cells and as contributors to "their desquamation" (Ross 1977, p.676): "Chronic hyperlipidemia (Ross 1976-b), various chemical factors such as homocystine (Harker 1974; Harker 1976), uremia, metabolites, infections, immunologic injury (Minick 1973; Hardin 1973), and mechanical factors (Ross 1973, + Stemerman 1972, + Bjorkerud 1971, + Moore 1973, + Helin 1971). Mechanical injury may occur at particular anatomic sites as a result of the increased shear stress applied to the endothelial cells from the flow of the blood at these sites (Glagov 1972, + Caro 1973)."

Ross's Update in 1993

In 1993, updating the "response-to-injury hypothesis of atherosclerosis," Russell Ross writes (Ross 1993, p.801): "Atherosclerosis is not merely a disease in its own right, but a process that is the principal contributor to the pathogenesis of myocardial and cerebral infarction, gangrene and loss of function in the extremities. The process, in normal circumstances a protective response to insults to the endothelium and smooth muscle cells of the wall of the artery, consists of the formation of fibrofatty and fibrous lesions, preceded and accompanied by inflammation. The advanced lesions of atherosclerosis, which, when excessive, become the disease and which may occlude the artery concerned, result from an excessive inflammatory-fibroproliferative response to numerous different forms of insult." And (Ross 1993, p.804):

"Central to the response-to-injury hypothesis (Ross 1986 + 1973 + 1976 + 1976-b + 1981) is the proposal that the different risk factors somehow lead to endothelial dysfunction, which can elicit a series of cellular interactions that culminate in the lesions of atherosclerosis. The initial injurious events do not necessarily lead to endothelial denudation. In fact, with better modes of tissue preservation, it is clear that the early lesions develop at sites of morphologically intact endothelium (Faggiotto 1984 + 1984-b; Masuda 1990 + 1990-b; Rosenfeld 1987 + 1987-b)."

2b. Some Recent Insights about the Endothelium

In 1997, Parmley noted (p.11) that "The endothelial cells lining vessels such as the coronary arteries were once thought to be relatively inert, functioning primarily as a barrier between the blood and the underlying vessel wall." Those days are past. Munro 1988 comments (p.251):

"Recent research on vascular endothelium, sparked to a considerable degree by the response to injury hypothesis, has established endothelial cells as active cells which perform a variety of critical functions." Both Parmley and Munro list such functions (and Munro provides numerous references). We will use a similar list of functions from Ross (Ross 1993, p.804):

(1) provision of a nonthrombogenic surface; (2) a permeability barrier through which there is exchange and active transport of substances into the artery wall; (3) maintenance of vascular tone by release of small molecules such as nitric oxide, prostacyclin, and endothelin that modulate vasodilation or vasoconstriction, respectively; (4) formation and secretion of growth-regulatory molecules and cytokines; (5) maintenance of the basement membrane collagen and proteoglycans upon which they rest; (6) provision of a nonadherent surface for leukocytes; and (7) ability to modify (oxidize) lipoproteins as they are transported into the artery wall." And (Ross 1993, p.804-805):

"Changes in one or more of these properties may represent the earliest manifestations of endothelial dysfunction. Evidence has accumulated to suggest that oxLDL [oxidized Low-Density Lipoprotein] is a key component in endothelial injury (Cathcart 1985; Rosenfeld 1990; Steinbrecher 1990; Steinberg 1991). Once formed by the endothelium, oxLDL may directly injure the endothelium and play an initial role in the increased adherence and migration of monocytes and T lymphocytes into the subendothelial space ..."

Parmley offers the opinion (Parmley 1997, p.11) that "It is clear that the endothelial cells have a major role in determining vascular reactivity, the potential for thrombosis, and perhaps, represent the key element in the development of atherosclerosis."

Variable permeability of the endothelial layer to solutes in the bloodstream --- to lipoproteins, for instance --- is an issue to which we return in Part 5 of this chapter.

2c. Injury by Viral Agents: The 1990 Concept of Melnick et al

The hypothesis, that viruses may play a causal role in Ischemic Heart Disease, has been around for quite a while --- with work centered on the herpes simplex virus and the cytomegalovirus (see, for example, Fabricant 1978, + Gyorkey 1984, + Adam 1987, + Hajjar 1988, + Yamashiroya 1988, + McDonald 1989).

In 1990, Joseph L. Melnick, Ervin Adam, and Michael E. DeBakey, at the Baylor College of Medicine in Houston, Texas in 1990, produced a "special communication" in the Journal of the

American Medical Association in which they reviewed existing evidence on this hypothesis (Melnick 1990). Near the outset of their review, they state (Melnick 1990, p.2204):

"Recent work in this area has been stimulated by the finding that atherosclerotic lesions, strikingly similar to those in human disease, were reproducibly induced in chickens by an avian herpesvirus ... Our studies at Baylor College of Medicine (Melnick 1983, + Petrie 1987, + Adam 1987, + Gyorkey 1984, + Petrie 1988) were stimulated by the avian model (Fabricant 1978). We looked for evidence of involvement of cytomegalovirus (CMV) and herpes simplex virus (HSV) in human arterial tissue. These agents belong to the same herpesvirus family. They are very common and cause primary infections in infants and young children in every country ..." On the same page, Melnick et al note that "... in the United States, the prevalence of antibodies to CMV is about 10% to 15% in the adolescent population, rises to more than 50% by age 35 years, and exceeds 70% in adults older than 64 years." Describing part of their own work, Melnick et al write (Melnick 1990, p.2204):

"In the arterial samples we studied, we were unable to detect markers of HSV infection. However, CMV-infected cells were detected in samples taken from a number of lesions taken from patients with atherosclerosis and also in biopsy specimens of apparently uninvolved aorta taken from such patients. Infectious virus was not recovered, but CMV antigen and viral DNA were detected. In addition, epidemiologic studies were conducted in which viral antibodies were measured in patients with atherosclerosis and in matched control patients. High levels of antibodies to CMV were associated with clinically manifest atherosclerotic disease."

The body of their paper goes into detail about the research at Baylor, and describes related findings at the University of Washington, the University of Illinois, the University of Limburg, Stanford University, and the University of Minnesota. Near the end of their paper, they conclude (Melnick 1990, p.2207):

"There is solid evidence that a member of the herpesvirus family can cause atherosclerosis in chickens ... The evidence for involvement of one or more members of the herpesvirus family in human atherosclerosis is much more circumstantial, but it is increasing. The findings of CMV and HSV antigens and nucleic acid sequences in arterial smooth muscle cells suggest that virus infection of the arterial wall may be common in patients with severe atherosclerosis. Although certainly suggestive, these findings by themselves do not demonstrate a role of viruses in the pathogenesis of atherosclerosis, but they lead to an attractive working hypothesis of the steps involved (Table 3). Of special importance are the recent findings that heart transplant recipients who are immunosuppressed and become infected with CMV are particularly prone to develop severe atherosclerosis in the transplanted organ." (They refer to Grattan 1989 and to McDonald 1989.)

The Melnick Recommendation

The final sentence in Melnick 1990 is: "... the studies reviewed herein should provide a basis for further investigation of the role of viruses in human atherogenesis and of their control by means of vaccination or chemotherapy" (p.2207).

Speir et al on Inhibition of Normal p53 Protein by a CMV Product

In 1994, Edith Speir, of the National Institutes of Health, and several co-workers published work entitled "Potential Role of Human Cytomegalovirus and p53 Interaction in Coronary Restenosis" (Speir 1994). The p53 protein (encoded by the p53 gene) is known as a "tumor suppressor" because one of its normal functions is to activate a gene which blocks a cell from completing division. The abstract of Speir 1994 is short (Speir 1994, p.391):

"A subset of patients who have undergone coronary angioplasty develop restenosis, a vessel renarrowing characterized by excessive proliferation of smooth muscle cells (SMCs). Of 60 human restenosis lesions examined, 23 (38 percent) were found to have accumulated high amounts of the tumor suppressor protein p53, and this correlated with the presence of human cytomegalovirus (HCMV) in the lesions. SMCs grown from the lesions expressed HCMV protein IE84 and high amounts of p53. HCMV infection of cultured SMCs enhanced p53 accumulation, which correlated temporally with IE84 expression. IE84 also bound to p53 and abolished its ability to transcriptionally activate a reporter gene. Thus, HCMV, and IE84-mediated inhibition of p53 function, may contribute to the development of restenosis."

We note that Speir and co-workers established to their satisfaction that the p53 protein in their positive restenosis samples was normal (Speir 1994, p.392). In other words, they are not proposing that the virus is mutating the p53 gene which codes for the p53 protein. They are proposing that the p53 protein is disabled by binding to the IE84 protein produced by the re-activated virus. Comments on their paper and related work include Epstein 1994, + Finkel 1995, + Zhou 1996, + Smith 1997, + Kol 1997, + Zhou 1997.

Speir and co-workers also examined specimens from 20 primary atherosclerotic lesions, taken from atherectomy patients who had never had angioplasty (Speir 1994, p.391). None of the 20 primary specimens were found p53-immunopositive (p.392). Their work as of 1994 appears not to be applicable to the pathogenesis of primary atherosclerotic lesions.

2d. Injury by Bacterial Agents: Chlamydia and Helicobacter Pylori

During the past decade, multiple investigating teams have looked at a variety of Chlamydia bacteria and Helicobacter pylori as possible atherogens. For example:

Thom et al, 1992
In 1992, Thom et al reported on a case-control study designed to evaluate "the association between prior infection with Chlamydia pneumoniae, as measured by IgG antibody, and coronary artery disease" (Thom 1992, p.68). To qualify as a CAD patient, individuals had to have "at least one coronary artery lesion occupying 50% or more of the luminal diameter" (p.68), ascertained by diagnostic coronary angiography. A positive association was found only in cigarette smokers (Thom 1992, p.68, p.71, p.72). Association has been reported, too, by Davidson (Davidson 1998).

Mendall et al, 1994
In 1994, Mendall et al reported on their pilot study, undertaken "to determine whether Helicobacter pylori, a childhood acquired chronic bacterial infection, is associated with an increased risk of coronary heart disease later in life" (Mendall 1994, p.437). In this case-control study of white males, Mendall et al found a positive association between documented Coronary Heart Disease (CHD) and seropositivity for H-pylori-specific IgG antibodies (Odds Ratio = 2.15, p = 0.03), after adjustment for other variables. They speculate that "H pylori could influence the development of CHD through the systemic effects of chronic active inflammation in a large viscus" or by infection-induced alterations in levels of fibrinogen and apolipoprotein-a (p.438-439). They conclude (Mendall 1994, p.439): "The association between H pylori and CHD reported here needs confirmation by other more rigorously controlled epidemiological studies ... A causal link between H pylori and CHD would be of major public health importance because the infection can be eradicated with a single course of antibiotics with little chance of reinfection."

Patel et al, 1995
In 1995, Patel, Mendall and co-workers reported their findings from a larger study of white males "to investigate the relation between seropositivity to chronic infections with Helicobacter pylori and Chlamydia pneumoniae and both coronary heart disease and cardiovascular risk factors" (Patel 1995, p.711). They found the Odds Ratios for electrocardiographic evidence of ischemia or infarction to be 3.82 (95% confidence interval 1.60 to 9.10) and 3.06 (1.33 to 7.01) "in men seropositive for H pylori and C pneumoniae, respectively, after adjustment for a range of socioeconomic indicators and risk factors for coronary heart disease ... Possible mechanisms [for the relationship] include an increase in risk factor levels due to a low grade chronic inflammatory response" (Patel 1995, p.711).

Patel 1995 reports, "Cardiovascular risk factors that were independently associated with seropositivity to H pylori included fibrinogen concentration and total leukocyte count. Seropositivity to C pneumoniae was independently associated with raised fibrinogen and malondialdehyde concentrations" (p.711). The authors conclude (Patel 1995, p.714): "We have shown an association of the prevalence of coronary heart disease with potentially treatable infections which are common in the general population. Our figures imply that between one third and one half of current coronary heart disease in this population was statistically attributable to either or both infections. What are required now are well conducted prospective studies and eradication trials to evaluate the causal relation of these infections to haemostatic function, progression of atherosclerosis, and cardiovascular morbidity and mortality." A related study is reported by Gupta (Gupta 1997).

Muhlestein et al, 1996
In 1996, Muhlestein et al reported the results from their study in which they tested for "an association between Chlamydia and atherosclerosis by comparing the incidence of the pathogen found within atherosclerotic plaques in patients undergoing directional coronary atherectomy with a variety of control specimens and comparing the clinical features between the groups ... Coronary specimens from

90 symptomatic patients undergoing coronary atherectomy were tested for the presence of Chlamydia species using direct immunofluorescence. Control specimens from 24 subjects without atherosclerosis (12 normal coronary specimens and 12 coronary specimens from cardiac transplant recipients with subsequent transplant-induced coronary disease) were also examined" (Muhlestein 1996, p.1555). The abstract states the results well (Muhlestein 1996, p.1555):

"Coronary atherectomy specimens were definitely positive in 66 (73%) and equivocally positive in 5 (6%), resulting in 79% of specimens showing evidence for the presence of Chlamydia species within the atherosclerotic tissue. In contrast, only 1 (4%) of 24 nonatherosclerotic coronary specimens showed any evidence of Chlamydia. The statistical significance of this difference is a p-value <0.001. Transmission electron microscopy was used to confirm the presence of appropriate organisms in three of five positive specimens. No clinical factors except the presence of a primary non-restenotic lesion (odds ratio 3.0, p=0.057) predicted the presence of Chlamydia. CONCLUSIONS. This high incidence of Chlamydia only in coronary arteries diseased by atherosclerosis suggests an etiologic role for Chlamydia infection in the development of coronary atherosclerosis that should be further studied."

Toward the end of Muhlestein 1996 (p.1559), the authors acknowledge that Chlamydia might be present in the lesions only as an "innocent bystander, finding fertile ground to grow within the diseased atherosclerotic wall." They reason that "it should also be found within the walls of arteries diseased by processes other than atherosclerosis," and they argue that its ABSENCE "within diseased coronary segments of transplanted hearts appears to increase the likelihood that Chlamydia plays an active role in the pathogenesis of natural atherosclerosis." Returning to this topic (on p.1560), they state: "It is still possible that atherosclerotic plaque is merely a more fertile ground for Chlamydia to be deposited and grow. If this were the case, the presence of the pathogens would be a result rather than a cause."

Meier et al, 1999

In early 1999, Meier et al reported findings consistent with an association between certain bacterial infectious agents and risk of acute myocardial infarction (Meier 1999). Commenting in the same JAMA issue on the hypothesis of a causal role for bacterial infections in myocardial infarction, Folsom writes (Folsom 1999, p.461): "Hypothesized mechanisms by which infection could cause vascular disease include direct infection of the arterial wall, systemic infection leading to endotoxin injury of the endothelium, autoimmunity ..., systemic inflammation, or increases in inflammatory mediators, such as C-reactive protein, fibrinogen, and white blood cell count (Danesh 1997, + Libby 1997)."

● Part 3. The "Lipid Hypothesis": Gradual Acceptance of a Causal Role

Almost wholly, lipids (including free cholesterol, esterified cholesterol, fatty acids, triglycerides, and phospholipids) are transported in the bloodstream by combination with proteins --- hence, "lipoproteins."

In contrast to the "Response to Injury Hypothesis," the "Lipid Hypothesis" of Atherogenesis proposes that the entrance of certain types of plasma lipoproteins, into the intima from the circulating blood, is the initial step in the whole atherosclerotic process, and that atherosclerosis develops more quickly and more seriously when blood concentrations of these lipoproteins are high than when they are low. The lipoproteins have no physiologic function between the endothelium and the internal elastic membrane. They are "out of place," and unless they exit rapidly while soluble, they become a "foreign substance" which elicits efforts at removal or isolation --- the inflammatory response.

3a. What Is Our Goal with Respect to the Lipid Hypothesis?

During 1949-1950, our group at the Donner Laboratory (University of California, Berkeley) showed that numerous species of lipoproteins exist (Part 3b, below). The immediate question became: "Which lipoprotein species (if any) have causal roles in atherosclerosis?" In terms of the Lipid Hypothesis, one requirement was that the offending lipoproteins must be able to penetrate into the intima. The implication: Size would probably matter.

With respect to disease, a prime reason to STUDY causation is to PRACTICE prevention. Therefore, our group had many questions:

● (A) Which lipoprotein species (if any) are atherogenic, and can we identify ALL of the independently atherogenic species? (Parts 3h and 3i). If one misses half of the independently atherogenic species, one may miss a large share of the opportunities to prevent the disease.

● (B) Can reducing the concentration of SOME lipoprotein species, in the blood, cause elevation in the plasma concentrations of OTHER species on a steady-state basis? The answer is yes (Nichols 1957, + Gofman 1958).

● (C) Do certain regimes, which successfully reduce the blood-levels of some ATHEROGENIC species, elevate levels of OTHER atherogenic species? By 1957, our evidence --- from altered diets --- provided an affirmative answer (Appendix F). We continue to worry that some medically-prescribed low-fat, high-carbohydrate diets may even produce a net INCREASE in atherosclerosis risk for a possibly large segment of the population.

● (D) Are some other regimes converting NON-atherogenic lipoproteins into ATHEROGENIC species? For example, do some substances which seem to reduce the blood-measurement called "total triglyceride" do so by hydrolyzing a component of a non-atherogenic species --- and does this action produce a net increase in the blood-concentrations of atherogenic species? This question deserves resolution.

● (E) Is it possible to rank the atherogenic types of lipoproteins by the degree of their menace? My own opinion: Not yet.

3b. What We Never Expected: Such a Variety of Plasma Lipoproteins

By 1949, chemical analysis had divided lipoproteins into two parts: Alpha and Beta (Cohn 1946, + Oncley 1947, + Gurd 1949). But in 1950, there was only "limited interest in plasma lipoproteins," as Donald Fredrickson has reported (Fredrickson 1993, p.III-1).

In 1950, my colleagues and I wrote (Gofman 1950-b, p.161): "For many years, it has been suspected that the blood lipids might in some way be related to the pathogenesis of human atherosclerosis. All the major blood lipid constituents, including cholesterol and its esters, phospholipids and neutral fats, have been investigated without leading to definitive conclusions. Cholesteremia itself has received much attention..." And (Gofman 1950-b, p.161):

"... the low relationship manifested between the extent of atherosclerosis and serum cholesterol level has thus cast some doubt in the minds of many investigators upon the significance of serum cholesterol level, per se, in the pathogenesis of this disease."

The inconclusive, and ostensibly conflicting, results which existed, regarding the relationship between blood lipids and human atherosclerosis, were partly or largely a result of measuring only the lipid components of lipoproteins (Gofman 1950-a, p.171, p.186). Prior to 1949, the extensive diversity of the lipoprotein molecules was not known.

That situation changed abruptly in 1949 and 1950. At the Donner Laboratory, my colleagues and I were able to discover that Beta lipoprotein really consisted of a great many species, distinct in size, density, weight, and chemical composition. Not only do the lipid components vary, but the protein components vary too. We were entering a new world of diverse lipoproteins (Gofman 1949, + Lindgren 1950, + Gofman 1950-a, + Gofman 1950-b, + Lindgren 1951). Some views on these breakthroughs include Fredrickson 1993, + Gofman 1996-b (FASEB's "Milestones" series).

A Series of "Giant Molecules" : Boxes 1 and 2.

Entry into this new world occurred thanks to two ultracentrifuges (the analytic and the preparative), engineered and made available to humanity through the genius of Dr. Edward G. Pickels (Pickels 1942 + 1943; Gofman 1990-b). The variety of lipoproteins could be visualized in the optical diagrams made during the ultracentrifugation of serum (or plasma). By adjusting the density of the solution, we were able to segregate various classes of molecules from each other, in nearly their native states. The optical (schlieren) diagrams were useful in quantifying the concentration of various lipoprotein classes in the circulating blood plasma.

We were not expecting to find the large number of species of lipoproteins, all undergoing flotation in a solution of density, 1.063 grams per milliliter. Consider going from "Beta lipoprotein" to at least 9 discrete species and possibly as many as 100 or more either discrete species or a continuum of species. We initially referred to the entire complex of species as "Low-Density Lipoproteins (LDL)," because all underwent flotation in a solution of 1.063 gms/ml. As the work progressed, a number of separate classes of Low-Density Lipoproteins were identified, differing from each other either by their flotation rates (in a solution of density 1.063 grams/ml) or by their buoyant density (Boxes 1 and 2).

We described the lipoproteins as a series of macro-molecules ("giants"), because even the smallest ones are very large (Box 1).

3c. The Sf Unit: A Major Lipoprotein Identifier

For the lipoproteins of density < 1.063 gms/ml, it was found that the larger the lipoprotein size, the lower is the density. The very largest and least dense lipoproteins are called the chylomicrons.

Density and size together determined the speed of migration, occurring against centrifugal force in a solution of density 1.063 gms/ml --- with speed measured in Svedberg flotation units, or Sf units (Gofman 1950-a, p.168). The Svedberg unit, and the Standardized Svedberg Sf unit (Std Sf unit), are defined in Box 3 of this chapter.

The lower the density of the lipoprotein, the faster the migration (Box 2). At equal molecular size, the higher the density of the lipoprotein, the slower the migration.

Flotation rates, for the lipoproteins of density < 1.063 gms/ml, range from near Sf 0 up through approximately Sf 40,000 (the chylomicrons). The following entities are commonly discussed in the literature: Sf 0-12, Sf 12-20, Sf 20-100, and Sf 100-400. Combinations are also commonly discussed, e.g., the plasma concentration of Sf 0-20 (the combined concentration of Sf 0-12 and Sf 12-20 lipoprotein classes), or Sf 20-400 (the combined concentration of Sf20-100 and Sf 100-400 lipoprotein classes). And a particular investigator might use other cutting points, for reasons indigenous to his or her own studies.

These divisions of flotation rates are not arbitrary. Each group embraces lipoproteins which generally share some distinct chemical, biochemical, and metabolic properties (Box 1, bottom). Moreover, flotation rates are very highly correlated with the relative size of these molecules, in their native state. The body itself is not likely to "overlook" the DISTINCTIONS, in size and in the various size-correlated properties of these lipoproteins, in their native state. So we, too, heed the distinctions, as we endeavor to learn which lipoproteins are atherogenic and which are not (Part 3h).

3d. LDL, IDL, and VLDL, with Their Corresponding "Sf" Ranges

Currently, Sf values are combined under the following names (Krauss 1987, p.62):

- LDL: Std Sf 0-12 lipoproteins are described as "Low-Density Lipoproteins."
- IDL: Std Sf 12-20 lipoproteins are described as "Intermediate-Density Lipoproteins."
- VLDL: Std Sf 20-400 lipoproteins are described as "Very-Low Density Lipoproteins."

The lipoproteins of still higher flotation rates are also referred to as "Very-Low Density" by some investigators. We suggest that the use of flotation rates rather than the term "very-low density," for these Std Sf 400-40,000 classes, will prevent potential confusion with the Std Sf 20-400 lipoproteins.

3e. HDL: High-Density Lipoproteins and "Small, Dense LDL Particles"

The smallest and most dense lipoproteins are called High-Density Lipoproteins: HDL-1, HDL-2, HDL-3. Their hydrated densities (below) are from DeLalla 1954-b, p.333.

HDL-2 (density 1.075 g/ml) and HDL-3 (density 1.145 g/ml) belong roughly in the category formerly called "Alpha" lipoprotein (Lindgren 1951, p.83). Some unesterfied fatty acids are bound in protein complexes of even higher density than the HDL-3 lipoproteins. Because HDL-2 and HDL-3 sink (rather than float) in a salt solution of 1.063 g/ml, they acquired the name "High-Density" Lipoproteins.

HDL (HDL 2 + 3) is often called "the good cholesterol" in popular media, because of its purported protective effect against atherosclerosis. Appendix-G discusses that claim.

HDL-1 lipoproteins have a density (1.05 g/ml) just below the density of a salt solution of 1.063 g/ml. The flotation rate of HDL-1 in such a solution is the lowest encountered in our work. On the basis of flotation in such a solution, HDL-1 lipoproteins are definitely LOW-density lipoproteins (approximately Sf 0-2), and they make a small contribution to the measured plasma concentration of the Sf 0-12 lipoproteins. Nonetheless, they acquired the name "HDL-1" because their own plasma concentration is best ascertained if they are centrifuged along with the High-Density Lipoproteins > 1.063 g/ml. Quantitative recovery of HDL-1 for analysis requires this special handling (Appendix-H, Parts 2 + 4b).

Currently, much attention is directed at the HDL-1 lipoproteins, under another name: Small, dense LDL particles. Appendix-H discusses the claim that they are especially atherogenic.

3f. Discrete Entities, vs. a Continuum

The High-Density Lipoproteins appear to be discrete or nearly discrete macromolecular species. In the Low-and-Intermediate Density spectrum (Std Sf 0-20), extensive studies by Frank Lindgren, Alex Nichols, and Norman Freeman led to identification of numerous nearly discrete entities on a density scale and flotation-rate scale. For lipoproteins with flotation rates above Std Sf 20, it is difficult to speak to the issue of discrete macromolecular entities versus a continuum of entities (Jones 1951, p.359). Between approximately Sf 400 and Sf 40,000, there is an apparent continuum, and as the flotation rates progressively increase, the densities progressively decrease. The largest molecules might be compared with protein sacks which can be filled with variable amounts of lipid. Triglyceride is the dominant cargo.

3g. "Cholesterol-Rich" vs. "Triglyceride-Rich" Lipoproteins: Box 1

Cholesterol is a constituent of all species of HDL, LDL, IDL and VLDL, but the amount of cholesterol per macromolecule, and its status as free or esterified cholesterol, differs extensively among the macromolecular lipoprotein entities. In addition to free or esterified cholesterol, the triglycerides (a variety of different ones), phospholipids, and fatty acids are constituents of the various lipoprotein entities, in varying amounts (Box 1, bottom).

The Std Sf 0-20 lipoproteins are commonly referred to as the "major cholesterol-bearing lipoproteins," or the "cholesterol-rich lipoproteins." And the Std Sf 20-400 lipoproteins are the "major triglyceride-bearing lipoproteins," or "triglyceride-rich lipoproteins." These are valid characterizations, but there is room for ambiguity with respect to the term triglyceride-rich. The lipids of the chylomicrons (not of the Sf 20-400 lipoproteins) are the richest in triglycerides, when richness is expressed as fraction of the lipid constituents within a macromolecule. However, in those blood samples where chylomicrons are scarce, the Sf 20-400 lipoproteins are the main source of the measured levels of triglyceride.

In general, in moving along the lipoprotein spectrum from Sf 400 toward Sf 0, the proportion of triglyceride versus other lipids per macromolecule decreases, and the proportion of cholesterol versus other lipids per macromolecule increases (Jones 1951, p.360, + Gofman 1952-c, p.282). As a result, high levels of TOTAL serum-cholesterol or TOTAL serum-triglyceride are, by themselves, uninformative about WHICH of the lipoproteins are elevated. Measurements which destroy the molecules, without first measuring them in their physiologically NATIVE state, do not identify the species (Lindgren 1951, p.80).

3h. The Atherogenic Lipoproteins, and the Non-Atherogenic Ones

Many of the historical and quantitative details, of how we come to know what we know (and what we do not yet know) about the Lipid Hypothesis, are provided in Appendices E, F, G, H, and I of this book.

Back in 1949-1950, we soon realized that it would be a monumental task to sort out any relationship of plasma "lipids" with atherosclerosis, because initially there was no overt reason to exclude ANY part of the vast lipoprotein spectrum --- fatty acid-protein complexes, HDL, Sf 0-40,000 lipoproteins.   Appendix-E provides details about some of the studies which have led to our conclusion that the following lipoprotein classes have independent causal roles in atherogenesis:

- Standard Sf 0-12 lipoproteins.
- Standard Sf 12-20 lipoproteins.
- Standard Sf 20-100 lipoproteins.
- Standard Sf 100-400 lipoproteins, with the caveat that the upper limit for atherogenicity may lie well below Sf 400. The evidence of independent atherogenicity is weaker for the Sf 100-400 range than for the three other ranges. The relative weakness may reflect an upper limit on the size of blood lipoproteins which can penetrate into the intima. Or it may reflect the greater variability in blood-level measurements for Sf 100-400, compared with Sf 0-100 (Appendix E, Part 3). Or both.

Our studies did not make a direct endeavor to ascertain any atherogenic properties of Sf 400-Sf40,000 lipoproteins. It is likely that atherogenicity for these classes is limited by physical size. There are some who might not agree that these very large lipoproteins fail to gain access to the arterial intima. At this time, we consider the evidence to be otherwise. But we are, of course, always open to new evidence.

The following lipoprotein classes are non-causal in atheroclerosis:

- High-Density Lipoproteins-Two (HDL-2).
- High-Density Lipoproteins-Three (HDL-3).

HDL-1, and its purported special atherogenicity, are separately addressed in Appendix-H.

Our Livermore Lipoprotein Study at 10 years provided the first prospective evidence that the HDL-2 and HDL-3 lipoproteins might be inversely related to atherogenesis (Gofman 1966, pp.686-687). However, we are unable to say that this represents evidence for a protective role against atherosclerosis. The strong inverse relationship in cross-sectional studies, of the HDL(2+3) with the atherogenic Std Sf 0-20 and 20-400 lipoproteins (data in Appendix G), is undoubtedly the cause of some of, and maybe most of, the observed APPARENT protective effect.

Some have argued that a protective role for HDL-2+3 is biologically plausible, but we are not impressed yet by that argument. We remain skeptical that the (HDL-2 + HDL-3) lipoproteins truly do provide a protective effect against atherogenesis (Appendix-G).

3i. Acceptance for Part of the Lipid Hypothesis: From "Gofman's Heresy" to Current Wisdom?

Our concept, that certain classes of lipoproteins have a CAUSAL role in atherosclerosis (and that others do not) met with considerable skepticism. It was even suggested that anything related to cholesterol could not be deadly, because the human body needs and synthesizes cholesterol on its own, regardless of diet.

Gradually, as other workers also came to the conclusion that the primary cholesterol-bearing low-density lipoproteins are causally associated with Coronary (Ischemic) Heart Disease, this part of the Lipid Hypothesis became generally accepted --- not universally, of course. By 1983, in the Lyman Duff Memorial Lecture, Henry C. McGill, Jr., commented: "Gofman's heresy, that some lipoproteins are atherogenic and others are not, is now dogma" (McGill 1984, p.443). In 1993, the former head of the National Institutes of Health, Donald S. Fredrickson, wrote: "Today, there is no quarrel over the measurement of LDL rather than total serum cholesterol as the more powerful indicator of coronary risk" (Fredrickson 1993, p.III-3).

By contrast, there is not yet general acceptance of our findings that the triglyceride-rich Sf 20-100 and some of the Sf 100-400 lipoproteins are also atherogenic. The lack of acceptance does not reflect an inability by others to replicate our findings. Instead, debate has been raging for 30 years over whether, or not, "triglyceride" is atherogenic. The controversy is linked with the erroneous assumption that, in studies of atherogenesis, a chemical triglyceride measurement is the equivalent of an ultracentrifugal measurement of Std Sf 20-100 lipoproteins, in their nearly native states. This error has persisted for decades, and is only now being corrected hesitantly in the medical literature. We comment on the "triglyceride" controversy (and why it is important to resolve it) in Part 4.

● Part 4. High-Carbohydrate Diets and Triglycerides: A Lipoprotein Pitfall?

In 1957-1958 (Nichols 1957, + Gofman 1958), we showed that a high-carbohydrate diet can INCREASE steady-state blood-levels of the "triglyceride-rich" lipoproteins --- Std Sf 20-400. Results are shown in Appendix-F. This finding is consistent with the fact that the metabolism of carbohydrates ("sugar") and the metabolism of lipids are closely linked. This linkage was not as well known in the past as it is today. A dramatic reminder, related to diabetes mellitus, is briefly provided in Part 4a.

4a. Intersection of Carbohydrate Metabolism and Lipid Metabolism

Diabetes mellitus is simultaneously a disease of carbohydrate metabolism and lipoprotein metabolism. In 1952, Dr. Felix Kolb, Dr. Oliver DeLalla, and I were lucky to have the opportunity to learn what happened to the spectrum of lipoproteins circulating in the blood of a diabetic patient in severe acidosis and nearly in coma, as the patient moved from essentially full decontrol of diabetes mellitus toward control of the crisis, by treatment with electrolyte-balancing and insulin.

Our observations are reported in the journal Metabolism (Kolb 1955) and are also presented in Box 3.

What occurred were orderly, non-random changes in the blood-concentrations of lipoprotein-classes. Initially (June 19, 1952), the plasma was exceedingly cloudy, almost certainly due to high concentrations of lipoproteins in the Sf 400 to chylomicron range. In the Sf 0-400 range, concentrations the Sf 100-400 lipoproteins were massively elevated: 3,739 mg per 100 ml. The dominant class in terms of concentration changed progressively over time to smaller lipoprotein molecules with lower triglyceride content. As the concentration of one class decreased, the concentration of progressively smaller, more dense lipoprotein molecules increased. And as these declined in concentration, the concentration of the even smaller macromolecules increased. The columns in Box 3 show clearly how it occurred.

We witnessed the essence of the steps in lipid metabolism, in relation to the change from decontrolled to controlled carbohydrate metabolism. As Box 3 indicates at the bottom, we also witnessed, on the patient's skin, the disappearance of the "eruptive" xanthomas --- lesions which are filled with lipids.

Of course, we do not imply that diabetics in acidosis are a model for the general population --- but we do emphasize that linkage, between carbohydrate metabolism and blood-levels of various lipoproteins, occurs within the general population, too.

4b. High-Carbohydrate Diets and the Unsettled "Triglyceride" Issue

For decades, high-carbohydrate diets have been commonly recommended as "heart healthy." At the same time, we know that such diets elevate blood-levels of "triglyceride-rich" lipoproteins (Std Sf 20-400 lipoproteins), for some segment of such dieters --- perhaps for most of them (Appendix F). A high-carbohydrate diet may turn out to be optimal for some people, but very unhealthy for others. Therefore, an urgent obligation exists finally to SETTLE the question:

Do elevated levels of some, most, or all, of the Std Sf 20-400 "triglyceride-rich" plasma lipoproteins make independent, causal contributions to atherogenesis?

In the next attempts to answer this question, the first principle should be to measure the levels of the lipoproteins at issue, instead of measuring something ELSE --- namely, the levels of triglyceride. Measurements of serum triglyceride-levels are simply NOT informative about the serum levels of the Std Sf 20-100 lipoproteins and the separate serum levels of the Std Sf 100-400 lipoproteins --- because EVERY lipoprotein in the spectrum, from HDL to chylomicroms, can contribute triglyceride to such a measurement (Box 1; also Appendix-E, Boxes 1 + 2). Under these circumstances, it is difficult to explain how anyone could expect to learn WHICH groups of serum lipoproteins are atherogenic, by studying serum-levels of "total triglyceride."

4c.   A "Shortcut" Which Fogs the Lens: A Self-Inflicted Handicap

Long ago, perhaps due to the time and cost involved in ultracentrifugal analysis of an individual's lipoprotein patterns, ultracentrifugal analysis of blood was virtually discarded in favor of chemical analysis. This is the "shortcut" which produces measurements like "total cholesterol," "total triglyceride," and a variety of chemical ratios. Clearly, "total cholesterol" and "total triglyceride" are measurements which do not reveal the distribution of the various lipoproteins which were carrying those lipids in the bloodstream.

To our amazement, the recent decades have produced study after study using only chemical determination of triglycerides, with no indication what percentages of the measurement come from (Std) Sf 0-20, Sf 20-100, Sf 100-400, and Sf 400-40,000 lipoproteins. Substituting "total triglyceride" measurements for ultracentrifugal analysis of serum lipoprotein patterns, in research into the atherogenicity of particular segments of the lipoprotein spectrum, has been like fogging the lens through which investigators look at such questions. It is a self-inflicted handicap which could be remedied at will.

No wonder that, 45 years beyond opening of the world of lipoproteins, contradictory findings exist on whether OR NOT elevated serum levels of "total triglyceride" (hypertriglyceridemia) are an independent cause of Ischemic Heart Disease, or just a "marker" for a real cause (for instance, see Bierman 1992, p.647, + Blankenhorn 1994, p.185, + Avins 1989, 1997, + Hokanson 1996, + Enas 1998, + Lamarche 1998). In 1996, succinctly describing the confusion, Meir Stampfer and co-workers wrote (Stampfer 1996, p.882; with the capitalization from the original):

"WHETHER TRIGLYCERIDE level is an independent risk factor has been controversial for many years."

4d.   Why It Is Important to Learn the Atherogenicity of Every Lipoprotein Class

Why is it important firmly to establish either the atherogenic potency or the protective potency of every class of lipoprotein?

● First, the knowledge is essential in order to guide progress in limiting Ischemic Heart Disease, by modifying the steady-state serum-levels of such classes. If the "triglyceride-rich" Std Sf 20-400 lipoproteins are atherogens and we fail to recognize them as such, then a major opportunity to limit IHD is ignored --- an abhorrent situation.

● Second, if the Std Sf 20-400 lipoproteins are atherogens and we fail to recognize them as such, then the possibility of causing net HARM with specific types of intervention is real (Part 3a) --- an intolerable development. We think that evidence of atherogenicity in the Std Sf 20-100 segment of the spectrum was already strong more than 30 years ago, and that the health of a great many people may have been harmed by failure to resolve the doubts.

It is heartening to observe a growing sense of urgency, expressed in the literature, about the need to settle the so-called "triglyceride" issue.

● Part 5.   Healthy Intima: Balanced Entry and Exit of Lipoproteins

Many of us who were working on the atherosclerosis problem in the 1950s considered it plausible that plasma lipoproteins filtered into the intima by penetration through an INTACT endothelial layer. The problem, as we saw it then, was getting the lipoproteins out of the arterial wall, before they underwent denaturation with loss of solubility. Wei Young and I wrote in 1963, "Clearly, a major effort is still required to understand the fate of lipoproteins in the region BETWEEN the endothelium and the internal elastic membrane" (Gofman 1963, p.217).

Major efforts were made and are still being made. Many are described in the 1988 review by Munro and Cotran. Today, workers on blood lipoproteins (for example, Young 1994) show prominent enthusiasm over the possibility that oxidation of lipoproteins, either before or after entry into the arterial intima, may be the key change in low-density lipoproteins to make them atherogenic or more atherogenic. (See also Ross 1993, p.805; + Fuster 1994, p.2128; + Libby 1995, p.2849.) Whether the oxidation story is correct or not, the lipoprotein-atherosclerosis causal relationship itself is unaffected. It has withstood plenty of scrutiny for a half-century --- because it is scientifically solid.

5a. Concept of a Constant Passage of Fluid and Solutes

In 1963, in contrast to claims that injured endothelium must be required to initiate atherosclerosis, Wei Young and I built on earlier work by others to describe a model for ingress and egress of lipoproteins with respect to the intima (Gofman 1963, p.198):

"Anitschkow 1933, following many years of highly productive investigation of atherosclerosis, held the opinion that the normal artery experiences a constant passage of fluid through its wall in the direction from arterial lumen to adventitia. Further, some of the constituents of such fluid were considered to pass on through, along with the fluid vehicle, thus leading to no pathological consequence whatever. Other constituents of the fluid vehicle ... tend to REMAIN WITHIN the wall of the artery ... Atheroma, it was reasoned, developed, its extent being conditioned by (a) the nature of the substances remaining behind in the arterial wall, and (b) the over-all 'responsiveness' of the arterial tissue to these substances. During the last decade, opportunity for critical evaluation of the Anitschkow concept has greatly increased."

Much of the work involved the passage of materials to extravascular spaces from the interior of intact CAPILLARIES (Pappenheimer 1953, + Kellner 1954, + Courtice and Garlick 1962). Courtice and Garlick found evidence supporting the concept that the transfer-rate of lipoprotein molecules across the capillary wall varied with variation in the size of such molecules. As of 1963, the mechanism of macromolecule transport across normal endothelium was still speculative. We wrote (Gofman 1963, p.208):

"It has been suggested that such passage may occur (a) via junction regions between endothelial cells (Chambers and Zweifach, 1947), (b) via some of the 'pores' which constitute part of the endothelial lining (Pappenheimer 1953), or (c) via vesicles of the endothelial cells, revealed by electron microscope studies (Palade 1956, + Bennett 1956) as an active transport process." As for egress of the plasma lipoproteins from the intima, we speculated (Gofman 1963, p.200):

"Presumably, in health, the fenestra provide an adequately competent path for removal of fluid and solutes, including such giant solutes as lipoprotein macromolecules. If, for any reason or set of reasons, the transfer of lipoproteins be impeded through such fenestra RELATIVE to their influx via the endothelial lining, the stage would be set for an accumulation of lipoproteins within the subintimal space (i.e., between the endothelium and internal elastic membrane). The entrapped lipoproteins could result in several of what are probably later manifestations of fully developed atherosclerotic lesions."

Below, we will refer to a balance of ingress and egress as the "equilibrium concept."

5b. Elspeth Smith on (a) Normal Endothelium and (b) Impeded Egress

By the 1970s, evidence was appearing which supports the concepts that (a) plasma lipoproteins pass from the lumen into the intima through uninjured endothelium, and (b) imbalance between ingress and normal egress is the key to lipid accumulation in the intima. For instance, Elspeth B. Smith, in the Department of Chemical Pathology at the University of Aberdeen (Scotland) was investigating these issues. Smith wrote (Smith 1977, p.672):

"Normal endothelium is PERMEABLE to plasma macromolecules even in healthy young animals (Bell 1974-a + 1974-b), but no appreciable amount seems to ACCUMULATE in the absence of diffuse intimal thickening (Smith 1974). Gelatinous lesions are characterized by a massive increase in the content of plasma macromolecules, but we do not know if they are initiated by increased endothelial permeability, so that the rate of entry exceeds the rate of clearance, or by a primary change in the retentiveness of the intima." And earlier in the same paper (Smith 1977, p.667):

"Both our steady-state concentration data (Smith 1974) and permeability data (Bell 1974-a + 1974-b) suggest that a rather constant volume of whole plasma enters normal intima. This concept of a PACKAGE OF PLASMA is also supported by the constancy of the relation between the concentrations of plasma proteins in intima when expressed in terms of plasma volumes (Smith 1976, + Smith 1975)." And (Smith 1977, p.667):

"In intima there is much greater accumulation of LDL [low-density lipoprotein] than of albumin relative to their concentrations in plasma (Smith 1974, + Smith 1972); together with the concept of a package of plasma, this suggests that the steady-state concentrations may be largely determined by rate of egress of the macromolecules, and there has been much speculation that glycosaminoglycans (GAG) are involved in retention of LDL either by formation of specific ionic complexes (Tracy 1965, + Bihari-Varga 1967, + Srinivasan 1972) or by molecular sieving in the GAG gel (Laurent 1963, + Iverius 1973)." The 1977 Smith paper also discusses other processes which may cause irreversible precipitation of lipoproteins in the intima, or may split the lipids off the proteins.

5c. Some Current Wisdom on Endothelial Injury: Munro, Ross, Brown

The concept that extracellular lipid in atherosclerotic plaques arrives within the intima from the lumen, without endothelial injury, is now clearly accepted by workers such as Munro and Cotran (Munro 1988, p.257, Figure 2), Ross, and Brown. In Ross's current model (Ross 1993, p.804-805), the endothelial cells have the "ability to modify (oxidize) lipoproteins as they are transported into the artery wall" (Ross 1993, p.804). See also Part 2a, above.

B. Greg Brown and colleagues, referring to mature plaques, comment (Brown 1993, p.1785): "Lipid may enter the core region of the fibrous plaque by transmural flux (Fry 1987) of its more mobile forms --- lipoprotein particles, droplets, and vesicles (Guyton 1985, + Smith 1967) --- or it may be deposited there during foam cell necrosis (Stary 1987, + Haust 1971)."

5d. Some Current Wisdom on Equilibrium: Fry, Munro, Fuster

With respect to the equilibrium concept, the concept is fundamental to an interesting mathematical model presented by Donald L. Fry in "Mass Transport, Atherogenesis, and Risk" (Fry 1987). Fry's model allows him to "play" with various factors which are thought to affect the balance between ingress and egress of atherogens in the intimal wall: "Endothelial gap fractional area" (increased endothelial permeability), elevated blood pressure, elevated serum concentration of the atherogen, internal elastica fenestration, and pre-existing intimal thickening. It is worth noting that his model, like some earlier ones, predicts that an "extraordinary increase in the apparent diffusive permeability of the endothelial surface ... occurs with only slight opening of the endothelial junctions" (Fry 1987, p.93).

Munro and Cotran explicitly endorse the equilibrium concept with respect to cholesterol in the atherosclerotic lesions (Munro 1988, p.257): "Cholesterol accumulation in the plaque should be viewed as reflecting imbalance between influx and efflux ..." The equilibrium concept is also embraced by Valentin Fuster, for instance, in his Lewis A. Conner Memorial Lecture at the American Heart Association's 1993 National Meeting (Fuster 1994). Discussing the progression of atherosclerosis, from early lesions to the troublesome lesions, he states (Fuster 1994, p.2127):

"The potential for clinical problems, however, begins when the process continues. That is, if the influx of lipids and/or accumulation into the vessel wall continues and is more significant than their efflux, then the process evolves with a continuously slow progression of these lesions."

● Part 6. Evidence of Fewer Acute IHD Events after Lipid-Lowering

In contrast to chronic Ischemic Heart Disease of various degrees, the ACUTE syndromes or "events" are myocardial infarction (not always fatal), unstable angina, and ischemic sudden death.

The causal role of the low-density lipoproteins in the acute IHD events is apparently affirmed by recent demonstrations that dramatically fewer acute IHD events occur after various regimes which lower the serum levels of LDL cholesterol and total triglyceride. Several lipid-lowering trials are referenced in Box 4.

Such evidence has caused expert panels in the United States (ExpertPan 1993) and in Europe (TaskForce 1994) to recommend use of regimes to reduce serum levels of LDL cholesterol "as one of the fundamental preventive measures to reduce mortality from coronary heart disease" (Pedersen 1995, p.1350). In the same editorial of the New England Journal of Medicine, Pedersen adds (p.1351): "The benefits of reducing cholesterol are now established beyond any reasonable doubt." A year later, in a review article entitled "Advances in Coronary Angioplasty," John W. Bittl ends by recommending "intensive efforts to lower lipid levels" (Bittl 1996, p.1300).

But dissent remains. Box 4 includes some critics of the Lipid Hypothesis --- such as Uffe Ravnskov (1991, 1992, 1993, 1994, 1995-a, 1995-b), who suggests that the positive clinical benefits are due to some properties of the regimes OTHER THAN their lipid-lowering properties.

Of all the lipid-lowering studies in Box 4, we shall describe only three here. All three involve large randomized trials with a "statin" drug. Simvastatin and pravastatin are HMG-CoA reductase inhibitors (HMG-CoA: 3-hydroxy-3-methylglutaryl co-enzyme A).

6a. Scandinavian Simvastatin Survival Study (4S), 1994

The first is the "4S Study" --- the Scandinavian Simvastatin Survival Study, published in the Lancet (Scandinavian 1994).

In this randomized double-blind study, 2,223 patients were in the placebo group and 2,221 were in the simvastatin group (p.1384). The patients enrolled were patients (81% males; 51% age 60+) with angina pectoris or previous myocardial infarction and initial serum cholesterol levels in the range 5.5 to 8.0 mmol/L on a lipid-lowering diet. All the baseline characteristics of the two groups are shown in the paper's Table 1. The median follow-up period was 5.4 years. The primary endpoint of the study was total mortality. The secondary endpoint was major coronary events, "which comprised coronary deaths, definite or probable hospital-verified non-fatal acute MI [myocardial infarction], resuscitated cardiac arrest, and definite silent MI verified by electrocardiogram" (p.1384).

"Over the whole course of the study, in the simvastatin group, the mean changes from baseline in Total, LDL, and HDL cholesterol, and serum triglycerides, were -25%, -35%, +8%, and -10% respectively. The corresponding values in the placebo group were +1%, +1%, +1%, and +7% respectively" (p.1385).

Results (p.1385). The relative risk of death from any cause was 0.70 with simvastatin (p = 0.0003). "The relative risk of coronary death was 0.58 (95% confidence interval, 0.46 to 0.73) with simvastatin. This 42% reduction in the risk of coronary death accounts for the improvement in survival." And (p.1387): "With the exclusion of silent MI, the risk of coronary death plus nonfatal MI was reduced by 37% over the whole study, by 26% in the first 2 years, and by 46% thereafter." And (p.1388): "The improvement in survival produced by simvastatin was achieved without any suggestion of an increase in non-CHD mortality, including deaths due to violence and cancer, which have raised concern in some overviews of cholesterol-lowering trials (Oliver 1992, + Smith 1992, + Muldoon 1990, + Rossouw 1990, + Ravnskov 1992)."

6b. West of Scotland Coronary Prevention Study (WOSCOP), 1995

"WOSCOP" (the West Scotland Coronary Prevention Study, published in the New England Journal of Medicine) is entitled "Prevention of Coronary Heart Disease with Pravastatin in Men with Hypercholesterolemia" (Shepherd 1995).

In this randomized, double-blind study, 3,293 men were in the placebo group and 3,302 in the pravastatin group (p.1302). The patients enrolled were men (average age, 55 years) "with moderate hypercholesterolemia and no history of myocardial infarction" (p.1301). All the baseline characteristics of the two groups are shown in the paper's Table 1. The average follow-up period was 4.9 years. The primary endpoint of the study was the occurrence of nonfatal myocardial infarction or death from coronary heart disease as a first event (p.1302).

"When the data were analyzed according to the treatment actually received, pravastatin was found to have lowered plasma levels of cholesterol by 20 percent, LDL cholesterol by 26 percent, and triglycerides by 12 percent, whereas HDL cholesterol was increased by 5 percent. There were no such changes with placebo" (p.1304).

Results (p.1303, Table 2). Pravastatin produced a 31% reduction in the risk of definite nonfatal myocardial infarction and CHD death (95% confidence interval, 17 to 43 percent; p < 0.001). For definite CHD death by itself, there was a 28% reduction (95% confidence interval of -10 to 52; p = 0.13). And (p.1306) "the benefit of pravastatin therapy with respect to fatal coronary events and the absence of any increase in the number of deaths from other causes led to a 22 percent reduction in the relative risk of death from any cause (p = 0.051)." The pravastatin group and placebo group showed no significant difference during the study in deaths from cancer, suicide, or trauma (p.1305). In their discussion, the authors also comment (Shepherd 1995, p.1306):

"The relative reductions in risk attributable to pravastatin therapy were not affected by age (<55 years vs. >= 55 years) or smoking status. Furthermore, a significant treatment effect was seen in the subgroup without multiple risk factors and the subgroup without pre-existing vascular disease. Thus, it is possible to conclude that in the subjects who might be considered to fall strictly into the primary prevention category, pravastatin therapy produced a significant reduction in the relative risk of a coronary event."

6c. Cholesterol and Recurrent Events Study (CARE), 1996

"CARE" (the Cholesterol And Recurrent Events Study, published in the New England Journal of Medicine) is entitled "Effect of Pravastatin on Coronary Events after Myocardial Infarction in Patients with Average Cholesterol Levels" (Sacks 1996).

In this study, 2,078 patients (290 females) were in the placebo group, and 2,081 (286 females) in the treatment group. The median follow-up time was 5 years. After treatment, comparison of the treated group with the placebo group showed decreases in the pravastatin group of 20% in total cholesterol, 28% in LDL cholesterol, 14% in triglycerides, and an increase of 5% in HDL cholesterol (Sacks 1996, p.1003). Coronary deaths were reduced 20% (P = 0.10), fatal plus nonfatal coronary events were reduced by 24% (P = 0.003), and fatal plus nonfatal strokes were reduced by 31% (P = 0.03) (from Sacks 1996, p.1003, Table 2). Hebert and co-workers (Hebert 1997, pp.318-319) suggest that the reduction in stroke may be due to the reduction in myocardial infarction, a risk-factor for stroke. The 4S study found a significant 28% reduction in stroke in the simvastatin-treated patients, too.

A Notable Result in the CARE Study: Excess Breast Cancer

Hebert and co-workers did a meta-analysis of lipid-lowering trials involving statin therapy ALONE (Hebert 1997). They looked at fatal cancers and total cancers in the 13 trials which reported these events, and found a relative risk in the treated groups combined of 0.95 for cancer deaths, and 1.03 for all cancer. The confidence intervals suggest probably no difference between treated and placebo groups (Hebert 1997, p.320, Table 8).

But notable is that, in the CARE Study, there were 12 cases of BREAST cancer (all nonfatal) in the pravastatin group, compared with only 1 case (a fatal recurrence) in the placebo group of equal size. Of the 12 cases in the pravastatin group, "3 occurred in patients who had previously had breast cancer, 1 was ductal carcinoma in situ, and 1 occurred in a patient who took pravastatin for only six weeks" (Sacks 1996, p.1006).

The authors write (Sacks 1996, p.1007): "In evaluating this finding, it should be noted that although there was one case of breast cancer among the women given placebo, five cases would have been expected on the basis of the rate of breast cancer in the general population for women of similar race and age." In other words, the small-numbers problem may account for much of the difference. We note that the same study showed an excess of colo-rectal cancer in the placebo group (21 cases) compared with the pravastatin group (12 cases). The authors also report (Sacks 1996, p.1007):

"Importantly, interim results of the Long-Term Intervention with Pravastatin in Ischemic Disease trial (LIPID 1995), from four years of treatment of 1,508 women, show no increase in breast cancer (Barter P., Safety and Data Monitoring Committee, LIPID Study; personal communication).

The totality of evidence suggests that these findings [excess breast cancer] in the CARE trial could be an anomaly ..." Let us hope so. Additional follow-up is imperative.

6d. Large Clinical Benefits, but Small Decrease in Stenosis

Today, with respect to acute IHD events, B. Greg Brown and colleagues have provided evidence that the large clinical benefits, from lowering the serum low-density lipoproteins, are associated with only minimal reduction of stenosis in the angiographically measured plaques. Brown et al analyzed results from nine lipid-lowering trials, all of which demonstrated a benefit from treatment, and conclude (Brown 1993, p.1784):

"As seen in Table 2, averaged estimates of disease severity per patient worsened (progressed) by about 3% stenosis among the control subjects, whereas they improved (regressed) by 1-2% stenosis among the treated patients. When these results are expressed in terms of absolute change in arterial narrowing, they appear to be remarkably small." In view of such evidence, various analysts reason that lipid-lowering must achieve its benefits by "stabilizing" atherosclerotic plaques (for example, Fuster 1994, p.2138, + Libby 1995, p.2849). Parmley comments (Parmley 1997, p.12):

"As one looks at all of the lipid-lowering studies where quantitative coronary arteriography has been used to judge efficacy, it is clear that there are only minor regressive changes in lipid lesions over a several year period. On the other hand, there is a dramatic reduction in clinical events, suggesting that somehow, lipid-lowering has stabilized the plaque so that it is not as easy to break down and cause unstable angina or acute myocardial infarction."

The recent observations, of major health benefits from only minor reductions in degree of stenosis, resemble a prediction we made in 1969 about Coronary Heart Disease: "A marked drop in CHD risk and incidence might be achieved if degree of coronary atherosclerosis is kept to values just below the steeply rising portion of the CHD risk curve" (Gofman 1969, p.38). Our view was based on our analysis of preliminary results from the International Atherosclerosis Project (Strong 1966), from which we had concluded (Gofman 1969, pp.37-38) that the curve --- relating risk of Coronary Heart Disease with extent of coronary atherosclerosis --- must be sigmoid rather than linear (x = degree of atherosclerosis, y = risk of IHD).

While the analysis in 1969 correctly predicted large benefits from only small reductions in degree of atherosclerosis, we did not imagine then where some of the explanation might lie --- a topic discussed in Part 7.

● Part 7. Plaque Rupture: Proximal Cause of a Large Share of IHD Deaths

Today, many leading figures in IHD research regard thrombus formation DUE TO PLAQUE RUPTURE OR FISSURE as the proximal cause of an important share of acute IHD events, including death. See, for instance: Davies 1985, p.364, p.370, + Brown 1993, p.1785, p.1787, + Nicod 1993, p.1749, + Blankenhorn 1994, p.178, + Fuster 1994, p.2129, + Libby 1995, p.2848, + Parmley 1997, p.12, + Burke 1999, Table 1, p.922. Plaque erosion, instead of rupture, may characterize sudden cardiac death in premenopausal women (Burke 1998). In addition, there are investigators who regard plaque ruptures which are nonfatal --- and which are followed by fibrotic organization of the rupture-induced thrombus and its incorporation as part of the plaque --- as the explanation for the sometimes very rapid and unpredictable growth, angiographically, of specific plaques. (See, for instance: Davies 1985, p.366, + Brown 1993, p.1785, + Fuster 1994, p.2126, p.2127, 2128, 2132, + Libby 1995, p.2844).

Awareness of plaque rupture, fissure, ulceration, and erosion, is not new, and papers by both Davies 1985 and Fuster 1994 include fine historical context.

Below, we begin by quoting from a particularly lucid presentation by Peter Libby, M.D., on the important role of plaque rupture in the acute IHD events.

7a. Reassessment of a "Central Dogma in Clinical Cardiology": Severe Stenosis

"From Bench to Bedside" is the mini-title for the paper entitled "Molecular Bases of the Acute

Coronary Syndromes," by Peter Libby, M.D., of the Vascular Medicine and Atherosclerosis Unit, Department of Medicine, at Boston's Brigham and Women's Hospital (Libby 1995). Near the outset, Libby states (Libby 1995, p.2844):

"Much of the basis of contemporary cardiology and cardiac surgery rests on the axiom: the greater the stenosis, the greater the risk of a clinical event such as myocardial infarction or unstable angina pectoris. However, data emerging from clinical and pathological studies over the past decade have occasioned a reassessment of this central dogma of clinical cardiology (Fuster 1994)." And (Libby 1995, p.2844):

"First, the use of thrombolytic therapy in acute myocardial infarction became widespread in the wake of the GISSI study in 1986 (Gruppo 1986). Angiographic studies performed after thrombolysis during acute myocardial infarction led to the surprising finding that the atherosclerotic lesions that gave rise to the occlusive thrombus did not cause high-grade stenoses in many cases. Assessment of angiograms obtained before acute myocardial infarctions and those obtained during the infarction corroborated this concept that the lesions most likely to precipitate an infarct-provoking thrombosis often did not appear highly stenotic by angiography (Hackett 1988, + Ambrose 1988, + Nobuyoshi 1991, + Giroud 1992)." And (Libby 1995, p.2844):

"Another line of clinical evidence suggested dissociation between the degree of stenosis and coronary events. In the past decade, a number of 'regression trials' tested the hypothesis that lipid-lowering regimens would reduce the degree of high-grade coronary stenoses. The angiographic results showed disappointingly minimal effects on established stenotic lesions. Yet these studies revealed a consistent and resounding decrease in acute clinical coronary events (Blankenhorn 1994, + Brown 1993)." And (Libby 1995, p.2844):

"Meanwhile, state-of-the-art pathological studies using perfusion fixation of freshly obtained material provided new evidence buttressing the concept that rupture of atherosclerotic plaques precipitates the formation of the occluding thrombus that causes acute myocardial infarction (Davies 1985). The elegant pathological studies of Davies and colleagues (Davies 1985 + 1990) also sought evidence for plaque disruption in hearts from patients dying of noncardiac causes. They documented evidence for plaque disruption in these patients even without overt symptoms of coronary disease or acute myocardial infarction. These results suggested that not all disruptions of atherosclerotic plaques lead to clinically apparent or symptomatic events. Such subclinical episodes of plaque disruption with local thrombin activation and subsequent healing may indeed represent a major pathway for progression of atherosclerotic lesions."

And Libby's Key Summary (Libby 1995, p.2844-45)

"Taken together, these new results suggest that while angiographically severe coronary artery disease clearly correlates with the propensity to develop or succumb from acute myocardial infarction, the presence of the severe stenoses may merely serve as a marker for the presence of angiographically modest or even inapparent, non-critically stenotic plaques more prone to precipitate acute myocardial infarction." And at this point, Libby reminds readers that plaques grow away from the lumen for a long time before they encroach on the lumen itself (see Chapter 43, Part 3e).

The evidence seems convincing that acute IHD events are most often precipitated by plaques associated (not long before the event) with mild to moderate degrees of stenosis, rather than with severe degrees of stenosis. (Some details from Brown 1993 are provided in Chapter 46, Part 3.) Among the figures who either generate or embrace such evidence are Ambrose 1986, 1988, + Brown 1986, 1993, + Little 1988, + Blankenhorn 1994, + Fuster 1994, + Libby 1995, + Bittl 1996, + Parmley 1997.

It is worth noting that the relationship of stenotic degree with acute IHD events is not identical with the issue of plaque rupture as the proximal cause of acute events (Parts 7b and 7c).

7b. "Thrombogenic Lipid-Cores" as Catastrophes-in-Waiting

Libby (1995, p.2845) reminds readers that the lipid-imbibing foam cells in the core of the "typical [atherosclerotic] lesion" can "produce large amounts of tissue factor, a powerful pro-coagulant that potently stimulates thrombus formation when in contact with blood (Wilcox 1989)." Fuster emphatically agrees (Fuster 1994, p.2133):

"Fernandez-Ortiz et al (1994) studied the thrombogenicity of various human atherosclerotic plaques, including fatty streaks (lesions types II and III), atheromatous plaques or lipid-rich plaques with abundant cholesterol crystals (lesions types IV and Va), and fibrotic plaques with collagen-rich matrix (lesions types Vb and Vc). The lipid core exposed in atheromatous lipid-rich plaques was the most thrombogenic, with thrombus formation fourfold to sixfold greater than that on all other substrates."

7c. Crucial Role of a Plaque's Fibrous Cap

Libby continues his exposition by explaining the crucial role of a plaque's fibrous cap (Libby 1995, p.2845):

"... the integrity of the fibrous cap overlying this lipid-rich core fundamentally determines the stability of an atherosclerotic plaque. Rupture-prone plaques tend to have thin, friable fibrous caps (Richardson 1989, + Loree 1992). Plaques not liable to precipitate acute myocardial events tend to have thicker caps that protect the blood compartment in the arterial lumen from potentially disastrous contact with the underlying thrombogenic lipid core ..." And (Libby 1995, p.2845):

"Biomechanical analyses demonstrate maxima of circumferential stress at sites of plaques prone to rupture (Richardson 1989, + Cheng 1993). Thus, mechanical forces concentrate on the fibrous cap, which must resist these high stresses to avoid rupture and the attendant risk of developing an acute coronary event. This stress-laden fibrous cap is all that stands between the blood and the thrombogenic lipid core of the lesion."

B. Greg Brown and colleagues report (1993, p.1789): "Fissuring is predicted by a large accumulation of core lipid in the plaque and by a high density of lipid-laden macrophages in its thinned fibrous cap. Lesions with these characteristics constitute only 10-20% of the overall lesion population but account for 80-90% of the acute clinical events."

According to Celermajer (1998, p.2014), Magnetic Resonance Imaging (MRI) may hold promise --- despite cardiac and respiratory motion --- for in vivo visualization of the fibrous cap, core, and other features which might permit identification of the rupture-prone plaques in coronary arteries, prior to rupture.

7d. Massive Shift in Thinking about Smooth Muscle Cells in Acute Events

We credit Libby with helping to create a massive shift in our own thinking --- and probably in the thinking of others --- about the role of smooth muscle cells in the acute and deadly events of IHD.

For instance, in his paper, "Molecular Bases of the Acute Coronary Syndromes," Libby identifies a molecule (interferon-gamma) which can inhibit smooth muscle cells in plaque from producing collagen. After describing several lines of evidence, Libby states (Libby 1995, p.2846):

"Taken together, these results concordantly suggest that chronic immune stimulation within atheroma leads to elaboration of interferon-gamma from T cells, inhibiting collagen synthesis in vulnerable regions of the plaque's fibrous cap. This mechanism provides a molecular explanation for impaired maintenance and repair of the collagenous meshwork in vulnerable plaques, rendering it [meshwork] weak and prone to rupture in the critical region of the plaque. Intact ability to synthesize collagen may sustain the ability of the fibrous cap to resist the concentration of mechanical forces in STABLE plaques." (Emphasis added.)

Libby's T-cell explanation for plaque rupture seems reasonable, but incomplete. Chronic immune stimulation and T cells are typically present in plaques, but only some plaques rupture. On the other hand, some contribution by T-cells to plaque rupture seems reasonable. We continue quoting Libby on another very important point (Libby 1995, pp.2846-2847):

"Curiously, proliferation of smooth muscle cells has dominated our thinking about the pathogenesis of atherosclerosis for decades. Smooth muscle cell growth may indeed contribute importantly to earlier phases of lesion development. Yet the present data, summarized above, suggested that the aspects of the biology of atheroma that actually lead to ACUTE clinical manifestations, depend on IMPAIRED smooth muscle cell growth and matrix elaboration, rather than the contrary." (Emphasis added.) And Libby warns (Libby 1995, p.2847):

"This concept warrants consideration by those embarking on therapeutic quests seeking inhibitors of smooth muscle cell proliferation as treatments for atherosclerosis, on the basis of relatively short-term experiments using simple animal models. Inhibition of smooth muscle cell proliferation in human patients might produce the undesired effect of destabilizing vulnerable regions of atherosclerotic plaques by the mechanisms described above."

7e. Plaque Rupture and Exercise: "Black's Crack in the Plaque"

In 1997, the New England Journal of Medicine carried an important exchange entitled "More on Coronary-Plaque Rupture Triggered by Snow Shoveling," in the correspondence section. Both paragraphs by Paul D. Thompson merit presentation here (Thompson 1997, p.1678):

"Hammoudeh and Haft (Hammoudeh 1996) report on 15 patients in whom acute coronary syndromes developed during or immediately after shoveling snow. They suggest that evidence of rupture of coronary plaque and acute thrombosis associated with physical exertion has not previously been observed in living patients. On the contrary, Ciampricotti et al (1989), using coronary angiography, found irregular coronary lesions 'consistent with' plaque rupture in 8 of 13 patients who had myocardial infarction or cardiac arrest within an hour of vigorous athletic activity." And (Thompson 1997, p.1678):

"Furthermore, in 1975 Black et al (Black 1975) reported autopsy, angiographic, or clinical evidence of acute plaque rupture in 13 patients with acute coronary syndromes related to vigorous exertion. In what now reads like a very clairvoyant discussion, Black et al suggested that the increased 'twisting and bending' of coronary arteries during vigorous exertion increased the frequency of plaque rupture and that 'Black's crack in the plaque' was responsible for most exertion-related acute coronary events. Once again, we are reminded that what seems new in medicine is often a rediscovery of what we knew, but forgot."

Haft and Hammoudeh, in reply (Haft 1997, pp.1678-79), begin with a recommendation: "Men and women in the age group that places them at increased risk of coronary events should be warned against shoveling snow ..." They also write that the concept of plaque rupture as a proximal cause of myocardial infarction "has been with us for some time (Daoud 1963), but its importance and frequency have been appreciated only since modern high-resolution coronary arteriography became available (Ambrose 1985)." Haft and Hammoudeh end with a comment (p.1679): "As with many pathophysiologic phenomena that originally appear to be rare, plaque rupture is emerging as the common mechanism in triggered myocardial infarction and other acute coronary events."

Others might not go that far. The frequency of fatal plaque rupture seems far lower at rest than after exercise (Burke 1999, p.922). In a New England Journal of Medicine editorial, Attilio Maseri (Maseri 1997, p.1015) states his view that fissuring of plaque is one of possibly "multiple causes" of acute ischemic events: "The search for the multiple pathogenic components of acute ischemia is a major challenge."

Meanwhile (during the search for additional proximal causes), we shall accept the existing evidence that disruption of atherosclerotic plaque explains some large share --- not yet quantified --- of deaths from Ischemic Heart Disease.

7f. Coming Full Circle: Lipids from Start to Finish?

According to the American Heart Association (1995, p.9), "In 48 percent of men and 63 percent of women who died suddenly of coronary heart disease, there were no previous symptoms of this disease."

This is consistent with the fact that atherosclerotic plaques can already be large before they ever intrude into the lumen, and with the observations described above (Part 7a) that the lesions most likely to precipitate an infarct-provoking thrombus often do not appear highly stenotic by angiography.

Libby's paper expounds extremely well about several molecules which could weaken the fibrous cap of a plaque --- any plaque. At the end, Libby comes to the ultimate issue: How can the risk of acute myocardial infarction be reduced? Citing the review paper by MacIsaac et al (1993), Libby writes (p.2849): "To reduce the risk of acute myocardial infarction, one must stabilize lesions to prevent their disruption, particularly the less stenotic plaques." Then Libby asks (Libby 1995, p.2849):

"How might one achieve such a goal? The results of recent lipid-lowering trials provide a hint. The reduction in clinical events without substantial change in the degree of luminal stenosis could reflect a stabilization of the non-critically stenotic lesions (Blankenhorn 1994, + Brown 1993). This stabilization might result from reducing the inflammatory stimuli provided by modified lipoproteins that could contribute to activation of lesional foam cells and T lymphocytes ..." (See also Buja 1994, + DeLorgeril 1994-b, + Fuster 1994, + Moreno 1994, + Van de Wal 1994.)

And so, we "come full circle," with the blood lipoproteins having a key causal role at the beginning, middle, and thrombogenic end of the Ischemic Heart Disease story. Therefore, if Hypothesis-2 is also valid, it must be consistent with the lipoprotein facts.

● **Part 8.   Introduction of a Tumor Hypothesis by the Benditts in 1973**

Earl P. Benditt makes a very important statement near the outset of the same paper from which we quoted in Chapter 43, Part 3a. He states (Benditt 1976, p.96):

"Frequently, we tend to confuse the lesions induced in animals with the real lesion. Our ultimate goal is the answer to the question, 'What is the nature and origin of HUMAN atherosclerosis?' Because the lesion has a mass of cells as a major feature, we can rephrase the question as follows: What is the nature of the cellular proliferation involved in the 'new formation,' the atherosclerotic plaque?" And (Benditt 1976, p.97):

"Specifically, we can ask, 'Is it of multicellular or of monoclonal origin?' The importance of this distinction lies in the fact that many neoplasms have been found to be of monoclonal origin. [A clone is a group of genetically identical cells descended from the same progenitor cell.] On the other hand, ordinary cell proliferations seen in embryogenesis, maintenance, and repair seem to be multicellular in character." And (Benditt 1976, p.97):

"It becomes immediately apparent, when one asks the question in the form indicated, that on the answer depends the direction of our search for factors that are responsible for the disease. Phrased this way, the question takes on a new significance, because there is now a basis for obtaining an answer, and the methods involved are applicable to human tissues and to human lesions."

8a.   Female Mosaicism: The Basis of Benditt's Method

Benditt summarized the method as follows (Benditt 1976, p.97):

"The method of analysis, applicable to study of proliferated masses of cells and for distinguishing the origin of these cells from one or from many precursor cells, requires individual organisms that are mosaics, that is, that comprise a mixture of two or more distinctive cell types. According to the concept of Lyon (1968), all human females are mosaics, composed of two phenotypically distinct cell types. This situation is due to the fact that early in embryonic development, there is a random inactivation in each somatic cell of one or the other of the two X-chromosomes." And (Benditt 1976, p.97):

"Once inactivation has occurred, each cell reproduces true to type, and all daughter cells of a particular cell exhibit the activity of the single same X-chromosomal genes. The stability of this state has been shown by cultivating single cells from connective tissues of human mosaic donors (Davidson 1963). Given a stable mosaic population, it is possible to ask questions with regard to cell population origins in embryogenesis and as to whether a pathologic new formation is derived from one or from many cells." And (Benditt 1976, p.97):

"The enzyme glucose 6-phosphate dehydrogenase (G-6-PD) is a polymorphic X-linked gene product. Particularly interesting ... are the common B form and the A form, which migrates more rapidly in an electric field. The A form of G-6-PD is present in a substantial proportion of the Black population. In our series, 40% of 115 cases of Black females are heterozygous, and their tissues comprise two cell types, each of which produces one or the other enzyme form (Beutler 1962). This property of females has been used to assess the origin of cell populations in several tumors (Fialkow 1974), including benign smooth muscle tumors [called leiomyoma] of the uterus (Linder 1965)."

8b. Findings of Benditt's Team and Pearson's Team

Benditt's first series (Benditt and Benditt 1973) involved three women, from whom 30 atherosclerotic plaques from the aorta were analyzed (table on Benditts' p.1754). In addition, the paper presented data on 59 samples of uninvolved (healthy) aorta. Writing in 1976, Benditt comments (p.97) that the plaques "revealed the startling fact that most discrete plaques appear to be monoclonal; that is, they comprise a monotypic cell population ... Pearson et al (1975) have recently published data on material examined in a similar way, and report almost the identical result." The Benditts' work was in the Department of Pathology at the University of Washington School of Medicine, and Pearson's team was in the Department of Pathology at the Johns Hopkins University School of Medicine.

Benditt presents his own results and Pearson's 1975 results in his Table 1 (Benditt 1976, p.97), which we reproduce nearby. Samples in the AB column have both cell-types present. Samples in the A column are monotypic; they have A-producing cells, but not B-producers. Samples in the B column are also monotypic; they have B-producing cells, but not A-producers.

Table 1 from Benditt 1976, p.97: "G-6-PD Phenotypes of Cell Populations of Human Atherosclerotic Plaques Compared to NonPlaque Artery Samples."

	<---	NON-PLAQUES	--->		<---	PLAQUES	--->	
	AB	A	B	Total	AB	A	B	Total
Benditt: 4 cases	57	0	2	59	6	8	16	30
Pearson 16 cases	99	1	1	101	3	17	9	29
Total	156	1	3	160	9	25	25	59

The difference is dramatic, between non-plaque and plaque samples. Only 2.5 % of non-plaque samples are monotypic (4 out of 160). But 85 % of the plaque samples are monotypic (50 out of 59). Monotypic A-lesions and monotypic-B lesions occur separately within a single vessel (Benditt 1976, p.98). And (Benditt 1976, pp.97-98):

"The fact that the bulk of raised atherosclerotic lesions are monotypic with regard to cell type, as indicated by isoenzyme pattern, does not immediately yield the strong inference that lesions of atherosclerosis are monoclonal." Benditt evaluates three possibilities, other than monoclonality, which might explain the observed difference in his Table 1 between healthy samples and atherosclerotic samples. He concludes (Benditt 1976, p.99):

"When all of the data are considered together, the reasons are strong for believing that raised lesions are monoclonal."

8c. Twelve Years Later: Benditt's View on Monoclonality in 1988

Benditt did not change his mind. In 1988, he wrote a paper entitled, "Origins of Human Atherosclerotic Plaques," in which he discusses monoclonality again. With reference to his own findings (1973, 1974), he says (Benditt 1988, p.998) that "Two other laboratories have reported essentially the same findings," and he cites Pearson 1977 and Thomas 1979. Benditt discusses Thomas's dissenting interpretation (Thomas 1983), cell-proliferation in plaque, and Ross's hypothesis (Part 2) that platelet-derived growth factor plays a key role in such cell proliferation. And (Benditt 1988, p.999):

"The role of oncogenes and related growth factors has surged to the forefront of research in neoplasia, proliferation of cells in atherosclerotic plaques, and embryologic development. Platelet-derived growth factor (PDGF) has been proposed by Ross (Ross 1986) to be a key element in inducing proliferation in atherosclerotic lesions ... The clonality of the atherosclerotic lesions raises the possibility that there is a change in the pattern of growth factor or oncogene expression characteristic of the smooth-muscle cell populations of plaques. We [his group at the University of Washington] have been examining this possibility." At the end of his paper, Benditt refers to "benign smooth muscle tumor" and he concludes (Benditt 1988, pp.999-1000):

"... while it is quite reasonable to look for alternatives to the inference of monoclonality, no substantiated alternative has appeared ... At the moment, our belief is that abnormal expression of these growth-related genes does not account for the abnormal cell growth in either the benign smooth muscle tumor or the plaque cells."

8d. Benditt on Potential CAUSES of Arterial Tumors: Silence on Radiation

In Benditt's 1976 paper (p.99), he speculates about various "factors" which may be "the causes of monoclonal proliferation." He says, "Prominent among these factors is the role of chemical mutagens (or premutagens) derived from the environment. Another possibility is the role of viral agents, such as that which causes the common wart, a monoclonal lesion." Benditt properly regards cigarette smoking as an environmental factor, meaning an exogenous factor. He continues (Benditt 1976, pp.99-100):

"Cigarette smoking is well established as a 'risk factor' [for IHD]. Burning of cigarettes produces aryl hydrocarbons, and some of these (benzo[a]pyrene) hydrocarbons are well-known precarcinogens. It is easy to believe that hydrocarbons from cigarette smoke affect the lung; can they also affect distant arteries? ... Aryl hydrocarbon hydroxylase, an inducible enzyme system, is elevated enormously in the placentas of women who are heavy smokers and to an intermediate extent in women of average smoking habits, when compared with enzyme levels in nonsmokers (Juchau 1974). Clearly, the route must be from the lung via the blood to the placenta." And (Benditt 1976, p.100):

"In addition to this finding, it has been shown that low-density lipoproteins are the main carriers of benzo[a]pyrene in the bloodstream (Margolis 1969). Coupling this with the now established fact (Bierman 1976) that low-density lipoproteins are preferentially taken up by smooth muscle cells derived from the arterial wall, we then have a system entirely compatible with the initiation of cellular alteration that leads to monoclonal growth."

Benditt's comments on smoking are fully compatible with the second part of Hypothesis-2 --- that medical radiation induces mutations in cells of the arterial wall. With respect not only to cancer but also to monoclonal mini-tumors in coronary arteries, medical xrays and chemical carcinogens can work as co-actors. Indeed, it is very likely that no single mutation in a somatic cell is clinically potent as a tumor-maker unless it receives "help" at some time from additional mutations and promoting agents in the same cell.

In 1976, the Benditt paper is silent about ionizing radiation as a mutagen. It mentions only chemicals and viral agents. By contrast, near the end of a paper in 1985, Benditt and colleagues write: "The origin of certain atheromata may be linked to the presence of chemical and physical mutagens and viruses in our diet or environment" (Majesky 1985, p.3453). Ionizing radiation is a "physical mutagen," of course. In 1988, ionizing radiation is mentioned neither indirectly nor by name, when Benditt and colleagues state that "chemical mutagens or viruses" are potential causes of the monoclonal plaque cells (Benditt 1988, p.999).

An undisputed fact deserves repeating: Ionizing radiation is a potent mutagen (Chapter 42, Part 2).

8e. Some of the Additional Contributions by Pearson et al, 1978 and 1983

In 1978, Pearson and colleagues contributed additional evidence of monoclonality in atherosclerotic plaques (Pearson 1978-b). At the outset (pp.93-94), they explain that, in heterozygous women: "Normal tissue will consist of a mosaic of patches of cells expressing the same isoenzyme [either A-type or B-type]. However, these patches are so minute that the assay of even very small bits of normal tissue will yield both isoenzymes. Monoclonal lesions, on the other hand, will contain only one isoenzyme type in the heterozygous female (Linder 1965, + Fialkow 1974)."

The new evidence in Pearson 1978-b comes from 10 aortic plaques (from 6 deceased heterozygous women), which Pearson and co-workers divided into 45 portions. In this inquiry, they compared the upper and lower layers of the plaques. Pearson et al found (Pearson 1978-b, pp.96-97):
- 15 (33%) were monoclonal in both layers.
- 9 (20%) were monoclonal in one layer (upper).
- 9 (20%) were monoclonal in the other layer (lower).
- 12 (27%) were not monoclonal in either layer.

Pearson and colleagues state: "The monoclonal nature of atherosclerotic fibrous plaques has been demonstrated by independent investigators with unequivocal results (Benditt 1973, + Pearson 1975, + Pearson 1977). Although the laboratory observations have been firmly established, a variety of interpretations have been made as to the mechanism by which monoclonal populations arise" (Pearson 1978-b, p.99, followed by their discussion).

In 1983, Thomas and Kim state (Thomas 1983, p.247) that the Benditts' observations "have been amply confirmed by Pearson et al (1975, 1978) in Baltimore and by Thomas et al (1979) in Albany ... However, Thomas et al ... have serious doubts regarding the validity of the monoclonal theory of origin of atherosclerotic lesions." And (Thomas 1983, p.247): "To date, we have studied more than 600 lesions [from 44 aortas]. Normal aortic tissue from these aortas showed both A and B forms of G-6-PD (ditypism). Among 25 of these aortas, one or more samples of atherosclerotic lesions showed a single G-6-PD type (monotypism). Among the 469 lesions sampled in these aortas, 160 [34%] yielded at least one monotypic sample. All remaining lesions (440 of the original 600) yielded only ditypic samples. Furthermore, when multiple samples were taken from the lesions with one or more monotypic samples, ditypic samples were also obtained, frequently in greater numbers than monotypic samples."

Monoclonal Characteristics of Fatty Streaks and Thickened Intima

In 1983, Pearson and co-workers published a study in which they looked for monoclonality in human samples of NON-plaque segments of normal aortic media (237 samples), thickened intima (133 samples), and fatty streaks (58 samples) --- all from 13 heterozygous black females. The observed frequency of monoclonality in these NON-plaque samples of (Pearson 1983-a, p.36):

In media, zero; in thickened intima, 2 out of 133 (1.5%); in fatty streaks, 1 out of 58 (1.7%). All of these frequencies are dramatically distinct from the frequency found in atherosclerotic plaques (Parts 8b and directly above). An interesting observation is that Pearson et al found the distribution of isoenzyme values from fatty streaks to be "markedly different" from the distribution in the thickened intima and in media --- as measured by "percent B Isoenzyme."

8f. An Excellent Review-Paper in 1990

In 1990, Mutation Research presented a fine review by Arthur Penn of the Benditts' "monoclonal hypothesis." Entitled "Mutational Events in the Etiology of Arteriosclerotic Plaques," Penn's paper begins (Penn 1990, p.149): "In this review, evidence is provided in support of the 'monoclonal' hypothesis of arteriosclerotic plaque formation. Experimental and clinical data, collected over the last 16 years, are presented that are consistent with the view that environmental mutagens, including viruses and chemical carcinogens, play a key role in plaque etiology."

We shall return to some of Penn's own work in Part 9c of this chapter. As indicated above, he, too, overlooks ionizing radiation as a proven mutagen to which people are commonly exposed.

● Part 9. Other Thinkers about Benditt's Tumor Hypothesis

The Benditts' proposal, of monoclonality and a tumor etiology for the process of atherosclerosis, came in June 1973. In July 1975, Martell linked (a) the Benditt tumor hypothesis, (b) the clear tumor-producing power of alpha-particle ionizing radiation, and (c) the evidence that smoking is a cause of coronary heart disease, to produce his own hypothesis about a cause of atherosclerosis.

9a. Martell's Hypothesis on a Cause of Atherogenesis

Martell states (Martell 1975, p.404):

"Alpha interactions with chromosomes of cells surrounding insoluble radioactive smoke particles may cause cancer and contribute to early atherosclerosis development in cigarette smokers."

There is no doubt whatsoever that cigarette smoking is a cause of coronary heart disease. In 1975, Martell is proposing WHY it is. He is proposing that the alpha-particle emitters in mainstream smoke eventually get to the arterial wall through the blood and lymph stream, after their initial deposition in the lungs.

It is clear, thanks to the work of Martell and others, that insoluble particles of radioactive lead-210 are present in cigarette smoke (our Chapter 48, Part 1c). Lead-210 (Pb-210) is a member of the uranium decay-series, and has a radioactive half-life of about 20 years. Lead-210 is a beta-emitting radio-nuclide which decays into radioactive bismuth-210, which is another beta-emitting radionuclide with a radioactive half-life of about 5 days. Bismuth-210 decays into polonium-210, which is an alpha-emitting radio-nuclide with a half-life of about 138 days. And finally, polonium-210 decays into stable lead-206. Lead-210 atoms retained in the body will continuously convert in this way to polonium-210 atoms, which are the alpha-emitters discussed by Martell (references in Chapter 48, Part 1c).

Martell cites the papers of Arthur Elkeles (1961, 1966, 1968) in support of his view that alpha particles are present at unusually high concentration in atherosclerotic plaques (Martell 1975, pp.409-410):

"The possibility that alpha radiation may be the mutagenic agent in atherosclerosis plaque formation is indicated by the results of Elkeles, who found anomalously high concentrations of alpha activity at the calcified plaque sites of atherosclerosis victims ... The high incidence of early coronaries among cigarette smokers may conceivably be explained by the accumulation of insoluble radioactive smoke particles at the plaque sites. Such a possibility should be experimentally evaluated."

Our View of the Martell Hypothesis --- and of Martell

We have no independent reason to question the observations by Elkeles, but we should point out that they may have an alternative interpretation.

Calcification is a process which can be contaminated with ions other than Ca++ within the insoluble crystalline material of the calcified lesion. So it is possible that lead-210, retained in the body from cigarette smoke, is co-precipitating in arterial lesions which are calcifying --- but this may not be evidence in favor of the polonium-210 as a CAUSE of arterial tumors. Unless there is contrary evidence of which we are unaware, we entertain the possibility that arrival of the lead-210 in the plaque, and the subsequent production there of extra alpha activity, may be a late and incidental event. Calcification, which often occurs in tubercular and syphilitic lesions too (not only in atherosclerotic lesions), and in breast lesions, is one of the body's common RESPONSES to a menacing lesion, rather than the lesion's cause.

On the other hand, Martell may have been on the right track. Alpha particles may well be one of the mutagens which INITIATE multiple mini-tumors in the coronary arterial walls. We think the idea deserves further attention.

Martell was a superb radio-chemist and fearless original thinker. His untimely death in July 1995 is a great loss for science and humanity. During his last year of life, I know from personal communications with him that he was absolutely convinced that alpha-emitters play an important role in the genesis of atherosclerosis.

9b. Trosko and Chang on the Role of Mutagens in Atherosclerosis and Cancer

In a 1980 issue of Medical Hypotheses, J.E. Trosko and C-c. Chang took note of Benditt's monoclonal hypothesis of cell-proliferation in human atherosclerosis, in their paper "An Integrative Hypothesis Linking Cancer, Diabetes and Atherosclerosis: The Role of Mutations and Epigenetic

Changes" (Trosko 1980, at p.456 and p.460). Unlike Benditt (1973, 1976, 1988), Trosko and Chang mention "radiations" as as one of the potential mutagens. Indeed, "radiations" are included even in the abstract (Trosko 1980, p.455):

"It appears that the disease states of cancer, atherosclerosis and diabetes might share a common etiology. These chronic diseases appear to be multi-staged in their progression, with genetic, nutritional, psycho-social, environmental and viral factors influencing their appearance. We offer a hypothesis (a "mutation theory of disease"), stating that these diseases can be described by initiation and promotion phases; initiation being the result of the production of mutated cells after unrepaired damaged DNA is replicated; promotion being the selective proliferation of the initiated cells to form clones of mutated cells. It is further postulated that promotion affects cell proliferation by altering a membrane-Ca++ regulatory system. Depending on the nature of the mutation in the clone of cells, specific disease states would result. The roles of radiations, chemicals, viruses, genes, nutrition and psycho-social stress are related to either the initiation (mutation production) or the promotion (cell proliferation) phase of these diseases."

Trosko and Chang hold the view that "mutagenesis is a necessary, but insufficient, step" for the pathogenesis of atherosclerosis (Trosko 1980, p.460). They repeatedly stress the need for an outside stimulus (promoter) to cause a mutated cell to divide, which of course it must do, if it is going to form a clone or tumor.

Like Benditt, Martell, and some others, Trosko and Chang deserve much credit for proposing --- long ago --- the likelihood that mutagens have a key role in the pathogenesis of atherosclerosis.

9c. Penn on Experimental Evidence for Benditt's Tumor Hypothesis

Like Martell, and Trosko, and Chang, Arthur Penn has appreciated the importance of the Benditts' monoclonal hypothesis, discussed in Part 8 above. Looking back, Penn writes in 1989 (p.190):

"An early prediction that arose from the monoclonal hypothesis was that viruses and chemical mutagens would be expected to play critical roles in plaque formation and development, just as they do in tumorigenesis. The earliest attempt at verifying this prediction came from Roy Albert and his co-workers at New York University's Department of Environmental Medicine. Weekly injections of the polycyclic aromatic hydrocarbon (PAH) carcinogens 7,12-dimethylbenz[a]anthracene (DMBA) or benzo[a]pyrene (BaP) resulted in large, proliferating plaques in the abdominal aorta in cockerels (Albert 1977). Plaques increased in size in a dose-dependent fashion (Penn 1981-a, with Bastatini + Albert) ..."

As techniques of molecular biology advanced, Penn and co-workers returned to exploring the tumor hypothesis in various ways. Among their many interesting findings, Penn and co-workers have shown (a) that DNA from the coronary-artery plaques of some patients can transform NIH 3T3 fibroblasts, which thereby acquire the power to produce tumors in nude mice (Penn 1986; plus similar findings based on arterial plaque from experimental animals, Penn 1989 + Penn 1991), and (b) that injection of experimental animals, with a variety of established chemical carcinogens and mutagens, promotes expansion of arterial plaques in such animals (Penn 1981-a + Penn 1988), and (c) that inhalation of cigarette smoke by cockerels accelerates development of arteriosclerotic plaques (Penn 1993, 1994, 1996).

Penn 1990 on Compatibility of Monoclonal and Injury Hypotheses

Penn et al have not been the only ones producing experimental evidence pertinent to the tumor hypothesis in development of arterial plaques. Benditt's group at the University of Washington, and others (for example, Yew 1989 + Ahmed 1990), also have done some work --- work which is referenced in Penn's various papers. Here, we limit ourselves to abbreviating Penn's ideas about the COMPATIBILITY of the "response to injury" and the "tumor" hypotheses of cell proliferation in human coronary-artery plaque. Penn writes in Mutation Research (Penn 1990, p.158):

"Although the 'injury' and 'monoclonal' hypotheses are generally regarded as being mutually exclusive, the available evidence points to possible roles for both injury and transformation in plaque etiology. The 'synthesis' ... can be summarized as follows: The primary events in plaque etiology involve DNA damage and its fixation. Elaboration of this damage in a clinically or experimentally significant way may require injury to the arterial wall." And later (Penn 1990, p.160):

"As reviewed earlier in this article, there is abundant evidence that injury to the arterial wall can stimulate smooth-muscle cell [SMC] proliferation ... If genetically altered SMCs are the progenitor cells of plaques, if these cells are distributed randomly throughout the artery wall, as seems reasonable, and if injury plays a role in plaque formation, then the focal nature of plaques points to localized injury as a possible stimulus for the proliferation of the already transformed SMCs."

9d. The Benditts' Monoclonal Hypothesis: Still Alive and Respected

The Benditts' monoclonal tumor hypothesis has been neither discredited nor widely explored yet. In 1988, Munro and Cotran singled out the hypothesis as meriting respectful attention (Munro 1988, p.255). Indeed, the monoclonal hypothesis appears to have inspired recent work at the National Institutes of Health by Speir et al (Speir 1994). However, the work reported in Speir 1994 turned out to be no test whatsoever of the Benditts' monoclonal tumor hypothesis (details in Marx 1994, p.320).

● Part 10. Homocysteine as a Risk-Factor for Cardiovascular Mortality

The idea, that elevated blood-levels of homocysteine (an amino acid) may be causally involved in cardiovascular mortality, is credited by Nygard et al (Nygard 1997) to K.S. McCully (McCully 1969). The idea arose because premature vascular disease is characteristic of individuals with homocystinuria --- an inborn metabolic error which results in high blood-levels of homocysteine (>100 micromol/liter) and elevated levels in the urine. Untreated, about 50% of such individuals have thrombo-embolic events, and about 20% are dead before the age of 30, according to Nygard (1997, p.230). Thrombosis rather than atherosclerosis is the basis of their cardiovascular complications. Nygard comments (Nygard 1997, p.230):

"There is increasing evidence that homocysteine may affect the coagulation system and the resistance of endothelium to thrombosis (Malinow 1994) and that it may interfere with the vasodilator and anti-thrombotic functions of nitric oxide (Stamler 1996)."

Nygard cites five prospective, nested case-control studies which explore the relation between total homocysteine levels and the frequency of vascular disease: Alfthan 1994, + Stampfer 1992, + Arnesen 1995, + Perry 1995, + Verhoef 1994. According to Nygard, all except the Alfthan Study found a relationship. (Additional studies of interest include Ridker 1999.)

Nygard and colleagues undertook a prospective study of the relationship between plasma total homocysteine levels and mortality among 587 patients with angiographically confirmed coronary artery disease. Of these patients, 94 had single-vessel disease, 172 patients had 2-vessel disease, and 321 patients had 3-vessel disease at the outset (Nygard 1997, p.233, Table 1). At the outset, 337 of the 587 had already experienced a myocardial infarction, and 64 had already undergone coronary-artery bypass grafting (Nygard 1997, p.232).

Baseline measurements of homocysteine and other risk-factors for coronary heart disease were made in 1991 or 1992. Subsequently, 318 patients had bypass surgery, 120 patients had angioplasty, and 149 were treated medically (Nygard 1997, p.230). After a median follow-up period of 4.6 years, 64 patients had died --- 50 of them from cardiovascular disease, including 26 deaths from myocardial infarction (Nygard 1997, p.231).

Analysis shows a clear, strong, and positive dose-response between levels of plasma homocysteine at baseline, and relative risk of death (all causes combined). Patients were classed into four levels of "dose," measured in micromol/liter. The 130 patients with homocysteine below 9.0 served as the control group. After adjustment for age and sex, the results in terms of mortality ratios are:

Patients	Homo. Level	All causes	CVD only
n = 130	<9.0	1.00	1.00
n = 372	9.0-14.9	2.35	3.30
n = 59	15.0-19.9	5.75	6.30
n = 26	=> 20.0	7.04	9.90

As shown above, when death from cardiovascular disease (instead of from any cause) is evaluated, the dose-response is even stronger (Nygard 1997, p.234). By contrast, "The lipid-related factors showed either no relation or a much weaker relation to mortality [all causes] than the total homocysteine level" (Nygard 1997, p.234). The paper appears to be silent about what fraction of patients were treated with lipid-lowering regimes after the baseline measurements. Nygard and colleagues conclude their paper (Nygard 1997, p.236):

"This prospective study does not prove a causal relation between total homocysteine and mortality, but our results should serve as an additional strong incentive to the initiation of intervention trials with homocysteine-lowering therapy." We concur.

● Part 11. Other Aspects of IHD Etiology

In this chapter, we have barely mentioned, if at all, such established risk factors for Ischemic Heart Disease as heredity, age, gender, smoking, hypertension, diabetes mellitus, overweight, physical inactivity, and others. Of course they, too, are important aspects of IHD etiology. Here, however, we have focused on the concepts which have helped to shape our thinking in Chapters 45 and 46.

>>>>>>>>>

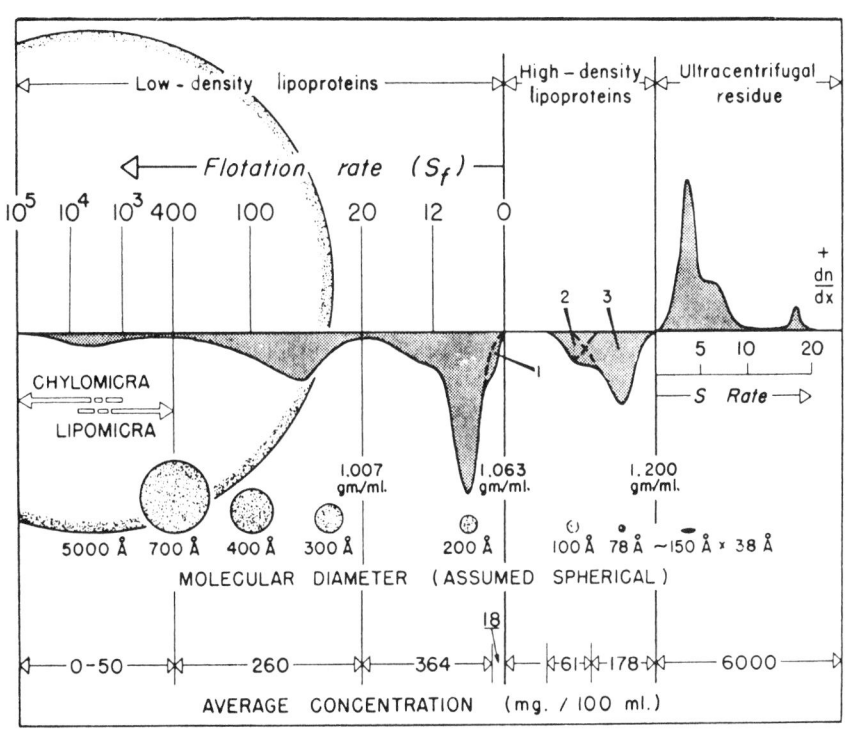

Related text = Parts 3b + 3c.

- The figure above shows the ultracentrifugal composition of human serum (reproduced from Lindgren 1956; also in Lindgren 1957 + 1959). During the 1950s, the term "Low-Density Lipoproteins" embraced all lipoproteins which floated during ultracentrifugation in a solution of density 1.063 gm/ml. In the figure, "average concentrations" in mg/dl are for 45-year-old males at that time. The size and area of the gray inverted peaks (see Box 2) result from the different concentrations of lipoproteins having various Sf values (e.g., the Sf 20 lipoproteins are in relatively lower concentration than adjacent varieties). In the Low-Density continuum, the Sf 6-8 lipoproteins are relatively small in size. Nonetheless, their estimated molecular weights are 3 to 6 million units. (By comparison, serum albumin has a molecular weight in the neighborhood of 70 thousand.)

- In the lipoprotein segment called "High-Density," the figure indicates that HDL-3 is more abundant than HDL-2. HDLs also float --- when in solutions which have densities adequately greater than the HDLs' own densities. The "residue" segment on the right consists of the much denser plasma proteins (e.g., albumin, globulins) and "20" components --- which sink instead of floating in a solution of 1.20 gm/ml. (Their plot is on 30-fold reduced dn/dx scale.)

Protein & Lipid: Approximate Shares per Molecule		
	Percent Protein	Percent Lipid
Chylo.	2	98
Sf 20-400	13	87
Sf 0-20	25	75
HDL-1	30	70
HDL-2,3	55	45

Lindgren 1955 + 1956.
Bragdon 1956.

Composition of the Lipid Part: Avg Percentage.
Source: Appendix E, Box 2.

Cholesterol ester — Unesterified cholesterol — Glyceride — Phospholipide — Unesterified fatty acid

$S_{f\,20-400}$: 15 | 8 | 55 | 20 | 2

$S_{f\,0-20}$: 46 | 14 | 14 | 25 | 1

$HDL_{2,3}$: 28 | 6 | 17 | 44 | 6

● Below are optical (schlieren) flotation diagrams which illustrate the distinctly different behavior of the Sf 13 and the Sf 6 lipoproteins during ultra-centrifugation (from Lindgren 1951, p.86).

● Each strip of photographs was made while the rotor was spinning at full speed (52,640 revolutions per minute), by passing a light-beam through the solution to expose a film cassette. From left to right, the frames of each strip were taken at 0, 8, 12, 22, 30, and 38 minutes after the rotor attained full speed. The cell used had a fluid column of 12 mm.

● The direction of centrifugal force is from left to right. If a lipoprotein floats against this force, it will move progressively from the right side of the first frame, toward the left side in subsequent frames.

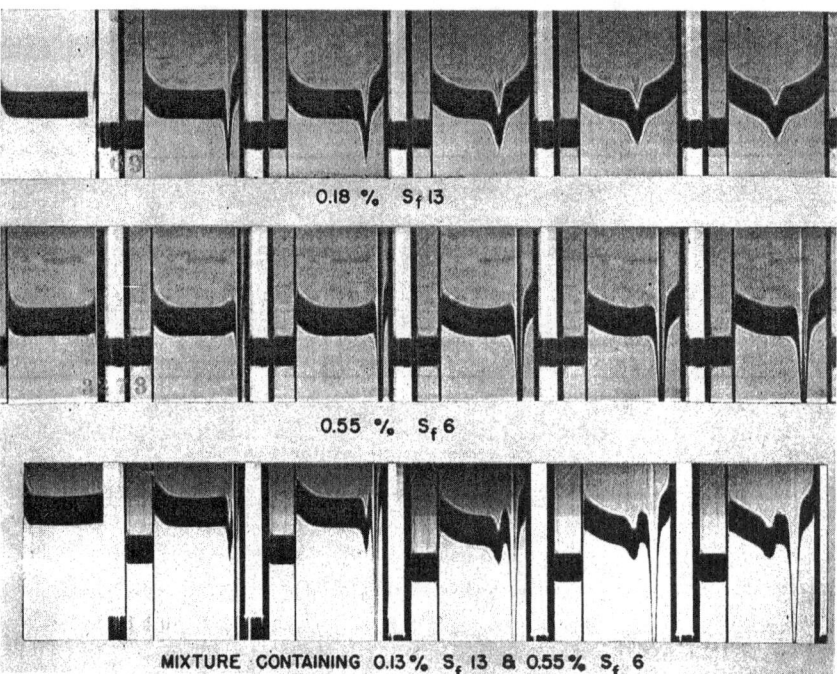

O min.   8 min.   12 min.   22 min.   30 min.   38 min.

0.18 % S_f 13

0.55 % S_f 6

MIXTURE CONTAINING 0.13 % S_f 13 & 0.55% S_f 6

● In the FIRST strip, depicting behavior of Sf 13 lipoproteins, an inverted peak is noted at 8 minutes. This peak results from movement of the Sf 13 lipoproteins from right to left, against the centrifugal force. At time = 38 minutes, the inverted peak is far from the righthand edge of the frame.

● In the SECOND strip, depicting behavior of the Sf 6 lipoproteins, the movement is very clearly slower than in the Sf 13 strip. At equal time-intervals, the Sf 13 lipoproteins have traveled (floated) farther from the right side of the frame than have the Sf 6 lipoproteins. The Sf 13 lipoproteins move faster because they are larger and even less dense than the Sf 6 lipoproteins (see continuum in Box 1). The area of the peak under its baseline is proportional to concentration. The area is clearly larger for the Sf 6 lipoproteins, whose initial concentration was higher than the Sf 13's concentration.

● In the THIRD strip, a mixed solution of Sf 13 (0.13% concentration) and Sf 6 (0.55% concentration) is spun. Their mixture causes two distinct peaks, floating at different rates. The experiment indicates that the molecules are stable and retain their physical identities when mixed.

Related text = Parts 3b + 3c.

See text, Part 4a.

Source: Felix O. Kolb + Oliver F. DeLalla + John W. Gofman, 1955, "The Hyperlipemias in Disorders of Carbohydrate Metabolism: Serial Lipoprotein Studies in Diabetic Acidosis with Xanthomatosis and in Glycogen Storage Disease." (Metabolism Vol.IV, No.4: 310–317, July 1955). Data are from Table 1, page 312.

• The four Low–Density Lipoprotein classes studied here are Sf 0–12, Sf 12–20, Sf 20–100, and Sf 100–400 . Sf 0–12 and Sf 12–20 are often combined as Sf 0–20. The Sf 20–100 and Sf 100–400 are often combined as Sf 20–400. The two High–Density Lipoprotein classes are HDL–2 and HDL–3. The HDL–1 class is special –– its flotation rate is the lowest encountered in solutions of density 1.063 gms/ml. So, while it is designated as a high–density lipoprotein, it actually is truly in the low–density class, defined by flotation in a solution of density, 1.063 gms/ml. A migration rate of 10^{-13} cm per second per unit field of force equals a rate of one Svedberg: In sedimentation, one S unit; in flotation, one Sf unit. When all adjustments are made for temperature and concentration effects, the flotation rates are said to be in "Standard Sf units " (Std Sf units). The Svedberg unit honors Thé Svedberg, the great Swedish physical chemist who pioneered the ultracentrifuge itself (Svedberg 1940).

• The measurements below show the metabolic inter–relations of various lipoprotein classes in a 37–year–old diabetic woman, during treatment of acidosis. The general pattern of lipoprotein conversions is as follows:

$$\text{Sf } 100\text{–}400 \quad \text{---> } \quad \text{Sf } 20\text{–}100$$
$$\text{Sf } 20\text{–}100 \quad \text{---> } \quad \text{Sf } 12\text{–}20$$
$$\text{Sf } 12\text{–}20 \quad \text{---> } \quad \text{Sf } 0\text{–}12$$

with concomitant, progressive loss of triglyceride catalyzed by Lipoprotein Lipase hydrolysis. Sugar–level in the urine falls from 4+ to 1+. The lipoprotein "spectrum" moves toward normality as the defective carbohydrate metabolism comes toward control.

• As the massive Sf 100–400 buildup declines (see second row) to 1530 from 3739---undoubtedly with hydrolysis of triglyceride--- the Sf 20–100 levels rise to a peak value of 1942, while the Sf 12–20 and Sf 0–12 also rise, but do not yet peak. With further therapy (see 3rd row), the Sf 100–400 levels fall even more to 685, the Sf 20–100 levels begine to decline, the Sf 12–20 levels reach their peak, and the Sf 0–12 levels are still building up . The Sf 0–12 levels are the last to peak. At the outset (Row 1), the HDL–1 massive elevation (168 mg per 100 ml is very high for HDL–1) is the same as that noted in chronic "essential hyperlipemia."

• The Lipoprotein and Cholesterol concentration measures are in units of milligrams per 100 ml.

Date of Study	Urinalysis Sugar	Acetone	Sf 0–12	Sf 12–20	Sf 20–100	Sf 100–400	HDL–1	HDL–2	HDL–3	Total Cholesterol
6/19/52	4+	4+	195	155	1120	3739	168	29	167	1535
6/22/52	3+	1+	444	352	1942	1530	141	7	179	1280
6/28/52	2+	0	744	428	1277	685	66	7	179	806
7/03/52	2+	0	939	338	670	139	34	14	172	668
7/18/52	3+	0	614	134	493	228	41	33	203	377
8/01/52	2+	0	531	148	432	132	25	33	219	342
8/08/52	2+	0	616	150	332	152	19	19	192	377
8/14/52	2+	0	549	108	150	31	----	----	----	320
9/12/52	2+	0	444	186	668	423	----	----	----	----
10/22/52	1+	0	452	237	988	461	52	29	157	398

6/24/52   "Tuberose" and "Eruptive" Xanthomata of approximately 3 months duration.

8/14/52   Partial Clearing of the Xanthomata.

2/27/53   Complete Clearing of the Xanthomata.

Related text = Part 4a.

This list is provided as a convenience for readers; it is not meant to be complete.
C = Clinical Findings. D = Discussion. Our Ref.Entry = Entry in our Reference List.

Year	Our Ref. Entry	Popular Abbreviation, Topic, or Title.
1978-C.	ComPrinci 1978.	Committee of Principal Investigators ... Clofibrate Trial in Primary Prevention.
1984-C.	Brensike 1984.	NHLBI: Natl. Heart, Lung, Blood Institute ... Cholestyramine Therapy.
1984-C.	LipidRC 1984.	Lipid Research Clinics Coronary Primary Prevention ... Cholesterol-Lowering.
1985-D.	Consensus 1985.	Concensus Conference on Lowering Blood Cholesterol to Prevent Heart Disease.
1987-C.	Blankenhorn 1987.	CLAS: Cholesterol-Lowering Atherosclerosis Study (Cholestipol Niacin Therapy).
1987-C.	Frick 1987.	Helsinki Heart Study: Primary Prevention Trial (with Gemfibrozil).
1987-D.	European 1987.	Policy Statement of the European Atherosclerosis Society.
1990-C.	Brown 1990.	FATS: Familial Atherosclerosis Treatment Study ... Lipid-Lowering Therapy.
1990-C.	Buchwald 1990.	POSCH: Program on Surgical Control of the Hyperlipidemias.
1990-C.	Cashin-Hemphill 1990.	CLAS II: CLAS 1987, for 2 more years.
1990-C.	Kane 1990.	UC-SCOR: Univ.Calif.Specialized Ctr. of Research (Familial HyperCholesterolemia).
1990-C.	Ornish 1990.	Lifestyle: The Lifestyle Heart Trial.
1990-D.	Holme 1990.	An Analysis of Random Trials on the Effect of Cholesterol Reduction.
1990-D.	Joint 1990.	Joint Statement by the AHA and NHLBI: Cholesterol Facts.
1991-D.	Oliver 1991.	Might Treatment ... Increase Non-Cardiac Mortality?
1992-C.	Quinn 1992.	SCRIP: Stanford Coronary Risk Intervention Project.
1992-C.	Schuler 1992.	Heidelberg Study (Low-Fat Diet and Exercise).
1992-C.	Watts 1992.	STARS: St. Thomas Atherosclerosis Regression Study (Lipid-Lowering).
1992-D.	Smith (G.D.) 1992.	"Should There Be a Moratorium on Use of Cholesterol-Lowering Drugs?"
1992-D.	Hulley 1992.	"Health Policy on Blood Cholesterol: Time to Change Directions."
1993-C.	Blankenhorn 1993.	MARS: Monitored Atherosclerosis Regression Study (with Lovastatin).
1993-C.	Pravastatin 1993.	Pravastatin Multinational Study Group.
1993-D.	Brown 1993.	Lipid-Lowering and Plaque Regression.
1993-D.	ExpertPanel 1993.	Expert Panel on ... Evaluation & Treatment of High Blood Cholesterol.
1993-D.	Pearson 1993.	Rapid Reduction in Cardiac Events with Lipid-Lowering Therapy.
1993-D.	Ravnskov 1993.	"Reducing Serum Cholesterol ... of Doubtful Benefit to Anyone."
1994-C.	Furberg 1994.	Effect of Lovastatin ... on Cardiovascular Events.
1994-C.	Haskell 1994.	SCRIP: Stanford Coronary Risk Intervention Project.
1994-C.	MAAS 1994.	MAAS: Multicentre Anti-Atheroma Study with Simvastatin.
1994-C.	Pedersen 1994.	4S: Scandinavian Simvastatin Survival Study, on Cholesterol-Lowering.
1994-C.	Waters 1994.	CCAIT: Canadian Coronary Atherosclerosis Intervention Trial.
1994-D.	Blankenhorn 1994.	Arterial Imaging and Atherosclerosis Reversal.
1994-D.	TaskForce 1994.	European Task Force: Recommendations on Prevention of IHD.
1995-C.	Jukema 1995.	REGRESS: Regression Growth Evaluation Statin Study, with Pravastatin.
1995-C.	Shepherd 1995.	Prevention of CHD with Pravastatin ...
1995-C.	Pitt 1995.	PLAC I: Pravastatin Limitation of Atherosclerosis ...
1995-D.	Gould 1995.	"Cholesterol Reduction Yields Clinical Benefit: New Look at Old Data."
1995-D.	Havel 1995.	"Management of Primary Hyperlipidemia."
1995-D.	Pedersen 1995.	"Lowering Cholesterol with Drugs and Diet" (Editorial).
1995-D.	Ravnskov 1995-a.	"Beneficial Effects of Simvastatin May Be Due to Non-Lipid Actions."
1996-C.	Sacks 1996.	CARE: Cholesterol And Recurrent Events Trial: Persons w. Avg. Cholesterol.
1997-C.	PostCABG 1997.	Post CABG: Post Coronary Artery Bypass Graft & Lipid-Lowering.
1997-D.	Oliver 1997.	"The Low-Fat, Low Cholesterol Diet Is Ineffective."
1998-C.	Ornish 1998.	Intensive Lifestyle Changes in Reversal of Coronary Heart Disease.
1998-C.	Downs 1998.	Primary Prevention of Acute Coronary Events w. Lovastatin: Persons w. Avg. Chol.
1998-C.	Rosenson 1998.	Anti-Atherothrombotic Properties of Statins ... and CV Event-Reduction.
1998-D.	Pearson 1998.	"Lipid-Lowering Therapy in Low-Risk Patients" (Commentary).

Related text = Part 6.

Part 1. Introduction: Who Needs Another Model?
Part 2. Lifelong "Foreign-Body Wars" in the Arteries: Role of Smooth Muscle Cells (SMCs)
Part 3. Start of a "Whole New Ballgame": Dysfunctional Clones of SMCs
Part 4. Some "Wartime Jobs" of SMCs, and the Meaning of Dysfunction
Part 5. Consequences of SMC Dysfunction: Plaques and Plaque Rupture

● Part 1. Introduction: Who Needs Another Model?

In this chapter, we present a Unified Model of Atherogenesis and of Acute IHD Death which seems to us to account for several otherwise unexplained and important aspects of Ischemic Heart Disease. For brevity, we will refer to it as our "Unified Model." We think the model is consistent with the observations reported by others, as well as with the new observations contributed by this book.

Who needs another model?

In the July 1, 1992 issue of the Journal of the American Medical Association, David H. Thom et al write (Thom 1992, p.68): "Established risk factors for coronary heart disease (CHD) such as serum lipids, smoking, hypertension, diabetes, age, gender, and family history, are usually estimated to account for less than half the variation in CHD."

Five years later, in the November 6, 1997 issue of the New England Journal of Medicine, in the Shattuck Lecture ("Special Article"), Eugene Braunwald states (p.1364): "Although much has been learned about the causes of coronary heart disease, the gaps in knowledge are noteworthy; for example, fully half of all patients with this condition do not have any of the established coronary risk factors (hypertension, hypercholesterolemia, cigarette smoking, diabetes mellitus, marked obesity, and physical inactivity)."

If it is true that so many cases lack any explanation, then "another model" should be welcome.

1a. Invitation in the New England Journal, 1997

A comprehensive theory should include the ACUTE events as well as the preceding atherogenesis, and so a model must deal also with Peter Libby's astute question (Libby 1995, p.2845): "What is it about certain plaques that renders them particularly susceptible to disruption and the ensuing acute manifestations such as myocardial infarction or unstable angina?"

In a recent editorial in the New England Journal of Medicine, Attilio Maseri issues an invitation for innovative research and new hypotheses (Maseri 1997, p.1014): "The weak relation between [acute] ischemic events and flow-limiting stenoses in the coronary and carotid arteries leaves room for innovative research (Maseri 1995). However, it is easier to study the details of accepted paradigms than it is to develop new hypotheses ..."

The search today for another PREVENTABLE cause of the acute ischemic events is emphasized by Valentin Fuster (Fuster 1994 p.2129): "Identification of vulnerable or unstable lesions prone to rupture, and measures for reversal of this pathological process, currently are subjects of worldwide investigation."

1b. A New "Smoking Gun": Stimulus for a Unified Model

This book has uncovered a new "smoking gun" with respect to causation of mortality from Ischemic Heart Disease. The irrefutable, strong, and positive dose-response uncovered here between PhysPop and IHD mortality demands an explanation. The dose-response is a clue of "smoking gun" magnitude, if ever there was one --- and it suggests a Unified Model of Atherogenesis and Acute

IHD Death. Our model offers an answer to Libby's question in Part 1a, and to several other questions. In addition, it points to a ready path to more PREVENTION of IHD deaths.

It should be self-evident that our Unified Model owes a lot to the work of others, already described in Chapter 44.

● Part 2. Lifelong "Foreign-Body Wars" in the Arteries: Role of Smooth Muscle Cells (SMCs)

As we see it, all our arterial beds (coronary, cerebral, and other) are in a lifelong process of clearing plasma lipoproteins out of the intimal layer after their influx (Chapter 44, Part 5). In that location, they are "foreign bodies," since normal metabolism would not use them in that place. We think that massive quantities of lipoproteins are processed in a lifetime in innumerable arterial walls. And obviously, most of this is handled successfully, since we rarely (if ever) see massive atherosclerosis all over the body.

2a. Diffuse Intimal Thickening

What we do observe, beginning in childhood, is gradual THICKENING of the intimal layer, and the arrival of smooth muscle cells (SMCs) --- capable of synthesizing connective tissue as needed (Chapter 44, Part 1a). McGill, citing earlier work (Geer 1968), concludes "as most others have" that the thickening of coronary intimas "represents normal arterial development, and is not a pathologic response to injury" (McGill 1984, p.447). Munro and Cotran also describe the thickening phenomenon as normal (Munro 1988, pp.250-251):

"In infancy, the endothelium of large and medium-sized arteries is generally directly apposed to [next to] the underlying media, but as the individual grows, these two structures become separated by accumulating extracellular matrix [connective tissue] and cells. The matrix contains collagen, elastin, proteoglycans, and fibronectins. Smooth muscle cells, macrophages, and T lymphocytes are present within the matrix, presumably having arrived from the media and arterial lumen. The smooth muscle cells can then be regarded as an intrinsic intimal population, and are often designated myointimal cells. After a few years of life the intima, now being composed of endothelium and underlying cells and matrix, continues to increase in height and is said to show diffuse intimal thickening. In general, this thickening can be taken as normal, presumably the result of hemodynamic or other stresses to the artery wall over time."

We would add: ... presumably the result of the continuous "Foreign-Body Wars" in the arterial intima --- the residue from mild inflammatory responses invoked when needed to help clear out the lipoproteins (and other foreign bodies). Of course, the intima's endothelial cells have roles in this inflammatory response, too.

2b. Development of Fatty Streaks

Also observed is widespread development, within the intimal layer, of fatty streaks --- in which smooth muscle cells have roles again. Fatty streaks are present "in most people under the age of 30 years regardless of their country of origin" (Fuster 1994, p.2126). Fatty streaks are found "in the coronary arteries of half of the autopsy specimens from children aged 10 to 14," according to Ross (Ross 1993, p.801). For a very long time, the relationship of fatty streaks and atherosclerotic plaques was hotly debated.

Under current classifications (mentioned in Chapter 43, Part 2c), Type I intimal lesions are those with isolated "foam cells" (macrophages containing lipid droplets); Type II lesions (flat fatty streaks) have foam cells of both the macrophage and SMC type; Type III lesions (raised fatty streaks) have "multiple but small extracellular lipid cores as well as lipid droplets in foam cells and an increasing number of smooth muscle cells" (Fuster 1994, p.2127). Writing about fatty streaks somewhat earlier, Munro and Cotran hold the view that (Munro 1988, p.250):

"On balance, it appears that fatty streaks are of universal occurrence and distribution, and most --- especially those in the aorta --- either disappear or remain harmless. In certain locations, (e.g., in coronary arteries) and especially in the predisposed individual, these streaks may conceivably evolve into fibrous plaques."

2c. Atherosclerotic Plaque: A Complex "Hive of Activity"

An atherosclerotic plaque can be a complex biochemical entity, humming with activity.

For instance, there is the continuing ingress and egress --- balanced or not --- of native and denatured lipoproteins. There is activity by monocytes, macrophages, T-cells, and platelets, all of which can secrete other molecules of great consequence (see, for instance, Munro 1988 or Libby 1995). In a large plaque, a micro-vasculature is growing to supply it with blood. Monocytes are transforming themselves into macrophages, which are trying to engulf the lipids. Such foam-cells produce large amounts of "tissue factor," a powerful stimulus of blood clotting. A highly thrombogenic lipid-pool can accumulate (Chapter 44, Part 7b).

Within this hive of activity, the smooth muscle cells need to provide high-quality collagen, elastin, and proteoglycans --- in the right quantity and at the correct speed. In advanced plaques, the products from SMCs are crucial to create, repair, and maintain a strong fibrous cap over the deadly lipid pool (Chapter 44, Part 7c).

● Part 3. Start of a "Whole New Ballgame": Dysfunctional Clones of Smooth Muscle Cells

When a failure occurs in the lifelong "Foreign-Body Wars," it is localized. Lipids start serious accumulation at a particular site in the intima --- not everywhere. Only particular patches become atherosclerotic plaques, surrounded by grossly normal tissue. Why does a plaque develop where it does? Is this totally random? We do not think so.

How do we visualize the change from routine processing of lipoproteins throughout the arteries, to the growth of a life-threatening and localized atherosclerotic plaque?

We envision it by paying attention to the new epidemiologic finding of this book with respect to mortality rates from Ischemic Heart Disease. We have uncovered a strong, positive dose-response between medical radiation and IHD death, in a huge prospective study (study-population of 150,000,000 people) --- a study based on neutral, objective data sources (Chapters 40 and 41). And we ask:

What can ionizing radiation do which could CAUSE a dose-dependent increase in death from Ischemic Heart Disease?

The answer is that ionizing radiation is a proven and potent mutagen (Chapter 2). It is capable --- most likely in concert with other agents --- of converting an individual smooth muscle cell of the artery into a benign mini-tumor consisting of clonal mutated cells which are LESS THAN COMPETENT at performing their crucial functions in the lifelong "Foreign-Body Wars" with intimal lipoproteins (Parts 4 and 5, below). Such mini-tumors would consist of DYSFUNCTIONAL clones. (A clone is a group of genetically identical cells descended from the same progenitor cell.)

● We consider that a "whole new ballgame" begins in a coronary artery wherever ionizing radiation causes this new entity, namely a clone of dysfunctional smooth muscle cells --- which gradually replace a small patch of non-tumorous tissue. We propose that some of these benign tumors have the behavior which leads to the local production of major atheromata.

● In regions adjacent to the mini-tumor, smooth muscle cells continue to perform their functions with competence --- even though they, too, were irradiated --- because exposure to a mutagen transforms only a very small share of cells into dysfunctional clones (the share rising with the dosage). Discussion in Part 4b.

Although we have mentioned only smooth muscle cells, we do not rule out the possibility that ionizing radiation induces some dysfunctional clones also among other types of arterial cells. So far, however, those who have investigated monoclonality in atherosclerotic plaques, implicate arterial smooth muscle cells (Benditt 1988, p.1000. Context in Chapter 44, Part 8).

● Part 4. Some "Wartime Jobs" of SMCs, and the Meaning of Dysfunction

So that the consequences of DYSFUNCTION can be appreciated, we will briefly describe some of the jobs, performed by normal arterial smooth muscle cells (SMCs), in the "Foreign-Body Wars" against lipoproteins in the intima.

4a. Several Functions of Normal Intimal Smooth Muscle Cells

In arterial inflammation, SMCs play the general role which fibroblasts play in other tissues (Chapter 44, Part 1a). In response to biochemical signals, smooth muscle cells synthesize and then secrete ("deposit") collagen, elastic tissue, and proteoglycans, to create interstitial connective tissue (extracellular matrix) as needed for the inflammatory response.

Another function, not mentioned in Chapter 44, is that smooth muscle cells are one of the cell-types which can produce a growth factor having the power to induce both SMC migration and proliferation (Munro 1988, p.255).

And perhaps of real importance, human vascular smooth muscle cells are also one of the cell-types which (A) can either INHIBIT degradation of the needed interstitial connective tissue, or (B) can PROMOTE such degradation when less connective tissue is needed. Peter Libby and co-workers have contributed years of laboratory work which has helped to uncover the multiple sources and complex interactions of the enzymes which control these processes. A very lucid description is presented in Libby 1995 (pp.2846-2847), from which we will extract what smooth muscle cells can contribute:

(A) Smooth muscle cells can express TIMPS --- Tissue INHIBITORS of MetalloProteinases. The metalloproteinases are a family of enzymes (proteins) which specialize in the DEGRADATION of extracellular matrix. This enzyme-family includes interstitial collagenase, gelatinase, and stromelysin. By expressing TIMPS, smooth muscle cells can help protect the extracellular matrix.

(B) Smooth muscle cells can also do the opposite: They can express some matrix-degrading enzymes. In response to biochemical signals, SMCs will express collagenase, gelatinase, and stromelysin.

Another job of normal SMCs, in the "Foreign-Body Wars," is probably to get out of the way when they are no longer needed --- a process which can involve apoptosis (cell suicide).

4b. What Qualifies as a Dysfunctional SMC Mini-Tumor?

Beginning in childhood, arterial smooth muscle cells participate in the mild inflammatory responses associated with successful clearance of most lipoproteins from the intima.

As the years pass, the arterial smooth muscle cells accumulate a rising exposure to mutagens (including ionizing radiation from natural sources and especially from medical radiation) --- and a rising frequency of cells which have acquired one or more mutations. Reminder: A mutation refers to permanent genetic change, and reflects damage which was never perfectly repaired. Of course, not all mutations are consequential. The specific consequences (if any) depend on which segment of the genome is altered.

A person's population of mutated, arterial smooth muscle cells must include some or all of the following categories:

● The Non-Clonal SMC. For this type of mutated cell, the accumulated mutations do not confer any proliferative advantage, when its neighborhood of smooth muscle cells receives mitotic signals. Thus, any acquired dysfunction is irrelevant because the dysfunction is not magnified by a burgeoning clone (tumor) of descendants.

● The Clonal but Otherwise Competent SMC. The accumulated mutations give this type of cell a proliferative advantage, but its burgeoning clone of descendants is innocuous because the descendants all are competent at their jobs. Such cells seem to correspond with the Benditts' monoclonal hypothesis (Chapter 44, Part 8).

● The Clonal AND Dysfunctional SMC. The accumulated mutations give such cells a proliferative advantage AND cause them to be incompetent in some degree at performing one or more of their jobs. These are the cells which become qualified as SMC Mini-Tumors consisting of DYSFUNCTIONAL clones --- corresponding with the second part of our Hypothesis-2. And because of their proliferative advantage, such dysfunctional cells gradually REPLACE competent cells, at a localized patch of artery.

● Part 5. Consequences of SMC Dysfunction: Plaques and Plaque Rupture

With respect to the lifelong "Foreign-Body Wars" against lipoproteins in the intima, we propose that plaques develop and that some of them subsequently rupture because mutation-induced mini-tumors (consisting of dysfunctional clones of smooth muscle cells) are unable to do some of their crucial wartime jobs adequately --- jobs required by the early battles and later by the final battles which cause mortality from Ischemic Heart Disease.

5a. Innumerable Genetic Pathways to Dysfunction

Within an arterial smooth muscle cell, there are INNUMERABLE genes involved in completing all the wartime jobs assigned to it. For example, many dozens of enzymes and other proteins are typically required, directly and indirectly, for the synthesis and secretion of a single strand of collagen or elastin or other molecules. Dozens of other enzymes and proteins can also be required to transmit a particular type of signal through the cell's outer membrane all the way to the nucleus of a cell, in order to activate or inactivate various genes. And numerous regulatory genes control the relative dominance of opposing tendencies in a cell.

A disabling mutation, of even ONE of these numerous genes in a clone's progenitor, can cause the clone to perform inadequately.

Ionizing radiation inflicts mutations at RANDOM locations of a cell's genome, and there is no part of the genome which is inaccessible to radiation-induced mutation (Chapter 2, Part 4). Consequently, it is highly improbable that any two dysfunctional clones have identical disabling mutations. For example:

One mini-tumor may evolve from a cell where the xyz-gene is mutated at a segment where damage impairs only slightly the function of the xyz-protein which the gene encodes. In another mini-tumor, the same xyz-gene can be mutated at a segment where the functional consequences are severe. And in yet another mini-tumor, the mutation may remove (delete) the ENTIRE xyz-gene, if the mutation is an unrepaired double-strand chromosome break. Such examples of variation are applicable to every gene required for SMC mini-tumors to do their wartime jobs. On the average, clones having multiple mutations (either in one gene or in several different genes) will be more dysfunctional than clones having a single mutation.

The particular nature of a clone's mutations will determine the changes in its biochemistry and the severity of its resulting dysfunction. The net result of disabling mutations will vary. For example:

● - One dysfunctional clone (patch) of smooth muscle cells might be slow in producing collagen --- producing too little per unit time. RATE MATTERS --- a fact well known to wartime commanders and biochemists alike.

● - Another dysfunctional clone might produce enough collagen, but it might be defective collagen which does not adequately fill its structural function, either in "early" lesions or in full-grown plaques. Quantity fine, but quality inadequate. Other clones might be unable to produce TIMP in the right quantity or quality (Part 4a), or at the NEEDED RATE.

● - Other dysfunctional clones might have their main problem in receiving signals --- with a net result of either under-responses or over-responses. Suppose that the SMCs in an advanced plaque end up over-responding to an apoptosis (suicide) signal. Then SMCs in the fibrous cap will become too scarce, and maintainance of the plaque's fibrous cap will be impaired (Chapter 44, Part 7d). Or suppose that the SMCs in an advanced plaque end up under-responding to the signal to produce TIMP (Part 4a). The result might well be inappropriate degradation of the collagen structure of the plaque's fibrous cap, and rupture of this weakened cap.

Thus, our Unified Model helps to answer a fundamental question: WHY do plaques of a single person differ from each other, for instance in the amount and form of accumulated lipid, the density of smooth muscle cells, the amount of fibrotic material, and the severity of the inflammatory response and its various other constituents? Plaques differ from each other because dysfunctional clones, of arterial smooth muscle cells, suffer from different sets of mutations.

5b. Libby's Question: Why Are CERTAIN Plaques Prone to Rupture?

We return to Libby's question (Libby 1995, p.2845): "What is it about certain plaques that renders them particularly susceptible to disruption and the ensuing acute manifestations such as myocardial infarction or unstable angina?"

Our Unified Model proposes an answer:

At different sites, the consequences of dysfunction accumulate in various degrees, because no two plaques share all the same mutations. Some dysfunctional clones will produce fibrous caps which are inherently more vulnerable than other caps to rupture. Moreover, an inferior cap which would suffice at a relatively quiet site may readily rupture, if it is located where the coronary arteries experience much greater twisting and mechanical stress (blood pressure, turbulence) than at other locations. Our model predicts that the plaques which DO rupture are determined jointly by the degree of their gene-based weakness and the degree of their external stress.

At sites of equal mechanical stress, plaques differ in their degree of vulnerability to "disruption" because they arise from mini-tumors of smooth muscle cells, and those mini-tumors suffer from DIFFERENT degrees of gene-based dysfunction. Particular mixtures of mutations result in fibrous caps which are more vulnerable to rupturing than the caps which result from other mixtures of mutations.

And Hypothesis-2 proposes that the main cause of these arterial mini-tumors is medical radiation. Were it not for these radiation-induced tumors, the normal inflammatory response, operating in its full majesty, would usually triumph in the lifelong "Foreign-Body Wars" of the arteries.

5c. What about Cases of IHD with No History of Medical Radiation?

We are well aware that Ischemic Heart Disease develops in some people who really have no history at all of medical irradiation. Such cases are consistent with our model.

To be clear: Our model does NOT "predict" that there was no Ischemic Heart Disease in humans before the introduction of xray machines. There are two crucial reminders. The first reminder is that every human being everywhere receives lifelong exposure to ionizing radiation from natural sources. The second reminder is that the coronary arteries are exposed to some OTHER mutagens besides ionizing radiation.

Our Unified Model of Atherogenesis and Acute IHD Death readily embraces a role for non-radiation mutagens also. Any mutagen qualifies if it can convert an arterial smooth muscle cell into the progenitor of a dysfunctional clone of smooth muscle cells. Whatever the source of acquired mutations (acquired at any interval after conception), the key issue is repairability --- because perfectly repaired gene-damage has no consequences.

It is worth noting that infants can acquire mutations during their gestation. With respect to xray-induced mutations, pelvimetry by xray examination was introduced soon after the 1896 discovery of the xray. How commonly was pelvimetry used? We can offer an estimate only for the 1947-1970 period. During those years, approximately 7.5% of mothers in the USA had such pre-delivery xray examinations --- so about 1 infant in every 13.5 experienced pre-birth medical radiation (Chapter 2, Part 2d).

In future research, on the role of acquired mutations in Ischemic Heart Disease, are there good reasons to focus on medical radiation, instead of on mutagens in general? Yes.

• The first reason: Medical radiation is a proven mutagen which has an irrefutable and positive dose-response with IHD mortality. That dose-response is staring at us from Chapters 40 and 41 --- and demanding "Explain me!"

• The second reason: Medical radiation is a mutagen to which vast numbers of people actually receive exposure during a lifetime.

• The third reason: Ionizing radiation has some unique and undisputed properties which give it access to every gene in every type of cell, and which enable it to cause complex double-strand chromosomal injuries --- some of which are never correctly repaired (Chapter 2, Part 4, and Appendices-B+C). The doubling-dose for xray-induced structural chromosomal mutations is very low (Chapter 2, Part 4b). Indeed, the xray doses accumulated from some common, current, nontherapeutic medical procedures are equivalent to such doubling-doses (Chapter 2, Part 7e). Medical radiation may well be the single most menacing mutagen to which people are routinely exposed.

CHAPTER 46

How the Unified Model Helps to Explain, or Is Consistent with, Established Observations

Part 1. Observation: Lipid–Lowering Reduces Acute IHD Events
Part 2. Observation: Lipid–Lowering Benefit Occurs with Little Change in Stenosis
Part 3. Observation: "Culprit Plaques" Are Often the Less Stenotic Ones
Part 4: Observations: Atherosclerotic Plaques Are Localized; Distribution Varies
Part 5: Observations about Endothelium, Homocysteine, Hypertension, Smoking, Diet
Part 6: Some Wise Words from Earl P. Benditt

Box 1. Illustrative Relationship between the Unified Model and Benefit of Lipid–Lowering.

● Part 1. Observation: Lipid–Lowering Reduces Acute IHD Events

Lipid–lowering regimes achieve marked reduction in acute IHD events, while causing only minimal reduction of stenosis in the angiographically measured plaques (Chapter 44, Parts 6d and 7a). Attempts to reconcile these two facts have elicited various conjectures.

1a. Reduced Inflammatory Response and/or Better Function of Endothelium

It has been proposed (for instance, by Brown 1993, p.1788, p.1789, + Libby 1995, p.2848, Figure 3) that reducing the influx–rate of lipoproteins into the intima stabilizes plaques against rupture because the reduced influx–rate decreases all aspects of the inflammatory response, including the number of lesional T cells and macrophages which may release matrix–degrading enzymes inappropriately.

We might add that reduced influx of lipoprotein into the intima should also lessen the clinical CONSEQUENCES of a plaque rupture, because reduced accumulation of lipids should reduce the frequency of foam cells ––– which produce thrombogenic tissue factor (Chapter 44, Part 7b).

Some other analysts postulate a role for the endothelium, in the reduced frequency of acute IHD events after lipid–lowering. For example, Thomas A. Pearson (citing Flavahan 1992) writes that "LDL cholesterol reduction" may restore the ability of endothelial cells to produce EDRF ––– endothelium–derived relaxing factor ––– and such restoration "may be an important mechanism" to explain the rapidly resulting decline in acute IHD events (Pearson 1993, p.1073). Referring to the reduced rate of cardiac events, John Bittl (1996, p.1300) writes that "The mechanism for this benefit has been attributed to minor regression of fixed stenoses, generalized improvement in endothelial function, and a decreased risk of plaque rupture during lipid–lowering therapy (Treasure 1995; Anderson 1995)." William Parmley offers a similar summary, when he says there are two hypotheses for the benefit (Parmley 1997, p.12):

"The first is that normalization of endothelial function is highly protective against clinical events. The second is that removal of the lipid pool plaques and lipid laden macrophages, particularly at the margins of plaques, may stabilize them so that they are less likely to break down. Perhaps both of these mechanisms are playing an important role in the dramatic reduction in clinical events which occurs from lipid lowering."

We agree that several mechanisms may contribute to the observed clinical benefit from reduced serum levels of atherogenic lipoproteins ––– and we certainly do not dismiss a role for the endothelium.

1b. How Our Unified Model Helps to Explain the Observations

Acquired mutations (which are permanent) are a fundamental part of our Unified Model, and lipid–lowering does not reverse such mutations (by definition). But since the Unified Model proposes that the COMBINATION of certain lipoproteins and acquired mutations is necessary for atherosclerosis, the model predicts effects from changing the frequency of either partner.

(A) HOLDING RADIATION DOSE CONSTANT.   At equal accumulated radiation dose (a potent mutagen), individuals with the higher plasma levels of atherogenic lipoproteins will fare worse. With greater influx of atherogenic lipoproteins into the intima per unit time, the urgency is greater for smooth muscle cells (SMCs) to do all their jobs well (Chapter 45, Part 4a).   While a higher lipid-load does not increase the absolute number and type of radiation-induced dysfunctional SMC clones, the higher lipid-load means that a growing FRACTION of the clones are severely inadequate to cope with the elevated demand on their performance.   Thus, the atherosclerotic process PROGRESSES at a larger fraction of the mini-tumors, and a growing fraction of fibrous caps become inadequate to contain their thrombogenic lipid-pools, and more plaques rupture per unit time.   Part 1c and Box 1 explain this fraction-concept in much more detail.

(B) HOLDING ATHEROGENIC LIPOPROTEINS CONSTANT.   At equal plasma levels of atherogenic lipoproteins, individuals with the higher accumulated radiation dose will fare worse.   On the average, they will have more mini-tumors of dysfunctional SMCs, and therefore, more plaques. The mini-tumors will produce plaques with various degrees of dysfunction and of vulnerability to rupture, because no two radiation-induced mini-tumors have the SAME mutated genome (Chapter 45, Part 5a).   A higher frequency of radiation-induced mini-tumors raises the frequency of both mildly dysfunctional and severely dysfunctional varieties.

1c. Box 1: Effect of a DECREASE in Lipid-Load

Scenario "A" in Part 1b means that if two individuals have the same number of radiation-induced dysfunctional SMC mini-tumors, the individual with the LOWER plasma levels of atherogenic lipoproteins will fare BETTER, all other things being equal.   Likewise, if ONE person lowers his/her lipid-load, the individual earns a lower risk of fatal Ischemic Heart Disease.   Box 1 provides an illustration of how this happens.

Because mini-tumors of dysfunctional smooth muscle cells will have various DEGREES of dysfunction, we can rank them in Box 1 (Column A) on a scale of competence.   In our hypothetical illustration, dysfunctional SMCs of Class-A clones are the most nearly competent.   SMCs of Class-K clones are the least competent.

To quantify their variation in competence, we say that Class A can handle a lipid-load of 90 arbitrary units every month without progression of prior atherosclerosis (Box 1, Column D).   By contrast, Class K can handle only 2 arbitrary units/month.   Rates matter (Chapter 45, Part 5a).   In the text and in Box 1, "lipid-load" refers only to the lipoproteins which are atherogenic (not to all lipoproteins).

We make the reasonable approximation that Class A mini-tumors are the most nearly competent BECAUSE they have accumulated the fewest mutations for dysfunction (Chapter 45, Part 5a), whereas, at equal radiation dose, we approximate that Class K mini-tumors are the least competent because they have acquired the MOST mutations for dysfunction.   Accumulation of MANY mutations for dysfunction in the same cell-nucleus is less likely to occur than accumulation of a few such mutations, from equal accumulated radiation doses.   So it follows that Class K mini-tumors occur less frequently than Class A mini-tumors, at equal radiation doses.   We arbitrarily establish a factor of about 2-fold disparity in frequency, in our illustration (Box 1, Columns B and C).

Box 1 uses our illustrative input to show how a reduction in lipid-load (from 60 arbitrary units per month to 40 arbitrary units per month) reduces the fraction of sites where the atherosclerotic process progresses.   Because of the reduced lipid-load, a higher fraction of the 773 dysfunctional clones becomes able to stop progression, and thus fewer sites ever develop large thrombogenic lipid-pools with incompetent containment --- the proximate cause of so many acute IHD events.

1d. Distinction between Comments in Part 1a and the Unified Model

The comments above closely resemble some of the comments in Part 1a about the inflammatory response.   Indeed they should.   The distinction, between such comments and our Unified Model, is the model's addition of the mutational component, which proposes how certain acquired mutations and atherogenic lipoproteins operate together.

When levels of atherogenic lipoproteins are reduced, a SMALLER fraction of dysfunctional SMC clones (dysfunctional due to acquired mutations) will be severely inadequate at their tasks of handling the lipid-load, per unit time, and a larger fraction will be ABLE to cope with the lesser load per unit time. In the lesions which are already advanced when the lipid-load is reduced, coping means especially producing, repairing, and maintaining an adequate fibrous cap. At a reduced plasma level of atherogenic lipoproteins, the rate of influx into the intima per unit time is lower, and the rate of efflux may be adequate to STABILIZE various lipid-pools at sizes with which more of the mini-tumors can cope. Although the mutation-induced dysfunctions persist, the frequency of fatal consequences can be reduced by lowering plasma levels of the atherogenic lipoproteins.

Box 1 illustrates how lipid-lowering can halt PROGRESSION of the atherosclerotic process at an increased fraction of sites. This halt will involve not only the stenotic sites, but also the many pre-stenotic sites (Chapter 43, Parts 3e and 4). Progression of atherosclerotic lesions predicts subsequent cardiac events (Blankenhorn 1994, pp.183-184, for instance). Less progression --- as in our Box 1 --- means fewer acute IHD events.

● **Part 2.   Observation: Lipid-Lowering Benefit Occurs with Little Change in Stenosis**

An analogy, between an atherosclerotic plaque and an office building, may help to elucidate the observation that lipid-lowering regimes achieve marked reduction in acute IHD events, while causing only minimal reduction of stenosis in the angiographically measured plaques (Chapter 44, Parts 6d and 7a).

The plaque's "office building" consists of the extracellular matrix (the structural proteins and proteoglycans), and in more advanced plaques, the microvasculature, crystallized cholesterol, and sometimes deposited calcium. The plaque's "population" consists of the extracellular lipids, and the smooth muscle cells, macrophages, foam cells, T cells, platelets, and other relatively mobile constituents of a plaque. Their interactions constitute a complex "hive of activity" (Chapter 45, Part 2c) --- the inflammatory response.

If a plaque's population is growing, the structure must expand too. A new "wing" is added ... with more floors. Normal smooth muscle cells, adjacent to the dysfunctional patch, also may grow into the plaque --- especially if the dysfunctional SMC clones are becoming senescent.

And what happens when the inflammatory response becomes less intense, so that the plaque's population is gradually DECLINING?

Just as a population can evacuate an office building while leaving the edifice intact, so could a plaque's inhabitants vacate some of the offices while leaving the edifice intact. As the plasma levels of atherogenic lipoproteins decline, the population in the plaque can also decline ... and do so, without tearing down the building. So it does not surprise us that a large clinical benefit occurs in lipid-lowering trials, while hardly any change (angiographically measurable "regression") occurs in the size of the edifice. It is often not the EDIFICE, but rather, contact between the thrombogenic lipid pool and the bloodstream, which causes the acute IHD events (Chapter 44, Part 7b). Reducing the frequency of thrombogenic lipid-pools can produce clinical benefits, even though the plaque's edifice remains.

● **Part 3.   Observation: "Culprit Plaques" Are Often the Less Stenotic Ones**

The plaques which trigger acute IHD events (unstable angina, myocardial infarction, sudden death) are often the less stenotic plaques (Chapter 44, Parts 6d and 7a).

3a. What Is Meant by "Often?"

What is meant by "often"? In 1986, Brown and co-workers reported some data on frequencies, summarized in Brown 1993, p.1786-1787:

"Acute ischemic syndromes are most commonly precipitated when mild or moderate coronary lesions [which they define as <70% stenosis] become disruptively transformed into severely obstructive culprit lesions. Such disruption usually involves fissuring of the fibrous cap of the atheroma, often with intramural hemorrhage and mural or occlusive thrombus ... Among patients undergoing thrombolytic therapy for acute myocardial infarction, the severity of the atherosclerotic stenosis underlying the thrombotic occlusion was measured at <50% diameter stenosis in one third of cases and between 50% and 60% stenosis in another third (Brown 1986)."

It is worth noting that, in the remaining third of those patients with acute myocardial infarction, the culprit lesions must have been those with >60% stenosis. So very stenotic lesions also cause acute troubles. Do they cause more than their share, or less? The answer is not at hand. We would need to know the ratio of (culprit mild-moderate lesions over total mild-moderate lesions), to compare with the ratio of (culprit severe lesions over total severe lesions).

In 1988, Little and co-workers reported frequencies from their study of 29 patients with myocardial infarction who each had had a coronary angiogram beforehand --- at an average of 706 days beforehand (Little 1988, p.1159, Table 1). Little and co-workers found that in 66% of the 29 patients, the most severe pre-infarction stenosis existing in the infarct-related artery was less than 50% luminal diameter narrowing. In 97% of the 29 patients, the most severe pre-infarction lesion in the relevant artery was less than 70% diameter stenosis (Little 1988, p.1159). These are important data.

Nonetheless, such data are not informative about how those culprit lesions might have evolved during the nearly two years, on the average, between pre-infarction measurement and the infarction. Major progression at a particular plaque-site can occur very rapidly --- for instance, due to silent rupture ("subclinical episodes," Chapter 44, Part 7 at its outset and Part 7a).

3b. Why Are Less Stenotic Lesions "Much More Numerous"?

From these studies and others, it seems established that the majority of acute IHD events are precipitated by atherosclerotic lesions which were, fairly recently, only mildly to moderately stenotic. Brown and colleagues propose a sensible explanation for why this is so: The mild to moderate lesions are "much more numerous." Here is the context (Brown 1993, p.1787):

"Although a given severe (>=70%) lesion is more likely to progress or totally occlude than a given mild or moderate lesion, clinical events are more frequently precipitated by lesions initially of the less severe type because these are much more numerous in the patient's anatomy (Brown 1989) and also because the majority of occlusions of severe stenoses occur without an event (Webster 1990)."

As far as it goes, we like the Brown 1993 explanation. Our Unified Model takes another step and also explains WHY less severely stenotic lesions are "much more numerous" and WHY some of the most severe stenoses continue growing "without an event."

The Unified Model interprets greater stenosis as usually signaling a greater degree of dysfunction in the SMC mini-tumor at that site. (The extra incompetence of the site's smooth muscle cells, at doing all of their jobs right, causes extra inflammation and an extra large "edifice" at that site.) The most severely incompetent mini-tumors generally evolve from cells which have accumulated multiple mutations. Since the frequency of cells with many mutations is lower than the frequency of cells with fewer mutations, the frequency of severely stenotic sites is relatively low.

The fact that SOME plaques can occlude a lumen without having "an event" (a known rupture) is consistent with the model, too. The model says that SMC mini-tumors differ in their mutations, and therefore they differ in their types and degree of dysfunction (Chapter 45, Part 5). It follows that some mini-tumors which are severely dysfunctional --- and produce very large plaques --- may nonetheless be adequate with respect to producing strong fibrous caps. They are the ones whose plaques can continue to grow, without having a rupture which kills the host.

3c. Observation: Progression of Lesions Is Not Uniform

The statement directly above relates to the observation by Valentin Fuster and others that all plaques do not inexorably follow the same path of behavior. For instance, Fuster presents text and a diagram showing a common path for the least severe lesions (types I, II, III), after which the path of a particular lesion is unpredictable. Fuster 1994 (p.2127):

"The type IV lesion has a predominance of extracellular lipid, mainly diffuse, and the type Va lesion has a high lipid content, mainly localized, and a very thin capsule. Types IV and Va lesions can evolve at an intermediate rate (months to a few years) into the more stenotic and fibrotic types Vb and Vc lesions. However, it appears that more often and acutely, these lesions rupture, and then a change in their geometry and subsequent thrombus formation may lead to the type VI complicated lesions ..." Also he notes that when stenosis evolves GRADUALLY instead of abruptly, the process all the way to occlusive stenosis can be silent (asymptomatic), because the resulting ischemia stimulates the growth of protective collateral circulation.

Such variation in the evolution of atherosclerotic plaques is exactly what our Unified Model predicts --- because the SMC mini-tumors have different mutations, and therefore they differ in types and degree of dysfunction (Chapter 45, Part 5).

● **Part 4: Observations: Atherosclerotic Plaques Are Localized; Distribution Varies**

Munro and Cotran (Munro 1988, p.251) comment that a comprehensive theory of atherogenesis should include an explanation for "the focal nature of the lesions and their general distribution."

The Focal Nature

In our model, the focal nature of the lesions, adjacent to plaque-free tissue, results from mutation-induced mini-tumors of dysfunctional SMC clones.

Quite obviously, routine processing of lipoproteins in the arterial wall does NOT result in atherosclerotic plaques everywhere. Why do we find atheromatous lesions highly localized, with apparently normal arterial tissue adjacent in the same artery? Because induction of dysfunctional clones (mini-tumors) of smooth muscle cells is a random and rare occurrence. Despite exposure to ionizing radiation (or other mutagens), the overwhelming share of arterial smooth muscle cells do not acquire the combination of mutations required to become both clonal and incompetent (Chapter 45, Part 4b).

The General Distribution

Munro and Cotran write (Munro 1988, p.250): "The distribution of plaques is important to explain. The most common site is the lower descending aorta, predominantly around the ostia of the major branches, followed by the coronary arteries, usually within the first six centimeters, the arteries of the lower extremities, descending thoracic aorta, the internal carotids, and the Circle of Willis. Some of this localization can be explained by hemodynamic factors, such as shear stress or disturbed flow (Glagov 1972; Ku 1985), but certainly not all of it."

Much of "the rest of it" may be explained by medical radiation, which exposes all of the sites named by Munro 1988. Munro and Cotran continue (Munro 1988, p.250): "Particularly vexing is the problem of determining the reason behind the heterogeneous distribution between and within individual subjects."

Our model predicts that some or most of the answer lies in the differing radiation histories of different cases.

Testing of this prediction will become possible someday, but not yet. The cases occurring today arise from radiation accumulated over a lifetime. For many people over age 50 today, a great deal of irradiation could have occurred during childhood, and some of it, even in-utero. Records are poor, and memories far worse. Unfortunately, even today, medical records seldom report how long a fluoroscopic xray beam is used and which organs are actually in the beam.

● **Part 5: Observations about Endothelium, Homocysteine, Hypertension, Smoking, Diet**

We will discuss briefly how our Unified Model relates to many of the established and proposed risk-factors for Ischemic Heart Disease, and to some of the current regimes for primary and secondary prevention.

5a. Proposed and Established Risk-Factors for IHD

● - DYSFUNCTIONAL ENDOTHELIUM IS A RISK-FACTOR FOR ATHEROSCLEROSIS. We presented several aspects of this hypothesis in Chapter 44, Part 2. In view of the endothelium's active roles in the passage of lipoproteins into the intimal layer, in the inflammatory response, in thrombogenesis, and in other functions, we have no trouble believing that --- at equal plasma levels of atherogenic lipoproteins and equal accumulated radiation dose --- individuals with an appropriately responding endothelium will fare better on the average than individuals with a dysfunctional endothelium. And this would be true regardless of the specific causes of endothelial dysfunction --- whether the causes are poor nutrition, infections, mutations (either acquired or inherited), or other injurious agents.

● - HIGH BLOOD-PRESSURE IS A RISK-FACTOR FOR ISCHEMIC HEART DISEASE. This is not in doubt. It follows that --- at equal plasma lipoprotein-loads and equal accumulated radiation dose --- individuals without hypertension will fare better on the average than individuals with hypertension. Without hypertension, we expect that the intima will have to handle less lipoprotein per unit time. Great progress has been made since 1940 in controlling hypertension (AHA 1995). Today, specific TYPES of hypertension which confer the most risk are being identified.

● - SMOKING IS A RISK-FACTOR FOR IHD. This seems to be well established. It follows that --- at equal plasma lipoprotein-loads and equal accumulated radiation dose from non-tobacco sources --- individuals who do not smoke will fare better on the average than individuals who smoke. The mechanism which makes smoking a cause of IHD is a separate issue. Benditt, Martell, Trosko, and Chang propose that the mechanism is delivery of mutagens, from smoke, into the coronary arteries (Chapter 44, Parts 8d, 9a, 9b).

● - OBESITY AND PHYSICAL INACTIVITY ARE RISK-FACTORS FOR IHD. In general, these conditions correlate with each other and with elevated plasma levels of atherogenic lipoproteins. (The weight-lipoprotein relationships can complicate prospective research; Appendix I, Parts 3+4). Obesity and inactivity are said to contribute to IHD risk in additional ways. If they do, it follows that --- at equal plasma lipoprotein-loads and equal accumulated radiation dose --- individuals who are slim and active will fare better on the average than individuals who are not.

● - HIGH PLASMA LEVELS OF ENDOGENOUS HOMOCYSTEINE MAY BE A RISK-FACTOR FOR ACUTE IHD EVENTS. We have presented some of the interesting observations on this issue in Chapter 44, Part 10. For cardiovascular diseases, this risk-factor seems potent. With respect specifically to Ischemic Heart Disease, we would have no trouble believing that --- at equal plasma lipoprotein-loads and equal accumulated radiation dose --- individuals who have lower levels of plasma homocysteine would fare better on the average than individuals with higher plasma levels of homocysteine.

● - LOW LEVELS OF ENDOGENOUS HEPARIN MAY BE AN INDEPENDENT RISK-FACTOR FOR IHD. In a recent review-article, Hyman Engelberg points to numerous lines of evidence for a possible causal role of low endogenous heparin activity in atherosclerosis (Engelberg 1996, p.84, Table 1), and concludes: "Ultimately, experimental observations must be validated by clinical studies in humans." Affirmation of this hypothesis would be fully compatible with our Hypothesis-2, by which we mean that at equal plasma lipoprotein-loads and equal accumulated radiation doses, individuals who have elevated levels of endogenous heparin would fare better on the average than individuals who do not.

● - DIABETES MELLITUS IS A RISK-FACTOR FOR IHD. There is simply no doubt that Ischemic Heart Disease (Coronary Heart Disease), now and for a long time, has been a frequent companion of Diabetes Mellitus --- as noted in Chapter 29, Box 1. Not all the reasons have been identified (discussion in Bierman 1992, for instance). Our Unified Model implies that, at equal dose-levels of medical radiation, diabetics will fare worse than non-diabetics, because diabetics on the average have higher blood-levels of atherogenic lipoproteins.

● - HYPERINSULINEMIA MAY BE AN INDEPENDENT RISK FACTOR FOR IHD. It has not been clear whether the association, between elevated blood-levels of insulin in fasting individuals and subsequent development of Ischemic Heart Disease, means that hyperinsulinemia is a MARKER for several established risk-factors with which hyperinsulinemia correlates (such as unfavorable lipoprotein levels, hypertension, obesity, etc.), or whether the association means that hyperinsulinemia independently confers a risk above-and-beyond its correlation with such other risk-factors. In April 1996, Jean-Pierre Despres and co-workers discussed the results of several earlier studies and reported the results of their own study (Despres 1996). In their own study, they conclude that hyperinsulinemia makes an independent contribution to the risk (Despres 1996, p.955). Affirmation of their work would be fully compatible with our Hypothesis-2, by which we mean that at equal plasma lipoprotein-loads and equal accumulated radiation doses, individuals who do not have elevated fasting levels of insulin would fare better on the average than individuals who do.

● - PARTICULAR NUTRITIONAL DEFICITS MAY BE RISK-FACTORS FOR IHD. In this group, for instance, one might include below-optimal levels of vitamins B-6, folic acid, and E, or too little absorbable magnesium, or too little consumption of omega-3 fatty acids. Other authors might mention a deficit of various hormone precursors and herbs and roots in the diet. One must expect that ANY substance (dietary or not) which appreciably worsens the problem of atherogenic lipids, or reduces the fraction of gene-injuries which are repaired correctly, or interferes with the body's optimal response to ischemia, thrombus formation, or to myocardial infarction (for instance), should show up as a risk-factor for IHD mortality. For each nutrition-based risk-factor which has been validated, searches for the mechanisms of harm should follow, of course. Appendix-F discusses dietary carbohydrates and fats, especially omega-3 fatty acids and the "Mediterranean Diet."

● - ADVANCING AGE IS A RISK-FACTOR FOR IHD. Accumulated radiation dose to the coronary arteries can only grow as age advances --- from natural, medical, and occupational sources, and from nuclear pollution. Indeed, radiation exposure from medical procedures rises steeply with advancing age, on the average. In addition, serum levels of atherogenic lipids generally start rising in the teen years --- much more for males than females, whose levels catch up when they are in their 50s (Glazier 1954). Some other risk-factors (such as body-weight, on the average, and reduced estrogen levels on the average) also rise with age. Moreover, atherosclerotic lesions take some time (variable) to develop into their life-threatening stages. Our Unified Model predicts that advancing age must be a risk-factor.

● - INHERITED DISORDERS WHICH CAUSE EXCESSIVE BLOOD-LEVELS OF ATHEROGENIC LIPOPROTEINS ARE A RISK-FACTOR FOR IHD. There is simply no doubt that some individuals inherit mutations (de novo or parental) which cause them to experience extremely high blood-levels of various atherogenic lipoproteins. By ages 20 or 30, they sometimes experience angina pectoris and/or xanthomata. Indeed, xanthoma tendinosum has occasionally been observed at birth. Are individuals who have inborn lipid disorders an exception to our Unified Model? Probably not. The fact that such individuals develop atherosclerotic lesions only at a finite number of sites is consistent with a requirement for dysfunctional clones caused by mutations which are acquired AFTER conception. Even mildly dysfunctional clones (Class A, in Box 1's Column A) are probably unable to cope with the enormously high "lipid-load" in the blood-stream of such individuals.

5b. Benefits from Pharmaceuticals Other Than Lipid-Lowerers

Valentin Fuster (1994, pp.2137-2140, with lots of references) presents a useful discussion of how (if known) various non-lipid-lowering pharmaceuticals achieve their benefits against mortality from Ischemic Heart Disease.

● - BETA-BLOCKERS. The beta-blockers reduce the rate of acute IHD events by reducing blood pressure and heart rate (Fuster 1994, p.2138).

● - ACE INHIBITORS. Angiotensin-Converting Enzyme Inhibitors reduce the rate of acute IHD events, including myocardial infarction, but "the mechanism of such reduction is uncertain," according to Fuster (p.2138). "Theoretically, it may be due to a decrease in plaque stress caused by lower blood pressure or reduced levels of neurohumoral activation (Francis 1989), thus reducing the possibility of plaque rupture. However, the decrease in blood pressure of these three trials [showing benefit] was relatively too small to suggest this hypothesis." Fuster favors the concept that ACE Inhibitors increase vasodilation of the micro-circulation and thus help to protect the myocardium from infarction during thrombotic episodes. William Parmley (1997. pp.12-13, with many references) mentions that ACE inhibitors not only increase vasodilation, but also may help block the thrombotic activity of Angiotensin II in vessel walls. Parmley reports that several large trials with ACE inhibitors are still underway to explore benefits apparently unrelated to reduced blood pressure.

● - ANTI-THROMBOTIC AGENTS. Such agents include platelet inhibitors and anti-coagulants. Fuster (1994, p.2139, citing many references) reports their effectiveness in preventing acute coronary events. Clot-formation appears to have a role in increasing the size of atherosclerotic plaques (Chapter 44, Part 7 at the outset and Part 7a). Therefore, Fuster reasons that anti-thrombotic agents "may offer some promise in preventing the progression of small coronary atherosclerotic plaques (Chesebro 1992)."

Fuster (1994, p.2139) points out that ASPIRIN is "the best-suited, least toxic, and most widely used antithrombotic agent in acute and chronic coronary artery disease," with benefits demonstrated in both primary and secondary prevention. Since aspirin and other antithrombotic agents each block only one of multiple pathways to thrombus formation, several ongoing trials (listed by Fuster) are trying to establish if use of a combination of agents will produce additional benefits, without unacceptable side-effects.

● - ESTROGEN-REPLACEMENT THERAPY. Such therapy for post-menopausal women appears to produce a large benefit in reduced mortality from cardiovascular disease --- "in part related to reduction in myocardial infarction" (Fuster 1994, p.2139, with references). Fuster reports lines of evidence that estrogens have direct anti-ischemic effects, perhaps acting as coronary artery vasodilators, for instance. Fuster also reports that the benefit comes partly from a resulting 15% decrease in LDL cholesterol, a 15% increase in HDL cholesterol, and inhibited influx of cholesterol through the endothelium into the intima. Citing Rosano 1993, Fuster states that "favorable alterations in lipid metabolism" probably do not explain ALL of the benefit, since the benefit is observed also in women who do not show the measured alterations.

In this context, we remind readers that many "alterations in lipid metabolism" go WITHOUT measurement. For example, the common short-cut of measuring "total triglycerides" --- instead of quantifying their lipoprotein sources, which differ in size and behavior --- may CREATE a number of unexplained results whose explanation is hidden primarily by the short-cut (Chapter 44, Part 4).

● - THYROID-HORMONE THERAPY. For hypo-thyroid individuals, adjustment of thyroid-hormone levels to more favorable values generally results in more favorable (less atherogenic) lipoprotein patterns.

● - Verified benefits from pharmaceutical agents are certainly compatible with our Unified Model, even when such agents achieve the benefit via co-actors in this disease OTHER THAN atherogenic lipoproteins and medical radiation. Obviously we expect that, at equal initial levels of atherogenic lipoproteins and equal accumulated radiation dose, individuals who take favorable pharmaceuticals should fare better on the average, with respect to IHD, than individuals who do not receive such help.

● Part 6: Some Wise Words from Earl P. Benditt

This chapter has explored how our Unified Model either helps to explain, or is consistent with, a great variety of observations about Ischemic Heart Disease. There is much, much more to be learned about the specific roles of acquired mutations in this disease. We would like to associate ourselves with the wise words of Earl P. Benditt, who wrote in 1988 (Benditt 1988, pp.1000-1001):

"Progress in science depends on consideration of various hypotheses as we search for the best explanations of natural phenomena. Recent progress with genetic analysis of cell function and disease aberrations clearly indicates the vast complexity yet to be uncovered."

The findings in Chapters 40 and 41 very strongly suggest that mutations, acquired in coronary arteries from medical radiation, are an important cause of mortality from Ischemic Heart Disease. Our Unified Model proposes HOW acquired mutations could cause such mortality. Much remains "yet to be uncovered."

Meanwhile, it deserves emphasis again that the powerful, positive dose-response uncovered in Chapters 40 and 41 between PhysPop and IHD MortRates, for both males and female, is a fact --- not an hypothesis.

>>>>>>>>>>

Please see text in Chapter 46, Part 1c. All entries below are arbitrary, for illustration only. "Lipid-load" is in arbitrary units, and of course refers only to the lipoproteins which are atherogenic.

(A) Mini-Tumor Class: Most Nearly Competent = A. Least = K.	(B) Frequency in a Universe of 773 Mini-Tumors.	(C) Percent of Universe.	(D) Lipid-Load/Month at Which Atherosclerosis Is Stabilized.
Class A	95	12.29%	90*
Class B	90	11.64%	80*
Class C	87	11.25%	70*
Class D	80	10.35%	60
Class E	76	9.83%	50
Class F	70	9.06%	40
Class G	65	8.41%	30
Class H	60	7.76%	20
Class I	55	7.12%	10
Class J	50	6.47%	5
Class K	45	5.82%	2
Sum	773	100.00%	

At a lipid-load (Col.D) of 60 units/month:

(95+90+87+80), or 352 mini-tumors (Col.B) are coping well enough to mean that the atherosclerotic process is not progressing at their plaque-sites. 352/773 = 0.46

(76+70+65+60+55+50+45), or 421 mini-tumors (Col.B) are NOT coping well enough to prevent progression of the atherosclerotic process at their plaque-sites. 421/773 = 0.54

Reduce lipid-load to 40 units/month:

(95+90+87+80+76+70), or 498 mini-tumors are coping well enough to mean that the atherosclerotic process is not progressing at their plaque-sites. 498/773 = 0.64

(65+60+55+50+45), or 275 mini-tumors are NOT coping well enough to prevent progression of the atherosclerotic process at their plaque-sites. 275/773 = 0.36

* Mini-tumors capable of coping with loads > 60 can also cope with 60 units, of course.

SUMMARY: Progression of the atherosclerotic process predicts cardiac events. Box 1 illustrates how a reduction in the serum levels of atherogenic lipoproteins halts progression at Class E and Class F sites. They become "stabilized." One could "play" with the arbitrary numbers in this box, but the point would remain.

Related text = Part 1c.

Part 1. Goal: To Evaluate Fractional Causation in the Post-1940 Decades
Part 2. Must Fractional Causation Change, If MortRates Rise or Fall?
Part 3. Post-1940 Behavior of PhysPop, and the Three "Trios"
Part 4. Cigarette-Smoking: Another Twentieth Century Carcinogen

Table 47-A. Averaged PhysPops, 1940-1990, and Ranking of the Census Divisions.
Table 47-B. Change-Factors for Averaged (Mean) PhysPops, by Census Trios.

● Part 1. Goal: To Evaluate Fractional Causation in the Post-1940 Decades

Studies in Section Five of this book will cover the SECOND half of the Twentieth Century. Section Two of this book uncovered extremely high and positive correlations between PhysPop and the 1940 cancer mortality-rates, by Census Divisions. Section Four of this book uncovered equally strong and positive correlations between the 1940-1950 PhysPops and the 1950 Ischemic Heart Disease MortRates. In dramatic contrast, Section Three uncovered an INVERSE correlation between PhysPop and all NonCancer NonIHD causes of death combined.

1a. Three Sets of Facts ... and a Single Explanation

Those three sets of findings are facts --- not interpretations --- and we have said that such facts "demand" an explanation.

We say the explanation is the proportionality of PhysPop with average accumulated per capita dose of MEDICAL RADIATION, by Census Divisions. In other words, the correlations are causal in nature --- they are dose-RESPONSES. Cancer is well-established as a disease inducible by ionizing radiation, at low doses as well as at high doses (Chapter 2, Parts 4c, 5b, 6). Our findings provide the first powerful epidemiologic evidence that Ischemic Heart Disease is also a radiation-inducible disease. Other than Cancer, there are very few well-established radiation-inducible causes of death (Chapter 23, Part 1). This being the case, we would expect NOT to find a positive correlation between PhysPop and NonCancer NonIHD causes of death --- and the real-world data support this expectation, by a relationship which is even inverse between PhysPop and such MortRates.

1b. Fractional Causation by Medical Radiation 1940-1950, and Beyond

Sections Two and Four of the book have used the Constants, generated by the mid-century dose-responses, to estimate Fractional Causation by medical radiation of the 1940 Cancer MortRates and of the 1950 IHD MortRates. By Fractional Causation, we mean the percentage of the MortRates which would have been absent in the absence of medical radiation. We emphatically do not mean that medical radiation was the only cause contributing to such cases (Introduction, Parts 4 and 5; Chapter 6, Part 6). Best estimates of Fractional Causation by medical radiation:

Male All-Cancer Mortality, 1940:	90%
Female All-Cancer Mortality, 1940:	58%
Male IHD Mortality, 1950:	79%
Female IHD Mortality, 1950:	97%

Now, Section Five evaluates Fractional Causation in subsequent decades. Our analyses indicate that very high Fractional Causations occur in every decade, including the decade ending in approximately 1990. With respect to Cancer, these findings are fully consistent with other evidence cited in Chapter 2:

(a) Ionizing radiation is a proven cause of almost every major type of human Cancer (Chapter 2, Part 4), and the carcinogenic properties of ionizing radiation persist even at the lowest possible dose and dose-rates (Chapter 2, Part 6, and Appendix-B).

(b) Medical xrays (including CT scans and fluoroscopy) are at least as carcinogenic per dose-unit as higher-energy gamma rays --- and probably are several times MORE carcinogenic per dose-unit (Chapter 2, Part 7).

(c) Radiation exposure of mixed-age populations causes radiation-induced Cancer in such populations for at least the subsequent forty-five years (Chapter 2, Part 8) --- probably longer.

(d) Nearly 93% of the 1990 cancer MortRate (USA) occurred in people old enough to have experienced now-obsolete practices in medical irradiation (Chapter 2, Part 3).

(e) Radiation doses from various current practices --- especially CT scans, fluoroscopy, and some nuclear medical procedures --- are very far from negligible, especially if repeated (Chapter 2, Parts 3 and 7e).

● Part 2. Must Fractional Causation Change, If MortRates Rise or Fall?

Fractional Causation of a disease by medical radiation is not automatically altered by changes in the National age-adjusted MortRate. Suppose that Fractional Causation of cancer mortality in 1940 is 75%. Then suppose, for illustrative purposes, that better treatments for Cancer, or better underlying health, cut the National age-adjusted MortRate in HALF by the year 2000. Medical radiation could still account for 75% of the cancer deaths which DO occur in the year 2000. Change (up or down) in a National MortRate, by itself, reveals nothing about Fractional Causation.

By contrast, there is a phenomenon which would definitely mean a decline in Fractional Causation, by medical radiation, of Cancer and IHD MortRates. If all use of ionizing radiation in medicine had CEASED in 1941 forever, then Fractional Causation of Cancer and IHD mortality by medical radiation would be truly NEGLIGIBLE in the year 2030 --- for nearly everyone who could have received medical radiation, before its 1941 cessation, would have died before the year 2030. In reality, of course, use of radiation in medicine has never ceased. While several aspects of its use have decreased since 1940, several other aspects of its use since 1940 have increased (Chapter 2, Part 3).

Fractional Causation by medical radiation of a National MortRate (Cancer, IHD) is determined by the relative size of the MortRate's two fractions: The radiation-induced cases (cases which would not have occurred in the absence of medical radiation), and the cases which occurred without help from radiation. The sum of the two fractions, of course, can not exceed 1.0 (100%). Thus, if either fraction increases between 1940 and 1990, the other must decrease.

● Part 3. Post-1940 Behavior of PhysPop, and the Three "Trios"

As we initiated our analyses of post-1940 Fractional Causation, we were mindful that no law of history establishes that the path leading to 1990 must be just like the path leading to 1940. And there is certainly no biological law requiring that the 1940 set of nine PhysPop values must be well correlated with the SUBSEQUENT sets for a half-century. Chaotic changes in post-1940 PhysPop values could eradicate meaningful differences, between populations of the Nine Census Divisions, in accumulated per capita radiation dosage from medical procedures (Chapter 3, Part 2c).

Existence of reliable dose-differences is, of course, the first requirement for dose-response studies. And when dose is accumulating over many decades in nine populations, such studies require reasonably STEADY proportions and rankings between the nine dose-levels. So, as we begin the post-1940 analyses, we need to demonstrate how well the PhysPop data meet this requirement.

3a. Calculation of the Averaged PhysPops: Table 47-A

Exposures to medical radiation in 1940, 1950, 1960, 1970, and 1980 all contribute to radiation-induced cancer-deaths in 1990 (Chapter 2, Part 8), and exposures received between 1980 and 1989 make some contribution too (Chapter 5, Part 4). Therefore, we will regress the 1990 cancer MortRates (or the very similar 1988 rates) upon the Averaged 1940-1990 PhysPops ... the 1980 cancer MortRates upon the Averaged 1940-1980 PhysPops ... and so forth.

Table 47-A, Part 1, presents the Averaged PhysPops for use in many subsequent chapters.

In the A-Bomb Survivor Study, 22% of all the fatal bomb-induced solid Cancers occurred during 1986 through 1990 --- 40 to 45 years after the Hiroshima-Nagasaki bombings (according to Pierce et al, 1996, as already stated in our Chapter 2, Part 8). In view of such evidence, it would be unreasonable to assume that the carcinogenic impact of radiation exposure disappears abuptly at 46 years post-irradiation. So, in our own work, when we regress the 1988 or 1990 U.S. cancer MortRates on the Averaged 1940-1990 U.S. PhysPops, we do not hesitate to assume that exposure to radiation in 1940 still contributes 48 to 50 years later to the 1988 or 1990 cancer MortRates.

3b. How Good Is the Correlation between Mean 1940-1990 and 1940 PhysPops?

The correlation is very good indeed between the Averaged (Mean) 1940-1990 PhysPops and the 1940 PhysPops. The R-Squared value is 0.91, as demonstrated below and as stated in Table 47-A, Part 1.

	x = PP 1940	y = PP 1940-90	Regression Output:	
Pacific	159.72	191.97	Constant	14.6705
NewEngland	161.55	208.20	Std Err of Y Est	11.2065
WestNoCentral	123.14	141.14	R Squared	0.9090
Mid-Atlantic	169.76	204.72	No. of Observations	9
EastNoCentral	133.36	146.19	Degrees of Freedom	7
Mountain	119.89	145.91		
WestSoCentral	103.94	126.28	X Coefficient(s)	1.1128
EastSoCentral	85.83	113.28	Std Err of Coef.	0.1331
SouthAtlantic	100.74	142.93	X-Coef / S.E. =	8.3616

3c. Census-Division Rankings by Their Mean PhysPop Values

The strong correlation found in Part 3b is in harmony with the rankings in Part 2 of Table 47-A. There, we show that the rankings of the Census Divisions, by their Averaged PhysPops, are quite steady over the half-century from 1940 to 1990. For instance:

● The Mid-Atlantic, New England, and Pacific Census Divisions are ALWAYS in the "trio" which has the highest Averaged PhysPops. We can call this the TopTrio, meaning Top-Dose Trio.

● Until inclusion of the 1990 PhyPops, the East North Central, West North Central, and Mountain Divisions are ALWAYS in the "trio" which has the mid-size Averaged PhysPops. We can call this the MidTrio, meaning Mid-Dose Trio.

● Until inclusion of the 1990 PhysPops, the West South Central, South Atlantic, and East South Central Divisions are ALWAYS in the "trio" which has the lowest Averaged PhysPops. We can call this the LowTrio, meaning Low-Dose Trio.

Inclusion of the 1990 PhysPops causes West North Central to move from MidTrio to LowTrio, while South Atlantic moves from LowTrio to MidTrio.

3d. "Trio Sequence" in Most Future Listings

Because of the observation described in Part 3c, we will allow one change in our Standard Sequence of the Census Divisions --- which is shown in Table 47-A, Part 1. If we list Mid-Atlantic ABOVE West North Central (instead of below it), the Nine Census Divisions will be grouped like the "Trios" described above. Of course, when Mid-Atlantic's name moves up in the listing by one position, the move has no effect on regression output because Mid-Atlantic's PhysPop and MortRate values move too, along with its name.

We will call the revised sequence of Census Divisions "the Trio Sequence." It is first shown in Chapter 48, Box 3. There, it is obvious that Mid-Atlantic is listed in third position (because of the exchanged position with West North Central). The exchange provides a visual reminder that Mid-Atlantic belongs to the TopTrio --- meaning Top-Dose Trio. However, Trio Sequence does not mean that the three Census Divisions WITHIN each Trio are listed in order of descending Mean

PhysPop Values. For example, Trio Sequence does not put Mid-Atlantic into the Trio's first position, even though Mid-Atlantic has the highest Averaged PhysPops until 1990 (Table 47-A, Part 2).

3e. Table 47-B: Change-Factors in Mean PhysPops over Time, by Trios

It is evident from Table 47-A, Part 1, that the Mean 1940-1990 PhysPop values are appreciably higher in every Census Division than the corresponding PhysPop values of 1940. If the 1940 values change by the SAME FACTOR in all three Trios, increments in MortRates will remain proportional to increments in Mean PhysPops regardless of these changes (Chapter 5, Part 6c) --- provided that the Census Divisions are well-matched for co-actors, of course.

Table 47-B examines the change-factors in Mean PhysPops over time. Column C shows the change-factor for each individual Census Division. Column D shows the average change-factor for each TRIO.

Then Table 47-B COMPARES the average change-factors in the MidTrio and the LowTrio with the average change-factor in the TopTrio. These ratios are summarized in the lower righthand corner of Table 47-B. We name these ratios "PhysPop Adjustment" factors (abbreviated "ppAdju"). No "ppAdju" is very different from 1.00, which means that the Mean PhysPop values change over time by very nearly the SAME FACTOR in all three Trios.

Summary of Part 3: Preservation of Dose-Differences

As we undertake the post-1940 analyses, the findings in Parts 3b, 3c, and 3e provide assurance that differences in accumulated radiation dose, between the Census Divisions, are quite well preserved between 1940 and 1990.

● Part 4. Cigarette-Smoking: Another Twentieth Century Carcinogen

At the very start of the Twentieth Century, a new carcinogen called medical radiation was widely introduced into the lives of males and females. Soon thereafter, a second new carcinogen (and a contributing cause of Ischemic Heart Disease) became very popular among males: Cigarette smoking. Females joined later.

Has cigarette-smoking occurred with equal intensity in all Nine Census Divisions?

As every epidemiologist well knows, the true correlation between a dose and its effect can be obscured --- or even appear to be a negative correlation when it is truly positive --- due to poor matching of the dose-groups for other agents which also contribute to the disease-rate. We illustrated the potential consequences of poor matching in Chapter 5, Part 7 (especially Figures 5-D and 5-E).

By contrast with Figures 5-D and 5-E, the very high and positive correlations of PhysPop, with the 1940 cancer MortRates, indicate that the Nine Census Divisions were probably well matched for the 1940 impact of carcinogenic co-actors. It should be noted that, during a period of time such as 1900 to 1940, exposure to any co-actor can be similar in all Census Divisions, even while such exposure is rising alike in all the Divisions or falling alike in all the Divisions.

But there is no law of history, we repeat, which guarantees that the path leading to 1990 must be just like the path leading to 1940. And there is certainly no law guaranteeing that the approximate matching of Census Divisions, for the impact of any carcinogenic co-actor up to 1940, will persist until 1990.

Cigarette smoking is a proven and powerful carcinogen. If its post-1940 impact is NOT well matched across the Census Divisions, cigarette smoking could degrade the observed post-1940 correlations between PhysPop and cancer MortRates. So, at the start of our post-1940 analyses, we needed to ascertain whether or not the post-1940 impact of cigarette smoking occurred with approximately equal intensity in all Nine Census Divisions. It did not. Therefore, cigarette smoking is the topic of the next chapter.

Table 47-A.
Averaged PhysPops, 1940–1990, and Ranking of the Census Divisions.

● Part 1. The Averaged PhysPops over Various Decades

The input data come from Table 3-A, and are presented at the bottom of this page for convenience.

Standard Sequence	1940 PhysPop	Mean1940 thru1950 PhysPop	Mean1940 thru1960 PhysPop	Mean1940 thru1970 PhysPop	Mean1940 thru1980 PhysPop	Mean1940 thru1990 PhysPop
Pacific	159.72	154.16	155.69	162.72	177.35	191.97
New England	161.55	162.03	162.81	168.74	185.86	208.20
West North Central	123.14	121.60	118.15	119.56	128.82	141.14
Mid-Atlantic	169.76	169.24	167.04	173.28	186.11	204.72
East North Central	133.36	128.53	123.87	124.70	133.71	146.19
Mountain	119.89	119.64	117.40	122.37	133.45	145.91
West South Central	103.94	102.64	102.31	105.03	114.66	126.28
East South Central	85.83	84.44	85.63	89.44	99.46	113.28
South Atlantic	100.74	99.91	101.72	108.97	124.62	142.93

NOTE: When the nine Mean 1940-1990 PhysPops of Part 1 are regressed on the nine 1940 PhysPops (or vice versa), they are correlated with an R-Squared value of 0.91.

● Part 2. Census Divisions, in shifting order, sorted by descending Averaged PhysPops.

	1940 PhysPop	Mean1940 thru1950 PhysPop	Mean1940 thru1960 PhysPop	Mean1940 thru1970 PhysPop	Mean1940 thru1980 PhysPop	Mean1940 thru1990 PhysPop
TopTrio	MidAtl NewEng Pac	MidAtl NewEng Pac	MidAtl NewEng Pac	MidAtl NewEng Pac	MidAtl NewEng Pac	NewEng MidAtl Pac
MidTrio	ENoCen WNoCen Mtn	ENoCen WNoCen Mtn	ENoCen WNoCen Mtn	ENoCen Mtn WNoCen	ENoCen Mtn WNoCen	ENoCen Mtn SoAtl
LowTrio	WSoCen SoAtl ESoCen	WSoCen SoAtl ESoCen	WSoCen SoAtl ESoCen	SoAtl WSoCen ESoCen	SoAtl WSoCen ESoCen	WNoCen WSoCen ESoCen

NOTE: Until the addition of the 1990 PhysPops, each "trio" of Census Divisions contains the same membership, decade after decade.

Below are the input-data used for calculating Part 1, above.

By taking the average of each horizontal row separately, for the appropriate number of columns, we obtained the Mean PhysPops shown in Part 1.

From Table 3-A.	1940 PhysPop	1950 PhysPop	1960 PhysPop	1970 PhysPop	1980 PhysPop	1990 PhysPop
Pacific	159.72	148.60	158.74	183.83	235.84	265.09
New England	161.55	162.51	164.37	186.51	254.37	319.88
West North Central	123.14	120.06	111.25	123.77	165.86	202.78
Mid-Atlantic	169.76	168.71	162.65	192.00	237.41	297.79
East North Central	133.36	123.69	114.56	127.17	169.79	208.54
Mountain	119.89	119.38	112.93	137.27	177.76	208.20
West South Central	103.94	101.34	101.65	113.20	153.18	184.34
East South Central	85.83	83.05	88.00	100.89	139.51	182.42
South Atlantic	100.74	99.07	105.36	130.70	187.22	234.48

Related text = Parts 3a, 3b, 3c.

===
Table 47-B
Change-Factors for Averaged (Mean) PhysPops, by Census Trios
===

Trio- Sequence	Col.A 1940 PhysPops Tab3-A	Col.B 1940-50 PhysPops Tab 47-A	Col.C Ratio Col.B /Col.A	Col.D Input from Col.C		Col.A 1940 PhysPops Tab3-A	Col.B 1940-60 PhysPops Tab 47-A	Col.C Ratio Col.B /Col.A	Col.D Input from Col.C
Pacific	159.72	154.16	0.965	Avg Chg		159.72	155.69	0.975	Avg Chg
NewEngland	161.55	162.03	1.003	TopTrio		161.55	162.81	1.008	TopTrio
Mid-Atlantic	169.76	169.24	0.997	0.988		169.76	167.04	0.984	0.989
WestNoCent.	123.14	121.60	0.987	Avg Chg		123.14	118.15	0.959	Avg Chg
EastNoCent.	133.36	128.53	0.964	MidTrio		133.36	123.87	0.929	MidTrio
Mountain	119.89	119.64	0.998	0.983		119.89	117.40	0.979	0.956
WestSoCent.	103.94	102.64	0.987	Avg Chg		103.94	102.31	0.984	Avg Chg
EastSoCent.	85.83	84.44	0.984	LowTrio		85.83	85.63	0.998	LowTrio
SoAtlantic	100.74	99.91	0.992	0.988		100.74	101.72	1.010	0.997

1950 ppAdju	Col.D: MidTrio/TopTrio =		0.99		1960 ppAdju	Col.D: Mid/Top =		0.97
	Col.D: LowTrio/TopTrio =		1.00			Col.D: Low/Top		1.01

Trio- Sequence	Col.A 1940 PhysPops Tab 3-A	Col.B 1940-70 PhysPops Tab 47-A	Col.C Ratio Col.B /Col.A	Col.D Input from Col.C		Col.A 1940 PhysPops Tab 3-A	Col.B 1940-80 PhysPops Tab 47-A	Col.C Ratio Col.B /Col.A	Col.D Input from Col.C
Pacific	159.72	162.72	1.019	Avg Chg		159.72	177.35	1.110	Avg Chg
NewEngland	161.55	168.74	1.045	TopTrio		161.55	185.86	1.150	TopTrio
Mid-Atlantic	169.76	173.28	1.021	1.028		169.76	186.11	1.096	1.119
WestNoCent.	123.14	119.56	0.971	Avg Chg		123.14	128.82	1.046	Avg Chg
EastNoCent.	133.36	124.70	0.935	MidTrio		133.36	133.71	1.003	MidTrio
Mountain	119.89	122.37	1.021	0.976		119.89	133.45	1.113	1.054
WestSoCent.	103.94	105.03	1.010	Avg Chg		103.94	114.66	1.103	Avg Chg
EastSoCent.	85.83	89.44	1.042	LowTrio		85.83	99.46	1.159	LowTrio
SoAtlantic	100.74	108.97	1.082	1.045		100.74	124.62	1.237	1.166

1970 ppAdju	Col.D: MidTrio/TopTrio =		0.95		1980 ppAdju	Col.D: Mid/Top =		0.94
	Col.D: LowTrio/TopTrio =		1.02			Col.D: Low/Top		1.04

Trio- Sequence	Col.A 1940 PhysPops Tab 3-A	Col.B 1940-90 PhysPops Tab 47-A	Col.C Ratio Col.B /Col.A	Col.D Input from Col.C
Pacific	159.72	191.97	1.202	Avg Chg
NewEngland	161.55	208.20	1.289	TopTrio
Mid-Atlantic	169.76	204.72	1.206	1.232
WestNoCent.	123.14	141.14	1.146	Avg Chg
EastNoCent.	133.36	146.19	1.096	MidTrio
Mountain	119.89	145.91	1.217	1.153
WestSoCent.	103.94	126.28	1.215	Avg Chg
EastSoCent.	85.83	113.28	1.320	LowTrio
SoAtlantic	100.74	142.93	1.419	1.318

1990 ppAdju	Col.D: MidTrio/TopTrio =		0.94
	Col.D: LowTrio/TopTrio =		1.07

PhysPop Adjustment ("ppAdju" in
subsequent tables and boxes)
quantifies change in MidTrio PP
and in LowTrio PP, relative
to change in TopTrio PP.

Year	ppAdju MidTrio	ppAdju LowTrio
1950	0.99	1.00
1960	0.97	1.01
1970	0.95	1.02
1980	0.94	1.04
1990	0.94	1.07

Related text = Part 3e.

CHAPTER 48

Cigarette Smoking: When, Who, How Much, and Especially Where

● Part 1. Recognition of Smoking as a Cause of Cancer and Ischemic Heart Disease

Cigarette smoking is established as a proven and very important cause of Respiratory-System Cancers, and is suspect as a contributing cause of many additional kinds of Cancer. When did this evidence develop?

1a. Warnings Which Preceded the 1964 "Surgeon General's Report"

In June 1956, at the instigation of the Surgeon General of U.S. Public Health Service, "a scientific Study Group [on relationships between smoking and health] was established jointly by the National Cancer Institute, the National Heart Institute, the American Cancer Society, and the American Heart Association. After appraising 16 independent studies carried on in five countries over a period of 18 years, this group concluded that there is a causal relationship between excessive smoking of cigarettes and lung cancer" (from pages 6-7 of the famous "Surgeon General's Report," which is in our Reference List as SurgeonGen 1964).

On July 12, 1957, after reviewing the report of the Study Group and other new evidence, the U.S. Surgeon General, Leroy E. Burney, issued a public warning: "The Public Health Service feels the weight of the evidence is increasingly pointing in one direction; that excessive smoking is one of the causative factors in lung cancer" (quoted from SurgeonGen 1964, p.7). In the November 28, 1959 issue of the Journal of the American Medical Association, Burney stated the belief of the Public Health Service that "The weight of the evidence at present implicates smoking as the principal factor in the increased incidence of lung cancer," and "Cigarette smoking particularly is associated with an increased chance of developing lung cancer" (quoted from SurgeonGen 1964, p.7).

Early in 1962, in London, a report was issued entitled "Smoking and Health: Summary and Report of the Royal College of Physicians of London on Smoking in Relation to Cancer of the Lung and Other Diseases." Its main conclusions: "Cigarette smoking is a cause of lung cancer and bronchitis, and probably contributes to the development of coronary heart disease and various less common diseases. It delays healing of gastric and duodenal ulcers" (quoted from SurgeonGen 1964, p.8).

1b. Principal Findings of the 1964 "Surgeon General's Report"

In 1964, the U.S. Public Service issued the 387-page "Surgeon General's Report" from which we have been quoting. It is formally entitled, "Smoking and Health: Report of the Advisory Committee to the Surgeon General of the Public Health Service," (PHS Publication Number 1103). The report's "Principal Findings" are summarized near its outset (abbreviated by us from pp.31-32):

● "Cigarette smoking is causally related to lung cancer in men ... The data for women, though less extensive, point in the same direction. The risk of developing lung cancer increases with duration of smoking and the number of cigarettes smoked per day, and is diminished by discontinuing smoking. In comparison with non-smokers, average male smokers of cigarettes have approximately a 9- to 10-fold risk of developing lung cancer, and heavy smokers at least a 20-fold risk. The risk of developing cancer of the lung for the combined group of pipe smokers, cigar smokers, and pipe and cigar smokers is greater than for non-smokers, but much less than for cigarette smokers."

● "Cigarette smoking is the most important of the causes of chronic bronchitis in the United States, and increases the risk of dying from chronic bronchitis and emphysema ... [For emphysema] it has not been established that the relationship is causal."

● "It is established that male cigarette smokers have a higher death rate from coronary artery disease than non-smoking males. Although the causative role of cigarette smoking in deaths from coronary artery disease is not proven, the Committee considers it more prudent from the public health viewpoint to assume that the established association has causative meaning than to suspend judgment until no uncertainty remains."

● "Pipe smoking appears to be causally related to lip cancer. Cigarette smoking is a significant factor in the causation of cancer of the larynx. The evidence supports the belief that an association exists between tobacco use and cancer of the esophagus, and between cigarette smoking and cancer of the urinary bladder in men, but the data are not adequate to decide whether these relationships are causal. Data on an association between smoking and cancer of the stomach are contradictory and incomplete."

The Requirement for Co-Action among Causes

Just before the summary above, the Report comments on co-action among causes (SurgeonGen 1964, p.31):

"It is recognized that no simple cause-and-effect relationship is likely to exist between a complex product like tobacco smoke and a specific disease in the variable human organism. It is also recognized that often the co-existence of several factors is required for the occurrence of a disease, and that one of the factors may play a determinant role; that is, without it, the other factors (such as genetic susceptibility) seldom lead to the occurrence of the disease."

1c. Is Smoke's Primary Carcinogen Really Alpha-Particle Radiation?

The carcinogenic agents from cigarette smoke may be chemical, and they may also be physical --- namely, ionizing radiation in the form of alpha particles, emitted by radioactive decay of polonium-210.

The late Dr. Edward A. Martell was a pioneer in pursuing the hypothesis that cigarette-induced Lung-Cancer results primarily from cigarette smoke's radioactive particles --- specifically from insoluble particles large enough for deposition in the bronchi, where the radioactive atoms of polonium (210, 212, 214) subsequently decay by alpha-particle emission (Martell 1974 + 1975 + 1982-a + 1982-b + 1982-c + 1983-a and 1983-b).

The delivery of polonium-210 to the lungs by cigarette smoking is a fact NOT IN DISPUTE. It has been reported for decades (for example, see Radford 1964, + Little 1965, + Hill 1965, + Holtzman 1966, + Blanchard 1967, + Radford 1977, + Winters 1982-a and Winters 1982-b, + NCRP 1984). In 1990, the BEIR-5 Report of the National Research Council acknowledged that portions of the bronchial epithelium of smokers receive a "relatively high dose (up to 0.2 Sv per year)" of radiation from this source (BEIR 1990, p.19). 0.2 Sv is equivalent to 20 rems, as stated in our Appendix A.

The role of alpha-particle radiation in smoking-induced Lung-Cancer is a very important and neglected issue --- but an issue outside the scope of this book.

1d. Cigar Smoking: Also a Carcinogen

In mid-April 1998, the National Cancer Institute (USA) released a 232-page report entitled

"Cigars: Health Effects and Trends. Monograph 9 on Smoking and Tobacco Control" (NCI 1998 in our Reference List). Monograph 9 is the work of 50 scientists, and reviews an extensive literature. The report warns:

● – Cigar smoking can cause oral, esophageal, laryngeal, and lung cancers. Regular cigar smokers who inhale, particularly those who smoke several cigars per day, have an increased risk of coronary heart disease and chronic obstructive pulmonary disease.

● – Cigar use in the USA has increased dramatically since 1993.

● – The Director of the NCI, Richard D. Klausner, M.D., comments in the Preface of Monograph 9 (pp. ii-iii): "We believe an accurate statement is that the risks of tobacco smoke exposure are similar for all sources of tobacco smoke, and the magnitude of the risks experienced by cigar smokers is proportionate to the nature and intensity of their exposure." And "To those cigarette smokers who are thinking of switching to cigars, don't be misled. Unless you substantially reduce your exposure to smoke, your risks will remain unchanged."

● **Part 2. Cigarette Smoking: Growth and Decline over Time (USA, UK)**

In the year 1900, cigarette smoking was very rare, both in the USA and Britain.

2a.   Changes in Per-Capita Use of Cigarettes per Year, 1900-1994

In the United States, changes in the annual use of cigarettes per capita of population (smokers + nonsmokers) are shown for 1900-1994 in the list below. The source is the CDC's Morbidity and Mortality Weekly Report (MMWR), November 18, 1994, Vol.43, No.SS-3, pp.6-7, Table 1, by Gary A. Giovino et al (MMWR 1994). The data are not provided by gender (see Part 3, below).

Year		Cigarettes Used Annually per Capita (males + females, smokers + nonsmokers combined, age 18 or older)	
1900		54	~1 per week
1910		151	
1920		665	
1930		1,485	
1940		1,976	
1950		3,552	
1960		4,171	
1963	Peak	4,345	11.9 per day
1970		3,985	
1980		3,849	
1990		2,817	
1994		2,493	

Figure 48-A: Cigarettes per Day in the UK, by Gender

The growth of cigarette smoking in the United Kingdom was also spectacular, according to a 1983 report from the Royal College of Physicians of London entitled "Health or Smoking? Follow-Up Report of the Royal College of Physicians" (Royal College 1983). In our Figure 48-A, we reproduce Figure 1.1 from that report. It shows a big difference between males and females in cigarette smoking.

Rapid Benefit for Physicians Who Quit Smoking

"Health or Smoking?" includes the following comment about causality (Royal College 1983, p.3): "The conclusion that cigarette smoking was responsible for this epidemic [of male Lung-Cancer mortality in Britain] was dramatically confirmed by looking at a group of the population that was giving up smoking --- doctors. Between 1954 and 1971, the proportion of male doctors smoking cigarettes halved (43% to 21%), while that for all men in England and Wales remained about the same. Over this period, the death rate in men from lung cancer fell by 25 percent in doctors while in the general population it increased by 26 percent (Doll 1976)."

2b. Figures 48-B, 48-C: Rates of Cigarette-Use, Lung-Cancer and IHD over Time

The tabulation in Part 2a shows the dramatic decline after 1963 of per capita cigarette-use in the USA. Our Figure 48-B depicts on a single graph (a) the growth and decline in annual per capita cigarette consumption in the USA (smokers and nonsmokers combined, genders combined), and (b) age-adjusted National Lung-Cancer MortRates for males, USA, back to 1930 (although Texas was not yet reporting in 1930; our Chapter 4).

The key point to note in Figure 48-B is that growth in per capita cigarette-consumption predicts growth in male Lung-Cancer mortality about 20 years LATER. Moreover, about 20 years after the decline began in cigarette-consumption, the male National Lung-Cancer MortRate appears to respond --- by ceasing its growth.

Quite different is the relationship of two curves in our Figure 48-C. In that figure, we plot (again) the nation's history of per capita cigarette consumption, this time with male MortRates from Ischemic Heart Disease. Both curves peak at the same time.

2c. Who Quits Smoking? Behavior and Formal Education: Box 1

Decline in cigarette consumption does not occur at random. The same issue of MMWR (Nov. 18, 1994, Table 2) presents compelling evidence that the greater the years of formal education, the greater is the decline between 1966 and 1991 in percentage of adults, >= 25 years of age, who are current smokers. Those data are presented in our Box 1.

● Part 3. Males, Females: Differences in Past Smoking Behavior

Box 2 shows the National MortRates in each decade from All-Cancers, Respiratory Cancers, and Difference-Cancers, and it calculates the growing percentages of All-Cancers contributed by Respiratory Cancers. Between the genders, there are marked differences, both in the Respiratory-Cancer MortRates and in the percentages of All-Cancers. The much lower rates and percentages for females are not surprising, in view of other data (below) which indicate that SMOKING-behavior in the past has been considerably less intense for females than for males in the USA.

● 1959-1960. In 1961, a paper entitled "Smoking Habits of Men and Women" by Hammond and Garfinkel (Hammond 1961) was published in the Journal of the National Cancer Institute. It is based on questionnaires answered in 1959-1960 by 43,000 adult Americans (age 30 or older) in 1,121 counties of 25 states. Among the findings: "Exposure to cigarette smoke is far less in the female than in the male population, as indicated by percent of heavy cigarette smokers [Table 4], degree of inhalation, nicotine and tar content of cigarettes, and age at which smoking was begun [Table 3]" (Hammond 1961, p.419). Hammond's Table 4 reports on "Current Regular Cigarette-Smoking by Number of Cigarettes per Day." For all ages combined, the gender-difference is shown below. Each percentage refers to the TOTAL sample (smokers + nonsmokers):

Current Cigarette Smoking, by Gender, 1959-1960		Males, Percent	Females, Percent
People currently NOT smoking regularly -->		53.3	72.7
Total who smoke cigarettes regularly -->		46.7	27.3
Number cigarettes smoked / day:	1-9	5.4	7.1
	10-19	8.8	8.2
	20	17.3	8.6
	21-39	9.2	2.4
	40	4.6	0.8
	41+	1.0	0.1
	Uncertain	0.3	--

● 1986. In an article entitled "Cigarette Smoking in the United States, 1986," the Morbidity and Mortality Weekly Report provides estimates for the 1986 prevalence of current smoking in persons age 17 or older. "Current cigarette smokers are defined as persons who have smoked at least 100 cigarettes in their lifetime and who are currently smoking cigarettes" (MMWR 1987, Vol.36, No.35, p.581, September 11, 1987). Results (from pp.582-583), based on survey by telephone of 13,031 respondents, indicate that the male-female difference narrowed a great deal after 1960:

Gender	Percent current smokers	Mean cigarettes / day
Male	29.5	22.8
Female	23.8	19.1

The same article presented estimates by gender, back to 1944. For males, the peak estimate of 54.2% for "percent current smokers" occurred in 1955, whereas for females, the peak estimate of 36% occurred in 1944 (a Gallup Poll). While not all the percentages are reliable, one can probably believe that a much lower percentage of females than males has EVER smoked cigarettes, if percents in all decades are averaged.

Such an inference is well supported by inspection of the female MortRates from Respiratory-System Cancers, in Box 2, Column B. In every decade from 1940 through 1988, the female rates are always much lower than the male rates. Indeed, the fact that the female rates are 3.3 in 1940, when the male rates are 11.0, is consistent with the likelihood that female smokers in the USA, like female smokers in the UK, adopted the smoking habit later and less intensely than males (Figure 48-A).

● 1995. In an article entitled "State-Specific Prevalence of Cigarette Smoking --- United States, 1995," the Morbidity and Mortality Weekly Report presents estimates for the 1995 prevalence of current smoking in persons age 18 or older, by states. Current cigarette smokers are defined as described above. Median values (from MMWR, Vol.45, No.44, November 8, 1996, p.963) are:

	Nat'l	Kentucky = highest	Utah = lowest
Male	24.7%	28.8%	16.4%
Female	20.9%	26.9%	10.0%

● Part 4. What Past Smoking-Data Are Available by States and Gender?

In order to ascertain whether or not the Nine Census Divisions have been approximately alike in smoking-intensity, we tried to acquire data back to 1930 (or earlier), by states and by gender. State-by-state data (which could be combined appropriately into Census Divisions) could quantify the distribution of this carcinogen among the Census Divisions, by decades. Additionally, gender-specific data would be extremely valuable because of the evidence that, as of 1960, fewer females than males were cigarette smokers and that females smoked a lot less intensely than male smokers (Part 3, above).

4a. Non-Existence of the Data We Sought

When our search at the medical library of the University of California at San Francisco did not yield data of the types we sought, we requested advice from the Office on Smoking and Health at the U.S. Centers for Disease Control and Prevention ("CDC"). Dr. Alyssa Easton responded with the following news, with respect to state-by-state estimates of cigarette-smoking prevalence:

1) The Behavioral Risk Factor Surveillance System (BRFSS) includes state-specific estimates of smoking prevalence. But it has been conducted only since 1984. At that time, only 15 states participated. The survey has been conducted annually, but participation by all 50 states did not occur until 1995.

2) The Current Population Survey began data-collection in 1985, with tobacco supplements conducted in 1985, 1989, and 1992-1993. "There is no pre-1985 information."

Indeed, it was June 1996 when the Council of State and Territorial Epidemiologists made a recommendation discussed in MMWR November 8, 1996, Vol.45, No.44, p.962:

"State-specific surveillance of the prevalence of cigarette smoking can be used to direct and evaluate public health interventions to reduce smoking and the burden of smoking-related diseases on society. In June 1996, the Council of State and Territorial Epidemiologists (CSTE) recommended that cigarette smoking be added to the list of conditions designated as reportable by States to CDC. This report [MMWR, November 8, 1996] responds to the CSTE recommendation and summarizes state-specific prevalences of cigarette smoking by U.S. adults in 1995."

Figure 48-D: Inverse Relationship for Smoking Prevalence 1995, PhysPop 1990

Figure 48-D regresses the 1995 smoking prevalences (male) by Census Divisions on 1990 PhysPops, and depicts the regression-input (boxy symbols) and line of best-fit. The correlation is inverse, with an R-squared value of 0.3568 and a ratio of -1.97 for X-Coef/SE. The relationship for the females (not shown) also is negative, but the R-squared value of 0.1237 from the female data has no significance. Since smoking habits in Census Divisions do not change "overnight," the inverse relationships in these recent data may indicate that relationships were inverse in earlier decades too. It is very disappointing that data by states and gender do not exist for the earlier decades.

Prevalence of Smoking: Not Informative about Intensity of Smoking

The prevalence-surveys of the Behavioral Risk Factor Surveillance System reveal nothing about the intensity of smoking among the smokers. The procedure is a state-based, random-digit-dialed telephone survey of the non-institutionalized U.S. population aged >= 18 years. Respondents are asked, "Have you smoked at least 100 cigarettes in your entire life?" and "Do you smoke cigarettes now?" Persons who answer yes to both questions are designated as "Current Smokers" (MMWR 1996, p.962).

4b. Bottom Line: Applying Reason to the Available Data

Despite the non-existence of the specific data which we would have liked to acquire, we are not helpless. In Part 5, we apply some reasoning to the two types of data which we DO have: PhysPops and MortRates.

● Part 5. No Reasonable Doubt: Smoking and PhysPop Become Inversely Related by Census Divisions

In Chapter 47, we established that the ranking of the Nine Census Divisions, by Averaged PhysPops, is quite steady during the 1940-1990 period. Indeed, Part 2 of Table 47-A shows that:

● The TopTrio (the three Census Divisions with the HIGHEST Mean PhysPops) always consisted of Mid-Atlantic, New England, and Pacific.

● The MidTrio always consisted of East North Central, West North Central, and Mountain --- until the 1990 PhysPops demoted West North Central to the LowTrio.

● The LowTrio always consisted of West South Central, South Atlantic, and East South Central --- until the 1990 PhysPops elevated South Atlantic into the MidTrio.

5a. The Relationship between MortRate Changes and PhysPop-Levels

Now we turn attention to the MortRate data for Respiratory-System Cancers in males, since females have a much less intense history of cigarette smoking (Part 3, above). In dramatic contrast to the post-1940 behavior of male MortRates for any other set of cancers, the National male MortRate for RESPIRATORY-SYSTEM Cancers rose from 11.0 in 1940, to 59.4 in 1980, and 59.7 in 1988 (Box 2). During the same period, male MortRates for All-Cancers EXCEPT Respiratory (that is, Difference Cancers) remained steady, in the range of 104.0 to 111.2 (also Box 2).

Box 2 obscures a key fact, however, because it is limited to the National rates. Box 3 shows that the spectacular rise in male Respiratory Cancers was VERY UNEVENLY distributed across the Census Divisions by 1988.

Box 3 compares the Top, Mid, and Low Trios for the MAGNITUDE OF CHANGE since 1940 in their Respiratory-Cancer MortRates. Change in a MortRate can be (and commonly is) expressed in either of two ways: As a ratio ("The new rate is 2.3 times higher than the old rate"), or as a difference

("The new rate is higher by 50 per 100,000"). Box 3 expresses change in both ways. Box 3 looks at 1960 as well as 1988.

    ● Comparison of Column A with Column G in Box 3 shows that the male MortRates from Respiratory Cancer increased enormously in EVERY Census Division, between 1940 and 1988.

    ● Column J measures the changes by subtraction (the 1988 rates minus the 1940 rates). Column K presents the average difference which developed in each Census Trio by 1988.

    ● Column H measures the changes by ratios (the 1988 rates divided by 1940 rates). Column I (Eye) shows the average ratio which developed in each Census Trio by 1988.

    ● In the Census Divisions where Mean PhysPop values are lowest (LowTrio), the average growth-ratio and growth-difference for Respiratory Cancers are highest.

    ● In the Census Divisions where Mean PhysPop values are highest (TopTrio), the average growth-ratio and growth-difference for Respiratory Cancers are lowest.

5b. Conclusion from These Facts: PhysPop and Smoking Inversely Related

The findings in Box 3 seem beyond challenge. What do they mean? They clearly mean that some cause of male Respiratory Cancers became much more intense in the LowTrio Census Divisions than in the HighTrio Census Divisions.

The identity of "some cause" can NOT be medical radiation. Mean PhysPop values have been persistently the lowest in the LowTrio Census Divisions (Table 47-A, Part 2). Mean PhysPop values grew in ALL the Trios between 1940 and 1988, but the growth-factor in the LowTrio was a mere 7% higher by 1988 than in the TopTrio (Table 47-B). This 7% disparity alone certainly can NOT explain why the Respiratory MortRate rose by a factor of 11.1 in the LowTrio, while rising by a factor of 3.9 in the Top Trio (Box 3, Column Eye). The explanation has to be that males in the LowTrio experienced some OTHER cause of Respiratory Cancers more intensely than did males in the TopTrio.

The identity of "some other cause" is almost surely cigarette smoking. After all, it is a PROVEN cause of Respiratory Cancer. And the time-frame is consistent with Figure 48-B. While the explanation of the facts in Box 3 MIGHT not be cigarette smoking, what matters is the evidence in Box 3 that, "beyond a reasonable doubt," SOME co-actor other than medical radiation has operated with greater intensity in the LowTrio Census Divisions than in the TopTrio Census Divisions. The name of this co-actor is not the issue. From here on, we will name it "smoking," because we think it is. But what really MATTERS is this:

A carcinogenic co-actor for Respiratory Cancers (which become a large constituent of All-Cancers by 1988) becomes INVERSELY related with the variable, PhysPop, whose correlation with Cancer we intend to analyze from 1950 to 1988. The inverse relationship of these two co-actors will result in false "findings" (Chapter 5, Part 7), if we fail to make appropriate adjustments.

In dose-response studies, appropriate adjustments are those which yield a reasonable approximation of what WOULD have been observed, if all variables (except the variable under study) had been WELL MATCHED across the dose-groups. "Adjusted data" are routinely used in the biomedical literature. Indeed, many studies make different adjustments in their data for three, four, five or more variables. Generally, readers are told only that adjustments have been made, but papers in journals rarely explain what was done. Readers who want to check the transformations, of observed data into adjusted data, must request aid from the paper's authors.

By contrast, we will make the necessary smoking-adjustment in full view. The next chapter explains each step. Many readers will skip over such steps, but all readers will be able easily to compare the "before and after" MortRate values, each time we make a MortRate adjustment in any chapter. Although showing our routine adjustment adds numerous pages to this part of the book, we feel strongly that real-world observations should not be adjusted "in the dark."

>>>>>>>>>>

● - The data (for males and females combined) come from interviews of people age 25 and older. The entries (rates per 100) represent the percentage who qualified as current smokers. The data below come from Morbidity and Mortality Weekly Report (MMWR), November 18, 1994, Vol. 43, No.SS-3, Table 2.

Year of Interview	<12	12	13-15	>=16
1965	--	--	--	--
1966	41.7	44.7	44.8	35.3
1970	37.5	39.3	38.7	28.8
1974	37.8	38.8	37.9	28.8
1978	35.7	37.0	34.3	24.2
1979	35.1	35.3	35.2	23.7
1980	35.1	35.4	33.9	24.5
1983	34.7	34.9	32.1	20.6
1985	34.2	33.4	30.6	19.0
1987	34.2	32.9	28.2	16.6
1988	32.9	32.7	28.1	16.3
1990	30.8	30.1	24.6	13.9
1991	31.4	30.6	25.5	13.9
Change from 1965 through 1991, converted to %	(41.7-31.4) / 41.7 = 0.247 Down by 24.7%	(44.7-30.6) / 44.7 = 0.315 Down by 31.5%	(44.8-25.5) / 44.8 = 0.431 Down by 43.1%	(35.3-13.9) / 35.3 = 0.606 Down by 60.6%

Number of years of education

Related text = Part 2c.

MALES, NATIONAL

Col. C entries are Col.A entries minus Col.B entries.
Col.D entries are Col.B entries divided by Col.A entries, then converted to percents.

	Col.A AllCancer MortRate. Table 6-B	Col.B RespSystCa MortRate. Table 16-B	Col.C Diff-Cancer MortRate. Table 18-B	Col.D Share of All Cancers from Respiratory System (Col.B / Col.A)
1940	115.0	11.0	104.0	9.57%
1950	132.8	21.6	111.2	16.27%
1960	145.7	35.2	110.5	24.16%
1970	155.1	47.3	107.8	30.50%
1980	164.5	59.4	105.1	36.11%
1988	162.7	59.7	103.0	36.69%

FEMALES, NATIONAL

1940	126.1	3.3	122.8	2.62%
1950	123.2	4.6	118.6	3.73%
1960	114.9	5.3	109.6	4.61%
1970	111.7	11.7	100.0	10.47%
1980	108.5	18.0	90.5	16.59%
1988	111.3	24.5	86.8	22.01%

● - At the same time when age-adjusted MortRates from Respiratory-System Cancers were soaring in each sex, the MortRates for all other types of cancer combined in Column C were either flat (males) or decreasing (females).

Related text = Part 3.

```
===================================================================================================
                                      Box 3, Chap. 48
                  Respiratory-System Cancers, Males:  Post-1940 Change in MortRates by Census Trios

1960 vs. 1940, by Trios:  Col.D expresses change by ratios.  Col.F expresses change by subtraction.
1988 vs. 1940, by Trios:  Col.I expresses change by ratios.  Col.K expresses change by subtraction.
MRs change inversely with PP.  High-PP Trio has lowest growth-factor.  Low-PP Trio has highest growth-factor.
```

	• 1940		>>> • Compare 1960 with 1940 • <<<				>>> • Compare 1988 with 1940 • <<<			
	Col.A	Col.B	Col.C	Col.D	Col.E	Col.F	Col.G	Col.H	Col.I	Col.J	Col.K
	1940	1960	Ratio	Input	Diff:	Input	1988	Ratio	Input	Diff:	Input
	MortRate	MortRate	Col.B	from	Col.B	from	MortRate	Col.G	from	Col.G	from
	Tab 16-A	Tab 16-A	/Col.A	Col.C	minus A	Col.E	Tab 16-A	/Col.A	Col.H	minus A	Col.J
Pacif	12.0	34.9	2.908	Avg Chg	22.9	Avg Chg	50.7	4.225	Avg Chg	38.7	Avg Chg
NewE	13.5	38.1	2.822	TopTrio	24.6	TopTrio	56.3	4.170	TopTrio	42.8	TopTrio
MidAtl	17.1	40.6	2.374	2.702	23.5	23.7	57.5	3.363	3.919	40.4	40.6
WNoCen	7.7	28.4	3.688	Avg Chg	20.7	Avg Chg	56.2	7.299	Avg Chg	48.5	Avg Chg
ENoCen	10.6	35.7	3.368	MidTrio	25.1	MidTrio	62.3	5.877	MidTrio	51.7	MidTrio
Mtn	7.8	25.5	3.269	3.442	17.7	21.2	44.2	5.667	6.281	36.4	45.5
WSoCen	7.6	34.9	4.592	Avg Chg	27.3	Avg Chg	67.9	8.934	Avg Chg	60.3	Avg Chg
ESoCen	4.9	29.0	5.918	LowTrio	24.1	LowTrio	79.1	16.143	LowTrio	74.2	LowTrio
SoAtl	8.3	35.7	4.301	4.937	27.4	26.3	68.5	8.253	11.110	60.2	64.9

```
===================================================================================================
```

Related text = Part 5a.

The notes below apply to Box 3 above and also to every Box 1 in Chapters 49 through 65.

MR = MortRate (mortality rate).   PP = PhysPop (physicians per 100,000 population).

High–PP Trio (TopTrio) = Three Census Divisions with the highest average accumulated doses from medical radiation (Table 47–A). These are Pacific, New England, Mid–Atlantic.

Low–PP Trio (LowTrio) = Three Census Divisions with the lowest average accumulated doses from medical radiation (Table 47–A). These are West South Central, East South Central, and South Atlantic.

• – Columns A, B, and G = Annual MortRates per 100,000 males, age-adjusted to the 1940 population distribution, from Table 16-A.

• – Col.C = The ratios of the 1960 MortRates divided by the 1940 MortRates. Col.H presents the ratio for 1988 MRs / 1940 MRs.

• – Col.D = The average value of Col.C, for each Trio of Census Divisions. Example: The 1960/1940 MortRate-ratios in the TopTrio Census Divisions were 2.908, 2.822, and 2.374, whose simple average is 2.702. In other words, on the average, the 1960 MortRates in the TopTrio were 2.702 times their 1940 values. The value, 2.702, is the "growth-factor" or "change-factor" for the TopTrio, 1960 vs. 1940. Col.I shows 1988 vs. 1940.

• – Col.E = The 1960 MortRates minus the 1940 MortRates. Col.E shows the difference, which is positive. Col. J compares 1988 with 1940.

• – Col.F = The average value of Col.E, for each Trio of Census Divisions. Example: The differences (1960 MR minus 1940 MR) in the TopTrio Census Divisions were 22.9, 24.6, and 23.5, whose simple average is 23.7. In other words, on the average, the 1960 MortRates in the TopTrio differed by +23.7 (per 100,000 population) from their 1940 values. Col.K compares 1988 with 1940.

Figure 48-A.
Growth of Smoking by Gender, 1890–1980, in the United Kingdom.

● Source: Chapter One of the 1983 report, "Health or Smoking? Follow-Up Report of the Royal College of Physicians of London" (Royal College 1983).

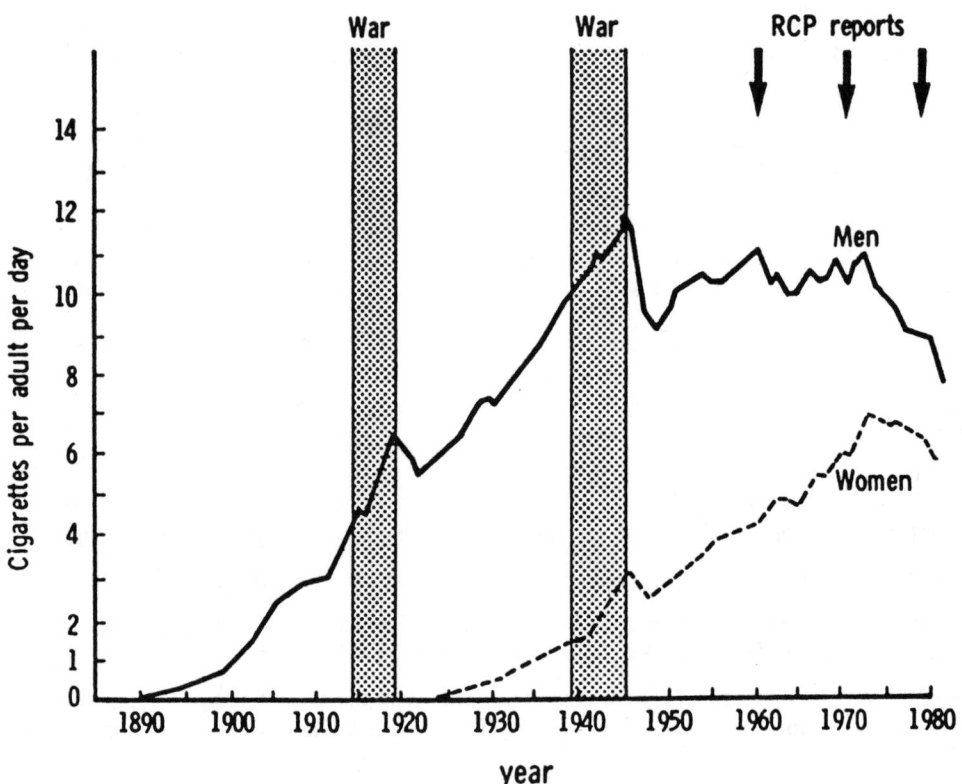

Related text = Part 2a.

Figure 1.1. Tobacco consumption in the UK 1890 to 1981, given as average number of cigarettes per adult per day for men and women separately, irrespective of whether they smoke or not. The arrows indicate the dates of the three previous Royal College of Physicians reports. Data from Tobacco Research (now Advisory) Council [reference 1 and unpublished data reproduced with permission]

Reference 1, above, is entered as Tobacco 1976 in our Reference List.

● The 16 diamond-like symbols depict male lung-cancer MortRates per 100,000 population. The rates are age-adjusted to 1970 because they were calculated by the American Cancer Society. We obtained these rates off the ACS graph at page 17 of Landis 1998. Except for the years before 1930, Figure 48-B depicts rates of lung-cancer mortality and rates of per capita cigarette-consumption for the same years.

● The 22 boxy symbols depict annual use of cigarettes per capita (smokers + nonsmokers, genders combined, USA), from 1900 to 1994. Source is MMWR, Nov. 18, 1994, Vol.43, No.SS-3, pp.6-7, Table 1. Because we have a single set of values on the vertical axis, cigarette-use is depicted at 1/50 of its actual rate. Example: The boxy symbol for 1930 is at about 30 on the vertical scale. This means the rate is (30 x 50), or 1,500 cigarettes per capita per year --- in harmony with the value of 1,485 shown in the tabulation of our text, Part 2a. The list on the left shows the value of each boxy symbol.

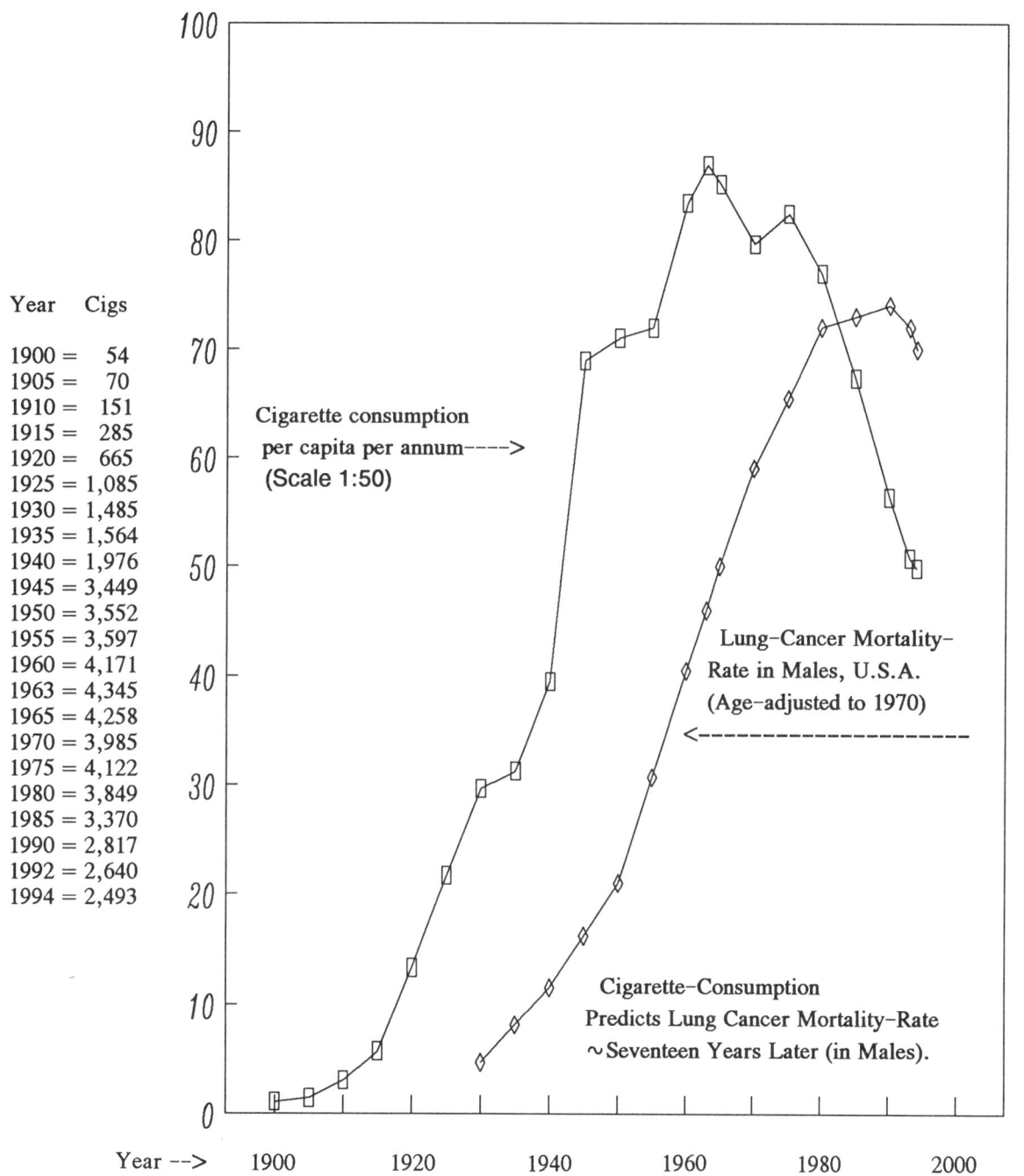

Year   Cigs

1900 =   54
1905 =   70
1910 =   151
1915 =   285
1920 =   665
1925 = 1,085
1930 = 1,485
1935 = 1,564
1940 = 1,976
1945 = 3,449
1950 = 3,552
1955 = 3,597
1960 = 4,171
1963 = 4,345
1965 = 4,258
1970 = 3,985
1975 = 4,122
1980 = 3,849
1985 = 3,370
1990 = 2,817
1992 = 2,640
1994 = 2,493

Cigarette consumption
per capita per annum---->
(Scale 1:50)

Lung-Cancer Mortality-
Rate in Males, U.S.A.
(Age-adjusted to 1970)
<----------------------

Cigarette-Consumption
Predicts Lung Cancer Mortality-Rate
~Seventeen Years Later (in Males).

Year -->   1900   1920   1940   1960   1980   2000

Related text = Part 2b.

Figure 48-C.
Ischemic Heart Disease: Per Capita Cigarette-Use and Male IHD MortRates.

● The 11 cross-like symbols depict male IHD MortRates per 100,000 population, age-adjusted to 1940, for the period 1950-1994. The rates come from our Table 40-A, except for the peak year (1963).

● With no change from our Figure 48-B, the 22 boxy symbols depict annual use of cigarettes per capita (smokers + nonsmokers, genders combined, USA), from 1900 to 1994. Source is MMWR, Nov. 18, 1994, Vol.43, No.SS-3, pp.6-7, Table 1. Because we have a single set of values on the vertical axis, cigarette-use is depicted at 1/10 of its actual rate. Example: The boxy symbol for 1930 is at about 150 on the vertical scale. This means the rate is (150 x 10), or 1,500 cigarettes per capita per year --- in harmony with the value of 1,485 shown in the tabulation of our text, Part 2a. The list on the left shows the value of each boxy symbol.

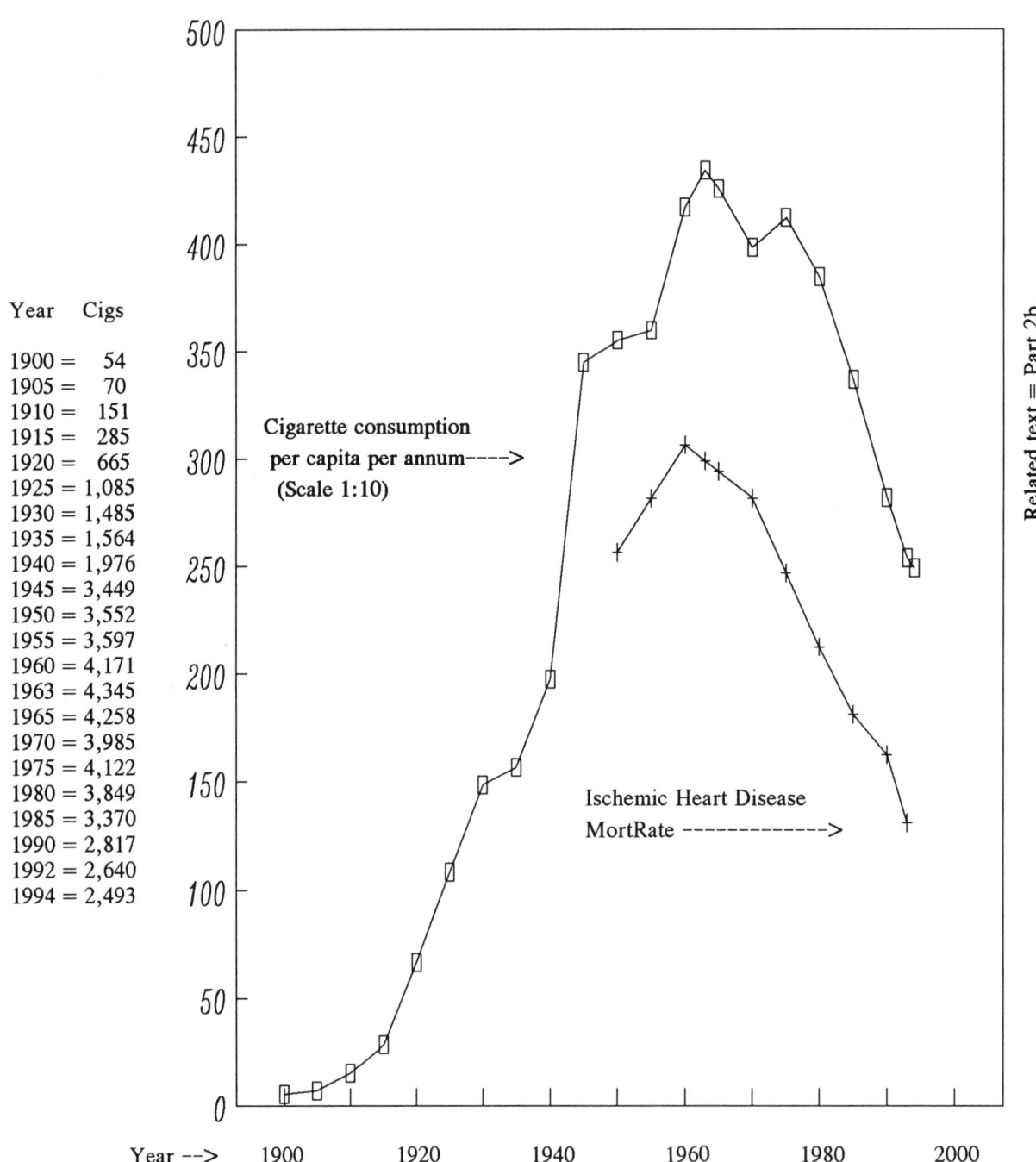

Year	Cigs
1900 =	54
1905 =	70
1910 =	151
1915 =	285
1920 =	665
1925 =	1,085
1930 =	1,485
1935 =	1,564
1940 =	1,976
1945 =	3,449
1950 =	3,552
1955 =	3,597
1960 =	4,171
1963 =	4,345
1965 =	4,258
1970 =	3,985
1975 =	4,122
1980 =	3,849
1985 =	3,370
1990 =	2,817
1992 =	2,640
1994 =	2,493

Cigarette consumption per capita per annum---->
(Scale 1:10)

Ischemic Heart Disease
MortRate ------------->

Year -->

Related text = Part 2b.

o Source: Male 1995 smoking prevalence, by Census Divisions, was calculated by us from the state–by–state data provided in CDC's Morbidity and Mortality Weekly Report (MMWR) Vol. 45, No. 44, Nov. 8, 1996. The PhysPop Values come from our own Universal PhysPop Table 3–A.

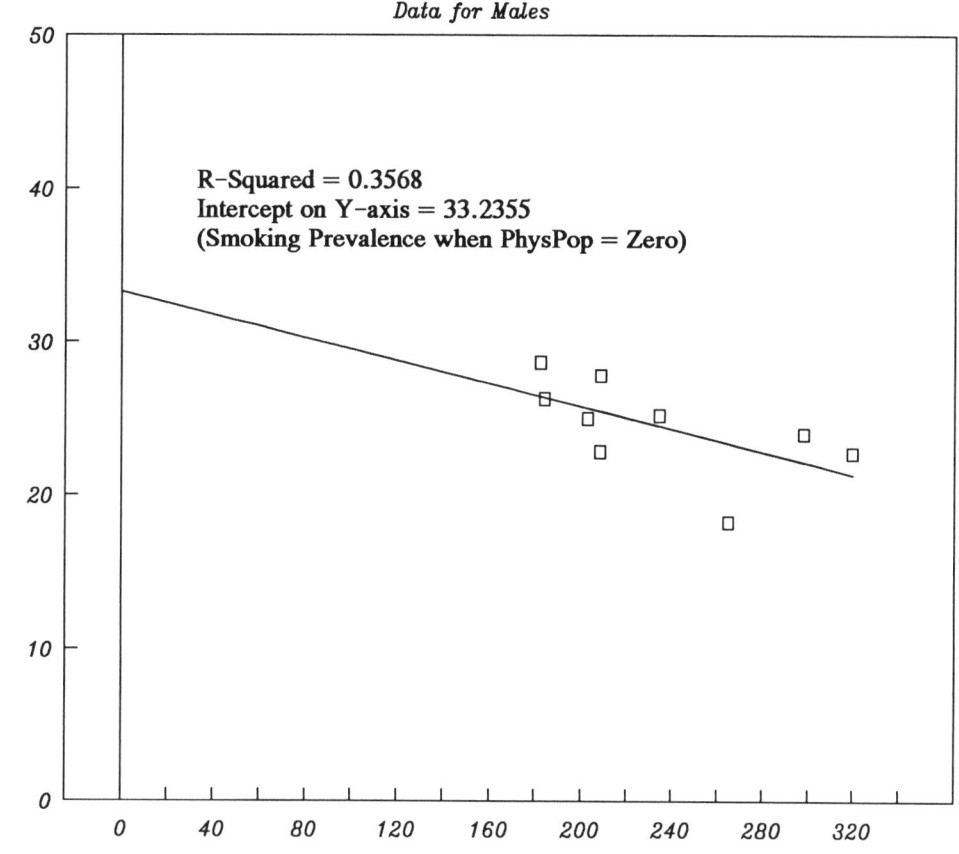

Smoking Prevalence vs. PhysPop

Data for Males

R–Squared = 0.3568
Intercept on Y–axis = 33.2355
(Smoking Prevalence when PhysPop = Zero)

Smoking Prevalence 1995

Physicians per 100,000 Population 1990

Related text = Part 4a.

Census Divisions	Males PhysPop 1990	Males SmokPrev 1995		
Pacific	265.09	18.24	Regression Output:	
New England	319.88	22.71	Constant	33.2355
West North Central	202.78	24.97	Std Err of Y Est	2.6669
Midatlantic	297.79	23.96	R Squared	0.3568
East North Central	208.54	27.81	No. of Observations	9
Mountain	208.20	22.82	Degrees of Freedom	7
West South Central	184.34	26.29		
East South Central	182.42	28.63	X Coefficient(s)	–0.0373
South Atlantic	234.48	25.18	Std Err of Coef.	0.0189
			Coeff. / S.E.	–1.9705

Equation of Best Fit: (–0.0373) * (PhysPop) + 33.2355

Figure 48–E

Per–Capita Consumption of Different Forms of Tobacco
in the USA, 1880–1997.

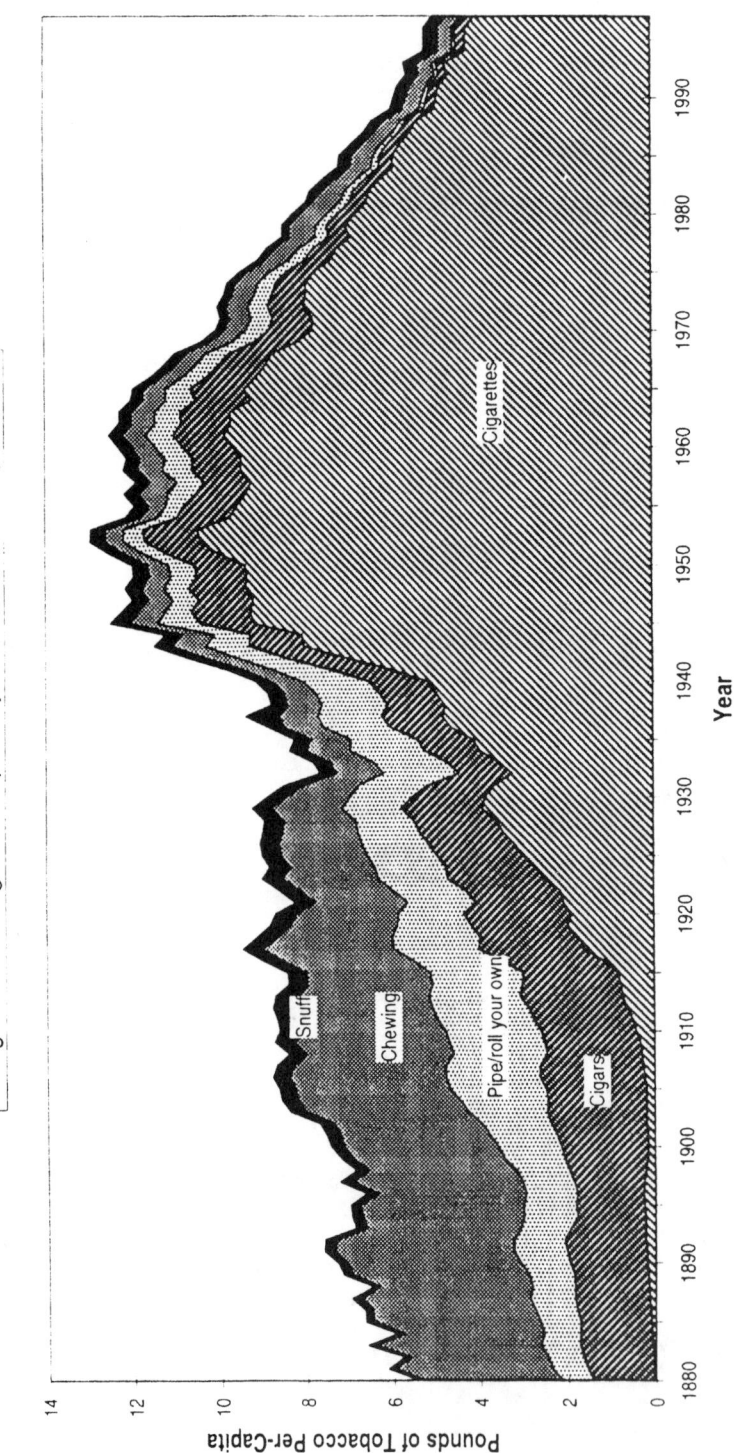

The figure above, and its title, are reproduced from a publication of the National Cancer Institute (NCI 1998, p.22,
Figure 1) entitled Cigars: Health Effects and Trends —– Monograph 9 on Smoking and Tobacco Control.

Related text = Part 2a.

All-Cancers, Males, 1940-1988: Fractional Causation by Medical Radiation

The boxes and tables in Chapter 49 provide the model for the subsequent chapters where we evaluate the post-1940 Fractional Causation, by medical radiation, of mortality from cancer and Ischemic Heart Disease.

● Table A summarizes the findings, by decade, including the 1940 percentage from Section 2 of the book. The sources for Table A are shown in its Column G. In several chapters, the table for 1970 is not shown in order to reduce pages. 1970 was chosen because the 1970 MortRates are interpolations (Chapter 4, Part 2b).

● Box 1 determines whether or not a post-1940 adjustment is obligatory, in order to achieve matching of the Census Divisions for smoking (Chapter 48, Part 5b). If adjustment is required, Box 2 calculates the appropriate adjustment factors. Chapter 49 shows the steps.

Because male All-Cancers include male Respiratory-System Cancers, it is no surprise that Box 1, below, shows results consistent with the results in Box 3 of Chapter 48. Below, Columns D and I, as well as Columns F and K, indicate clearly that a carcinogenic co-actor (smoking), which can contribute to male MortRates from All-Cancers, is operating more strongly in the LowTrio than in the TopTrio. We must match the Census Divisions for smoking.

Table 49-A
All-Cancers, Males: Fractional Causation by Medical Radiation over Time

Year	Col.A Natl MR	Col.B Frac.C		Col.C R-Sq	Col.D X-Coef	Col.E StdErr	Col.F Coef/SE	Col.G Source
1940	115.0	90%		0.9508	0.7557	0.0650	11.6276	Chap.6
1950	132.8	84%		0.9330	0.8462	0.0857	9.8703	Tab 49-B
1960	145.7	83%		0.9407	0.9251	0.0878	10.5397	Tab 49-C
1970	155.1	79%		0.9415	0.9073	0.0855	10.6122	Tab 49-D
1980	164.5	75%		0.9386	0.8480	0.0820	10.3418	Tab 49-E
1988	162.7	74%		0.9348	0.7488	0.0748	10.1056	Tab 49-F

```
===============================================================================================
                                Box 1, Chap. 49
                  All-Cancers, Males:  Post-1940 Change in MortRates by Census Trios

1960 vs. 1940, by Trios:  Col.D expresses change by ratios.  Col.F expresses change by subtraction.
1988 vs. 1940, by Trios:  Col.I expresses change by ratios.  Col.K expresses change by subtraction.
High-PhysPop Trio shows the lowest growth-ratio.  Low-PhysPop Trio shows the highest growth-ratio.


          ● 1940  |    >>>  ● Compare 1960 with 1940 ●  <<<  |    >>>  ● Compare 1988 with 1940 ●  <<<

          Col.A   |  Col.B    Col.C    Col.D    Col.E    Col.F   |  Col.G    Col.H    Col.I    Col.J    Col.K
          1940    |  1960     Ratio    Input    Diff:    Input   |  1988     Ratio    Input    Diff:    Input
          MortRate|  MortRate Col.B    from     Col.B    from    |  MortRate Col.G    from     Col.G    from
          Tab 6-A |  Tab 6-A  /Col.A   Col.C    minus A  Col.E   |  Tab 6-A  /Col.A   Col.H    minus A  Col.J

Pacif     122.9   |  140.7    1.145    Avg Chg   17.8    Avg Chg |  148.5    1.208    Avg Chg   25.6    Avg Chg
NewE      135.5   |  164.6    1.215    TopTrio   29.1    TopTrio |  167.1    1.233    TopTrio   31.6    TopTrio
MidAtl    140.9   |  164.0    1.164    1.175     23.1     23.3   |  168.4    1.195    1.212     27.5     28.2

WNoCen    110.9   |  135.6    1.223    Avg Chg   24.7    Avg Chg |  155.9    1.406    Avg Chg   45.0    Avg Chg
ENoCen    119.6   |  150.7    1.260    MidTrio   31.1    MidTrio |  171.2    1.431    MidTrio   51.6    MidTrio
Mtn        99.8   |  118.7    1.189    1.224     18.9     24.9   |  139.1    1.394    1.410     39.3     45.3

WSoCen     86.9   |  133.8    1.540    Avg Chg   46.9    Avg Chg |  172.9    1.990    Avg Chg   86.0    Avg Chg
ESoCen     73.6   |  125.1    1.700    LowTrio   51.5    LowTrio |  188.2    2.557    LowTrio  114.6    LowTrio
SoAtl      88.9   |  137.1    1.542    1.594     48.2     48.9   |  175.8    1.978    2.175     86.9     95.8
===============================================================================================
```

Part 1. Overview, and Purpose of Box 1
Part 2. Co-Action among Carcinogens, Such as Xrays and Cigarettes
Part 3. The Essence of the Smoking Adjustment
Part 4. Explanation of Box 2
Part 5. Explanation of Table 49-B and Every Similar Table

● Part 1. Overview, and Purpose of Box 1

Although Box 1 and Table 49-A appear on the first page of this chapter, Box 2 and Tables 49-B through 49-F are located in the usual place --- AFTER the text, at the end of the chapter.

1a. Chapter 49 as the Model for Subsequent Chapters

The text will explain how we make the Smoking Adjustment and how we then calculate Fractional Causation, by medical radiation, of the post-1940 Observed National MortRates for Cancer and Ischemic Heart Disease. The same steps will be used without further explanation in Chapters 50 through 65. Each chapter has its own Table-A, in which Column B summarizes the findings on Fractional Causation by medical radiation. The findings very strongly support Hypotheses 1+2. Chapter 66 summarizes the findings from Chapters 49 - 65, for easy comparison.

1b. The Purpose of Box 1

Box 1 of this chapter follows the model of Box 3 in Chapter 48, where its columns are explained in detail. Here, in Box 1 of Chapter 49, we look first at the findings in Columns D and I:

● - Col.D: The 1960 MortRates in the TopTrio of Census Divisions are 1.175 times their values in 1940, whereas the 1960 MortRates in the LowTrio of Census Divisions are 1.594 times their values in 1940.
● - Col.I: The 1988 MortRates in the TopTrio are 1.212 times their values in 1940, whereas the 1988 MortRates in the LowTrio are 2.175 times their values in 1940.
● - Columns F and K provide confirmation that a carcinogenic co-actor (smoking), which can contribute to male MortRates from All-Cancers, is operating more strongly in the LowTrio than in the TopTrio (Chapter 48, Part 5b). We must match the Census Divisions for smoking.

In the chapters which follow, Smoking Adjustments are made only if Box 1 demonstrates that they are required. The requirement is not assumed. For example, Box 1 of Chapter 62 does NOT produce a clear indication that a Smoking Adjustment is required.

● Part 2. Co-Action among Carcinogens, Such as Xrays and Cigarettes

Co-action among carcinogenic agents has been discussed already in the Introduction (Parts 4 + 5), and in Chapter 6 (Part 6). Co-action is a widely (but not universally) accepted expectation, which is supported by several lines of evidence.

Cigarette smoking and exposure to ionizing radiation are each well established causes of Cancer. And smoking is specifically identified as a CO-ACTOR with radiation by the National Research Council's BEIR-5 and BEIR-6 Committees (BEIR 1990, p.152, + BEIR 1999, p.33) --- meaning that smoking and radiation can make necessary contributions to the same fatal cases of Cancer and that the two carcinogens modify each other's carcinogenic potency.

2a. The 1990 BEIR Report's List of Co-Actors with Radiation

For the readers' convenience, we repeat BEIR 1990's list of co-actors with ionizing radiation (from BEIR 1990, p.152; already presented in our Chapter 6, Part 6): "As discussed in the preceding section, the carcinogenic process includes the successive stages of initiation and promotion. The latter phase, promotion, appears to be particularly susceptible to modulation, with cigarette smoking being a conspicuous example of a modulating factor. Susceptibility to the carcinogenic effects of radiation can thus be affected by a number of factors, such as genetic constitution, sex, age at initiation, physiological state, smoking habits, drugs, and various other physical and chemical agents (UNSCEAR 1982)." (BEIR 1990 is citing the 1982 UNSCEAR Report in support of these statements.) Some of the confirmatory experimental evidence, of alteration of radiation's per-rad potency by other agents, is presented in BEIR 1990 also (pp.145-147).

The 1999 BEIR Committee (BEIR-6), which has almost no overlapping membership with the 1990 BEIR Committee (BEIR-5), embraces the same expectation of co-action among carcinogenic (and anti-carcinogenic) agents: "Radiation carcinogenesis, in common with any other form of cancer induction, is likely to be a complex multi-step process that can be influenced by other agents and genetic factors at each step" (BEIR 1999, p.5).

The BEIR 1990 list, of co-actors with ionizing radiation, embraces co-actors which may themselves be mutagens and explicitly embraces inherited mutations ("genetic constitution"). The list also calls the co-actors "factors" --- meaning that co-actors modify the carcinogenic per-rad potency of ionizing radiation by multiplication.

2b. Two Illustrations: How Co-Actors Can Biologically Alter the Potency of Xrays

How could exposure to non-xray co-actors (for instance, smoking-induced co-actors) multiply the carcinogenic potency of medical radiation, per rad of dose? Two illustrations suffice here. One way would be by interfering with correct repair of xray-induced damage to the genetic molecules. The result would be a higher frequency of xray-induced mutations per rad of dose. A different way would be by intensifying the carcinogenic CONSEQUENCES of xray-induced mutations --- for example, by blocking a signal for apoptosis (cell suicide), and leaving the cancer-prone cell alive. (Chapter 67, Part 2b, provides an additional type of illustration.) There is no reason to imagine that medical radiation is the only carcinogenic agent whose potency is affected by other agents.

2c. Implications of Co-Action for Changes in Cancer MortRates over Time

In the monograph's Introduction (Part 5), our illustrations reflect the expectation that all carcinogenic co-actors can interact with each other, and that each case of Cancer may have multiple co-actors. Therefore, if exposures to various co-actors change over time, the subsequent observed cancer MortRates reflect the NET effect of upward pressure from some agents and downward pressure from others.

The impact of nonradiation agents, upon the per-rad potency of radiation, means that we should expect that per-rad potency of xrays and gamma rays may vary from organ to organ. Although xrays and gamma rays have access to every cell of every organ, the chemical and infectious co-actors may not have "universal access" and, even when they have access, they may not have identical activity in cells of different types. Therefore, rising (or falling) exposure to a specific non-xray co-actor can be irrelevant, with respect to the subsequent cancer MortRates of any organ where that particular co-actor has little access or little activity.

2d. How Co-Action Determines the Approach to a Smoking Adjustment

Our Smoking Adjustment applies the expectations of co-action among carcinogenic agents, and modulation of radiation's per-rad potency by interaction with smoking and other carcinogenic agents. In our opinion, it would be irrational for anyone to propose that virtually all carcinogenic agents EXCEPT smoking co-act with each other.

Nonetheless, rejection of co-action is the assumption in Appendix-M, which is included in this monograph in order to show that support for Hypothesis-1 does not depend on co-action. In Appendix-M, we make the post-1940 Smoking Adjustment by assuming that smoking and medical radiation do not contribute to causing the same cases of Cancer. The rejection of co-action implies that, if smoking had been matched across the Census Divisions (emphasis on "if"), smoking would have simply added the same number of post-1940 cancer-deaths in every Census Division to the baseline cancer MortRate (the Constant). Such post-1940 additions do not degrade the tight and positive linear dose-responses, observed in 1940, between PhysPop and cancer MortRates (discussion in Chapter 5, Part 6a). The Fractional Causations calculated in Appendix-M support Hypothesis-1 for All-Cancers-Combined and Hypothesis-2 for Ischemic Heart Disease. But since we do NOT embrace the denial of co-action assumed in Appendix-M, we proceed in Part 3 (below) with a Smoking Adjustment which incorporates the premises of co-action and modulation of radiation's potency by co-actors. Additional implications, of the Smoking Adjustment employed in Chapters 49 through 65, are discussed in Appendix-M, Parts 3 and 4..

● **Part 3. The Essence of the Smoking Adjustment**

The goal of the Smoking Adjustment is not to eliminate all of the post-1940 impact of cigarette smoking upon the MidTrio and LowTrio cancer MortRates. Rather, the goal is to eliminate only the post-1940 impact of EXTRA smoking in the MidTrio and LowTrio Census Divisions, compared with the TopTrio Census Divisions. Note: We use the phrase "post-1940 impact" as a reminder that exposure to a carcinogen or atherogen may precede some of the impact (the fatal consequences) by decades. All of Part 3 applies also to Chapters 64 and 65 (Ischemic Heart Disease).

3a. An Uncomplicated Method

In each decade after 1940, the average All-Cancer MortRate in the TopTrio changed, relative to its 1940 value. For each decade, we can ascertain by what factor the 1940 cancer MortRate changed in the TopTrio (Examples from Box 1, Columns D and I: 1960 change-factor = 1.175, and 1988 change-factor = 1.212). Then --- one decade at a time --- we can multiply the Observed 1940 Cancer MortRates, of the MidTrio and LowTrio Census Divisions, by the SAME change-factor which is observed in the TopTrio during each post-1940 decade. That is the method.

The resulting MidTrio and LowTrio MortRate values, for each decade, are reasonable approximations of the post-1940 cancer MortRates which would have been observed in the MidTrio and LowTrio Census Divisions, if there had not been EXTRA smoking (relative to the TopTrio) in the MidTrio and LowTrio Census Divisions. Part 3b explains what makes them reasonable approximations. The adjusted values become even better approximations, logically, after we apply a second change-factor for the post-1940 behavior of PhysPop itself (Part 4b).

3b. Why We Should Use the Same Change-Factor for All Census Divisions

Part 3a states that we multiply the Observed 1940 Cancer MortRates in the MidTrio and LowTrio by SAME change-factor observed in the TopTrio. Why are we entitled to say that the SAME change-factor would operate in all Nine Census Divisions, if the post-1940 impact of the cigarette co-actor had been alike (matched) across Census Divisions?

(A) We start with the observation that, in 1940, the impact of cigarettes and other nonxray co-actors with medical radiation was well matched across the Nine Census Divisions. The evidence for this statement consists of the very strong and positive LINEAR dose-responses observed between cancer MortRates and PhysPop (dose) in 1940. By definition (Chapter 5, Part 5a), the nearly perfect linear dose-responses mean that potency per dose-unit (PhysPop-unit) was nearly equal in all Nine Census Divisions. Such equality requires well matched co-actors (potency-modulators).

(B) In the ABSENCE of degraded matching during the post-1940 decades, we would necessarily expect that the nearly perfect and positive linear correlations observed in 1940, between medical radiation (PhysPop) and cancer MortRates, would persist indefinitely --- even if the slopes and National Cancer MortRates were to change (as many of them did) during the post-1940 decades. To expect otherwise (i.e., to expect 1940's tight, positive linear dose-responses to disappear) would be to expect repeal of the well-established causal relationship between ionizing radiation and cancer induction.

(C) The requirement here, for transferring the tight positive linear correlation observed in 1940 to the post-1940 decades, is to multiply the Observed 1940 Cancer MortRates by the SAME change-factor in all Nine Census Divisions. This operation is illustrated in Chapter 5, Part 6c. In regression #3 of that illustration, the original correlation persists unchanged AFTER every MortRate has been multiplied by the same change-factor. If the change-factor were 1.2 (for example) instead of 0.8, then the equation for the linear dose-response ($y = mx + c$) would become $1.2y = 1.2mx + 1.2c$. The operation is valid not only for perfect correlations, but also for R-squared values below 1.0. Biologically, these relationships make sense. When smoking increases all the MortRates by the same factor, it is because smoking has increased radiation's potency per dose-unit ("m") and also has raised the Constant (because smoking will increase the potency of the non-xray co-actors interacting with each other, too).

(D) When our Smoking Adjustment ensures that Post-1940 Cancer MortRates in all Nine Census Divisions change by the SAME change-factor, the adjustment ensures that post-1940 dose-responses, between PhysPop and Adjusted Cancer MortRates, will retain a tight, positive linear dose-response (Paragraph B). This is the rational expectation in the absence of extra smoking in the MidTrio and LowTrio Census Divisions, relative to the TopTrio. By making the post-1940 MortRate change-factor the same in all Nine Census Divisions, the adjustment ensures equal potency per dose-unit in all Nine Census Divisions (Paragraph C), and thus eliminates the unequal potency per dose-unit which results from extra smoking in the MidTrio and LowTrio Census Divisions.

Summary: Our Smoking Adjustment provides reasonable sets of Post-1940 Cancer MortRates because it (a) retains all the impact of smoking upon the TopTrio's MortRates, whose Post-1940 Observed Cancer MortRates are retained without any adjustment at all, and (b) adjusts the MidTrio and LowTrio MortRates to the values they would have had, if there had not been EXTRA smoking in those Census Divisions. After we obtain Post-1940 Adjusted MortRates for the MidTrio and LowTrio Census Divisions, we proceed to calculate Post-1940 Fractional Causations by medical radiation (Part 5c).

● Part 4. Explanation of Box 2

Because the Census Divisions were NOT properly matched for smoking, a Smoking Adjustment for the cancer MortRates is required. We must ascertain by WHAT factors the TopTrio's 1940 All-Cancer MortRates changed in subsequent decades. Then we can apply these reality-based factors to the Observed 1940 All-Cancer MortRates of the MidTrio and LowTrio.

4a. Part 1 of Box 2: TopTrio's Population-Weighted MortRates

Box 1 already provides 1.175 as the MortRate change-factor in the TopTrio for 1960 vs. 1940, and provides 1.212 as the factor for 1988 vs. 1940. But these factors were calculated, as an approximation, without obtaining population-weighted MortRates for the Top Trio. Now Part 1 of Box 2 provides average MortRates which are population-weighted for the TopTrio.

Example (1940), from the upper-left corner of Box 2, Part 1: Col.A presents the Observed MortRates from Table 6-A for the Pacific, New England, and Mid-Atlantic Census Divisions. Col.B presents the population of each Division from Table 3-B. The sum (45,710,039) is the 1940 population of the TopTrio. Then Col.C divides Col.B by 45,710,039, in order to find out what fraction of the TopTrio's population is contributed by each Census Division. That share is the weighting factor. Col.D multiplies each Observed MortRate by its own weighting factor, and the sum (136.070) is the population-weighted All-Cancer MortRate for TopTrio males in 1940. Box 2 follows the same procedure in all six sections of its Part 1.

4b. Part 2 of Box 2: Obtaining the Adjustment Factors

In Part 2 of Box 2, the Col.A presents the population-weighted MortRates for 1950-1988, obtained in Part 1. Col.B presents the 1940 population-weighted MortRate, also obtained in Part 1. Then Col.C divides each MortRate in Col.A by the 1940 MortRate, to obtain the factor by which the 1940 MortRates changed in each subsequent decade.

The change-factors in Col.C describe what was observed only in the TopTrio --- not in the MidTrio or LowTrio.

How different are results in Box 2, compared with the approximations used in Box 1? The population-weighted factor for 1960 is 1.151 (Box 2, Part 2, Col.C), whereas the non-weighted factor for 1960 is 1.175 (Box 1, Col.D) --- a ratio of 0.98 (for PopWeighted / NonWeighted). For 1988, the comparison is 1.174 (Box 2) versus 1.212 (Box 1, Col.I) --- a ratio of 0.97 (for PopWeighted / NonWeighted). Although the differences between population-weighted and non-weighted values turn out to be negligible here, the determination makes population-weighted values available, and so we use them.

Col.C: The Change-Factors in the Top Trio, during 50 Years

In Col.C of Part 2, the change-factors contain no assumptions whatsoever. They report on the MortRate changes which were observed in the TopTrio (high-dose) Census Divisions over time.

If we multiplied each of the 1940 MortRates, observed in the six Census Divisions of the MidTrio and LowTrio, by 1.085 (from Col.C, for the year 1950), we would obtain six 1950 Adjusted MortRates which would no longer be elevated by the post-1940 impact from EXTRA smoking in those six Census Divisions. The 1950 Adjusted MortRates would incorporate approximately the SAME smoking effect observed in the Top Trio --- but they would not yet register some of the PhysPop effect, as explained below.

In eliminating the effect of unmatched cigarette smoking, we must NOT eliminate the effect of post-1940 growth (or decline) in Averaged PhysPops in the MidTrio and LowTrio, relative to the TopTrio. So, we must consult Table 47-B, which quantifies what happened after 1940. There are no assumptions in Table 47-B.

Column D: The PhysPop Adjustment

Col.D of Part 2 presents the ratios (from Table 47-B) which show that, relative to Averaged PhysPops in the TopTrio, Averaged PhysPop values fell a little in the MidTrio and rose a little in the LowTrio during the 1940-1990 years. Because radiation-induced cancer MortRates are proportional to radiation dose, we must also adjust the 1940 MortRates for these slightly uneven post-1940 changes in Averaged PhysPop over time.

Therefore, we plan to multiply the MidTrio and LowTrio 1940 MortRates by TWO adjustment factors (from Col.C and Col.D). Of course the sequence makes no difference. So we can multiply Col.C by Col.D ahead of time, and obtain a combined Adjustment Factor in Col.E. The Col.E Adjustment Factors for the year 1950 get used in Table 49-B, the 1960 Adjustment Factors get used in Table 49-C, and so forth.

● Part 5. Explanation of Table 49-B and Every Similar Table

There are three parts in Table 49-B and similar tables:
- ● - Calculation of the Adjusted MortRates for MidTrio and LowTrio.
- ● - Linear regression-analysis: MortRates upon Averaged PhysPops.
- ● - Calculation of estimated Fractional Causation, by medical radiation, of the Observed National MortRates for Cancer (Chapters 49 - 63) and for Ischemic Heart Disease (Chapters 64, 65).

5a. Part 1 of Table 49-B: Adjusted MortRates for the Year 1950

COLUMN C. Col.C is included just as an easy error-check on the entries in Col.A and Col.B. If the sum of Col.C is NOT a close match for the Observed 1950 National MortRate from Table 6-B, then we know that we have made an entry-error somewhere.

COLUMN F. Although this column is labeled "Adjusted MortRates," the MortRates for the TopTrio are not adjusted in any manner. They are identical with the Observed 1950 MortRates in Col.B. By contrast, the six entries for the MidTrio and LowTrio are their Observed 1940 MortRates (in Col.D) times the Adjustment Factor in Col.E (which comes from Box 2, Part 2, Col.E). This set of Adjusted MortRates retains all of the smoking-effect on the TopTrio MortRates, and retains a comparable amount of smoking-effect on the MidTrio and Low Trio MortRates. Only the effect of EXTRA smoking has been eliminated from the MidTrio and LowTrio MortRates. Note: Column F of Table 49-B differs in one small way from the illustration discussed in Part 3b of the text. The 1950 cancer Mortrates in Col. F, for the three census divisions of TopTrio, are the cancer MortRates OBSERVED in 1950, directly from Table 6-A. If they were the observed 1940 rates times the 1950 change-factor of 1.085 (from Box 2, Part 2, Col.C, then the top three entries in Col.F would be calculated from Table 6-A as follows: Pacific = 122.9 * 1.085 = 133.3; and New England = 135.5 * 1.085 = 147.0; and Mid-Atlantic = 140.9 * 1.085 = 152.9.

COLUMN G.   In order to obtain a population-weighted National Adjusted MortRate, we multiply the values in Col.F by their own population-fraction in Col.A.

Return to a Promise Made at the End of Chapter 48

We promised, at the end of Chapter 48, that readers would be able easily to compare the "before and after" MortRate values, each time we make a MortRate adjustment in any chapter.   The "before" rates are always in Col.B, and the "after" rates are always in Col.F.   As noted above, the TopTrio MortRates are not adjusted at all --- and are the same in Col.B and in Col.F.

Eliminating the effect of EXTRA smoking on MortRates, in the MidTrio and LowTrio, always produces Adjusted MortRates which are lower than the Observed MortRates in those six Census Divisions.   And, consequently, the National Adjusted MortRate is always lower than the National Observed MortRate.   The difference between the two rates quantifies the part of the Observed National MortRate which results from the co-action, of the UNMATCHED share of smoking, with medical radiation and with other carcinogenic agents.

5b. Part 2 of Table 49-B: Linear Regression Analysis

COLUMN C.   This is the output from linear regression analysis, so familiar from earlier chapters.   The very high R-squared value in Column C matches expectation and confirms that the Adjustment achieves the tight linear correlation which would occur if co-actors were matched (text, Part 3c).   INPUT: The Averaged PhysPop values come from Table 47-A.   The 1950 Adjusted MortRates come from the same Table 49-B: Part 1, Col.F.

COLUMN E.   This regression-output comes from substituting the 1940 PhysPop values (Col.D) for the 1940-1950 Averaged PhysPops.   We do this as another easy error-check.   The PhysPop rankings are so stable over time that, even in 1990, the correlation between Mean 1940-1990 PhysPops and 1940 PhysPops has an R-squared value of 0.91 (Chapter 47, Part 3b).   Therefore, we must expect that this extra regression will produce similar output to the main regression in Col.C.   If it does NOT, then we know that we have made an entry-error somewhere.

5c. Part 3-A of Table 49-B: Fractional Causation

Here, the calculation of post-1940 Fractional Causation is essentially the same as the calculation in Chapters 6 through 21, and Chapters 40 and 41.   Here, too, the radiation-induced MortRate is divided by the entire OBSERVED National Cancer MortRate.

The Smoking Adjustment permits us to estimate what the national radiation-induced Cancer MortRate would have been, if Cancer MortRates in the MidTrio and LowTrio had not been elevated by EXTRA smoking.   Then we ask, "And what share does the estimated radiation-induced MortRate contribute to the ENTIRE Observed National Cancer MortRate, which INCLUDES the consequences of extra smoking in the MidTrio and LowTrio?"   Paragraph 3 of Part 3-A provides the answer (with Part 3-B as an error-check):

Medical radiation accounts for about 84 percent of the entire Observed National All-Cancer MortRate for males in 1950.

5d. Approximations: A Characteristic of Epidemiology

Our post-1940 Smoking Adjusted MortRates in the MidTrio and LowTrio Census Divisions are necessarily approximations --- as are all the statistical adjustments for matching which fill the peer-reviewed epidemiological literature.

A reminder here is appropriate.   Approximations can suffice, in epidemiology, in answering some questions definitively.   Here, the question is: Did medical radiation cease, after mid-century, to be a major cause of mortality rates for Cancer and Ischemic Heart Disease in the USA, or did it CONTINUE to be a necessary co-actor in a very large share of the national mortality rates for these two diseases?   The estimated post-1940 Fractional Causations, from the tables in Chapters 49 through 65, provide a clear answer.

```
=============================================================================================
                              Box 2, Chap. 49
                  All-Cancers, Males:  Calculation of Adjustment Factor

                    This adjustment is discussed fully in Chapter 49.
  ● Part1:  Calculate average population-weighted MortRate for the combined TopTrio Census Divs.
```

Census Div.	Col.A 1940 MR Tab 6-A	Col.B 1940 Pop'n Tab 3-B	Col.C 1940 Popn /45,710,039	Col.D Col.A * Col.C	Census Div.	Col.A 1950 MR Tab 6-A	Col.B 1950 Pop'n Tab 3-B	Col.C 1950 Popn /53,964,513	Col.D Col.A * Col.C
Pacific	122.9	9,733,262	0.2129	26.17	Pacific	127.2	14,486,527	0.2684	34.15
NewEng	135.5	8,437,290	0.1846	25.01	NewEng	152.4	9,314,453	0.1726	26.30
Mid-Atl	140.9	27,539,487	0.6025	84.89	Mid-Atl	156.0	30,163,533	0.5590	87.20
1940		Sum TopTrio 45,710,039	Sum 1.0000	TopTrio 136.070	1950		Sum TopTrio 53,964,513	Sum 1.0000	TopTrio 147.647

Census Div.	Col.A 1960 MR Tab 6-A	Col.B 1960 Pop'n Tab 3-B	Col.C 1960 Popn /65,875,863	Col.D Col.A * Col.C	Census Div.	Col.A 1970 MR Tab 6-A	Col.B 1970 Pop'n Tab 3-B	Col.C 1970 Popn /75,017,000	Col.D Col.A * Col.C
Pacific	140.7	21,198,044	0.3218	45.28	Pacific	147.2	26,087,000	0.3477	51.19
NewEng	164.6	10,509,367	0.1595	26.26	NewEng	167.5	11,781,000	0.1570	26.30
Mid-Atl	164.0	34,168,452	0.5187	85.06	Mid-Atl	167.9	37,149,000	0.4952	83.15
1960		Sum TopTrio 65,875,863	Sum 1.0000	TopTrio 156.598	1970		Sum TopTrio 75,017,000	Sum 1.0000	TopTrio 160.639

Census Div.	Col.A 1980 MR Tab 6-A	Col.B 1980 Pop'n Tab 3-B	Col.C 1980 Popn /80,615,000	Col.D Col.A * Col.C	Census Div.	Col.A 1988 MR Tab 6-A	Col.B 1990 Pop'n Tab 3-B	Col.C 1990 Popn /88,495,000	Col.D Col.A * Col.C
Pacific	153.7	31,523,000	0.3910	60.10	Pacific	148.5	37,837,000	0.4276	63.49
NewEng	170.3	12,322,000	0.1528	26.03	NewEng	167.1	12,998,000	0.1469	24.54
Mid-Atl	171.8	36,770,000	0.4561	78.36	Mid-Atl	168.4	37,660,000	0.4256	71.66
1980		Sum TopTrio 80,615,000	Sum 1.0000	TopTrio 164.493	1988		Sum TopTrio 88,495,000	Sum 1.0000	TopTrio 159.701

```
  ● Part 2:  Take ratios of these TopTrio MortRates, with 1940 as the denominator of each ratio.
      Col.D modifies Col.C by separate PhysPop adjustments for MidTrio and LowTrio Census Divisions.
```

	Col.A TopTrio Mean MR	Col.B 1940 TopTrio Mean MR	Col.C = Col.A / Col.B	Col.D ppAdju Tab 47-B	Col.E = Col.C * Col.D	ALL CANCERS. Males.
				MidTrio		
1950	147.647	136.070	1.085	0.99	1.07	= MidTrio Adjustment Factor, 1950
1960	156.598	136.070	1.151	0.97	1.12	= MidTrio Adjustment Factor, 1960
1970	160.639	136.070	1.181	0.95	1.12	= MidTrio Adjustment Factor, 1970
1980	164.493	136.070	1.209	0.94	1.14	= MidTrio Adjustment Factor, 1980
1988	159.701	136.070	1.174	0.94	1.10	= MidTrio Adjustment Factor, 1988
				LowTrio		
1950	147.647	136.070	1.085	1.00	1.09	= LowTrio Adjustment Factor, 1950
1960	156.598	136.070	1.151	1.01	1.16	= LowTrio Adjustment Factor, 1960
1970	160.639	136.070	1.181	1.02	1.20	= LowTrio Adjustment Factor, 1970
1980	164.493	136.070	1.209	1.04	1.26	= LowTrio Adjustment Factor, 1980
1988	159.701	136.070	1.174	1.07	1.26	= LowTrio Adjustment Factor, 1988

```
=============================================================================================
```

===

Table 49-B
All Cancers, Males: Fractional Causation in 1950

===

Part 1.
Calculation of the 6 Adjusted MortRates (Col.F) and the National Adjusted MortRate (Col.G).
The last six entries in Part 1, Col.F, are the products of (Col.D * Col.E), as discussed in Chap. 49.

Trio-Sequence	Col.A 1950 PopFrac Tab 3-B	Col.B 1950 Obs MR Tab 6-A	Col.C A * B	Col.D 1940 MR Mid,Low Tab 6-A	Col.E AdjuFact Bx2,Pt2 Col.E	Col.F 1950 Adju MortRates	Col.G A * F
Pacific	0.0961	127.2	12.224			127.2	12.224
New England	0.0618	152.4	9.418			152.4	9.418
Mid-Atlantic	0.2002	156.0	31.231			156.0	31.231
WestNoCentral	0.0933	125.3	11.690	110.9	1.07	118.66	11.071
EastNoCentral	0.2017	138.3	27.895	119.6	1.07	127.97	25.812
Mountain	0.0337	108.1	3.643	99.8	1.07	106.79	3.599
WestSoCentral	0.0965	112.7	10.876	86.9	1.09	94.72	9.141
EastSoCentral	0.0762	104.7	7.978	73.6	1.09	80.22	6.113
SouthAtlantic	0.1406	116.3	16.352	88.9	1.09	96.90	13.624

Sum = 131.3

1950 Observed Natl MR from Table 6-B =   132.8   1950 Natl Adjusted MR =   122.2333

Part 2. --

Trio-Seq.	Col.A Mean1940 thru1950 PPs from Tab 47-A x'	Col.B 1950 Adju MRs from Col.F Part 1	Col.C All Cancers, Males: 1950 Adjusted MortRates regressed on Mean 1940 thru 1950 PPs Regression Output:		Col.D 1940 PPs from Table 3-A (TrioSeq) x''	Col.E All Cancers, Males: 1950 Adjusted MortRates regressed on 1940 PhysPops Regression Output:	
Pac	154.16	127.2	Constant	10.4866	159.72	Constant	10.7576
NewEng	162.03	152.4	Std Err of Y Est	7.1588	161.55	Std Err of Y Est	7.9010
MidAtl	169.24	156.0	R Squared	0.9330	169.76	R Squared	0.9183
WNoCen	121.60	118.663	No. of Observation	9	123.14	No. of Observation	9
ENoCen	128.53	127.972	Degrees of Freedom	7	133.36	Degrees of Freedom	7
Mtn	119.64	106.786			119.89		
WSoCen	102.64	94.721	X Coefficient(s)	0.8462	103.94	X Coefficient(s)	0.8326
ESoCen	84.44	80.224	Std Err of Coef.	0.0857	85.83	Std Err of Coef.	0.0938
SoAtl	99.91	96.901	XCoef / S.E. =	9.8703	100.74	XCoef / S.E.	8.8729

Part 3-A. | Part 3-B.
Calculation of Fractional Causation | Calculation of Fractional Causation
from Averaged PhysPops | from 1940 PhysPops
                                                       |
1. Nonradiation rate is Adjusted | 1. Nonradiation rate is Adjusted
    Constant (Part 2, Col.C) =   10.4866 | Constant (Part 2, Col.E) =   10.7576
                                                       |
2. Radiation rate is Natl Adjusted | 2. Radiation rate is Natl Adjusted
    MortRate (Part 1, Col.G = 122.2333) | MortRate (Part 1, Col.G = 122.2333)
    minus Nonradiation rate (10.4866) =   111.7467 | minus Nonradiation rate (10.7576) =   111.4757
                                                       |
3. 1950 Fractional Causation is radiation | 3. 1950 Fractional Causation is radiation
    rate (111.7467) divided by OBSERVED | rate (111.4757) divided by OBSERVED
    Natl MR Part 1,Col.C=   132.8   =   0.84 | Natl MR Part 1, Col.C=   132.8   =   0.84

```
=========================================================================================================
                                        Table 49-C
                      All Cancers, Males:  Fractional Causation in 1960
=========================================================================================================
```

Part 1.
Calculation of the 6 Adjusted MortRates (Col.F) and the National Adjusted MortRate (Col.G).
The last six entries in Part 1, Col.F, are the products of (Col.D * Col.E), as discussed in Chap. 49.

Trio-Sequence	Col.A 1960 PopFrac Tab 3-B	Col.B 1960 Obs MR Tab 6-A	Col.C A * B	Col.D 1940 MR Mid,Low Tab 6-A	Col.E AdjuFact Bx2,Pt2 Col.E	Col.F 1960 Adju MortRates	Col.G A * F
Pacific	0.1182	140.7	16.631			140.7	16.631
New England	0.0586	164.6	9.646			164.6	9.646
Mid-Atlantic	0.1905	164.0	31.242			164.0	31.242
WestNoCentral	0.0858	135.6	11.634	110.9	1.12	124.21	10.657
EastNoCentral	0.2020	150.7	30.441	119.6	1.12	133.95	27.058
Mountain	0.0382	118.7	4.534	99.8	1.12	111.78	4.270
WestSoCentral	0.0945	133.8	12.644	86.9	1.16	100.80	9.526
EastSoCentral	0.0672	125.1	8.407	73.6	1.16	85.38	5.737
SouthAtlantic	0.1448	137.1	19.852	88.9	1.16	103.12	14.932

```
                                        Sum =  145.0                                          Sum =
            1960 Observed Natl MR from Table 6-B =  145.7          1960 Natl Adjusted MR =    129.6991
```

Part 2. --

Trio-Seq.	Col.A Mean1940 thru1960 PPs from Tab 47-A x'	Col.B 1960 Adju MRs from from Col.F Part 1	Col.C All Cancers, Males: 1960 Adjusted MortRates regressed on Mean 1940 thru 1960 PPs Regression Output:		Col.D 1940 PPs from Table 3-A (TrioSeq) x''	Col.E All Cancers, Males: 1960 Adjusted MortRates regressed on 1940 PhysPops Regression Output:	
Pac	155.69	140.7	Constant	8.7654	159.72	Constant	8.1440
NewEng	162.81	164.6	Std Err of Y Est	7.2600	161.55	Std Err of Y Est	6.9237
MidAtl	167.04	164.0	R Squared	0.9407	169.76	R Squared	0.9461
WNoCen	118.15	124.208	No. of Observation	9	123.14	No. of Observation	9
ENoCen	123.87	133.952	Degrees of Freedom	7	133.36	Degrees of Freedom	7
Mtn	117.40	111.776			119.89		
WSoCen	102.31	100.804	X Coefficient(s)	0.9251	103.94	X Coefficient(s)	0.9113
ESoCen	85.63	85.376	Std Err of Coef.	0.0878	85.83	Std Err of Coef.	0.0822
SoAtl	101.72	103.124	XCoef / S.E. =	10.5397	100.74	XCoef / S.E.	11.0832

Part 3-A.
Calculation of Fractional Causation
from Averaged PhysPops

1. Nonradiation rate is Adjusted
   Constant (Part 2, Col.C) =   8.7654

2. Radiation rate is Natl Adjusted
   MortRate (Part 1, Col.G = 129.6991)
   minus Nonradiation rate (8.7654) =   120.9337

3. 1960 Fractional Causation is radiation
   rate (120.9337) divided by OBSERVED
   Natl MR Part 1,Col.C=   145.7   =   0.83

Part 3-B.
Calculation of Fractional Causation
from 1940 PhysPops

1. Nonradiation rate is Adjusted
   Constant (Part 2, Col.E) =   8.1440

2. Radiation rate is Natl Adjusted
   MortRate (Part 1, Col.G = 129.6991)
   minus Nonradiation rate (8.1440) =   121.5551

3. 1960 Fractional Causation is radiation
   rate (121.5551) divided by OBSERVED
   Natl MR Part 1, Col.C=   145.7   =   0.83

● Table 49-D is not included. Its results, for 1970, are shown in Table 49-A (p.375).

```
===============================================================================================
                                        Table 49-E
                        All Cancers, Males:  Fractional Causation in 1980
===============================================================================================
Part 1.
Calculation of the 6 Adjusted MortRates (Col.F) and the National Adjusted MortRate (Col.G).
The last six entries in Part 1, Col.F, are the products of (Col.D * Col.E), as discussed in Chap. 49.

                 Col.A    Col.B    Col.C        Col.D    Col.E   Col.F    Col.G
                 1980     1980                  1940 MR  AdjuFact 1980
                 PopFrac  Obs MR   A * B        Mid,Low  Bx2,Pt2  Adju     A * F
Trio-Sequence    Tab 3-B  Tab 6-A               Tab 6-A  Col.E   MortRates
     Pacific     0.1398   153.7    21.487                        153.7    21.487
     New England 0.0546   170.3    9.298                         170.3    9.298
     Mid-Atlantic 0.1630  171.8    28.003                        171.8    28.003
     WestNoCentral 0.0759 152.0    11.537       110.9    1.14    126.43   9.596
     EastNoCentral 0.1846 169.5    31.290       119.6    1.14    136.34   25.169
     Mountain    0.0502   134.7    6.762        99.8     1.14    113.77   5.711
     WestSoCentral 0.1049 162.9    17.088       86.9     1.26    109.49   11.486
     EastSoCentral 0.0646 174.1    11.247       73.6     1.26    92.74    5.991
     SouthAtlantic 0.1624 171.4    27.835       88.9     1.26    112.01   18.191

                            Sum =   164.5                         Sum =
        1980 Observed Natl MR from Table 6-B =  164.5     1980 Natl Adjusted MR =  134.9330

Part 2.  ---------------------------------------------------------------------------------------
          Col.A    Col.B                   Col.C                Col.D            Col.E
          Mean1940  1980           All Cancers, Males:          1940      All Cancers, Males:
          thru1980 Adju MRs        1980 Adjusted MortRates      PPs from   1980 Adjusted MortRates
Trio-     PPs from from Col.F          regressed on            Table 3-A        regressed on
Seq.      Tab 47-A Part 1         Mean 1940 thru 1980 PPs      (TrioSeq)     1940 PhysPops
             x'                      Regression Output:           x''         Regression Output:
Pac       177.35    153.7         Constant          10.8567     159.72      Constant          12.5043
NewEng    185.86    170.3         Std Err of Y Est  7.4657      161.55      Std Err of Y Est  5.9915
MidAtl    186.11    171.8         R Squared         0.9386      169.76      R Squared         0.9604
WNoCen    128.82    126.43        No. of Observation 9          123.14      No. of Observation 9
ENoCen    133.71    136.34        Degrees of Freedom 7          133.36      Degrees of Freedom 7
Mtn       133.45    113.77                                      119.89
WSoCen    114.66    109.49        X Coefficient(s)  0.8480      103.94      X Coefficient(s)  0.9276
ESoCen    99.46     92.74         Std Err of Coef.  0.0820      85.83       Std Err of Coef.  0.0712
SoAtl     124.62    112.01        XCoef / S.E. =    10.3418     100.74      XCoef / S.E.      13.0356


----------------------------------------------------------------------------------------------
Part 3-A.                                        | Part 3-B.
Calculation of Fractional Causation              | Calculation of Fractional Causation
from Averaged PhysPops                           | from 1940 PhysPops
                                                 |
1.  Nonradiation rate is Adjusted                | 1.  Nonradiation rate is Adjusted
    Constant (Part 2, Col.C) =        10.8567    |     Constant (Part 2, Col.E) =       12.5043
                                                 |
2.  Radiation rate is Natl Adjusted              | 2.  Radiation rate is Natl Adjusted
    MortRate (Part 1, Col.G = 134.9330)          |     MortRate (Part 1, Col.G = 134.9330)
    minus Nonradiation rate (10.8567) = 124.0763 |     minus Nonradiation rate (12.5043) = 122.4287
                                                 |
3.  1980 Fractional Causation is radiation       | 3.  1980 Fractional Causation is radiation
    rate (124.0763) divided by OBSERVED          |     rate (122.4287) divided by OBSERVED
    Natl MR Part 1,Col.C=   164.5    =    0.75   |     Natl MR Part 1, Col.C=   164.5    =   0.74
----------------------------------------------------------------------------------------------
```

===

Table 49-F
All Cancers, Males: Fractional Causation in 1988

===

Part 1.
Calculation of the 6 Adjusted MortRates (Col.F) and the National Adjusted MortRate (Col.G).
The last six entries in Part 1, Col.F, are the products of (Col.D * Col.E), as discussed in Chap. 49.

Trio-Sequence	Col.A 1990 PopFrac Tab 3-B	Col.B 1988 Obs MR Tab 6-A	Col.C A * B	Col.D 1940 MR Mid,Low Tab 6-A	Col.E AdjuFact Bx2,Pt2 Col.E	Col.F 1988 Adju MortRates	Col.G A * F
Pacific	0.1535	148.5	22.795			148.5	22.795
New England	0.0527	167.1	8.806			167.1	8.806
Mid-Atlantic	0.1527	168.4	25.715			168.4	25.715
WestNoCentral	0.0721	155.9	11.240	110.9	1.10	121.99	8.795
EastNoCentral	0.1713	171.2	29.327	119.6	1.10	131.56	22.536
Mountain	0.0543	139.1	7.553	99.8	1.10	109.78	5.961
WestSoCentral	0.1087	172.9	18.794	86.9	1.26	109.49	11.902
EastSoCentral	0.0621	188.2	11.687	73.6	1.26	92.74	5.759
SouthAtlantic	0.1725	175.8	30.325	88.9	1.26	112.01	19.322

Sum = 166.2
1988 Observed Natl MR from Table 6-B = 162.7

Sum =
1988 Natl Adjusted MR = 131.5917

Part 2. ---

Trio-Seq.	Col.A Mean1940 thru1990 PPs from Tab 47-A x'	Col.B 1988 Adju MRs from Col.F Part 1	Col.C All Cancers, Males: 1988 Adjusted MortRates regressed on Mean 1940 thru 1990 PPs Regression Output:		Col.D 1940 PPs from Table 3-A (TrioSeq) x''	Col.E All Cancers, Males: 1988 Adjusted MortRates regressed on 1940 PhysPops Regression Output:	
Pac	191.97	148.5	Constant	10.8756	159.72	Constant	16.4305
NewEng	208.20	167.1	Std Err of Y Est	7.3475	161.55	Std Err of Y Est	7.1662
MidAtl	204.72	168.4	R Squared	0.9348	169.76	R Squared	0.9379
WNoCen	141.14	121.99	No. of Observation	9	123.14	No. of Observation	9
ENoCen	146.19	131.56	Degrees of Freedom	7	133.36	Degrees of Freedom	7
Mtn	145.91	109.78			119.89		
WSoCen	126.28	109.49	X Coefficient(s)	0.7488	103.94	X Coefficient(s)	0.8754
ESoCen	113.28	92.74	Std Err of Coef.	0.0748	85.83	Std Err of Coef.	0.0851
SoAtl	142.93	112.01	XCoef / S.E. =	10.0156	100.74	XCoef / S.E.	10.2865

Part 3-A.
Calculation of Fractional Causation
from Averaged PhysPops

1. Nonradiation rate is Adjusted
    Constant (Part 2, Col.C) =   10.8756

2. Radiation rate is Natl Adjusted
    MortRate (Part 1, Col.G = 131.5917)
    minus Nonradiation rate (10.8756) =   120.7161

3. 1988 Fractional Causation is radiation
    rate (120.7161) divided by OBSERVED
    Natl MR Part 1,Col.C=   162.7   =   0.74

Part 3-B.
Calculation of Fractional Causation
from 1940 PhysPops

1. Nonradiation rate is Adjusted
    Constant (Part 2, Col.E) =   16.4305

2. Radiation rate is Natl Adjusted
    MortRate (Part 1, Col.G = 131.5917)
    minus Nonradiation rate (16.4305) =   115.1612

3. 1988 Fractional Causation is radiation
    rate (115.1612) divided by OBSERVED
    Natl MR Part 1,Col.C=   162.7   0.71

CHAPTER 50

All-Cancers, Females, 1940-1988: Fractional Causation by Medical Radiation

- Table 50-A, Column A, shows that the female National All-Cancer MortRate is falling through time —- despite the rising female MortRate from Respiratory-System Cancers. The net decline, in the female's National All-Cancer MortRate, receives big assistance from the steep post-1940 declines in the National MortRate for female Genital Cancers (Table 14-B) and female Digestive- System Cancers (Table 10-B).

- Box 1 includes ratios below 1.00 in Columns D and I, and negative numbers in Columns F and K. These findings reflect the fact that, after 1940, the female populations in the TopTrio and MidTrio enjoy a DECREASE in their own 1940 All-Cancer MortRates. A falling rate produces a ratio (fraction) below 1.0, of course. By 1988, female All-Cancer MortRates in the TopTrio are down to 0.835 (83.5%) of their 1940 value. In the MidTrio, rates are down to 89.1% (Column I). Meanwhile, the LowTrio population experiences an 8% INCREASE in its own 1940 rates (ratio = 1.081, in Column I). The facts in Box 1 indicate clearly that a carcinogenic co-actor (smoking) is operating more strongly in the LowTrio than in the TopTrio (Chapter 48, Part 5b). We must match the Census Divisions for smoking, before evaluating Fractional Causation by medical radiation.

- Besides Chapter 50, Chapter 56 also permits evaluation of Fractional Causation of female All-Cancer MortRates by medical radiation. The two chapters are in very satisfactory accord.

	Table 50-A All-Cancers, Females: Fractional Causation by Medical Radiation over Time						
Year	Col.A Natl MR	Col.B Frac.C	Col.C R-Sq	Col.D X-Coef	Col.E StdErr	Col.F Coef/SE	Col.G Source
1940	126.1	58%	0.8608	0.5279	0.0802	6.5801	Chap.7
1950	123.2	53%	0.8644	0.4894	0.0733	6.6803	Tab 50-B
1960	114.9	54%	0.8689	0.4661	0.0684	6.8105	Tab 50-C
1970	111.7	52%	0.8799	0.4285	0.0598	7.1600	Tab 50-D
1980	108.5	52%	0.8839	0.3857	0.0528	7.3005	Tab 50-E
1988	111.3	50%	0.8703	0.3393	0.0495	6.8536	Tab 50-F

```
================================================================================================
                                    Box 1, Chap. 50
                  All-Cancers, Females:  Post-1940 Change in MortRates by Census Trios

1960 vs. 1940, by Trios:  Col.D expresses change by ratios.  Col.F expresses change by subtraction.
1988 vs. 1940, by Trios:  Col.I expresses change by ratios.  Col.K expresses change by subtraction.
High-PhysPop Trio shows the lowest growth-ratio.  Low-PhysPop Trio shows the highest growth-ratio.


         • 1940   |   >>> • Compare 1960 with 1940 • <<<   |   >>> • Compare 1988 with 1940 • <<<

         Col.A    |  Col.B    Col.C    Col.D    Col.E    Col.F  |  Col.G    Col.H    Col.I    Col.J    Col.K
         1940     |  1960     Ratio    Input    Diff:    Input  |  1988     Ratio    Input    Diff:    Input
         MortRate |  MortRate Col.B    from     Col.B    from   |  MortRate Col.G    from     Col.G    from
         Tab 7-A  |  Tab 7-A  /Col.A   Col.C    minus A  Col.E  |  Tab 7-A  /Col.A   Col.H    minus A  Col.J
                  |                                             |
Pacif    127.4    |  110.1    0.864    Avg Chg  -17.3    Avg Chg|  111.5    0.875    Avg Chg  -15.9    Avg Chg
NewE     145.3    |  122.4    0.842    TopTrio  -22.9    TopTrio|  116.4    0.801    TopTrio  -28.9    TopTrio
MidAtl   142.9    |  127.4    0.892    0.866    -15.5    -18.6  |  118.6    0.830    0.835    -24.3    -23.0
                  |                                             |
WNoCen   120.1    |  109.3    0.910    Avg Chg  -10.8    Avg Chg|  106.8    0.889    Avg Chg  -13.3    Avg Chg
ENoCen   131.4    |  119.8    0.912    MidTrio  -11.6    MidTrio|  116.5    0.887    MidTrio  -14.9    MidTrio
Mtn      111.8    |  101.0    0.903    0.908    -10.8    -11.1  |  100.4    0.898    0.891    -11.4    -13.2
                  |                                             |
WSoCen    99.8    |  102.9    1.031    Avg Chg    3.1    Avg Chg|  109.8    1.100    Avg Chg   10.0    Avg Chg
ESoCen   102.5    |  104.8    1.022    LowTrio    2.3    LowTrio|  112.7    1.100    LowTrio   10.2    LowTrio
SoAtl    106.9    |  107.4    1.005    1.019      0.5     2.0   |  111.6    1.044    1.081      4.7     8.3
                  |                                             |
================================================================================================
```

===
Box 2, Chap. 50
All-Cancers, Females: Calculation of Adjustment Factor

This adjustment is discussed fully in Chapter 49.
● Part 1: Calculate average population-weighted MortRate for the combined TopTrio Census Divs.

Census Div.	Col.A 1940 MR Tab 7-A	Col.B 1940 Pop'n Tab 3-B	Col.C 1940 Popn /45,710,039	Col.D Col.A * Col.C		Census Div.	Col.A 1950 MR Tab 7-A	Col.B 1950 Pop'n Tab 3-B	Col.C 1950 Popn /53,964,513	Col.D Col.A * Col.C
Pacific	127.4	9,733,262	0.2129	27.13		Pacific	117.7	14,486,527	0.2684	31.60
NewEng	145.3	8,437,290	0.1846	26.82		NewEng	132.1	9,314,453	0.1726	22.80
Mid-Atl	142.9	27,539,487	0.6025	86.09		Mid-Atl	137.0	30,163,533	0.5590	76.58
1940		Sum TopTrio 45,710,039	Sum 1.0000	TopTrio 140.043		1950		Sum TopTrio 53,964,513	Sum 1.0000	TopTrio 130.973

Census Div.	Col.A 1960 MR Tab 7-A	Col.B 1960 Pop'n Tab 3-B	Col.C 1960 Popn /65,875,863	Col.D Col.A * Col.C		Census Div.	Col.A 1970 MR Tab 7-A	Col.B 1970 Pop'n Tab 3-B	Col.C 1970 Popn /75,017,000	Col.D Col.A * Col.C
Pacific	110.1	21,198,044	0.3218	35.43		Pacific	110.2	26,087,000	0.3477	38.32
NewEng	122.4	10,509,367	0.1595	19.53		NewEng	119.4	11,781,000	0.1570	18.75
Mid-Atl	127.4	34,168,452	0.5187	66.08		Mid-Atl	122.4	37,149,000	0.4952	60.61
1960		Sum TopTrio 65,875,863	Sum 1.0000	TopTrio 121.035		1970		Sum TopTrio 75,017,000	Sum 1.0000	TopTrio 117.686

Census Div.	Col.A 1980 MR Tab 7-A	Col.B 1980 Pop'n Tab 3-B	Col.C 1980 Popn /80,615,000	Col.D Col.A * Col.C		Census Div.	Col.A 1988 MR Tab 7-A	Col.B 1990 Pop'n Tab 3-B	Col.C 1990 Popn /88,495,000	Col.D Col.A * Col.C
Pacific	110.4	31,523,000	0.3910	43.17		Pacific	111.5	37,837,000	0.4276	47.67
NewEng	116.4	12,322,000	0.1528	17.79		NewEng	116.4	12,998,000	0.1469	17.10
Mid-Atl	117.5	36,770,000	0.4561	53.59		Mid-Atl	118.6	37,660,000	0.4256	50.47
1980		Sum TopTrio 80,615,000	Sum 1.0000	TopTrio 114.556		1988		Sum TopTrio 88,495,000	Sum 1.0000	TopTrio 115.241

● Part 2: Take ratios of these TopTrio MortRates, with 1940 as the denominator of each ratio.
  Col.D modifies Col.C by separate PhysPop adjustments for MidTrio and LowTrio Census Divisions.

	Col.A TopTrio Mean MR	Col.B 1940 TopTrio Mean MR	Col.C = Col.A / Col.B	Col.D ppAdju Tab 47-B	Col.E = Col.C * Col.D	ALL CANCERS. Females.
				MidTrio		
1950	130.973	140.043	0.935	0.99	0.93	= MidTrio Adjustment Factor, 1950
1960	121.035	140.043	0.864	0.97	0.84	= MidTrio Adjustment Factor, 1960
1970	117.686	140.043	0.840	0.95	0.80	= MidTrio Adjustment Factor, 1970
1980	114.556	140.043	0.818	0.94	0.77	= MidTrio Adjustment Factor, 1980
1988	115.241	140.043	0.823	0.94	0.77	= MidTrio Adjustment Factor, 1988
				LowTrio		
1950	130.973	140.043	0.935	1.00	0.94	= LowTrio Adjustment Factor, 1950
1960	121.035	140.043	0.864	1.01	0.87	= LowTrio Adjustment Factor, 1960
1970	117.686	140.043	0.840	1.02	0.86	= LowTrio Adjustment Factor, 1970
1980	114.556	140.043	0.818	1.04	0.85	= LowTrio Adjustment Factor, 1980
1988	115.241	140.043	0.823	1.07	0.88	= LowTrio Adjustment Factor, 1988

===

```
================================================================================================
                                        Table 50-B
                    All Cancers, Females:  Fractional Causation in 1950
================================================================================================
```

Part 1.
Calculation of the 6 Adjusted MortRates (Col.F) and the National Adjusted MortRate (Col.G).
The last six entries in Part 1, Col.F, are the products of (Col.D * Col.E), as discussed in Chap. 49.

Trio-Sequence	Col.A 1950 PopFrac Tab 3-B	Col.B 1950 Obs MR Tab 7-A	Col.C A * B	Col.D 1940 MR Mid,Low Tab 7-A	Col.E AdjuFact Bx2,Pt2 Col.E	Col.F 1950 Adju MortRates	Col.G A * F
Pacific	0.0961	117.7	11.311			117.7	11.311
New England	0.0618	132.1	8.164			132.1	8.164
Mid-Atlantic	0.2002	137.0	27.427			137.0	27.427
WestNoCentral	0.0933	117.1	10.925	120.1	0.93	111.69	10.421
EastNoCentral	0.2017	127.5	25.717	131.4	0.93	122.20	24.648
Mountain	0.0337	106.0	3.572	111.8	0.93	103.97	3.504
WestSoCentral	0.0965	109.3	10.547	99.8	0.94	93.81	9.053
EastSoCentral	0.0762	110.3	8.405	102.5	0.94	96.35	7.342
SouthAtlantic	0.1406	113.3	15.930	106.9	0.94	100.49	14.128

```
                                      Sum =    122.0                              Sum =
            1950 Observed Natl MR from Table 7-B =    123.2        1950 Natl Adjusted MR =    115.9982
```

Part 2. --

Trio-Seq.	Col.A Mean1940 thru1950 PPs from Tab 47-A x'	Col.B 1950 Adju MRs from Col.F Part 1	Col.C All Cancers, Females: 1950 Adjusted MortRates regressed on Mean 1940 thru 1950 PPs Regression Output:		Col.D 1940 PPs from Table 3-A (TrioSeq) x''	Col.E All Cancers, Females: 1950 Adjusted MortRates regressed on 1940 PhysPops Regression Output:	
Pac	154.16	117.7	Constant	50.6985	159.72	Constant	50.6257
NewEng	162.03	132.1	Std Err of Y Est	6.1181	161.55	Std Err of Y Est	6.2790
MidAtl	169.24	137.0	R Squared	0.8644	169.76	R Squared	0.8572
WNoCen	121.60	111.69	No. of Observation	9	123.14	No. of Observation	9
ENoCen	128.53	122.20	Degrees of Freedom	7	133.36	Degrees of Freedom	7
Mtn	119.64	103.97			119.89		
WSoCen	102.64	93.81	X Coefficient(s)	0.4894	103.94	X Coefficient(s)	0.4834
ESoCen	84.44	96.35	Std Err of Coef.	0.0733	85.83	Std Err of Coef.	0.0746
SoAtl	99.91	100.49	XCoef / S.E. =	6.6803	100.74	XCoef / S.E. =	6.4819

```
------------------------------------------------------------------------------------------------
```

Part 3-A. | Part 3-B.
Calculation of Fractional Causation | Calculation of Fractional Causation
from Averaged PhysPops | from 1940 PhysPops
                                                   |
1. Nonradiation rate is Adjusted | 1. Nonradiation rate is Adjusted
    Constant (Part 2, Col.C) =   50.6985 | Constant (Part 2, Col.E) =   50.6257
                                                   |
2. Radiation rate is Natl Adjusted | 2. Radiation rate is Natl Adjusted
    MortRate (Part 1, Col.G = 115.9982) | MortRate (Part 1, Col.G = 115.9982)
    minus Nonradiation rate (50.6985) = 65.2997 | minus Nonradiation rate (50.6257) = 65.3726
                                                   |
3. 1950 Fractional Causation is radiation | 3. 1950 Fractional Causation is radiation
    rate (65.2997) divided by OBSERVED | rate (65.3726) divided by OBSERVED
    Natl MR Part 1,Col.C=   123.2   =   0.53 | Natl MR Part 1,Col.C=   123.2   =   0.53
```
----------------------------------------------------------------------------------------------
```

===
Table 50-C
All Cancers, Females: Fractional Causation in 1960
===

Part 1.
Calculation of the 6 Adjusted MortRates (Col.F) and the National Adjusted MortRate (Col.G).
The last six entries in Part 1, Col.F, are the products of (Col.D * Col.E), as discussed in Chap. 49.

Trio-Sequence	Col.A 1960 PopFrac Tab 3-B	Col.B 1960 Obs MR Tab 7-A	Col.C A * B	Col.D 1940 MR Mid,Low Tab 7-A	Col.E AdjuFact Bx2,Pt2 Col.E	Col.F 1960 Adju MortRates	Col.G A * F
Pacific	0.1182	110.1	13.014			110.1	13.014
New England	0.0586	122.4	7.173			122.4	7.173
Mid-Atlantic	0.1905	127.4	24.270			127.4	24.270
WestNoCentral	0.0858	109.3	9.378	120.1	0.84	100.88	8.656
EastNoCentral	0.2020	119.8	24.200	131.4	0.84	110.38	22.296
Mountain	0.0382	101.0	3.858	111.8	0.84	93.91	3.587
WestSoCentral	0.0945	102.9	9.724	99.8	0.87	86.83	8.205
EastSoCentral	0.0672	104.8	7.043	102.5	0.87	89.18	5.993
SouthAtlantic	0.1448	107.4	15.552	106.9	0.87	93.00	13.467

Sum =   114.2   Sum =
1960 Observed Natl MR from Table 7-B =   114.9   1960 Natl Adjusted MR =   106.6598

Part 2. --

Trio-Seq.	Col.A Mean1940 thru1960 PPs from Tab 47-A x'	Col.B 1960 Adju MRs from Col.F Part 1	Col.C All Cancers, Females: 1960 Adjusted MortRates regressed on Mean 1940 thru 1960 PPs Regression Output:		Col.D 1940 PPs from Table 3-A (TrioSeq) x''	Col.E All Cancers, Females: 1960 Adjusted MortRates regressed on 1940 PhysPops Regression Output:	
Pac	155.69	110.1	Constant	45.0231	159.72	Constant	44.6563
NewEng	162.81	122.4	Std Err of Y Est	5.6609	161.55	Std Err of Y Est	5.5178
MidAtl	167.04	127.4	R Squared	0.8689	169.76	R Squared	0.8754
WNoCen	118.15	100.88	No. of Observation	9	123.14	No. of Observation	9
ENoCen	123.87	110.38	Degrees of Freedom	7	133.36	Degrees of Freedom	7
Mtn	117.40	93.91			119.89		
WSoCen	102.31	86.83	X Coefficient(s)	0.4661	103.94	X Coefficient(s)	0.4596
ESoCen	85.63	89.18	Std Err of Coef.	0.0684	85.83	Std Err of Coef.	0.0655
SoAtl	101.72	93.00	XCoef / S.E. =	6.8105	100.74	XCoef / S.E.	7.0134

Part 3-A. | Part 3-B.
Calculation of Fractional Causation | Calculation of Fractional Causation
from Averaged PhysPops | from 1940 PhysPops
                                                        |
1. Nonradiation rate is Adjusted | 1. Nonradiation rate is Adjusted
    Constant (Part 2, Col.C) =   45.0231 | Constant (Part 2, Col.E) =   44.6563
                                                        |
2. Radiation rate is Natl Adjusted | 2. Radiation rate is Natl Adjusted
    MortRate (Part 1, Col.G = 106.6598) | MortRate (Part 1, Col.G = 106.6598)
    minus Nonradiation rate (45.0231) =   61.6368 | minus Nonradiation rate (44.6563) =   62.0036
                                                        |
3. 1960 Fractional Causation is radiation | 3. 1960 Fractional Causation is radiation
    rate (61.6368) divided by OBSERVED | rate (62.0036) divided by OBSERVED
    Natl MR Part 1,Col.C=   114.9   =   0.54 | Natl MR Part 1, Col.C=   114.9   =   0.54

```
========================================================================================================
                                           Table 50-E
                       All Cancers, Females:  Fractional Causation in 1980
========================================================================================================
```

Part 1.
Calculation of the 6 Adjusted MortRates (Col.F) and the National Adjusted MortRate (Col.G).
The last six entries in Part 1, Col.F, are the products of (Col.D * Col.E), as discussed in Chap. 49.

Trio-Sequence	Col.A 1980 PopFrac Tab 3-B	Col.B 1980 Obs MR Tab 7-A	Col.C A * B	Col.D 1940 MR Mid,Low Tab 7-A	Col.E AdjuFact Bx2,Pt2 Col.E	Col.F 1980 Adju MortRates	Col.G A * F
Pacific	0.1398	110.4	15.434			110.4	15.434
New England	0.0546	116.4	6.355			116.4	6.355
Mid-Atlantic	0.1630	117.5	19.153			117.5	19.153
WestNoCentral	0.0759	101.0	7.666	120.1	0.77	92.48	7.019
EastNoCentral	0.1846	112.0	20.675	131.4	0.77	101.18	18.677
Mountain	0.0502	94.9	4.764	111.8	0.77	86.09	4.322
WestSoCentral	0.1049	100.1	10.500	99.8	0.85	84.83	8.899
EastSoCentral	0.0646	103.2	6.667	102.5	0.85	87.13	5.628
SouthAtlantic	0.1624	105.0	17.052	106.9	0.85	90.87	14.756

```
                                    Sum =   108.3                              Sum =
          1980 Observed Natl MR from Table 7-B =   108.5      1980 Natl Adjusted MR =   100.2433
```

Part 2. --

Trio-Seq.	Col.A Mean1940 thru1980 PPs from Tab 47-A x'	Col.B 1980 Adju MRs from Col.F Part 1	Col.C All Cancers, Females: 1980 Adjusted MortRates regressed on Mean 1940 thru 1980 PPs Regression Output:		Col.D 1940 PPs from Table 3-A (TrioSeq) x''	Col.E All Cancers, Females: 1980 Adjusted MortRates regressed on 1940 PhysPops Regression Output:	
Pac	177.35	110.4	Constant	43.5132	159.72	Constant	45.3316
NewEng	185.86	116.4	Std Err of Y Est	4.8101	161.55	Std Err of Y Est	5.1053
MidAtl	186.11	117.5	R Squared	0.8839	169.76	R Squared	0.8692
WNoCen	128.82	92.48	No. of Observation	9	123.14	No. of Observation	9
ENoCen	133.71	101.18	Degrees of Freedom	7	133.36	Degrees of Freedom	7
Mtn	133.45	86.09			119.89		
WSoCen	114.66	84.83	X Coefficient(s)	0.3857	103.94	X Coefficient(s)	0.4136
ESoCen	99.46	87.13	Std Err of Coef.	0.0528	85.83	Std Err of Coef.	0.0606
SoAtl	124.62	90.87	XCoef / S.E. =	7.3005	100.74	XCoef / S.E.	6.8210

Part 3-A.
Calculation of Fractional Causation
from Averaged PhysPops

1. Nonradiation rate is Adjusted
    Constant (Part 2, Col.C) =   43.5132

2. Radiation rate is Natl Adjusted
    MortRate (Part 1, Col.G = 100.2433)
    minus Nonradiation rate (43.5132) =   56.7300

3. 1980 Fractional Causation is radiation
    rate (56.7300) divided by OBSERVED
    Natl MR Part 1,Col.C=   108.5   =   0.52

Part 3-B.
Calculation of Fractional Causation
from 1940 PhysPops

1. Nonradiation rate is Adjusted
    Constant (Part 2, Col.E) =   45.3316

2. Radiation rate is Natl Adjusted
    MortRate (Part 1, Col.G = 100.2433)
    minus Nonradiation rate (45.3316) =   54.9117

3. 1980 Fractional Causation is radiation
    rate (54.9117) divided by OBSERVED
    Natl MR Part 1, Col.C=   108.5   =   0.51

```
=====================================================================================
                                  Table 50-F
                   All Cancers, Females:  Fractional Causation in 1988
=====================================================================================
```

Part 1.

Calculation of the 6 Adjusted MortRates (Col.F) and the National Adjusted MortRate (Col.G).

The last six entries in Part 1, Col.F, are the products of (Col.D * Col.E), as discussed in Chap. 49.

	Col.A 1990 PopFrac Tab 3-B	Col.B 1988 Obs MR Tab 7-A	Col.C A * B	Col.D 1940 MR Mid,Low Tab 7-A	Col.E AdjuFact Bx2,Pt2 Col.E	Col.F 1988 Adju MortRates	Col.G A * F
Pacific	0.1535	111.5	17.115			111.5	17.115
New England	0.0527	116.4	6.134			116.4	6.134
Mid-Atlantic	0.1527	118.6	18.110			118.6	18.110
WestNoCentral	0.0721	106.8	7.700	120.1	0.77	92.48	6.668
EastNoCentral	0.1713	116.5	19.956	131.4	0.77	101.18	17.332
Mountain	0.0543	100.4	5.452	111.8	0.77	86.09	4.674
WestSoCentral	0.1087	109.8	11.935	99.8	0.88	87.82	9.546
EastSoCentral	0.0621	112.7	6.999	102.5	0.88	90.20	5.601
SouthAtlantic	0.1725	111.6	19.251	106.9	0.88	94.07	16.227

```
                                    Sum =  112.7                               Sum =
        1988 Observed Natl MR from Table 7-B =  111.3      1988 Natl Adjusted MR =   101.4089
```

Part 2. --

Trio-Seq.	Col.A Mean1940 thru1990 PPs from Tab 47-A x'	Col.B 1988 Adju MRs from Col.F Part 1	Col.C All Cancers, Females: 1988 Adjusted MortRates regressed on Mean 1940 thru 1990 PPs Regression Output:		Col.D 1940 PPs from Table 3-A (TrioSeq) x''	Col.E All Cancers, Females: 1988 Adjusted MortRates regressed on 1940 PhysPops Regression Output:	
Pac	191.97	111.5	Constant	46.2557	159.72	Constant	50.5238
NewEng	208.20	116.4	Std Err of Y Est	4.8659	161.55	Std Err of Y Est	5.8211
MidAtl	204.72	118.6	R Squared	0.8703	169.76	R Squared	0.8144
WNoCen	141.14	92.48	No. of Observation	9	123.14	No. of Observation	9
ENoCen	146.19	101.18	Degrees of Freedom	7	133.36	Degrees of Freedom	7
Mtn	145.91	86.09			119.89		
WSoCen	126.28	87.82	X Coefficient(s)	0.3393	103.94	X Coefficient(s)	0.3831
ESoCen	113.28	90.20	Std Err of Coef.	0.0495	85.83	Std Err of Coef.	0.0691
SoAtl	142.93	94.07	XCoef / S.E. =	6.8536	100.74	XCoef / S.E.	5.5419

Part 3-A.

Calculation of Fractional Causation from Averaged PhysPops

1. Nonradiation rate is Adjusted
   Constant (Part 2, Col.C) =   46.2557

2. Radiation rate is Natl Adjusted
   MortRate (Part 1, Col.G = 101.4089)
   minus Nonradiation rate (46.2557) =   55.1532

3. 1988 Fractional Causation is radiation
   rate (55.1532) divided by OBSERVED
   Natl MR Part 1,Col.C=   111.3   =   0.50

Part 3-B.

Calculation of Fractional Causation from 1940 PhysPops

1. Nonradiation rate is Adjusted
   Constant (Part 2, Col.E) =   50.5238

2. Radiation rate is Natl Adjusted
   MortRate (Part 1, Col.G = 101.4089)
   minus Nonradiation rate (50.5238) =   50.8851

3. 1988 Fractional Causation is radiation
   rate (50.8851) divided by OBSERVED
   Natl MR Part 1, Col.C=   111.3   =   0.46

CHAPTER 51

Respiratory-System Cancers, Males, 1940-1988: Fractional Causation by Medical Radiation

● Table 51-A, Column A, shows the dramatic post-1940 rise in the male National MortRate from Respiratory-System Cancers. The relationship with cigarette smoking is reviewed in Chapter 48. Although male Respiratory Cancer MortRates rose in every Census Division, the observations in Box 1 clearly indicate that smoking-intensity and PhysPop are inversely related (Chapter 48, Part 5b). In order to match the Census Divisions for smoking, an adjustment is obligatory. Box 2 calculates the appropriate adjustment factors.

● Chapter 51 contains a second set of tables (51-AA through 51-FF), located after Table 51-F. In Tables 51-BB through FF, the regressions produce positive Constants --- in contrast with the negative Constants produced in Tables 51-B through 51-F. The discussion is located with Table 51-AA.

Table 51-A							
Respiratory Cancers, Males: Fractional Causation by Medical Radiation over Time							
Year	Col.A Natl MR	Col.B Frac.C	Col.C R-Sq	Col.D X-Coef	Col.E StdErr	Col.F Coef/SE	Col.G Source
1940	11.0	~100%	0.8673	0.1169	0.0173	6.7640	Chap.16
1950	21.6	84%	0.9007	0.2070	0.0260	7.9692	Tab 51-B
1960	35.2	77%	0.9375	0.3312	0.0323	10.2463	Tab 51-C
1970	47.3	71%	0.9417	0.3997	0.0376	10.6295	Tab 51-D
1980	59.4	67%	0.9447	0.4498	0.0411	10.9368	Tab 51-E
1988	59.7	65%	0.9445	0.4042	0.0370	10.9188	Tab 51-F

```
==================================================================================================
                                    Box 1, Chap. 51
          Respiratory-System Cancers, Males:  Post-1940 Change in MortRates by Census Trios

1960 vs. 1940, by Trios:  Col.D expresses change by ratios.  Col.F expresses change by subtraction.
1988 vs. 1940, by Trios:  Col.I expresses change by ratios.  Col.K expresses change by subtraction.
MRs change inversely with PP.  High-PP Trio has lowest growth-factor.  Low-PP Trio has highest growth-factor.

          Col.A   |  Col.B    Col.C    Col.D    Col.E    Col.F  |  Col.G    Col.H    Col.I    Col.J    Col.K
          1940    |  1960     Ratio    Input    Diff:    Input  |  1988     Ratio    Input    Diff:    Input
          MortRate|  MortRate Col.B    from     Col.B    from   |  MortRate Col.G    from     Col.G    from
          Tab 16-A|  Tab 16-A /Col.A   Col.C    minus A  Col.E  |  Tab 16-A /Col.A   Col.H    minus A  Col.J
                  |                                             |
Pacif     12.0    |  34.9     2.908    Avg Chg  22.9     Avg Chg|  50.7     4.225    Avg Chg  38.7     Avg Chg
NewE      13.5    |  38.1     2.822    TopTrio  24.6     TopTrio|  56.3     4.170    TopTrio  42.8     TopTrio
MidAtl    17.1    |  40.6     2.374    2.702    23.5     23.7   |  57.5     3.363    3.919    40.4     40.6
                  |                                             |
WNoCen    7.7     |  28.4     3.688    Avg Chg  20.7     Avg Chg|  56.2     7.299    Avg Chg  48.5     Avg Chg
ENoCen    10.6    |  35.7     3.368    MidTrio  25.1     MidTrio|  62.3     5.877    MidTrio  51.7     MidTrio
Mtn       7.8     |  25.5     3.269    3.442    17.7     21.2   |  44.2     5.667    6.281    36.4     45.5
                  |                                             |
WSoCen    7.6     |  34.9     4.592    Avg Chg  27.3     Avg Chg|  67.9     8.934    Avg Chg  60.3     Avg Chg
ESoCen    4.9     |  29.0     5.918    LowTrio  24.1     LowTrio|  79.1     16.143   LowTrio  74.2     LowTrio
SoAtl     8.3     |  35.7     4.301    4.937    27.4     26.3   |  68.5     8.253    11.110   60.2     64.9
==================================================================================================
```

```
=====================================================================================================
                                      Box 2, Chap. 51
                    Respiratory-System Cancers, Males:  Calculation of Adjustment Factor

                        This adjustment is discussed fully in Chapter 49.
 ● Part 1:  Calculate average population-weighted MortRate for the combined TopTrio Census Divs.


            Col.A      Col.B      Col.C      Col.D  |            Col.A      Col.B      Col.C      Col.D
 Census    1940 MR    1940 Pop'n  1940 Popn  Col.A * | Census   1950 MR    1950 Pop'n  1950 Popn  Col.A *
 Div.      Tab 16-A   Tab 3-B    /45,710,039 Col.C  | Div.      Tab 16-A   Tab 3-B    /53,964,513 Col.C
                                                    |
 Pacific   12.0       9,733,262   0.2129     2.56   | Pacific   21.1       14,486,527  0.2684     5.66
 NewEng    13.5       8,437,290   0.1846     2.49   | NewEng    23.6       9,314,453   0.1726     4.07
 Mid-Atl   17.1       27,539,487  0.6025     10.30  | Mid-Atl   28.4       30,163,533  0.5590     15.87

   1940              Sum TopTrio        Sum TopTrio |   1950              Sum TopTrio        Sum TopTrio
                      45,710,039   1.0000  15.350   |                      53,964,513   1.0000  25.612
 ---------------------------------------------------+-------------------------------------------------
            Col.A      Col.B      Col.C      Col.D  |            Col.A      Col.B      Col.C      Col.D
 Census    1960 MR    1960 Pop'n  1960 Popn  Col.A * | Census   1970 MR    1970 Pop'n  1970 Popn  Col.A *
 Div.      Tab 16-A   Tab 3-B    /65,875,863 Col.C  | Div.      Tab 16-A   Tab 3-B    /75,017,000 Col.C
                                                    |
 Pacific   34.9       21,198,044  0.3218     11.23  | Pacific   44.2       26,087,000  0.3477     15.37
 NewEng    38.1       10,509,367  0.1595     6.08   | NewEng    47.7       11,781,000  0.1570     7.49
 Mid-Atl   40.6       34,168,452  0.5187     21.06  | Mid-Atl   49.5       37,149,000  0.4952     24.51

   1960              Sum TopTrio        Sum TopTrio |   1970              Sum TopTrio        Sum TopTrio
                      65,875,863   1.0000  38.367   |                      75,017,000   1.0000  47.374
 ---------------------------------------------------+-------------------------------------------------
            Col.A      Col.B      Col.C      Col.D  |            Col.A      Col.B      Col.C      Col.D
 Census    1980 MR    1980 Pop'n  1980 Popn  Col.A * | Census   1988 MR    1990 Pop'n  1990 Popn  Col.A *
 Div.      Tab 16-A   Tab 3-B    /80,615,000 Col.C  | Div.      Tab 16-A   Tab 3-B    /88,495,000 Col.C
                                                    |
 Pacific   53.5       31,523,000  0.3910     20.92  | Pacific   50.7       37,837,000  0.4276     21.68
 NewEng    57.3       12,322,000  0.1528     8.76   | NewEng    56.3       12,998,000  0.1469     8.27
 Mid-Atl   58.4       36,770,000  0.4561     26.64  | Mid-Atl   57.5       37,660,000  0.4256     24.47

   1980              Sum TopTrio        Sum TopTrio |   1988              Sum TopTrio        Sum TopTrio
                      80,615,000   1.0000  56.316   |                      88,495,000   1.0000  54.416
 ---------------------------------------------------+-------------------------------------------------
```

● Part 2: Take ratios of these TopTrio MortRates, with 1940 as the denominator of each ratio.
     Col.D modifies Col.C by separate PhysPop adjustments for MidTrio and LowTrio Census Divisions.

```
            Col.A      Col.B      Col.C   Col.D     Col.E
            TopTrio   1940 TopTrio = Col.A  ppAdju   = Col.C                    RESPIRATORY CANCERS.
            Mean MR    Mean MR    / Col.B Tab 47-B  * Col.D                          Males.
                                          MidTrio
 1950       25.612     15.350     1.669   0.99       1.65   = MidTrio Adjustment Factor, 1950
 1960       38.367     15.350     2.500   0.97       2.42   = MidTrio Adjustment Factor, 1960
 1970       47.374     15.350     3.086   0.95       2.93   = MidTrio Adjustment Factor, 1970
 1980       56.316     15.350     3.669   0.94       3.45   = MidTrio Adjustment Factor, 1980
 1988       54.416     15.350     3.545   0.94       3.33   = MidTrio Adjustment Factor, 1988
 -------------------------------------------------  LowTrio  -------------------------------------------
 1950       25.612     15.350     1.669   1.00       1.67   = LowTrio Adjustment Factor, 1950
 1960       38.367     15.350     2.500   1.01       2.52   = LowTrio Adjustment Factor, 1960
 1970       47.374     15.350     3.086   1.02       3.15   = LowTrio Adjustment Factor, 1970
 1980       56.316     15.350     3.669   1.04       3.82   = LowTrio Adjustment Factor, 1980
 1988       54.416     15.350     3.545   1.07       3.79   = LowTrio Adjustment Factor, 1988
=====================================================================================================
```

```
=========================================================================================================
                                         Table 51-B
                    Respiratory Cancers, Males:  Fractional Causation in 1950
=========================================================================================================
```

Part 1.

Calculation of the 6 Adjusted MortRates (Col.F) and the National Adjusted MortRate (Col.G).

The last six entries in Part 1, Col.F, are the products of (Col.D * Col.E), as discussed in Chap. 49.

Trio-Sequence	Col.A 1950 PopFrac Tab 3-B	Col.B 1950 Obs MR Tab 16-A	Col.C A * B	Col.D 1940 MR Mid,Low Tab 16-A	Col.E AdjuFact Bx2,Pt2 Col.E	Col.F 1950 Adju MortRates	Col.G A * F
Pacific	0.0961	21.1	2.028			21.1	2.028
New England	0.0618	23.6	1.458			23.6	1.458
Mid-Atlantic	0.2002	28.4	5.686			28.4	5.686
WestNoCentral	0.0933	16.5	1.539	7.7	1.65	12.705	1.185
EastNoCentral	0.2017	21.8	4.397	10.6	1.65	17.490	3.528
Mountain	0.0337	16.7	0.563	7.8	1.65	12.870	0.434
WestSoCentral	0.0965	19.0	1.834	7.6	1.67	12.692	1.225
EastSoCentral	0.0762	14.7	1.120	4.9	1.67	8.183	0.624
SouthAtlantic	0.1406	19.8	2.784	8.3	1.67	13.861	1.949

```
                                  Sum =   21.4                                  Sum =
      1950 Observed Natl MR from Table 16-B =   21.6        1950 Natl Adjusted MR =   18.1159
```

Part 2. --

Trio-Seq.	Col.A Mean1940 thru1950 PPs from Tab 47-A x'	Col.B 1950 Adju MRs from Col.F Part 1	Col.C Respiratory Ca, Males: 1950 Adjusted MortRates regressed on Mean 1940 thru 1950 PPs Regression Output:		Col.D 1940 PPs from Table 3-A (TrioSeq) x''	Col.E Respiratory Ca, Males: 1950 Adjusted MortRates regressed on 1940 PhysPops Regression Output:	
Pac	154.16	21.1	Constant	-9.5039	159.72	Constant	-9.4898
NewEng	162.03	23.6	Std Err of Y Est	2.1691	161.55	Std Err of Y Est	2.2817
MidAtl	169.24	28.4	R Squared	0.9007	169.76	R Squared	0.8901
WNoCen	121.60	12.705	No. of Observation	9	123.14	No. of Observation	9
ENoCen	128.53	17.490	Degrees of Freedom	7	133.36	Degrees of Freedom	7
Mtn	119.64	12.870			119.89		
WSoCen	102.64	12.692	X Coefficient(s)	0.2070	103.94	X Coefficient(s)	0.2041
ESoCen	84.44	8.183	Std Err of Coef.	0.0260	85.83	Std Err of Coef.	0.0271
SoAtl	99.91	13.861	XCoef / S.E. =	7.9692	100.74	XCoef / S.E.	7.5313

Part 3-A.

Calculation of Fractional Causation
from Averaged PhysPops

1. Nonradiation rate is Adjusted
    Constant (Part 2, Col.C) = NEG =   0.0

2. Radiation rate is Natl Adjusted
    MortRate (Part 1, Col.G = 18.1159)
    minus Nonradiation rate (0.0) =   18.1159

3. 1950 Fractional Causation is radiation
    rate (18.1159) divided by OBSERVED
    Natl MR Part 1,Col.C=   21.6   =   0.84

Part 3-B.

Calculation of Fractional Causation
from 1940 PhysPops

1. Nonradiation rate is Adjusted
    Constant (Part 2, Col.E) = NEG =   0.0

2. Radiation rate is Natl Adjusted
    MortRate (Part 1, Col.G = 18.1159)
    minus Nonradiation rate (0.0) =   18.1159

3. 1950 Fractional Causation is radiation
    rate (18.1159) divided by OBSERVED
    Natl MR Part 1,Col.C=   21.6   =   0.84

```
===================================================================================================
                                         Table 51-C
                   Respiratory Cancers, Males:  Fractional Causation in 1960
===================================================================================================
Part 1.
Calculation of the 6 Adjusted MortRates (Col.F) and the National Adjusted MortRate (Col.G).
The last six entries in Part 1, Col.F, are the products of (Col.D * Col.E), as discussed in Chap. 49.
```

	Col.A 1960 PopFrac Tab 3-B	Col.B 1960 Obs MR Tab 16-A	Col.C A * B	Col.D 1940 MR Mid,Low Tab 16-A	Col.E AdjuFact Bx2,Pt2 Col.E	Col.F 1960 Adju MortRates	Col.G A * F
Pacific	0.1182	34.9	4.125			34.9	4.125
New England	0.0586	38.1	2.233			38.1	2.233
Mid-Atlantic	0.1905	40.6	7.734			40.6	7.734
WestNoCentral	0.0858	28.4	2.437	7.7	2.42	18.634	1.599
EastNoCentral	0.2020	35.7	7.211	10.6	2.42	25.652	5.182
Mountain	0.0382	25.5	0.974	7.8	2.42	18.876	0.721
WestSoCentral	0.0945	34.9	3.298	7.6	2.52	19.152	1.810
EastSoCentral	0.0672	29.0	1.949	4.9	2.52	12.348	0.830
SouthAtlantic	0.1448	35.7	5.169	8.3	2.52	20.916	3.029

```
                                   Sum =    35.1                                 Sum =
          1960 Observed Natl MR from Table 16-B =    35.2      1960 Natl Adjusted MR =    27.2620
```

```
Part 2.  ------------------------------------------------------------------------------------------
```

Trio-Seq.	Col.A Mean1940 thru1960 PPs from Tab 47-A x'	Col.B 1960 Adju MRs from Col.F Part 1	Col.C Respiratory Ca, Males: 1960 Adjusted MortRates regressed on Mean 1940 thru 1960 PPs Regression Output:		Col.D 1940 PPs from Table 3-A (TrioSeq) x''	Col.E Respiratory Ca, Males: 1960 Adjusted MortRates regressed on 1940 PhysPops Regression Output:	
Pac	155.69	34.9	Constant	-16.2889	159.72	Constant	-15.7619
NewEng	162.81	38.1	Std Err of Y Est	2.6735	161.55	Std Err of Y Est	3.2174
MidAtl	167.04	40.6	R Squared	0.9375	169.76	R Squared	0.9095
WNoCen	118.15	18.634	No. of Observation	9	123.14	No. of Observation	9
ENoCen	123.87	25.652	Degrees of Freedom	7	133.36	Degrees of Freedom	7
Mtn	117.40	18.876			119.89		
WSoCen	102.31	19.152	X Coefficient(s)	0.3312	103.94	X Coefficient(s)	0.3204
ESoCen	85.63	12.348	Std Err of Coef.	0.0323	85.83	Std Err of Coef.	0.0382
SoAtl	101.72	20.916	XCoef / S.E. =	10.2463	100.74	XCoef / S.E. =	8.3861

```
------------------------------------------------------------------------------------------------
Part 3-A.                                      | Part 3-B.
Calculation of Fractional Causation            | Calculation of Fractional Causation
from Averaged PhysPops                          | from 1940 PhysPops
                                                |
1.  Nonradiation rate is Adjusted              | 1.  Nonradiation rate is Adjusted
    Constant (Part 2, Col.C) = NEG =      0.0  |     Constant (Part 2, Col.E) = NEG =      0.0
                                                |
2.  Radiation rate is Natl Adjusted            | 2.  Radiation rate is Natl Adjusted
    MortRate (Part 1, Col.G = 27.2620)          |     MortRate (Part 1, Col.G = 27.2620)
    minus Nonradiation rate (0.0) =   27.2620  |     minus Nonradiation rate (0.0) =   27.2620
                                                |
3.  1960 Fractional Causation is radiation     | 3.  1960 Fractional Causation is radiation
    rate (27.2620) divided by OBSERVED          |     rate (27.2620) divided by OBSERVED
    Natl MR Part 1,Col.C=   35.2   =     0.77  |     Natl MR Part 1,Col.C=   35.2   =     0.77
------------------------------------------------------------------------------------------------
```

===

Table 51-E

Respiratory Cancers, Males: Fractional Causation in 1980

===

Part 1.

Calculation of the 6 Adjusted MortRates (Col.F) and the National Adjusted MortRate (Col.G).
The last six entries in Part 1, Col.F, are the products of (Col.D * Col.E), as discussed in Chap. 49.

Trio-Sequence	Col.A 1980 PopFrac Tab 3-B	Col.B 1980 Obs MR Tab 16-A	Col.C A * B	Col.D 1940 MR Mid,Low Tab 16-A	Col.E AdjuFact Bx2,Pt2 Col.E	Col.F 1980 Adju MortRates	Col.G A * F
Pacific	0.1398	53.5	7.479			53.5	7.479
New England	0.0546	57.3	3.129			57.3	3.129
Mid-Atlantic	0.1630	58.4	9.519			58.4	9.519
WestNoCentral	0.0759	53.7	4.076	7.7	3.45	26.565	2.016
EastNoCentral	0.1846	61.4	11.334	10.6	3.45	36.570	6.751
Mountain	0.0502	43.6	2.189	7.8	3.45	26.910	1.351
WestSoCentral	0.1049	62.8	6.588	7.6	3.82	29.032	3.045
EastSoCentral	0.0646	70.8	4.574	4.9	3.82	18.718	1.209
SouthAtlantic	0.1624	65.2	10.588	8.3	3.82	31.706	5.149

                                        Sum =   59.5   Sum =
       1980 Observed Natl MR from Table 16-B =   59.4   1980 Natl Adjusted MR =   39.6488

Part 2. --

Trio-Seq.	Col.A Mean1940 thru1980 PPs from Tab 47-A x'	Col.B 1980 Adju MRs from Col.F Part 1	Col.C Respiratory Ca, Males: 1980 Adjusted MortRates regressed on Mean 1940 thru 1980 PPs		Col.D 1940 PPs from Table 3-A (TrioSeq) x''	Col.E Respiratory Ca, Males: 1980 Adjusted MortRates regressed on 1940 PhysPops	
			Regression Output:			Regression Output:	
Pac	177.35	53.5	Constant	-26.5400	159.72	Constant	-22.8215
NewEng	185.86	57.3	Std Err of Y Est	3.7445	161.55	Std Err of Y Est	5.4754
MidAtl	186.11	58.4	R Squared	0.9447	169.76	R Squared	0.8818
WNoCen	128.82	26.565	No. of Observation	9	123.14	No. of Observation	9
ENoCen	133.71	36.570	Degrees of Freedom	7	133.36	Degrees of Freedom	7
Mtn	133.45	26.910			119.89		
WSoCen	114.66	29.032	X Coefficient(s)	0.4498	103.94	X Coefficient(s)	0.4699
ESoCen	99.46	18.718	Std Err of Coef.	0.0411	85.83	Std Err of Coef.	0.0650
SoAtl	124.62	31.706	XCoef / S.E. =	10.9368	100.74	XCoef / S.E.	7.2261

Part 3-A. | Part 3-B.
Calculation of Fractional Causation | Calculation of Fractional Causation
from Averaged PhysPops | from 1940 PhysPops
                                                         |
1. Nonradiation rate is Adjusted | 1. Nonradiation rate is Adjusted
    Constant (Part 2, Col.C) = NEG =   0.0 | Constant (Part 2, Col.E) = NEG =   0.0
                                                         |
2. Radiation rate is Natl Adjusted | 2. Radiation rate is Natl Adjusted
    MortRate (Part 1, Col.G = 39.6488) | MortRate (Part 1, Col.G = 39.6488)
    minus Nonradiation rate (0.0) =   39.6488 | minus Nonradiation rate (0.0) =   39.6488
                                                         |
3. 1980 Fractional Causation is radiation | 3. 1980 Fractional Causation is radiation
    rate (39.6488) divided by OBSERVED | rate (39.6488) divided by OBSERVED
    Natl MR Part 1,Col.C=   59.4   =   0.67 | Natl MR Part 1, Col.C=   59.4   =   0.67

```
==================================================================================================
                                        Table 51-F
                    Respiratory Cancers, Males:  Fractional Causation in 1988
==================================================================================================
Part 1.
Calculation of the 6 Adjusted MortRates (Col.F) and the National Adjusted MortRate (Col.G).
The last six entries in Part 1, Col.F, are the products of (Col.D * Col.E), as discussed in Chap. 49.


                      Col.A   Col.B   Col.C        Col.D   Col.E   Col.F   Col.G
                      1990    1988                 1940 MR AdjuFact 1988
                      PopFrac Obs MR  A * B        Mid,Low Bx2,Pt2  Adju    A * F
Trio-Sequence         Tab 3-B Tab 16-A             Tab 16-A Col.E  MortRates
        Pacific       0.1535  50.7    7.782                        50.7    7.782
        New England   0.0527  56.3    2.967                        56.3    2.967
        Mid-Atlantic  0.1527  57.5    8.780                        57.5    8.780
        WestNoCentral 0.0721  56.2    4.052        7.7     3.33    25.641  1.849
        EastNoCentral 0.1713  62.3   10.672        10.6    3.33    35.298  6.047
        Mountain      0.0543  44.2    2.400        7.8     3.33    25.974  1.410
        WestSoCentral 0.1087  67.9    7.381        7.6     3.79    28.804  3.131
        EastSoCentral 0.0621  79.1    4.912        4.9     3.79    18.571  1.153
        SouthAtlantic 0.1725  68.5   11.816        8.3     3.79    31.457  5.426

                              Sum =   60.8                                Sum =
        1988 Observed Natl MR from Table 16-B =   59.7   1988 Natl Adjusted MR =   38.5459


Part 2. ------------------------------------------------------------------------------------------
          Col.A   Col.B                        Col.C              Col.D                   Col.E
          Mean1940  1988            Respiratory Ca, Males:         1940         Respiratory Ca, Males:
          thru1990  Adju MRs        1988 Adjusted MortRates        PPs from     1988 Adjusted MortRates
Trio-     PPs from  from Col.F         regressed on               Table 3-A        regressed on
Seq.      Tab 47-A  Part 1          Mean 1940 thru 1990 PPs       (TrioSeq)      1940 PhysPops
            x'                        Regression Output:            x''           Regression Output:
Pac       191.97   50.7            Constant        -27.1037       159.72       Constant        -21.3904
NewEng    208.20   56.3            Std Err of Y Est  3.6381       161.55       Std Err of Y Est   5.6760
MidAtl    204.72   57.5            R Squared         0.9445       169.76       R Squared          0.8650
WNoCen    141.14   25.641          No. of Observation     9       123.14       No. of Observation      9
ENoCen    146.19   35.298          Degrees of Freedom     7       133.36       Degrees of Freedom      7
Mtn       145.91   25.974                                         119.89
WSoCen    126.28   28.804          X Coefficient(s)  0.4042       103.94       X Coefficient(s)   0.4515
ESoCen    113.28   18.571          Std Err of Coef.  0.0370        85.83       Std Err of Coef.   0.0674
SoAtl     142.93   31.457          XCoef / S.E. =   10.9188       100.74       XCoef / S.E.        6.6973


-----------------------------------------------------|------------------------------------------------
Part 3-A.                                            | Part 3-B.
Calculation of Fractional Causation                  | Calculation of Fractional Causation
from Averaged PhysPops                               | from 1940 PhysPops
                                                     |
1.  Nonradiation rate is Adjusted                    | 1.  Nonradiation rate is Adjusted
    Constant (Part 2, Col.C) = NEG =        0.0      |     Constant (Part 2, Col.E) = NEG =        0.0
                                                     |
2.  Radiation rate is Natl Adjusted                  | 2.  Radiation rate is Natl Adjusted
    MortRate (Part 1, Col.G = 38.5459)               |     MortRate (Part 1, Col.G = 38.5459)
    minus Nonradiation rate (0.0) =        38.5459   |     minus Nonradiation rate (0.0) =        38.5459
                                                     |
3.  1988 Fractional Causation is radiation           | 3.  1988 Fractional Causation is radiation
    rate (38.5459) divided by OBSERVED               |     rate (38.5459) divided by OBSERVED
    Natl MR Part 1,Col.C=   59.7    =       0.65     |     Natl MR Part 1,Col.C=   59.7    =       0.65
-----------------------------------------------------|------------------------------------------------
```

● Special Section: Tables 51-AA through 51-FF

At the outset of this chapter, we mentioned the negative Constants produced by the regressions in Tables 51-B through 51-F. Their production is not surprising, because (a) the 1940 PhysPop-MortRate dose-response for male Respiratory-System Cancers also produces a negative Constant of appreciable magnitude relative to the 1940 National MortRate, and (b) the 1940 dose-response is the foundation for the Smoking Adjustment which precedes the post-1940 regressions.

Biologically, MortRates below zero do not exist, as we stressed in Chapter 22 (Part 3). Nonetheless, we must expect that some negative Constants will show up when we subdivide the 1940 All-Cancer MortRates to make 16 separate examinations (Chapters 8 through 21). A negative Constant reflects a best-fit slope which is too steep to be biologically realistic. Such a slope may well be the result of a few 'outliers' --- datapoints which are 'way out of line' in a series of observations. In epidemiology, a few outliers are no justification for disbelieving the bigger picture.

Two Questions Answered by the Special Tabulation (Next Page)

We asked, "Which analyses of 1940 MortRates produced negative Constants which exceeded about 20% of the corresponding National Cancer MortRate?" There were only two: Male Respiratory-System Cancers and male Urinary-System Cancers, as shown in Column E of the Special Tabulation on the next page. It is notable that the corresponding female MortRates did NOT produce the aberration --- so there is no reason to suspect that there is something special about Respiratory-System Cancers or Urinary-System Cancers. Moreover, from Column D of the tabulation, we note that these two Cancers account for only (11.0 + 7.4) / (115.0), or 16% of the male All-Cancer MortRate in 1940. In other words, appreciable negative Constants do NOT characterize the "bigger picture" in 1940 for either males or females.

Also we asked: "Is the too-steep slope in 1940, for male Respiratory-System and male Urinary-System Cancers, predictable from their 1940 High-5/Low-4 MortRate Ratios?" The answer is yes, as demonstrated by Column B of the tabulation. Those two entities, having Hi5/Lo4 ratios of 1.70 and 1.77 respectively, are the ones having values far above the center of the distribution of such ratios (approximateley 1.45). Their large Hi5/Lo4 ratios, for their 1940 MortRates, identify them as the outliers.

We have demonstrated to ourselves (the work is not included here) that the too-steep slopes and resulting negative Constants, in 1940, account for the too-steep slopes and negative Constants which show up in our post-1940 Smoking Adjusted MortRates. If the highest 1940 MortRate is reduced somewhat and the lowest 1940 MortRate is raised somewhat, not only does the negative Constant in 1940 become positive, but also the post-1940 Constants become positive.

Obtaining Positive Constants in Tables 51-BB through 51-FF

Because we know that the slopes are unrealistically steep for male Respiratory-System Cancers in Tables 51-B through 51-F, we have produced more realistic slopes in Tables 51-BB through 51-FF. How? Here, we do NOT disturb the Observed 1940 MortRates in Column D of Table 51-BB. Instead, we change the adjustment factor in Column E. For Table 51-BB, we multiplied the adjustment factors from Column E of Table 51-B by 1.4. Thus, the adjustment factors used in Column E of Table 51-BB are 2.31 (which is 1.65 * 1.4) and 2.338 (which is 1.67 * 1.4). As a result, the six adjusted MortRates and the National Adjusted MortRate are higher in Table 51-BB than in Table 51-B. With these y-values in the new regression, the slope (X-Coefficient) is lower in Table 51-BB than in Table 51-B, and the Constant becomes positive.

We followed the same procedure for Tables 51-CC, 51-DD (not shown), 51-EE, and 51-FF: We multiplied the adjustment factors in Column E of the first set of tables by 1.4. As a result, all the Constants become positive --- an outcome which suggests that the second set of adjustment factors is more realistic. We also note that the R-squared values in this second set of tables are lower (although still very strong) than they were in the first set. Table 51-AA (next page) summarizes the results. Readers can make their own comparison of Table 51-AA with Table 51-A.

Table 51-AA
Respiratory Cancers, Males: Fractional Causation by Medical Radiation over Time

Year	Col.A Natl MR	Col.B Frac.C	Col.C R–Sq	Col.D X–Coef	Col.E StdErr	Col.F Coef/SE	Col.G Source
1940	11.0	~100%	0.8673	0.1169	0.0173	6.7640	Chap.16
1950	21.6	89%	0.7168	0.1403	0.0333	4.2087	Tab 51-BB
1960	35.2	86%	0.7917	0.2241	0.0435	5.1581	Tab 51-CC
1970	47.3	79%	0.7951	0.2677	0.0514	5.2118	Tab 51 DD
1980	59.4	71%	0.7948	0.2984	0.0573	5.2076	Tab 51-EE
1988	59.7	74%	0.7816	0.2652	0.0530	5.0050	Tab 51-FF

Special Tabulation for Chap. 51

Col.A		Col.B: 1940 MortRate Hi5/Lo4 Ratio	Col.C: 1940 Constant	Col.D 1940 Natl MortRate	Col.E: Ratio of ColC/ColD
All Cancers Combined	Male	1.44	11.6	115.0	0.10
Chaps. 6+7	Fem	1.27	53.0	126.1	0.42
Breast Cancer, Chap.8	Fem	1.55	-2.2	23.3	-0.09
Digestive-Syst. Cancer	Male	1.52	1.9	60.4	0.03
Chaps. 9+10	Fem	1.39	10.2	50.1	0.20
Urinary-Syst. Cancers	Male	1.77	-2.8	7.4	-0.38
Chaps. 11+12	Fem	1.40	0.6	4.0	0.15
Genital Cancers	Male	1.32	3.2	15.2	0.21
Chaps. 13+14	Fem	1.04	29.1	32.1	0.91
Buccal-Pharynx, Ch15	Male	1.56	-0.2	5.1	-0.04
Respiratory-Syst. Canc	Male	1.70	-5.1	11.0	-0.46
Chaps. 16+17	Fem	1.46	0.1	3.3	0.03
"Difference" Cancers	Male	1.42	16.7	104.0	0.16
Chaps. 18+19	Fem	1.42	52.9	122.8	0.43
All-Except-Genital Ca	Male	1.46	6.4	99.8	0.06
Chap. 20	Fem	1.36	23.9	94.0	0.25
All-Exc-Gen+Resp Ca	Male	1.44	11.5	88.8	0.13
Chap. 21	Fem	1.36	23.8	90.7	0.26

>>>>>>>>>>

===

Table 51-BB
Respiratory Cancers, Males: Fractional Causation in 1950

===

Part 1.
Calculation of the 6 Adjusted MortRates (Col.F) and the National Adjusted MortRate (Col.G).
The last six entries in Part 1, Col.F, are the products of (Col.D * Col.E), as discussed in Chap. 49.

Trio-Sequence	Col.A 1950 PopFrac Tab 3-B	Col.B 1950 Obs MR Tab 16-A	Col.C A * B	Col.D 1940 MR Mid,Low Tab 16-A	Col.E AdjuFact Bx2,Pt2 Col.E	Col.F 1950 Adju MortRates	Col.G A * F
Pacific	0.0961	21.1	2.028			21.1	2.028
New England	0.0618	23.6	1.458			23.6	1.458
Mid-Atlantic	0.2002	28.4	5.686			28.4	5.686
WestNoCentral	0.0933	16.5	1.539	7.7	2.31	17.787	1.660
EastNoCentral	0.2017	21.8	4.397	10.6	2.31	24.486	4.939
Mountain	0.0337	16.7	0.563	7.8	2.31	18.018	0.607
WestSoCentral	0.0965	19.0	1.834	7.6	2.338	17.769	1.715
EastSoCentral	0.0762	14.7	1.120	4.9	2.338	11.456	0.873
SouthAtlantic	0.1406	19.8	2.784	8.3	2.338	19.405	2.728

Sum = 21.4

1950 Observed Natl MR from Table 16-B = 21.6

Sum =

1950 Natl Adjusted MR = 21.6935

Part 2. ---

Trio-Seq.	Col.A Mean1940 thru1950 PPs from Tab 47-A x'	Col.B 1950 Adju MRs from Col.F	Col.C Respiratory Ca, Males: 1950 Adjusted MortRates regressed on Mean 1940 thru 1950 PPs Regression Output:		Col.D 1940 PPs from Table 3-A (TrioSeq) x''	Col.E Respiratory Ca, Males: 1950 Adjusted MortRates regressed on 1940 PhysPops Regression Output:	
Pac	154.16	21.1	Constant	2.4168	159.72	Constant	2.3293
NewEng	162.03	23.6	Std Err of Y Est	2.7841	161.55	Std Err of Y Est	2.7873
MidAtl	169.24	28.4	R Squared	0.7168	169.76	R Squared	0.7161
WNoCen	121.60	17.787	No. of Observation	9	123.14	No. of Observation	9
ENoCen	128.53	24.486	Degrees of Freedom	7	133.36	Degrees of Freedom	7
Mtn	119.64	18.018			119.89		
WSoCen	102.64	17.769	X Coefficient(s)	0.1403	103.94	X Coefficient(s)	0.1391
ESoCen	84.44	11.456	Std Err of Coef.	0.0333	85.83	Std Err of Coef.	0.0331
SoAtl	99.91	19.405	XCoef / S.E. =	4.2087	100.74	XCoef / S.E.	4.2018

Part 3-A.
Calculation of Fractional Causation
from Averaged PhysPops

1. Nonradiation rate is Adjusted
    Constant (Part 2, Col.C) =   2.4168

2. Radiation rate is Natl Adjusted
    MortRate (Part 1, Col.G = 21.6935)
    minus Nonradiation rate (2.4168) =   19.2766

3. 1950 Fractional Causation is radiation
    rate (19.2766) divided by OBSERVED
    Natl MR Part 1,Col.C=   21.6   =   0.89

Part 3-B.
Calculation of Fractional Causation
from 1940 PhysPops

1. Nonradiation rate is Adjusted
    Constant (Part 2, Col.E) =   2.3293

2. Radiation rate is Natl Adjusted
    MortRate (Part 1, Col.G = 21.6935)
    minus Nonradiation rate (2.3293) =   19.3642

3. 1950 Fractional Causation is radiation
    rate (19.3642) divided by OBSERVED
    Natl MR Part 1,Col.C=   21.6   =   0.90

===

Table 51-CC
Respiratory Cancers, Males: Fractional Causation in 1960

===

Part 1.
Calculation of the 6 Adjusted MortRates (Col.F) and the National Adjusted MortRate (Col.G).
The last six entries in Part 1, Col.F, are the products of (Col.D * Col.E), as discussed in Chap. 49.

Trio-Sequence	Col.A 1960 PopFrac Tab 3-B	Col.B 1960 Obs MR Tab 16-A	Col.C A * B	Col.D 1940 MR Mid,Low Tab 16-A	Col.E AdjuFact Bx2,Pt2 Col.E	Col.F 1960 Adju MortRates	Col.G A * F
Pacific	0.1182	34.9	4.125			34.9	4.125
New England	0.0586	38.1	2.233			38.1	2.233
Mid-Atlantic	0.1905	40.6	7.734			40.6	7.734
WestNoCentral	0.0858	28.4	2.437	7.7	3.388	26.088	2.238
EastNoCentral	0.2020	35.7	7.211	10.6	3.388	35.913	7.254
Mountain	0.0382	25.5	0.974	7.8	3.388	26.426	1.009
WestSoCentral	0.0945	34.9	3.298	7.6	3.528	26.813	2.534
EastSoCentral	0.0672	29.0	1.949	4.9	3.528	17.287	1.162
SouthAtlantic	0.1448	35.7	5.169	8.3	3.528	29.282	4.240

Sum =   35.1
1960 Observed Natl MR from Table 16-B =   35.2   1960 Natl Adjusted MR =   Sum = 32.5299

Part 2.

Trio-Seq.	Col.A Mean1940 thru1960 PPs Tab 47-A x'	Col.B 1960 Adju MRs from Col.F Part 1	Col.C Respiratory Ca, Males: 1960 Adjusted MortRates regressed on Mean 1940 thru 1960 PPs Regression Output:		Col.D 1940 PPs from Table 3-A (TrioSeq) x''	Col.E Respiratory Ca, Males: 1960 Adjusted MortRates regressed on 1940 PhysPops Regression Output:	
Pac	155.69	34.9	Constant	2.3462	159.72	Constant	1.9609
NewEng	162.81	38.1	Std Err of Y Est	3.5939	161.55	Std Err of Y Est	3.4375
MidAtl	167.04	40.6	R Squared	0.7917	169.76	R Squared	0.8094
WNoCen	118.15	26.088	No. of Observation	9	123.14	No. of Observation	9
ENoCen	123.87	35.913	Degrees of Freedom	7	133.36	Degrees of Freedom	7
Mtn	117.40	26.426			119.89		
WSoCen	102.31	26.813	X Coefficient(s)	0.2241	103.94	X Coefficient(s)	0.2226
ESoCen	85.63	17.287	Std Err of Coef.	0.0435	85.83	Std Err of Coef.	0.0408
SoAtl	101.72	29.282	XCoef / S.E. =	5.1581	100.74	XCoef / S.E.	5.4528

Part 3-A.
Calculation of Fractional Causation
from Averaged PhysPops

1. Nonradiation rate is Adjusted
    Constant (Part 2, Col.C) =   2.3462

2. Radiation rate is Natl Adjusted
    MortRate (Part 1, Col.G = 32.5299)
    minus Nonradiation rate (2.3462) =   30.1838

3. 1960 Fractional Causation is radiation
    rate (30.1838) divided by OBSERVED
    Natl MR Part 1,Col.C=   35.2   =   0.86

Part 3-B.
Calculation of Fractional Causation
from 1940 PhysPops

1. Nonradiation rate is Adjusted
    Constant (Part 2, Col.E) =   1.9609

2. Radiation rate is Natl Adjusted
    MortRate (Part 1, Col.G = 32.5299)
    minus Nonradiation rate (1.9609) =   30.5690

3. 1960 Fractional Causation is radiation
    rate (30.5690) divided by OBSERVED
    Natl MR Part 1,Col.C=   35.2   =   0.87

```
===============================================================================================
                                      Table 51-EE
                   Respiratory Cancers, Males:  Fractional Causation in 1980
===============================================================================================
Part 1.
Calculation of the 6 Adjusted MortRates (Col.F) and the National Adjusted MortRate (Col.G).
The last six entries in Part 1, Col.F, are the products of (Col.D * Col.E), as discussed in Chap. 49.
```

	Col.A 1980 PopFrac Tab 3-B	Col.B 1980 Obs MR Tab 16-A	Col.C A * B	Col.D 1940 MR Mid,Low Tab 16-A	Col.E AdjuFact Bx2,Pt2 Col.E	Col.F 1980 Adju MortRates	Col.G A * F
Pacific	0.1398	53.5	7.479			53.5	7.479
New England	0.0546	57.3	3.129			57.3	3.129
Mid-Atlantic	0.1630	58.4	9.519			58.4	9.519
WestNoCentral	0.0759	53.7	4.076	7.7	4.83	37.191	2.823
EastNoCentral	0.1846	61.4	11.334	10.6	4.83	51.198	9.451
Mountain	0.0502	43.6	2.189	7.8	4.83	37.674	1.891
WestSoCentral	0.1049	62.8	6.588	7.6	5.348	40.645	4.264
EastSoCentral	0.0646	70.8	4.574	4.9	5.348	26.205	1.693
SouthAtlantic	0.1624	65.2	10.588	8.3	5.348	44.388	7.209

```
                                   Sum =   59.5                            Sum =
          1980 Observed Natl MR from Table 16-B =   59.4    1980 Natl Adjusted MR =   47.4574
```

```
Part 2. ---------------------------------------------------------------------------------------
        Col.A    Col.B                           Col.C                Col.D                      Col.E
        Mean1940   1980              Respiratory Ca, Males:            1940           Respiratory Ca, Males:
        thru1980  Adju MRs           1980 Adjusted MortRates           PPs from       1980 Adjusted MortRates
Trio-   PPs from  from Col.F            regressed on                  Table 3-A          regressed on
Seq.    Tab 47-A  Part 1            Mean 1940 thru 1980 PPs           (TrioSeq)        1940 PhysPops
        x'                            Regression Output:               x''               Regression Output:
Pac     177.35    53.5      Constant              2.5975             159.72       Constant              3.6939
NewEng  185.86    57.3      Std Err of Y Est      5.2167             161.55       Std Err of Y Est      5.2342
MidAtl  186.11    58.4      R Squared             0.7948             169.76       R Squared             0.7935
WNoCen  128.82    37.191    No. of Observation       9               123.14       No. of Observation       9
ENoCen  133.71    51.198    Degrees of Freedom       7               133.36       Degrees of Freedom       7
Mtn     133.45    37.674                                             119.89
WSoCen  114.66    40.645    X Coefficient(s)      0.2984             103.94       X Coefficient(s)      0.3223
ESoCen   99.46    26.205    Std Err of Coef.      0.0573              85.83       Std Err of Coef.      0.0622
SoAtl   124.62    44.388    XCoef / S.E. =        5.2076             100.74       XCoef / S.E.          5.1857
```

```
-----------------------------------------------------------------------------------------------
Part 3-A.                                         |   Part 3-B.
Calculation of Fractional Causation               |   Calculation of Fractional Causation
from Averaged PhysPops                            |   from 1940 PhysPops
                                                  |
1.  Nonradiation rate is Adjusted                 |   1.  Nonradiation rate is Adjusted
    Constant (Part 2, Col.C) =        2.5975      |       Constant (Part 2, Col.E) =        3.6939
                                                  |
2.  Radiation rate is Natl Adjusted               |   2.  Radiation rate is Natl Adjusted
    MortRate (Part 1, Col.G = 47.4574)            |       MortRate (Part 1, Col.G = 47.4574)
    minus Nonradiation rate (2.5975) =  44.8600   |       minus Nonradiation rate (3.6939) =  43.7635
                                                  |
3.  1980 Fractional Causation is radiation        |   3.  1980 Fractional Causation is radiation
    rate (44.8600) divided by OBSERVED            |       rate (43.7635) divided by OBSERVED
    Natl MR Part 1,Col.C=   59.4    =    0.71     |       Natl MR Part 1, Col.C=   59.4    =    0.74
-----------------------------------------------------------------------------------------------
```

===
Table 51-FF
Respiratory Cancers, Males: Fractional Causation in 1988
===

Part 1.
Calculation of the 6 Adjusted MortRates (Col.F) and the National Adjusted MortRate (Col.G).
The last six entries in Part 1, Col.F, are the products of (Col.D * Col.E), as discussed in Chap. 49.

Trio-Sequence	Col.A 1990 PopFrac Tab 3-B	Col.B 1988 Obs MR Tab 16-A	Col.C A * B	Col.D 1940 MR Mid,Low Tab 16-A	Col.E AdjuFact Bx2,Pt2 Col.E	Col.F 1988 Adju MortRates	Col.G A * F
Pacific	0.1535	50.7	7.782			50.7	7.782
New England	0.0527	56.3	2.967			56.3	2.967
Mid-Atlantic	0.1527	57.5	8.780			57.5	8.780
WestNoCentral	0.0721	56.2	4.052	7.7	4.662	35.897	2.588
EastNoCentral	0.1713	62.3	10.672	10.6	4.662	49.417	8.465
Mountain	0.0543	44.2	2.400	7.8	4.662	36.364	1.975
WestSoCentral	0.1087	67.9	7.381	7.6	5.306	40.326	4.383
EastSoCentral	0.0621	79.1	4.912	4.9	5.306	25.999	1.615
SouthAtlantic	0.1725	68.5	11.816	8.3	5.306	44.040	7.597

Sum =   60.8

1988 Observed Natl MR from Table 16-B =   59.7

Sum =

1988 Natl Adjusted MR =   46.1524

Part 2. ---

Trio-Seq.	Col.A Mean1940 thru1990 PPs from Tab 47-A x'	Col.B 1988 Adju MRs from Col.F Part 1	Col.C Respiratory Ca, Males: 1988 Adjusted MortRates regressed on Mean 1940 thru 1990 PPs Regression Output:		Col.D 1940 PPs from Table 3-A (TrioSeq) x''	Col.E Respiratory Ca, Males: 1988 Adjusted MortRates regressed on 1940 PhysPops Regression Output:	
Pac	191.97	50.7	Constant	2.1988	159.72	Constant	4.7678
NewEng	208.20	56.3	Std Err of Y Est	5.2078	161.55	Std Err of Y Est	5.4503
MidAtl	204.72	57.5	R Squared	0.7816	169.76	R Squared	0.7608
WNoCen	141.14	35.897	No. of Observation	9	123.14	No. of Observation	9
ENoCen	146.19	49.417	Degrees of Freedom	7	133.36	Degrees of Freedom	7
Mtn	145.91	36.364			119.89		
WSoCen	126.28	40.326	X Coefficient(s)	0.2652	103.94	X Coefficient(s)	0.3054
ESoCen	113.28	25.999	Std Err of Coef.	0.0530	85.83	Std Err of Coef.	0.0647
SoAtl	142.93	44.040	XCoef / S.E. =	5.0050	100.74	XCoef / S.E.	4.7183

Part 3-A. Calculation of Fractional Causation from Averaged PhysPops	Part 3-B. Calculation of Fractional Causation from 1940 PhysPops
1. Nonradiation rate is Adjusted Constant (Part 2, Col.C) = 2.1988	1. Nonradiation rate is Adjusted Constant (Part 2, Col.E) = 4.7678
2. Radiation rate is Natl Adjusted MortRate (Part 1, Col.G = 46.1524) minus Nonradiation rate (2.1988) = 43.9536	2. Radiation rate is Natl Adjusted MortRate (Part 1, Col.G = 46.1524) minus Nonradiation rate 4.7678 = 41.3846
3. 1988 Fractional Causation is radiation rate (43.9536) divided by OBSERVED Natl MR Part 1,Col.C= 59.7 = 0.74	3. 1988 Fractional Causation is radiation rate (41.3846) divided by OBSERVED Natl MR Part 1,Col.C= 59.7 = 0.69

CHAPTER 52

Respiratory-System Cancers, Females, 1940-1988

● Table 52-A, Column A, shows the dramatic post-1940 rise in the female National MortRate from Respiratory-System Cancers. The relationship with cigarette smoking is reviewed in Chapter 48.

● Box 1 indicates in Columns D and I that a carcinogenic co-actor (smoking), which can contribute to female MortRates from Respiratory-System Cancers, is operating more strongly in the LowTrio than in the TopTrio. However, when post-1940 change is expressed by subtraction (Column K), the change is nearly the same in all Trios. By contrast, the comparable Column K for females in Chapters 50, 54, 55, 56, 58, 60, and 65, all clearly support the conclusion that a co-actor, which can contribute to female MortRates from cancer and IHD, is operating more strongly in the LowTrio than in the TopTrio. Therefore, despite Column K, below, we are convinced that female Respiratory Cancers are no exception.

● With respect to the LowTrio MortRates in Columns A and G of Box 1, below, we remind readers that these entries are not errors. Please see note in Table 17-A.

		Table 52-A						
		Respiratory Cancers, Females: Fractional Causation by Medical Radiation over Time						
Year	Col.A Natl MR	Col.B Frac.C		Col.C R-Sq	Col.D X-Coef	Col.E StdErr	Col.F Coef/SE	Col.G Source
1940	3.3	97%		0.9625	0.0238	0.0018	13.4046	Chap.17
1950	4.6	76%		0.9420	0.0341	0.0032	10.6614	Tab 52-B
1960	5.3	85%		0.9521	0.0346	0.0029	11.7954	Tab 52-C
1970	11.7	83%		0.8987	0.0721	0.0091	7.8795	Tab 52-D
1980	18.0	81%		0.8624	0.1005	0.0152	6.6231	Tab 52-E
1988	24.5	83%		0.8975	0.1265	0.0162	7.8277	Tab 52-F

1950: The anomalous value in Col.B is probably related to the anomaly noted for 1950 in Table 17-A.

```
================================================================================
                            Box 1, Chap. 52
            Respiratory Cancer, Females:  Post-1940 Change in MortRates by Census Trios

1960 vs. 1940, by Trios:  Col.D expresses change by ratios.  Col.F expresses change by subtraction.
1988 vs. 1940, by Trios:  Col.I expresses change by ratios.  Col.K expresses change by subtraction.
MRs change inversely with PP.  High-PP Trio has lowest growth-factor.  Low-PP Trio has highest growth-factor.
```

	Col.A 1940 MortRate Tab 17-A		Col.B 1960 MortRate Tab 17-A	Col.C Ratio Col.B /Col.A	Col.D Input from Col.C	Col.E Diff: Col.B minus A	Col.F Input from Col.E		Col.G 1988 MortRate Tab 17-A	Col.H Ratio Col.G /Col.A	Col.I Input from Col.H	Col.J Diff: Col.G minus A	Col.K Input from Col.J
Pacif	3.8		5.9	1.553	Avg Chg	2.1	Avg Chg		27.8	7.316	Avg Chg	24.0	Avg Chg
NewE	4.1		5.6	1.366	TopTrio	1.5	TopTrio		26.9	6.561	TopTrio	22.8	TopTrio
MidAtl	4.2		6.0	1.429	1.449	1.8	1.8		25.8	6.143	6.673	21.6	22.8
WNoCen	3.1		4.4	1.419	Avg Chg	1.3	Avg Chg		23.1	7.452	Avg Chg	20.0	Avg Chg
ENoCen	3.2		5.1	1.594	MidTrio	1.9	MidTrio		26.4	8.250	MidTrio	23.2	MidTrio
Mtn	2.9		4.1	1.414	1.476	1.2	1.5		22.2	7.655	7.786	19.3	20.8
WSoCen	2.4		5.2	2.167	Avg Chg	2.8	Avg Chg		26.6	11.083	Avg Chg	24.2	Avg Chg
ESoCen	2.4		4.7	1.958	LowTrio	2.3	LowTrio		26.6	11.083	LowTrio	24.2	LowTrio
SoAtl	2.4		5.0	2.083	2.069	2.6	2.6		26.6	11.083	11.083	24.2	24.2

```
================================================================================
```

```
===================================================================================================
                                        Box 2, Chap. 52
                  Respiratory-System Cancers, Females:  Calculation of Adjustment Factor

                      This adjustment is discussed fully in Chapter 49.
● Part 1: Calculate average population-weighted MortRate for the combined TopTrio Census Divs.
```

Census Div.	Col.A 1940 MR Tab 17-A	Col.B 1940 Pop'n Tab 3-B	Col.C 1940 Popn /45,710,039	Col.D Col.A * Col.C	Census Div.	Col.A 1950 MR Tab 17-A	Col.B 1950 Pop'n Tab 3-B	Col.C 1950 Popn /53,964,513	Col.D Col.A * Col.C
Pacific	3.8	9,733,262	0.2129	0.81	Pacific	4.4	14,486,527	0.2684	1.18
NewEng	4.1	8,437,290	0.1846	0.76	NewEng	4.1	9,314,453	0.1726	0.71
Mid-Atl	4.2	27,539,487	0.6025	2.53	Mid-Atl	5.0	30,163,533	0.5590	2.79
1940		Sum TopTrio 45,710,039	Sum 1.0000	TopTrio 4.096	1950		Sum TopTrio 53,964,513	Sum 1.0000	TopTrio 4.684

Census Div.	Col.A 1960 MR Tab 17-A	Col.B 1960 Pop'n Tab 3-B	Col.C 1960 Popn /65,875,863	Col.D Col.A * Col.C	Census Div.	Col.A 1970 MR Tab 17-A	Col.B 1970 Pop'n Tab 3-B	Col.C 1970 Popn /75,017,000	Col.D Col.A * Col.C
Pacific	5.9	21,198,044	0.3218	1.90	Pacific	13.6	26,087,000	0.3477	4.73
NewEng	5.6	10,509,367	0.1595	0.89	NewEng	12.1	11,781,000	0.1570	1.90
Mid-Atl	6.0	34,168,452	0.5187	3.11	Mid-Atl	12.3	37,149,000	0.4952	6.09
1960		Sum TopTrio 65,875,863	Sum 1.0000	TopTrio 5.904	1970		Sum TopTrio 75,017,000	Sum 1.0000	TopTrio 12.721

Census Div.	Col.A 1980 MR Tab 17-A	Col.B 1980 Pop'n Tab 3-B	Col.C 1980 Popn /80,615,000	Col.D Col.A * Col.C	Census Div.	Col.A 1988 MR Tab 17-A	Col.B 1990 Pop'n Tab 3-B	Col.C 1990 Popn /88,495,000	Col.D Col.A * Col.C
Pacific	21.2	31,523,000	0.3910	8.29	Pacific	27.8	37,837,000	0.4276	11.89
NewEng	18.5	12,322,000	0.1528	2.83	NewEng	26.9	12,998,000	0.1469	3.95
Mid-Atl	18.5	36,770,000	0.4561	8.44	Mid-Atl	25.8	37,660,000	0.4256	10.98
1980		Sum TopTrio 80,615,000	Sum 1.0000	TopTrio 19.556	1988		Sum TopTrio 88,495,000	Sum 1.0000	TopTrio 26.817

```
● Part 2:  Take ratios of these TopTrio MortRates, with 1940 as the denominator of each ratio.
   Col.D modifies Col.C by separate PhysPop adjustments for MidTrio and LowTrio Census Divisions.
```

	Col.A TopTrio Mean MR	Col.B 1940 TopTrio Mean MR	Col.C = Col.A / Col.B	Col.D ppAdju Tab 47-B	Col.E = Col.C * Col.D	RESPIRATORY CANCERS. Females.
				MidTrio		
1950	4.684	4.096	1.143	0.99	1.13	= MidTrio Adjustment Factor, 1950
1960	5.904	4.096	1.441	0.97	1.40	= MidTrio Adjustment Factor, 1960
1970	12.721	4.096	3.105	0.95	2.95	= MidTrio Adjustment Factor, 1970
1980	19.556	4.096	4.774	0.94	4.49	= MidTrio Adjustment Factor, 1980
1988	26.817	4.096	6.546	0.94	6.15	= MidTrio Adjustment Factor, 1988
				LowTrio		
1950	4.684	4.096	1.143	1.00	1.14	= LowTrio Adjustment Factor, 1950
1960	5.904	4.096	1.441	1.01	1.46	= LowTrio Adjustment Factor, 1960
1970	12.721	4.096	3.105	1.02	3.17	= LowTrio Adjustment Factor, 1970
1980	19.556	4.096	4.774	1.04	4.96	= LowTrio Adjustment Factor, 1980
1988	26.817	4.096	6.546	1.07	7.00	= LowTrio Adjustment Factor, 1988

```
===================================================================================================
```

```
===================================================================================================
                                      Table 52-B
                   Respiratory Cancers, Females:  Fractional Causation in 1950
===================================================================================================
```

Part 1.
Calculation of the 6 Adjusted MortRates (Col.F) and the National Adjusted MortRate (Col.G).
The last six entries in Part 1, Col.F, are the products of (Col.D * Col.E), as discussed in Chap. 49.

Trio-Sequence	Col.A 1950 PopFrac Tab 3-B	Col.B 1950 Obs MR Tab 17-A	Col.C A * B	Col.D 1940 MR Mid,Low Tab 17-A	Col.E AdjuFact Bx2,Pt2 Col.E	Col.F 1950 Adju MortRates	Col.G A * F
Pacific	0.0961	4.4	0.423			4.4	0.423
New England	0.0618	4.1	0.253			4.1	0.253
Mid-Atlantic	0.2002	5.0	1.001			5.0	1.001
WestNoCentral	0.0933	4.8	0.448	3.1	1.13	3.50	0.327
EastNoCentral	0.2017	4.5	0.908	3.2	1.13	3.62	0.729
Mountain	0.0337	4.2	0.142	2.9	1.13	3.28	0.110
WestSoCentral	0.0965	4.3	0.415	2.4	1.14	2.74	0.264
EastSoCentral	0.0762	4.7	0.358	2.4	1.14	2.74	0.208
SouthAtlantic	0.1406	4.7	0.661	2.4	1.14	2.74	0.385

```
                                   Sum =    4.6                                  Sum =
            1950 Observed Natl MR from Table 17-B =   4.6      1950 Natl Adjusted MR =   3.7010
```

Part 2. --

Trio-Seq.	Col.A Mean1940 thru1950 PPs from Tab 47-A x'	Col.B 1950 Adju MRs from Col.F Part 1	Col.C Respiratory Ca, Females: 1950 Adjusted MortRates regressed on Mean 1940 thru 1950 PPs Regression Output:		Col.D 1940 PPs from Table 3-A (TrioSeq) x''	Col.E Respiratory Ca, Females: 1950 Adjusted MortRates regressed on 1940 PhysPops Regression Output:	
Pac	154.16	4.4	Constant	0.2183	159.72	Constant	0.2144
NewEng	162.03	4.1	Std Err of Y Est	0.2669	161.55	Std Err of Y Est	0.2327
MidAtl	169.24	5.0	R Squared	0.9420	169.76	R Squared	0.9270
WNoCen	121.60	3.50	No. of Observation	9	123.14	No. of Observation	9
ENoCen	128.53	3.62	Degrees of Freedom	7	133.36	Degrees of Freedom	7
Mtn	119.64	3.28			119.89		
WSoCen	102.64	2.74	X Coefficient(s)	0.0341	103.94	X Coefficient(s)	0.0261
ESoCen	84.44	2.74	Std Err of Coef.	0.0032	85.83	Std Err of Coef.	0.0028
SoAtl	99.91	2.74	XCoef / S.E. =	10.6614	100.74	XCoef / S.E.	9.4294

Part 3-A. | Part 3-B.
Calculation of Fractional Causation | Calculation of Fractional Causation
from Averaged PhysPops | from 1940 PhysPops
                                                             |
1. Nonradiation rate is Adjusted | 1. Nonradiation rate is Adjusted
    Constant (Part 2, Col.C) =   0.2183 | Constant (Part 2, Col.E) =   0.2144
                                                             |
2. Radiation rate is Natl Adjusted | 2. Radiation rate is Natl Adjusted
    MortRate (Part 1, Col.G = 3.7010) | MortRate (Part 1, Col.G = 3.7010)
    minus Nonradiation rate (0.2183) =   3.4828 | minus Nonradiation rate (0.2144) =   3.4866
                                                             |
3. 1950 Fractional Causation is radiation | 3. 1950 Fractional Causation is radiation
    rate (3.4828) divided by OBSERVED | rate (3.4866) divided by OBSERVED
    Natl MR Part 1,Col.C=   4.6   =   0.76 | Natl MR Part 1,Col.C=   4.6   =   0.76

```
=====================================================================================
                                    Table 52-C
                  Respiratory Cancers, Females:  Fractional Causation in 1960
=====================================================================================
```

Part 1.
Calculation of the 6 Adjusted MortRates (Col.F) and the National Adjusted MortRate (Col.G).
The last six entries in Part 1, Col.F, are the products of (Col.D * Col.E), as discussed in Chap. 49.

Trio-Sequence	Col.A 1960 PopFrac Tab 3-B	Col.B 1960 Obs MR Tab 17-A	Col.C A * B	Col.D 1940 MR Mid,Low Tab 17-A	Col.E AdjuFact Bx2,Pt2 Col.E	Col.F 1960 Adju MortRates	Col.G A * F
Pacific	0.1182	5.9	0.697			5.9	0.697
New England	0.0586	5.6	0.328			5.6	0.328
Mid-Atlantic	0.1905	6.0	1.143			6.0	1.143
WestNoCentral	0.0858	4.4	0.378	3.1	1.40	4.34	0.372
EastNoCentral	0.2020	5.1	1.030	3.2	1.40	4.48	0.905
Mountain	0.0382	4.1	0.157	2.9	1.40	4.06	0.155
WestSoCentral	0.0945	5.2	0.491	2.4	1.46	3.50	0.331
EastSoCentral	0.0672	4.7	0.316	2.4	1.46	3.50	0.235
SouthAtlantic	0.1448	5.0	0.724	2.4	1.46	3.50	0.507

```
                                      Sum =     5.3                        Sum =
           1960 Observed Natl MR from Table 17-B =    5.3     1960 Natl Adjusted MR =    4.6749
```

Part 2. --

Trio-Seq.	Col.A Mean1940 thru1960 PPs from Tab 47-A x'	Col.B 1960 Adju MRs from Col.F Part 1	Col.C Respiratory Ca, Females: 1960 Adjusted MortRates regressed on Mean 1940 thru 1960 PPs Regression Output:		Col.D 1940 PPs from Table 3-A (TrioSeq) x''	Col.E Respiratory Ca, Females: 1960 Adjusted MortRates regressed on 1940 PhysPops Regression Output:	
Pac	155.69	5.9	Constant	0.1825	159.72	Constant	0.1694
NewEng	162.81	5.6	Std Err of Y Est	0.2426	161.55	Std Err of Y Est	0.2401
MidAtl	167.04	6.0	R Squared	0.9521	169.76	R Squared	0.9531
WNoCen	118.15	4.34	No. of Observation	9	123.14	No. of Observation	9
ENoCen	123.87	4.48	Degrees of Freedom	7	133.36	Degrees of Freedom	7
Mtn	117.40	4.06			119.89		
WSoCen	102.31	3.50	X Coefficient(s)	0.0346	103.94	X Coefficient(s)	0.0340
ESoCen	85.63	3.50	Std Err of Coef.	0.0029	85.83	Std Err of Coef.	0.0029
SoAtl	101.72	3.50	XCoef / S.E. =	11.7954	100.74	XCoef / S.E. =	11.9244

Part 3-A. | Part 3-B.
Calculation of Fractional Causation | Calculation of Fractional Causation
from Averaged PhysPops | from 1940 PhysPops

1. Nonradiation rate is Adjusted | 1. Nonradiation rate is Adjusted
    Constant (Part 2, Col.C) =   0.1825 | Constant (Part 2, Col.E) =   0.1694

2. Radiation rate is Natl Adjusted | 2. Radiation rate is Natl Adjusted
    MortRate (Part 1, Col.G = 4.6749) | MortRate (Part 1, Col.G = 4.6749)
    minus Nonradiation rate (0.1825) =   4.4925 | minus Nonradiation rate (0.1694) =   4.5055

3. 1960 Fractional Causation is radiation | 3. 1960 Fractional Causation is radiation
    rate (4.4925) divided by OBSERVED | rate (4.5055) divided by OBSERVED
    Natl MR Part 1,Col.C=   5.3   =   0.85 | Natl MR Part 1,Col.C=   5.3   =   0.85

```
===================================================================================================
                                         Table 52-E
                    Respiratory Cancers, Females:  Fractional Causation in 1980
===================================================================================================
Part 1.
Calculation of the 6 Adjusted MortRates (Col.F) and the National Adjusted MortRate (Col.G).
The last six entries in Part 1, Col.F, are the products of (Col.D * Col.E), as discussed in Chap. 49.


                  Col.A    Col.B    Col.C        Col.D    Col.E    Col.F    Col.G
                  1980     1980                  1940 MR  AdjuFact 1980
                  PopFrac  Obs MR   A * B        Mid,Low  Bx2,Pt2  Adju     A * F
Trio-Sequence     Tab 3-B  Tab 17-A             Tab 17-A Col.E    MortRates
      Pacific     0.1398   21.2     2.964                          21.2     2.964
      New England 0.0546   18.5     1.010                          18.5     1.010
      Mid-Atlantic 0.1630  18.5     3.016                          18.5     3.016
      WestNoCentral 0.0759 15.0     1.138        3.1      4.49      13.92    1.056
      EastNoCentral 0.1846 18.1     3.341        3.2      4.49      14.37    2.652
      Mountain    0.0502   14.9     0.748        2.9      4.49      13.02    0.654
      WestSoCentral 0.1049 17.3     1.815        2.4      4.96      11.90    1.249
      EastSoCentral 0.0646 17.0     1.098        2.4      4.96      11.90    0.769
      SouthAtlantic 0.1624 17.9     2.907        2.4      4.96      11.90    1.933

                           Sum =    18.0                                    Sum =
        1980 Observed Natl MR from Table 17-B =  18.0   1980 Natl Adjusted MR =    15.3027

Part 2. --------------------------------------------------------------------------------------------
        Col.A    Col.B                        Col.C            Col.D                       Col.E
        Mean1940 1980              Respiratory Ca, Females:    1940            Respiratory Ca, Females:
        thru1980 Adju MRs          1980 Adjusted MortRates     PPs from        1980 Adjusted MortRates
Trio-   PPs from from Col.F            regressed on            Table 3-A           regressed on
Seq.    Tab 47-A Part 1           Mean 1940 thru 1980 PPs      (TrioSeq)        1940 PhysPops
        x'                             Regression Output:      x''                  Regression Output:
Pac     177.35   21.2             Constant          0.6797     159.72         Constant          1.0980
NewEng  185.86   18.5             Std Err of Y Est  1.3822     161.55         Std Err of Y Est  1.4193
MidAtl  186.11   18.5             R Squared         0.8624     169.76         R Squared         0.8549
WNoCen  128.82   13.92            No. of Observation   9       123.14         No. of Observation   9
ENoCen  133.71   14.37            Degrees of Freedom   7       133.36         Degrees of Freedom   7
Mtn     133.45   13.02                                         119.89
WSoCen  114.66   11.90            X Coefficient(s)  0.1005     103.94         X Coefficient(s)  0.1082
ESoCen   99.46   11.90            Std Err of Coef.  0.0152      85.83         Std Err of Coef.  0.0169
SoAtl   124.62   11.90            XCoef / S.E. =    6.6231     100.74         XCoef / S.E.      6.4216


--------------------------------------------------------- -----------------------------------------------
Part 3-A.                                                | Part 3-B.
Calculation of Fractional Causation                     | Calculation of Fractional Causation
from Averaged PhysPops                                   | from 1940 PhysPops
                                                        |
1.  Nonradiation rate is Adjusted                       | 1.  Nonradiation rate is Adjusted
    Constant (Part 2, Col.C) =            0.6797         |     Constant (Part 2, Col.E) =            1.0980
                                                        |
2.  Radiation rate is Natl Adjusted                     | 2.  Radiation rate is Natl Adjusted
    MortRate (Part 1, Col.G = 15.3027)                  |     MortRate (Part 1, Col.G = 15.3027)
    minus Nonradiation rate (0.6797) =    14.6230        |     minus Nonradiation rate (1.0980) =    14.2048
                                                        |
3.  1980 Fractional Causation is radiation              | 3.  1980 Fractional Causation is radiation
    rate (14.623) divided by OBSERVED                   |     rate (14.2048) divided by OBSERVED
    Natl MR Part 1,Col.C=    18.0    =     0.81          |     Natl MR Part 1, Col.C=    18.0    =     0.79
--------------------------------------------------------- -----------------------------------------------
```

===
Table 52-F
Respiratory Cancers, Females: Fractional Causation in 1988
===

Part 1.
Calculation of the 6 Adjusted MortRates (Col.F) and the National Adjusted MortRate (Col.G).
The last six entries in Part 1, Col.F, are the products of (Col.D * Col.E), as discussed in Chap. 49.

Trio-Sequence	Col.A 1990 PopFrac Tab 3-B	Col.B 1988 Obs MR Tab 17-A	Col.C A * B	Col.D 1940 MR Mid,Low Tab 17-A	Col.E AdjuFact Bx2,Pt2 Col.E	Col.F 1988 Adju MortRates	Col.G A * F
Pacific	0.1535	27.8	4.267			27.8	4.267
New England	0.0527	26.9	1.418			26.9	1.418
Mid-Atlantic	0.1527	25.8	3.940			25.8	3.940
WestNoCentral	0.0721	23.1	1.666	3.1	6.15	19.07	1.375
EastNoCentral	0.1713	26.4	4.522	3.2	6.15	19.68	3.371
Mountain	0.0543	22.2	1.205	2.9	6.15	17.84	0.968
WestSoCentral	0.1087	26.6	2.891	2.4	7.00	16.80	1.826
EastSoCentral	0.0621	26.6	1.652	2.4	7.00	16.80	1.043
SouthAtlantic	0.1725	26.6	4.589	2.4	7.00	16.80	2.898

Sum = 26.1
1988 Observed Natl MR from Table 17-B =   24.5   1988 Natl Adjusted MR =   Sum = 21.1062

Part 2. --

Trio-Seq.	Col.A Mean1940 thru1990 PPs from Tab 47-A x'	Col.B 1988 Adju MRs from Col.F Part 1	Col.C Respiratory Ca, Females: 1988 Adjusted MortRates regressed on Mean 1940 thru 1990 PPs Regression Output:		Col.D 1940 PPs from Table 3-A (TrioSeq) x''	Col.E Respiratory Ca, Females: 1988 Adjusted MortRates regressed on 1940 PhysPops Regression Output:	
Pac	191.97	27.8	Constant	0.8622	159.72	Constant	2.0364
NewEng	208.20	26.9	Std Err of Y Est	1.5884	161.55	Std Err of Y Est	1.7301
MidAtl	204.72	25.8	R Squared	0.8975	169.76	R Squared	0.8784
WNoCen	141.14	19.07	No. of Observation	9	123.14	No. of Observation	9
ENoCen	146.19	19.68	Degrees of Freedom	7	133.36	Degrees of Freedom	7
Mtn	145.91	17.84			119.89		
WSoCen	126.28	16.80	X Coefficient(s)	0.1265	103.94	X Coefficient(s)	0.1461
ESoCen	113.28	16.80	Std Err of Coef.	0.0162	85.83	Std Err of Coef.	0.0205
SoAtl	142.93	16.80	XCoef / S.E. =	7.8277	100.74	XCoef / S.E.	7.1096

Part 3-A. | Part 3-B.
Calculation of Fractional Causation | Calculation of Fractional Causation
from Averaged PhysPops | from 1940 PhysPops
                                                     |
1. Nonradiation rate is Adjusted | 1. Nonradiation rate is Adjusted
    Constant (Part 2, Col.C) =   0.8622 | Constant (Part 2, Col.E) =   2.0364
                                                     |
2. Radiation rate is Natl Adjusted | 2. Radiation rate is Natl Adjusted
    MortRate (Part 1, Col.G = 21.1062) | MortRate (Part 1, Col.G = 21.1062)
    minus Nonradiation rate (0.8622) =   20.2440 | minus Nonradiation rate (2.0364)) =   19.0699
                                                     |
3. 1988 Fractional Causation is radiation | 3. 1988 Fractional Causation is radiation
    rate (20.2440) divided by OBSERVED | rate (19.0699) divided by OBSERVED
    Natl MR Part 1,Col.C=   24.5   =   0.83 | Natl MR Part 1, Col.C=   24.5   0.78

CHAPTER 53

Difference–Cancers, Males, 1940–1988

- Difference–Cancers are, of course, All–Cancers Minus Respiratory Cancers. Table 53–A, Column A, shows that male National MortRates for Difference–Cancers are approximately steady in the 1940–1988 period. Box 1 looks at the rates by Census Divisions.

- Box 1 shows that, while Difference–Cancer MortRates are GROWING in the LowTrio compared with 1940, they are FALLING in the TopTrio compared with 1940. A falling rate produces a ratio (fraction) below 1.0, as noted in Chapter 50. For the TopTrio, we find ratios below 1.00 in Columns D and I, and negative numbers in Columns F and K. By contrast, ratios in the LowTrio are above 1.0 and values in Columns F and K are positive. The facts in Box 1 mean that a carcinogenic co–actor which can contribute to male MortRates, from Difference Cancers, is operating more strongly in the LowTrio than in the TopTrio (Chapter 48, Part 5b). Our opinion is that the identity of this co–actor is cigarette smoke.

- Abundant studies implicate cigarette smoking in several types of cancer outside the respiratory system, including adult leukemia, colo–rectal cancer, breast cancer, and male bladder cancer. Some of these studies have been characterized as inconclusive. Box 1 in this chapter, and Boxes 1 in the subsequent chapters, lend support to the strong suspicion that cigarette smoke elevates mortality from many non–respiratory types of cancer. However, we need not "settle" the issue here, because we must match the Census Divisions for the co–actor, regardless of its identity.

		Table 53–A					
		Difference–Cancers, Males: Fractional Causation by Medical Radiation over Time					
Year	Col.A Natl MR	Col.B Frac.C	Col.C R–Sq	Col.D X–Coef	Col.E StdErr	Col.F Coef/SE	Col.G Source
1940	104.0	84%	0.9342	0.6388	0.0641	9.9695	Chap.18
1950	111.2	80%	0.9099	0.6722	0.0799	8.4103	Tab 53–B
1960	110.5	78%	0.9153	0.6603	0.0759	8.6991	Tab 53–C
1970	107.8	75%	0.9167	0.5975	0.0681	8.7784	Tab 53–D
1980	105.1	75%	0.9113	0.4858	0.0573	8.4805	Tab 53–E
1988	103.0	72%	0.9158	0.4622	0.0530	8.7250	Tab 53–F

===

Box 1, Chap. 53
Difference-Cancers, Males: Post-1940 Change in MortRates by Census Trios

1960 vs. 1940, by Trios: Col.D expresses change by ratios. Col.F expresses change by subtraction.
1988 vs. 1940, by Trios: Col.I expresses change by ratios. Col.K expresses change by subtraction.
MRs change inversely with PP. High-PP Trio has lowest growth-factor. Low-PP Trio has highest growth-factor.

	Col.A 1940 MortRate Tab 18-A	Col.B 1960 MortRate Tab 18-A	Col.C Ratio Col.B /Col.A	Col.D Input from Col.C	Col.E Diff: Col.B minus A	Col.F Input from Col.E	Col.G 1988 MortRate Tab 18-A	Col.H Ratio Col.G /Col.A	Col.I Input from Col.H	Col.J Diff: Col.G minus A	Col.K Input from Col.J
Pacif	110.9	105.8	0.954	Avg Chg	-5.1	Avg Chg	97.8	0.882	Avg Chg	-13.1	Avg Chg
NewE	122.0	126.5	1.037	TopTrio	4.5	TopTrio	110.8	0.908	TopTrio	-11.2	TopTrio
MidAtl	123.8	123.4	0.997	0.996	-0.4	-0.3	110.9	0.896	0.895	-12.9	-12.4
WNoCen	103.2	107.2	1.039	Avg Chg	4.0	Avg Chg	99.7	0.966	Avg Chg	-3.5	Avg Chg
ENoCen	109.0	115.0	1.055	MidTrio	6.0	MidTrio	108.9	0.999	MidTrio	-0.1	MidTrio
Mtn	92.0	93.2	1.013	1.036	1.2	3.7	94.9	1.032	0.999	2.9	-0.2
WSoCen	79.3	98.9	1.247	Avg Chg	19.6	Avg Chg	105.0	1.324	Avg Chg	25.7	Avg Chg
ESoCen	68.7	96.1	1.399	LowTrio	27.4	LowTrio	109.1	1.588	LowTrio	40.4	LowTrio
SoAtl	80.6	101.4	1.258	1.301	20.8	22.6	107.3	1.331	1.414	26.7	30.9

===

===
Box 2, Chap. 53
Difference-Cancers, Males: Calculation of Adjustment Factor

This adjustment is discussed fully in Chapter 49.
● Part 1: Calculate average population-weighted MortRate for the combined TopTrio Census Divs.

Census Div.	Col.A 1940 MR Tab 18-A	Col.B 1940 Pop'n Tab 3-B	Col.C 1940 Popn /45,710,039	Col.D Col.A * Col.C		Census Div.	Col.A 1950 MR Tab 18-A	Col.B 1950 Pop'n Tab 3-B	Col.C 1950 Popn /53,964,513	Col.D Col.A * Col.C
Pacific	110.9	9,733,262	0.2129	23.61		Pacific	106.1	14,486,527	0.2684	28.48
NewEng	122.0	8,437,290	0.1846	22.52		NewEng	128.8	9,314,453	0.1726	22.23
Mid-Atl	123.8	27,539,487	0.6025	74.59		Mid-Atl	127.6	30,163,533	0.5590	71.32
1940		Sum TopTrio 45,710,039	Sum 1.0000	TopTrio 120.721		1950		Sum TopTrio 53,964,513	Sum 1.0000	TopTrio 122.036

Census Div.	Col.A 1960 MR Tab 18-A	Col.B 1960 Pop'n Tab 3-B	Col.C 1960 Popn /65,875,863	Col.D Col.A * Col.C		Census Div.	Col.A 1970 MR Tab 18-A	Col.B 1970 Pop'n Tab 3-B	Col.C 1970 Popn /75,017,000	Col.D Col.A * Col.C
Pacific	105.8	21,198,044	0.3218	34.05		Pacific	103.0	26,087,000	0.3477	35.82
NewEng	126.5	10,509,367	0.1595	20.18		NewEng	119.8	11,781,000	0.1570	18.81
Mid-Atl	123.4	34,168,452	0.5187	64.01		Mid-Atl	118.4	37,149,000	0.4952	58.63
1960		Sum TopTrio 65,875,863	Sum 1.0000	TopTrio 118.231		1970		Sum TopTrio 75,017,000	Sum 1.0000	TopTrio 113.265

Census Div.	Col.A 1980 MR Tab 18-A	Col.B 1980 Pop'n Tab 3-B	Col.C 1980 Popn /80,615,000	Col.D Col.A * Col.C		Census Div.	Col.A 1988 MR Tab 18-A	Col.B 1990 Pop'n Tab 3-B	Col.C 1990 Popn /88,495,000	Col.D Col.A * Col.C
Pacific	100.2	31,523,000	0.3910	39.18		Pacific	97.8	37,837,000	0.4276	41.82
NewEng	113.0	12,322,000	0.1528	17.27		NewEng	110.8	12,998,000	0.1469	16.27
Mid-Atl	113.4	36,770,000	0.4561	51.72		Mid-Atl	110.9	37,660,000	0.4256	47.19
1980		Sum TopTrio 80,615,000	Sum 1.0000	TopTrio 108.177		1988		Sum TopTrio 88,495,000	Sum 1.0000	TopTrio 105.284

● Part 2: Take ratios of these TopTrio MortRates, with 1940 as the denominator of each ratio.
   Col.D modifies Col.C by separate PhysPop adjustments for MidTrio and LowTrio Census Divisions.

	Col.A TopTrio Mean MR	Col.B 1940 TopTrio Mean MR	Col.C = Col.A / Col.B	Col.D ppAdju Tab 47-B	Col.E = Col.C * Col.D	DIFFERENCE CANCERS. Males.
				MidTrio		
1950	122.036	120.721	1.011	0.99	1.00	= MidTrio Adjustment Factor, 1950
1960	118.231	120.721	0.979	0.97	0.95	= MidTrio Adjustment Factor, 1960
1970	113.265	120.721	0.938	0.95	0.89	= MidTrio Adjustment Factor, 1970
1980	108.177	120.721	0.896	0.94	0.84	= MidTrio Adjustment Factor, 1980
1988	105.284	120.721	0.872	0.94	0.82	= MidTrio Adjustment Factor, 1988
				LowTrio		
1950	122.036	120.721	1.011	1.00	1.01	= LowTrio Adjustment Factor, 1950
1960	118.231	120.721	0.979	1.01	0.99	= LowTrio Adjustment Factor, 1960
1970	113.265	120.721	0.938	1.02	0.96	= LowTrio Adjustment Factor, 1970
1980	108.177	120.721	0.896	1.04	0.93	= LowTrio Adjustment Factor, 1980
1988	105.284	120.721	0.872	1.07	0.93	= LowTrio Adjustment Factor, 1988

===

```
=============================================================================================
                                      Table 53-B
                  Difference Cancers, Males:  Fractional Causation in 1950
=============================================================================================
Part 1.
Calculation of the 6 Adjusted MortRates (Col.F) and the National Adjusted MortRate (Col.G).
The last six entries in Part 1, Col.F, are the products of (Col.D * Col.E), as discussed in Chap. 49.
```

	Col.A 1950 PopFrac Tab 3-B	Col.B 1950 Obs MR Tab 18-A	Col.C A * B	Col.D 1940 MR Mid,Low Tab 18-A	Col.E AdjuFact Bx2,Pt2 Col.E	Col.F 1950 Adju MortRates	Col.G A * F
Trio-Sequence							
Pacific	0.0961	106.1	10.196			106.1	10.196
New England	0.0618	128.8	7.960			128.8	7.960
Mid-Atlantic	0.2002	127.6	25.546			127.6	25.546
WestNoCentral	0.0933	108.8	10.151	103.2	1.00	103.20	9.629
EastNoCentral	0.2017	116.5	23.498	109.0	1.00	109.00	21.985
Mountain	0.0337	91.4	3.080	92.0	1.00	92.00	3.100
WestSoCentral	0.0965	93.7	9.042	79.3	1.01	80.09	7.729
EastSoCentral	0.0762	90.0	6.858	68.7	1.01	69.39	5.287
SouthAtlantic	0.1406	96.5	13.568	80.6	1.01	81.41	11.446

```
                                      Sum =    109.9                             Sum =
          1950 Observed Natl MR from Table 18-B =    111.2       1950 Natl Adjusted MR =   102.8778
```

```
Part 2. ------------------------------------------------------------------------------------
```

	Col.A Mean1940 thru1950 PPs from Tab 47-A x'	Col.B 1950 Adju MRs from Col.F Part 1	Col.C Difference Cancers, Males: 1950 Adjusted MortRates regressed on Mean 1940 thru 1950 PPs Regression Output:		Col.D 1940 PPs from Table 3-A (TrioSeq) x''	Col.E Difference Ca, Males: 1950 Adjusted MortRates regressed on 1940 PhysPops Regression Output:	
Trio-Seq.							
Pacific	154.16	106.1	Constant	14.4291	159.72	Constant	14.6584
NewEng	162.03	128.8	Std Err of Y Est	6.6738	161.55	Std Err of Y Est	7.1929
MidAtl	169.24	127.6	R Squared	0.9099	169.76	R Squared	0.8954
WNoCen	121.60	103.20	No. of Observation	9	123.14	No. of Observation	9
ENoCen	128.53	109.00	Degrees of Freedom	7	133.36	Degrees of Freedom	7
Mtn	119.64	92.00			119.89		
WSoCen	102.64	80.09	X Coefficient(s)	0.6722	103.94	X Coefficient(s)	0.6612
ESoCen	84.44	69.39	Std Err of Coef.	0.0799	85.83	Std Err of Coef.	0.0854
SoAtl	99.91	81.41	XCoef / S.E. =	8.4103	100.74	XCoef / S.E.	7.7407

```
--------------------------------------------------------------------------------------------
Part 3-A.                                              |  Part 3-B.
Calculation of Fractional Causation                    |  Calculation of Fractional Causation
from Averaged PhysPops                                  |  from 1940 PhysPops
                                                       |
1.  Nonradiation rate is Adjusted                      |  1.  Nonradiation rate is Adjusted
    Constant (Part 2, Col.C) =         14.4291         |      Constant (Part 2, Col.E) =         14.6584
                                                       |
2.  Radiation rate is Natl Adjusted                    |  2.  Radiation rate is Natl Adjusted
    MortRate (Part 1, Col.G = 102.8778)                |      MortRate (Part 1, Col.G = 102.8778)
    minus Nonradiation rate (14.4291) =   88.4487      |      minus Nonradiation rate (14.6584) =   88.2194
                                                       |
3.  1950 Fractional Causation is radiation             |  3.  1950 Fractional Causation is radiation
    rate (88.4487) divided by OBSERVED                 |      rate (88.2194) divided by OBSERVED
    Natl MR Part 1,Col.C=   111.2    =     0.80        |      Natl MR Part 1,Col.C=   111.2    =     0.79
--------------------------------------------------------------------------------------------
```

```
================================================================================
                              Table 53-C
              Difference Cancers, Males:  Fractional Causation in 1960
================================================================================
```

Part 1.

Calculation of the 6 Adjusted MortRates (Col.F) and the National Adjusted MortRate (Col.G).

The last six entries in Part 1, Col.F, are the products of (Col.D * Col.E), as discussed in Chap. 49.

Trio-Sequence	Col.A 1960 PopFrac Tab 3-B	Col.B 1960 Obs MR Tab 18-A	Col.C A * B	Col.D 1940 MR Mid,Low Tab 18-A	Col.E AdjuFact Bx2,Pt2 Col.E	Col.F 1960 Adju MortRates	Col.G A * F
Pacific	0.1182	105.8	12.506			105.8	12.506
New England	0.0586	126.5	7.413			126.5	7.413
Mid-Atlantic	0.1905	123.4	23.508			123.4	23.508
WestNoCentral	0.0858	107.2	9.198	103.2	0.95	98.04	8.412
EastNoCentral	0.2020	115.0	23.230	109.0	0.95	103.55	20.917
Mountain	0.0382	93.2	3.560	92.0	0.95	87.40	3.339
WestSoCentral	0.0945	98.9	9.346	79.3	0.99	78.51	7.419
EastSoCentral	0.0672	96.1	6.458	68.7	0.99	68.01	4.570
SouthAtlantic	0.1448	101.4	14.683	80.6	0.99	79.79	11.554

```
                                        Sum =  109.9                              Sum =
        1960 Observed Natl MR from Table 18-B =  110.5       1960 Natl Adjusted MR =  99.6373
```

Part 2. --

Trio-Seq.	Col.A Mean1940 thru1960 PPs from Tab 47-A x'	Col.B 1960 Adju MRs from Col.F Part 1	Col.C Difference Cancers, Males: 1960 Adjusted MortRates regressed on Mean 1940 thru 1960 PPs Regression Output:		Col.D 1940 PPs from Table 3-A (TrioSeq) x''	Col.E Difference Ca, Males: 1960 Adjusted MortRates regressed on 1940 PhysPops Regression Output:	
Pacific	155.69	105.8	Constant	13.5357	159.72	Constant	12.9994
NewEng	162.81	126.5	Std Err of Y Est	6.2782	161.55	Std Err of Y Est	6.0029
MidAtl	167.04	123.4	R Squared	0.9153	169.76	R Squared	0.9226
WNoCen	118.15	98.04	No. of Observation	9	123.14	No. of Observation	9
ENoCen	123.87	103.55	Degrees of Freedom	7	133.36	Degrees of Freedom	7
Mtn	117.40	87.40			119.89		
WSoCen	102.31	78.51	X Coefficient(s)	0.6603	103.94	X Coefficient(s)	0.6512
ESoCen	85.63	68.01	Std Err of Coef.	0.0759	85.83	Std Err of Coef.	0.0713
SoAtl	101.72	79.79	XCoef / S.E. =	8.6991	100.74	XCoef / S.E.	9.1340

Part 3-A.

Calculation of Fractional Causation
from Averaged PhysPops

1. Nonradiation rate is Adjusted
   Constant (Part 2, Col.C) =   13.5357

2. Radiation rate is Natl Adjusted
   MortRate (Part 1, Col.G = 99.6373)
   minus Nonradiation rate (13.5357) =   86.1016

3. 1960 Fractional Causation is radiation
   rate (86.1016) divided by OBSERVED
   Natl MR Part 1,Col.C=   110.5   =   0.78

Part 3-B.

Calculation of Fractional Causation
from 1940 PhysPops

1. Nonradiation rate is Adjusted
   Constant (Part 2, Col.E) =   12.9994

2. Radiation rate is Natl Adjusted
   MortRate (Part 1, Col.G = 99.6373)
   minus Nonradiation rate (12.9994) =   86.6379

3. 1960 Fractional Causation is radiation
   rate (86.6379) divided by OBSERVED
   Natl MR Part 1, Col.C=   110.5   =   0.78

===

Table 53-E
Difference Cancers, Males: Fractional Causation in 1980

===

Part 1.
Calculation of the 6 Adjusted MortRates (Col.F) and the National Adjusted MortRate (Col.G).
The last six entries in Part 1, Col.F, are the products of (Col.D * Col.E), as discussed in Chap. 49.

Trio-Sequence	Col.A 1980 PopFrac Tab 3-B	Col.B 1980 Obs MR Tab 18-A	Col.C A * B	Col.D 1940 MR Mid,Low Tab 18-A	Col.E AdjuFact Bx2,Pt2 Col.E	Col.F 1980 Adju MortRates	Col.G A * F
Pacific	0.1398	100.2	14.008			100.2	14.008
New England	0.0546	113.0	6.170			113.0	6.170
Mid-Atlantic	0.1630	113.4	18.484			113.4	18.484
WestNoCentral	0.0759	98.3	7.461	103.2	0.84	86.69	6.580
EastNoCentral	0.1846	108.1	19.955	109.0	0.84	91.56	16.902
Mountain	0.0502	91.1	4.573	92.0	0.84	77.28	3.879
WestSoCentral	0.1049	100.1	10.500	79.3	0.93	73.75	7.736
EastSoCentral	0.0646	103.3	6.673	68.7	0.93	63.89	4.127
SouthAtlantic	0.1624	106.2	17.247	80.6	0.93	74.96	12.173

Sum = 105.1

1980 Observed Natl MR from Table 18-B = 105.1

Sum = 90.0598

1980 Natl Adjusted MR = 90.0598

Part 2. --

Trio-Seq.	Col.A Mean1940 thru1980 PPs from Tab 47-A x'	Col.B 1980 Adju MRs from Col.F Part 1	Col.C Difference Cancers, Males: 1980 Adjusted MortRates regressed on Mean 1940 thru 1980 PPs Regression Output:		Col.D 1940 PPs from Table 3-A (TrioSeq) x''	Col.E Difference Ca, Males: 1980 Adjusted MortRates regressed on 1940 PhysPops Regression Output:	
Pacific	177.35	100.2	Constant	11.6156	159.72	Constant	13.6840
NewEng	185.86	113.0	Std Err of Y Est	5.6305	161.55	Std Err of Y Est	4.0882
MidAtl	186.11	113.4	R Squared	0.9113	169.76	R Squared	0.9532
WNoCen	128.82	86.69	No. of Observation	9	123.14	No. of Observation	9
ENoCen	133.71	91.56	Degrees of Freedom	7	133.36	Degrees of Freedom	7
Mtn	133.45	77.28			119.89		
WSoCen	114.66	73.75	X Coefficient(s)	0.4858	103.94	X Coefficient(s)	0.5800
ESoCen	99.46	63.89	Std Err of Coef.	0.0573	85.83	Std Err of Coef.	0.0486
SoAtl	124.62	74.96	XCoef / S.E. =	8.4805	100.74	XCoef / S.E.	11.9456

Part 3-A.
Calculation of Fractional Causation
from Averaged PhysPops

1. Nonradiation rate is Adjusted
    Constant (Part 2, Col.C) =   11.6156

2. Radiation rate is Natl Adjusted
    MortRate (Part 1, Col.G = 90.0598)
    minus Nonradiation rate (11.6156) =   78.4442

3. 1980 Fractional Causation is radiation
    rate (78.4442) divided by OBSERVED
    Natl MR Part 1,Col.C=   105.1   =   0.75

Part 3-B.
Calculation of Fractional Causation
from 1940 PhysPops

1. Nonradiation rate is Adjusted
    Constant (Part 2, Col.E) =   13.6840

2. Radiation rate is Natl Adjusted
    MortRate (Part 1, Col.G = 90.0598)
    minus Nonradiation rate (13.6840) =   76.3758

3. 1980 Fractional Causation is radiation
    rate (76.3758) divided by OBSERVED
    Natl MR Part 1, Col.C=   105.1   =   0.73

```
=================================================================================================
                                        Table 53-F
                        Difference Cancers, Males:  Fractional Causation in 1988
=================================================================================================
```

Part 1.
Calculation of the 6 Adjusted MortRates (Col.F) and the National Adjusted MortRate (Col.G).
The last six entries in Part 1, Col.F, are the products of (Col.D * Col.E), as discussed in Chap. 49.

Trio-Sequence	Col.A 1990 PopFrac Tab 3-B	Col.B 1988 Obs MR Tab 18-A	Col.C A * B	Col.D 1940 MR Mid,Low Tab 18-A	Col.E AdjuFact Bx2,Pt2 Col.E	Col.F 1988 Adju MortRates	Col.G A * F
Pacific	0.1535	97.8	15.012			97.8	15.012
New England	0.0527	110.8	5.839			110.8	5.839
Mid-Atlantic	0.1527	110.9	16.934			110.9	16.934
WestNoCentral	0.0721	99.7	7.188	103.2	0.82	84.62	6.101
EastNoCentral	0.1713	108.9	18.655	109.0	0.82	89.38	15.311
Mountain	0.0543	94.9	5.153	92.0	0.82	75.44	4.096
WestSoCentral	0.1087	105.0	11.414	79.3	0.93	73.75	8.017
EastSoCentral	0.0621	109.1	6.775	68.7	0.93	63.89	3.968
SouthAtlantic	0.1725	107.3	18.509	80.6	0.93	74.96	12.930

```
                                        Sum =    105.5                          Sum =
          1988 Observed Natl MR from Table 18-B =  103.0      1988 Natl Adjusted MR =    88.2089
```

Part 2. ---

Trio-Seq.	Col.A Mean1940 thru1990 PPs from Tab 47-A x'	Col.B 1988 Adju MRs from Col.F Part 1	Col.C Difference Cancers, Males: 1988 Adjusted MortRates regressed on Mean 1940 thru 1990 PPs		Col.D 1940 PPs from Table 3-A (TrioSeq) x''	Col.E Difference Ca, Males: 1988 Adjusted MortRates regressed on 1940 PhysPops	
				Regression Output:			Regression Output:
Pac	191.97	97.8	Constant	13.8764	159.72	Constant	16.4360
NewEng	208.20	110.8	Std Err of Y Est	5.2069	161.55	Std Err of Y Est	4.3199
MidAtl	204.72	110.9	R Squared	0.9158	169.76	R Squared	0.9420
WNoCen	141.14	84.62	No. of Observation	9	123.14	No. of Observation	9
ENoCen	146.19	89.38	Degrees of Freedom	7	133.36	Degrees of Freedom	7
Mtn	145.91	75.44			119.89		
WSoCen	126.28	73.75	X Coefficient(s)	0.4622	103.94	X Coefficient(s)	0.5472
ESoCen	113.28	63.89	Std Err of Coef.	0.0530	85.83	Std Err of Coef.	0.0513
SoAtl	142.93	74.96	XCoef / S.E. =	8.7250	100.74	XCoef / S.E.	10.6659

Part 3-A.
Calculation of Fractional Causation
from Averaged PhysPops

1. Nonradiation rate is Adjusted
    Constant (Part 2, Col.C) =   13.8764

2. Radiation rate is Natl Adjusted
    MortRate (Part 1, Col.G = 88.2089)
    minus Nonradiation rate (13.8764) = 74.3325

3. 1988 Fractional Causation is radiation
    rate (74.2335) divided by OBSERVED
    Natl MR Part 1,Col.C=   103   =   0.72

Part 3-B.
Calculation of Fractional Causation
from 1940 PhysPops

1. Nonradiation rate is Adjusted
    Constant (Part 2, Col.E) =   16.4360

2. Radiation rate is Natl Adjusted
    MortRate (Part 1, Col.G = 88.2089)
    minus Nonradiation rate (16.4360) = 71.7729

3. 1988 Fractional Causation is radiation
    rate (71.7729) divided by OBSERVED
    Natl MR Part 1, Col.C=   103   0.70

CHAPTER 54

Difference-Cancers, Females, 1940-1988

● Difference Cancers are, of course, All-Cancers Minus Respiratory Cancers.

● Unlike the males (Chapter 53), the females enjoy a dramatic decline in post-1940 National MortRates from Difference Cancers, as shown in Table 54-A, Column A. But this decline does not happen with equal impact in all Census Divisions.

● Box 1 shows that the ratios in Columns D and I are below 1.00 in every Trio, because every Trio experiences a fall in the MortRates of 1940. In Column I, the lowest ratio (0.660) occurs in the TopTrio, while the highest fraction (0.843) occurs in the LowTrio. The 1988 MortRates in the TopTrio are down to 0.660 of their 1940 values, while the 1988 MortRates in the LowTrio are still at 0.843 of their 1940 values. Column K shows that the 1940 rates fell by 45.8 per 100,000 in the TopTrio and by only 15.9 per 100,000 in the LowTrio. The facts in Box 1 mean that a carcinogenic co-actor which can contribute to female MortRates, from Difference Cancers, is operating more strongly in the LowTrio than in the TopTrio (Chapter 48, Part 5b). We think this co-actor is cigarette smoke. Please see the text in Chapter 53.

	Col.A	Col.B		Col.C	Col.D	Col.E	Col.F	Col.G
Year	Natl MR	Frac.C		R-Sq	X-Coef	StdErr	Coef/SE	Source
1940	122.8	57%		0.8528	0.5041	0.0792	6.3682	Chap.19
1950	118.6	53%		0.8601	0.4711	0.0718	6.5597	Tab 54-B
1960	109.6	52%		0.8528	0.4279	0.0672	6.3677	Tab 54-C
1970	100.1	53%		0.8499	0.3573	0.0567	6.2968	Tab 54-D
1980	90.5	51%		0.8561	0.3143	0.0487	6.4521	Tab 54-E
1988	86.8	48%		0.8437	0.2536	0.0413	6.1462	Tab 54-F

Table 54-A
Difference-Cancers, Females: Fractional Causation by Medical Radiation over Time

==

```
                                 Box 1, Chap. 54
              Difference-Cancers, Females: Post-1940 Change in MortRates by Census Trios

1960 vs. 1940, by Trios:  Col.D expresses change by ratios.  Col.F expresses change by subtraction.
1988 vs. 1940, by Trios:  Col.I expresses change by ratios.  Col.K expresses change by subtraction.
MRs change inversely with PP.  High-PP Trio has lowest growth-factor.  Low-PP Trio has highest growth-factor.
```

	Col.A 1940 MortRate Tab 19-A		Col.B 1960 MortRate Tab 19-A	Col.C Ratio Col.B /Col.A	Col.D Input from Col.C	Col.E Diff: Col.B minus A	Col.F Input from Col.E		Col.G 1988 MortRate Tab 19-A	Col.H Ratio Col.G /Col.A	Col.I Input from Col.H	Col.J Diff: Col.G minus A	Col.K Input from Col.J
Pacif	123.6	|	104.2	0.843	Avg Chg	-19.4	Avg Chg		83.7	0.677	Avg Chg	-39.9	Avg Chg
NewE	141.2	|	116.8	0.827	TopTrio	-24.4	TopTrio		89.5	0.634	TopTrio	-51.7	TopTrio
MidAtl	138.7	|	121.4	0.875	0.849	-17.3	-20.4		92.8	0.669	0.660	-45.9	-45.8
WNoCen	117.0	|	104.9	0.897	Avg Chg	-12.1	Avg Chg		83.7	0.715	Avg Chg	-33.3	Avg Chg
ENoCen	128.2	|	114.7	0.895	MidTrio	-13.5	MidTrio		90.1	0.703	MidTrio	-38.1	MidTrio
Mtn	108.9	|	96.9	0.890	0.894	-12.0	-12.5		78.2	0.718	0.712	-30.7	-34.0
WSoCen	97.4	|	97.7	1.003	Avg Chg	0.3	Avg Chg		83.2	0.854	Avg Chg	-14.2	Avg Chg
ESoCen	100.1	|	100.1	1.000	LowTrio	0.0	LowTrio		86.1	0.860	LowTrio	-14.0	LowTrio
SoAtl	104.5	|	102.4	0.980	0.994	-2.1	-0.6		85.0	0.813	0.843	-19.5	-15.9

==

```
===================================================================================================
                                      Box 2, Chap. 54
                 Difference-Cancers, Females:  Calculation of Adjustment Factor

                      This adjustment is discussed fully in Chapter 49.
  • Part 1: Calculate average population-weighted MortRate for the combined TopTrio Census Divs.
```

Census Div.	Col.A 1940 MR Tab 19-A	Col.B 1940 Pop'n Tab 3-B	Col.C 1940 Popn /45,710,039	Col.D Col.A * Col.C	Census Div.	Col.A 1950 MR Tab 19-A	Col.B 1950 Pop'n Tab 3-B	Col.C 1950 Popn /53,964,513	Col.D Col.A * Col.C
Pacific	123.6	9,733,262	0.2129	26.32	Pacific	113.3	14,486,527	0.2684	30.41
NewEng	141.2	8,437,290	0.1846	26.06	NewEng	128.0	9,314,453	0.1726	22.09
Mid-Atl	138.7	27,539,487	0.6025	83.56	Mid-Atl	132.0	30,163,533	0.5590	73.78
1940		Sum TopTrio 45,710,039	Sum 1.0000	TopTrio 135.946	1950		Sum TopTrio 53,964,513	Sum 1.0000	TopTrio 126.290

Census Div.	Col.A 1960 MR Tab 19-A	Col.B 1960 Pop'n Tab 3-B	Col.C 1960 Popn /65,875,863	Col.D Col.A * Col.C	Census Div.	Col.A 1970 MR Tab 19-A	Col.B 1970 Pop'n Tab 3-B	Col.C 1970 Popn /75,017,000	Col.D Col.A * Col.C
Pacific	104.2	21,198,044	0.3218	33.53	Pacific	96.7	26,087,000	0.3477	33.63
NewEng	116.8	10,509,367	0.1595	18.63	NewEng	107.4	11,781,000	0.1570	16.87
Mid-Atl	121.4	34,168,452	0.5187	62.97	Mid-Atl	110.2	37,149,000	0.4952	54.57
1960		Sum TopTrio 65,875,863	Sum 1.0000	TopTrio 115.131	1970		Sum TopTrio 75,017,000	Sum 1.0000	TopTrio 105.066

Census Div.	Col.A 1980 MR Tab 19-A	Col.B 1980 Pop'n Tab 3-B	Col.C 1980 Popn /80,615,000	Col.D Col.A * Col.C	Census Div.	Col.A 1988 MR Tab 19-A	Col.B 1990 Pop'n Tab 3-B	Col.C 1990 Popn /88,495,000	Col.D Col.A * Col.C
Pacific	89.2	31,523,000	0.3910	34.88	Pacific	83.7	37,837,000	0.4276	35.79
NewEng	97.9	12,322,000	0.1528	14.96	NewEng	89.5	12,998,000	0.1469	13.15
Mid-Atl	99.0	36,770,000	0.4561	45.16	Mid-Atl	92.8	37,660,000	0.4256	39.49
1980		Sum TopTrio 80,615,000	Sum 1.0000	TopTrio 95.000	1988		Sum TopTrio 88,495,000	Sum 1.0000	TopTrio 88.424

```
  • Part 2: Take ratios of these TopTrio MortRates, with 1940 as the denominator of each ratio.
    Col.D modifies Col.C by separate PhysPop adjustments for MidTrio and LowTrio Census Divisions.
```

	Col.A TopTrio Mean MR	Col.B 1940 TopTrio Mean MR	Col.C = Col.A / Col.B	Col.D ppAdju Tab 47-B	Col.E = Col.C * Col.D	DIFFERENCE CANCERS. Females.
				MidTrio		
1950	126.290	135.946	0.929	0.99	0.92	= MidTrio Adjustment Factor, 1950
1960	115.131	135.946	0.847	0.97	0.82	= MidTrio Adjustment Factor, 1960
1970	105.066	135.946	0.773	0.95	0.73	= MidTrio Adjustment Factor, 1970
1980	95.000	135.946	0.699	0.94	0.66	= MidTrio Adjustment Factor, 1980
1988	88.424	135.946	0.650	0.94	0.61	= MidTrio Adjustment Factor, 1988
				LowTrio		
1950	126.290	135.946	0.929	1.00	0.93	= LowTrio Adjustment Factor, 1950
1960	115.131	135.946	0.847	1.01	0.86	= LowTrio Adjustment Factor, 1960
1970	105.066	135.946	0.773	1.02	0.79	= LowTrio Adjustment Factor, 1970
1980	95.000	135.946	0.699	1.04	0.73	= LowTrio Adjustment Factor, 1980
1988	88.424	135.946	0.650	1.07	0.70	= LowTrio Adjustment Factor, 1988

```
===================================================================================================
```

```
===================================================================================
                                   Table 54-B
                  Difference Cancers, Females:  Fractional Causation in 1950
===================================================================================
```

Part 1.
Calculation of the 6 Adjusted MortRates (Col.F) and the National Adjusted MortRate (Col.G).
The last six entries in Part 1, Col.F, are the products of (Col.D * Col.E), as discussed in Chap. 49.

Trio-Sequence	Col.A 1950 PopFrac Tab 3-B	Col.B 1950 Obs MR Tab 19-A	Col.C A * B	Col.D 1940 MR Mid,Low Tab 19-A	Col.E AdjuFact Bx2,Pt2 Col.E	Col.F 1950 Adju MortRates	Col.G A * F
Pacific	0.0961	113.3	10.888			113.3	10.888
New England	0.0618	128.0	7.910			128.0	7.910
Mid-Atlantic	0.2002	132.0	26.426			132.0	26.426
WestNoCentral	0.0933	112.3	10.478	117.0	0.92	107.64	10.043
EastNoCentral	0.2017	123.0	24.809	128.2	0.92	117.94	23.789
Mountain	0.0337	101.8	3.431	108.9	0.92	100.19	3.376
WestSoCentral	0.0965	105.0	10.133	97.4	0.93	90.58	8.741
EastSoCentral	0.0762	105.6	8.047	100.1	0.93	93.09	7.094
SouthAtlantic	0.1406	108.6	15.269	104.5	0.93	97.19	13.664

```
                                  Sum =  117.4                        Sum =
   1950 Observed Natl MR from Table 19-B =  118.6    1950 Natl Adjusted MR =  111.9324
```

Part 2. ---

Trio-Seq.	Col.A Mean1940 thru1950 PPs from Tab 47-A x'	Col.B 1950 Adju MRs from Col.F Part 1 y	Col.C Difference Ca., Females: 1950 Adjusted MortRates regressed on Mean 1940 thru 1950 PPs Regression Output:		Col.D 1940 PPs from Table 3-A (TrioSeq) x''	Col.E Difference Ca., Females: 1950 Adjusted MortRates regressed on 1940 PhysPops Regression Output:	
Pac	154.16	113.3	Constant	49.0882	159.72	Constant	49.0527
NewEng	162.03	128.0	Std Err of Y Est	5.9977	161.55	Std Err of Y Est	6.1704
MidAtl	169.24	132.0	R Squared	0.8601	169.76	R Squared	0.8519
WNoCen	121.60	107.64	No. of Observation	9	123.14	No. of Observation	9
ENoCen	128.53	117.94	Degrees of Freedom	7	133.36	Degrees of Freedom	7
Mtn	119.64	100.19			119.89		
WSoCen	102.64	90.58	X Coefficient(s)	0.4711	103.94	X Coefficient(s)	0.4650
ESoCen	84.44	93.09	Std Err of Coef.	0.0718	85.83	Std Err of Coef.	0.0733
SoAtl	99.91	97.19	XCoef / S.E. =	6.5597	100.74	XCoef / S.E. =	6.3458

Part 3-A.
Calculation of Fractional Causation
from Averaged PhysPops

1. Nonradiation rate is Adjusted
   Constant (Part 2, Col.C) =   49.0882

2. Radiation rate is Natl Adjusted
   MortRate (Part 1, Col.G = 111.9324)
   minus Nonradiation rate (49.0882) =   62.8442

3. 1950 Fractional Causation is radiation
   rate (62.8442) divided by OBSERVED
   Natl MR Part 1,Col.C=   118.6   =   0.53

Part 3-B.
Calculation of Fractional Causation
from 1940 PhysPops

1. Nonradiation rate is Adjusted
   Constant (Part 2, Col.E) =   49.0527

2. Radiation rate is Natl Adjusted
   MortRate (Part 1, Col.G = 111.9324)
   minus Nonradiation rate (49.0527) =   62.8798

3. 1950 Fractional Causation is radiation
   rate (62.8798) divided by OBSERVED
   Natl MR Part 1,Col.C=   118.6   =   0.53

```
=============================================================================================================
                                          Table 54-C
                   Difference Cancers, Females:  Fractional Causation in 1960
=============================================================================================================
Part 1.
Calculation of the 6 Adjusted MortRates (Col.F) and the National Adjusted MortRate (Col.G).
The last six entries in Part 1, Col.F, are the products of (Col.D * Col.E), as discussed in Chap. 49.
```

Trio-Sequence	Col.A 1960 PopFrac Tab 3-B	Col.B 1960 Obs MR Tab 19-A	Col.C A * B	Col.D 1940 MR Mid,Low Tab 19-A	Col.E AdjuFact Bx2,Pt2 Col.E	Col.F 1960 Adju MortRates	Col.G A * F
Pacific	0.1182	104.2	12.316			104.2	12.316
New England	0.0586	116.8	6.844			116.8	6.844
Mid-Atlantic	0.1905	121.4	23.127			121.4	23.127
WestNoCentral	0.0858	104.9	9.000	117.0	0.82	95.94	8.232
EastNoCentral	0.2020	114.7	23.169	128.2	0.82	105.12	21.235
Mountain	0.0382	96.9	3.702	108.9	0.82	89.30	3.411
WestSoCentral	0.0945	97.7	9.233	97.4	0.86	83.76	7.916
EastSoCentral	0.0672	100.1	6.727	100.1	0.86	86.09	5.785
SouthAtlantic	0.1448	102.4	14.828	104.5	0.86	89.87	13.013

```
                                              Sum =    108.9                              Sum =
              1960 Observed Natl MR from Table 19-B =   109.6          1960 Natl Adjusted MR =   101.8794
```

```
Part 2.  ----------------------------------------------------------------------------------------------------
```

Trio-Seq.	Col.A Mean1940 thru1960 PPs from Tab 47-A x'	Col.B 1960 Adju MRs from Col.F Part 1 y	Col.C Difference Ca., Females: 1960 Adjusted MortRates regressed on Mean 1940 thru 1960 PPs Regression Output:		Col.D 1940 PPs from Table 3-A (TrioSeq) x''	Col.E Difference Ca., Females: 1960 Adjusted MortRates regressed on 1940 PhysPops Regression Output:	
Pac	155.69	104.2	Constant	45.2239	159.72	Constant	45.0348
NewEng	162.81	116.8	Std Err of Y Est	5.5577	161.55	Std Err of Y Est	5.5244
MidAtl	167.04	121.4	R Squared	0.8528	169.76	R Squared	0.8545
WNoCen	118.15	95.94	No. of Observation	9	123.14	No. of Observation	9
ENoCen	123.87	105.12	Degrees of Freedom	7	133.36	Degrees of Freedom	7
Mtn	117.40	89.30			119.89		
WSoCen	102.31	83.76	X Coefficient(s)	0.4279	103.94	X Coefficient(s)	0.4207
ESoCen	85.63	86.09	Std Err of Coef.	0.0672	85.83	Std Err of Coef.	0.0656
SoAtl	101.72	89.87	XCoef / S.E. =	6.3677	100.74	XCoef / S.E.	6.4127

```
------------------------------------------------------------|-------------------------------------------------------
Part 3-A.                                                   |  Part 3-B.
Calculation of Fractional Causation                         |  Calculation of Fractional Causation
from Averaged PhysPops                                      |  from 1940 PhysPops
                                                            |
1.  Nonradiation rate is Adjusted                           |  1.  Nonradiation rate is Adjusted
    Constant (Part 2, Col.C) =            45.2239           |      Constant (Part 2, Col.E) =            45.0348
                                                            |
2.  Radiation rate is Natl Adjusted                         |  2.  Radiation rate is Natl Adjusted
    MortRate (Part 1, Col.G = 101.8794)                     |      MortRate (Part 1, Col.G = 101.8794)
    minus Nonradiation rate (45.2239) =    56.6554          |      minus Nonradiation rate (45.0348) =    56.8446
                                                            |
3.  1960 Fractional Causation is radiation                  |  3.  1960 Fractional Causation is radiation
    rate (56.6554) divided by OBSERVED                      |      rate (56.8446) divided by OBSERVED
    Natl MR Part 1,Col.C=  109.6    =       0.52            |      Natl MR Part 1, Col.C=  109.6    =       0.52
------------------------------------------------------------|-------------------------------------------------------
```

```
=========================================================================================
                                    Table 54-E
                 Difference Cancers, Females:  Fractional Causation in 1980
=========================================================================================
Part 1.
Calculation of the 6 Adjusted MortRates (Col.F) and the National Adjusted MortRate (Col.G).
The last six entries in Part 1, Col.F, are the products of (Col.D * Col.E), as discussed in Chap. 49.


                    Col.A    Col.B    Col.C      Col.D    Col.E    Col.F    Col.G
                    1980     1980                1940 MR  AdjuFact 1980
                    PopFrac  Obs MR   A * B      Mid,Low  Bx2,Pt2  Adju     A * F
Trio-Sequence       Tab 3-B  Tab 19-A           Tab 19-A Col.E    MortRates
      Pacific       0.1398   89.2     12.470                      89.2     12.470
      New England   0.0546   97.9     5.345                       97.9     5.345
      Mid-Atlantic  0.1630   99.0     16.137                      99.0     16.137
      WestNoCentral 0.0759   86.0     6.527      117.0    0.66     77.22    5.861
      EastNoCentral 0.1846   93.9     17.334     128.2    0.66     84.61    15.619
      Mountain      0.0502   80.0     4.016      108.9    0.66     71.87    3.608
      WestSoCentral 0.1049   82.8     8.686      97.4     0.73     71.10    7.459
      EastSoCentral 0.0646   86.2     5.569      100.1    0.73     73.07    4.721
      SouthAtlantic 0.1624   87.1     14.145     104.5    0.73     76.29    12.389

                              Sum =    90.2                                 Sum =
          1980 Observed Natl MR from Table 19-B =   90.5      1980 Natl Adjusted MR =   83.6087


Part 2.  ------------------------------------------------------------------------------------------
           Col.A    Col.B                        Col.C                 Col.D              Col.E
           Mean1940 1980              Difference Ca., Females:          1940       Difference Ca., Females:
           thru1980 Adju MRs          1980 Adjusted MortRates           PPs from    1980 Adjusted MortRates
Trio-      PPs from from Col.F        regressed on                     Table 3-A   regressed on
Seq.       Tab 47-A Part 1           Mean 1940 thru 1980 PPs           (TrioSeq)   1940 PhysPops
             x'       y              Regression Output:                 x''        Regression Output:
Pac        177.35   89.2            Constant          37.4145          159.72      Constant          38.9905
NewEng     185.86   97.9            Std Err of Y Est   4.4348          161.55      Std Err of Y Est   4.7021
MidAtl     186.11   99.0            R Squared          0.8561          169.76      R Squared          0.8382
WNoCen     128.82   77.22           No. of Observation    9            123.14      No. of Observation    9
ENoCen     133.71   84.61           Degrees of Freedom    7            133.36      Degrees of Freedom    7
Mtn        133.45   71.87                                               119.89
WSoCen     114.66   71.10           X Coefficient(s)   0.3143          103.94      X Coefficient(s)   0.3362
ESoCen      99.46   73.07           Std Err of Coef.   0.0487           85.83      Std Err of Coef.   0.0558
SoAtl      124.62   76.29           XCoef / S.E. =     6.4521          100.74      XCoef / S.E. =     6.0213


------------------------------------------------------------------------------------------------------
Part 3-A.                                             |  Part 3-B.
Calculation of Fractional Causation                   |  Calculation of Fractional Causation
from Averaged PhysPops                                |  from 1940 PhysPops
                                                      |
                                                      |
1.  Nonradiation rate is Adjusted                     |  1.  Nonradiation rate is Adjusted
    Constant (Part 2, Col.C) =          37.4145       |      Constant (Part 2, Col.E) =          38.9905
                                                      |
2.  Radiation rate is Natl Adjusted                   |  2.  Radiation rate is Natl Adjusted
    MortRate (Part 1, Col.G = 83.6087)                |      MortRate (Part 1, Col.G = 83.6087)
    minus Nonradiation rate (37.4145) =  46.1942      |      minus Nonradiation rate (38.9905) =  44.6183
                                                      |
3.  1980 Fractional Causation is radiation            |  3.  1980 Fractional Causation is radiation
    rate (46.1942) divided by OBSERVED                |      rate (44.6183) divided by OBSERVED
    Natl MR Part 1,Col.C=   90.5    =      0.51       |      Natl MR Part 1, Col.C=   90.5    =      0.49
------------------------------------------------------------------------------------------------------
```

```
================================================================================
                                  Table 54-F
              Difference Cancers, Females:  Fractional Causation in 1988
================================================================================
Part 1.
Calculation of the 6 Adjusted MortRates (Col.F) and the National Adjusted MortRate (Col.G).
The last six entries in Part 1, Col.F, are the products of (Col.D * Col.E), as discussed in Chap. 49.
```

	Col.A	Col.B	Col.C	Col.D	Col.E	Col.F	Col.G
	1990	1988		1940 MR	AdjuFact	1988	
	PopFrac	Obs MR	A * B	Mid,Low	Bx2,Pt2	Adju	A * F
Trio-Sequence	Tab 3-B	Tab 19-A		Tab 19-A	Col.E	MortRates	
Pacific	0.1535	83.7	12.848			83.7	12.848
New England	0.0527	89.5	4.717			89.5	4.717
Mid-Atlantic	0.1527	92.8	14.171			92.8	14.171
WestNoCentral	0.0721	83.7	6.035	117.0	0.61	71.37	5.146
EastNoCentral	0.1713	90.1	15.434	128.2	0.61	78.20	13.396
Mountain	0.0543	78.2	4.246	108.9	0.61	66.43	3.607
WestSoCentral	0.1087	83.2	9.044	97.4	0.70	68.18	7.411
EastSoCentral	0.0621	86.1	5.347	100.1	0.70	70.07	4.351
SouthAtlantic	0.1725	85.0	14.663	104.5	0.70	73.15	12.618

```
                                        Sum =  86.5                    Sum =
        1988 Observed Natl MR from Table 19-B =  86.8    1988 Natl Adjusted MR =  78.2649
```

```
Part 2. ------------------------------------------------------------------------
```

	Col.A	Col.B	Col.C		Col.D	Col.E	
	Mean1940	1988	Difference Ca., Females:		1940	Difference Ca., Females:	
	thru1990	Adju MRs	1988 Adjusted MortRates		PPs from	1988 Adjusted MortRates	
Trio-	PPs from	from Col.F	regressed on		Table 3-A	regressed on	
Seq.	Tab 47-A	Part 1	Mean 1940 thru 1990 PPs		(TrioSeq)	1940 PhysPops	
	x'	y	Regression Output:		x''	Regression Output:	
Pac	191.97	83.7	Constant	37.0164	159.72	Constant	40.3826
NewEng	208.20	89.5	Std Err of Y Est	4.0551	161.55	Std Err of Y Est	4.7895
MidAtl	204.72	92.8	R Squared	0.8437	169.76	R Squared	0.7819
WNoCen	141.14	71.37	No. of Observation	9	123.14	No. of Observation	9
ENoCen	146.19	78.20	Degrees of Freedom	7	133.36	Degrees of Freedom	7
Mtn	145.91	66.43			119.89		
WSoCen	126.28	68.18	X Coefficient(s)	0.2536	103.94	X Coefficient(s)	0.2850
ESoCen	113.28	70.07	Std Err of Coef.	0.0413	85.83	Std Err of Coef.	0.0569
SoAtl	142.93	73.15	XCoef / S.E. =	6.1462	100.74	XCoef / S.E.	5.0097

```
--------------------------------------------------------------------------------
Part 3-A.                                     | Part 3-B.
Calculation of Fractional Causation           | Calculation of Fractional Causation
from Averaged PhysPops                         | from 1940 PhysPops

1.  Nonradiation rate is Adjusted             | 1.  Nonradiation rate is Adjusted
    Constant (Part 2, Col.C) =     37.0164    |     Constant (Part 2, Col.E) =     40.3826
                                              |
2.  Radiation rate is Natl Adjusted           | 2.  Radiation rate is Natl Adjusted
    MortRate (Part 1, Col.G = 78.2649)        |     MortRate (Part 1, Col.G = 78.2649)
    minus Nonradiation rate (37.0164) = 41.2485|     minus Nonradiation rate (40.3826) = 37.8823
                                              |
3.  1988 Fractional Causation is radiation    | 3.  1988 Fractional Causation is radiation
    rate (41.2485) divided by OBSERVED        |     rate (37.8823) divided by OBSERVED
    Natl MR Part 1,Col.C=  86.8    =   0.48   |     Natl MR Part 1, Col.C=  86.8    =   0.44
--------------------------------------------------------------------------------
```

CHAPTER 55

Breast-Cancers, Females, 1940-1990

- Table 55-A provides the summary on Fractional Causation by medical radiation, by decades starting with 1940. Column A shows that age–adjusted National Breast Cancer MortRates, per 100,000 women, have barely changed since 1940 (although incidence rates have risen dramatically). Breast Cancer MortRates by separate age-bands are shown, by decades from 1950 through 1990, in Chapter 4 (Box 3). For some age–groups, the trend has been up --- and for others, the trend has been down.

- Although National Breast Cancer MortRates are steady in the 1940-1988 period, they are not steady everywhere. Box 1 shows that MortRates FELL in the TopTrio of Census Divisions, while simultaneously RISING in the LowTrio of Census Divisions. The facts in Box 1 mean that a carcinogenic co–actor which can contribute to female MortRates, from Breast Cancers, is operating more strongly in the LowTrio than in the TopTrio (Chapter 48, Part 5b). We must match the Census Divisions for this co–actor, regardless of its identity. We believe its identity is cigarette smoke.

Table 55-A
Breast-Cancers, Females: Fractional Causation by Medical Radiation over Time

Year	Col.A Natl MR	Col.B Frac.C		Col.C R-Sq	Col.D X-Coef	Col.E StdErr	Col.F Coef/SE	Col.G Source
1940	23.3	~100%		0.9153	0.1906	0.0219	8.6965	Chap.8
1950	22.5	93%		0.9098	0.1572	0.0187	8.4021	Tab 55-B
1960	22.9	90%		0.9043	0.1569	0.0193	8.1309	Tab 55-C
1970	23.1	87%		0.8875	0.1467	0.0197	7.4317	Tab 55-D
1980	22.6	85%		0.8717	0.1335	0.0194	6.8964	Tab 55-E
1990	23.1	83%		0.8924	0.1187	0.0156	7.6213	Tab 55-F

```
================================================================================================
                              Box 1, Chap. 55
          Breast Cancer, Females:  Post-1940 Change in MortRates by Census Trios

1960 vs. 1940, by Trios:  Col.D expresses change by ratios.  Col.F expresses change by subtraction.
1990 vs. 1940, by Trios:  Col.I expresses change by ratios.  Col.K expresses change by subtraction.
MRs change inversely with PP.  In high-PP Trio, rates decline.  In Low-PP Trio, rates grow.

          Col.A  | Col.B     Col.C     Col.D     Col.E     Col.F  | Col.G     Col.H     Col.I     Col.J     Col.K
          1940   | 1960      Ratio     Input     Diff:     Input  | 1990      Ratio     Input     Diff:     Input
          MortRate| MortRate  Col.B     from      Col.B     from   | MortRate  Col.G     from      Col.G     from
          Tab 8-A | Tab 8-A   /Col.A    Col.C     minus A   Col.E  | Tab 8-A   /Col.A    Col.H     minus A   Col.J
                 |                                               |
Pacif     26.7   | 23.3      0.873     Avg Chg   -3.4      Avg Chg| 22.7      0.850     Avg Chg   -4.0      Avg Chg
NewE      28.8   | 25.9      0.899     TopTrio   -2.9      TopTrio| 24.3      0.844     TopTrio   -4.5      TopTrio
MidAtl    27.8   | 26.8      0.964     0.912     -1.0      -2.4   | 25.8      0.928     0.874     -2.0      -3.5
                 |                                               |
WNoCen    22.6   | 22.8      1.009     Avg Chg   0.2       Avg Chg| 22.6      1.000     Avg Chg   0.0       Avg Chg
ENoCen    24.3   | 24.3      1.000     MidTrio   0.0       MidTrio| 24.1      0.992     MidTrio   -0.2      MidTrio
Mtn       18.6   | 20.3      1.091     1.033     1.7       0.6    | 21.0      1.129     1.040     2.4       0.7
                 |                                               |
WSoCen    15.1   | 17.8      1.179     Avg Chg   2.7       Avg Chg| 20.8      1.377     Avg Chg   5.7       Avg Chg
ESoCen    15.1   | 17.6      1.166     LowTrio   2.5       LowTrio| 21.4      1.417     LowTrio   6.3       LowTrio
SoAtl     18.3   | 19.4      1.060     1.135     1.1       2.1    | 22.6      1.235     1.343     4.3       5.4
================================================================================================
```

===

Box 2, Chap. 55
Breast-Cancers, Females: Calculation of Adjustment Factor

This adjustment is discussed fully in Chapter 49.
● Part 1: Calculate average population-weighted MortRate for the combined TopTrio Census Divs.

Census Div.	Col.A 1940 MR Tab 8-A	Col.B 1940 Pop'n Tab 3-B	Col.C 1940 Popn /45,710,039	Col.D Col.A * Col.C		Census Div.	Col.A 1950 MR Tab 8-A	Col.B 1950 Pop'n Tab 3-B	Col.C 1950 Popn /53,964,513	Col.D Col.A * Col.C
Pacific	26.7	9,733,262	0.2129	5.69		Pacific	23.8	14,486,527	0.2684	6.39
NewEng	28.8	8,437,290	0.1846	5.32		NewEng	25.8	9,314,453	0.1726	4.45
Mid-Atl	27.8	27,539,487	0.6025	16.75		Mid-Atl	26.5	30,163,533	0.5590	14.81
1940		Sum TopTrio 45,710,039	Sum 1.0000	TopTrio 27.750		1950		Sum TopTrio 53,964,513	Sum 1.0000	TopTrio 25.654

Census Div.	Col.A 1960 MR Tab 8-A	Col.B 1960 Pop'n Tab 3-B	Col.C 1960 Popn /65,875,863	Col.D Col.A * Col.C		Census Div.	Col.A 1970 MR Tab 8-A	Col.B 1970 Pop'n Tab 3-B	Col.C 1970 Popn /75,017,000	Col.D Col.A * Col.C
Pacific	23.3	21,198,044	0.3218	7.50		Pacific	22.3	26,087,000	0.3477	7.75
NewEng	25.9	10,509,367	0.1595	4.13		NewEng	25.3	11,781,000	0.1570	3.97
Mid-Atl	26.8	34,168,452	0.5187	13.90		Mid-Atl	26.4	37,149,000	0.4952	13.07
1960		Sum TopTrio 65,875,863	Sum 1.0000	TopTrio 25.530		1970		Sum TopTrio 75,017,000	Sum 1.0000	TopTrio 24.801

Census Div.	Col.A 1980 MR Tab 8-A	Col.B 1980 Pop'n Tab 3-B	Col.C 1980 Popn /80,615,000	Col.D Col.A * Col.C		Census Div.	Col.A 1990 MR Tab 8-A	Col.B 1990 Pop'n Tab 3-B	Col.C 1990 Popn /88,495,000	Col.D Col.A * Col.C
Pacific	21.2	31,523,000	0.3910	8.29		Pacific	22.7	37,837,000	0.4276	9.71
NewEng	24.7	12,322,000	0.1528	3.78		NewEng	24.3	12,998,000	0.1469	3.57
Mid-Atl	25.9	36,770,000	0.4561	11.81		Mid-Atl	25.8	37,660,000	0.4256	10.98
1980		Sum TopTrio 80,615,000	Sum 1.0000	TopTrio 23.879		1990		Sum TopTrio 88,495,000	Sum 1.0000	TopTrio 24.254

● Part 2: Take ratios of these TopTrio MortRates, with 1940 as the denominator of each ratio.
  Col.D modifies Col.C by separate PhysPop adjustments for MidTrio and LowTrio Census Divisions.

	Col.A TopTrio Mean MR	Col.B 1940 TopTrio Mean MR	Col.C = Col.A / Col.B	Col.D ppAdju Tab 47-B	Col.E = Col.C * Col.D	BREAST CANCERS. Females.
				MidTrio		
1950	25.654	27.750	0.924	0.99	0.92	= MidTrio Adjustment Factor, 1950
1960	25.530	27.750	0.920	0.97	0.89	= MidTrio Adjustment Factor, 1960
1970	24.801	27.750	0.894	0.95	0.85	= MidTrio Adjustment Factor, 1970
1980	23.879	27.750	0.860	0.94	0.81	= MidTrio Adjustment Factor, 1980
1990	24.254	27.750	0.874	0.94	0.82	= MidTrio Adjustment Factor, 1990
				LowTrio		
1950	25.654	27.750	0.924	1.00	0.92	= LowTrio Adjustment Factor, 1950
1960	25.530	27.750	0.920	1.01	0.93	= LowTrio Adjustment Factor, 1960
1970	24.801	27.750	0.894	1.02	0.91	= LowTrio Adjustment Factor, 1970
1980	23.879	27.750	0.860	1.04	0.89	= LowTrio Adjustment Factor, 1980
1990	24.254	27.750	0.874	1.07	0.94	= LowTrio Adjustment Factor, 1990

===

```
===================================================================================================
                                         Table 55-B
                        Breast Cancers, Females:  Fractional Causation in 1950
===================================================================================================
```

Part 1.
Calculation of the 6 Adjusted MortRates (Col.F) and the National Adjusted MortRate (Col.G).
The last six entries in Part 1, Col.F, are the products of (Col.D * Col.E), as discussed in Chap. 49.

Trio-Sequence	Col.A 1950 PopFrac Tab 3-B	Col.B 1950 Obs MR Tab 8-A	Col.C A * B	Col.D 1940 MR Mid,Low Tab 8-A	Col.E AdjuFact Bx2,Pt2 Col.E	Col.F 1950 Adju MortRates	Col.G A * F
Pacific	0.0961	23.8	2.287			23.8	2.287
New England	0.0618	25.8	1.594			25.8	1.594
Mid-Atlantic	0.2002	26.5	5.305			26.5	5.305
WestNoCentral	0.0933	22.6	2.109	22.6	0.92	20.79	1.940
EastNoCentral	0.2017	23.5	4.740	24.3	0.92	22.36	4.509
Mountain	0.0337	18.8	0.634	18.6	0.92	17.11	0.577
WestSoCentral	0.0965	16.6	1.602	15.1	0.92	13.89	1.341
EastSoCentral	0.0762	16.6	1.265	15.1	0.92	13.89	1.059
SouthAtlantic	0.1406	18.4	2.587	18.3	0.92	16.84	2.367

```
                                        Sum =    22.1                              Sum =
          1950 Observed MR from Table 8-B        22.5        1950 Natl Adjusted MR =     20.9790
```

Part 2. ---

Trio-Seq.	Col.A Mean1940 thru1950 PPs from Tab 47-A x'	Col.B 1950 Adju MRs from Col.F Part 1 y	Col.C Breast Cancers, Females: 1950 Adjusted MortRates regressed on Mean 1940 thru 1950 PPs Regression Output:		Col.D 1940 PPs from Table 3-A (TrioSeq) x''	Col.E Breast Cancers, Females: 1950 Adjusted MortRates regressed on 1940 PhysPops Regression Output:	
Pac	154.16	23.8	Constant	0.1607	159.72	Constant	-0.0134
NewEng	162.03	25.8	Std Err of Y Est	1.5622	161.55	Std Err of Y Est	1.5088
MidAtl	169.24	26.5	R Squared	0.9098	169.76	R Squared	0.9159
WNoCen	121.60	20.79	No. of Observation	9	123.14	No. of Observation	9
ENoCen	128.53	22.36	Degrees of Freedom	7	133.36	Degrees of Freedom	7
Mtn	119.64	17.11			119.89		
WSoCen	102.64	13.89	X Coefficient(s)	0.1572	103.94	X Coefficient(s)	0.1564
ESoCen	84.44	13.89	Std Err of Coef.	0.0187	85.83	Std Err of Coef.	0.0179
SoAtl	99.91	16.84	XCoef / S.E. =	8.4021	100.74	XCoef / S.E.	8.7284

Part 3-A. | Part 3-B.
Calculation of Fractional Causation | Calculation of Fractional Causation
from Averaged PhysPops | from 1940 PhysPops

1. Nonradiation rate is Adjusted | 1. Nonradiation rate is Adjusted
    Constant (Part 2, Col.C) =   0.1607 | Constant (Part 2, Col.E) = Negative   0.0

2. Radiation rate is Natl Adjusted | 2. Radiation rate is Natl Adjusted
    MortRate (Part 1, Col.G = 20.9790) | MortRate (Part 1, Col.G = 20.9790)
    minus Nonradiation rate (0.1607) = 20.8182 | minus Nonradiation rate (0.0) =   20.9790

3. 1950 Fractional Causation is radiation | 3. 1950 Fractional Causation is radiation
    rate (20.8182) divided by OBSERVED | rate (20.9790) divided by OBSERVED
    Natl MR Part 1,Col.C=   22.5   =   0.93 | Natl MR Part 1,Col.C=   22.5   =   0.93

```
=================================================================================================
                                        Table 55-C
                        Breast Cancers, Females:  Fractional Causation in 1960
=================================================================================================
Part 1.
Calculation of the 6 Adjusted MortRates (Col.F) and the National Adjusted MortRate (Col.G).
The last six entries in Part 1, Col.F, are the products of (Col.D * Col.E), as discussed in Chap. 49.
```

	Col.A	Col.B	Col.C	Col.D	Col.E	Col.F	Col.G
	1960	1960		1940 MR	AdjuFact	1960	
	PopFrac	Obs MR	A * B	Mid,Low	Bx2,Pt2	Adju	A * F
Trio-Sequence	Tab 3-B	Tab 8-A		Tab 8-A	Col.E	MortRates	
Pacific	0.1182	23.3	2.754			23.3	2.754
New England	0.0586	25.9	1.518			25.9	1.518
Mid-Atlantic	0.1905	26.8	5.105			26.8	5.105
WestNoCentral	0.0858	22.8	1.956	22.6	0.89	20.11	1.726
EastNoCentral	0.2020	24.3	4.909	24.3	0.89	21.63	4.369
Mountain	0.0382	20.3	0.775	18.6	0.89	16.55	0.632
WestSoCentral	0.0945	17.8	1.682	15.1	0.93	14.04	1.327
EastSoCentral	0.0672	17.6	1.183	15.1	0.93	14.04	0.944
SouthAtlantic	0.1448	19.4	2.809	18.3	0.93	17.02	2.464

```
                                    Sum =   22.7                          Sum =
            1960 Observed MR from Table 8-B  22.9     1960 Natl Adjusted MR =   20.8391
```

```
Part 2.  ------------------------------------------------------------------------------------------
         Col.A   Col.B                      Col.C               Col.D              Col.E
         Mean1940  1960        Breast Cancers, Females:          1940        Breast Cancers, Females:
         thru1960 Adju MRs     1960 Adjusted MortRates         PPs from      1960 Adjusted MortRates
Trio-    PPs from from Col.F        regressed on               Table 3-A          regressed on
Seq.     Tab 47-A  Part 1      Mean 1940 thru 1960 PPs         (TrioSeq)        1940 PhysPops
           x'       y               Regression Output:           x''          Regression Output:
```

	x'	y	(regression)		x''	(regression)	
Pac	155.69	23.3	Constant	0.1508	159.72	Constant	-0.0073
NewEng	162.81	25.9	Std Err of Y Est	1.5963	161.55	Std Err of Y Est	1.5108
MidAtl	167.04	26.8	R Squared	0.9043	169.76	R Squared	0.9142
WNoCen	118.15	20.11	No. of Observation	9	123.14	No. of Observation	9
ENoCen	123.87	21.63	Degrees of Freedom	7	133.36	Degrees of Freedom	7
Mtn	117.40	16.55			119.89		
WSoCen	102.31	14.04	X Coefficient(s)	0.1569	103.94	X Coefficient(s)	0.1550
ESoCen	85.63	14.04	Std Err of Coef.	0.0193	85.83	Std Err of Coef.	0.0179
SoAtl	101.72	17.02	XCoef / S.E. =	8.1309	100.74	XCoef / S.E.	8.6381

```
----------------------------------------------------------------------------------------------------
Part 3-A.                                       |  Part 3-B.
Calculation of Fractional Causation             |  Calculation of Fractional Causation
from Averaged PhysPops                          |  from 1940 PhysPops
                                                |
1.  Nonradiation rate is Adjusted               |  1.  Nonradiation rate is Adjusted
    Constant (Part 2, Col.C) =        0.1508    |      Constant (Part 2, Col.E) = Negative    0.0
                                                |
2.  Radiation rate is Natl Adjusted             |  2.  Radiation rate is Natl Adjusted
    MortRate (Part 1, Col.G = 20.8391)          |      MortRate (Part 1, Col.G = 20.8391)
    minus Nonradiation rate (0.1508) =  20.6883 |      minus Nonradiation rate (0.0) =   20.8391
                                                |
3.  1960 Fractional Causation is radiation      |  3.  1960 Fractional Causation is radiation
    rate (20.6883) divided by OBSERVED          |      rate (20.8391) divided by OBSERVED
    Natl MR Part 1,Col.C=  22.9   =     0.90    |      Natl MR Part 1,Col.C=  22.9   =     0.91
----------------------------------------------------------------------------------------------------
```

==
Table 55-E
Breast Cancers, Females: Fractional Causation in 1980
==
Part 1.
Calculation of the 6 Adjusted MortRates (Col.F) and the National Adjusted MortRate (Col.G).
The last six entries in Part 1, Col.F, are the products of (Col.D * Col.E), as discussed in Chap. 49.

Trio-Sequence	Col.A 1980 PopFrac Tab 3-B	Col.B 1980 Obs MR Tab 8-A	Col.C A * B	Col.D 1940 MR Mid,Low Tab 8-A	Col.E AdjuFact Bx2,Pt2 Col.E	Col.F 1980 Adju MortRates	Col.G A * F
Pacific	0.1398	21.2	2.964			21.2	2.964
New England	0.0546	24.7	1.349			24.7	1.349
Mid-Atlantic	0.1630	25.9	4.222			25.9	4.222
WestNoCentral	0.0759	21.7	1.647	22.6	0.81	18.31	1.389
EastNoCentral	0.1846	24.0	4.430	24.3	0.81	19.68	3.633
Mountain	0.0502	20.3	1.019	18.6	0.81	15.07	0.756
WestSoCentral	0.1049	18.9	1.983	15.1	0.89	13.44	1.410
EastSoCentral	0.0646	19.6	1.266	15.1	0.89	13.44	0.868
SouthAtlantic	0.1624	21.0	3.410	18.3	0.89	16.29	2.645

| | | | Sum = 22.3 | | | | Sum = |
| 1980 Observed MR from Table 8-B | | | 22.6 | 1980 Natl Adjusted MR = | | | 19.2362 |

Part 2. --

Trio-Seq.	Col.A Mean1940 thru1980 PPs from Tab 47-A x'	Col.B 1980 Adju MRs from Col.F Part 1 y	Col.C Breast Cancers, Females: 1980 Adjusted MortRates regressed on Mean 1940 thru 1980 PPs Regression Output:		Col.D 1940 PPs from Table 3-A (TrioSeq) x''	Col.E Breast Cancers, Females: 1980 Adjusted MortRates regressed on 1940 PhysPops Regression Output:	
Pac	177.35	21.2	Constant	-0.3834	159.72	Constant	-0.0638
NewEng	185.86	24.7	Std Err of Y Est	1.7630	161.55	Std Err of Y Est	1.6597
MidAtl	186.11	25.9	R Squared	0.8717	169.76	R Squared	0.8863
WNoCen	128.82	18.31	No. of Observation	9	123.14	No. of Observation	9
ENoCen	133.71	19.68	Degrees of Freedom	7	133.36	Degrees of Freedom	7
Mtn	133.45	15.07			119.89		
WSoCen	114.66	13.44	X Coefficient(s)	0.1335	103.94	X Coefficient(s)	0.1456
ESoCen	99.46	13.44	Std Err of Coef.	0.0194	85.83	Std Err of Coef.	0.0197
SoAtl	124.62	16.29	XCoef / S.E. =	6.8964	100.74	XCoef / S.E.	7.3870

Part 3-A. | Part 3-B.
Calculation of Fractional Causation | Calculation of Fractional Causation
from Averaged PhysPops | from 1940 PhysPops
                                                   |
1. Nonradiation rate is Adjusted | 1. Nonradiation rate is Adjusted
    Constant (Part 2, Col.C) = Negative   0.0 | Constant (Part 2, Col.E) = Negative   0.0
                                                   |
2. Radiation rate is Natl Adjusted | 2. Radiation rate is Natl Adjusted
    MortRate (Part 1, Col.G = 19.2362) | MortRate (Part 1, Col.G = 19.2362)
    minus Nonradiation rate (0.0) =   19.2362 | minus Nonradiation rate (0.0) =   19.2362
                                                   |
3. 1980 Fractional Causation is radiation | 3. 1980 Fractional Causation is radiation
    rate (19.2362) divided by OBSERVED | rate (19.2362) divided by OBSERVED
    Natl MR Part 1,Col.C=   22.6   =   0.85 | Natl MR Part 1, Col.C=   22.6   =   0.85

===
Table 55-F
Breast Cancers, Females: Fractional Causation in 1990
===
Part 1.
Calculation of the 6 Adjusted MortRates (Col.F) and the National Adjusted MortRate (Col.G).
The last six entries in Part 1, Col.F, are the products of (Col.D * Col.E), as discussed in Chap. 49.

Trio-Sequence	Col.A 1990 PopFrac Tab 3-B	Col.B 1990 Obs MR Tab 8-A	Col.C A * B	Col.D 1940 MR Mid,Low Tab 8-A	Col.E AdjuFact Bx2,Pt2 Col.E	Col.F 1990 Adju MortRates	Col.G A * F
Pacific	0.1535	22.7	3.484			22.7	3.484
New England	0.0527	24.3	1.281			24.3	1.281
Mid-Atlantic	0.1527	25.8	3.940			25.8	3.940
WestNoCentral	0.0721	22.6	1.629	22.6	0.82	18.53	1.336
EastNoCentral	0.1713	24.1	4.128	24.3	0.82	19.93	3.413
Mountain	0.0543	21.0	1.140	18.6	0.82	15.25	0.828
WestSoCentral	0.1087	20.8	2.261	15.1	0.94	14.19	1.543
EastSoCentral	0.0621	21.4	1.329	15.1	0.94	14.19	0.881
SouthAtlantic	0.1725	22.6	3.899	18.3	0.94	17.20	2.967

```
                                      Sum =    23.1                          Sum =
      1990 Observed MR from Table 8-B          23.1    1990 Natl Adjusted MR =    19.6741
```

Part 2. --

Trio-Seq.	Col.A Mean1940 thru1990 PPs from Tab 47-A x'	Col.B 1990 Adju MRs from Col.F Part 1 y	Col.C Breast Cancers, Females: 1990 Adjusted MortRates regressed on Mean 1940 thru 1990 PPs Regression Output:		Col.D 1940 PPs from Table 3-A (TrioSeq) x''	Col.E Breast Cancers, Females: 1990 Adjusted MortRates regressed on 1940 PhysPops Regression Output:	
Pac	191.97	22.7	Constant	0.3898	159.72	Constant	1.2779
NewEng	208.20	24.3	Std Err of Y Est	1.5304	161.55	Std Err of Y Est	1.5142
MidAtl	204.72	25.8	R Squared	0.8924	169.76	R Squared	0.8947
WNoCen	141.14	18.53	No. of Observation	9	123.14	No. of Observation	9
ENoCen	146.19	19.93	Degrees of Freedom	7	133.36	Degrees of Freedom	7
Mtn	145.91	15.25			119.89		
WSoCen	126.28	14.19	X Coefficient(s)	0.1187	103.94	X Coefficient(s)	0.1387
ESoCen	113.28	14.19	Std Err of Coef.	0.0156	85.83	Std Err of Coef.	0.0180
SoAtl	142.93	17.20	XCoef / S.E. =	7.6213	100.74	XCoef / S.E.	7.7128

Part 3-A. | Part 3-B.
Calculation of Fractional Causation | Calculation of Fractional Causation
from Averaged PhysPops | from 1940 PhysPops
                                              |
1. Nonradiation rate is Adjusted | 1. Nonradiation rate is Adjusted
    Constant (Part 2, Col.C) =   0.3898 | Constant (Part 2, Col.E) =   1.2779
                                              |
2. Radiation rate is Natl Adjusted | 2. Radiation rate is Natl Adjusted
    MortRate (Part 1, Col.G = 19.6741) | MortRate (Part 1, Col.G = 19.6741)
    minus Nonradiation rate (0.3898) = 19.2843| minus Nonradiation rate (1.2779) = 18.3962
                                              |
3. 1990 Fractional Causation is radiation | 3. 1990 Fractional Causation is radiation
    rate (19.2843) divided by OBSERVED | rate (18.3962) divided by OBSERVED
    Natl MR Part 1,Col.C=   23.1   =   0.83 | Natl MR Part 1, Col.C=   23.1   =   0.80

CHAPTER 56

All-Cancers-Except-Genital, Females, 1940-1980

- All-Cancers-Except-Genital is a category which we examined in Chapter 20, because Chapter 14 revealed no detectable dose-response between PhysPop and the female MortRates in 1940 from Genital Cancers.

- In Table 56-A, below, Column A shows that the National female MortRate (All-Cancers-Except-Genital) barely changed from 1940 to 1980. However, rates were not steady everywhere. Box 1 shows that they FELL in the TopTrio while simultaneously RISING in the LowTrio. The facts in Box 1 mean that a carcinogenic co-actor which can contribute to female MortRates, from All-Cancers-Except-Genital, is operating more strongly in the LowTrio than in the TopTrio (Chapter 48, Part 5b). We must match the Census Divisions for this co-actor, regardless of its identity. We believe that its identity is smoking.

- Tables 56-B through 56-E have an extra calculation at the bottom, which yields Fractional Causation by medical radiation of ALL-Cancers, female. These results are discussed on a special page after Table 56-E, where they are compared with the results obtained in Chapter 50. The two sets of estimates are in very satisfactory agreement.

	Year	Col.A Natl MR	Col.B Frac.C		Col.C R-Sq	Col.D X-Coef	Col.E StdErr	Col.F Coef/SE	Col.G Source
	1940	94.0	75%		0.8675	0.5087	0.0751	6.7698	Chap.20
	1950	96.0	69%		0.8946	0.5012	0.0650	7.7074	Tab 56-B
	1960	92.5	68%		0.9084	0.4794	0.0575	8.3301	Tab 56-C
	1970	93.7	67%		0.9201	0.4658	0.0519	8.9763	Tab 56-D
	1980	94.8	66%		0.9290	0.4292	0.0448	9.5711	Tab 56-E
	1988								None

Table 56-A
All-Cancers-Except-Genital, Females: Fractional Causation by Medical Radiation over Time

```
=====================================================================================================
                                    Box 1, Chap. 56
            All-Cancer-Except-Genital, Females: Post-1940 Change in MortRates by Census Trios

1960 vs. 1940, by Trios:  Col.D expresses change by ratios.  Col.F expresses change by subtraction.
1980 vs. 1940, by Trios:  Col.I expresses change by ratios.  Col.K expresses change by subtraction.
MRs change inversely with PP.  High-PP Trio has lowest growth-factor.  Low-PP Trio has highest growth-factor.

          Col.A  |  Col.B    Col.C    Col.D    Col.E    Col.F  |  Col.G    Col.H    Col.I    Col.J    Col.K
          1940   |  1960     Ratio    Input    Diff:    Input  |  1980     Ratio    Input    Diff:    Input
        MortRate | MortRate  Col.B    from     Col.B    from   | MortRate  Col.G    from     Col.G    from
        Tab 20-A | Tab 20-A  /Col.A   Col.C    minus A  Col.E  | Tab 20-A  /Col.A   Col.H    minus A  Col.J
               |
Pacif   94.3   |   90.1     0.955   Avg Chg    -4.2   Avg Chg  |  97.1     1.030   Avg Chg     2.8   Avg Chg
NewE   112.5   |  100.7     0.895   TopTrio   -11.8   TopTrio  | 103.0     0.916   TopTrio    -9.5   TopTrio
MidAtl 110.2   |  105.2     0.955    0.935     -5.0    -7.0    | 103.2     0.936    0.961     -7.0    -4.6
               |                                              |
WNoCen  91.7   |   89.0     0.971   Avg Chg    -2.7   Avg Chg  |  87.7     0.956   Avg Chg    -4.0   Avg Chg
ENoCen  98.2   |   95.6     0.974   MidTrio    -2.6   MidTrio  |  97.5     0.993   MidTrio    -0.7   MidTrio
Mtn     84.0   |   82.6     0.983    0.976     -1.4    -2.2    |  83.2     0.990    0.980     -0.8    -1.8
               |                                              |
WSoCen  69.8   |   80.8     1.158   Avg Chg    11.0   Avg Chg  |  87.6     1.255   Avg Chg    17.8   Avg Chg
ESoCen  69.3   |   80.1     1.156   LowTrio    10.8   LowTrio  |  88.9     1.283   LowTrio    19.6   LowTrio
SoAtl   74.4   |   83.4     1.121    1.145      9.0    10.3    |  91.5     1.230    1.256     17.1    18.2
=====================================================================================================
```

===
Box 2, Chap. 56
All Cancers Except Genital, Females: Calculation of Adjustment Factor

This adjustment is discussed fully in Chapter 49.
● Part 1: Calculate average population-weighted MortRate for the combined TopTrio Census Divs.

Census Div.	Col.A 1940 MR Tab 20-A	Col.B 1940 Pop'n Tab 3-B	Col.C 1940 Popn /45,710,039	Col.D Col.A * Col.C		Census Div.	Col.A 1950 MR Tab 20-A	Col.B 1950 Pop'n Tab 3-B	Col.C 1950 Popn /53,964,513	Col.D Col.A * Col.C
Pacific	94.3	9,733,262	0.2129	20.08		Pacific	92.2	14,486,527	0.2684	24.75
NewEng	112.5	8,437,290	0.1846	20.77		NewEng	107.0	9,314,453	0.1726	18.47
Mid-Atl	110.2	27,539,487	0.6025	66.39		Mid-Atl	109.8	30,163,533	0.5590	61.37
1940		Sum TopTrio 45,710,039	Sum 1.0000	TopTrio 107.239		1950		Sum TopTrio 53,964,513	Sum 1.0000	TopTrio 104.592

Census Div.	Col.A 1960 MR Tab 20-A	Col.B 1960 Pop'n Tab 3-B	Col.C 1960 Popn /65,875,863	Col.D Col.A * Col.C		Census Div.	Col.A 1970 MR Tab 20-A	Col.B 1970 Pop'n Tab 3-B	Col.C 1970 Popn /75,017,000	Col.D Col.A * Col.C
Pacific	90.1	21,198,044	0.3218	28.99		Pacific	93.6	26,087,000	0.3477	32.55
NewEng	100.7	10,509,367	0.1595	16.06		NewEng	101.9	11,781,000	0.1570	16.00
Mid-Atl	105.2	34,168,452	0.5187	54.57		Mid-Atl	104.2	37,149,000	0.4952	51.60
1960		Sum TopTrio 65,875,863	Sum 1.0000	TopTrio 99.623		1970		Sum TopTrio 75,017,000	Sum 1.0000	TopTrio 100.153

Census Div.	Col.A 1980 MR Tab 20-A	Col.B 1980 Pop'n Tab 3-B	Col.C 1980 Popn /80,615,000	Col.D Col.A * Col.C		Census Div.	Col.A 1988 MR Tab 20-A	Col.B 1990 Pop'n Tab 3-B	Col.C 1990 Popn /88,495,000	Col.D Col.A * Col.C
Pacific	97.1	31,523,000	0.3910	37.97		Pacific	--	37,837,000	0.4276	0.00
NewEng	103.0	12,322,000	0.1528	15.74		NewEng	--	12,998,000	0.1469	0.00
Mid-Atl	103.2	36,770,000	0.4561	47.07		Mid-Atl	--	37,660,000	0.4256	0.00
1980		Sum TopTrio 80,615,000	Sum 1.0000	TopTrio 100.784		1988		Sum TopTrio 88,495,000	Sum 1.0000	TopTrio 0.000

● Part 2: Take ratios of these TopTrio MortRates, with 1940 as the denominator of each ratio.
   Col.D modifies Col.C by separate PhysPop adjustments for MidTrio and LowTrio Census Divisions.

	Col.A TopTrio Mean MR	Col.B 1940 TopTrio Mean MR	Col.C = Col.A / Col.B	Col.D ppAdju Tab 47-B	Col.E = Col.C * Col.D	ALL CANCERS, EXCEPT GENITAL. Females.
				MidTrio		
1950	104.592	107.239	0.975	0.99	0.97	= MidTrio Adjustment Factor, 1950
1960	99.623	107.239	0.929	0.97	0.90	= MidTrio Adjustment Factor, 1960
1970	100.153	107.239	0.934	0.95	0.89	= MidTrio Adjustment Factor, 1970
1980	100.784	107.239	0.940	0.94	0.88	= MidTrio Adjustment Factor, 1980
1988	--	107.239	--	0.94	--	= MidTrio Adjustment Factor, 1988
				LowTrio		
1950	104.592	107.239	0.975	1.00	0.98	= LowTrio Adjustment Factor, 1950
1960	99.623	107.239	0.929	1.01	0.94	= LowTrio Adjustment Factor, 1960
1970	100.153	107.239	0.934	1.02	0.95	= LowTrio Adjustment Factor, 1970
1980	100.784	107.239	0.940	1.04	0.98	= LowTrio Adjustment Factor, 1980
1988	--	107.239	--	1.07	--	= LowTrio Adjustment Factor, 1988

===

```
=====================================================================================================
                                        Table 56-B
                 All-Cancers-Except-Genital, Females:  Fractional Causation in 1950
=====================================================================================================
```

Part 1.
Calculation of the 6 Adjusted MortRates (Col.F) and the National Adjusted MortRate (Col.G).
The last six entries in Part 1, Col.F, are the products of (Col.D * Col.E), as discussed in Chap. 49.

	Col.A 1950 PopFrac Tab 3-B	Col.B 1950 Obs MR Tab 20-A	Col.C A * B		Col.D 1940 MR Mid,Low Tab 20-A	Col.E AdjuFact Bx2,Pt2 Col.E	Col.F 1950 Adju MortRates	Col.G A * F
Trio-Sequence								
Pacific	0.0961	92.2	8.860				92.2	8.860
New England	0.0618	107.0	6.613				107.0	6.613
Mid-Atlantic	0.2002	109.8	21.982				109.8	21.982
WestNoCentral	0.0933	93.7	8.742		91.7	0.97	88.95	8.299
EastNoCentral	0.2017	99.2	20.009		98.2	0.97	95.25	19.213
Mountain	0.0337	82.4	2.777		84.0	0.97	81.48	2.746
WestSoCentral	0.0965	81.9	7.903		69.8	0.98	68.40	6.601
EastSoCentral	0.0762	80.6	6.142		69.3	0.98	67.91	5.175
SouthAtlantic	0.1406	83.4	11.726		74.4	0.98	72.91	10.251

```
                                  Sum =   94.8                                      Sum =
        1950 Observed MR from Table 20-A   96.0            1950 Natl Adjusted MR = 89.7400
```

Part 2. --

	Col.A Mean1940 thru1950 PPs from Tab 47-A	Col.B 1950 Adju MRs from Col.F Part 1	Col.C (All Minus Genital) Ca. Females: 1950 Adjusted MortRates regressed on Mean 1940 thru 1950 PPs Regression Output:		Col.D 1940 PPs from Table 3-A (TrioSeq)	Col.E (All Minus Gen) Ca. Females 1950 Adjusted Mort-Rates regressed on 1940 PhysPops Regression Output:
Trio-Seq.	x'	y			x''	
Pac	154.16	92.2	Constant	23.4999	159.72	Constant	23.5299
NewEng	162.03	107.0	Std Err of Y Est	5.4298	161.55	Std Err of Y Est	5.6911
MidAtl	169.24	109.8	R Squared	0.8946	169.76	R Squared	0.8842
WNoCen	121.60	88.95	No. of Observation	9	123.14	No. of Observation	9
ENoCen	128.53	95.25	Degrees of Freedom	7	133.36	Degrees of Freedom	7
Mtn	119.64	81.48			119.89		
WSoCen	102.64	68.40	X Coefficient(s)	0.5012	103.94	X Coefficient(s)	0.4941
ESoCen	84.44	67.91	Std Err of Coef.	0.0650	85.83	Std Err of Coef.	0.0676
SoAtl	99.91	72.91	XCoef / S.E. =	7.7074	100.74	XCoef / S.E.	7.3107

Part 3-A. Calculation of Fractional Causation from Averaged PhysPops	Part 3-B. Calculation of Fractional Causation from 1940 PhysPops
1. Nonradiation rate is Adjusted Constant (Part 2, Col.C) = 23.4999	1. Nonradiation rate is Adjusted Constant (Part 2, Col.E) = 23.5299
2. Radiation rate is Natl Adjusted MortRate (Part 1, Col.G = 89.7400) minus Nonradiation rate (23.4999) = 66.2401	2. Radiation rate is Natl Adjusted MortRate (Part 1, Col.G = 89.7400) minus Nonradiation rate (23.5299) = 66.2101
3. 1950 Fractional Causation is radiation rate (66.2401) divided by OBSERVED Natl MR Part 1,Col.C= 96.0 = 0.69	3. 1950 Fractional Causation is radiation rate (66.2101) divided by OBSERVED Natl MR Part 1, Col.C= 96.0 = 0.69

----

4. Calculation for Table 56-AA at the end of this chapter:
   (Radiation rate of 66.2401 / Obs. Natl. female All-Cancer MortRate of 123.2) = 0.54 .

```
===============================================================================================
                                       Table 56-C
                All-Cancers-Except-Genital, Females:  Fractional Causation in 1960
===============================================================================================
Part 1.
Calculation of the 6 Adjusted MortRates (Col.F) and the National Adjusted MortRate (Col.G).
The last six entries in Part 1, Col.F, are the products of (Col.D * Col.E), as discussed in Chap. 49.
```

	Col.A 1960 PopFrac Tab 3-B	Col.B 1960 Obs MR Tab 20-A	Col.C A * B	Col.D 1940 MR Mid,Low Tab 20-A	Col.E AdjuFact Bx2,Pt2 Col.E	Col.F 1960 Adju MortRates	Col.G A * F
Trio-Sequence							
Pacific	0.1182	90.1	10.650			90.1	10.650
New England	0.0586	100.7	5.901			100.7	5.901
Mid-Atlantic	0.1905	105.2	20.041			105.2	20.041
WestNoCentral	0.0858	89.0	7.636	91.7	0.90	82.53	7.081
EastNoCentral	0.2020	95.6	19.311	98.2	0.90	88.38	17.853
Mountain	0.0382	82.6	3.155	84.0	0.90	75.60	2.888
WestSoCentral	0.0945	80.8	7.636	69.8	0.94	65.61	6.200
EastSoCentral	0.0672	80.1	5.383	69.3	0.94	65.14	4.378
SouthAtlantic	0.1448	83.4	12.076	74.4	0.94	69.94	10.127

```
                                Sum =    91.8                                    Sum =
          1960 Observed MR from Table 20-A    92.5          1960 Natl Adjusted MR = 85.1178
```

```
Part 2.  ------------------------------------------------------------------------------------
```

Trio-Seq.	Col.A Mean1940 thru1960 PPs from Tab 47-A x'	Col.B 1960 Adju MRs from Col.F Part 1 y	Col.C (All Minus Genital) Ca. Females: 1960 Adjusted MortRates regressed on Mean 1940 thru 1960 PPs Regression Output:		Col.D 1940 PPs from Table 3-A (TrioSeq) x''	Col.E (All Minus Gen) Ca. Females 1960 Adjusted Mort-Rates regressed on 1940 PhysPops Regression Output:	
Pac	155.69	90.1	Constant	22.1426	159.72	Constant	21.4743
NewEng	162.81	100.7	Std Err of Y Est	4.7599	161.55	Std Err of Y Est	4.3352
MidAtl	167.04	105.2	R Squared	0.9084	169.76	R Squared	0.9240
WNoCen	118.15	82.53	No. of Observation	9	123.14	No. of Observation	9
ENoCen	123.87	88.38	Degrees of Freedom	7	133.36	Degrees of Freedom	7
Mtn	117.40	75.60			119.89		
WSoCen	102.31	65.61	X Coefficient(s)	0.4794	103.94	X Coefficient(s)	0.4749
ESoCen	85.63	65.14	Std Err of Coef.	0.0575	85.83	Std Err of Coef.	0.0515
SoAtl	101.72	69.94	XCoef / S.E. =	8.3301	100.74	XCoef / S.E. =	9.2246

```
-----------------------------------------------------------------------------------------------
Part 3-A.                                        |  Part 3-B.
Calculation of Fractional Causation              |  Calculation of Fractional Causation
from Averaged PhysPops                           |  from 1940 PhysPops
                                                 |
1.  Nonradiation rate is Adjusted                |  1.  Nonradiation rate is Adjusted
    Constant (Part 2, Col.C) =       22.1426     |      Constant (Part 2, Col.E) =       21.4743
                                                 |
2.  Radiation rate is Natl Adjusted              |  2.  Radiation rate is Natl Adjusted
    MortRate (Part 1, Col.G = 85.1178)           |      MortRate (Part 1, Col.G = 85.1178)
    minus Nonradiation rate (22.1426) = 62.9752  |      minus Nonradiation rate (21.4743) = 63.6435
                                                 |
3.  1960 Fractional Causation is radiation       |  3.  1960 Fractional Causation is radiation
    rate (62.9752) divided by OBSERVED           |      rate (63.6435) divided by OBSERVED
    Natl MR Part 1,Col.C=   92.5    =    0.68     |      Natl MR Part 1,Col.C=   92.5    =    0.69
-----------------------------------------------------------------------------------------------
4.  Calculation for Table 56-AA at the end of this chapter:
    (Radiation rate of 62.9752 / Obs. Natl. female All-Cancer MortRate of 114.9) = 0.55 .
```

```
===============================================================================================
                                      Table 56-E
              All-Cancers-Except-Genital, Females:  Fractional Causation in 1980
===============================================================================================
```

Part 1.
Calculation of the 6 Adjusted MortRates (Col.F) and the National Adjusted MortRate (Col.G).
The last six entries in Part 1, Col.F, are the products of (Col.D * Col.E), as discussed in Chap. 49.

Trio-Sequence	Col.A 1980 PopFrac Tab 3-B	Col.B 1980 Obs MR Tab 20-A	Col.C A * B	Col.D 1940 MR Mid,Low Tab 20-A	Col.E AdjuFact Bx2,Pt2 Col.E	Col.F 1980 Adju MortRates	Col.G A * F
Pacific	0.1398	97.1	13.575			97.1	13.575
New England	0.0546	103.0	5.624			103.0	5.624
Mid-Atlantic	0.1630	103.2	16.822			103.2	16.822
WestNoCentral	0.0759	87.7	6.656	91.7	0.88	80.70	6.125
EastNoCentral	0.1846	97.5	17.999	98.2	0.88	86.42	15.952
Mountain	0.0502	83.2	4.177	84.0	0.88	73.92	3.711
WestSoCentral	0.1049	87.6	9.189	69.8	0.98	68.40	7.176
EastSoCentral	0.0646	88.9	5.743	69.3	0.98	67.91	4.387
SouthAtlantic	0.1624	91.5	14.860	74.4	0.98	72.91	11.841

```
                                    Sum =    94.6                          Sum =
          1980 Observed MR from Table 20-A   94.8       1980 Natl Adjusted MR = 85.2117
```

Part 2. --

Trio-Seq.	Col.A Mean1940 thru1980 PPs from Tab 47-A x'	Col.B 1980 Adju MRs from Col.F y	Col.C (All Minus Genital) Ca. Females: 1980 Adjusted MortRates regressed on Mean 1940 thru 1980 PPs Regression Output:		Col.D 1940 PPs from Table 3-A (TrioSeq) x''	Col.E (All Minus Gen) Ca. Females 1980 Adjusted Mort-Rates regressed on 1940 PhysPops Regression Output:	
Pac	177.35	97.1	Constant	22.4922	159.72	Constant	23.4300
NewEng	185.86	103.0	Std Err of Y Est	4.0831	161.55	Std Err of Y Est	3.5151
MidAtl	186.11	103.2	R Squared	0.9290	169.76	R Squared	0.9474
WNoCen	128.82	80.70	No. of Observation	9	123.14	No. of Observation	9
ENoCen	133.71	86.42	Degrees of Freedom	7	133.36	Degrees of Freedom	7
Mtn	133.45	73.92			119.89		
WSoCen	114.66	68.40	X Coefficient(s)	0.4292	103.94	X Coefficient(s)	0.4687
ESoCen	99.46	67.91	Std Err of Coef.	0.0448	85.83	Std Err of Coef.	0.0417
SoAtl	124.62	72.91	XCoef / S.E. =	9.5711	100.74	XCoef / S.E.	11.2269

Part 3-A.
Calculation of Fractional Causation
from Averaged PhysPops

1. Nonradiation rate is Adjusted
    Constant (Part 2, Col.C) =   22.4922

2. Radiation rate is Natl Adjusted
    MortRate (Part 1, Col.G = 85.2117)
    minus Nonradiation rate (22.4922) =   62.7195

3. 1980 Fractional Causation is radiation
    rate (62.7195) divided by OBSERVED
    Natl MR Part 1,Col.C=   94.8   =   0.66

Part 3-B.
Calculation of Fractional Causation
from 1940 PhysPops

1. Nonradiation rate is Adjusted
    Constant (Part 2, Col.E) =   23.4300

2. Radiation rate is Natl Adjusted
    MortRate (Part 1, Col.G = 85.2117)
    minus Nonradiation rate (23.4300) =   61.7817

3. 1980 Fractional Causation is radiation
    rate (61.7817) divided by OBSERVED
    Natl MR Part 1,Col.C=   94.8   =   0.65

----

4. Calculation for Table 56-AA at the end of this chapter:
    (Radiation rate of 62.7195 / Obs. Natl. female All-Cancer MortRate of 108.5) = 0.58 .

Female All-Cancers: Another "Path" to Fractional Causation

Female Genital Cancers are the single cancer-group for which there is no apparent dose-response with PhysPop (Chapters 14 and 62). This finding may be "taken at face value" (accepted without challenge), or it might be misleading --- for example, if the Census Divisions are badly matched for an important carcinogenic co-actor which is specific for female Genital Cancers.

If we take the lack of dose-response at face value, we can divide the 1950 radiation rate of 66.24 for All-Except-Genital (Table 56-B, Part 3-A) by the 1950 National MortRate for All-Cancers (123.2) instead of by the National MortRate for All-Except-Genital (96.0). We can make this substitution because we assume that no female Genital Cancers are induced by medical radiation. Thus, when Table 56-B uncovers a 1950 rate of 66.24 radiation-induced cancer-deaths/100K, we "know" that inclusion of female Genital Cancers would add nothing to that rate of radiation-induced cancers.

In such a case, 66.24/100K should still be the radiation rate when we study female All-Cancers. And indeed, in Table 50-B (Part 3-A), the radiation rate for female All-Cancers is extremely close: 65.3/100K. In Tables 50-C and 56-C, in Tables 50-D and 56-D (not shown), and in Tables 50-E and 56-E, radiation rates also are in very reasonable agreement with each other.

In Table 56-AA, below, we collect the estimates of Fractional Causation by medical radiation of female All-Cancers, located as (4) at the very bottom of Tables 56-B through 56-E. Table 56-AA also lists the corresponding estimates from Chapter 50, for easy comparison. The two sets of estimates are quite similar.

	Table 56-AA	
All-Cancers, Females: Fractional Causation by Medical Radiation over Time		
Year	Col.A Estimates via the "All-Except-Genital" Path	Col.B Estimates from Chap.50, Table 50-A
1940	56%	58%
1950	54%	53%
1960	55%	54%
1970	57%	52%
1980	58%	52%
1988	--	50%

CHAPTER 57

Digestive–System Cancers, Males, 1940–1988

- In Table 57-A, Column A shows that the National MortRate from Digestive–System Cancers in males fell appreciably in the 1940-1988 period. Box 1 shows that such MortRates fell much more in the TopTrio than in the LowTrio. These facts mean that a carcinogenic co–actor which can contribute to male MortRates, from Digestive Cancers, is operating more strongly in the LowTrio than in the TopTrio (Chapter 48, Part 5b). We must match the Census Divisions for this co–actor, whatever its identity. We believe that its identity is smoking.

- Of course, matching the Census Divisions for smoking does not "abolish" the steady decline in male MortRates from Digestive–System Cancers in Tables 57-B through 57-F. (Nor did the smoking adjustment "abolish" the simultaneous INCREASE in male MortRates from Respiratory–System Cancers, in Tables 51-B through 51-FF). The steady net decline in Digestive–System Cancers, for both males and females, almost certainly reflects gradual reduction of one or more carcinogenic co–actors for Digestive–System Cancers --- possibly nitrates in the diet, and/or Helicobacter pylori infections in the stomach (Munoz 1994 + WHO 1996, p.59).

Table 57-A
Digestive–System Cancers, Males: Fractional Causation by Medical Radiation over Time

Year	Col.A Natl MR	Col.B Frac.C		Col.C R-Sq	Col.D X-Coef	Col.E StdErr	Col.F Coef/SE	Col.G Source
1940	60.4	97%		0.9078	0.4262	0.0513	8.3009	Chap.9
1950	55.4	93%		0.8591	0.3874	0.0593	6.5325	Tab 57-B
1960	49.7	92%		0.8673	0.3488	0.0516	6.7647	Tab 57-C
1970	45.7	89%		0.8709	0.2986	0.0434	6.8722	Tab 57-D
1980	41.7	86%		0.8718	0.2454	0.0356	6.9002	Tab 57-E
1988	38.8	82%		0.8745	0.1984	0.0284	6.9825	Tab 57-F

```
==================================================================================================
                                   Box 1, Chap. 57
           Digestive-System Cancers, Males:  Post-1940 Change in MortRates by Census Trios

1960 vs. 1940, by Trios:  Col.D expresses change by ratios.  Col.F expresses change by subtraction.
1988 vs. 1940, by Trios:  Col.I expresses change by ratios.  Col.K expresses change by subtraction.
MRs change inversely with PP.  High-PP Trio has lowest growth-factor.  Low-PP Trio has highest growth-factor.

          Col.A  |  Col.B    Col.C    Col.D    Col.E    Col.F  |  Col.G    Col.H    Col.I    Col.J    Col.K
          1940   |  1960     Ratio    Input    Diff:    Input  |  1988     Ratio    Input    Diff:    Input
          MortRate| MortRate  Col.B    from     Col.B    from   |  MortRate  Col.G    from     Col.G    from
          Tab 9-A | Tab 9-A   /Col.A   Col.C    minus A  Col.E  |  Tab 9-A   /Col.A   Col.H    minus A  Col.J
                  |                                             |
Pacif     63.4   |  46.4     0.732    Avg Chg  -17.0    Avg Chg |  36.3     0.573    Avg Chg  -27.1    Avg Chg
NewE      71.7   |  58.9     0.821    TopTrio  -12.8    TopTrio |  42.1     0.587    TopTrio  -29.6    TopTrio
MidAtl    74.7   |  60.1     0.805    0.786    -14.6    -14.8   |  43.3     0.580    0.580    -31.4    -29.4
                  |                                             |
WNoCen    59.9   |  46.3     0.773    Avg Chg  -13.6    Avg Chg |  35.8     0.598    Avg Chg  -24.1    Avg Chg
ENoCen    64.9   |  53.0     0.817    MidTrio  -11.9    MidTrio |  40.2     0.619    MidTrio  -24.7    MidTrio
Mtn       52.1   |  38.9     0.747    0.779    -13.2    -12.9   |  33.0     0.633    0.617    -19.1    -22.6
                  |                                             |
WSoCen    42.3   |  40.6     0.960    Avg Chg   -1.7    Avg Chg |  36.5     0.863    Avg Chg   -5.8    Avg Chg
ESoCen    38.2   |  39.4     1.031    LowTrio    1.2    LowTrio |  38.0     0.995    LowTrio   -0.2    LowTrio
SoAtl     43.4   |  43.1     0.993    0.995     -0.3    -0.3    |  38.5     0.887    0.915     -4.9    -3.6
==================================================================================================
```

===

Box 2, Chap. 57
Digestive-System Cancers, Males: Calculation of Adjustment Factor

This adjustment is discussed fully in Chapter 49.

● Part 1: Calculate average population-weighted MortRate for the combined TopTrio Census Divs.

Census Div.	Col.A 1940 MR Tab 9-A	Col.B 1940 Pop'n Tab 3-B	Col.C 1940 Popn /45,710,039	Col.D Col.A * Col.C	Census Div.	Col.A 1950 MR Tab 9-A	Col.B 1950 Pop'n Tab 3-B	Col.C 1950 Popn /53,964,513	Col.D Col.A * Col.C
Pacific	63.4	9,733,262	0.2129	13.50	Pacific	50.8	14,486,527	0.2684	13.64
NewEng	71.7	8,437,290	0.1846	13.23	NewEng	66.3	9,314,453	0.1726	11.44
Mid-Atl	74.7	27,539,487	0.6025	45.01	Mid-Atl	67.7	30,163,533	0.5590	37.84
1940		Sum TopTrio 45,710,039	Sum 1.0000	TopTrio 71.740	1950		Sum TopTrio 53,964,513	Sum 1.0000	TopTrio 62.922

- -

Census Div.	Col.A 1960 MR Tab 9-A	Col.B 1960 Pop'n Tab 3-B	Col.C 1960 Popn /65,875,863	Col.D Col.A * Col.C	Census Div.	Col.A 1970 MR Tab 9-A	Col.B 1970 Pop'n Tab 3-B	Col.C 1970 Popn /75,017,000	Col.D Col.A * Col.C
Pacific	46.4	21,198,044	0.3218	14.93	Pacific	42.9	26,087,000	0.3477	14.92
NewEng	58.9	10,509,367	0.1595	9.40	NewEng	52.5	11,781,000	0.1570	8.24
Mid-Atl	60.1	34,168,452	0.5187	31.17	Mid-Atl	54.2	37,149,000	0.4952	26.84
1960		Sum TopTrio 65,875,863	Sum 1.0000	TopTrio 55.500	1970		Sum TopTrio 75,017,000	Sum 1.0000	TopTrio 50.003

- -

Census Div.	Col.A 1980 MR Tab 9-A	Col.B 1980 Pop'n Tab 3-B	Col.C 1980 Popn /80,615,000	Col.D Col.A * Col.C	Census Div.	Col.A 1988 MR Tab 9-A	Col.B 1990 Pop'n Tab 3-B	Col.C 1990 Popn /88,495,000	Col.D Col.A * Col.C
Pacific	39.3	31,523,000	0.3910	15.37	Pacific	36.3	37,837,000	0.4276	15.52
NewEng	46.0	12,322,000	0.1528	7.03	NewEng	42.1	12,998,000	0.1469	6.18
Mid-Atl	48.3	36,770,000	0.4561	22.03	Mid-Atl	43.3	37,660,000	0.4256	18.43
1980		Sum TopTrio 80,615,000	Sum 1.0000	TopTrio 44.429	1988		Sum TopTrio 88,495,000	Sum 1.0000	TopTrio 40.131

- -

● Part 2: Take ratios of these TopTrio MortRates, with 1940 as the denominator of each ratio.
     Col.D modifies Col.C by separate PhysPop adjustments for MidTrio and LowTrio Census Divisions.

	Col.A TopTrio Mean MR	Col.B 1940 TopTrio Mean MR	Col.C = Col.A / Col.B	Col.D ppAdju Tab 47-B	Col.E = Col.C * Col.D	DIGESTIVE-SYSTEM CANCERS. Males.
				MidTrio		
1950	62.922	71.740	0.877	0.99	0.87	= MidTrio Adjustment Factor, 1950
1960	55.500	71.740	0.774	0.97	0.75	= MidTrio Adjustment Factor, 1960
1970	50.003	71.740	0.697	0.95	0.66	= MidTrio Adjustment Factor, 1970
1980	44.429	71.740	0.619	0.94	0.58	= MidTrio Adjustment Factor, 1980
1988	40.131	71.740	0.559	0.94	0.53	= MidTrio Adjustment Factor, 1988
				LowTrio		
1950	62.922	71.740	0.877	1.00	0.88	= LowTrio Adjustment Factor, 1950
1960	55.500	71.740	0.774	1.01	0.78	= LowTrio Adjustment Factor, 1960
1970	50.003	71.740	0.697	1.02	0.71	= LowTrio Adjustment Factor, 1970
1980	44.429	71.740	0.619	1.04	0.64	= LowTrio Adjustment Factor, 1980
1988	40.131	71.740	0.559	1.07	0.60	= LowTrio Adjustment Factor, 1988

===

==
Table 57-B
Digestive-System Cancers, Males: Fractional Causation in 1950
==

Part 1.

Calculation of the 6 Adjusted MortRates (Col.F) and the National Adjusted MortRate (Col.G).

The last six entries in Part 1, Col.F, are the products of (Col.D * Col.E), as discussed in Chap. 49.

Trio-Sequence	Col.A 1950 PopFrac Tab 3-B	Col.B 1950 Obs MR Tab 9-A	Col.C A * B	Col.D 1940 MR Mid,Low Tab 9-A	Col.E AdjuFact Bx2,Pt2 Col.E	Col.F 1950 Adju MortRates	Col.G A * F
Pacific	0.0961	50.8	4.882			50.8	4.882
New England	0.0618	66.3	4.097			66.3	4.097
Mid-Atlantic	0.2002	67.7	13.554			67.7	13.554
WestNoCentral	0.0933	51.9	4.842	59.9	0.87	52.11	4.862
EastNoCentral	0.2017	60.0	12.102	64.9	0.87	56.46	11.389
Mountain	0.0337	43.7	1.473	52.1	0.87	45.33	1.528
WestSoCentral	0.0965	42.9	4.140	42.3	0.88	37.22	3.592
EastSoCentral	0.0762	41.3	3.147	38.2	0.88	33.62	2.562
SouthAtlantic	0.1406	45.2	6.355	43.4	0.88	38.19	5.370

```
                                   Sum =    54.6                          Sum =
        1950 Observed MR from Table 9-B     55.4        1950 Natl Adjusted MR =    51.8345
```

Part 2. ---

Trio-Seq.	Col.A Mean1940 thru1950 PPs from Tab 47-A x'	Col.B 1950 Adju MRs from Col.F Part 1 y	Col.C Digestive Sys. Cancers, Males: 1950 Adjusted MortRates regressed on Mean 1940 thru 1950 PPs Regression Output:		Col.D 1940 PPs from Table 3-A (TrioSeq) x''	Col.E Digestive Cancers, Males: 1950 Adjusted MortRates regressed on 1940 PhysPops Regression Output:	
Pac	154.16	50.8	Constant	0.5831	159.72	Constant	0.8384
NewEng	162.03	66.3	Std Err of Y Est	4.9522	161.55	Std Err of Y Est	5.2586
MidAtl	169.24	67.7	R Squared	0.8591	169.76	R Squared	0.8411
WNoCen	121.60	52.11	No. of Observation	9	123.14	No. of Observation	9
ENoCen	128.53	56.46	Degrees of Freedom	7	133.36	Degrees of Freedom	7
Mtn	119.64	45.33			119.89		
WSoCen	102.64	37.22	X Coefficient(s)	0.3874	103.94	X Coefficient(s)	0.3802
ESoCen	84.44	33.62	Std Err of Coef.	0.0593	85.83	Std Err of Coef.	0.0625
SoAtl	99.91	38.19	XCoef / S.E. =	6.5325	100.74	XCoef / S.E.	6.0872

Part 3-A. | Part 3-B.
Calculation of Fractional Causation | Calculation of Fractional Causation
from Averaged PhysPops | from 1940 PhysPops
                                                   |
1. Nonradiation rate is Adjusted | 1. Nonradiation rate is Adjusted
      Constant (Part 2, Col.C) =   0.5831 | Constant (Part 2, Col.E) =   0.8384
                                                   |
2. Radiation rate is Natl Adjusted | 2. Radiation rate is Natl Adjusted
      MortRate (Part 1, Col.G = 51.8345) | MortRate (Part 1, Col.G = 51.8345)
      minus Nonradiation rate (0.5831) =   51.2513 | minus Nonradiation rate (0.8384) = 50.9961
                                                   |
3. 1950 Fractional Causation is radiation | 3. 1950 Fractional Causation is radiation
      rate (51.2513) divided by OBSERVED | rate (50.9961) divided by OBSERVED
      Natl MR Part 1,Col.C=   55.4   =   0.93 | Natl MR Part 1, Col.C=   55.4   =   0.92

```
=======================================================================================================
                                           Table 57-C
                    Digestive-System Cancers, Males:  Fractional Causation in 1960
=======================================================================================================
Part 1.
Calculation of the 6 Adjusted MortRates (Col.F) and the National Adjusted MortRate (Col.G).
The last six entries in Part 1, Col.F, are the products of (Col.D * Col.E), as discussed in Chap. 49.


                    Col.A    Col.B    Col.C        Col.D    Col.E    Col.F    Col.G
                    1960     1960                  1940 MR  AdjuFact 1960
                    PopFrac  Obs MR   A * B        Mid,Low  Bx2,Pt2  Adju     A * F
Trio-Sequence       Tab 3-B  Tab 9-A              Tab 9-A  Col.E    MortRates
        Pacific     0.1182   46.4     5.484                         46.4     5.484
        New England 0.0586   58.9     3.452                         58.9     3.452
        Mid-Atlantic 0.1905  60.1     11.449                        60.1     11.449
        WestNoCentral 0.0858 46.3     3.973        59.9     0.75     44.93    3.855
        EastNoCentral 0.2020 53.0     10.706       64.9     0.75     48.68    9.832
        Mountain    0.0382   38.9     1.486        52.1     0.75     39.08    1.493
        WestSoCentral 0.0945 40.6     3.837        42.3     0.78     32.99    3.118
        EastSoCentral 0.0672 39.4     2.648        38.2     0.78     29.80    2.002
        SouthAtlantic 0.1448 43.1     6.241        43.4     0.78     33.85    4.902


                             Sum =    49.3                          Sum =
        1960 Observed MR from Table 9-B   49.7    1960 Natl Adjusted MR =     45.5866

Part 2.  -------------------------------------------------------------------------------------------
            Col.A    Col.B                Col.C                Col.D                Col.E
            Mean1940  1960      Digestive Sys. Cancers, Males:   1940      Digestive Cancers, Males:
            thru1960  Adju MRs   1960 Adjusted MortRates        PPs from   1960 Adjusted MortRates
Trio-       PPs from  from Col.F    regressed on                Table 3-A       regressed on
Seq.        Tab 47-A  Part 1    Mean 1940 thru 1960 PPs         (TrioSeq)    1940 PhysPops
            x'        y          Regression Output:             x''          Regression Output:
Pac         155.69    46.4      Constant        -0.1152         159.72     Constant        -0.6028
NewEng      162.81    58.9      Std Err of Y Est  4.2648        161.55     Std Err of Y Est  4.0169
MidAtl      167.04    60.1      R Squared        0.8673         169.76     R Squared        0.8823
WNoCen      118.15    44.93     No. of Observation    9         123.14     No. of Observation    9
ENoCen      123.87    48.68     Degrees of Freedom    7         133.36     Degrees of Freedom    7
Mtn         117.40    39.08                                     119.89
WSoCen      102.31    32.99     X Coefficient(s)  0.3488        103.94     X Coefficient(s)  0.3456
ESoCen       85.63    29.80     Std Err of Coef.  0.0516         85.83     Std Err of Coef.  0.0477
SoAtl       101.72    33.85     XCoef / S.E. =    6.7647        100.74     XCoef / S.E. =    7.2439


-----------------------------------------------------------------------------------------------------
Part 3-A.                                          |  Part 3-B.
Calculation of Fractional Causation                |  Calculation of Fractional Causation
from Averaged PhysPops                              |  from 1940 PhysPops Negative
                                                   |
1. Nonradiation rate is Adjusted                   |  1. Nonradiation rate is Adjusted
   Constant (Part 2, Col.C) = NEGATIVE     0.0     |     Constant (Part 2, Col.E) = NEGATIVE     0.0
                                                   |
2. Radiation rate is Natl Adjusted                 |  2. Radiation rate is Natl Adjusted
   MortRate (Part 1, Col.G = 45.5866)              |     MortRate (Part 1, Col.G = 45.5866)
   minus Nonradiation rate (0.0) =      45.5866    |     minus Nonradiation rate (0.0) =      45.5866
                                                   |
3. 1960 Fractional Causation is radiation          |  3. 1960 Fractional Causation is radiation
   rate (45.5866) divided by OBSERVED              |     rate (45.5866) divided by OBSERVED
   Natl MR Part 1,Col.C=   49.7    =    0.92       |     Natl MR Part 1, Col.C=   49.7    =    0.92
-----------------------------------------------------------------------------------------------------
```

```
=====================================================================================================
                                      Table 57-E
                  Digestive-System Cancers, Males:  Fractional Causation in 1980
=====================================================================================================
```

Part 1.
Calculation of the 6 Adjusted MortRates (Col.F) and the National Adjusted MortRate (Col.G).
The last six entries in Part 1, Col.F, are the products of (Col.D * Col.E), as discussed in Chap. 49.

Trio-Sequence	Col.A 1980 PopFrac Tab 3-B	Col.B 1980 Obs MR Tab 9-A	Col.C A * B	Col.D 1940 MR Mid,Low Tab 9-A	Col.E AdjuFact Bx2,Pt2 Col.E	Col.F 1980 Adju MortRates	Col.G A * F
Pacific	0.1398	39.3	5.494			39.3	5.494
New England	0.0546	46.0	2.512			46.0	2.512
Mid-Atlantic	0.1630	48.3	7.873			48.3	7.873
WestNoCentral	0.0759	38.5	2.922	59.9	0.58	34.74	2.637
EastNoCentral	0.1846	43.7	8.067	64.9	0.58	37.64	6.949
Mountain	0.0502	33.7	1.692	52.1	0.58	30.22	1.517
WestSoCentral	0.1049	37.0	3.881	42.3	0.64	27.07	2.840
EastSoCentral	0.0646	38.4	2.481	38.2	0.64	24.45	1.579
SouthAtlantic	0.1624	40.1	6.512	43.4	0.64	27.78	4.511

```
                                           Sum =   41.4                              Sum =
              1980 Observed MR from Table 9-B       41.7        1980 Natl Adjusted MR =   35.9112
```

Part 2. --

Trio-Seq.	Col.A Mean1940 thru1980 PPs from Tab 47-A x'	Col.B 1980 Adju MRs from Col.F Part 1 y	Col.C Digestive Sys. Cancers, Males: 1980 Adjusted MortRates regressed on Mean 1940 thru 1980 PPs Regression Output:		Col.D 1940 PPs from Table 3-A (TrioSeq) x''	Col.E Digestive Cancers, Males: 1980 Adjusted MortRates regressed on 1940 PhysPops Regression Output:	
Pac	177.35	39.3	Constant	0.0426	159.72	Constant	-0.2819
NewEng	185.86	46.0	Std Err of Y Est	3.2382	161.55	Std Err of Y Est	2.3234
MidAtl	186.11	48.3	R Squared	0.8718	169.76	R Squared	0.9340
WNoCen	128.82	34.74	No. of Observation	9	123.14	No. of Observation	9
ENoCen	133.71	37.64	Degrees of Freedom	7	133.36	Degrees of Freedom	7
Mtn	133.45	30.22			119.89		
WSoCen	114.66	27.07	X Coefficient(s)	0.2454	103.94	X Coefficient(s)	0.2747
ESoCen	99.46	24.45	Std Err of Coef.	0.0356	85.83	Std Err of Coef.	0.0276
SoAtl	124.62	27.78	XCoef / S.E. =	6.9002	100.74	XCoef / S.E.	9.9539

Part 3-A. | Part 3-B.
Calculation of Fractional Causation | Calculation of Fractional Causation
from Averaged PhysPops | from 1940 PhysPops Negative
                                                       |
1. Nonradiation rate is Adjusted | 1. Nonradiation rate is Adjusted
    Constant (Part 2, Col.C) =   0.0426 | Constant (Part 2, Col.E) = Neg.   0.0
                                                       |
2. Radiation rate is Natl Adjusted | 2. Radiation rate is Natl Adjusted
    MortRate (Part 1, Col.G = 35.9112) | MortRate (Part 1, Col.G = 35.9112)
    minus Nonradiation rate (0.0426) =   35.8687 | minus Nonradiation rate (0.0) =   35.9112
                                                       |
3. 1980 Fractional Causation is radiation | 3. 1980 Fractional Causation is radiation
    rate (35.8687) divided by OBSERVED | rate (35.9112) divided by OBSERVED
    Natl MR Part 1,Col.C=   41.7   =   0.86 | Natl MR Part 1, Col.C=   41.7   =   0.86

==

Table 57-F
Digestive-System Cancers, Males: Fractional Causation in 1988

==

Part 1.
Calculation of the 6 Adjusted MortRates (Col.F) and the National Adjusted MortRate (Col.G).
The last six entries in Part 1, Col.F, are the products of (Col.D * Col.E), as discussed in Chap. 49.

Trio-Sequence	Col.A 1990 PopFrac Tab 3-B	Col.B 1988 Obs MR Tab 9-A	Col.C A * B	Col.D 1940 MR Mid,Low Tab 9-A	Col.E AdjuFact Bx2,Pt2 Col.E	Col.F 1988 Adju MortRates	Col.G A * F
Pacific	0.1535	36.3	5.572			36.3	5.572
New England	0.0527	42.1	2.219			42.1	2.219
Mid-Atlantic	0.1527	43.3	6.612			43.3	6.612
WestNoCentral	0.0721	35.8	2.581	59.9	0.53	31.75	2.289
EastNoCentral	0.1713	40.2	6.886	64.9	0.53	34.40	5.892
Mountain	0.0543	33.0	1.792	52.1	0.53	27.61	1.499
WestSoCentral	0.1087	36.5	3.968	42.3	0.60	25.38	2.759
EastSoCentral	0.0621	38.0	2.360	38.2	0.60	22.92	1.423
SouthAtlantic	0.1725	38.5	6.641	43.4	0.60	26.04	4.492

Sum = 38.6

1988 Observed MR from Table 9-B   38.8   1988 Natl Adjusted MR =   Sum = 32.7572

Part 2. --

Trio-Seq.	Col.A Mean1940 thru1990 PPs from Tab 47-A x'	Col.B 1988 Adju MRs from Col.F Part 1 y	Col.C Digestive Sys. Cancers, Males: 1988 Adjusted MortRates regressed on Mean 1940 thru 1990 PPs Regression Output:		Col.D 1940 PPs from Table 3-A (TrioSeq) x''	Col.E Digestive Cancers, Males: 1988 Adjusted MortRates regressed on 1940 PhysPops Regression Output:	
Pac	191.97	36.3	Constant	0.8879	159.72	Constant	1.3466
NewEng	208.20	42.1	Std Err of Y Est	2.7922	161.55	Std Err of Y Est	1.9620
MidAtl	204.72	43.3	R Squared	0.8745	169.76	R Squared	0.9380
WNoCen	141.14	31.75	No. of Observation	9	123.14	No. of Observation	9
ENoCen	146.19	34.40	Degrees of Freedom	7	133.36	Degrees of Freedom	7
Mtn	145.91	27.61			119.89		
WSoCen	126.28	25.38	X Coefficient(s)	0.1984	103.94	X Coefficient(s)	0.2398
ESoCen	113.28	22.92	Std Err of Coef.	0.0284	85.83	Std Err of Coef.	0.0233
SoAtl	142.93	26.04	XCoef / S.E. =	6.9825	100.74	XCoef / S.E. =	10.2915

Part 3-A.
Calculation of Fractional Causation
from Averaged PhysPops

1. Nonradiation rate is Adjusted
    Constant (Part 2, Col.C) =   0.8879

2. Radiation rate is Natl Adjusted
    MortRate (Part 1, Col.G = 32.7572)
    minus Nonradiation rate (0.8879) =   31.8694

3. 1988 Fractional Causation is radiation
    rate (31.8694) divided by OBSERVED
    Natl MR Part 1,Col.C=   38.8   =   0.82

Part 3-B.
Calculation of Fractional Causation
from 1940 PhysPops

1. Nonradiation rate is Adjusted
    Constant (Part 2, Col.E) =   1.3466

2. Radiation rate is Natl Adjusted
    MortRate (Part 1, Col.G = 32.7572)
    minus Nonradiation rate (1.3466) =   31.4106

3. 1988 Fractional Causation is radiation
    rate (31.4106) divided by OBSERVED
    Natl MR Part 1,Col.C=   38.8   =   0.81

CHAPTER 58

Digestive–System Cancers, Females, 1940–1988

• In Table 58–A, Column A shows that the female National MortRate from Digestive–System Cancers fell in half in the 1940–1988 period. Box 1 shows that such rates fell much more in the TopTrio than in the LowTrio. Please see the text in Chapter 57.

• It is noteworthy that in the 1940–1988 period, female National MortRates decline for Digestive–System Cancers, rise for Respiratory–System Cancers, and remain steady for Breast Cancers. These very different behaviors are consistent with high Fractional Causation by medical radiation for all three groups of cancers, throughout the entire period. High Fractional Causation simply means that medical radiation has been a NECESSARY co–actor for most of the fatal cases, whether the MortRate was rising or falling or flat.

• An additional concept is relevant. The impact on MortRates of a carcinogen, per unit, almost certainly varies with the levels of its co–actors. If the co–actors with medical radiation rise for Respiratory Cancers, while other co–actors for Digestive Cancers fall, then each unit of medical radiation would become more likely than previously to produce Respiratory Cancers, and less likely than previously to produce Digestive Cancers ––– "all other things being equal."

				Table 58–A			
			Digestive–System Cancers, Females: Fractional Causation by Medical Radiation over Time				
Year	Col.A Natl MR	Col.B Frac.C	|	Col.C R–Sq	Col.D X–Coef	Col.E StdErr	Col.F Coef/SE	Col.G Source
1940	50.1	80%	|	0.7550	0.2895	0.0623	4.6442	Chap.10
1950	42.4	76%	|	0.7707	0.2431	0.0501	4.8506	Tab 58–B
1960	35.8	75%	|	0.7985	0.2048	0.0389	5.2675	Tab 58–C
1970	31.0	73%	|	0.8365	0.1668	0.0279	5.9850	Tab 58–D
1980	26.2	70%	|	0.8547	0.1271	0.0198	6.4177	Tab 58–E
1988	23.5	68%	|	0.8637	0.0999	0.0150	6.6597	Tab 58–F

===

Box 1, Chap. 58
Digestive–System Cancers, Females: Post–1940 Change in MortRates by Census Trios

1960 vs. 1940, by Trios: Col.D expresses change by ratios. Col.F expresses change by subtraction.
1988 vs. 1940, by Trios: Col.I expresses change by ratios. Col.K expresses change by subtraction.
MRs change inversely with PP. High–PP Trio has lowest growth–factor. Low–PP Trio has highest growth–factor.

	Col.A 1940 MortRate Tab 10–A	|	Col.B 1960 MortRate Tab 10–A	Col.C Ratio Col.B /Col.A	Col.D Input from Col.C	Col.E Diff: Col.B minus A	Col.F Input from Col.E	|	Col.G 1988 MortRate Tab 10–A	Col.H Ratio Col.G /Col.A	Col.I Input from Col.H	Col.J Diff: Col.G minus A	Col.K Input from Col.J
Pacif	46.8	|	32.5	0.694	Avg Chg	-14.3	Avg Chg	|	22.8	0.487	Avg Chg	-24.0	Avg Chg
NewE	61.3	|	40.7	0.664	TopTrio	-20.6	TopTrio	|	24.7	0.403	TopTrio	-36.6	TopTrio
MidAtl	60.2	|	42.9	0.713	0.690	-17.3	-17.4	|	26.0	0.432	0.441	-34.2	-31.6
WNoCen	49.7	|	34.1	0.686	Avg Chg	-15.6	Avg Chg	|	21.8	0.439	Avg Chg	-27.9	Avg Chg
ENoCen	53.1	|	38.5	0.725	MidTrio	-14.6	MidTrio	|	24.2	0.456	MidTrio	-28.9	MidTrio
Mtn	47.7	|	30.5	0.639	0.684	-17.2	-15.8	|	21.1	0.442	0.446	-26.6	-27.8
WSoCen	34.5	|	29.6	0.858	Avg Chg	-4.9	Avg Chg	|	21.5	0.623	Avg Chg	-13.0	Avg Chg
ESoCen	36.3	|	29.6	0.815	LowTrio	-6.4	LowTrio	|	23.3	0.642	LowTrio	-13.0	LowTrio
SoAtl	37.3	|	30.6	0.820	0.834	-6.7	-6.0	|	22.8	0.611	0.625	-14.5	-13.5

===

```
=======================================================================================
                              Box 2, Chap. 58
              Digestive-System Cancers, Females:  Calculation of Adjustment Factor

                   This adjustment is discussed fully in Chapter 49.
• Part 1:  Calculate average population-weighted MortRate for the combined TopTrio Census Divs.


           Col.A      Col.B        Col.C     Col.D  |            Col.A      Col.B         Col.C     Col.D
Census     1940 MR    1940 Pop'n   1940 Popn Col.A * | Census     1950 MR    1950 Pop'n    1950 Popn Col.A *
Div.       Tab 10-A   Tab 3-B      /45,710,039 Col.C | Div.       Tab 10-A   Tab 3-B       /53,964,513 Col.C
                                                     |
Pacific    46.8       9,733,262    0.2129    9.97   | Pacific    37.3       14,486,527    0.2684    10.01
NewEng     61.3       8,437,290    0.1846    11.31  | NewEng     48.9       9,314,453     0.1726    8.44
Mid-Atl    60.2       27,539,487   0.6025    36.27  | Mid-Atl    51.1       30,163,533    0.5590    28.56
                                                     |
  1940               Sum TopTrio      Sum  TopTrio  |   1950               Sum TopTrio      Sum  TopTrio
                     45,710,039    1.0000   57.550  |                      53,964,513    1.0000   47.016
----------------------------------------------------------------------------------------
           Col.A      Col.B        Col.C     Col.D  |            Col.A      Col.B         Col.C     Col.D
Census     1960 MR    1960 Pop'n   1960 Popn Col.A * | Census     1970 MR    1970 Pop'n    1970 Popn Col.A *
Div.       Tab 10-A   Tab 3-B      /65,875,863 Col.C | Div.       Tab 10-A   Tab 3-B       /75,017,000 Col.C
                                                     |
Pacific    32.5       21,198,044   0.3218    10.46  | Pacific    28.9       26,087,000    0.3477    10.05
NewEng     40.7       10,509,367   0.1595    6.49   | NewEng     34.7       11,781,000    0.1570    5.45
Mid-Atl    42.9       34,168,452   0.5187    22.25  | Mid-Atl    36.5       37,149,000    0.4952    18.08
                                                     |
  1960               Sum TopTrio      Sum  TopTrio  |   1970               Sum TopTrio      Sum  TopTrio
                     65,875,863    1.0000   39.202  |                      75,017,000    1.0000   33.574
----------------------------------------------------------------------------------------
           Col.A      Col.B        Col.C     Col.D  |            Col.A      Col.B         Col.C     Col.D
Census     1980 MR    1980 Pop'n   1980 Popn Col.A * | Census     1988 MR    1990 Pop'n    1990 Popn Col.A *
Div.       Tab 10-A   Tab 3-B      /80,615,000 Col.C | Div.       Tab 10-A   Tab 3-B       /88,495,000 Col.C
                                                     |
Pacific    25.4       31,523,000   0.3910    9.93   | Pacific    22.8       37,837,000    0.4276    9.75
NewEng     28.7       12,322,000   0.1528    4.39   | NewEng     24.7       12,998,000    0.1469    3.63
Mid-Atl    30.1       36,770,000   0.4561    13.73  | Mid-Atl    26.0       37,660,000    0.4256    11.06
                                                     |
  1980               Sum TopTrio      Sum  TopTrio  |   1988               Sum TopTrio      Sum  TopTrio
                     80,615,000    1.0000   28.048  |                      88,495,000    1.0000   24.441
----------------------------------------------------------------------------------------
```

• Part 2: Take ratios of these TopTrio MortRates, with 1940 as the denominator of each ratio.
   Col.D modifies Col.C by separate PhysPop adjustments for MidTrio and LowTrio Census Divisions.

```
           Col.A      Col.B        Col.C   Col.D     Col.E
           TopTrio    1940 TopTrio = Col.A ppAdju    = Col.C      DIGESTIVE-SYSTEM CANCERS.
           Mean MR    Mean MR      / Col.B Tab 47-B  * Col.D          Females.
                                          MidTrio
1950       47.016     57.550       0.817   0.99       0.81  = MidTrio Adjustment Factor, 1950
1960       39.202     57.550       0.681   0.97       0.66  = MidTrio Adjustment Factor, 1960
1970       33.574     57.550       0.583   0.95       0.55  = MidTrio Adjustment Factor, 1970
1980       28.048     57.550       0.487   0.94       0.46  = MidTrio Adjustment Factor, 1980
1988       24.441     57.550       0.425   0.94       0.40  = MidTrio Adjustment Factor, 1988
-------------------------------------------  LowTrio  ------------------------------------------
1950       47.016     57.550       0.817   1.00       0.82  = LowTrio Adjustment Factor, 1950
1960       39.202     57.550       0.681   1.01       0.69  = LowTrio Adjustment Factor, 1960
1970       33.574     57.550       0.583   1.02       0.60  = LowTrio Adjustment Factor, 1970
1980       28.048     57.550       0.487   1.04       0.51  = LowTrio Adjustment Factor, 1980
1988       24.441     57.550       0.425   1.07       0.45  = LowTrio Adjustment Factor, 1988
=======================================================================================
```

```
===================================================================================================
                                        Table 58-B
                   Digestive-System Cancers, Females:  Fractional Causation in 1950
===================================================================================================
```

Part 1.
Calculation of the 6 Adjusted MortRates (Col.F) and the National Adjusted MortRate (Col.G).
The last six entries in Part 1, Col.F, are the products of (Col.D * Col.E), as discussed in Chap. 49.

Trio-Sequence	Col.A 1950 PopFrac Tab 3-B	Col.B 1950 Obs MR Tab 10-A	Col.C A * B	Col.D 1940 MR Mid,Low Tab 10-A	Col.E AdjuFact Bx2,Pt2 Col.E	Col.F 1950 Adju MortRates	Col.G A * F
Pacific	0.0961	37.3	3.585			37.3	3.585
New England	0.0618	48.9	3.022			48.9	3.022
Mid-Atlantic	0.2002	51.1	10.230			51.1	10.230
WestNoCentral	0.0933	40.4	3.769	49.7	0.81	40.26	3.756
EastNoCentral	0.2017	44.7	9.016	53.1	0.81	43.01	8.675
Mountain	0.0337	34.8	1.173	47.7	0.81	38.64	1.302
WestSoCentral	0.0965	33.3	3.213	34.5	0.82	28.29	2.730
EastSoCentral	0.0762	34.5	2.629	36.3	0.82	29.77	2.268
SouthAtlantic	0.1406	34.9	4.907	37.3	0.82	30.59	4.300

```
                                    Sum =   41.5                            Sum =
    1950   Observed MR from Table 10-B     42.4       1950 Natl Adjusted MR =    39.8687
```

Part 2. --

Trio-Seq.	Col.A Mean1940 thru1950 PPs from Tab 47-A x'	Col.B 1950 Adju MRs from Col.F Part 1 y	Col.C Digestive Cancers, Females: 1950 Adjusted MortRates regressed on Mean 1940 thru 1950 PPs Regression Output:		Col.D 1940 PPs from Table 3-A (TrioSeq) x''	Col.E Digestive Cancers, Females: 1950 Adjusted MortRates regressed on 1940 PhysPops Regression Output:	
Pac	154.16	37.3	Constant	7.7959	159.72	Constant	8.1574
NewEng	162.03	48.9	Std Err of Y Est	4.1853	161.55	Std Err of Y Est	4.4162
MidAtl	169.24	51.1	R Squared	0.7707	169.76	R Squared	0.7447
WNoCen	121.60	40.26	No. of Observation	9	123.14	No. of Observation	9
ENoCen	128.53	43.01	Degrees of Freedom	7	133.36	Degrees of Freedom	7
Mtn	119.64	38.64			119.89		
WSoCen	102.64	28.29	X Coefficient(s)	0.2431	103.94	X Coefficient(s)	0.2370
ESoCen	84.44	29.77	Std Err of Coef.	0.0501	85.83	Std Err of Coef.	0.0524
SoAtl	99.91	30.59	XCoef / S.E. =	4.8506	100.74	XCoef / S.E.	4.5188

Part 3-A.
Calculation of Fractional Causation
from Averaged PhysPops

1. Nonradiation rate is Adjusted
   Constant (Part 2, Col.C) =   7.7959

2. Radiation rate is Natl Adjusted
   MortRate (Part 1, Col.G = 39.8687)
   minus Nonradiation rate (7.7959) =   32.0728

3. 1950 Fractional Causation is radiation
   rate (32.0728) divided by OBSERVED
   Natl MR Part 1,Col.C=   42.4   =   0.76

Part 3-B.
Calculation of Fractional Causation
from 1940 PhysPops

1. Nonradiation rate is Adjusted
   Constant (Part 2, Col.E) =   8.1574

2. Radiation rate is Natl Adjusted
   MortRate (Part 1, Col.G = 39.8687)
   minus Nonradiation rate (8.1574) =   31.7112

3. 1950 Fractional Causation is radiation
   rate (31.7112) divided by OBSERVED
   Natl MR Part 1, Col.C=   42.4   =   0.75

```
=====================================================================================================
                                        Table 58-C
                Digestive-System Cancers, Females:  Fractional Causation in 1960
=====================================================================================================
Part 1.
Calculation of the 6 Adjusted MortRates (Col.F) and the National Adjusted MortRate (Col.G).
The last six entries in Part 1, Col.F, are the products of (Col.D * Col.E), as discussed in Chap. 49.
```

	Col.A	Col.B	Col.C	Col.D	Col.E	Col.F	Col.G
	1960	1960		1940 MR	AdjuFact	1960	
	PopFrac	Obs MR	A * B	Mid,Low	Bx2,Pt2	Adju	A * F
Trio-Sequence	Tab 3-B	Tab 10-A		Tab 10-A	Col.E	MortRates	
Pacific	0.1182	32.5	3.842			32.5	3.842
New England	0.0586	40.7	2.385			40.7	2.385
Mid-Atlantic	0.1905	42.9	8.172			42.9	8.172
WestNoCentral	0.0858	34.1	2.926	49.7	0.66	32.80	2.814
EastNoCentral	0.2020	38.5	7.777	53.1	0.66	35.05	7.079
Mountain	0.0382	30.5	1.165	47.7	0.66	31.48	1.203
WestSoCentral	0.0945	29.6	2.797	34.5	0.69	23.81	2.250
EastSoCentral	0.0672	29.9	2.009	36.3	0.69	25.05	1.683
SouthAtlantic	0.1448	30.6	4.431	37.3	0.69	25.74	3.727

```
                                    Sum =    35.5                          Sum =
     1960    Observed MR from Table 10-B     35.8     1960 Natl Adjusted MR =    33.1547
```

```
Part 2.  ---------------------------------------------------------------------------------------------
        Col.A    Col.B                      Col.C                Col.D                      Col.E
        Mean1940   1960        Digestive Cancers, Females:        1940        Digestive Cancers, Females:
        thru1960  Adju MRs     1960 Adjusted MortRates           PPs from     1960 Adjusted MortRates
Trio-   PPs from  from Col.F      regressed on                   Table 3-A       regressed on
Seq.    Tab 47-A  Part 1      Mean 1940 thru 1960 PPs           (TrioSeq)     1940 PhysPops
           x'      y                  Regression Output:           x''            Regression Output:
Pac     155.69   32.5         Constant           6.4099         159.72        Constant           6.1060
NewEng  162.81   40.7         Std Err of Y Est   3.2153         161.55        Std Err of Y Est   3.0943
MidAtl  167.04   42.9         R Squared          0.7985         169.76        R Squared          0.8134
WNoCen  118.15   32.80        No. of Observation    9           123.14        No. of Observation    9
ENoCen  123.87   35.05        Degrees of Freedom    7           133.36        Degrees of Freedom    7
Mtn     117.40   31.48                                          119.89
WSoCen  102.31   23.81        X Coefficient(s)   0.2048         103.94        X Coefficient(s)   0.2030
ESoCen   85.63   25.05        Std Err of Coef.   0.0389          85.83        Std Err of Coef.   0.0367
SoAtl   101.72   25.74        XCoef / S.E. =     5.2675         100.74        XCoef / S.E. =     5.5243
---------------------------------------------------------------------------------------------------
```

```
Part 3-A.                                          |  Part 3-B.
Calculation of Fractional Causation                |  Calculation of Fractional Causation
from Averaged PhysPops                             |  from 1940 PhysPops
                                                   |
1.  Nonradiation rate is Adjusted                  |  1.  Nonradiation rate is Adjusted
    Constant (Part 2, Col.C) =          6.4099     |      Constant (Part 2, Col.E) =          6.1060
                                                   |
2.  Radiation rate is Natl Adjusted                |  2.  Radiation rate is Natl Adjusted
    MortRate (Part 1, Col.G = 33.1547)             |      MortRate (Part 1, Col.G = 33.1547)
    minus Nonradiation rate (6.4099) =   26.7448   |      minus Nonradiation rate (6.1060) =   27.0487
                                                   |
3.  1960 Fractional Causation is radiation         |  3.  1960 Fractional Causation is radiation
    rate (26.7448) divided by OBSERVED             |      rate (27.0487) divided by OBSERVED
    Natl MR Part 1,Col.C=    35.8    =    0.75     |      Natl MR Part 1, Col.C=   35.8    =    0.76
---------------------------------------------------------------------------------------------------
```

```
=====================================================================================================
                                        Table 58-E
                  Digestive-System Cancers, Females:  Fractional Causation in 1980
=====================================================================================================
Part 1.
Calculation of the 6 Adjusted MortRates (Col.F) and the National Adjusted MortRate (Col.G).
The last six entries in Part 1, Col.F, are the products of (Col.D * Col.E), as discussed in Chap. 49.


                   Col.A    Col.B    Col.C          Col.D    Col.E    Col.F    Col.G
                   1980     1980                    1940 MR  AdjuFact 1980
                   PopFrac  Obs MR   A * B          Mid,Low  Bx2,Pt2  Adju     A * F
Trio-Sequence      Tab 3-B  Tab 10-A               Tab 10-A Col.E    MortRates
       Pacific     0.1398   25.4     3.551                                    25.4     3.551
       New England 0.0546   28.7     1.567                                    28.7     1.567
       Mid-Atlantic 0.1630  30.1     4.906                                    30.1     4.906
       WestNoCentral 0.0759 24.6     1.867          49.7     0.46     22.86    1.735
       EastNoCentral 0.1846 27.1     5.003          53.1     0.46     24.43    4.509
       Mountain    0.0502   22.2     1.114          47.7     0.46     21.94    1.101
       WestSoCentral 0.1049 23.6     2.476          34.5     0.51     17.60    1.846
       EastSoCentral 0.0646 24.4     1.576          36.3     0.51     18.51    1.196
       SouthAtlantic 0.1624 24.4     3.963          37.3     0.51     19.02    3.089

                            Sum =    26.0                                     Sum =
       1980 Observed MR from Table 10-B    26.2         1980 Natl Adjusted MR =       23.5010

Part 2. ---------------------------------------------------------------------------------------------
          Col.A    Col.B                        Col.C                Col.D              Col.E
          Mean1940 1980            Digestive Cancers, Females:       1940      Digestive Cancers, Females:
          thru1980 Adju MRs        1980 Adjusted MortRates           PPs from  1980 Adjusted MortRates
Trio-     PPs from from Col.F          regressed on                  Table 3-A     regressed on
Seq.      Tab 47-A Part 1         Mean 1940 thru 1980 PPs            (TrioSeq)     1940 PhysPops
          x'       y                   Regression Output:            x''           Regression Output:
Pac       177.35   25.4          Constant          5.0370           159.72    Constant          4.8931
NewEng    185.86   28.7          Std Err of Y Est  1.8034           161.55    Std Err of Y Est  1.3933
MidAtl    186.11   30.1          R Squared         0.8547           169.76    R Squared         0.9133
WNoCen    128.82   22.86         No. of Observation     9           123.14    No. of Observation     9
ENoCen    133.71   24.43         Degrees of Freedom     7           133.36    Degrees of Freedom     7
Mtn       133.45   21.94                                            119.89
WSoCen    114.66   17.60         X Coefficient(s)  0.1271           103.94    X Coefficient(s)  0.1421
ESoCen    99.46    18.51         Std Err of Coef.  0.0198            85.83    Std Err of Coef.  0.0165
SoAtl     124.62   19.02         XCoef / S.E. =    6.4177           100.74    XCoef / S.E.      8.5865


----------------------------------------------------------------------------------------------------
Part 3-A.                                          | Part 3-B.
Calculation of Fractional Causation                | Calculation of Fractional Causation
from Averaged PhysPops                              | from 1940 PhysPops
                                                   |
1.  Nonradiation rate is Adjusted                  | 1.  Nonradiation rate is Adjusted
    Constant (Part 2, Col.C) =         5.0370      |     Constant (Part 2, Col.E) =          4.8931
                                                   |
2.  Radiation rate is Natl Adjusted                | 2.  Radiation rate is Natl Adjusted
    MortRate (Part 1, Col.G = 23.5010)             |     MortRate (Part 1, Col.G = 23.5010)
    minus Nonradiation rate (5.0370) =  18.4639    |     minus Nonradiation rate (4.8931) =  18.6079
                                                   |
3.  1980 Fractional Causation is radiation         | 3.  1980 Fractional Causation is radiation
    rate (18.4639) divided by OBSERVED             |     rate (18.6079) divided by OBSERVED
    Natl MR Part 1,Col.C=   26.2     =    0.70     |     Natl MR Part 1, Col.C=   26.2     =    0.71
----------------------------------------------------------------------------------------------------
```

===
```
                                        Table 58-F
                   Digestive-System Cancers, Females:  Fractional Causation in 1988
```
===
Part 1.
Calculation of the 6 Adjusted MortRates (Col.F) and the National Adjusted MortRate (Col.G).
The last six entries in Part 1, Col.F, are the products of (Col.D * Col.E), as discussed in Chap. 49.

Trio-Sequence	Col.A 1990 PopFrac Tab 3-B	Col.B 1988 Obs MR Tab 10-A	Col.C A * B	Col.D 1940 MR Mid,Low Tab 10-A	Col.E AdjuFact Bx2,Pt2 Col.E	Col.F 1988 Adju MortRates	Col.G A * F
Pacific	0.1535	22.8	3.500			22.8	3.500
New England	0.0527	24.7	1.302			24.7	1.302
Mid-Atlantic	0.1527	26.0	3.970			26.0	3.970
WestNoCentral	0.0721	21.8	1.572	49.7	0.40	19.88	1.433
EastNoCentral	0.1713	24.2	4.145	53.1	0.40	21.24	3.638
Mountain	0.0543	21.1	1.146	47.7	0.40	19.08	1.036
WestSoCentral	0.1087	21.5	2.337	34.5	0.45	15.53	1.688
EastSoCentral	0.0621	23.3	1.447	36.3	0.45	16.34	1.014
SouthAtlantic	0.1725	22.8	3.933	37.3	0.45	16.79	2.895

```
                                  Sum =    23.4                              Sum =
          1988 Observed MR from Table 10-B  23.5       1988 Natl Adjusted MR =      20.4769
```

Part 2. ---

Trio-Seq.	Col.A Mean1940 thru1990 PPs from Tab 47-A x'	Col.B 1988 Adju MRs from Col.F Part 1 y	Col.C Digestive Cancers, Females 1988 Adjusted MortRates regressed on Mean 1940 thru 1990 PPs Regression Output:		Col.D 1940 PPs from Table 3-A (TrioSeq) x''	Col.E Digestive Cancers, Females 1988 Adjusted MortRates regressed on 1940 PhysPops Regression Output:	
Pac	191.97	22.8	Constant	4.4992	159.72	Constant	4.6537
NewEng	208.20	24.7	Std Err of Y Est	1.4736	161.55	Std Err of Y Est	1.0129
MidAtl	204.72	26.0	R Squared	0.8637	169.76	R Squared	0.9356
WNoCen	141.14	19.88	No. of Observation	9	123.14	No. of Observation	9
ENoCen	146.19	21.24	Degrees of Freedom	7	133.36	Degrees of Freedom	7
Mtn	145.91	19.08			119.89		
WSoCen	126.28	15.53	X Coefficient(s)	0.0999	103.94	X Coefficient(s)	0.1213
ESoCen	113.28	16.34	Std Err of Coef.	0.0150	85.83	Std Err of Coef.	0.0120
SoAtl	142.93	16.79	XCoef / S.E. =	6.6597	100.74	XCoef / S.E. =	10.0845

```
Part 3-A.                                      |  Part 3-B.
Calculation of Fractional Causation            |  Calculation of Fractional Causation
from Averaged PhysPops                          |  from 1940 PhysPops
                                                |
1.  Nonradiation rate is Adjusted              |  1.  Nonradiation rate is Adjusted
      Constant (Part 2, Col.C) =       4.4992  |        Constant (Part 2, Col.E) =       4.6537
                                                |
2.  Radiation rate is Natl Adjusted            |  2.  Radiation rate is Natl Adjusted
      MortRate (Part 1, Col.G = 20.4769)       |        MortRate (Part 1, Col.G = 20.4769)
      minus Nonradiation rate (4.4992) = 15.9777 |      minus Nonradiation rate (4.6537) = 15.8232
                                                |
3.  1988 Fractional Causation is radiation     |  3.  1988 Fractional Causation is radiation
      rate (15.9777) divided by OBSERVED       |        rate (15.8232) divided by OBSERVED
      Natl MR Part 1,Col.C=   23.5   =   0.68  |        Natl MR Part 1,Col.C=   23.5   =   0.67
```

CHAPTER 59

Urinary–System Cancers, Males, 1940–1980

● Box 1, below, shows the familiar pattern revealed in the previous Boxes 1. Therefore, Box 2 evaluates an adjustment factor, in order to match the Census Divisions for the unmatched co–actor.

● Table 59–A deviates from the model of Chapter 49. We have added a second set of entries (Tables 59–BB through 59–EE) because the regressions in Tables 59–B through 59–E produce negative constants, whose magnitude is not trivial relative to the male National MortRate for Urinary–System Cancers. We regard this as a signal that something is not REALISTIC about the adjustment factor used in those tables for the MidTrio and LowTrio MortRates. This happened once before. Please refer to the text in Chapter 51, preceding Table 51–AA. Here, in Chapter 59, the factor which abolishes the negative sign on Constants is 1.35, and it is used in the same way the factor of 1.4 was used to produce Table 51–BB through 51–FF.

Table 59–A							
Urinary–System Cancers, Males: Fractional Causation by Medical Radiation over Time							
Year	Col.A Natl MR	Col.B Frac.C	Col.C R–Sq	Col.D X–Coef	Col.E StdErr	Col.F Coef/SE	Col.G Source
1940	7.4	~100	0.9208	0.0750	0.0083	9.0208	Chap.11
1950	8.1	93%	0.9329	0.0832	0.0084	9.8635	Tab 59–B
1960	8.5	86%	0.9170	0.0833	0.0095	8.7960	Tab 59–C
1970	8.35	84%	0.9220	0.0761	0.0084	9.0993	Tab 59–D
1980	8.2	79%	0.9275	0.0676	0.0071	9.4602	Tab 59–E
1950	8.1	99%	0.5967	0.0601	0.0187	3.2180	Tab 59–BB
1960	8.5	91%	0.5901	0.0585	0.0184	3.1746	Tab 59–CC
1970	8.35	86%	0.6018	0.0524	0.0161	3.2526	Tab 59–DD
1980	8.2	83%	0.6089	0.0458	0.0139	3.3016	Tab 59–EE

```
===================================================================================================
                                     Box 1, Chap. 59
                  Urinary-System Cancers, Males:  Post-1940 Change in MortRates by Census Trios

1960 vs. 1940, by Trios:  Col.D expresses change by ratios.  Col.F expresses change by subtraction.
1980 vs. 1940, by Trios:  Col.I expresses change by ratios.  Col.K expresses change by subtraction.
MRs change inversely with PP.  High-PP Trio has lowest growth-factor.  Low-PP Trio has highest growth-factor.

          Col.A  |  Col.B    Col.C    Col.D    Col.E    Col.F  |  Col.G    Col.H    Col.I    Col.J    Col.K
          1940   |  1960     Ratio    Input    Diff:    Input  |  1980     Ratio    Input    Diff:    Input
          MortRate| MortRate  Col.B    from     Col.B    from   |  MortRate  Col.G    from     Col.G    from
          Tab 11-A| Tab 11-A  /Col.A   Col.C    minus A  Col.E  |  Tab 11-A  /Col.A   Col.H    minus A  Col.J
                 |
Pacif      8.1   |  8.2      1.012    Avg Chg   0.1     Avg Chg |  7.7      0.951    Avg Chg  -0.4     Avg Chg
NewE       9.1   |  10.7     1.176    TopTrio   1.6     TopTrio |  9.5      1.044    TopTrio   0.4     TopTrio
MidAtl     10.2  |  10.2     1.000    1.063     0.0      0.6    |  9.2      0.902    0.966    -1.0     -0.3
                 |                                             |
WNoCen     6.7   |  8.3      1.239    Avg Chg   1.6     Avg Chg |  7.9      1.179    Avg Chg   1.2     Avg Chg
ENoCen     8.1   |  9.4      1.160    MidTrio   1.3     MidTrio |  8.7      1.074    MidTrio   0.6     MidTrio
Mtn        6.5   |  7.8      1.200    1.200     1.3      1.4    |  7.0      1.077    1.110     0.5      0.8
                 |                                             |
WSoCen     4.3   |  6.6      1.535    Avg Chg   2.3     Avg Chg |  7.0      1.628    Avg Chg   2.7     Avg Chg
ESoCen     3.0   |  5.2      1.733    LowTrio   2.2     LowTrio |  7.3      2.433    LowTrio   4.3     LowTrio
SoAtl      5.3   |  6.9      1.302    1.523     1.6      2.0    |  7.8      1.472    1.844     2.5      3.2
===================================================================================================
```

```
=========================================================================================================
                                        Box 2, Chap. 59
                    Urinary-System Cancers, Males:  Calculation of Adjustment Factor

                        This adjustment is discussed fully in Chapter 49.
● Part 1: Calculate average population-weighted MortRate for the combined TopTrio Census Divs.


           Col.A      Col.B        Col.C      Col.D   |            Col.A      Col.B        Col.C      Col.D
Census     1940 MR    1940 Pop'n   1940 Popn  Col.A * | Census     1950 MR    1950 Pop'n   1950 Popn  Col.A *
Div.       Tab 11-A   Tab 3-B      /45,710,039 Col.C  | Div.       Tab 11-A   Tab 3-B      /53,964,513 Col.C

Pacific    8.1        9,733,262    0.2129     1.72    | Pacific    8.4        14,486,527   0.2684     2.25
NewEng     9.1        8,437,290    0.1846     1.68    | NewEng     10.5       9,314,453    0.1726     1.81
Mid-Atl    10.2       27,539,487   0.6025     6.15    | Mid-Atl    10.5       30,163,533   0.5590     5.87

  1940                Sum TopTrio       Sum  TopTrio  |   1950                Sum TopTrio       Sum  TopTrio
                      45,710,039     1.0000  9.550    |                       53,964,513     1.0000  9.936
---------------------------------------------------------------------------------------------------------
           Col.A      Col.B        Col.C      Col.D   |            Col.A      Col.B        Col.C      Col.D
Census     1960 MR    1960 Pop'n   1960 Popn  Col.A * | Census     1970 MR    1970 Pop'n   1970 Popn  Col.A *
Div.       Tab 11-A   Tab 3-B      /65,875,863 Col.C  | Div.       Tab 11-A   Tab 3-B      /75,017,000 Col.C

Pacific    8.2        21,198,044   0.3218     2.64    | Pacific    8.0        26,087,000   0.3477     2.78
NewEng     10.7       10,509,367   0.1595     1.71    | NewEng     10.1       11,781,000   0.1570     1.59
Mid-Atl    10.2       34,168,452   0.5187     5.29    | Mid-Atl    9.7        37,149,000   0.4952     4.80

  1960                Sum TopTrio       Sum  TopTrio  |   1970                Sum TopTrio       Sum  TopTrio
                      65,875,863     1.0000  9.636    |                       75,017,000     1.0000  9.172
---------------------------------------------------------------------------------------------------------
           Col.A      Col.B        Col.C      Col.D   |            Col.A      Col.B        Col.C      Col.D
Census     1980 MR    1980 Pop'n   1980 Popn  Col.A * | Census     1988 MR    1990 Pop'n   1990 Popn  Col.A *
Div.       Tab 11-A   Tab 3-B      /80,615,000       4| Div.       Tab 11-A   Tab 3-B      /88,495,000 Col.C

Pacific    7.7        31,523,000   0.3910     3.01    | Pacific    --         37,837,000   0.4276     --
NewEng     9.5        12,322,000   0.1528     1.45    | NewEng     --         12,998,000   0.1469     --
Mid-Atl    9.2        36,770,000   0.4561     4.20    | Mid-Atl    --         37,660,000   0.4256     --

  1980                Sum TopTrio       Sum  TopTrio  |   1988                Sum TopTrio       Sum  TopTrio
                      80,615,000     1.0000  8.659    |                       88,495,000     1.0000  --
---------------------------------------------------------------------------------------------------------
```

● Part 2: Take ratios of these TopTrio MortRates, with 1940 as the denominator of each ratio.
    Col.D modifies Col.C by separate PhysPop adjustments for MidTrio and LowTrio Census Divisions.

```
           Col.A        Col.B        Col.C     Col.D    Col.E
           TopTrio      1940 TopTrio = Col.A   ppAdju   = Col.C      URINARY-SYSTEM CANCERS.
           Mean MR      Mean MR      / Col.B   Tab 47-B * Col.D           Males.
                                               MidTrio
1950       9.936        9.550        1.040     0.99      1.03    = MidTrio Adjustment Factor, 1950
1960       9.636        9.550        1.009     0.97      0.98    = MidTrio Adjustment Factor, 1960
1970       9.172        9.550        0.960     0.95      0.91    = MidTrio Adjustment Factor, 1970
1980       8.659        9.550        0.907     0.94      0.85    = MidTrio Adjustment Factor, 1980
1988       --           9.550        --        0.94      0.00    = MidTrio Adjustment Factor, 1988
-------------------------------------------------    LowTrio    --------------------------------------
1950       9.936        9.550        1.040     1.00      1.04    = LowTrio Adjustment Factor, 1950
1960       9.636        9.550        1.009     1.01      1.02    = LowTrio Adjustment Factor, 1960
1970       9.172        9.550        0.960     1.02      0.98    = LowTrio Adjustment Factor, 1970
1980       8.659        9.550        0.907     1.04      0.94    = LowTrio Adjustment Factor, 1980
1988       --           9.550        --        1.07      0.00    = LowTrio Adjustment Factor, 1988
=========================================================================================================
```

```
===============================================================================================
                                        Table 59-B
                    Urinary-System Cancers, Males:  Fractional Causation in 1950
===============================================================================================
```

Part 1.
Calculation of the 6 Adjusted MortRates (Col.F) and the National Adjusted MortRate (Col.G).
The last six entries in Part 1, Col.F, are the products of (Col.D * Col.E), as discussed in Chap. 49.

Trio-Sequence	Col.A 1950 PopFrac Tab 3-B	Col.B 1950 Obs MR Tab 11-A	Col.C A * B	Col.D 1940 MR Mid,Low Tab 11-A	Col.E AdjuFact Bx2,Pt2 Col.E	Col.F 1950 Adju MortRates	Col.G A * F
Pacific	0.0961	8.4	0.807			8.4	0.807
New England	0.0618	10.5	0.649			10.5	0.649
Mid-Atlantic	0.2002	10.5	2.102			10.5	2.102
WestNoCentral	0.0933	7.2	0.672	6.7	1.03	6.90	0.644
EastNoCentral	0.2017	8.6	1.735	8.1	1.03	8.34	1.683
Mountain	0.0337	6.1	0.206	6.5	1.03	6.70	0.226
WestSoCentral	0.0965	5.8	0.560	4.3	1.04	4.47	0.432
EastSoCentral	0.0762	5.0	0.381	3.0	1.04	3.12	0.238
SouthAtlantic	0.1406	6.1	0.858	5.3	1.04	5.51	0.775

```
                                      Sum =    8.0                           Sum =
        1950   Observed MR from Table 11-B     8.1        1950 Natl Adjusted MR =   7.5548
```

Part 2.

Trio-Seq.	Col.A Mean1940 thru1950 PPs from Tab 47-A x'	Col.B 1950 Adju MRs from Col.F Part 1 y	Col.C Urinary Syst. Ca. Males: 1950 Adjusted MortRates regressed on Mean 1940 thru 1950 PPs Regression Output:		Col.D 1940 PPs from Table 3-A (TrioSeq) x''	Col.E Urinary Syst. Ca. Males: 1950 Adjusted MortRates regressed on 1940 PhysPops Regression Output:	
Pac	154.16	8.4	Constant	-3.4038	159.72	Constant	-3.4043
NewEng	162.03	10.5	Std Err of Y Est	0.7047	161.55	Std Err of Y Est	0.7548
MidAtl	169.24	10.5	R Squared	0.9329	169.76	R Squared	0.9230
WNoCen	121.60	6.90	No. of Observation	9	123.14	No. of Observation	9
ENoCen	128.53	8.34	Degrees of Freedom	7	133.36	Degrees of Freedom	7
Mtn	119.64	6.70			119.89		
WSoCen	102.64	4.47	X Coefficient(s)	0.0832	103.94	X Coefficient(s)	0.0821
ESoCen	84.44	3.12	Std Err of Coef.	0.0084	85.83	Std Err of Coef.	0.0090
SoAtl	99.91	5.51	XCoef / S.E. =	9.8635	100.74	XCoef / S.E. =	9.1609

Part 3-A.
Calculation of Fractional Causation
from Averaged PhysPops

1. Nonradiation rate is Adjusted
    Constant (Part 2, Col.C) = NEGATIVE   0.0

2. Radiation rate is Natl Adjusted
    MortRate (Part 1, Col.G = 7.5548)
    minus Nonradiation rate (0.0) =   7.5548

3. 1950 Fractional Causation is radiation
    rate (7.5548) divided by OBSERVED
    Natl MR Part 1,Col.C=   8.1   =   0.93

Part 3-B.
Calculation of Fractional Causation
from 1940 PhysPops

1. Nonradiation rate is Adjusted
    Constant (Part 2, Col.E) = NEGATIVE   0.0

2. Radiation rate is Natl Adjusted
    MortRate (Part 1, Col.G = 7.5548)
    minus Nonradiation rate (0.0) =   7.5548

3. 1950 Fractional Causation is radiation
    rate (7.5548) divided by OBSERVED
    Natl MR Part 1, Col.C=   8.1   =   0.93

```
=================================================================================
                                  Table 59-C
            Urinary-System Cancers, Males:  Fractional Causation in 1960
=================================================================================
Part 1.
Calculation of the 6 Adjusted MortRates (Col.F) and the National Adjusted MortRate (Col.G).
The last six entries in Part 1, Col.F, are the products of (Col.D * Col.E), as discussed in Chap. 49.


                    Col.A    Col.B    Col.C        Col.D    Col.E   Col.F    Col.G
                    1960     1960                  1940 MR  AdjuFact 1960
                    PopFrac  Obs MR   A * B        Mid,Low  Bx2,Pt2  Adju     A * F
Trio-Sequence       Tab 3-B  Tab 11-A             Tab 11-A Col.E    MortRates
        Pacific     0.1182    8.2     0.969                          8.2      0.969
        New England 0.0586   10.7     0.627                         10.7      0.627
        Mid-Atlantic 0.1905  10.2     1.943                         10.2      1.943
        WestNoCentral 0.0858  8.3     0.712         6.7     0.98     6.57     0.563
        EastNoCentral 0.2020  9.4     1.899         8.1     0.98     7.94     1.603
        Mountain     0.0382   7.8     0.298         6.5     0.98     6.37     0.243
        WestSoCentral 0.0945  6.6     0.624         4.3     1.02     4.39     0.414
        EastSoCentral 0.0672  5.2     0.349         3.0     1.02     3.06     0.206
        SouthAtlantic 0.1448  6.9     0.999         5.3     1.02     5.41     0.783


                              Sum =    8.4                           Sum =
      1960  Observed MR from Table 11-B    8.5     1960 Natl Adjusted MR =      7.3524

Part 2. -------------------------------------------------------------------------
        Col.A   Col.B                 Col.C                Col.D               Col.E
        Mean1940 1960          Urinary Syst. Ca. Males:    1940         Urinary Syst. Ca. Males:
        thru1960 Adju MRs      1960 Adjusted MortRates     PPs from     1960 Adjusted MortRates
Trio-   PPs from from Col.F        regressed on            Table 3-A        regressed on
Seq.    Tab 47-A Part 1       Mean 1940 thru 1960 PPs      (TrioSeq)      1940 PhysPops
        x'       y                Regression Output:        x''            Regression Output:
Pac     155.69   8.2          Constant         -3.5171     159.72      Constant         -3.5565
NewEng  162.81  10.7          Std Err of Y Est  0.7830     161.55      Std Err of Y Est  0.7719
MidAtl  167.04  10.2          R Squared         0.9170     169.76      R Squared         0.9194
WNoCen  118.15   6.57         No. of Observation     9     123.14      No. of Observation     9
ENoCen  123.87   7.94         Degrees of Freedom     7     133.36      Degrees of Freedom     7
Mtn     117.40   6.37                                      119.89
WSoCen  102.31   4.39         X Coefficient(s)  0.0833     103.94      X Coefficient(s)  0.0819
ESoCen   85.63   3.06         Std Err of Coef.  0.0095      85.83      Std Err of Coef.  0.0092
SoAtl   101.72   5.41         XCoef / S.E. =    8.7960     100.74      XCoef / S.E.      8.9342
```

Part 3-A.
Calculation of Fractional Causation
from Averaged PhysPops

1. Nonradiation rate is Adjusted
    Constant (Part 2, Col.C) = NEGATIVE   0.0

2. Radiation rate is Natl Adjusted
    MortRate (Part 1, Col.G = 7.3524)
    minus Nonradiation rate (0.0) =   7.3524

3. 1960 Fractional Causation is radiation
    rate (7.3524) divided by OBSERVED
    Natl MR Part 1,Col.C=   8.5   =   0.86

Part 3-B.
Calculation of Fractional Causation
from 1940 PhysPops

1. Nonradiation rate is Adjusted
    Constant (Part 2, Col.E) = NEGATIVE   0.0

2. Radiation rate is Natl Adjusted
    MortRate (Part 1, Col.G = 7.3524)
    minus Nonradiation rate (0.0) =   7.3524

3. 1960 Fractional Causation is radiation
    rate (7.3524) divided by OBSERVED
    Natl MR Part 1, Col.C=   8.5   =   0.86

```
==============================================================================================
                                    Table 59-E
                Urinary-System Cancers, Males:  Fractional Causation in 1980
==============================================================================================
Part 1.
Calculation of the 6 Adjusted MortRates (Col.F) and the National Adjusted MortRate (Col.G).
The last six entries in Part 1, Col.F, are the products of (Col.D * Col.E), as discussed in Chap. 49.
```

	Col.A 1980 PopFrac Tab 3-B	Col.B 1980 Obs MR Tab 11-A	Col.C A * B	Col.D 1940 MR Mid,Low Tab 11-A	Col.E AdjuFact Bx2,Pt2 Col.E	Col.F 1980 Adju MortRates	Col.G A * F
Trio-Sequence							
Pacific	0.1398	7.7	1.076			7.7	1.076
New England	0.0546	9.5	0.519			9.5	0.519
Mid-Atlantic	0.1630	9.2	1.500			9.2	1.500
WestNoCentral	0.0759	7.9	0.600	6.7	0.85	5.70	0.432
EastNoCentral	0.1846	8.7	1.606	8.1	0.85	6.89	1.271
Mountain	0.0502	7.0	0.351	6.5	0.85	5.52	0.277
WestSoCentral	0.1049	7.0	0.734	4.3	0.94	4.04	0.424
EastSoCentral	0.0646	7.3	0.472	3.0	0.94	2.82	0.182
SouthAtlantic	0.1624	7.8	1.267	5.3	0.94	4.98	0.809

```
                                    Sum =    8.1                              Sum =
        1980 Observed MR from Table 11-B     8.2        1980 Natl Adjusted MR =    6.4906
```

```
Part 2.  ----------------------------------------------------------------------------------
        Col.A   Col.B                              Col.C              Col.D                    Col.E
      Mean1940    1980      Urinary Syst. Ca. Males:                   1940      Urinary Syst. Ca. Males:
      thru1980 Adju MRs       1980 Adjusted MortRates               PPs from      1980 Adjusted MortRates
 Trio- PPs from from Col.F         regressed on                     Table 3-A          regressed on
 Seq.  Tab 47-A  Part 1    Mean 1940 thru 1980 PPs                  (TrioSeq)       1940 PhysPops
          x'       y             Regression Output:                    x''          Regression Output:
Pac     177.35    7.7      Constant              -3.3843             159.72      Constant              -3.2198
NewEng  185.86    9.5      Std Err of Y Est       0.6507             161.55      Std Err of Y Est       0.5795
MidAtl  186.11    9.2      R Squared              0.9275             169.76      R Squared              0.9425
WNoCen  128.82    5.70     No. of Observation        9               123.14      No. of Observation        9
ENoCen  133.71    6.89     Degrees of Freedom        7               133.36      Degrees of Freedom        7
Mtn     133.45    5.52                                                119.89
WSoCen  114.66    4.04     X Coefficient(s)       0.0676             103.94      X Coefficient(s)       0.0737
ESoCen   99.46    2.82     Std Err of Coef.       0.0071              85.83      Std Err of Coef.       0.0069
SoAtl   124.62    4.98     XCoef / S.E. =         9.4602             100.74      XCoef / S.E.          10.7079
```

```
--------------------------------------------------------------------------------------------
Part 3-A.                                         |  Part 3-B.
Calculation of Fractional Causation               |  Calculation of Fractional Causation
from Averaged PhysPops                             |  from 1940 PhysPops
                                                  |
1.  Nonradiation rate is Adjusted                 |  1.  Nonradiation rate is Adjusted
    Constant (Part 2, Col.C) = NEGATIVE    0.0     |      Constant (Part 2, Col.E) = NEGATIVE    0.0
                                                  |
2.  Radiation rate is Natl Adjusted               |  2.  Radiation rate is Natl Adjusted
    MortRate (Part 1, Col.G = 6.4906)             |      MortRate (Part 1, Col.G = 6.4906)
    minus Nonradiation rate (0.0) =     6.4906    |      minus Nonradiation rate (0.0) =     6.4906
                                                  |
3.  1980 Fractional Causation is radiation        |  3.  1980 Fractional Causation is radiation
    rate (6.4906) divided by OBSERVED             |      rate (6.4906) divided by OBSERVED
    Natl MR Part 1,Col.C=    8.2    =    0.79     |      Natl MR Part 1, Col.C=    8.2    =    0.79
--------------------------------------------------------------------------------------------
```

```
==================================================================================================
                                        Table 59-BB
                     Urinary-System Cancers, Males:  Fractional Causation in 1950
==================================================================================================
Part 1.
Calculation of the 6 Adjusted MortRates (Col.F) and the National Adjusted MortRate (Col.G).
The last six entries in Part 1, Col.F, are the products of (Col.D * Col.E), as discussed in Chap. 49.


                  Col.A    Col.B    Col.C        Col.D    Col.E    Col.F    Col.G
                  1950     1950                  1940 MR  AdjuFact  1950
                  PopFrac  Obs MR   A * B        Mid,Low  Bx2,Pt2  Adju     A * F
Trio-Sequence     Tab 3-B  Tab 11-A             Tab 11-A  Col.E   MortRates
      Pacific     0.0961    8.4     0.807                           8.4     0.807
      New England 0.0618   10.5     0.649                          10.5     0.649
      Mid-Atlantic 0.2002  10.5     2.102                          10.5     2.102
      WestNoCentral 0.0933  7.2     0.672          6.7     1.39     9.32    0.869
      EastNoCentral 0.2017  8.6     1.735          8.1     1.39    11.26    2.272
      Mountain    0.0337    6.1     0.206          6.5     1.39     9.04    0.305
      WestSoCentral 0.0965  5.8     0.560          4.3     1.40     6.04    0.583
      EastSoCentral 0.0762  5.0     0.381          3.0     1.40     4.21    0.321
      SouthAtlantic 0.1406  6.1     0.858          5.3     1.40     7.44    1.046

                            Sum =   8.0                            Sum =
   1950   Observed MR from Table 11-B   8.1     1950 Natl Adjusted MR =    8.9536

Part 2.  -----------------------------------------------------------------------------------------
         Col.A    Col.B                                                              Col.E
         Mean1940  1950                          Col.C           Col.D     Urinary Syst. Ca. Males:
         thru1950 Adju MRs        Urinary Syst. Ca. Males:       1940      1950 Adjusted MortRates
Trio-    PPs from from Col.F      1950 Adjusted MortRates        PPs from        regressed on
Seq.     Tab 47-A Part 1              regressed on               Table 3-A    1940 PhysPops
           x'       y           Mean 1940 thru 1950 PPs        (TrioSeq)
                                     Regression Output:             x''         Regression Output:
Pac      154.16    8.4          Constant          0.8998       159.72     Constant          0.8522
NewEng   162.03   10.5          Std Err of Y Est  1.5587       161.55     Std Err of Y Est  1.5568
MidAtl   169.24   10.5          R Squared         0.5967       169.76     R Squared         0.5977
WNoCen   121.60    9.32         No. of Observation    9        123.14     No. of Observation    9
ENoCen   128.53   11.26         Degrees of Freedom    7        133.36     Degrees of Freedom    7
Mtn      119.64    9.04                                        119.89
WSoCen   102.64    6.04         X Coefficient(s)  0.0601       103.94     X Coefficient(s)  0.0596
ESoCen    84.44    4.21         Std Err of Coef.  0.0187        85.83     Std Err of Coef.  0.0185
SoAtl     99.91    7.44         XCoef / S.E. =    3.2180       100.74     XCoef / S.E.      3.2248


---------------------------------------------------        ---------------------------------------
Part 3-A.                                            |      Part 3-B.
Calculation of Fractional Causation                  |      Calculation of Fractional Causation
from Averaged PhysPops                               |      from 1940 PhysPops
                                                     |
1.  Nonradiation rate is Adjusted                    |      1.  Nonradiation rate is Adjusted
    Constant (Part 2, Col.C) =           0.8998      |          Constant (Part 2, Col.E) =           0.8522
                                                     |
2.  Radiation rate is Natl Adjusted                  |      2.  Radiation rate is Natl Adjusted
    MortRate (Part 1, Col.G = 8.9536)                |          MortRate (Part 1, Col.G = 8.9536)
    minus Nonradiation rate (0.8998) =   8.0538      |          minus Nonradiation rate (0.8522) =   8.1014
                                                     |
3.  1950 Fractional Causation is radiation           |      3.  1950 Fractional Causation is radiation
    rate (8.0538) divided by OBSERVED                |          rate (8.1014) divided by OBSERVED
    Natl MR Part 1,Col.C=    8.1    =     0.99       |          Natl MR Part 1, Col.C=    8.1    =    1.00
-----------------------------------------------------------------------------------------------------
```

```
===================================================================================================
                                        Table 59-CC
                    Urinary-System Cancers, Males:  Fractional Causation in 1960
===================================================================================================
Part 1.
Calculation of the 6 Adjusted MortRates (Col.F) and the National Adjusted MortRate (Col.G).
The last six entries in Part 1, Col.F, are the products of (Col.D * Col.E), as discussed in Chap. 49.


                  Col.A    Col.B    Col.C         Col.D    Col.E    Col.F    Col.G
                  1960     1960                   1940 MR  AdjuFact 1960
                  PopFrac  Obs MR   A * B         Mid,Low  Bx2,Pt2  Adju     A * F
  Trio-Sequence   Tab 3-B  Tab 11-A               Tab 11-A Col.E    MortRates
        Pacific   0.1182    8.2     0.969                            8.2     0.969
        New England 0.0586 10.7     0.627                           10.7     0.627
        Mid-Atlantic 0.1905 10.2    1.943                           10.2     1.943
        WestNoCentral 0.0858 8.3    0.712          6.7     1.32      8.86     0.761
        EastNoCentral 0.2020 9.4    1.899          8.1     1.32     10.72     2.165
        Mountain    0.0382   7.8     0.298          6.5     1.32      8.60     0.329
        WestSoCentral 0.0945 6.6    0.624          4.3     1.38      5.92     0.560
        EastSoCentral 0.0672 5.2    0.349          3.0     1.38      4.13     0.278
        SouthAtlantic 0.1448 6.9    0.999          5.3     1.38      7.30     1.057


                           Sum =   8.4                               Sum =
  1960    Observed MR from Table 11-B  8.5        1960 Natl Adjusted MR =     8.6870

Part 2.  --------------------------------------------------------------------------------------
         Col.A    Col.B                      Col.C                Col.D                    Col.E
         Mean1940  1960          Urinary Syst. Ca. Males:          1940        Urinary Syst. Ca. Males:
         thru1960 Adju MRs       1960 Adjusted MortRates          PPs from     1960 Adjusted MortRates
  Trio-  PPs from from Col.F        regressed on                  Table 3-A        regressed on
  Seq.   Tab 47-A  Part 1       Mean 1940 thru 1960 PPs           (TrioSeq)     1940 PhysPops
           x'       y               Regression Output:              x''          Regression Output:
  Pac     155.69    8.2       Constant          0.9177            159.72      Constant          0.5824
  NewEng  162.81   10.7       Std Err of Y Est  1.5241            161.55      Std Err of Y Est  1.4247
  MidAtl  167.04   10.2       R Squared         0.5901            169.76      R Squared         0.6418
  WNoCen  118.15    8.86      No. of Observation    9             123.14      No. of Observation    9
  ENoCen  123.87   10.72      Degrees of Freedom    7             133.36      Degrees of Freedom    7
  Mtn     117.40    8.60                                          119.89
  WSoCen  102.31    5.92      X Coefficient(s)  0.0585            103.94      X Coefficient(s)  0.0599
  ESoCen   85.63    4.13      Std Err of Coef.  0.0184             85.83      Std Err of Coef.  0.0169
  SoAtl   101.72    7.30      XCoef / S.E. =    3.1746            100.74      XCoef / S.E.      3.5416


---------------------------------------------------|------------------------------------------------
Part 3-A.                                          | Part 3-B.
Calculation of Fractional Causation                | Calculation of Fractional Causation
from Averaged PhysPops                             | from 1940 PhysPops
                                                   |
1.  Nonradiation rate is Adjusted                  | 1.  Nonradiation rate is Adjusted
    Constant (Part 2, Col.C) =          0.9177     |     Constant (Part 2, Col.E) =          0.5824
                                                   |
2.  Radiation rate is Natl Adjusted                | 2.  Radiation rate is Natl Adjusted
    MortRate (Part 1, Col.G = 8.6870)              |     MortRate (Part 1, Col.G = 8.6870)
    minus Nonradiation rate (0.9177) =    7.7693   |     minus Nonradiation rate (0.5824) =    8.1046
                                                   |
3.  1960 Fractional Causation is radiation         | 3.  1960 Fractional Causation is radiation
    rate (7.7693) divided by OBSERVED              |     rate (8.1046) divided by OBSERVED
    Natl MR Part 1,Col.C=    8.5    =      0.91     |     Natl MR Part 1, Col.C=    8.5    =      0.95
---------------------------------------------------|------------------------------------------------
```

```
==================================================================================================
                                        Table 59-EE
                      Urinary-System Cancers, Males:  Fractional Causation in 1980
==================================================================================================
Part 1.
Calculation of the 6 Adjusted MortRates (Col.F) and the National Adjusted MortRate (Col.G).
The last six entries in Part 1, Col.F, are the products of (Col.D * Col.E), as discussed in Chap. 49.


                  Col.A    Col.B    Col.C        Col.D    Col.E    Col.F    Col.G
                  1980     1980                  1940 MR  AdjuFact 1980
                  PopFrac  Obs MR   A * B        Mid,Low  Bx2,Pt2  Adju     A * F
   Trio-Sequence  Tab 3-B  Tab 11-A             Tab 11-A Col.E    MortRates
        Pacific   0.1398   7.7      1.076                          7.7      1.076
        New England 0.0546 9.5      0.519                          9.5      0.519
        Mid-Atlantic 0.1630 9.2     1.500                          9.2      1.500
        WestNoCentral 0.0759 7.9    0.600        6.7     1.15      7.69     0.584
        EastNoCentral 0.1846 8.7    1.606        8.1     1.15      9.29     1.716
        Mountain  0.0502   7.0      0.351        6.5     1.15      7.46     0.374
        WestSoCentral 0.1049 7.0    0.734        4.3     1.27      5.46     0.572
        EastSoCentral 0.0646 7.3    0.472        3.0     1.27      3.81     0.246
        SouthAtlantic 0.1624 7.8    1.267        5.3     1.27      6.73     1.092

                            Sum =    8.1                                    Sum =
           1980 Observed MR from Table 11-B     8.2        1980 Natl Adjusted MR =    7.6791

Part 2. -----------------------------------------------------------------------------------------
        Col.A    Col.B                            Col.C            Col.D                    Col.E
        Mean1940  1980               Urinary Syst. Ca. Males:      1940      Urinary Syst. Ca. Males:
        thru1980 Adju MRs            1980 Adjusted MortRates       PPs from  1980 Adjusted MortRates
   Trio- PPs from from Col.F            regressed on               Table 3-A    regressed on
   Seq.  Tab 47-A Part 1             Mean 1940 thru 1980 PPs       (TrioSeq) 1940 PhysPops
          x'       y                           Regression Output:    x''              Regression Output:
   Pac   177.35   7.7      Constant              0.8918            159.72    Constant              0.5316
   NewEng 185.86  9.5      Std Err of Y Est      1.2629            161.55    Std Err of Y Est      1.0819
   MidAtl 186.11  9.2      R Squared             0.6089            169.76    R Squared             0.7130
   WNoCen 128.82  7.69     No. of Observation    9                 123.14    No. of Observation    9
   ENoCen 133.71  9.29     Degrees of Freedom    7                 133.36    Degrees of Freedom    7
   Mtn   133.45   7.46                                             119.89
   WSoCen 114.66  5.46     X Coefficient(s)      0.0458            103.94    X Coefficient(s)      0.0536
   ESoCen 99.46   3.81     Std Err of Coef.      0.0139            85.83     Std Err of Coef.      0.0128
   SoAtl 124.62   6.73     XCoef / S.E. =        3.3016            100.74    XCoef / S.E.          4.1703


-----------------------------------------------------------------------------------------------
Part 3-A.                                        |  Part 3-B.
Calculation of Fractional Causation              |  Calculation of Fractional Causation
from Averaged PhysPops                           |  from 1940 PhysPops
                                                 |
1.  Nonradiation rate is Adjusted                |  1.  Nonradiation rate is Adjusted
    Constant (Part 2, Col.C) =         0.8918    |      Constant (Part 2, Col.E) =         0.5316
                                                 |
2.  Radiation rate is Natl Adjusted              |  2.  Radiation rate is Natl Adjusted
    MortRate (Part 1, Col.G = 7.6791)            |      MortRate (Part 1, Col.G = 7.6791)
    minus Nonradiation rate (0.8918) =  6.7873   |      minus Nonradiation rate (0.5316) =  7.1475
                                                 |
3.  1980 Fractional Causation is radiation       |  3.  1980 Fractional Causation is radiation
    rate (6.7873) divided by OBSERVED            |      rate (7.1475) divided by OBSERVED
    Natl MR Part 1,Col.C=   8.2   =      0.83    |      Natl MR Part 1,Col.C=   8.2   =      0.87
-----------------------------------------------------------------------------------------------
```

CHAPTER 60

Urinary-System Cancers, Females, 1940-1980

● Table 60-A, Column A, shows the female National MortRates from Urinary-System Cancers. Relative to most other sets of cancers studied in this book, the rates per 100,000 female population are very small numbers in every decade. Ordinarily, one should be wary of believing that a change from 4.0 to 3.0 is real, and we are wary. Nonetheless, the change occurs with such steadiness over the 1940-1980 period, and comes out of such a huge database, that a decline in the National MortRate is probably real.

● Box 1 shows that, by 1960 (Columns D and F), the MortRates are falling in the TopTrio and MidTrio, while rising in the LowTrio. By 1980 (Columns I and K), the MortRates are falling in all three Trios, but more in the TopTrio than in the LowTrio. These observations mean that a carcinogenic co-actor which can contribute to female MortRates, from Urinary-System Cancers, is operating more strongly in the LowTrio than in the TopTrio (Chapter 48, Part 5b). We must match the Census Divisions for this co-actor, whatever its identity. We believe that its identity is smoking.

Table 60-A Urinary-System Cancers, Females: Fractional Causation by Medical Radiation over Time							
Year	Col.A Natl MR	Col.B Frac.C		Col.C R-Sq	Col.D X-Coef	Col.E StdErr	Col.F Coef/SE	Col.G Source
1940	4.0	86%		0.9395	0.0247	0.0024	10.4305	Chap.12
1950	3.9	76%		0.9508	0.0223	0.0019	11.6295	Tab 60-B
1960	3.6	77%		0.9346	0.0211	0.0021	10.0025	Tab 60-C
1970	3.3	77%		0.9263	0.0189	0.0020	9.3816	Tab 60-D
1980	3.0	78%		0.9112	0.0161	0.0019	8.4777	Tab 60-E

```
===================================================================================================
                                    Box 1, Chap. 60
            Urinary-System Cancers, Females:  Post-1940 Change in MortRates by Census Trios

1960 vs. 1940, by Trios:  Col.D expresses change by ratios.  Col.F expresses change by subtraction.
1980 vs. 1940, by Trios:  Col.I expresses change by ratios.  Col.K expresses change by subtraction.
MRs change inversely with PP.  High-PP Trio has lowest growth-factor.  Low-PP Trio has highest growth-factor.


           Col.A   |  Col.B    Col.C    Col.D    Col.E    Col.F   |  Col.G    Col.H    Col.I    Col.J    Col.K
           1940    |  1960     Ratio    Input    Diff:    Input   |  1980     Ratio    Input    Diff:    Input
           MortRate|  MortRate Col.B    from     Col.B    from    |  MortRate Col.G    from     Col.G    from
           Tab 12-A|  Tab 12-A /Col.A   Col.C    minus A  Col.E   |  Tab 12-A /Col.A   Col.H    minus A  Col.J
                   |                                              |
Pacif      4.1     |  3.3      0.805    Avg Chg  -0.8     Avg Chg  |  2.8      0.683    Avg Chg  -1.3     Avg Chg
NewE       4.7     |  3.9      0.830    TopTrio  -0.8     TopTrio  |  3.4      0.723    TopTrio  -1.3     TopTrio
MidAtl     4.9     |  4.0      0.816    0.817    -0.9     -0.8     |  3.2      0.653    0.686    -1.7     -1.4
                   |                                              |
WNoCen     3.7     |  3.3      0.892    Avg Chg  -0.4     Avg Chg  |  3.0      0.811    Avg Chg  -0.7     Avg Chg
ENoCen     4.1     |  3.9      0.951    MidTrio  -0.2     MidTrio  |  3.0      0.732    MidTrio  -1.1     MidTrio
Mtn        3.5     |  3.4      0.971    0.938    -0.1     -0.2     |  2.5      0.714    0.752    -1.0     -0.9
                   |                                              |
WSoCen     3.1     |  3.2      1.032    Avg Chg  0.1      Avg Chg  |  2.8      0.903    Avg Chg  -0.3     Avg Chg
ESoCen     2.7     |  3.0      1.111    LowTrio  0.3      LowTrio  |  2.8      1.037    LowTrio  0.1      LowTrio
SoAtl      3.0     |  3.3      1.100    1.081    0.3      0.2      |  2.9      0.967    0.969    -0.1     -0.1
===================================================================================================
```

===
Box 2, Chap. 60
Urinary-System Cancers, Females: Calculation of Adjustment Factor

This adjustment is discussed fully in Chapter 49.
● Part 1: Calculate average population-weighted MortRate for the combined TopTrio Census Divs.

Census Div.	Col.A 1940 MR Tab 12-A	Col.B 1940 Pop'n Tab 3-B	Col.C 1940 Popn /45,710,039	Col.D Col.A * Col.C	Census Div.	Col.A 1950 MR Tab 12-A	Col.B 1950 Pop'n Tab 3-B	Col.C 1950 Popn /53,964,513	Col.D Col.A * Col.C
Pacific	4.1	9,733,262	0.2129	0.87	Pacific	3.9	14,486,527	0.2684	1.05
NewEng	4.7	8,437,290	0.1846	0.87	NewEng	3.9	9,314,453	0.1726	0.67
Mid-Atl	4.9	27,539,487	0.6025	2.95	Mid-Atl	4.5	30,163,533	0.5590	2.52
1940		Sum TopTrio 45,710,039	Sum 1.0000	TopTrio 4.693	1950		Sum TopTrio 53,964,513	Sum 1.0000	TopTrio 4.235

Census Div.	Col.A 1960 MR Tab 12-A	Col.B 1960 Pop'n Tab 3-B	Col.C 1960 Popn /65,875,863	Col.D Col.A * Col.C	Census Div.	Col.A 1970 MR Tab 12-A	Col.B 1970 Pop'n Tab 3-B	Col.C 1970 Popn /75,017,000	Col.D Col.A * Col.C
Pacific	3.3	21,198,044	0.3218	1.06	Pacific	3.1	26,087,000	0.3477	1.08
NewEng	3.9	10,509,367	0.1595	0.62	NewEng	3.7	11,781,000	0.1570	0.58
Mid-Atl	4.0	34,168,452	0.5187	2.07	Mid-Atl	3.6	37,149,000	0.4952	1.78
1960		Sum TopTrio 65,875,863	Sum 1.0000	TopTrio 3.759	1970		Sum TopTrio 75,017,000	Sum 1.0000	TopTrio 3.442

Census Div.	Col.A 1980 MR Tab 12-A	Col.B 1980 Pop'n Tab 3-B	Col.C 1980 Popn /80,615,000	Col.D Col.A * Col.C	Census Div.	Col.A 1988 MR Tab 12-A	Col.B 1990 Pop'n Tab 3-B	Col.C 1990 Popn /88,495,000	Col.D Col.A * Col.C
Pacific	2.8	31,523,000	0.3910	1.09	Pacific	--	37,837,000	0.4276	--
NewEng	3.4	12,322,000	0.1528	0.52	NewEng	--	12,998,000	0.1469	--
Mid-Atl	3.2	36,770,000	0.4561	1.46	Mid-Atl	--	37,660,000	0.4256	--
1980		Sum TopTrio 80,615,000	Sum 1.0000	TopTrio 3.074	1988		Sum TopTrio 88,495,000	Sum 1.0000	TopTrio --

----

● Part 2: Take ratios of these TopTrio MortRates, with 1940 as the denominator of each ratio.
   Col.D modifies Col.C by separate PhysPop adjustments for MidTrio and LowTrio Census Divisions.

	Col.A TopTrio Mean MR	Col.B 1940 TopTrio Mean MR	Col.C = Col.A / Col.B	Col.D ppAdju Tab 47-B	Col.E = Col.C * Col.D	URINARY-SYSTEM CANCERS. Females.
				MidTrio		
1950	4.235	4.693	0.903	0.99	0.89	= MidTrio Adjustment Factor, 1950
1960	3.759	4.693	0.801	0.97	0.78	= MidTrio Adjustment Factor, 1960
1970	3.442	4.693	0.733	0.95	0.70	= MidTrio Adjustment Factor, 1970
1980	3.074	4.693	0.655	0.94	0.62	= MidTrio Adjustment Factor, 1980
1988	--	4.693	--	0.94	--	= MidTrio Adjustment Factor, 1988
				LowTrio		
1950	4.235	4.693	0.903	1.00	0.90	= LowTrio Adjustment Factor, 1950
1960	3.759	4.693	0.801	1.01	0.81	= LowTrio Adjustment Factor, 1960
1970	3.442	4.693	0.733	1.02	0.75	= LowTrio Adjustment Factor, 1970
1980	3.074	4.693	0.655	1.04	0.68	= LowTrio Adjustment Factor, 1980
1988	--	4.693	--	1.07	--	= LowTrio Adjustment Factor, 1988

===

```
=============================================================================================
                                       Table 60-B
                Urinary-System Cancers, Females:  Fractional Causation in 1950
=============================================================================================
Part 1.
Calculation of the 6 Adjusted MortRates (Col.F) and the National Adjusted MortRate (Col.G).
The last six entries in Part 1, Col.F, are the products of (Col.D * Col.E), as discussed in Chap. 49.
```

	Col.A 1950 PopFrac Tab 3-B	Col.B 1950 Obs MR Tab 12-A	Col.C A * B	Col.D 1940 MR Mid,Low Tab 12-A	Col.E AdjuFact Bx2,Pt2 Col.E	Col.F 1950 Adju MortRates	Col.G A * F
Trio-Sequence							
Pacific	0.0961	3.9	0.375			3.9	0.375
New England	0.0618	3.9	0.241			3.9	0.241
Mid-Atlantic	0.2002	4.5	0.901			4.5	0.901
WestNoCentral	0.0933	3.6	0.336	3.7	0.89	3.29	0.307
EastNoCentral	0.2017	4.2	0.847	4.1	0.89	3.65	0.736
Mountain	0.0337	3.5	0.118	3.5	0.89	3.12	0.105
WestSoCentral	0.0965	3.4	0.328	3.1	0.90	2.79	0.269
EastSoCentral	0.0762	3.6	0.274	2.7	0.90	2.43	0.185
SouthAtlantic	0.1406	3.6	0.506	3.0	0.90	2.70	0.380

```
                                     Sum =     3.9                            Sum =
        1950 Observed MR from Table 12-B       3.9       1950 Natl Adjusted MR =     3.4989
```

```
Part 2. -------------------------------------------------------------------------------------
       Col.A    Col.B                          Col.C                Col.D                        Col.E
       Mean1940   1950        Urinary Sys Ca. Females:              1940         Urinary Sys Ca. Females:
       thru1950 Adju MRs       1950 Adjusted MortRates             PPs from       1950 Adjusted MortRates
Trio-  PPs from from Col.F        regressed on                     Table 3-A          regressed on
Seq.   Tab 47-A Part 1       Mean 1940 thru 1950 PPs              (TrioSeq)         1940 PhysPops
         x'       y            Regression Output:                    x''           Regression Output:
Pac    154.16    3.9       Constant            0.5306              159.72       Constant            0.5040
NewEng 162.03    3.9       Std Err of Y Est    0.1603              161.55       Std Err of Y Est    0.1475
MidAtl 169.24    4.5       R Squared           0.9508              169.76       R Squared           0.9583
WNoCen 121.60    3.29      No. of Observation      9               123.14       No. of Observation      9
ENoCen 128.53    3.65      Degrees of Freedom      7               133.36       Degrees of Freedom      7
Mtn    119.64    3.12                                              119.89
WSoCen 102.64    2.79      X Coefficient(s)    0.0223              103.94       X Coefficient(s)    0.0222
ESoCen  84.44    2.43      Std Err of Coef.    0.0019               85.83       Std Err of Coef.    0.0018
SoAtl   99.91    2.70      XCoef / S.E. =     11.6295              100.74       XCoef / S.E.       12.6873
```

```
-----------------------------------------------------------|-----------------------------------------------------
Part 3-A.                                                  | Part 3-B.
Calculation of Fractional Causation                        | Calculation of Fractional Causation
from Averaged PhysPops                                     | from 1940 PhysPops
                                                           |
1. Nonradiation rate is Adjusted                           | 1. Nonradiation rate is Adjusted
   Constant (Part 2, Col.C) =            0.5306            |    Constant (Part 2, Col.E) =            0.5040
                                                           |
2. Radiation rate is Natl Adjusted                         | 2. Radiation rate is Natl Adjusted
   MortRate (Part 1, Col.G = 3.4989)                       |    MortRate (Part 1, Col.G = 3.4989)
   minus Nonradiation rate (0.5306) =    2.9684            |    minus Nonradiation rate (0.5040) =    2.9949
                                                           |
3. 1950 Fractional Causation is radiation                  | 3. 1950 Fractional Causation is radiation
   rate (2.9684) divided by OBSERVED                       |    rate (2.9949) divided by OBSERVED
   Natl MR Part 1,Col.C=   3.9    =       0.76             |    Natl MR Part 1, Col.C=   3.9    =      0.77
-----------------------------------------------------------|-----------------------------------------------------
```

===
Table 60-C
Urinary-System Cancers, Females: Fractional Causation in 1960
===

Part 1.

Calculation of the 6 Adjusted MortRates (Col.F) and the National Adjusted MortRate (Col.G).

The last six entries in Part 1, Col.F, are the products of (Col.D * Col.E), as discussed in Chap. 49.

	Col.A 1960 PopFrac Tab 3-B	Col.B 1960 Obs MR Tab 12-A	Col.C A * B		Col.D 1940 MR Mid,Low Tab 12-A	Col.E AdjuFact Bx2,Pt2 Col.E	Col.F 1960 Adju MortRates	Col.G A * F
Trio-Sequence								
Pacific	0.1182	3.3	0.390				3.3	0.390
New England	0.0586	3.9	0.229				3.9	0.229
Mid-Atlantic	0.1905	4.0	0.762				4.0	0.762
WestNoCentral	0.0858	3.3	0.283		3.7	0.78	2.89	0.248
EastNoCentral	0.2020	3.9	0.788		4.1	0.78	3.20	0.646
Mountain	0.0382	3.4	0.130		3.5	0.78	2.73	0.104
WestSoCentral	0.0945	3.2	0.302		3.1	0.81	2.51	0.237
EastSoCentral	0.0672	3.0	0.202		2.7	0.81	2.19	0.147
SouthAtlantic	0.1448	3.3	0.478		3.0	0.81	2.43	0.352

Sum =   3.6   Sum =
1960 Observed MR from Table 12-B   3.6   1960 Natl Adjusted MR =   3.1146

Part 2. --

	Col.A Mean1940 thru1960 PPs from Tab 47-A x'	Col.B 1960 Adju MRs from Col.F Part 1 y	Col.C Urinary Sys Ca. Females: 1960 Adjusted MortRates regressed on Mean 1940 thru 1960 PPs Regression Output:		Col.D 1940 PPs from Table 3-A (TrioSeq) x''	Col.E Urinary Sys Ca. Females: 1960 Adjusted MortRates regressed on 1940 PhysPops Regression Output:
Trio-Seq.						
Pac	155.69	3.3	Constant 0.3602		159.72	Constant 0.3525
NewEng	162.81	3.9	Std Err of Y Est 0.1742		161.55	Std Err of Y Est 0.1731
MidAtl	167.04	4.0	R Squared 0.9346		169.76	R Squared 0.9355
WNoCen	118.15	2.89	No. of Observation 9		123.14	No. of Observation 9
ENoCen	123.87	3.20	Degrees of Freedom 7		133.36	Degrees of Freedom 7
Mtn	117.40	2.73			119.89	
WSoCen	102.31	2.51	X Coefficient(s) 0.0211		103.94	X Coefficient(s) 0.0207
ESoCen	85.63	2.19	Std Err of Coef. 0.0021		85.83	Std Err of Coef. 0.0021
SoAtl	101.72	2.43	XCoef / S.E. = 10.0025		100.74	XCoef / S.E. 10.0723

Part 3-A.

Calculation of Fractional Causation
from Averaged PhysPops

1. Nonradiation rate is Adjusted
    Constant (Part 2, Col.C) =   0.3602

2. Radiation rate is Natl Adjusted
    MortRate (Part 1, Col.G = 3.1146)
    minus Nonradiation rate (0.3602) =   2.7544

3. 1960 Fractional Causation is radiation
    rate (2.7544) divided by OBSERVED
    Natl MR Part 1,Col.C=   3.6   =   0.77

Part 3-B.

Calculation of Fractional Causation
from 1940 PhysPops

1. Nonradiation rate is Adjusted
    Constant (Part 2, Col.E) =   0.3525

2. Radiation rate is Natl Adjusted
    MortRate (Part 1, Col.G = 3.1146)
    minus Nonradiation rate (0.3525) =   2.7621

3. 1960 Fractional Causation is radiation
    rate (2.7621) divided by OBSERVED
    Natl MR Part 1, Col.C=   3.6   =   0.77

==

Table 60-D

Urinary-System Cancers, Females: Fractional Causation in 1970

==

Part 1.

Calculation of the 6 Adjusted MortRates (Col.F) and the National Adjusted MortRate (Col.G).

The last six entries in Part 1, Col.F, are the products of (Col.D * Col.E), as discussed in Chap. 49.

Trio-Sequence	Col.A 1970 PopFrac Tab 3-B	Col.B 1970 Obs MR Tab 12-A	Col.C A * B	Col.D 1940 MR Mid,Low Tab 12-A	Col.E AdjuFact Bx2,Pt2 Col.E	Col.F 1970 Adju MortRates	Col.G A * F
Pacific	0.1293	3.1	0.401			3.1	0.401
New England	0.0584	3.7	0.216			3.7	0.216
Mid-Atlantic	0.1842	3.6	0.663			3.6	0.663
WestNoCentral	0.0805	3.2	0.258	3.7	0.70	2.59	0.208
EastNoCentral	0.1993	3.5	0.698	4.1	0.70	2.87	0.572
Mountain	0.0408	3.0	0.122	3.5	0.70	2.45	0.100
WestSoCentral	0.0948	3.0	0.284	3.1	0.75	2.33	0.220
EastSoCentral	0.0631	2.9	0.183	2.7	0.75	2.03	0.128
SouthAtlantic	0.1496	3.1	0.464	3.0	0.75	2.25	0.337

Sum =   3.3

1970 Observed MR from Table 12-B   3.3   1970 Natl Adjusted MR =   Sum = 2.8453

Part 2. --

Trio- Seq.	Col.A Mean1940 thru1970 PPs from Tab 47-A x'	Col.B 1970 Adju MRs from Col.F Part 1 y	Col.C Urinary Sys Ca. Females: 1970 Adjusted MortRates regressed on Mean 1940 thru 1970 PPs Regression Output:		Col.D 1940 PPs from Table 3-A (TrioSeq) x''	Col.E Urinary Sys Ca. Females: 1970 Adjusted MortRates regressed on 1940 PhysPops Regression Output:	
Pac	162.72	3.1	Constant	0.3018	159.72	Constant	0.2882
NewEng	168.74	3.7	Std Err of Y Est	0.1726	161.55	Std Err of Y Est	0.1679
MidAtl	173.28	3.6	R Squared	0.9263	169.76	R Squared	0.9303
WNoCen	119.56	2.59	No. of Observation	9	123.14	No. of Observation	9
ENoCen	124.70	2.87	Degrees of Freedom	7	133.36	Degrees of Freedom	7
Mtn	122.37	2.45			119.89		
WSoCen	105.03	2.33	X Coefficient(s)	0.0189	103.94	X Coefficient(s)	0.0193
ESoCen	89.44	2.03	Std Err of Coef.	0.0020	85.83	Std Err of Coef.	0.0020
SoAtl	108.97	2.25	XCoef / S.E. =	9.3816	100.74	XCoef / S.E.	9.6658

Part 3-A. | Part 3-B.

Calculation of Fractional Causation | Calculation of Fractional Causation

from Averaged PhysPops | from 1940 PhysPops

1. Nonradiation rate is Adjusted | 1. Nonradiation rate is Adjusted
    Constant (Part 2, Col.C) =   0.3018 | Constant (Part 2, Col.E) =   0.2882

2. Radiation rate is Natl Adjusted | 2. Radiation rate is Natl Adjusted
    MortRate (Part 1, Col.G = 2.8453) | MortRate (Part 1, Col.G = 2.8453)
    minus Nonradiation rate (0.3018) =   2.5434 | minus Nonradiation rate (0.2882) =   2.5571

3. 1970 Fractional Causation is radiation | 3. 1970 Fractional Causation is radiation
    rate (2.5434) divided by OBSERVED | rate (2.5571) divided by OBSERVED
    Natl MR Part 1,Col.C=   3.3   =   0.77 | Natl MR Part 1,Col.C=   3.3   =   0.77

```
===================================================================================================
                                        Table 60-E
                    Urinary-System Cancers, Females:  Fractional Causation in 1980
===================================================================================================
Part 1.
Calculation of the 6 Adjusted MortRates (Col.F) and the National Adjusted MortRate (Col.G).
The last six entries in Part 1, Col.F, are the products of (Col.D * Col.E), as discussed in Chap. 49.


                   Col.A    Col.B    Col.C        Col.D    Col.E    Col.F    Col.G
                   1980     1980                  1940 MR  AdjuFact 1980
                   PopFrac  Obs MR   A * B        Mid,Low  Bx2,Pt2  Adju     A * F
    Trio-Sequence  Tab 3-B  Tab 12-A              Tab 12-A Col.E    MortRates
         Pacific   0.1398   2.8      0.391                          2.8      0.391
         New England 0.0546 3.4      0.186                          3.4      0.186
         Mid-Atlantic 0.1630 3.2     0.522                          3.2      0.522
         WestNoCentral 0.0759 3.0    0.228        3.7      0.62     2.29     0.174
         EastNoCentral 0.1846 3.0    0.554        4.1      0.62     2.54     0.469
         Mountain  0.0502   2.5      0.126        3.5      0.62     2.17     0.109
         WestSoCentral 0.1049 2.8    0.294        3.1      0.68     2.11     0.221
         EastSoCentral 0.0646 2.8    0.181        2.7      0.68     1.84     0.119
         SouthAtlantic 0.1624 2.9    0.471        3.0      0.68     2.04     0.331

                            Sum =    3.0                            Sum =
         1980 Observed MR from Table 12-B   3.0       1980 Natl Adjusted MR =   2.5220


Part 2.  ------------------------------------------------------------------------------------------
           Col.A    Col.B                      Col.C              Col.D              Col.E
           Mean1940 1980           Urinary Sys Ca. Females:       1940        Urinary Sys Ca. Females:
           thru1980 Adju MRs       1980 Adjusted MortRates        PPs from     1980 Adjusted MortRates
    Trio-  PPs from from Col.F          regressed on              Table 3-A        regressed on
    Seq.   Tab 47-A Part 1        Mean 1940 thru 1980 PPs         (TrioSeq)      1940 PhysPops
           x'       y                  Regression Output:         x''            Regression Output:
    Pac    177.35   2.8     Constant            0.1957            159.72    Constant            0.2578
    NewEng 185.86   3.4     Std Err of Y Est    0.1725            161.55    Std Err of Y Est    0.1764
    MidAtl 186.11   3.2     R Squared           0.9112            169.76    R Squared           0.9072
    WNoCen 128.82   2.29    No. of Observation   9                123.14    No. of Observation   9
    ENoCen 133.71   2.54    Degrees of Freedom   7                133.36    Degrees of Freedom   7
    Mtn    133.45   2.17                                          119.89
    WSoCen 114.66   2.11    X Coefficient(s)    0.0161            103.94    X Coefficient(s)    0.0173
    ESoCen 99.46    1.84    Std Err of Coef.    0.0019            85.83     Std Err of Coef.    0.0021
    SoAtl  124.62   2.04    XCoef / S.E. =      8.4777            100.74    XCoef / S.E.        8.2714


    ---------------------------------------------------|--------------------------------------------
Part 3-A.                                              | Part 3-B.
Calculation of Fractional Causation                    | Calculation of Fractional Causation
from Averaged PhysPops                                 | from 1940 PhysPops
                                                       |
1.  Nonradiation rate is Adjusted                      | 1.  Nonradiation rate is Adjusted
    Constant (Part 2, Col.C) =           0.1957        |     Constant (Part 2, Col.E) =           0.2578
                                                       |
2.  Radiation rate is Natl Adjusted                    | 2.  Radiation rate is Natl Adjusted
    MortRate (Part 1, Col.G = 2.522)                   |     MortRate (Part 1, Col.G = 2.522)
    minus Nonradiation rate (0.1957) =   2.3263        |     minus Nonradiation rate (0.2578) =   2.2642
                                                       |
3.  1980 Fractional Causation is radiation             | 3.  1980 Fractional Causation is radiation
    rate (2.3887) divided by OBSERVED                  |     rate (2.2642) divided by OBSERVED
    Natl MR Part 1,Col.C=   3.0    =      0.78         |     Natl MR Part 1,Col.C=   3.0    =      0.75
    ---------------------------------------------------|--------------------------------------------
```

CHAPTER 61

Genital Cancers, Males, 1940–1990

- Male Genital Cancers include prostate and testis cancers. For the cancers combined, Column A of Table 61-A shows about a 16% increase in the age-adjusted National MortRate, between 1960 and 1990. Measured separately, age-adjusted MortRates from Prostate Cancers rose during the 1973–1994 period, while age-adjusted MortRates from Testis Cancer fell (SEER 1997, p.45).

- Box 1 shows that while MortRates fell or were steady in the TopTrio, the MortRates INCREASED appreciably in the LowTrio. The observations in Box 1 mean that a carcinogenic co-actor which can contribute to male MortRates, from Genital Cancers, is operating more strongly in the LowTrio than in the TopTrio (Chapter 48, Part 5b). We must match the Census Divisions for this co-actor, whatever its identity. We believe that its identity is smoking.

			Table 61-A				
		Genital Cancers, Males:	Fractional Causation by Medical Radiation over Time				
Year	Col.A Natl MR	Col.B Frac.C	Col.C R-Sq	Col.D X-Coef	Col.E StdErr	Col.F Coef/SE	Col.G Source
1940	15.2	79%	0.7754	0.0932	0.0190	4.9160	Chap.13
1950	14.9	58%	0.7241	0.0676	0.0158	4.2865	Tab 61-B
1960	14.6	55%	0.7486	0.0628	0.0137	4.5658	Tab 61-C
1970	14.8	52%	0.7840	0.0585	0.0116	5.0402	Tab 61-D
1980	15.0	50%	0.8044	0.0517	0.0096	5.3656	Tab 61-E
1990	16.9	47%	0.7921	0.0498	0.0096	5.1650	Tab 61-F

```
=====================================================================================================
                                  Box 1, Chap. 61
              Genital Cancers, Males:  Post-1940 Change in MortRates by Census Trios

1960 vs. 1940, by Trios:  Col.D expresses change by ratios.  Col.F expresses change by subtraction.
1990 vs. 1940, by Trios:  Col.I expresses change by ratios.  Col.K expresses change by subtraction.
MRs change inversely with PP.  High-PP Trio has lowest growth-ratio.  Low-PP Trio has highest growth-ratio.

          Col.A  |  Col.B    Col.C    Col.D    Col.E    Col.F  |  Col.G    Col.H    Col.I    Col.J    Col.K
          1940   |  1960     Ratio    Input    Diff:    Input  |  1990     Ratio    Input    Diff:    Input
          MortRate| MortRate  Col.B    from     Col.B    from   |  MortRate  Col.G    from     Col.G    from
          Tab 13-A| Tab 13-A  /Col.A   Col.C    minus A  Col.E  |  Tab 13-A  /Col.A   Col.H    minus A  Col.J
                 |                                             |
Pacif     17.2   |  13.4     0.779    Avg Chg   -3.8    Avg Chg |  15.9     0.924    Avg Chg   -1.3    Avg Chg
NewE      18.2   |  15.7     0.863    TopTrio   -2.5    TopTrio |  16.6     0.912    TopTrio   -1.6    TopTrio
MidAtl    15.8   |  13.6     0.861    0.834     -2.2    -2.8    |  16.8     1.063    0.967      1.0    -0.6
                 |                                             |
WNoCen    16.5   |  15.4     0.933    Avg Chg   -1.1    Avg Chg |  16.3     0.988    Avg Chg   -0.2    Avg Chg
ENoCen    15.8   |  15.1     0.956    MidTrio   -0.7    MidTrio |  17.2     1.089    MidTrio    1.4    MidTrio
Mtn       15.8   |  15.2     0.962    0.950     -0.6    -0.8    |  16.6     1.051    1.042      0.8     0.7
                 |                                             |
WSoCen    11.6   |  14.6     1.259    Avg Chg    3.0    Avg Chg |  16.7     1.440    Avg Chg    5.1    Avg Chg
ESoCen    10.4   |  15.9     1.529    LowTrio    5.5    LowTrio |  17.5     1.683    LowTrio    7.1    LowTrio
SoAtl     12.8   |  14.6     1.141    1.309      1.8     3.4    |  18.6     1.453    1.525      5.8     6.0
=====================================================================================================
```

```
==========================================================================================================
                                         Box 2, Chap. 61
                       Genital Cancers, Males:  Calculation of Adjustment Factor

                        This adjustment is discussed fully in Chapter 49.
  ● Part 1:  Calculate average population-weighted MortRate for the combined TopTrio Census Divs.


            Col.A       Col.B       Col.C     Col.D  |            Col.A       Col.B       Col.C     Col.D
  Census    1940 MR    1940 Pop'n  1940 Popn  Col.A * | Census    1950 MR    1950 Pop'n  1950 Popn  Col.A *
  Div.      Tab 13-A   Tab 3-B     /45,710,039 Col.C  | Div.      Tab 13-A   Tab 3-B     /53,964,513 Col.C
                                                     |
  Pacific    17.2      9,733,262    0.2129    3.66   | Pacific    14.0      14,486,527    0.2684    3.76
  NewEng     18.2      8,437,290    0.1846    3.36   | NewEng     16.6       9,314,453    0.1726    2.87
  Mid-Atl    15.8     27,539,487    0.6025    9.52   | Mid-Atl    14.2      30,163,533    0.5590    7.94
                                                     |
   1940                Sum TopTrio      Sum  TopTrio |  1950                Sum TopTrio      Sum  TopTrio
                       45,710,039    1.0000  16.541  |                      53,964,513    1.0000  14.561
  --------------------------------------------------------------------------------------------------------
            Col.A       Col.B       Col.C     Col.D  |            Col.A       Col.B       Col.C     Col.D
  Census    1960 MR    1960 Pop'n  1960 Popn  Col.A * | Census    1970 MR    1970 Pop'n  1970 Popn  Col.A *
  Div.      Tab 13-A   Tab 3-B     /65,875,863 Col.C  | Div.      Tab 13-A   Tab 3-B     /75,017,000 Col.C
                                                     |
  Pacific    13.4     21,198,044    0.3218    4.31   | Pacific    13.9      26,087,000    0.3477    4.83
  NewEng     15.7     10,509,367    0.1595    2.50   | NewEng     15.3      11,781,000    0.1570    2.40
  Mid-Atl    13.6     34,168,452    0.5187    7.05   | Mid-Atl    14.2      37,149,000    0.4952    7.03
                                                     |
   1960                Sum TopTrio      Sum  TopTrio |  1970                Sum TopTrio      Sum  TopTrio
                       65,875,863    1.0000  13.871  |                      75,017,000    1.0000  14.268
  --------------------------------------------------------------------------------------------------------
            Col.A       Col.B       Col.C     Col.D  |            Col.A       Col.B       Col.C     Col.D
  Census    1980 MR    1980 Pop'n  1980 Popn  Col.A * | Census    1990 MR    1990 Pop'n  1990 Popn  Col.A *
  Div.      Tab 13-A   Tab 3-B     /80,615,000 Col.C  | Div.      Tab 13-A   Tab 3-B     /88,495,000 Col.C
                                                     |
  Pacific    14.4     31,523,000    0.3910    5.63   | Pacific    15.9      37,837,000    0.4276    6.80
  NewEng     14.8     12,322,000    0.1528    2.26   | NewEng     16.6      12,998,000    0.1469    2.44
  Mid-Atl    14.8     36,770,000    0.4561    6.75   | Mid-Atl    16.8      37,660,000    0.4256    7.15
                                                     |
   1980                Sum TopTrio      Sum  TopTrio |  1990                Sum TopTrio      Sum  TopTrio
                       80,615,000    1.0000  14.644  |                      88,495,000    1.0000  16.386
  --------------------------------------------------------------------------------------------------------
  ● Part 2:  Take ratios of these TopTrio MortRates, with 1940 as the denominator of each ratio.
     Col.D modifies Col.C by separate PhysPop adjustments for MidTrio and LowTrio Census Divisions.


            Col.A        Col.B       Col.C   Col.D      Col.E
            TopTrio    1940 TopTrio  = Col.A  ppAdju    = Col.C      GENITAL CANCERS.
            Mean MR      Mean MR     / Col.B Tab 47-B   * Col.D         Males.
                                            MidTrio
  1950      14.561       16.541      0.880   0.99        0.87    = MidTrio Adjustment Factor, 1950
  1960      13.871       16.541      0.839   0.97        0.81    = MidTrio Adjustment Factor, 1960
  1970      14.268       16.541      0.863   0.95        0.82    = MidTrio Adjustment Factor, 1970
  1980      14.644       16.541      0.885   0.94        0.83    = MidTrio Adjustment Factor, 1980
  1990      16.386       16.541      0.991   0.94        0.93    = MidTrio Adjustment Factor, 1990
  ------------------------------------------------      LowTrio   ------------------------------------------
  1950      14.561       16.541      0.880   1.00        0.88    = LowTrio Adjustment Factor, 1950
  1960      13.871       16.541      0.839   1.01        0.85    = LowTrio Adjustment Factor, 1960
  1970      14.268       16.541      0.863   1.02        0.88    = LowTrio Adjustment Factor, 1970
  1980      14.644       16.541      0.885   1.04        0.92    = LowTrio Adjustment Factor, 1980
  1990      16.386       16.541      0.991   1.07        1.06    = LowTrio Adjustment Factor, 1990
==========================================================================================================
```

```
===================================================================================================
                                         Table 61-B
                    Genital Cancers, Males:  Fractional Causation in 1950
===================================================================================================
Part 1.
Calculation of the 6 Adjusted MortRates (Col.F) and the National Adjusted MortRate (Col.G).
The last six entries in Part 1, Col.F, are the products of (Col.D * Col.E), as discussed in Chap. 49.


                    Col.A    Col.B   Col.C        Col.D   Col.E   Col.F   Col.G
                    1950     1950                 1940 MR AdjuFact 1950
                    PopFrac  Obs MR  A * B        Mid,Low Bx2,Pt2 Adju    A * F
Trio-Sequence       Tab 3-B  Tab 13-A            Tab 13-A Col.E  MortRates
        Pacific     0.0961   14.0    1.345                        14.0    1.345
        New England 0.0618   16.6    1.026                        16.6    1.026
        Mid-Atlantic 0.2002  14.2    2.843                        14.2    2.843
        WestNoCentral 0.0933 16.6    1.549        16.5    0.87    14.36   1.339
        EastNoCentral 0.2017 15.1    3.046        15.8    0.87    13.75   2.773
        Mountain    0.0337   13.6    0.458        15.8    0.87    13.75   0.463
        WestSoCentral 0.0965 13.3    1.283        11.6    0.88    10.21   0.985
        EastSoCentral 0.0762 14.7    1.120        10.4    0.88     9.15   0.697
        SouthAtlantic 0.1406 14.7    2.067        12.8    0.88    11.26   1.584

                             Sum =   14.7                         Sum =
     1950  Observed MR from Table 13-B   14.9    1950 Natl Adjusted MR =     13.0554

Part 2. --------------------------------------------------------------------------------------------
          Col.A    Col.B                    Col.C               Col.D                    Col.E
          Mean1940  1950            Genital Ca. Males            1940           Genital Ca. Males:
          thru1950 Adju MRs        1950 Adjusted MortRates      PPs from       1950 Adjusted MortRates
Trio-     PPs from from Col.F         regressed on             Table 3-A           regressed on
Seq.      Tab 47-A Part 1         Mean 1940 thru 1950 PPs      (TrioSeq)        1940 PhysPops
          x'       y               Regression Output:          x''              Regression Output:
Pac       154.16   14.0    Constant          4.4558            159.72    Constant          4.4912
NewEng    162.03   16.6    Std Err of Y Est  1.3162            161.55    Std Err of Y Est  1.3483
MidAtl    169.24   14.2    R Squared         0.7241            169.76    R Squared         0.7105
WNoCen    121.60   14.36   No. of Observation    9             123.14    No. of Observation    9
ENoCen    128.53   13.75   Degrees of Freedom    7             133.36    Degrees of Freedom    7
Mtn       119.64   13.75                                       119.89
WSoCen    102.64   10.21   X Coefficient(s)  0.0676            103.94    X Coefficient(s)  0.0664
ESoCen     84.44    9.15   Std Err of Coef.  0.0158             85.83    Std Err of Coef.  0.0160
SoAtl      99.91   11.26   XCoef / S.E. =    4.2865            100.74    XCoef / S.E.      4.1447


--------------------------------------------------------------------------------------------
Part 3-A.                                     | Part 3-B.
Calculation of Fractional Causation           | Calculation of Fractional Causation
from Averaged PhysPops                         | from 1940 PhysPops
                                              |
1.  Nonradiation rate is Adjusted             | 1.  Nonradiation rate is Adjusted
    Constant (Part 2, Col.C) =      4.4558    |     Constant (Part 2, Col.E) =      4.4912
                                              |
2.  Radiation rate is Natl Adjusted           | 2.  Radiation rate is Natl Adjusted
    MortRate (Part 1, Col.G = 13.0554)        |     MortRate (Part 1, Col.G = 13.0554)
    minus Nonradiation rate (4.4558) =  8.5996|     minus Nonradiation rate (4.4912) =  8.5642
                                              |
3.  1950 Fractional Causation is radiation    | 3.  1950 Fractional Causation is radiation
    rate (8.5996) divided by OBSERVED         |     rate (8.5642) divided by OBSERVED
    Natl MR Part 1,Col.C=   14.9   =   0.58   |     Natl MR Part 1, Col.C=  14.9   =   0.57
--------------------------------------------------------------------------------------------
```

===
Table 61-C
Genital Cancers, Males: Fractional Causation in 1960
===

Part 1.
Calculation of the 6 Adjusted MortRates (Col.F) and the National Adjusted MortRate (Col.G).
The last six entries in Part 1, Col.F, are the products of (Col.D * Col.E), as discussed in Chap. 49.

	Col.A	Col.B	Col.C		Col.D	Col.E	Col.F	Col.G
	1960	1960			1940 MR	AdjuFact	1960	
	PopFrac	Obs MR	A * B		Mid,Low	Bx2,Pt2	Adju	A * F
Trio-Sequence	Tab 3-B	Tab 13-A			Tab 13-A	Col.E	MortRates	
Pacific	0.1182	13.4	1.584				13.4	1.584
New England	0.0586	15.7	0.920				15.7	0.920
Mid-Atlantic	0.1905	13.6	2.591				13.6	2.591
WestNoCentral	0.0858	15.4	1.321		16.5	0.81	13.37	1.147
EastNoCentral	0.2020	15.1	3.050		15.8	0.81	12.80	2.585
Mountain	0.0382	15.2	0.581		15.8	0.81	12.80	0.489
WestSoCentral	0.0945	14.6	1.380		11.6	0.85	9.86	0.932
EastSoCentral	0.0672	15.9	1.068		10.4	0.85	8.84	0.594
SouthAtlantic	0.1448	14.6	2.114		12.8	0.85	10.88	1.575

		Sum =	14.6				Sum =	
1960	Observed MR from Table 13-B		14.6		1960 Natl Adjusted MR =			12.4167

Part 2. ---

	Col.A	Col.B		Col.C		Col.D		Col.E	
	Mean1940	1960			Genital Ca. Males	1940		Genital Ca. Males:	
	thru1960	Adju MRs			1960 Adjusted MortRates	PPs from		1960 Adjusted MortRates	
Trio-	PPs from	from Col.F			regressed on	Table 3-A		regressed on	
Seq.	Tab 47-A	Part 1			Mean 1940 thru 1960 PPs	(TrioSeq)		1940 PhysPops	
	x'	y			Regression Output:	x''		Regression Output:	
Pac	155.69	13.4		Constant	4.4476	159.72	Constant	4.3818	
NewEng	162.81	15.7		Std Err of Y Est	1.1370	161.55	Std Err of Y Est	1.1171	
MidAtl	167.04	13.6		R Squared	0.7486	169.76	R Squared	0.7574	
WNoCen	118.15	13.37		No. of Observation	9	123.14	No. of Observation	9	
ENoCen	123.87	12.80		Degrees of Freedom	7	133.36	Degrees of Freedom	7	
Mtn	117.40	12.80				119.89			
WSoCen	102.31	9.86		X Coefficient(s)	0.0628	103.94	X Coefficient(s)	0.0620	
ESoCen	85.63	8.84		Std Err of Coef.	0.0137	85.83	Std Err of Coef.	0.0133	
SoAtl	101.72	10.88		XCoef / S.E. =	4.5658	100.74	XCoef / S.E.	4.6744	

Part 3-A. | Part 3-B.
Calculation of Fractional Causation | Calculation of Fractional Causation
from Averaged PhysPops | from 1940 PhysPops
                                                       |
1. Nonradiation rate is Adjusted | 1. Nonradiation rate is Adjusted
    Constant (Part 2, Col.C) =   4.4476 | Constant (Part 2, Col.E) =   4.3818
                                                       |
2. Radiation rate is Natl Adjusted | 2. Radiation rate is Natl Adjusted
    MortRate (Part 1, Col.G = 12.4167) | MortRate (Part 1, Col.G = 12.4167)
    minus Nonradiation rate (4.4476) =   7.9692 | minus Nonradiation rate (4.3818) =   8.0350
                                                       |
3. 1960 Fractional Causation is radiation | 3. 1960 Fractional Causation is radiation
    rate (7.9692) divided by OBSERVED | rate (8.0350) divided by OBSERVED
    Natl MR Part 1,Col.C=   14.6   =   0.55 | Natl MR Part 1, Col.C=   14.6   =   0.55

===
Table 61-E
Genital Cancers, Males: Fractional Causation in 1980
===

Part 1.
Calculation of the 6 Adjusted MortRates (Col.F) and the National Adjusted MortRate (Col.G).
The last six entries in Part 1, Col.F, are the products of (Col.D * Col.E), as discussed in Chap. 49.

Trio-Sequence	Col.A 1980 PopFrac Tab 3-B	Col.B 1980 Obs MR Tab 13-A	Col.C A * B	Col.D 1940 MR Mid,Low Tab 13-A	Col.E AdjuFact Bx2,Pt2 Col.E	Col.F 1980 Adju MortRates	Col.G A * F
Pacific	0.1398	14.4	2.013			14.4	2.013
New England	0.0546	14.8	0.808			14.8	0.808
Mid-Atlantic	0.1630	14.8	2.412			14.8	2.412
WestNoCentral	0.0759	14.2	1.078	16.5	0.83	13.69	1.039
EastNoCentral	0.1846	15.3	2.824	15.8	0.83	13.11	2.421
Mountain	0.0502	14.5	0.728	15.8	0.83	13.11	0.658
WestSoCentral	0.1049	14.3	1.500	11.6	0.92	10.67	1.119
EastSoCentral	0.0646	15.4	0.995	10.4	0.92	9.57	0.618
SouthAtlantic	0.1624	16.4	2.663	12.8	0.92	11.78	1.912

	Sum =	15.0			Sum =	

1980 Observed MR from Table 13-B   15.0   1980 Natl Adjusted MR =   13.0022

Part 2. ---

Trio-Seq.	Col.A Mean1940 thru1980 PPs from Tab 47-A x'	Col.B 1980 Adju MRs from Col.F Part 1 y	Col.C Genital Ca. Males 1980 Adjusted MortRates regressed on Mean 1940 thru 1980 PPs Regression Output:		Col.D 1940 PPs from Table 3-A (TrioSeq) x''	Col.E Genital Ca. Males: 1980 Adjusted MortRates regressed on 1940 PhysPops Regression Output:	
Pac	177.35	14.4	Constant	5.5093	159.72	Constant	5.4226
NewEng	185.86	14.8	Std Err of Y Est	0.8769	161.55	Std Err of Y Est	0.7257
MidAtl	186.11	14.8	R Squared	0.8044	169.76	R Squared	0.8660
WNoCen	128.82	13.69	No. of Observation	9	123.14	No. of Observation	9
ENoCen	133.71	13.11	Degrees of Freedom	7	133.36	Degrees of Freedom	7
Mtn	133.45	13.11			119.89		
WSoCen	114.66	10.67	X Coefficient(s)	0.0517	103.94	X Coefficient(s)	0.0580
ESoCen	99.46	9.57	Std Err of Coef.	0.0096	85.83	Std Err of Coef.	0.0086
SoAtl	124.62	11.78	XCoef / S.E. =	5.3656	100.74	XCoef / S.E.	6.7273

Part 3-A. | Part 3-B.
Calculation of Fractional Causation | Calculation of Fractional Causation
from Averaged PhysPops | from 1940 PhysPops
                                                       |
1. Nonradiation rate is Adjusted | 1. Nonradiation rate is Adjusted
    Constant (Part 2, Col.C) =   5.5093 | Constant (Part 2, Col.E) =   5.4226
                                                       |
2. Radiation rate is Natl Adjusted | 2. Radiation rate is Natl Adjusted
    MortRate (Part 1, Col.G = 13.0022) | MortRate (Part 1, Col.G = 13.0022)
    minus Nonradiation rate (5.5093) =   7.4929 | minus Nonradiation rate (5.4226) =   7.5796
                                                       |
3. 1980 Fractional Causation is radiation | 3. 1980 Fractional Causation is radiation
    rate (7.4929) divided by OBSERVED | rate (7.5796) divided by OBSERVED
    Natl MR Part 1,Col.C=   15.0   =   0.50 | Natl MR Part 1, Col.C=   15.0   =   0.51

===
Table 61-F
Genital Cancers, Males: Fractional Causation in 1990
===

Part 1.
Calculation of the 6 Adjusted MortRates (Col.F) and the National Adjusted MortRate (Col.G).
The last six entries in Part 1, Col.F, are the products of (Col.D * Col.E), as discussed in Chap. 49.

Trio-Sequence	Col.A 1990 PopFrac Tab 3-B	Col.B 1990 Obs MR Tab 13-A	Col.C A * B	Col.D 1940 MR Mid,Low Tab 13-A	Col.E AdjuFact Bx2,Pt2 Col.E	Col.F 1990 Adju MortRates	Col.G A * F
Pacific	0.1535	15.9	2.441			15.9	2.441
New England	0.0527	16.6	0.875			16.6	0.875
Mid-Atlantic	0.1527	16.8	2.565			16.8	2.565
WestNoCentral	0.0721	16.3	1.175	16.5	0.93	15.35	1.106
EastNoCentral	0.1713	17.2	2.946	15.8	0.93	14.69	2.517
Mountain	0.0543	16.6	0.901	15.8	0.93	14.69	0.798
WestSoCentral	0.1087	16.7	1.815	11.6	1.06	12.30	1.337
EastSoCentral	0.0621	17.5	1.087	10.4	1.06	11.02	0.685
SouthAtlantic	0.1725	18.6	3.209	12.8	1.06	13.57	2.340

Sum =   17.0

1990 Observed MR from Table 13-B   16.9   1990 Natl Adjusted MR =   14.6638

Part 2. ---

Trio-Seq.	Col.A Mean1940 thru1990 PPs from Tab 47-A x'	Col.B 1990 Adju MRs from Col.F Part 1 y	Col.C Genital Ca. Males 1990 Adjusted MortRates regressed on Mean 1940 thru 1990 PPs Regression Output:		Col.D 1940 PPs from Table 3-A (TrioSeq) x''	Col.E Genital Ca. Males: 1990 Adjusted MortRates regressed on 1940 PhysPops Regression Output:	
Pac	191.97	15.9	Constant	6.6865	159.72	Constant	6.7256
NewEng	208.20	16.6	Std Err of Y Est	0.9476	161.55	Std Err of Y Est	0.7594
MidAtl	204.72	16.8	R Squared	0.7921	169.76	R Squared	0.8665
WNoCen	141.14	15.35	No. of Observation	9	123.14	No. of Observation	9
ENoCen	146.19	14.69	Degrees of Freedom	7	133.36	Degrees of Freedom	7
Mtn	145.91	14.69			119.89		
WSoCen	126.28	12.30	X Coefficient(s)	0.0498	103.94	X Coefficient(s)	0.0608
ESoCen	113.28	11.02	Std Err of Coef.	0.0096	85.83	Std Err of Coef.	0.0090
SoAtl	142.93	13.57	XCoef / S.E. =	5.1650	100.74	XCoef / S.E.	6.7400

Part 3-A. | Part 3-B.
Calculation of Fractional Causation | Calculation of Fractional Causation
from Averaged PhysPops | from 1940 PhysPops
                                                       |
1. Nonradiation rate is Adjusted | 1. Nonradiation rate is Adjusted
    Constant (Part 2, Col.C) =   6.6865 | Constant (Part 2, Col.E) =   6.7256
                                                       |
2. Radiation rate is Natl Adjusted | 2. Radiation rate is Natl Adjusted
    MortRate (Part 1, Col.G = 14.6638) | MortRate (Part 1, Col.G = 14.6638)
    minus Nonradiation rate (6.6865) =   7.9773 | minus Nonradiation rate (6.7256) =   7.9382
                                                       |
3. 1990 Fractional Causation is radiation | 3. 1990 Fractional Causation is radiation
    rate (7.9773) divided by OBSERVED | rate (7.9382) divided by OBSERVED
    Natl MR Part 1,Col.C=   16.9   =   0.47 | Natl MR Part 1,Col.C=   16.9   =   0.47

CHAPTER 62

Genital Cancers, Females, 1940-1980

● Female Genital Cancers are the single cancer-group for which there is no apparent dose-response with PhysPop (Chapters 14 and 62). This finding may be accepted without challenge, or it may mean that that the Census Divisions are badly matched for an important carcinogenic co-actor which is specific for female Genital Cancers (discussion in Chapter 14). Although it will be important for female health (and for insights into radiation carcinogenesis) to establish which is the correct explanation, resolution of the issue is beyond the scope of this book.

● Until the issue is resolved, we strongly caution against a hasty belief that female genital tissues are invulnerable to radiation carcinogenesis. If such a belief is NOT true, the belief could result in a great deal of harm to female health.

● In Chapter 14, when we regressed the 1940 MortRates for female Genital Cancers on PhysPop, we found no detectable dose-response. Before we regress the 1950, 1960, 1970, and 1980 MortRates on the appropriate PhysPops (next page), we must examine Box 1 to learn whether or not we are obliged to adjust the MortRates.

● We see in Box 1 that the 1940 MortRates for female Genital Cancers are almost equal in all Trios, and decline thereafter almost equally in all Trios. There is very little separation in the 1940 MortRates among the Census Divisions, and very little separation in 1960, and very little separation in 1980. This finding is, of course, consistent with Table 14-A, which shows that the High5/Low4 Ratio hovers close to unity for the entire period. By contrast, the separation of Mean PhysPop values, by Census Divisions, persists quite well during the same period (Chapter 47, Part 3).

● Even though Box 1 (Column D) suggests a slight "Trio-Effect," the effect disappears in Box 2 (not shown), where we use population-weighted MortRates and "ppAdju" from Table 47-B. Box 2 produces no evidence that a carcinogenic co-actor which can contribute to female MortRates, from Genital Cancers, is operating more strongly in the LowTrio than in the TopTrio. Therefore, we make no adjustments before doing the post-1940 regressions (next page).

● Those regressions confirm that no dose-response develops after 1940 --- a result which could be predicted from Columns B and G of Box 1. No dose-response means no detectable Fractional Causation by medical radiation. Thus, there is no Table 62-A in this chapter.

```
=================================================================================================
                                       Box 1, Chap. 62
              Genital Cancers, Females:  Post-1940 Change in MortRates by Census Trios

1960 vs. 1940, by Trios:  Col.D expresses change by ratios.  Col.F expresses change by subtraction.
1980 vs. 1940, by Trios:  Col.I expresses change by ratios.  Col.K expresses change by subtraction.
1940 MRs are almost equal in all Trios, and decline almost equally in all Trios by 1980.
```

	Col.A 1940 MortRate Tab 14-A	Col.B 1960 MortRate Tab 14-A	Col.C Ratio Col.B /Col.A	Col.D Input from Col.C	Col.E Diff: Col.B minus A	Col.F Input from Col.E	Col.G 1980 MortRate Tab 14-A	Col.H Ratio Col.G /Col.A	Col.I Input from Col.H	Col.J Diff: Col.G minus A	Col.K Input from Col.J
Pacif	33.1	20.0	0.604	Avg Chg	-13.1	Avg Chg	13.3	0.402	Avg Chg	-19.8	Avg Chg
NewE	32.8	21.7	0.662	TopTrio	-11.1	TopTrio	13.4	0.409	TopTrio	-19.4	TopTrio
MidAtl	32.7	22.2	0.679	0.648	-10.5	-11.6	14.3	0.437	0.416	-18.4	-19.2
WNoCen	28.4	20.3	0.715	Avg Chg	-8.1	Avg Chg	13.3	0.468	Avg Chg	-15.1	Avg Chg
ENoCen	33.2	24.2	0.729	MidTrio	-9.0	MidTrio	14.5	0.437	MidTrio	-18.7	MidTrio
Mtn	27.8	18.4	0.662	0.702	-9.4	-8.8	11.7	0.421	0.442	-16.1	-16.6
WSoCen	30.0	22.1	0.737	Avg Chg	-7.9	Avg Chg	12.5	0.417	Avg Chg	-17.5	Avg Chg
ESoCen	33.2	24.7	0.744	LowTrio	-8.5	LowTrio	14.3	0.431	LowTrio	-18.9	LowTrio
SoAtl	32.5	24.0	0.738	0.740	-8.5	-8.3	13.5	0.415	0.421	-19.0	-18.5

```
=================================================================================================
```

● Regression of Post-1940 MortRates on Mean PhysPops

       Below, the Mean PhysPops (x-values) come from Table 47-A, and the MortRates (y-values) from Table 14-A. Both are arranged in Trio-Sequence here.

	1940-50 Mean PP	1950 MortRate	Genital Cancers, Female: Regression Output:	
Pacific	154.16	25.5	Constant	31.3346
New England	162.03	25.1	Std Err of Y Est	2.3191
Mid-Atlantic	169.24	27.2	R Squared	0.1996
West North Central	121.60	23.4	No. of Observations	9
East North Central	128.53	28.3	Degrees of Freedom	7
Mountain	119.64	23.6		
West South Central	102.64	27.4	X Coefficient(s)	-0.0367
East South Central	84.44	29.7	Std Err of Coef.	0.0278
South Atlantic	99.91	29.9	Coefficient / S.E.	-1.3213

Trio-Seq.	1940-60 Mean PP	1960 MortRate	Genital Cancers, Female: Regression Output:	
Pacific	155.69	20.0	Constant	25.3185
New England	162.81	21.7	Std Err of Y Est	2.1123
Mid-Atlantic	167.04	22.2	R Squared	0.1348
West North Central	118.15	20.3	No. of Observations	9
East North Central	123.87	24.2	Degrees of Freedom	7
Mountain	117.40	18.4		
West South Central	102.31	22.1	X Coefficient(s)	-0.0267
East South Central	85.63	24.7	Std Err of Coef.	0.0255
South Atlantic	101.72	24.0	Coefficient / S.E.	-1.0445

Trio-Seq.	1940-70 Mean PP	1970 MortRate	Genital Cancers, Female: Regression Output:	
Pacific	162.72	16.7	Constant	19.0584
New England	168.74	17.6	Std Err of Y Est	1.5206
Mid-Atlantic	173.28	18.3	R Squared	0.0461
West North Central	119.56	16.8	No. of Observations	9
East North Central	124.70	19.4	Degrees of Freedom	7
Mountain	122.37	15.0		
West South Central	105.03	17.3	X Coefficient(s)	-0.0103
East South Central	89.44	19.5	Std Err of Coef.	0.0177
South Atlantic	108.97	18.8	Coefficient / S.E.	-0.5819

Trio-Seq.	1940-80 Mean PP	1980 MortRate	Genital Cancers, Female: Regression Output:	
Pacific	177.35	13.3	Constant	12.9516
New England	185.86	13.4	Std Err of Y Est	0.9611
Mid-Atlantic	186.11	14.3	R Squared	0.0138
West North Central	128.82	13.3	No. of Observations	9
East North Central	133.71	14.5	Degrees of Freedom	7
Mountain	133.45	11.7		
West South Central	114.66	12.5	X Coefficient(s)	0.0033
East South Central	99.46	14.3	Std Err of Coef.	0.0106
South Atlantic	124.62	13.5	Coefficient / S.E.	0.3125

>>>>>>>>>>

CHAPTER 63

Buccal-Pharynx Cancers, Males, 1940-1980

• Box 1, below, shows the familiar pattern revealed in the previous Boxes 1. Therefore, Box 2 evaluates an adjustment factor, in order to match the Census Divisions for the unmatched co-actor.

• Table 63-A deviates from the model of Chapter 49. We have added a second set of entries (Table 63-BB through 63-EE) because the regressions in Tables 63-B through 63-E produce negative Constants, whose magnitude is not trivial relative to the male National MortRate for Buccal-Pharynx Cancers. We regard this as a signal that something is not REALISTIC about the adjustment factor used in those tables for the MidTrio and LowTrio MortRates. This has happened before. Please refer to the text in Chapter 51, preceding Table 51-AA. Here, in Chapter 63, the factor which abolishes the negative sign on Constants is 1.15. and it is used in the same way the factor of 1.4 was used to produce Tables 51-BB through 51-FF.

		Col.A Natl MR	Col.B Frac.C		Col.C R-Sq	Col.D X-Coef	Col.E StdErr	Col.F Coef/SE	Col.G Source
	Year								
	1940	5.1	~100%		0.7234	0.0382	0.0089	4.2782	Chap.15
	1950	5.0	89%		0.7137	0.0377	0.0090	4.1778	Tab 63-B
	1960	4.7	88%		0.7135	0.0389	0.0093	4.1754	Tab 63-C
	1970	4.65	86%		0.7101	0.0356	0.0086	4.1405	Tab 63-D
	1980	4.6	81%		0.7327	0.0315	0.0072	4.3806	Tab 63-E
	1950	5.0	81%		0.6094	0.0303	0.0092	3.3048	Tab 63-BB
	1960	4.7	88%		0.6129	0.0316	0.0095	3.3293	Tab 63-CC
	1970	4.65	84%		0.6003	0.0286	0.0088	3.2425	Tab 63-DD
	1980	4.6	81%		0.6177	0.0252	0.0075	3.3630	Tab 63-EE

Table 63-A
Buccal-Pharynx Cancers, Males: Fractional Causation by Medical Radiation over Time

```
=================================================================================
                              Box 1, Chap. 63
           Buccal-Pharynx Cancers, Males:  Post-1940 Change in MortRates by Census Trios

1960 vs. 1940, by Trios:  Col.D expresses change by ratios.   Col.F expresses change by subtraction.
1980 vs. 1940, by Trios:  Col.I expresses change by ratios.   Col.K expresses change by subtraction.
MRs change inversely with PP.  High-PP Trio has lowest growth-factor.  Low-PP Trio has highest growth-factor.


          Col.A  |  Col.B    Col.C    Col.D    Col.E    Col.F  |  Col.G    Col.H    Col.I    Col.J    Col.K
          1940   |  1960     Ratio    Input    Diff:    Input  |  1980     Ratio    Input    Diff:    Input
          MortRate| MortRate  Col.B    from     Col.B    from   |  MortRate  Col.G    from     Col.G    from
          Tab 15-A| Tab 15-A  /Col.A   Col.C    minus A  Col.E  |  Tab 15-A  /Col.A   Col.H    minus A  Col.J
                 |                                              |
Pacif     5.3    |  4.6      0.868    Avg Chg  -0.7     Avg Chg |  4.2      0.792    Avg Chg  -1.1     Avg Chg
NewE      6.4    |  6.6      1.031    TopTrio   0.2     TopTrio |  5.7      0.891    TopTrio  -0.7     TopTrio
MidAtl    6.9    |  5.4      0.783    0.894    -1.5     -0.7    |  5.1      0.739    0.807    -1.8     -1.2
                 |                                              |
WNoCen    4.6    |  4.0      0.870    Avg Chg  -0.6     Avg Chg |  3.5      0.761    Avg Chg  -1.1     Avg Chg
ENoCen    4.8    |  4.9      1.021    MidTrio   0.1     MidTrio |  4.6      0.958    MidTrio  -0.2     MidTrio
Mtn       2.8    |  2.6      0.929    0.940    -0.2     -0.2    |  2.9      1.036    0.918     0.1     -0.4
                 |                                              |
WSoCen    4.0    |  3.9      0.975    Avg Chg  -0.1     Avg Chg |  4.2      1.050    Avg Chg   0.2     Avg Chg
ESoCen    3.3    |  4.2      1.273    LowTrio   0.9     LowTrio |  4.4      1.333    LowTrio   1.1     LowTrio
SoAtl     4.3    |  4.5      1.047    1.098     0.2      0.3    |  5.0      1.163    1.182     0.7      0.7
=================================================================================
```

===

Box 2, Chap. 63
Buccal-Pharynx Cancers, Males: Calculation of Adjustment Factor

This adjustment is discussed fully in Chapter 49.

● Part 1: Calculate average population-weighted MortRate for the combined TopTrio Census Divs.

Census Div.	Col.A 1940 MR Tab 15-A	Col.B 1940 Pop'n Tab 3-B	Col.C 1940 Popn /45,710,039	Col.D Col.A * Col.C	Census Div.	Col.A 1950 MR Tab 15-A	Col.B 1950 Pop'n Tab 3-B	Col.C 1950 Popn /53,964,513	Col.D Col.A * Col.C
Pacific	5.3	9,733,262	0.2129	1.13	Pacific	4.7	14,486,527	0.2684	1.26
NewEng	6.4	8,437,290	0.1846	1.18	NewEng	6.5	9,314,453	0.1726	1.12
Mid-Atl	6.9	27,539,487	0.6025	4.16	Mid-Atl	5.9	30,163,533	0.5590	3.30
1940		Sum TopTrio 45,710,039	Sum 1.0000	TopTrio 6.467	1950		Sum TopTrio 53,964,513	Sum 1.0000	TopTrio 5.681

Census Div.	Col.A 1960 MR Tab 15-A	Col.B 1960 Pop'n Tab 3-B	Col.C 1960 Popn /65,875,863	Col.D Col.A * Col.C	Census Div.	Col.A 1970 MR Tab 15-A	Col.B 1970 Pop'n Tab 3-B	Col.C 1970 Popn /75,017,000	Col.D Col.A * Col.C
Pacific	4.6	21,198,044	0.3218	1.48	Pacific	4.4	26,087,000	0.3477	1.53
NewEng	6.6	10,509,367	0.1595	1.05	NewEng	6.2	11,781,000	0.1570	0.97
Mid-Atl	5.4	34,168,452	0.5187	2.80	Mid-Atl	5.3	37,149,000	0.4952	2.62
1960		Sum TopTrio 65,875,863	Sum 1.0000	TopTrio 5.334	1970		Sum TopTrio 75,017,000	Sum 1.0000	TopTrio 5.128

Census Div.	Col.A 1980 MR Tab 15-A	Col.B 1980 Pop'n Tab 3-B	Col.C 1980 Popn /80,615,000	Col.D Col.A * Col.C	Census Div.	Col.A 1990 MR Tab 15-A	Col.B 1990 Pop'n Tab 3-B	Col.C 1990 Popn /88,495,000	Col.D Col.A * Col.C
Pacific	4.2	31,523,000	0.3910	1.64	Pacific	--	37,837,000	0.4276	0.00
NewEng	5.7	12,322,000	0.1528	0.87	NewEng	--	12,998,000	0.1469	0.00
Mid-Atl	5.1	36,770,000	0.4561	2.33	Mid-Atl	--	37,660,000	0.4256	0.00
1980		Sum TopTrio 80,615,000	Sum 1.0000	TopTrio 4.840	1990		Sum TopTrio 88,495,000	Sum 1.0000	TopTrio 0.000

● Part 2: Take ratios of these TopTrio MortRates, with 1940 as the denominator of each ratio.
     Col.D modifies Col.C by separate PhysPop adjustments for MidTrio and LowTrio Census Divisions.

	Col.A TopTrio Mean MR	Col.B 1940 TopTrio Mean MR	Col.C = Col.A / Col.B	Col.D ppAdju Tab 47-B	Col.E = Col.C * Col.D	BUCCAL-PHARYNX CANCERS. Males.
				MidTrio		
1950	5.681	6.467	0.879	0.99	0.87	= MidTrio Adjustment Factor, 1950
1960	5.334	6.467	0.825	0.97	0.80	= MidTrio Adjustment Factor, 1960
1970	5.128	6.467	0.793	0.95	0.75	= MidTrio Adjustment Factor, 1970
1980	4.840	6.467	0.748	0.94	0.70	= MidTrio Adjustment Factor, 1980
1990	0.000	6.467	0.000	0.94	0.00	= MidTrio Adjustment Factor, 1990
				LowTrio		
1950	5.681	6.467	0.879	1.00	0.88	= LowTrio Adjustment Factor, 1950
1960	5.334	6.467	0.825	1.01	0.83	= LowTrio Adjustment Factor, 1960
1970	5.128	6.467	0.793	1.02	0.81	= LowTrio Adjustment Factor, 1970
1980	4.840	6.467	0.748	1.04	0.78	= LowTrio Adjustment Factor, 1980
1990	0.000	6.467	0.000	1.07	0.00	= LowTrio Adjustment Factor, 1990

===

```
==================================================================================================
                                        Table 63-B
                    Buccal-Pharynx Cancers, Males:  Fractional Causation in 1950
==================================================================================================
```

Part 1.

Calculation of the 6 Adjusted MortRates (Col.F) and the National Adjusted MortRate (Col.G).

The last six entries in Part 1, Col.F, are the products of (Col.D * Col.E), as discussed in Chap. 49.

Trio-Sequence	Col.A 1950 PopFrac Tab 3-B	Col.B 1950 Obs MR Tab 15-A	Col.C A * B	Col.D 1940 MR Mid,Low Tab 15-A	Col.E AdjuFact Bx2,Pt2 Col.E	Col.F 1950 Adju MortRates	Col.G A * F
Pacific	0.0961	4.7	0.452			4.7	0.452
New England	0.0618	6.5	0.402			6.5	0.402
Mid-Atlantic	0.2002	5.9	1.181			5.9	1.181
WestNoCentral	0.0933	4.5	0.420	4.6	0.87	4.002	0.373
EastNoCentral	0.2017	4.9	0.988	4.8	0.87	4.176	0.842
Mountain	0.0337	3.2	0.108	2.8	0.87	2.436	0.082
WestSoCentral	0.0965	4.3	0.415	4.0	0.88	3.520	0.340
EastSoCentral	0.0762	4.2	0.320	3.3	0.88	2.904	0.221
SouthAtlantic	0.1406	4.7	0.661	4.3	0.88	3.784	0.532

```
                                     Sum =    4.9                                 Sum =
        1950 Observed Natl MR from Table 15-B   5.0        1950 Natl Adjusted MR =   4.4253
```

Part 2. --

Trio-Seq.	Col.A Mean1940 thru1950 PPs from Tab 47-A x'	Col.B 1950 Adju MRs from Col.F Part 1 y	Col.C Buccal-Phar. Ca. Males: 1950 Adjusted MortRates regressed on Mean 1940 thru 1950 PPs		Col.D 1940 PPs from Table 3-A (TrioSeq) x''	Col.E Buccal-Phar. Ca. Males: 1950 Adjusted MortRates regressed on 1940 PhysPops	
			Regression Output:			Regression Output:	
Pac	154.16	4.7	Constant	-0.5725	159.72	Constant	-0.5304
NewEng	162.03	6.5	Std Err of Y Est	0.7538	161.55	Std Err of Y Est	0.7797
MidAtl	169.24	5.9	R Squared	0.7137	169.76	R Squared	0.6937
WNoCen	121.60	4.002	No. of Observation	9	123.14	No. of Observation	9
ENoCen	128.53	4.176	Degrees of Freedom	7	133.36	Degrees of Freedom	7
Mtn	119.64	2.436			119.89		
WSoCen	102.64	3.520	X Coefficient(s)	0.0377	103.94	X Coefficient(s)	0.0369
ESoCen	84.44	2.904	Std Err of Coef.	0.0090	85.83	Std Err of Coef.	0.0093
SoAtl	99.91	3.784	XCoef / S.E. =	4.1778	100.74	XCoef / S.E.	3.9820

```
--------------------------------------------------------------------------------------------------
```

Part 3-A. | Part 3-B.

Calculation of Fractional Causation | Calculation of Fractional Causation

from Averaged PhysPops | from 1940 PhysPops

1. Nonradiation rate is Adjusted | 1. Nonradiation rate is Adjusted
    Constant (Part 2, Col.C) = NEGATIVE   0.0 | Constant (Part 2, Col.E) = NEGATIVE   0.0

2. Radiation rate is Natl Adjusted | 2. Radiation rate is Natl Adjusted
    MortRate (Part 1, Col.G = 4.4253) | MortRate (Part 1, Col.G = 4.4253)
    minus Nonradiation rate (0.0) =   4.4253 | minus Nonradiation rate (0.0) =   4.4253

3. 1950 Fractional Causation is radiation | 3. 1950 Fractional Causation is radiation
    rate (4.4253) divided by OBSERVED | rate (4.4253) divided by OBSERVED
    Natl MR Part 1,Col.C=   5.0   =   0.89 | Natl MR Part 1, Col.C=   5.0   =   0.89

```
--------------------------------------------------------------------------------------------------
```

```
================================================================================
                                  Table 63-C
                Buccal-Pharynx Cancers, Males:  Fractional Causation in 1960
================================================================================
Part 1.
Calculation of the 6 Adjusted MortRates (Col.F) and the National Adjusted MortRate (Col.G).
The last six entries in Part 1, Col.F, are the products of (Col.D * Col.E), as discussed in Chap. 49.

                Col.A   Col.B   Col.C      Col.D    Col.E   Col.F    Col.G
                1960    1960               1940 MR  AdjuFact 1960
                PopFrac Obs MR  A * B      Mid,Low  Bx2,Pt2  Adju     A * F
Trio-Sequence   Tab 3-B Tab 15-A           Tab 15-A  Col.E   MortRates
     Pacific    0.1182   4.6    0.544                          4.6    0.544
     New England 0.0586  6.6    0.387                          6.6    0.387
     Mid-Atlantic 0.1905 5.4    1.029                          5.4    1.029
     WestNoCentral 0.0858 4.0   0.343       4.6     0.80      3.680   0.316
     EastNoCentral 0.2020 4.9   0.990       4.8     0.80      3.840   0.776
     Mountain    0.0382   2.6    0.099       2.8     0.80      2.240   0.086
     WestSoCentral 0.0945 3.9   0.369       4.0     0.83      3.320   0.314
     EastSoCentral 0.0672 4.2   0.282       3.3     0.83      2.739   0.184
     SouthAtlantic 0.1448 4.5   0.652       4.3     0.83      3.569   0.517

                        Sum =   4.7                                  Sum =
        1960 Observed Natl MR from Table 15-B  4.7    1960 Natl Adjusted MR =   4.1508


Part 2.  -----------------------------------------------------------------------
            Col.A   Col.B                Col.C               Col.D                      Col.E
            Mean1940  1960          Buccal-Phar. Ca. Males:   1940            Buccal-Phar. Ca. Males:
            thru1960 Adju MRs       1960 Adjusted MortRates   PPs from         1960 Adjusted MortRates
Trio-       PPs from from Col.F        regressed on           Table 3-A           regressed on
Seq.        Tab 47-A Part 1        Mean 1940 thru 1960 PPs    (TrioSeq)          1940 PhysPops
             x'       y            Regression Output:          x''             Regression Output:
Pac         155.69   4.6          Constant        -0.9113     159.72          Constant        -0.7373
NewEng      162.81   6.6          Std Err of Y Est  0.7715    161.55          Std Err of Y Est  0.8398
MidAtl      167.04   5.4          R Squared         0.7135    169.76          R Squared         0.6606
WNoCen      118.15   3.680        No. of Observation   9      123.14          No. of Observation   9
ENoCen      123.87   3.840        Degrees of Freedom   7      133.36          Degrees of Freedom   7
Mtn         117.40   2.240                                    119.89
WSoCen      102.31   3.320        X Coefficient(s)  0.0389    103.94          X Coefficient(s)  0.0368
ESoCen       85.63   2.739        Std Err of Coef.  0.0093     85.83          Std Err of Coef.  0.0100
SoAtl       101.72   3.569        XCoef / S.E. =    4.1754    100.74          XCoef / S.E.      3.6909


---------------------------------------------|----------------------------------------------
Part 3-A.                                    | Part 3-B.
Calculation of Fractional Causation          | Calculation of Fractional Causation
from Averaged PhysPops                        | from 1940 PhysPops
                                             |
1. Nonradiation rate is Adjusted             | 1. Nonradiation rate is Adjusted
   Constant (Part 2, Col.C) = NEGATIVE  0.0  |    Constant (Part 2, Col.E) = NEGATIVE  0.0
                                             |
2. Radiation rate is Natl Adjusted           | 2. Radiation rate is Natl Adjusted
   MortRate (Part 1, Col.G = 4.1508)         |    MortRate (Part 1, Col.G = 4.1508)
   minus Nonradiation rate (0.0) =   4.1508  |    minus Nonradiation rate (0.0) =   4.1508
                                             |
3. 1960 Fractional Causation is radiation    | 3. 1960 Fractional Causation is radiation
   rate (4.1508) divided by OBSERVED         |    rate (4.1508) divided by OBSERVED
   Natl MR Part 1,Col.C=  4.7   =    0.88    |    Natl MR Part 1,Col.C=  4.7   =    0.88
---------------------------------------------|----------------------------------------------
```

```
=====================================================================================================
                                         Table 63-E
                    Buccal-Pharynx Cancers, Males:  Fractional Causation in 1980
=====================================================================================================
Part 1.
Calculation of the 6 Adjusted MortRates (Col.F) and the National Adjusted MortRate (Col.G).
The last six entries in Part 1, Col.F, are the products of (Col.D * Col.E), as discussed in Chap. 49.


                    Col.A    Col.B    Col.C        Col.D    Col.E    Col.F    Col.G
                    1980     1980                  1940 MR  AdjuFact 1980
                    PopFrac  Obs MR   A * B        Mid,Low  Bx2,Pt2  Adju     A * F
Trio-Sequence       Tab 3-B  Tab 15-A             Tab 15-A  Col.E   MortRates
        Pacific     0.1398   4.2      0.587                          4.2      0.587
        New England 0.0546   5.7      0.311                          5.7      0.311
        Mid-Atlantic 0.1630  5.1      0.831                          5.1      0.831
        WestNoCentral 0.0759 3.5      0.266        4.6      0.70     3.220    0.244
        EastNoCentral 0.1846 4.6      0.849        4.8      0.70     3.360    0.620
        Mountain    0.0502   2.9      0.146        2.8      0.70     1.960    0.098
        WestSoCentral 0.1049 4.2      0.441        4.0      0.78     3.120    0.327
        EastSoCentral 0.0646 4.4      0.284        3.3      0.78     2.574    0.166
        SouthAtlantic 0.1624 5.0      0.812        4.3      0.78     3.354    0.545

                              Sum =    4.5                                    Sum =
        1980 Observed Natl MR from Table 15-B     4.6        1980 Natl Adjusted MR =    3.7310

Part 2. --------------------------------------------------------------------------------------------
        Col.A    Col.B                     Col.C                Col.D                      Col.E
        Mean1940 1980         Buccal-Phar. Ca. Males:            1940          Buccal-Phar. Ca. Males:
        thru1980 Adju MRs     1980 Adjusted MortRates           PPs from       1980 Adjusted MortRates
Trio-   PPs from from Col.F      regressed on                   Table 3-A         regressed on
Seq.    Tab 47-A Part 1      Mean 1940 thru 1980 PPs           (TrioSeq)        1940 PhysPops
        x'       y              Regression Output:             x''               Regression Output:
Pac     177.35   4.2      Constant        -0.8783              159.72       Constant        -0.5053
NewEng  185.86   5.7      Std Err of Y Est  0.6554             161.55       Std Err of Y Est  0.7520
MidAtl  186.11   5.1      R Squared         0.7327             169.76       R Squared         0.6481
WNoCen  128.82   3.220    No. of Observation    9              123.14       No. of Observation    9
ENoCen  133.71   3.360    Degrees of Freedom    7              133.36       Degrees of Freedom    7
Mtn     133.45   1.960                                         119.89
WSoCen  114.66   3.120    X Coefficient(s)  0.0315             103.94       X Coefficient(s)  0.0321
ESoCen   99.46   2.574    Std Err of Coef.  0.0072              85.83       Std Err of Coef.  0.0089
SoAtl   124.62   3.354    XCoef / S.E. =    4.3806             100.74       XCoef / S.E.      3.5908


-----------------------------------------------------|----------------------------------------------
Part 3-A.                                            | Part 3-B.
Calculation of Fractional Causation                  | Calculation of Fractional Causation
from Averaged PhysPops                                | from 1940 PhysPops
                                                     |
1.  Nonradiation rate is Adjusted                    | 1.  Nonradiation rate is Adjusted
    Constant (Part 2, Col.C) = NEGATIVE     0.0       |     Constant (Part 2, Col.E) = NEGATIVE     0.0
                                                     |
2.  Radiation rate is Natl Adjusted                  | 2.  Radiation rate is Natl Adjusted
    MortRate (Part 1, Col.G = 3.7310)                 |     MortRate (Part 1, Col.G = 3.7310)
    minus Nonradiation rate (0.0) =        3.7310     |     minus Nonradiation rate (0.0) =        3.7310
                                                     |
3.  1980 Fractional Causation is radiation           | 3.  1980 Fractional Causation is radiation
    rate (3.7310) divided by OBSERVED                 |     rate (3.7310) divided by OBSERVED
    Natl MR Part 1,Col.C=    4.6    =      0.81       |     Natl MR Part 1, Col.C=    4.6    =      0.81
-----------------------------------------------------|----------------------------------------------
```

```
===================================================================================================
                                        Table 63-BB
                    Buccal-Pharynx Cancers, Males:  Fractional Causation in 1950
===================================================================================================
```

Part 1.
Calculation of the 6 Adjusted MortRates (Col.F) and the National Adjusted MortRate (Col.G).
The last six entries in Part 1, Col.F, are the products of (Col.D * Col.E), as discussed in Chap. 49.

	Col.A	Col.B	Col.C		Col.D	Col.E	Col.F	Col.G
	1950	1950			1940 MR	AdjuFact	1950	
	PopFrac	Obs MR	A * B		Mid,Low	Bx2,Pt2	Adju	A * F
Trio-Sequence	Tab 3-B	Tab 15-A			Tab 15-A	Col.E	MortRates	
Pacific	0.0961	4.7	0.452				4.7	0.452
New England	0.0618	6.5	0.402				6.5	0.402
Mid-Atlantic	0.2002	5.9	1.181				5.9	1.181
WestNoCentral	0.0933	4.5	0.420		4.6	1.00	4.602	0.429
EastNoCentral	0.2017	4.9	0.988		4.8	1.00	4.802	0.969
Mountain	0.0337	3.2	0.108		2.8	1.00	2.801	0.094
WestSoCentral	0.0965	4.3	0.415		4.0	1.01	4.048	0.391
EastSoCentral	0.0762	4.2	0.320		3.3	1.01	3.340	0.254
SouthAtlantic	0.1406	4.7	0.661		4.3	1.01	4.352	0.612

```
                              Sum =    4.9                              Sum =
        1950 Observed Natl MR from Table 15-B    5.0       1950 Natl Adjusted MR =    4.7839
```

Part 2. --

	Col.A	Col.B	Col.C		Col.D	Col.E
	Mean1940	1950	Buccal-Phar. Ca. Males:		1940	Buccal-Phar. Ca. Males:
	thru1950	Adju MRs	1950 Adjusted MortRates		PPs from	1950 Adjusted MortRates
Trio-	PPs from	from Col.F	regressed on		Table 3-A	regressed on
Seq.	Tab 47-A	Part 1	Mean 1940 thru 1950 PPs		(TrioSeq)	1940 PhysPops
	x'	y	Regression Output:		x''	Regression Output:
Pac	154.16	4.7	Constant	0.7112	159.72	Constant	0.7444
NewEng	162.03	6.5	Std Err of Y Est	0.7664	161.55	Std Err of Y Est	0.7828
MidAtl	169.24	5.9	R Squared	0.6094	169.76	R Squared	0.5926
WNoCen	121.60	4.602	No. of Observation	9	123.14	No. of Observation	9
ENoCen	128.53	4.802	Degrees of Freedom	7	133.36	Degrees of Freedom	7
Mtn	119.64	2.801			119.89		
WSoCen	102.64	4.048	X Coefficient(s)	0.0303	103.94	X Coefficient(s)	0.0297
ESoCen	84.44	3.340	Std Err of Coef.	0.0092	85.83	Std Err of Coef.	0.0093
SoAtl	99.91	4.352	XCoef / S.E. =	3.3048	100.74	XCoef / S.E.	3.1907

Part 3-A.	Part 3-B.
Calculation of Fractional Causation	Calculation of Fractional Causation
from Averaged PhysPops	from 1940 PhysPops

1. Nonradiation rate is Adjusted Constant (Part 2, Col.C) = 0.7112	1. Nonradiation rate is Adjusted Constant (Part 2, Col.E) = 0.7444
2. Radiation rate is Natl Adjusted MortRate (Part 1, Col.G = 4.7839) minus Nonradiation rate (0.7112) = 4.0727	2. Radiation rate is Natl Adjusted MortRate (Part 1, Col.G = 4.7839) minus Nonradiation rate (0.7444) = 4.0395
3. 1950 Fractional Causation is radiation rate (4.0727) divided by OBSERVED Natl MR Part 1,Col.C= 5.0 = 0.81	3. 1950 Fractional Causation is radiation rate (4.0395) divided by OBSERVED Natl MR Part 1, Col.C= 5.0 = 0.81

```
===================================================================================================
                                        Table 63-CC
                   Buccal-Pharynx Cancers, Males:  Fractional Causation in 1960
===================================================================================================
```

Part 1.
Calculation of the 6 Adjusted MortRates (Col.F) and the National Adjusted MortRate (Col.G).
The last six entries in Part 1, Col.F, are the products of (Col.D * Col.E), as discussed in Chap. 49.

Trio-Sequence	Col.A 1960 PopFrac Tab 3-B	Col.B 1960 Obs MR Tab 15-A	Col.C A * B	Col.D 1940 MR Mid,Low Tab 15-A	Col.E AdjuFact Bx2,Pt2 Col.E	Col.F 1960 Adju MortRates	Col.G A * F
Pacific	0.1182	4.6	0.544			4.6	0.544
New England	0.0586	6.6	0.387			6.6	0.387
Mid-Atlantic	0.1905	5.4	1.029			5.4	1.029
WestNoCentral	0.0858	4.0	0.343	4.6	0.92	4.232	0.363
EastNoCentral	0.2020	4.9	0.990	4.8	0.92	4.416	0.892
Mountain	0.0382	2.6	0.099	2.8	0.92	2.576	0.098
WestSoCentral	0.0945	3.9	0.369	4.0	0.95	3.818	0.361
EastSoCentral	0.0672	4.2	0.282	3.3	0.95	3.150	0.212
SouthAtlantic	0.1448	4.5	0.652	4.3	0.95	4.104	0.594

```
                                   Sum =    4.7                              Sum =
        1960 Observed Natl MR from Table 15-B    4.7      1960 Natl Adjusted MR =    4.4795
```

Part 2. ---

Trio-Seq.	Col.A Mean1940 thru1960 PPs from Tab 47-A x'	Col.B 1960 Adju MRs from Col.F Part 1 y	Col.C Buccal-Phar. Ca. Males: 1960 Adjusted MortRates regressed on Mean 1940 thru 1960 PPs Regression Output:		Col.D 1940 PPs from Table 3-A (TrioSeq) x''	Col.E Buccal-Phar. Ca. Males: 1960 Adjusted MortRates regressed on 1940 PhysPops Regression Output:	
Pac	155.69	4.6	Constant	0.3338	159.72	Constant	0.4593
NewEng	162.81	6.6	Std Err of Y Est	0.7859	161.55	Std Err of Y Est	0.8263
MidAtl	167.04	5.4	R Squared	0.6129	169.76	R Squared	0.5721
WNoCen	118.15	4.232	No. of Observation	9	123.14	No. of Observation	9
ENoCen	123.87	4.416	Degrees of Freedom	7	133.36	Degrees of Freedom	7
Mtn	117.40	2.576			119.89		
WSoCen	102.31	3.818	X Coefficient(s)	0.0316	103.94	X Coefficient(s)	0.0300
ESoCen	85.63	3.150	Std Err of Coef.	0.0095	85.83	Std Err of Coef.	0.0098
SoAtl	101.72	4.104	XCoef / S.E. =	3.3293	100.74	XCoef / S.E. =	3.0593

Part 3-A. Calculation of Fractional Causation from Averaged PhysPops	Part 3-B. Calculation of Fractional Causation from 1940 PhysPops
1. Nonradiation rate is Adjusted Constant (Part 2, Col.C) = 0.3338	1. Nonradiation rate is Adjusted Constant (Part 2, Col.E) = 0.4593
2. Radiation rate is Natl Adjusted MortRate (Part 1, Col.G = 4.4795) minus Nonradiation rate (0.3338) = 4.1457	2. Radiation rate is Natl Adjusted MortRate (Part 1, Col.G = 4.4795) minus Nonradiation rate (0.4593) = 4.1508
3. 1960 Fractional Causation is radiation rate (4.1457) divided by OBSERVED Natl MR Part 1,Col.C= 4.7 = 0.88	3. 1960 Fractional Causation is radiation rate (4.1508) divided by OBSERVED Natl MR Part 1,Col.C= 4.7 = 0.86

```
===========================================================================================
                                     Table 63-EE
                 Buccal-Pharynx Cancers, Males:  Fractional Causation in 1980
===========================================================================================
Part 1.
Calculation of the 6 Adjusted MortRates (Col.F) and the National Adjusted MortRate (Col.G).
The last six entries in Part 1, Col.F, are the products of (Col.D * Col.E), as discussed in Chap. 49.
```

	Col.A 1980 PopFrac Tab 3-B	Col.B 1980 Obs MR Tab 15-A	Col.C A * B	Col.D 1940 MR Mid,Low Tab 15-A	Col.E AdjuFact Bx2,Pt2 Col.E	Col.F 1980 Adju MortRates	Col.G A * F
Trio-Sequence							
Pacific	0.1398	4.2	0.587			4.2	0.587
New England	0.0546	5.7	0.311			5.7	0.311
Mid-Atlantic	0.1630	5.1	0.831			5.1	0.831
WestNoCentral	0.0759	3.5	0.266	4.6	0.80	3.703	0.281
EastNoCentral	0.1846	4.6	0.849	4.8	0.80	3.864	0.713
Mountain	0.0502	2.9	0.146	2.8	0.80	2.254	0.113
WestSoCentral	0.1049	4.2	0.441	4.0	0.90	3.588	0.376
EastSoCentral	0.0646	4.4	0.284	3.3	0.90	2.960	0.191
SouthAtlantic	0.1624	5.0	0.812	4.3	0.90	3.857	0.626

```
                                    Sum =     4.5                          Sum =
            1980 Observed Natl MR from Table 15-B    4.6      1980 Natl Adjusted MR =    4.0312
```

```
Part 2.  -----------------------------------------------------------------------------------
```

| | Col.A
Mean1940
thru1980
PPs from
Tab 47-A
x' | Col.B
1980
Adju MRs
from Col.F
Part 1
y | Col.C
Buccal-Phar. Ca. Males:
1980 Adjusted MortRates
regressed on
Mean 1940 thru 1980 PPs
Regression Output: | | Col.D
1940
PPs from
Table 3-A
(TrioSeq)
x'' | Col.E
Buccal-Phar. Ca. Males:
1980 Adjusted MortRates
regressed on
1940 PhysPops
Regression Output: | |
Trio- Seq.							
Pac	177.35	4.2	Constant	0.3234	159.72	Constant	0.6049
NewEng	185.86	5.7	Std Err of Y Est	0.6814	161.55	Std Err of Y Est	0.7377
MidAtl	186.11	5.1	R Squared	0.6177	169.76	R Squared	0.5518
WNoCen	128.82	3.703	No. of Observation	9	123.14	No. of Observation	9
ENoCen	133.71	3.864	Degrees of Freedom	7	133.36	Degrees of Freedom	7
Mtn	133.45	2.254			119.89		
WSoCen	114.66	3.588	X Coefficient(s)	0.0252	103.94	X Coefficient(s)	0.0257
ESoCen	99.46	2.960	Std Err of Coef.	0.0075	85.83	Std Err of Coef.	0.0088
SoAtl	124.62	3.857	XCoef / S.E. =	3.3630	100.74	XCoef / S.E.	2.9356

```
-------------------------------------------------------------------------------------------
Part 3-A.                                      |  Part 3-B.
Calculation of Fractional Causation            |  Calculation of Fractional Causation
from Averaged PhysPops                          |  from 1940 PhysPops
                                               |
1.  Nonradiation rate is Adjusted              |  1.  Nonradiation rate is Adjusted
       Constant (Part 2, Col.C) =       0.3234 |         Constant (Part 2, Col.E) =       0.6049
                                               |
2.  Radiation rate is Natl Adjusted            |  2.  Radiation rate is Natl Adjusted
       MortRate (Part 1, Col.G = 4.0312)       |         MortRate (Part 1, Col.G = 4.0312)
       minus Nonradiation rate (0.3234) = 3.7078 |       minus Nonradiation rate (0.6049) = 3.4263
                                               |
3.  1980 Fractional Causation is radiation     |  3.  1980 Fractional Causation is radiation
       rate (3.7078) divided by OBSERVED       |         rate (3.4263) divided by OBSERVED
       Natl MR Part 1,Col.C=   4.6   =   0.81  |         Natl MR Part 1,Col.C=   4.6   =   0.74
-------------------------------------------------------------------------------------------
```

CHAPTER 64

Ischemic Heart Disease, Males, 1950–1993

- Table 64-A, Column A, shows a pattern of National MortRates unlike any of the patterns for cancer. The National MortRate from Ischemic Heart Disease rose until 1963 (a year not shown in our Table 40-B), and then began its dramatic decline.

- Box 1 reveals that these events occurred unevenly across the Census Divisions, however. The facts in Box 1 show that a co-actor (smoking), which can contribute to male MortRates from Ischemic Heart Disease, is operating more strongly in the LowTrio than in the TopTrio (Chapter 48, Part 5b). We must match the Census Divisions for smoking.

- Unlike the smoking adjustment for cancer, which is based on 1940 MortRates, the adjustment for IHD has to be based on 1950 MortRates --- which have been impacted by an extra decade of the inverse relationship between smoking and PhysPop. If we had been able to evaluate the 1940 dose-response between PhysPop and male IHD MortRates, the 1940 Fractional Causation by medical radiation would probably have been higher than 79%, because the impact of extra smoking in the LowTrio would probably have been less severe in 1940 than 1950. This conjecture is consistent with the finding that, for females, the 1950 Fractional Causation by medical radiation of IHD mortality is higher than for males --- and females had a history of less intense smoking (Chapter 48, Part 3).

Table 64-A
Ischemic Heart Disease, Males: Fractional Causation by Medical Radiation over Time

Year	Col.A Natl MR	Col.B Frac.C		Col.C R-Sq	Col.D X-Coef	Col.E StdErr	Col.F Coef/SE	Col.G Source
1940	No data available							No data
1950	256.4	79%		0.9475	1.4852	0.1321	11.2446	Chap. 40
1960	306.5	74%		0.8733	1.7259	0.2485	6.9463	Tab 64-B
1970	259.7	72%		0.8344	1.3710	0.2309	5.9387	Tab 64-C
1980	212.8	70%		0.7714	1.0117	0.2081	4.8608	Tab 64-D
1993	131.0	63%		0.7279	0.4969	0.1148	4.3271	Tab 64-E

```
=================================================================================================
                                   Box 1, Chap. 64
               Ischemic Heart Disease, Males:  Post-1940 Change in MortRates by Census Trios

1960 vs. 1950, by Trios:  Col.D expresses change by ratios.  Col.F expresses change by subtraction.
1993 vs. 1950, by Trios:  Col.I expresses change by ratios.  Col.K expresses change by subtraction.
MRs change inversely with PP.  High-PP Trio has lowest growth-factor.  Low-PP Trio has highest growth-factor.

          Col.A  |  Col.B    Col.C    Col.D    Col.E    Col.F  |  Col.G    Col.H    Col.I    Col.J    Col.K
          1950   |  1960     Ratio    Input    Diff:    Input  |  1993     Ratio    Input    Diff:    Input
          MortRate| MortRate  Col.B    from     Col.B    from   |  MortRate  Col.G    from     Col.G    from
          Tab 40-A| Tab 40-A  /Col.A   Col.C    minus A  Col.E  |  Tab 40-A  /Col.A   Col.H    minus A  Col.J
                 |
Pacif     283.2  |  284.2    1.004    Avg Chg    1.0    Avg Chg |  112.4    0.397    Avg Chg  -170.8   Avg Chg
NewE      297.1  |  347.1    1.168    TopTrio   50.0    TopTrio |  117.8    0.396    TopTrio  -179.3   TopTrio
MidAtl    310.3  |  355.0    1.144    1.105     44.7     31.9   |  147.9    0.477    0.423    -162.4   -170.8
                 |
WNoCen    228.4  |  284.1    1.244    Avg Chg   55.7    Avg Chg |  129.9    0.569    Avg Chg   -98.5   Avg Chg
ENoCen    258.9  |  320.8    1.239    MidTrio   61.9    MidTrio |  140.5    0.543    MidTrio  -118.4   MidTrio
Mtn       214.8  |  256.8    1.196    1.226     42.0     53.2   |  101.2    0.471    0.528    -113.6   -110.2
                 |
WSoCen    206.1  |  269.4    1.307    Avg Chg   63.3    Avg Chg |  137.6    0.668    Avg Chg   -68.5   Avg Chg
ESoCen    176.8  |  254.4    1.439    LowTrio   77.6    LowTrio |  145.8    0.825    LowTrio   -31.0   LowTrio
SoAtl     222.0  |  286.4    1.290    1.345     64.4     68.4   |  128.7    0.580    0.691    -93.3    -64.3
=================================================================================================
```

==

Box 2, Chap. 64
Ischemic Heart Disease, Males: Calculation of Adjustment Factor

This adjustment is discussed fully in Chapter 49.

● Part 1: Calculate average population-weighted MortRate for the combined TopTrio Census Divs.

1940 Ischemic Heart Disease MortRates	Census Div.	Col.A 1950 MR Tab 40-A	Col.B 1950 Pop'n Tab 3-B	Col.C 1950 Popn /53,964,513	Col.D Col.A * Col.C
not available by Census Divisions and gender.					
	Pacific	283.2	14,486,527	0.2684	76.02
	NewEng	297.1	9,314,453	0.1726	51.28
	Mid-Atl	310.3	30,163,533	0.5590	173.44
	1950		Sum TopTrio 53,964,513	Sum 1.0000	TopTrio 300.747

Census Div.	Col.A 1960 MR Tab 40-A	Col.B 1960 Pop'n Tab 3-B	Col.C 1960 Popn /65,875,863	Col.D Col.A * Col.C	Census Div.	Col.A 1970 MR Tab 40-A	Col.B 1970 Pop'n Tab 3-B	Col.C 1970 Popn /75,017,000	Col.D Col.A * Col.C
Pacific	284.2	21,198,044	0.3218	91.45	Pacific	231.0	26,087,000	0.3477	80.33
NewEng	347.1	10,509,367	0.1595	55.37	NewEng	285.3	11,781,000	0.1570	44.80
Mid-Atl	355.0	34,168,452	0.5187	184.13	Mid-Atl	300.8	37,149,000	0.4952	148.96
1960		Sum TopTrio 65,875,863	Sum 1.0000	TopTrio 330.957	1970		Sum TopTrio 75,017,000	Sum 1.0000	TopTrio 274.093

Census Div.	Col.A 1980 MR Tab 40-A	Col.B 1980 Pop'n Tab 3-B	Col.C 1980 Popn /80,615,000	Col.D Col.A * Col.C	Census Div.	Col.A 1993 MR Tab 40-A	Col.B 1990 Pop'n Tab 3-B	Col.C 1990 Popn /88,495,000	Col.D Col.A * Col.C
Pacific	177.7	31,523,000	0.3910	69.49	Pacific	112.4	37,837,000	0.4276	48.06
NewEng	223.5	12,322,000	0.1528	34.16	NewEng	117.8	12,998,000	0.1469	17.30
Mid-Atl	246.6	36,770,000	0.4561	112.48	Mid-Atl	147.9	37,660,000	0.4256	62.94
1980		Sum TopTrio 80,615,000	Sum 1.0000	TopTrio 216.127	1993		Sum TopTrio 88,495,000	Sum 1.0000	TopTrio 128.301

● Part 2: Take ratios of these TopTrio MortRates, with 1950 as the denominator of each ratio.
    Col.D modifies Col.C by separate PhysPop adjustments for MidTrio and LowTrio Census Divisions.

	Col.A TopTrio Mean MR	Col.B 1950 TopTrio Mean MR	Col.C = Col.A / Col.B	Col.D ppAdju Tab 47-B	Col.E = Col.C * Col.D	ISCHEMIC HEART DISEASE. Males.
				MidTrio		
1960	330.957	300.747	1.100	0.97	1.07	= MidTrio Adjustment Factor, 1960
1970	274.093	300.747	0.911	0.95	0.87	= MidTrio Adjustment Factor, 1970
1980	216.127	300.747	0.719	0.94	0.68	= MidTrio Adjustment Factor, 1980
1993	128.301	300.747	0.427	0.94	0.40	= MidTrio Adjustment Factor, 1993
				LowTrio		
1960	330.957	300.747	1.100	1.01	1.11	= LowTrio Adjustment Factor, 1960
1970	274.093	300.747	0.911	1.02	0.93	= LowTrio Adjustment Factor, 1970
1980	216.127	300.747	0.719	1.04	0.75	= LowTrio Adjustment Factor, 1980
1993	128.301	300.747	0.427	1.07	0.46	= LowTrio Adjustment Factor, 1993

==

```
=====================================================================================================
                                         Table 64-B
                     Ischemic Heart Disease, Males:  Fractional Causation in 1960
=====================================================================================================
Part 1.
Calculation of the 6 Adjusted MortRates (Col.F) and the National Adjusted MortRate (Col.G).
The last six entries in Part 1, Col.F, are the products of (Col.D * Col.E), as discussed in Chap. 49.
```

	Col.A	Col.B	Col.C	Col.D	Col.E	Col.F	Col.G
	1960	1960		1950 MR	AdjuFact	1960	
	PopFrac	Obs MR	A * B	Mid,Low	Bx2,Pt2	Adju	A * F
Trio-Sequence	Tab 3-B	Tab 40-A		Tab 40-A	Col.E	MortRates	
Pacific	0.1182	284.2	33.592			284.2	33.592
New England	0.0586	347.1	20.340			347.1	20.340
Mid-Atlantic	0.1905	355.0	67.628			355.0	67.628
WestNoCentral	0.0858	284.1	24.376	228.4	1.07	244.388	20.968
EastNoCentral	0.2020	320.8	64.802	258.9	1.07	277.023	55.959
Mountain	0.0382	256.8	9.810	214.8	1.07	229.836	8.780
WestSoCentral	0.0945	269.4	25.458	206.1	1.11	228.771	21.619
EastSoCentral	0.0672	254.4	17.096	176.8	1.11	196.248	13.188
SouthAtlantic	0.1448	286.4	41.471	222.0	1.11	246.420	35.682

```
                                   Sum =   304.6                                Sum =
        1960 Observed MR from Table 40-B    306.5        1960 Natl Adjusted MR =      277.7552
```

```
Part 2.  -------------------------------------------------------------------------------------------
        Col.A    Col.B                      Col.C                Col.D                     Col.E
        Mean1940  1960          Ischemic Ht. Dis. Males:          1940         Ischemic Ht. Dis. Males:
        thru1960  Adju MRs      1960 Adjusted MortRates          PPs from      1960 Adjusted MortRates
Trio-   PPs from  from Col.F        regressed on                 Table 3-A         regressed on
Seq.    Tab 47-A  Part 1       Mean 1940 thru 1960 PPs          (TrioSeq)      1940 PhysPops
        x'        y                Regression Output:            x''              Regression Output:
Pac     155.69    284.2        Constant          50.0790        159.72        Constant          52.9369
NewEng  162.81    347.1        Std Err of Y Est  20.5514        161.55        Std Err of Y Est  22.6343
MidAtl  167.04    355.0        R Squared          0.8733        169.76        R Squared          0.8463
WNoCen  118.15    244.388      No. of Observation     9         123.14        No. of Observation     9
ENoCen  123.87    277.023      Degrees of Freedom     7         133.36        Degrees of Freedom     7
Mtn     117.40    229.836                                       119.89
WSoCen  102.31    228.771      X Coefficient(s)   1.7259        103.94        X Coefficient(s)   1.6690
ESoCen   85.63    196.248      Std Err of Coef.   0.2485         85.83        Std Err of Coef.   0.2688
SoAtl   101.72    246.420      XCoef / S.E. =     6.9463        100.74        XCoef / S.E.       6.2089
```

```
-----------------------------------------------------------------------------------------------------
Part 3-A.                                         |  Part 3-B.
Calculation of Fractional Causation               |  Calculation of Fractional Causation
from Averaged PhysPops                            |  from 1940 PhysPops
                                                  |
1.  Nonradiation rate is Adjusted                 |  1.  Nonradiation rate is Adjusted
    Constant (Part 2, Col.C) =         50.0790    |      Constant (Part 2, Col.E) =         52.9369
                                                  |
2.  Radiation rate is Natl Adjusted               |  2.  Radiation rate is Natl Adjusted
    MortRate (Part 1, Col.G = 277.7552)           |      MortRate (Part 1, Col.G = 277.7552)
    minus Nonradiation rate (50.0790) =  227.6762 |      minus Nonradiation rate (52.9369) =  224.8183
                                                  |
3.  1960 Fractional Causation is radiation        |  3.  1960 Fractional Causation is radiation
    rate (227.6762) divided by OBSERVED           |      rate (224.8183) divided by OBSERVED
    Natl MR Part 1,Col.C=   306.5     =    0.74   |      Natl MR Part 1, Col.C=   306.5     =    0.73
-----------------------------------------------------------------------------------------------------
```

```
=======================================================================================================
                                          Table 64-C
                      Ischemic Heart Disease, Males:  Fractional Causation in 1970
=======================================================================================================
Part 1.
Calculation of the 6 Adjusted MortRates (Col.F) and the National Adjusted MortRate (Col.G).
The last six entries in Part 1, Col.F, are the products of (Col.D * Col.E), as discussed in Chap. 49.


                  Col.A    Col.B    Col.C        Col.D    Col.E    Col.F    Col.G
                  1970     1970                  1950 MR  AdjuFact 1970
                  PopFrac  Obs MR   A * B        Mid,Low  Bx2,Pt2  Adju     A * F
Trio-Sequence     Tab 3-B  Tab 40-A             Tab 40-A Col.E    MortRates
        Pacific   0.1293   231.0    29.868                        231.0    29.868
        New England 0.0584 285.3    16.662                        285.3    16.662
        Mid-Atlantic 0.1842 300.8   55.407                        300.8    55.407
        WestNoCentral 0.0805 245.1  19.731       228.4    0.87    198.708  15.996
        EastNoCentral 0.1993 274.1  54.628       258.9    0.87    225.243  44.891
        Mountain  0.0408   215.2    8.780        214.8    0.87    186.876= 7.625
        WestSoCentral 0.0948 232.0  21.994       206.1    0.93    191.673  18.171
        EastSoCentral 0.0631 236.8  14.942       176.8    0.93    164.424  10.375
        SouthAtlantic 0.1496 248.7  37.206       222.0    0.93    206.460  30.886

                           Sum =    259.2                                  Sum =
         1970 Observed MR from Table 40-B  259.7       1970 Natl Adjusted MR =   229.8808


Part 2.  -----------------------------------------------------------------------------------------------
         Col.A    Col.B                     Col.C                 Col.D                        Col.E
         Mean1940 1970            Ischemic Ht. Dis. Males:        1940            Ischemic Ht. Dis. Males:
         thru1970 Adju MRs        1970 Adjusted MortRates         PPs from        1970 Adjusted MortRates
Trio-    PPs from from Col.F          regressed on                Table 3-A           regressed on
Seq.     Tab 47-A Part 1         Mean 1940 thru 1970 PPs         (TrioSeq)        1940 PhysPops
         x'       y                  Regression Output:          x''                 Regression Output:
Pac      162.72   231.0       Constant          42.1983          159.72       Constant          45.1199
NewEng   168.74   285.3       Std Err of Y Est  19.7900          161.55       Std Err of Y Est  21.6442
MidAtl   173.28   300.8       R Squared         0.8344           169.76       R Squared         0.8019
WNoCen   119.56   198.708     No. of Observation     9           123.14       No. of Observation     9
ENoCen   124.70   225.243     Degrees of Freedom     7           133.36       Degrees of Freedom     7
Mtn      122.37   186.876                                        119.89
WSoCen   105.03   191.673     X Coefficient(s)  1.3710           103.94       X Coefficient(s)  1.3683
ESoCen    89.44   164.424     Std Err of Coef.  0.2309            85.83       Std Err of Coef.  0.2570
SoAtl    108.97   206.460     XCoef / S.E. =    5.9387           100.74       XCoef / S.E.      5.3232


-------------------------------------------------------------- -----------------------------------------
Part 3-A.                                                      | Part 3-B.
Calculation of Fractional Causation                           | Calculation of Fractional Causation
from Averaged PhysPops                                        | from 1940 PhysPops
                                                              |
1.  Nonradiation rate is Adjusted                             | 1.  Nonradiation rate is Adjusted
    Constant (Part 2, Col.C) =              42.1983           |     Constant (Part 2, Col.E) =              45.1199
                                                              |
2.  Radiation rate is Natl Adjusted                          | 2.  Radiation rate is Natl Adjusted
    MortRate (Part 1, Col.G = 229.8808)                       |     MortRate (Part 1, Col.G = 229.8808)
    minus Nonradiation rate (42.1983) =     187.6826          |     minus Nonradiation rate (45.1199) =     184.7609
                                                              |
3.  1970 Fractional Causation is radiation                   | 3.  1970 Fractional Causation is radiation
    rate (187.6826) divided by OBSERVED                       |     rate (184.7609) divided by OBSERVED
    Natl MR Part 1,Col.C=  259.7    =        0.72             |     Natl MR Part 1,Col.C=  259.7    =        0.71
-------------------------------------------------------------- -----------------------------------------
```

```
============================================================================================
                                     Table 64-D
                Ischemic Heart Disease, Males:  Fractional Causation in 1980
============================================================================================
Part 1.
Calculation of the 6 Adjusted MortRates (Col.F) and the National Adjusted MortRate (Col.G).
The last six entries in Part 1, Col.F, are the products of (Col.D * Col.E), as discussed in Chap. 49.
```

	Col.A 1980 PopFrac Tab 3-B	Col.B 1980 Obs MR Tab 40-A	Col.C A * B	Col.D 1950 MR Mid,Low Tab 40-A	Col.E AdjuFact Bx2,Pt2 Col.E	Col.F 1980 Adju MortRates	Col.G A * F
Trio-Sequence							
Pacific	0.1398	177.7	24.842			177.7	24.842
New England	0.0546	223.5	12.203			223.5	12.203
Mid-Atlantic	0.1630	246.6	40.196			246.6	40.196
WestNoCentral	0.0759	206.1	15.643	228.4	0.68	155.312	11.788
EastNoCentral	0.1846	227.4	41.978	258.9	0.68	176.052	32.499
Mountain	0.0502	173.6	8.715	214.8	0.68	146.064	7.332
WestSoCentral	0.1049	194.5	20.403	206.1	0.75	154.575	16.215
EastSoCentral	0.0646	219.2	14.160	176.8	0.75	132.600	8.566
SouthAtlantic	0.1624	210.9	34.250	222.0	0.75	166.500	27.040

```
                                      Sum =    212.4                            Sum =
          1980 Observed MR from Table 40-B     212.8         1980 Natl Adjusted MR =     180.6816
```

```
Part 2.  -----------------------------------------------------------------------------------
```

	Col.A Mean1940 thru1980 PPs from Tab 47-A x'	Col.B 1980 Adju MRs from Col.F Part 1 y	Col.C Ischemic Ht. Dis. Males: 1980 Adjusted MortRates regressed on Mean 1940 thru 1980 PPs Regression Output:		Col.D 1940 PPs from Table 3-A (TrioSeq) x''	Col.E Ischemic Ht. Dis. Males: 1980 Adjusted MortRates regressed on 1940 PhysPops Regression Output:	
Trio- Seq.							
Pac	177.35	177.7	Constant	31.0994	159.72	Constant	38.4779
NewEng	185.86	223.5	Std Err of Y Est	18.9493	161.55	Std Err of Y Est	20.5756
MidAtl	186.11	246.6	R Squared	0.7714	169.76	R Squared	0.7305
WNoCen	128.82	155.3	No. of Observation	9	123.14	No. of Observation	9
ENoCen	133.71	176.1	Degrees of Freedom	7	133.36	Degrees of Freedom	7
Mtn	133.45	146.1			119.89		
WSoCen	114.66	154.6	X Coefficient(s)	1.0117	103.94	X Coefficient(s)	1.0645
ESoCen	99.46	132.6	Std Err of Coef.	0.2081	85.83	Std Err of Coef.	0.2444
SoAtl	124.62	166.5	XCoef / S.E. =	4.8608	100.74	XCoef / S.E. =	4.3563

```
    --------------------------------------------------------------------------------------
```

Part 3-A. Calculation of Fractional Causation from Averaged PhysPops	Part 3-B. Calculation of Fractional Causation from 1940 PhysPops

1. Nonradiation rate is Adjusted
     Constant (Part 2, Col.C) =   31.0994
	1. Nonradiation rate is Adjusted
	     Constant (Part 2, Col.E) = 38.4779

2. Radiation rate is Natl Adjusted
     MortRate (Part 1, Col.G = 180.6816)
     minus Nonradiation rate (31.0994) =   149.5822
	2. Radiation rate is Natl Adjusted
	     MortRate (Part 1, Col.G = 180.6816)
	     minus Nonradiation rate (38.4779) = 142.2037

3. 1980 Fractional Causation is radiation
     rate (149.5822) divided by OBSERVED
     Natl MR Part 1,Col.C=   212.8   =   0.70
	3. 1980 Fractional Causation is radiation
	     rate (142.2037) divided by OBSERVED
	     Natl MR Part 1,Col.C= 212.8 = 0.67

===
Table 64-E
Ischemic Heart Disease, Males: Fractional Causation in 1993
===

Part 1.

Calculation of the 6 Adjusted MortRates (Col.F) and the National Adjusted MortRate (Col.G).

The last six entries in Part 1, Col.F, are the products of (Col.D * Col.E), as discussed in Chap. 49.

Trio-Sequence	Col.A 1990 PopFrac Tab 3-B	Col.B 1993 Obs MR Tab 40-A	Col.C A * B	Col.D 1950 MR Mid,Low Tab 40-A	Col.E AdjuFact Bx2,Pt2 Col.E	Col.F 1993 Adju MortRates	Col.G A * F
Pacific	0.1535	112.4	17.253			112.4	17.253
New England	0.0527	117.8	6.208			117.8	6.208
Mid-Atlantic	0.1527	147.9	22.584			147.9	22.584
WestNoCentral	0.0721	129.9	9.366	228.4	0.40	91.360	6.587
EastNoCentral	0.1713	140.5	24.068	258.9	0.40	103.560	17.740
Mountain	0.0543	101.2	5.495	214.8	0.40	85.920	4.665
WestSoCentral	0.1087	137.6	14.957	206.1	0.46	94.806	10.305
EastSoCentral	0.0621	145.8	9.054	176.8	0.46	81.328	5.050
SouthAtlantic	0.1725	128.7	22.201	222.0	0.46	102.120	17.616

Sum =   131.2

1993 Observed MR from Table 40-B   131.0   1993 Natl Adjusted MR =   108.0097

Part 2. --

Trio-Seq.	Col.A Mean1940 thru1990 PPs from Tab 47-A x'	Col.B 1993 Adju MRs from Col.F Part 1 y	Col.C Ischemic Ht. Dis. Males: 1993 Adjusted MortRates regressed on Mean 1940 thru 1990 PPs Regression Output:		Col.D 1940 PPs from Table 3-A (TrioSeq) x''	Col.E Ischemic Ht. Dis. Males: 1993 Adjusted MortRates regressed on 1940 PhysPops Regression Output:	
Pac	191.97	112.4	Constant	25.6954	159.72	Constant	31.6182
NewEng	208.20	117.8	Std Err of Y Est	11.2868	161.55	Std Err of Y Est	12.0989
MidAtl	204.72	147.9	R Squared	0.7279	169.76	R Squared	0.6873
WNoCen	141.14	91.360	No. of Observation	9	123.14	No. of Observation	9
ENoCen	146.19	103.560	Degrees of Freedom	7	133.36	Degrees of Freedom	7
Mtn	145.91	85.920			119.89		
WSoCen	126.28	94.806	X Coefficient(s)	0.4969	103.94	X Coefficient(s)	0.5636
ESoCen	113.28	81.328	Std Err of Coef.	0.1148	85.83	Std Err of Coef.	0.1437
SoAtl	142.93	102.120	XCoef / S.E. =	4.3271	100.74	XCoef / S.E.	3.9225

Part 3-A.
Calculation of Fractional Causation
from Averaged PhysPops

1. Nonradiation rate is Adjusted
    Constant (Part 2, Col.C) =   25.6954

2. Radiation rate is Natl Adjusted
    MortRate (Part 1, Col.G = 108.0097)
    minus Nonradiation rate (25.6954) =   82.3143

3. 1993 Fractional Causation is radiation
    rate (82.3143) divided by OBSERVED
    Natl MR Part 1,Col.C=   131.0   =   0.63

Part 3-B.
Calculation of Fractional Causation
from 1940 PhysPops

1. Nonradiation rate is Adjusted
    Constant (Part 2, Col.E) =   31.6182

2. Radiation rate is Natl Adjusted
    MortRate (Part 1, Col.G = 108.0097)
    minus Nonradiation rate (31.6182) =   76.3915

3. 1993 Fractional Causation is radiation
    rate (76.3915) divided by OBSERVED
    Natl MR Part 1, Col.C=   131.0   =   0.58

CHAPTER 65

Ischemic Heart Disease, Females, 1950-1993

- Table 65-A, Column A, shows a pattern of National MortRates similar to the male pattern in Table 64-A, but at half the magnitude. If we assume (reasonably) that dosage of medical radiation is approximately the same for males and females, the gender-difference in magnitude of the IHD MortRates is another example of a concept we have emphasized throughout this book: Medical radiation has been a NECESSARY contributor to most fatal cases of cancer and IHD in the Twentieth Century, and its impact per unit dose on MortRates is modulated by the levels of its co-actors. For IHD, such co-actors with medical radiation include smoking, and unfavorable patterns of blood lipid-levels, and other agents.

- Box 1 shows that, by 1960, the 1950 MortRates increase most in the LowTrio and least in the TopTrio. MortRates peak in the mid-1960s. By 1993, the decline from the 1950 MortRates is greatest in the TopTrio and least in the LowTrio. Such observations indicate that smoking has been most intense in the LowTrio and least intense in the TopTrio (Chapter 48, Part 5b). We must match the Census Divisions for smoking.

- Although Tables 65-B, 65-C, and 65-D produce negative Constants, their values are such a small fraction of the Observed and Adjusted MortRates that they could readily have fallen upon the Origin or on its positive side --- as does the positive Constant in 1993.

	Table 65-A						
	Ischemic Heart Disease, Females: Fractional Causation by Medical Radiation over Time						
Year	Col.A Natl MR	Col.B Frac.C	Col.C R-Sq	Col.D X-Coef	Col.E StdErr	Col.F Coef/SE	Col.G Source
1940	No data available						No data
1950	126.5	97%	0.8669	0.9041	0.1339	6.7531	Chap. 41
1960	152.5	89%	0.8084	1.0346	0.1904	5.4353	Tab 65-B
1970	124.9	86%	0.7980	0.8074	0.1535	5.2593	Tab 65-C
1980	97.2	83%	0.7620	0.5620	0.1187	4.7340	Tab 65-D
1993	64.7	78%	0.6816	0.3025	0.0782	3.8710	Tab 65-E

===

```
                              Box 1, Chap. 65
          Ischemic Heart Disease, Females:  Post-1940 Change in MortRates by Census Trios

1960 vs. 1950, by Trios:  Col.D expresses change by ratios.  Col.F expresses change by subtraction.
1993 vs. 1950, by Trios:  Col.I expresses change by ratios.  Col.K expresses change by subtraction.
MRs change inversely with PP.  High-PP Trio has lowest growth-factor.  Low-PP Trio has highest growth-factor.
```

	Col.A 1950 MortRate Tab 41-A		Col.B 1960 MortRate Tab 41-A	Col.C Ratio Col.B /Col.A	Col.D Input from Col.C	Col.E Diff: Col.B minus A	Col.F Input from Col.E		Col.G 1993 MortRate Tab 41-A	Col.H Ratio Col.G /Col.A	Col.I Input from Col.H	Col.J Diff: Col.G minus A	Col.K Input from Col.J
Pacif	125.0	|	133.4	1.067	Avg Chg	8.4	Avg Chg	|	57.7	0.462	Avg Chg	-67.3	Avg Chg
NewE	153.2	|	176.3	1.151	TopTrio	23.1	TopTrio	|	55.7	0.364	TopTrio	-97.5	TopTrio
MidAtl	169.4	|	189.7	1.120	1.113	20.3	17.3	|	78.8	0.465	0.430	-90.6	-85.1
WNoCen	104.1	|	135.8	1.305	Avg Chg	31.7	Avg Chg	|	58.3	0.560	Avg Chg	-45.8	Avg Chg
ENoCen	124.2	|	162.2	1.306	MidTrio	38.0	MidTrio	|	70.2	0.565	MidTrio	-54.0	MidTrio
Mtn	96.2	|	118.9	1.236	1.282	22.7	30.8	|	46.3	0.481	0.536	-49.9	-49.9
WSoCen	94.0	|	123.9	1.318	Avg Chg	29.9	Avg Chg	|	66.5	0.707	Avg Chg	-27.5	Avg Chg
ESoCen	84.7	|	126.2	1.490	LowTrio	41.5	LowTrio	|	67.7	0.799	LowTrio	-17.0	LowTrio
SoAtl	103.4	|	132.4	1.280	1.363	29.0	33.5	|	61.6	0.596	0.701	-41.8	-28.8

===

==

Box 2, Chap. 65
Ischemic Heart Disease, Females: Calculation of Adjustment Factor

This adjustment is discussed fully in Chapter 49.

● Part 1: Calculate average population-weighted MortRate for the combined TopTrio Census Divs.

		Col.A 1950 MR	Col.B 1950 Pop'n	Col.C 1950 Popn	Col.D Col.A *
1940 Ischemic Heart Disease MortRates not available by Census Divisions and gender.	Census Div.	Tab 41-A	Tab 3-B	/53,964,513	Col.C
	Pacific	125.0	14,486,527	0.2684	33.56
	NewEng	153.2	9,314,453	0.1726	26.44
	Mid-Atl	169.4	30,163,533	0.5590	94.69
	1950		Sum TopTrio	Sum	TopTrio
			53,964,513	1.0000	154.685

	Col.A 1960 MR	Col.B 1960 Pop'n	Col.C 1960 Popn	Col.D Col.A *		Col.A 1970 MR	Col.B 1970 Pop'n	Col.C 1970 Popn	Col.D Col.A *
Census Div.	Tab 41-A	Tab 3-B	/65,875,863	Col.C	Census Div.	Tab 41-A	Tab 3-B	/75,017,000	Col.C
Pacific	133.4	21,198,044	0.3218	42.93	Pacific	107.4	26,087,000	0.3477	37.35
NewEng	176.3	10,509,367	0.1595	28.13	NewEng	138.0	11,781,000	0.1570	21.67
Mid-Atl	189.7	34,168,452	0.5187	98.39	Mid-Atl	154.9	37,149,000	0.4952	76.71
1960		Sum TopTrio	Sum	TopTrio	1970		Sum TopTrio	Sum	TopTrio
		65,875,863	1.0000	169.446			75,017,000	1.0000	135.728

	Col.A 1980 MR	Col.B 1980 Pop'n	Col.C 1980 Popn	Col.D Col.A *		Col.A 1993 MR	Col.B 1990 Pop'n	Col.C 1990 Popn	Col.D Col.A *
Census Div.	Tab 41-A	Tab 3-B	/80,615,000	Col.C	Census Div.	Tab 41-A	Tab 3-B	/88,495,000	Col.C
Pacific	81.4	31,523,000	0.3910	31.83	Pacific	57.7	37,837,000	0.4276	24.67
NewEng	99.6	12,322,000	0.1528	15.22	NewEng	55.7	12,998,000	0.1469	8.18
Mid-Atl	120.1	36,770,000	0.4561	54.78	Mid-Atl	78.8	37,660,000	0.4256	33.53
1980		Sum TopTrio	Sum	TopTrio	1993		Sum TopTrio	Sum	TopTrio
		80,615,000	1.0000	101.834			88,495,000	1.0000	66.386

● Part 2: Take ratios of these TopTrio MortRates, with 1950 as the denominator of each ratio.
  Col.D modifies Col.C by separate PhysPop adjustments for MidTrio and LowTrio Census Divisions.

	Col.A TopTrio Mean MR	Col.B 1950 TopTrio Mean MR	Col.C = Col.A / Col.B	Col.D ppAdju Tab 47-B	Col.E = Col.C * Col.D	ISCHEMIC HEART DISEASE. Females.
				MidTrio		
1960	169.446	154.685	1.095	0.97	1.06	= MidTrio Adjustment Factor, 1960
1970	135.728	154.685	0.877	0.95	0.83	= MidTrio Adjustment Factor, 1970
1980	101.834	154.685	0.658	0.94	0.62	= MidTrio Adjustment Factor, 1980
1993	66.386	154.685	0.429	0.94	0.40	= MidTrio Adjustment Factor, 1993
				LowTrio		
1960	169.446	154.685	1.095	1.01	1.11	= LowTrio Adjustment Factor, 1960
1970	135.728	154.685	0.877	1.02	0.89	= LowTrio Adjustment Factor, 1970
1980	101.834	154.685	0.658	1.04	0.68	= LowTrio Adjustment Factor, 1980
1993	66.386	154.685	0.429	1.07	0.46	= LowTrio Adjustment Factor, 1993

==

===

Table 65-B
Ischemic Heart Disease, Females: Fractional Causation in 1960

===

Part 1.
Calculation of the 6 Adjusted MortRates (Col.F) and the National Adjusted MortRate (Col.G).
The last six entries in Part 1, Col.F, are the products of (Col.D * Col.E), as discussed in Chap. 49.

Trio-Sequence	Col.A 1960 PopFrac Tab 3-B	Col.B 1960 Obs MR Tab 41-A	Col.C A * B	Col.D 1950 MR Mid,Low Tab 41-A	Col.E AdjuFact Bx2,Pt2 Col.E	Col.F 1960 Adju MortRates	Col.G A * F
Pacific	0.1182	133.4	15.768			133.4	15.768
New England	0.0586	176.3	10.331			176.3	10.331
Mid-Atlantic	0.1905	189.7	36.138			189.7	36.138
WestNoCentral	0.0858	135.8	11.652	104.1	1.06	110.346	9.468
EastNoCentral	0.2020	162.2	32.764	124.2	1.06	131.652	26.594
Mountain	0.0382	118.9	4.542	96.2	1.06	101.972	3.895
WestSoCentral	0.0945	123.9	11.709	94.0	1.11	104.340	9.860
EastSoCentral	0.0672	126.2	8.481	84.7	1.11	94.017	6.318
SouthAtlantic	0.1448	132.4	19.172	103.4	1.11	114.774	16.619

Sum =   150.6

1960 Observed MR from Table 41-B   152.5

Sum =
1960 Natl Adjusted MR =   134.9910

Part 2. --

Trio-Seq.	Col.A Mean1940 thru1960 PPs from Tab 47-A x'	Col.B 1960 Adju MRs from Col.F Part 1 y	Col.C Ischemic Ht. Dis. Females: 1960 Adjusted MortRates regressed on Mean 1940 thru 1960 PPs Regression Output:	Col.D 1940 PPs from Table 3-A (TrioSeq) x''	Col.E Ischemic Ht. Dis. Females: 1960 Adjusted MortRates regressed on 1940 PhysPops Regression Output:
Pac	155.69	133.4	Constant -1.9357	159.72	Constant 0.6663
NewEng	162.81	176.3	Std Err of Y Est 15.7446	161.55	Std Err of Y Est 17.1514
MidAtl	167.04	189.7	R Squared 0.8084	169.76	R Squared 0.7727
WNoCen	118.15	110.346	No. of Observation 9	123.14	No. of Observation 9
ENoCen	123.87	131.652	Degrees of Freedom 7	133.36	Degrees of Freedom 7
Mtn	117.40	101.972		119.89	
WSoCen	102.31	104.340	X Coefficient(s) 1.0346	103.94	X Coefficient(s) 0.9936
ESoCen	85.63	94.017	Std Err of Coef. 0.1904	85.83	Std Err of Coef. 0.2037
SoAtl	101.72	114.774	XCoef / S.E. = 5.4353	100.74	XCoef / S.E. 4.8779

Part 3-A. | Part 3-B.
Calculation of Fractional Causation | Calculation of Fractional Causation
from Averaged PhysPops | from 1940 PhysPops
                                                   |
1. Nonradiation rate is Adjusted | 1. Nonradiation rate is Adjusted
    Constant (Part 2, Col.C) = Negative   0.0 | Constant (Part 2, Col.E) =   0.6663
                                                   |
2. Radiation rate is Natl Adjusted | 2. Radiation rate is Natl Adjusted
    MortRate (Part 1, Col.G = 134.9910) | MortRate (Part 1, Col.G = 134.9910)
    minus Nonradiation rate (0.0) =   134.9910 | minus Nonradiation rate (0.6663) =   134.3247
                                                   |
3. 1960 Fractional Causation is radiation | 3. 1960 Fractional Causation is radiation
    rate (134.9910) divided by OBSERVED | rate (134.3247) divided by OBSERVED
    Natl MR Part 1,Col.C=   152.5   =   0.89 | Natl MR Part 1, Col.C=   152.5   =   0.88

===
Table 65-C
Ischemic Heart Disease, Females: Fractional Causation in 1970
===

Part 1.

Calculation of the 6 Adjusted MortRates (Col.F) and the National Adjusted MortRate (Col.G).

The last six entries in Part 1, Col.F, are the products of (Col.D * Col.E), as discussed in Chap. 49.

Trio-Sequence	Col.A 1970 PopFrac Tab 3-B	Col.B 1970 Obs MR Tab 41-A	Col.C A * B	Col.D 1950 MR Mid,Low Tab 41-A	Col.E AdjuFact Bx2,Pt2 Col.E	Col.F 1970 Adju MortRates	Col.G A * F
Pacific	0.1293	107.4	13.887			107.4	13.887
New England	0.0584	138.0	8.059			138.0	8.059
Mid-Atlantic	0.1842	154.9	28.533			154.9	28.533
WestNoCentral	0.0805	111.0	8.936	104.1	0.83	86.403	6.955
EastNoCentral	0.1993	134.5	26.806	124.2	0.83	103.086	20.545
Mountain	0.0408	96.6	3.941	96.2	0.83	79.846	3.258
WestSoCentral	0.0948	105.5	10.001	94.0	0.89	83.660	7.931
EastSoCentral	0.0631	110.7	6.985	84.7	0.89	75.383	4.757
SouthAtlantic	0.1496	111.6	16.695	103.4	0.89	92.026	13.767

                                           Sum =   123.8   Sum =
              1970 Observed MR from Table 41-B   124.9   1970 Natl Adjusted MR =   107.6915

Part 2. ---

Trio-Seq.	Col.A Mean1940 thru1970 PPs from Tab 47-A x'	Col.B 1970 Adju MRs from Col.F y	Col.C Ischemic Ht. Dis. Females: 1970 Adjusted MortRates regressed on Mean 1940 thru 1970 PPs		Col.D 1940 PPs from Table 3-A (TrioSeq) x''	Col.E Ischemic Ht. Dis. Females: 1970 Adjusted MortRates regressed on 1940 PhysPops	
Pac	162.72	107.4	Regression Output:		159.72	Regression Output:	
NewEng	168.74	138.0	Constant	-3.0913	161.55	Constant	-0.8291
MidAtl	173.28	154.9	Std Err of Y Est	13.1597	169.76	Std Err of Y Est	14.3762
WNoCen	119.56	86.403	R Squared	0.7980	123.14	R Squared	0.7590
ENoCen	124.70	103.086	No. of Observation	9	133.36	No. of Observation	9
Mtn	122.37	79.846	Degrees of Freedom	7	119.89	Degrees of Freedom	7
WSoCen	105.03	83.660			103.94		
ESoCen	89.44	75.383	X Coefficient(s)	0.8074	85.83	X Coefficient(s)	0.8016
SoAtl	108.97	92.026	Std Err of Coef.	0.1535	100.74	Std Err of Coef.	0.1707
			XCoef / S.E. =	5.2593		XCoef / S.E.	4.6949

Part 3-A. | Part 3-B.
Calculation of Fractional Causation | Calculation of Fractional Causation
from Averaged PhysPops | from 1940 PhysPops
                                                   |
1. Nonradiation rate is Adjusted | 1. Nonradiation rate is Adjusted
    Constant (Part 2, Col.C) = Negative   0.0 | Constant (Part 2, Col.E) = NEGATIVE   0.0
                                                   |
2. Radiation rate is Natl Adjusted | 2. Radiation rate is Natl Adjusted
    MortRate (Part 1, Col.G = 107.6915) | MortRate (Part 1, Col.G = 107.6915)
    minus Nonradiation rate (0.0) =   107.6915 | minus Nonradiation rate (0.0) =   107.6915
                                                   |
3. 1970 Fractional Causation is radiation | 3. 1970 Fractional Causation is radiation
    rate (107.6915) divided by OBSERVED | rate (107.6915) divided by OBSERVED
    Natl MR Part 1,Col.C= 124.9   =   0.86 | Natl MR Part 1, Col.C= 124.9   =   0.86

```
===============================================================================================
                                         Table 65-D
                    Ischemic Heart Disease, Females:  Fractional Causation in 1980
===============================================================================================
```

Part 1.
Calculation of the 6 Adjusted MortRates (Col.F) and the National Adjusted MortRate (Col.G).
The last six entries in Part 1, Col.F, are the products of (Col.D * Col.E), as discussed in Chap. 49.

	Col.A 1980 PopFrac Tab 3-B	Col.B 1980 Obs MR Tab 41-A	Col.C A * B		Col.D 1950 MR Mid,Low Tab 41-A	Col.E AdjuFact Bx2,Pt2 Col.E	Col.F 1980 Adju MortRates	Col.G A * F
Trio-Sequence								
Pacific	0.1398	81.4	11.380				81.4	11.380
New England	0.0546	99.6	5.438				99.6	5.438
Mid-Atlantic	0.1630	120.1	19.576				120.1	19.576
WestNoCentral	0.0759	86.1	6.535		104.1	0.62	64.542	4.899
EastNoCentral	0.1846	106.8	19.715		124.2	0.62	77.004	14.215
Mountain	0.0502	74.2	3.725		96.2	0.62	59.644	2.994
WestSoCentral	0.1049	87.1	9.137		94.0	0.68	63.920	6.705
EastSoCentral	0.0646	95.2	6.150		84.7	0.68	57.596	3.721
SouthAtlantic	0.1624	90.8	14.746		103.4	0.68	70.312	11.419

```
                                      Sum =   96.4                          Sum =
        1980 Observed MR from Table 41-B      97.2      1980 Natl Adjusted MR =    80.3466
```

Part 2. --

	Col.A Mean1940 thru1980 PPs from Tab 47-A x'	Col.B 1980 Adju MRs from Col.F Part 1 y	Col.C Ischemic Ht. Dis. Females: 1980 Adjusted MortRates regressed on Mean 1940 thru 1980 PPs		Col.D 1940 PPs from Table 3-A (TrioSeq) x''	Col.E Ischemic Ht. Dis. Females: 1980 Adjusted MortRates regressed on 1940 PhysPops
Trio-Seq.			Regression Output:			Regression Output:
Pac	177.35	81.4	Constant	-3.0605	159.72	Constant	0.6652
NewEng	185.86	99.6	Std Err of Y Est	10.8092	161.55	Std Err of Y Est	11.5409
MidAtl	186.11	120.1	R Squared	0.7620	169.76	R Squared	0.7287
WNoCen	128.82	64.542	No. of Observation	9	123.14	No. of Observation	9
ENoCen	133.71	77.004	Degrees of Freedom	7	133.36	Degrees of Freedom	7
Mtn	133.45	59.644			119.89		
WSoCen	114.66	63.920	X Coefficient(s)	0.5620	103.94	X Coefficient(s)	0.5943
ESoCen	99.46	57.596	Std Err of Coef.	0.1187	85.83	Std Err of Coef.	0.1371
SoAtl	124.62	70.312	XCoef / S.E. =	4.7340	100.74	XCoef / S.E. =	4.3359

Part 3-A. | Part 3-B.
Calculation of Fractional Causation | Calculation of Fractional Causation
from Averaged PhysPops | from 1940 PhysPops
                                                   |
1. Nonradiation rate is Adjusted | 1. Nonradiation rate is Adjusted
    Constant (Part 2, Col.C) = Negative   0.0 | Constant (Part 2, Col.E) =   0.6652
                                                   |
2. Radiation rate is Natl Adjusted | 2. Radiation rate is Natl Adjusted
    MortRate (Part 1, Col.G = 80.3466) | MortRate (Part 1, Col.G = 80.3466)
    minus Nonradiation rate (0.0) =   80.3466 | minus Nonradiation rate (0.6652) = 79.6814
                                                   |
3. 1980 Fractional Causation is radiation | 3. 1980 Fractional Causation is radiation
    rate (80.3466) divided by OBSERVED | rate (79.6814) divided by OBSERVED
    Natl MR Part 1,Col.C=   97.2   =   0.83 | Natl MR Part 1,Col.C=   97.2   =   0.82

```
===================================================================================================
                                        Table 65-E
                      Ischemic Heart Disease, Females:  Fractional Causation in 1993
===================================================================================================
Part 1.
Calculation of the 6 Adjusted MortRates (Col.F) and the National Adjusted MortRate (Col.G).
The last six entries in Part 1, Col.F, are the products of (Col.D * Col.E), as discussed in Chap. 49.
```

	Col.A	Col.B	Col.C	Col.D	Col.E	Col.F	Col.G
	1990	1993		1950 MR	AdjuFact	1993	
	PopFrac	Obs MR	A * B	Mid,Low	Bx2,Pt2	Adju	A * F
Trio-Sequence	Tab 3-B	Tab 41-A		Tab 41-A	Col.E	MortRates	
Pacific	0.1535	57.7	8.857			57.7	8.857
New England	0.0527	55.7	2.935			55.7	2.935
Mid-Atlantic	0.1527	78.8	12.033			78.8	12.033
WestNoCentral	0.0721	58.3	4.203	104.1	0.40	41.640	3.002
EastNoCentral	0.1713	70.2	12.025	124.2	0.40	49.680	8.510
Mountain	0.0543	46.3	2.514	96.2	0.40	38.480	2.089
WestSoCentral	0.1087	66.5	7.229	94.0	0.46	43.240	4.700
EastSoCentral	0.0621	67.7	4.204	84.7	0.46	38.962	2.420
SouthAtlantic	0.1725	61.6	10.626	103.4	0.46	47.564	8.205

```
                                Sum =      64.6                           Sum =
        1993 Observed Natl MR from Table 41-B =   64.7      1993 Natl Adjusted MR =     52.7515
```

```
Part 2. -------------------------------------------------------------------------------------------
        Col.A   Col.B                       Col.C                 Col.D                      Col.E
        Mean1940  1993           Ischemic Ht. Dis. Females:        1940        Ischemic Ht. Dis. Females:
        thru1990  Adju MRs       1993 Adjusted MortRates          PPs from     1993 Adjusted MortRates
Trio-   PPs from  from Col.F        regressed on                  Table 3-A       regressed on
Seq.    Tab 47-A  Part 1        Mean 1940 thru 1990 PPs           (TrioSeq)    1940 PhysPops
          x'       y                     Regression Output:         x''                    Regression Output:
Pac     191.97   57.7           Constant            2.4435        159.72       Constant             5.4867
NewEng  208.20   55.7           Std Err of Y Est    7.6810        161.55       Std Err of Y Est     7.9358
MidAtl  204.72   78.8           R Squared           0.6816        169.76       R Squared            0.6601
WNoCen  141.14   41.640         No. of Observation     9          123.14       No. of Observation      9
ENoCen  146.19   49.680         Degrees of Freedom     7          133.36       Degrees of Freedom      7
Mtn     145.91   38.480                                           119.89
WSoCen  126.28   43.240         X Coefficient(s)    0.3025        103.94       X Coefficient(s)     0.3475
ESoCen  113.28   38.962         Std Err of Coef.    0.0782        85.83        Std Err of Coef.     0.0942
SoAtl   142.93   47.564         XCoef / S.E. =      3.8710        100.74       XCoef / S.E.         3.6872
```

```
--------------------------------------------------------------------------------------------------
Part 3-A.                                          |  Part 3-B.
Calculation of Fractional Causation                |  Calculation of Fractional Causation
from Averaged PhysPops                             |  from 1940 PhysPops
                                                   |
1.  Nonradiation rate is Adjusted                  |  1.  Nonradiation rate is Adjusted
    Constant (Part 2, Col.C) =          2.4435     |      Constant (Part 2, Col.E) =          5.4867
                                                   |
2.  Radiation rate is Natl Adjusted                |  2.  Radiation rate is Natl Adjusted
    MortRate (Part 1, Col.G = 52.7515)             |      MortRate (Part 1, Col.G = 52.7515)
    minus Nonradiation rate (2.4435) =   50.3080   |      minus Nonradiation rate (5.4867) =   47.2648
                                                   |
3.  1993 Fractional Causation is radiation         |  3.  1993 Fractional Causation is radiation
    rate (50.3080) divided by OBSERVED             |      rate (47.2648) divided by OBSERVED
    Natl MR Part 1,Col.C=    64.7    =    0.78     |      Natl MR Part 1, Col.C=    64.7    =    0.73
--------------------------------------------------------------------------------------------------
```

Box 1, and Collection of the Tables "A"

Box 1 of this chapter summarizes the findings from two different sets of analyses: The earliest set and the most recent set. For Cancer, the earliest analyses are for the 1940 MortRates, and the most recent are usually for MortRates of 1988 or 1990 (a few for 1980). For Ischemic Heart Disease, the earliest analyses are for the 1950 MortRates, and the most recent are for the 1993 MortRates. Appendix-M presents a directly comparable Box 1, which summarizes the results from an alternative method of making the Smoking Adjustment. The results in Appendix-M are very similar to results summarized in this chapter, except for Respiratory-System Cancers.

In this chapter, following Box 1, we reproduce the more detailed Tables A or AA, from Chapters 49 through 65 --- so that they may be easily compared with each other. Reminder: "Natl MR" means "National Mortality-Rate per 100,000 Population, Age-Adjusted to the 1940 Reference Year."

When all the findings are assembled here, it is clear that they strongly support Hypotheses 1+2. Even VERY much lower values for Fractional Causation, by medical radiation, also would have supported these hypotheses.

The Key Datum from the Year 1896

Hypothesis-1 embraces the entire Twentieth Century: Medical radiation is a highly important cause (probably the principal cause) of cancer-mortality in the United States during the Twentieth Century.

Although complete Cancer MortRates for each state and gender are not available before 1940, we have a datum from 1896 which is both highly reliable and highly informative with respect to Hypothesis-1. Namely, we know that Fractional Causation by medical radiation of the Cancer (and IHD) MortRates in 1896 was zero --- because 1896 was the very first year of any use of xrays.

Fractional Causation by medical radiation rises from ZERO percent in 1896
- to 90% of the male Observed All-Cancer MortRates in 1940,
- to 58% of the female Observed All-Cancer MortRates in 1940,
- to 75% of the female Observed All-Cancer-Except-Genital MortRate in 1940,
- to 79% of the male Observed IHD MortRate in 1950, and
- to 97% of the female Observed IHD MortRate in 1950. These mid-century percentages would almost certainly be even higher (especially for males), if we were able to adjust the 1940 cancer MortRates and the 1950 IHD MortRates to eliminate the effect of EXTRA smoking in the LowTrio and MidTrio Census Divisions, relative to the TopTrio (Chapter 49, Part 1c).

The Change in Fractional Causation after Mid-Century

After 1940, Column C of Box 1 shows that Fractional Causation of Cancer and IHD mortality, by medical radiation, declines moderately. In other words, medical radiation is a required co-actor in a somewhat lower fraction of the fatal cases at the end of the Twentieth Century than at mid-century. We suspect (but can not demonstrate) that the disparity in Fractional Causation by medical radiation, mid-century versus late-century, would have been even smaller if the population had not embraced cigarette-smoking with a vengeance.

Box 1 of Chapter 66
Final Summary for Fractional Causation, by Medical Radiation, of Cancer and Ischemic Heart Disease.

● – The range of values below represents the earliest year and the most recent year named in Column A.
The values come from the "A" or "AA" tables in Chapters 49 – 65. "Diff–Ca" = All–Cancers except Respiratory.
● – Hypotheses 1+2 are strongly supported by all of the percentages in Column C.

Col.A: M = Male. F = Fem.	Col.B: Natl Age–Adjusted Mortality Rate	Col.C: Frac. Causation by Medical Radn	Col.D: R-squared	Col.E: X–Coefficient	Col.F: Ratio of XCoef/Std.Error
Ch49, 1940–88, All–Cancer: M	Big net rise. 115.0 --> 162.7	90% --> 74%	0.95 --> 0.93	0.76 --> 0.75	11.6 --> 10.1
Ch50, 1940–88, All–Cancer: F	Net decline. 126.1 --> 111.3	58% --> 50%	0.86 --> 0.87	0.53 --> 0.34	6.6 --> 6.9
Ch51, 1940–88, Resp'y Ca: M	Enormous rise. 11.0 --> 59.7	~100% --> 74%	0.87 --> 0.78	0.12 --> 0.27	6.8 --> 5.0
Ch52, 1940–88, Resp'y Ca: F	Enormous rise. 3.3 --> 24.5	97% --> 83%	0.96 --> 0.90	0.02 --> 0.13	13.4 --> 7.8
Ch53, 1940–88, Diff–Ca: M	Approx. flat. 104.0 --> 103.0	84% --> 72%	0.93 --> 0.92	0.64 --> 0.46	10.0 --> 8.7
Ch54, 1940–88, Diff–Ca: F	Big decline. 122.8 --> 86.8	57% --> 48%	0.85 --> 0.84	0.50 --> 0.25	6.3 --> 6.1
Ch55, 1940–90, Breast–Ca: F	Flat. 23.3 --> 23.1	~100% --> 83%	0.92 --> 0.89	0.19 --> 0.12	8.7 --> 6.7
Ch56, 1940–80, AllExcGen: F	Flat. 94.0 --> 94.8	75% --> 66%	0.87 --> 0.93	0.51 --> 0.43	6.8 --> 9.6
Ch57, 1940–88, Digest–Ca: M	Big decline. 60.4 --> 38.8	97% --> 82%	0.91 --> 0.87	0.43 --> 0.20	8.3 --> 7.0
Ch58, 1940–88, Digest–Ca: F	Big decline. 50.1 --> 23.5	80% --> 68%	0.76 --> 0.86	0.29 --> 0.10	4.6 --> 6.7
Ch59, 1940–80, Urinary–Ca: M	Approx. flat. 7.4 --> 8.2	~100% --> 83%	0.92 --> 0.61	0.08 --> 0.05	9.0 --> 3.3
Ch60, 1940–80, Urinary–Ca: F	Decline. 4.0 --> 3.0	86% --> 78%	0.94 --> 0.91	0.02 --> 0.02	10.4 --> 8.5
Ch61, 1940–90, Genital–Ca: M	Some rise. 15.2 --> 16.9	79% --> 47%	0.77 --> 0.79	0.09 --> 0.05	4.9 --> 5.2
Ch63, 1940–80, Buccal–Phar: M	Approx. flat. 5.1 --> 4.6	~100% --> 81%	0.72 --> 0.73	0.04 --> 0.03	4.3 --> 4.4
Ch64, 1950–93, IHD: M	Enormous fall. 256.4 --> 131.0	79% --> 63%	0.95 --> 0.73	1.49 --> 0.50	11.2 --> 4.3
Ch65, 1950–93, IHD: F	Enormous fall. 126.5 --> 64.7	97% --> 78%	0.87 --> 0.68	0.90 --> 0.30	6.8 --> 3.9

Table 49-A
All-Cancers, Males: Fractional Causation by Medical Radiation over Time

Year	Col.A Natl MR	Col.B Frac.C		Col.C R-Sq	Col.D X-Coef	Col.E StdErr	Col.F Coef/SE	Col.G Source
1940	115.0	90%		0.9508	0.7557	0.0650	11.6276	Chap.6
1950	132.8	84%		0.9330	0.8462	0.0857	9.8703	Tab 49-B
1960	145.7	83%		0.9407	0.9251	0.0878	10.5397	Tab 49-C
1970	155.1	79%		0.9415	0.9073	0.0855	10.6122	Tab 49-D
1980	164.5	75%		0.9386	0.8480	0.0820	10.3418	Tab 49-E
1988	162.7	74%		0.9348	0.7488	0.0748	10.1056	Tab 49-F

Table 50-A
All-Cancers, Females: Fractional Causation by Medical Radiation over Time

Year	Col.A Natl MR	Col.B Frac.C		Col.C R-Sq	Col.D X-Coef	Col.E StdErr	Col.F Coef/SE	Col.G Source
1940	126.1	58%		0.8608	0.5279	0.0802	6.5801	Chap.7
1950	123.2	53%		0.8644	0.4894	0.0733	6.6803	Tab 50-B
1960	114.9	54%		0.8689	0.4661	0.0684	6.8105	Tab 50-C
1970	111.7	52%		0.8799	0.4285	0.0598	7.1600	Tab 50-D
1980	108.5	52%		0.8839	0.3857	0.0528	7.3005	Tab 50-E
1988	111.3	50%		0.8703	0.3393	0.0495	6.8536	Tab 50-F

Table 51-AA
Respiratory Cancers, Males: Fractional Causation by Medical Radiation over Time

Year	Col.A Natl MR	Col.B Frac.C		Col.C R-Sq	Col.D X-Coef	Col.E StdErr	Col.F Coef/SE	Col.G Source
1940	11.0	~100%		0.8673	0.1169	0.0173	6.7640	Chap.16
1950	21.6	89%		0.7168	0.1403	0.0333	4.2087	Tab 51-BB
1960	35.2	86%		0.7917	0.2241	0.0435	5.1581	Tab 51-CC
1970	47.3	79%		0.7951	0.2677	0.0514	5.2118	Tab 51 DD
1980	59.4	71%		0.7948	0.2984	0.0573	5.2076	Tab 51-EE
1988	59.7	74%		0.7816	0.2652	0.0530	5.0050	Tab 51-FF

Table 52-A
Respiratory Cancers, Females: Fractional Causation by Medical Radiation over Time

Year	Col.A Natl MR	Col.B Frac.C		Col.C R-Sq	Col.D X-Coef	Col.E StdErr	Col.F Coef/SE	Col.G Source
1940	3.3	97%		0.9625	0.0238	0.0018	13.4046	Chap.17
1950	4.6	76%		0.9420	0.0341	0.0032	10.6614	Tab 52-B
1960	5.3	85%		0.9521	0.0346	0.0029	11.7954	Tab 52-C
1970	11.7	83%		0.8987	0.0721	0.0091	7.8795	Tab 52-D
1980	18.0	81%		0.8624	0.1005	0.0152	6.6231	Tab 52-E
1988	24.5	83%		0.8975	0.1265	0.0162	7.8277	Tab 52-F

Table 53-A
Difference-Cancers, Males: Fractional Causation by Medical Radiation over Time

Year	Col.A Natl MR	Col.B Frac.C		Col.C R-Sq	Col.D X-Coef	Col.E StdErr	Col.F Coef/SE	Col.G Source
1940	104.0	84%		0.9342	0.6388	0.0641	9.9695	Chap.18
1950	111.2	80%		0.9099	0.6722	0.0799	8.4103	Tab 53-B
1960	110.5	78%		0.9153	0.6603	0.0759	8.6991	Tab 53-C
1970	107.8	75%		0.9167	0.5975	0.0681	8.7784	Tab 53-D
1980	105.1	75%		0.9113	0.4858	0.0573	8.4805	Tab 53-E
1988	103.0	72%		0.9158	0.4622	0.0530	8.7250	Tab 53-F

Table 54-A
Difference-Cancers, Females: Fractional Causation by Medical Radiation over Time

Year	Col.A Natl MR	Col.B Frac.C		Col.C R-Sq	Col.D X-Coef	Col.E StdErr	Col.F Coef/SE	Col.G Source
1940	122.8	57%		0.8528	0.5041	0.0792	6.3682	Chap.19
1950	118.6	53%		0.8601	0.4711	0.0718	6.5597	Tab 54-B
1960	109.6	52%		0.8528	0.4279	0.0672	6.3677	Tab 54-C
1970	100.1	53%		0.8499	0.3573	0.0567	6.2968	Tab 54-D
1980	90.5	51%		0.8561	0.3143	0.0487	6.4521	Tab 54-E
1988	86.8	48%		0.8437	0.2536	0.0413	6.1462	Tab 54-F

Table 55-A
Breast-Cancers, Females: Fractional Causation by Medical Radiation over Time

Year	Col.A Natl MR	Col.B Frac.C		Col.C R-Sq	Col.D X-Coef	Col.E StdErr	Col.F Coef/SE	Col.G Source
1940	23.3	~100%		0.9153	0.1906	0.0219	8.6965	Chap.8
1950	22.5	93%		0.9098	0.1572	0.0187	8.4021	Tab 55-B
1960	22.9	90%		0.9043	0.1569	0.0193	8.1309	Tab 55-C
1970	23.1	87%		0.8875	0.1467	0.0197	7.4317	Tab 55-D
1980	22.6	85%		0.8717	0.1335	0.0194	6.8964	Tab 55-E
1990	23.1	83%		0.8924	0.1187	0.0156	7.6213	Tab 55-F

Table 56-A
All-Cancers-Except-Genital, Females: Fractional Causation by Medical Radiation over Time

Year	Col.A Natl MR	Col.B Frac.C		Col.C R-Sq	Col.D X-Coef	Col.E StdErr	Col.F Coef/SE	Col.G Source
1940	94.0	75%		0.8675	0.5087	0.0751	6.7698	Chap.20
1950	96.0	69%		0.8946	0.5012	0.0650	7.7074	Tab 56-B
1960	92.5	68%		0.9084	0.4794	0.0575	8.3301	Tab 56-C
1970	93.7	67%		0.9201	0.4658	0.0519	8.9763	Tab 56-D
1980	94.8	66%		0.9290	0.4292	0.0448	9.5711	Tab 56-E
1988								None

Table 57-A
Digestive-System Cancers, Males: Fractional Causation by Medical Radiation over Time

Year	Col.A Natl MR	Col.B Frac.C		Col.C R-Sq	Col.D X-Coef	Col.E StdErr	Col.F Coef/SE	Col.G Source
1940	60.4	97%		0.9078	0.4262	0.0513	8.3009	Chap.9
1950	55.4	93%		0.8591	0.3874	0.0593	6.5325	Tab 57-B
1960	49.7	92%		0.8673	0.3488	0.0516	6.7647	Tab 57-C
1970	45.7	89%		0.8709	0.2986	0.0434	6.8722	Tab 57-D
1980	41.7	86%		0.8718	0.2454	0.0356	6.9002	Tab 57-E
1988	38.8	82%		0.8745	0.1984	0.0284	6.9825	Tab 57-F

Table 58-A
Digestive-System Cancers, Females: Fractional Causation by Medical Radiation over Time

Year	Col.A Natl MR	Col.B Frac.C		Col.C R-Sq	Col.D X-Coef	Col.E StdErr	Col.F Coef/SE	Col.G Source
1940	50.1	80%		0.7550	0.2895	0.0623	4.6442	Chap.10
1950	42.4	76%		0.7707	0.2431	0.0501	4.8506	Tab 58-B
1960	35.8	75%		0.7985	0.2048	0.0389	5.2675	Tab 58-C
1970	31.0	73%		0.8365	0.1668	0.0279	5.9850	Tab 58-D
1980	26.2	70%		0.8547	0.1271	0.0198	6.4177	Tab 58-E
1988	23.5	68%		0.8637	0.0999	0.0150	6.6597	Tab 58-F

Table 59-A
Urinary-System Cancers, Males: Fractional Causation by Medical Radiation over Time

Year	Col.A Natl MR	Col.B Frac.C		Col.C R-Sq	Col.D X-Coef	Col.E StdErr	Col.F Coef/SE	Col.G Source
1940	7.4	~100		0.9208	0.0750	0.0083	9.0208	Chap.11
1950	8.1	93%		0.9329	0.0832	0.0084	9.8635	Tab 59-B
1960	8.5	86%		0.9170	0.0833	0.0095	8.7960	Tab 59-C
1970	8.35	84%		0.9220	0.0761	0.0084	9.0993	Tab 59-D
1980	8.2	79%		0.9275	0.0676	0.0071	9.4602	Tab 59-E
1950	8.1	99%		0.5967	0.0601	0.0187	3.2180	Tab 59-BB
1960	8.5	91%		0.5901	0.0585	0.0184	3.1746	Tab 59-CC
1970	8.35	86%		0.6018	0.0524	0.0161	3.2526	Tab 59-DD
1980	8.2	83%		0.6089	0.0458	0.0139	3.3016	Tab 59-EE

Table 60-A
Urinary-System Cancers, Females: Fractional Causation by Medical Radiation over Time

Year	Col.A Natl MR	Col.B Frac.C		Col.C R-Sq	Col.D X-Coef	Col.E StdErr	Col.F Coef/SE	Col.G Source
1940	4.0	86%		0.9395	0.0247	0.0024	10.4305	Chap.12
1950	3.9	76%		0.9508	0.0223	0.0019	11.6295	Tab 60-B
1960	3.6	77%		0.9346	0.0211	0.0021	10.0025	Tab 60-C
1970	3.3	77%		0.9263	0.0189	0.0020	9.3816	Tab 60-D
1980	3.0	78%		0.9112	0.0161	0.0019	8.4777	Tab 60-E

Table 61-A
Genital Cancers, Males: Fractional Causation by Medical Radiation over Time

Year	Col.A Natl MR	Col.B Frac.C		Col.C R-Sq	Col.D X-Coef	Col.E StdErr	Col.F Coef/SE	Col.G Source
1940	15.2	79%		0.7754	0.0932	0.0190	4.9160	Chap.13
1950	14.9	58%		0.7241	0.0676	0.0158	4.2865	Tab 61-B
1960	14.6	55%		0.7486	0.0628	0.0137	4.5658	Tab 61-C
1970	14.8	52%		0.7840	0.0585	0.0116	5.0402	Tab 61-D
1980	15.0	50%		0.8044	0.0517	0.0096	5.3656	Tab 61-E
1990	16.9	47%		0.7921	0.0498	0.0096	5.1650	Tab 61-F

We uncovered no dose-response between PhysPop and Female Genital Cancers.

Table 63-A
Buccal-Pharynx Cancers, Males: Fractional Causation by Medical Radiation over Time

Year	Col.A Natl MR	Col.B Frac.C		Col.C R-Sq	Col.D X-Coef	Col.E StdErr	Col.F Coef/SE	Col.G Source
1940	5.1	~100%		0.7234	0.0382	0.0089	4.2782	Chap.15
1950	5.0	89%		0.7137	0.0377	0.0090	4.1778	Tab 63-B
1960	4.7	88%		0.7135	0.0389	0.0093	4.1754	Tab 63-C
1970	4.65	86%		0.7101	0.0356	0.0086	4.1405	Tab 63-D
1980	4.6	81%		0.7327	0.0315	0.0072	4.3806	Tab 63-E
1950	5.0	81%		0.6094	0.0303	0.0092	3.3048	Tab 63-BB
1960	4.7	88%		0.6129	0.0316	0.0095	3.3293	Tab 63-CC
1970	4.65	84%		0.6003	0.0286	0.0088	3.2425	Tab 63-DD
1980	4.6	81%		0.6177	0.0252	0.0075	3.3630	Tab 63-EE

Table 64-A
Ischemic Heart Disease, Males: Fractional Causation by Medical Radiation over Time

Year	Col.A Natl MR	Col.B Frac.C		Col.C R-Sq	Col.D X-Coef	Col.E StdErr	Col.F Coef/SE	Col.G Source
1940	No data available							No data
1950	256.4	79%		0.9475	1.4852	0.1321	11.2446	Chap. 40
1960	306.5	74%		0.8733	1.7259	0.2485	6.9463	Tab 64-B
1970	259.7	72%		0.8344	1.3710	0.2309	5.9387	Tab 64-C
1980	212.8	70%		0.7714	1.0117	0.2081	4.8608	Tab 64-D
1993	131.0	63%		0.7279	0.4969	0.1148	4.3271	Tab 64-E

Table 65-A
Ischemic Heart Disease, Females: Fractional Causation by Medical Radiation over Time

Year	Col.A Natl MR	Col.B Frac.C		Col.C R-Sq	Col.D X-Coef	Col.E StdErr	Col.F Coef/SE	Col.G Source
1940	No data available							No data
1950	126.5	97%		0.8669	0.9041	0.1339	6.7531	Chap. 41
1960	152.5	89%		0.8084	1.0346	0.1904	5.4353	Tab 65-B
1970	124.9	86%		0.7980	0.8074	0.1535	5.2593	Tab 65-C
1980	97.2	83%		0.7620	0.5620	0.1187	4.7340	Tab 65-D
1993	64.7	78%		0.6816	0.3025	0.0782	3.8710	Tab 65-E

Part 1. Do Our Findings Conflict with High Fractional Causations by Smoking and Diet?
Part 2. If All Cancer-Types Share Xrays as a Key Cause, Why Not Share One MortRate Trend?
Part 3. Did Cancer MortRates Rise Enough after the Year 1900?
Part 4. Did Cancer & IHD MortRates Rise Enough after PhysPop's Post-1965 Rise?
Part 5. Are the Findings on Cancer Compatible with Existing Data on Xray Dose?
Part 6. Is the New Finding about IHD in Conflict with the A-Bomb Study?

Box 1. Century Begins: Year-1900 Age-Specific and Age-Adjusted All-Cancer MortRates (USA).
Box 2. Years 1900 through 1990: Age-Specific All-Cancer Death-Rates (USA) across Time.
Figure 67-A. The 1930-1988 Trends in MortRates for Nine Types of Cancer.
Figure 67-B. The 1973-1994 Trends in Cancer Incidence and Mortality.

If two sets of findings in science are inherently contradictory, then at least one set must be untrue (and possibly both sets are untrue). Among scientific truths, harmony is a requirement. So, the last stage of checking our own work for errors is to ask: Are our estimates of Fractional Causation by medical radiation, of mortality rates from Cancer and Ischemic Heart Disease, incompatible with any OTHER evidence which we and others properly regard as incontrovertible?

This chapter tests some "general wisdom" and some incontrovertible facts, for potential conflicts with our own findings. Earlier chapters have already examined several aspects of the issue.

● Part 1. Do Our Findings Conflict with High Fractional Causations by Smoking and Diet?

Part 1 addresses the question: Do our findings, of high Fractional Causations by medical radiation, conflict with estimates of high Fractional Causations by smoking and by unfavorable nutrition?

A piece of "general wisdom" is that about 30% of cancer deaths are attributable to tobacco and another 35% are attributable to unfavorable nutrition (AICR 1997). In addition, there are estimates that at least 10% of cancer deaths are caused by occupational exposure to carcinogens in the workplace (Landrigan 1996, p.67, citing Landrigan 1989). Additional estimates attribute 5% to 10% of cancer deaths to inherited mutations.

Even if the estimates above were incontrovertible, such estimates would be fully compatible with our findings that Fractional Causation by MEDICAL RADIATION exceeds 50%. Reminder: Explanation of the compatibility occurs in this monograph's Introduction (Part 5).

1a. The Meaning of "Cause" When Cases Have More Than One Cause

A statement, that two or more co-actors each contribute to the same fatal case of a disease, means that each contribution is necessary --- for if the death would have happened as it did in the ABSENCE of one, then that one did not really contribute anything to the outcome. The concept of necessary co-actors was stated very nicely by the authors of the famous 1964 "Surgeon General's Report" on smoking (SurgeonGen 1964, p.31): "It is recognized that often the co-existence of several factors is required for the occurrence of a disease, and that one of the factors may play a dominant role; that is, without it, the other factors (such as genetic susceptibility) seldom lead to the occurrence of the disease."

Among current cancer biologists, the concept --- that more than one cause per case is generally required --- is widely embraced as part of the initiation-promotion, multi-step, multi-mutation models of carcinogenesis. Such co-action has been discussed in the Introduction (Parts 4 + 5), Chapter 6 (Part 6), and Chapter 49 (Part 2).

We know of virtually no one, today, who thinks that each case of fatal Cancer has a single cause. When estimates are made about the percentage of Cancer caused by smoking, or by unfavorable nutrition, or by medical radiation, such estimates do not imply that such causes act alone. Such estimates refer to the share of Cancer which would be absent (prevented) by the absence of that particular cause. Example:

Our work indicates that Fractional Causation of the cancer MortRate (male) in 1988 was about 74% by medical radiation. The work also indicates that Fractional Causation of the male MortRate from Ischemic Heart Disease in 1993 was about 63% by medical radiation. These findings refer to the percentages in which medical radiation was a necessary co-actor --- and thus, the percentages which would not have occurred in the absence of exposure to medical radiation.

1b. A Distinction between Our Cancer-Findings and our IHD-Findings

The findings summarized in Chapter 66 show that, during the Twentieth Century, medical radiation has become a necessary co-actor in a very large share of the mortality rate from Cancer (USA), and the findings very strongly indicate that the same is true for Ischemic Heart Disease. The distinction lies in the fact that medical radiation is already a PROVEN cause of Cancer, whereas the evidence in this book is the FIRST evidence that medical radiation is a cause also of Ischemic Heart Disease.

With respect to Cancer, our findings do not conflict at all with the "general wisdom" about the importance of tobacco, nutrition, workplace (and environmental) carcinogens, and of inherited mutations in cancer-causation. Regarding inherited mutations, we suspect that they have a predisposing role in probably the majority of cancer-deaths (Gofman 1994, Chapter 7, Part 2b). With respect to Ischemic Heart Disease, our findings do not conflict at all with the very important roles, already well established, for various non-xray co-actors in that disease, including elevated blood-levels of atherogenic lipoproteins, high blood-pressure, smoking, diabetes, and many others (Chapters 44, 45, 46).

● Part 2. If All Cancer-Types Share Xrays as a Key Cause, Why Not Share One MortRate Trend?

Part 2 addresses the question: If all kinds of Cancer share medical radiation as a key cause, why do they NOT share the same trend in their MortRates? (Part 2a describes the trends.) Why have MortRates increased for some types of Cancer and simultaneously decreased for others? We think that the question's correct answer is co-action --- multiple causes per fatal case --- as illustrated in Part 2b.

2a. Some Nearly "Incontrovertible" Facts: Figure 67-A

We consider the simultaneous rise and fall of MortRates, for different types of Cancer, to be "incontrovertible." We have called attention to this phenomenon, not only in Chapter 22 (Part 1c) and in Chapter 58, but also by featuring such changes in every Table "A" of Chapters 49 through 63.

Figure 67-A depicts the contrast in MortRate-trends among nine types of Cancer, for the period 1930-1988. For the MortRates which decline in those graphs, the explanation is almost certainly NOT a rising fraction of cured cases --- because the declining MortRates embrace years when there were very few effective treatments for Cancer. The 1930 MortRates in Figure 67-A necessarily omit a few states (Chapter 4, Part 1), but the 1930 age-adjusted rates may be reliable enough to say that the trend was DOWNWARD between 1930-1940 for the following three types of Cancers in Figure 67-A (and may have been downward long before 1930):

Females: Uterus (cervix + corpus), Stomach, Liver.
Males: Stomach, Liver.

We note a common feature among these three types of Cancer, which appear to have had declining MortRates between 1930 (or earlier) and 1940: Infectious agents are regarded as extremely important causes. Cervical Cancer is causally associated with the Human Papilloma Virus (Chapter 14); Stomach Cancer, with the bacterium, Helicobacter pylori (Chapter 57); Liver Cancer (Hepatocellular Carcinoma), with infection by the Hepatitis Virus, usually type-B (WHO 1996, p.59).

2b. At Constant Xray Dosage: Opposite MortRate Trends

In any decade, the Observed Cancer MortRates are the result of co-action among carcinogens --- the presence of which has been rising for some co-actors, constant for other co-actors, and falling for other co-actors. Unlike xrays, which have access to every organ of the body, chemical and infectious carcinogens do not necessarily have "universal access", nor do they necessarily have equal activity within cells of different organs. Thus if exposure decreases to some non-xray co-actors for Liver Cancer, while exposure simultaneously increases to some other non-xray co-actors for Respiratory-System Cancers, the outcome (at constant exposure to xrays) is very likely to be MortRates which simultaneously decline for Liver Cancer and rise for Respiratory-System Cancers. And yet Fractional Causation, by medical radiation, can remain high for these two Cancers --- which have OPPOSITE trends in their MortRates.

How can this happen? An uncomplicated model can suffice here, although additional models also are consistent with our findings. (Chapter 49, Part 2b, provides two related illustrations.)

We can suppose that Liver Cancer requires the presence of two carcinogenic mutations in the same cell, and that by mid-century, medical radiation has become the source of one (sometimes both) of the mutations in most cases of Liver Cancer.   Also, we can suppose that the U.S. population consists of 1,000,000 persons, and that they have a combined total 1,000 "liver-cells" which have acquired, or will acquire, one of the necessary mutations from MEDICAL XRAYS.

Lowering the cells' exposure, to non-xray mutagenic co-actors, reduces the fraction of those 1,000 cells which ever acquire both of the required mutations. Result: A falling MortRate from Liver Cancer. By contrast, increasing the cells' exposure, to non-xray mutagenic co-actors, increases the fraction of those 1,000 cells which ultimately acquire both of the required mutations. Result: A rising MortRate from Liver Cancer. Yet, in either case (a falling or a rising MortRate from Liver Cancer), medical radiation can be a necessary co-actor in most of the fatal cases of Liver Cancer which DO occur.

If one repeats the illustration for Respiratory-System Cancers, it is clear that livers and lungs, which may share the SAME history of medical radiation, can develop OPPOSITE trends in their Cancer MortRates.

● Part 3. Did Cancer MortRates RISE Enough after the Year 1900?

Part 3 addresses the question: Did Cancer MortRates RISE enough after the year 1900 to be consistent with our findings?

3a. The Solid Basis for an Affirmative Answer

The answer is, yes. Indeed, even a post-1900 decline in the age-adjusted National All-Cancer MortRate would be compatible with our findings of high Fractional Causation by medical radiation. Fractional Causation refers to the age-adjusted National Cancer MortRates which DID OCCUR in post-1900 decades, and is independent of whether the trend in post-1900 Cancer MortRates was rising, flat, or falling (a point made in Chapter 22, Part 1c).

Introduction of medical radiation in 1896 requires that the RADIATION-INDUCED SHARE of All-Cancer MortRates must rise from zero percent in 1896 to some positive fraction in subsequent decades --- fractions whose quantification has been the goal of our research. But introduction of medical radiation does not automatically require a net rise in post-1900 All-Cancer MortRates. After all, specific non-xray causes may have declined after 1900. So, the direction of post-1900 All-Cancer MortRates can not be predicted --- the direction can be learned only by observation.

3b. Age-Adjusted All-Cancer MortRates: Post-1900 Trends

The available records indicate that the age-adjusted All-Cancer MortRate rose dramatically in the USA between 1900 and 1940. For males and females combined, including all "races" without any exclusions, the age-adjusted rate rose from 79.6 per 100,000 population in the year 1900, to 120.3 per 100,000 in the year 1940 --- an upward change by a factor of 1.5 --- an increase by 50%.

Box 1 shows the sources of these data, and shows how the crude rate in 1900 (which was 64.0 per 100,000) converts to the age-adjusted rate in 1900 of 79.6 per 100,000. The "general wisdom," that age-adjusted Cancer MortRates increased by a large factor between 1900 and 1940, is probably solid (perhaps even "incontrovertible"). However, we have little faith that the specific factor of 1.5 is reliable.

One source of unreliability in the factor of 1.5 is the fact that the 1900 records include only the ten "registration states" (Chapter 4, Part 1), whereas the 1940 data include all 48 states. So this is a comparison of "apples with oranges." Would complete data from 1900 increase the factor of 1.5, or lower it? It will never be known. A second source of unreliability is present if there was greater under-reporting of Cancer in 1900 than in 1940. Approximately 38% of deaths which occurred in 1900 "had no cause listed," according to Patterson 1987 (p.80), who states also that it was acceptable to report "old age" as a cause of death. How many cancer deaths in 1900 were never reported as Cancer? This, too, will never be known. Even in 1940, there was a stigma associated with having a cancer death in the family, and the stigma --- absurd though it was --- may have been stronger in 1900. These considerations (and some others) make the upward change-factor of 1.5 unreliable for the age-adjusted National All-Cancer MortRates. As for Ischemic Heart Disease, there are no records back to 1900. However, for the reasons stated in Part 3a, the magnitude of the change-factors, for Cancer and IHD MortRates between 1900 and mid-century, is essentially irrelevant to the validity of our findings on Fractional Causation by medical radiation --- as is the magnitude of the change-factor between 1940 and 1990.

3c. National Cancer MortRates (Box 2): Trends from 1900 to 1990

For everyone's convenience, we present a comparison right here of the National All-Cancer MortRates (USA) for various decades. The intervening decades are shown in the bottom section of Box 2. These rates are for males and females combined, per 100,000 population, and the non-1940 rates are age-adjusted to the population's age-distribution in 1940. The entries, prior to 1940, do not include all the states (Chapter 4, Part 1). The year 1900 includes the fewest states (only ten).

1900 = 79.6   1920 = 104.9   1940 = 120.3   1960 = 129.1   1990 = 135.0

In contrast with the bottom section of Box 2, the upper section of Box 2 presents the age-SPECIFIC All-Cancer MortRates (USA) from the year 1900 through 1990. Because each row displays rates per 100,000 persons of the SAME age, these entries can be directly compared across time. Illustration: Per 100,000 persons who REACH age 60, the row for ages 55-64 gives directly comparable cancer death rates across nine decades. With respect to the early decades, the warnings given in Part 3b apply also to Box 2.

For cancer death below age 45, rates peaked around 1950, and then declined. For cancer death at about age 50, the peak was in 1970. For cancer death above age 54, the age-specific rates have been climbing ever since the year 1900. In the mid-1990s (not shown), a slight decline may have occurred. We hope that the decline endures --- but we note that the decline in 1970 for ages 85+ did NOT endure.

The rates in Box 2, like the rates in Box 1, are fully consistent with our findings about Fractional Causation by medical radiation (Part 3a).

● Part 4. Did Cancer and IHD MortRates Rise Enough after PhysPop's Post-1965 Rise?

Part 4 addresses the question: Did National MortRates from Cancer and Ischemic Heart Disease rise enough to be consistent with the spectacular post-1965 rise in PhysPop values?

There is very little doubt that PhysPop values rose dramatically in all Nine Census Divisions after 1965 (after the enactment of Medicare and related legislation). Readers can see this for themselves in Chapter 3, PhysPop Table 3-A and in the top half of Figure 3-A. Although the National Average PhysPop values in Table 3-A are not weighted by the populations of the Census Divisions, the National Averages there are good approximations. The National Average was 133.01 in 1965, and was 233.72 in 1990 --- an upward change-factor of 1.76.

Concurrently, the National All-Cancer MortRate (Table 6-B) rose, from 129.1 in 1960, to

135.0 in 1988 --- an upward change-factor of 1.05. The National Ischemic Heart Disease MortRate (Table 40-B) fell, from 190.0 in 1960, to 94.9 in 1993 --- a downward change-factor of 0.50.

Should these virtually incontrovertible facts raise any doubt about the importance of medical radiation in causation of both diseases? Not at all. Hypotheses 1+2 say that medical radiation is a required co-actor in most of the fatal cases which DO occur, but Hypotheses 1+2 certainly do NOT require cancer and IHD MortRates to rise by 1.76-fold if PhysPop rises by 1.76-fold. Why not?

● (1) Physpop is PROPORTIONAL to dose in rads from medical radiation, but PhysPop is not EQUAL to such dose. Even while annual PhyPop values rise by 1.76-fold , average annual per capita dose from medical radiation could actually decline, if dosage per procedure declines sufficiently, for instance. Indeed, although there is no way to prove it, we believe that an unquantifiable net decline HAS occurred since 1960 in average annual per capita dose from medical radiation (Chapter 2, Part 3g) --- and most people in medicine seem to make the same assumption. Whatever the change, it is reasonable to assume that the change-factor has been similar in all Nine Census Divisions.

● (2) Even if annual per capita xray dosage had grown by 1.76-fold between 1965 and 1990 (which almost certainly did not occur), still there would be no reason to expect a simultaneous 1.76-fold increase in the MortRates from Cancer and IHD. The very gradual delivery of radiation-induced Cancer, in a mixed-age population, means that the full effect of any new dose-level would not occur until at least 45 years later (Chapter 2, Part 8, and Chapter 5, Part 1). And even if a 1.76-fold elevated dose-level were maintained for at least 45 years beyond 1990, still one should not expect MortRates from Cancer and IHD to rise 1.76-fold by the year 2035 (1990 + 45 years) --- unless exposure to co-actors had been constant (Point 3, below).

● (3) The per-rad potency of ionizing radiation --- the X-coefficient --- is modulated by co-actors (Chapter 49, Part 2). If exposure to co-actors is not constant over time, the X-coefficient for dose (PhysPop) will not be the same in every decade. Abundant evidence, of change in exposure to non-xray co-actors over time, is another reason for NOT expecting PhysPop values and MortRates for Cancer and IHD to change by the same factor (1.76, for instance). Cigarettes provide a noteworthy example of change in a co-actor. When PhysPop values experienced their big increase after 1965 in the USA, cigarette consumption simultaneously experienced a big decline in the USA (Chapter 48, Part 2a). Hypertension is another important co-actor for Ischemic Heart Disease. Death-rates from Hypertension (age-adjusted to the 1940 reference year) fell from about 68 per 100,000 population in 1940 to about 4 per 100,000 in 1990 --- as approximated from Figure 40-B. This dramatic decline almost certainly reflects a reduction also in the average severity of non-fatal cases of Hypertension. With respect to Cancer, Figure 67-A reflects big changes in levels of some unspecified co-actors (discussion in Part 2, above).

● (4) There is a fourth reason, for NOT expecting PhysPop values and MortRates (from Cancer, IHD) necessarily to change by the same factor in the same direction. It is the expectation that better underlying health and better treatments can prevent a growing share of radiation-induced cases from becoming fatal.

● Part 5. Are the Findings on Cancer Compatible with Existing Data on Xray Dose?

Part 5 addresses the question: Are the new findings, that medical radiation accounts for a very large share of all cancer deaths in the USA, compatible with existing data on dosage from medical radiation?

5a. A Solution to the Permanent Uncertainty about Dosage in Rads

We expect that this new work will face the following claim from some people: Doses from medical radiation have never been high enough to cause such a Roentgen Tragedy. Such assertions, about past and current dosage, would not be based on any "well established" evidence --- as readers of Chapter 2 (Part 3) already know.

Of course, we do not expect such claims from anyone who has READ this work. Readers will understand that Fractional Causation in this work is not tied to specific dose-levels in rads. It is tied to the RELATIVE size of per capita medical doses in the Nine Census Divisions, as indicated by the relative number of physicians per 100,000 population.

And this tie, to the RELATIVE magnitude of per capita doses, is one of the important scientific strengths of our method. Our analysis has tested (and validated) Hypothesis-1 without resorting to any questionable estimates, in medical RADS, of average annual per capita doses from medical radiation. Estimates in rads could easily be wrong by factors like 3 to 10.

A related strength of our method is that it employs no estimates of risk per medical rad. While the evidence is incontrovertible that medical xrays induce cancer, the NUMBER of cases induced per medical rad is far from being a settled issue (Chapter 2, Part 7c) --- which is inevitable when the doses are so very uncertain in most of the studies which produce the risk-per-rad estimates.

Instead of depending on controversial estimates of doses and of risk per unit dose, we have let the RELATIVE sizes of medical doses, in the Nine Census Divisions, directly reveal the magnitude of Fractional Causation of cancer death-rates by medical radiation.

5b. Breast Cancer: Two Analyses Tied to Estimated Doses and Risk/Rad

By contrast with this book, our previous monograph (Gofman 1995/96) examines the induction of only Breast Cancer by medical radiation. In Gofman 1995/96, we undertook a scholarly effort to estimate average annual per capita breast-dose received in the USA by females of various ages, during forty relevant years (1920 - 1960). We determined an estimate of such dose, and then multiplied dose by the number of Breast Cancers (per 10,000 females) expected per rad of breast-irradiation --- based largely on our earlier work (Gofman 1990). The resulting number of radiation-induced Breast Cancers, divided by the total annual incidence of 182,000 Breast Cancers (USA, in 1995), yielded a Fractional Causation by medical radiation of 75% of current annual Breast Cancer incidence.

Reviewing our book in the Journal of the American Medical Association, the American Cancer Society's Clark W. Heath, Jr., began as follows (Heath 1995, p.657): "Although breast tissue is particularly sensitive to the carcinogenic effects of ionizing radiation, especially at young ages, it is likely that less than 5% of all Breast Cancer in the United States can be attributed to radiation and less than 1% to medical uses of radiation (Evans 1986)." There is a 75-fold difference in these two estimates (Evans compared with Gofman).

The Input on Estimated Dose to Breasts

The 1986 Evans' estimate, conspicuously embraced by Heath, was based on an average lifetime dose to the breasts of 2.1 rad "from diagnostic radiographs" (Evans 1986, p.812) of the 1970-1980 variety, whose types and frequency were assessed primarily from the 1977 records of Blue Cross, Medicare, and Medicaid patients in Maine (McNeil 1985), and whose presumed doses per exam were calculated with phantoms and Monto Carlo simulations. Breast irradiation received during interventional fluoroscopic procedures was omitted entirely. Estimated doses received during diagnostic fluoroscopy were included, but in the related paper (McNeil 1985, p.55), fluoroscopic doses were acknowledged to be uncertain due to "marked variations in fluoroscopy times and technique."

In making their estimate of Fractional Causation, Evans and co-workers used the 1980 Mortality Rates from Breast Cancer (Evans 1986, p.812). Breast irradiation occurring in 1977 had very little impact on the 1980 MortRates from Breast Cancer. For Breast-Cancer deaths in 1980, the average annual breast-dose during the 1920-1960 period is certainly much more relevant, but Evans and co-workers did not evaluate such doses.

By contrast, our careful study of practices in the 1920-1960 period yielded 27.9 rads as the relevant lifetime average breast-dose from diagnostic, interventional, and non-cancer therapeutic radiologic procedures (Gofman 1995/96, p.267, Col.T).

The Input on Estimated Risk of Breast Cancer per Rad

It is clear that dose-estimates account for about a 13.3-fold difference (Evans compared with Gofman), in estimates of Fractional Causation of recent Breast Cancer by medical radiation. Because 13.3 times 5.64 = 75, a disparity of only about 5.64-fold in estimated risk of Breast Cancer, PER RAD, would account for the 75-fold difference in estimated Fractional Causation.

When Evans et al chose the risk/rad estimates they used, they correctly acknowledged (Evans

1986, p.811): "The limitations of the available data and the reliance on unverified assumptions introduce uncertainty into any estimates of risk." We, too, explicitly emphasized the inherent uncertainties about risk per rad (Gofman 1995/96, p.273, 276, 281, 285). Indeed, available evidence explicitly on xray-induced Breast Cancer yields a 6-fold range in estimated risk per rad (BEIR 1990, p.255). The Evans paper uses risk/rad estimates which exclude the 6-fold higher values from the Nova Scotia evidence (Evans 1986, p.811), and employs risk models prepared by Gilbert for the Nuclear Regulatory Commission (Evans 1986, p.810; see specifically Gilbert 1985).

No Conflict between Our Findings and Well-Established Facts

A difference in estimates of Fractional Causation, traceable to openly acknowledged uncertainties, clearly does not amount to a conflict between any of our findings and well-established, scientifically-solid facts. There is no mystery why estimates of Fractional Causation will continue to vary so dramatically, if analysts continue to depend exclusively on two key inputs whose values are so uncertain. By contrast, the method used in this book is independent of uncertain estimates of dose and of risk per dose-unit --- as emphasized in Part 5a.

5c. Two Routes --- but Arrival at a Single Answer

Because of claims like Clark Heath's (Part 5b), it is interesting to compare our results for Breast Cancer in the 1995/96 monograph, with the results for Breast Cancer in this monograph --- where the results are derived from a method and from data which are completely independent from the earlier analysis.

Gofman 1995/96:   75% of Breast-Cancer Cases in 1995 induced by medical radiation.
This analysis: 83% of Breast-Cancer Deaths in 1988 induced by medical radiation.

Naturally, we are pleased that both sets of findings are compatible --- indeed, nearly identical. But if someone asks which estimate we trust the most, the answer will be that we regard the results produced in this monograph as the more reliable --- due to the superior method.

Indeed, for several reasons, we doubt that there will ever be a MORE reliable test of Hypothesis-1 than the testing in this book. The first reason is that this testing avoids the pitfalls described in Part 5a. The second reason is that our testing "enrolled" the entire U.S. population in nine dose-cohorts (the Census Divisions) --- and a large prospective study is more likely to be correct than a small one, all other things being equal. The third reason, already discussed in Chapter 3 (Part 1c), is that this testing has been done by a straightforward, replicable method using utterly neutral data --- data collected long ago for other purposes and thus free from any conceivable bias with respect to Hypothesis-1.

● Part 6. Is the New Finding about IHD in Conflict with the A-Bomb Study?

Part 6 addresses the question: Is our finding, that medical radiation even at low and moderate doses is an important cause of Ischemic Heart Disease, in conflict with any well-established findings from the Atomic-Bomb Survivor Study?

There appears to be no conflict. We say "appears" because we have not analyzed the Japanese IHD data ourselves, using the "constant-cohort, dual dosimetry" method described in Chapter 2 (Part 5c). Between what OTHERS have recently reported from the A-Bomb Study regarding Ischemic Heart Disease, and what we report from our PhysPop studies in this monograph, there is no conflict.

6a. Positive Dose-Response in People below Age 40 during the Bombings

What do analysts at RERF report? Yukiko Shimizu, Hiroo Kato, William Schull, and Donald Hoel present the findings on Coronary Heart Disease (Ischemic Heart Disease) within the Life-Span Study, in a 1992 paper (Shimizu 1992, Table 6, p.255). The IHD results in their Table 6 are limited to the 1966-1985 time-period, and to IHD deaths among participants who were under age 40 at the time of bombing in 1945. The doses are entrance doses, not internal organ-doses (Shimizu, p.250), and they are bomb-rads, not medical rads. To convert bomb-rads to medical rads, the bomb-rads need multiplication by a factor of about 0.375 (details in Chapter 2, Part 7d). The total deaths from

Coronary Heart Disease in the study-sample are 363 deaths, distributed as follows (from Shimizu 1992, Table 6, p.255):

IHD deaths = 136.	Dose = 0.	Relative risk = 1.0.
IHD deaths = 190.	Entrance dose = 1-49 rads.	Relative risk = 1.35.
IHD deaths = 19.	Entrance dose = 50-99 rads.	Relative risk = 1.25.
IHD deaths = 8.	Entrance dose = 100-199 rads.	Relative risk = 0.92.
IHD deaths = 6.	Entrance dose = 200-299 rads.	Relative risk = 2.57.
IHD deaths = 4.	Entrance dose = 300-600 rads.	Relative risk = 2.38.

Although Shimizu's Table 6 does not show the number of participants in each dose-group, Shimizu's calculation of relative risk is, of course, the observed number of IHD deaths per 100 participants in an exposed group, divided by the observed number of IHD deaths per 100 participants in the zero-dose group.

We note the anomalous relative risk value of 0.92 in the middle of a positive dose-response trend. Biologically, the anomaly is improbable, while statistically, it is rather probable --- since it arises from only 8 cases.

6b. An Interesting Comment on CardioVascular Disease by Shimizu et al

The 1992 Shimizu paper also reports (Shimizu 1992, p.254): "Mortality from circulatory disease [cardio-vascular disease] in the years 1950-1985 shows a significant association with dose." Commenting later in their paper on this finding, Shimizu and co-workers seem to entertain the possibility that radiation-induced genetic mutations play a role in atherosclerosis. In their comment, below, they allude to the work of Arthur Penn (our Chapter 44, Part 9c) and to the concepts of stochastic and non-stochastic phenomenon. If a radiation-induced health disorder has a stochastic basis, its probability of OCCURRING in an exposed population is a function of dose; if it has a non-stochastic basis, the SEVERITY of radiation-induced cases will be a function of dose. Here is their comment, in full (Shimizu 1992, p.260):

"It is not unreasonable to assume that the effects of radiation on cancer induction (which is presumed to be a stochastic phenomenon) and on noncancer mortality differ, and that the latter may follow a nonstochastic process with a threshold dose. However, given the recent evidence of a transforming gene in the DNA from an atheromatous plaque (Penn 1986), the increase in cardiovascular disease is a particularly intriguing finding, and may suggest, if the association is real, that the effect of ionizing radiation on atherosclerosis should be treated as a stochastic phenomenon. Further data will be especially interesting in this regard."

Our own findings in this monograph, which arise from enormous databases (males and females separately), demonstrate a powerful dose-response between medical radiation and mortality from Ischemic Heart Disease. Such findings provide the first very strong epidemiologic evidence that ionizing radiation has a causal role in Ischemic Heart Disease. Moreover, the role of medical radiation is large, not small. In the USA, Fractional Causation indicates that medical radiation has been and continues to be an extremely important cause of fatal cases of Ischemic Heart Disease.

>>>>>>>>>>

Box 1 of Chapter 67

Century Begins: Year–1900 Age–Specific and Age–Adjusted All–Cancer MortRates (USA).

Related text = Part 3b.

● This box (for the year 1900) is directly comparable to Box 4 of Chapter 4 (for the year 1990).

● Col.A entries are the observed age–specific 1900 All–Cancer MortRates per 100,000 [100K] population of each age–band, for both sexes combined. The data cover only the ten "Death–Registration States" (Chapter 4, Part 1). Source: Linder 1947, Table 14, p.250. The crude rate in 1900, before adjustment to the 1940 reference year, is 64.0 per 100,000 population (Linder 1947, Table 14, p.250).

● Col.B is the weighting factor, from the "Standard Million Population, 1940" (see Chapter 4, Box 1, "Fraction of Total").

● Col.C is the product of Col.A times Col.B. The SUM of Col.C is the 1900 age–adjusted MortRate, adjusted to the 1940 reference year: 79.5788 per 100,000 population. This age–adjusted rate, of 80 per 100,000 population, is confirmed in Grove 1968, Figure 21, p.87.

● Col.D divides each entry in Col.C by 79.5788, and thus determines the fraction of the 1900 Age–Adjusted Rate (79.5788 per 100,000) contributed by each age–band. The sum of the fractions for age 45 and older = 0.86550, or 86.5% .

Age–Band	Col.A AgeSpecific Rate/100K	Col.B Weighting Factor	Col.C = A times B	Col.D = Col.C / 79.5788
Under 1 yr	3.2	0.015343	0.04910	0.00062
1–4	2.9	0.064718	0.18768	0.00236
5–14	1.8	0.170355	0.30664	0.00385
15–24	3.2	0.181678	0.58137	0.00731
25–34	14.0	0.162065	2.26891	0.02851
35–44	52.5	0.139237	7.30994	0.09186
45–54	139.1	0.117811	16.38749	0.20593
55–64	260.9	0.080294	20.94861	0.26324
65–74	421.0	0.048426	20.38726	0.25619
75–84	544.7	0.017304	9.42536	0.11844
85 and up	623.2	0.002770	1.72640	0.02169
CANCER Sums		1.000000	79.5788	1.0000

For All–Cancer MortRates per 100,000 population (USA), this box presents age–SPECIFIC rates in its upper section, and age–ADJUSTED rates in its lower section, for males and females combined. Prior to 1933, not all states reported. Entries for 1900 are based on the fewest states (only ten).

Sources are (1) Vital Statistics Rates in the United States 1900–1940 (Linder 1947 in our Reference List, p.250, Table 14, "Death Rates" for "Cancer and Other Malignant Tumors, 45–55") and (2) Health: United States 1995 (PHS 1995 in our Reference List, p.132, Table 39, "Death Rates for Malignant Neoplasms").

Because entries in the upper section are age–specific rates, per 100,000 population of each age–band, different decades can be directly compared without age–adjusting.

Age	1900	1910	1920	1930	1940	1950	1960	1970	1980	1990
Under 1 year	3.2	3.1	3.2	3.1	4.4	8.7	7.2	4.7	3.2	2.3
Ages 1–4	2.9	3.3	3.1	4.1	4.8	11.7	10.9	7.5	4.5	3.5
Ages 5–14	1.8	1.5	1.6	2.0	3.0	6.7	6.8	6.0	4.3	3.1
Ages 15–24	3.2	3.5	3.8	4.2	5.4	8.6	8.3	8.3	6.3	4.9
Ages 25–34	14.0	14.1	14.7	16.7	17.3	20.0	19.5	16.5	13.7	12.6
Ages 35–44	52.5	55.6	56.0	58.9	61.1	62.7	59.7	59.5	48.6	43.3
Ages 45–54	139.1	156.7	155.1	159.6	168.8	175.1	177.0	182.5	180.0	158.9
Ages 55–64	260.9	322.4	341.2	355.6	369.6	392.9	396.8	423.0	436.1	449.6
Ages 65–74	421.0	541.7	607.7	677.1	695.2	692.5	713.9	751.2	817.9	872.3
Ages 75–84	544.7	749.9	900.0	1,019.7	1,161.0	1,153.3	1,127.4	1,169.2	1,232.3	1,348.5
Ages 85+	623.2	810.9	1,017.5	1,196.3	1,319.0	1,451.0	1,450.0	1,320.7	1,594.6	1,752.9

Entries in the three rows below are the age–adjusted National Cancer Mortality Rates per 100,000 population. Cancer–deaths at all ages are included in these age–adjusted rates. Because each entry is adjusted to the same "standard population" (1940), the entries in this row are directly comparable with each other --- despite changes over the century in infant mortality, average lifespan, etc. The male–female difference after 1940 results largely from the males' much higher rate of Respiratory–System Cancers.

	1900	1910	1920	1930	1940	1950	1960	1970	1980	1990
Both Sexes	79.6	97.0	104.9	113.4	120.3	127.7	129.1	129.8	131.9	135.0
Males	--	--	--	--	115.0	132.8	145.7	155.1	164.5	162.7
Females	--	--	--	--	126.1	123.2	114.9	111.7	108.5	111.3

Related text = Part 3c.

Figure 67-A

Related text = Part 2a.

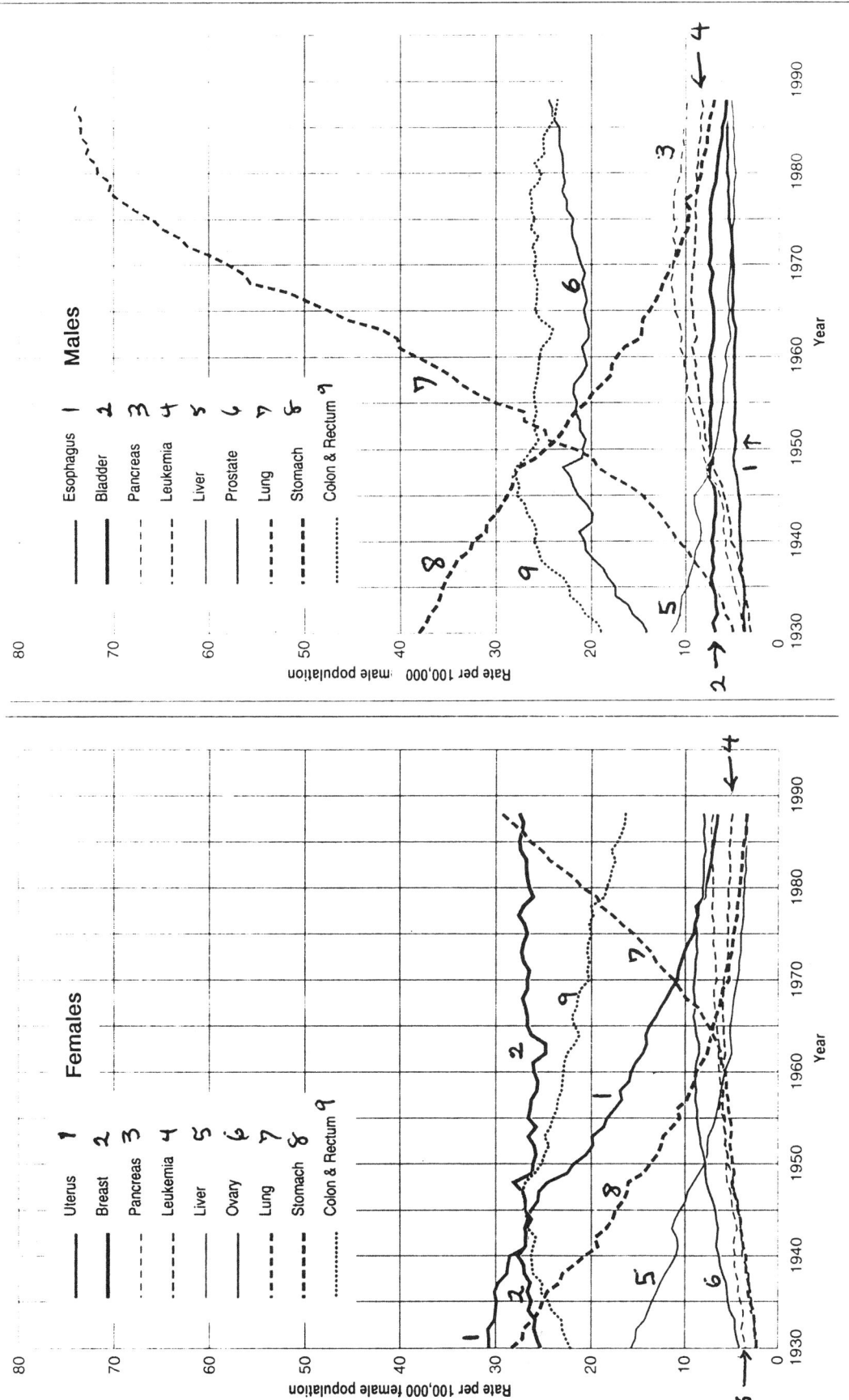

The two graphs above depict "Age–Adjusted Cancer Death Rates for Selected Sites, United States, 1930–1988." Except for our by–hand addition of the numerical labels, both graphs are reproduced from the January–February 1992 issue of the American Cancer Society's publication, CA – A Cancer Journal for Clinicians, Vol.42, No.1: pp.28–29. The age–adjustment by the ACS is to the age distribution of the 1970 U.S. Census. Sources of data: U.S. National Ctr. for Health Statistics and U.S. Bureau of the Census. (Reminder from Chap.4: Not every state reported Cancer MortRates in 1930.)

Figure 67–B

Related text = Part 2.

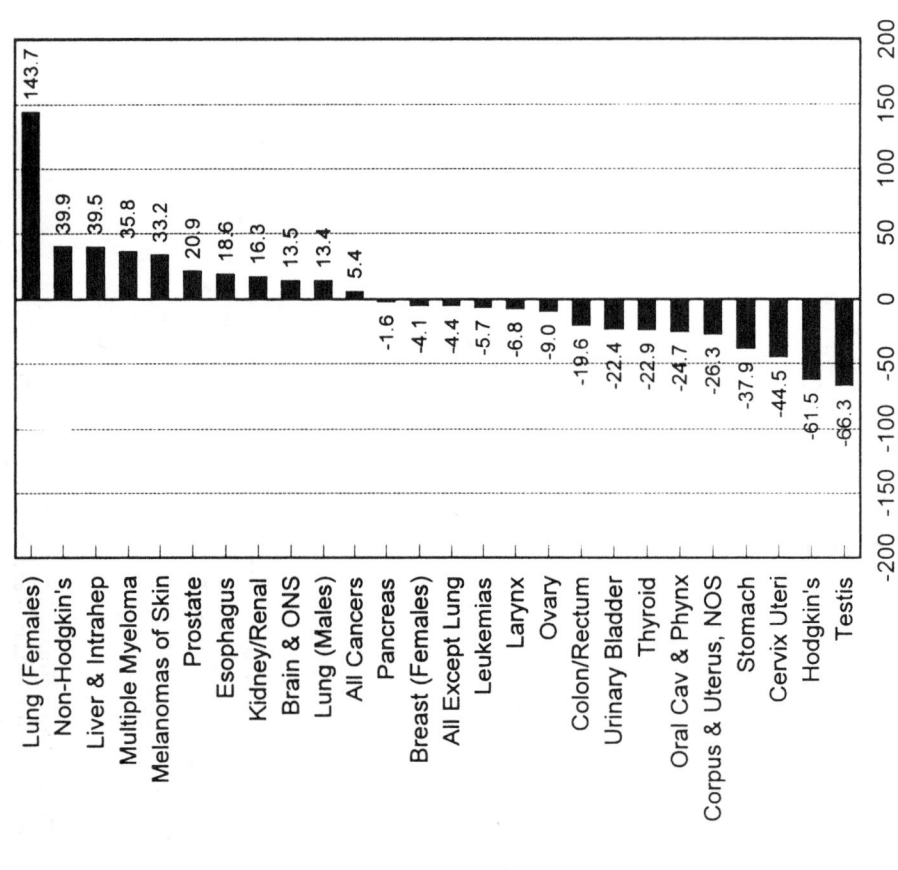

Trends in U.S. Cancer Mortality Rates

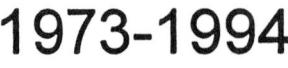

1973-1994

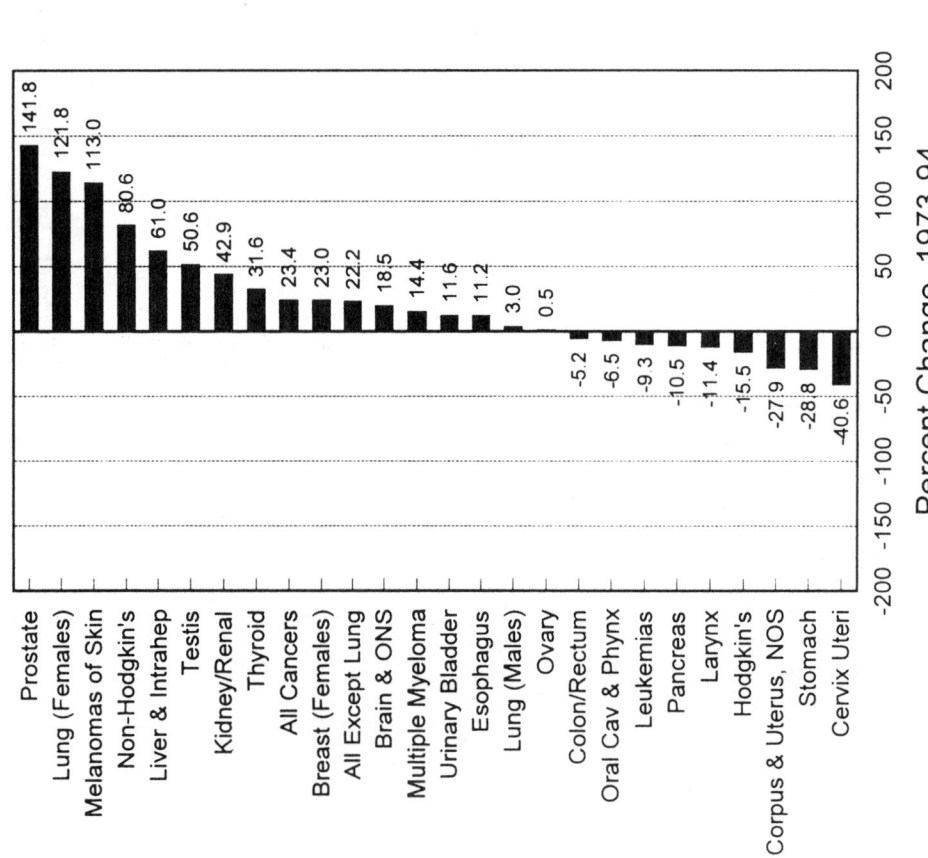

Trends in SEER Incidence Rates

The two bar-graphs above depict "Trends in SEER Incidence and U.S. Mortality Rates by Primary Cancer Site, 1973–1994." They are reproduced from Figure I-4 at page 45 of SEER Cancer Statistics Review, 1973-1994 (SEER 1994 in our Reference List). SEER abbreviates Surveillance, Epidemiology, and End Results. SEER is a program, initiated in 1973 under the National Cancer Institute, to evaluate cancer incidence-trends in the USA. The trends in Incidence Rates are based on five states and four to six metropolitan areas (SEER 1994, p.1). Trends in the Mortality Rates are based on the entire USA.

CHAPTER 68

Is There a Reasonable Non-Radiation Explanation for the Observations?

Part 1. High-Density of Physicians and High Density of Sick People?
Part 2. Some Other Carcinogenic/Mutagenic Activity of Physicians?
Part 3. A Causal Role for RadioTherapy and ChemoTherapy in IHD MortRates?
Part 4. Pressure Not to Report Cancer in Low-PhysPop Census Divisions?
Part 5. A Consequence of Autopsy Rates?
Part 6. An Aspect of Urbanization as the Explanation?
Part 7. What about Migration between High and Low PhysPop Census Divisions?
Part 8. A Reminder about the Law of Minimum Hypotheses

The key observations in this book are the strong, positive dose-responses, by Census Divisions, between Physicians-per-100,000 Population (PhysPop) and 1940 Cancer death-rates, and the strong, positive dose-responses in 1950 between PhysPop and Ischemic Heart Disease death-rates (age-adjusted to the 1940 reference year). So of course we asked, could reasons OTHER than medical radiation cause these tight, linear, positive correlations between PhysPop and each disease? This chapter presents our reasoning, about various alternative explanations.

● Part 1. High-Density of Physicians and High Density of Sick People?

WHY do the relationships, described above, occur? Are these observations explicable by proposing that the Census Divisions with the highest PhysPop values also had the highest densities of very sick people per 100,000 population --- in other words, by proposing that "doctors and sick people attract each other to the same location"? The explanation fails, because we also uncovered a significant NEGATIVE correlation between PhysPop and the 1950 MortRates from all NonCancer NonIHD causes combined (Chapter 25, Box 1).

And so we explored the proposition, that the strong POSITIVE correlations (PhysPop with Cancer MortRates and with IHD MortRates) are due to the combined force of the "attraction" speculation above PLUS the lack of effective treatments for Cancer and IHD at mid-century. But this modified proposition still provides no reasonable substitute for medical radiation as the correct explanation --- because several OTHER major diseases, which also lacked effective treatments at mid-century, do NOT have strong positive correlations with PhysPop. Reminder: A negative sign on the ratio (Xcoef/SE) reflects a DOWNWARD slope.

Chap.	Disease	Male R-sq.	Male Xcoef/SE	Fem R-Sq	Fem Xcoef/SE
27	CNS Vascular (Stroke)	0.40	-2.16	0.29	-1.68
28	Chronic Nephritis	0.46	-2.42	0.27	-1.60
30	Hypertensive Disease (1950)	0.35	-1.92	0.19	-1.28
31	Influenza + Pneumonia	0.83	-5.94	0.88	-7.34
34	RheumHeart + RheumFever	0.002	+0.12	0.06	+0.64
35	Syphilis + Sequelae	0.33	-1.85	--	--
36	Tuberculosis	0.21	-1.35	0.64	-3.51
29	Diabetes Mellitus is omitted due to the reason described in Chap. 29, Box 1.				

It is likely that the PhysPop-MortRates correlations for Cancer in 1940 would resemble the OTHERS on the list above, if it were not for the fact that physicians can CAUSE Cancer by causing medical irradiation. What distinguishes Cancer with certainty, from all the other diseases above, is Cancer's proven inducibility by ionizing radiation --- not its lack of effective treatment. So the "attraction" speculation does not explain the real-world observations --- whereas medical irradiation, caused by physicians, explains the observed correlation between Cancer MortRates and PhysPop very, very reasonably.

● **Part 2. Some Other Carcinogenic/Mutagenic Activity of Physicians?**

Besides causing irradiation of patients, did physicians also behave very commonly in some OTHER way, during the first half of the Twentieth Century, which could cause the extremely strong and positive correlations between PhysPop and 1940 Cancer MortRates, by Census Divisions? And between PhysPop and age-adjusted 1950 IHD MortRates, by Census Divisions? Although such an explanation can not be absolutely ruled out, it would be irrational to choose to explain the correlations by an imaginary carcinogen/mutagen, instead of explanation by a proven, potent, identified carcinogen/mutagen, very commonly dispensed by physicians to both male and female patients --- namely, medical radiation.

Of course, both radiotherapy and chemotherapy for Cancer are themselves carcinogens/mutagens, dispensed by physicians. But neither type of therapy can explain the mid-century dose-responses between PhysPop and Cancer MortRates, by Census Divisions. Both types of therapy are a RESULT of Cancer, and can not explain its occurrence in the first place. IHD is considered in Part 3.

● **Part 3. A Causal Role for RadioTherapy and ChemoTherapy in IHD MortRates?**

For both males and females, a positive correlation exists, by Census Divisions, between age-adjusted MortRates for Cancer and age-adjusted MortRates for IHD. The details are in Appendix-N. To a good approximation, the IHD MortRates are high in the same Census Divisions where the Cancer MortRates are high, and are low in other Census Divisions where Cancer MortRates are low. Smoking is an inadequate explanation (Appendix-N).

3a. Is RadioTherapy the Explanation?

In light of the well-known heart damage by high-dose medical radiation (Appendix-J), we asked ourselves: If deaths due to high-dose cardiac damage have been routinely mis-reported as deaths from IHD, could radioTHERAPY for Cancer explain the correlations between Cancer MortRates and IHD MortRates? Are IHD MortRates high where Cancer MortRates are high, simply because radiation-treated cancer patients are switched --- by their radiation treatment --- from the Cancer death-stream into the "Heart Disease" death-stream? No. We will explain how we reach that conclusion.

High-dose radiogenic heart lesions require radiation therapy to a Cancer existing near the HEART. Otherwise, the heart will not receive the very high doses used in cancer therapy. The two frequent cancers in the heart-region are Respiratory-System Cancers and Breast Cancer. But Respiratory-System Cancers are explicitly excluded from Difference Cancers. By definition, Difference Cancers are All-Cancers-Combined EXCEPT respiratory-system cancers (Chapter 18). That leaves, for males, no high-frequency chest cancer within the Difference-Cancer MortRates. Hodgkin's disease, lymphomas, and leukemia are too infrequent to enter these considerations.

Therefore, if a high positive correlation exists between male MortRates for Difference-Cancers and for IHD, by Census Divisions, we can conclude reasonably that radiotherapy to the chest is not the explanation. Such correlation does exist. Based on the Observed MortRates (from Chapters 18 and 40), the R-squared values are: 0.81 in 1950; and 0.96 in 1960; and 0.94 in 1970; and 0.76 in 1980; and 0.53 in 1990. So we can say with assurance that radiotherapy to the chest does not explain the MortRate-MortRate (Cancer,IHD) correlations for males. And if radiation therapy is not the explanation for males, it is very unlikely to be an adequate explanation for females --- despite radiotherapy for some Breast-Cancers.

RadioTherapy for Breast-Cancer: The Cuzick Study, 1994; the Clarke Report, 1995

A reminder here may be helpful: "Cardiac-related" deaths and IHD deaths are not the same entity. When someone reports an excess of "cardiac-related" deaths, it is possible that NONE of those deaths are IHD deaths. For instance, in 1994, Cuzick and co-workers reported an elevated death-rate from "cardiac-related" causes among breast-cancer patients who received surgical and radiation therapies (circa 1950 to 1976), by comparison with breast-cancer patients in the same studies who received only surgical therapy (P < 0.001) (Cuzick 1994, p.447; pp.451-452). Cuzick specified that "cardiac-related" causes included all the entities with ICD/9 codes 390-429 --- a range which includes many more entities than IHD. Fortunately for breast-cancer patients, current techniques of radiation therapy result in lower radiation doses to the heart than such therapy in the past (Cuzick 1994, p.451; see also "Radiotherapy for Breast Cancer" on the Internet at <http://oncolink.upenn.edu/cancernet/ >).

Regarding radiotherapy for "Early Breast Cancer," a 1995 meta-analysis involving 10-year survival --- for 17,273 women treated surgically for "Early Breast Cancer" --- produced no significant difference between women who had both surgery and radiotherapy and women who had only surgery (p = 0.3; Clarke 1995, p.1444, pp.1445-1447). Radiotherapy was associated with a lower risk of death from Breast Cancer (Odds Ratio = 0.94) but a higher risk of death from other causes (Odds Ratio = 1.24) which were not specified in the report.

3b. Radiation-Conversion of IHD Patients to Cancer Deaths

There is no doubt that IHD patients receive a lot of xray procedures (diagnostic and interventional) to parts of the chest.

For male IHD patients, xray procedures associated with IHD are undoubtedly converting some patients from IHD death to Respiratory-Cancer death. But since Difference-Cancers exclude Respiratory-Cancers, such conversions can not explain the persistent correlations, by Census Divisions, between male IHD MortRates and male Difference-Cancer MortRates.

For female IHD patients, xray procedures associated with IHD are undoubtedly converting some female IHD patients from IHD deaths into Breast-Cancer deaths, which ARE included with female Difference-Cancers. But we doubt very, very much that such conversions suffice to explain the following positive correlations between female MortRates for Difference-Cancers and for IHD, by Census Divisions (MortRates from Chapters 19 and 41): 0.90 in 1950; and 0.98 in 1960; and 0.94 in 1970; and 0.72 in 1980; and 0.59 in 1990.

3c. Does CHEMOtherapy for Cancer Explain the IHD-Cancer Correlations?

The second part of Hypothesis-2 proposes that multiple mini-tumors, induced by ionizing radiation in the walls of the coronary arteries, probably constitute the mechanism by which very-low to moderate doses of medical radiation become a cause of IHD. Of course, the coronary arteries are exposed to non-radiation mutagens too. Some of the chemical agents used in cancer therapy are mutagens. Since these chemicals circulate in the blood-stream, a patient's Cancer need not be located in the CHEST in order for the therapeutic chemicals to have a tumorigenic impact on the coronary arteries.

So we asked ourselves: Could CHEMOtherapy for Cancer (rather than low-and-moderate doses of medical radiation) be the explanation for the the observed MortRate-MortRate correlations (IHD with Cancer, by Census Divisions)?

There is an overwhelming reason to answer "No." It is the timing. We have shown a strong correlation in 1950 (Appendix-N), the earliest year for which IHD data are available. Chemotherapy was very rarely used for Cancer in 1950, so it could not possibly explain the 1950 correlation between MortRates for IHD and for Cancer, by Census Divisions. Indeed, the first hundred cancer patients who underwent chemotherapy did so in 1943 (Moss 1995, p.18). Between 1945 and the present time, use of chemotherapy against various Cancers has grown dramatically, but by 1960 or 1970, it was still not common enough to explain the extremely strong correlations (IHD MortRates and Cancer MortRates, by Census Divisions) observed in 1960 and 1970.

"I'll eat my hat" if a good share of IHD deaths in 1960 or 1970 had a prior history of chemotherapy for Cancer. If chemotherapy for Cancer cannot explain the 1950, 1960, or 1970 observations, why ask chemotherapy to explain the 1980 and 1990 correlations? By contrast, if Hypothesis-2 is valid, then all ten sets of observations (1950, 1960, 1970, 1980, 1990, for males and females separately) can be explained by exposure to low-and-moderate doses of medical radiation --- a phenomenon whose effects have been in operation since 1896.

● Part 4. Pressure Not to Report Cancer in Low-PhysPop Census Divisions?

        At mid-century, the general wisdom which I heard in medical school was that having a
cancer-death in the family was, at least among some families, regarded as somehow shameful. No one
can possibly evaluate how common this attitude may have been, nor can anyone evaluate how often a
family pleaded successfully with a physician to report a cancer-death as something else. Still, we
asked ourselves: Could this attitude explain why age-adjusted Cancer MortRates in 1940 were low in
the Census Divisions where PhysPop was low, and high where PhysPop was high?

        To take this speculation seriously, we would need to believe firmly that the frequency of the
attitude was proportional to PhysPop, and also that physicians routinely helped to cover-up cancer
deaths when asked. And what about Ischemic Heart Disease? An explanation is required also for
IHD's irrefutable and strong positive correlation with PhysPop in 1950, by Census Divisions. So we
would need not only a second stigma --- but also one whose frequency was proportional to PhysPop.
And what about tuberculosis-death? It has a NEGATIVE relationship with PhysPop at mid-century.
So we would also need to believe that stigma was ABSENT for tuberculosis-death and all other causes
of death which had negative relationships with PhysPop at mid-century (Chapter 38, Box 1). The
"stigma" speculation can not compete, in evidence or reasonableness, with medical radiation as the
correct explanation for our findings.

● Part 5. A Consequence of Autopsy Rates?

        We consider it likely that wherever PhysPop was higher, the autopsy-rate was higher. In
Chapter 4, Part 3a, we related reports which attempt to quantify how often Cancer was first discovered
at autopsy to be the cause of a death. The statements of Hill (1992, p.48) suggest that hospital
autopsies in the first half of the Twentieth Century produced a net increase of about 25% in deaths
assigned to Cancer, compared to clinical pre-autopsy counts. Additionally, Hill relates (p.51) that at
mid-century in the USA, more than half of patients dying IN HOSPITALS were autopsied. We cannot
independently verify either piece of information, but we can use both pieces to explore some
implications of an "autopsy" speculation.

        Suppose that in the Census Division with the lowest 1940 PhysPop, no autopsies at all were
performed. And suppose that in the Census Division with the highest 1940 PhysPop, 100% of all
deaths were autopsied. We know that both suppositions are preposterous, but we are just exploring.
In these unrealistic circumstances, we could expect that the 100% autopsy-rate would increase the
reported Cancer MortRate in the high-PhysPop Census Division to a value about 25% higher than the
reported Cancer MortRate in the low-PhysPop no-autopsy Census Division. But we know that a 25%
increment would be a highly exaggerated expectation, because there never really was a 1940
autopsy-rate of zero in any Census Division, and there never really was an autopsy-rate of 100% in
any Census Division (many people died AT HOME). So --- still assuming that autopsy-rates were
higher where PhysPop values were higher --- we will approximate that higher autopsy-rates could
cause reported cancer MortRates to be 13% higher in the the Census Division with the highest PhysPop
than in the Census Division with the lowest PhysPop. Thirteen percent still may be quite an
exaggerated expectation. Then we make a reality-check:

        Males, All-Cancers (data from Chapter 6, Part 2j): For males in 1940, the Cancer MortRate is
73.6 where PhysPop is lowest. Therefore, due to a higher autopsy-rate where PhysPop is highest, the
expected Cancer MortRate might be (73.6 times 1.13), or 83.2. In reality, the observed Cancer
MortRate is 140.9 where PhysPop is highest.

        Females, All-Cancers (data from Chapter 7, Part 2j): For females in 1940, the Cancer
MortRate is 102.5 where PhysPop is lowest. Therefore, due to a higher autopsy-rate where PhysPop
is highest, the expected Cancer MortRate might be (102.5 times 1.13), or 115.8. In reality, the
observed Cancer MortRate is 142.9 where PhysPop is highest.

        Our reality-check on a potential "autopsy effect" indicates that, in all likelihood, it is far from
adequate to account for the observed increment in either the male or female 1940 Cancer MortRate,
between the lowest PhysPop Census Division and the highest PhysPop Census Division. In other
words, the "autopsy" speculation does not explain the observed, tight, positive correlations between
PhysPop and 1940 All-Cancer MortRates. Something else was at work. The superior explanation is

medical radiation --- which is a proven cause of Cancer and is an explanation which ALSO accounts for the observation that PhysPop shows a dramatically different relationship with the NonCancer NonIHD causes of death.

● **Part 6. An Aspect of Urbanization as the Explanation?**

We know a peer-reviewer who is fond of invoking "urbanization" as a possible explanation for many physical and mental afflictions. So we asked ourselves: What would be required, for urbanization to explain our mid-century findings?

First, it would require evidence that urbanization and physician-density per 100,000 population were very highly and positively correlated in the decades before mid-century. And then it would require evidence that city-life in the years 1900-1950 was a potent risk factor both for Cancer and for IHD. With such EVIDENCE, it would become reasonable to propose that PhysPop was a surrogate, in our mid-century regression analyses, for the accumulated health effects of city-life.

And then, if such evidence were at hand, we would have to ask ourselves: What ASPECT of city-living might elevate the MortRates from Cancer and IHD? With respect particularly to Cancer --- because ionizing radiation is a proven carcinogen --- a reasonable answer would be that the extra medical radiation administered by the extra city-based physicians would be a very reasonable explanation.

● **Part 7. What about Migration between High and Low PhysPop Census Divisions?**

There is no doubt that, both before and after mid-century, migration from one Census Division to another has occurred. And when people receive, say, half of their medical irradiation in one Census Division, and then move to another Census Division, they carry with them the xray-induced mutations already inscribed upon their genetic molecules.

If 20 million people (illustrative number) move from the three High-PhysPop Census Divisions to the three Low-PhysPop Census Divisions, where they die, the effect would be to elevate the Cancer MortRate in the Low-PhysPop Divisions without any increase in PhysPop values --- even though the migrants from High-PhysPop Divisions had caused an increase in the average accumulated per capita dose of medical radiation in the population of the Low-PhysPop Census Divisions. And likewise, if 20 million other people migrate from the three Low-PhysPop Census Divisions to the three High-PhysPop Census Divisions, where they die, the effect would be to lower the Cancer MortRate in the High-PhysPop Divisions without any decrease in PhysPop values --- even though the migrants from Low-PhysPop Divisions had caused a decrease in the average accumulated per capita dose of medical radiation in the population of the High-PhysPop Census Divisions.

Under the above circumstances, the TRUE difference between High-PhysPop and Low-PhysPop Census Divisions, in average accumulated per capita dose of medical radiation, would be less than indicated by the ratio of their PhysPop values. Because of less dose-difference, the Cancer MortRates would become more similar in High-PhysPop and Low-PhysPop Census Divisions.

In Chapters 49 through 65 (exception: female Genital Cancers), the Boxes 1 always reveal that after mid-century, something caused more upward pressure on the Cancer and IHD MortRates in the Low-PhysPop Census Divisions than in the High-PhysPop Census Divisions. Even when National MortRates fell relative to 1940 or 1950, they fell by a smaller factor in the Low-PhysPop Divisions. Sometimes MortRates fell by just a LITTLE in the High-PhysPop Divisions while RISING in the Low-PhysPop Divisions. The Boxes 1 are consistent with inspection of Tables 6-A and 7-A (and subsequent MortRates tables, including 40-A and 41-A), with attention to the row marked "Ratio, Hi5/Lo4." It is obvious that, over time, Cancer MortRates grew more similar among the Census Divisions, and so did IHD MortRates.

We had to ask ourselves: Are these observations more consistent with extra smoking in the Low-PhysPop Divisions, or with a net migration from Higher-PhysPop Divisions to Lower-PhysPop Divisions?

Regression-Output without the Smoking Adjustment

We answered the question by studying the regression-outputs (not shown) of post-1940 "raw" Cancer and IHD MortRates ("raw" meaning without a smoking-adjustment) regressed upon Mean PhysPop values (Table 47-A). Those R-squared values decline with every decade --- and for MALES, a few of the X-coefficients turn negative. By the time one arrives at raw 1988 Cancer MortRates (1993 for IHD), regressed upon 1940-1990 Mean PhysPop values, the R-squared values range all over the place, even though most R-squared values were in a tight range of 0.85 to 0.95 in 1940 (for IHD, in 1950) --- with only positive slopes.

	Male R-sq	Xcoef/SE	Fem R-sq	Xcoef/SE
All-Cancers 1988	0.11	-0.91	0.22	+1.41
Respiratory 1988	0.31	-1.78	0.06	+0.70
Diff-Cancers 1988	0.03	+0.43	0.24	+1.48
Digestive-Ca 1988	0.32	+1.79	0.42	+2.23
Breast Cancer 1990	--	--	0.57	+3.07
IHD 1993	0.06	-0.67	0.0011	+0.09

The Better Explanation: The Inverse Relationship of PhysPop with Smoking

In the preceding tabulation, the opposite slopes and wide range of R-squared values are not indicative of migration as the explanation. If migration were the explanation, the correlations should degrade about equally for all types of Cancer, per gender. For example, migrants from one Census Division to another can not decide to carry their xray-induced mutations for Digestive-System Cancers WITH them, but choose to leave BEHIND their xray-induced mutations for Respiratory-System Cancers --- or vice versa.

By contrast, the tabulated results are very reasonably explained by the rising post-1940 impact of the inverse relationship between PhysPop (the x-variable) and smoking-intensity. By 1988-1993, the inverse PhysPop-Smoking relationship not only conceals the true PhysPop-MortRate relationships, but does so UNEQUALLY --- with the most intense impacts exactly where they are expected: Upon the PhysPop correlations with Respiratory-Cancer and IHD MortRates.

Of course, we also asked ourselves: Could the explanation of the tabulated values be that medical radiation just lost its importance, relative to some new and independent cause of Cancer? No. We have already demonstrated (Chapter 5, Part 6a) that a high correlation between PhysPop and Cancer MortRates, by Census Divisions, would NOT be lowered just by addition of a new and independent cause of Cancer --- if the intensity of the new cause were MATCHED across the Census Divisions.

● Part 8.　A Reminder about the Law of Minimum Hypotheses

In logic and science, the Law of Minimum Hypotheses (Ockham's Razor) advises: To explain a phenomenon, invoke only as many explanations as required. This monograph has revealed certain phenomena:

● For Cancer, powerful positive correlations exist between PhysPop values and Cancer MortRates, by Census Divisions, for males and females separately.

● For Ischemic Heart Disease, powerful positive correlations exist between PhysPop values and IHD MortRates, by Census Divisions, for males and females separately.

● For NonCancer NonIHD causes of death, powerful positive correlations with PhysPop do NOT exist across the Census Divisions; instead, the relationships are NEGATIVE.

These three facts "demand" an explanation. This chapter has reviewed various explanations which do NOT appear reasonable. By contrast, medical radiation is almost certainly the correct and sufficient explanation, for the reasons which we have presented earlier. From this it follows (by the logic demonstrated in this book) that medical radiation is a highly important cause of both Cancer and Ischemic Heart Disease, in the USA, for most of the Twentieth Century --- including the present day.

CHAPTER 69

Conclusion: Making Sense of Three Sets of Irrefutable Correlations

Part 1. Newly Uncovered: Three Remarkable Sets of Facts
Part 2. Basis for Correctly Identifying the Explanation
Part 3. Are High Fractional Causations Hard to Believe?

At the conclusion of a work like this, it can be helpful to review what is based on irrefutable observation and what is based on logic.

● Part 1. Newly Uncovered: Three Remarkable Sets of Facts

This final chapter focuses on the findings that death-rates at mid-century from Cancer and Ischemic Heart Disease each have a very strong positive correlation with physician-density (PhysPop), by Census Divisions, while death-rates from NonCancer NonIHD Causes have a signficant inverse correlation.

These three sets of relationships are irrefutable facts, not interpretations. And they are remarkable. What EXPLAINS the extreme similarity in the relationship of Cancer and of Ischemic Heart Disease (IHD) with PhysPop, and the unambiguously different relationship of NonCancer NonIHD Causes of Death with PhysPop?

By investigating the major subsets of All-Cancers, we have confirmed that the positive correlation is not just the net result of some positive and some negative correlations. And by investigating the major subsets of All NonCancer NonIHD Causes, we have established that the negative correlation is overwhelmingly supported by the subsets. The mid-century results are easily compared with each other in Chapter 38's Box 1 (also reproduced in Chapter 1). The columns of that single page present a mountain of irrefutable facts, without any interpretation.

It is startling, indeed, to realize that the 1940 National All-Cancer Death-Rates are quite well predicted from PhysPop values 10 years earlier --- even 20 years earlier (Chapter 22, Box 4). And the 1950 National Death-Rates from Ischemic Heart Disease are also quite well predicted from the PhysPop values 20 years earlier.

● Part 2. Basis for Correctly Identifying the Explanation

While the correlations we uncovered are irrefutable observations, correlations alone can never PROVE causation. However, strong correlations are properly regarded as one of the "gold standards" when it comes to establishing causation, because they provide such strong circumstantial evidence.

And when strong correlations are supported by supplemental information and logic, the combination can establish causality beyond reasonable doubt. We are on extremely solid ground when we assert that the correlations we have uncovered are causal in nature. First, we consider the strong positive correlation uncovered between PhysPop and MortRates from Cancer, by Census Divisions.

The Cancer-Response to PhysPop

Chapter 3 shows, with evidence and logic, that PhysPop can be regarded as approximately proportional to the average per capita accumulated dose from medical radiation. This relationship constitutes information supplemental to the PhysPop-MortRate correlations. Additionally, Chapter 2 points to a large body of evidence that ionizing radiation (including the xray) is a well-established cause of Cancer. Such evidence is also a supplement to the PhysPop-MortRate correlations. Under these circumstances, it is highly reasonable to conclude that the strong positive correlations, between PhysPop and cancer MortRates, are causal --- a dose-response between medical radiation and cancer MortRates, by Census Divisions.

Even if there were NO pre-existing evidence that medical radiation is a cause of Cancer, the strong positive correlation would be presumed to be causal --- in the absence (Chapter 68) of a better explanation.

The IHD Response to PhysPop

Next, we consider the strong positive correlation uncovered between PhysPop and MortRates from Ischemic Heart Disease, by Census Divisions. This finding was an enormous surprise to us. But the correlation is an irrefutable fact, and it "demands" explanation. Within the same database, medical radiation is the cause of the cancer-response, so it would be irrational to assume (with no basis) a DIFFERENT cause of the IHD-response. Supplemental support is provided by pre-existing evidence on the role of acquired mutations in atherogenesis (Chapter 44, Parts 8+9) and on the structure of atherosclerotic plaques and the activity within them (Chapter 44, Part 7). Such evidence combines with logic (Chapter 45) to indicate that xray-induced mutations are the reasonable explanation for the observed positive dose-response between PhysPop and MortRates from Ischemic Heart Disease, by Census Divisions.

In research, Ockham's Razor is an important admonition: To explain a phenomenon, invoke only as many explanations as required. Avoid fabricating many explanations if one suffices. Of course, the explanation MUST be consistent with other well-established observations. Radiation-induction of Ischemic Heart Disease, via induction of mutations in the coronary arteries, is fully consistent with other well-established observations, and indeed, helps to explain some of them (Chapters 45, 46, and Appendix-N).

The NonCancer NonIHD Response to PhysPop

Lastly, we consider the ABSENCE of a strong positive dose-response between PhysPop and NonCancer NonIHD Causes of Death, by Census Divisions. This finding is another irrefutable fact --- a fact which is consistent with the "general wisdom" that ionizing radiation is not a cause of such deaths (Chapter 23). Indeed, the significant negative correlation, produced by the evidence, is also in harmony with common sense: Physician-activity, including the use of medical radiation, helps to PREVENT such deaths.

The unambiguous difference, between the cancer-response to Phys-Pop and the NonCancer NonIHD response to PhysPop, constitutes very strong confirmation that PhysPop is reliable as a surrogate for medical radiation. And the observed behavior of PhysPop values, over the 1921-1990 period, supports their use as surrogates for average ACCUMULATED doses from medical radiation, by Census Divisions (Chapters 3 and 47).

In short: Medical Radiation Explains the Irrefutable Facts

Thus, the foundation is solid for saying that MEDICAL RADIATION explains the irrefutable facts set forth in Part 1:

1) The very strong positive correlation which exists between PhysPop and Cancer MortRates, by Census Divisions.

2) The very strong positive correlation which exists between PhysPop and IHD MortRates, by Census Divisions.

3) The dramatically different relationship which exists between PhysPop and NonCancer NonIHD MortRates, by Census Divisions.

● Part 3. Are High Fractional Causations Hard to Believe?

The importance of medical radiation, in the etiology of both Cancer and Ischemic Heart Disease, has been evaluated in terms of Fractional Causation of their National Mortality Rates, decade by decade, from 1940-50 to 1988-93. For the reasons discussed above and additional reasons discussed in the text (especially Chapters 1, 2, 6, 48, 49, 67, 68) we have a high level of confidence that the observed correlations and the applied logic are sound.

The resulting estimates of Fractional Causation, by medical radiation, are summarized in Chapters 1 and 66. The high values are neither hard to believe nor inconsistent with well-established facts, if one remembers that:

● Xrays are a uniquely potent mutagen, able to induce virtually every kind of mutation in the cells of every organ. Moreover, the xray doubling-dose for structural chromosomal mutations is very low. Xrays are also an established cause of genomic instability.

● Although past and current exposure to medical radiation is extremely common in the USA, the magnitudes of accumulated doses are simply not known --- due to a persistent lack of routine measurements. One result is great uncertainty about the true risk per dose-unit. Nonetheless, our method is able to produce credible estimates of the consequences of such exposure, because PhysPop values are credible measures of the RELATIVE magnitude of accumulated doses in the Nine Census Divisions.

● A very high Fractional Causation --- such as 80% of a MortRate by medical radiation --- does not mean that medical radiation is the only cause of 80% of the MortRate. It means that medical radiation is a NECESSARY co-actor (with other causes) in 80% of the MortRate, and that approximately 80% of the MortRate would be absent, in the absence of medical radiation.

But nothing in this book argues against the use of medical radiation, which can have undeniable benefits. The findings in this book do argue, strongly, that we will safely prevent much of the future mortality from Cancer and Ischemic Heart Disease, if we eliminate uselessly high doses of medical radiation. Techniques (already proven) exist for obtaining every benefit of medical radiation at much lower doses, without eliminating a single radiation procedure (Chapter 1, Parts 3, 9, and 10).

The findings in this book also argue, strongly, that epidemiologists may obtain seriously flawed results in studies of non-xray causes of Cancer and Ischemic Heart Disease, if their studies have cohorts which are poorly matched for accumulated exposure to medical radiation.

The irrefutable observations presented by this work are at variance with some comfortable assumptions. The benefits for human health, of discarding comfortable, but erroneous assumptions, will be very large. Whether the field is biomedicine or astro-physics or engineering, some words from a successful pioneer may have merit. Orville Wright, one of the two brothers who had to discard some prevailing wisdom in order to develop the first operable airplane, wrote: "If we all worked on the assumption, that what is accepted as true really is true, there would be little hope of advance."

Part 1. Where Xrays Fit, among a Variety of Ionizing Radiations
Part 2. The Relatively Reliable Dose-Units: Rad (Centi-Gray) and Roentgen
Part 3. The Less Reliable Dose-EQUIVALENT Unit: Rem (Centi-Sievert)
Part 4. The Least Reliable Dose-Unit: EFFECTIVE Dose Equivalent
Part 5. Dose-Levels: What is Low ... Moderate ... High?
Part 6. Possible Leveling and Decline of Dose-Response at High Doses
Part 7. A Contrast: Natural Background Dose vs. Promptly Lethal Dose

● **Part 1. Where Xrays Fit, among a Variety of Ionizing Radiations**

Ionizing radiation is radiation with enough energy not only to "kick" electrons out of their normal atomic orbits, but also to endow these liberated electrons with kinetic energy which sets them into high-speed linear travel (details in Gofman 1990, Chapters 20 + 32). Ionizing radiation can be divided into two classes.

● One class begins as photons (xrays, gamma rays), which spend their energy on kicking electrons out of orbit and endowing them with the kinetic energy for high-speed linear travel. As they travel, these high-speed high-energy electrons drop portions of their energy, like small bombs or grenades, upon various molecules in their paths. These energy-deposits do the damage.

● The other class begins as high-speed high-energy particles (such as alpha particles, beta particles, positrons, etc.), which also drop portions of their energy upon various molecules in their paths. Usually, the source of these high-speed high-energy "bullets" is the natural decay of unstable (radioactive) atomic nuclei. Our monograph says little else about this second category, because XRAYS are certainly the main source of exposure from diagnostic and interventional uses of medical radiation. Exposure to alpha-radiation is discussed elsewhere (Gofman 1976 + 1981 + 1994).

● **Part 2. The Relatively Reliable Dose-Units: Rad (Centi-Gray) and Roentgen**

● Rad abbreviates "radiation absorbed dose." The rad is defined as the following amount of energy deposited by ionizing radiation PER GRAM OF TISSUE:

10^{-5} joule. ($\char94$ is our sign for exponent.) Or the equivalent:
$6.24 * 10^{10}$ KeV. (* is our sign for multiplication.) Or the equivalent:
100 ergs.

In the way that the price of candy is a ratio of dollars per gram, the rad is a ratio of energy per gram. 100 rads means that the ratio rises by 100-fold --- e.g., 100 rads would be 10^{-3} joule per gram of tissue, or $6.24 * 10^{12}$ KeV per gram, or 10,000 ergs per gram.

● The milli-rad is one-thousandth of a rad: 0.001 rad or 10^{-3} rad.

● Centi-gray (cGy) is a more recent name for rad. The two dose-units are identical. We (and many others) prefer the shorter term. There are 100 centi-grays (or 100 rads) per Gray. Thus a dose of 0.2 Gray (Gy) is the same as 20 rads. A dose of 0.2 milli-gray (mGy) is the same as 20 milli-rads (mrads).

● Roentgen (abbreviated R, or r) is a dose-unit for xrays and gamma rays which, in the energy range of 100 to 3,000 KeV, produces 0.96 rad in tissue (Shapiro 1990, p.48). In other words, rad and roentgen can be considered as nearly equivalent, for most xrays and gamma rays.

Two Billion Photons ... 675 Million Cells

If a gram of tissue has received a dose of 1 rad, from medical xrays with an average energy of

30 KeV per photon, it turns out that about TWO BILLION photons have delivered all of their energy within that gram of tissue (from Gofman 1990, Table 20-C). It also turns out that, on the average, every cell-nucleus in that gram has been traversed 1.33 times by a primary high-speed high-energy electron (Appendix-B, Part 1b). There are an estimated 675 million cell-nuclei, on the average, in a gram of human tissue (Gofman 1990, Chapter 20, Part 2).

● **Part 3. The Less Reliable Dose-EQUIVALENT Unit: Rem (Centi-Sievert)**

There is little doubt that, at EQUAL rads to the same tissue, some types of ionizing radiation do more damage than other types. Alpha radiation is particularly violent. Thus, it is correct to say that radiations differ in their "Relative Biological Effectiveness" (RBE) --- with "effectiveness" as a euphemism for harmfulness. The type of harmfulness which relates to this book is mutagenic potency. In Chapter 2 (Part 7), we discussed the evidence that, at EQUAL rads, medical xrays are 2 to 4 times more harmful than the high-energy gamma rays of the Hiroshima-Nagasaki bombs. Because of this distinction, we spoke of "bomb rads" and "medical rads" (Chapter 2, Part 7d).

For decades, the differences in RBE have been attributed largely to differences in LET (Linear Energy Transfer) --- which means: Attributed to differences in the average amount of energy lost, by a high-speed high-energy particle, per unit of its track-length (BEIR 1990, p.395). The higher the LET, the higher is the harmfulness. In other words, when the energy-deposits are larger and/or closer to each other, the harmfulness rises. The greater size and closer proximity of these energy "grenades" probably cause damage which is more complex and harder for a cell to repair correctly.

For electrons, the higher is the initial energy, the lower is the Linear Energy Transfer (BEIR 1972, p.215). This is consistent with the observation that lower-energy xrays are more harmful, per rad, than high-energy gamma rays from the Hiroshima-Nagasaki bombs.

● Rem is a dose-unit introduced in the hope of establishing a system to handle RBE. A rem is numerically equal to the absorbed dose in rads, multiplied by a factor equal to the applicable RBE --- with the reference value for RBE being 250 KvP xrays (BEIR 1972, p.216). Thus, if alpha radiation is estimated to be 10-fold more mutagenic per rad than 250 KvP xrays, its RBE for mutagenicity would be 10, and someone exposed to 2 rads of alpha radiation would be said to have received a dose-equivalent of 20 rems. However, if the source does not identify the chosen RBE as 10, the reader is left to wonder. Twenty rems could mean 1 rad at RBE = 20, or 2 rads at RBE = 10, or 4 rads at RBE = 5.

● Centi-sievert (cSv) is a more recent name for rem. There are 100 centi-sieverts (or 100 rems) per Sievert. Thus a dose of 0.2 Sievert (Sv) is the same as 20 rems. A dose of 0.2 milli-sievert (mSv) is the same as 20 milli-rems (mrems).

We must regard these dose-equivalent units as "less reliable" than the rad (or cGy) and the Roentgen, because the RBE values are still so uncertain. It is much easier to measure energy per gram of tissue than to establish the Relative Biological Effectiveness in humans. Moreover, the latter effort must start with reliable dose-measurements, and so it is hardly helped by the failure to make good measurements in routine medical practice.

● **Part 4. The Least Reliable Dose-Unit: EFFECTIVE Dose Equivalent**

The least reliable of all dose-measurements is reported, in rems or sieverts, as an EFFECTIVE dose-equivalent. The effective dose-equivalent incorporates not only significant assumptions about RBE values, but many additional assumptions. UNSCEAR describes the input as follows (UNSCEAR 1993, p.12/69):

"The various organs and tissues in the body differ in their response to radiation. To allow for this, a further quantity, effective dose, is used. The equivalent dose [in rems or sieverts] in each tissue or organ is multiplied by a tissue weighting factor, and the sum of these products over the whole body is called the effective dose. The effective dose is an indicator of the total detriment due to stochastic effects in the exposed individual and his or her descendants." And what is the input for those "tissue weighting factors"? UNSCEAR explains (1993, p.13/77):

The International Commission on Radiological Protection "takes account of the attributable probability of fatal cancer in different organs, of the additional detriment from non-fatal cancer and hereditary disorders, and of the different latency periods for cancers of different kinds. All these features are included in the selection of weighting factors for converting equivalent dose into effective dose."

The concept of evaluating "total detriment" is attractive. The big problem is the currently poor quality of the evidence required to do it. Because the quantitative evidence, needed for such tissue weighting factors, is thin to really non-existent, we regard the "effective dose equivalent" as a step very likely to introduce needless ERRORS into this field at this time.

● Part 5.   Dose-Levels: What is Low ... Moderate ... High?

There are no formal cutting points, on the continuum of dosage, between very low, low, moderate, high, very high. Different analysts in different decades mean different things by such terms. What do WE mean, when we incorporate the phrase "even at very low and moderate doses" into our Hypothesis-2?

First: We Mean Absorbed Organ-Doses

First, we mean absorbed organ-doses, which can be quite a bit lower than the xray dose measured when the beam first touches the skin. The following ratios come from Gofman 1985 (p.404), where the beam quality and other conditions are specified:

Example 1: If the skin receives 1.0 rad of entrance dose from an xray beam, traveling from front to back, the heart may receive an average dose of 0.50 rad. Now: Reverse the direction of the xray beam. If the skin receives 1.0 rad of entrance dose from an xray beam, traveling from back to front, the heart may receive an average dose of about 0.06 rad.

Example 2: Change heart to kidneys. From 1.0 rad of entrance dose, with the xray beam traveling front to back, the kidneys receive an average dose of about 0.07 rad. But reverse the beam, so that it travels back to front, and the kidneys receive an average dose of about 0.75 rad.

Example 3: Change kidneys to breasts. From 1.0 rad of entrance dose, with the beam traveling from front to back, the breasts receive an average dose of about 0.70 rad. Reverse the beam, and they receive about 0.04 rad.

Second: We Mean Accumulated Organ-Dose

The existing evidence is that radiation-induced mutations endure for decades, and probably for the entire subsequent lifespan --- unless the mutated cell dies without reproducing itself. To a good "first approximation," we have to say that carcinogenic and atherogenic mutations accumulate with each additional dose. Therefore, Hypothesis-2 refers to accumulated dosage from medical radiation.

Division of the Dose-Spectrum into Levels

The doses below refer to increments above the accumulated background dose from natural sources. Reminder: Any divisions are arbitrary, and they vary from author to author and vary with the RBE of the radiation.

- Very low-dose accumulated increment: Up to 5 rads to any organ from medical xrays.

- Low-dose accumulated increment: 5 to 10 rads to any organ from medical xrays.

- Moderate-dose accumulated increment: 10 to 50 rads to any organ from medical xrays.

- High-dose accumulated increment: 50 to 200 rads to any organ from medical xrays.

- Very high-dose accumulated increment: Over 200 rads to any organ from medical xrays. Appendix-J discusses the non-atherogenic coronary effects of heart-irradiation by a thousand or more rads.

● **Part 6.   Possible Leveling and Decline of Dose-Response at High Doses**

The evidence from experiments with cells and with animals is that the dose-response, for carcinogenesis and mutagenesis, frequently levels off at high doses and then declines at very high doses (NCRP 1980, p.17; discussion in Gofman 1990, Chapter 23). Such findings are a warning that estimates of risk-per-rad at very low dose-levels probably will be underestimated, if their basis includes much high-dose evidence.

● **Part 7.   A Contrast: Natural Background Dose vs. Promptly Lethal Dose**

The Accumulated Dose from Natural Background Radiation

At sea-level, people accumulate radiation exposure from natural radiation sources at an average rate of about 100 milli-rems or 0.1 rem per year --- about 7.5 rems of whole-body dose by age 75 (excluding the "effective dose equivalent" from radon daughter-products; BEIR 1990, p.18). Natural background dose varies with altitude and other factors. "Natural" radiation, which puts ionization tracks through cell-nuclei, is surely responsible for a share of Inherited Afflictions (Gofman 1998) and Cancer and --- according to the evidence uncovered in this book --- Ischemic Heart Disease.

The Promptly Lethal Whole-Body Dose

The high-speed high-energy particles from ionizing radiation can damage EVERY kind of molecule in their pathways --- not only the genetic molecules. As dose rises, the level of mayhem and chaos rises. For some critical systems, there will be a dose-level at which too many cells become so damaged, in so many ways, that they do not recover quickly enough to perform their essential functions. The person dies promptly --- not from Cancer or Ischemic Heart Disease, but from "acute radiation syndrome." Some of the workers who tried to extinguish the fire at the Chernobyl nuclear power plant, in 1986, died of that syndrome.

For HALF of the humans exposed, promptly-lethal doses have been estimated by the radiation community at around 300 or 400 internal-organ rads, accumulated in one week or less by the whole body (NCRP 1989-b, p.70, p.73). Radiotherapy for Cancer delivers total doses far above 300 or 400 rads, but (a) only a small section of the body is irradiated, and (b) the total dose is delivered in fractions, with time for some recovery between doses.

>>>>>>>>>>

The Safe-Dose Fallacy: Summary of Three Remarkably Similar Reports

Part 1. Gofman 1990: Proof That There Is No Threshold Dose or Dose-Rate
Part 2. UNSCEAR 1993: "Sometimes Misrepair Can Occur"
Part 3. NRPB 1995: Evidence "Falls Decisively" against a Threshold
Part 4. Alpha Particles, Xrays, and the Major Medical Journals

By "Safe-Dose Fallacy," we refer to the mistaken idea that no cancer-risk occurs from ionizing radiation if a dose is below a certain level (below a "threshold dose"). Appendix-B expands on the very brief discussion in Chapter 2, Part 6.

The threshold hypothesis, with respect to radiation carcinogenesis, has been invalidated in three major reports: Gofman 1990, UNSCEAR 1993, and NRPB 1995. (UNSCEAR is the United Nations Scientific Committee on the Effects of Atomic Radiation. NRPB is Britain's National Radiological Protection Board.)

● Part 1. Gofman 1990: Proof There Is No Threshold Dose or Dose-Rate

The no-risk speculation about low-dose radiation has been tied for a long time to the fact that cell-nuclei have massive capacity to repair DNA damage (Part 1c). Once upon a time, nearly everyone (myself included) hoped that carcinogenic lesions might invariably be repaired --- correctly --- whenever the repair-system was not overwhelmed by "too much" radiation-induced damage all at once.

In the 1970s, however, it was already clear that perfect repair of injured human chromosomes did NOT occur, even when low total doses of radiation were received very slowly from weapons-testing fallout or chronic occupational exposures. And some evidence was already solid that radiation-induced human CANCER is associated with very low doses and dose-rates. But might there be a safe dose (no-risk dose) at even lower levels?

Between 1970 and 1990, it was frequently asserted that the safe-dose issue could never be settled, because of the limits of epidemiology. In Gofman 1990, however, we were able to prove, by any reasonable standard of biomedical proof, that no safe dose or dose-rate exists with respect to radiation carcinogenesis.

The key breakthrough lies in recognizing that the relevant way to define the lowest possible dose and dose-rate of radiation is NOT in fractions of a rad. The RELEVANT definition occurs in "tracks" per cell (Gofman 1971, pp.275-276; Gofman 1981, pp.405-411; Gofman 1986, pp.6-14). We will show why, by explaining "tracks" in Section 1a, below.

1a. The Least Possible Amount of Damage to Repair

● - (1) "The dose from low-LET ionizing radiation is delivered by high-speed electrons, traveling through human cells and creating primary ionization tracks" (Gofman 1990, p.18-2). Low-LET radiation includes xrays, gamma rays, and beta particles.

● - (2) When genetic molecules are damaged by ionizing radiation, each cell-nucleus attempts to un-do the damage by repair. The damage done by a single primary ionization track is the LEAST POSSIBLE damage which the repair-system ever can face. "Fractional tracks do not exist. Either a track traverses a nucleus somewhere (one nuclear track) or it does not (zero nuclear track)" (1990, p.19-2).

● - (3) "For disproof of any safe dose or dose-rate, it is more important to establish the dose in terms of the average number of tracks per nucleus, than to establish it in terms of rads. The reason is that the lowest conceivable dose or dose-rate with respect to repair is not a millionth or any other tiny fraction of a rad or centi-gray. The lowest conceivable dose or dose-rate is one track per nucleus plus sufficient time to repair it" (1990, p.18-3,4).

● - (4) "Because the minimal event in dose-delivery of ionizing radiation is a single track, we can define the least possible disturbance to a single cell-nucleus: It is the traversal of the nucleus by just one primary ionization track" (1990, p.19-1). The traversal is complete in a tiny fraction of one second.

● - (5) "Single, primary ionization-tracks, acting independently of each other, are never innocuous with respect to creating carcinogenic injuries in the cells which they traverse. Every track --- without help from any other track --- has a chance of inducing cancer by creating such injuries" (1990, p.18-2).

● - (6) "... Any lesion which can be inflicted in a nucleus by a PAIR of tracks, can also be inflicted by a single track acting ALONE ... The earlier parts of this chapter leave no doubt that events [injuries] at multiple, separate sites are certainly producible by a single track, acting alone" (1990, p.19-8).

1b. What Dose in Rads Delivers an Average of ONE Track / Nucleus?

● - (7) Because a single primary track represents the least possible challenge to the repair-system in a cell-nucleus, we wanted to find out if there is solid human evidence of radiation-induced Cancer as a result of doses which deliver just one track or a few tracks per nucleus. If such evidence exists, it indicates that repair is not always perfect, even when the challenge is about as low as it can ever get. In other words, it would be DIRECT evidence that the hypothesis of a no-risk dose is false, with respect to radiation-induced cancer.

● - (8) So a necessary step in our analysis was figuring out what dose in rads (cGy) delivers an average of ONE primary track per cell-nucleus. Chapters 20, 32, and 33 in Gofman 1990 show how such doses were derived, step-by-step. The doses vary with the diameter of the cell-nucleus and with the energy of the radiation.

● - (9) The values in the box apply to cell-nuclei with an average diameter of 7.1 micrometers (p.20-3). The heading "Medical Xrays" refers to diagnostic xrays with an average energy of 30 KeV, generated when the peak kilovoltage across the xray tube is 90 KeV. The heading "596 KeV Gammas" refers to gamma rays from radium-226 and daughters. Several additional sources of radiation are evaluated in Tables 20-M and 20-"0" of Gofman 1990.

Radiation	Average number of tracks per nucleus	Tissue-dose in rads (centi-grays)
Medical Xrays	1 track	0.75 rad
	10 tracks	7.48 rads
	134 tracks	100.00 rads
596 KeV Gammas	1 track	0.34 rad
	10 tracks	3.40 rads
	294 tracks	100.00 rads

From Gofman 1990, Table 20-M.

● - (10) When the AVERAGE number of primary tracks per nucleus is one, then:

   37 percent of cell-nuclei experience no primary track at all;
   37 percent of cell-nuclei experience one primary track;
   18 percent of cell-nuclei experience two primary tracks;
    6 percent of cell-nuclei experience three primary tracks;

1.5 percent of cell-nuclei experience four primary tracks;
Half-percent of cell-nuclei experience more than four primary tracks.
(From Table 20-N of Gofman 1990).

1c. How Many Tracks at Once Can Overwhelm the Repair System?

● – (11) In our 1990 analysis, we reviewed the existing experimental evidence on what radiation doses are required to overwhelm the repair-system for genetic molecules. In Gofman 1990, p.18-4, we quote Albrecht Kellerer, one of the leading experts on the issue:

"There is, at present, no experimental evidence for a reduction of the repair capacity or the rate of repair at doses of a few gray [a few hundred rads] which are relevant to cellular radiation effects" (Kellerer 1987, p.346). And: "There is little or no evidence for an impairment of enzymatic repair processes at doses of a few gray. Studies, for example by Virsik et al on chromosome aberrations, have established characteristic repair times that are substantially constant up to 10 Gy [1,000 rads], that is, up to the highest doses investigated" (Kellerer 1987, p.358).

● – (12) We also reviewed the existing evidence on the time required to finish repair (Gofman 1990, Chapter 18). Numerous studies indicate that cell-nuclei finish whatever repair they can perform on genetic molecules within 3 to 6 hours, even after doses of 100 to 400 rads.

● – (13) "The dazzling speed of repair has an extremely important implication for settling the threshold issue. It means that certain HIGH-dose evidence can reveal a great deal, as we will explain" (Gofman 1990, p.18-5).

1d. Existing Human Evidence of Cancer from Minimal Doses

● – (14) The relevant high-dose evidence comes from studies of breast-cancer rates among women who received serial fluoroscopies in the course of pneumo-thorax treatment for tuberculosis (see entries in the Reference list for Boice 1977, Boice 1978, Boice 1981, Howe 1984, Hrubec 1989, MacKenzie 1965, Miller 1989, Myrden + Hiltz 1969).

Because the women had so many fluoroscopic exams over months and years of treatment, their breasts accumulated radiation doses ranging from about 150 rads to over 1,000 rads (Gofman 1990, Chapter 21). But each exposure delivered single doses of 1.5 to 7.5 rads at a time. Such doses deliver, respectively, an average of just 2 or 10 tracks per cell-nucleus, as we see from paragraph 9 above.

● – (15) These are very nearly the lowest POSSIBLE doses and dose-rates, with respect to challenging the repair-system in a cell-nucleus. If the repair-capacity of cell-nuclei is not overwhelmed by the tracks from hundreds of simultaneous rads (paragraphs 11 and 12, above), we can regard 10 tracks per nucleus, on the average, as nearly minimal.

● – (16) Referring to the Nova Scotia Fluoroscopy Study of female tuberculosis patients, we wrote (1990, p.21-2):

"If carcinogenic injury was produced in the irradiated women at their first fluoroscopy exposure-session, but if repair-systems were able to perform flawless repair afterwards, then that particular exposure-session would have left no residual harm, in terms of any increased risk of future breast-cancer." And:

"Similar carcinogenic injury inflicted at EVERY subsequent fluoroscopy session would also have been without residual harm, if a flawless repair-system operated at a total dose per exposure-session of 7.5 rads. And thus, after accumulating 850 rads in this fashion, the irradiated women would have had NO radiation-induced breast-cancer." And:

"The Nova Scotia Study is certainly not a high-dose study; at every critical step along the way, it is a test of how perfectly the repair-system can un-do carcinogenic injury produced by 7.5 rads, or 10 nuclear tracks on the average --- a LOW dose and dose-rate." Between exposures, ample time elapsed for completion of repair-work (paragraph 12).

● - (17) The repair-system FAILED the test, conclusively, not only in the Nova Scotia series of women, but also in additional pneumo-thorax series in Canada and in Massachusetts. The evidence of excess Breast Cancer in the fluoroscoped women is very solid, and shows a positive dose-response. This evidence of radiation-induced human cancer is widely acknowledged and cited, but not many people recognize that it shows REPAIR-FAILURE even after a challenge which was MINIMAL.

● - (18) Our disproof of any threshold dose or dose-rate includes six additional studies from the mainstream literature which show radiation-induced cancer when the average number of tracks per cell-nucleus ranged from 0.3 track to 12 tracks (Gofman 1990, Table 21-A). They are the Israeli Scalp-Irradiation Study (Modan 1977, 1989); the Stewart In-Utero Studies (1956, 1958, 1970); MacMahon's In-Utero Study (1962); the British Luminizer Study (Baverstock 1981, 1983, 1987); Harvey's In-Utero Study of Twins (1985); Modan's Study of Breast-Cancer in the Scalp Irradiation Study (1989). The evidence against any threshold embraces infants in-utero, children, adolescents, young women, high-energy gamma rays, medical x-rays, acute single doses, acute serial doses, and chronic occupational doses.

● - (19) "In recent years, it has been fashionable to suggest that epidemiologic investigations can not usefully address the low-dose radiation question. The epidemiologic studies described here make it apparent that this is incorrect ... When the effort is made to evaluate the doses in such studies, in terms of tracks-per-nucleus, then it becomes evident that studies whose doses are not 'next-to-zero' are nonetheless studies of truly minimal doses and dose-rates" (Gofman 1990, p.21-19).

1e. Failure of Repair: "The Troublesome Trio"

● - (20) It is the COMBINATION of epidemiology with track-analysis which reveals that we already know that (a) repair has failures even when the repair-system has the least possible challenge, and (b) the failure has CANCER consequences. We do not need impossible-to-obtain studies at doses like 10 milli-rads or 10 micro-rads --- because the least possible challenge to the repair-system occurs at much higher doses.

● - (21) "One can look with awe, humility, and gratitude at a system of repair with the capacities demonstrated by the DNA repair-system. But an independent analyst, or a realist of any stripe, does not casually dismiss the troublesome trio: Unrepaired lesions. Unrepairable lesions. Misrepaired lesions" (1990, p.18-6). And:

"One cannot fault the repair-system in cell-nuclei for leaving a relatively small number of injuries unrepaired, or misrepaired, or for having some inherent inability to repair every conceivable type of injury inflicted at random by the tracks of high-speed electrons ... " (1990, p.18-6).

● - (22) "... the human epidemiological evidence on dose versus cancer-response provides no support for the speculation that repair makes each rad less carcinogenic as dose falls. If that were the net result of repair, the shape of dose-response would be concave-UPWARD. But what is seen in the A-Bomb Study and in others is NOT concavity-upward. The finding is either supra-linearity or linearity --- both of which are inconsistent with the speculation that repair processes make each rad less carcinogenic as dose and dose-rate fall" (1990, p.18-6, 18-7).

● - (23) "Our entire experience with human radiation carcinogenesis should have made it evident that the problem we might be facing is that --- regardless of dose-level --- some fraction of radiation injury to nuclei is unrepaired ... some fraction is unrepairable ... and some fraction is misrepaired" (1990, p.18-7).

1f. Not "Hypothetical": Fatal Cancers from Minimal Doses

● - (24) "The radiation-induced cancers arising from the unrepaired lesions at low doses do not wear a little flag identifying them as any different from cancers induced by higher doses of radiation, or induced by causes entirely unrelated to radiation. Therefore, threshold proponents cannot argue that the cancers arising from the lowest conceivable doses of radiation will somehow be eliminated by the immune system or any other bodily defenses against cancer. Such an argument would require the elimination of cancer in general by such defenses. Instead, we observe that cancer is a major killer ... So the proposition would lead to a non-credible consequence, and must be rejected" (Gofman 1990, p.18-2).

● - (25) What about the speculation that low radiation doses may induce a net health benefit, by stimulating DNA repair or by stimulating the immune system? "When excess fatal cancer is observed in humans after such exposures [minimal doses and dose-rates], the excess has occurred DESPITE any possible stimulation of the repair- and immune-responses by low-doses. The NET result is injury, not benefit. I wish it were otherwise" (1990, p.18-2).

● - (26) "By reasonable standards of proof, the safe-dose hypothesis is not merely implausible --- it is disproven ... We conclude with a warning: Disproof of any safe dose or dose-rate means that fatal cancers from minimal doses and dose-rates of ionizing radiation are not imaginary. They are really occurring in exposed populations. Proposals, to declare that they need not be considered, have health implications extending far beyond the radiation issue ..." (1990, p.18-18).

● Part 2. UNSCEAR 1993: "Sometimes Misrepair Can Occur"

UNSCEAR 1993, written by the United Nations Scientific Committee on the Effects of Atomic Radiation, is a 922-page report (with no index) which presents a lot of valuable information and analysis. Pagination in the report is consecutive from beginning to end, but paragraph numbers start over with each annex. Below, we will separate the page number and the paragraph number by a slash.

Although authors of its nine big sections (called "annexes") are not identified, the total international membership of the Committee is identified on page 29. The biggest delegations are from Canada (9), China (7), France (9), Germany (7), Japan (11), Russian Federation (12), United States (11). Staff and consultants are identified on page 30.

● - (27) In its introduction, the report states: "The combination of epidemiology and radiobiology, particularly at the molecular and cellular levels, is a useful tool for elucidating the consequences of low doses of radiation" (1993, p.27/184). That very combination is the essence of our proof, above, that there is no threshold dose with respect to radiation carcinogenesis.

● - (28) UNSCEAR also affirms our premise in paragraph 24, when it states: "Epidemiological studies of human groups exposed to low-LET radiation show that a range of neoplasms are represented in excess and, broadly, that these do not differ markedly from those arising spontaneously in the population ... no unique neoplastic signature of human radiation exposure is, as yet, apparent" (p.578/153).

2a. The Smallest Possible "Insult" at the Cellular Level

● - (29) UNSCEAR 1993, like Gofman, recognizes the importance of using an APPROPRIATE definition of the lowest possible radiation dose or dose-rate. And it embraces our "microdosimetric approach to defining low doses and low dose rates" (p.680/321):

"Photons deposit energy in cells in the form of tracks, comprising ionizations and excitations from energetic electrons, and the smallest insult each cell can receive is the energy deposited from one electron entering or being set in motion within a cell." See paragraphs 1-4 above.

● - (30) The only conversion offered by UNSCEAR between tracks and dose in rads (centi-grays) is for cobalt-60, which produces a far more energetic gamma ray than the 596 KeV gammas presented above in our paragraph 9. Says UNSCEAR (p.680/321):

"For cobalt-60 gamma rays and a spherical cell (or nucleus) assumed to be 8 micrometers in diameter, there is an average of one track per cell (or nucleus) when the absorbed dose is about 1 mGy [0.1 cGy or rad]. The dose, corresponding to one track per cell, on average, varies inversely with volume and is also dependent on radiation quality, being much larger for high-LET radiation."

● - (31) At page 696, UNSCEAR supplies Table 17, "Proportion of a cell population traversed by tracks at various levels of track density." It is like Table 20-N in Gofman 1990. For instance, it shows what percentage of cells experience 0, 1, 2, 3, 4, and more tracks per cell-nucleus, when the average track density is ONE track per cell-nucleus. The percentages are the same as we show in paragraph 10, above.

● – (32) The UNSCEAR authors define the region of "definite" single-track action as the dose-region where not more than TWO PERCENT of the cell-nuclei experience more than a single track. "In this dose-region, there are so few radiation tracks that a single cell (or nucleus) is very unlikely to be traversed by more than one track" (p.628/42). For cobalt-60, the two-percent criterion means a tissue-dose of 0.2 mGy. Two percent is an arbitrary choice which seems completely unrelated to the repair-issue --- even though UNSCEAR agrees with us that the repair-issue is a critical part of the threshold-issue, as we will show. However, after choosing cobalt-60 and a dose of only 0.2 mGy (20 milli-rads), the UN authors are correct in saying that there are no corresponding human or animal data (p.628/42).

2b. UNSCEAR: The Carcinogenic Potency of a Single Track

● – (33) "The most basic, although not sufficient, condition for a true dose threshold is that any single track of the radiation should be totally unable to produce the effect" (p.630/54).

● – (34) "Radiation is able to induce a diversity of genomic lesions, ranging from damage to single bases to gross DNA deletions and rearrangements" (p.578/153).

And: "Biophysical analyses based on Monte Carlo simulations of track structure show clearly that all types of ionizing radiation should be capable of producing, by single-track action, a variety of damage to DNA, including double-strand breaks alone or in combination with associated damage to the DNA and adjacent proteins" (p.632/63).

And: "In all these mechanistic models, a single radiation track from any radiation is capable of producing the full damage and hence the cellular effect" (p.632/64).

● – (35) "There is compelling evidence that most, if not all, cancers originate from damage to single cells … Point mutations and chromosomal damage play roles in the initiation of neoplasia" (p.8/37).

And: "Single changes in the cell genetic code are usually insufficient to result in a fully transformed cell capable of leading to cancer; a series of several mutations (perhaps two to seven) is required … The whole process is called multi-stage carcinogenesis" (p.8/38). And: "It is possible that radiation acts at several stages in multi-stage carcinogenesis, but its principal role seems to be in the initial conversion of normal stem cells to an initiated, pre-neoplastic state" (p.8/39).

● – (36) "… the majority of neoplasms originate from damage to single cells. In principle, therefore, the traversal of a single target cell by one ionizing track from radiation has a finite probability, albeit low, of initiating neoplastic change" (p.556/26).

● – (37) Our topic here is real-world human evidence relating to the threshold-issue for radiation-induced cancer. We omit unrelated references by UNSCEAR to dose-response curves induced in various experiments, although we are interested in such experiments (see Gofman 1990, Chapter 23). With respect to the threshold-issue, we quote UNSCEAR:

"Multi-stage models of carcinogenesis could lead to expectations of a dose threshold, or a response with no linear term, under particular, highly restricted sets of assumptions" (p.636/84). But, "it would be difficult to conclude on theoretical grounds that a true threshold should be expected even from multi-stage mechanisms of carcinogenesis, unless there were clear evidence that it was necessary for more than one time-separated change to be caused by radiation alone" (p.633/69).

2c. UNSCEAR: "Sometimes Misrepair Can Occur"

A threshold-dose for radiation-induced cancer is a dose below which there is NO risk of radiation-induced cancer. A safe dose.

● – (38) As long as there are any primary tracks at all occurring in a biological tissue, a radiation dose is occurring. UNSCEAR acknowledges that "the dose and dose-rate region of main practical relevance in radiation protection (0-50 mSv per year) [0-5 rems per year] is characterized by small average numbers of tracks per cell with long intervals of time between them. Effects are, therefore, likely to be dominated by individual tracks, acting alone" (p.628/43). This is precisely the point made in Gofman 1990, p.20-7.

● - (39) "Cells are able to repair both single- and double-strand breaks in DNA over a period of a few hours, but sometimes misrepair can occur" (p.625/28).

● - (40) "The extent to which radiation-induced DNA damage may be correctly repaired at very low doses and very low dose rates is beyond the resolution of current experimental techniques. If DNA double-strand breaks are critical lesions determining a range of cellular responses, including perhaps neoplastic transformation, then it may be that wholly accurate cellular repair is unlikely even at the very low lesion abundance expected after low dose and low-dose-rate irradiation" (p.634/74).

● - (41) "It is highly unlikely that a dose threshold exists for the initial molecular damage to DNA, because a single track from any ionizing radiation has a finite probability of producing a sizable cluster of atomic damage directly in, or near, the DNA. Only if the resulting molecular damage, plus any additional associated damage from the same track, were always repaired with total efficiency could there be any possibility of a dose threshold for consequent cellular effects" (p.636/84).

● - (42) "Biological effects are believed to arise predominantly from residual DNA changes that originate from radiation damage to chromosomal DNA. It is the repair response of the cell that determines its fate. The majority of damage is repaired, but it is the remaining unrepaired or misrepaired damage that is then considered responsible for cell killing, chromosomal aberrations, mutations, transformations and cancerous changes" (p.680-681/323).

● Part 3. NRPB 1995: Evidence "Falls Decisively" against a Threshold

In October 1995, Britain's National Radiological Protection Board released a 77-page report entitled "Risk of Radiation-Induced Cancer at Low Doses and Dose Rates for Radiation Protection Purposes" (NRPB 1995). Its five authors are Cox, Muirhead, Stather, Edwards, and Little.

● - (43) Chapter 2 of NRPB 1995 reviews the existing human epidemiologic evidence and concludes (p.25/61): "It is important to note that the studies of low-LET exposure considered in this chapter are consistent with a linear trend in cancer risks at low doses without threshold." This statement embraces the pneumothorax-fluoroscopy studies (p.13/23).

● - (44) Chapter 5 of NRPB 1995 reviews "Cellular and molecular mechanisms of radiation tumorigenesis." There, the authors also state the now-familiar definition of the lowest possible dose and dose-rate from ionizing radiation:

"It may be argued ... that a single radiation track (the lowest dose and dose rate possible) traversing the nucleus of an appropriate target cell, has a finite probability, albeit low, of generating the specific damage that will result in tumour-initiating mutation" p.58/27).

● - (45) The authors consider existing evidence relating to the reduction of radiation risk by so-called cellular "adaptive" responses and immune-system responses. In particular, they discuss issues raised in UNSCEAR 1993 and in UNSCEAR 1994 (Annex B). The authors reach the same conclusion that we do: Such cellular responses do not provide any threshold dose with respect to post-repair genetic damage. NRPB concludes (p.75/21):

"Whilst adaptive responses or other protective mechanisms may influence the risk of tumour development, they do not provide a sound basis for judgement that tumorigenic response at low doses and low dose rates of radiation is likely to have a non-linear component which might result in a dose threshold below which the risk may approach zero."

3a. NRPB on Special Difficulties in Repairing Radiation Damage

The NRPB authors understand very well that imperfect repair is the key to the absence of any threshold dose. The following excerpts from their 1995 report show they understand that ionizing radiation has the power to induce some UNREPAIRABLE damage to chromosomes and DNA, and that a difference exists between action by primary ionization tracks, and action by the free radicals which are produced by normal cellular metabolism (see Appendix-C of this book).

● – (46) "Radiation-induced damage to DNA nucleotide bases and to the sugar-phosphate backbone on one strand of the DNA duplex closely resembles the cellular damage that occurs through normal endogenous metabolic processes" (p.59/28).

"It is generally accepted that, in the absence of exogenous agents, each cell in the human body sustains 5,000 to 10,000 DNA damage events per hour [they cite Ames 1989 and Billen 1990], principally as a consequence of thermodynamic instability and attack by chemical radicals produced via endogenous biochemical reactions; this damage is believed to contribute to natural cancer risk" (p.59/29).

● – (47) "On this basis, arguments have been made [they cite Billen 1990 and Abelson 1994] that the small increment of additional cellular DNA damage resulting from low dose radiation exposure will have an insignificant effect on the frequency of gene and chromosomal mutations, and by implication, on cancer risk. This would be a valid hypothesis if the DNA damage resulting from spontaneous endogenous processes were to be IDENTICAL with that induced by ionising radiation. There is, however, strong evidence that this is not the case and, consequently, that the hypothesis lacks credibility" (p.59/30).

● – (48) "The vast majority of endogenous DNA lesions takes the form of DNA base damage, base losses, and breaks to one of the sugar-phosphate backbone strands of the duplex. Such single-strand DNA damage may be reconstituted rapidly in an error-free fashion by cellular repair processes ..." (p.59/31).

● – (49) "In contrast, although a single ionising track of radiation will also induce single-strand damage when an energy-loss event takes place in close proximity to one DNA strand, a cluster of such loss events within the diameter of the DNA duplex, of about 2 nanometers, has a significant probability of simultaneously inducing coincident damage to both strands. In support of this, an approximately linear dose-response for double-strand break induction by low-LET radiation is observed, confirming that breakage of BOTH STRANDS of the duplex may be achieved by the traversal of a SINGLE IONISING TRACK and does not demand multiple-track action ..." (p.59/32). And:

"There is also evidence that a proportion of radiation-induced double-strand breaks are complex and involve local multiply damaged sites --- LMDS [they cite Ward 1991-a] ..." (p.59/32).

● – (50) "A given fraction of radiation-inducible double-strand damage will be repaired efficiently and correctly, but error-free repair of all such damage even at the low abundance expected after low dose exposure should not be anticipated" (p.60/33). And:

"Unlike damage to a SINGLE-strand of the DNA duplex, a proportion of double-strand lesions --- perhaps that component represented by LMDS --- will result in loss of DNA coding from BOTH strands. Such losses are inherently difficult to repair correctly, and it is believed that misrepair of such DNA double-strand lesions is the crucial factor underlying the induction of chromosomal aberrations and gene deletions that represent the principal hallmarks of stable mutations induced by ionising radiation of various qualities" (p.60/33). And:

"Double-strand DNA losses may in principle be repaired correctly by DNA recombination, but there is evidence that radiation-induced DNA damage may be subject to error-prone illegitimate DNA recombination which can result in the forms of gene and chromosomal mutations that are known to characterise malignant development" (p.60/33).

● – (51) "The importance of DNA double-strand damage and its repair for the radiation response of cells is further supported by studies indicating, firstly, that the repair of such damage is the principal determinant of dose and dose-rate effects after low-LET radiation and, secondly, that genetically determined cellular radiosensitivity is predominantly associated with deficiencies in DNA double-strand break repair. Finally, there is evidence that it is the difference in the QUALITY and not the QUANTITY of induced DNA double-strand lesions that principally provide for the increased biological effectiveness of high-LET radiation such as alpha particles compared with low-LET radiation such as xrays and gamma rays; these observations are best explained by experimental and computational data indicating that, overall, DNA double-strand lesions in cells induced by high-LET radiation are more complex and less likely to be repaired correctly than those induced by low-LET radiation ..." (p.60/34).

● – (52) "In summary, a coherent argument may be assembled that at low doses and low dose rates of low-LET radiation, DNA single-strand damage either is repaired in an error-free fashion or is an insignificant component of tumour risk. For double-strand DNA damage, there is good reason to believe that repair has an error-prone mutagenic component irrespective of damage-abundance and, by implication, will, even at very low doses, contribute to tumour risk" (p.60/36).

● – (53) "It may be concluded ... that existing data from both in vitro and in vivo [radiation] studies support a linear rather than a threshold-type response for neoplasia-initiating gene mutations" (p.61/38).

3b. NRPB's Conclusion on a Threshold Dose

● – (54) "It is concluded ... that data relating to the role of gene mutations in tumorigenesis, the monoclonal origin of tumours, and the relationship between DNA damage repair, gene/chromosomal mutation and neoplasia are well established and broadly consistent with the thesis that, at low doses and low dose rates, the risk of induced neoplasia rises as a simple function of dose and does not have a DNA damage or DNA repair related threshold-like component" (p.75/21). And:

● – (55) The following statement by the NRPB authors is remarkably similar to paragraph (26):

"In consideration of a broad body of relevant cellular and molecular data, it is concluded that the weight of the evidence, in respect of the induction of the majority of common human tumours, falls decisively in favor of the thesis that, at low doses and low dose rates, tumorigenic risk rises as a simple function of dose without a low dose interval within which risk may be discounted" (p.68/80).

● Part 4. Alpha Particles, Xrays, and the Major Medical Journals

The facts and logic in Parts 1, 2, and 3 above are applicable not only to xrays and other low-LET ionizing radiation, but are applicable also to high-LET ionizing radiation, such as alpha particles (see Appendix-A). Therefore, it should surprise no one that, in 1997, Hei and co-workers demonstrated that traversal of human-hamster hybrid cells, by a single alpha particle per cell, can induce structural chromosomal mutations (Hei 1997; commentary by Little 1997; see also Riches 1997). From some elegant experimental work, Hei et al report (Hei 1997, p.3765):

"Although single-particle traversal was only slightly cyto-toxic to [these] cells (survival fraction $\sim$ 0.82), it was highly mutagenic, and the induced mutant fraction averaged 110 mutants per 100,000 survivors ... These data provide direct evidence that a single alpha particle traversing a nucleus will have a high probability of resulting in a mutation and highlight the need for radiation protection at low doses."

While one chance in 1,000 per cell may not sound like "a high-probability," one must remember the PER CELL part of the finding. There are approximately 600 million typical cells in one cubic centimeter of human tissue (calculation in Gofman 1990, p.20-5). Mutation-induction by alpha-particles is explored further in Wu 1999.

Underway at NCRP: Evaluation of the Linear NonThreshold Dose-Response Model

The National Council on Radiation Protection and Measurements (NCRP is described in our Reference List) has undertaken an evaluation of the threshold issue, under its Scientific Committee 1-6 (Arthur C. Upton, Chair). Here, we do not quote from the online draft report (October 1998) because draft reports are subject to change or non-publication.

Comment: Time for a Policy-Change

In view of Parts 1, 2, and 3 of this Appendix, we urge that editors and reviewers at the major medical journals challenge any submitted paper which uncritically incorporates the safe-dose fallacy. With respect to mutagenesis and carcinogenesis by xrays and other classes of ionizing radiation, the epidemiologic and experimental evidence "falls decisively against" any safe dose (risk-free dose).

APPENDIX-C
The Free-Radical Fallacy about Ionizing Radiation:
Demonstration that a Popular Comparison Is Senseless

Part 1. Does "Just Living" Hurt DNA More Seriously Than Ionizing Radiation?
Part 2. The Relative Frequency of DNA Damage-Events
Part 3. Reality-Check for the "Same-Nature" Assumption
Part 4. The Unique Power of Ionizing Radiation

Free radicals are highly reactive molecules possessing an unpaired electron. In cells, such radicals can do injury (for instance, oxidative damage) to proteins and other molecules --- including injury to the DNA molecules which encode the human genes.

● **Part 1. Does "Just Living" Hurt DNA More Seriously than Ionizing Radiation?**

1a ● In some peer-review journals and various interviews in the media, what we call the Free-Radical Fallacy has been employed in order to belittle the health menace of low-dose xrays, gamma rays, and beta particles (low-LET ionizing radiation). We will demonstrate the nature of the fallacy in Part 3, below.

1b ● There is no doubt that routine metabolic chemistry in each cell produces, every hour, legions of free-radicals and consequent DNA damage-events in the process of "just living." And there is no doubt that exposure, to a small dose of low-LET ionizing radiation, adds relatively a very small number of DNA damage-events to irradiated cells.

1c ● In 1990, Dr. Daniel Billen of the Oak Ridge Associated Universities proposed: "It would seem reasonable to conclude that, due to common oxidizing radicals, many of the qualitative changes in DNA are quite similar for radiation-induced or spontaneous DNA damage" (Billen 1990, p.243). Having assumed a qualitative equivalence, Billen concentrated on comparing the NUMBER of DNA damage-events per cell caused by natural intrinsic processes ("just living") versus the much lower number caused by small doses of low-LET ionizing radiation. And this type of comparison has become a refrain which is frequently incorporated, these days, into attempts to calm concerns about medical radiation and nuclear pollution.

1d ● Part 3 demonstrates why such comparisons are fatally flawed and senseless. In short, a reality-check demonstrates that the nature of DNA damage from ionizing radiation and the nature of DNA damage from intrinsic processes cannot possibly be qualitatively equivalent.

● **Part 2. The Relative Frequency of DNA Damage-Events**

2a ● Billen (1990, p.242) cites various mainstream sources for two estimates: (1) "Approximately 10,000 measurable DNA modification events occur per hour in each mammalian cell due to intrinsic causes," and (2) "About 100 (or fewer) measurable DNA alterations occur per centi-Gray of low-LET radiation per mammalian cell."   These two values are made comparable in Part 2d, below.

2b ● The goodness of both estimates, above, will surely improve a great deal with future methods of measurement, but neither Billen's presentation nor refutation of its key assumption depends on precision in these two values.

2c ● Billen states his conclusion (p.242): "Therefore, every HOUR, human and other mammalian cells undergo at least 50-100 times as much spontaneous or natural DNA damage as would result from exposure to 1 centi-Gray of ionizing radiation." Centi-Gray and "rad" are two names for the same amount of radiation exposure. How much is one rad of exposure?

2d ● On the average, it takes about 10 years for a person to accumulate one rad of whole-body exposure from natural background radiation. So Billen's numbers mean that the ratio of damage-events PER UNIT OF TIME (per hour, or per day, or per year) may be as large as **8.8 million** endogenous damage-events for each damage-event due to natural background radiation. A very large difference ... but is it meaningful?

2e ● The estimates presented by Billen of "DNA modifications" and "DNA alterations" are estimated numbers PRIOR to repair-work by the cell. Here (and elsewhere in the literature), the term "damage-events" is preferred, to signal that the event is not necessarily an unrepairable PERMANENT mutation of the DNA.

2f ● Billen's arithmetic is correct, but a reality-check is needed for his assumption that the nature of DNA damage-events is the same from routine cellular metabolism and from ionizing radiation.

● **Part 3. Reality-Check for the "Same-Nature" Assumption**

3a ● According to Billen (2a), a rad (centi-Gray) causes about 100 or fewer measurable DNA damage-events per cell.

3b ● According to Billen (2b), the number of comparable damage-events from intrinsic causes per cell, every HOUR, is 50 to 100 times higher, which means 5,000 to 10,000 damage-events every hour from intrinsic causes, per cell. (Bruce Ames 1995, p.5259, provides an estimate per DAY, not per hour: "The number of oxidative hits to DNA per cell per day is estimated to be about 100,000 in the rat and roughly ten times fewer in the human." We will include this estimate in Point 3e.)

3c ● It follows from Billen that per DAY, the DNA damage-events per cell from endogenous causes are either:

(5,000 events/hr) x (24 hr/day) = 120,000 events/day, or:
(10,000 events/hr) x (24 hr/day) = 240,000 events/day ... in each cell.

3d ● And something else follows from Billen's assumption that there is no important difference between the endogenous and the radiation-induced damage-events. If correct, then the DNA-based consequences from a radiation dose which delivers 120,000 or 240,000 damage-events each day, per cell, should be the same as from 120,000 or 240,000 such events per cell each day, from endogenous sources.

3e ● The whole-body radiation dose per day required (by Billen's numbers) to deliver 120,000 to 240,000 such DNA damage-events per cell, each day, would be either:

(120,000 events) x (1 rad/100 events) = 1,200 rads, or:
(240,000 events) x (1 rad/100 events) = 2,400 rads. And:

If we substitute Ames' figure (from Point 3b), we would calculate (10,000 events) x (1 rad/100 events) = 100 whole-body rads per day to deliver DNA damage equivalent to daily damage from intrinsic causes.

Bottom Line of the Reality-Check

3f ● If there were equivalence between DNA damage from normal, intrinsic processes and DNA damage from ionizing radiation, then whole-body doses of 100 rads to 2,400 rads per day EVERY day would be easily tolerated. Instead, such doses are promptly LETHAL.

3g ● For half the humans exposed, promptly-lethal doses are estimated by the radiation community at 300 or 400 whole-body internal-organ rads accumulated in one week or less (NCRP 1989-b, p.70, p.73).

3h ● There is an additional observation worth noting. Unrepaired and misrepaired chromosomal (DNA) injuries are widely accepted as a cause of Cancer. The background rate of clinical Cancer is estimated (largely on the basis of the Atomic-Bomb Study) to be doubled by extra radiation doses of a few hundred whole-body rads of non-xray exposure. Suppose (for illustrative purposes) that 300 whole-body rads were required in order to double the background rate of clinical Cancer. According to Billen (2a), 300 rads of low-LET radiation would cause about 30,000 or fewer DNA damage-events per cell. But 30,000 damage-events per cell would be far exceeded by intrinsic processes in a single day (Billen, 3c) or in ten days (Ames: 10,000 per day * ten days). If DNA damage from intrinsic causes and from low-LET ionizing radiation were equivalent, it is hard to see how anyone could escape having MULTIPLE clinical Cancers from intrinsic processes.

3i • From these two reality-based observations (acute lethal doses and doubling-doses for radiation-induced Cancer), we have demonstrated that the nature of damage caused by ionizing radiation CANNOT POSSIBLY BE THE SAME as it is from normal metabolic processes and oxidative damage. Without an equivalence, the Billen argument and its variations collapse. The Free-Radical Refrain is a Free-Radical Fallacy.

● Part 4. The Unique Power of Ionizing Radiation

4a • The difference between free-radical damage from routine metabolism and from ionizing radiation almost surely lies in REPAIRABILITY. If DNA damage is perfectly repaired by a cell, such damage has no health consequences. It is inconsequential. The consequences arise only from injuries which are non-repairable or mis-repaired.

4b • The demonstration in Part 3 supports other evidence (and vice versa) that ionizing radiation can induce the special kinds of complex DNA damage which CANNOT BE PERFECTLY REPAIRED. A leading figure in this research is John F. Ward; see Reference List.

4c • The power of ionizing radiation to induce the complex injuries is not in dispute. Billen himself appears to acknowledge it, but then to ignore it (Billen 1991, p.388).

4d • The power of ionizing radiation to induce particularly complex and unrepairable genetic injuries is surely related to a UNIQUE PROPERTY of this agent. Ionizing radiation instantly unloads biologically abnormal amounts of energy at random in an irradiated cell. Biochemical reactions in a cell generally involve net energy-transfers in the ballpark of 10 electron-volts and below. By contrast, Ward reports (1988, p.103) that the average energy-deposit from low-LET ionizing radiation is thought to be about 60 electron-volts, all within an area having a diameter of only 4 nanometers. (The diameter of the DNA double-helix is 2 nanometers). In other words, ionizing radiation produces violent energy-transfers of a type simply absent in a cell's natural biochemistry.

4e • Because of its unique property, ionizing radiation is a unique menace to our DNA and chromosomes. This fact deserves wide recognition, as mankind learns that FAR more health problems are mutation-based than anyone could prove 15 years ago.

>>>>>>>>>>

Part 1. Genomic Instability: "One of the Hallmarks of the Cancer Cell"
Part 2. A Deep Insight from 1914, Slowly Confirmed
Part 3. Ionizing Radiation (Xrays Included): A Cause of the Instability
Part 4. Implications: Curing versus Preventing Cancer
Part 5. Five Key Facts and Three Moderate Comments

Box 1. Mini-Glossary

● Part 1. Genomic Instability: "One of the Hallmarks of the Cancer Cell"

Genomic instability --- also called "genetic instability" and "chromosomal instability" --- refers to abnormally high rates (possibly accelerating rates) of genetic change occurring serially and spontaneously in cell-populations, as they descend from the same ancestral cell. Some additional terms are discussed at the end of this Appendix, Box 1.

By contrast, normal cells maintain genomic STABILITY by operation of elaborate systems which ensure accurate duplication and distribution of DNA to progeny-cells (Cheng 1993, p.124), and which prevent duplication of genetically abnormal cells. These systems ("metabolic pathways") involve an estimated 100 genes (Cheng 1993, p.142).

Why is genomic instability so important? Many (not all) cancer biologists now believe that genomic instability "not only initiates carcinogenesis, but also allows the tumor cell to become metastatic and evade drug toxicity" (Tlsty 1993, p.645), and "The loss of stability of the genome is becoming accepted as one of the most important aspects of carcinogenesis" (Morgan 1996, p.247), and "One of the hallmarks of the cancer cell is the inherent instability of its genome" (Morgan 1996, p.254).

Although such observations are far from new (Part 2), they certainly did not receive the attention which they merit until recently.

● Part 2. A Deep Insight from 1914, Slowly Confirmed

It was the year 1956 when the normal number of human chromosomes per cell was firmly established as 46. Soon thereafter, it became clear that cells of advanced Cancers have often evolved an abnormal number of chromosomes ("aneuploidy").

1914: Theodor Boveri's Great Insight

Such observations were consistent with the prediction of Theodor Boveri (Boveri 1914), a great German embryologist who postulated that malignancy is the result of inappropriate balance of instructions (genetic information) in the tumor cells. Such "imbalance" can result not only from numerical chromosome aberrations, but also from structural alterations within the 46 chromosomes. As a leading cause of structural chromosome aberrations (deletions, acentric fragments, translocations, inversions, dicentrics, etc.), ionizing radiation is well-established.

When my colleagues and I initiated a research program in 1963 (at the Atomic Energy Commission's Livermore National Laboratory), to test Boveri's hypothesis, there was very little interest in the concept. Although the techniques for detecting structural chromosome aberrations were extremely crude then, compared with current techniques, we were making gradual progress (Minkler 1970, + Minkler 1971). However, the Atomic Energy Commission became angry with me after a paper I presented at an IEEE Symposium (Gofman 1969-b), and canceled our funding in the early 1970s (Seaborg 1993, Chapter 8, "Challenge from Within," + Terkel 1995, pp.406-408).

1976: Peter C. Nowell's Classic Paper on Tumor Evolution

In October 1976, the journal Science published Peter C. Nowell's classic paper entitled, "The Clonal Evolution of Tumor Cell Populations" --- a paper almost always cited by today's analysts of genomic instability. Among other things, Nowell's 1976 paper discussed evidence, from various analysts, indicating that as tumor cells become increasingly aneuploid, the malignancy becomes increasingly aggressive (Nowell, p.25). Reasoning from the available evidence at that time, Nowell proposed the following model of multi-step carcinogenesis:

Tumor initiation occurs by an induced change in a single, previously normal cell, which makes the cell "neoplastic" (partially liberated from normal growth controls) and provides the cell with a selective growth advantage over adjacent normal cells (Nowell, p.23).

"From time to time, as a result of genetic instability in the expanding tumor population, mutant cells are produced ... Nearly all of these variants are eliminated, because of metabolic disadvantage or immunologic destruction ... but occasionally one has an additional selective advantage with respect to the original tumor cells as well as normal cells, and this mutant becomes the precursor of a new predominant subpopulation" (Nowell, p.23). And:

"Over time, there is sequential selection by an evolutionary process of sub-lines which are increasingly abnormal, both genetically and biologically ... Ultimately, the fully developed malignancy as it appears clinically has a unique, aneuploid karyotype associated with aberrant metabolic behavior and specific antigenic properties, and it also has the capability of continued variation as long as the tumor persists" (Nowell, p.23). And:

"The major contention of this article is that the biological events recognized in tumor progression represent (i) the effects of acquired genetic instability in the neoplastic cells, and (ii) the sequential selection of variant subpopulations produced as a result of that genetic instability" (Nowell, p.25).

The recent surge of interest in genomic instability reflects the recognition that the cancer process represents a trip (or set of trips) from the stable genome to the genome with diverse deviations. It has been a long wait for Boveri.

● Part 3. Ionizing Radiation (Xrays Included): A Cause of Genomic Instability

Today, laboratory researchers are performing reality-checks on this logic: Genomic instability can be initiated and intensified by any type of genetic mutation (including chromosome aberrations), when such mutation alters some of the DNA which maintains genomic STABILITY. Of course, such DNA includes the numerous DNA segments which govern DNA synthesis, cell-division, and also the routine REPAIR of the genome --- the "repair genes" (Cheng 1993, p.131; Morgan 1996, p.248).

When a mutagen has induced genomic instability in a cell, some of the cell's descendants will experience new and unrepaired genetic abnormalities at an excessive rate, even though the descendants themselves received no exposure to the mutagen used in the experiment. This occurs because such cells have inherited a genome which was injured with respect to maintaining genomic STABILITY.

Aneuploidy, Deletions, Gene-Amplifications

Very recently, a technique has been developed for efficiently detecting three of the types of chromosome aberrations which are very prominent in genomic instability: Aneuploidy (wrong number of chromosomes), deletions (permanent removal of DNA segments, long or short), and gene-amplifications (extra copies of specific DNA segments). This technique, called Comparative Genomic Hybridization, was first described by Kallioniemi (1992, in Science). However, such a technique does not detect many other kinds of mutations.

The nature of the genetic code is such that mutations need not be gross in order to have gross biological consequences. For instance, permanent removal of a single nucleotide (a micro-deletion) can totally garble much of a gene's code, by causing what is called a "frame-shift." Then this non-functional gene can be the phenomenon which wrecks part of the system which would otherwise maintain genetic STABILITY.

Amplification (instead of injury), of the crucial genes in the stability-system, also can permit a cell to escape the controls which otherwise prevent duplication of cells with injured genomes. Evidence is developing that gene amplification is associated with dicentric chromosomes and circular acentric fragments called "double minutes" (DiLeonardo 1993, p.656) --- very well-known products among the consequences of ionizing radiation.

The sequence, in which various mutations accumulate in tumor cells, may or may not matter. "For example, one or more pre-cancerous mutations might lie dormant until additional mutations create an environment in which the prior changes confer a selective advantage" (DiLeonardo 1993, p.655, citing Kemp 1993, + Fearon 1990, + Temin 1988).

Confirmations: Ionizing Radiation Is a Cause of Genomic Instability

The fact, that ionizing radiation is a mutagen capable of causing all known types of genetic mutation --- from micro to gross, at any DNA location along any chromosome --- made it utterly predictable that ionizing radiation would be a cause of genomic instability. Indeed, one of the last projects completed by our research group at the Livermore Laboratory, before the Atomic Energy Commission shut down our work, was a demonstration which showed that ionizing radiation can induce genomic instability. Our experiments used gamma rays and cultured human fibroblasts (Minkler 1971).

During recent years, multiple experiments have confirmed the fact that ionizing radiation can cause genomic instability. Such results have been observed after both low-LET radiation (such as xrays and gamma rays) and high-LET radiation (such as alpha particles). Among numerous papers, see, for instance:

Kadhim 1992;
Holmberg 1993 (who cites Minkler 1971);
Marder 1993 (especially p.6674);
Mendonca 1993;
Kadhim 1994;
Kronenberg 1994 (radiation dose-response, p.605);
Kadhim 1995;
Morgan 1996 (review).

No Discontinuity between Cause and Effect

In the mass media, some writers have expressed astonishment that radiation-induced genomic instability is not detected until several cell-divisions have occurred after the radiation exposure. They seem to imagine that the delay reflects a mysterious discontinuity between cause and effect. There is NO discontinuity, of course --- a point made explicitly in Kadhim 1992 (p.739). With current techniques, and with uncertainties about where to search closely among a billion nucleotides, it is just not possible to detect every intermediate step.

● Part 4. Implications: Curing vs. Preventing Cancer

The induction of genomic instability in a cell does not guarantee that it will become malignant. Genomic instability increases the RATE of mutation in that cell and its descendants, and with this higher rate, the cells each have a higher PROBABILITY that at least one of them will accumulate all the genetic powers of a killer-cancer. These powers include the ability to thrive BETTER than normal cells, to invade inappropriate tissue, to adapt to the new conditions there, to recruit a blood supply, to fool the immune system, and many other properties.

No one claims, yet, that genomic instability must precede every case of Cancer. However, genomic instability helps to explain why Cancer is sometimes called "at least a hundred different diseases." Indeed, genomic instability means that each case of Cancer may develop a genome like no other case. Is it any wonder that individual tumors often differ in behavior from each other?

Nowell's 1976 paper was certainly not the last one to observe that Cancers become increasingly deviant in their genomes, as they "advance." Tlsty 1993 (p.645) cites several more recent papers. Near the end of his paper, Nowell wrote (p.27):

"The fact that most human malignancies are aneuploid and individual in their cytogenetic alterations is somewhat discouraging with respect to therapeutic considerations ... With variants being continually produced, and even increasing in frequency with tumor progression, the neoplasm possesses a marked capacity for generating mutant sub-lines, resistant to whatever therapeutic modality the physician introduces ... The same capacity for variation and selection which permitted the evolution of a malignant population [of cells] from the original aberrant cell, also provides the opportunity for the tumor to adapt successfully to the inimical environment of therapy, to the detriment of the patient."

And Some Lessons:

(A) • Recognition, that genomic instability constitutes a serious obstacle to curing Cancer, has stimulated strategies to evade the problem --- perhaps by preventing a tumor from acquiring its necessary blood-supply. For some 30 years now, Judah Folkman has been an admirable pioneer in researching this aspect of angiogenesis.

(B) • The quickest path to less cancer-misery in the future would be a policy of reducing exposure to carcinogens.

(C) • Ionizing radiation is almost certainly the most potent carcinogen to which vast numbers of people are actually exposed (see Part 5).

• Part 5. Five Key Facts and Three Moderate Comments

(1) • Ionizing radiation is a mutagen having special properties which make some radiation-induced genetic injuries complex and impossible for a cell to repair correctly --- quite unlike the routine damage from endogenous free radicals (Appendix-B, Part 3a, + Appendix-C).

(2) • Ionizing radiation is a mutagen which undeniably can cause every known kind of mutation, at any DNA location along any chromosome. The body does not always eliminate cells having harmful mutations. If it did, there would be very little Cancer --- and no inherited afflictions.

(3) • Ionizing radiation is a mutagen known to induce genomic instability (Parts 2 and 3, above).

(4) • Ionizing radiation is a human carcinogen at every dose-level, not just at high doses; there is no threshold dose. A single photon or a single high-speed particle can result in unrepairable genetic damage (Appendix-B).

(5) • Ionizing radiation is a mutagen observed to induce virtually every kind of human Cancer (Chapter 2, Part 4c).

And the Comments:

(1) • In view of all the five facts above, it would be inappropriate to doubt the health-menace of low-dose ionizing radiation.

(2) • And in view of all the five facts, it is strange --- in studies which attempt to explain a difference in cancer-rates between two groups --- that the question is so seldom asked: How do the radiation histories differ between the groups? In view of the five facts above, it should be the FIRST question.

(3) • And in view of the five facts, it might be appropriate for the American Medical Association, the National Cancer Institute, the American Cancer Society, the American College of Radiology, the American Society of Radiologic Technologists, and dozens of similar organizations, to insist that uselessly high exposures to ionizing radiation, in pre-cancer medical procedures, be eliminated. The ways are known (Chapters 1 and 2).

Today, the two largest sources of voluntary radiation exposure are (i) pre-cancer medical procedures, including CT scans and fluoroscopy (NCRP 1987, p.59, + NCRP 1989, p.69) and (ii) cigarette smoking --- which delivers appreciable alpha-particle radiation to the lungs (Chapter 48, Part 1c). As for involuntary exposures accumulated from nuclear pollution, they have been poorly ascertained --- to put it in a kindly fashion.

Box 1 of Appendix-D

Mini-Glossary

● GENOME. A person's genome is one set of his (or her) genes. The human genes, which control a cell's structure, operation, and division, are located in the cell's nucleus. The full human genome (estimated at 50,000 to 100,000 genes) is present in every cell-nucleus, even though many genes are inactive in cells which have specialized functions (the "differentiated" cells).

● GENES AND CHROMOSOMES. Genes are composed of segments of DNA. In normal cell-nuclei, the DNA is distributed among 46 chromosomes (23 inherited at conception from a person's father, and 23 from the mother). Each chromosome consists of one very long strand of DNA and numerous proteins, which are required for successful management of the long DNA molecule. The longest chromosomes each "carry" thousands of genes. Every time a cell divides, the cell must duplicate the 46 chromosomes and must distribute one copy of each to the two resulting cells.

● THE CODE. The DNA of each chromosome is composed of units --- "nucleotides" of four different types (A, T, G, C). These nucleotides are linked to each other in linear fashion. The sequence of the four types of nucleotides is critical, because the sequence produces the "code" which (a) determines the function of each particular gene, (b) identifies the gene's start-point and stop-point along the DNA strand, and (c) permits certain regulatory functions. The code of the human genome consists of more than a billion nucleotides.

● THE MITOCHONDRIAL DNA (mtDNA). Outside the nucleus, human cells also have some "foreign" DNA located in structures called the mitochondria. This small and separate set of DNA does not participate in the 46 human chromosomes, and is not part of "the genomic DNA." The mitochondria are inherited from the mother.

Part 1.   Some Remarkable Rabbits Who Made a Lasting Impression
Part 2.   An Encouraging Piece of Luck, As We Began Human Case-Control Studies
Part 3.   How Do Sf 0-400 Measurements Vary with Food Intake?
Part 4.   From Person to Person, Do the Sf 6-8 Lipoproteins Have the Same Lipid Constituents?
Part 5.   Molecules in a Metabolic Chain? The Steady-State Concentrations
Part 6.   How Do Lipoproteins Move from an Artery's Lumen to Its Intima?
Part 7.   Why We Should Identify Non-IHD Disorders with Aberrant Lipoprotein Patterns
Part 8.   A "Smoking Gun": The Reversal of Xanthomas by Lipid-Lowering Regimes
Part 9.   Early Evidence of Independent Atherogenicity, Sf 12-20 vs. Sf 20-100 Lipoproteins
Part 10. Segregation of Infarcts from Normals: Sf-Class vs. Total Cholesterol
Part 11. Do Elevated Serum Lipoproteins Precede or Follow an Infarct?
Part 12. Inclusion of Std Sf 0-12 and Std Sf 100-400 in Two Prospective Studies

Box 1.   Comparison of 5 Individuals, Regarding Lipid Composition of Sf 6-8 Lipoproteins.
Box 2.   Comparison of 9 Individuals: Lipid Composition of Sf 0-20 and Sf 20-400 Lipoproteins.
Box 3.   Aberrant Lipoprotein Levels in Xanthoma, Nephrosis, Biliary Obstruction, Myxedema, etc.
Figure E-1. Evidence of the Independent Atherogenicity of Sf 12-20 and Sf 20-100 Lipoproteins.
Figure E-2. Segregation of Infarcts from Normals, by Sf 12-20 Levels Independent of Cholesterol.
Table E-1. The Matched-Series Method of Assessing Indpendent Contribution to Atherosclerosis.

The Lipid Hypothesis of Atherosclerosis and IHD proposes that entrance of certain types of plasma lipoproteins into the intima, from the circulating blood, is the initial step in the whole atherosclerotic process. Between the endothelium and the internal elastic membrane, the plasma lipoproteins (oxidized, or not oxidized) have no physiologic function. They are "out of place," and unless they exit rapidly, they become a "foreign substance" which elicits efforts at removal or isolation --- the inflammatory response (Chapter 44, Part 3).

The hypothesis suggests that, "all other things being equal," the higher is the concentration of the pathogenic lipoproteins in the bloodstream, the greater will probably be their infiltration of the intima. This, then, suggests that the important measure at issue is the blood-level of the culprit lipoproteins, "operating" over time. Of course, back in 1949, all this remained to be established.

● Part 1. Some Remarkable Rabbits Who Made a Lasting Impression

In medical school, I had been quite impressed by the work of Anitschkow (1933) on the cholesterol-fed rabbit. Dietary cholesterol is an exotic substance to rabbits, who are herbivorous animals. The work showed induction of massive elevation in the blood-levels of cholesterol, AND it showed induction of an atherosclerosis-like picture in the rabbit's large arterial blood vessels. In academic circles, the results were disparaged by some professors. How, they asked, could feeding cholesterol to rabbits teach us anything useful, when the human body needs and synthesizes cholesterol on its own (regardless of diet)?

They seemed unable to look past the FEEDING aspect of the work. But if one looks past WHY those rabbits had high blood-levels of cholesterol, one is staring at strong evidence that elevated blood-levels appeared to cause DEPOSITION of cholesterol in the rabbits' arteries.

It seemed quite unreasonable to me that anyone would refuse to look further into the question: "How does the blood level of cholesterol come to be related to the atherosclerotic-like deposits in arterial walls of rabbits, and could a similar mechanism be causing atherosclerosis in humans?"

At that time (1947), two classes of plasma lipoproteins (alpha and beta) had been identified, and there was also some human evidence that in certain families, arterial disease was related to the level of blood cholesterol. But for the general population, the findings were said to be very uncertain (Chapter 44, Part 3b).

Two new sets of rabbits suddenly made the problem irresistible.

In 1948 and 1949, G. Lyman Duff and Gardner McMillan reported on the development of atherosclerosis in rabbits with alloxan-induced diabetes. The results were remarkable: Compared with non-diabetic rabbits which were fed a diet equivalently high in cholesterol, the alloxan-diabetic rabbits developed equal or higher blood-levels of cholesterol, but markedly LESS atherosclerosis. This contrast was intriguing in its own right --- and all the more so because, among humans, the diabetics are MORE prone to atherosclerosis than non-diabetics.

Duff and McMillan made the signal observation that an excessive, visible lipemia (producing a cloudy, creamy look) characterized the blood of the alloxan-diabetic rabbits. These investigators surmised that the physical state of the cholesterol in such blood might have been altered in the alloxanized animals, and that the alteration might account for the lower degree of atherosclerosis.

Atherogenicity: Indication that Size Matters, among Lipoproteins

At the Donner Laboratory, we pursued their speculation --- with our new capability of identifying the spectrum of lipoproteins in their NATIVE states (Chapter 44, Parts 3b + 3g). In the early 1950s, with new sets of rabbits, our group at Donner demonstrated that aortic atherogenesis, at least in the rabbit, was strongly related to the serum level of certain lipoproteins of a limited size-range, and NOT to lipoproteins above or below that size-range (Gofman 1950-a, + Jones 1951, + Pierce 1952).

Even massive elevation, in concentration of lipoproteins larger than a certain size, resulted in little or no incremental atherosclerosis in rabbits. Our interpretation: Probably the larger molecule's size prevented entry into the rabbit's aortic intima. With additional sets of rabbits, Frank Pierce (1952, Tables 2 and 4) confirmed that alloxan-diabetic, cholesterol-fed rabbits on the average developed a much lesser degree of atherosclerosis than normal non-alloxanized non-diabetic rabbits, fed a comparable high-cholesterol diet.

MOST IMPORTANTLY, Pierce (1952, Table 3) showed that the diabetic cholesterol-fed rabbits transported their massive concentrations of cholesterol mainly in large lipoproteins having Sf values above 100. By contrast, normal non-diabetic cholesterol-fed rabbits who develop atherosclerosis, carry the cholesterol mostly in smaller lipoproteins of the Sf 12-30 range.

The various sets of rabbits left us with a strong expectation of a causal relationship in HUMANS between some classes, but not all classes, of plasma lipoproteins and atherosclerosis.

"Man Is Not a Rabbit!" The Necessity of Human Data

Of course, lipid-induced atherosclerosis in rabbits can never establish causal relationships for HUMANS. After all, "Man is not a rabbit!" From the start, it was evident to us that such questions had to be studied in humans. Prospective studies are the most reliable, but by their very nature, they require years for their completion. Meanwhile, our group undertook case-control studies. A question demanded intense investigation by all possible routes:

> "Could phenomena, similar to the observed phenomena in
> rabbits, explain atherogenesis in humans in the coronary, carotid,
> peripheral, and cerebral arteries?"

● Part 2. An Encouraging Piece of Luck, As We Began Human Case-Control Studies

Very early in our work at Donner, we began to analyze samples of blood from clinically healthy humans. It was clear that the lipoprotein spectrum in humans was VASTLY larger than the range in normal rabbits on their normal diet. Such rabbits have serum lipoproteins in the Sf 5-8 range (Gofman 1950-a, p.168). Analysis of the blood from healthy young persons (18-30 years of age) revealed lipoproteins in the Sf 6-8 range --- very much like the normal rabbits. But at older ages, the human spectrum grew very much broader.

Where, in the lipoprotein spectrum of humans, should we BEGIN our studies of a possibly causal role of lipoproteins in atherogenesis? We could not simultaneously develop reliable laboratory techniques with the new methods, and also study everything at once.

In those early days at Donner, we knew (from the work of others) that feeding cholesterol to normal non-diabetic rabbits would produce atherosclerosis in a few months. As our Donner rabbits progressed in the cholesterol feeding program, the blood levels of Sf 5-8 lipoproteins increased. After an even longer time on cholesterol-feeding, an additional peak appeared in their ultracentrifugal flotation patterns (Chapter 44, Box 2). The rabbits' additional peak, in the Sf 10-20 region, announced the development of an additional set of lipoproteins circulating in the rabbits' blood --- while atherosclerosis was developing (confirmed at autopsy).

Luck: Our First Look at the Blood of a Heart-Attack Survivor

Soon after the Sf 10-20 lipoproteins had developed in our rabbits, we decided to find out what the blood looked like in at least one proven case of myocardial infarction. We obtained a blood sample from a lady several weeks after her heart attack. We spun her serum sample ... with immense curiosity.

Her serum lipoprotein pattern looked just like the pattern in rabbits, after several weeks of cholesterol-feeding: A fairly high concentration of lipoproteins of the Sf 0-10 class AND an appreciable elevation of lipoproteins of the Sf 10-20 class.

As we would soon learn, NOT all heart-attack survivors have this pattern. But the fact, that the very first blood-sample from a human heart-attack survivor matched blood from the rabbits who were developing experimental atherosclerosis (confirmed at autopsy), was a piece of good luck which encouraged us to believe that our investigations were definitely worth pursuing.

2a. Results from Our First Case-Control Studies

As a result of this event, we chose serum concentrations of the Sf 10-20 segment of the lipoprotein spectrum, as the parameter for comparison in our first case-control study of humans. In a series of 104 cases of confirmed myocardial infarction, we found elevated concentrations of the Sf 10-20 lipoproteins, compared with controls who had no overt coronary disease (Gofman 1950-a).

By this time also, we had analyzed enough blood samples from clinically healthy males and females, from ages 20 to 40 and from 40 to 70, to perceive the significantly higher mean concentrations of the Sf 10-20 lipoproteins in males than in females before age 40, and to perceive that beyond age 40, concentrations increase in both sexes --- with the females catching up to the males. We noted (Gofman 1950-a, pp.170-171) that these findings --- together with the observation that men (on the average) have a higher fequency of atherosclerosis before age 40 than do women, and that the frequency of atherosclerosis increases with age in both sexes --- were consistent with the hypothesis that the Sf 10-20 lipoproteins are atherogenic.

The parallelism of our rabbit and early human studies created the suspicion that there might be something very special about the Sf 10-20 lipoproteins, and that whatever those special properties might be, they might also cause atherogenesis.

Studying More of the Spectrum

As soon as we were able, we added the Sf 20-100 segment of lipoproteins to our case-control studies. By 1951, comparison of patients having coronary disease, versus normals, indicated that the Sf 20-100 lipoproteins in humans are also and independently atherogenic (Jones 1951, + Gofman 1952-a, pp.125-126 + Gofman 1952-c, p.286). By mid-1953, such findings were based largely on serum lipoprotein measurements of 239 males with clinical Coronary Heart Disease, versus 740 males of corresponding ages (40-59 years) without overt CHD (Gofman 1953).

2b. Correction of an Error Concerning the Sf 0-12 Lipoproteins

From ultracentrifugal work by others, we were aware that migration-rates can be affected by the concentrations of various molecules in a sample (Gofman 1950-a, p.168). Concentration affects rates in two ways: By self-slowing of molecules and by interaction between molecules.

Before long, we realized that our flotation-technique underestimated the concentration of especially the Sf 0-12 lipoproteins --- and increasingly so, with increase in their true serum levels. The result was that the net difference in Sf 0-12 lipoprotein levels, between persons with overt CHD and persons without overt CHD, was underestimated. Hence, we erroneously reported that the Sf 0-12 segment appeared NOT to be atherogenic (Gofman 1952-a, p.122).

During 1952, this error was eliminated from our work by making adjustments which took into account the effects of concentration on the flotation rates, not only of the Sf 0-12 lipoproteins but also on the segments of the lipoprotein spectrum above Sf 12 (technical details in DeLalla 1954-a; discussion in Co-op 1956, p.696, pp.714-715). The improved methodology became our routine practice in 1952 and thereafter, as did the use of "Std Sf" (Standard Sf or, in certain journals, Sf with a superscript of zero) to indicate that measurements were properly adjusted for the effects of concentrations.

With the error corrected, the Std Sf 0-12 lipoproteins were also recognized as atherogenic (Gofman 1953).

Decision Not to Pursue Our Ultracentrifugal Studies above Sf 400

Of course, the entire lipoprotein spectrum needed evaluation, for any relationship with atherogenesis in humans. However, for technical and biological reasons, we soon decided to limit our ultracentrifugal studies of the "Low-Density Lipoproteins" to the Sf 0-400 segment. High variability of lipoprotein findings near and above Sf 400, with respect to meals, convinced us that lipoproteins above Sf 400 would need to be studied in a different manner (Glazier 1954, p.396).

Evaluation of the Lipid Hypothesis in a Rigorous Manner

Case-control studies, excellent for guidance in some general directions, do not suffice for evaluation of the Lipid Hypothesis in a rigorous manner. A prospective study was initiated in 1950 (Part 11, below). And many additional lines of inquiry were required --- some of which are described in Parts 3 through 10, below.

● Part 3. How Do Sf 0-400 Measurements Vary with Food Intake?

If one studies the parameter, "blood levels of the Sf 0-12, 12-20, 20-100, and 100-400 lipoproteins," in relationship with Coronary Heart Disease, one would like to know the nature and extent of variation in those measurements within every individual studied, both acutely over daily cycles and chronically during decades of life. Although this goal is not attainable, one does what one can. Indeed, like many investigators, we began with measurements of the most easily controlled persons: Ourselves. The team at Donner had some pretty sore arms from sampling our own blood, day and night, in relationship with various diets.

From our measurements for the first prospective study (Part 11 of this Appendix), soon we also had data from 1,231 employes of the Los Angeles Civil Service and from 2,105 participants in the Framingham Heart Study. The Los Angeles data are for fasting blood samples; the Framingham data are for non-fasting samples. While we cannot be positive that the two population samples are identical in all other respects, we have taken data from these two population surveys to represent fasting (Los Angeles) and non-fasting (Framingham) lipoprotein levels. For ages below 30, we made non-fasting measurements of 222 members of the university population at U.C. Berkeley, and of 15 clinically healthy children of a Pediatric Clinic.

The measurements, fasting vs. non-fasting by gender and age, are tabulated and graphed in Glazier 1954. Below, we present the major findings just for age 40 and higher.

Mean Values of Lipoprotein Levels: Std Sf 0-12, Std Sf 12-20, Std Sf 20-100, Std Sf 100-400

For Males	Fasting	vs	Non-Fasting

Std Sf 0-12   Fasting samples average about 11 % higher than non-fasting
Std Sf 12-20   Fasting samples average about 14 % higher than non-fasting
Std Sf 20-100   Fasting samples average about 5% LOWER than non-fasting

Std Sf 100-400   Fasting samples are, on the average, about HALF the level of non-fasting

For Females
Std Sf 0-12   Fasting samples average about 7 % higher than non-fasting
Std Sf 12-20   Fasting samples average about 2-3 % higher than non-fasting
Std Sf 20-100   Fasting samples average about 20 % LOWER than non-fasting
Std Sf 100-400   Fasting samples are, on the average, about HALF the level of non-fasting

● **Part 4. From Person to Person, Do the Sf 6-8 Lipoproteins Have the Same Lipid Constituents?**

Having opened up the vast spectrum of plasma lipoproteins for study, the team at Donner Laboratory needed to learn whether a specific class of ultracentrifugally-defined lipoproteins --- Sf 6-8, for example --- had a constant chemical meaning from one person to another, and for one person at different times.

Frank T. Lindgren, Alex V. Nichols, Bernard Shore, Virgie Shore, Thomas L. Hayes, Norman K. Freeman, and Gary Nelson attacked this large and difficult problem with admirable hard work and success, during much of the first decade of the work at Donner. Some of their major findings are brought together in an excellent overview paper in the Annals of the New York Academy of Sciences: "Structure and Homogeneity of the Low-Density Serum Lipoproteins" (Lindgren 1959).

Boxes 1 and 2 of this appendix indicate that, WITHIN a specific Sf-range, the lipid composition of the LDL, IDL, and VLDL classes is remarkably similar from person to person.

The constancy of composition was a most welcome feature, as we proceeded with various case-control and prospective studies of the atherogenicity of various Sf-classes of lipoproteins, and proceeded to study the metabolism of these molecules.

● **Part 5. Molecules in a Metabolic Chain? The Steady-State Concentrations**

By 1951, we had developed the view "that all the molecules from Sf 40,000 down to Sf 4 (and possibly into the high density class) represent a sequence of molecules in a metabolic chain involved in the ultimate utilization of glyceryl esters and/or fatty acids" (Jones 1951, p.360). And (also p.360): "From tracer studies now in progress, it appears that the average lifetime of all the lipoprotein species of the low density group is of the order of several hours. Therefore, the most probable explanation of a given concentration of a particular lipoprotein is that it is the steady-state concentration at which the influx and utilization rates are equal." (See also Gofman 1952-c, + McGinley 1952, + Pierce 1954-a, + Lindgren 1956, + and Lindgren 1959.)

● **Part 6. How Do Lipoproteins Move from an Artery's Lumen to Its Intima?**

If certain plasma lipoproteins are the source of the lipids observed in atherosclerotic lesions, we (and others) wanted to learn how the lipoproteins moved from the artery's lumen into its intima.   At the outset, we had no preconceived notions. For instance, we did not start in 1949 with the infiltration concept discussed in Chapter 44, Part 5.

In 1949, it was proposed that the lipid material present in atheromatous lesions might be carried there by cells such as macrophages (Leary 1949). In 1949, Kuntz and Sulkin tested the idea by injecting hypercholesterolemic rabbits with dyes. The dyes ended up in visceral phagocytes but not in the lipid-filled foam cells of atherosclerotic lesions (Kuntz 1949). We felt that their experiment did not settle the issue, however, because the dyes had not been injected BEFORE cholesterol-feeding. At the Donner Lab, John Simonton undertook to inject rabbits with a radioisotope which was demonstrated to concentrate in visceral macrophages. Then cholesterol-feeding was initiated. The results confirmed that macrophage migration into the intima was not a significant source of the lipid-filled foam cells in the rabbits' subsequent atherosclerotic lesions (Simonton 1951).

● **Part 7. Why We Should Identify Non-IHD Disorders with Aberrant Lipoprotein Patterns**

The Donner team gave considerable attention to identifying non-IHD disorders and conditions which are characterized by aberrant concentrations of plasma lipoproteins. These include diabetes (Engelberg 1952-a + Engelberg 1952-b, + Kolb 1955), infectious hepatitis (Pierce 1954-b), infectious mononucleosis (Rubin 1954), biliary cirrhosis (McGinley 1952), obesity (Gofman 1952-b), xanthoma tendinosum, xanthoma tuberosum, nephrotic syndrome, chronic biliary obstruction, myxedema, "essential hyperlipemia," and possibly others (Gofman 1954-e).

When persons have disorders and conditions which can cause unusual concentrations of various classes of lipoproteins, these persons necessarily influence the AVERAGE concentrations measured in population samples. This can cause false conclusions, if the frequency of such persons is appreciably higher in one group than in a second group with which the first group is compared. We and others needed to be watchful for these potential pitfalls, when comparing average concentrations of serum lipoprotein classes in population samples with clinical Ischemic Heart Disease and samples without overt IHD.

Box 3 (from Gofman 1954-e) constituted a warning, about the patterns of extraordinary serum lipoprotein levels which may characterize certain disorders.

● Part 8. A "Smoking Gun": Reversal of Xanthomas by Lipid-Lowering Regimes

In the early 1950s, the Lipid Hypothesis of Atherogenesis was VERY far from accepted. Among the clues ("smoking guns") which convinced the Donner group that our work was well worth pursuing were our clinical observations with respect to xanthomas --- lipid-filled lesions visible on the outside of people with the heritable disorders called xanthoma tuberosum and xanthoma tendinosum.

We soon developed the opinion that visible xanthomatous lesions, of the skin and tendon sheath, are the equivalent of atheromas, and that these lesions --- like atheromas of the arteries --- have their genesis in the deposition of lipoprotein molecules, from the circulating blood, into tissue-sites where deposition should not occur.

Quite early in our work, also, we detected a striking difference in the serum lipoprotein patterns of people with xanthoma tuberosum versus xanthoma tendinosum (McGinley 1952, + Gofman 1954-e). The mean levels, tabulated below, are in mg per 100 ml, and the numbers in parentheses are for matched controls (from Gofman 1954-a, p.432, 434). Because the two disorders typically differ in the distribution patterns of the lesions on the body, we postulated that different serum lipoproteins deposited generally in different tissues because of the molecules' different SIZES. And we wondered why this should not be the case for deposition in arterial walls. too.

	Sf 0-12	Sf 12-20	Sf 20-100	Sf 100-400
X.Tendinosum (18 cases)	793 (336)	150 (65)	128 (92)	36 (56)
X.Tuberosum (23 cases)	206 (358)	128 (74)	616 (105)	650 (72)

Most exciting of all, when we were able to reduce serum lipoprotein levels pharmacologically or by diet, the patients stopped developing new lesions, and we could watch the skin lesions gradually involute and disappear, unless they had been of long standing and were fibrotic (Gofman 1952-d, + Gofman 1958, + Gofman 1959 at pp.210-211).

● Part 9. Early Evidence of Independent Atherogenicity, Sf 12-20 vs. Sf 20-100 Lipoproteins

From the outset, we knew the importance of finding out WHICH segments (if any) of the plasma lipoprotein spectrum are atherogenic (Chapter 44, Part 3a). We faced the possibility that some segments may have elevated concentrations in cases of Coronary Heart Disease (CHD), only because of potentially tight positive correlations with an elevated concentration in IHD cases of one TRULY atherogenic segment of the spectrum. So we began looking at this issue in our early case-control studies (presented in Jones 1951, + Gofman 1952-a, + Gofman 1952-c).

Figure E-1 (which is adapted from Gofman 1952-c, p.286) is one way in which we presented our findings. The plasma lipoprotein measurements on 253 "normals" (males, ages 41-50, NOT having overt CHD) are compared with such measurements in 72 male patients (ages 41-50) having overt CHD. The comparison is based on the lines of best fit from linear regression. Figure E-1 demonstrates that, at any serum Sf 12-20 lipoprotein level (in mg/100 ml, or mg%), the Sf 20-100 lipoprotein level is higher in CHD cases than in normals.

This finding represents evidence for an INDEPENDENT atherogenic contribution by Sf 20-100 lipoproteins. If the presence of Sf 20-100 lipoproteins contributed nothing to atherogenic risk beyond the contribution from the Sf 12-20 lipoproteins, the two lines of best fit (CHD cases and normals) would fall on top of each other, within experimental error.

Although we regard this type of evidence as a strong indication that the triglyceride-rich Sf 20-100 lipoproteins are independently atherogenic, we have pointed out that the issue remains unsettled today, some 47 years later (Chapter 44, Parts 3i + 4).

Separately, we approached the independence issue by using linear regression to evaluate how strongly or weakly blood-levels of various segments of the lipoprotein spectrum correlate with each other, in clinically healthy persons by gender and age (Gofman 1954-a, Tables 8, 9, 10, + DeLalla 1961, Tables 40 through 46). We reasoned that if blood-levels of two segments of the lipoprotein spectrum are not strongly correlated in clinically healthy people, AND if these two segments both are above normal levels in CHD cases, then each segment may be contributing, INDEPENDENTLY, to atherogenic risk. Such findings create the obligation to find out --- but it is surely not easy to do so.

● **Part 10. Segregation of Infarcts from Normals: Sf-Class vs. Total Cholesterol**

Not long after the Donner team revealed the vast spectrum of plasma lipoproteins, there were voices at some scientific meetings suggesting that serum measurements of the ultracentrifugally-identified lipoproteins would add nothing useful about atherogenesis beyond what could be determined from the simpler chemical measurement of total serum cholesterol.

Their prediction deserved testing as soon as feasible. By 1951, the Donner team published its first results --- from comparing male survivors of Myocardial Infarcts with male "normals."

Results are shown in our Figure E-2 (adapted from Jones 1951, Figure 2, p.364). The results for males ages 41-50 (based on 64 survivors of Myocardial Infarcts and 273 "normals") revealed that there was significant segregation of infarcts from normals by the Sf 12-20 measurement, independent of serum cholesterol. Not shown here are the results for males ages 51-60 (based on 92 survivors of Myocardial Infarcts and 126 "normals"). Those results also revealed that there was significant segregation of infarcts from normals by the Sf 12-20 measurement, independent of serum cholesterol (Jones 1951, p.364).

Our analysis revealed, too, that for males ages 41-50 (but not for ages 51-60), there was independent segregation of infarcts from normals by the serum cholesterol measurement, independent of the Sf 12-20 lipoprotein measurement.

Our Table E-1 (adapted from Jones 1951, Table 3, p.366) demonstrates this finding by the "matched-series" method, for males ages 41-50. Our interpretation is that the independence of the serum cholesterol measurement (independent from the Sf 12-20 measurement) resided in the contributions of extra serum cholesterol from atherogenic lipoproteins of the cholesterol-rich Sf 0-12 segment, as well as some cholesterol from the glyceride-rich Sf 20-400 lipoproteins. However, for the males of ages 51-60, the cholesterol measurement did not independently segregate infarcts from normals, while the Sf 12-20 measurement successfully did so (Jones 1951, pp.366-367).

Of course, the merits of the ultracentrifugal and chemical measurements would not be established by a single study. There was much more to be done.

● **Part 11. Do Elevated Serum Lipoproteins Precede or Follow an Infarct?**

The initial findings from our case-control data (Gofman 1950-a) created considerable interest, but case-control data could never prove that the observed elevation of Sf 12-20 lipoproteins in survivors of myocardial infarction did not occur AFTER (and due to) the infarction. Only a prospective study could establish which came first.

11a. An Answer from the First Prospective Study

Before the end of 1950 and under a grant from the National Heart Institute, a prospective study was underway at four laboratories: Donner Laboratory at U.California at Berkeley, the Cleveland Clinic Foundation, the Dept. of Biophysics at U. Pittsburgh, and the Dept. of Nutrition at Harvard School of Public Health. The effort was called a "Cooperative Study of Lipoproteins and Atherosclerosis" (Co-op 1956 in our Reference List).

The study did establish that, relative to the base population in which they started, the participants who developed definite new "events" (myocardial infarction, coronary thrombosis, ECG abnormality associated with coronary artery disease, angina pectoris) had above-average serum lipoproteins BEFOREHAND. Total serum cholesterol, transported only by lipoproteins, was of course elevated BEFOREHAND, too.

The total follow-up time was cut off at 2 years or less. Among the 4,914 males, ages 40-59 at entry to the study, 82 "events" developed within that time-frame, according to the Review Committee. These 82 events included 27 "Myocardial Infarctions, definite" and 38 "Myocardial Infarctions, definite, by ECG only" (Co-op 1956, p.709).

Because one goal of the Co-op Study was "to determine the range of values of the Sf 12-20 fraction (lipoprotein) in 'normal' persons of all ages and both sexes," the four laboratories actually measured the serum lipids of approximately 15,000 persons in a period of about 3 years (Co-op 1956, p.692, p.693). Measurements of the total serum cholesterol and the Sf 20-100 lipoproteins were made, too (Co-op 1956, p.695-696).

11b. Disagreement over Converting to the More Accurate Std Sf Measurement

Among the four research teams, major differences of opinion --- much of which is presented in Co-op 1956 --- developed concerning several aspects of the studies themselves and of the results. There was some merit on both sides of the arguments which ensued.

The central disagreement centered on the whether or not to convert the Sf measurements to Standard Sf measurements (Part 2d of this Appendix) --- a conversion which would have been fully consistent with the study's "blinding" protocol. The more refined Std Sf measurements made very substantial differences in levels of both the Sf 12-20 and Sf 20-100 segment of the lipoprotein spectrum --- as can be seen by comparing Exhibit D with Exhibit G in Co-op 1956 (p.735 and p.738). Fortunately, use of the more refined measurement-procedure resulted in no loss of previously made measurements, because the original film could be re-evaluated by the refined procedure (Co-op 1956, p.715).

We at Donner argued vigorously for using the more accurate technique, but we were not in charge and we did not prevail. Therefore, in order that measurements from all four laboratories be comparable, the Donner team agreed to continue providing the less accurate measurements. However, we also provided the more accurate measurements (Appendix-A of Co-op 1956).

Describing the outcome of the Co-op Study, there is a concluding statement by Professor E. Cowles Andrus (Johns Hopkins University) in the report at pp.713-714:

"The participants in this study concur in the presentation to this point. They agree in finding that atherosclerosis, as manifested by clinical signs of coronary artery disease, is associated with a disorder of lipid metabolism and that there is some predictive value in the various lipid measurements examined. However, because of clear divergence of opinion between the Eastern Laboratories and the Donner group with regard to the degree and specificity of this predictive value, and indeed with regard to the significance of certain data in relation thereto, it was agreed that the discussion and conclusions would be prepared independently and presented separately by [the] two groups ..."

11c. Some Valuable Lessons from the Co-operative Study

Not too bad, as an outcome, when we consider that such a project was probably launched prematurely --- only about one year after any technique for separating and measuring serum lipoproteins ultracentrifugally had been initiated. We were still learning about its complexity.

Nonetheless, four separate laboratories undertook this complex new technology, and managed to carry through the measurements with satisfactory and provable reproducibility. And all four teams had some exceedingly valuable education in two areas of great importance:

(1) We received advice generously from Dr. J. Franklin Yeager of the National Heart Institute (Grants and Training Branch) on how to work together and not to be scientific prima-donnas. And:

(2) We received invaluable advice and firmness from Felix E. Moore of the National Heart Institute (Biometric Research Section) on the rules of credible, blinded epidemiologic research --- rules which did characterize the actual conduct of the project. Sadly, bio-medical research sometimes still permits (when it should disdain) violations of such fundamental rules as "Never tamper with the input data, once the results of the follow-up are known."

● **Part 12.   Inclusion of Std Sf 0-12 and Std Sf 100-400 in Two Prospective Studies**

The Co-operative Study (1956), which did not include consideration of the Sf 0-12 or the Sf 100-400 serum lipoproteins, clearly could not address the question: Are these segments of the lipoprotein spectrum also predictive of de novo Ischemic Heart Disease? Additional prospective inquiry would be required to find out.

12a.   1966: Results from the Framingham Study

During the Co-operative Study, the Donner team had made ultracentrifugal measurements of serum lipoproteins (Std Sf 0-12, 12-20, 20-100, 100-400) and chemical measurements of total serum cholesterol for "townspeople" who were then new entrants to the Framingham Heart Study. We measured the bloods of 2,022 males, ages 30-69 at entry, and 2,487 women, ages 30-69 at entry (Gofman 1966, p.685). Our measurements were permanently "set in concrete," when they were contemporaneously filed by Donner with both Framingham and with Felix Moore at the National Heart Institute.

In 1965, Thomas R. Dawber, M.D. (then Director of the Framingham Study) and his colleagues generously provided us at Donner with a listing of the 319 de novo cases of Ischemic Heart Disease which had occurred during the intervening 12 years among those 4,509 Framingham entrants measured by Donner. In other words, the list represented results for about a 12-year follow-up. In toto, there were 221 de novo cases among the men, and 98 cases among the women. The results for all males combined, all females combined, and for subsets by age, were fully reported in Gofman 1966 (pp.681-685).

Compared with the male base population, all ages combined, the de novo IHD cases had very significantly (p < 0.001) higher serum levels of Std. Sf. 0-12, 12-20, 20-100, 100-400 lipoproteins and total cholesterol (Gofman 1966, p.682, Table 1). In other words, EACH of these five measurements was predictive of clinical Ischemic Heart Disease. When examined by age-groups, the differences between base and cases declined with age at entry, and none was significant in the group which had been 56-62 years old at entry (Gofman 1966, p.683, Table 6). This decline in segregation, with advancing age at measurement, had also been observed in our case-control studies (Gofman 1954-d, p.593).

For the females who had been ages 40-69 at entry, the de novo cases had higher levels of each of the five measurements than did the base, but the findings were statistically significant only for the Std Sf 0-12, Std Sf 20-100, and total serum cholesterol (Gofman 1966, p.682, Table 2).

In Appendix-G, Part 4, we show the mean measurements, de novo cases versus base, for the males ages 30-39 at entry.

12b.   1986: Some Later Results from the Framingham Study

In 1986, William P. Castelli, M.D. (then Director of the Framingham Heart Study) authored "The Triglyceride Issue: A View from Framingham" in the American Heart Journal (Castelli 1986). In that paper, he briefly discussed Sf 0-20 and Sf 20-400 measurements made on males entering into the Framingham Study (Castelli 1986, pp.434-435).

His Figure 4 is entitled "CHD risk according to lipid level on entry into the Framingham Study:

Men aged 30-62." It shows clearly that the Morbidity Ratio rises in a steady linear relationship with rising serum levels of the cholesterol-rich Sf 0-20 lipoproteins, and separately, with rising levels of the triglyceride-rich Sf 20-400 lipoproteins.

12c. 1966: The Livermore Lipoprotein Study

In 1954, the Livermore (Weapons) Laboratory asked me to organize its industrial medical facility, where employes at all levels (from the very top to the bottom) would receive complete medical examinations at intervals of approximately 1.5 years.

Nature of This Database

During the years 1954-1957, the Donner Laboratory analyzed the serum lipoproteins and total serum cholesterol from the non-fasting bloods of this population of workers. Oliver F. DeLalla described the study as follows (DeLalla 1958, p.18):

"The healthy subjects for this investigation were employes of the University of California Radiation Laboratory at Livermore, California, all of whom are periodically given routine physical examinations. Venous blood samples --- 30 ml from each subject --- were taken between the hours of 8 a.m. and 3 p.m., without any restriction of diet ... There were certain restrictions, however; no individual was included in the study whose medical records showed any of the following conditions." List follows, verbatim:

Acute infection at the time of sampling.
Poliomyelitis with residual deformity.
Surgery involving removal of part or all of any organ.
Any condition requiring that the subject take medications such as thyroid, steroids, etc.
Cardiovascular disease history.
Cancer history.
Multiple sclerosis.
Pregnant women and persons following special diets were also excluded.

"The total number of subjects qualifying was 2,297, of which 1,961 were males and 336 were females. Their ages ranged from 17 to 65 years. The 2,297 samples were collected and analyzed at a rate of about 20 per week over a 3-year period. There were no delays in the analysis of any sample. Whenever the samples were collected, they were immediately put through the complete analysis, which normally takes from 7 to 10 days." The more detailed distribution is as follows (from DeLalla 1961, p.139):

Ages	Males	Females
17-29	585	190
30-39	834	99
40-49	399	37
50-65	143	10
Sums	1,961	336

The database for each person includes ultracentrifugal measurements of serum lipoproteins Std Sf 0-12, 12-20, 20-100, 100-400, HDL-1, HDL-2, HDL-3, plus the Atherogenic Index, total cholesterol, systolic blood pressure, diastolic blood pressure, relative weight. Persons studied twice = 374.

Follow-Up Findings in 1966

In 1966, we conducted a follow-up of the males in this population, which we reported in Gofman 1966. Out of the base population of 1,961 subjects, 38 cases of de novo Ischemic Heart Disease had developed.

Compared with the male base population, all ages combined, the de novo IHD cases had very significantly ($p < 0.001$) higher serum levels of Std Sf 0-12 and Std Sf 12-20 and total cholesterol, significantly higher ($p < 0.01$) higher levels of Std Sf 20-100, and higher (but not significantly) Std Sf 100-400 lipoproteins (Gofman 1966, p.684, Table 10). In other words, EACH of these five measurements was elevated BEFORE the development of clinical Ischemic Heart Disease. The findings about HDL are discussed in Appendix-G.

12d. Prospective Studies: A Generic Problem

In 1978, we examined another aspect of the Livermore database: The 374 participants for whom we had two measurements about 1.5 years apart. The findings left us greatly impressed that the Framingham and Livermore Studies in 1966 had been capable of discerning statistically significant differences between de novo cases and base populations. The relationship between elevation of certain serum lipoproteins and Ischemic Heart Disease must be very strong indeed, when it remained detectable despite the phenomenon which we quantified in 1978. The 1978 analysis is summarized in Appendix I (eye), because it remains applicable today to prospective studies in many fields of bio-medical inquiry.

>>>>>>>>>>

CSE = cholesterol esters. UCS = unesterified cholesterol. G = glycerides. PL = phospholipids.

● The figure below compares the lipid composition of Sf 6-8 serum lipoproteins from 5 individuals (reproduced from Lindgren 1959, p.834).

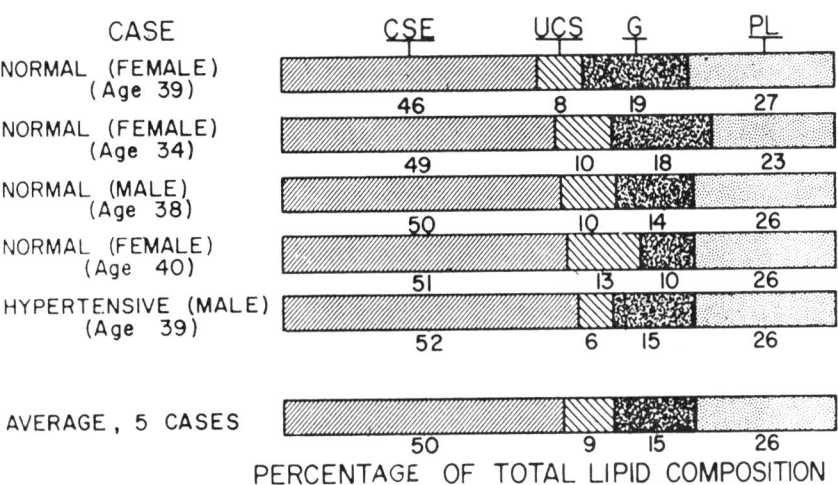

PERCENTAGE OF TOTAL LIPID COMPOSITION

Related Text = Part 4.

● Having opened up the vast spectrum of plasma lipoproteins for study, the team at Donner Laboratory needed to learn whether a specific class of ultracentrifugally-defined lipoproteins (say, Std Sf 12-20) had a constant chemical meaning. In short: Is the chemical internal structure the same for Std Sf 12-20 lipoproteins, for one person from time to time, and for one person compared with another person?

● If, for a specific Sf-range of lipoproteins, one could NOT count on internal structure being reasonably constant from person to person, it would be an obstacle to understanding lipoprotein metabolism --- for instance, the apparent conversion of molecules of high Sf values into molecules of lower Sf values (text, Part 3).

● Box 2 shows results, comparable with the figure above, for the combined Sf 0-20 lipoproteins and the combined Sf 20-400 lipoproteins, from the fasting sera of 9 individuals.

● Boxes 1 and 2 indicate that, WITHIN a specific Sf-range, the lipid composition of the lipoproteins is remarkably similar, quantitatively, for both sexes in health and for several major lipoprotein disorders, and across a range of ages. Moreover, the distribution of lipids is distinctly different in the Sf 0-20 lipoproteins vs. the Sf 20-400 lipoproteins --- as already indicated in Chapter 44, Box 1, lower right corner.

● The constancy of composition was a most welcome feature, as the team at Donner proceeded with various case-control and prospective studies of the atherogenicity of various Sf-classes of lipoproteins.

• The figures below compare the findings from the blood lipoproteins of 9 individuals (fasting measurements). Reproduced from Lindgren 1956; also in Lindgren 1957 + 1959.
• Please see text in Box 1 of this Appendix.

CSE = cholesterol esters. UCS = unesterified cholesterol. G = glycerides. PL = phospholipids.
UFA = unesterified fatty acids. Far right: Avg flotation rate (Sf) in a solution of 1.063 gm/ml.

Sf 0–20 Lipoproteins: Lipid Composition, Expressed as Percent of Total Lipid.

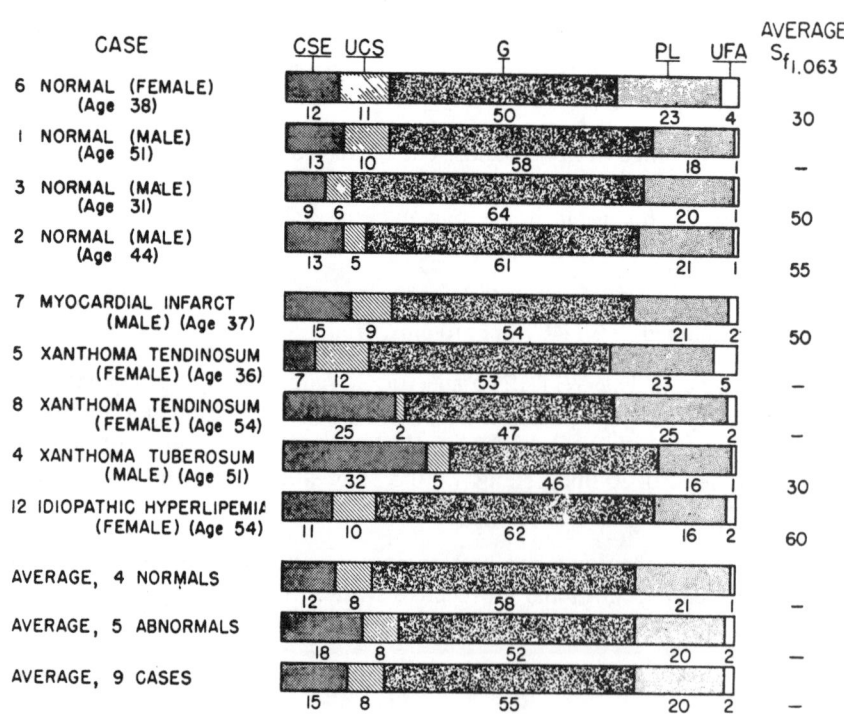

Related text = Part 7. Reproduced from Gofman 1954-e, p.518.

Hyperlipoproteinemia—*Gofman et al.*   518

Related text = Part 7.

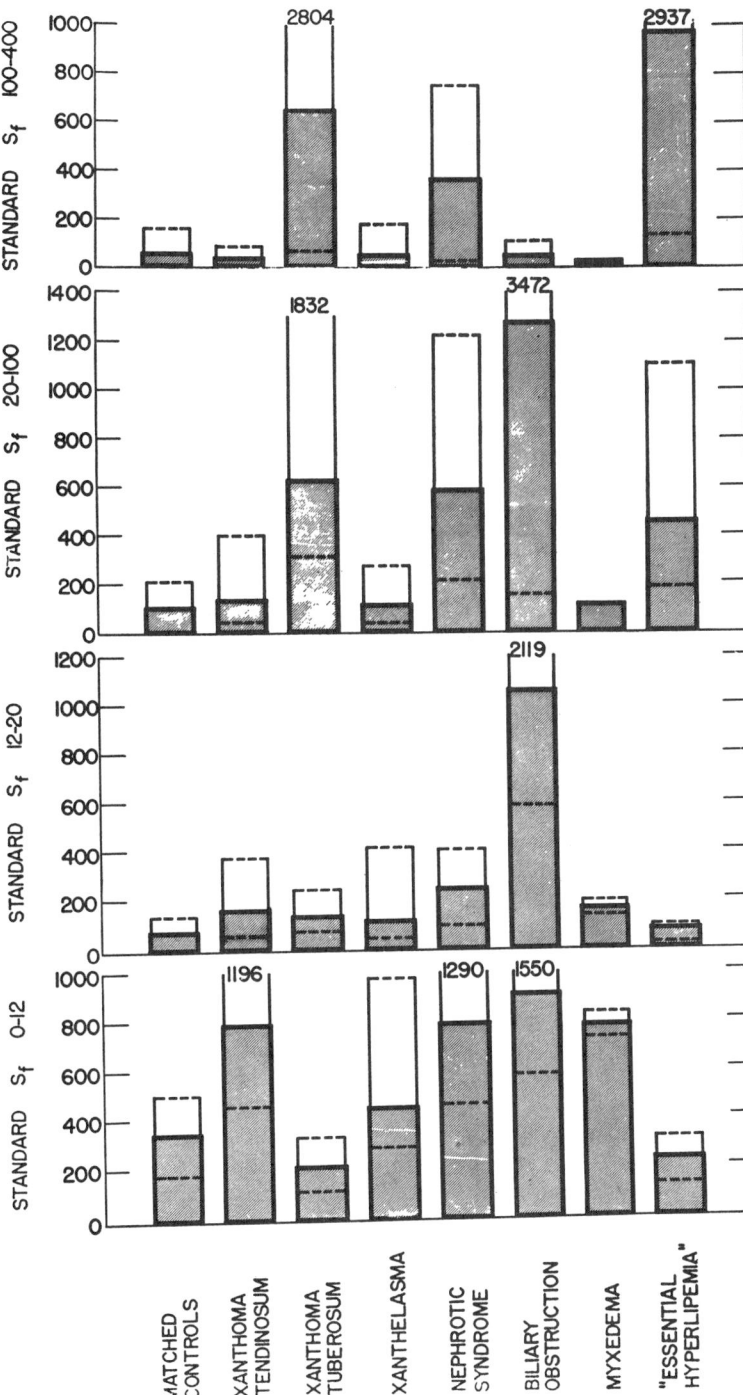

FIG. 1. The mean levels of serum lipoprotein concentration in the various disease states and a control group matched for age and sex for the four lipoprotein classes are shown by the hatched areas. The entire ranges of values found in the patients described here are shown by the interrupted lines. In some disease categories the highest level of lipoproteins found was off the scale of the graph; these high values are written in the appropriate space. For the matched control group the interrupted lines show the levels at two standard deviations from the control mean values.

Figure E-1

**Evidence of the Independent Atherogenicity of the Triglyceride-Rich Sf 20-100 Lipoproteins.
Serum Levels of Sf 20-100 Lipoproteins Regressed upon Sf 12-20 Lipoproteins.**

- The figure below is adapted from Gofman 1952-c, p.286 (case-control data).
- Line A in the graph is the regression line for normals (253 men without overt Coronary Heart Disease).
- Line B in the graph is the regression line for 72 patients with CHD.

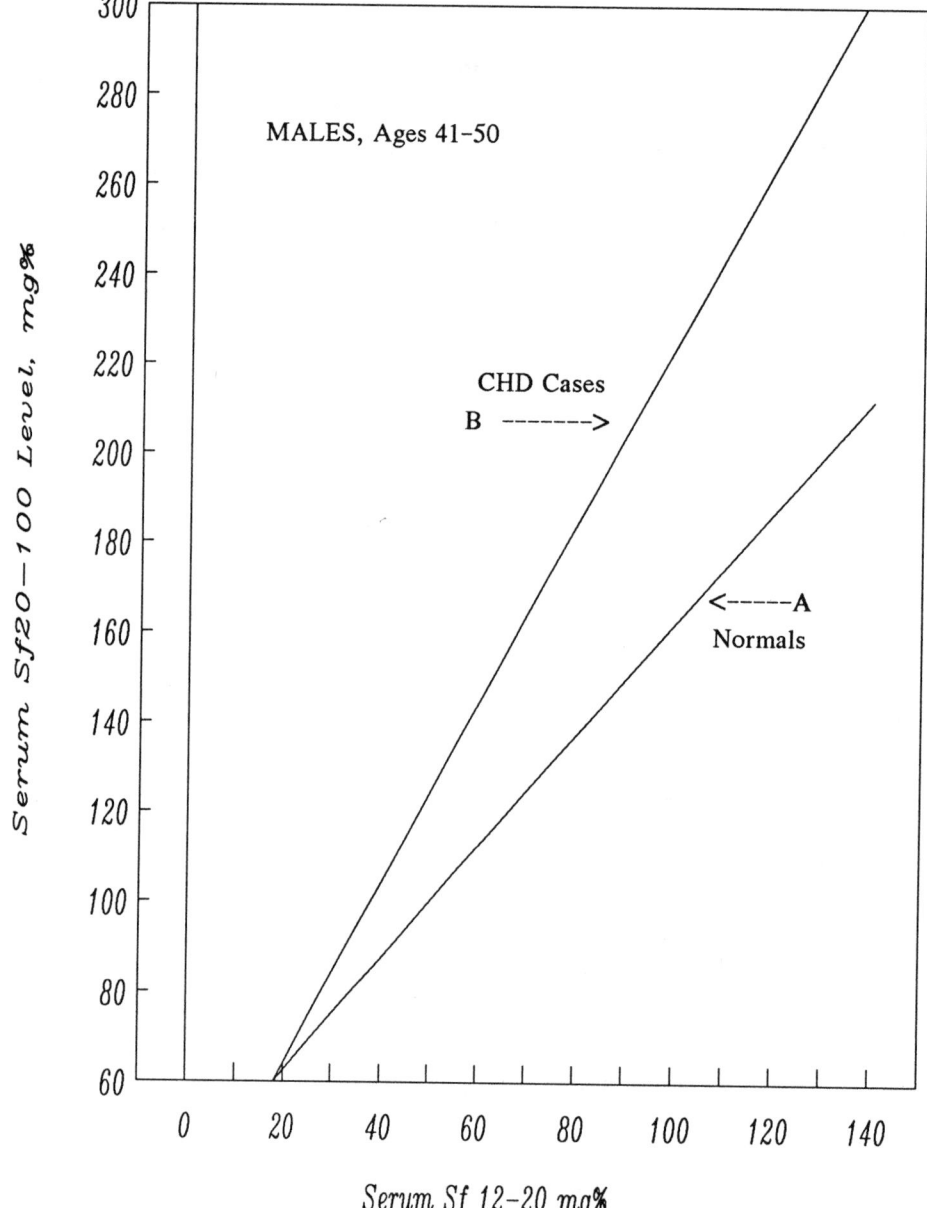

- At any serum Sf 12-20 level, the mean Sf 20-100 level (in mg/100 ml) can be read for either normals or for CHD cases from the appropriate regression line.
- At every serum Sf 12-20 level, the corresponding Sf 20-100 level is higher, on the average, in CHD cases than in normals.

- This finding represents evidence for an INDEPENDENT atherogenic contribution by the triglyceride-rich Sf 20-100 lipoproteins. If the presence of the Sf 20-100 lipoproteins contributed nothing to atherogenic risk beyond the contribution from the Sf 12-20 lipoproteins, the two lines of best fit (for CHD cases and normals) would fall on top of each other, within experimental error.

Figure E-2

Segregation of Infarcts from Normals, by Sf 12-20 Levels Independent of Cholesterol Levels. Serum Sf 12-20 Lipoprotein Levels Regressed upon Serum Cholesterol Levels.

- The figure below is adapted from Jones 1951, p.364 (case-control data).
- Line A in the graph is the regression line for normals (273 males without overt Coronary Heart Disease).
- Line B in the graph is the regression line for 64 survivors of Myocardial Infarction.

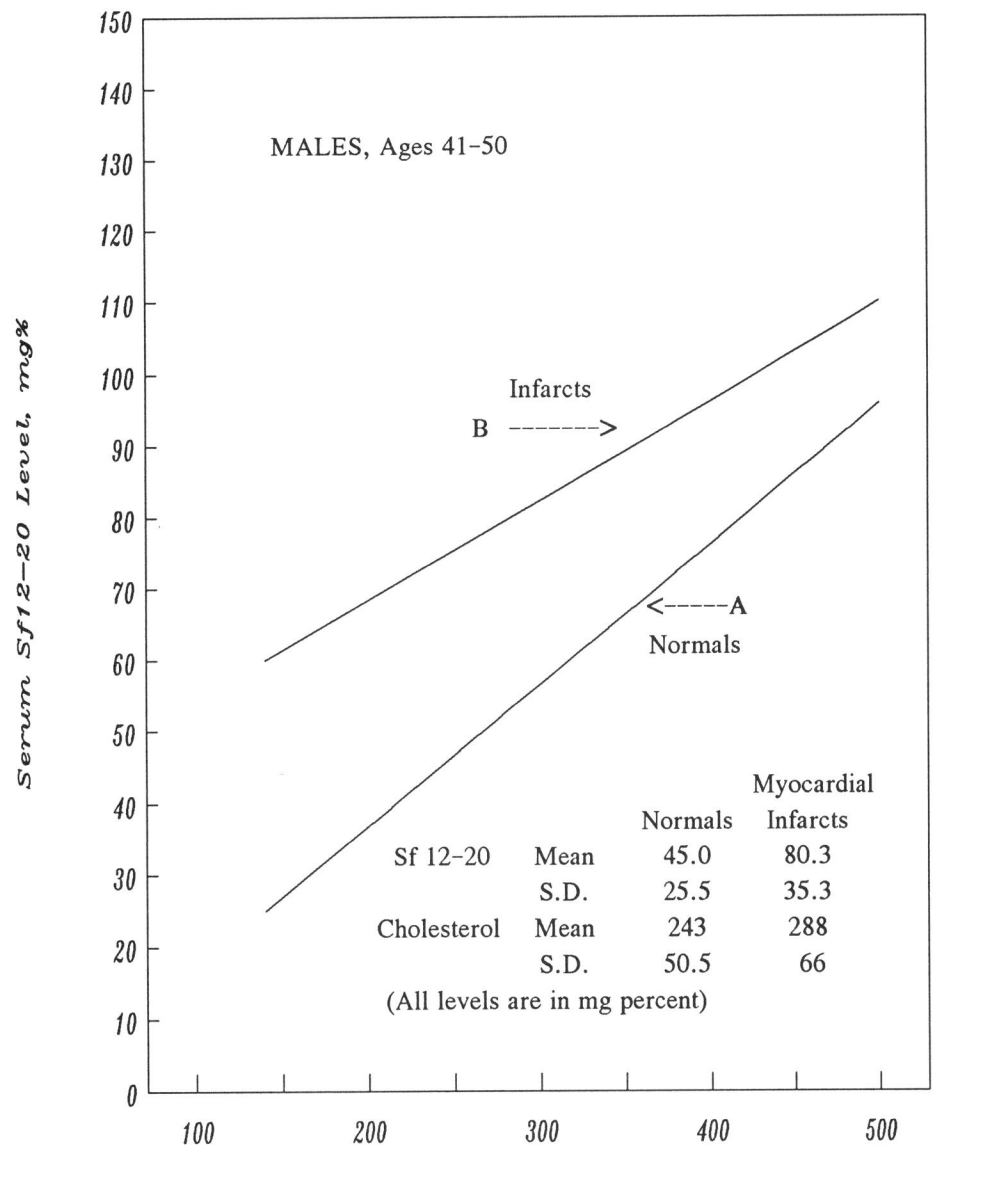

		Normals	Myocardial Infarcts
Sf 12-20	Mean	45.0	80.3
	S.D.	25.5	35.3
Cholesterol	Mean	243	288
	S.D.	50.5	66

(All levels are in mg percent)

MALES, Ages 41-50

Related text = Part 10.

- At any serum cholesterol level, the mean Sf 12-20 level (in mg/100 ml) can be read for either normals or for Infarct survivors from the appropriate regression line.
- At every serum cholesterol level, the corresponding Sf 12-20 level is higher, on the average, in the Infarct survivors than in normals.

- This finding represents evidence for an INDEPENDENT atherogenic contribution by the Sf 12-20 lipoproteins, beyond their positive correlation with serum cholesterol. If the levels of the Sf 12-20 lipoproteins contributed nothing to CHD risk beyond the contribution measured by total serum cholesterol, the two lines of best fit (for Infarcts and for normals) would fall on top of each other, within experimental error.

Table E-1
The Matched-Series Method of Assessing Independent Contribution to Atherogenesis.

● The tabulations below, based on case-control data, are adapted from Jones 1951, Table 3, p.366. Tabulations 1 and 2 are based on males ages 41-50. Lipoprotein and cholesterol determinations were done on aliquots of the same serum sample from each individual in the study.

● 1. Myocardial Infarcts Matched with Normals by Serum Cholesterol Levels.

Group	Number of Cases	Mean Serum Cholesterol	Mean Sf 12-20 Level
Infarcts	55	291.3 mg. %	73.8 mg. %
Normals	55	291.5 mg. %	53.7 mg. %

Difference in Sf 12-20 = 20.1 mg. % (+,- 5.5). Significance: $p < 1\%$.

● 2. Myocardial Infarcts Matched with Normals by Sf 12-20 Levels.

Group	Number of Cases	Mean Serum Cholesterol	Mean Sf 12-20 Level
Infarcts	55	292.3 mg. %	74.3 mg. %
Normals	55	273.2 mg. %	74.3 mg. %

Difference in Serum Cholesterol = 19.1 mg. % (+,- 6.4). Significance: $p < 1\%$.

● Quoting from Jones 1951 (p.365): "Matched series [above, in Part 1] have been used to assess the relationship of Sf 12-20 lipoproteins to atherosclerosis, while nullifying the effect of serum cholesterol PER SE, and conversely [in Part 2, above] to assess the relationship of serum cholesterol to atherosclerosis, while nullifying the effect of Sf 12-20 lipoproteins." And (p.365):

● "Thus [for Part 1], the cases of myocardial infarction were listed by cholesterol levels. From the normal data, cases were matched at random at corresponding cholesterol levels. Then the Sf 12-20 levels for the myocardial infarctions were compared with those for the normals that have been matched with these infarcts at the same age and serum cholesterol." And (p.365):

● "Similarly [for Part 2], the infarct cases were listed by Sf 12-20 levels and matched at random with normals having the same Sf 12-20 levels. Then the serum cholesterol levels were compared for the two groups." And (Jones p.365, referring to Table 3 on p.366):

● "The data in Table 3 may be interpreted as follows: [1] For the 41-50 year age group, at the same serum cholesterol, the myocardial infarctions average 20.1 mg. per cent higher in Sf 12-20 levels than normals. [2] At the same Sf 12-20 levels, the myocardial infarctions average 19.1 mg. per cent higher in serum cholesterol than do normals."

● There is significant segregation of infarcts from normals by Sf 12-20 measurement, independent of the serum cholesterol. And in this age-band, there is also independent segregation of infarcts from normals by serum cholesterol levels. However, for the males ages 51-60, the cholesterol measurement did not independently segregate infarcts from normals, while the Sf 12-20 measurement successfully did so.

Related text = Part 10.

The new evidence, presented earlier in this monograph, has identified medical radiation as a very important cause of Ischemic Heart Disease (Chapters 40, 41, 64, 65). That finding is fully consistent with the Lipid Hypothesis of Atherogenesis. Indeed, our Unified Model of Atherogenesis and Acute IHD Events (Chapter 45) proposes that medical radiation and atherogenic serum lipoproteins BOTH are necessary co-actors in most cases of fatal IHD in the United States. This monograph has identified some practical ways to reduce exposure to medical radiation (Chapter 1), and so it should also identify some practical (dietary) ways to reduce exposure to atherogenic serum lipoproteins. That is the purpose of Appendix-F.

● Part 1. Open Exasperation in the Literature over Dietary Advice

"Ninety years of research should permit the scientific community to speak with a consistent voice about diet and the prevention of coronary disease," wrote Dr. William E. Connor and Sonja L. Connor (M.S., R.D.) in a recent "Clinical Debate" in the New England Journal of Medicine (Connor 1997, p.566). The topic was "Should a Low-Fat, High-Carbohydrate Diet Be Recommended for Everyone?" The Connors' apparent exasperation is over the LACK of consensus.

The Carbohydrate Effect: Elevated Serum Triglyceride
Elsewhere (not in that particular "Clinical Debate"), the so-called "triglyceride" issue is an intimate part of the debate over LowFat HighCarbo Diets. Such diets very commonly --- although not in every case --- ELEVATE blood levels of the triglyceride-rich Std Sf 20-400 segment of the serum lipoproteins (Chapter 44, Part 4). Not surprisingly, total serum triglyceride also is often found to be elevated by such diets (Knopp 1997).

Elevation of Std Sf 20-400 serum lipoproteins has serious negative implications for health, if some or much of that lipoprotein spectrum is atherogenic. In our opinion, the evidence was already strong 40 years ago that the Std Sf 20-100 lipoproteins and some portion of the Std Sf 100-400 lipoprotein spectrum ARE independently atherogenic, and now additional evidence exists (some of it is mentioned in Part 9).

Here, as promised at the outset of Chapter 44, Part 4, we present some compelling experimental evidence (from consenting humans), on the effects of dietary changes upon serum lipoprotein levels. It is particularly compelling evidence because (a) only ONE dietary variable was evaluated per experiment, and (b) the experiments were not encumbered by the simultaneous administration of a variety of pharmaceutical preparations which could render results uninterpretable.

The evidence in Parts 3, 4, and 5 was first presented in reports by Nichols, Dobbin + Gofman in the journal Geriatrics (Nichols 1957) and by Gofman in the American Journal of Cardiology (Gofman 1958).

The findings and statements issued in 1958 are still highly germane to the dietary debates in 1998. Moreover, the evidence argues strongly for making measurements TODAY which are capable of revealing whether or not certain dietary advice results in a harmful conversion of non-atherogenic lipoproteins into atherogenic varieties (Parts 6 + 8).

The Omega-3 Fatty Acids: Cardio-Protective Effects
"Heart healthy" dietary advice today is also intimately involved with the issue of a possible protective effect from ingestion of omega-3 (n-3) fatty acids, listed in Part 7. In Parts 8 + 10, we will consider some of the recent, exciting work on this issue by Connor, Harris, DeLorgeril, Renaud, and others.

The Weight-Gain Effect: Elevated Triglyceride-Rich Serum Lipoproteins
Lastly (Part 12), we offer evidence which leads us to wonder if much of the unhealthy increase, in atherogenic serum lipoproteins with advancing age in the USA, could be eliminated if we would just maintain the lower weights we had at around age 20.

● Part 2. Dietary Management of Blood Lipoproteins: Goals and Pitfalls

Goals and pitfalls, of dietary management of blood lipoproteins, have hardly changed in some 40 years. The paper "Diet in the Prevention and Treatment of Myocardial Infarction" (Gofman 1958) presented (a) the discussion which follows below, and (b) the experimental evidence which is shown in Parts 3 and 4 (based on Nichols 1957). What follows is taken from Gofman 1958 (pp.271-273). We have subdivided paragraphs and added some subtitles here.

2a. No Single Dietary Regime Is Beneficial for Everyone; Potential Harm

"Historically, the earliest interest in blood lipids in relationship to coronary disease centered around the blood cholesterol level, and certainly such interest provided an important stepping stone to our present position of knowledge." And:

"However, today, attention to the blood cholesterol alone provides only the most naive approach to the problem of clinical management of myocardial infarction. Indeed, if preventive and therapeutic measures are considered only in the light of the blood cholesterol level, serious errors of management will eventuate and many patients will be denied effective therapy." And:

"Cholesterol is only one of several chemical lipid constituents of the blood, some of the other major chemicals being triglyceride, fatty acids, and phospholipids. None of these chemical entities has an existence in the blood stream as such, but all are instead building blocks of a series of very large lipid-protein complexes designated as lipoproteins. Cholesterol is a constituent of essentially all the lipoproteins of blood, being quite abundant in some but at very low abundance in others. Similarly, triglyceride or phospholipid is to be found in essentially all the lipoproteins, but at differing abundance in the various lipoprotein classes." And:

"Thus, if cholesterol represents only 5 per cent of the composition of one class of lipoproteins whereas it represents 30 per cent of the composition of some other class of lipoproteins, it becomes evident that one milligram of cholesterol measured in the blood could mean the presence of 20 mg of the first class of lipoproteins but it could mean only 3 1/3 mg of the second class of lipoproteins. Similar considerations apply to the measurements of phospholipids and of triglycerides. The study of the blood lipoproteins in relationship to coronary arteriosclerosis and myocardial infarction has demonstrated that the desired information is the actual level of certain of the lipoprotein classes that circulate in the blood (Gofman 1950-b)." And:

"In addition it has been shown that the various lipoprotein classes involved in coronary disease have differing responses to dietary measures being considered for preventive and therapeutic purposes, indeed differing to the extent that a particular diet may depress the blood level of one important lipoprotein class while at the same time it elevates the level of one of the other important lipoprotein classes (Nichols 1957). This carries with it the implication that a particular dietary regimen designed for a patient with an elevation of one lipoprotein class may in fact be quite harmful for a patient whose lipoprotein elevation is largely in some other class."

2b. Targeting the Specific Lipoprotein Classes Which Are Elevated

"The most extensive detailed characterization of the lipoproteins as they actually circulate in the blood is available through the technique of ultracentrifugation (DeLalla 1954-a). Indeed, at present no other technique reveals the lipoprotein differences that are so crucial in the design of preventive and therapeutic regimens for myocardial infarction ... Four lipoprotein classes have been identified to be associated with coronary intimal arteriosclerosis and to have predictive implications for the development of myocardial infarction. These are the Standard Sf 0-12, Standard Sf 12-20, Standard Sf 20-100, and Standard Sf 100-400 lipoprotein classes ... All four of these lipoprotein classes are involved in the development of coronary heart disease." And:

"However, the extent to which one of these lipoprotein classes is elevated in a particular person, as compared with the extent to which the other three classes are elevated, is quite different from individual to individual. Thus marked elevation of the Std Sf 0-12 lipoproteins may exist in an individual with moderate or even LOW levels of Std Sf 12-20, Std Sf 20-100, and Std Sf 100-400 lipoproteins. In this individual, the marked elevation of Std Sf 0-12 lipoproteins is the lipid factor predisposing to myocardial infarction and hence preventive or therapeutic efforts would be centered around the modification of the blood level of this particular class of lipoproteins. In some other patient the efforts would be directed against elevation of some other class of lipoproteins."

2c. Patients with Xanthoma: Visible Evidence of Response to Targeted Measures

"The general philosophy underlying such an approach to coronary disease is exceedingly simple and direct. If high levels of certain lipoproteins increase the risk of coronary disease, prevention would appear to lie in the direction of lowering the blood levels of such lipoproteins. There is ample experimental evidence in humans [with their consent] to support this approach to the problem." And:

"Thus in cases of xanthoma tuberosum and more recently in xanthoma tendinosum, each associated with extreme specific lipoprotein-class elevation, we have been able to demonstrate consistently that dietary and pharmacologic measures which lower the levels of the involved classes of lipoproteins result in diminution in size of xanthomatous lesions or in their complete disappearance. Further, in such individuals the development of new lesions is inhibited when the lipoprotein levels are maintained at lowered values. It cannot be proved that the intimal arteriosclerotic lesions will behave precisely as does the xanthomatous lesion, but there is every reason to consider that the same type of processes will occur, although at perhaps a different rate and to a greater or lesser extent."

2d. Not Speculative: Lipoprotein Classes Respond Differently to the Same Therapy

"If the factors which control the levels of the various classes of lipoproteins were identical and if the responses to therapeutic measures were identical, it would be of little moment which particular classes of lipoproteins contributed to the coronary disease risk in a patient, since the management would be the same for all types of lipoprotein derangement. SUCH, unfortunately, IS DISTINCTLY NOT THE CASE." [Emphasis is like that in the original]. And:

"Our studies have proven very conclusively that measures which modify the Std Sf 0-12 lipoproteins favorably may have absolutely no effect on the Std Sf 100-400 lipoproteins or may even have a deleterious effect of raising the level of such lipoproteins. Conversely, factors which may favorably modify elevated levels of the Std Sf 20-400 lipoproteins may have no effect whatever on elevated levels of Std Sf 0-12 or Std Sf 12-20 lipoproteins."

2e. Clinical Implications: What Measurements the Physician Must Obtain

"The implication of these facts for the prevention and management of acute myocardial infarction is extremely important to the clinician. In order to manage the problem intelligently in a particular patient, the physician must know which of the lipoproteins are elevated if he is to advise preventive or therapeutic measures. Without such knowledge, a dietary or pharmacologic approach is 'hit or miss'; it may fail to provide certain patients with the correct measures, and it may result in the advice of positively harmful measures in other cases." And:

"As will be pointed out later in this paper, the measurement of the serum cholesterol level will not serve as a substitute for a lipoprotein analysis, for it will fail to reveal which [atherogenic] lipoprotein classes are predominantly elevated and hence in need of correction." Part 5a and Box 3 of this Appendix provide a real-world example.

2f. "We All Eat Too Much Fat" --- A Dangerous Over-Generalization

"Several years ago, when the author and other workers proposed dietary measures for the management of lipoprotein elevations in the effort to prevent and treat myocardial infarction, there existed considerable skepticism ... about the possible efficacy of diet in lowering lipoprotein levels. Today the pendulum has swung so far in the other direction that the problem no longer resides in skepticism concerning dietary measures, but rather in the irrational blanket-use of dietary measures in all persons irrespective of the indications in a particular individual. Thus we are today bombarded by generalizations such as 'we all eat too much fat' or 'we all eat too much animal fat'." And:

"There is some element of truth in such statements, but on the other hand there is a considerable element of falsehood in them. Action based upon these generalizations may do almost as much medical harm as good, and in individual cases we can be certain that more harm than good will result. That the medical profession and the public have become receptive to the concept, that perhaps coronary disease can be modified by dietary means, is welcome. It is a corollary to this that valid and sane advice be provided as to how to put such dietary measures into intelligent use."

● Part 3. Experimental Dietary Evidence: Vegetable-Fat vs. Animal-Fat (Box 1)

Nichols, Dobbin, and Gofman (1957) evaluated the effect of both the quantity and origin (vegetable vs. animal) of dietary fats upon the blood level of all four classes of serum lipoproteins involved in coronary disease, namely the Std Sf 0-12, Sf 12-20, Sf 20-100, and Sf 100-400 classes. As noted in Part 1, these studies gave careful attention to evaluation of one dietary variable at a time, and the studies were not encumbered by the simultaneous administration of a variety of pharmaceutical preparations which could render results uninterpretable.

3a. Conduct of the Dietary Experiments

The effects of three different diets were evaluated in five males, ranging in age from 20 to 49 years. The same men followed each of the three diets (Nichols 1957, p.9, Table 5). Here, for brevity, we shall call these diets the Vegetable-Fat Diet (duration = 10.5 weeks), Animal-Fat Diet (duration = 11 weeks), and HighCarbo-LowFat Diet (total duration = 24 weeks).

Breakfast was standard throughout all the dietary periods and was eaten at home by the subjects. The average breakfast consisted of fruit juice, toast or cereal, and skim milk or coffee. Such breakfasts contained on the average about 13 grams of protein, 2 grams of fat, 54 grams of carbohydrate, and negligible quantities of cholesterol. Lunch and dinner were served at the Cowell Memorial Hospital of the University of California (Berkeley) under the supervision of Chief Dietitian, Mrs. Virginia Dobbin.

Boxes 1 and 2, adapted from Gofman 1958, present the composition of all three diets, including the breakfast. The diets were designed so that the total daily intake of calories was very nearly the same in all three regimes.

During the study, lipoprotein levels were determined at weekly intervals on all subjects by the ultracentrifugal technique (DeLalla 1954-a). It became apparent in these studies that, with respect to the dietary alterations of lipoprotein levels, the Std Sf 0-12 and Std Sf 12-20 classes of lipoproteins behaved similarly. Hence their measurements were combined as the Std Sf 0-20 class. The Std Sf 20-100 and Std Sf 100-400 classes behaved similarly to each other, but very differently from the lipoproteins of the Std Sf 0-20 class. The results for the Std Sf 20-100 and Std Sf 100-400 lipoproteins are presented together as the Std Sf 20-400 class.

Boxes 1 and 2 show the mean values of these two lipoprotein classes, individually for the five participants and for the five participants combined.

3b. Box 1: Results & Clinical Implications --- Veg-Fat vs. Animal-Fat

Box 1 shows that, at equal daily intake of fat and unequal daily intake of cholesterol:

• 1. The cholesterol-rich Std Sf 0-20 lipoprotein levels were consistently and highly significantly elevated during the Animal-Fat Diet, compared with the Veg-Fat Diet. Conversely, the Std Sf 0-20 lipoprotein levels were very much lower in all five participants during the Veg-Fat Diet than during the Animal-Fat Diet.

• 2. The triglyceride-rich Std Sf 20-400 lipoprotein levels showed no consistent trend and no significant difference in mean level, when the two dietary periods are compared.

At the time, we wrote: "Why there should exist this remarkable difference in behavior of the two lipoprotein classes, with respect to the origin of dietary fat, is not at all clear from present evidence. The fact of the existence of the difference is however solidly established, and as such has important bearing on the problem of prevention and treatment of myocardial infarction" (Gofman 1958, pp.274-275) And (p.275):

"It should be clear that where a patient has an elevation of the Std Sf 0-20 lipoproteins, one can expect to accomplish a striking reduction in lipoprotein levels by a shift from the ingestion of animal fat to the use of vegetable oil (unhydrogenated cottonseed oil in these studies), even without any restriction upon the TOTAL fat intake. However, if a patient is already low in the Std Sf 0-20 lipoproteins but high in the Std Sf 20-400 lipoproteins, essentially nothing would be accomplished by such a dietary shift. This would be one illustration of the folly of such a nonscientific blanket generalization as 'too much animal fat is the problem'."

3c. A Puzzle: What Explains Such a Remarkable Result?

The remarkable difference --- with respect to effect on the serum levels of Std Sf 0-20 lipoproteins --- between a diet high in fat of vegetable origin and one high in fat of animal origin, posed several questions, including (Gofman 1958, p.275):

(1) Is the different effect caused by some noxious substance, present in animal fat, which raises the Std Sf 0-20 lipoprotein levels?

(2) Is the different effect caused by some substance, present in the vegetable fat, which lowers the Std Sf 0-20 lipoprotein levels?

(3) If there is something noxious in animal fat, could its effect be overcome by the addition of vegetable oil to the diet without removal of the animal fat from the diet?

In seeking answers to questions (1) and (2), we undertook the comparison described in Part 4.

• Part 4. More Evidence: Vegetable-Fat Diet vs. HighCarbo-LowFat Diet (Box 2)

We reasoned that, if vegetable fat provides a positively beneficial substance which lowers the Std Sf 0-20 lipoprotein levels, then Std Sf 0-20 lipoprotein levels would RISE on a diet from which we removed that kind of fat, since the hypothetical protective substance would be absent. So we removed most of the vegetable fat from the Veg-Fat Diet and replaced it mainly by carbohydrate. This is the diet called HighCarbo-LowFat in Box 2, and it was followed by the same five participants who had followed the Veg-Fat and Animal-Fat Diets (Box 1).

We anticipated, also, that if the animal fat contains a possible noxious substance which raises the Std Sf 0-20 lipoprotein levels, then there would be little difference in levels of the Std Sf 0-20 lipoproteins in a comparison of the Veg-Fat and HighCarbo-LowFat diets, because animal fat would be virtually absent in both diets --- as shown in the "Composition" section of Box 2.

4a. Results: Std Sf 0-20 Lipoprotein Levels on the HighCarbo-LowFat Diet

The Std Sf 0-20 lipoprotein levels were not shifted in a consistent direction, nor was there any significant change in their mean levels when the Veg-Fat and HighCarbo-LowFat Diets are compared (Box 2). These results provided no basis for believing that vegetable fat contains any positively beneficial agent capable of actively lowering the Std Sf 0-20 lipoprotein levels. The results seemed to suggest that the elevation of Std Sf 0-20 lipoprotein levels observed on the Animal-Fat Diet (Box 1) was due to the presence in such fats of a factor or factors capable of elevating them.

Our suspicions were not limited to animal-fats. We cited the work of Ahrens 1955 for the extension of suspicion to saturated and hydrogenated fats of any origin, and we wrote (Gofman 1958, p.278): "It appears at present that the natural fats of animal origin such as dairy fat, meat fat, and egg fat are least favorable with respect to content of the noxious agents which affect the blood lipids [Std Sf 0-20], that saturated vegetable oils, either naturally occurring or produced by hydrogenation, are unfavorable but not to the extent of animal fats, and that the unsaturated oils, while not BENEFICIAL with respect to maintaining low blood lipid levels [Std Sf 0-20], are at least neutral in this regard."

4b. Results: Std Sf 20-400 Lipoprotein Levels on the HighCarbo-LowFat Diet

The Std Sf 20-400 lipoprotein levels were consistently and highly significantly elevated during the HighCarbo-LowFat Diet, compared with the Veg-Fat Diet (Box 2). We had not anticipated such a strong effect.

The "Carbohydrate Effect," as it came to be called in many circles, is the unwelcome effect of high carbohydrate intake on RAISING the levels of the atherogenic triglyceride-rich Std Sf 20-400 lipoprotein levels.

4c. Knopp 1997: The "Unwanted" Carbohydrate Effect

By now, there are many references in the literature to the fact that high carbohydrate intake can elevate serum levels of total triglyceride --- a measurement which reflects elevation in levels of the triglyceride-rich lipoproteins, without revealing which part of the Sf 20-40,000 segment is the source.

Here, we cite only a recent example (Knopp 1997) which recognizes elevated serum triglyceride (hypertriglyceridemia) to be an "unwanted" effect. Knopp and co-workers studied the effect of 4 fat-restricted diets (followed for 1 year) on two groups of men having high serum levels of LDL-cholesterol. One group (HC for Hyper-Cholesterolemic) had only the LDL-cholesterol elevated, pre-diet. The other group (CHL for Combined Hyperlipidemic) ALSO had serum total triglyceride elevated, pre-diet. One of their concluding comments is this (Knopp 1997, p.1514):

"Among the unwanted effects of an aggressively low-fat diet were the plasma TG [triglyceride] increases of 22% and 39% on diets 3 and 4 in HC subjects, even after 1 year. This elevation was confirmed when TG levels were examined among subjects with intakes of carbohydrate consistently higher than 60% (Retzlaff 1995). These findings indicate that hypertriglyceridemia is induced by an aggressively fat-restricted, high-carbohydrate diet and that the elevation persists longterm. The failure of plasma palmitic acid content to fall, despite progressive dietary saturated fat restriction, suggests that fatty acid synthesis may increase as fat intake is restricted and carbohydrate intake is increased. This mechanism may explain both the induction of hypertriglyceridemia and the lack of further LDL-C lowering with extreme dietary fat restriction." (Note: The mean LDC-cholesterol reductions on the 4 diets were modest --- ranging from 2.8% to 13.4%; Knopp, p.1509). In 1998, Hegsted cites a report (Antonis 1961) that the duration of the carbohydrate effect on plasma triglyceride is "temporary" (Hegsted 1998, p.918). In response, Hu et al cite evidence (Mensink 1992, + Knopp 1997) that the effect is NOT "transient" (Hu 1998, p.919).

● Part 5. Infarct Case + Xanthoma Case: Responses to LOW-Carbo Diets (Boxes 3 + 4)

Boxes 3 and 4 illustrate --- forcefully --- how the Carbohydrate Effect can be clinically applied, in reverse, to LOWER the blood levels of the atherogenic Std Sf 20-400 lipoproteins in persons who need such therapy.

5a. Infarct Survivor with Very-High Sf 20-400 & Below-Avg. Sf 0-20 Levels

There exist many persons in whom the hazard of premature atherosclerosis and myocardial infarction is causally related to a marked elevation of Std Sf 20-100 lipoproteins, or to Std Sf 100-400 lipoproteins, or to a combination of these two classes. We do not consider it reasonable to re-prove these facts. In such individuals, the Std Sf 0-20 lipoproteins may be quite low, and hence contribute little to the coronary disease risk.

Box 3 presents such a case: A 65-year-old male survivor of a Myocardial Infarction, studied well beyond the acute phase of his infarct. On his usual diet, before he followed a very LOW-Carbohydrate diet, his blood levels of the Std Sf 0-20 lipoproteins were BELOW the average of his day, and his 20-400 levels were FAR ABOVE the average for his day. These comparisons can be made by looking at average levels in Box 5.

In such cases, the problem of management necessarily resides in efforts to reduce the elevated Std Sf 20-400 levels. An obvious application of the findings reported in Part 4 would be to recommend a LOW-CARBOHYDRATE diet. The diet advised for this case contained only 100 g of carbohydrate per day, and the diet used vegetable oil to replace calories lost from carbohydrate. No appreciable change in body-weight occurred during this controlled diet (duration = 60 days).

Box 3 presents the favorable response: A massive fall occurred in the plasma Std Sf 20-400 lipoprotein levels, without a significant increase in the Std Sf 0-20 lipoprotein levels. Notably, the total serum cholesterol measurement did not change during the diet. Thus, the cholesterol measurement revealed nothing about the major improvement in what we consider to be major plasma atherogens.

5b. A Case of Xanthoma Tuberosum with Exceedingly High Sf 20-400 Levels

This same principle of carbohydrate restriction has been applied successfully in several types of extreme derangement of lipoprotein levels. For example:

Box 4 presents the lipoprotein and cholesterol measurements of a patient with Xanthoma Tuberosum, pre-diet and during 36 days of an extremely LOW-carbohydrate diet. Pre-diet, her Std Sf 20-400 levels were exceedingly high (Box 5 provides a comparison with "normals" of that time). After five weeks on the controlled diet, her mean levels of Std Sf 20-400 lipoproteins were dramatically down (with no dramatic rise in Std Sf 0-20 levels) --- although, to be sure, there was quite a lot of room for improvement beyond what was achieved. In fact, this patient's xanthomata underwent considerable regression during further periods of carbohydrate restriction on a home diet. And she was not an exception (Gofman 1959, p.211).

● Part 6. Could a Declining Level of Total Serum Triglyceride Be Unhealthy?

We have worried a lot that certain preventive or therapeutic regimes might transform non-atherogenic lipoprotein molecules into atherogenic ones, without recognition (Chapter 44, Part 3a).

Specifically, we have worried that some pharmaceutical and dietary regimes which reduce hypertriglyceridemia --- as measured by total serum triglyceride (TG) --- might actually increase the atherogenic Std Sf 20-400 lipoproteins. We think patients deserve evidence that this is NOT happening. In our own work, we had such assurance because we measured the responses to therapy of the Std Sf 0-12, 12-20, 20-100, and 100-400 lipoproteins in their native state, and we saw levels of the Sf 20-400 lipoproteins DECLINE --- for instance, during LowCarbo-High(Veg)Fat diets.

How might it be possible to INCREASE the triglyceride-rich Sf 20-400 lipoproteins while simultaneously REDUCING the total serum triglyceride? Hydrolysis of triglyceride transforms it into something else (fatty acids), and thus TG hydrolysis results in a lower level of total serum triglyceride. Suppose that a therapeutic regime promotes TG hydrolysis. Suppose that the lower TOTAL serum triglyceride conceals an unhealthy redistribution of residual post-hydrolysis triglyceride --- i.e., conceals a lower frequency of the Sf 400-40,000 lipoproteins and a HIGHER frequency (concentration) of the atherogenic Sf 20-400 lipoproteins.

This would represent a serious increase in atherogenic risk --- and no warning would be provided by the chemical measurement which shows a reduced amount of total serum triglyceride. By itself, the chemical triglyceride analysis does not provide the crucial information. If proper studies are done, either by ultracentrifugal analysis or by some equivalent alternative, we can really KNOW the answer --- an answer which we are obligated to know.

Today, there exists some potentially good news on this very issue, within the elegant work of William S. Harris and colleagues on omega-3 fatty acids and levels of blood lipoproteins (Part 8b). Before describing it, we must set forth a few terms (Part 7).

● **Part 7. The Healthful-Oils Hypothesis: Nomenclature and Sources**

A great deal of evidence has been published indicating that the distinction, between fats of animal vs. plant origin, is not adequate to separate harmful dietary fats from healthful ones. Certain "marine oils" from fatty fish (clearly an animal source) and from some other marine animals may be very healthful, while coconut oil (a plant source) has a composition very similar to beef-fat and butter-fat (animal sources). Soon after the lipoprotein spectrum was revealed, the specific fatty-acid composition of dietary fats was studied at many labs, including Donner (Freeman 1952 + 1953). Ongoing research today indicates that anti-atherogenic benefits may depend upon EXACTLY which kinds of fatty acids are in the dietary fat. Before discussing evidence which may be of great benefit to persons trying to avoid IHD, Part 7 defines certain terms and abbreviations.

7a. Fatty Acids, Triglycerides, and Esters

● FATTY ACID. Fatty acids have a methyl group at one end, a carbon-carbon chain of variable length, and a carboxyl terminus (COOH --- which makes them "carboxylic" acids).

● TRIGLYCERIDE. Triglyceride (sometimes called "neutral fat" or just "fat") is any molecule formed when the 3-carbon alcohol (glycerol) combines with 3 fatty acids. The combination of glycerol with only 1 fatty acid yields a monoglyceride; with 2 fatty acids, a diglyceride.

● ESTER, ETHYL ESTER, and ESTERIFIED. Esters are organic compounds which are formed by the splitting out of water between an alcohol (e.g., glycerol, ethanol, cholesterol) and a carboxylic acid. Triglyceride is a fatty acid glyceryl ester. When the alcohol is ethanol, the combination is an "ethyl ester." The "cholesterol esters" are a prominent feature in Appendix-E, Boxes 1 + 2. When an alcohol and carboxylic acid combine, the alcohol is "esterified."

7b. Major Long-Chain Fatty Acids of Dietary Importance

Because fatty acids differ in the length of their carbon-carbon chains and in the number of double-bonds in that chain, a notation is commonly used, to indicate "what's what." For example, 12:0 denotes the fatty acid which has 12 carbons in its chain and zero double-bonds in that chain: Lauric Acid.

By definition, saturated fatty acids have no double-bonds in the carbon-carbon chain, so the entry after the colon is ZERO for all of the saturated fatty acids.

By contrast, the UNsaturated fatty acids have at least one double-bond in the carbon-carbon chain. The position of a double-bond in the carbon chain is denoted either as its omega-position or its n-position. Thus, omega-3 is the same as n-3. When a chain has multiple double-bonds, usually the position only of the FIRST double-bond is named.

● SATURATED FATTY ACIDS.

Lauric Acid	12:0
Myristic Acid	14:0
Palmitic Acid	16:0
Stearic Acid	18:0

● MONO-UNSATURATED FATTY ACIDS.

Oleic Acid	18:1 n-9	One double-bond, at the 9 position in the chain.

● POLY-UNSATURATED FATTY ACIDS.

Linoleic Acid	18:2 n-6	Two double bonds, one at n-6 and one at n-9.
a-Linolenic Acid	18:3 n-3	Three double bonds: n-3, n-6, n-9.

EicosaPentaenoic Acid (also referred to as EPA). Eicosa signals the 20; Penta, the 5.
          20:5 n-3   Five double bonds: n-3, n-6, n-9, n-12, n-15.

DocosaHexaenoic Acid (also referred to as DHA). Docosa signals the 22; Hexa, the 6.
          22:6 n-3   Six double bonds: n-3, n-6, n-9, n-12, n-15, n-18.

Some Rich Dietary Sources of the UnSaturated Fatty Acids

● Oleic Acid (n-9): Olive oil.
● Linoleic Acid (n-6): Certain oils, especially safflower, sunflower, walnut, soybean, and corn.
● Alpha-Linolenic Acid (n-3): Certain oils, especially flaxseed oil, walnut oil, canola (rapeseed) oil.
● EPA and DHA (both n-3): Certain fish or "marine oils," including salmon, tuna, herring, mackerel. However, the body can and does produce these fatty acids from their precursor: Alpha-Linolenic Acid.

● The dietary examples above are much expanded in books such as Simopoulos 1998, and others.

● Part 8. Encouraging Evidence about the n-3 Marine Oils: The 1989 Review by W.S. Harris

For about two decades, William S. Harris, William E. Connor, D. Roger Illingworth, B.E. Phillipson, and co-workers have been leading figures in the effort to elucidate the effects of dietary omega-3 fatty acids on lipid metabolism and on management of serum lipoprotein levels, especially with respect to Ischemic Heart Disease. Harris provided a superb review of the "fish oil" evidence in his 1989 paper "Fish Oils and Plasma Lipid and Lipoprotein Metabolism in Humans: A Critical Review" (Journal of Lipid Research). Harris begins (Harris 1989, p.785):

"The purpose of this report is to review the effects of the long-chain, n-3 fatty acids (also referred to as omega-3 fatty acids) found in fish oils on human plasma lipids and lipoproteins." His magnificent Table 1 lists the results from 68 dietary experiments which increased the intake of fish oils, and reported upon the observed changes in (a) total serum cholesterol, (b) total serum triglyceride, (c) LDL cholesterol, and (d) HDL cholesterol. Altogether, there were 928 participants, including 596 persons whose pre-diet plasma levels were considered "normal," + 37 "Type II-a patients" (hypercholesterolemia), + 194 "Type II-b patients" (combined hyperlipidemia), + 101 patients with isolated hypertriglyceridemia (Types IV and V).

8a. Overall Results on the Harris Meta-Analysis

Harris' Figure 1 (p.790) summarizes the results of the meta-analysis, in percent change. All four types of participants experienced large decreases in total serum triglyceride, and small increases in HDL cholesterol. Changes in total cholesterol and LDL cholesterol were negligible, with one exception: The patients with isolated hypertriglyceridemia (who experienced a 52% decrease in total serum triglyceride) experienced a 30% increase in LDL cholesterol (mostly in the Type-IV patients).

Also notable, in terms of fish oil in the potential prevention of IHD, are the responses specifically of the participants with pre-diet "normal" measurements. They "responded to n-3 fatty acids with essentially no change in total or LDL cholesterol, a 25% decrease in triglyceride levels, and a slight rise in HDL-C (+3%)" (Harris 1989, p.787).

We note that these results are echoed in the 1995 review paper by Martijn B. Katan and colleagues, who cite work by Harris and others about EPA and DHA: "These very-long-chain (n-3) polyunsaturates do not share the LDL-lowering effect of linoleic acid (18:2 n-6, n-9). On the contrary, several studies have shown that fish oils raise LDL and apoprotein B, and similar results have been reported for fatty fish. Fish oil and fatty fish do, however, have favorable effects on serum triglycerides and very-low-density lipoprotein (VLDL), which can be reduced by intake of a few grams of fish oil per day. Whether this effect is responsible for the lower incidence of Coronary Heart Disease, observed in fish-eating populations in epidemiologic studies, remains uncertain; fish oils also can modulate many other physiological processes, including blood platelet function" (Katan 1995, p.1371-S). Katan and colleagues also discuss the "carbohydrate effect".

The parts of the 1989 Harris paper, discussed below, relate (a) to our worry that falling levels of total serum triglyceride might conceal an unrecognized increase in blood levels of the atherogenic Std Sf 20-400 lipoproteins, and (b) to the growing recognition that such lipoproteins are atherogenic. We recommend also all parts of the Harris paper NOT discussed below, including "Fish Oils and Lipoprotein Metabolism," "Fish versus Fish Oil Supplements," "Linolenic Acid versus Fish Oils," and "Fish Oils in the Treatment of Hyperlipidemia."

8b. Does Extra Fish Oil Increase the Atherogenic Std Sf 20-400 Lipoproteins?

Harris 1989 uses the following terminology at page 794:
- VLDL-1 refers to large lipoproteins in the Sf 100-400 range.
- VLDL-2 refers to smaller lipoproteins in the Sf 20-60 range.

Harris, citing Sullivan 1986, describes the response of four patients (Type-IV) with isolated hypertriglyceridemia. They took a 15 gram dose of MaxEPA daily for 2 weeks. MaxEPA is described (in Simopoulos 1998, p.353) as concentrated fish-body oils having (17.8 grams EPA + 11.6 grams DHA) per 100 grams of total oil. The result of the Sullivan experiment is possibly worrisome. From Harris 1989, p.793-794:

"The distribution of light and heavy LDL particles, the percent apolipoprotein B, and the cholesterol to protein ratios were not different before and after fish oil in the Type IV patients. However, the size distribution of VLDL changed significantly, with a decrease in the large particles (Sf 100-400) and an increase in the small particles (Sf 20-60) with fish oil supplementation. The average diameter of the VLDL particles decreased by 22%."

Although the findings are worrisome, it is a real pleasure to see use of an appropriate technique (ultracentrifugation) to identify the VLDLs under investigation. From the limited evidence Harris gives us on the Sullivan experiments, we wonder whether there was a net increase in atherogenic risk, as a result of some decrease in Sf 100-400 and some increase in the Sf 20-60 VLDLs. These data leave open the possibility of decrease overall, no change overall, or increase overall in the atherogenic impact of the remaining mix of VLDLs.

In addition, Harris provides relevant evidence from his own laboratory, and it is NOT ambiguous on this important issue. He reports (Harris 1989, p.794): "In a preliminary study from our laboratory (Inagaki 1988), five Type IV patients were given 6 grams of n-3 fatty acid ethyl esters daily for 4 weeks. Before and after the supplements, the plasma lipoproteins were separated by gel filtration chromatography according to Rudel, Marzetta, and Johnson (Rudel 1986) and analyzed for compositional changes (Table 2). There appeared to be greater reduction in the large VLDL-1 particles than in the small ones (VLDL-2 and IDL) after the fish oil treatment. LDL and HDL cholesterol and protein contents both increased."

These results indicate that serum levels decreased for BOTH groups of VLDL (Sf 20-60 and Sf 100-400). This study (small though it is) provides encouragement that fish oil supplements may be able to lower the IHD risk from both groups of VLDL at the same time, without causing a HIDDEN increase in atherogenic species --- hidden by a reduced "total triglyceride" measurement which might conceal a net INCREASE in levels of Std Sf 20-400 lipoproteins (Part 6).

Of course, we do not dismiss the non-hidden increase in LDL cholesterol in the patients who had Type IV hypertriglyceridemia. We note, however, that the LDL cholesterol rise was absent or small in the other three types of participants described in Part 8a. Nonetheless, there are many other types of persons whose responses we do not know. Meanwhile, we consider the "fish oil" frontier to be an extremely important area in efforts to help prevent IHD.

● **Part 9.   Why Reduce Serum Triglyceride Levels?   Harris's "Good Reasons"**

In Harris's section on "Fish Oils in the Treatment of Hyperlipidemia," he asks (Harris 1989, p.801):

"Since elevated triglyceride levels are not generally regarded as independent risk factors for CHD, why be concerned about them?   There are several good reasons why VLDL in hypertriglyceridemic patients should be reduced."

9a.   The "Good Reasons" Listed by Harris (Harris 1989, p.801)

● "Data from the Framingham Heart study have shown that 90% of hypertriglyceridemic individuals are, in fact, at increased risk for CHD; the exception being the 10% of patients with low total to HDL cholesterol ratios (Castelli 1986)."   And (p.801):

● "In women over age 50, triglyceride levels are independently correlated with CHD risk and are even better predictors than LDL-C levels (Reardon 1985)."   And (p.801):

● "Carlson, Bottiger and Ahfeldt in the Stockholm Prospective Study found that serum triglyceride levels were independent predictors of subsequent myocardial infarction in both sexes (Carlson 1979)."   And (p.801):

● "Gemfibrozil was recently shown to lower coronary risk in the Helsinki Heart Study (Frick 1987).   This drug lowered VLDL levels by 40%, raised HDL-C levels by 8%, and lowered LDL-C by only about 9%.   Since gemfibrozil and fish oils appear to produce similar changes in VLDL and HDL, one might expect that n-3 fatty acids may be anti-atherogenic as well."

A Recommendation from Harris for Any Patient at Increased Risk for CHD

Several paragraphs later, Harris concludes the section (p.801):   "Finally, a case could be made for providing low levels of n-3 fatty acids (0.3 to 1 gram/day, or one or two capsules/day) to any patient at increased risk for CHD, especially if that patient cannot or will not increase fish intake.   This level of supplementation would provide as much n-3 fatty acid as consuming three 100-g servings of salmon per week, and would surpass the n-3 fatty acid intake associated with reduced coronary events in the study of Kromhout et al (Kromhout 1985).   Although this low dose may have little, if any, measurable effect on plasma lipid levels, it may, in the long run, slow the progression of atherosclerosis, and the potential risk to the patient from this intake is essentially nil."

9b.   "The Triglyceride Issue: A View from Framingham" (Castelli 1986)

Directly from William Castelli's 1986 paper, we provide some details not mentioned by Harris, above.

Castelli presents two bar graphs showing "CHD Risk According to Lipid Level on Entry into the Framingham Study; Men Aged 30-62" (Castelli 1986, p.434, Figure 4).   These graphs show lipoprotein level versus CHD morbidity ratio, separately for the cholesterol-rich Sf 0-20 lipoproteins and for the triglyceride-rich Sf 20-400 lipoproteins.   Each graph displays an almost perfect positive relationship between rising level of serum lipoprotein and morbidity ratio.   Castelli states (p.434):

"Figure 4, which shows lipid levels in men on entry into the Framingham Study, indicates that the higher the VLDL or triglyceride level, measured as Sf 20-400, the higher the subsequent CHD rate.   This is the univariate association of triglyceride levels with the risk of CHD."   He continues (p.435):

"Early in the study, however, it was observed that if one adjusted the relationship of the Sf 20-400 by the concomitant cholesterol level, the impact of triglyceride on risk disappeared.   It was then concluded that triglycerides were not an independent risk factor, at least in men.   In women the story is quite different ..."   In women, the triglyceride levels "were better than LDL in predicting subsequent CHD."   Returning to the men, Castelli states (p.435):

"More recent analyses by Abbott et al (Abbott 1984) ... indicated that triglycerides do play an independent role in CHD risk in men over the age of 50, and that this role has been masked previously by the association of triglycerides with other risk factors (e.g., HDL) that are themselves related to CHD. When one untangles such relationships, triglycerides emerge as an important risk for CHD in men (Table 2)." Our Appendix-G presents some striking graphs and data on the inverse relationship between levels of EACH class of serum lipoproteins (Std Sf 0-12, 12-20, 20-100, 100-400) and levels of HDL. Castelli concludes (pp.436-437):

"The final view from Framingham is that individuals who have high triglyceride levels should be considered at high risk for CHD. The only exceptions to this judgment are those with a very low total/HDL cholesterol ratio. Since only about 10% of these high-triglyceride patients will have a low total/HDL cholesterol ratio (<3.5), 9 of 10 patients with high triglyeride levels are at a high risk for premature CHD, even though almost half of these patients will have a total cholesterol less than 250 mg/dl."

● **Part 10. A Beneficial "Mediterranean Diet": The Lyon Diet Heart Study**

With respect to reducing mortality rates from Ischemic Heart Disease, we are impressed by results reported in the Lyon Diet Heart Study, conducted by Michel de Lorgeril, Serge Renaud, and co-workers. The relationship of their results with the Lipid Hypothesis, and with our Unified Model of Atherogenesis and IHD Death, will be discussed too (Part 11). Papers presenting the Lyon Diet Heart Study and its results include:

● DeLorgeril et al, 1994 in the Lancet 343: 1454-1459, "Mediterranean Alpha-Linolenic-Acid-Rich Diet in Secondary Prevention of Coronary Heart Disease."

● Renaud et al, 1995 in the American Journal of Clinical Nutrition 61 (suppl): 1360S-1367S, "Cretan Mediterranean Diet for Prevention of Coronary Heart Disease."

● DeLorgeril et al, 1996 in the Journal of the American College of Cardiology 28: 1103-1108, "Effect of a Mediterranean Type of Diet on the Rate of Cardiovascular Complications in Patients with Coronary Artery Disease; Insights into the Cardio-protective Effect of Certain Nutriments." See also DeLorgeril 1997 and 1998 in our Reference List.

● DeLorgeril et al, 1999 in the journal Circulation 99: 779-785, "... Final Report of the Lyon Diet Heart Study." Commentary: Leaf 1999 in our Reference List.

10a. A Comparison of the Test-Diet and the Control-Diet

The purpose of the Lyon Diet Heart Study was "to test in France the hypothesis that a Mediterranean diet, especially the one consumed on the island of Crete in the early 1960s, could have more of a protective effect against Coronary Heart Disease than the prudent diet that is often recommended for coronary patients" (Renaud 1995, p.1360S; see also pp.1364-65S).

In the Lyon Diet Heart Study, "the control patients received no dietary advice apart from that of hospital dietitians or attending physicians. Patients in the experimental group were advised by the research cardiologist and dietitian, during a one-hour-long session, to adopt a Mediterranean-type diet: More bread, more root vegetables and green vegetables, more fish, less meat (beef, lamb, and pork to be replaced by poultry), no day without fruit, and butter and cream to be replaced with margarine supplied by the study ... The oils recommended for salads and food preparation were rapeseed [canola] and olive oils exclusively. Moderate alcohol consumption in the form of wine was allowed at meals. At each subsequent visit of the experimental patients, a dietary survey and further counseling were done by the research dietitian. Diet evaluation comprised a 24-hour recall and a frequency questionnaire" (DeLorgeril 1994, p.1455). The margarine supplied, free of cost, was based on rapeseed (canola) oil. It had a composition very similar to olive oil (DeLorgeril 1994, p.1455), and contained only 6% trans fatty acids (Renaud 1995, p.1365S).

A comparison of the average control-diet and average test-diet, after 1 to 4 years of follow-up, is provided in DeLorgeril 1994 (p.1456, Table 5) in terms of daily intake (g/day) of various food-types. Omitting the standard errors and p-values, we list the mean values in g/day from Table 5, with the control diet listed first and then the experimental diet:

Bread 145, 167. Cereals 99.4, 94.0. Legumes 9.9, 19.9. Vegetables 288, 316. Fruits 203, 251. Delicatessen (ham, sausage, offal) 13.4, 6.4. Meat 60.4, 40.8. Poultry 52.8, 57.8. Cheese 35.0, 32.2. Butter + cream 16.6, 2.8. Margarine 5.1, 19.0. Oil 16.5, 15.7. Fish 39.5, 46.5. DeLorgeril 1994 (p.1454) reports:

"The experimental group consumed significantly less lipids, saturated fat, cholesterol, and linoleic acid but more oleic and alpha-linolenic acids confirmed by measurements in plasma. Serum lipids, blood pressure, and body mass index remained similar in the 2 groups. In the experimental group, plasma levels of albumin, vitamin E, and vitamin C were increased, and granulocyte count decreased." And (p.1457):

"After 52 weeks, there were higher [plasma] concentrations of oleic, alpha-linolenic, and eicosaPentaenoic [EPA] acids and reduced concentrations of stearic, linoleic, and arachidonic acids in the experimental group (Table 4). The increase in eicosaPentaenoic acids was probably related to alpha-linolenic acid intake since intake of fish was not significantly increased (Table 5)."

10b. Who Was Enrolled, and What Were the Clinical Outcomes?

The Lyon Diet Heart Study was a prospective, randomized, single-blinded, multi-clinic, secondary prevention trial which enrolled (between March 1988 and March 1992) mostly male patients under age 70 who had survived a first myocardial infarction six or fewer months beforehand. After randomization, the control and experimental groups were quite well matched in infarction history and treatment, age (mean = 53.5 years), and "main cardiovascular risk factors" (DeLorgeril 1994, Tables 2 + 6). At the outset, there were 303 patients in the control group and 302 in the experimental group.

"After the randomization visit, patients of both groups were scheduled to be seen 2 months later and then annually at the Research Unit. These visits did not replace their regular visits to the attending physicians, who were responsible for all aspects of treatment, including use of medication and of invasive diagnostic and therapeutic procedures" (DeLorgeril 1996, p.1104). Mean follow-up time was nearly the same: 27.1 months (594 person-years) in the control group and 26.9 months (606 person-years) in the experimental group (DeLorgeril 1994, p.1456, and p.1457, Table 6).

The primary endpoints evaluated were cardiovascular death and nonfatal acute myocardial infarction. Six major secondary endpoints and ten minor secondary endpoints also were evaluated (DeLorgeril 1996, p.1105, Table 2).

With respect to the primary endpoints, there were 33 in the control group, and 8 in the experimental group. With respect to the secondary endpoints, there were 37 in the control group and 6 in the study group. With respect to the minor secondary endpoints, there were 84 in the control group and 58 in the experimental group (DeLorgeril 1996, p.1105, Table 2). These are striking differences.

We also wish to mention that, in this study, use of aspirin (about 250 mg/day) was the only pharmaceutical which showed a significant and inverse relationship with the primary endpoints (DeLorgeril 1996, p.1105).

10c. DeLorgeril and Colleagues Ask: Are the Benefits "Plausible"?

Although one can not rule out the possibility, in such studies, that the result was "built-in at the start" (i.e., that pure bad luck distributed more extremely sick patients into the control group than into the experimental group), it would be irresponsible to dismiss beneficial results merely because such a possibility exists. More studies either will or will not confirm this very promising work.

DeLorgeril and co-workers themselves address several questions about possible bias in the study. For example, with respect to possibly different TREATMENT (other than diet) within the two groups, they note that "no difference between the two groups in the use of medication was detectable" (DeLorgeril 1996, p.1106).

According to Renaud (as quoted in Jones 1996, p.63), "The experiment had to be stopped. It would have been unethical to continue" because too many of the patients on the regular diet were dying compared with those on the Mediterranean Diet. "You cannot continue. You are watching them die."

Convinced that the benefits of the Mediterranean Diet are real, DeLorgeril, Renaud, and colleagues want to know WHICH elements of the diet explain those benefits (DeLorgeril 1996, p.1107):

"To evaluate the plausibility of the results of the trial, it is important to try to identify which biologic factors, modified by the Mediterranean Diet, may have been cardio-protective. Two major biologic factors were modified by the intervention: 1) anti-oxidant vitamins, alpha-tocopherol and ascorbic acid, which were increased in the plasma of the study [experimental] patients; 2) the plasma fatty acid profile, with a noticeable increase in omega-3 fatty acids and a decrease in omega-6 fatty acids in the study group. Other factors such as the anti-oxidant flavonoids and minerals, arginine, glutamine and methionine and vitamins of the B-group including folic acid probably played important roles but were not measured in the study."

DeLorgeril and co-workers point out that their results are consistent with other sets of clinical observations linked to diets which reduce the ratio of n-6 to n-3 fatty acids to the neighborhood of about 5:1 (see especially Renaud 1995, pp.1365-66 S; also DeLorgeril 1996, p.1103). They discuss:

1) The lower mortality rate from Coronary Heart Disease in southern Europe compared with northern Europe (Keys 1984), and particularly lower CHD mortality in Crete, where there is a relatively low ratio of n-6 to n-3 fatty acids in serum cholesterol esters (Sandker 1993); 2) the very low MortRates from CHD in the Japanese of Kohama Island, where plasma levels are high in the n-3 alpha-linolenic acid (Kardinal 1993); 3) the reduced rate of death and myocardial re-infarction in the DART Study which had an increased intake of fish and n-3 fish oil (Burr 1989); 4) "results comparable to the Lyon study" in the diet-study reported by Singh et al (Singh 1992-a + 1992-b).

DeLorgeril, Renaud and co-workers also point out that there are several lines of other evidence which may explain WHY the Mediterranean Diet would have the striking benefits which it appears to have. Part 11, below, describes some of them.

● Part 11. How the Lyon Study May Relate to the Lipid Hypothesis and Our "Unified Model"

The Final Report of the Lyon Diet Heart Study states: "The protective effect of the Mediterranean dietary pattern was maintained up to 4 years after the first infarction, confirming previous intermediate analyses. Major traditional risk factors, such as high blood cholesterol and blood pressure, were shown to be independent and joint predictors of recurrence, indicating that the Mediterranean dietary pattern did not alter, at least qualitatively, the usual relationships between major risk factors and recurrence" (DeLorgeril 1999, p.779).

The Lyon Diet Heart Study did not measure serum lipoprotein levels. It measured total serum cholesterol, LDL-cholesterol, HDL cholesterol, and total serum triglyceride in control patients and in experimental patients, at the outset of the study, at week 8, at week 52 in about 80% of the initial participants, and at week 104 in about 57% of the initial participants (DeLorgeril 1994, p.1457, Table 6).

The two groups were well matched in the mean measurements at the outset, and they REMAINED alike. Such measurements can not reveal whether changes occurred in the lipoprotein pattern. However, if one considers "total triglyceride" (TG), it is clear that the appreciable decreases observed during diets supplemented by EPA and DHA fish oils (Harris review, in Part 8, above) did NOT occur in the Lyon Diet Heart Study. Is this an instance of conflicting observations? Probably not. Rather, the results may reflect an effect of dosage. In the Lyon Study, the amount of ingested fish oil was almost certainly far lower than in the diets described by Harris in Part 8.

If the benefits observed in the Lyon Diet Heart Study are real, and if we suppose that they occur without reducing blood-levels of the atherogenic lipoproteins, then what is their explanation?

11a. Four of the Explanations Proposed by DeLorgeril, Renaud, and Colleagues

In lucid and well-documented discussions, DeLorgeril 1994, Renaud 1995, and DeLorgeril 1996 present lines of evidence that cardio-protective benefits of the Mediterranean-Type Diet could arise from four different routes.

1) ANTI-INFLAMMATORY EFFECT. DeLorgeril 1996 (p.1107) reports that "Recent studies in humans (Rapp 1991, + Felton 1994) have shown a direct influence of dietary fatty acids on the fatty acid composition of arterial lesions. Rapp et al reported the incorporation of dietary omega-3 fatty acids in obstructive arterial lesions within some days after starting supplementation ... Omega-3 fatty acids may have an anti-inflammatory and stabilizing effect on the lipid-rich lesions because they have been shown in various animal models [two references] and humans [four references] to interfere with the many secretory and pro-inflammatory properties of leukocytes ... Thus, loading plaque with omega-3 fatty acids, as occurs in patients with high intake and high plasma levels of omega-3 fatty acids (Rapp 1991, + Felton 1994), can induce local anti-inflammatory activity." Reduction in certain aspects of the inflammatory response (Chapter 44, Part 7b) may indeed help to make the lipid-pools of plaques less thrombogenic --- when such plaques have not been prevented in the first place.

2) ANTI-OXIDANT EFFECT. DeLorgeril and Renaud cite observations that there are lower rates of CHD in persons with high intake of the anti-oxidant Vitamin E (for instance, Rimm 1993, + Stampfer 1993) and high intake of oleic acid, which is "remarkably resistant to oxidation" compared with some other fatty acids (DeLorgeril 1996, p.1107, citing 2 references, and citing Witztum 1991 for the view that "oxidized lipids are also thought to play a major role in arterial complications by stimulating macrophages, injuring endothelial cells and promoting leukocyte coagulant activity and platelet reactivity"). Both Vitamin-E and oleic acid intake were elevated in the Lyon Diet Heart Study.

3) ANTI-THROMBOTIC EFFECT. Renaud 1995 (p.1365S) states that the intake of 18:3 n-3 [Alpha-Linolenic Acid] is associated with inhibitory effects on the clotting activity of platelets, on their response to thrombin [2 references], and on the regulation of arachidonic acid metabolism (Budowski 1985)," and that "n-3 fatty acids give rise to a different family of prostanoids than do n-6 fatty acids, with major consequences for thrombogenesis (Knapp 1986)." In addition, he cites two studies which report that a high intake of 18:2 n-6 [Linoleic Acid] reduces the conversion of 18:3 n-3 (Alpha-Linolenic Acid] to the longer-chain n-3 fatty acids, EPA and DHA. DeLorgeril 1996 (p.1107) states that the Lyon Diet produced a ratio of arachidonic acid to eicosaPentanoic acid [EPA] in the plasma which "was also extremely favorable for obtaining an anti-thrombotic effect through an improved balance in the generation of prostacyclin and thromboxane [2 references]."

4) ANTI-ARRHYTHMIC EFFECT. DeLorgeril 1994 (p.1459) ends the discussion section as follows: "The fact that no sudden death occurred in the experimental group [of the Lyon Diet Heart Study] against 8 in the control group, suggests a possible additional anti-arrhythmic effect, consistent with observations in man (Burr 1989, + Riemersma 1989) and animals (McLennan 1993) indicating that n-3 fatty acids, especially alpha-linolenic acid, markedly reduced the incidence of lethal arrhythmias."

11b. How Do These Possible Mechanisms Relate to the Lipid Hypothesis?

The four possible mechanisms described above, for explaining the cardio-protective effects of the Mediterranean-Type Diet, are fully consistent with the Lipid Hypothesis and with our Unified Model of Atherogenesis and Acute IHD Events (Chapter 45). The Unified model proposes that atherosclerotic plaques in the coronary arteries and most acute IHD events occur due to the interaction of atherogenic blood lipoproteins and atherogenic mutations in those arteries.

When people have any atherosclerotic plaques in the their coronary arteries, we would expect such persons to fare better with respect to outcome if they have help from anti-inflammatory, anti-thrombotic, anti-arrhythmic, and anti-oxidative agents than if they lack such help (Chapter 46, Part 5). Levels of these helpers may well be increased by the Mediterranean-Type Diet. We think that the contributions by DeLorgeril, Renaud, and their colleagues look very promising indeed --- with one possible caveat: A 1999 multi-variate analysis (incorporating adjustments for over a dozen variables) reports "an increased risk of breast cancer associated with omega-3 fat from fish" (Holmes 1999, p.919; Table 2 at p.916 shows MultiVariate Relative Risk = 1.09).

● Part 12. Effects of Weight-Changes on the Atherogenic Lipoproteins

So far, Appendix-F has described effects from manipulating the composition of diets, while maintaining calorie-intakes at levels which do not alter weight. Now we must mention the observed effects of weight-changes on the blood levels of atherogenic lipoproteins, for no chapter about dietary prevention and management of Ischemic Heart Disease should fail to mention this important aspect of diet.

Box 5 shows that, between the ages of about 20 to 60 years in both males and females (USA), the serum levels of the cholesterol-rich Std Sf 0-20 lipoproteins and the triglyceride-rich Std Sf 20-400 lipoproteins rise. During these years, Americans are typically gaining weight.

12a. At Equal Age, Do Average Lipoprotein Levels Rise with Weight?

If we hold age constant, do we see a difference in average levels of the atherogenic lipoproteins as weight increases?

In order to explore this question, we examine 834 males, all in the age-range of 30-39 years. The data below come from the Livermore Lipoprotein Study (Appendix-E, Part 12-c). We sort the 834 men by their relative weight, divide them into six groups of ascending weight from 0.86 to 1.37, and calculate for each group the mean serum levels of the Std Sf 0-20, 20-400, and 0-400 lipoproteins. On their habitual U.S. diet, these men --- all within a fairly narrow age-band --- experience major increases of Std Sf 20-400 serum lipoproteins with increase in relative weight.

Number of Subjects	Relative weight (Means)	LDL + IDL Std Sf 0-20	VLDL Std Sf 20-400	Total Std Sf 0-400
97	0.86	382	105	487
219	0.95	397	122	519
249	1.05	407	140	547
168	1.14	422	165	587
75	1.23	416	191	607
26	1.37	425	203	627

12b. Do Average Lipoprotein Levels Decrease with Weight-Loss?

Below are data from the study of 28 women who participated in a weight-reduction program under medical guidance (Nichols 1957, p.11). The diet prescribed was a 1,000 calorie diet, low in animal fat, and necessarily low in carbohydrate. Over a 2-month period, the women lost an average of 14 pounds each. The changes in serum lipoproteins levels are tabulated below (all concentrations are in mg/dl).

	LDL + IDL Std Sf 0-20	VLDL Std Sf 20-400	Combination Std Sf 0-400	Weight (pounds)
Initial Values	465	158	623	212
Final Values	387	99	486	198
Change in Values	-78	-59	-137	-14

For the Std Sf 0-400 combination, the change is -9.8 mg/dl per pound of weight-loss in 2 months.

The data presented here show that a 1,000 calorie weight-reduction diet resulted in appreciable falls both in the Std Sf 0-20 lipoproteins and the Std Sf 20-400 lipoproteins. The magnitude of the effects observed is such that the lowering of Std Sf 0-20 lipoprotein levels could be attributed to the reduced animal-fat intake of the 1,000 calorie diet, and that the lowering of the Std Sf 20-400 lipoprotein levels could be attributed to the reduced carbohydrate intake of this diet. Thus, the major effects of weight reduction upon blood lipids in these women could be assigned to mechanisms known to operate in the absence of weight reduction.

12c. How Helpful Would It Be, to Maintain Our Age-20 Weights?

These are only two brief illustrations from a vast literature which shows that a relationship exists between serum lipoprotein levels and weight or weight-change. Appendix-I presents additional evidence for U.S. adults on the habitual American diet of the 1950s, namely:

● Positive caloric balance (weight-gain) is associated with RISING levels of Std Sf 0-400 lipoproteins.

● Negative caloric balance (weight-loss) is associated with FALLING levels of Std Sf 0-400 lipoproteins.

● Caloric equilibrium (stable weight) is associated with STABLE levels of Std Sf 0-400 lipoproteins. This observation is fully consistent with the observations that, at caloric equilibrium, certain changes in the COMPOSITION of the diet will change lipoprotein profiles (for instance, Boxes 1 + 2).

On the average, caloric equilibrium at a higher weight yields higher Std Sf 0-400 levels than caloric equilibrium at a lower weight (Part 12a). Thus, once a period of weight-gain is over, one continues indefinitely to "pay the price" of having gained the weight.

Such observations, combined with the tendency in the USA to gain weight as age advances from 20 toward 60, suggests that weight-gain may explain nearly all of the observed RISE in the average levels of atherogenic Std Sf 0-400 lipoproteins during the adult years (Box 5). There is one highly probable exception: The small percent of the population having Std Sf 0-100 (Type-II) hyperlipoproteinemias. These disorders are hereditarily controlled and are not determined by the same forces affecting the population at large.

Although the typical American diet and average serum lipoprotein levels may have changed since the 1950s, we doubt very much that those changes have repealed the weight-effect. We expect that, even on truly heart-healthy diets, the lipoprotein profile becomes less favorable if caloric balance is positive.

Indeed, we have been wondering for about 20 years (Gofman 1978) if a large share of the unhealthy increase in atherogenic serum lipoproteins, which occurs in the USA with advancing age, could be prevented if we would just maintain the lower weights we had at around age 20.

>>>>>>>>>>

Box 1 of Appendix-F
Comparison: Impact of Vegetable-Fat Diet vs. Animal-Fat Diet on Lipoprotein Levels.
This box is adapted from Nichols 1957 and Gofman 1958, pp.274-275.

● The same five males (ages 20-49) followed the three diets described in Boxes 1 + 2.
● Ultracentrifugal measurement of serum lipoprotein levels was done weekly during the diet-periods, and the means are presented below. Measurements are mg/100 ml serum.

● Sf 0-20: Mean Measurements for the Cholesterol-Rich Std Sf 0-20 Lipoproteins:

Individual Case	Vegetable-Fat Diet Duration = 10.5 weeks	Animal-Fat Diet Duration = 11 weeks
No.1	337	448
No.2	386	563
No.3	346	571
No.4	292	373
No.5	352	430
Grand Mean	343	477

● Sf 20-400: Mean Measurements for the Triglyceride-Rich Std Sf 20-400 Lipoproteins:

Individual Case	Vegetable-Fat Diet	Animal-Fat Diet
No.1	277	251
No.2	236	241
No.3	157	201
No.4	48	57
No.5	208	193
Grand Mean	185	189

● Composition of the IsoCaloric Vegetable-Fat and Animal-Fat Diets:

● For diet-composition, entries below do not match the tabulated entries in Gofman 1958 because the entries here include the standard breakfast, whereas breakfast grams were only in the text of Gofman 1958 (p.274). Our error-check (by calories) yields just trivial discrepancies.

	Veg-Fat Diet: Daily Intake = ~ 2,360 calories		Animal-Fat Diet: Daily Intake = ~ 2,345 calories	
Calories				
from fat	~ 40%	944 cal.	~ 40%	945 cal.
from protein	~ 16%	378 cal.	~ 16%	375 cal.
from carbo.	~ 44%	1038 cal.	~ 44%	1032 cal.
....	....		....	
Vegetable Fat	87 g	783 cal.	9 g	81 cal.
Animal Fat	15 g	135 cal.	96 g	864 cal.
Protein	91 g	364 cal.	91 g	364 cal.
Carbohydrate	254 g	1016 cal.	254 g	1016 cal.

● The Vegetable-Fat Diet is the same in Boxes 1 and 2. In the Vegetable-Fat Diet, unhydrogenated cottonseed oil (Wesson) was the main source of the vegetable fat. The small amount of fat of animal origin was almost wholly meat fat. Cholesterol intake/day was about 0.5 g.

● In the Animal-Fat Diet, the distribution of animal fat was 30 g from meat, 23 g from dairy products, 37 g from egg origin. Cholesterol intake/day was about 2.2 g, almost wholly from 7 egg-yolks per day.

Box 2 of Appendix-F
Comparison: Impact of Veg-Fat Diet vs. HighCarbo-LowFat Diet on Lipoprotein Levels.
This box is adapted from Nichols 1957 and Gofman 1958, pp.275-276.

- The same five males (ages 20-49) followed the three diets described in Boxes 1 + 2.
- Ultracentrifugal measurement of serum lipoprotein levels was done weekly during
the diet-periods, and the means are presented below. Measurements are mg/100 ml serum.

- Sf 0-20: Mean Measurements for the Cholesterol-Rich Std Sf 0-20 Lipoproteins:

Individual Case	Vegetable-Fat Diet Duration = 10.5 weeks	HighCarbo-LowFat Diet Duration = 24 weeks
No.1	337	391
No.2	386	395
No.3	346	358
No.4	292	292
No.5	352	317
Grand Mean	343	351

- Sf 20-400: Mean Measurements for the Triglyceride-Rich Std Sf 20-400 Lipoproteins:

Individual Case	Vegetable-Fat Diet	HighCarbo-LowFat Diet
No.1	277	477
No.2	236	310
No.3	157	234
No.4	48	89
No.5	208	215
Grand Mean	185	265

- Composition of the IsoCaloric Vegetable-Fat and HighCarbo-LowFat Diets:

- For diet-composition, entries below do not match the tabulated entries in Gofman 1958 because the entries here include the standard breakfast, whereas breakfast grams/calories were only in the text of Gofman 1958 (p.274). Our error-check (by calories) yields only trivial discrepancies.

	Veg-Fat Diet: Daily Intake = ∼ 2,360 calories		HighCarbo LowFat Diet: = ∼ 2,370 calories/day	
Calories				
from fat	∼ 40%	944 cal.	∼ 8%	190 cal.
from protein	∼ 16%	378 cal.	∼ 18%	427 cal.
from carbo.	∼ 44%	1038 cal.	∼ 74%	1754 cal.
Vegetable Fat	87 g	783 cal.	10 g	90 cal.
Animal Fat	15 g	135 cal.	10 g	90 cal.
Protein	91 g	364 cal.	108 g	436 cal.
Carbohydrate	254 g	1016 cal.	439 g	1756 cal.

- The Vegetable-Fat Diet is the same in Boxes 1 and 2. In the Vegetable-Fat Diet, unhydrogenated cottonseed oil (Wesson) was the main source of the vegetable fat. The small amount of fat of animal origin was almost wholly meat fat. Cholesterol intake/day was about 0.5 g.

- The High-Carbohydrate Low-Fat Diet differs from the Vegetable-Fat Diet mainly in the substitution of carbohydrate for vegetable fat. The number of calories/day are virtually the same in both diets.

Box 3 of Appendix-F
Infarct Case: Effect of a LOW-Carbohydrate HighFat Diet on Lipoprotein Levels.
This box is adapted from Gofman 1958 (p.279).

- The patient was a 65-year-old male survivor of a past Myocardial Infarction.
- The post-infarct low-carbohydrate diet (only 100 g/day of carbohydrate) did not cause appreciable weight-change during the 60-day period of observation and measurement (August 20-Oct.19).

- The tabulation below reveals the very different kind of information conveyed by the Std Sf 20-400 measurement, which fell to about HALF of its pre-diet level, and the total cholesterol measurement, which did not change with the diet. The total cholesterol measurement provided no hint of the dramatic decrease which occurred in the level of the atherogenic triglyceride-rich Std Sf 20-400 lipoproteins. The mean level of cholesterol-rich Std Sf 0-20 lipoproteins rose by a factor of 1.08 --- (308/286).

- Composition of the LowCarbo-HighFat diet was as follows:
  - Carbohydrate .... 100 g/day
  - Animal fat .... 60 g/day
  - Vegetable oil (unhydrogenated cottonseed oil) .... 80 g/day
  - Protein .... 91 g/day
  - Total calories .... 2,024 cal/day

Pre-diet levels (on the patient's usual home diet)

Date	Std Sf 0-20 (mg/100 ml)	Std Sf 20-400 (mg/100 ml)	Cholesterol (mg/100 ml)
7/18	347	482	---
8/11	226	585	186
Mean Values (pre-diet)	286	533	

Values on low-carbohydrate diet (100 g carbohydrate daily, started 8/20)

Date	Std Sf 0-20	Std Sf 20-400	Cholesterol
8/31	320	225	209
9/7	281	150	170
9/14	279	245	172
9/21	297	253	190
9/28	271	230	166
10/5	399	245	206
10/12	297	237	186
10/19	321	234	186

Mean values during the 60 days of low-carbohydrate diet

308	227	186

Box 4 of Appendix-F
Xanthoma Case: Effect of a LOW-Carbohydrate HighFat Diet on Xanthomata & Lipoprotein Levels.
This box is adapted from Gofman 1958 (p.280).

● The patient was a retired female school teacher with long-standing Xanthoma Tuberosum --- a disorder typically associated with extremely high serum levels of the Std Sf 20-400 lipoproteins.

● From January 24-March 1, the patient followed a very LOW-carbohydrate diet (details are below), during which the second set of measurements was made.

● By comparison with mean measurements on the patient's usual home diet, mean measurements during the 36-days on the LowCarbo HighFat Diet showed that Std Sf 0-20 lipoprotein levels rose by a factor of 1.07, that Std Sf 20-400 lipoprotein levels fell by a factor of 0.52, and that serum cholesterol fell by a factor of 0.85.

● Subsequent to the 36-day period, the patient underwent further periods of carbohydrate restriction on a home diet, and her xanthomata underwent considerable regression.

● Composition of the LowCarbo-HighFat Diet was as follows:

Carbohydrate	100 g/day
Animal fat	100 g/day
Vegetable oil (unhydrogenated cottonseed oil)	20 g/day
Protein	75 g/day
Total calories	1,780 cal/day

Pre-diet levels (on the patient's usual home diet)

Date	Std Sf 0-20 (mg/100 ml)	Std Sf 20-400 (mg/100 ml)	Cholesterol (mg/100 ml)
1/5	361	915	408
1/24	411	767	418
Mean Values (pre-diet)	386	841	413

Values on low-carbohydrate diet (100 g carbohydrate daily, started 1/24)

2/1	512	452	388
2/8	404	401	333
2/15	392	334	320
2/21	352	465	342
3/1	400	550	383

Mean values during the 36 days of low-carbohydrate diet

412	440	353

● The plots demonstrate the rise in mean serum lipoprotein levels beyond about 20 years of age until --- especially for males --- the rising slope flattens and even declines a bit. The rise after age 20 coincides with the tendency (USA) to gain weight during adulthood. The text provides data on the effect of weight-change on serum lipoprotein levels.

● The scale of values is different for the three plots. All measurements, made by the Donner Lab, come from Glazier 1954 and are listed below the graphs. From age 30 upward, the population sample consisted of clinically healthy entrants into the Framingham Heart Study, and below age 30, samples are from clinically healthy children and persons of the U.C. Berkeley campus (details in Appendix-E, Part 3).

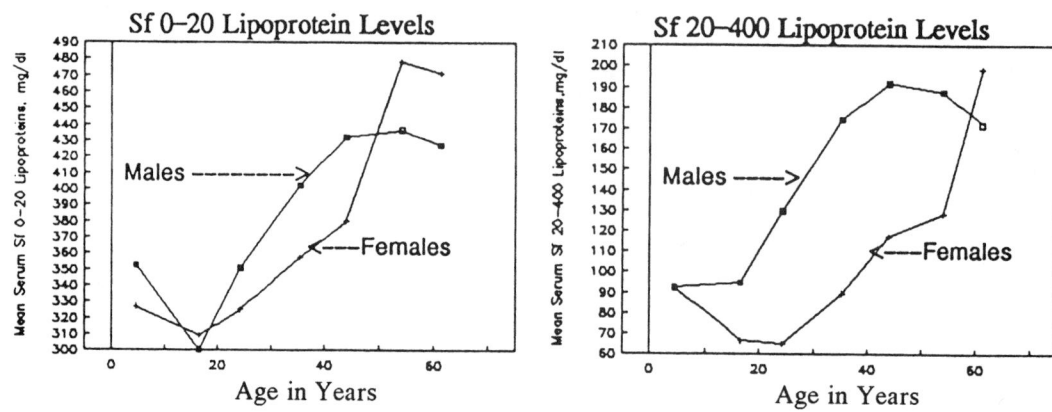

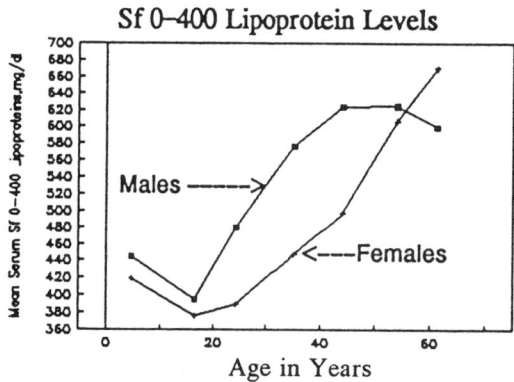

Related text = Parts 5a + 12.

Mean Lipoprotein Concentrations (mg/dl) in Clinically Healthy Persons as a Function of Age:

Mean Age (Years)	Persons: Males	Std Sf 0-20	Std Sf 20-400	Std Sf 0-400
4.3	9	352.4	92.3	444.7
16.2	29	300.4	94.7	395.1
25.0	75	350.6	129.5	480.1
35.1	358	402.1	174.5	576.6
44.2	313	432.1	191.8	623.9
54.1	228	436.6	187.7	624.3
61.1	43	427.4	172.0	599.4
	Females			
5.0	6	326.9	92.0	418.9
16.9	32	309.1	66.5	375.6
23.5	86	325.0	65.0	390.0
35.2	452	357.5	89.4	446.9
44.0	399	379.7	117.4	497.1
53.9	269	478.2	128.4	606.6
61.5	43	471.3	198.7	670.0

● **Part 1. Some Background: Controversy over the Goodness of "Good Cholesterol"**

We shall use the term HDL(2+3) to signify the combination of the true High-Density Lipoproteins, HDL-2 and HDL-3 (Chapter 44, Part 3e).

During the 1950s, it became evident that plasma levels of HDL(2+3) are inversely related to plasma levels of the Sf 0-400 lipoproteins, in clinically healthy populations (details in Gofman 1954-a, + DeLalla 1958, + DeLalla 1961). Additionally, HDL(2+3) levels are depressed in a number of clinical entities where there is a marked elevation of lipoproteins of the Std Sf 0-400 classes --- for instance, Xanthoma Tendinosum, Active Nephrotic Syndrome, Chronic Biliary Obstruction, Glycogen Storage Disease (details in Gofman 1954-a), and Acute Hepatitis (Pierce 1954-b, p.235). During that same period, evidence was accumulating that the cholesterol-rich lipoproteins (Sf 0-20) and the triglyceride-rich lipoproteins (Sf 20-400) are each independently atherogenic (Appendix-E).

In the 1965 Lyman Duff Memorial Lecture (Gofman 1966), we presented results from two prospective studies: Framingham at about 12 years of follow-up, and Livermore at about 10 years of follow-up (details in Appendix E, Part 12c). The Livermore Study, which included measurements of HDL-2 and HDL-3, provided the first PROSPECTIVE confirmation that their plasma concentrations might be inversely related to de novo cases of Ischemic Heart Disease (Gofman 1966, pp.686-687). By contrast, the HDL-1 concentrations were virtually identical in the base population and in the de novo IHD cases. The various Livermore findings were based on 38 de novo IHD cases which grew out of a base-population of 1,961 men, with average age of 43.7 years at entry to the study. We wrote (Gofman 1966, p.687):

"From these data, it is not possible to conclude whether or not the observed lowerings of HDL-2 and HDL-3 in Ischemic Heart Disease are in excess of those anticipated from the inverse correlations [with levels of Sf 0-400]. This, again, would ultimately be desirable information, since if there is any lowering beyond that expected from interclass correlations, the possibility of a protective role of High-Density Lipoproteins would require consideration."

The possibility, of an anti-atherogenic effect from the HDL(2+3) lipoproteins, was a topic of numerous studies in the 1960s and 1970s. In 1978 (Gofman 1978, pp.14-18), I explained my skepticism that the existing evidence supported an independent anti-atherogenic role for HDL(2+3). And some 20 years later, I am still a skeptic. This appendix describes, in Part 5, the kind of testing which I believe would be required to settle the issue.

We have not been alone in our doubts.

Is there any goodness in "good cholesterol"? The terms "good cholesterol" and "bad cholesterol" are everywhere, now. This suggests that the cholesterol transported by the purportedly protective High-Density Lipoproteins is "good," and the cholesterol transported by the atherogenic

Low-Density Lipoproteins is "bad" --- a concept which was explicitly challenged in the New England Journal of Medicine during 1989 by Gordon + Rifkind (Part 2).

● Part 2.   Important Issues Raised by Gordon and Rifkind

        David J. Gordon and Basil M. Rifkind (Gordon 1989) are the authors of "High-Density Lipoprotein --- the Clinical Implications of Recent Studies," in the NEJM (Gordon 1989). We can associate ourselves with several of their doubts and comments. For instance, they state (Gordon 1989, p.1314):

        "The association, of lower HDL levels with higher rates of coronary disease within populations in observational epidemiologic studies, has given rise to the hypothesis that interventions that raise low levels of HDL cholesterol will reduce coronary disease rates. However, neither our present understanding of lipid metabolism nor these epidemiologic observations can provide assurance that low levels of HDL cholesterol are a causative rather than a coincidental factor in coronary disease, or that intervention would be beneficial."

        Gordon and Rifkind point out (at p.1312) that "It has been hypothesized that HDL is involved in the 'reverse transport' of cholesterol from peripheral tissues to the liver." About this idea, Gordon and Rifkind have the following relevant observations (Gordon 1989, p.1312):

        "At least three caveats should be kept in mind. First, the relevance of these reverse-transport pathways to the rate of deposition (or removal) of cholesterol in atherosclerotic plaques has yet to be established. Second, the complex interrelation of cholesterol and triglyceride metabolism and the many lipoproteins involved may make it misleading to consider any single component of this system in isolation. Low plasma levels of HDL are often found in conjunction with high plasma levels of atherogenic, triglyceride-rich lipoproteins, and it is difficult to determine whether low levels of HDL cholesterol have a direct etiologic role in atherogenesis or serve only as a marker of a more fundamental disorder. Finally, the popular designation of HDL as 'the good cholesterol' is misleading, because the anti-atherogenic role that has been hypothesized for it pertains not to any unique property of its cholesterol but to the direction in which it transports that cholesterol."

Emphasis Added --- By the Explicit Data in Our Figure G-1

        From the preceding paragraph, we shall repeat, underline, and comment upon the following sentence:

        "... Low plasma levels of HDL are often found in conjunction with high plasma levels of atherogenic, triglyceride-rich lipoproteins, and it is difficult to determine whether low levels of HDL cholesterol have a direct etiologic role in atherogenesis or serve only as a marker of a more fundamental disorder ..."

        These words in 1989 suggest that little progress had occurred on the problem we described in 1966 (Part 1, above): The need to determine whether LOW levels of HDL(2+3) are an independent cause of Ischemic Heart Disease, or whether such levels are "automatically" low when the levels of the atherogenic Sf 0-400 lipoproteins are high.

        Not only are low plasma levels of HDL(2+3) "often" found in conjunction with high plasma levels of atherogenic, triglyceride-rich lipoproteins, but we can show that this relationship is PROMINENT in a sample of 891 American males, ages 30-39, whose lipoproteins were measured in our Livermore Lipoprotein Study (Appendix-E). I would be extremely surprised if a similar inverse relationship failed to exist in other (non-Livermore-Lab) institutions in the United States. Our Figure G-1 depicts the strong inverse relationship between HDL(2+3) and the combined Sf 0-400 lipoproteins --- details in Part 3.

● Part 3.   Data from a Livermore Population: Inverse Correlations

        In our Livermore Lipoprotein Study, 891 male participants were in the age-band 30-39 years old when we enrolled them into the database and measured their plasma lipoproteins, during the years 1954-1957. With 891 persons, this age-band constituted over half of the 1,961 males in the study.

Figure G-1: HDL(2+3) Regressed on Std. Sf 0-400 Lipoproteins

To prepare Figure G-1, we sorted the 891 records in ascending order by their plasma concentrations (milligrams per deciliter) of the combined Std Sf 0-400 lipoproteins. Then we divided the database into deciles, with each of the first nine having 89 persons and with the tenth having 90 persons. For each decile, we calculated the average concentrations of the Std Sf 0-400 and the HDL(2+3) lipoproteins. The ten resulting pairs of Observed Values are tabulated in Figure G-1, and shown as boxy symbols within the graph.

The Observed HDL(2+3) values are regressed linearly on the Observed Std Sf 0-400 values. The regression output is shown to the right, in Figure G-1. Then, following the steps described in Chapter 6, Part 3, we write the Equation of Best Fit and calculate the third column of values --- the Calculated Best-Fit HDL(2+3) values, including the two "extensions." The Line of Best Fit in the graph reflects the pairing of the Observed Std Sf 0-400 values with the Calculated Best-Fit HDL(2+3) values.

Next, we examine each of the four major segments, within the atherogenic band of Std Sf 0-400 lipoproteins. Each is in a demonstrably inverse relationship with the HDL(2+3) in this population sample. However, the following point deserves emphasis:

By themselves, these inverse relationships with the atherogenic lipoproteins are NOT evidence that High-Density Lipoproteins are anti-atherogenic --- as Part 5 shows.

Figures G-2, G-3, + G-4: HDL vs. Segments of the Std Sf 0-400 Lipoproteins

Figures G-2, G-3, and G-4 are prepared in the manner described for Figure G-1, except that the 891 records were sorted by the indicated SEGMENTS of the Std Sf 0-400 spectrum. Additionally, when there are no low values on the horizontal axis, the scale of that axis does not start at zero. This can cause the mistaken impression that the y-intercept would not match the Constant --- an illusion which vanishes if one widens those graphs so that the scale begins at zero.

All segments of the Std Sf 0-400 lipoproteins are in an inverse relationship with HDL(2+3). The relationships for the Std Sf 20-100 and 100-400 segments of the spectrum have a steep component and a flatter component, making their relationships less linear and more complex than the overall relationship in Figure G-1. (We note that observations in Rubins 1995 appear consistent with our 1957 data.)

● **Part 4. Data from a Framingham Population: Effects of Selective Pressure**

Part 4 illustrates the effects of selective pressure in an epidemiologic study --- the development of an "outgrowth" population from a base population.

During the 1950s, our group at the Donner Laboratory measured the Std Sf 0-12, 12-20, 20-100, and 100-400 lipoproteins on several thousand entrants to the Framingham Heart Study (Appendix-E, Part 12). These Framingham entrants included 687 men in the 30-39 year age-band --- which is the same age-band evaluated in Part 3 above from a Livermore population. For the Livermore population (but not the Framingham population), the HDL-1, HDL-2, and HDL-3 measurements were made in addition to the Std Sf 0-12, 12-20, 20-100, and 100-400 lipoproteins.

In 1965, Dr. Thomas R. Dawber (then Director of the Framingham Study) provided a listing of the 319 de novo cases of IHD which had occurred during the intervening years among the entrants measured about 12 years earlier by Donner, and we reported the results in our Lyman Duff Memorial Lecture (Gofman 1966). Below are the results for the 687 males, ages 30-39 when measured, from Gofman 1966 (p.683, Table 3, which includes standard deviations of the means). All lipoprotein and cholesterol measurements are in mg/dl.

Measures (Mean)	De Novo IHD	Base Population	Difference	Significance test
Std Sf 0-12	390.2	341.9	48.3	p=0.001
Std Sf 12-20	75.4	62.1	13.3	p=0.01
Std Sf 20-100	139.7	102.3	37.4	p<0.001

Std Sf 100-400	145.8	77.3	68.5	$p < 0.001$
Std Sf 0-400	751.1	583.6	167.5	$p < 0.001$
Atherogenic Index	102.1	76.6	25.5	$p < 0.001$
Cholesterol	267.7	222.6	45.1	$p < 0.001$
Systolic B.P.	136.2	130.3	5.9	$p \sim 0.05$
Diastolic B.P.	91.6	84.6	7.0	$p = 0.001$
Relative Weight	119.3	111.9	7.4	$p < 0.01$

These Framingham results illustrate the role of SELECTIVE PRESSURES in an epidemiological study. What grows out of this base population (of 687 males at Framingham) depends upon such pressures. Since the four Standard Sf classes are atherogens, we know that their levels will --- as a result of the selective pressure --- be elevated in the group which develops de novo ISCHEMIC HEART DISEASE, compared with levels in the base population from which the cases grew out. And that is precisely what is observed in the results which are tabulated above.

Outgrowth of De Novo IHD from a Base Population-Sample

The base population-samples, in the prospective Framingham and Livermore Studies, are not assumed to be "risk-free" with respect to future manifestation of clinical Ischemic Heart Disease. Rather, these base populations are assumed to have a distribution of persons with various risk-factors for future manifestation of clinical IHD. If a particular biochemical variable is suspected of contributing to that risk, a prospective study (having adequate size of sample and duration of follow-up time) is expected to show that the variable is indeed associated with IHD evolution in some members of the base population.

The Framingham results tabulated above show that plasma lipoprotein levels in each segment of the Sf 0-400 lipoprotein spectrum were significantly higher in the "outgrowth" population (the 32 persons who later developed overt de novo IHD) than in the base population. Such findings are consistent with a causal relationship of those lipoproteins with the development of IHD --- and with the Lipid Hypothesis of Atherosclerosis.

HDL(2+3) in the IHD Cases Above vs. the Base Population

Although the Framingham base population above does not have HDL(2+3) measurements, it has the same age-band and gender as the Livermore base population which shows HDL(2+3) concentration having a strong, inverse relationship with the concentration of Sf 0-400 lipoproteins (Figure G-1). Therefore, in the absence of contrary evidence, it seems likely that the outgrowth sample (above) in the Framingham Study would have shown lower mean concentrations of HDL(2+3) than its base population --- because the outgrowth sample showed higher mean concentrations of Sf 0-400 lipoproteins than its base population.

Using the entire base population of males in the Livermore Study, our 10-year follow-up showed statistically significant reductions in mean HDL-2 and HDL-3 concentrations in the outgrowth sample of 38 de novo IHD cases, compared with the base population's mean concentrations (Gofman 1966, p.686, Table 14). The same outgrowth sample of de novo IHD cases also showed statistically significant elevations in mean values of Sf 0-12, Sf 12-20, and Sf 20-100 lipoproteins (Gofman 1966, p.684, Table 10) --- as reported in Appendix-E.

● Part 5. Testing a "Protective" Effect for HDL(2+3): Three Possibilities

Three possibilities exist for HDL(2+3):

Case 1: The HDLs themselves are neutral with respect to atherogenesis and IHD development.
Case 2: The HDLs themselves are independent anti-atherogens (protective against IHD).
Case 3: The HDLs themselves are independent atherogens.

5a. Case 1: Outgrowth De Novo Sample and Base Population on Same Figure

Suppose HDL(2+3) themselves are neutral with respect to atherogenesis and IHD development. And suppose we did a prospective study with a very large number of persons in the base population and a large outgrowth of de novo IHD cases. Then what would we expect to see if we made a new and expanded Figure G-1 from the results?

On the new Figure G-1, we would plot the values and best-fit line (or curve, if curvature exists) not only for the base population, but also separately for the outgrowth de-novo-IHD population sample.

Neutrality of HDL(2+3) with respect to atherogenesis and IHD would mean that there would be NO SELECTIVE PRESSURE for or against HDL(2+3) in the cohort of de novo IHD cases which would grow out of the base population. Therefore, we would expect to find that the line of best fit for the de novo IHD cases would lie directly over (within experimental error) the line of best fit for the base population.

At EQUAL concentrations of Sf 0-400 lipoproteins, the concentrations of HDL(2+3) would not differ significantly between the base and the outgrowth populations. Such a result would be consistent, of course, with higher MEAN Sf 0-400 and lower MEAN HDL(2+3) levels among the IHD cases, than among the base population.

5b. Case 2: The HDLs Themselves Are Independently Protective against IHD

If we would go through the same exercise for Case 2 as we did for Case 1, we would find that the line (or curve) for the de novo IHD cases would lie BENEATH the line for the base population in a revised Figure G-1 --- if HDL(2+3) have an independent anti-atherogenic effect.

Why BENEATH? If HDL(2+3) have an independent protective effect against IHD, above and beyond their inverse relationship with the Sf 0-400 lipoproteins, then there would be SELECTIVE PRESSURE to prevent the HDLs from getting into the IHD outgrowth group. This is the expectation for an anti-atherogen which protects against IHD development. The de novo IHD sample would grow out of the base population partly BECAUSE it is impoverished in the protective HDL(2+3). So, the protective HDL would necessarily "stay behind," and the de novo IHD cohort would be LESS RICH in HDL(2+3) than the base population. For each value of Std Sf 0-400 among the de novo IHD cases, the HDL (2+3) would lie below the value which would have obtained for "neutrality." This is the implication of claims that HDL(2+3) have a protective effect against IHD.

5c. Case 3: The HDLs Themselves Are Independent Atherogens

If we would go through the same exercise for Case 3 as we did for Case 1, we would find that the line (or curve) for the de novo IHD cases would lie ABOVE the line for the base population in a revised Figure G-1 --- if HDL(2+3) are independent atherogens.

Why ABOVE? If HDL(2+3) are independent atherogens, the SELECTIVE PRESSURE would make the IHD outgrowth cohort ENRICHED in the HDL(2+3) compared with the base population. Therefore, at each point along the Std Sf 0-400 line (or curve), the HDL(2+3) values would be higher in the IHD cohort than in the base population.

● Part 6. Does the Existing Evidence Pass the Test in Part 5b?

We do not rule out the existence of HDL anti-atherogens. Either the existing evidence can pass the test for Case 2 described above (or an equivalent test), or it can not pass the test. Unless one becomes convinced that existing evidence has already passed such a test, there seems to be little basis for considering that any HDL entity truly merits to be called "protective" or "the good cholesterol."

>>>>>>>>>>

Figure G-1

Regression of HDL (2+3) on Std Sf 0-400

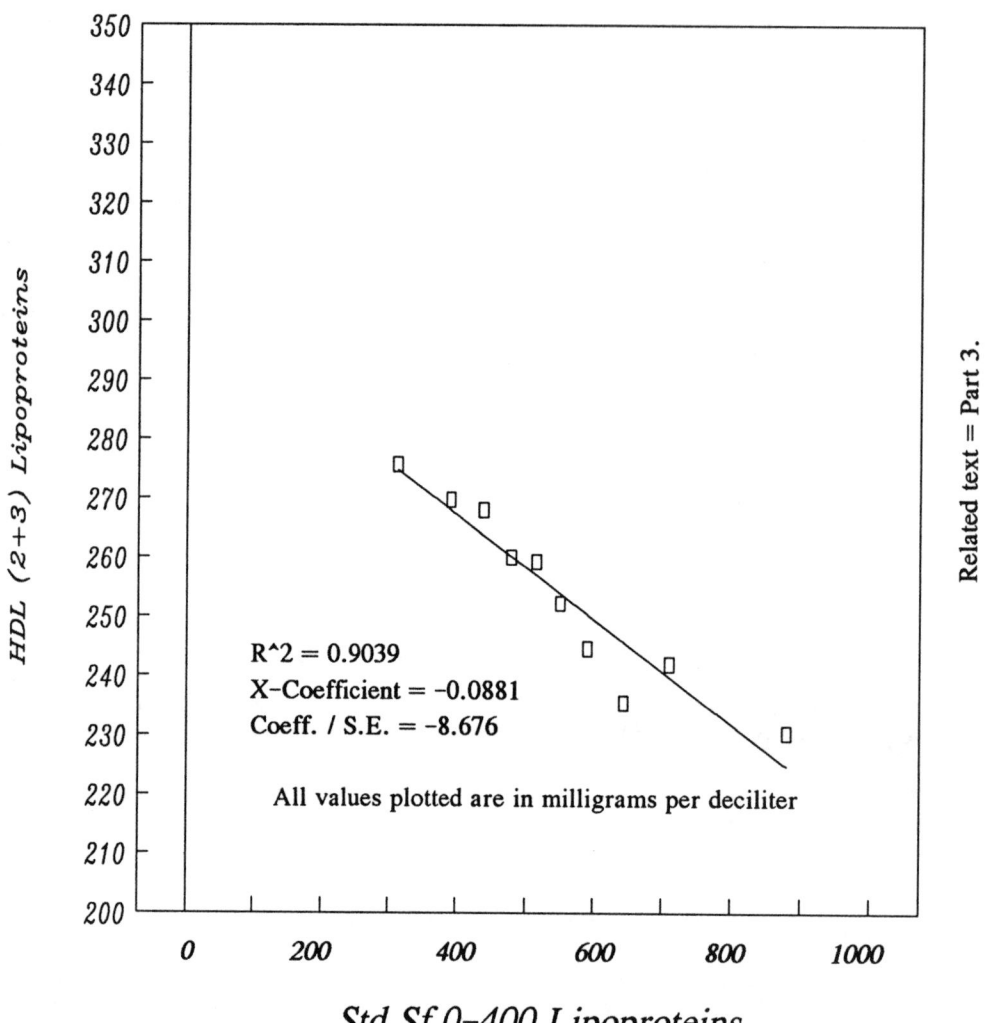

R^2 = 0.9039
X-Coefficient = -0.0881
Coeff. / S.E. = -8.676

All values plotted are in milligrams per deciliter

Related text = Part 3.

Std Sf 0-400 Lipoproteins

Data for Plotting (HDL(2+3) vs. Std Sf 0-400

Deciles	STD Sf 0-400	HDL(2+3) Obs.	HDL(2+3) Calc.
Decile 1	312.3	275.7	274.8
Decile 2	390.8	269.7	267.9
Decile 3	439.3	267.9	263.6
Decile 4	479.1	260.0	260.1
Decile 5	515.7	259.2	256.9
Decile 6	551.1	252.1	253.7
Decile 7	591.3	244.5	250.2
Decile 8	643.8	235.5	245.6
Decile 9	710.8	242.0	239.7
Decile 10	880.3	230.4	224.8
Extension	900.0		223.0
Extension	925.0		220.8

Regression of HDL(2+3) on Std Sf0-400
Regression Output:

Constant	302.2803
Std Err of Y Est	5.0322
R Squared	0.9039
No. of Observations	10
Degrees of Freedom	8
X Coefficient(s)	-0.0881
Std Err of Coef.	0.0102
Coeff. / S.E.	-8.6764

Equation of best fit:

HDL(2+3) = (-0.0881 * Std Sf 0-400) + 302.2803

Figure G-2

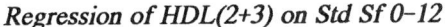

Regression of HDL(2+3) on Std Sf 0-12

Regression of HDL (2+3) on Std Sf 12-20

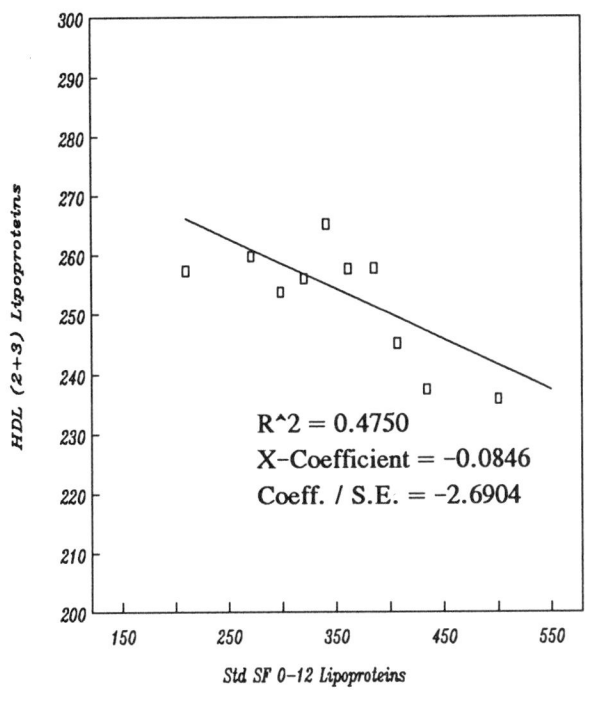

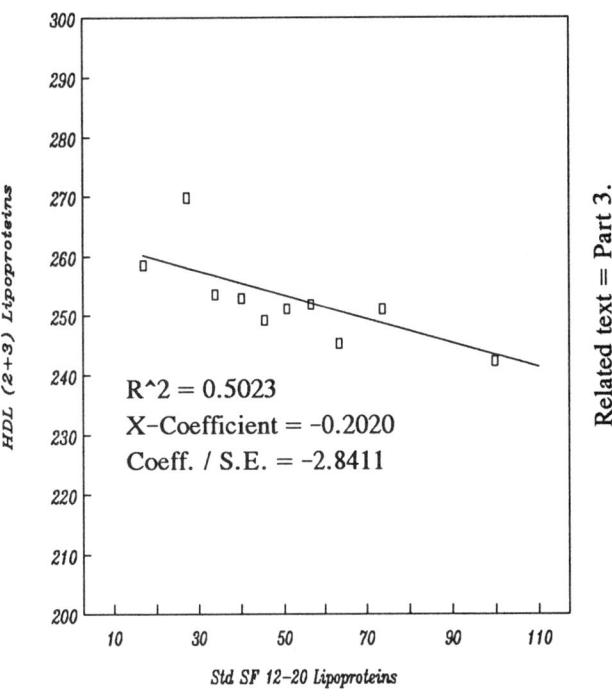

Related text = Part 3.

All values plotted are in milligrams per deciliter
Data for Plotting (HDL(2+3) vs. Std Sf 0-12

Deciles	STD Sf 0-12	HDL(2+3) Obs.	HDL(2+3) Calc.
Decile 1	210.0	257.4	266.1
Decile 2	271.5	259.8	260.9
Decile 3	298.3	256.8	258.7
Decile 4	320.3	258.4	256.8
Decile 5	341.2	265.2	255.0
Decile 6	361.6	257.7	253.3
Decile 7	385.3	264.2	251.3
Decile 8	406.8	247.7	249.5
Decile 9	434.1	237.4	247.2
Decile 10	500.4	235.8	241.6
Extension	525.0		239.5
Extension	550.0		237.4

Regression Output:

Constant	283.9044
Std Err of Y Est	7.9335
R Squared	0.4750
No. of Observations	10
Degrees of Freedom	8
X Coefficient(s)	−0.0846
Std Err of Coef.	0.0315
Coeff./ S.E.	−2.6904

Equation of best fit:
HDL(2+3) = (−0.0846 * Std Sf 0-12) + 283.904

All values plotted are in milligrams per deciliter
Data for Plotting (HDL(2+3) vs. Std Sf 12-20

Deciles	STD Sf 12-20	HDL(2+3) Obs.	HDL(2+3) Calc.
Decile 1	16.9	255.3	260.2
Decile 2	27.1	269.8	258.1
Decile 3	34.0	253.6	256.7
Decile 4	40.0	257.7	255.5
Decile 5	45.2	251.1	254.5
Decile 6	50.7	251.1	253.4
Decile 7	56.6	251.9	252.2
Decile 8	63.5	247.1	250.8
Decile 9	73.7	251.1	248.7
Decile 10	99.8	244.6	243.4
Extension	100.0		243.4
Extension	110.0		241.4

Regression Output:

Constant	263.5890
Std Err of Y Est	5.1742
R Squared	0.5023
No. of Observations	10
Degrees of Freedom	8
X Coefficient(s)	−0.2020
Std Err of Coef.	0.0711
Coeff./ S.E.	−2.8411

Equation of best fit:
HDL(2+3) = (−0.2020 * Std Sf 12-20) + 263.589

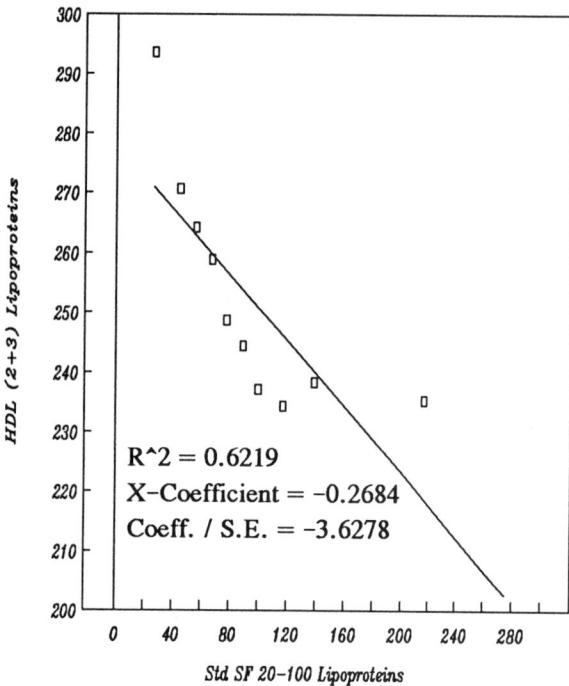

Regression of HDL (2+3) on Std Sf 20-100

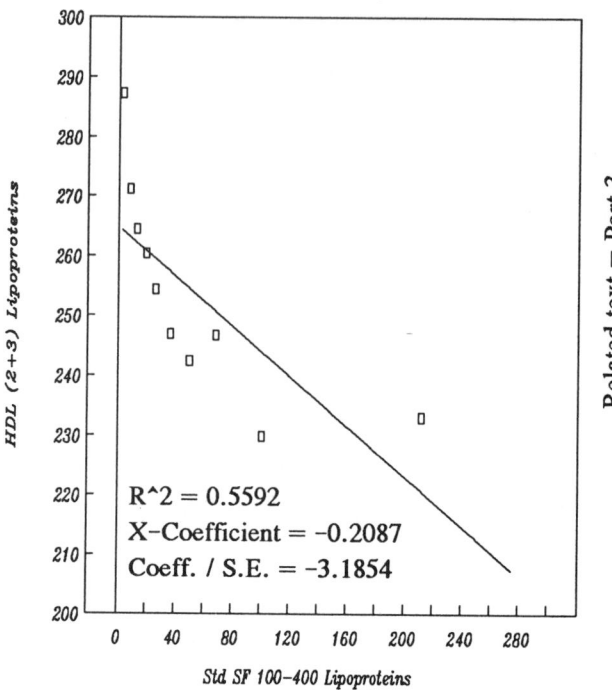

Regression of HDL (2+3) on Std Sf 100-400

Related text = Part 3.

All values plotted are in milligrams per deciliter
Data for Plotting (HDL(2+3) vs. Std Sf 20-100)

Deciles	STD Sf 20-100	HDL(2+3) Obs.	HDL(2+3) Calc.
Decile 1	26.6	293.6	271.7
Decile 2	44.6	270.7	266.8
Decile 3	55.9	266.2	263.8
Decile 4	67.6	258.8	260.7
Decile 5	77.7	250.9	258.0
Decile 6	89.1	244.9	254.9
Decile 7	100.0	239.7	252.0
Decile 8	117.2	237.1	247.3
Decile 9	138.9	238.3	241.5
Decile 10	216.8	237.0	220.6
Extension	250.0		211.7
Extension	275.0		205.0

Regression Output:

Constant	278.7990
Std Err of Y Est	12.1795
R Squared	0.6219
No. of Observations	10
Degrees of Freedom	8
X Coefficient(s)	-0.2684
Std Err of Coef.	0.0740
Coeff./ S.E.	-3.6278

Equation of best fit:
HDL(2+3) = (-0.2684 * Std Sf 20-100) + 278.799

All values plotted are in milligrams per deciliter
Data for Plotting (HDL(2+3) vs. Std Sf 100-400)

Deciles	STD Sf 100-400	HDL(2+3) Obs.	HDL(2+3) Calc.
Decile 1	2.9	287.2	264.3
Decile 2	8.3	271.2	263.1
Decile 3	12.9	264.4	262.2
Decile 4	19.5	260.3	260.8
Decile 5	26.2	254.3	259.4
Decile 6	36.7	246.9	257.2
Decile 7	49.8	242.4	254.5
Decile 8	68.1	246.7	250.6
Decile 9	101.2	229.8	243.7
Decile 10	212.2	233.0	220.6
Extension	250.0		212.7
Extension	275.0		207.5

Regression Output:

Constant	264.8587
Std Err of Y Est	12.4490
R Squared	0.5592
No. of Observations	10
Degrees of Freedom	8
X Coefficient(s)	-0.2087
Std Err of Coef.	0.0655
Coeff. / S.E.	-3.1854

Equation of best fit:
HDL(2+3) = (-0.2087 * Std Sf 100-400) + 264.8587

Regression of HDL (2+3) on Std Sf 0–20

Regression of HDL (2+3) on Std Sf 20–400

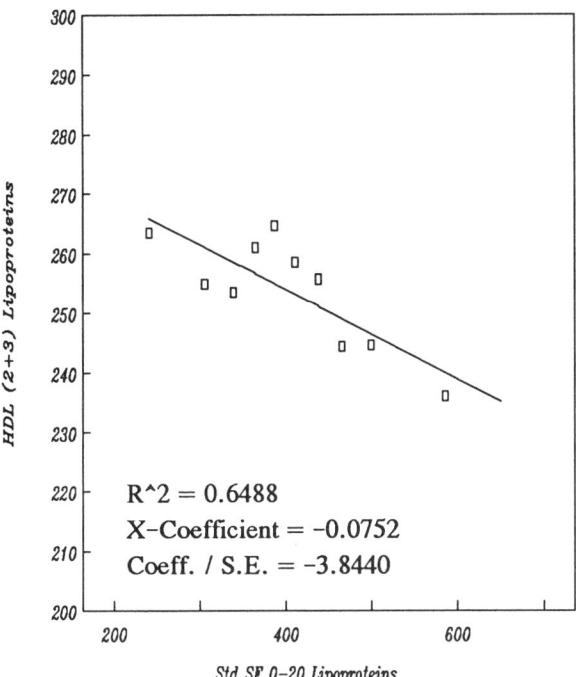

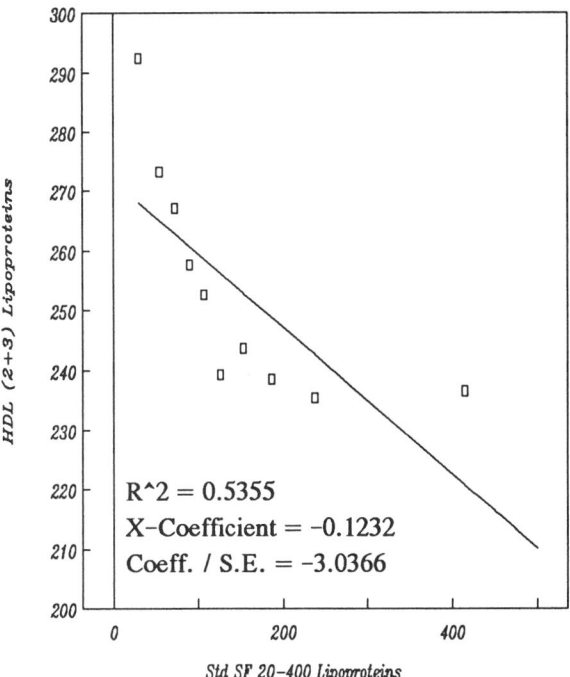

All values plotted are in milligrams per deciliter
Data for Plotting (HDL(2+3) vs. Std Sf 0–20)

Deciles	STD Sf 0–20	HDL(2+3) Obs.	HDL(2+3) Calc.
Decile 1	241.4	263.5	265.9
Decile 2	305.6	254.9	261.1
Decile 3	339.6	253.5	258.5
Decile 4	364.5	261.0	256.6
Decile 5	386.9	264.7	254.9
Decile 6	410.7	258.6	253.1
Decile 7	438.2	255.7	251.1
Decile 8	465.3	244.4	249.0
Decile 9	499.7	244.6	246.5
Decile 10	585.7	236.1	240.0
Extension	600.0		238.9
Extension	650.0		235.2

Regression Output:

Constant	284.0316
Std Err of Y Est	5.8424
R Squared	0.6488
No. of Observations	10
Degrees of Freedom	8
X Coefficient(s)	−0.0752
Std Err of Coef.	0.0196
Coeff. /S.E.	−3.8440

Equation of best fit:
HDL(2+3) = (−0.0752 * Std Sf 0–20) + 284.0316

All values plotted are in milligrams per deciliter
Data for Plotting (HDL(2+3) vs. Std Sf 20–400)

Deciles	STD Sf 20–400	HDL(2+3) Obs.	HDL(2+3) Calc.
Decile 1	30.5	292.4	268.0
Decile 2	54.8	273.2	265.0
Decile 3	72.4	267.1	262.8
Decile 4	89.8	257.7	260.7
Decile 5	106.8	252.7	258.6
Decile 6	126.3	239.2	256.2
Decile 7	154.0	243.6	252.8
Decile 8	186.2	238.5	248.8
Decile 9	236.8	235.4	242.6
Decile 10	415.0	236.5	220.6
Extension	450.0		216.3
Extension	500.0		210.2

Regression Output:

Constant	271.7754
Std Err of Y Est	13.7250
R Squared	0.5355
No. of Observations	10
Degrees of Freedom	8
X Coefficient(s)	−0.1232
Std Err of Coef.	0.0406
Coeff. /S.E.	−3.0366

Equation of best fit:
HDL(2+3) = (−0.1232 * Std Sf 20–400) + 271.7754

Are "Small, Dense LDL Particles" Especially Atherogenic? A Basis for Strong Doubt

Part 1. Evidence, from the Livermore Study, Which Challenges a Current Hypothesis
Part 2. What Are These "Small, Dense Low-Density Lipoprotein Particles"?
Part 3. Average LDL Diameters: IHD Cases vs. Controls --- from Gardner, Stampfer
Part 4. The Livermore Study: The "Massive HDL-1 Hyperlipoproteinemia Syndrome"
Part 5. The Very Notable Distribution of Elevated HDL-1 among 891 Livermore Males
Part 6. Definition and Demonstration of the RATIO Used in Our Livermore Analysis
Part 7. Results of Our Livermore Analysis, Regarding LDL Diameter and IHD
Part 8. Conclusion

Box 1. Calculation of the Mean RATIO, for Decile 1 of the 891 Livermore Males.

● **Part 1. Evidence, from the Livermore Study, Which Challenges a Current Hypothesis**

During the 1990s, a large literature has accumulated around the hypothesis that the smallest, most dense molecules within the Low-Density class of serum lipoproteins are the key atherogenic species. There is clearly renewed interest in the fact that lipoprotein species WITHIN the Std Sf 0-12 segment of the spectrum differ from each other in density, size, and physical behavior --- a fact solidly demonstrated in the early 1950s (Chapter 44, Part 3f, and Chapter 44, Box 2) but de-emphasized by reliance on LDL-cholesterol measurements.

The purpose of Appendix-H is to present some striking evidence from the Livermore Lipoprotein Study --- evidence which persuades us that the purported high atherogenicity of the "small, dense" LDL molecules reflects, instead, the elevated mean level of the triglyceride-rich Std Sf 20-400 serum lipoproteins in Ischemic Heart Disease.

<u>Some Resources in the Literature</u>

Within the following papers, readers can find the history and details of the Hypothesis of Small, Dense LDL Particles:

● 1994: Ronald M. Krauss, "Heterogeneity of Plasma Low-Density Lipoproteins and Atherosclerosis Risk," (Review), CURRENT OPINION IN LIPIDOLOGY Vol.5: 339-349.

● 1994: A.H. Slyper, "Low-Density Lipoprotein Density and Atherosclerosis: Unraveling the Connection," JOURNAL OF THE AMERICAN MEDICAL ASSN. Vol.272: 305-308.

● 1996: Christopher D. Gardner + Stephen P. Fortmann + Ronald M. Krauss, "Association of Small Low-Density Lipoprotein Particles with the Incidence of Coronary Artery Disease in Men and Women," JOURNAL OF THE AMERICAN MEDICAL ASSN. Vol.276, No.11: 875-881.

● 1996: Meir J. Stampfer + Ronald Krauss + Jing Ma + 4 co-workers, "A Prospective Study of Triglyceride Level, Low-Density Lipoprotein Particle Diameter, and Risk of Myocardial Infarction," JAMA Vol.276, No.11: 882-888.

● 1996: Josef Coresh + Peter O. Kwiterovich, Jr., "Small, Dense Low-Density Lipoprotein Particles and Coronary Heart Disease Risk: A Clear Association with Uncertain Implications," (Editorial), JAMA Vol.276, No.11: 914-915.

● 1997: Benoit Lamarche + Andre Tchernof + Sital Moorjani + 4 co-workers, "Small, Dense Low-Density Lipoprotein Particles as a Predictor of the Risk of Ischemic Heart Disease in Men: Prospective Results from the Quebec CardioVascular Study," CIRCULATION Vol.95, No.1: 69-75.

● **Part 2. What Are These "Small, Dense Low-Density Lipoprotein Particles"?**

Gardner and colleagues have recently stated (Gardner 1996, p.875):

"LOW-DENSITY lipoprotein (LDL) particles are heterogeneous in size, density and composition. Using different methods, including gradient gel electrophoresis, density gradient ultracentrifugation, and analytical ultracentrifugation, various investigators have identified and defined LDL heterogeneity as consisting of 2 to 15 different fractions, patterns, types, diameter ranges, size intervals, subspecies, or peak flotation rates [references provided]. Despite these multiple approaches for defining LDL subclasses, they all similarly differentiate relatively smaller, denser, and lipid-depleted particles from those that are larger, more buoyant, and lipid enriched."

Appendix-H focuses on evidence concerning one such "small, dense lipoprotein" in the LDL group of lipoproteins of density less than 1.063 gms/ml. That lipoprotein, having a density of 1.05 g/ml, was described as HDL-1 over 40 years ago by DeLalla 1954-b (details in Chapter 44, Part 3e). It was always clear that HDL-1, by its density and ability to float in a salt solution of 1.063 g/ml, is truly a member of the LDL group (Std Sf 0-12 lipoproteins). Indeed, HDL-1 is labeled "low density" in the figure from 1956 which is reproduced in Chapter 44, Box 1. It is the smallest and most dense species which we could identify in the Std Sf 0-12 segment of serum lipoproteins. HDL-1 can be regarded as approximately Sf 0-2, in the Sf 0-12 spectrum. Nonetheless, it acquired the name "HDL-1" at Donner Lab because it proved effective to quantify its concentration during runs made to isolate what we named the HDL-2 class of High-Density Lipoproteins.

There is no doubt that HDL-1 must constitute a large share of the "small, dense LDL lipoproteins" to which Gardner and others have been referring.

● Part 3. Average LDL Diameters: IHD Cases vs. Controls --- from Gardner, Stampfer

Both Gardner 1996 and Stampfer 1996 report that the average diameter of serum LDL molecules is smaller in patients who have Ischemic Heart Disease (Coronary Artery Disease) than in controls. The mean measurements in Part 3a come from their papers: Gardner's Table 3 (p.878) and Stampfer's Table 1 (p.884).

3a. Differences in Diameter of LDL Molecules, in Nanometers

● Gardner 1996, Cases and Controls.   Difference: Cases minus Controls.

	No. of Pairs	Cases	Controls		
Men	90	26.05	26.58	-0.52	p value <0.001
Women	34	26.46	26.94	-0.48	p value = 0.06
Both	124	26.17	26.68	-0.51	p value <0.001

● Stampfer 1996.   Difference: Cases minus Controls.

Men, 266 Cases	25.6 nm		
Men, 308 Controls	25.9 nm	-0.3 nm	p value <0.001

While the statistical significance is high, the differences between cases and controls seem quite small in terms of percentage. For the Gardner report, the difference in average LDL diameters is (0.51/26.68), or 1.9 percent. For the Stampfer report, the difference in LDL diameters is even smaller: (0.3/25.9), or 1.2 percent. We will demonstrate (Part 7b) why such differences can be expected --- wholly aside from any special atherogenicity related to reduced LDL diameters in the small, dense range of LDL.

3b. The Views of Gardner and Co-Workers, 1996

Gardner and colleagues state (1996, p.875):

"A large and growing body of epidemiologic evidence shows a consistent association between small, dense LDL particles and prevalent coronary artery disease (CAD) in case-control studies. In each of these studies, the case-control difference in LDL size or subclass concentration was statistically more significant than the difference in LDL cholesterol (LDC) levels. This epidemiologic evidence is supported by mechanistic evidence from human, animal, and in vitro studies that suggest the smaller, more dense LDL Particles are relatively more atherogenic than larger, more buoyant particles." And later (Gardner 1996, p.880):

"The findings reported herein add to the accumulating evidence that small, dense LDL particles meet many of the criteria for being an important CAD risk factor. Angiography and MI survivor case-control studies have consistently found an association between small, dense LDL and CAD. The present study and a recent report from the Physicians' Health Study demonstrate that the presence of small LDL particles precedes CAD. Our data suggest that the relationship with CAD is graded. Similar to our own findings, other investigators have reported that the relative risk of CAD among individuals with small LDL particles is strong ... The association between LDL size and CAD thus appears to be consistent, prospective, graded, strong, and biologically plausible."

We think the association has a different explanation. And we do not consider any size-effect to be biologically plausible, for LDL molecules in this part of the lipoprotein spectrum.

● **Part 4. The Livermore Study: The "Massive HDL-1 Hyperlipoproteinemia Syndrome"**

Long ago, the Donner team found the HDL-1 class to be massively elevated in the blood of (a) Diabetics during acidosis and decontrol (Gofman 1952-d, + Kolb 1955), (b) Persons with Glycogen Storage Disease (Kolb 1955), and (c) Persons with "Essential Hyperlipemia" who have "creamy serum" but have no overt clinical troubles (Gofman 1954-a).

4a. A Search for HLD-1 "Out-Liers" in the Livermore Lipoprotein Study

In 1993, we examined the records of the 2,297 participants in the Livermore Lipoprotein Study, whose database of clinically healthy adults is described in Appendix-E, Part 12c. During this examination, we searched for persons with serum HDL-1 levels equal to or greater than 100 mg/dl. A level of "only" 100 mg/dl would represent Massive HDL-1 Hyperlipoproteinemia, since it would be some four times above the mean and median HDL-1 levels in the overall database (approximately 6 Standard Scores above the mean). Such persons would be real HDL-1 "out-liers." We found twelve persons in the entire database with this syndrome --- half of one percent. The frequencies, by age and gender, are indicated below:

Males (ages)	Total Persons	HDL-1 "Outliers"	Females Age Band	Total Persons	HDL-1 "Outliers"
17-29	585	ZERO	17-29	190	ZERO
30-39	834	8	30-39	99	ZERO
40-49	399	3	40-49	37	ZERO
50-65	143	1	50-65	10	ZERO

Male frequency = (12/1961) = 0.006. The absence of such cases in the 17-29 year-old males suggests that this is an inherited abnormality which becomes expressed after age 30 in persons who appear to be "normal" before age 30. The absence of any such persons among the females may just mean that the Livermore population sample of females, age 30 and older, is too small.

4b. What ELSE Is Characteristic of Massive HDL-1 Hyperlipoproteinemia?

For the 12 men identified with Massive HDL-1 Hyperlipoproteinemia, the tabulation below shows age at entry (which is the same as age at measurement) and the measured levels in mg/dl of

Case	Age at Entry	HDL-1	Std Sf 0-12	Std Sf 12-20	Std Sf 20-100	Std Sf 100-400	HDL-2	HDL-3
1	32	126	227	76	314	415	19	185
2	33	195	349	101	417	237	22	223
3	36	186	338	67	231	320	0	201
4	37	167	310	38	297	116	42	187
5	38	111	208	76	332	215	13	235
6	38	149	318	43	217	320	24	179
7	38	216	296	92	327	379	16	205
8	39	147	336	65	235	264	10	184
9	41	219	293	47	222	412	10	184
10	44	156	172	47	383	1030	64	205
11	49	117	166	56	177	90	0	139
12	54	264	311	74	367	562	10	221
Averages	39.9	171	277	65	293	363	19	196

HDL-1 and other serum lipoproteins. These measurements were made, ultracentrifugally, in 1954-1957. The AVERAGE values, for this group of 12 men, are shown at the tabulation's bottom.

For age, the mean is 39.9 years; for HDL-1, the mean level is 171 mg/dl in these 12 men. By contrast, the mean HDL-1 level is only 24 for the 891 Livermore males in the age-band 30-39 years (shown in Part 5, below). Although the age-match is not perfect, it is close enough to make meaningful comparisons.

How do the non-HDL-1 measurements in the 12 cases of Massive HDL-1 compare with the measurements in the cohort of 891 Livermore males (ages 30-39)? We can make the comparison by taking means, for the cohort of 891 males, from the measurements listed in Appendix-G, Figures G-2 and G-3.

The tabulated comparison, which follows, shows that the 12 males with Massive HDL-1 Hyperlipoproteinemia ALSO have massively elevated levels of the triglyceride-rich Std Sf 20-100 and 100-400 lipoproteins. From here on, our analysis in this Appendix will use the combined Std Sf 20-100 and 100-400 measurements: Std Sf 20-400.

	12 Men w. Massive HDL-1	891 Men, Ages 30-39
Std Sf 0-12 *	277 mg/dl	353 mg/dl
Std Sf 12-20:	65 mg/dl	51 mg/dl
Std Sf 20-100:	293 mg/dl	93 mg/dl
Std Sf 100-400:	363 mg/dl	54 mg/dl

* Note: The Std Sf 0-12 lipoprotein spectrum includes HDL-1, and Std Sf 0-12 measurements include HDL-1. Although the usual 1.063 gm/ml flotation runs recover somewhat less than 100% of the relatively slow-moving HDL-1 into the top fraction of the preparative ultracentrifugal run, those preparative runs do recover a high percentage of the HDL-1. Because the percentage is not 100%, the subsequent analytical runs routinely underestimate the serum concentration of the total Std Sf 0-12 lipoproteins by a small percentage.

● **Part 5.   The Very Notable Distribution of Elevated HDL-1 among 891 Livermore Males**

The next tabulation shows the distribution of mean HDL-1 measurements when the 891 Livermore males, ages 30-39, are sorted by ascending serum levels of Std Sf 20-400 lipoproteins. After the sort, the cohort is divided into deciles, with 89 persons in each of the first nine deciles and 90 persons in the tenth. Column F shows the mean levels of Std Sf 20-400 serum lipoproteins per decile, and Column C shows the corresponding mean levels of HDL-1. Column D shows the number of persons in each decile who have HDL-1 values equal to or greater than 40 mg/dl, and Column E shows the frequency of such persons (HDL-1 => 40 mg/dl) per 1,000 persons. The last part of the tabulation has divided the tenth decile into its lower and upper halves.

Col.A Decile	Col.B Number of Men	Col.C Mean HDL-1 (mg/dl)	Col.D Cases w. HDL-1 => 40 mg/dl	Col.E HDL-1 => 40 mg/dl: Rate/1000	Col.F Mean Std Sf 20-400 (mg/dl)
1	89	22.303	1	11.2	30.47
2	89	23.460	1	11.2	54.80
3	89	22.663	1	11.2	72.45
4	89	23.247	2	22.5	89.81
5	89	23.225	2	22.5	106.8
6	89	22.326	0	0.0	126.3
7	89	22.416	2	22.5	154.0
8	89	21.663	1	11.2	186.2
9	89	21.022	0	0.0	236.8
10	90	35.177	10	111.1	415.0
	Avg = 23.750		20 = Sum		Avg = 147.263
Tenth:low half	45	20.511	0	0.0	303.133
Tenth:top half	45	49.844	10	222.2	526.933

Column C makes it obvious that the elevation in average HDL-1 levels occurs in the tenth decile of persons, ranked by ascending levels of the atherogenic triglyceride-rich Std Sf 20-400 serum lipoproteins. Column E confirms that the frequency of persons with HDL-1 elevation (=> 40 mg/dl) is 5-fold to 10-fold higher in the tenth decile (111.1 per 1,000 persons) than in the first nine deciles.

After the tenth decile is split into its lower and upper halves, it becomes clear that the upper FIVE PERCENT of the population, ranked by ascending levels of the atherogenic triglyceride-rich Std Sf 20-400 lipoproteins, account for the real elevation of mean HDL-1 levels (Column C). The top five percent of these 891 men have a frequency of elevated HDL-1 which is 222.2 per 1,000 persons --- a rate ten to twenty times the frequency in the other 95% of the sample (Column E). The same 45 men have a mean concentration of Std Sf 20-400 lipoproteins of 527 mg/dl, compared with a mean value of 147 mg/dl in the 891 men considered as a whole.

The distribution-patterns demonstrated in Parts 4 and 5 are striking --- and important. For reasons set forth in Parts 6 and 7, the distributions explain very nicely why mean LDL particle-size is lower in Cases of Ischemic Heart Disease than in Controls.

● Part 6.   Definition and Demonstration of the RATIO Used in Our Analysis

The LDL or Low-Density Lipoprotein spectrum is defined, ultracentrifugally, by flotation rates of Sf 0-12 (Chapter 44, Part 3d). LDL comprises several distinct species, one of which is called High-Density Lipoprotein-1 (HDL-1) --- as discussed in Part 2. HDL-1 is the smallest and most dense Low-Density Lipoprotein which we were able to distinguish ultracentrifugally at Donner Lab. It has the correct properties of size and of density to constitute all, or nearly all, of the "small, dense LDL particles" referred to by Gardner, Stampfer, and others (Part 2). It is correct to say:

LDL = (HDL-1) + (all of the Std Sf 0-12 lipoproteins OTHER THAN HDL-1).

6a.   Definition of "the RATIO" Which Is Calculated in Box 1

In Box 1 and in the next two tabulations, we refer to "the RATIO," which is this:

RATIO =   (HDL-1 Concentration) / (Std Sf 0-12 Concentration Minus HDL-1 Concentration).

We can characterize the ratio as follows:

$$\text{RATIO} = \frac{\text{The HDL-1 part (small, dense part) of the LDL (Std Sf 0-12) lipoproteins}}{\text{The larger and less-dense part of the LDL (Std Sf 0-12) Lipoproteins}}$$

● The higher this ratio is, the SMALLER will be the AVERAGE size of LDL (Std Sf 0-12) lipoproteins. And:

● The lower this ratio is, the LARGER will be the AVERAGE Size of LDL (Std Sf 0-12) lipoproteins.

6b.   The RATIO for Each of 891 Livermore Males (Ages 30-39)

In Part 5, above, we sorted the 891 Livermore males (ages 30-39) by ascending levels of Std Sf 20-400 serum lipoproteins, and divided them into deciles. Here, we are going to calculate the RATIO, as defined in Part 6a, for each individual in each of those same deciles --- 891 separate RATIOS.

Box 1 demonstrates the 89 calculations for Decile 1, and the arrival at the MEAN ratio for Decile 1. The other nine deciles are handled in the same way. The results are tabulated below. From here on, we will refer to this real-world cohort of 891 Livermore males as the "Control Group" or the "Controls." Part 7 explains why.

Tabulation of Mean Values for the Controls

Mean values in Columns A, B, D, and E in this tabulation match Part 5's tabulation. Columns C and F, below, present additional mean values. One purpose of this new tabulation is to ascertain the single average RATIO for the entire group of 891 controls. It is the last entry in Column G, below:

Col.A Decile	Col.B Number of Men	Col.C Std Sf 0-12	Col.D HDL-1	Col.E Std Sf 20-400	Col.F RATIO	Col.G Cases times RATIO
1	89	308.88	22.303	30.472	0.0815	7.254
2	89	325.27	23.461	54.798	0.0812	7.227
3	89	342.37	22.663	72.449	0.0736	6.550
4	89	362.10	23.247	89.809	0.0722	6.426
5	89	354.07	23.225	106.753	0.0734	6.533
6	89	369.42	22.326	126.337	0.0817	7.271
7	89	367.56	22.416	154.045	0.0669	5.954
8	89	374.73	21.663	186.236	0.0647	5.758
9	89	383.60	21.022	236.832	0.0599	5.331
10	90	343.48	35.177	415.033	0.1968	17.712

Sum of Col.G =   76.016

Avg. RATIO = (76.016 / 891) =   0.085

We are now in a position to examine, in Part 7, why Ischemic Heart Disease will have an association with LDL-size.

● Part 7. Results of Our Livermore Analysis, Regarding LDL Diameter and IHD

Suppose that we regard the 891 clinically healthy Livermore males as a Control Group. Appendix-E, Part 12c, describes the nature of this population sample. The mean level of Std Sf 20-400 serum lipoproteins in this Control Group is 147 mg/dl (from above, Part 5, Column F). By contrast, in a matched set of 891 Cases of ISCHEMIC HEART DISEASE, the mean levels of the triglyceride-rich Std Sf 20-400 lipoproteins would be higher than 147 mg/dl --- as demonstrated by the statistically very significant difference (105.9 mg/dl) observed between IHD cases and the baseline population of males aged 30-39 years in the prospective Framingham Study (Gofman 1966, p.683, Table 3).

7a. A Simulated Set of 891 Cases of IHD

For illustrative purposes, we will simulate a set of 891 Cases of IHD, identical with the Control Group in Part 6b except for two modifications. The first modification: We eliminate Decile 1, which is the least likely to contribute any IHD cases. The second modification: We add a duplicate of Decile 10, which is the most likely to contribute IHD cases. These modifications assure that the mean level of Std Sf 20-400 lipoproteins in the Case Group will be greater than in the Control Group. (And this expectation is confirmed: The average of the ten Column-E entries below = 185.7.) In making these two modifications, we re-name old Decile 2 as Decile 1, etc. The new tabulation follows.

Tabulation of Mean Values for the IHD Cases (Simulated Set)

Col.A NEW Decile	Col.B Number of Men	Col.C Std Sf 0-12	Col.D HDL-1	Col.E Std Sf 20-400	Col.F RATIO	Col.G Cases times RATIO
1, new	89	325.27	23.461	54.798	0.0812	7.227
2, new	89	342.37	22.663	72.449	0.0736	6.550
3, new	89	362.10	23.247	89.809	0.0722	6.426
4, new	89	354.07	23.225	106.753	0.0734	6.533
5, new	89	369.42	22.326	126.337	0.0817	7.271
6, new	89	367.56	22.416	154.045	0.0669	5.954
7, new	89	374.73	21.663	186.236	0.0647	5.758
8, new	89	383.60	21.022	236.832	0.0599	5.331
9, new	89	343.48	35.178	415.033	0.1968	17.515
10, new	90	343.48	35.178	415.033	0.1968	17.712

Average, Col. E->   185.733   Sum, Col.G =   86.278

Avg. RATIO = (86.278 / 891) =   0.097

Reminder: The Case Group's average RATIO of 0.097 means 97 mg/dl of HDL-1 per 1,000 mg/dl of (the Sf 0-12 lipoproteins minus HDL-1).

7b. Meaning of a Higher RATIO in the IHD Cases than in the Controls

We can now compare the average RATIO of the IHD Case Group from Part 7a, with the average RATIO of the Control Group from Part 6b:

- RATIO in IHD Case Group: 0.097
- RATIO in the Control Group: 0.085
- The relative value is (0.097 / 0.085) = 1.14 .

The IHD Cases have an average RATIO which is 14 % higher than the RATIO for the controls.

$$\text{RATIO} = \frac{\text{The HDL-1 part (small, dense part) of the LDL (Std Sf 0-12 lipoproteins)}}{\text{The larger and less-dense part of the LDL (Std Sf 0-12 lipoproteins)}}$$

- The higher this RATIO is, the SMALLER will be the AVERAGE size of LDL particles (Std Sf 0-12 lipoproteins). And:

- The lower this RATIO is, the LARGER will be the AVERAGE size of LDL particles (Std Sf 0-12 lipoproteins).

- It follows, from their higher RATIO, that the cases of Ischemic Heart Disease (Coronary Artery Disease) on the average will be found to have LDL lipoproteins with a SMALLER average size than the LDL lipoproteins in the Controls.

- **Part 8. Conclusion**

Point 1: There is no doubt that the overall Std Sf 0-12 lipoprotein spectrum is atherogenic. What we do not presently accept is the hypothesis that the smallest, most dense species in the 0-12 spectrum are ESPECIALLY atherogenic.

Point 2: Average LDL diameter has been reported by Stampfer and Gardner to be 1.2% to 1.9% smaller in IHD Cases than in Controls (Part 3a). Our own comparison, of Cases vs. Controls (Parts 7a and 7b), used serum concentrations to establish that a shift toward smaller average LDL diameters in IHD Cases vs. Controls will occur inevitably as the BY-PRODUCT of there being higher mean serum concentrations of triglyceride-rich Std Sf 20-400 lipoproteins in IHD Cases than in Controls. Therefore, the observation of smaller average LDL size, in IHD Cases than in Controls, does not necessarily indicate that the small, dense LDL particles are at all atherogenic.

>>>>>>>>>>

A peer-reviewer has suggested that, in Part 6b, the RATIO for the Control Group could also be calculated by dividing the sum of the 891 HDL-1 concentration values by the sum of the 891 concentration values for the quantity (total Std Sf 0-12 lipoproteins minus HDL-1 lipoproteins). That calculation produces a RATIO of 0.0721. A similar calculation in Part 7b for the IHD Case Group produces a RATIO of 0.0756. Then Cases/Controls have a ratio of 1.048. This alternate calculation also supports the meaning and conclusions described in Parts 7b and 8.

John W. Gofman 2001

Box 1 of Appendix-H
Calculation of the Mean RATIO for Decile 1 of the 891 Livermore Males.

Col.A HDL-1	Col.B STD (Sf 0-12 Minus HDL-1)	Col.C STD (Sf 0-12)	RATIO Col.D HDL-1 / (Sf 0-12 Minus HDL-1)	Continuation HDL-1	STD (Sf 0-12 Minus HDL-1)	STD (Sf 0-12)	RATIO HDL-1 / (Sf 0-12 Minus HDL-1)
30	198	228	0.1515	24	316	340	0.0759
27	358	385	0.0754	27	378	405	0.0714
21	248	269	0.0847	18	448	466	0.0402
15	274	289	0.0547	42	193	235	0.2176
12	333	345	0.0360	24	316	340	0.0759
27	172	199	0.1570	21	340	361	0.0618
24	325	349	0.0738	21	225	246	0.0933
24	361	385	0.0665	30	295	325	0.1017
24	265	289	0.0906	18	378	396	0.0476
24	339	363	0.0708	24	265	289	0.0906
24	308	332	0.0779	21	149	170	0.1409
15	191	206	0.0785	12	373	385	0.0322
21	203	224	0.1034	27	240	267	0.1125
34	289	323	0.1176	24	256	280	0.0938
30	281	311	0.1068	9	262	271	0.0344
21	228	249	0.0921	24	316	340	0.0759
36	365	401	0.0986	18	305	323	0.0590
24	252	276	0.0952	26	324	350	0.0802
36	296	332	0.1216	18	246	264	0.0732
18	202	220	0.0891	14	259	273	0.0541
36	399	435	0.0902	18	258	276	0.0698
17	346	363	0.0491	18	347	365	0.0519
18	269	287	0.0669	18	150	168	0.1200
21	198	219	0.1061	18	345	363	0.0522
24	269	293	0.0892	15	182	197	0.0824
30	422	452	0.0711	18	172	190	0.1047
39	418	457	0.0933	15	274	289	0.0547
24	294	318	0.0816	23	192	215	0.1198
18	363	381	0.0496	23	230	253	0.1000
16	353	369	0.0453	27	327	354	0.0826
24	274	298	0.0876	18	338	356	0.0533
12	228	240	0.0526	27	299	326	0.0903
21	290	311	0.0724	24	386	410	0.0622
24	238	262	0.1008	27	255	282	0.1059
27	265	292	0.1019	30	405	435	0.0741
27	387	414	0.0698	18	331	349	0.0544
21	239	260	0.0879	18	302	320	0.0596
39	405	444	0.0963	15	207	222	0.0725
15	263	278	0.0570	15	276	291	0.0543
21	293	314	0.0717	21	241	262	0.0871
30	243	273	0.1235	15	205	220	0.0732
12	163	175	0.0736	18	287	305	0.0627
18	246	264	0.0732	17	397	414	0.0428
30	275	305	0.1091	18	343	361	0.0525
18	244	262	0.0738				

Sum of RATIOS for 45 men---> 3.8357 | Sum of RATIOS for 44 men----> 3.4150

Sum of RATIOS for 89 men = 3.8357 + 3.4150 = 7.2507
For Decile 1, Mean RATIO, all 89 men = 7.2507 / 89 = 0.0815

RATIO, defined in text:

$$\text{RATIO} = \frac{\text{Concentration of HDL-1 Lipoproteins}}{\text{Concentration of Std Sf 0-12 Lipoproteins MINUS Concentration of HDL-1 Lipoproteins}}$$

Related text = Part 6b.

"Snapshot" Epidemiology: Why Lipid Levels Only APPEARED Less Important at Older Ages

Part 1. Emergence of a Clinical Disappointment and a Scientific Puzzle
Part 2. "Snapshot" Epidemiology: Nature Is Not Obliged to Be Convenient
Part 3. Remarkably Frequent Changes in Weight and Lipoprotein-Levels: Livermore Data
Part 4. Why Lipoprotein Levels Only APPEAR Less Important at Older Ages

● Part 1. Emergence of a Clinical Disappointment and a Scientific Puzzle

In the early 1950's, the Donner team advanced evidence --- initially from case-control studies --- that all members of the Std Sf 0-400 serum lipoproteins (LDL, IDL, and VLDL) are significantly and independently elevated in Ischemic Heart Disease (Appendix-E). We proposed that arterial exposure, over time, to elevated levels of these serum lipoproteins causes acceleration of and greater amounts of coronary artery atherosclerosis. During that same period, prospective studies were launched, including one at the Lawrence Livermore Laboratory --- a study which included all members of the entire lipoprotein spectrum, low and high density, except for the Std Sf 400+ lipoproteins.

1a. Confirmations --- Including a Perplexing Observation

By 1965, the prospective studies provided some major and extensive confirmation of the case-control findings, but there were some major unresolved questions, too (Gofman 1966). The powerful picture which had emerged by 1965, from two independent prospective studies --- Livermore and Framingham, both of which were based upon ultracentrifugal analysis --- was of an extremely strong relationship of Std Sf 0-400 lipoproteins with Ischemic Heart Disease (IHD) for men entering the study at 30-39 years of age. Moreover, the same lipoprotein classes, Std Sf 0-20 and Std Sf 20-400, which had been proven to be independently associated with IHD risk in the case-control studies, were shown prospectively to be strongly associated with IHD risk in men (more details in Appendix-E, Part 12).

Both prospective studies also confirmed a perplexing observation from the early case-control studies --- namely, the IHD-lipoprotein association appeared to be weaker in the older groups than in the younger groups of such studies.

1b. What Made This Observation "Perplexing" for Many Years?

The observed decline, in the IHD-lipoprotein relationship with advancing age, represented both a scientific puzzle and a clinical disappointment. It hardly made sense that a group of circulating lipoproteins, measured at relatively young ages, could be causally related to development of coronary atherosclerosis --- and NOT be causally related if they were measured, instead, at ages beyond 55 years. The clinically disappointing aspect was that it appeared no medical benefit could be achieved by attention to lipoprotein levels beyond 55 years of age, since IHD risk appeared so weakly associated with Std Sf 0-400 lipoprotein level.

Further consideration of the puzzle identified a reason for the apparent age-related decline in association of Std Sf 0-400 lipoprotein levels with IHD risk. The word "apparent" must be stressed, since our subsequent analysis (Gofman 1978) led us to conclude that, most probably, the Std Sf 0-400 lipoproteins are causally important for coronary atherogenesis and risk of clinical IHD at all ages, including advanced ages. It is the purpose of Appendix-I to summarize that analysis, and to reflect on the limits of "snapshot epidemiology" in general.

● Part 2. "Snapshot Epidemiology": Nature Is Not Obliged to Be Convenient

Whether one studies how a disease relates to some aspect of the blood in a population, or to some other parameters, the easy parameters to evaluate are those which are essentially constant, both acutely and over decades. In such a simplified situation, individuals can be ranked according to the parameter at one point in life, and be relied upon to remain in that rank order throughout the decades during which a process such as atherosclerosis develops.

The usual prospective study takes a "snapshot" (a single measurement) of a parameter at one point in time, and neglects its past history as well as its future behavior. If such studies deal with a stable parameter, one "snapshot" can suffice.

2a. A Severe Inconvenience in Medical Research

But Nature is not compelled to be kind to medical investigators, and only a very few parameters are fixed diurnally and over decades. Diurnal variation deserves special mention. The vague medical notion, that stable parameters are more important for disease development than are those with substantial diurnal variation, has no basis. Nature is under no obligation to prevent a highly variable parameter from being the key element in a disease process, even at the loss of convenience for the medical investigator.

If the parameter under study is not fixed, the investigators in a "snapshot" study can hope that the non-measured exposures do not change the relative RANKING (established by the "snapshot" measurement) of a study's participants upon that parameter. The investigators start their study with an appreciable gradient in measurements, from low to high. But without their knowing it, 10 years after the "snapshot" or 10 years before the "snapshot," many participants may have OTHER places in the ranking (due to changes in the parameter which were never measured). The "snapshot" rank-order may be a poor (even misleading) indicator of differences in lifetime exposure to the parameter, if the parameter is quite unstable over time.

It appears that a proper prospective study would start early in childhood, measure crucial parameters at least annually (and with a check on diurnal variation), and would follow this procedure for the full lifespans of the participants.

2b. The Serum Lipoprotein Levels: Not a Stable Parameter

Ischemic Heart Disease (Coronary Artery Disease) has two phases in time: (1) Coronary atherogenesis, which is a silent pathological process occurring over decades, and (2) Clinical Ischemic Heart Disease, which is a relatively or extremely abrupt announcement of the pathological process --- a process which may well continue, if its announcement was not fatal.

The Lipid Hypothesis means that risk of a fatal outcome is higher in persons with higher average exposure to atherogenic lipoproteins over TIME (all other co-actors being equal). Therefore, variation of exposure-level matters, daily and over the lifespan.

Nonetheless, virtually all reported prospective studies, on the relationship of serum lipoprotein levels and Ischemic Heart Disease, have failed to find out how serum concentrations of the Std Sf 0-12, 12-20, 20-100, and 100-400 lipoproteins VARIED in each participant, acutely over daily cycles and chronically over decades of life. We must fault the magnificent Framingham Study and our own Livermore Study on both counts.

It turns out that the Livermore Study contains evidence which WARNS that a remarkably large share of American adults change their lipoprotein measurements even during a brief period of 18-months. Part 3 presents that evidence.

● Part 3. Remarkably Frequent Changes in Weight and Lipoprotein-Levels: Livermore Data

By 1952, it was already evident that obesity was associated with unfavorable serum elevation of certain atherogenic lipoproteins (Gofman 1952-b, p.514). The fact, that serum lipoprotein levels rise with weight-gain and fall with weight-loss, was illustrated in Appendix-F, Part 12, by some specific evidence.

Here, we add to that evidence with some interesting measurements from the Livermore

Lipoprotein Study. Of the 2,297 persons in the Livermore database, 374 individuals were examined twice, with the average interval between examinations being 1.5 years. The findings were as follows for this interval of 1.5 years:

Number whose weight changed by less than 5 pounds: 164 persons.

Number who lost five pounds or MORE: 73 persons.
Average loss of weight = 10.7 pounds.
Avg change in Std Sf 0-400 lipoprotein level = decline by 5.0 mg/dl for each pound of weight lost.

Number who gained five pounds or MORE: 103 persons.
Average gain of weight = 9.6 pounds.
Avg change in Std Sf 0-400 lipoprotein level = rise by 5.0 mg/dl for each pound of weight gained.

Number who did not change weight: 34 persons.
Stable weight was associated with stable Std Sf 0-400 lipoprotein level.

There were 176 persons (73+103) who either gained an average of about ten pounds (4.54 kilograms) in the 1.5 year interval or lost about ten pounds in that interval. None of the individuals had been advised, requested, or instructed by the Livermore Laboratory to change weight. Since there was a total of 374 individuals who were studied twice, it follows that (176 / 374), or 47 percent of the Livermore population sample showed large changes in weight and Std Sf 0-400 lipoprotein levels in the brief span of 1.5 years. This remarkable fact is ordinarily obscured by the AVERAGE weight-gain trend during the decades of adulthood in the United States.

How does this information relate to "snapshot" epidemiology? One illustration can suffice for endless variations.

Suppose that two individuals, A and B, have the same age, and have equal levels of Std Sf 0-400 serum lipoproteins up to a given time. Both have followed the mean trend of rising serum lipoprotein levels with advancing age (shown in Appendix-F, Box 5). Now, suppose that during the 1.5 years which follow the "given time," person A gains 10 pounds of weight (as did 27.5% of the Livermore employes) and person B loses 10 pounds of weight (as did 19.5% of the Livermore employes).

Then we come along and enroll them in our "snapshot" prospective study about the relationship of serum lipoprotein levels with Ischemic Heart Disease. At this point, A has appreciably higher levels than B, although the levels had been the same until 1.5 years earlier. In such a study, having no prior lipoprotein measurements of A and B, we in effect assume (but mistakenly) that A and B have differed in their lifetime exposure to the atherogenic lipoproteins. And we assume that they will continue to differ. In truth, however, A and B have had nearly equal INTEGRATED exposure, prior to our "snapshot." Moreover, the data above certainly shatter the assumption that their "snapshot" measurements will remain constant during the follow-up years.

Part 4 combines the information In Part 3 with the information from Appendix-F, Box 5, to address the APPARENT weakening of the male IHD-lipoprotein relationship at older ages.

● **Part 4. Why Lipoprotein Levels Only APPEAR Less Important at Older Ages**

The data and graphs in Appendix-F, Box 5, show that average serum levels of the Std Sf 0-20 and 20-400 lipoproteins rise steeply in males from about age 18 to age 42. Then the increase flattens out or is already declining, on the average. Our interpretation of the curving-over of these plots is that it results from an increase in the number of men who shift, from positive caloric balance or from caloric equilibrium, into negative caloric balance as they age. It can be anticipated that in culturally different populations, these curvatures might be different, if customary weight-behavior is different.

The same Box 5 (Appendix-F) shows that levels for females during adulthood are lower than levels for males --- until the women catch up with the men at about age 50 for the Std Sf 0-20 lipoproteins and at about age 59 for the Std Sf 20-100. After these ages, the female levels exceed the male levels. Box 5 makes it evident that the average female has a lower INTEGRATED exposure to the atherogenic lipoproteins than the average male. Their age-adjusted MortRates from Ischemic Heart Disease have always been much lower than male rates, too (Table 41-B).

4a. Age at "Snapshot": Males 30-39, and Females

Despite the rising, non-constant lipoprotein levels so obvious in Box 5, investigators necessarily hope in "snapshot" prospective studies that individual participants do, at least, retain their relative RANK ORDER with respect to Std Sf 0-400 lipoprotein levels. For men in the age-band of about the 20-35 years, we are lucky, simply because relatively few individuals (relative to older age-bands) "fall out of order" during the steep upward trend in average weight (positive caloric balance being the rule).

Thus, at the end of a 12-year follow-up, we observed a very strong positive association of Std Sf 0-400 lipoprotein level with IHD risk, for those men who entered the "snapshot" prospective study between 30 and 39 years of age (Gofman 1966, p.683, Table 3). We suggest that the association would have been stronger yet, if not for perturbations in lipoprotein levels caused by relatively acute fluctuations in weight (Part 3). Moreover, at 12 years after the "snapshot," the surviving men had an average age of about 48 years, which means they had spent little time beyond the age when peak-exposure to Std Sf 0-400 lipoproteins changes to its downward direction.

What about the women of ages 30-49 years at their "snapshot" measurement? Their average age was 44.4 years then, and about 56 years at follow-up. Because of the much lower rates of IHD in females than males, there were still too few cases of de novo IHD (16 cases in 1,635 women) to obtain much statistical power (Gofman 1966, p.683, Table 7). The statistically strongest IHD-lipoprotein associations occurred in the female cohort of ages 50-55 (average age of about 65 at follow-up). In this group, there were 39 de novo cases among 468 women (Gofman 1966, p.683, Table 8). But for the oldest female cohort (having ages 56-69 at "snapshot" and a mean age of about 71 years at follow-up), the statistically significant IHD-lipoprotein associations appear to have vanished (Gofman 1966, p.684, Table 9). So, we do not know to what extent --- if any --- the discussion here is applicable to females.

4b. The Mis-Match between True Rank-Order and Ostensible Rank-Order

By contrast with the men ages 30-39 at "snapshot," men who were ages 40-49, or 50-59, or 60-69 at the time of their "snapshot" measurement, spent most or all of their follow-up time at ages when the AVERAGE level of Std Sf 0-400 lipoproteins was declining (Appendix-F, Box 5).

The observed change, from average rise in lipoprotein level to average decline, can have two possible explanations. Either all the men are changing in this manner, or some are still rising in lipoprotein level (and in weight-level) while an increasing proportion are leveling out or declining in lipoprotein level (and in weight-level) with increasing age. We suggest that experience favors the second interpretation.

Under the latter interpretation, the change from an average rise in lipoprotein level to an average decline, offers extra opportunities for individual participants to experience major, unrecorded changes in their lipoprotein RANK ORDER within their age-band.

The men who were ages 40-49, at their "snapshot" measurement, have experienced not only more years of acute lipoprotein fluctuations (Part 3) than the 30-39 year cohort, but a growing share of men in the 40-49 age-group shift from increasing levels to decreasing levels of serum lipoproteins. In the 40-49 age-group, these circumstances cause the rank order of Std Sf 0-400 lipoprotein levels to become materially perturbed (if not massively so), relative to the rank order assigned by the "snapshot" measurements.

A mis-match is developing, between the true rank order and the ostensible rank order.

The result: The association of Std Sf 0-400 level with IHD risk in males appears to be markedly weakened. The IHD-lipoprotein association necessarily appears to weaken even more, as age at the "snapshot" measurement increases. When 58 is the average age at "snapshot" (age 70 at follow-up), all of the IHD-lipoprotein association appears to be lost for males in a prospective study (Gofman 1966, p.683, Table 6).

The REASON is that the "snapshot" Std Sf 0-400 level tells us less and less about the integrated Std Sf 0-400 lipoprotein exposure, as age-at-entry to such a study becomes higher. It is INTEGRATED exposure which relates to risk of IHD (all other co-actors being equal).

4c. The Need to Find Out the True Rank Order

This problem in "snapshot" epidemiology is not going to go away, and the problem is not peculiar to the study of the IHD-lipoprotein relationship. It applies to many etiologic co-actors in Ischemic Heart Disease and in other disorders which build up over decades. What is needed is an appreciable increase in the frequency of measuring the key variables, if one hopes to reduce the frequency of FALSE conclusions.

Parts 4a and 4b suggest that "snapshot" epidemiology came close to concealing the now well-confirmed IHD-lipoprotein relationship --- and might have done so, were it not for the inclusion of the men ages 30-39 at their "snapshot" measurement. Parts 3 and 4 help to explain why this group was able to avoid the pitfall.

The nature of the pitfall indicates that the APPARENT loss of the IHD-lipoprotein relationship, in men measured only at advanced ages, was probably not a REAL absence of the relationship. Recognition of this pitfall moves us to warn that serum lipoproteins which are proven atherogens at one age almost certainly remain atherogenic at even the most advanced ages.

>>>>>>>>>>

● **Part 1. Two Distinct Types of Radiation-Induced Damage**

There are reports as early as 1899 to 1909 which examine the types of damage inflicted by high-dose xrays to small blood vessels, especially in the skin (Part 7). By 1922, reports began appearing on damage to the heart from high-dose medical radiation (Part 7). By 1942, occurrence of such damage was well established (Part 7). Starting in the 1960s, numerous reports related xray therapy --- for various Cancers in the chest --- to xray-induced damage to the heart and its vessels (Parts 3, 4, 5, and 6). Also, there have been some animal studies (rabbits, rats) which report synergism between xray exposure and elevated plasma-levels of atherogenic lipoproteins (Part 5c, for example).

The evidence described in this appendix has firmly established that exposure of the heart and the coronary arteries, to very high-dose medical radiation, can result in damage typically having results very different from atherosclerosis. What are such results?

The answer requires explanation of a few more terms. The heart muscle itself (the myocardium) is surrounded by a sac of tissue whose outer side (adjacent to other organs) is called the pericardium, and whose inner layer is called the epicardium. The lining of the four chambers of the heart is called endocardium.

1a. Types of "Non-Atherogenic High-Dose Radiogenic Damage"

Among various patients who have received a few thousand rads of radiation involving the chest, injuries of virtually every component of the heart and its coronary arteries have been observed. These include severe pericarditis (both of the pericardium and the epicardium), myocardial fibrosis, myocardial infarction, endocardial fibrosis, and a variety of injuries to the coronary arteries. When very high-dose radiation has injured a coronary artery, the observors typically (but not always) report that the resulting lesions DIFFER from the usual atherosclerotic plaques. Parts 3, 4, 5 and 6 provide the observations in great detail.

We can refer to all types of radiation-induced non-atherosclerotic injuries, to the heart and its vessels, as "non-atherogenic high-dose radiogenic damage." We found no reliable estimate of its frequency, per 1,000 patients whose hearts have received very high-dose irradiation.

An undisputed point deserves emphasis, here at the outset: Not every type of radiation-induced injury to the heart itself, and to the coronary arteries, is part of the ATHEROSCLEROTIC process --- the process which underlies Ischemic Heart Disease (Coronary Heart Disease, Coronary Artery Disease) in the human.

Nonetheless, the kinds of non-atherogenic damage described above have been called "heart disease" or "coronary heart disease" or "coronary artery disease" in the biomedical literature (Parts 4, 5, and 6). Unfortunately, these labels are almost certainly the basis of some confusion.

1b. The Atherogenic Damage from Xrays: Radiation-Induced Mutations

In contrast to the well-established "non-atherogenic high-dose radiogenic damage" described in this appendix, we have proposed that ionizing radiation is also a cause of atherogenic damage and fatal Ischemic Heart Disease --- as a result of radiation's undisputed power to cause mutations of virtually every type in the coronary arteries. Chapter 45 introduces our Unified Model of Atherogenesis and Acute IHD Death.

● Part 2. RadioTherapy and Angioplasty: Range of Xray Dosage

High-dose irradiation, of all or part of the heart and coronary arteries, can be unavoidable during radiation therapy for certain malignancies (for example, breast cancer, Hodgkin's disease or other lymphomas). Such therapies, received in a series of doses over a period of weeks, can amount to internal radiation doses (all exposures combined) in the range of 1,000 to 5,000 rads, and even higher, in parts of the chest (Part 4a). These are very high doses (Appendix-A). Tissues located adjacent to an xray beam (or "field") receive some dose too, from internally scattered photons. Of course, the amount of scattered radiation declines as distance from the edge of the field increases.

Cancer therapy is not the only cause, in modern medicine, of high-dose heart irradiation. Heart patients who receive MULTIPLE cardiac catheterizations and angioplasties can accumulate skin doses from fluoroscopic xrays in the range of 1,000 to 3,500 rads (Lichtenstein 1996), which could mean accumulated doses of a few hundred rads --- even 1,000 rads --- to parts of the heart and the coronary vessels. These are very high doses, too. Under ordinary circumstances, however, the estimated average dose from angioplasty is about 60 rads per procedure to the skin (less to the heart) if one stenosis is dilated, and about 130 rads to the skin if two are dilated (NCRP 1989, p.31). Because of self-shielding, the side of the heart which is proximal, to the source of the xray beam, receives a higher dose than the distal side of the heart.

● Part 3. Autopsy Reports in 1965 on Two Very Interesting Patients

On July 31, 1965, the Lancet published a three-paragraph letter entitled "Myocardial Infarction Following Radiation," by Dollinger, Lavine, and Foye of the San Francisco Veterans Adminstration Hospital (Dollinger 1965). They reported on what "appears to be the first case of acute myocardial infarction secondary to radiation therapy to be studied histologically" [by pathologists]. The next year, they reported at length on the case in JAMA, under the title "Myocardial Infarction due to Post-Irradiation Fibrosis of the Coronary Arteries" (Dollinger 1966).

3a. Lesions Observed by Dollinger: Not the Usual Atherosclerosis

The case was a man diagnosed with Hodgkin's disease at age 23. However, he did not receive radiation therapy until he was age 31. Two months later, he experienced an acute but nonfatal myocardial infarction. In subsequent years, he developed angina pectoris, congestive heart failure, and several other afflictions. He died of pulmonary impairment in 1961, eighteen years after his initial diagnosis. When he died, autopsy revealed "extensive infarction of the right ventricle and posterior septum" (1965, p246). And how did his coronary arteries look? Dollinger reports (1966, p.317):

"The left coronary arteries showed only a few plaques and were widely patent [unobstructed]. However, the right coronary artery, 2 cm distal to its origin, abruptly changed to a pliable fibrous cord with a pinpoint lumen. This obliteration extended the full length of the right circumflex coronary artery down into the beginning of the posterior descending coronary artery, which was patent and thin-walled ... Multiple sections of the right coronary artery showed a marked degree of fibro-muscular proliferation interior to a well-preserved internal elastic membrane. This fibrous tissue was rather dense and contained a small number of fibroblasts with very scanty cytoplasm. No foam cells were present, and there was no evidence of cholesterol clefting, which was found to a minimal degree in areas of the left coronary artery ... The small branches of the right coronary artery in this area also showed intimal fibrous proliferation. The aorta and its major branches showed only minimal atherosclerosis." And in the "Comments" section (p.318):

"... it appears that the fibrous muscular proliferation of the right coronary artery was due to radiation. The patient was 31 years of age at the time of his myocardial infarction, two months after completion of radiation therapy. There was no significant atheromatous change in the coronary arteries or other major arteries, there was no family history of coronary artery disease, and the serum cholesterol value was normal."

Dollinger and co-workers characterize the radiation-induced damage as "fibrosis of the coronary arteries" (their title), they never describe the lesions as atherosclerotic, and they specifically emphasize that "there was no atheromatous change in the coronary arteries." So the high-dose radiogenic lesions did not look to THEM like the "usual" atherosclerotic lesions.

3b. Lesions Observed by Prentice: Typical Atheromatous Change

Only three weeks after Dollinger's 1965 letter, Lancet published a comment on it, provided by R.T.W. Prentice of the Western Infirmary in Glasgow. Prentice had a case of his/her own (Prentice 1965). In the Glasgow hospital, a young man had recently died at age 19 from "acute left ventricular failure." At age 15, he had received chest irradiation (3,250 Roentgens) after diagnosis of probable lymphoma of the Hodgkin group. For a few years, he seemed "generally well" until he developed angina pectoris, which could not be relieved. "There was no family history of premature coronary-artery disease ... His serum-cholesterol was within normal limits." After his death, his coronary arteries were studied:

"Histological examination confirmed the presence of typical atheromatous change in all three main branches of the coronary vessels; there was no evidence of deposits of Hodgkin's tissue in the vessels, serial sectioning of which showed extensive fibrin deposition in the intima of the left descending artery, but no evidence of ante-mortem thrombus formation. There was recent extensive myocardial infarction of the left ventricle and septum. There was no atheroma of the vessels elsewhere." And Prentice added that the histological findings fail to resemble "those described by Dr. Dollinger and his colleagues ..." And Prentice ends by saying:

"While it is not possible to postulate a direct causal relation between radiotherapy and severe coronary-artery disease in this case, it is felt that the association of these two events in a young man of this age should be reported." Indeed.

3c. One Agent and Two Different Responses

Dollinger's case appears to have developed severe "non-atherogenic high-dose radiogenic damage" in his right coronary artery, and no apparent radiation-induced atherosclerosis. Prentice's case appears to have developed radiation-induced atherosclerosis in the three main branches of the coronary arteries, and no severe "non-atherogenic high-dose radiogenic damage." These two cases are consistent with our point (Part 1) that ionizing radiation can induce two DIFFERENT kinds of heart-related damage.

● Part 4. Distinctions Found and Emphasized by Cohn, Fajardo, Stewart

In 1967 and 1968, the California team of Cohn (medicine), Fajardo (pathology), and Stewart (radiology) published three papers on "radiation-induced heart disease" (Cohn 1967; Stewart 1967; Fajardo 1968). These papers report on some or all of 25 patients in whom "significant heart disease followed radiation therapy to the chest for a variety of malignant tumors" (Stewart 1967, p.302). Those tumors included cancers of the lung, breast, esophagus, thymus, and Hodgkin's disease. None of the patients had evidence of "heart disease" before they received radiation therapy.

4a. Radiation Dosage, and Its Consequences

Dose to all or part of the heart ranged from 3,000 to 9,800 rads per patient, administered at 1,100 rads per week (Fajardo 1968, p.512). The interval of time, between initiation of radiation and taking of the tissue for study, ranged from 6 to 84 months (Fajardo 1967, tabulation on p.513 for 16 patients). The discussion in Stewart 1967 begins (p.307):

"Scattered case reports have described radiation-induced heart disease of essentially all the types seen in this series. Acute pericarditis (Jones 1960; Portioli 1963; Spodick 1959), chronic pericardial and myocardial fibrosis (Rubin 1963; Gimlette 1959; Hurst 1959), and myocardial infarction (Prentice 1965) have been noted following thoracic irradiation. The present series of cases from a single institution [Stanford University School of Medicine] is unique in that the number of reported cases is high ...It is our contention that many similar cases could be found in other institutions where large numbers of patients are treated with irradiation." And the summary on all 25 patients states (Stewart 1967, p.308):

"Acute pericarditis, often with pericardial effusion [presence of escaped fluid] and tamponade [acute compression of the heart], chronic pericardial effusion, and chronic constrictive pericarditis sometimes associated with myocardial and/or endocardial fibrosis or fibroelastosis were most commonly seen. Five of these cases were fatal. An additional patient [Case 21] died at the age of fifteen of myocardial infarction sixteen months after irradiation. In the majority of patients, detectable heart disease developed after a delay of several months. Most had received at least 4,000 rads to a sizable portion of the heart, and the more severe cases occurred among patients who had received the highest radiation doses to the heart (up to 8,950 rads) ... Radiation-induced heart disease, particularly pericarditis, is apparently more frequent and important than is commonly appreciated ..."

Reporting on 16 of the 25 patients, the Fajardo 1968 paper begins (p.512):

"Connective tissue proliferation is the hallmark of these late radiation lesions wherever located in the heart. Organizing pericarditis with extensive fibrosis is the most frequent alteration ... Diffuse interstitial myocardial fibrosis is commonly present. Endocardial and vascular changes are much less frequent. The diffuse nature of the myocardial fibrosis makes it difficult to detect grossly, and pathologists should be aware of its deceivingly innocent appearance."

4b. One Anomalous Case in the Series

In both Stewart 1967 and Fajardo 1968, the authors emphatically differentiate between one non-typical case in their series (Case 21) and all the other cases. Stewart 1967 describes the "spectrum of heart disease" observed in the 25 cases, and when the authors reach "Coronary Artery Disease and Myocardial Infarction," only ONE patient fits the category (p.305):

"A 15-year-old boy died of myocardial infarction sixteen months after receiving 4,000 rads to the 'mantle' for Hodgkin's disease. There was no known familial or metabolic predisposition to atherosclerosis, and an autopsy showed intimal proliferation and atheromatous deposits in the coronary arteries with a fresh myocardial infarction." In Fajardo 1968, the authors organize the results in anatomical groups, and when they reach "Vessels," they report (p.516):

"The vessels did not show consistent changes. However, proliferation of endothelial cells was observed in six of the 12 cases where large, medium, and small vessels could be examined. Case 21 was a remarkable exception. This 15-year-old boy, who absorbed 4,400 rads in the heart during treatment for Hodgkin's disease, developed severe coronary atherosclerosis. There was marked thickening of the intima by fibroblasts, collagen, endothelial cells, and histiocytes, some with large foamy cytoplasm. This thickening resulted in severe narrowing of the lumen of the left anterior descending branch to less than one eighth of its expected cross sectional area. Thrombosis was found at necropsy [autopsy] with an extensive area of ischemic necrosis of myocardium. No evidence of atherosclerosis was found in other vessels at necropsy and there was no history of familial hyperlipemia or severe juvenile atherosclerosis."

So, these investigators find that the damage from very high-dose radiation typically is NOT atherogenic damage --- which is why they comment so clearly on this single, apparent exception.

4c. Myocardial Fibrosis: Due to Radiation-Damaged Micro-Circulation

In 1973 and 1977, Fajardo proposed that closure of small irradiated vessels, resulting in "insufficient micro-circulation," is the cause of the myocardial fibrosis observed after high-dose irradiation of the heart (Fajardo 1973 and 1977).

In 1984, Stewart and Fajardo jointly proposed that existing evidence is "consistent with the following explanation of the pathogenesis of radiation-induced myocardial fibrosis. Radiation injures capillary endothelial cells leading to destruction or obstruction of capillaries. A compensatory mechanism of endothelial cell renewal is set in motion, but is inadequate to reconstruct the damaged capillary network. Ischemia [deficiency of blood supply] results from the insufficient micro-circulation and leads finally to fibrosis."

4d. Harmony between Hypothesis-2 and the California Studies

In 1977, Fajardo wrote an editorial in the journal Chest, in which he says (Fajardo 1977, p.564): "As far as I know, typical coronary atherosclerosis, of the type seen in man, has not been produced experimentally [in animals] by the administration of radiation alone."

We are not surprised. Our Unified Model of Atherogenesis and Acute IHD Death requires BOTH mutation-induced dysfunctional clones in the coronary arteries AND elevated levels of atherogenic lipoproteins in flux through the intima of those arteries (Chapter 45; see also Part 5c, below, on synergism).

In 1984, Stewart and Fajardo published a 21-page paper entitled "Radiation-Induced Heart Disease: An Update," in which they again draw the distinction between Coronary Artery Disease (CAD) and the heart troubles they observe after very high-dose radiation. Unfortunately, because they say "radiation" without specifying "very high-dose," their text may appear to be in conflict with Hypothesis-2 --- although there is no conflict at all. For example (Stewart 1984, p.175):

"... one must conclude that the evidence for an important role of radiation in the pathogenesis of CAD is weak (Stewart 1978)."

What they studied was damage to the heart and its vessels after very high-dose radiation. If they had written what they meant --- which was surely that "the role of VERY HIGH-DOSE radiation in the pathogenesis of CAD is weak" --- then our position would be like theirs. Such a statement would be fully compatible with Hypothesis-2 and with a highly consequential role of medical radiation in the etiology of fatal Ischemic Heart Disease. After all, the number of people whose hearts ever accumulate very high-dose radiation, is very small by comparison with the number whose hearts accumulate low to moderate doses of radiation from medical procedures.

On another point, agreement clearly exists between Stewart, Fajardo, and us. We agree that the so-called "radiation-induced heart disease" on which they report, with elegance, is NOT THE SAME DISORDER as atherosclerosis --- the disorder which causes the overwhelming share of Ischemic Heart Disease.

● Part 5. Distinctions Found and Emphasized by McReynolds

Before describing the McReynolds' case, we will describe the Huff case, which was reported four years earlier.

5a. The Huff Case, 1972: Infarction Followed by Cardiogenic Shock

Harry Huff and E. Max Sanders reported (Huff 1972, p.780) on a patient, age 21, who had an acute myocardial infarction about nine months after receiving 3,500 Roentgens to the left cervical region (of the heart) for Stage 1a Hodgkin's disease. Huff and Sanders do not report whether any of the heart was directly in the xray field.

Following complications of the infarct, the patient died of cardiogenic shock (circulatory failure due to sudden decrease in cardiac output). The patient's prior records showed "normal average serum cholesterol," normal blood pressure and fasting blood sugars. He lacked a family history of arteriosclerotic heart disease, hyperlipidemia, or hypertension. They state (Huff 1972):

"At autopsy, adhesive pericarditis with loculated areas of pericardial fluid and an anterior infarction with definite extension into the ventricular septum were found. The proximal 2 cm of the left anterior descending coronary artery was completely occluded by organizing thrombus superimposed on an atherosclerotic plaque. The right coronary, left coronary and posterior circumflex arteries were grossly normal, but microscopical observation revealed atheromatous changes here as well. No residual Hodgkin's disease was found." And (Huff 1972):

"Myocardial and pericardial disease secondary to radiation is widely acknowledged. Since many patients subject to radiation therapy are also candidates for coronary-vessel occlusion, the independent effect of radiation on the coronary vessels is difficult to evaluate clinically ... We should be interested in knowing whether other cases exist ..."

5b. The McReynolds' Case 1976: "Adventitial Scarring" of the Coronaries

In January 1976, McReynolds and co-workers began a paper as follows (McReynolds 1976, p.39):

"Attention is called to the development of coronary heart disease in two patients several years after they received mediastinal irradiation [mid-chest irradiation] for Hodgkin's disease. One patient, a 33-year old man, died suddenly eight years after irradiation; necropsy disclosed marked narrowing of all three major coronary arteries. In addition to severe intimal fibrous thickening, there also was considerable adventitial scarring of the coronary arteries. This type of coronary sclerosis is different from that seen in the usual patient with coronary heart disease." They present no autopsy data on the second patient, age 42, because he was RECOVERING in 1976 from his second acute myocardial infarction (p.42).

Details of the Observed Differences

In the "Comments" section of the paper, McReynolds and co-workers say (McReynolds 1976, p.43):

"Several factors support the view that the coronary disease in Case 1 [deceased] was the result of mediastinal irradiation rather than the result of atherosclerosis of the usual type: (1) The patient had received large doses of irradiation to the mediastinal tissues eight years earlier. (2) He was young (age 33 years) at the time of sudden coronary death. (3) The patient had other lesions which were readily attributable to the effects of irradiation, including pericardial adhesions and thickening, interstitial myocardial fibrosis of the right ventricle and focal thickening of the mural endocardium of both ventricles. (4) The patient had coronary arterial lesions of a type uncommonly observed in the usual type of coronary atherosclerosis, including severe coronary adventitial fibrosis, usually in continuity with overlying epicardial fibrous tissue, and marked paucity of lipid in the intimal lesions." And (McReynolds 1976, p.43):

"In our experience (Roberts 1972, 1973, 1974, 1975) the dominant component of atherosclerotic plaques large enough to be fatal is fibrous tissue (collagen), but lipid, primarily extracellular, is present in most plaques and constitutes a significant proportion of them (at least 25 per cent, occasionally much more). In the present patient, the amount of lipid in the plaques examined was minute, and this probably is particularly unusual for such a young patient. Furthermore, the collection of cells in the adventitia of the coronary arteries included large numbers of plasma cells and fewer lymphocytes. In the usual type of coronary atherosclerosis, lymphocytes are dominant and plasma cells are infrequent ('adventitial lymphocytosis'). (5) The patient had relatively 'low level' risk factors to the development of accelerated atherosclerosis ... (6) He had virtually no atherosclerosis in other systemic arteries, including the aorta."

5c. Observations of Synergism between Radiation and Cholesterol

In the same paper, McReynolds and co-workers allude to the experimental animal evidence on synergism between very high-dose xrays ("ballpark" of 2,000 rads in these studies) and cholesterol (McReynolds 1976, p.44-45):

"Experimental studies also support the thesis that therapeutic levels of irradiation can produce or hasten coronary atherosclerosis (Gold 1961 and 1962; Amromin 1964; Lamberts 1964). The effects of radiation alone, however, are minimal, but when irradiation is given to animals (rabbits or rats) on high cholesterol diets, severe coronary atherosclerosis results, far more severe degrees of atherosclerosis than that resulting from the hypercholesterolemia alone. Irradiation and hypercholesterolemia appear to act synergistically to produce considerably more atherosclerosis than that produced by either radiation or hypercholesterolemia alone."

Very similar comments on synergism, citing the same sources, are found also in Fajardo 1977 and Stewart 1984 (who adds Artom 1965 to the list of sources for such observations).

● **Part 6. Distinctions Found and Emphasized by Brosius and by Dunsmore**

6a. The work of Brosius, Waller, and Roberts

In 1981, Brosius and co-workers (at the National Heart, Lung, and Blood Institute) published a study entitled "Radiation Heart Disease: Analysis of 16 Young (Aged 15 to 33 Years) Necropsy Patients Who Received over 3,500 Rads to the Heart." Their study reported in parallel manner on findings from 10 control patients. "Although only six of our 16 patients had clinical evidence of cardiac dysfunction, at necropsy, all had anatomic evidence of cardiac abnormality. All but one had fibrous thickening of the pericardium ..." (p.526). We will omit the now-familiar details. The authors themselves say (Brosius 1981, p.527):

"The most important contribution of the present study is the detailed information regarding the epicardial [extramural] coronary arteries. Damage to these vessels by high-dose irradiation has been reported previously [many references cited], but the extent of, and the type of, damage to these arteries have not been described. It is now clear that high-dose irradiation can cause damage to the epicardial coronary arteries and allow intimal proliferation of mainly fibrous tissue to produce luminal narrowing." And (Brosius 1981, p.527-528):

"Of our 16 patients, 16 of the 64 major (right, left main, left anterior descending and left circumflex) epicardial coronary arteries were narrowed >75 percent in cross-sectional area, primarily by fibrous plaques. In contrast, of 10 control subjects of similar age and sex, only one of 40 major epicardial coronary arteries was similarly narrowed ... Thus, our study-patients clearly had more coronary narrowing than did the control subjects, and none had pronounced recognized risk factors to premature atherosclerosis." And (Brosius 1981, p.528):

"The dominant component of the atherosclerotic plaques in both our study patients and in the control subjects was fibrous tissue; very little lipid was present in either." Even though this is not the "usual" atherosclerotic plaque, the authors call it by the usual name. They continue by describing additional non-typical findings (Brosius 1981, p.528):

"The study patients, however, had a striking loss of smooth muscle cells from the media (76 percent of sections versus 10 percent of sections in the controls). Furthermore, adventitial fibrosis was noted in 49 percent of the coronary sections in the study patients and in only 3 percent of the sections in the control subjects. Thus, high dose radiation causes coronary luminal narrowing ... adventitial scarring and damage to the smooth muscle cells in the media."

6b. Case Reported by the Dunsmores, 1986 and 1996

In 1986, Dunsmore, LoPonte, and Dunsmore authored a paper entitled "Radiation-Induced Coronary Artery Disease" (Dunsmore 1986). It is inevitable that a large proportion of physicians reading that title would assume the disorder under study to be no different from atherosclerotic disease of the coronary arteries. But in 1996, the Dunsmores made a special effort to correct that assumption, as we shall show.

The 1986 paper reports on three patients with Hodgkin's disease who received radiation therapy at ages 19, 19, and 32, respectively --- exposing "large areas of the heart" to 4,000 rads of dosage in 1965, 1966, and 1970 respectively. None received adjunctive chemotherapy. All developed "coronary artery disease," after 8, 12, and 4 years, respectively. All "succumbed" to their heart problems before age 40 (Dunsmore 1996). In Dunsmore 1986, there are autopsy findings only from Case 2.

Case 2, male, was age 19 when he was diagnosed with Stage IIB Hodgkin's disease in 1974. He received 4,000 rads of mediastinal midplane radiation dose. In 1978, he suffered two acute myocardial infarcts, but survived. At that time, he quit smoking two packs of cigarettes per day. In 1983, he was hospitalized with a large pleural effusion. He had 80% stenosis of the left mainstem coronary artery, and "the left anterior descending artery was totally occluded near its midpoint and the distal segment was thin and irregular and not bypassable. The main circumflex artery was occluded.

The right coronary artery had extensive luminal irregularity without critical obstruction." Coronary bypass surgery was performed. "The patient had a stormy post-operative course and died 48 hours later" (p.240). The Dunsmore paper presents four photos of coronary artery samples from Case 2, with the following captions:

Figure 1: "Case 2. Coronary artery showing compromise of the lumen by thickened intima. Intima and media are replaced by dense hyalinized tissue with only a peripheral rim of adventitia discernible." Hyalinization means conversion into amorphous tissue lacking definite and specialized structure.

Figure 2. "Case 2. Medium power view of a coronary artery showing dense areas of sclerotic hyalinization with focal calcification replacing the bulk of the coronary wall."

Figure 3. "Case 2. Medium power view of the coronary artery wall showing clusters of fibroblasts in dense hyalinized tissue."

Figure 4. "Close-up of wall of coronary artery showing fibroblasts in an area of dense fibrotic hyalinization."

In the summary of Dunsmore 1986 (p.243), the authors state: "... a striking difference in the pathologic features of the coronary arteries has been noted in those patients exposed to [therapeutic] radiation, as in our Case 2, compared with the typical atherosclerotic lesions of non-irradiated patients."

1996: Emphasis Again on Distinction from "Typical" Atherosclerosis

On February 27, 1996, the Wall Street Journal (p.B-1, B-6, "Can Radiation Help Fight Heart Disease?") reported on some small trials in which coronary angioplasty patients receive radioactive stents, as a potential method for preventing re-stenosis. The article elicited a cautionary response from Doctors Richard and Lillian Dunsmore, whose letter to the editor was printed on April 1, 1996. They warn:

"... the long-term results of coronary-artery radiation may terminate in further heart damage, inability to perform bypass surgery and even untimely deaths in patients treated with radiation. For decades the teaching had been that the heart was resistant to radiation. However, in 1986, we presented data suggesting to the contrary ... Our data showed that coronary arteries, exposed to radiation, resulted in fibrosis of the walls of the vessels with resultant narrowing in contrast to the atherosclerosis present in the typical cases of coronary-artery disease."

The Dunsmores clearly emphasize that the arterial fibrosis induced by the patient's high-dose irradiation DIFFERS from "the atherosclerosis present in typical cases of coronary-artery disease."

● Part 7. A Little History: Some Interesting Observations in 1899-1909

In 1942 and 1943, the eminent pathologist, Shields Warren, published a series of articles, in the Archives of Pathology, covering the effects of ionizing radiation on normal tissues. This was quite a comprehensive set of publications. The December 1942 article (Warren 1942, p.1070) is entitled "Effects of Radiation on the Cardiovascular System." It contains a section entitled "The Heart" which begins as follows (Warren 1942, p.1070):

"Relatively little attention has been paid to the effects of radiation on the heart." After reviewing some studies (earliest: 1922), Warren summarizes (p.1074): "The various forms of cardiac damage secondary to radiation therapy cannot be recognized as specific in themselves, but the aseptic necrosis, hyaline fibrosis, and obliterative vascular changes combine to form a fairly characteristic lesion."

The next section is entitled "The Blood Vessels." The evidence cited therein deals largely with blood vessels of the skin, and Warren draws no inferences about radiogenic responses in the coronary arteries. Nonetheless, the very old observations cited by Warren may be relevant to the explanation proposed by Fajardo and Stewart for radiation-induced myocardial fibrosis (Part 4c, above).

Unfortunately, the magnitude of the skin-doses in the old work is unknown, but we doubt that such doses could have been comparable with heart-doses in the Fajardo and Stewart series.

Warren documents that, very quickly after discovery of the xray ("roentgen ray"), researchers had ascertained that high doses injured the small cutaneous vessels (blood and lymph vessels in the skin). Indeed, xray-induced skin damage was rather common in the years before xray operators understood the need to control dosage (Chapter 2, Part 2b). Some very detailed studies of damaged skin were published between 1899 and 1909. (However, one should be doubtful whether, in those years, the investigators could really distinguish reliably between the types of cells they observed. Even today, it is a matter of sophistication to prove that certain cells are really certain types.)

7a. Narrowing, and Even Closure, of Small Irradiated Vessels

A prominent consequence, in vessels which received high-dose radiation, was proliferation of cells and of connective tissue, resulting in the narrowing and sometimes permanent CLOSURE of the small irradiated vessels. Warren relates the following (1942, p.1074):

"The early observations (Baermann 1904, + Gassmann 1899, 1904) established that although no element of the blood vessel is immune to radiation injury, the endothelium is the most susceptible; consequently, the major changes are seen in those vessels in which the endothelium makes up a proportionately large part of the wall. Injury to large vessels is rare with doses below 500 Roentgens..." And (Warren 1942, p.1075):

"As early as 1899 the intimal thickening, the swelling and proliferation of the endothelium and the vacuolation of the smooth muscle were noted by Gassmann, who stated these alterations would lead to 'starvation of the surrounding tissues' and hence explained the intractable character of the roentgen ulcer [xray-induced skin-ulcer]. He later (1904) treated rabbits with roentgen rays and sectioned the resulting cutaneous ulcers one month after their development. He found obliteration of lymphatics [lymphatic channels] by endothelial proliferation as well as endothelial proliferation in the arteries and vacuolation of their smooth muscle cells ..." And (Warren 1942, p.1075):

7b. Delayed Consequences: "The Progressive Character of Vascular Lesions"

"In human skin treated with roentgen rays of low voltage, Linser (1904) noted, at the end of four days, fissuring of the media of vessels, occlusion of vessels by thrombi and slight perivascular round cell infiltration [round cells are lymphocytes], reaching its peak at eight days. After twenty days, intimal thickening by connective tissue was marked, with obliteration of some vessels ..." And (Warren 1942, p.1076):

"The importance of vascular changes in the cutaneous radiation effects was clearly presented by Wolbach (1909). He gave detailed descriptions of the walls of the blood vessels in the later radiation changes and defined the changes occurring in the endothelium and the supporting tissues of the walls. If the endothelium was not killed, proliferation often recurred, even to the point of obliteration of the capillaries. Sometimes the swollen or vacuolated endothelial cells formed tufts projecting into the lumen. In the veins and arteries subintimal fibrosis, with the collagen often showing some degree of hyalinization, resulted in thickening of the wall at the expense of the lumen." And (Warren 1942, p.1076):

"In the media, the elastic tissue degenerated and the smooth muscle cells showed vacuolation, hyalinization or atrophy. This coat was thickened as well by the presence of large, sometimes branching fibroblasts with abundant collagen. The degeneration of elastic lamellas was sometimes complete with substitution of fibrous tissue or bands of hyalinized collagen. He [Wolbach 1909] also emphasized the progressive character of the vascular lesions and cited the proliferation of fibroblasts in the media of arteries as late as four years after the last exposure to roentgen rays. However, even in severe damage, some normal blood vessels may be seen."

● Part 8: Summary, and a Recommendation

Existence of severe injury to the heart and its coronary arteries, inflicted by medical irradiation at very high doses, has been acknowledged for at least five decades. However, the number of people who accumulate only low and moderate doses, of heart irradiation, far exceeds the number of cancer patients whose hearts receive very high-dose irradiation.   This was true in the past, and is likely to remain so in the future.

The dose-response evidence in Chapters 40 and 41, combined with collateral circumstantial evidence, is persuasive that routine medical radiation was and continues to be a highly consequential "player" in the etiology of coronary atherosclerosis and fatal Ischemic Heart Disease.

In light of such evidence and the pathologic evidence in this appendix, we are confident that xrays can cause TWO types of heart-related damage: a) Atherogenic lesions and b) Non-atherogenic high-dose radiogenic damage to the heart and its coronary arteries.

We have found no dispute over the evidence that the damage to the heart and the coronary arteries, from very high-dose xrays, typically (not always) DIFFERS from atherosclerotic lesions. Therefore, we urge that distinct labels be used in the literature, whenever lesions have different characteristics and different etiologies.

If investigators and journal editors persist in referring to non-atherogenic high-dose radiogenic damage as "heart disease," or "atherosclerotic plaque," or "Coronary Artery Disease," the practice can badly obscure what is really going on, and even can introduce unnecessary bickering among investigators, to no good end.

>>>>>>>>>>

Mid-Century: Average Annual Per Capita Dose from Diagnostic Medical Xrays

Part 1. An Important Illustration of Uncertainty
Part 2. How Many Xray Examinations Occurred per Year?
Part 3. What Might the Average Dose Have Been, per Xray Exam?
Part 4. The Average Annual Per Capita Whole-Body Organ-Dose in 1950

Box 1. Physicians in 1950 Owning Their Own Roentgen-Ray Units, by Type of Practice.
Box 2. Independent Checks on Estimated Number of Annual Xray Exams at Mid-Century.

● Part 1. An Important Illustration of Uncertainty

In the second half of the Twentieth Century, some developments in medicine tended to reduce the average annual per capita dose from medical radiation --- while other developments tended to increase such doses (Chapter 2, Part 3). What might have been the magnitude of the average annual per capita dose at mid-century?

Appendix-K offers an approximate ("ballpark") answer. The process of obtaining an answer illustrates an important fact: All such estimates are highly uncertain and will necessarily remain so. For making the 1950 estimate, our two key references are:

1951: S.W. Donaldson, M.D. (Director of the Professional Bureau of the American College of Radiology), "The Practice of Radiology in the United States: Facts and Figures," American Journal of Roentgenology Vol.66, No.4: 929-946. December 1951.

1953: Dade W. Moeller, M.S. (Public Health Service) + James G. Terrill, Jr., C.E., M.B., + Samuel C. Ingraham, II, M.D., M.P.H., "Radiation Exposure in the United States," Public Health Reports Vol.68, No.1: 57-65. January 1953.

● Part 2. How Many Xray Examinations Occurred per Year?

According to Donaldson (1951, p.935, Table 1), there were 151,267 practicing physicians (USA) in 1950. Within this total, there were 3,000 certified radiologists devoting full time to the specialty, plus 500 specialists devoting most of their time to radiology, plus 600 second and third year residents in radiology --- a total of 4,100 active practitioners in radiology (Donaldson p.931, and p.937, Table 7).

Donaldson estimates that the typical radiologist administered 20 non-therapeutic radiologic examinations per day, worked 306 days per year, and thus administered about 6,000 examinations per year (Donaldson 1951, p.932, p.937, and p.945, Table 34.)

2a. Radiologists: The Number of Annual Xray Exams

Using these figures, Moeller reasonably estimates that 4,100 radiologists, each giving about 6,000 examinations per year, gave a total of about 25 million xray examinations per year (Moeller 1953, p.58).

In 1950, the U.S. population was 150 million persons. It follows that, every year, the equivalent of one sixth of the entire population was receiving an xray examination FROM A RADIOLOGIST. But radiologists do not constitute the whole "story." Far from it. A few reminders from Chapter 2 (Part 2c) are appropriate:

In 1923, Dr. Preston Hickey reported to the American Roentgen Ray Society that "It is interesting to note also the large number of internists who have placed fluoroscopes in their offices, not with the idea of specializing in xray work, but simply wishing to have conveniently at hand an xray control of their physical findings" (Hickey 1923).

In 1937, Dr. Eugene Leddy of the Mayo Clinis reported, "In fact, roentgenologic methods of diagnosis are so important that no investigation of a patient is considered complete without roentgenologic examinations, which generally include roentgenoscopy [fluoroscopy]. These studies are often carried out by a general practitioner or surgeon in his office because of lack of facilities for expert study nearby or because the physician sees no need to refer the patient to a roentgenologist" (Leddy 1937, p.924).

By 1940 (perhaps much earlier), some pediatricians (not all) performed fluoroscopic examinations as part of the routine monthly check-ups for problem-free babies (Buschke 1942).

2b. Non-Radiologists: The Number of Annual Xray Exams

If radiologists in 1950 gave 25 million xray examinations per year, how many additional xray exams were given by the non-radiologists?

In the absence of the data which we would like, we can begin with some other estimates from Donaldson 1951. He writes (p.931): "Information was obtained from experienced investigators in the field of the costs of medical care, who estimate that the 150,000 practicing physicians render annually 750,000,000 medical services to the 150,000,000 persons in the United States." Out of 150,000,000 persons, approximately 430,000 (about 0.3%) were "confined to general hospitals" on the average day (p.931). There were 4,761 registered general hospitals (Donaldson p.931).

If the estimate above is nearly correct, then on the average, every man, women and child received 5 medical services from physicians each year --- some receiving none at all, and others receiving many more than 5.

How many of the estimated 750,000,000 annual services included an xray examination rendered by a NON-radiologist? Important additional information, confirming appreciable xray activity by non-radiologists, is presented in our Box 1 (from Donaldson's Tables 25 and 27).

● Lesson One: General practitioners owned even more xray units (20,000) than the total number located in hospitals (13,000).

● Lesson Two: Out of a total of 50,000 xray equipment "units," approximately 31,000 or more were owned by non-radiologists.

But 31,000 is an underestimate because Donaldson does not consider equipment owned by osteopaths and chiropractors. Moeller reports that there were 11,000 osteopaths and chiropractors, plus 67,000 dentists (Moeller 1953, p.57). We assume that many or most of these 78,000 persons were using xrays.

The Issue of Dental Xrays

At mid-century, xrays beams were allowed to expose much more area than needed, and we suspect (but do not know) that dental xrays may have irradiated significant segments of the head, neck and even some of the chest. Nonetheless, we exclude dental xrays from our considerations, unless some reliable information becomes available.

Number of Annual Exams: A 2-to-1 Ratio (Box 1)

Having excluded xray units dedicated to dental xrays, we will approximate (a) that about 38,000 xray units were under the control of non-radiologists (~31,000 from Box 1 plus 7,000 assigned by us to osteopaths and chiropractors), and (b) that about 16,000 units were under the control of radiologists (3,000 directly and 13,000 located in hospitals). This is an equipment ratio of 2 to 1, non-radiologists to radiologists.

How often did non-radiologists use their xray equipment? We (and others) are left to speculate --- which is one of the reasons that estimates of annual average per capita xray dosage are inherently unreliable. Leddy's comments (Part 2a) might suggest "every visit."

Would non-radiologic physicians take the time? Not many minutes are consumed in giving the patient a fluoroscopic "once over." Moreover, fluoroscopy provides instant information, without any delay and without the expense and expertise required for proper film exposure and chemical processing. In offices which did make xray films (roentgenograms or radiographs), the person who did it was often a technician, nurse or clerk. Xrays enjoyed a good reputation, and some anecdotal evidence supports the expectation that patients (and parents of children) were pleased --- even pressing --- to have xray exams.

In view of the distribution of xray units (2:1), we will explore the proposition that non-radiologists (combined) administered twice as many xray exams each year as did radiologists (combined).

2c. Total Yearly Xray Exams, by (Radiologists + Non-Radiologists)

If non-radiologists administered twice as many xray exams per year at mid-century as did radiologists (who administered ∿ 25 million exams per year), then the total xray exams in 1950 would be 50 million plus 25 million, or 75 million xray exams. In 1950, the U.S. population was 150 million persons, so 75 million xray exams/year would be an annual rate of 1 exam per 2 persons, or 500 exams per 1,000 population.

This estimate for mid-century excludes dental xrays and therapeutic uses of xrays. A radiological examination is not an xray treatment.

For comparison with our 1950 estimated rate, we consider an estimated rate for 1964. Excluding dental xrays, the "estimated total number of diagnostic xray procedures" during 1964 is presented as 109 million, or 580 xray exams per 1,000 population (NCRP 1989, p.15, Table 3.7 --- taken from Mettler 1987).

● Part 3. What Might the Average Dose Have Been, per Xray Exam in 1950?

Now we pile more uncertainties upon the considerable uncertainties identified in Part 2.

3a. Relative Shares: Films, PhotoFluorograms, Fluoroscopies

Donaldson provides a Table entitled "Comparative Distribution of Roentgen-Ray Examinations" (Donaldson 1951, p.944, Table 30). It is based on the experience of the University of Minnesota Hospitals with 58,497 radiologic patients:

Roentgenographic examinations (30,355)	30,355	51.88%
Photofluorograms (19,677)	19,677	33.64%
Roengenoscopic examinations (8,465)	8,465	14.48%

A photofluorogram is a photograph taken of an image while the image is present on a fluoroscopy screen.

Total patients	58,497	100.00%

This is the distribution, at ONE set of hospitals, of examinations generally performed by or for radiologists. It is uncertain whether or not the distribution was TYPICAL for hospitals. There is very little reason to assume that the distribution was typical for non-radiologists. Indeed, it is likely that many non-radiologists used fluoroscopy even more often than did radiologists, for the reasons mentioned in Part 2b. Nonetheless, this distribution is what we have to work with, and we will treat it as applicable to all 75 million annual xray examinations (est.) at mid-century.

3b. Weighted Average Entrance Dose per Xray Exam

Moeller embraces the distribution provided by Donaldson (above). Then Moeller assigns each type of examination an average dose in roentgens (almost comparable to rads; see Appendix-A). These dose-estimates are entrance doses at the skin, where the xray beam enters the body. Moeller (1953, pp.58-59) provides the following dose-estimates in Column A, borrows the distribution in Column B from Donaldson, and arrives at a weighted average entrance dose per exam in Column C.

Type of examination	(A) Avg. dosage (roentgens)	(B) Share of Exams	(C) A times B
Radiographic	2.7	0.5188	1.401
Photofluorographic	1.0	0.3364	0.336
Fluoroscopic	65.0	0.1448	9.412
Weighted Average Dose per Xray Examination ----->			11.149

Are the Doses in Column A Credible?

Moeller's Reference List, which provides no sources for the values in Column A, includes the notation that "a complete bibliography of the source material for this article" is available upon request. That was in 1953. A request 45 years later might be impossible to honor.

The values in Column A appear credible to us for radiologic practice, in view of certain surveys conducted in the 1970s (discussed in Gofman 1985) and in view of certain reports about fluoroscopy machines in the 1930s and 1940s (discussed in Gofman 1995/96).

However, if Moeller's figure of 2.7 roentgens per radiographic exam is based on the practice of well-trained radiologists and radiologic technologists, and if twice as many exams were given by poorly trained personnel, then that average dose/exam could be quite an underestimate.

With respect to fluoroscopy, if some readers think that 65 roentgens "must be too high" for the TYPICAL fluoroscopic exam, they might consider this: In the 1930s and 1940s, the dose-rate from many fluoroscopy machines employed in medical practice was in the range of 25 to 35 roentgens per minute (Buschke 1942, p.525, p.527). Carl B. Braestrup, of the New York City Department of Hospitals, called such machines "a lethal diagnostic weapon" (Braestrup 1942, pp.210-211). Depending on the milli-amperes during operation, the dose-rate could exceed even 100 roentgens per minute --- without the operator realizing it (Buschke 1942, p.525). In fact, the Wappler Fluoroscope produced 125-150 roentgens per minute at the panel (Braestrup 1969). Some mobile units, operating close to the skin at bedside and during surgery, could deliver 1,000 roentgens per minute (Braestrup 1942, p.213).

Such considerations make 65 roentgens of entrance dose, during a typical fluoroscopic exam at mid-century, seem credible. Fluoroscopes in current medical practice typically operate in the dose-range of 2 to 20 rads of skin-dose per minute (FDA 1994, pp.2-3) --- and there are several surgical procedures during which a fluoroscope operates for 50 minutes or more (Chapter 2, Part 3d).

3c. An Independent Check on the Number of Annual Xray Examinations: Box 2

If there were 75 million xray exams per year at mid-century (Part 2c), and if 51.88% were radiographs, 33.64% were photofluorographs, and 14.48% were fluoroscopies (Part 3a), then:

Total radiographic exams, 1950 =	~39,910,000
Total photofluorographic exams, 1950 =	~25,230,000
Total fluoroscopic exams, 1950 =	~10,860,000

We can make an independent check, on the reasonableness of the estimate of 75 million, if we assume that the AGES of patients receiving fluoroscopy was about the same in 1950 as it was in the early 1970s. We have done that work in Box 2. It suggests that the estimate of 75 million xray exams per year might be a bit low.

● Part 4. The Average Annual Per Capita Whole-Body Organ-Dose in 1950

The dose-estimate which is relevant to induction of Cancer of ALL types, in the U.S. population, is the Average Per-Capita Whole-Body Internal Organ-Dose. To arrive at such an estimate, we must employ the estimates in Parts 2 and 3 plus some additional approximations.

4a. Radiographic Exams: Est. Annual Per Capita Whole-Body Organ-Dose

	Dose
● A) Per radiographic exam, entrance dose in roentgens =	2.700
● B) Per CAPITA entrance dose = (Row A) * (39.910 million exams from Part 3c) / (150 million persons) =	0.718
● C) Average dose to irradiated organs in the xray beam = (Row B * 0.4) because we will apply an average conversion-factor of 0.4 rads of average internal organ-dose per roentgen of entrance dose (details in Gofman 1985, p.404, Table C) =	0.287
● D) Average dose to ALL the internal organs, both inside and outside the xray beam = (Row C * 0.6), because we approximate that during the typical mid-century radiographic exam, about 60% of the body (excluding the limbs) was in the beam =	0.172

● SUMMARY: On the basis of these approximations, the average annual per capita "whole-body" internal organ-dose from diagnostic medical radiographs was about 0.172 rad at mid-century.

4b. Photofluorograms: Est. Annual Per Capita Whole-Body Organ-Dose

	Dose
● A) Per photofluorogram, entrance dose in roentgens =	1.000
● B) Per CAPITA entrance dose = (Row A) * (25.230 million exams from Part 3c) / (150 million persons) =	0.168
● C) Average dose to irradiated organs in the xray beam = (Row B * 0.334) because we will apply an average conversion-factor of 0.334 rads of average internal organ-dose per roentgen of entrance dose (this factor differs from Part 4a because photofluorograms are largely lung procedures) =	0.056
● D) Average dose to ALL the internal organs, both inside and outside the xray beam = (Row C * 0.6), because we approximate that during the typical mid-century photofluorogram, about 20% of the body (excluding the limbs) was in the beam =	0.011

● SUMMARY: On the basis of these approximations, the average annual per capita "whole-body" internal organ-dose from diagnostic medical photofluorograms was about 0.011 rad at mid-century.

4c. Fluoroscopies: Est. Annual Per Capita Whole-Body Organ-Dose

	Dose
● A) Per fluoroscopic exam, entrance dose in roentgens =	65.000
● B) Per CAPITA entrance dose = (Row A) * (10.860 million exams from Part 3c) / (150 million persons) =	4.706
● C) Average dose to irradiated organs in the xray beam = (Row B * 0.5) because we will apply an average conversion-factor of 0.5 rads of average internal organ-dose per roentgen of entrance dose (all beam-directions combined) =	2.353
● D) Average dose to ALL the internal organs, both inside and outside the xray beam = (Row C * 0.2), because we approximate that during the typical mid-century fluoroscopic exam, about 20% of the body (excluding the limbs) was in the beam =	0.471

● SUMMARY: On the basis of these approximations, the average annual per capita "whole-body" internal organ-dose from fluoroscopic xrays was about 0.471 rad at mid-century.

4d. Est. TOTAL Annual Per Capita Whole-Body Internal Organ-Dose

Now the estimates from 4a, 4b, and 4c need to be combined:

From diagnostic medical radiographs:	0.172 rad
From diagnostic photofluorograms:	0.011 rad
From fluoroscopies:	0.471 rad

Average annual per capita internal organ-dose from non-therapeutic
use of xrays in medicine, mid-century: 0.654 rad

Many consequential approximations are incorporated into the estimate. For what it covers, the estimate might be either too low or too high. It covers no contribution from dental xrays and no contribution from therapeutic uses of xrays, which were still used at mid-century to treat a variety of non-malignant diseases (Chapter 2, Part 2).

When readers are offered such estimates, for either past or current values, they need to keep such uncertainties in mind.

>>>>>>>>>>

Box 1 of Appendix-K.
Estimated Distribution of Physicians in the United States Owning
Their Own Roentgen-Ray Units, by Type of Practice (Modern Medicine 1949).

The title and reference above, and the entries in Columns A, B, C, and E below, are reproduced from Donaldson 1951, p.943, Table 25. We have inserted Column D, and the "All Combined" row, and the comments.

Col. A	Col.B	Col.C	Col.D	Col.E
		Est. number	Percent	Percent
		Owning	of Each	of the
	Total	Xray	Sub-set	32,250
Type of Practice	Physicians	Equipment	(ColC/ColB)	Owners
General practitioners	73,079	19,680	26.9%	61.0%
General surgery	19,976	3,000	15.0%	9.3%
Internal medicine	12,079	2,200	18.2%	6.8%
Radiology	3,559	1,300	36.5%	4.0%
Eye, ear, nose, throat	10,788	1,090	10.1%	3.4%
Dermatology	2,110	1,050	49.8%	3.3%
Industrial medicine	2,122	840	39.6%	2.6%
Orthopedic surgery	2,532	810	32.0%	2.5%
Pediatrics	6,321	750	11.9%	2.3%
Urology	2,869	700	24.4%	2.2%
Cardiology	916	290	31.7%	0.9%
Gastro-enterology & TB	2,277	200	8.8%	0.6%
Obstetrics & Gynecology	6,798	180	2.6%	0.6%
Allergy	703	160	22.8%	0.5%
All combined	146,129	32,250	22.1%	1

● - For Radiology, the 36.5% entry in Column D is a reminder that physicians of all types were able to use and share xray equipment which they did not personally own.

● - Among the 32,250 owners, 61% were general practitioners. Clearly, such physicians dominated the ownership at mid-century. The approximately 20,000 units owned by general practitioners exceeded even the estimated 13,000 units at the hospitals (see below).

● - Fluoroscopes (roentgenoscopes) were among the xray equipment commonly used by various non-radiologists (text, Part 2a).

● - The estimated total number of "Roentgen-Ray Equipment Units" was 50,000, located as follows (from Donaldson p.944, Table 27):

      20,000   units with 19,600 General Practitioners (1 unit each).
      13,000   units in 5,200 Hospitals (∼3 units each --- some owned by radiologists, p.942).
      12,000   units with ∼12,000 non-radiologic specialists (1 unit each).
       3,000   units with 1,300 Radiologists (∼2.5 units each).
       2,000   units in 1,000 Clinics and other groups (2 units each).

Sum -->   50,000

Box 2 of Appendix-K.

Independent Checks on Est. Number of Annual Xray Examinations at Mid-Century.

- In Col.B, the population entries for 1950 come from Grove 1968 (p.789).
- In Col.C, the rates per 100 persons come from Shleien 1977, Figure 2, and are estimates from sampling in the early 1970s. Shleien provides no entries above age 74.
- The upper tabulation is for the annual number of medical radiographs, with the exclusion (by Shleien) of examinations of the extremities. The lower tabulation is for the annual number of fluoroscopic examinations, including spot films and plates. Shleien provides no separate rates for "photofluorograms."
- Unless the AGES at which patients received examinations changed a great deal between 1950 and the early 1970s, the entries in Col.C permit an independent check on our estimates in the text, Part 2c.

Col.A Age-band	Col.B 1950 Population	Col.C Annual Number of Radiographs per 100 Persons		Col.D Absolute Number of Exams/Year = (B*C)/100
Under 1 yr	3,146,948	16		503,512
1-4 yrs	13,016,623	16		2,082,660
5-14 yrs	24,318,953	16		3,891,032
15-24 yrs	22,098,426	42		9,281,339
25-34 yrs	23,759,267	56		13,305,190
35-44 yrs	21,450,359	65		13,942,733
45-54 yrs	17,342,653	72		12,486,710
55-64 yrs	13,369,520	73		9,759,750
65-74 yrs	8,339,960	73		6,088,171
75-84 yrs	3,277,751	73	assumed	2,392,758
85 and over	576,901	73	assumed	421,138
Sums --->	150,697,361	Radiographs/yr ---> excluding extremities		74,154,992

Col.A	Col.B 1950 Population	Col.C Annual Number of Fluoroscopies per 100 Persons		Col.D Absolute Number of Exams/Year = (B*C)/100
Under 1 yr	3,146,948	1		31,469
1-4 yrs	13,016,623	1		130,166
5-14 yrs	24,318,953	1		243,190
15-24 yrs	22,098,426	3		662,953
25-34 yrs	23,759,267	5		1,187,963
35-44 yrs	21,450,359	9		1,930,532
45-54 yrs	17,342,653	12		2,081,118
55-64 yrs	13,369,520	13		1,738,038
65-74 yrs	8,339,960	15		1,250,994
75-84 yrs	3,277,751	15	assumed	491,663
85 and over	576,901	15	assumed	86,535
Sums --->	150,697,361	Fluoroscopies/yr --->		9,834,621

Estimated Yearly (Radiographs + Fluoroscopies), excluding
examinations of the extremities, at mid-century   ----->   83,989,614

Radiation "Hormesis" : How an Illusion Can Arise from "Perfectly Good Data"

Part 1. The Meaning and Attraction of Radiation "Hormesis"
Part 2. A Demonstrable Cause of a Spurious J-Shaped Dose-Response
Part 3. The Illusion of Hormesis --- due to Unmatched Medical Care

Figure L-1. How the Hormetic Illusion Can Arise from "Perfectly Good Data."

● Part 1. The Meaning and Attraction of Radiation "Hormesis"

Hormesis is the phenomenon when a poison or harmful agent is good for health at low doses, even when it is manifestly deadly at higher doses. The dose-response for hormesis is a J-shaped curve, when the poisonous agent is on the horizontal axis, and the mortality rate is on the vertical axis. The MortRate DECLINES between zero dose and some small dose of the poison, and thereafter, as dose rises, so does the MortRate.

Obviously, radiation hormesis would be a welcome phenomenon, if real. Such an attractive concept inspires hope and speculation.

Unfortunately, we and others have shown (in Appendix-B, for example) that there is NO dose of ionizing radiation which is either harmless or beneficial with respect to carcinogenesis (and mutagenesis). However, Cancer causes "only" 23% of the deaths in the USA, and Ischemic Heart Disease causes "only" 22% (Chapter 39, Part 4).

Suppose that OTHER causes of death were, somehow, reduced by exposure to low-dose radiation? Such speculations often refer to the fact (not in dispute) that radiation exposure stimulates measurable responses at the cellular level. For instance, concentrations of some enzymes temporarily increase and others decrease. In the search for radiation hormesis, a large literature has developed around such "adaptive" responses --- for example, discussions in Gofman 1990 (Chapter 35) and in UNSCEAR 1994 (its "Annex" B).

● Part 2. A Demonstrable Cause of a Spurious J-Shaped Dose-Response

We expect to see some epidemiologic studies showing the J-shaped curve, with radiation exposure on the horizontal axis and MortRates on the vertical axis. If a study produces several curves including the J-shaped type, ALL must be published and none hidden. We do not pre-judge a J-shaped curve to be spurious, but --- like every other wished-for curve --- it deserves especially rigorous scrutiny with respect to potential pitfalls.

One potential pitfall is a weak or non-existent "blinding" procedure. Exposure-status and health-status should not be known by the SAME decision-makers, when radiation doses (which are likely to be rough estimates) are assigned to participants, or when radiation doses are "revised" for certain participants, or when certain participants are thrown out of the study completely.

But the topic of this appendix is a wholly separate pitfall, which also could produce a spurious J-shaped dose-reponse.

Medical Care: Control-Group vs. Exposed-Groups

Any variable which causes the control group to have a higher MortRate, than the baseline MortRate of the study-population, can result in a J-shaped curve. Can this happen, even though proper matching is one of the fundamental rules of epidemiologic research?

Yes it can, despite the best intentions. Part 3 will demonstrate how it can happen, by using "Medical Care" as the variable which may turn out NOT to be properly matched.

Medical Care and MortRates

We begin with the proposition that medical services do reduce some types of mortality dramatically. This is almost beyond dispute.

When I was a senior in medical school and an intern in medicine (1946-47), the revolution in managing a number of very prominent infectious diseases was evident before our eyes. Pneumonias which were considered to be "in the lap of the gods" with respect to recovery, became treatable with full recovery. Syphilis became successfully treatable beyond any expectations. Tuberculosis first became more treatable (by lung collapse), and later highly successfully treatable with antibiotics. Tuberculosis sanitoria closed their doors all over the country. Dreaded streptococcal and meningococcal infections yielded to a variety of antibiotics. Rheumatic fever deaths and deaths from rheumatic heart disease became vastly more rare, as the underlying infections were managed with antibiotic prophylaxis and therapy.

It was during my internship period that some of the dreaded subacute infectious diseases began yielding to prolonged therapy with penicillin. Subacute bacterial endocarditis --- previously 90-95% fatal --- became curable, if caught before appreciable valvular damage had occurred. My wife was cured in 1948 of a post-partum fulminant bacteroides funduliformis infection, by ten million units of penicillin daily --- at a time when the medical experts pronounced that the disease was 95% fatal if it did not respond to 500,000 units of penicillin daily.

Fatal, infectious complications of trauma and surgery, for otherwise non-fatal disorders, became less common. For diabetics, too, the bacterial infections which had been such a common and often fatal complication of the disease, became more manageable and less of a menace. Meanwhile, surgery began to develop the ability to manage, increasingly, many congenital heart malformations and other disorders --- aided in no small measure by advances in the use of blood and other fluids.

However, medical advances ALONE do not determine the death-rate in a population-sample. AVAILABILITY of medical care --- and on a timely basis --- is another factor which affects specific MortRates. Long before work-related health insurance became common, there were some corporations and institutions which established "health preservation" programs, and even installed medical facilities right at their "plants." Indeed, in 1954, I organized the industrial medical facility at the Lawrence Livermore Laboratory. We instituted a policy of complete medical examinations for everyone, at intervals of approximately 1.5 years. For those engaged in particularly hazardous work, the exam frequency was greater. There is no doubt that such examinations did pick up manageable diseases, for which therapy made a difference in outlook.

Overall, with respect to non-malignant disease, there is little doubt that the more medical care exists for a U.S. population sample, the lower will be the death rate on an age-adjusted basis. Indeed, the proposition is consistent with the studies in Section 3 of this book (Chapters 23-38).

● Part 3. The Illusion of Hormesis --- due to Unmatched Medical Care

Suppose that an epidemiologic study is undertaken to assess the effect of radiation exposure upon mortality rate (from all causes combined). For such a study, investigators generally seek plenty of participants who have doses appreciably above normal, from occupational or medical irradiation.

For our illustrative demonstration, we can gloss over the many difficulties in such a study. For example, we will assume that reliable and comparable estimates of accumulated dose (medical + non-medical) exist for all participants, individually. The participants with the lowest accumulated doses (no one has zero dose) are designated to be the control-group, and the remaining participants are designated as the study-group. The study-group, sorted by individual accumulated radiation dose, can be divided along its dose-continuum into ten groups of progressively higher dose.

Somehow (a word which reflects more "glossing-over"), the participants in each dose-group and in the control-group are well-matched with each other for age, gender, occupation, smoking habits, nutrition, and other variables which can affect the MortRates --- except for one variable: MEDICAL CARE. Suppose it turns out that the control-group has less medical care, or worse medical care, than the study-group, but the analysts do not realize it. What can happen?

Illustrative Numbers and a J-Shaped Curve in Figure L-1

For purposes of comparison, we will call the All-Cause MortRate observed in the control-group "unity" (1.00). Thus, a MortRate 15% below the control-group's MortRate would be 0.85, and a MortRate 15% higher than the control-group's MortRate would be 1.15. In the tabulation below Figure L-1, the control-group is called dose-level 0, and its observed MortRate (in Column D) = 1.00 .

Now, thanks to the "extra" medical care, each part of the study-group will have an All-Cause MortRate which is 15% lower than the MortRate of the control-group. The value of 15% is purely illustrative. In Column C below Figure L-1, we enter 0.85 as the effect of "extra" medical care upon the MortRate in the study-group.

And thanks to the extra accumulated doses of ionizing radiation, each part of the study-group will have a higher MortRate from Cancer and Ischemic Heart Disease than the control-group. To illustrate the point, suppose that in the first dose-group, radiation is sufficient to increase the All-Cause MortRate by 10% above the MortRate observed in the control-group, and by an additional 10% in each successively higher dose-group --- as shown in Column B of the tabulation presented below Figure L-1. Although the 10% value is purely illustrative, the expectation of an increase in the study-group --- due to fatal cases of radiation-induced Cancer and radiation-induced Ischemic Heart Disease --- is based on real-world evidence.

The NET observed All-Cause MortRate is the product of two factors: The elevation by radiation, and the decrement by better medical care. Column D calculates the product of Column B times Column C. The results are depicted by the graph of Figure L-I.

Hormetic Illusions: A Realistic Pitfall in Epidemiology

There it is: The J-shaped dose-response. A little extra radiation appears to make the All-Cause MortRate fall BELOW the rate in the control-group --- a spurious hormetic effect, because the harm from even the lowest level of extra radiation is present in Column B. The hormetic illusion is produced by the fact that the control group and the study-group are not matched for medical care.

The same problem --- less, or less effective, medical care in the control group --- could also produce a J-shaped dose-response in a similar study concerning death only from Cancer (or IHD). Moreover, the problem could produce a dose-response with the SAME MortRate in the control-group and at dose-level 1 --- which would be a threshold illusion instead of an hormetic illusion.

If the hormetic illusion can arise out of "perfectly good data" having nearly IDEAL conditions for an epidemiologic study, then the hormetic illusion is a realistic pitfall in the ACTUAL world of epidemiology --- where matching for socio-economic factors may not assure comparable medical care, where reliable and comparable estimates of accumulated radiation dose (medical + non-medical) almost never exist, and where a general population serves rather often as the control-group for a relatively particular study-group.

● Below the graph, "perfectly good data" match the circumstances described in the Text, Part 3. The control-group and the irradiated study-group are perfectly matched with each other on all variables --- except medical care. The study-group is divided into ten dose-levels. The horizontal axis shows the dose-levels from Col.A, where 0 is the control-group. The vertical axis shows the net observed death-rates from Col.D, where unity (1.00) is the death-rate observed in the control group.

● The J-shaped dose-response depicts a spurious hormetic effect. The harm from even the lowest extra dose of radiation is present in Col.B, below. The hormetic illusion is produced by a failure to match the control-group and the exposed groups for medical care.

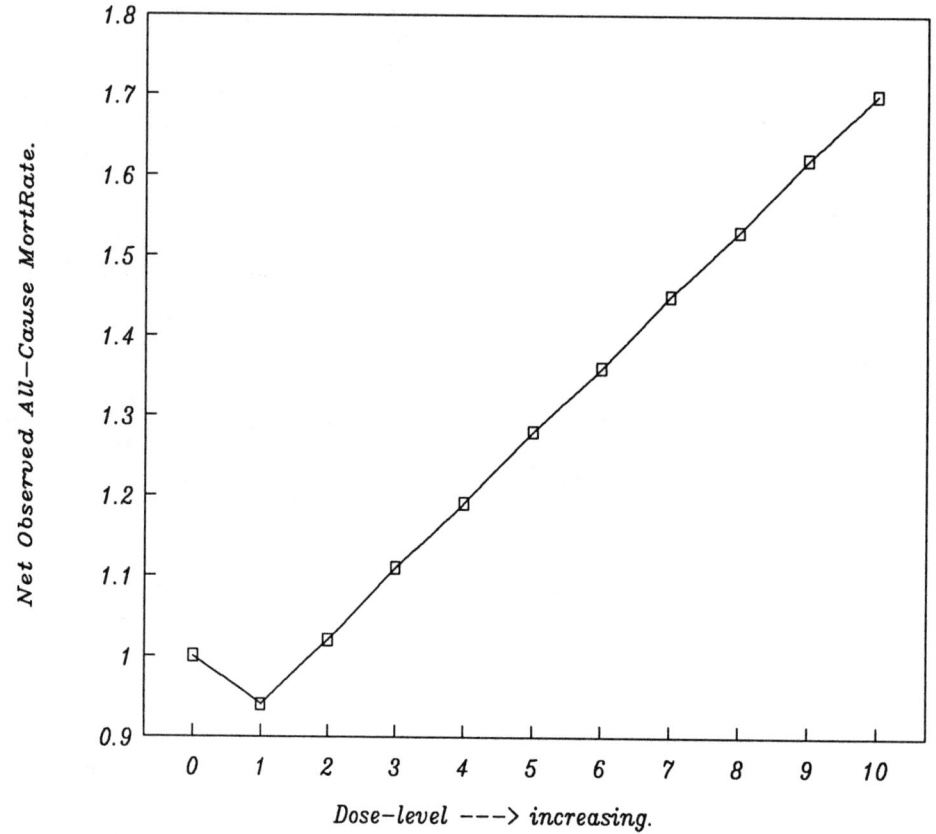

Related text = Part 3.

Col.A Dose- Level	Col.B Change in MortRate due to Extra Radiation	Col.C Change in MortRate due to Extra Medical Care	Col.D Net Obs. MortRate = B times C
0	1.00	1.00	1.00
1	1.10	0.85	0.94
2	1.20	0.85	1.02
3	1.30	0.85	1.11
4	1.40	0.85	1.19
5	1.50	0.85	1.28
6	1.60	0.85	1.36
7	1.70	0.85	1.44
8	1.80	0.85	1.53
9	1.90	0.85	1.62
10	2.00	0.85	1.70

Fractional Causation, 1980-1993, after an Alternative Smoking Adjustment

Part 1. Purpose of Appendix-M, and Overview of Results
Part 2. Explanation of Table M-1: All-Cancers, Males, 1988
Part 3. Biological Premises: Factor vs. Difference Methods of Smoking Adjustment
Part 4. Additional Features Which Recommend the Factor Method

Box 1. Difference Method: Fractional Causation, by Medical Radiation, of Cancer + IHD

● Part 1. Purpose of Appendix-M, and Overview of Results

        Appendix-M uses an alternative method to calculate Smoking Adjusted MortRates. This alternative method can be called the Difference Method, to distinguish it from the method used in Chapters 49 through 65, which we can call the Factor Method. Appendix-M uses the alternative set of Adjusted MortRates in calculating an alternative set of Fractional Causations.

        The results are summarized in Box 1 of this Appendix. Like the results summarized in Box 1 of Chapter 66, the results here strongly support Hypotheses One and Two. Our opinion is that the Factor Method is more reasonable, biologically, than the Difference Method for making the Smoking Adjustment. But we wondered if validation of Hypothesis-1 depends on that opinion. Appendix-M shows that Hypothesis-1 does NOT depend on that opinion.

● Part 2. Explanation of Table M-1: All-Cancers, Males, 1988

        Column G in Table M-1 is the feature which distinguishes the Difference Method from the Factor Method. Readers who have studied the Factor Method will easily comprehend the distinction.

2a. Column G: The Essential Distinction between Methods

        The header of Column G says that Col.G adds the value of +28.2 to every value in Col.F. The value of +28.2 comes from Chapter 49, for All-Cancers, males. Specifically, +28.2 is the value found for the TopTrio in Box 1, Column K, of Chapter 49. (In the successive tables of Appendix-M, one visits the comparable place in Chapter 50, 51, 52, etc.)

        What is +28.2? It is the average DIFFERENCE in MortRate per 100,000 population, if one moves from the TopTrio's 1940 All-Cancer MortRates to the TopTrio's 1988 All-Cancer MortRates. The Smoking Adjustment in Table M-1 is based on the DIFFERENCE per 100K which developed between the TopTrio's Observed 1940 MortRates and its Observed MortRates in subsequent decades.

        The Difference Method permits the post-1940 MortRates in the MidTrio and LowTrio Census Divisions to increase (or decrease) relative to their Observed 1940 values, but the change (in cases/100K population) will be adjusted to have the same size and direction in the MidTrio and LowTrio as in the TopTrio. The Difference Method and Factor Method each leave the Observed Post-1940 MortRates in the TopTrio intact. Thus both methods are designed to eliminate only the effect of EXTRA smoking in the MidTrio and LowTrio, relative to the TopTrio.

        In Table M-1, Column G adds exactly the same value (+28.2) to the MortRates in Col.F, regardless of the PhysPop level. This is equivalent to doing what we discussed in Chapter 5, Part 6a --- where we added +20 cases/100K to MortRates which were perfectly proportional to PhysPop. In Figure 5-C, all 20 of the additional cases became part of the Constant --- the "non-radiation rate." Likewise, when one uses the Difference Method to make the Smoking Adjustment in Table M-1, the underlying assumption is that medical radiation is NOT a co-actor in the 28.2 fatal cases/100K added in Col.G. Such cases should not be multiplied by the PhysPop Adjustment from Table 47-B. Therefore, Col.E (the PhysPop Adjustment) occurs BEFORE addition of the 28.2 cases in Col.G.

The other steps in Table M-1 are exactly like the steps in Table 49-B. We have done Difference Tables only for the most recent years, because we are confident that the most recent years must be the LEAST favorable to Hypothesis-1. Box 1 of Appendix-M summarizes the results from 16 tables, in exactly the same format as Box 1 of Chapter 66, to facilitate comparisons.

2b. Two Comments on the Output of the M-Tables

ONE. Box 1, Column C, shows a series of very high Fractional Causations for the 1980-1993 period --- with one expected exception. The relatively low Fractional Causations for Respiratory-System Cancers (male, female) are expected because the largest impact of smoking is upon that system. Thus, Table M-3 adds 40.6 cases/100K (Col.G) to 1940 rates which are in the ballpark of 8 cases/100K (Col.D). Table M-4 adds 22.8 cases/100K to 1940 rates which are in the ballpark of 3 cases/100K. By contrast, the change for other Cancers does not even approach the magnitude of their 1940 levels. Moreover, the Difference Method assumes (mistakenly, in our opinion) that the entire impact of smoking should be distributed to the Constant (meaning no co-action between smoking and xrays). So for Respiratory-System Cancers, the Constants rise to become a large share of the 1988 National Observed MortRates, and the calculations yield relatively low Fractional Causations by medical radiation for Respiratory-System Cancers.

TWO. We note that the Smoking Adjustment by the Difference Method produces negative Constants which are large fractions of the National MortRates for two important entities: Digestive-System Cancers in 1988 (Tables M-9 and M-10) and of the National MortRates for Ischemic Heart Disease in 1993 (Tables M-15 and M-16). This occurs because both entities experience steep NET declines in their TopTrio MortRates, despite the impact of smoking. Thus, the Difference Method requires subtraction of large numbers of fatal cases/100K from the mid-century observed MortRates in the MidTrio and LowTrio. The result is a slope so steep that it intersects the y-axis below zero.

In our opinion, some negative Constants are to be expected on the basis of occasional anomalies in the observations (Chapter 22, Part 3). The presumed anomaly occurs in the observations for the TopTrio, which supplies the change-factor for the Factor Method and supplies the change-difference for the Difference Method. In Table M-15, a relatively small modification --- reducing the change-difference from 170.8 to 130.0 in Col.G --- would eliminate the negative Constant. (Also, the modification would raise Fractional Causation to 80%, reduce R-squared to 0.6464, and reduce Xcoef/SE to 3.5770). Because we do not accept the premise of the Difference Method (which denies co-action), we do not devote any pages to modified M-tables.

● Part 3. Biological Premises: Factor vs. Difference Methods of Smoking Adjustment

In a dose-response study of medical radiation (PhysPop), the dose-cohorts (the populations of the Nine Census Divisions) must be matched for all non-xray carcinogens. If matching exists at the outset of a 50-year study like ours, various non-xray carcinogens can subsequently increase or decrease in intensity, and yet the matching persists PROVIDED that all Nine Census Divisions experience the SAME changes in the non-xray carcinogens.

Both the Factor Method and the Difference Method, for making the Smoking Adjustment, address the question: What would the post-1940 MortRates have been, if there had not been EXTRA smoking in the LowTrio and MidTrio, relative to the TopTrio? To eliminate the effect of EXTRA smoking on the LowTrio and MidTrio post-1940 MortRates, each method evaluates how much the Observed 1940 MortRates in the TopTrio changed during subsequent decades, and then applies the SAME change to the Observed 1940 MortRates of the LowTrio and MidTrio. But the Factor Method measures "change" in the TopTrio by the ratio of a later MortRate over the 1940 MortRate, while the Difference Method measures "change" by finding the difference in cases/100,000 in moving from 1940 to later years. The distinction has biological implications.

3a. The Factor Method

The underlying biological premise of the Factor Method is that an elevated level of co-actors can make each rad of medical radiation more potent --- and that a diminished level of co-actors can make each rad of medical radiation less potent. Milieu matters. The premise that carcinogenic co-actors modulate (regulate) each other's potency is not exotic. Its foundations are presented in the

Introduction (Part 4), Chapter 6 (Part 6), Chapter 49 (Part 2), and Chapter 67 (Part 2b).

If milieu matters, how does it affect expectations regarding the introduction of cigarette smoking?

Illustration: Let the Census Divisions be matched for non-xray co-actors, and let no one smoke cigarettes. Then let everything be held constant --- except that cigarette smoking joins the mix of non-xray co-actors, and joins at equal intensity per capita in all Census Divisions. Expectation: In this new milieu, each rad of medical radiation (each PhysPop unit) would become more potent in every Census Division. Therefore, the new milieu would cause cancer MortRates to rise by a greater absolute number in the high-dose Census Divisions (TopTrio) than in the low-dose Census Divisions (LowTrio).

Revision of Circumstances: Let the smoking-intensity NOT be matched across the Census Divisions. Let it be greater in the LowTrio than in the TopTrio --- which actually happened (Chapter 48). This would cause the potency of each medical rad to DIFFER across the Census Divisions --- to be higher in the LowTrio than in the TopTrio. The LowTrio would still have fewer rads than the TopTrio, but would develop more Cancer and IHD deaths than the TopTrio PER RAD (per PhysPop unit). The difference in MortRates would diminish, between TopTrio and LowTrio. For male Respiratory-System Cancers in 1940, the TopTrio MortRates very clearly exceeded the LowTrio MortRates (Figure 16-A). By about 1970, a reversal had occurred: Table 16-A shows that LowTrio MortRates actually exceed TopTrio MortRates by then. (Reminder: Mid-Atlantic is in the TopTrio; Chapter 3, Box 1, Part 2).

3b. The Difference Method

By contrast with the Factor Method, the underlying biological premise of the Difference Method is that an elevated or diminished level of non-xray carcinogens has no effect on the potency of each rad of medical radiation. Milieu does NOT matter.

Expectation: Again, let the Census Divisions be matched for non-xray carcinogens, and let no one smoke cigarettes. Then let everything be held constant --- except that cigarette smoking (matched across the Census Divisions) joins the mix. The new carcinogenic milieu remains matched across the Census Divisions. If we say (for illustrative purposes) that addition of smoking to the milieu adds +20 fatal cancers/100,000 population in the TopTrio, then addition of smoking would add +20 cases/100K in every Census Division --- in the Difference Method. This outcome is depicted in Chapter 5 by comparison of Figure 5-B with Figure 5-C. The Difference Method reflects what we regard as a biologically improbable premise: That carcinogens co-produce the cancer MortRate of each Census Division in an exclusively additive way, without co-action.

3c. Co-Action: Cigarettes Modulate Xray Potency, and Xrays Modulate Cigarette Potency

If we hold ALL non-xray carcinogenic co-actors, cigarette smoking included, constant at some level (any level), matched in all Census Divisions, while ONLY the number of PhysPop units (dose of medical radiation) differs per Census Division, then co-actors modulate the potency of each PhysPop unit by the same force in all Census Divisions. (Reminder: Per-PATIENT dose is very similar in all Census Divisions --- Chapter 5, Part 5d.) Because matching gives co-actors equal modulating force in all the Census Divisions, the potency per PhysPop unit is the SAME in all the Census Divisions --- which results in a tight linear and positive dose-response between medical radiation and cancer MortRates, by Census Divisions.

On the other hand, if the dose of medical radiation and dose of other non-smoking carcinogens are the SAME (matched) across the Census Divisions, while the dose ONLY of smoking-induced co-actors differs among the Census Divisions, one would see a positive dose-response between SMOKING and cancer MortRates. And the xray-induced mutations --- present at EQUAL frequency in all the Census Divisions --- would modulate the carcinogenic potency of each cigarette.

● Part 4. Additional Features Which Recommend the Factor Method

Our prediction, that evidence will firmly establish that carcinogenic co-actors modulate each other's potency, is one reason that we consider the Factor Method to be superior to the Difference Method of making the Smoking Adjustment. There are two additional features (described in Parts 4a + 4b) which recommend the Factor Method.

4a. Accommodation of the "Drive" toward Equilibrium

The Factor Method accommodates the likelihood that equilibrium has not yet been reached in 1940. Chapter 5, Part 2, explains the equilibrium concept. A new carcinogen, medical radiation, is introduced into human experience in 1896. If annual exposure to this new carcinogen AND exposure to carcinogenic co-actors occur at steady levels over time, ultimately the annual PRODUCTION-rate of radiation-induced Cancers will equal the annual DELIVERY-rate of radiation-induced Cancers (on an age-adjusted basis). This is equilibrium. Until equilibrium is attained, the radiation-induced cancer MortRate is continuously increasing (Figure 5-A).

Equilibrium is a useful concept, even though we doubt very much that annual xray doses have been steady since 1896 (Part 4b, below), and we know that exposure to carcinogenic co-actors has not been steady since 1896 (cigarette smoking constitutes a prime example).

It is likely that an approximation of equilibrium was NOT reached by 1940, because it is very likely that the carcinogenic effect from radiation-induced mutations lasts for a population's remaining lifespan (Chapter 2, Parts 8b and 8c). If so, then deliveries of Cancer --- from mutations which were induced by xrays in 1896 --- contribute to the annual cancer MortRate until virtually everyone who received medical radiation in 1896 has died. Such deliveries occur in proportion to PhysPop values, because the xray-induced mutations occur in proportion to PhysPop values.

4b. Accommodation of Changes in Averge Rads per Capita

The Factor Method also accommodates the likelihood that average annual per capita xray dose (in rads) has changed somewhat since 1940.

PhysPop is approximately proportional to average per capita dose from medical radiation. Table 47-A reveals that the Averaged PhysPops of the Nine Census retained their 1940 proportions very well over the subsequent 50 years. Those proportions were only minimally affected by the dramatic rise, after the mid-1960s, in the absolute number of physicians per 100,000 population. But steady PhysPop PROPORTIONS do not rule out the likelihood that the average per capita radiation dose (in rads per year), caused per physician, changed from its 1940 level. Some post-1940 forces would help to lower the average annual population dose per capita, and others would help to raise it (Chapter 2, Part 3). It is not possible to quantify the net change. Whatever the net result (probably somewhat downward), the resulting cancer MortRates would remain proportional to PhysPop values.

4c. Proportionality with the Nine PhysPop Values

Parts 4a and 4b indicate that it is likely that NOT ALL of the observed change in the TopTrio's cancer MortRate, 1988 compared with 1940, is due to post-1940 changes in various non-xray co-actors. The post-1940 changes in the TopTrio's 1940 MortRate are probably the NET effect of (a) upward pressure from the "drive" toward equilibrium, (b) downward pressure from a somewhat reduced annual per capita dose from medical radiation, (c) upward pressure from smoking-induced co-actors, and (d) pressures both upward and downward from changes in other non-xray co-actors.

Nonetheless, the observed change-FACTORS in the TopTrio's post-1940 cancer MortRates (relative to the TopTrio's 1940 MortRates) are reasonable guides to what would have happened to cancer MortRates in the LowTrio and MidTrio --- if non-xray co-actors had been matched with the TopTrio. In that case, the effects of (a), (b), (c), and (d) in every Census Division would each be proportional to the Division's PhysPop value. Thus, when we use the Factor Method to adjust the MidTrio and LowTrio MortRates --- by multiplying their Observed 1940 MortRates by the same change-factor observed in the TopTrio --- we appropriately accommodate (a), (b), (c) and (d) --- at the same time that we eliminate the effect of EXTRA smoking in the MidTrio and LowTrio.

● – The range of values below represents the earliest year and the most recent year named in Col. A.

● – Again, Column C strongly supports the validity of Hypotheses-1 & 2 --- as it does in Chapter 66, Box 1.

Col. A: M = Male. F = Fem.	Col. B: Nat'l Age-Adjusted Mortality Rate	Col. C: Frac. Causation by Medical Radn	Col. D: R-squared	Col. E: X-Coefficient	Col. F: Ratio of XCoef/Std. Error
Ch49, 1940-88, All-Cancer: M	Big net rise. 115.0 --> 162.7	90% --> 58%	0.95 --> 0.91	0.76 --> 0.58	11.6 --> 8.3
Ch50, 1940-88, All-Cancer: F	Net decline. 126.1 --> 111.3	58% --> 56%	0.86 --> 0.86	0.53 --> 0.38	6.6 --> 6.6
Ch51, 1940-88, Resp'y Ca: M	Enormous rise. 11.0 --> 59.7	~100% --> 27%	0.87 --> 0.81	0.12 --> 0.10	6.8 --> 5.5
Ch52, 1940-88, Resp'y Ca: F	Enormous rise. 3.3 --> 24.5	97% --> 11%	0.96 --> 0.52	0.02 --> 0.02	13.4 --> 2.7
Ch53, 1940-88, Diff-Ca: M	Approx. flat. 104.0 --> 103.0	84% --> 75%	0.93 --> 0.90	0.64 --> 0.48	10.0 --> 8.1
Ch54, 1940-88, Diff-Ca: F	Big decline. 122.8 --> 86.8	57% --> 69%	0.85 --> 0.85	0.50 --> 0.37	6.3 --> 6.3
Ch55, 1940-90, Breast-Ca: F	Flat. 23.3 --> 23.1	~100% --> 82%	0.92 --> 0.89	0.19 --> 0.14	8.7 --> 7.7
Ch56, 1940-80, AllExcGen: F	Flat. 94.0 --> 94.8	75% --> 66%	0.87 --> 0.92	0.51 --> 0.43	6.8 --> 8.8
Ch57, 1940-88, Digest-Ca: M	Big decline. 60.4 --> 38.8	97% --> 73%	0.91 --> 0.87	0.43 --> 0.32	8.3 --> 6.9
Ch58, 1940-88, Digest-Ca: F	Big decline. 50.1 --> 23.5	80% --> 69%	0.76 --> 0.86	0.29 --> 0.21	4.6 --> 6.4
Ch59, 1940-80, Urinary-Ca: M	Approx. flat. 7.4 --> 8.2	~100% --> 81%	0.92 --> 0.89	0.08 --> 0.07	9.0 --> 7.6
Ch60, 1940-80, Urinary-Ca: F	Decline. 4.0 --> 3.0	86% --> 79%	0.94 --> 0.92	0.02 --> 0.02	10.4 --> 9.0
Ch61, 1940-90, Genital-Ca: M	Some rise. 15.2 --> 16.9	79% --> 53%	0.77 --> 0.84	0.09 --> 0.06	4.9 --> 6.1
Ch63, 1940-80, Buccal-Phar: M	Approx. flat. 5.1 --> 4.6	~100% --> 79%	0.72 --> 0.70	0.04 --> 0.03	4.3 --> 4.1
Ch64, 1950-93, IHD: M	Enormous fall. 256.4 --> 131.0	79% --> 62%	0.95 --> 0.87	1.49 --> 1.18	11.2 --> 6.9
Ch65, 1950-93, IHD: F	Enormous fall. 126.5 --> 64.7	97% --> 57%	0.87 --> 0.84	0.90 --> 0.69	6.8 --> 6.2

```
=================================================================================================
                                    Table M-1
          All-Cancers, Males, 1988:  Alternative Smoking Adju and Fractional Causation
=================================================================================================
```

This table is like Table 49-F, except for the Smoking Adjustment. Here in Part 1, we do not multiply the 1940 MidTrio and LowTrio MortRates by a factor; we adjust them by the DIFFERENCE (+28.2 cases) which we obtain from Chap.49, Box 1, Col.K. Of course, Col.E below keeps the PP Adju (see Tab 47-B & Chap.49, Box 2, Part 2, Col.D).

Part 1. Calculation of the Alternative Smoking Adjustment (Col.G) and the National Adjusted MortRate (Col.I).

	Col.A	Col.B	Col.C		Col.D =	Col.E=	Col.F=	Col.G=	Col.H=	Col.I =
	1990	1988			1940 MRs	PP Adju	D * E	Col.F	Adju	A * H
Trio-	PopFrac	Obs MR	A * B		Mid,Low	Factor		+ 28.2	MortRate	
Seq.	Tab 3-B	Tab 6-A			Tab 6-A	Tab 47-B		(TopTrio,		
								Bx1,ColK)		
Pac	0.1535	148.5	22.795						148.5	22.795
NewEng	0.0527	167.1	8.806						167.1	8.806
MidAtl	0.1527	168.4	25.715						168.4	25.715
WNoCen	0.0721	155.9	11.240		110.9	0.94	104.2	132.4	132.446	9.549
ENoCen	0.1713	171.2	29.327		119.6	0.94	112.4	140.6	140.624	24.089
Mtn	0.0543	139.1	7.553		99.8	0.94	93.8	122.0	122.012	6.625
WSoCen	0.1087	172.9	18.794		86.9	1.07	93.0	121.2	121.183	13.173
ESoCen	0.0621	188.2	11.687		73.6	1.07	78.8	107.0	106.952	6.642
SoAtl	0.1725	175.8	30.325		88.9	1.07	95.1	123.3	123.323	21.273
Weighted avg. Col.C =		166.2							Sum =	
1988 Obs.Natl MR, Tab 6-B=		162.7				1988 Natl Adju MR =			138.6666	

Part 2. ---

	Col.A	Col.B		Col.C		Col.D		Col.E	
	Mean1940	1988		All-Cancers, Males:		1940		All-Cancers, Males:	
	thru1990	AdjuMRs		1988 Adjusted MortRates		PPs from		1988 Adjusted MortRates	
Trio-	PPs from	Col.H		regressed on		Table 3-A		regressed on	
Seq.	Tab 47-A	Part 1		Mean 1940 thru 1990 PhysPops		TrioSeq.		1940 PhysPops	
	x'			Regression Output:		x''		Regression Output:	
Pac	191.97	148.5		Constant	44.6445	159.72		Constant	48.3344
NewEng	208.20	167.1		Std Err of Y Est	6.9167	161.55		Std Err of Y Est	6.2743
MidAtl	204.72	168.4		R Squared	0.9075	169.76		R Squared	0.9239
WNoCen	141.14	132.446		No. of Observation	9	123.14		No. of Observation	9
ENoCen	146.19	140.624		Degrees of Freedom	7	133.36		Degrees of Freedom	7
Mtn	145.91	122.012				119.89			
WSoCen	126.28	121.183		X Coefficient(s)	0.5834	103.94		X Coefficient(s)	0.6870
ESoCen	113.28	106.952		Std Err of Coef.	0.0704	85.83		Std Err of Coef.	0.0745
SoAtl	142.93	123.323		XCoef / S.E. =	8.2894	100.74		XCoef / S.E.	9.2201

```
-----------------------------------------------------------|-----------------------------------------------------------
Part 3-A.                                                  | Part 3-B.
Calculation of Fractional Causation                        | Calculation of Fractional Causation
from Averaged PhysPops                                     | from 1940 PhysPops
                                                           |
1.  Nonradiation rate is Adjusted                          | 1.  Nonradiation rate is Adjusted
    Constant (Part 2, Col.C) =            44.6445          |     Constant (Part 2, Col.E) =            48.3344
                                                           |
2.  Radiation rate is Natl Adjusted                        | 2.  Radiation rate is Natl Adjusted
    MortRate (Part 1, Col.I = 138.6666)                    |     MortRate (Part 1, Col.I = 138.6666)
    minus Nonradiation rate (44.6445) =    94.0221         |     minus Nonradiation rate (48.3344) =    90.3322
                                                           |
3.  1988 Fractional Causation is radiation                 | 3.  1988 Fractional Causation is radiation
    rate (94.0221) divided by OBSERVED                     |     rate (90.3322) divided by OBSERVED
    Natl MR Part 1,Col.C=    162.7     =       0.58        |     Natl MR Part 1, Col.C=    162.7           0.56
4.  Comparable est. = 0.74 from Table 49-F.                |
-----------------------------------------------------------|-----------------------------------------------------------
```

===
Table M-2
All-Cancers, Females, 1988: Alternative Smoking Adju and Fractional Causation
===

This table is like Table 50-F, except for the Smoking Adjustment. Here in Part 1, we do not multiply the 1940
MidTrio and LowTrio MortRates by a factor; we adjust them by the DIFFERENCE (-23.0 cases) which we obtain from
Chap.50, Box 1, Col.K. Of course, Col.E below keeps the PP Adju (see Tab 47-B & Chap.50, Box 2, Part 2, Col.D).

Part 1. Calculation of the Alternative Smoking Adjustment (Col.G) and the National Adjusted MortRate (Col.I).

Trio-Seq.	Col.A 1990 PopFrac Tab 3-B	Col.B 1988 Obs MR Tab 7-A	Col.C A * B	Col.D = 1940 MRs Mid,Low Tab 7-A	Col.E= PPAdju Factor Tab47-B	Col.F= D * E	Col.G= Col.F - 23.0 (TopTrio, Bx1,ColK)	Col.H= Adju MortRate	Col.I = A * H
Pac	0.1535	111.5	17.115					111.5	17.115
NewEng	0.0527	116.4	6.134					116.4	6.134
MidAtl	0.1527	118.6	18.110					118.6	18.110
WNoCen	0.0721	106.8	7.700	120.1	0.94	112.9	89.894	89.894	6.481
ENoCen	0.1713	116.5	19.956	131.4	0.94	123.5	100.516	100.516	17.218
Mtn	0.0543	100.4	5.452	111.8	0.94	105.1	82.092	82.092	4.458
WSoCen	0.1087	109.8	11.935	99.8	1.07	106.8	83.786	83.786	9.108
ESoCen	0.0621	112.7	6.999	102.5	1.07	109.7	86.675	86.675	5.383
SoAtl	0.1725	111.6	19.251	106.9	1.07	114.4	91.383	91.383	15.764
Weighted avg. Col.C =		112.7						Sum =	
1988 Obs.Natl MR, Tab 7-B=		111.3			1988 Natl Adju MR =				99.7707

Part 2. ---

Trio-Seq.	Col.A Mean1940 thru1990 PPs from Tab 47-A x'	Col.B 1988 AdjuMRs Col.H Part 1	Col.C All-Cancers, Females: 1988 Adjusted MortRates regressed on Mean 1940 thru 1990 PhysPops Regression Output:		Col.D 1940 PPs from Table 3-A TrioSeq. x''	Col.E All-Cancers, Females: 1988 Adjusted MortRates regressed on 1940 PhysPops Regression Output:	
Pac	191.97	111.5	Constant	37.3965	159.72	Constant	41.8295
NewEng	208.20	116.4	Std Err of Y Est	5.6859	161.55	Std Err of Y Est	6.5349
MidAtl	204.72	118.6	R Squared	0.8624	169.76	R Squared	0.8182
WNoCen	141.14	89.894	No. of Observation	9	123.14	No. of Observation	9
ENoCen	146.19	100.516	Degrees of Freedom	7	133.36	Degrees of Freedom	7
Mtn	145.91	82.092			119.89		
WSoCen	126.28	83.786	X Coefficient(s)	0.3831	103.94	X Coefficient(s)	0.4356
ESoCen	113.28	86.675	Std Err of Coef.	0.0579	85.83	Std Err of Coef.	0.0776
SoAtl	142.93	91.383	XCoef / S.E. =	6.6225	100.74	XCoef / S.E.	5.6126

Part 3-A. | Part 3-B.
Calculation of Fractional Causation | Calculation of Fractional Causation
from Averaged PhysPops | from 1940 PhysPops
                                                              |
1. Nonradiation rate is Adjusted | 1. Nonradiation rate is Adjusted
    Constant (Part 2, Col.C) =   37.3965 | Constant (Part 2, Col.E) =   41.8295
                                                              |
2. Radiation rate is Natl Adjusted | 2. Radiation rate is Natl Adjusted
    MortRate (Part 1, Col.I = 99.7707) | MortRate (Part 1, Col.I = 99.7707)
    minus Nonradiation rate (37.3965) = 62.3743 | minus Nonradiation rate (41.8295) = 57.9412
                                                              |
3. 1988 Fractional Causation is radiation | 3. 1988 Fractional Causation is radiation
    rate (62.3743) divided by OBSERVED | rate (57.9412) divided by OBSERVED
    Natl MR Part 1,Col.C=   111.3   =   0.56 | Natl MR Part 1, Col.C=   111.3   0.52
4. Comparable est. = 0.50 from Table 50-F. |

```
=====================================================================================================
                                         Table M-3
            Respiratory-System Cancers, Males, 1988:  Alternative Smoking Adju and Fractional Causation
=====================================================================================================
```

This table is like Table 51-FF, except for the Smoking Adjustment. Here in Part 1, we do not multiply the 1940 MidTrio and LowTrio MortRates by a factor; we adjust them by the DIFFERENCE (+40.6 cases) which we obtain from Chap.51, Box 1, Col.K. Of course, Col.E below keeps the PP Adju (see Tab 47-B & Chap.51, Box 2, Part 2, Col.D).

Part 1. Calculation of the Alternative Smoking Adjustment (Col.G) and the National Adjusted MortRate (Col.I).

Trio-Seq.	Col.A 1990 PopFrac Tab 3-B	Col.B 1988 Obs MR Tab 16-A	Col.C A * B	Col.D = 1940 MRs Mid,Low Tab 16-A	Col.E = PP Adju Factor Tab 47-B	Col.F = D * E	Col.G = Col.F + 40.6 (TopTrio, Bx1,ColK)	Col.H = Adju MortRate	Col.I = A * H
Pac	0.1535	50.7	7.782					50.7	7.782
NewEng	0.0527	56.3	2.967					56.3	2.967
MidAtl	0.1527	57.5	8.780					57.5	8.780
WNoCen	0.0721	56.2	4.052	7.7	0.94	7.2	47.8	47.838	3.449
ENoCen	0.1713	62.3	10.672	10.6	0.94	10.0	50.6	50.564	8.662
Mtn	0.0543	44.2	2.400	7.8	0.94	7.3	47.9	47.932	2.603
WSoCen	0.1087	67.9	7.381	7.6	1.07	8.1	48.7	48.732	5.297
ESoCen	0.0621	79.1	4.912	4.9	1.07	5.2	45.8	45.843	2.847
SoAtl	0.1725	68.5	11.816	8.3	1.07	8.9	49.5	49.481	8.535

Weighted avg. Col.C =   60.8   Sum =
1988 Obs.Natl MR, Tab 16-B=   59.7   1988 Natl Adju MR =   50.9226

Part 2.
```
            Col.A     Col.B                                    Col.D                                    Col.E
            Mean1940   1988          Col.C                      1940              Col.E
            thru1990  AdjuMRs   Respiratory-Ca, Males:          PPs from      Respiratory-Ca, Males:
  Trio-     PPs from   Col.H    1988 Adjusted MortRates         Tab 3-A       1988 Adjusted MortRates
  Seq.      Tab 47-A  Part 1       regressed on                TrioSeq          regressed on
               x'                Mean 1940 thru 1990 PhysPops      x''          1940 PhysPops
  Pac       191.97    50.7                    Regression Output:                     Regression Output:
  NewEng    208.20    56.3     Constant            34.5305     159.72     Constant            36.1389
  MidAtl    204.72    57.5     Std Err of Y Est     1.8044     161.55     Std Err of Y Est     2.1820
  WNoCen    141.14    47.838   R Squared            0.8135     169.76     R Squared            0.7273
  ENoCen    146.19    50.564   No. of Observation        9    123.14     No. of Observation        9
  Mtn       145.91    47.932   Degrees of Freedom        7    133.36     Degrees of Freedom        7
  WSoCen    126.28    48.732                                  119.89
  ESoCen    113.28    45.843   X Coefficient(s)     0.1014    103.94     X Coefficient(s)     0.1120
  SoAtl     142.93    49.481   Std Err of Coef.     0.0184     85.83     Std Err of Coef.     0.0259
                               XCoef / S.E. =       5.5257    100.74     XCoef / S.E.         4.3206
```

Part 3-A. Calculation of Fractional Causation from Averaged PhysPops	Part 3-B. Calculation of Fractional Causation from 1940 PhysPops
1. Nonradiation rate is Adjusted Constant (Part 2, Col.C) = 34.5305	1. Nonradiation rate is Adjusted Constant (Part 2, Col.E) = 36.1389
2. Radiation rate is Natl Adjusted MortRate (Part 1, Col.I = 50.9226) minus Nonradiation rate (34.5305) = 16.3922	2. Radiation rate is Natl Adjusted MortRate (Part 1, Col.I = 50.9226) minus Nonradiation rate (36.1389) = 14.7838
3. 1988 Fractional Causation is radiation rate (16.3922) divided by OBSERVED Natl MR Part 1,Col.C= 59.7 = 0.27	3. 1988 Fractional Causation is radiation rate (14.7838) divided by OBSERVED Natl MR Part 1, Col.C= 59.7 0.25
4. Comparable est. = 0.74 from Table 51-FF.	

```
===================================================================================
                                  Table M-4
       Respiratory-System Cancers, Females, 1988:  Alternative Smoking Adju and Fractional Causation
===================================================================================
```

This table is like Table 52-F, except for the Smoking Adjustment. Here in Part 1, we do not multiply the 1940
MidTrio and LowTrio MortRates by a factor; we adjust them by the DIFFERENCE (+22.8 cases) which we obtain from
Chap.52, Box 1, Col.K. Of course, Col.E below keeps the PP Adju (see Tab 47-B & Chap.52, Box 2, Part 2, Col.D).

Part 1. Calculation of the Alternative Smoking Adjustment (Col.G) and the National Adjusted MortRate (Col.I).

Trio-Seq.	Col.A 1990 PopFrac Tab 3-B	Col.B 1988 Obs MR Tab 17-A	Col.C A * B	Col.D = 1940 MRs Mid,Low Tab 17-A	Col.E= PPAdju Factor Tab47-B	Col.F= D * E	Col.G= Col.F + 22.8 (TopTrio, Bx1,ColK)	Col.H= Adju MortRate	Col.I = A * H
Pac	0.1535	27.8	4.267					27.8	4.267
NewEng	0.0527	26.9	1.418					26.9	1.418
MidAtl	0.1527	25.8	3.940					25.8	3.940
WNoCen	0.0721	23.1	1.666	3.1	0.94	2.9	25.714	25.714	1.854
ENoCen	0.1713	26.4	4.522	3.2	0.94	3.0	25.808	25.808	4.421
Mtn	0.0543	22.2	1.205	2.9	0.94	2.7	25.526	25.526	1.386
WSoCen	0.1087	26.6	2.891	2.4	1.07	2.6	25.368	25.368	2.758
ESoCen	0.0621	26.6	1.652	2.4	1.07	2.6	25.368	25.368	1.575
SoAtl	0.1725	26.6	4.589	2.4	1.07	2.6	25.368	25.368	4.376

```
Weighted avg. Col.C =            26.1                                              Sum =
1988 Obs.Natl MR, Tab 17-B=      24.5                     1988 Natl Adju MR =      25.9944
```

Part 2. --

Trio-Seq.	Col.A Mean1940 thru1990 PPs from Tab 47-A x'	Col.B 1988 AdjuMRs Col.H Part 1	Col.C Respiratory-Ca, Females: 1988 Adjusted MortRates regressed on Mean 1940 thru 1990 PhysPops Regression Output:		Col.D 1940 PPs from Table 3-A TrioSeq. x''	Col.E Respiratory-Ca, Females: 1988 Adjusted MortRates regressed on 1940 PhysPops Regression Output:	
Pac	191.97	27.8	Constant	23.2229	159.72	Constant	23.3763
NewEng	208.20	26.9	Std Err of Y Est	0.6223	161.55	Std Err of Y Est	0.6274
MidAtl	204.72	25.8	R Squared	0.5175	169.76	R Squared	0.5095
WNoCen	141.14	25.714	No. of Observation	9	123.14	No. of Observation	9
ENoCen	146.19	25.808	Degrees of Freedom	7	133.36	Degrees of Freedom	7
Mtn	145.91	25.526			119.89		
WSoCen	126.28	25.368	X Coefficient(s)	0.0173	103.94	X Coefficient(s)	0.0201
ESoCen	113.28	25.368	Std Err of Coef.	0.0063	85.83	Std Err of Coef.	0.0075
SoAtl	142.93	25.368	XCoef / S.E. =	2.7400	100.74	XCoef / S.E. =	2.6964

Part 3-A.
Calculation of Fractional Causation
from Averaged PhysPops

1. Nonradiation rate is Adjusted
    Constant (Part 2, Col.C) =   23.2229

2. Radiation rate is Natl Adjusted
    MortRate (Part 1, Col.I = 25.9944)
    minus Nonradiation rate (23.2229) =   2.7715

3. 1988 Fractional Causation is radiation
    rate (2.7715) divided by OBSERVED
    Natl MR Part 1,Col.C=   24.5   =   0.11
4. Comparable est. = 0.83 from Table 52-F.

Part 3-B.
Calculation of Fractional Causation
from 1940 PhysPops

1. Nonradiation rate is Adjusted
    Constant (Part 2, Col.E) =   23.3763

2. Radiation rate is Natl Adjusted
    MortRate (Part 1, Col.I = 25.9944)
    minus Nonradiation rate (23.3763) =   2.6181

3. 1988 Fractional Causation is radiation
    rate (2.6181) divided by OBSERVED
    Natl MR Part 1, Col.C=   24.5   0.11

```
===================================================================================================
                                        Table M-5
          Difference-Cancers, Males, 1988:  Alternative Smoking Adju and Fractional Causation
===================================================================================================
```

This table is like Table 53-F, except for the Smoking Adjustment. Here in Part 1, we do not multiply the 1940
MidTrio and LowTrio MortRates by a factor; we adjust them by the DIFFERENCE (-12.4 cases) which we obtain from
Chap.53, Box 1, Col.K. Of course, Col.E below keeps the PP Adju (see Tab 47-B & Chap.53, Box 2, Part 2, Col.D).

Part 1. Calculation of the Alternative Smoking Adjustment (Col.G) and the National Adjusted MortRate (Col.I).

Trio-Seq.	Col.A 1990 PopFrac Tab 3-B	Col.B 1988 Obs MR Tab 18-A	Col.C A * B	Col.D = 1940 MRs Mid,Low Tab 18-A	Col.E= PPAdju Factor Tab47-B	Col.F= D * E	Col.G= Col.F -12.4 (TopTrio, Bx1,ColK)	Col.H= Adju MortRate	Col.I = A * H
Pac	0.1535	97.8	15.012					97.8	15.012
NewEng	0.0527	110.8	5.839					110.8	5.839
MidAtl	0.1527	110.9	16.934					110.9	16.934
WNoCen	0.0721	99.7	7.188	103.2	0.94	97.0	84.608	84.608	6.100
ENoCen	0.1713	108.9	18.655	109.0	0.94	102.5	90.06	90.060	15.427
Mtn	0.0543	94.9	5.153	92.0	0.94	86.5	74.08	74.080	4.023
WSoCen	0.1087	105.0	11.414	79.3	1.07	84.9	72.451	72.451	7.875
ESoCen	0.0621	109.1	6.775	68.7	1.07	73.5	61.109	61.109	3.795
SoAtl	0.1725	107.3	18.509	80.6	1.07	86.2	73.842	73.842	12.738

Weighted avg. Col.C =   105.5   Sum =
1988 Obs.Natl MR, Tab 18-B=   103.0   1988 Natl Adju MR =   87.7440

Part 2. --

Trio-Seq.	Col.A Mean1940 thru1990 PPs from Tab 47-A x'	Col.B 1988 AdjuMRs Col.H Part 1	Col.C Difference-Ca, Males: 1988 Adjusted MortRates regressed on Mean 1940 thru 1990 PhysPops Regression Output:		Col.D 1940 PPs from Table 3-A TrioSeq. x''	Col.E Difference-Ca, Males: 1988 Adjusted MortRates regressed on 1940 PhysPops Regression Output:	
Pac	191.97	97.8	Constant	10.1140	159.72	Constant	12.1956
NewEng	208.20	110.8	Std Err of Y Est	5.8803	161.55	Std Err of Y Est	4.4832
MidAtl	204.72	110.9	R Squared	0.9026	169.76	R Squared	0.9434
WNoCen	141.14	84.608	No. of Observation	9	123.14	No. of Observation	9
ENoCen	146.19	90.060	Degrees of Freedom	7	133.36	Degrees of Freedom	7
Mtn	145.91	74.080			119.89		
WSoCen	126.28	72.451	X Coefficient(s)	0.4819	103.94	X Coefficient(s)	0.5751
ESoCen	113.28	61.109	Std Err of Coef.	0.0598	85.83	Std Err of Coef.	0.0532
SoAtl	142.93	73.842	XCoef / S.E. =	8.0547	100.74	XCoef / S.E.	10.8008

Part 3-A. | Part 3-B.
Calculation of Fractional Causation | Calculation of Fractional Causation
from Averaged PhysPops | from 1940 PhysPops
                                                       |
1. Nonradiation rate is Adjusted | 1. Nonradiation rate is Adjusted
   Constant (Part 2, Col.C) =   10.1140 | Constant (Part 2, Col.E) =   12.1956
                                                       |
2. Radiation rate is Natl Adjusted | 2. Radiation rate is Natl Adjusted
   MortRate (Part 1, Col.I = 87.7440) | MortRate (Part 1, Col.I = 87.7440)
   minus Nonradiation rate (10.1140) =   77.6300 | minus Nonradiation rate (12.1956) =   75.5484
                                                       |
3. 1988 Fractional Causation is radiation | 3. 1988 Fractional Causation is radiation
   rate (77.6300) divided by OBSERVED | rate (75.5484) divided by OBSERVED
   Natl MR Part 1,Col.C=   103.0   =   0.75 | Natl MR Part 1, Col.C=   103.0   0.73
4. Comparable est. = 0.72 from Table 53-F. |

```
================================================================================================
                                      Table M-6
           Difference-Cancers, Females, 1988: Alternative Smoking Adju and Fractional Causation
================================================================================================
```

This table is like Table 54-F, except for the Smoking Adjustment. Here in Part 1, we do not multiply the 1940
MidTrio and LowTrio MortRates by a factor; we adjust them by the DIFFERENCE (-45.8 cases) which we obtain from
Chap.54, Box 1, Col.K. Of course, Col.E below keeps the PP Adju (see Tab 47-B & Chap.54, Box 2, Part 2, Col.D).

Part 1. Calculation of the Alternative Smoking Adjustment (Col.G) and the National Adjusted MortRate (Col.I).

Trio-Seq.	Col.A 1990 PopFrac Tab 3-B	Col.B 1988 Obs MR Tab 19-A	Col.C A * B	Col.D = 1940 MRs Mid,Low Tab 19-A	Col.E= PPAdju Factor Tab47-B	Col.F= D * E	Col.G= Col.F -45.8 (TopTrio, Bx1,ColK)	Col.H= Adju MortRate	Col.I = A * H
Pac	0.1535	83.7	12.848					83.7	12.848
NewEng	0.0527	89.5	4.717					89.5	4.717
MidAtl	0.1527	92.8	14.171					92.8	14.171
WNoCen	0.0721	83.7	6.035	117.0	0.94	110.0	64.180	64.180	4.627
ENoCen	0.1713	90.1	15.434	128.2	0.94	120.5	74.708	74.708	12.797
Mtn	0.0543	78.2	4.246	108.9	0.94	102.4	56.566	56.566	3.072
WSoCen	0.1087	83.2	9.044	97.4	1.07	104.2	58.418	58.418	6.350
ESoCen	0.0621	86.1	5.347	100.1	1.07	107.1	61.307	61.307	3.807
SoAtl	0.1725	85.0	14.663	104.5	1.07	111.8	66.015	66.015	11.388

```
Weighted avg. Col.C =        86.5                                              Sum =
1988 Obs.Natl MR, Tab 19-B=  86.8                        1988 Natl Adju MR =       73.7763
```

Part 2. --

Trio-Seq.	Col.A Mean1940 thru1990 PPs from Tab 47-A x'	Col.B 1988 AdjuMRs Col.H Part 1	Col.C Difference-Ca, Females: 1988 Adjusted MortRates regressed on Mean 1940 thru 1990 PhysPops Regression Output:		Col.D 1940 PPs from Table 3-A TrioSeq. x''	Col.E Difference-Ca, Females: 1988 Adjusted MortRates regressed on 1940 PhysPops Regression Output:	
Pac	191.97	83.7	Constant	14.1736	159.72	Constant	18.4532
NewEng	208.20	89.5	Std Err of Y Est	5.6649	161.55	Std Err of Y Est	6.4698
MidAtl	204.72	92.8	R Squared	0.8519	169.76	R Squared	0.8069
WNoCen	141.14	64.180	No. of Observation	9	123.14	No. of Observation	9
ENoCen	146.19	74.708	Degrees of Freedom	7	133.36	Degrees of Freedom	7
Mtn	145.91	56.566			119.89		
WSoCen	126.28	58.418	X Coefficient(s)	0.3658	103.94	X Coefficient(s)	0.4155
ESoCen	113.28	61.307	Std Err of Coef.	0.0576	85.83	Std Err of Coef.	0.0768
SoAtl	142.93	66.015	XCoef / S.E. =	6.3460	100.74	XCoef / S.E.	5.4076

Part 3-A. | Part 3-B.
Calculation of Fractional Causation | Calculation of Fractional Causation
from Averaged PhysPops | from 1940 PhysPops

1. Nonradiation rate is Adjusted | 1. Nonradiation rate is Adjusted
   Constant (Part 2, Col.C) =   14.1736 | Constant (Part 2, Col.E) =   18.4532
                                                       |
2. Radiation rate is Natl Adjusted | 2. Radiation rate is Natl Adjusted
   MortRate (Part 1, Col.I = 73.7763) | MortRate (Part 1, Col.I = 73.7763)
   minus Nonradiation rate (14.1736) =   59.6028 | minus Nonradiation rate (18.4532) =   55.3231
                                                       |
3. 1988 Fractional Causation is radiation | 3. 1988 Fractional Causation is radiation
   rate (59.6028) divided by OBSERVED | rate (55.3231) divided by OBSERVED
   Natl MR Part 1,Col.C=   86.8   =   0.69 | Natl MR Part 1, Col.C=   86.8   0.64
4. Comparable est. = 0.48 from Table 54-F. |

```
==================================================================================================
                                         Table M-7
              Breast-Cancers, Females, 1990:  Alternative Smoking Adju and Fractional Causation
==================================================================================================
```

This table is like Table 55-F, except for the Smoking Adjustment. Here in Part 1, we do not multiply the 1940
MidTrio and LowTrio MortRates by a factor; we adjust them by the DIFFERENCE (-3.5 cases) which we obtain from
Chap.45, Box 1, Col.K. Of course, Col.E below keeps the PP Adju (see Tab 47-B & Chap.45, Box 2, Part 2, Col.D).

Part 1. Calculation of the Alternative Smoking Adjustment (Col.G) and the National Adjusted MortRate (Col.I).

Trio-Seq.	Col.A 1990 PopFrac Tab 3-B	Col.B 1990 Obs MR Tab 8-A	Col.C A * B	Col.D = 1940 MRs Mid,Low Tab 8-A	Col.E= PPadju Factor Tab47-B	Col.F= D * E	Col.G= Col.F -3.5 (TopTrio, Bx1,ColK)	Col.H= Adju MortRate	Col.I = A * H
Pac	0.1535	22.7	3.484					22.7	3.484
NewEng	0.0527	24.3	1.281					24.3	1.281
MidAtl	0.1527	25.8	3.940					25.8	3.940
WNoCen	0.0721	22.6	1.629	22.6	0.94	21.2	17.744	17.744	1.279
ENoCen	0.1713	24.1	4.128	24.3	0.94	22.8	19.342	19.342	3.313
Mtn	0.0543	21.0	1.140	18.6	0.94	17.5	13.984	13.984	0.759
WSoCen	0.1087	20.8	2.261	15.1	1.07	16.2	12.657	12.657	1.376
ESoCen	0.0621	21.4	1.329	15.1	1.07	16.2	12.657	12.657	0.786
SoAtl	0.1725	22.6	3.899	18.3	1.07	19.6	16.081	16.081	2.774

```
Weighted avg. Col.C =      23.1                                                    Sum =
1990 Obs.Natl MR, Tab 8-B= 23.1                       1990 Natl Adju MR =       18.9925
```

Part 2. --

Trio-Seq.	Col.A Mean1940 thru1990 PPs from Tab 47-A x'	Col.B 1990 AdjuMRs Col.H Part 1	Col.C Breast Cancer, Females: 1990 Adjusted MortRates regressed on Mean 1940 thru 1990 PhysPops Regression Output:		Col.D 1940 PPs from Table 3-A TrioSeq. x''	Col.E Breast Cancer, Females: 1990 Adjusted MortRates regressed on 1940 PhysPops Regression Output:	
Pac	191.97	22.7	Constant	-3.1287	159.72	Constant	-2.1643
NewEng	208.20	24.3	Std Err of Y Est	1.7378	161.55	Std Err of Y Est	1.6787
MidAtl	204.72	25.8	R Squared	0.8944	169.76	R Squared	0.9015
WNoCen	141.14	17.744	No. of Observation	9	123.14	No. of Observation	9
ENoCen	146.19	19.342	Degrees of Freedom	7	133.36	Degrees of Freedom	7
Mtn	145.91	13.984			119.89		
WSoCen	126.28	12.657	X Coefficient(s)	0.1362	103.94	X Coefficient(s)	0.1595
ESoCen	113.28	12.657	Std Err of Coef.	0.0177	85.83	Std Err of Coef.	0.0199
SoAtl	142.93	16.081	XCoef / S.E. =	7.7005	100.74	XCoef / S.E.	8.0026

Part 3-A. Calculation of Fractional Causation from Averaged PhysPops	Part 3-B. Calculation of Fractional Causation from 1940 PhysPops
1. Nonradiation rate is Adjusted Constant (Part 2, Col.C) = NEG = 0.0	1. Nonradiation rate is Adjusted Constant (Part 2, Col.E) = NEG = 0.0
2. Radiation rate is Natl Adjusted MortRate (Part 1, Col.I = 18.9925) minus Nonradiation rate (0.0) = 18.9925	2. Radiation rate is Natl Adjusted MortRate (Part 1, Col.I = 18.9925) minus Nonradiation rate (0.0) = 18.9925
3. 1990 Fractional Causation is radiation rate (18.9925) divided by OBSERVED Natl MR Part 1,Col.C= 23.1 = 0.82	3. 1990 Fractional Causation is radiation rate (18.9925) divided by OBSERVED Natl MR Part 1, Col.C= 23.1 0.82
4. Comparable est. = 0.83 from Table 55-F.	

```
==================================================================================
                                   Table M-8
          All-Cancers-Except-Genital, Females, 1980:  Alternative Smoking Adju and Fractional Causation
==================================================================================
```

This table is like Table 56-E, except for the Smoking Adjustment. Here in Part 1, we do not multiply the 1940
MidTrio and LowTrio MortRates by a factor; we adjust them by the DIFFERENCE (-4.6 cases) which we obtain from
Chap.56, Box 1, Col.K. Of course, Col.E below keeps the PP Adju (see Tab 47-B & Chap.56, Box 2, Part 2, Col.D).

Part 1. Calculation of the Alternative Smoking Adjustment (Col.G) and the National Adjusted MortRate (Col.I).

Trio-Seq.	Col.A 1980 PopFrac Tab 3-B	Col.B 1980 Obs MR Tab 20-A	Col.C A * B	Col.D = 1940 MRs Mid,Low Tab 20-A	Col.E= PPAdju Factor Tab47-B	Col.F= D * E	Col.G= Col.F -4.6 (TopTrio, Bx1,ColK)	Col.H= Adju MortRate	Col.I = A * H
Pac	0.1398	97.1	13.575					97.1	13.575
NewEng	0.0546	103.0	5.624					103.0	5.624
MidAtl	0.1630	103.2	16.822					103.2	16.822
WNoCen	0.0759	87.7	6.656	91.7	0.94	86.2	81.598	81.598	6.193
ENoCen	0.1846	97.5	17.999	98.2	0.94	92.3	87.708	87.708	16.191
Mtn	0.0502	83.2	4.177	84.0	0.94	79.0	74.360	74.360	3.733
WSoCen	0.1049	87.6	9.189	69.8	1.04	72.6	67.992	67.992	7.132
ESoCen	0.0646	88.9	5.743	69.3	1.04	72.1	67.472	67.472	4.359
SoAtl	0.1624	91.5	14.860	74.4	1.04	77.4	72.776	72.776	11.819

```
Weighted avg. Col.C =        94.6                                       Sum =
1980 Obs.Natl MR, Tab 20-A=  94.8              1980 Natl Adju MR =      85.4469
```

Part 2. --

Trio-Seq.	Col.A Mean1940 thru1980 PPs from Tab 47-A x'	Col.B 1980 AdjuMRs Col.H Part 1	Col.C All-Except-Genital, Fems: 1980 Adjusted MortRates regressed on Mean 1940 thru 1980 PhysPops Regression Output:		Col.D 1940 PPs from Table 3-A TrioSeq. x''	Col.E All-Except-Genital, Fems: 1980 Adjusted MortRates regressed on 1940 PhysPops Regression Output:	
Pac	177.35	97.1	Constant	22.5894	159.72	Constant	23.0656
NewEng	185.86	103.0	Std Err of Y Est	4.4334	161.55	Std Err of Y Est	3.4487
MidAtl	186.11	103.2	R Squared	0.9176	169.76	R Squared	0.9501
WNoCen	128.82	81.598	No. of Observation	9	123.14	No. of Observation	9
ENoCen	133.71	87.708	Degrees of Freedom	7	133.36	Degrees of Freedom	7
Mtn	133.45	74.360			119.89		
WSoCen	114.66	67.992	X Coefficient(s)	0.4298	103.94	X Coefficient(s)	0.4729
ESoCen	99.46	67.472	Std Err of Coef.	0.0487	85.83	Std Err of Coef.	0.0410
SoAtl	124.62	72.776	XCoef / S.E. =	8.8271	100.74	XCoef / S.E.	11.5470

Part 3-A.
Calculation of Fractional Causation
from Averaged PhysPops

1. Nonradiation rate is Adjusted
    Constant (Part 2, Col.C) =   22.5894

2. Radiation rate is Natl Adjusted
    MortRate (Part 1, Col.I = 85.4469)
    minus Nonradiation rate (22.5894) =   62.8575

3. 1980 Fractional Causation is radiation
    rate (62.8575) divided by OBSERVED
    Natl MR Part 1,Col.C=   94.8   =   0.66

4. Comparable est. = 0.66 from Table 56-E.

Part 3-B.
Calculation of Fractional Causation
from 1940 PhysPops

1. Nonradiation rate is Adjusted
    Constant (Part 2, Col.E) =   23.0656

2. Radiation rate is Natl Adjusted
    MortRate (Part 1, Col.I = 85.4469)
    minus Nonradiation rate (23.0656) =   62.3813

3. 1980 Fractional Causation is radiation
    rate (62.3813) divided by OBSERVED
    Natl MR Part 1,Col.C=   94.8   0.66

===
Table M-9
Digestive-System Cancers, Males, 1988: Alternative Smoking Adju and Fractional Causation
===

This table is like Table 57-F, except for the Smoking Adjustment. Here in Part 1, we do not multiply the 1940
MidTrio and LowTrio MortRates by a factor; we adjust them by the DIFFERENCE (-29.4 cases) which we obtain from
Chap.57, Box 1, Col.K. Of course, Col.E below keeps the PP Adju (see Tab 47-B & Chap.57, Box 2, Part 2, Col.D).

Part 1. Calculation of the Alternative Smoking Adjustment (Col.G) and the National Adjusted MortRate (Col.I).

Trio-Seq.	Col.A 1990 PopFrac Tab 3-B	Col.B 1988 Obs MR Tab 9-A	Col.C A * B	Col.D = 1940 MRs Mid,Low Tab 9-A	Col.E= PPAdju Factor Tab47-B	Col.F= D * E	Col.G= Col.F -29.4 (TopTrio, Bx1,ColK)	Col.H= Adju MortRate	Col.I = A * H
Pac	0.1535	36.3	5.572					36.3	5.572
NewEng	0.0527	42.1	2.219					42.1	2.219
MidAtl	0.1527	43.3	6.612					43.3	6.612
WNoCen	0.0721	35.8	2.581	59.9	0.94	56.3	26.906	26.906	1.940
ENoCen	0.1713	40.2	6.886	64.9	0.94	61.0	31.606	31.606	5.414
Mtn	0.0543	33.0	1.792	52.1	0.94	49.0	19.574	19.574	1.063
WSoCen	0.1087	36.5	3.968	42.3	1.07	45.3	15.861	15.861	1.724
ESoCen	0.0621	38.0	2.360	38.2	1.07	40.9	11.474	11.474	0.713
SoAtl	0.1725	38.5	6.641	43.4	1.07	46.4	17.038	17.038	2.939

Weighted avg. Col.C =   38.6   Sum =
1988 Obs.Natl MR, Tab 9-B=   38.8   1988 Natl Adju MR =   28.1952

Part 2. ---

Trio-Seq.	Col.A Mean1940 thru1990 PPs from Tab 47-A x'	Col.B 1988 AdjuMRs Col.H Part 1	Col.C Digestive-Ca, Males: 1988 Adjusted MortRates regressed on Mean 1940 thru 1990 PhysPops Regression Output:		Col.D 1940 PPs from Table 3-A TrioSeq. x''	Col.E Digestive-Ca, Males: 1988 Adjusted MortRates regressed on 1940 PhysPops Regression Output:	
Pac	191.97	36.3	Constant	-23.0444	159.72	Constant	-22.9761
NewEng	208.20	42.1	Std Err of Y Est	4.5486	161.55	Std Err of Y Est	2.5486
MidAtl	204.72	43.3	R Squared	0.8708	169.76	R Squared	0.9594
WNoCen	141.14	26.906	No. of Observation	9	123.14	No. of Observation	9
ENoCen	146.19	31.606	Degrees of Freedom	7	133.36	Degrees of Freedom	7
Mtn	145.91	19.574			119.89		
WSoCen	126.28	15.861	X Coefficient(s)	0.3179	103.94	X Coefficient(s)	0.3894
ESoCen	113.28	11.474	Std Err of Coef.	0.0463	85.83	Std Err of Coef.	0.0303
SoAtl	142.93	17.038	XCoef / S.E. =	6.8682	100.74	XCoef / S.E.	12.8668

Part 3-A. | Part 3-B.
Calculation of Fractional Causation | Calculation of Fractional Causation
from Averaged PhysPops | from 1940 PhysPops
                                                      |
1. Nonradiation rate is Adjusted | 1. Nonradiation rate is Adjusted
    Constant (Part 2, Col.C) = NEG =   0.0 | Constant (Part 2, Col.E) = NEG =   0.0
                                                      |
2. Radiation rate is Natl Adjusted | 2. Radiation rate is Natl Adjusted
    MortRate (Part 1, Col.I = 28.1952) | MortRate (Part 1, Col.I = 28.1952)
    minus Nonradiation rate (0.0) =   28.1952 | minus Nonradiation rate (0.0) =   28.1952
                                                      |
3. 1988 Fractional Causation is radiation | 3. 1988 Fractional Causation is radiation
    rate (28.1952) divided by OBSERVED | rate (28.1952) divided by OBSERVED
    Natl MR Part 1,Col.C=   38.8   =   0.73 | Natl MR Part 1, Col.C=   38.8   0.73
4. Comparabe est. = 0.82 from Table 57-F. |

```
=====================================================================================
                                    Table M-10
        Digestive-System Cancers, Females, 1988:  Alternative Smoking Adju and Fractional Causation
=====================================================================================
```

This table is like Table 58-F, except for the Smoking Adjustment. Here in Part 1, we do not multiply the 1940 MidTrio and LowTrio MortRates by a factor; we adjust them by the DIFFERENCE (-31.6 cases) which we obtain from Chap.58, Box 1, Col.K. Of course, Col.E below keeps the PP Adju (see Tab 47-B & Chap.58, Box 2, Part 2, Col.D).

Part 1. Calculation of the Alternative Smoking Adjustment (Col.G) and the National Adjusted MortRate (Col.I).

Trio-Seq.	Col.A 1990 PopFrac Tab 3-B	Col.B 1988 Obs MR Tab 10-A	Col.C A * B	Col.D = 1940 MRs Mid,Low Tab 10-A	Col.E= PP Adju Factor Tab47-B	Col.F= D * E	Col.G= Col.F -31.6 (TopTrio, Bx1,ColK)	Col.H= Adju MortRate	Col.I = A * H
Pac	0.1535	22.8	3.500					22.8	3.500
NewEng	0.0527	24.7	1.302					24.7	1.302
MidAtl	0.1527	26.0	3.970					26.0	3.970
WNoCen	0.0721	21.8	1.572	49.7	0.94	46.7	15.118	15.118	1.090
ENoCen	0.1713	24.2	4.145	53.1	0.94	49.9	18.314	18.314	3.137
Mtn	0.0543	21.1	1.146	47.7	0.94	44.8	13.238	13.238	0.719
WSoCen	0.1087	21.5	2.337	34.5	1.07	36.9	5.315	5.315	0.578
ESoCen	0.0621	23.3	1.447	36.3	1.07	38.8	7.241	7.241	0.450
SoAtl	0.1725	22.8	3.933	37.3	1.07	39.9	8.311	8.311	1.434

Weighted avg. Col.C =   23.4
1988 Obs.Natl MR, Tab 10-B=   23.5   1988 Natl Adju MR =   16.1788 (Sum =)

Part 2. --

Trio-Seq.	Col.A Mean1940 thru1990 PPs from Tab 47-A x'	Col.B 1988 AdjuMRs Col.H Part 1	Col.C Digestive-Ca, Females: 1988 Adjusted MortRates regressed on Mean 1940 thru 1990 PhysPops		Col.D 1940 PPs from Table 3-A TrioSeq. x''	Col.E Digestive-Ca, Females: 1988 Adjusted MortRates regressed on 1940 PhysPops	
			Regression Output:			Regression Output:	
Pac	191.97	22.8	Constant	-17.0424	159.72	Constant	-17.0334
NewEng	208.20	24.7	Std Err of Y Est	3.1585	161.55	Std Err of Y Est	1.9494
MidAtl	204.72	26.0	R Squared	0.8559	169.76	R Squared	0.9451
WNoCen	141.14	15.118	No. of Observation	9	123.14	No. of Observation	9
ENoCen	146.19	18.314	Degrees of Freedom	7	133.36	Degrees of Freedom	7
Mtn	145.91	13.238			119.89		
WSoCen	126.28	5.315	X Coefficient(s)	0.2072	103.94	X Coefficient(s)	0.2542
ESoCen	113.28	7.241	Std Err of Coef.	0.0321	85.83	Std Err of Coef.	0.0232
SoAtl	142.93	8.311	XCoef / S.E. =	6.4489	100.74	XCoef / S.E.	10.9799

Part 3-A.
Calculation of Fractional Causation
from Averaged PhysPops

1. Nonradiation rate is Adjusted
   Constant (Part 2, Col.C) = NEG =   0.0

2. Radiation rate is Natl Adjusted
   MortRate (Part 1, Col.I = 16.1788)
   minus Nonradiation rate (0.0) =   16.1788

3. 1988 Fractional Causation is radiation
   rate (16.1788) divided by OBSERVED
   Natl MR Part 1,Col.C=   23.5   =   0.69

4. Comparable est. = 0.68 from Table 58-F.

Part 3-B.
Calculation of Fractional Causation
from 1940 PhysPops

1. Nonradiation rate is Adjusted
   Constant (Part 2, Col.E) = NEG =   0.0

2. Radiation rate is Natl Adjusted
   MortRate (Part 1, Col.I = 16.1788)
   minus Nonradiation rate (0.0) =   16.1788

3. 1988 Fractional Causation is radiation
   rate (16.1788) divided by OBSERVED
   Natl MR Part 1, Col.C=   23.5   0.69

```
==================================================================================================
                                       Table M-11
          Urinary-System Cancers, Males, 1980:  Alternative Smoking Adju and Fractional Causation
==================================================================================================
```

This table is like Table 59-EE, except for the Smoking Adjustment. Here in Part 1, we do not multiply the 1940 MidTrio and LowTrio MortRates by a factor; we adjust them by the DIFFERENCE (-0.3 case) which we obtain from Chap.59, Box 1, Col.K. Of course, Col.E below keeps the PP Adju (see Tab 47-B & Chap.59, Box 2, Part 2, Col.D).

Part 1. Calculation of the Alternative Smoking Adjustment (Col.G) and the National Adjusted MortRate (Col.I).

Trio-Seq.	Col.A 1980 PopFrac Tab 3-B	Col.B 1980 Obs MR Tab 11-A	Col.C A * B	Col.D = 1940 MRs Mid,Low Tab 11-A	Col.E= PPAdju Factor Tab47-B	Col.F= D * E	Col.G= Col.F -0.3 (TopTrio, Bx1,ColK)	Col.H= Adju MortRate	Col.I = A * H
Pac	0.1398	7.7	1.076					7.7	1.076
NewEng	0.0546	9.5	0.519					9.5	0.519
MidAtl	0.1630	9.2	1.500					9.2	1.500
WNoCen	0.0759	7.9	0.600	6.7	0.94	6.3	5.998	5.998	0.455
ENoCen	0.1846	8.7	1.606	8.1	0.94	7.6	7.314	7.314	1.350
Mtn	0.0502	7.0	0.351	6.5	0.94	6.1	5.810	5.810	0.292
WSoCen	0.1049	7.0	0.734	4.3	1.04	4.5	4.172	4.172	0.438
ESoCen	0.0646	7.3	0.472	3.0	1.04	3.1	2.820	2.820	0.182
SoAtl	0.1624	7.8	1.267	5.3	1.04	5.5	5.212	5.212	0.846

```
Weighted avg. Col.C =          8.1                                                    Sum =
1980 Obs.Natl MR, Tab 11-B=    8.2                           1980 Natl Adju MR =             6.6581
```

Part 2. --

Trio-Seq.	Col.A Mean1940 thru1980 PPs from Tab 47-A x'	Col.B 1980 AdjuMRs Col.H Part 1	Col.C Urinary-System Ca: Males 1980 Adjusted MortRates regressed on Mean 1940 thru 1980 PhysPops Regression Output:		Col.D 1940 PPs from Table 3-A TrioSeq. x''	Col.E Urinary-System Ca: Males 1980 Adjusted MortRates regressed on 1940 PhysPops Regression Output:	
Pac	177.35	7.7	Constant	-2.9136	159.72	Constant	-2.8529
NewEng	185.86	9.5	Std Err of Y Est	0.7785	161.55	Std Err of Y Est	0.6428
MidAtl	186.11	9.2	R Squared	0.8931	169.76	R Squared	0.9271
WNoCen	128.82	5.998	No. of Observation	9	123.14	No. of Observation	9
ENoCen	133.71	7.314	Degrees of Freedom	7	133.36	Degrees of Freedom	7
Mtn	133.45	5.810			119.89		
WSoCen	114.66	4.172	X Coefficient(s)	0.0654	103.94	X Coefficient(s)	0.0720
ESoCen	99.46	2.820	Std Err of Coef.	0.0086	85.83	Std Err of Coef.	0.0076
SoAtl	124.62	5.212	XCoef / S.E. =	7.6461	100.74	XCoef / S.E. =	9.4354

Part 3-A. Calculation of Fractional Causation from Averaged PhysPops	Part 3-B. Calculation of Fractional Causation from 1940 PhysPops
1. Nonradiation rate is Adjusted Constant (Part 2, Col.C) = NEG = 0.0	1. Nonradiation rate is Adjusted Constant (Part 2, Col.E) = NEG = 0.0
2. Radiation rate is Natl Adjusted MortRate (Part 1, Col.I = 6.6581) minus Nonradiation rate (0.0) = 6.6581	2. Radiation rate is Natl Adjusted MortRate (Part 1, Col.I = 6.6581) minus Nonradiation rate (0.0) = 6.6581
3. 1980 Fractional Causation is radiation rate (6.6581) divided by OBSERVED Natl MR Part 1,Col.C= 8.2 = 0.81	3. 1980 Fractional Causation is radiation rate (6.6581) divided by OBSERVED Natl MR Part 1, Col.C= 8.2 0.81

4. Comparable est. = 0.83 from Table 59-EE.

```
===============================================================================================
                                          Table M-12
            Urinary-System Cancers, Females, 1980:  Alternative Smoking Adju and Fractional Causation
===============================================================================================
```

This table is like Table 60-E, except for the Smoking Adjustment. Here in Part 1, we do not multiply the 1940
MidTrio and LowTrio MortRates by a factor; we adjust them by the DIFFERENCE (-1.4 case) which we obtain from
Chap.60, Box 1, Col.K. Of course, Col.E below keeps the PP Adju (see Tab 47-B & Chap.60, Box 2, Part 2, Col.D).

Part 1. Calculation of the Alternative Smoking Adjustment (Col.G) and the National Adjusted MortRate (Col.I).

Trio-Seq.	Col.A 1980 PopFrac Tab 3-B	Col.B 1980 Obs MR Tab 12-A	Col.C A * B	Col.D = 1940 MRs Mid,Low Tab 12-A	Col.E= PPAdju Factor Tab47-B	Col.F= D * E	Col.G= Col.F -1.4 (TopTrio, Bx1,ColK)	Col.H= Adju MortRate	Col.I = A * H
Pac	0.1398	2.8	0.391					2.8	0.391
NewEng	0.0546	3.4	0.186					3.4	0.186
MidAtl	0.1630	3.2	0.522					3.2	0.522
WNoCen	0.0759	3.0	0.228	3.7	0.94	3.5	2.078	2.078	0.158
ENoCen	0.1846	3.0	0.554	4.1	0.94	3.9	2.454	2.454	0.453
Mtn	0.0502	2.5	0.126	3.5	0.94	3.3	1.890	1.890	0.095
WSoCen	0.1049	2.8	0.294	3.1	1.04	3.2	1.824	1.824	0.191
ESoCen	0.0646	2.8	0.181	2.7	1.04	2.8	1.408	1.408	0.091
SoAtl	0.1624	2.9	0.471	3.0	1.04	3.1	1.720	1.720	0.279

```
Weighted avg. Col.C =          3.0                                                   Sum =
1980 Obs.Natl MR, Tab 12-B=    3.0                          1980 Natl Adju MR =            2.3659
```

Part 2. --

Trio-Seq.	Col.A Mean1940 thru1980 PPs from Tab 47-A x'	Col.B 1980 AdjuMRs Col.H Part 1	Col.C Urinary-System Ca: Females 1980 Adjusted MortRates regressed on Mean 1940 thru 1980 PhysPops Regression Output:		Col.D 1940 PPs from Table 3-A TrioSeq. x''	Col.E Urinary-System Ca: Females 1980 Adjusted MortRates regressed on 1940 PhysPops Regression Output:	
Pac	177.35	2.8	Constant	-0.6480	159.72	Constant	-0.6026
NewEng	185.86	3.4	Std Err of Y Est	0.2105	161.55	Std Err of Y Est	0.1852
MidAtl	186.11	3.2	R Squared	0.9199	169.76	R Squared	0.9380
WNoCen	128.82	2.078	No. of Observation	9	123.14	No. of Observation	9
ENoCen	133.71	2.454	Degrees of Freedom	7	133.36	Degrees of Freedom	7
Mtn	133.45	1.890			119.89		
WSoCen	114.66	1.824	X Coefficient(s)	0.0207	103.94	X Coefficient(s)	0.0226
ESoCen	99.46	1.408	Std Err of Coef.	0.0023	85.83	Std Err of Coef.	0.0022
SoAtl	124.62	1.720	XCoef / S.E. =	8.9631	100.74	XCoef / S.E.	10.2876

Part 3-A. Calculation of Fractional Causation from Averaged PhysPops	Part 3-B. Calculation of Fractional Causation from 1940 PhysPops
1. Nonradiation rate is Adjusted Constant (Part 2, Col.C) = NEG = 0.0	1. Nonradiation rate is Adjusted Constant (Part 2, Col.E) = NEG = 0.0
2. Radiation rate is Natl Adjusted MortRate (Part 1, Col.I = 2.3659) minus Nonradiation rate (0.0) = 2.3659	2. Radiation rate is Natl Adjusted MortRate (Part 1, Col.I = 2.3659) minus Nonradiation rate (0.0) = 2.3659
3. 1980 Fractional Causation is radiation rate (2.3659) divided by OBSERVED Natl MR Part 1,Col.C= 3.0 = 0.79	3. 1980 Fractional Causation is radiation rate (2.3659) divided by OBSERVED Natl MR Part 1,Col.C= 3.0 0.79
4. Comparable est. = 0.78 from Table 60-E.	

```
===================================================================================================
                                         Table M-13
              Genital Cancers, Males, 1990: Alternative Smoking Adju and Fractional Causation
===================================================================================================
```

This table is like Table 61-F, except for the Smoking Adjustment. Here in Part 1, we do not multiply the 1940
MidTrio and LowTrio MortRates by a factor; we adjust them by the DIFFERENCE (-0.6 case) which we obtain from
Chap.61, Box 1, Col.K. Of course, Col.E below keeps the PP Adju (see Tab 47-B & Chap.61, Box 2, Part 2, Col.D).

Part 1. Calculation of the Alternative Smoking Adjustment (Col.G) and the National Adjusted MortRate (Col.I).

Trio-Seq.	Col.A 1990 PopFrac Tab 3-B	Col.B 1990 Obs MR Tab 13-A	Col.C A * B	Col.D = 1940 MRs Mid,Low Tab 13-A	Col.E= PP Adju Factor Tab47-B	Col.F= D * E	Col.G= Col.F -0.6 (TopTrio, Bx1,ColK)	Col.H= Adju MortRate	Col.I = A * H
Pac	0.1535	15.9	2.441					15.9	2.441
NewEng	0.0527	16.6	0.875					16.6	0.875
MidAtl	0.1527	16.8	2.565					16.8	2.565
WNoCen	0.0721	16.3	1.175	16.5	0.94	15.5	14.91	14.910	1.075
ENoCen	0.1713	17.2	2.946	15.8	0.94	14.9	14.252	14.252	2.441
Mtn	0.0543	16.6	0.901	15.8	0.94	14.9	14.252	14.252	0.774
WSoCen	0.1087	16.7	1.815	11.6	1.07	12.4	11.812	11.812	1.284
ESoCen	0.0621	17.5	1.087	10.4	1.07	11.1	10.528	10.528	0.654
SoAtl	0.1725	18.6	3.209	12.8	1.07	13.7	13.096	13.096	2.259

```
Weighted avg. Col.C =          17.0                                                    Sum =
1990 Obs.Natl MR, Tab 13-B=    16.9                        1990 Natl Adju MR =        14.3679
```

Part 2. --

Trio-Seq.	Col.A Mean1940 thru1990 PPs from Tab 47-A x'	Col.B 1990 AdjuMRs Col.H Part 1	Col.C Genital-Ca, Males: 1990 Adjusted MortRates regressed on Mean 1940 thru 1990 PhysPops Regression Output:		Col.D 1940 PPs from Table 3-A TrioSeq. x''	Col.E Genital-Ca, Males: 1990 Adjusted MortRates regressed on 1940 PhysPops Regression Output:	
Pac	191.97	15.9	Constant	5.3635	159.72	Constant	5.4998
NewEng	208.20	16.6	Std Err of Y Est	0.9080	161.55	Std Err of Y Est	0.7170
MidAtl	204.72	16.8	R Squared	0.8411	169.76	R Squared	0.9009
WNoCen	141.14	14.910	No. of Observation	9	123.14	No. of Observation	9
ENoCen	146.19	14.252	Degrees of Freedom	7	133.36	Degrees of Freedom	7
Mtn	145.91	14.252			119.89		
WSoCen	126.28	11.812	X Coefficient(s)	0.0562	103.94	X Coefficient(s)	0.0679
ESoCen	113.28	10.528	Std Err of Coef.	0.0092	85.83	Std Err of Coef.	0.0085
SoAtl	142.93	13.096	XCoef / S.E. =	6.0861	100.74	XCoef / S.E. =	7.9766

Part 3-A. | Part 3-B.
Calculation of Fractional Causation | Calculation of Fractional Causation
from Averaged PhysPops | from 1940 PhysPops

1. Nonradiation rate is Adjusted | 1. Nonradiation rate is Adjusted
    Constant (Part 2, Col.C) =   5.3635 | Constant (Part 2, Col.E) =   5.4998

2. Radiation rate is Natl Adjusted | 2. Radiation rate is Natl Adjusted
    MortRate (Part 1, Col.I = 14.3679) | MortRate (Part 1, Col.I = 14.3679)
    minus Nonradiation rate (5.3635) =   9.0044 | minus Nonradiation rate (5.4998) =   8.8681

3. 1990 Fractional Causation is radiation | 3. 1990 Fractional Causation is radiation
    rate (9.0044) divided by OBSERVED | rate (8.8681) divided by OBSERVED
    Natl MR Part 1,Col.C=   16.9   =   0.53 | Natl MR Part 1, Col.C=   16.9   0.52
4. Comparable est. = 0.47 from Table 61-F. |

```
===============================================================================
                                  Table M-14
       Buccal-Pharynx Cancers, Males, 1980:  Alternative Smoking Adju and Fractional Causation
===============================================================================
```

This table is like Table 63-EE, except for the Smoking Adjustment. Here in Part 1, we do not multiply the 1940
MidTrio and LowTrio MortRates by a factor; we adjust them by the DIFFERENCE (-1.2 case) which we obtain from
Chap.63, Box 1, Col.K. Of course, Col.E below keeps the PP Adju (see Tab 47-B & Chap.63, Box 2, Part 2, Col.D).

Part 1. Calculation of the Alternative Smoking Adjustment (Col.G) and the National Adjusted MortRate (Col.I).

Trio-Seq.	Col.A 1980 PopFrac Tab 3-B	Col.B 1980 Obs MR Tab 15-A	Col.C A * B	Col.D = 1940 MRs Mid,Low Tab 15-A	Col.E= PPAdju Factor Tab47-B	Col.F= D * E	Col.G= Col.F -1.2 (TopTrio, Bx1,ColK)	Col.H= Adju MortRate	Col.I = A * H
Pac	0.1398	4.2	0.587					4.2	0.587
NewEng	0.0546	5.7	0.311					5.7	0.311
MidAtl	0.1630	5.1	0.831					5.1	0.831
WNoCen	0.0759	3.5	0.266	4.6	0.94	4.3	3.124	3.124	0.237
ENoCen	0.1846	4.6	0.849	4.8	0.94	4.5	3.312	3.312	0.611
Mtn	0.0502	2.9	0.146	2.8	0.94	2.6	1.432	1.432	0.072
WSoCen	0.1049	4.2	0.441	4.0	1.04	4.2	2.960	2.960	0.311
ESoCen	0.0646	4.4	0.284	3.3	1.04	3.4	2.232	2.232	0.144
SoAtl	0.1624	5.0	0.812	4.3	1.04	4.5	3.272	3.272	0.531

```
Weighted avg. Col.C =          4.5                                              Sum =
1980 Obs.Natl MR, Tab 15-B=    4.6               1980 Natl Adju MR =           3.6361
```

Part 2. --

Trio-Seq.	Col.A Mean1940 thru1980 PPs from Tab 47-A x'	Col.B 1980 AdjuMRs Col.H Part 1	Col.C Buccal-Pharynx Ca: Males 1980 Adjusted MortRates regressed on Mean 1940 thru 1980 PhysPops Regression Output:		Col.D 1940 PPs from Table 3-A TrioSeq. x''	Col.E Buccal-Pharynx Ca: Males 1980 Adjusted MortRates regressed on 1940 PhysPops Regression Output:	
Pac	177.35	4.2	Constant	-1.4889	159.72	Constant	-1.1135
NewEng	185.86	5.7	Std Err of Y Est	0.7775	161.55	Std Err of Y Est	0.8659
MidAtl	186.11	5.1	R Squared	0.7039	169.76	R Squared	0.6327
WNoCen	128.82	3.124	No. of Observation	9	123.14	No. of Observation	9
ENoCen	133.71	3.312	Degrees of Freedom	7	133.36	Degrees of Freedom	7
Mtn	133.45	1.432			119.89		
WSoCen	114.66	2.960	X Coefficient(s)	0.0348	103.94	X Coefficient(s)	0.0357
ESoCen	99.46	2.232	Std Err of Coef.	0.0085	85.83	Std Err of Coef.	0.0103
SoAtl	124.62	3.272	XCoef / S.E. =	4.0797	100.74	XCoef / S.E.	3.4728

```
----------------------------------------------------------------------------------
Part 3-A.                                        | Part 3-B.
Calculation of Fractional Causation              | Calculation of Fractional Causation
from Averaged PhysPops                            | from 1940 PhysPops
                                                  |
1. Nonradiation rate is Adjusted                  | 1. Nonradiation rate is Adjusted
   Constant (Part 2, Col.C) = NEG =        0.0    |    Constant (Part 2, Col.E) = NEG =        0.0
                                                  |
2. Radiation rate is Natl Adjusted                | 2. Radiation rate is Natl Adjusted
   MortRate (Part 1, Col.I = 3.6361)              |    MortRate (Part 1, Col.I = 3.6361)
   minus Nonradiation rate (0.0) =      3.6361    |    minus Nonradiation rate (0.0) =      3.6361
                                                  |
3. 1980 Fractional Causation is radiation         | 3. 1980 Fractional Causation is radiation
   rate (3.6361) divided by OBSERVED              |    rate (3.6361) divided by OBSERVED
   Natl MR Part 1,Col.C=    4.6     =     0.79    |    Natl MR Part 1,Col.C=    4.6            0.79
4. Comparable est. = 0.81 from Table 63-EE.       |
----------------------------------------------------------------------------------
```

```
===============================================================================================
                                    Table M-15
          Ischemic Heart Disease, Males, 1993:  Alternative Smoking Adju and Fractional Causation
===============================================================================================
```

This table is like Table 64-E, except for the Smoking Adjustment. Here in Part 1, we do not multiply the 1940
MidTrio and LowTrio MortRates by a factor; we adjust them by the DIFFERENCE (-170.8 cases) which we obtain from
Chap.64, Box 1, Col.K. Of course, Col.E below keeps the PP Adju (see Tab 47-B & Chap.64, Box 2, Part 2, Col.D).

Part 1. Calculation of the Alternative Smoking Adjustment (Col.G) and the National Adjusted MortRate (Col.I).

Trio-Sequence	Col.A 1990 PopFrac Tab 3-B	Col.B 1993 Obs MR Tab 40-A	Col.C A * B	Col.D 1950 MR Mid,Low Tab 40-A	Col.E= PP Adju Factor Tab 47-B	Col.F= D * E	Col.G= Col.F -170.8 (TopTrio, Bx1,ColK)	Col.H= Adju MortRate	Col.I = A * G
Pacific	0.1535	112.4	17.253					112.4	17.253
New England	0.0527	117.8	6.208					117.8	6.208
Mid-Atlantic	0.1527	147.9	22.584					147.9	22.584
WestNoCentral	0.0721	129.9	9.366	228.4	0.94	214.696	43.90	43.896	3.165
EastNoCentral	0.1713	140.5	24.068	258.9	0.94	243.366	72.57	72.566	12.431
Mountain	0.0543	101.2	5.495	214.8	0.94	201.912	31.11	31.112	1.689
WestSoCentral	0.1087	137.6	14.957	206.1	1.07	220.527	49.73	49.727	5.405
EastSoCentral	0.0621	145.8	9.054	176.8	1.07	189.176	18.38	18.376	1.141
SouthAtlantic	0.1725	128.7	22.201	222.0	1.07	237.540	66.74	66.740	11.513
		Sum =	131.2						Sum
1993 Observed Natl MR from Table 40-B =			131.0			1993 Natl Adjusted MR =			81.3898

Part 2. --

Trio-Seq.	Col.A Mean1940 thru1990 PPs from Tab 47-A x'	Col.B 1993 Adju MRs from Col.H Part 1	Col.C IHD, Males: 1993 Adjusted MortRates regressed on Mean 1940 thru 1990 PPs		Col.D 1940 PPs from Table 3-A (TrioSeq) x''	Col.E IHD, Males: 1993 Adjusted MortRates regressed on 1940 PhysPops	
			Regression Output:			Regression Output:	
Pac	191.97	112.4	Constant	-112.3211	159.72	Constant	-97.0234
NewEng	208.20	117.8	Std Err of Y Est	16.7476	161.55	Std Err of Y Est	20.3378
MidAtl	204.72	147.9	R Squared	0.8720	169.76	R Squared	0.8112
WNoCen	141.14	43.896	No. of Observation	9	123.14	No. of Observation	9
ENoCen	146.19	72.566	Degrees of Freedom	7	133.36	Degrees of Freedom	7
Mtn	145.91	31.112			119.89		
WSoCen	126.28	49.727	X Coefficient(s)	1.1765	103.94	X Coefficient(s)	1.3245
ESoCen	113.28	18.376	Std Err of Coef.	0.1704	85.83	Std Err of Coef.	0.2415
SoAtl	142.93	66.740	XCoef / S.E. =	6.9045	100.74	XCoef / S.E. =	5.4839

Part 3-A.
Calculation of Fractional Causation
from Averaged PhysPops

1. Nonradiation rate is Adjusted
    Constant (Part 2, Col.C) = NEG. =   0.0000

2. Radiation rate is Natl Adjusted
    MortRate (Part 1, Col.I = 81.3898)
    minus Nonradiation rate (0.0) =   81.3898

3. 1993 Fractional Causation is radiation
    rate (81.3898) divided by OBSERVED
    Natl MR Part 1,Col.C=   131.0   =   0.62
4. Comparable est. = 0.63 from Table 64-E.

Part 3-B.
Calculation of Fractional Causation
from 1940 PhysPops

1. Nonradiation rate is Adjusted
    Constant (Part 2, Col.E) = NEG. =   0.0

2. Radiation rate is Natl Adjusted
    MortRate (Part 1, Col.I = 81.3898)
    minus Nonradiation rate (0.0) =   81.3898

3. 1993 Fractional Causation is radiation
    rate (81.3898) divided by OBSERVED
    Natl MR Part 1, Col.C=   131.0   0.62

```
=====================================================================================================
                                          Table M-16
            Ischemic Heart Disease, Females, 1993:  Alternative Smoking Adju and Fractional Causation
=====================================================================================================
```

This table is like Table 65-E, except for the Smoking Adjustment. Here in Part 1, we do not multiply the 1940
MidTrio and LowTrio MortRates by a factor; we adjust them by the DIFFERENCE (-85.1 cases) which we obtain from
Chap.65, Box 1, Col.K. Of course, Col.E below keeps the PP Adju (see Tab 47-B & Chap.65, Box 2, Part 2, Col.D).

Part 1. Calculation of the Alternative Smoking Adjustment (Col.G) and the National Adjusted MortRate (Col.I).

Trio-Sequence	Col.A 1990 PopFrac Tab 3-B	Col.B 1993 Obs MR Tab 41-A	Col.C A * B	Col.D 1950 MR Mid,Low Tab 41-A	Col.E= PP Adju Factor Tab 47-B	Col.F= D * E	Col.G= Col.F -85.1 (TopTrio, Bx1,ColK)	Col.H= Adju MortRate	Col.I = A * G
Pacific	0.1535	57.7	8.857					57.7	8.857
New England	0.0527	55.7	2.935					55.7	2.935
Mid-Atlantic	0.1527	78.8	12.033					78.8	12.033
WestNoCentral	0.0721	58.3	4.203	104.1	0.94	97.854	12.754	12.754	0.920
EastNoCentral	0.1713	70.2	12.025	124.2	0.94	116.748	31.648	31.648	5.421
Mountain	0.0543	46.3	2.514	96.2	0.94	90.428	5.328	5.328	0.289
WestSoCentral	0.1087	66.5	7.229	94.0	1.07	100.580	15.480	15.480	1.683
EastSoCentral	0.0621	67.7	4.204	84.7	1.07	90.629	5.529	5.529	0.343
SouthAtlantic	0.1725	61.6	10.626	103.4	1.07	110.638	25.538	25.538	4.405
		Sum =	64.6						Sum =
1993 Observed Natl MR from Table 41-B =			64.7				1993 Natl Adjusted MR =		36.8866

Part 2. --

Trio-Seq.	Col.A Mean1940 thru1990 PPs from Tab 47-A x'	Col.B 1993 Adju MRs from Col.H Part 1	Col.C IHD, Females: 1993 Adjusted MortRates regressed on Mean 1940 thru 1990 PPs Regression Output:		Col.D 1940 PPs from Table 3-A (TrioSeq) x''	Col.E IHD, Females: 1993 Adjusted MortRates regressed on 1940 PhysPops Regression Output:	
Pac	191.97	57.7	Constant	-77.5231	159.72	Constant	-68.8802
NewEng	208.20	55.7	Std Err of Y Est	11.0499	161.55	Std Err of Y Est	12.7956
MidAtl	204.72	78.8	R Squared	0.8449	169.76	R Squared	0.7920
WNoCen	141.14	12.754	No. of Observation	9	123.14	No. of Observation	9
ENoCen	146.19	31.648	Degrees of Freedom	7	133.36	Degrees of Freedom	7
Mtn	145.91	5.328			119.89		
WSoCen	126.28	15.480	X Coefficient(s)	0.6942	103.94	X Coefficient(s)	0.7845
ESoCen	113.28	5.529	Std Err of Coef.	0.1124	85.83	Std Err of Coef.	0.1520
SoAtl	142.93	25.538	XCoef / S.E. =	6.1745	100.74	XCoef / S.E. =	5.1625

Part 3-A. | Part 3-B.
Calculation of Fractional Causation | Calculation of Fractional Causation
from Averaged PhysPops | from 1940 PhysPops
                                                          |
1. Nonradiation rate is Adjusted | 1. Nonradiation rate is Adjusted
    Constant (Part 2, Col.C) = NEG. =   0.0 | Constant (Part 2, Col.E) = NEG. =   0.0
                                                          |
2. Radiation rate is Natl Adjusted | 2. Radiation rate is Natl Adjusted
    MortRate (Part 1, Col.I = 36.8866) | MortRate (Part 1, Col.I = 36.8866)
    minus Nonradiation rate (0.0) =   36.8866 | minus Nonradiation rate (0.0) =   36.8866
                                                          |
3. 1993 Fractional Causation is radiation | 3. 1993 Fractional Causation is radiation
    rate (36.8866) divided by OBSERVED | rate (36.8866) divided by OBSERVED
    Natl MR Part 1,Col.C=   64.7   =   0.57 | Natl MR Part 1, Col.C=   64.7   0.57
4. Comparable est. = 0.78 from Table 65-E. |

Various findings in this work stand alone, without any interpretation, as irrefutable. We will add one new set of such findings here.

If medical radiation is an important cause of both Cancer and Ischemic Heart Disease (IHD), then we "predict" that the age-adjusted MortRates for the two diseases should show a PERSISTENT positive correlation with each other over time, by Census Divisions --- and should simultaneously show a distinctly different relationship with MortRates for NonCancer NonIHD causes of death, which are NOT inducible by ionizing radiation.

We start with the knowledge that cigarette smoking is ALSO an important cause of both Cancer and IHD. We return to that in Part 2.

● Part 1. The Findings from Both Smoking-Adjusted and from "Raw" MortRates

Our expectation is very well met, as shown by the nearby Tabulations 1,2,3,4. The regressions were done in the manner shown hundreds of times in this book, using the MortRates from our prior tables. One regression is shown below as a sample, to emphasize that PhysPop values do not participate in any of these regressions. m = male, and f = female.

Census Div. Trio- Sequence	x= 1988 AllCa-m Tab 49-F Col.F	y= 1993 IHD-m Tab 64-E Col.F	1993 IHD MortRates, Male, regressed on 1988 All-Cancer MortRates, Male Regression Output:	
Pacific	148.5	112.4	Constant	17.5269
New England	167.1	117.8	Std Err of Y Est	9.7711
Mid-Atlantic	168.4	147.9	R Squared	0.7961
WestNoCentral	122.0	91.4	No. of Observations	9
EastNoCentral	131.6	103.6	Degrees of Freedom	7
Mountain	109.8	85.9		
WestSoCentral	109.5	94.8	X Coefficient(s)	0.6710
EastSoCentral	92.7	81.3	Std Err of Coef.	0.1284
SouthAtlantic	112.0	102.1	Ratio, Xcoef/SE =	5.2272

A positive correlation between Cancer MortRates and IHD MortRates, by Census Divisions, implies no causation of one disease by the other, of course. It implies that the MortRates of the two diseases each have an independent and positive correlation with a cause which they share in common.

In all of the tabulations, a negative sign on the ratio (X-Coefficient / Std. Error) reflects a negative X-coefficient --- an inverse relationship between the two sets of MortRates. The abbreviation "NonNon" refers to "NonCancer NonIHD." The "NonNon" MortRates come from Table 25-A (where 1980 is the most recent entry). The "year 1990" means, as usual in this monograph, 1988 for Cancer MortRates and 1993 for IHD MortRates (Chapter 4, Part 2b).

Tabulations 1+2: The Smoking Adjusted MortRates for All-Cancers come from Chapters 49,50, for Difference-Cancers from Chapters 53,54, and for Ischemic Heart Disease from Chapters 64,65. We start with 1960, because 1960 is the earliest year for which we have Smoking Adjusted MortRates for IHD. The Smoking Adjusted MortRates mean that smoking-level does NOT vary across the Census Divisions --- but PhysPop-level DOES vary across the Census Divisions. The results are shown in Tabulations 1+2.

Tabulations 3+4: Because we can not rule out a smoking effect on NonCancer NonIHD MortRates, and because we have no Smoking Adjusted MortRates for NonCancer NonIHD causes of death, we also did the entire test with the "raw" MortRates for All-Cancers (from Chapters 6,7), Difference-Cancers (from Chapters 18,19), and IHD (from Chapters 40,41) --- starting with 1950. The results are shown in Tabulations 3+4.

Tabulations 1 through 4

● **Tabulation 1, with All-Cancers. MortRate-MortRate Correlations, by Census Divisions.**

SmoAdju (Ca,IHD)		Year	R-sq	Ratio		Year	R-sq	Ratio
x	y							
AllCa-m	IHD-m	1960	0.93	+9.98		1990	0.80	+5.23
AllCa-f	IHD-f	1960	0.91	+8.61		1990	0.79	+5.16
NonNon-m	IHD-m	1960	0.65	-3.62		1980	0.45	-2.38
NonNon-f	IHD-f	1960	0.35	-1.94		1980	0.32	-1.82
NonNon-m	AllCa-m	1960	0.84	-5.98		1980	0.71	-4.16
NonNon-f	AllCa-f	1960	0.49	-2.61		1980	0.48	-2.52

● **Tabulation 2, with Diff-Cancers. MortRate-MortRate Correlations, by Census Divisions.**

SmoAdju (Ca,IHD)		Year	R-sq	Ratio		Year	R-sq	Ratio
x	y							
Diff-m	IHD-m	1960	0.92	+8.78		1990	0.76	+4.76
Diff-f	IHD-f	1960	0.93	+9.67		1990	0.83	+5.90
NonNon-m	IHD-m	1960	0.65	-3.62		1980	0.45	-2.38
NonNon-f	IHD-f	1960	0.35	-1.94		1980	0.32	-1.82
NonNon-m	Diff-m	1960	0.85	-6.37		1980	0.75	-4.59
NonNon-f	Diff-f	1960	0.46	-2.44		1980	0.49	-2.57

● **Tabulation 3, with All-Cancers. MortRate-MortRate Correlations, by Census Divisions.**

"Raw" MortRates		Year	R-sq	Ratio		Year	R-sq	Ratio
x	y							
AllCa-m	IHD-m	1950	0.86	+6.63		1990	0.66	+3.72
AllCa-f	IHD-f	1950	0.89	+7.60		1990	0.57	+3.06
NonNon-m	IHD-m	1950	0.71	-4.13		1980	0.06	-0.67
NonNon-f	IHD-f	1950	0.37	-2.03		1980	0.06	-0.68
NonNon-m	AllCa-m	1950	0.66	-3.72		1980	0.01	+0.32
NonNon-f	AllCa-f	1950	0.23	-1.47		1980	0.25	-1.55

● **Tabulation 4, with Diff-Cancers. MortRate-MortRate Correlations, by Census Divisions.**

"Raw" MortRates		Year	R-sq	Ratio		Year	R-sq	Ratio
x	y							
Diff-m	IHD-m	1950	0.81	+5.54		1990	0.53	+2.81
Diff-f	IHD-f	1950	0.90	+7.79		1990	0.59	+3.16
NonNon-m	IHD-m	1950	0.71	-4.13		1980	0.06	-0.67
NonNon-f	IHD-f	1950	0.37	-2.03		1980	0.06	-0.68
NonNon-m	Diff-m	1950	0.70	-4.04		1980	0.12	-0.97
NonNon-f	Diff-f	1950	0.25	-1.52		1980	0.34	-1.91

Text continues on page 644 --->'

In all four tabulations, the MortRates for Cancer and for IHD show a persistent, significant, positive correlation with each OTHER, by Census Divisions, over time --- and simultaneously show a distinctly DIFFERENT relationship with NonCancer NonIHD causes of death, which are NOT inducible by ionizing radiation.

● **Part 2. What Is a Reasonable Interpretation of These Additional Observations?**

The findings in Part 1 are free from interpretation. Tabulations 3 and 4 do not even use smoking-adjusted MortRates. These findings are facts which "demand" an explanation. The Law of Minimum Hypotheses says that the explanation is staring at us: Medical radiation is an important cause not only of Cancer, but also of Ischemic Heart Disease.

Although cigarette smoking is ALSO a cause shared by Cancer and IHD, smoking is not an adequate explanation for these correlations. If we seek the entries (in the tabulations) which would reflect the LEAST possible impact from cigarette smoking, we must look at Tabulations 2 and 4, because Difference-Cancers exclude Respiratory-System Cancers. And in Tabulations 2 and 4, we must look at the earliest entries for FEMALES, because (a) the percentage of female-smokers used to be far lower than for males, and the smoking-intensity by females used to be much lower than for males (Chapter 48, Part 3), (b) the cancer-impact of smoking appears to be far lower upon females than upon males (Chapter 48, Box 2), and (c) females did not used to smoke cigars or chew tobacco.

For females in 1950, Tabulation 4 shows the correlation between female MortRates from Difference-Cancers and female MortRates from Ischemic Heart Disease, by Census Divisions. The R-squared value is 0.90 (with +7.79 as the ratio of X-Coefficient over its Standard Error) --- a very high positive correlation, indeed.

WHY? What is the explanation? Such things do not happen by accident. Such evidence points to medical radiation, rather than smoking, as the correct explanation. And if medical radiation is the correct explanation for the 1950 female correlation (1950 MortRates for Diff-Ca with 1950 MortRates for IHD), it would be irrational to assume --- in the absence of evidence --- that something ELSE explains the later correlations and the male correlations.

>>>>>>>>>>

● As promised in Chapter 2, dots in the left margin identify sources with extensive bibliographies which reflect the vast body of evidence establishing that ionizing radiation is a cause of almost all kinds of human Cancer. Medical xrays are the source of much of the evidence. [Dotted entries: BEIR, Gofman, ICRP, NAS, NRPB, UNSCEAR.]

● When the same last name appears several times, entries are sequenced from the earliest year to the most recent, regardless of the author's first name.

Abbott 1984. R.D. Abbott + R. Carroll, "Interpreting Multiple Logistic Regression Coefficients in Prospective Observational Studies," AMERICAN JOURNAL OF EPIDEMIOLOGY Vol.119: 830+. 1984.

Abelson 1994. Philip H. Abelson, "Risk Assessments of Low-Level Exposures," (editorial), SCIENCE Vol.265: 1507. September 9, 1994.

ACS (general): The American Cancer Society is a non-profit organization (USA) whose headquarters are at 1599 Clifton Road, N.E., Atlanta, Georgia 30329. Website: < www.cancer.org >. Tel: 404-320-3333.

ACS 1992. American Cancer Society, "Guidelines for the Wise Use of Medical Xrays." Item Number 2900 on the ACS "Cancer Response System" (Telephone: 1-800-ACS-2345). "Date reviewed" by ACS: January 10, 1992.

ACS-CA (general). This abbreviates the American Cancer Society's peer-reviewed publication CA - A CANCER JOURNAL FOR CLINICIANS. From Jan. 1996, text is available online: www.ca-journal.org

ACS-CA 1992. Catherine C. Boring + Teresa S. Squires + Tony Tong, "Cancer Statistics, 1992," CA - A CANCER JOURNAL FOR CLINICIANS Vol.42, No.1: 19-43. Jan/Feb 1992.

ACS-CA 1997. Sheryl L. Parker + Tony Tong + Sherry Bolden + Phyllis A. Wingo, "Cancer Statistics, 1997," CA - A CANCER JOURNAL FOR CLINICIANS Vol.47, No.1: 5-27. Jan/Feb 1997.

Adam 1987. E. Adam + J.L. Melnick et al, "High Levels of Cytomegalovirus Antibody in Patients Requiring Vascular Surgery for Atherosclerosis," LANCET: 291-293. 1987.

Adler 1986. Yolanda T. Adler, book review of Xrays: Health Effects of Common Exams, by John W. Gofman (1985), in JOURNAL OF THE AMERICAN MEDICAL ASSOCIATION Vol.256, No.21. Dec. 5, 1986.

AHA (general): The American Heart Association is a non-profit organization (USA) located at 7272 Greenville Ave., Dallas, TX 75231. Website: < www.americanheart.org >. Among its publications is CIRCULATION.

AHA 1995. American Heart Association, HEART AND STROKE FACTS: 1996 STATISTICAL SUPPLEMENT. 23 pages. This publication uses 1993 provisional and 1992 final mortality statistics. Copyright 1995.

AHA 1996. American Heart Association, 1997 HEART AND STROKE STATISTICAL UPDATE. 29 pages. This publication uses 1994 provisional and 1993 final mortality statistics. Copyright 1996.

Ahmed 1990. Aftab J. Ahmed + Bert W. O'Malley + Frank M. Yatsu, "Presence of a Putative Transforming Gene in Human Atherosclerotic Plaques," (abstract only), ARTERIOSCLEROSIS Vol.10: 755a. 1990.

Ahrens 1955. E.H. Ahrens + T.T. Tsaltas + J. Hirsch + W. Insull, Jr., "Effect of Dietary Fats on the Serum Lipids of Human Subjects," JOURNAL OF CLINICAL INVESTIGATION Vol.34: 918+. 1955.

AICR 1997. American Institute for Cancer Research, "The Cancer Process" 1986. Provides the estimate that ~ 35% of Cancer is due to unfavorable diet. Oct. 1997 and April 1999, AICR confirmed this as the current estimate. AICR is co-sponsor of a 660-page report, issued Sept. 1997, entitled FOOD, NUTRITION, and the PREVENTION OF CANCER: A GLOBAL PERSPECTIVE, prepared by a panel of 16 scholars, chaired by Prof. John D. Potter, Fred Hutchinson Cancer Center in Seattle. (AICR, 1759 "R" Street, NW, Washington DC 20009. Tel: 1-800-843-8114. Website: www.aicr.org).

Albert 1977. R.E. Albert et al, "Effect of Carcinogens on Chicken Atherosclerosis," CANCER RESEARCH Vol.37: 2232-2235. 1977.

Alfthan 1994. G. Alfthan + J. Pekkanen + M. Jauhiainen et al, "Relation of Serum Homocysteine and Lipoprotein(a) Concentrations to Atherosclerotic Disease in a Prospective Finnish Population Based Study," ATHEROSCLEROSIS Vol.106: 9-19. 1994.

AMA (general): The American Medical Association is a non-profit organization of physicians (USA), with its headquarters at 515 North State St., Chicago IL 60610. Tel: 312-464-5000. Website: www.ama-assn.org/ Among AMA's publications is JAMA, the Journal of the American Medical Assn.

AMA 1950. American Medical Assn., AMERICAN MEDICAL DIRECTORY, 1950: A REGISTER OF PHYSICIANS OF THE UNITED STATES ... 18th Edition. Edited by Frank V. Cargill. 2,912 pages. 1950.

AMA 1965. American Medical Assn., DISTRIBUTION OF PHYSICIANS IN THE U.S. by State, Region, District, and County / 1965. Prepared by Christ N. Theodore, Gerald E. Sutter, Grant M. Osborn, Raymond L. White. PhysPop data are "as of April 6, 1964." UCSF Library: RA410.7, D614, 1965.

AMA 1982.　American Medical Assn., PHYSICIAN CHARACTERISTICS AND DISTRIBUTION IN THE U.S., 1982 Edition.　1982.

AMA 1986.　American Medical Assn., PHYSICIAN CHARACTERISTICS AND DISTRIBUTION IN THE U.S., 1986 Edition.　1986.

AMA 1990.　Gene Roback + Lillian Randolph + Bradley Seidman, PHYSICIAN CHARACTERISTICS & DISTRIBU- TION in the U.S., 1990 Edition.　278 pages.　ISBN 0-89970-403-4.　American Medical Assn., Physician Data Services, Division of Survey and Data Resources.　1990.　UCSF Library: RA410.7, D614, 1990.

AMA 1993.　Gene Roback + Lillian Randolph + Bradley Seidman, PHYSICIAN CHARACTERISTICS AND DISTRIBUTION IN THE U.S., 1993 Edition.　284 pages.　ISBN 0-89970-492-1.　American Medical Assn., Dept. of Physician Data Services, Division of Survey and Data Resources.　1993.　UCSF Library: RA410.7, D614, 1993.

AMA 1994.　Gene Roback + Lillian Randolph + Bradley Seidman + Thomas Pasko, PHYSICIAN CHARACTERISTICS AND DISTRIBUTION IN THE U.S., 1994 Edition.　Data up to Jan. 1, 1993.　American Medical Assn., Dept. of Data Services, Division of Survey and Data Resources.　1994.　UCSF Library: RA410.7, D614, 1994.

Ambrose 1985.　J.A. Ambrose et al, "Coronary Angiographic Morphology in Myocardial Infarction: A Link between the Pathogenesis of Unstable Angina and Myocardial Infarction," JOURNAL OF THE AMERICAN COLLEGE OF CARDIOLOGY Vol.6: 1233-1238.　1985.

Ambrose 1988.　J.A. Ambrose + 8 co-workers (including V. Fuster), "Angiographic Progression of Coronary Artery Disease and the Development of Myocardial Infarction," JOURNAL OF THE AMERICAN COLLEGE OF CARDIOLOGY Vol.12: 56-62.　1988.

American Cancer Society.　Please see ACS (general), above.

American Heart Association.　Please see AHA (general), above.

Ames 1989.　Bruce N. Ames, "Endogenous DNA Damage as Related to Cancer and Aging," MUTATION RESEARCH Vol.214: 41-46.　1989.

Ames 1995.　Bruce N. Ames, "The Causes and Prevention of Cancer," PROCEEDINGS OF THE NATIONAL ACADEMY OF SCIENCES (USA) Vol.92: 5258-5265.　1995.

Amromin 1964.　G.D. Amromin + H.L. Gildenhorn + R.D. Solomon + B.B. Nadkrni + M.L. Jacobs, "The Synergism of X-Irradiation and Cholesterol-Fat Feeding on the Development of Coronary Artery Lesions," J. ATHEROSCLER. RES. Vol.4: 325-334.　1964.

Anderson 1985.　K. Anderson + O. Mattsson, "Critical Analysis of Dose Reduction Trends with Special Reference to Procedures Involved in Fluoroscopy," BRITISH J. OF RADIOLOGY Vol.18 (Suppl.): 46-49.　1985.

Anderson 1989.　Robert E. Anderson + Rolla B. Hill + C.R. Key, "The Sensitivity and Specificity of Clinical Diagnostics during Five Decades: Toward an Understanding of Necessary Fallibility," JOURNAL of the AMERICAN MEDICAL ASSN. Vol.261: 1610-1617.　1989.

Anderson 1990.　Robert E. Anderson + Rolla B. Hill + F. Gorstein, "A Model for the Autopsy-Based Quality Assessment of Medical Diagnostics," HUMAN PATHOLOGY Vol.21: 174-181.　1990.

Anderson 1995.　T.J. Anderson + I.T. Meredith + A.C. Yeung + B. Frei + Andrew P. Selwyn + Peter Ganz, "The Effect of Cholesterol-Lowering and Anti-Oxidant Therapy on Endothelium-Dependent Coronary Vasomotion," NEW ENGLAND JOURNAL OF MEDICINE Vol.332: 488-493.　1995.

Anitschkow 1933.　N. Anitschkow, "Experimental Arteriosclerosis in Animals," in ARTERIOSCLEROSIS, edited by E.V. Cowdry.　(McMillan Company, New York.)　1933.

Antonis 1961.　A. Antonis + I. Bersohn, "The Influence of Diet on Serum-Triglycerides in South African White and Bantu Prisoners," LANCET 1: 3-9.　1961.

Apple 1994.　Raymond J. Apple + 5 co-workers, "HLA DR-DQ Associations with Cervical Carcinoma Show Papilloma Virus-type Specificity," NATURE GENETICS Vol.6: 157-162.　February 1994.

Apple 1995.　Raymond J. Apple + Thomas M. Becker + Cosette M. Wheeler + Henry A. Ehrlich, "Comparison of Human Leukocyte Antigen DR-DQ Disease Associations Found with Cervical Dysplasia and Invasive Cervical Carcinoma," JOURNAL OF THE NATL CANCER INSTITUTE Vol.87, No.6: 427-436.　March 15, 1995.

Arbeit 1996.　Jeffrey M. Arbeit + Peter M. Howley + Douglas Hanahan, "Chronic Estrogen-Induced Cervical and Vaginal Squamous Carcinogenesis in Human PapillomaVirus Type-16 Transgenic Mice," PROCEEDINGS of the NATL. ACADEMY OF SCIENCES USA 93: 2930-2935.　April 1996.　Online journal: www.pnas.org

Armstrong 1989.　M.L. Armstrong + M.B. Megan + D.D. Heistad, "Adaptive Responses of the Artery Wall as Human Atherosclerosis Develops," pp.469-480 in PATHOBIOLOGY OF THE HUMAN ATHEROSCLEROTIC PLAQUE, edited by S. Glagov + W.P. Newman + S.A. Schaffer.　(New York: Springer-Verlag.)　1989.

Arnesen 1995.　E. Arnesen + H. Refsum et al, "Serum Total Homocysteine and Coronary Artery Disease," INTERNATIONAL JOURNAL OF EPIDEMIOLOGY Vol.24: 704-709.　1995.

Artom 1965. C. Artom + H.B. Loftand Jr. + T.B. Clarkson, "Ionizing Radiation Atherosclerosis and Lipid Metabolism in Pigeons," RADIATION RESEARCH Vol.26: 165–177. 1965.

Avins 1989. Andrew L. Avins + R.J. Haber + S.B. Hulley, "The Status of Hypertriglyceridemia as a Risk Factor for Coronary Heart Disease," CLIN. LAB. MED. Vol.9: 153–168. 1989.

Avins 1997. Andrew L. Avins, "The Management of High Blood Triglycerides," pp.155-168 in syllabus of 25TH ANNUAL ADVANCES IN INTERNAL MEDICINE, presented by the Dept. of Medicine, University of California San Francisco School of Medicine. (Prof. Avins, UCSF Box 0656, San Francisco CA 94143-0656). June 23, 1997.

Baermann 1904. G. Baermann + P. Linser, "Uber die Lokale und Allgemeine Wirkung der Rontgenstrahlen," MUNCHEN MED. WOCHENSCHR. Vol.51: 996–999. 1904.

Bailar 1997. John C. Bailar III + Heather L. Gornik, "Cancer Undefeated," (Special Article), NEW ENGLAND JOURNAL OF MEDICINE Vol.336, No.22: 1569-1574. May 29, 1997.

Baverstock 1981. Keith F. Baverstock et al, "Risk of Radiation at Low Dose Rates," LANCET 1: 430–433. February 21, 1981.

Baverstock 1983. Keith F. Baverstock, "A Note on Radium Body Content and Breast Cancers in U.K. Radium Luminizers," HEALTH PHYSICS Vol.44, No.1 Suppl.: 575-577, 1983.

Baverstock 1987. Keith F. Baverstock, "The U.K. Luminizer Survey," BRITISH JOURNAL OF RADIOLOGY Supplemental BIR Report 21, pp.71-76. BIR = Brit. Inst. of Radiology. 1987.

Baverstock 1991. Response to Billen 1990. Keith F. Baverstock, "Comments on the Commentary by D. Billen," (letter) RADIATION RESEARCH 126: 383-384. 1991.

Baverstock 1992. Keith Baverstock + 4 co-workers, "Thyroid Cancer after Chernobyl," (correspondence) NATURE Vol.359: 21-22. September 3, 1992.

● BEIR 1972. Committee on the Biological Effects of Ionizing Radiation, THE EFFECTS ON POPULATIONS OF EXPOSURE TO LOW LEVELS OF IONIZING RADIATION (also known as the BEIR-1 Report). 217 pages. (National Academy of Sciences, National Research Council, Washington DC 20418.) November 1972.

● BEIR 1980. Committee on the Biological Effects of Ionizing Radiation, THE EFFECTS ON POPULATIONS OF EXPOSURE TO LOW LEVELS OF IONIZING RADIATION (also known as the BEIR-3 Report). (National Academy of Sciences, National Research Council, Washington DC 20418.) 1980.

● BEIR 1990. Committee on the Biological Effects of Ionizing Radiation, HEALTH EFFECTS OF EXPOSURE TO LOW LEVELS OF IONIZING RADIATION (also known as the BEIR-5 Report). National Research Council. Contract from the federal government's Office of Science and Technology. 421 pages. ISBN 0-309-03995-9. (National Academy Press, Washington DC 20418.) 1990.

● BEIR 1999. Committee on the Biological Effects of Ionizing Radiation, HEALTH EFFECTS OF EXPOSURE TO RADON (also known as the BEIR-6 Report). Committee on Health Risks of Exposure to Radon. National Research Council. Prepared under a grant from the Environmental Protection Agency. 500 pages. ISBN 0-309-05645-4. LCCN 98-25503. (National Academy Press, Washington DC 20418.) 1999.

Bell 1974-a. F.P. Bell + I.L. Adamson + C.J. Schwartz, "Aortic Endothelial Permeability to Albumin: Focal and Regional Patterns of Uptake and Transmural Distribution of I-131-Albumin in the Young Pig," EXP. MOL. PATHOL. Vol.20: 57-68. 1974.

Bell 1974-b. F.P. Bell + A.S. Gallus + C.J. Schwartz, " Focal and Regional Patterns of Uptake and the Transmural Distribution of I-131-Fibrinogen in the Pig Aorta in Vivo," EXP. MOL. PATHOL. Vol.20: 281-292. 1974.

Benditt 1973. Earl P. Benditt + John M. Benditt, "Evidence for a Monoclonal Origin of Human Atherosclerotic Plaques," PROCEEDINGS OF THE NATIONAL ACADEMY OF SCIENCES Vol.70, No.6: 1753-1756. June 1973.

Benditt 1974. Earl P. Benditt, "Evidence for a Monoclonal Origin of Human Atherosclerotic Plaques and Some Implications," CIRCULATION Vol.50: 650-652. 1974.

Benditt 1976. Earl P. Benditt, "Implications of the Monoclonal Character of Human Athersclerotic Plaques," ANNALS OF THE NEW YORK ACADEMY OF SCIENCES Vol.275: 96-100. 1976.

Benditt 1977. Earl P. Benditt, "The Origin of Atherosclerosis," SCIENTIFIC AMERICAN Vol.236: 74-85. 1977.

Benditt 1988. Earl P. Benditt, "Origins of Human Atherosclerotic Plaques," ARCHIVES OF PATHOLOGY AND LABORATORY MEDICINE Vol.112: 997-1001. October 1988.

Bennett 1956. H.S. Bennett, in J. BIOPHYS. BIOCHEM. CYTOL. Vol.2 (Suppl.): 99-103. 1956.

Beutler 1962. E. Beutler + M. Yeh + V.F. Fairbanks, "The Normal Human Female as a Mosaic of X–Chromosome Activity ...," PROCEEDINGS OF THE NATIONAL ACADEMY OF SCIENCES (USA) Vol.48: 9-16. 1962.

Bierman 1976. E.L. Bierman + J.J. Albers, in ANNALS OF THE NEW YORK ACADEMY OF SCIENCES Vol.275. 1976

Bierman 1992. Edwin L. Bierman, "Atherogenesis in Diabetes," (George Lyman Duff Memorial Lecture), ARTERIOSCLEROSIS AND THROMBOSIS Vol.12, No.6: 647-656. June 1992.

Bihari-Varga 1967. M. Bihari-Varga + M. Vegh, "Quantitative Studies on the Complexes Formed between Aortic
          Mucopolysaccharides and Serum Lipoproteins," BIOCHIM. BIOPHYS. ACTA Vol.144: 202-210. 1967.

Billen 1990. Daniel Billen, "Spontaneous DNA Damage and Its Significance for the 'Negligible Dose' Controversy in Radiation
          Protection," (Commentary), RADIATION RESEARCH Vol.124: 242-245. 1990. See also Baverstock 1991,
          Ward 1991.

Billen 1991. Daniel Billen "Response to Comments of K.F. Baverstock and J.F. Ward," (letter), RADIATION
          RESEARCH Vol.126: 388-389.

Bishop 1922. Louis Faugeres Bishop, "Fluoroscope in the Diagnosis of Diseases of the Heart," MEDICAL RECORD
          Vol.101, No.12: 489-491. March 25, 1922.

Bittl 1996. John A. Bittl, "Advances in Coronary Angioplasty," (review article), NEW ENGLAND JOURNAL OF MEDICINE
          Vol.335, No.17: 1290-1302. October 24, 1996.

Bjorkerud 1971. S. Bjorkerud + G. Bondjers, "Arterial Repair and Atherosclerosis after Mechanical Injury ..."
          ATHEROSCLEROSIS Vol.13: 355-363. 1971.

Black 1975. A. Black + M.M. Black + G. Gensini, "Exertion and Acute Coronary Injury," ANGIOLOGY
          Vol.26: 759-783. 1975.

Blanchard 1967. R.L. Blanchard, "Concentrations of Pb-210 and Po-210 in Human Soft Tissues," HEALTH PHYSICS Vol.13:
          625-632. 1967.

Blankenhorn 1987. (CLAS Study.) David H. Blankenhorn + S.A. Nessim + R.L. Johnson + M.E. Sanmarco + S.P. Azen + L.
          Cashin-Hamphill, "Beneficial Effects of Colestipol Niacin Therapy on Coronary Atherosclerosis and Coronary
          Venous Bypass Grafts," JOURNAL OF THE AMERICAN MEDICAL ASSN. Vol.257: 3233-3240. 1987.

Blankenhorn 1993. (MARS Study.) David H. Blankenhorn + S.P. Azen + D.M. Kramsch + W.J. Mack + L. Cashin-Hemphill +
          Howard N. Hodis + the MARS research group, "The Monitored Atherosclerosis Regression Study (MARS):
          Coronary Angiographic Changes with Lovastatin Therapy," ANNALS OF INTERNAL MEDICINE Vol.119:
          969-976. 1993.

Blankenhorn 1994. David H. Blankenhorn + Howard N. Hodis (1992 Lyman Duff Memorial Lecture), "Arterial Imaging and
          Atherosclerosis Reversal," ARTERIOSCLEROSIS AND THROMBOSIS Vol.14, No.2: 177-192. February 1994.

Blatz 1970. Hanson W. Blatz, "Regulatory Changes for Effective Programs," in SECOND ANNUAL NATIONAL
          CONFERENCE ON RADIATION CONTROL, U.S. Dept. of Health, Education, and Welfare Report
          BRH/ORO 70-5. Please see Shapiro 1990, at page 421.

Boice 1977. John D. Boice, Jr. + Richard R. Monson, "Breast Cancer in Women after Repeated Fluoroscopic Examinations
          of the Chest," JOURNAL OF THE NATIONAL CANCER INSTITUTE Vol.59: 823-832. 1977.

Boice 1978. John D. Boice, Jr. + Marvin Rosenstein + E. Dale Trout, "Estimation of Breast Doses and Breast Cancer Risk
          Associated with Repeated Fluoroscopic Chest Examinations of Women with Tuberculosis," RADIATION
          RESEARCH Vol.73: 373-390. 1978.

Boice 1981. John D. Boice, Jr. + Richard R. Monson + Marvin Rosenstein, "Cancer Mortality in Women after Repeated
          Fluoroscopic Examinations of the Chest," JOURNAL OF THE NATIONAL CANCER INSTITUTE Vol.66:
          863-867. 1981. See also update: Boice 1991.

Boice 1991. John D. Boice, Jr. + Dale Preston + Faith G. Davis + Richard R. Monson, "Frequent Chest Xray Fluoroscopy and
          Breast Cancer Incidence among Tuberculosis Patients in Massachusetts," RADIATION RESEARCH Vol.125:
          214-222. 1991.

Bond 1978. Victor P. Bond + Charles B. Meinhold + H.H. Rossi, "Low-Dose RBE and Q for X Ray Compared to Gamma Ray
          Radiation," HEALTH PHYSICS Vol.34: 433-438. 1978.

Borek 1983. C. Borek + E.J. Hall + M. Zaider, "Xrays May Be Twice as Potent as Gamma for Malignant Transformation at Low
          Doses," NATURE Vol.301: 156-158. 1983.

Bosch 1995. F.X. Bosch et al, "Prevalence of Human PapillomaVirus in Cervical Cancer: A Worldwide Perspective,"
          JOURNAL OF THE NATIONAL CANCER INST. (USA) Vol.87: 796-802. 1995.

Boveri 1914. Theodore Boveri, THE ORIGIN OF MALIGNANT TUMORS. First issued in 1914 in the German language.
          English-language edition published in 1929 by Williams and Wilkins, Baltimore, Maryland, USA. 1914.

Boyer 1996. Herb Boyer, co-founder of Genentech, interviewed by Carl T. Hall in "Biotechnology Revolution, 20 Years
          Later: Genentech's Founders Pioneered It All," pp.B-1 + B-2, SAN FRANCISCO CHRONICLE, May 28, 1996.

Braestrup 1942. Carl. B. Braestrup, "Xray Protection in Diagnostic Radiology," RADIOLOGY Vol.38:
          207-216. 1942.

Braestrup 1969. Carl B. Braestrup, PAST AND PRESENT STATUS OF RADIATION PROTECTION. A COMPARISON.
          Report, Seminar Paper 005. (U.S. Department of Health, Education, and Welfare, Consumer Protection and
          Environmental Control Administration, Washington, DC.) 1969. Please see Shapiro 1990, at page 379.

Bragdon 1956. J.H. Bragdon + R.J. Havel + Boyle, in JOURNAL OF LABORATORY AND CLINICAL MEDICINE
          Vol.48: 36+. 1956.

Braunwald 1997. Eugene Braunwald, "Cardiovascular Medicine at the Turn of the Millennium: Triumphs, Concerns, and Opportunities," Shattuck Lecture (special article) NEW ENGLAND JOURNAL OF MEDICINE Vol.337, No.19: 1360-1369. November 6, 1997.

Brenner 1989. D.J. Brenner + H.I. Amols, "Enhanced Risk from Low-Energy Screen-Film Mammography Xrays," BRITISH JOURNAL OF RADIOLOGY Vol.62: 910-914. 1989.

Brensike 1984. (NHLBI Study.) J.F. Brensiki + 8 co-workers, "Effects of Therapy with Cholestyramine on Progression of Coronary Atherosclerosis: Results of the NHLBI Type II Coronary Intervention Study," CIRCULATION Vol.69: 313-324. 1984. NHLBI = Natl. Heart, Lung, Blood Inst.

Brewen 1973. J.G. Brewen + R.J. Preston + K.P. Jones + D.G. Gosslee, "Genetic Hazards of Ionizing Radiations: Cytogenetic Extrapolations from Mouse to Man," MUTATION RESEARCH Vol.17: 245-254. 1973.

Brosius 1981. Frank C. Brosius III + Bruce F. Waller + William C. Roberts, "Radiation Heart Disease: Analysis of 16 Young (Aged 15 to 33 Years) Necropsy Patients Who Received Over 3,500 Rads to the Heart," AMERICAN JOURNAL OF MEDICINE Vol.70: 519-530. March 1981.

Brown 1986. B. Greg Brown + 7 co-workers, "Incomplete Lysis of Thrombus in the Moderate Underlying Atherosclerotic Lesion during Intracoronary Infusion of Streptokinase for Acute Myocardial Infarction: Quantitative Angiographic Observations," CIRCULATION Vol.73: 653-661. 1986.

Brown 1989. B. Greg Brown + 6 co-workers, "Progression of Coronary Atherosclerosis in Patients with Probable Familial Hypercholesterolemia: Quantitative Arteriographic Assessment of Patients in NHLBI Type II Study," ARTERIOSCLEROSIS Vol.9 (suppl.1): 81-90. 1989. NHLBI = Natl. Heart, Lung, Blood Inst.

Brown 1990. (FATS Study.) B. Greg Brown + John J. Albers et al, "Regression of Coronary Artery Disease as a Result of Intensive Lipid-Lowering Therapy in Men with High Levels of Apolipoprotein-B, " NEW ENGLAND JOURNAL OF MEDICINE Vol.323: 1289-1298. 1990.

Brown 1993. B. Greg Brown + Xue-Qiao Zhao + Dianne E. Sacco + John J. Albers, "Lipid Lowering and Plaque Regression: New Insights into Prevention of Plaque Disruption and Clinical Events in Coronary Disease," (Clinical Progress Series), CIRCULATION Vol.87, No.6: 1781-1791. June 1993.

Buchwald 1990. (POSCH Study.) H. Buchwald + R.L. Varco + J.P. Matts + J.M. Long + L.L. Fitch + G.S. Campbell, "Effect of Partial Ileal Bypass on Mortality and Morbidity from Coronary Heart Disease in Patients with Hypercholesterolemia: Report of the Program on Surgical Control of the Hyperlipidemias (POSCH), NEW ENGLAND JOURNAL OF MEDICINE Vol.323: 946-955. 1990.

Bucky 1927. Gustav Bucky, "'Grenz' (Infra-Roentgen) Ray Therapy," AMERICAN JOURNAL OF ROENTGENOLOGY AND RADIUM THERAPY Vol.17, No.6: 645-650. 1927.

Budowski 1985. P. Budowski + M.A. Crawford, "Alpha-Linolenic Acid as a Regulator of the Metabolism of Arachidonic Acid: Dietary Implications of the Ratio, n-6:n-3 Fatty Acids," PROCEEDINGS OF THE NUTRITION SOCIETY Vol.44: 221-229. 1985.

Buja 1994. L.M. Buja + J.T. Willerston, "Role of Inflammation in Coronary Plaque Disruption," CIRCULATION Vol. 89: 503-503. 1994.

Burke 1998. A.P. Burke + G.T. Malcom + Y. Liang + J. Smialek + Renu Virmani, "Effect of Risk Factors on the Mechanism of Acute Thrombosis and Sudden Coronary Death in Women," CIRCULATION Vol.97, No.21: 2110-2116. June 2, 1998.

Burke 1999. Allen P. Burke + 5 co-workers, "Plaque Rupture and Sudden Death Related to Exertion in Men with Coronary Artery Disease," JOURNAL OF THE AMERICAN MEDICAL ASSN Vol.281, No.10: 921-926. March 10, 1999.

Burr 1989. M.L. Burr + A.M. Fehily + J.F. Gilbert et al, "Effects of Changes in Fat, Fish, and Fibre Intakes on Death and Myocardial Infarction: Diet and Reinfarction Trial (DART), LANCET Vol.334: 757-761. 1989.

Burton 1998. Elizabeth C. Burton + Dana A. Troxclair + William P. Newman III, "Autopsy Diagnoses of Malignant Neoplasms: How Often Are Clinical Diagnoses Incorrect?", JOURNAL OF THE AMERICAN MEDICAL ASSN. Vol.280: 1245-1248. 1998.

Buschke 1942. Franz Buschke + Herbert M. Parker, "Possible Hazards of Repeated Fluoroscopies in Infants," JOURNAL OF PEDIATRICS Vol.21: 524-533. October 1942.

Campbell 1988. Gordon R. Campbell + Julie H. Campbell + John A. Manderson + Sophie Horrigan + Robyn E. Rennick, "Arterial Smooth Muscle: A Multifunctional Mesenchymal Cell," ARCHIVES OF PATHOLOGY AND LABORATORY MEDICINE Vol.112: 977-986. October 1988.

Carlson 1979. L.A. Carlson + L.E. Bottiger + P-E. Ahlfeldt, "Risk Factors for Myocardial Infarction in the Stockholm Perspective Study," ACTA MEDICA SCANDINAVIA 206: 351-360, 1979.

Caro 1971. C. Caro + J. Fitz-Gerald + R. Schroter, "Atheroma and Arterial Wall Shear: Observation, Correlation and Proposal of a Shear-Dependent Mass Transfer Mechanism for Atherogenesis," PROCEEDINGS OF THE ROYAL SOCIETY (London), B177, 109-159. 1971.

Cashin-Hemphill 1990. (CLAS II Study.) L. Cashin-Hemphill + W.J. Mack + M.J. Pogoda + M.E. Sanmarco + S.P. Azen + David H. Blankenhorn, "Beneficial Effects of Colestipol-Niacin on Coronary Atherosclerosis," JOURNAL OF THE AMERICAN MEDICAL ASSN. Vol.264: 3013-3017. 1990.

Castelli 1986. William P. Castelli, "The Triglyceride Issue: A View from Framingham," AMERICAN HEART JOURNAL Vol.112, No.2: 432–437. August 1986.

Cathcart 1985. M.K. Cathcart + D.W. Morel + G.M. Chisholm, in J. LEUK. BIOL. Vol.38: 341–350. 1985.

Celermajer 1998. David S. Celermajer, "NonInvasive Detection of Atherosclerosis," (Editorial), NEW ENGLAND JOURNAL OF MEDICINE Vol.339, No.27: 2014–2015. Dec. 31, 1998.

Census Bureau 1951. Bureau of the Census, U.S. Dept. of Commerce. "Intercensal Estimates of the Population of Regions, Divisions, and States: July 1, 1940 to 1949. CURRENT POPULATION REPORTS, POPULATION ESTIMATES, Series P–25, No.47, March 9, 1951.

Census Bureau 1959. Bureau of the Census, U.S. Dept. of Commerce, CURRENT POPULATION REPORTS, POPULATION ESTIMATES, Series P–25, No.210, December 27, 1959. State populations: July 1, 1959.

Census Bureau 1994. Bureau of the Census, Economics and Statistics Administration, U.S. Dept. of Commerce, STATISTICAL ABSTRACT OF THE UNITED STATES, 114th Edition, The National Data Book. 1994.

Chambers + Zweifach 1947. R. Chambers + B.W. Zweifach, in PHYSIOL. REVIEW Vol.27: 436–463. 1947.

Cheng 1993. (Please compare with next entry.) G.C. Cheng + H.M. Loree + R.D. Kamm + M.C. Fishbein + R.T. Lee, "Distribution of Circumferential Stress in Ruptured and Stable Atherosclerotic Lesions: A Structural Analysis with Histopathologic Correlation," CIRCULATION Vol.87: 1179–1187. 1993.

Cheng 1993. (Please compare with preceding entry.) Keith C. Cheng + Lawrence A. Loeb, "Genomic Instability and Tumor Progression: Mechanistic Considerations," ADVANCES IN CANCER RESEARCH 60: 121–156. 1993.

Chesebro 1992. J.H. Chesebro + M.W.I. Webster + P. Zoldhelyi + P.C. Roche + L. Badimon + J.J. Badimon, "Antithrombotic Therapy and Progression of Coronary Artery Disease," CIRCULATION Vol.86 (suppl.3): 100–111. 1992.

Ciampricotti 1989. R. Ciampricotti + M.E. el Gamal + J.J. Bonnier + T.H. Relik, "Myocardial Infarction and Sudden Death after Sport: Acute Angiographic Findings," CATHET. CARDIOVASC. DIAGN. Vol.17: 193–197. 1989.

Clarke 1995. Mike Clarke + Rory Collins + Jon Godwin + Richard Gray + Richard Peto (writing for the Early Breast Cancer Trialists' Collaborative Group) "Effects of RadioTherapy and Surgery in Early Breast Cancer: An Overview of Randomized Trials," NEW ENGLAND J OF MEDICINE Vol.333, No.22: 1444–1455. Nov. 30, 1995.

Clarkson 1994. T.B. Clarkson + R.W. Prichard + 3 co-workers, "Remodeling of Coronary Arteries in Human and NonHuman Primates," JOURNAL of the AMERICAN MEDICAL ASSN. Vol.271: 289–294. 1994.

Cohen 1985. M. Cohen. "Quality Assurance as an Optimising Procedure in Diagnostic Radiology, "BRITISH JOURNAL OF RADIOLOGY Vol.18 (Suppl.): 134–141. 1985.

Cohn 1946. Edwin J. Cohn + 5 co-workers, "Preparation and Properties of Serum and Plasma Proteins: IV. System for Separation into Fractions of Protein and Lipoprotein Components of Biological Tissues and Fluids," in JOURNAL OF THE AMERICAN CHEMICAL SOCIETY Vol.68: 459–475. 1946.

Cohn 1967. Keith E. Cohn + J. Robert Stewart + Luis F. Fajardo + E. William Hancock + Henry S. Kaplan, "Heart Disease following Radiation," MEDICINE Vol.46: 281–298. May 1967.

ComPrinci 1978. Report from the Committee of Principal Investigators, "A Co-operative Trial in the Primary Prevention of Ischaemic Heart Disease Using Clofibrate," BRITISH HEART JOURNAL Vol.40: 1069–1118. 1978.

Connor 1986. William E. Connor, 1986 "Hypolipidemic Effects of Dietary Omega-3 Fatty Acids in Seafoods." In HEALTH EFFECTS OF POLYUNSATURATED FATTY ACIDS IN SEAFOODS, A.P. Simopoulos + R.R. Kifer + R.E. Martin, editors. ACADEMIC PRESS. Orlando, Florida. 1986.

Connor 1997. William E. Connor + Sonja L. Connor, "Should a Low-Fat, High-Carbohydrate Diet Be Recommended for Everyone?" (Clinical Debate, vs. Martijn B. Katan, Scott M. Grundy, Walter C. Willett), NEW ENGLAND JOURNAL OF MEDICINE Vol.337, No.8: 562–567. August 21, 1997.

Consensus 1985. Consensus Conference, "Lowering Blood Cholesterol to Prevent Heart Disease: Consensus Conference," JOURNAL OF THE AMERICAN MEDICAL ASSN. Vol.253: 2080–2090. 1985.

Cook 1997. Linda S. Cook + Mary L. Kamb + Noel S. Weiss, "Perineal Powder Exposure and the Risk of Ovarian Cancer," AMERICAN JOURNAL OF EPIDEMIOLOGY Vol.145, No.5: 459–465. 1997.

Co-op 1956. Co-operative Study of Lipoproteins and Atherosclerosis, performed by 4 teams: Donner Laboratory at the Univ. of Calif., Berkeley + Cleveland Clinic Foundation + Dept. of Biophysics at the Univ. of Pittsburgh + Dept. of Nutrition at Harvard School of Public Health. Title: "Evaluation of Serum Lipoprotein and Cholesterol Measurements as Predictors of Clinical Complications of Atherosclerosis," CIRCULATION Vol.14, No.4, Part Two: 691–741. October 1956.

Coresh 1996. Josef Coresh + Peter O. Kwiterovich, Jr., "Small, Dense Low-Density Lipoprotein Particles and Coronary Heart Disease Risk: A Clear Association with Uncertain Implications, " JAMA Vol. 276, No. 11. 914–915. September 18, 1996.

Courtice + Garlick 1962. F.C. Courtice + D.G. Garlick, in QUARTERLY J. OF EXPERIMENTAL PHYSIOL. Vol.47: 211–220. 1962.

Cralley 1968. L.J. Cralley + 4 co-workers, "Fibrous and Mineral Content of Cosmetic Talcum Products," AMERICAN INDUSTRIAL HYGIENE ASSOC. JOURNAL 1968, pp.350–354.

Cramer 1982. Daniel W. Cramer + William R. Welch + Robert E. Scully + Carol A. Wojciechowski, "Ovarian Cancer and Talc: A Case–Control Study," CANCER Vol.50, No.2: 372–376. July 15, 1982.

Cuzick 1994. J. Cuzick + H. Stewart + L. Rutqvist + 10 co-workers, "Cause–Specific Mortality in Long–Term Survivors of Breast Cancer Who Participated in Trials of Radiotherapy," (rapid publication), JOURNAL OF CLINICAL ONCOLOGY Vol.12, No.3: 447–453. March 1994.

Danesh 1997. J. Danesh + R. Collins + R. Peto, "Chronic Infections and Coronary Heart Disease: Is There a Link?" LANCET Vol.350: 430–436. 1997.

Daoud 1963. A.S. Daoud + 4 co-workers, "Possible Relationship of Coronary Thrombosis and Rupture of Arteriosclerotic Plaque," LAB. INVEST. Vol.12: 863. 1963.

Davidson 1963. R.G. Davidson + H.M. Nitowsky + B. Childs, "Demonstration of Two Populations of Cells in the Human Female Heterozygous for Glucose–6–Phosphate Dehydrogenase Variants," PROCEEDINGS OF THE NATIONAL ACADEMY OF SCIENCES (USA) Vol.50: 481–485. 1963.

Davidson 1998. Michael Davidson et al, "Confirmed Previous Infection with Chlamydia Pneumoniae (TWAR) and Its Presence in Early Coronary Atherosclerosis," CIRCULATION Vol.98: 628–633. 1998.

Davies 1985. Michael J. Davies + Anthony C. Thomas, "Plaque Fissuring --- the Cause of Acute Myocardial Infarction, Sudden Ischaemic Death, and Crescendo Angina," (review), BRITISH HEART JOURNAL Vol.53: 363–373. April 1985.

Davies 1990. Michael J. Davies, "A Macro and Micro View of Coronary Vascular Insult in Ischemic Heart Disease," CIRCULATION Vol.82 Suppl.2: 38–46. 1990.

DeLalla 1954-a. Oliver F. DeLalla + John W. Gofman, "Ultracentrifugal Analysis of Human Serum Lipoproteins," in METHODS OF BIOCHEMICAL ANALYSIS Vol.1, edited by D. Glick. (Interscience, New York City.) 1954.

DeLalla 1954-b. Oliver F. DeLalla + Harold A. Elliott + John W. Gofman, "Ultracentrifugal Studies of High Density Serum Lipoproteins in Clinically Healthy Adults," THE AMERICAN JOURNAL OF PHYSIOLOGY Vol.179, No. 2: 333–337. 1954.

DeLalla 1958. Oliver Francis DeLalla, BIOLOGY AND ULTRACENTRIFUGAL METHODOLOGY OF THE HIGH–DENSITY LIPOPROTEINS AND THE PROTEINS OF HUMAN SERUM. 91 pages. (Thesis.) University of California, Lawrence Radiation Laboratory, Berkeley, CA. UCRL Report 8550. November 25, 1958. Available (then) from U.S. Dept. of Commerce, Office of Technical Services, Washington 25, D.C.

DeLalla 1961. Oliver F. DeLalla + John W. Gofman, "On the Population Distributions of Both the Low–Density Lipoproteins and the High–Density Lipoproteins and the Serum Proteins: A) Biological Distributions. B) Lipoprotein and Protein Interclass Relationships," (chapter) pp.137–151 in LIPIDE UND LIPOPROTEIDE IM BLUTPLASMA, Edited by Fritz Pezold. (Springer Verlag, Berlin.) 1961.

DeLorgeril 1994. Michel deLorgeril + Serge Renaud + Nicole Mamelle + Patricia Salen, + Jeana-Louis Martin + Isabelle Monjaud + Jeannine Guidollet + Paul Touboul + Jacques Delaye, "Mediterranean Alpha–Linolenic Acid–Rich Diet in Secondary Prevention of Coronary Heart Disease," THE LANCET Vol. 343 June 11 1454–59, 1994.

DeLorgeril 1994-b. Michel deLorgeril + R. Forrat + R. Ferrara + others, "Chronic Immune–Induced Inflammation in Genesis of Arterial Manifestations," J. IMMUNOPHARMACOLOGY Vol.14: 53–58. 1994.

DeLorgeril 1996. Michel deLorgeril + Patricia Salen + Jean-Louis Martin + Nicole Mamelle + Isabelle Monjaud + Paul Touboul + Jacques Delaye, "Effect of a Mediterranean Type of Diet on the Rate of Cardiovascular Complications in Patients with Coronary Artery Disease," JOURNAL OF THE AMERICAN COLLEGE OF CARDIOLOGY Vol. 28: 1103–1108. 1996.

DeLorgeril 1997. Michel DeLorgeril et al, "Control of Bias in Dietary Trial to Prevent Coronary Recurrences: The Lyon Diet Heart Study," EUROPEAN JOURNAL OF CLINICAL NUTRITION Vol.51: 116–122. 1997.

DeLorgeril 1998. Michel deLorgeril + Patricia Salen + Jen-Louis Martin + Isabelle Monjaud + Philippe Boucher + Nicole Mamelle, "Mediterranean Dietary Pattern in a Randomized Trial: Prolonged Survival and Possible Reduced Cancer Rate," ARCHIVES OF INTERNAL MEDICINE Vol.158: 1181–1187. 1998.

DeLorgeril 1999. Michel deLorgeril + Patricia Salen + Jean-Louis Martin + Isabelle Monjaud + Jacques Delaye + Nicole Mamelle, "Mediterranean Diet, Traditional Risk Factors, and the Rate of CardioVascular Complications after Myocardial Infarction: Final Report of the Lyon Diet Heart Study," CIRCULATION Vol.99: 779–785. February 16, 1999.

Despres 1996. Jean-Pierre Despres + 6 co-workers, "HyperInsulinemia as an Independent Risk Factor for Ischemic Heart Disease," NEW ENGLAND J. OF MEDICINE Vol.334, No.15: 952–957. April 11, 1996.

Dewing 1965. Stephen B. Dewing, RADIOTHERAPY OF BENIGN DISEASE. 311 pages. Library of Congress Catalog Number 65–11686. (Publisher: Charles C. Thomas, Springfield, Illinois, USA). 1965.

DiLeonardo 1993. A. DiLeonardo + S.P. Linke + Y. Yin + G.M. Wahl, "Cell Cycle Regulation of Gene Amplification," COLD SPRING HARBOR SYMPOSIA ON QUANTITATIVE BIOLOGY 58: 655–667. 1993.

Dobson 1976. R.L. Dobson + T.C. Kwan, "The RBE of Tritium Radiation Measured in Mouse Oocytes: Increase at Low Exposure Levels," RADIATION RESEARCH Vol.66: 615–625. 1976.

Doll 1976. R. Doll + R. Peto, "Mortality in Relation to Smoking: Twenty Years' Observations of British Doctors,"
          BRITISH MEDICAL JOURNAL Vol.4: 1525-1536. 1976.

Dollinger 1965. Malin R. Dollinger + Daniel M. Lavine + Laurance V. Foye, Jr., "Myocardial Infarction Following
          Radiation," LANCET Vol.2: 246. July 31, 1965.

Dollinger 1966. Malin R. Dollinger + Daniel M. Lavine + Laurance V. Foye, Jr., "Myocardial Infarction due to
          Post-Irradiation Fibrosis of the Coronary Arteries," JOURNAL OF THE AMERICAN
          MEDICAL ASSN Vol.195, No.4: 176-179. January 24, 1966.

Donaldson 1951. S.W. Donaldson, "The Practice of Radiology in the United States: Facts and Figures,"
          AMERICAN JOURNAL OF ROENTGENOLOGY Vol.66, No.6: 929-946. December 1951.

Doud 1964. A. Doud + J. Jarmolych et al, " 'Pre-Atheroma' Phase of Coronary Atherosclerosis in Man,"
          EXP. MOL. PATHOLOGY Vol.3: 465-484. 1964.

Downs 1998. John R. Downs + Michael Clearfield + Stephen Weis + others: All for the AFCAPS/TexCAPS Research Group,
          "Primary Prevention of Acute Coronary Events with Lovostatin in Men and Women with Average Cholesterol
          Levels: Results of AFCAPS/TexCAPS," J. AMERICAN MEDICAL ASSN Vol.279: 1615-1622. May 27, 1998.

Duff 1948. G. Lyman Duff + Gardner C. McMillan, "Inhibition of Experimental Cholesterol Atherosclerosis by Alloxan
          Diabetes in the Rabbit," (abstract), AMERICAN HEART JOURNAL Vol.36: 469. 1948.

Duff 1949. G. Lyman Duff + Gardner C. McMillan, "Effect of Alloxan Diabetes on Experimental Cholesterol Atherosclerosis in
          the Rabbit," JOURNAL OF EXPERIMENTAL MEDICINE Vol.89: 611+. 1949.

Dunsmore 1986. Lillian D. Dunsmore + Marie A. LoPonte + Richard A. Dunsmore, "Radiation-Induced Coronary Artery
          Disease," JOURNAL AMER. COLLEGE OF CARDIOLOGY Vol.8, No.1: 239-244. July 1986.

Dunsmore 1996. Richard A. Dunsmore + Lillian D. Dunsmore, "Coronary Radiation Could Damage Heart," letter in WALL
          STREET JOURNAL. April 1, 1996.

Elkeles 1961. Arthur Elkeles, "Radioactivity in Calcified Atherosclerosis," BRITISH JOURNAL OF
          RADIOLOGY Vol.34: 602-605. 1961.

Elkeles 1966. Arthur Elkeles, "Atherosclerosis and Radioactivity," JOURNAL OF THE AMERICAN
          GERIATRICS SOCIETY Vol.14: 895. 1966.

Elkeles 1968. Arthur Elkeles, "Alpha-Ray Activity in Coronary Artery Disease," JOURNAL OF THE
          AMERICAN GERIATRICS SOCIETY Vol.16, No.5: 576-583. 1968.

Elkeles 1969. Arthur Elkeles, "Alpha-Ray Activity in the Large Human Arteries," NATURE Vol.221:
          662-664. 1969.

Elkeles 1977. Arthur Elkeles, "Metabolic Behavior of Alpha-Ray Activity in Large Human Arteries: Relationship to
          Atherosclerosis," JOURNAL OF THE AMERICAN GERIATRICS SOCIETY Vol.25: 179-182.
          (Editors mistakenly interchanged the radio-autographs for Figures 1 and 2.) 1977.

Enas 1998. Enas A. Enas, "Triglycerides and Small, Dense Low-Density Lipoproteins," (letter), JOURNAL OF
          THE AMERICAN MEDICAL ASSN Vol.280, No.23: 1990. Dec. 16, 1998.

Engelberg 1952-a. Hyman Engelberg + John W. Gofman + Hardin B. Jones, "Serum Lipids and Lipoproteins in Diabetic
          Glomerulosclerosis," METABOLISM Vol.1: 300-306. 1952.

Engelberg 1952-b. Hyman Engelberg + John W. Gofman + Hardin B. Jones, "Serum Lipids and Lipoproteins in Diabetic
          Glomerulosclerosis: Preliminary Observations of the Effect of Heparin upon the Disease," DIABETES
          Vol.1, No.6: 425-433. Nov-Dec. 1952.

Engelberg 1996. Hyman Engelberg, "Endogenous Heparin Activity Deficiency and Atherosclerosis," (review), CLIN.
          APPL. THROMBOSIS/HEMOSTASIS Vol.2, No.2: 83-93. 1996.

Epstein 1994. Stephen E. Epstein + Edith Speir + Ellis F. Unger + Raul J. Guzman + Toren Finkel, "The Basis of Molecular
          Strategies for Treating Coronary Restenosis after Angioplasty," J. AMER. COLLEGE OF CARDIOLOGY
          Vol.23, No.6: 1278-1288. May 1994. Has a bibliography of 189 references. Related paper: Finkel 1995.

European 1987. European Atherosclerosis Society, "Strategies for the Prevention of Coronary Heart Disease: Policy Statement
          of the European Atherosclerosis Society," EUROPEAN HEART JOURNAL Vol.8: 77-88. 1987.

Evans 1978. H.J. Evans + David C. Lloyd (editors), MUTAGEN-INDUCED CHROMOSOME DAMAGE IN MAN.
          (University Press, Edinburgh, Scotland.) 1978.

Evans 1979. H.J. Evans + K.E. Buckton + G.E. Hamilton + A. Carothers, "Radiation-Induced Chromosome Aberrations in
          Nuclear-Dockyard Workers," NATURE Vol.277: 531-534. February 15, 1979.

Evans 1986. John S. Evans + John E. Wennberg + Barbara J. McNeil, "The Influence of Diagnostic Radiography on the Incidence
          of Breast Cancer and Leukemia," NEW ENGL. J. MED Vol.315, No.13: 810-815. Sept. 25, 1986.

Evens 1995. Ronald G. Evens, "Roentgen Retrospective: One Hundred Years of a Revolutionary Technology,"
          J. AMERICAN MEDICAL ASSN. Vol.274, No.11: 912-916. September 20, 1995.

Expert Panel 1993. Expert Panel on Detection, Evaluation, and Treatment of High Blood Cholesterol in Adults. "Summary
          of the Second Report of the National Cholesterol Education Program (NCEP), Adult Treatment Panel II,"
          JOURNAL OF THE AMERICAN MEDICAL ASSN. Vol.269: 3015-3023. 1993.

Fabricant 1978. C.G. Fabricant + J. Fabricant + M.M. Litrenta + C.R. Minick, "Virus-Induced Atherosclerosis,"
          JOURNAL OF EXPERIMENTAL MED. Vol.148: 335-340. 1978.

Faggiotto 1984. A. Faggiotto + Russell Ross + Laurence Harker, ARTERIOSCLEROSIS Vol.4: 323–340. 1984.

Faggiotto 1984–b. Faggiotto + Russell Ross, ARTERIOSCLEROSIS Vol.4: 341–356. 1984.

Fajardo 1968. Luis F. Fajardo + J. Robert Stewart + Keith E. Cohn, "Morphology of Radiation–Induced Heart Disease," ARCHIVES OF PATHOLOGY Vol.86: 512–519. November 1968.

Fajardo 1973. Luis F. Fajardo + J. Robert Stewart, "Pathogenesis of Radiation–Induced Myocardial Fibrosis," LAB. INVEST. Vol.29: 244–257. 1973.

Fajardo 1977. Luis F. Fajardo, "Radiation–Induced Coronary Artery Disease," (editorial), CHEST Vol.71, No.5: 563–564. May 1977.

FDA 1992. Food and Drug Administration, "HANDBOOK OF SELECTED TISSUE DOSES FOR THE UPPER GASTRO–INTESTINAL FLUOROSCOPIC EXAMINATION. By Marvin Rosenstein + 4 co–workers. 27 pages. HHS Publication FDA 92–8282. (U.S. Food and Drug Admin., Center for Devices and Radiological Health, Rockville MD 20857.) June 1992.

FDA 1994. Food and Drug Administration, Public Health Advisory, "Avoidance of Serious Xray–Induced Skin Injuries to Patients during Fluoroscopically–Guided Procedures." 6 pages. (U.S. Food and Drug Administration, Center for Devices and Radiological Health, Rockville, Maryland USA 20857.) September 9, 1994.

FDA 1995. Food and Drug Administration, HANDBOOK OF SELECTED TISSUE DOSES FOR FLUOROSCOPIC AND CINE–ANGIOGRAPHIC EXAMINATION OF THE CORONARY ARTERIES. By Stanley Stern + 3 co–workers. 27 pages. HHS Publication FDA 95–8288. (U.S. Food and Drug Admin., Center for Devices and Radiological Health; see FDA 1994.) September 1995.

Fearon 1990. E.R. Fearon + Bert Vogelstein, "A Genetic Model for Colorectal Tumorigenesis," CELL 61: 759+. 1990.

Felton 1994. C.V. Felton + D. Crook + M.J. Davies + M.F. Oliver, "Dietary PolyUnsaturated Fatty Acids and Composition of Human Aortic Plaques," LANCET Vol.344: 1195–1196. 1994.

Fernando–Ortiz 1994. A. Fernandez–Ortiz + J.J. Badimon + E. Falk + Valentin Fuster + 5 others, "Characterization of Relative Thrombogenicity of Atherosclerotic Plaque Components: Implications for Consequences of Plaque Rupture," J. AMERICAN COLLEGE OF CARDIOLOGY Vol.23: 1564–1569. 1994.

Fialkow 1974. P.J. Fialkow, "The Origin and Development of Human Tumors Studied with Cell Markers," NEW ENGLAND JOURNAL OF MEDICINE Vol.291: 26–35. 1974.

Finkel 1995. Toren Finkel + Stephen E. Epstein, "Gene Therapy for Vascular Disease," FASEB JOURNAL (Federation of Amer. Societies for Experimental Biology) Vol.9: 843–851. July 1995.

Flavahan 1992. N.A. Flavahan, "Atherosclerosis or Lipid–Induced Endothelial Dysfunction; Potential Mechanisms Underlying Reduction in EDRF/Nitric–Oxide Activity," CIRCULATION Vol.85: 1927–1938. 1992.

Folsom 1999. Aaron R. Folsom, "AntiBiotics for Prevention of Myocardial Infarction? Not Yet!" (editorial), JOURNAL OF THE AMERICAN MEDICAL ASSN Vol.281, No.5: 461–462. Feb. 3, 1999.

Francis 1989. G.S. Francis, "The Relationship of the Sympathetic Nervous System and the Renin–Angiotensin System in Congestive Heart Failure," AMERICAN HEART JOURNAL Vol.118: 642–648. 1989.

Fredrickson 1993. Donald S. Fredrickson, "Phenotyping: On Reaching Base Camp (1950–1975)," pp.1–15 in DYSLIPOPROTEINEMIA: FROM PHENOTYPES TO GENOTYPES ... A REMARKABLE QUARTER CENTURY. Supplement to CIRCULATION Vol.87, No.4: April 1993.

Freeman 1952. Norman K. Freeman, "Infrared Spectrum of Branched Long–Chain Fatty Acids," JOURNAL OF THE AMERICAN CHEMICAL SOCIETY 74: 2523–2528. 1952

Freeman 1953. Norman K. Freeman + Frank T. Lindgren + Yook C. Ng + Alex V. Nichols, "Infra–Red Spectra of some Lipoproteins and Related Lipids," J. BIOLOGICAL CHEMISTRY, Vol. 203: 293–304. 1953.

Frick 1987. (Helsinki Heart Study.) M.H. Frick + O. Elo + K. Haapa et al, "Helsinki Heart Study: Primary Prevention Trial with Gemfibrozil in Middle–Aged Men with Dyslipidemia: Safety of Treatment, Changes in Risk Factors, and Incidence of Coronary Heart Disease," NEW ENGLAND J. MED. Vol.317: 1237–1245. 1987.

Frigerio 1976. N.A. Frigerio + R.S. Stowe, "Carcinogenic and Genetic Hazard from Background Radiation," pp.385–393 in BIOLOGICAL AND ENVIRONMENTAL EFFECTS OF LOW–LEVEL RADIATION. (International Atomic Energy Agency, Vienna, Austria.) 1976.

Fry 1968. Donald L. Fry, "Acute Vascular Endothelial Changes Associated with Increased Blood Velocity Gradients," CIRCULATION RESEARCH Vol.22: 165–197. 1968.

Fry 1987. Donald L. Fry, "Mass Transport, Atherogenesis, and Risk," ARTERIOSCLEROSIS Vol.7, No.1: 88–100. Jan–Feb. 1987.

Furberg 1994. C.D. Furberg et al, "Effect of Lovastatin on Early Carotid Atherosclerosis and Cardiovascular Events," CIRCULATION Vol.90: 1679–1687. 1994.

Fuster 1992. Valentin Fuster + L. Badimon + J.J. Badimon + J.H. Chesebro, "The Pathogenesis of Coronary Artery Disease and the Acute Coronary Syndromes," NEW ENGLAND J. MEDICINE Vol.326: 242–250, 310–318. 1992.

Fuster 1994. Valentin Fuster, "Mechanisms Leading to Myocardial Infarction: Insights from Studies of Vascular Biology," (Lewis A. Conner Memorial Lecture), CIRCULATION Vol.90: 2126–2146. Oct. 1994. (Erratum: Vol.91: 256.)

Fuster 1995. Valentin Fuster + J.J. Badimon, "Regression or Stabilization of Atherosclerosis Means Regression or Stabilization of What We Don't See in the Arteriogram," EUROPEAN HEART JOURNAL Vol.16, suppl.E: 6–12. 1995.

Fuster 1997. Valentin Fuster + David A. Vorchheimer, "Prevention of Atherosclerosis in Coronary–Artery Bypass Grafts," (editorial), NEW ENGLAND JOURNAL OF MEDICINE Vol.336, No.3: 212–213. 1997.

Gabay 1999. Cem Gabay + Irving Kushner, "Acute–Phase Proteins and Other Systemic Responses to Inflammation," (Mechanisms of Disease), NEW ENGLAND JOURNAL OF MEDICINE Vol.340, No.6: 448–454. Feb. 11, 1999.

Gallini 1985. R. Gallini + S. Belletti + U. Giugni, "Cost Benefit Evaluation in a Quality Control Programme for Conventional Radiodiagnosis," BRITISH JOURNAL OF RADIOLOGY Vol. 18 (Suppl.): 49–50. 1985.

Gardner 1996. Christopher D. Gardner + Stephen P. Fortman + Ronald M. Krauss, "Association of Small Low–Density Lipoprotein Particles with the Incidence of Coronary Artery Disease in Men and Women." JOURNAL OF THE AMERICAN MEDICAL ASSN. Vol.276, No.11. 875–881. September 18, 1996.

Gassmann 1899. A. Gassmann, "Zur Histologie der Rontgenulcere," in FORTSCHR. a. d. GEBIETE d. ROENTGENSTRAHLEN Vol.2: 199–207. 1899.

Gassmann 1904. A. Gassmann, in ARCH FUER DERMAT. UND SYPH. Vol.70: 97. 1904.

Geer 1961. J.C. Geer + Henry C. McGill, Jr. + J.P. Strong, "The Fine Structure of Human Atherosclerotic Lesions," AMERICAN JOURNAL OF PATHOLOGY Vol.38: 263–287. 1961.

Geer 1968. J.C. Geer + Henry C. McGill, Jr. + W.B. Robertson + J.P. Strong, "Histologic Characteristics of Coronary Artery Fatty Streaks," LAB. INVEST. Vol.18: 565–570. 1968.

Gilbert 1985. Ethel Gilbert, "Late Somatic Effects," in HEALTH EFFECTS MODELS FOR NUCLEAR POWER PLANT ACCIDENT CONSEQUENCE ANALYSIS. NUREG/CR–4214. No index. The report was prepared by the Harvard School of Public Health under sub–contract to the Sandia National Laboratories. Principal investigators: John S. Evans + Dade W. Moeller + Douglas W. Cooper. (U.S. Nuclear Regulatory Commission, Division of Risk Analysis and Operations.) 1985.

Gimlette 1959. T.M. Gimlette, "Constrictive Pericarditis," BRITISH HEART JOURNAL Vol.21: 9–16. January 1959.

Giovino 1994.

Please see MMWR 1994.

Giroud 1992. D. Giroud et al, "Relation of the Site of Acute Myocardial Infarction to the Most Severe Coronary Arterial Stenosis at Prior Angiography," AMERICAN JOURNAL OF CARDIOLOGY Vol.69: 729–732. 1992.

Glagov 1972. S. Glagov, "Hemodynamic Risk Factors: Mechanical Stress, Mural Architecture, Medial Nutrition, and the Vulnerability of Arteries to Atherosclerosis," in THE PATHOGENESIS OF ATHEROSCLEROSIS, edited by R.W. Wissler + J.C. Geer. (Publisher: Williams and Wilkins, Baltimore, Maryland.) 1972.

Glagov 1987. S. Glagov + E. Weisenberg + 3 co–workers, "Compensatory Enlargement of Human Atherosclerotic Coronary Arteries," NEW ENGLAND JOURNAL OF MEDICINE Vol.316: 371–375. 1987.

Glazier 1954. Frank W. Glazier + Arthur R. Tamplin + Beverly Strisower + Oliver F. DeLalla + John W. Gofman + Thomas R. Dawber + Edward Phillips, "Human Serum Lipoprotein Concentrations," JOURNAL OF GERONTOLOGY Vol.9: 395–402. October 1954.

Gofman 1949. John W. Gofman + Frank T. Lindgren + Harold A. Elliott, "Ultracentrifugal Studies of Lipoproteins of Human Serum," JOURNAL OF BIOLOGICAL CHEMISTRY Vol.179: 973–979. 1949.

Gofman 1950–a. John W. Gofman + Frank T. Lindgren + Harold A. Elliott + William Mantz + John Hewitt + Beverly Strisower + Virgil Herring + Thomas P. Lyon, "The Role of Lipids and Lipoproteins in Atherosclerosis," SCIENCE Vol.111, No.2877: 166–171 & 186. Feb. 17, 1950.

Gofman 1950–b. John W. Gofman + Hardin B. Jones + Frank T. Lindgren + Thomas P. Lyon + Harold A. Elliott + Beverly Strisower, "Blood Lipids and Human Atherosclerosis," CIRCULATION Vol.2, No.2: 161–178. August 1950.

Gofman 1951–a. John W. Gofman + Frank T. Lindgren + Hardin B. Jones + Thomas P. Lyon + Beverly Strisower, "Lipoproteins and Atherosclerosis," JOURNAL OF GERONTOLOGY, Vol.6, No.2: 105–119. April 1951.

Gofman 1951–b. John W. Gofman, "Biophysical Approaches to Atherosclerosis," in BIOLOGICAL and MEDICAL PHYSICS, Vol.2: 269–280. 1951.

Gofman 1952–a. John W. Gofman + Hardin B. Jones + Thomas P. Lyon + Frank T. Lindgren + Beverly Strisower + David Colman + Virgil Herring. "Blood Lipids and Human Atherosclerosis," CIRCULATION Vol.5, No.1: 110–134. Jan. 1952.

Gofman 1952–b. John W. Gofman + Hardin B. Jones, "Obesity, Fat Metabolism and Cardiovascular Disease," CIRCULATION Vol.5, No.4: 514–517. April 1952.

Gofman 1952–c. John W. Gofman, "Diet and Lipotrophic Agents in Atherosclerosis." BULLETIN OF THE NEW YORK ACADEMY of MEDICINE, Vol. 28, No.5: 279–293. May 1952.

Gofman 1952–d. John W. Gofman, "Lipoprotein Transformations in Health and Disease," EDUCATIONAL PROCEEDINGS OF THE PERMANENTE HOSPITALS (Kaiser) Vol.2: 174+. 1952.

Gofman 1953. John W. Gofman + Beverly Strisower + Oliver DeLalla + Arthur Tamplin + Hardin B. Jones + Frank
     Lindgren, "Index of Coronary Artery Atherogenesis," MODERN MEDICINE, June 15, 1953, Pages 119–140.

Gofman 1954–a. John W. Gofman + Oliver DeLalla + Frank Glazier + Norman K. Freeman + Frank T. Lindgren + Alex V.
     Nichols + Beverly Strisower + Arthur R. Tamplin, "The Serum Lipoprotein Transport System in Health, Metabolic
     Disorders, Atherosclerosis and Coronary Heart Disease," PLASMA (Milano, Italy) Vol.2, No.4: 413–484. 1954.

Gofman 1954–b. John W. Gofman, "The Nature of the Relationship of Disturbance in Blood Lipid Transport with the Evolution
     of Clinical Coronary Heart Disease," TRANSACTIONS OF THE AMERICAN COLLEGE OF CARDIOLOGY
     Vol.4: 198–220. 1954.

Gofman 1954–c. John W. Gofman + Arthur Tamplin + Beverly Strisower, "Relation of Fat and Caloric Intake to
     Atherosclerosis," JOURNAL OF THE AMER. DIETETIC ASSN Vol.30, No.4: 317–326. April 1954.

Gofman 1954–d. John W. Gofman + Frank Glazier + Arthur Tamplin + Beverly Strisower + Oliver DeLalla, "Lipoproteins,
     Coronary Heart Disease, and Atherosclerosis," PHYSIOLOGICAL REVIEWS, Vol.34, No. 2: 589–607.
     July 1954.

Gofman 1954–e. John W. Gofman + Leonard Rubin + James P. McGinley + Hardin B. Jones, "Hyperlipoproteinemia,"
     AMERICAN JOURNAL OF MEDICINE Vol.17, No.4: 514–520. October 1954.

Gofman 1955. John W. Gofman + Frank T. Lindgren + Beverly Strisower + Oliver F. DeLalla + Frank Glazier + Arthur R.
     Tamplin, "Cigarette Smoking, Serum Lipoproteins, and Coronary Heart Disease." GERIATRICS Vol. 10: 349–354.
     August 1955.

Gofman 1957. John W. Gofman, WHAT WE DO KNOW ABOUT HEART ATTACKS. (G.P. Putnam's Sons, publisher, New
     York City.) 1957.

Gofman 1958. John W. Gofman, "Diet in the Prevention and Treatment of Myocardial Infarction," THE AMERICAN
     JOURNAL OF CARDIOLOGY Vol.1, No.2: 271–283. February 1958.

Gofman 1958–b. John W. Gofman + Alex V. Nichols + E. Virginia Dobbin, DIETARY PREVENTION AND TREATMENT
     OF HEART DISEASE. 256 pages. LCCN 58–10072. (G.P. Putnam's Sons, publisher, New York City.) 1958.

Gofman 1959. John W. Gofman, CORONARY HEART DISEASE. 353 pages. LCCN 58–14073. (Charles C. Thomas
     Publishers, Springfield, Illinois.) 1959.

Gofman 1963. John W. Gofman + Wei Young, "The Filtration Concept of Atherosclerosis and Serum Lipids in the
     Diagnosis of Atherosclerosis," Chapter 6 in ATHEROSCLEROSIS AND ITS ORIGINS. (Academic
     Press Inc., New York City.) 1963.

Gofman 1966. John W. Gofman + Wei Young + Robert Tandy, "Ischemic Heart Disease, Atherosclerosis, and Longevity,"
     (the George Lyman Duff Memorial Lecture), CIRCULATION Vol.34: 679–697. October 1966.

Gofman 1969. John W. Gofman, "The Quantitative Nature of the Relationship of Coronary Artery Atherosclerosis and
     Coronary Heart Disease Risk," CARDIOLOGY DIGEST Vol.4, No.2: 28–38. February 1969.

Gofman 1969–b. John W. Gofman + Arthur R. Tamplin, "Low Dose Radiation and Cancer," paper presented Oct. 29, 1969 at
     the IEEE Nuclear Science Symposium, San Francisco. In IEEE TRANSACTIONS ON NUCLEAR SCIENCE
     Vol.NS–17, No.1: 1–9, February 1970. (Institute of Electrical and Electronics Engineering, New York City.)

● Gofman 1971. John W. Gofman + Arthur R. Tamplin, "Epidemiologic Studies of Carcinogenesis by Ionizing Radiation,"
     presented July 20, 1971. Pages 235–277 in PROCEEDINGS OF THE SIXTH BERKELEY SYMPOSIUM ON
     MATHEMATICAL STATISTICS AND PROBABILITY: VOL. 6: EFFECTS OF POLLUTION ON HEALTH.
     Edited by L.M. Lecam + J. Neyman + E. Scott. 1972. (University of California Press, Berkeley CA 94720.)

Gofman 1976. John W. Gofman, "The Plutonium Controversy," JOURNAL OF THE AMERICAN MEDICAL ASSN.
     Vol.236, No.3: 284–286. July 19, 1976.

Gofman 1978. John W. Gofman, "Serum Lipoproteins, Coronary Atherosclerosis, and Ischemic Heart Disease," paper given at
     the San Francisco symposium "New Frontiers on the Relationships of Lipids, Lipoproteins, and the Arterial Wall in
     Cardiovascular Disease," jointly sponsored by the University of California San Francisco School of Medicine,
     Extended Programs in Medical Education, and U.C. Berkeley Lifelong Learning Program. January 28, 1978.

● Gofman 1981. John W. Gofman, RADIATION AND HUMAN HEALTH. 908 pages. ISBN 0–87156–275–8. LCCN
     80–26484. (Sierra Club Books, San Francisco.) 1981. [A Japanese–language edition (1991) is
     available from Shakai Shiso Sha Ltd., ISBN 4–390–50185–2.]

Gofman 1985. John W. Gofman + Egan O'Connor, XRAYS: HEALTH EFFECTS OF COMMON EXAMS. 440 pages.
     ISBN 0–87156–838–1. LCCN 84–23527. (Sierra Club Books, San Francisco.) 1985.

Gofman 1986. John W. Gofman, "Assessing Chernobyl's Cancer–Consequences," presentation as a panelist at the
     SYMPOSIUM ON LOW–LEVEL RADIATION, 192nd National Meeting of the American Chemical Society, held in
     Anaheim, California. Sept. 9, 1986.

Gofman 1988. John W. Gofman, "A Proposal Concerning 'the New Dosimetry'," (letter), HEALTH PHYSICS Vol.55,
     No.1: 580–581. September 1988.

● Gofman 1990. John W. Gofman, RADIATION–INDUCED CANCER FROM LOW–DOSE EXPOSURE: AN INDEPENDENT
     ANALYSIS. 480 pages. ISBN 0–932682–89–8. LCCN 89–62431. (Committee for Nuclear Responsibility
     Books. San Francisco.) 1990. Also available online at www.ratical.org/radiation/CNR/

[A Russian–language edition (1995) is available from the Socio–Ecological Union, Lydia Popova, Center for Nuclear Ecology, Moscow. Email: seulydia@glas.apc.org   Website: www.ecoline.ru]

Gofman 1990–b. Recollections of the "impossible" early events in the Donner group's lipoprotein–atherosclerosis research. Presented informally at "The Donner Dinner," June 22, 1990, in honor of Virgie Shore and Frank T. Lindgren.

Gofman 1992. John W. Gofman, "Bio–Medical 'Un–Knowledge' and Nuclear Pollution: A Common–Sense Proposal," presented Dec. 9, 1992 on the occasion of the Right Livelihood Award, Stockholm. (Right Livelihood Award Foundation, Stockholm, Sweden). 1992.

Gofman 1994. John W. Gofman, CHERNOBYL ACCIDENT: RADIATION CONSEQUENCES FOR THIS AND FUTURE GENERATIONS. 574 pages, in the Russian language. Translated from English by Professors Emanuel I. Volmyansky and Olga A. Volmyanskaya. ISBN 5–339–00869–X. (Vysheishaya Shkola Publishing House, 11 Masharov Avenue, Minsk 220048, Belarus.) 1994.

Gofman 1995. John W. Gofman, PREVENTING BREAST CANCER: THE STORY OF A MAJOR, PROVEN, PREVENTABLE CAUSE OF THIS DISEASE, First Edition. 339 pages. LCCN 94–69129. ISBN 0–932682–94–4. (Committee for Nuclear Responsibility Books, San Francisco.) 1995.

Gofman 1995/96. This notation refers to the identical parts of Gofman 1995 and 1996 ––– namely, Chapters 1 through 42.

Gofman 1996. John W. Gofman, PREVENTING BREAST CANCER: THE STORY OF A MAJOR, PROVEN, PREVENTABLE CAUSE OF THIS DISEASE, Second Edition. (Second Edition adds Section 5, covering critiques of the First Edition.) 422 pages. LCCN 96–2453. ISBN 0–932682–96–0. (Committee for Nuclear Responsibility Books, San Francisco.) 1996. Also available online: www.ratical.org/radiation/CNR/

Gofman 1996–b. John W. Gofman, "Atherosclerotic Heart Disease and Cancer: Looking for the 'Smoking Guns'," (in the "Milestones in Biological Research" series of the Federation of American Societies for Experimental Biology), FASEB JOURNAL Vol.10: 661–663. April 1996.

Gofman 1998. John W. Gofman, "Asleep at the Wheel: The Special Menace of Inherited Afflictions from Ionizing Radiation," CNR PUBLICATION 9810. (Committee for Nuclear Responsibility, San Francisco.) 1998. Also online at www.ratical.org/radiation/CNR/

Gold 1961. Harold Gold, "Production of Arteriosclerosis in the Rat: Effect of X–Ray and a High–Fat Diet," ARCHIVES OF PATHOLOGY Vol.71: 268–273. March 1961.

Gold 1962. Harold Gold, "Atherosclerosis in the Rat: Effect of X–Ray and a High Fat Diet," PROC. SOC. EXP. BIOL. MED. Vol.111: 593–595. 1962.

Goldman 1983. L. Goldman + R. Sayson + S. Robbins + co–workers, "The Value of the Autopsy in Three Medical Areas," NEW ENGLAND JOURNAL OF MEDICINE Vol.308: 1000–1005. 1983.

Gordon 1989. David J. Gordon + Basil M. Rifkind, "High–Density Lipoprotein ––– the Clinical Implications of Recent Studies," (Medical Intelligence), NEW ENGLAND J. OF MEDICINE Vol.321, No.19: 1311–1316. Nov. 9, 1989.

Gould 1995. A.L. Gould + J.E. Rossouw + N.C. Santanello + J.G. Heyse + C.D. Furberg, "Cholesterol Reduction Yields Clinical Benefit: A New Look at Old Data," CIRCULATION Vol.91: 2274–2282. 1995.

Graham 1967. J. Graham + R. Graham, "Ovarian Cancer and Asbestos," ENVIRONMENTAL RESEARCH Vol.1: 115–128. 1967.

Grattan 1989. M.T. Grattan and 5 co–workers, "Cytomegalovirus Infection Is Associated with Cardiac Allograft Rejection and Atherosclerosis," JOURNAL OF THE AMERICAN MEDICAL ASSN. Vol.261: 3561–3566. 1989.

Gray 1983. Joel E. Gray + Alan D. Hoffman + Hamlet A. Peterson, "Reduction of Radiation Exposure during Radiography for Scoliosis," JOURNAL OF BONE and JOINT SURGERY 65–A: 5–12. Jan.1983.

Greenfield 1986. Maurice M. Greenfield, book review of Xrays: Health Effects of Common Exams, by Gofman + O'Connor (1985), in NEW ENGLAND JOURNAL OF MEDICINE Vol.314, No.6: 393. Feb. 6, 1986.

Grove 1968. Robert D. Grove + Alice M. Hetzel, VITAL STATISTICS RATES IN THE UNITED STATES 1940–1960. 881 pages. No index. Public Health Service Publication 1677. Public Health Service, National Center for Health Statistics (Division of Vital Statistics), U.S. Dept. of Health, Education, and Welfare. U.S Government Printing Office. 1968. UCSF Library: HA211, A3, 1940/60.

Gruppo 1986. Gruppo Italiano per lo Studio della Streptochinasi nell'Infarto Miocardico (GISSI), "Effectiveness of Intravenous Thrombolytic Treatment in Acute Myocardial Infarction," LANCET 1: 397–402. 1986.

Gupta 1997. S. Gupta + E.W. Leatham + D. Carrington + M.A. Mendall + J.C. Kaski + A.J. Camm, "Elevated Chlamydia Pneumoniae Antibodies, CardioVascular Events, and Azithromycin in Male Survivors of Myocardial Infarction," CIRCULATION Vol.96: 404–407. 1997.

Gurd 1949. F.R.N. Gurd + J.L. Oncley + John T. Edsall + Edwin J. Cohn, "The Lipo–Proteins of Human Plasma," FARADAY DISCUSS. CHEM. SOCIETY Vol.6: 70. 1949.

Guyton 1985. J.R. Guyton + T.M.A. Bocan, "Human Aortic Fibrolipid Lesions: Progenitor Lesions for Fibrous Plaques, Exhibiting Early Formation of the Cholesterol–Rich Core," AMER. J. OF PATH. Vol.120: 193–206. 1985.

Gyorkey 1984. F. Gyorkey + J.L. Melnick + G.A. Guinn + P. Gyorkey + M.E. DeBakey, "Herpesviridae in the Endothelial and Smooth Muscle Cells of the Proximal Aorta in Atherosclerosis Patients," EXPERIMENTAL MOLECULAR PATHOLOGY Vol.40: 328–339. 1984.

Hackett 1988. D. Hackett + G. Davies + Attilio Maseri, "Pre-Existing Coronary Stenoses in Patients with First Myocardial Infarction Are Not Necessarily Severe," EUROPEAN HEART JOURNAL Vol.9: 1317-1323. 1988.

Haft 1997. Jacob I. Haft + Ayman J. Hammoudeh, "More on Coronary-Plaque Rupture Triggered by Snow Shoveling," (letter), NEW ENGLAND JOURNAL OF MEDICINE Vol.336, No.23: 1678-1679. June 5, 1997.

Hajjar 1988. D.P. Hajjar + C. Fabricant + C.R. Minick + J. Fabricant, "Herpesvirus Infection Alters Aortic Cholesterol Metabolism and Accumulation," AMERICAN JOURNAL OF PATHOLOGY Vol.122: 62-70. 1988.

Hall 1990. Judith G. Hall, "Genomic Imprinting: Review and Relevance to Human Diseases," AMERICAN JOURNAL OF HUMAN GENETICS Vol.46: 857-873. 1990. See also: Lyon 1968.

Hammerstein 1979. G.R. Hammerstein + D.W. Miller + D.R. White et al, "Absorbed Radiation Dose in Mammography," RADIOLOGY Vol. 130: 485-491. 1979.

Hammond 1961. E. Cuyler Hammond + Lawrence Garfinkel, "Smoking Habits of Men and Women," JOURNAL OF THE NATIONAL CANCER INSTITUTE Vol.27, No.2: 419-442. August 1961.

Hammoudeh 1996. Ayman J. Hammoudeh + Jacob I. Haft, "Coronary-Plaque Rupture in Acute Coronary Syndromses Triggered by Snow Shoveling," NEW ENGLAND JOURNAL OF MEDICINE Vol.335: 2001. 1996.

Hardin 1973. N.J. Hardin + C.R. Minick + G.E. Murphy, "Experimental Induction of Atherosclerosis by the Synergy of Allergic Injury to Arteries and Lipid-Rich Diet. III, Role of Earlier Acquired Fibromuscular Intimal Thickening in the Pathogenesis of Later Developing Atherosclerosis," AMERICAN JOURNAL OF PATHOLOGY Vol.73: 301-325. 1973. See also Minick 1973.

Harker 1974. L.A. Harker + S.J. Slichter + C.R. Scott + R. Ross, "Homocystinemia: Vascular Injury and Arterial Thrombosis," NEW ENGLAND JOURNAL OF MEDICINE Vol.291: 537-543. 1974.

Harker 1976. L. Harker et al, "Homocystine-Induced Arteriosclerosis: The Role of Endothelial Cell Injury and Platelet Response in Its Genesis," J. CLIN. INVEST. Vol.58: 731-741. 1976.

Harris 1989. William S. Harris, "Fish Oils and Plasma Lipid and Lipoprotein Metabolism in Humans: A Critical Review," JOURNAL OF LIPID RESEARCH Vol.30: 785-807. 1989.

Harvey 1985. Elizabeth B. Harvey + John D. Boice, Jr. et al, "Prenatal Xray Exposure and Childhood Cancer in Twins," NEW ENGLAND JOURNAL OF MEDICINE Vol.312, No.9: 541-545. 1985.

Haskell 1994. (SCRIP Study.) W. L. Haskell + E.L. Alderman + J.M. Fair et al, "Effects of Intensive Multiple Risk Factor Reduction on Coronary Atherosclerosis and Clinical Cardiac Events in Men and Women with Coronary Artery Disease. The Stanford Coronary Risk Intervention Project (SCRIP), CIRCULATION Vol.89: 975-990. 1994.

Haust 1960. M.D. Haust + R.H. More + H.Z. Movat, "The Role of Smooth Muscle Cells in the Fibrogenesis of Atherosclerosis," AMERICAN JOURNAL OF PATHOLOGY Vol.37: 377-389. 1960.

Haust 1971. M.D. Haust, "The Morphogenesis and Fate of Potential and Early Atherosclerotic Lesions in Man," HUMAN PATHOLOGY Vol.2: 1-29. 1971.

Havel 1995. R.J. Havel + E. Rapaport, "Management of Primary Hyperlipidemia," NEW ENGLAND JOURNAL OF MEDICINE Vol.332: 1491-1498. 1995.

Heath 1995. Clark W. Heath, Jr., book review of Preventing Breast Cancer: The Story of a Major, Proven, Preventable Cause of This Disease, by John W. Gofman (1995), in JOURNAL OF THE AMERICAN MEDICAL ASSN. Vol.274, No.8: 657. August 23/30, 1995.

Hebert 1997. Patricia R. Hebert + J. Michael Gaziano + Ki Sau Chan + Charles H. Hennekens, "Cholesterol Lowering with Statin Drugs, Risk of Stroke, and Total Mortality: An Overview of Randomized Trials," JOURNAL OF THE AMERICAN MEDICAL ASSN. Vol.278, No.4: 313-321. 1997.

Hegsted 1998. D.M. Hegsted, "Dietary Fat Intake and the Risk of Coronary Heart Disease in Women," (letter), NEW ENGLAND J OF MEDICINE Vol.338, No.13: 917-918. March 26, 1998.

Hei 1997. Tom K. Hei + Li-Jun Wu + Su-Xian Liu + Diane Vannais + Charles A. Waldren + Gerhard Randers-Pehrson, "Mutagenic Effects of a Single and an Exact Number of Alpha Particles in Mammalian Cells," PROCEEDINGS of the NATL. ACADEMY OF SCIENCES (USA) Vol.94: 3765-3770. April 1997.

Helin 1971. P. Helin et al, "Arteriosclerosis in Rabbit Aorta Induced by Mechanical Dilation ..." ATHEROSCLEROSIS Vol.13: 319-331. 1971.

Henderson 1971. W.J. Henderson + C.A.F. Joslin + A.C. Turnbull + K. Griffiths, "Talc and Carcinoma of the Ovary and Cervix," J. OF OBSTET. & GYNAECOL. BR. COMMONWEALTH Vol.78: 266-272. 1971.

Hickey 1923. Preston Hickey, "The Effect of the War on the Development of Roentgenology," AMERICAN JOURNAL OF ROENTGENOLOGY AND RADIUM THERAPY Vol.10: 70-75. 1923.

Hill 1965. C.R. Hill, "Polonium-210 in Man," NATURE Vol.208: 423-428. 1965.

Hill 1992. Rolla B. Hill + Robert E. Anderson, "The Autopsy in Oncology," (opinion article), CA - A CANCER JOURNAL FOR CLINICIANS Vol.42, No.1: 47-56. Jan/Feb. 1992.

Ho 1998. Gloria Y.F. Ho + 4 co-workers, "Natural History of CervicoVaginal PapillomaVirus Infection in Young Women," NEW ENGLAND JOURNAL OF MEDICINE Vol.338, No.7: 423-428. February 12, 1998.

Hokanson 1996.　J.E. Hokanson + M. Austin, "Plasma Triglyceride Level Is a Risk Factor for CardioVascular Disease Independent of High Density Lipoprotein Levels: A Meta-Analysis of Population-Based Prospective Studies," JOURNAL OF CARDIOVASCULAR RISK Vol.3: 213-229. 1996.

Holm 1988.　Lars-Erik Holm + 11 co-workers, "Thyroid Cancer after Diagnostic Doses of Iodine-131: A Retrospective Cohort Study," JOURNAL OF THE NATIONAL CANCER INSTITUTE (USA) Vol.80: 1132-1138. 1988.

Holmberg 1993.　Kerstin Holmberg + Susann Falt + Annelie Johansson + Bo Lambert, "Clonal Chromosome Aberrations and Genomic Instability in X-Irradiated Human T-Lymphocyte Cultures," MUTATION RESEARCH Vol.286: 321-330. 1993.

Holme 1990.　I. Holme, "An Analysis of Randomized Trials Evaluating the Effect of Cholesterol Reduction on Total Mortality and Coronary Heart Disease Incidence," CIRCULATION Vol.82: 1916-1924. 1990.

Holmes 1999.　Michelle D. Holmes + 7 co-workers, "Association of Dietary Intake of Fat and Fatty Acids with Risk of Breast Cancer," JOURNAL OF THE AMERICAN MEDICAL ASSN Vol.218, No.10: 914-920. March 10, 1999.

Holtzman 1966.　R.B. Holtzman + F.H. Ilcewicz, "Lead-210 and Polonium-210 in Tissues of Cigarette Smokers," SCIENCE Vol.153: 1259-1260. 1966. See also Holtzman 1967 in SCIENCE Vol.155: 607.

Howe 1984.　Geoffrey R. Howe, "Epidemiology of Radiogenic Breast Cancer," pp.119-129 in RADIATION CARCINOGENESIS EPIDEMIOLOGY AND BIOLOGICAL SIGNIFICANCE, edited by John D. Boice, Jr. + Joseph F. Fraumeni. (Raven Press, New York.) 1984.

Hrubec 1989.　Zdenek Hrubec + John D. Boice, Jr. + Richard R. Monson + Marvin Rosenstein, "Breast Cancer after Multiple Chest Fluoroscopies," CANCER RESEARCH Vol.49: 229-234. 1989.

Hsieh 1999.　Hsieh W.A. + six co-workers, "Alpha Coefficient of Dose-Response for Chromosome Translocations Measured by FISH in Human Lymphocytes Exposed to Chronic Cobalt-60 Gamma Rays at Body Temperature," INTERNATIONAL JOURNAL OF RADIATION BIOLOGY Vol.75, No.4: 435-439. 1999.

Hu 1998.　Frank B. Hu + Meir J. Stampfer + Walter C. Willett, "Dietary Fat Intake and the Risk of Coronary Heart Disease in Women," (letter), NEW ENGLAND J OF MEDICINE Vol.338, No.13: 918-919. March 26, 1998.

Huda 1984.　W. Huda, "Is Energy Imparted a Good Measure of the Radiation Risk Associated with CT Examination?" PHYS. MED. BIOL. Vol. 29: 1137-1142. 1984.

Huff 1972.　Harry Huff + E. Max Sanders, correspondence, "Coronary-Artery Occlusion after Radiation," NEW ENGLAND JOURNAL OF MEDICINE Vol.286, No.14: 780. April 6, 1972.

Hulley 1992.　S.B. Hulley + J.M.B. Walsh + T.B. Newman, "Health Policy on Blood Cholesterol: Time to Change Directions," CIRCULATION Vol.86: 1026-1029. 1992.

Hurst 1959.　D.W. Hurst, "Radiation Fibrosis of Pericardium with Cardiac Tamponade; Case Report with Post-Mortem Studies and Review of the Literature," CANADIAN MEDICAL ASSN. JOURNAL Vol.81: 377-380. September 1, 1959.

Husik 1926.　David N. Husik, "Thymic Death in an Adult during Tonsillectomy under Local Anesthesia," ATLANTIC MEDICAL JOURNAL 857-869. September 1926.

IARC 1995.　International Agency for Research on Cancer, HUMAN PAPILLOMAVIRUS. IARC Monograph Eval. Carcinogenic Risk Hum 64. 1995.

● ICRP 1990 (1991).　International Commission on Radiological Protection, RECOMMENDATIONS OF THE ICRP. Publication 60. Annals of the ICRP 21 (1-3). (Pergamon Press, Oxford, England.) 1991.

ICRU 1986.　International Commission on Radiation Units and Measurements, THE QUALITY FACTOR IN RADIATION PROTECTION. ICRU Report 40. 1986.

Inagaki 1988.　M. Inagaki + W.S. Harris + K. Arakawa, "Plasma Lipid and Apoprotein Levels in Hypertriglyceridemic Patients Taking Fish Oil," ARTERIOSCLEROSIS Vol.8: 634-A. 1988.

International Classification of Diseases (ICD).
　　　Please see WHO 1958 in this Reference List.

Isherwood 1978.　I. Isherwood + B.R. Pullan + R. Ritchings, "Radiation Dose in Neuroradiological Procedures," NEURORADIOLOGY Vol.16: 477-481. 1978.

Iverius 1973.　P-H Iverius, "Possible Role of the Glycosaminoglycans in the Genesis of Atherosclerosis," in ATHEROGENESIS: INITIATING FACTORS, Ciba Foundation Symposium, Vol.12 NS: 185-193. 1973.

Jankowski 1984.　J. Jankowski, "Organ Doses in Diagnostic X-Ray Procedures," HEALTH PHYSICS 46: 228-234. 1984.

Johnson + Goetz 1986.　David W. Johnson + Walter A. Goetz, "Patient Exposure Trends in Medical and Dental Radiography," HEALTH PHYSICS Vol.50, No.1: 107-116. 1986.

Joint 1990.　Joint Statement by the American Heart Assn. and the National Heart, Lung, and Blood Institute, "The Cholesterol Facts: A Summary of the Evidence Relating Dietary Fats, Serum Cholesterol, and Coronary Heart Disease," CIRCULATION Vol.81: 1721-1733. 1990.

Jonasson 1986.　L. Jonasson + J. Holm + O. Skalli et al, "Regional Accumulation of T Cells, Macrophages, and Smooth Muscle Cells in the Human Atherosclerotic Plaque," ARTERIOSCLEROSIS Vol.6: 131-138. 1986.

Jones 1951.　Hardin B. Jones + John W. Gofman + Frank T. Lindgren + Thomas P. Lyon + Dean M. Graham + Beverly Strisower + Alex V. Nichols , "Lipoproteins in Atherosclerosis," THE AMERICAN JOURNAL OF MEDICINE Vol.11, No.3: 358-380. September 1951.

Jones 1960. A. Jones + J. Wedgwood, "Effects of Radiations on the Heart," BRITISH JOURNAL OF RADIOLOGY Vol.33: 138-158. March 1960.

Jones 1996. Frank Jones, THE SAVE YOUR HEART WINE GUIDE. 244 pages. ISBN 0-312-14729-5. (A Thomas Dunne Book, an imprint of St. Martin's Press, New York City.) 1996.

Juchau 1974. M.R. Juchau + 5 co-workers, "Studies on Human Placenta Carbon Monoxide-Binding Cytochromes," DRUG METAB. DISPOSIT. Vol.2: 79-86. 1974.

Jukema 1995. (REGRESS Study.) J.W. Jukema + A.V.G. Bruschke + A.J. van Boven et al, "Effects of Lipid Lowering by Pravastatin on Progression and Regression of Coronary Artery Disease in Symptomatic Men with Normal to Moderately Elevated Serum Cholesterol Levels: The Regression Growth Evaluation Statin Study (REGRESS)," CIRCULATION Vol.91: 2528-2540. 1995.

Kadhim 1992. Munira A. Kadhim + D.A. MacDonald + Dudley T. Goodhead + Sally A. Lorimore + 2 co-workers, "Transmission of Chromosomal Instability after Plutonium Alpha-Particle Irradiation," NATURE Vol.355: 738-740. 1992.

Kadhim 1994. Munira A. Kadhim + Sally A. Lorimore + Mary D. Hepburn + Dudley T. Goodhead + 2 co-workers, "Alpha-Particle Induced Chromosomal Instability in Human Bone-Marrow Cells," LANCET Vol.344: 987-988. 1994.

Kadhim 1995. Munira A. Kadhim + Sally A. Lorimore + K.M.S. Townsend + Dudley T. Goodhead + 2 co-workers, "Radiation-Induced Genomic Instability: Delayed Cytogenetic Aberrations and Apoptosis in Primary Human Bone-Marrow Cells," INTERNATL. J. OF RADIATION BIOLOGY Vol.67: 287-293. 1995.

Kallioniemi 1992. Anne Kallioniemi + Olli-P. Kallioniemi + Damir Sudar + 4 co-workers, "Comparative Genomic Hybridization for Molecular Cytogenetic Analysis of Solid Tumors," SCIENCE Vol.258: 818-821. 1992.

Kane 1990. (UC-SCOR Study.) J.P. Kane + M.J. Malloy + T.A Ports + N.R. Phillips + J.C. Diehl + R.J. Havel, "Regression of Coronary Atherosclerosis during Treatment of Familial Hypercholesterolemia with Combined Drug Regimens," JOURNAL OF THE AMERICAN MEDICAL ASSN. Vol.264: 3007-3012. 1990.

Kardinal 1993. A.F.M. Kardinal + F.J. Kok + J. Ringstad, et al, "AntiOxidants in Adipose Tissue and Risk of Myocardial Infarction: The Euramic Study," LANCET Vol.342: 1379-1384. 1993.

Katan 1995. Martijn B. Katan + Peter L. Zock + Ronald Mensink, "Dietary Oils, Serum Lipoproteins, and Coronary Heart Disease," AMERICAN J. CLINICAL NUTRITION Vol.61 (Suppl.): 1368S-1373S. 1995.

Katan 1997. Martijn B. Katan, Scott M. Grundy, Walter C. Willett: Clinical Debate. Please see Connor 1997.

Kazakov 1992. Vasili S. Kazakov + Evgeni P. Demidchik + Larisa N. Astakova, "Thyroid Cancer after Chernobyl," (correspondence) NATURE Vol.359: 21. Sept. 3, 1992. See also Baverstock 1992.

Kellerer 1987. Albrecht M. Kellerer, "Models of Cellular Radiation Action," pp.305-375 in KINETICS OF NONHOMOGENEOUS PROCESSES, edited by Gordon R. Freeman. (John Wiley & Sons, New York.) 1987.

Kellner 1954. A. Kellner, pp.42-49 in SYMPOSIUM ON ATHEROSCLEROSIS, National Academy of Sciences, Washington, D.C. 1954.

Kelly 1975. Kevin M. Kelly + Dale A. Madden + Joseph Arcarese + Mark Bennett + Reynold F. Brown, "The Utilization and Efficacy of Pelvimetry," AMERICAN JOURNAL OF ROENTGENOLOGY AND RADIUM THERAPY Vol.125, No.1: 66-74. September 1975.

Kemp 1993. C.J. Kemp + L.A. Donehower + A. Bradley + A. Balmain, "Reduction of p53 Gene Dosage Does Not Increase Initiation or Promotion but Enhances Malignant Progression of Chemically Induced Skin Tumors," CELL Vol.74: 813+. 1993.

Keys 1984. A. Keys + A. Menotti + C. Aravanis et al, "The Seven Countries Study: 2289 Deaths in 15 Years," JOURNAL OF PREVENTIVE MEDICINE Vol.13: 141-154. 1984.

Knapp 1986. H.R. Knapp + I.A.G. Reilly + P. Alessandrini + G.A. Fitzgerald, "In Vivo Indexes of Platelet and Vascular Function during Fish Oil Administration in Patients with Atherosclerosis," NEW ENGL.J.MED. 31: 937-942. 1986.

Knopp 1997. Robert H. Knopp + Carolyn E. Walden + Barbara M. Retzlaff + Barbara S. McCann + Alice A. Dowdy + John J. Albers + George O. Gey + Manuel N. Cooper, "Long-Term Cholesterol-Lowering Effects of 4 Fat-Restricted Diets in Hypercholesterolemia and Combined Hyperlipidemic Men: The Dietary Alternatives Study," JOURNAL OF THE AMERICAN MEDICAL ASSN. Vol.278, No. 18: 1509-1515. November 12, 1997.

Kodama 1993. Yoshiaki Kodama + Mimako Nakano + Kazuo Ohtaki + Akio A. Awa + Joe N. Lucas + Tore Straume + D. Pinkel + J.W. Gray, "Biotechnology Contributes to Biological Dosimetry: Using Fluorescence In-Situ Hybridization to Detect Chromosome Translocations, Radiation- and Chemical-Induced Chromosome Changes That Can Be Identified Decades after Exposure," RERF UPDATE Vol.4: 6-7. Winter 1992-1993. (Radiation Effects Research Foundation, Hiroshima, Japan.) 1993.

Kodama 1993-b. Kazunori Kodama, "CardioVascular Disease in Atomic-Bomb Survivors," in RERF UPDATE Vol.5, No.4: 3-4. Winter 1993-94. (Rad'n Effects Research Fdn., Hiroshima.) Preceded by Shimizu 1992.

Kol 1997. Amir Kol + Giovanni Sperti + Attilio Maseri, "Association between Prior Cytomegalovirus Infections and the Risk of Restenosis...", (letter), NEW ENGLAND J. MED. Vol.336: 587-588. Feb. 20, 1997.

Kolb 1955. Felix O. Kolb + Oliver F. DeLalla + John W. Gofman, "The Hyperlipemias in Disorders of Carbohydrate Metabolism: Serial Lipoprotein Studies in Diabetic Acidosis with Xanthomatosis and in Glycogen Storage Disease," METABOLISM Vol.4, No.4: 310–317. July 1955.

Kramsch 1971. D.M. Kramsch + C. Franzblau + W. Hollander, "The Protein and Lipid Composition of Arterial Elastin and Its Relationship to Lipid Accumulation in the Atherosclerotic Plaque," J. CLIN. INVEST. Vol.50: 1666+. 1971.

Krauss 1982. Ronald M. Krauss + David J. Burke, "Identification of Multiple Subclasses of Plasma Low Density Lipoproteins in Normal Humans," JOURNAL OF LIPID RESEARCH Vol.23: 97–104. 1982.

Krauss 1987. Ronald M. Krauss + Paul T. Williams + John Brensike + Katherine M. Detre + Frank T. Lindgren et al, "Intermediate–Density Lipoproteins and Progression of Coronary Artery Disease In Hypercholesterolaemic Men," THE LANCET, July 11, 1987, pp.62–66.

Krauss 1994. Ronald M. Krauss, "Heterogeneity of Plasma Low–Density Lipoproteins and Atherosclerosis Risk," (review), CURRENT OPINION IN LIPIDOLOGY Vol.5: 339–349. 1994.

Krieger 1996. Lisa M. Krieger, "Consumers Rebel over HMO Practices," SAN FRANCISCO EXAMINER p.A–1, March 10, 1996.

Kromhout 1985. Daan Kromhout + E.B. Bosschieter + C.L. Coulnader, "The Inverse Relation between Fish Consumption and 20–Year Mortality from Coronary Heart Disease," NEW ENGLAND J. MED. Vol.312: 1205–1209. 1985.

Kronenberg 1994. A. Kronenberg, "Radiation–Induced Genomic Instability," INTERNATIONAL JOURNAL OF RADIATION BIOLOGY Vol.66: 603–609. 1994.

Ku 1985. D.N. Ku + D.P. Giddens + C.K. Zarins + S. Glagov, "Pulsatile Flow and Atherosclerosis in the Human Carotid Bifurcation; Positive Correlation between Plaque Location and Low and Oscillating Shear Stress," ARTERIOSCLEROSIS Vol.5: 293+. 1985.

Kucerova 1972. M. Kucerova + A.B.J. Anderson + K.E. Buckton + H.J. Evans, "Xray–Induced Chromosome Aberrations in Human Peripheral Blood Lymphocytes," INTERNATL. J. RADIATION BIOL. Vol.21: 389–396. 1972.

Kuhn 1985. H.F. Kuhn, "Methods for Reducing Patient Dose: Rare Earth Screens, Filtration, Spot–Film Technique and Digital Radiography," BRITISH JOURNAL OF RADIOLOGY Vol.18 (Suppl.): 37–39. 1985.

Kuntz 1949. A. Kuntz + N.M. Sulkin, "Lesions Induced in Rabbits by Cholesterol Feeding, with Special Reference to Their Origin," ARCHIVES OF PATHOLOGY Vol.47: 248+. 1949.

Lamarche 1997. Benoit Lamarche + 6 co–workers, "Small, Dense Low–Density Lipoprotein Particles as a Predictor of the Risk of Ischemic Heart Disease in Men: Prospective Results from the Quebec Cardiovascular Study," CIRCULATION Vol.95, No.1: 69–75. January 7, 1997.

Lamarche 1998. Benoit Lamarche + co–workers, "Triglycerides and Small, Dense Low–Density Lipoprotein," (letter), JOURNAL OF THE AMERICAN MEDICAL ASSN Vol.280, No.23: 1990–91. Dec. 16, 1998.

Lamberts 1964. H.B. Lamberts + W.G.R.M. de Boer, "Contributions to the Study of Immediate and Early X–Ray Reactions with Regard to Chemoprotection: IX. X–Ray–Induced Coronary Occlusion Leading to Heart Damage in Rabbits," INTERNATL. JOURNAL OF RADIATION BIOLOGY Vol.8, No.4: 359–365. 1964.

Landis 1998. Sarah H. Landis + Taylor Murray + Sherry Bolden + Phyllis A. Wingo, "Cancer Statistics, 1998," CA – A CANCER JOURNAL FOR CLINICIANS Vol.48, No.1: 6–29. January–February 1998.

Landrigan 1989. Philip J. Landrigan + S. Markowitz, "Current Magnitude of Occupational Disease in the United States: Estimates from New York State," ANNALS OF NEW YORK ACADEMY OF SCIENCES 572: 27–45. 1989.

Landrigan 1996. Philip J. Landrigan, "The Prevention of Occupational Cancer," guest editorial in CA – A CANCER JOURNAL FOR CLINICIANS Vol.46, No.2: 67–69. March/April 1996.

Laurent 1963. T.C. Laurent et al, "On the Interaction between Polysaccharides and Other Macromolecules," BIOCHIM. BIOPHYS. ACTA Vol.78: 351–359. 1963.

Laws 1980. Priscilla Laws + Marvin Rosenstein, "Quantitative Analysis of the Reduction in Organ Doses in Diagnostic Radiology by Means of Entrance Exposure Guidelines." U.S. Dept. of Health, Education and Welfare (HEW). Food and Drug Admin. (FDA). HEW (FDA) Publication 80–8107. 1980.

Leaf 1999. Alexander Leaf, "Dietary Prevention of Coronary Heart Disease: The Lyon Diet Heart Study," (editorial), CIRCULATION Vol.99: 733–735. February 16, 1999.

Leary 1949. T. Leary, "Crystalline Ester Cholesterol and Atherosclerosis," ARCHIVES OF PATHOLOGY Vol.47: 1+. 1949.

Leddy 1937. Eugene T. Leddy, "The Dangers of Roentgenoscopy: Summary and Recommendations," (editorial), AMERICAN JOURNAL OF ROENTGENOLOGY AND RADIUM THERAPY Vol.38: 924–927. 1937.

Leibovic 1983. S.J. Leibovic + W.J.H. Caldicott, "Gastrointestinal Fluoroscopy: Patient Dose and Methods for its Reduction," BRITISH JOURNAL OF RADIOLOGY Vol.56: 715–719. 1983.

Libby 1995. Peter Libby, "Molecular Bases of the Acute Coronary Syndromes: From Bench to Bedside," CIRCULATION Vol.91, No.11: 2844–2850. June 1, 1995.

Libby 1997. Peter Libby + Debra Egan + Sonia Skarlatos, "Roles of Infectious Agents in Atherosclerosis and Restenosis: An Assessment of the Evidence & Need for Future Research," CIRCULATION Vol.96: 4095–4103. Dec.2, 1997.

Lichtenstein 1996. Daniel A. Lichtenstein et al, "Chronic Radiodermatitis Following Cardiac Catheterization," ARCHIVES OF DERMATOLOGY Vol.132: 663–667. 1996.

Linder 1947.   Forrest E. Linder + Robert D. Grove, VITAL STATISTICS RATES IN THE UNITED STATES 1900–1940.
          1,051 pages. Government Printing Office, Washington, DC. 1947.

Linder 1965.   D. Linder + Stanley M. Gartler, "Glucose–6–Phosphate Dehydrogenase Mosaicism: Utilization as a Cell Marker
          in the Study of Leiomyomas," SCIENCE Vol.150: 67–69. 1965.

Lindgren 1950.   Frank T. Lindgren + Harold A. Elliott + John W. Gofman + Beverly Strisower, "The Ultracentrifugal Composi-
          tion of Normal Rabbit Serum," JOURNAL OF BIOLOGICAL CHEMISTRY Vol.182, No.1: 1–4. January 1950.

Lindgren 1951.   Frank T. Lindgren + Harold A. Elliott + John W. Gofman, "The Ultracentrifugal Characterization and Isolation
          of Human Blood Lipids and Lipoproteins, with Applications to the Study of Atherosclerosis,"
          JOURNAL OF PHYSICAL AND COLLOID CHEMISTRY Vol.55, No.1: 80–93. January 1951.

Lindgren 1955.   Frank T. Lindgren + Alex V. Nichols + Norman K. Freeman, "Physical and Chemical Composition Studies on the
          Lipoproteins of Fasting and Heparinized Human Sera," JOURNAL OF PHYSICAL CHEMISTRY Vol.59:
          930–938. 1955.

Lindgren 1956.   Frank T. Lindgren + Norman K. Freeman + Alex V. Nichols + John W. Gofman, "The Physical Chemistry of
          Lipoprotein Transformation," pp.224–242 in THE BLOOD LIPIDS AND THE CLEARING FACTOR,
          proceedings of the Third International Conference on Biochemical Problems of Lipids
          (Brussels), July 26–28, 1956.

Lindgren 1957.   Frank T. Lindgren + John W. Gofman, "The Role of Lipoproteins in Coronary Disease," BULLETIN OF THE
          SWISS ACADEMY OF MEDICAL SCIENCES Vol.13, Fasc.1–4: 152–178. 1957. From Symposium on
          Arteriosclerosis, in Basle, August 8–10, 1956.

Lindgren 1959.   Frank T. Lindgren + Alex V. Nichols + Thomas L. Hayes + Norman K. Freeman + John W. Gofman,
          "Structure and Homogeneity of the Low–Density Serum Lipoproteins," ANNALS OF THE NEW YORK
          ACADEMY OF SCIENCES Vol.72, Article 14: 826–844. June 16, 1959.

Lindgren 1960.   Frank T. Lindgren + Alex V. Nichols, "Structure and Function of Human Serum Lipoproteins," Chapter 11 (pp.
          1–58) in THE PLASMA PROTEINS Vol.2: Biosynthesis, Metabolism, Alterations in Disease, edited by Frank W.
          Putnam. LCCN 59–15756. (Academic Press, New York, London). UCSF Lib'y: QP 99.3, P7, P87, v.2 1960.

Linser 1904.   P. Linser, in FORTSCHR. a. d. GEB. d. ROENTGENSTRAHLEN Vol.8: 97. 1904.

Lipid 1995.   (Long–Term Intervention with Pravastatin, in Ischemic Disease.) Lipid Study Group, "Design Features
          and Baseline Characteristics of the LIPID Study: A Randomized Trial in Patients with Previous Acute Myocardial
          Infarction and/or Unstable Angina Pectoris," AMERICAN J. OF CARDIOLOGY Vol.76: 474–479. 1995.

LipidRC 1984.   "Lipid Research Clinics Coronary Primary Prevention Trial Results I. Reduction in Incidence of Coronary
          Heart Disease," and "Results II. The Relationship of Reduction in Incidence of Coronary Heart Disease to
          Cholesterol Lowering," J. OF THE AMER. MEDICAL ASSN. Vol.251: 351–374. 1984.

Little 1965.   J.B. Little + Edward P. Radford + H.L. McCombs + V.R. Hunt, "Distribution of Polonium–210 in Pulmonary
          Tissues of Cigarette Smokers," NEW ENGLAND J. OF MED. Vol.273: 1343–1351. 1965.

Little 1988.   William C. Little + Martin Constantinescu + Robert J. Applegate + Michael A. Kutcher + 3 colleagues, "Can
          Coronary Angiography Predict the Site of a Subsequent Myocardial Infarction in Patients with Mild–to–Moderate
          Coronary Artery Disease?" CIRCULATION Vol.78, No.5: 1157–1166. November 1988.

Little 1997.   John B. Little, "What Are the Risks of Low–Level Exposure to Alpha–Radiation from Radon?" (commentary),
          PROCEEDINGS OF THE NATL. ACADEMY OF SCIENCES (USA) Vol.94: 5996–5997. June 1997.

Liuzzo 1994.   G. Liuzzo et al, "The Prognostic Value of C–Reactive Protein and Serum Amyloid A Protein in Severe
          Unstable Angina," NEW ENGLAND J. MED. Vol.331: 417–424. 1994.

Lloyd 1986.   David C. Lloyd + A.A. Edwards + J.S. Prosser, "Chromosome Aberrations Induced in Human Lymphocytes by
          in Vitro Acute X and Gamma Radiation," RADIATION PROTECTION & DOSIMETRY Vol.15: 83–88. 1986.

Lloyd 1988.   David C. Lloyd + 8 co–workers, "Frequencies of Chromosomal Aberrations Induced in Human Blood Lymphocytes
          by Low Doses of Xrays," INTERNATIONAL J. OF RADIATION BIOLOGY Vol.53, No.1: 49–55. 1988.

Lloyd 1992.   David C. Lloyd + 10 co–workers, "Chromosomal Aberrations in Human Lymphocytes Induced in Vitro by Very
          Low Doses of Xrays," INTERNATIONAL JOURNAL OF RADIATION BIOLOGY Vol.61: 335–343. 1992.

Longo 1979.   D.L. Longo + R.C. Young, "Cosmetic Talc and Ovarian Cancer," LANCET ii: 349–351. 1979.

Loree 1992.   H.M. Loree + R.D. Kamm + R.G. Stringfellow + R.T. Lee, "Effects of Fibrous Cap Thickness on Peak
          Circumferential Stress in Model Atherosclerotic Vessels," CIRC. RESEARCH 71: 850–858. 1992.

Lorgeril.
          Please see DeLorgeril.

Lucas 1992.   Joe N. Lucas + Akio Awa + Tore Straume + M. Poggensee + Yoshiaki Kodama + 6 additional co–workers, "Rapid
          Translocation Frequency Analysis in Humans Decades after Exposure to Ionizing Radiation,"
          INTERNATIONAL JOURNAL OF RADIATION BIOLOGY Vol.62, No.1: 53–63. 1992.

Lucas 1995.   Joe N. Lucas + F. Hill + C. Burk + T. Fester + Tore Straume, "Dose–Response Curve for Chromosome
          Translocations Measured in Human Lymphocytes Exposed to Cobalt–60 Gamma Rays," HEALTH PHYSICS Vol.68,
          No.6: 761–765. June 1995.

Lucas 1997.   Joe N. Lucas, "Dose Reconstruction for Individuals Exposed to Ionizing Radiation Using Chromosome Painting," RADIATION RESEARCH Vol.148: S33–S38. 1997.

Lucas 1999.   Joe N. Lucas + 9 co-workers, "Background Ionizing Radiation Plays a Minor Role in the Production of Chromosome Translocations in a Control Population," INTERNATIONAL JOURNAL OF RADIATION BIOLOGY Vol.75: 819–827. 1999.

Lyon 1968.   M.F. Lyon, "Chromosomal and Subchromosomal Inactivation," ANNUAL REVIEW OF GENETICS Vol.2: 31–51. 1968. See also: Hall 1990.

MAAS 1994.   (MultiCentre Anti-Atheroma Study.) MAAS Investigators, "Effect of Simvastatin on Coronary Atheroma: The Multicentre Anti-Atheroma Study (MAAS)," LANCET Vol.344: 633–638. 1994. (See also Erratum, LANCET 344: 762. 1994.)

MacIsaac 1993.   Andrew I. MacIsaac + James D. Thomas + Eric J. Topol, "Toward the Quiescent Coronary Plaque," (review paper), JOURNAL of the AMER. COLLEGE OF CARDIOLOGY Vol.22: 1228–1241. Oct.1993.

MacKee 1922.   George M. MacKee + George C. Andrews, "The Value of Roentgen Therapy in Dermatology," AMERICAN JOURNAL OF ROENTGENOLOGY AND RADIUM THERAPY Vol.9: 241–246. 1922.

MacKee 1938.   George M. MacKee, XRAYS AND RADIUM IN THE TREATMENT OF DISEASES OF THE SKIN, Third Edition. 830 pages. Several chapters have co-authors. (Lea & Febiger, Malvern PA 19355 USA.) 1938.

MacKenzie 1965.   Ian MacKenzie, "Breast Cancer Following Multiple Fluoroscopies," BRITISH JOURNAL OF CANCER Vol.19: 1–8. March 1965.

MacMahon 1962.   Brian MacMahon, "PreNatal Xray Exposure and Childhood Cancer," JOURNAL OF THE NATIONAL CANCER INSTITUTE (USA) Vol.28: 1173–1191. 1962.

Mahan 1975.   Bruce H. Mahan, UNIVERSITY CHEMISTRY, Third Edition. 894 pages. LCCN 74–19696. ISBN 0–201–04405–6. (Addison-Wesley Publishing Co., Reading, Massachusetts.) 1975.

Majesky 1985.   Mark W. Majesky + Michael A. Reidy + Earl P. Benditt + Mont R. Juchau, "Focal Smooth Muscle Cell Proliferation in the Aortic Intima Produced by an Initiation-Promotion Sequence," PROCEEDINGS OF THE NATIONAL ACADEMY OF SCIENCES (USA), Vol.82: 3450–3454. May 1985.

Malinow 1994.   M.R. Malinow, "Homocyst(e)ine and Arterial Occlusive Diseases," JOURNAL OF INTERNAL MEDICINE Vol.236: 603–617. 1994.

Marder 1993.   Brad A. Marder + William F. Morgan, "Delayed Chromosomal Instability Induced by DNA Damage," MOLECULAR AND CELL BIOLOGY Vol.13: 6667–6677. 1993.

Margolis 1969.   S. Margolis, "Structure of Very Low and Low Density Lipoproteins," in STRUCTURAL AND FUNCTIONAL ASPECTS OF LIPOPROTEINS IN LIVING SYSTEMS, edited by E. Trea + A.M. Scanu. (Academic Press, New York City.) 1969.

Martell 1974.   Edward A. Martell, "Radioactivity of Tobacco Trichomes and Insoluble Cigarette Smoke Particles," NATURE Vol.249: 215–217. May 17, 1974.

Martell 1975.   Edward A. Martell, "Tobacco Radioactivity and Cancer in Smokers," AMERICAN SCIENTIST Vol.63: 404–412. July–August 1975.

Martell 1982-a.   Edward A. Martell, "Radioactivity in Cigarette Smoke," (letter), NEW ENGLAND J. OF MEDICINE Vol.307, No.5: 309. July 29, 1982.

Martell 1982-b.   Edward A. Martell, "The Natural Alpha Radiation Environment: A Preliminary Assessment," pp.121–130 in PROCEEDINGS, NATURAL RADIATION ENVIRONMENT, edited by K.G. Vohra et al. (Publisher: Wiley Eastern Ltd., New Delhi.) 1982.

Martell 1982-c.   Edward A. Martell + K.S. Sweder, "The Roles of Polonium Isotopes in the Etiology of Lung Cancer in Cigarette Smokers and Uranium Miners," Chapter 61 (pp.383–389) in RADIATION HAZARDS IN MINING, proceedings of a 1981 symposium edited by M. Gomez. (American Institute of Mining Engineers, New York.) 1982.

Martell 1983-a.   Edward A. Martell, "Alpha-Radiation Dose at Bronchial Bifurcations of Smokers from Indoor Exposure to Radon Progeny," PROC. OF THE NATL. ACADEMY OF SCIENCES (USA) Vol.80: 1285–1289. March 1983.

Martell 1983-b.   Edward A. Martell, "Bronchial Cancer Induction by Alpha Radiation: A New Hypothesis," paper C6–11 in PROCEEDINGS OF THE 7TH INTERNATIONAL CONGRESS ON RADIATION RESEARCH, edited by J.J. Broerse et al. (Publisher: Martinus Nijhoff, Amsterdam, Netherlands.) 1983

Marx 1994.   Jean Marx, "Coronary Artery Disease: CMV-p53 Interaction May Help Explain Clogged Arteries," SCIENCE Vol.265: 320. July 15, 1994.

Maseri 1997.   Attilio Maseri, "Inflammation, Atherosclerosis, and Ischemic Events --- Exploring the Hidden Side of the Moon," (editorial), NEW ENGLAND J. OF MEDICINE Vol.336, No.14: 1014–1016. April 3, 1997.

Mass 1890.   Secretary of the Commonwealth of Massachusetts, FORTY-NINTH REPORT TO THE LEGISLATURE OF MASSACHUSETTS RELATING TO THE REGISTRY AND RETURN OF BIRTHS, MARRIAGES, AND DEATHS IN THE COMMONWEALTH FOR THE YEAR ENDING DECEMBER 31, 1890. With editorial remarks by Samuel W. Abbott, M.D. ~417 pages. Public Document No.1., printed 1891. Available courtesy of Naomi Allen, librarian at the Massachusetts State Library, Room 341 State House, Boston, MA 02133, USA. Call number of this report: MR 614.1M3, S44R, 1890.

Masuda 1990. J. Masuda + Russell Ross, ARTERIOSCLEROSIS Vol.10: 164–177. 1990.

Masuda 1990–b. J. Masuda + Russell Ross, ARTERIOSCLEROSIS Vol.10: 178–187. 1990.

McCully 1969. K.S. McCully, "Vascular Pathology of Homocysteinemia: Implications for the Pathogenesis of
    Arteriosclerosis," AMERICAN JOURNAL OF PATHOLOGY Vol.56: 111–128. 1969.

McDonald 1989. K. McDonald + R.S. Rector et al, "Association of Coronary Artery Disease in Cardiac Transplant Recipients
    with Cytomegalovirus Infection," AMERICAN JOURNAL OF PATHOLOGY Vol.64: 359–362. 1989.

McGill 1984. Henry C. McGill, Jr., "Persistent Problems in the Pathogenesis of Atherosclerosis," (George Lyman Duff
    Memorial Lecture), ARTERIOSCLEROSIS Vol.4: 443–451. Sept–Oct. 1984.

McGinley 1952. James P. McGinley + Hardin B. Jones + John W. Gofman, "Lipoproteins and Xanthomatous Diseases,"
    JOURNAL OF INVESTIGATIVE DERMATOLOGY Vol.19, No.1:71–82, 1952.

McLennan 1993. P.L. McLennan, "Relative Effects of Dietary Saturated, MonoUnsaturated, and PolyUnsaturated Fatty Acids
    on Cardiac Arrhythmias in Rats," AMER. J. CLINICAL NUTRITION Vol.57: 207–212. 1993.

McNeil 1985. Barbara J. McNeil + Dennis Tihansky + John E. Wennberg, "Use of Medical Radiographs: Extent of Variation and
    Associated Active Bone Marrow Doses," RADIOLOGY Vol.156: 51–56. July 1985.

McReynolds 1976. Richard A. McReynolds + G. Lennard Gold + William C. Roberts, "Coronary Heart Disease after
    Mediastinal Irradiation for Hodgkin's Disease," AMERICAN JOURNAL OF MEDICINE Vol.60: 39–45. 1976.

Meier 1999. Christoph R. Meier + Laura E. Derby + Susan S. Jick + Catherine Vasilakis + Jershel Jick, "AntiBiotics and Risk
    of Subsequent First–Time Acute Myocardial Infarction," JOURNAL OF THE AMERICAN MEDICAL ASSN.
    Vol.218, No.5: 427–431. Feb. 3, 1999.

Melnick 1983. J.L. Melnick + B.L. Petrie et al, "Cytomegalovirus Antigen within Human Arterial Smooth Muscle Cells,"
    LANCET Vol.2: 644–647. 1983.

Melnick 1990. J.L. Melnick + E. Adam + M.E. DeBakey, "Possible Role of Cytomegalovirus in Atherogenesis,"
    JOURNAL OF THE AMERICAN MEDICAL ASSN Vol.263: 2204–2207. 1990.

Mendall 1994. Michael A. Mendall + 7 co–workers, "Relation of Helicobacter Pylori Infection and Coronary Heart Disease,"
    BRITISH HEART JOURNAL Vol.71: 437–439. 1994.

Mendonca 1993. Marc S. Mendonca + Ronald J. Antoniono + J. Leslie Redpath, "Delayed Heritable Damage and Epigenetics in
    Radiation–Induced Neoplastic Transformation of Human Hybrid Cells," RADIATION RESEARCH
    Vol.134: 209–216. 1993.

Mensink 1992. R.P. Mensink + M.B. Katan, "Effect of Dietary Fatty Acids on Serum Lipids and Lipoproteins: A
    Meta–Analysis of 27 Trials," ARTERIOSCLEROSIS AND THROMBOSIS Vol.12: 911–919. 1992.

Mettler 1987. Fred A. Mettler, Jr., "Diagnostic Radiology: Usage and Trends in the United States, 1964–1980," RADIOLOGY
    Vol.162: 263–266. January 1987.

Miller 1989. Anthony B. Miller + Geoffrey R. Howe + 6 co–workers, "Mortality from Breast Cancer after Irradiation during
    Fluoroscopic Examinations in Patients Being Treated for Tuberculosis," NEW ENGLAND JOURNAL OF
    MEDICINE Vol.321, No.19: 1285–1289. 1989.

Minick 1973. C.R. Minick + G.E. Murphy, "Experimental Induction of Atherosclerosis by the Synergy of Allergic Injury to
    Arteries and Lipid Rich Diet: II, Effect of Repeatedly Injected Foreign Protein in Rabbits Fed a Lipid–Rich,
    Cholesterol–Poor Diet," AMERICAN J. PATHOLOGY Vol.73: 265–300. 1973. See also Hardin 1973.

Minkler 1970. Jason L. Minkler + John W. Gofman + Robert K. Tandy, "A Specific Common Chromosomal Pathway for
    the Origin of Human Malignancy," BRITISH JOURNAL OF CANCER Vol.24: 726–740. 1970.

Minkler 1971. Jason L. Minkler + Dolores Piluso + John W. Gofman + Robert K. Tandy, "A Long–Term Effect of Radiation on
    Chromosomes of Cultured Human Fibroblasts," MUTATION RESEARCH Vol.13: 67–75. 1971.

MMWR (general). Morbidity and Mortality Weekly Report is published by the Epidemiology Program Office, U.S. Centers for
    Disease Control and Prevention (CDC), Atlanta, Georgia 30333.

MMWR 1987. "Cigarette Smoking in the United States, 1986," MORBIDITY AND MORTALITY WEEKLY REPORT
    Vol.36, No.35: 581–585. September 11, 1987.

MMWR 1988. "Autopsy Frequency --- United States, 1980–1985," MORBIDITY & MORTALITY
    WEEKLY REPORT Vol.37: 191–194. 1988.

MMWR 1994. Gary A. Giovino + 8 co–workers, "Surveillance for Selected Tobacco–Use Behaviors --- United States,
    1900–1994," MORBIDITY & MORTALITY WEEKLY REPORT Vol.43, No. SS–3: 1–43. Nov.18, 1994.

MMWR 1996. "State–Specific Prevalence of Cigarette Smoking --- United States, 1995," MORBIDITY & MORTALITY
    WEEKLY REPORT Vol.45, No.44: 962–974. November 8, 1996.

Modan 1977. Baruch Modan et al, "Thyroid Cancer Following Scalp Irradiation," RADIOLOGY Vol.123: 741–744.
    1977.

Modan 1989. Baruch Modan + Angela Chetrit + Esther Alfandary + Leah Katz, "Increased Risk of Breast Cancer after Low–Dose
    Irradiation," LANCET: 629–631. March 25, 1989.

Modern Medicine 1949. Modern Medicine, Minneapolis, Advertising Sales Pamphlet, 1949 --- cited by Donaldson 1951,
    Table 25.

Moeller 1953. Dade W. Moeller + James G. Terrill + Samuel C. Ingraham, "Radiation Exposure in the United States," PUBLIC HEALTH REPORTS Vol.68, No.1: 57–65. January 1953.

Montanara 1986. A. Montanara + R. Pani + R. Pellegrini et al, "The Radiation Dose to the Lens in Radiology of the Orbit," BRITISH JOURNAL OF RADIOLOGY Vol.59: 1171–1173. 1986

Monthly Vital Statistics Report, or MVS (general). MVS Reports are published by the National Center for Health Statistics. Please see NatCtrHS (general).

Moore 1973. S. Moore, "Thromboatherosclerosis in Normolipemic Rabbits; a Result of Continued Endothelial Damage," LABORATORY INVESTIGATION Vol.29: 478–487. 1973.

Morbidity and Mortality Weekly Report.
   Please see MMWR, above.

Moreno 1994. P.R. Moreno + E. Falk + I.F. Palacios + J.B. Newell + V. Fuster + J.T. Fallon, "Macrophage Infiltration in Acute Coronary Syndromes. Implications for Plaque Rupture," CIRCULATION Vol.90: 775–78. 1994.

Morgan 1971. Karl Z. Morgan, "Never Do Harm," ENVIRONMENT Vol.13: 28–38. 1971.

Morgan 1996. William F. Morgan + 4 co-workers, "Genomic Instability Induced by Ionizing Radiation," (review), RADIATION RESEARCH Vol.146: 247–258. 1996.

Morgan 1999. Karl Z. Morgan + Ken M. Peterson, THE ANGRY GENIE: ONE MAN'S WALK THROUGH THE NUCLEAR AGE. 218 pages. LCCN 98-34766. ISBN 0-8061-3122-5. (University of Oklahoma Press, Norman OK 73069, USA.) 1999.

Morris 1984. N. Morris + B. Young, "The Accuracy and Interpretation of Numbers for Practical Radiography," RADIOGRAPHER pp.107–109. 1984.

Moss 1995. Ralph W. Moss, QUESTIONING CHEMOTHERAPY. 214 pages. ISBN 1-881025-25-X. LCCN 95-11440. (Equinox Press, 144 St. John's Place, Brooklyn NY 11217). 1995.

Muhlestein 1996. Joseph B. Muhlestein + 7 co-workers, "Increased Incidence of Chlamydia Species within the Coronary Arteries of Patients with Symptomatic Atherosclerosis versus Other Forms of Cardiovascular Disease," JOURNAL OF THE AMERICAN COLLEGE OF CARDIOLOGY Vol.27, No.7: 1555–1561. June 1996.

Muldoon 1990. M.F. Muldoon + S.B. Manuck + K.A. Matthews, "Lowering Cholesterol Concentrations and Mortality: A Quantitative Review of Primary Prevention Trials," BRITISH MEDICAL JOURNAL Vol.301: 309–314. 1990.

Munoz 1994. Nubia Munoz, "Is Helicobacter Pylori a Cause of Gastric Cancer? An Appraisal of the Sero-Epidemiological Evidence," (review), CANCER EPIDEMIOLOGY, BIOMARKERS & PREVENTION Vol.3: 445–451. Jul–Aug 1994.

Munro 1988. J. Michael Munro + Ramzi S. Cotran, "Biology of Disease: The Pathogenesis of Atherosclerosis: Atherogenesis and Inflammation," LABORATORY INVESTIGATION Vol.58, No.3: 249–261. 1988.

Murata 1986. K. Murata + T. Motayama + C. Kotake, "Collagen Types in Various Layers of the Human Aorta and Their Changes with the Atherosclerotic Process," ATHEROSCLEROSIS Vol.60: 251+. 1986.

Mustafa 1985. A.A. Mustafa + K. Kouris, "Effective Dose Equivalent and Associated Risks from Mass Chest Radiography in Kuwait," HEALTH PHYSICS Vol. 49: 1147–1154. 1985.

MVS (general). The Monthly Vital Statistics Report is published by the National Center for Health Statistics. Please see NatCtrHS (general).

Myrden + Hiltz 1969. J.A. Myrden + J.E. Hiltz, "Breast Cancer Following Multiple Fluoroscopies during Artificial Pneumothorax Treatment of Pulmonary Tuberculosis," CANADIAN MEDICAL ASSOCIATION JOURNAL Vol.100: 1032–1034. 1969.

● NAS 1956. National Academy of Sciences, National Research Council, THE BIOLOGICAL EFFECTS OF RADIATION: A REPORT TO THE PUBLIC. 1956. This is one of a series issued by NAS Committees in 1956 on the Biological Effects of Atomic Radiation ... on Genetic Effects ... on Agriculture and Food Supplies ... and on Oceanography and Fisheries.

NatCtrHS (general). The National Center for Health Statistics is a subdivision of the (U.S.) Centers for Disease Control and Prevention (U.S. Dept. of Health & Human Services). The NatCtrHS is located at 6525 Belcrest Road, Hyattsville MD 20782. Vital Statistics, Mortality: Telephone 301-436-8884.

NatCtrHS 1980. National Center for Health Statistics, AVERAGE AGE-ADJUSTED DEATH RATES AND STANDARD ERRORS (SE) FOR MAJOR CAUSES, BY RACE AND SEX; UNITED STATES AND RANK FOR EACH STATE, 1979-1981. "Based on age-specific death rates per 100,000 population ... using as standard population the age distribution of total U.S. population as enumerated in 1940." These data were provided to us in 1995 by Jeff Maurer at the NatCtrHS, as a printout of over 150 pages.

NatCtrHS 1993. National Center for Health Statistics, ICD 410 TO 414.9, DEATH RATE/100,000, 1992-1994; ALL AGES; ALL RACES; BY STATE; STD. POP. 1940. This is a 7-page printout --- 3 pages for males, 3 pages for females, plus one page for both sexes combined, all states combined. Extracted by Jeff Maurer at the NatCtrHS from the Centers for Disease Control "wonder" database at <http://wonder.cdc.gov>.

National Center for Health Statistics.
   Please see NatCtrHS (general).

National Council on Radiation Protection and Measurements.
          Please see NCRP (general).
National Research Council (USA). This is an agency of the National Academy of Sciences. The Academy is a society of
          scholars chartered by Congress in 1863 to advise the federal government on scientific and technical matters.
          Please see BEIR.
NCI 1990. National Cancer Institute (USA). No author named. EVERYTHING DOESN'T CAUSE CANCER. Booklet, 12
          pages. NIH Publication No. 90-2039. (National Cancer Institute, National Institutes of Health, Public Health
          Service, U.S. Dept. of Health and Human Services.) March 1990.
NCI 1998. National Cancer Institute (USA). CIGARS: HEALTH EFFECTS AND TRENDS. Monograph 9. 232 pages.
          No index. This report includes work by "over 50 scientists both within and outside the Federal Government."
          Senior Scientific Editor: David M. Burns, Professor of Medicine, University of California School of Medicine, San
          Diego. NIH Publication No. 98-4302. (National Cancer Institute, Rockville, Maryland, USA 20892.) Feb.1998.
NCRP (general). The National Council on Radiation Protection and Measurements is a nonprofit organization. In each of its
          reports, NCRP publishes a list of radiation-related organizations from which it receives "generous support." A
          sample from the 1989 list: American College of Nuclear Physicians, American College of Radiology, American
          Dental Assn., American Medical Assn., American Nuclear Society, American Roentgen Ray Society, Federal
          Emergency Management Agency, Health Physics Society, Institute of Nuclear Power Operations, Radiological
          Society of North America, U.S. Airforce, Army, and Navy, U.S. Dept. of Energy, U.S. Nuclear Regulatory
          Commission. Address: 7910 Woodmont Avenue, Bethesda MD 20814, USA. Website: www.ncrp.com
NCRP 1980. National Council on Radiation Protection & Measurements, INFLUENCE OF DOSE AND ITS DISTRIBUTION
          IN TIME ON DOSE-RESPONSE RELATIONSHIPS FOR LOW-LET RADIATIONS. Report 64. 1980.
NCRP 1984. National Council on Radiation Protection & Measurements, EVALUATION OF OCCUPATIONAL AND
          ENVIRONMENTAL EXPOSURES TO RADON AND RADON DAUGHTERS. Report 78. 1984.
NCRP 1986. National Council on Radiation Protection & Measurements, MAMMOGRAPHY: A USER'S GUIDE. Report 85.
          1986.
NCRP 1987. National Council on Radiation Protection & Measurements, IONIZING RADIATION EXPOSURE
          OF THE POPULATION OF THE U.S. Report 93. 87 pages. ISBN 0-913392-91-X. LCCN 87-22062. 1987.
NCRP 1989. National Council on Radiation Protection & Measurements, EXPOSURE OF THE U.S. POPULATION
          FROM DIAGNOSTIC MEDICAL RADIATION. Report 100. 105 pages. LCCN 88-25316. 1989.
NCRP 1989-b. National Council on Radiation Protection & Measurements, GUIDANCE ON RADIATION RECEIVED IN
          SPACE ACTIVITIES. Report 98. 227 pages. ISBN 0-929600-04-5. LCCN 89-3023. 1989.
NCRP 1990. National Council on Radiation Protection & Measurements, THE RELATIVE BIOLOGICAL EFFECTIVENESS
          OF RADIATIONS OF DIFFERENT QUALITY. Report 104. 1990.
Neamiro 1983. E. Neamiro + G. Balode, "Photofluorography of the Thorax." (in the Russian Language). 1983.

Newlin 1978. N. Newlin, "Reduction in Radiation Exposure: The Rare Earth Screen," AMERICAN JOURNAL OF
          ROENTGENOLOGY Vol.130: 1195-1196. 1978.
Nichols 1957. Alex Nichols + Virginia Dobbin + John W. Gofman, "The Influence of Dietary Factors upon Human Serum
          Lipoprotein Concentrations," GERIATRICS 12: 7-17. 1957.
Nichols 1957-b. Alex V. Nichols + Frank T. Lindgren + John W. Gofman, "Estimation of Atherogenic Index and
          Lipoprotein Distribution in Men: Evaluation from Serum Gravimetric Total Lipid and Total Cholesterol
          Concentration," GERIATRICS: 130-138. February 1957.
Nicod 1993. Pascal Nicod + Urs Scherrer, "Explosive Growth of Coronary Angioplasty: Success Story of a Less-Than-
          Perfect Procedure," (editorial comment), CIRCULATION Vol.87: 1749-1751. May 1993.
Nobuyoshi 1991. M. Nobuyoshi + 8 co-workers, "Progression of Coronary Atherosclerosis: Is Coronary Spasm Related to
          Progression?" JOURNAL OF THE AMERICAN COLLEGE OF CARDIOLOGY Vol.18: 904-910. 1991.
Nowell 1976. Peter C. Nowell, "The Clonal Evolution of Tumor Cell Populations," SCIENCE Vol.194: 23-28.
          1976.
• NRPB 1995. National Radiological Protection Board (Britain), RISK OF RADIATION-INDUCED CANCER AT LOW
          DOSES AND LOW DOSE RATES FOR RADIATION PROTECTION PURPOSES. Prepared by Roger Cox (head
          of biomedical effects) + Colin Muirhead (head of epidemiology) + John W. Stather + A.A. Edwards + M.P.
          Little. 77 pages. ISBN 0-85951-386-6. Vol.6, No.1 in the series, Documents of the NRPB. (NRPB, Chilton,
          Didcot, Oxon OX11 ORQ, Britain.) October 1995.
Nygard 1997. Ottar Nygard + Jan Erik Nordrehaug + Helga Refsum + Per Magne Ueland + Mikael Farstad + Stein Emil
          Vollset, "Plasma Homocysteine Levels and Mortality in Patients with Coronary Artery Disease," NEW
          ENGLAND JOURNAL OF MEDICINE Vol.337, No.4: 230-236. July 24, 1997.
Oliver 1991. Michael F. Oliver, "Might Treatment of Hypercholesterolaemia Increase Non-Cardiac Mortality?"
          LANCET Vol.337: 1529-1531. 1991.
Oliver 1992. Michael F. Oliver, "Doubts about Preventing Coronary Heart Disease. Multiple Interventions in Middle-Aged
          Men May Do More Harm Than Good," BRITISH MEDICAL JOURNAL Vol.304: 393-394. 1992.

Oliver 1997. Michael F. Oliver + Laura A. Corr, "The Low Fat/Low Cholesterol Diet Is Ineffective," EUROPEAN
        HEART JOURNAL Vol.18: 18-22. 1997.

Oncley 1947. John L. Oncley + G. Scatchard + A. Brown, "Physical-Chemical Characteristics of Certain of the Proteins of
        Normal Human Plasma," JOURNAL OF PHYSIOLOGICAL CHEMISTRY Vol.51: 184-198. 1947.

Ornish 1990. (Lifestyle Heart Trial.) Dean Ornish + 9 co-workers, "Can Lifestyle Changes Reverse Coronary Heart Disease?"
        LANCET Vol.336: 129-133. 1990.

Ornish 1998. Dean Ornish + 10 co-workers, "Intensive Lifestyle Changes for Reversal of Coronary Heart Disease,"
        JOURNAL OF THE AMERICAN MEDICAL ASSN Vol.280, No.23: 2001-2007. Dec. 16, 1998.

Osler 1908. William Osler, "Diseases of the Arteries," pp.429-447 in MODERN MEDICINE: ITS PRACTICE AND THEORY,
        edited by William Osler. (Lea & Febiger, Philadelphia, Pennsylvania, USA.) 1908.

Palade 1956. G. Palade, in J. BIOPHYS. BIOCHEM. CYTOL. Vol.2 (Suppl.): 85-98. 1956.

Pappenheimer 1953. J.R. Pappenheimer, in PHYSIOL. REV. Vol.14: 404-481. 1953.

Parkes 1991. Joan Lee Parkes + Robert R. Cardell + Frank C. Hubbard, Jr. + Dale Hubbard + Alan Meltzer + Arthur Penn,
        "Cultured Human Atherosclerotic Plaque Smooth Muscle Cells Retain Transforming Potential and Display Enhanced
        Expression of the myc Proto-Oncogene," AMER. J. PATHOLOGY 138: 765-775. March 1991. See also: Penn.

Parmley 1997. Wm. W. Parmley, "Clinical Significance of Endothelial Dysfunction," pp. 11-17 in syllabus of 25TH ANNUAL
        ADVANCES IN MEDICINE, presented by the Department of Medicine, University of California San Francisco
        School of Medicine. (Prof. Parmley, UCSF Box 0656, San Francisco CA 94143-0656). June 23, 1997.

Patel 1995. P. Patel + Michael A. Mendall + 9 co-workers, "Association of Helicobacter Pylori and Chlamydia Pneumoniae
        Infections with Coronary Heart Disease and Cardiovascular Risk Factors," BRITISH MEDICAL JOURNAL
        Vol.311: 711-714. September 16, 1995.

Patterson 1987. James T. Patterson, THE DREAD DISEASE: CANCER AND MODERN AMERICAN CULTURE.
        380 pages. ISBN 0-674-21625-3. (Harvard University Press, Cambridge, Mass.) 1987.

Pearson 1975. Thomas A. Pearson + A. Wang + Kim Solez + Robert H. Heptinstall, "Clonal Characteristics of Fibrous Plaques
        and Fatty Streaks from Human Aortas," AMERICAN J. PATHOLOGY Vol.81, No.2: 379-387. Nov.1975.

Pearson 1977. Thomas A. Pearson + E.C. Kramer + Kim Solez + Robert H. Heptinstall, "The Human Atherosclerotic Plaque,"
        AMERICAN JOURNAL OF PATHOLOGY Vol.86, No.3: 657-664. March 1977.

Pearson 1978-a. Thomas A. Pearson + John M. Dillman + Kim Solez + Robert H. Heptinstall, "Clonal Markers in the Study
        of the Origin and Growth of Human Atherosclerotic Lesions," CIRCULATION RESEARCH Vol.43: 10-18. 1978.

Pearson 1978-b. Thomas A. Pearson + John M. Dillman + Kim Solez + Robert H. Heptinstall, "Clonal Characteristics in
        Layers of Human Atherosclerotic Plaques: A Study of the Selection Hypothesis of Monoclonality," AMERICAN
        JOURNAL OF PATHOLOGY Vol.93, No.1: 93-102. October 1978.

Pearson 1980. Thomas A. Pearson + John M. Dillman + Kim Solez + Robert H. Heptinstall, "Evidence for Two Populations of
        Fatty Streaks with Different Roles in the Atherogenic Process," LANCET 2: 496-498. 1980.

Pearson 1983-a. Thomas A. Pearson + John M. Dillman + Robert H. Heptinstall, "The Clonal Characteristics of Human Aortic
        Intima: Comparison with Fatty Streaks and Normal Media," AMER. J. PATHOLOGY Vol.113: 33-40. Oct. 1983.

Pearson 1983-b. Thomas A. Pearson et al, "Cholesterol-Induced Atherosclerosis: Clonal Characteristics of Arterial Lesions
        in the Hybrid Hare," ARTERIOSCLEROSIS Vol.3: 574-580. Nov-Dec. 1983.

Pearson 1993. Thomas A. Pearson + Herbert J. Marx, "Rapid Reduction in Cardiac Events with Lipid-Lowering Therapy:
        Mechanisms & Implications," (commentary), AMER. J. CARDIOLOGY Vol.72: 1072-1073. Nov.1, 1993.

Pearson 1998. Thomas A. Pearson, "Lipid-Lowering Therapy in Low-Risk Patients," (commentary), JOURNAL OF
        THE AMERICAN MEDICAL ASSN. Vol. 279, No.20: 1659-1661. May 27, 1998.

Pedersen 1994.
        Please see Scandinavian 1994.

Pedersen 1995. Terje R. Pedersen, "Lowering Cholesterol with Drugs and Diet," (editorial), NEW ENGLAND
        J. OF MEDICINE Vol.333, No.20: 1350-1351. November 16, 1995. See also Scandinavian 1994.

Penn 1980. Arthur Penn + G. Batastini + Roy Albert, "Age-Dependent Changes in Prevalence, Size, and Proliferation of
        Arterial Lesions in the Cockerel. I. Spontaneous Lesions," ARTERY Vol.7: 448-463. 1980.

Penn 1981-a. Arthur Penn + G. Batastini + J. Solomon + F. Burns + Roy Albert, "Dose-Dependent Size Increases of Aortic
        Lesions Following Chronic Exposure to 7,12-Dimethylbenz[a]Anthracene," CANCER RESEARCH 41: 588-92.
        1981.

Penn 1981-b. Arthur Penn et al, "Age-Dependent Changes in Prevalence, Size and Proliferation of Arterial Lesions in
        Cockerels. II. Carcinogen-Associated Lesions," ARTERY Vol.9, No.5: 382-393. 1981.

Penn 1986. Arthur Penn + Seymour J. Garte + Lisa Warren + Douglas Nesta + Bruce Mindich, "Transforming Gene in Human
        Atherosclerotic Plaque DNA," PROC. NATL. ACADEMY SCI. USA Vol.83: 7951-7955. Oct.1986.

Penn 1988. Arthur Penn + Carroll A. Snyder, "Arteriosclerotic Plaque Development Is 'Promoted' by Polynuclear Aromatic
        Hydrocarbons," CARCINOGENESIS Vol.9, No.12: 2185-2189. 1988.

Penn 1989. Arthur Penn, "Molecular Alterations Critical to the Development of Arteriosclerotic Plaques: A Role for Environmental Agents," ENVIRONMENTAL HEALTH PERSPECTIVES Vol.81: 189-192. 1989.

Penn 1990. Arthur Penn, "Mutational Events in the Etiology of Arteriosclerotic Plaques," (review), MUTATION RESEARCH Vol.239: 149-162. 1990.

Penn 1991. Arthur Penn + Frank C. Hubbard, Jr. + Joan Lee Parkes, "Transforming Potential Is Detectable in Arteriosclerotic Plaques of Young Animals," ARTERIOSCLEROSIS AND THROMBOSIS Vol.11, No.4: 1053-1058. July/August 1991. See also Parkes 1991.

Penn 1993. Arthur Penn + Carroll A. Snyder, "Inhalation of Sidestream Cigarette Smoke Accelerates Development of Arteriosclerotic Plaques," CIRCULATION Vol.88: 1820-1825. 1993.

Penn 1994. Arthur Penn + Lung-Chi Chen + Carroll A. Snyder, "Inhalation of Steady-State Sidestream Smoke from One Cigarette Promotes Arteriosclerotic Plaque Development," CIRCULATION Vol.90: 1363-1367. Sept.1994.

Penn 1996. Arthur Penn et al, "The Tar Fraction of Cigarette Smoke Does Not Promote Arteriosclerotic Plaque Development," ENVIRONMENTAL HEALTH PERSPECTIVES Vol.104, No.10: 1108-1113. October 1996.

Pennell 1952. Maryland Y. Pennell + Marion E. Altenderfer, HEALTH MANPOWER SOURCE BOOK I. PHYSICIANS. 70 pages. Public Health Service Publication No. 263. Section I, May 1952. Produced by the Federal Security Agency, Public Health Service, Division of Public Health Methods. Printed by U.S. Government Printing Office. UCSF Library: RA410.7, A1, A36, No.1, 1952.

Perry 1995. I.J. Perry + H. Refsum + R.W. Morris et al, "Prospective Study of Serum Homocysteine Concentration and Risk of Stroke in Middle-Aged British Men," LANCET Vol.346: 1395-1398. 1995.

Petrie 1987. B.L. Petrie + Joseph L. Melnick et al, "Nucleic Acid Sequences of Cytomegalovirus in Cells Cultured from Human Arterial Tissue," J. INFECT. DIS. Vol.155: 158-159. 1987.

Petrie 1988. B.L. Petrie + Ervin Adam + Joseph L. Melnick, "Association of Herpesvirus/CMV Infections with Human Atherosclerosis," PROG. MED. VIROL. Vol.35: 21-42. 1988.

PHS 1959. "Report to the Surgeon General, U.S. Public Health Service, on the Control of Radiation Hazards in the United States." Prepared by the National Advisory Committee on Radiation. 20 pages. March 1959. Committee Members: Dr. Victor P. Bond + Dr. Richard H. Chamberlain + Dr. James F. Crow + Dr. Herman E. Hilleboe + Dr. Hardin B. Jones + Dr. Edward B. Lewis + Dr. Berwyn F. Mattison + Dr. Russell H. Morgan (Chairman) + Mr. Lauriston Taylor + Dr. George W. Thorn + Dr. Abel Wolman + Dr. Arthur H. Wuehrmann.

PHS 1992. Public Health Service report, HEALTH: UNITED STATES 1992, AND HEALTHY PEOPLE 2000 REVIEW. 390 pages. UCSF Library: RA407.3, U576, 1992.

PHS 1995. Public Health Service report, HEALTH, UNITED STATES, 1995. 320+ pages. This is the twentieth report in a series submitted by the Secretary of Health and Human Services to the President and Congress of the United States. The report was prepared by the Centers for Disease Control, National Center for Health Statistics. The entire report is available on the Internet (as an Acrobat.pdf file) at http://www.cdc.gov/nchswww/nchshome.htm. Also, the entire report is available for purchase on a CD-ROM "Publications from the National Center for Health Statistics, Featuring HEALTH, UNITED STATES, 1995, Vol.2, No.1, June 1996," and its 148 tables can be purchased as Lotus 123 spreadsheet files.

Pickels 1942. Edward G. Pickels, "The Ultracentrifuge: Practical Aspects of the Ultracentrifugal Analysis of Proteins," CHEMICAL REVIEWS Vol.30: 341-355. 1942.

Pickels 1943. Edward G. Pickels, "Sedimentation in the Angle Centrifuge," JOURNAL OF GENERAL PHYSIOLOGY Vol.26: 341-360. 1943.

Pierce 1952. Frank T. Pierce, Jr., "The Relationship of Serum Lipoproteins to Atherosclerosis in the Cholesterol-Fed Alloxanized Rabbit," CIRCULATION Vol. V, No. 3: 401-407. March 1952.

Pierce 1954-a. Frank T. Pierce, Jr., "The Interconversion of Serum Lipoproteins in Vivo," METABOLISM Vol.3, No.2: 142-153. March 1954.

Pierce 1954-b. Frank T. Pierce, Jr. + Joe R. Kimmel + Thomas W. Burns, "Lipoproteins in Infectious Hepatitis," METABOLISM Vol.3: 228-239. 1954.

Pierce 1996-a. Donald A. Pierce + Dale L. Preston, "Risks from Low Doses of Radiation," (letter), SCIENCE Vol.272: 632-633. May 3, 1996.

Pierce 1996-b. Donald A. Pierce + Yukiko Shimizu + Dale L. Preston + 2 co-workers, "Studies of the Mortality of Atomic Bomb Survivors. Report 12, Part 1. Cancer: 1950-1990," RADIATION RESEARCH Vol. 146: 1-27. 1996.

Pifer 1963. James W. Pifer + Edward T. Toyooka + Robert W. Murray + Wendell R. Ames + Louis H. Hempelmann, "Neoplasms in Children Treated with Xrays for Thymic Enlargement. I. Neoplasms and Mortality," J. NATIONAL CANCER INSTITUTE (USA) Vol.31, No.6: 1333-1356. December 1963.

Pitt 1995. (PLAC-1 Study.) B. Pitt + B.J. Mancini + S.G. Ellis et al for the PLAC-1 Investigators, "Pravastatin Limitation of Atherosclerosis in the Coronary Arteries (PLAC-1): Reduction in Atherosclerosis Progression and Clinical Events," JOURNAL OF THE AMERICAN COLLEGE OF CARDIOLOGY Vol.26: 1133-1139. 1995.

Poretti 1985. G. Poretti, "Radiatiaon Exposure of a Population due to Diagnostic X-Ray Examinations: Some Critical Remarks," PHYS. MED. BIOL. Vol. 30: 1017-1027. 1985.

Portioli 1963. R.I. Portioli + G. Botti, "Pericardite acuta da raggi," FOLIA CARDIOL. Vol.22: 257–265.
         May–June 1963.
PostCABG 1997. Post Coronary Artery Bypass Graft Trial Investigators, "Effect of Aggressive Lowering of Low–Density
         Lipoprotein Cholesterol Levels and Low–Dose Anticoagulation on Obstructive Changes in Saphenous–Vein
         Coronary–Artery Bypass Grafts," NEW ENGLAND J. OF MEDICINE Vol.336, No.3: 153–162. January 16, 1997.
Pravastatin 1993. The Pravastatin Multinational Study Group for Cardiac Risk Patients, "Effects of Pravastatin in Patients
         with Serum Total Cholesterol Levels from 5.2 to 7.8 mmol/liter (200–300 mg/dl) plus Two Additional
         Atherosclerotic Risk Factors," AMERICAN J. OF CARDIOLOGY Vol.72: 1031–1037. 1993.
Prentice 1965. R.T.W. Prentice, "Myocardial Infarction following Radiation," (letter), LANCET Vol.2,
         No.7408: 388. August 21, 1995.
Preston 1997. Dale L. Preston, Chief, Department of Statistics, Radiation Effects Research Foundation, Hiroshima, Japan:
         Personal written communication, May 20, 1997. Quoted verbatim, with permission.
Prokopcyzk 1996. Bogdan Prokopczyk, quoted in SCIENCE NEWS Vol.149: 282. May 4, 1996.

Prokopcyzk 1997. Bogdan Prokopczyk + Jonathan E. Cox + Dietrich Hoffmann + Steven E. Waggoner, "Identification of
         Tobacco–Specific Carcinogen in the Cervical Mucus of Smokers and NonSmokers," JOURNAL OF THE
         NATIONAL CANCER INSTITUTE (USA) Vol.89, No.12: 868–873. June 18, 1997.
Properzio 1985. W.S. Properzio + R.L. Burkhart, "A Review of the Experience with Diagnostic X–Ray Quality Assurance
         in the United States." BRITISH JOURNAL OF RADIOLOGY Vol. 18 (Suppl.): 75–78. 1985.
Prosser 1983. J.S. Prosser + D.C. Lloyd + A.A. Edwards + J.W. Stather, "Induction of Chromosome Aberrations in Human
         Lymphocytes by Exposure to Tritiated Water in Vitro," RAD'N PROTECTION DOSIMETRY 4: 21–26. 1983.
Purrott 1977. R.J. Purrott + E.J. Reeder + S. Lowell, "Chromosome Aberration Yields Induced in Human Lymphocytes by
         15 MeV Electrons Given at a Conventional Dose–Rate and in Microsecond Pulses," INTERNATIONAL
         JOURNAL OF RADIATION BIOLOGY Vol.31: 251–256. 1977.
Quinn 1992. (SCRIP STudy.) T.G. Quinn + E. Alderman + A. McMillan + W. Haskell + SCRIP investigators, "Reduction
         in the Development of New Coronary Atherosclerotic Lesions during the Stanford Coronary Risk Intervention
         Project (SCRIP)," (abstract), CIRCULATION Vol.86 (Suppl.1): 62. 1992.
Radford 1964. Edward P. Radford + V.R. Hunt, "Polonium–210: A Volatile Radio–Element in Cigarettes," SCIENCE
         Vol.143: 247–249. 1964.
Radford 1977. Edward P. Radford + Edward A. Martell, "Polonium–210:Lead–210 Ratios as an Index of Residence Times
         of Insoluble Particles from Cigarette Smoke in Bronchial Epithelium," pp.567–580 in INHALED PARTICLES,
         Part 2, edited by W.H. Walton. (Pergamon Press, Oxford UK.) 1977.
Radford 1982. Edward P. Radford, in "A Roundtable: With Radiation, How Little Is Too Much?" A three–way discussion
         among Dr. Radford, Dr. John W. Gofman, and Prof. Edward W. Webster, in NEW YORK TIMES, Section 4:
         Week in Review, p.EY 19. September 26, 1982.
Radiation Effects Research Foundation (general). RERF is the successor of the ABCC (Atomic Bomb Casualty Commis-
         sion), and controls the A–Bomb Survivor databases. It is funded jointly by the U.S. and Japanese governments.
         5–2 Hijiyama Park, Minami–ku, Hiroshima, 732 Japan; or via Natl.Acad.Sci.USA. Website: www.rerf.or.jp.
Rapp 1991. J.H. Rapp + W.E. Connor + D.S. Lin + J.M. Porter, "Dietary EicosaPentaenoic Acid and DocosaHexaenoic Acid
         from Fish Oil: Their Incorporation into Advanced Human Atherosclerotic Plaques,"
         ARTERIOSCLER. THROMB. Vol.11: 903–911. 1991.
Ravnskov 1991. Uffe Ravnskov, "An Elevated Serum Cholesterol Is Secondary, Not Causal, in Coronary Heart Disease,"
         MEDICAL HYPOTHESES Vol.36: 238–241. 1991.
Ravnskov 1992. Uffe Ravnskov, "Cholesterol Lowering Trials in Coronary Heart Disease: Frequency of Citation and
         Outcome," BRITISH MEDICAL JOURNAL Vol.305: 15–19. 1992.
Ravnskov 1993. Uffe Ravnskov, "Reducing Serum Cholesterol. Lower Cholesterol of Doubtful Benefit to Anyone,"
         BRITISH MEDICAL JOURNAL Vol.307: 125. 1993.
Ravnskov 1994. Uffe Ravenskov, "Is Intake of Trans–Fatty Acids and Saturated Fat Causal in Coronary Heart Disease?",
         CIRCULATION Vol.90: 2568–69. 1994.
Ravnskov 1995–a. Uffe Ravnskov, "Beneficial Effects of Simvastatin May Be Due to Non–Lipid Actions," BRITISH
         MEDICAL JOURNAL Vol.311: 1436–37. 1995.
Ravnskov 1995–b. Uffe Ravnskov, "Quotation Bias in Reviews of the Diet–Heart Idea," JOURNAL OF CLINICAL
         EPIDEMIOLOGY Vol.48: 713–719. 1995.
Rayer 1823. P. Rayer, "Memoire sur l'Ossification Morbide, Consideree comme une Terminaison des Phlegmasies,"
         ARCH. GEN. de MED. (Paris) Vol.1: 313+. 1823.
Reardon 1985. M.F. Reardon + P.J. Nestel + I.H. Craig + R.W. Harper, "Lipoprotein Predictors of the Severity of Coronary
         Artery Disease in Men and Women," CIRCULATION Vol.71: 881–888. 1985.
Renaud 1995. Serge Renaud + Michel de Lorgeril + Jacques Delaye + Janine Guidollet + Franck Jacquard + Nicole Mamelle
         + Jean–Louis Martin + Isabelle Monjaud + Patricia Salen + Paul Toubol, "Cretan Mediterranean Diet for Prevention
         of Coronary Heart Disease," AMER. J. CLINICAL NUTRITION 61 (suppl): 1360S–1367S. 1995.

Retzlaff 1995. B.M. Retzlaff + C.W. Walden + A.A. Dowdy + M.S. McCann + K.A. Anderson + R.H. Knopp, "Changes in Plasma TriacylGlycerol Concentrations among Free-Living Hyperlipidemic Men Adopting Different Carbohydrate Intakes over 2 Years: Dietary Alternatives Study," AM. J. CLINICAL NUTRITION 62: 988-995. 1995.

Richardson 1989. P.D. Richardson + Michael J. Davies + G.V. Born, "Influence of Plaque Configuration and Stress Distribution on Fissuring of Coronary Atherosclerotic Plaques," LANCET 2: 941-944. 1989.

Riches 1997. A.C. Riches + Z. Herceg + P.E. Bryant + D.L. Stevens + D.T. Goodhead, "Radiation-Induced Transformation of SV40 Immortalized Human Thyroid Epithelial Cells by Single Exposure to Plutonium Alpha-Particles in Vitro," INTERNATL. J. RADN. BIOL. 72: 515-521. 1997. (Experiment includes gamma radiation.)

Ricketts 1951. William E. Ricketts + Walter L. Palmer, "Radiation Therapy in Peptic Ulcer," Chapter 34 in PEPTIC ULCER: CLINICAL ASPECTS-DIAGNOSIS-MANAGEMENT. Edited by David Sandweiss. (Saunders Company, Philadelphia, Pennsylvania, USA.) 1951.

Ridker 1997. Paul M. Ridker + Mary Cushman + Meir J. Stampfer + Russell P. Tracy + Charles H. Hennekens, "Inflammation, Aspirin, and the Risk of Cardiovascular Disease in Apparently Healthy Men," NEW ENGLAND J. MED Vol.336, No.14: 973-979. April 3, 1997. [Erratum: NEJM 1997, Vol.337: 356.]

Ridker 1998. Paul M. Ridker + 4 co-workers, "Prospective Study of C-Reactive Protein and the Risk of Future Cardiovascular Events among Apparently Healthy Women," (Brief Rapid Communication), CIRCULATION Vol.98: 731-733. 1998.

Ridker 1999. Paul M. Ridker + JoAnn E. Manson + Julie E. Buring + Jessie Shih + Matthew Matias + Charles H. Hennekens, "Homocysteine and Risk of Cardiovascular Disease among PostMenopausal Women," JOURNAL OF THE AMERICAN MEDICAL ASSN. Vol.281, No.19: 1817-1821. May 10, 1999.

Riemersma 1989. R.A. Riemersma + C.A. Sargent, "Dietary Fish Oil and Ischemic Arrhythmias," JOURNAL OF INTERNAL MEDICINE Vol.225: 111-116. 1989.

Rimkus 1984. D. Rimkus + N.A. Baily, "Patient Exposure Requirements for High Contrast Resolution in Digital Radiographic Systems," AMERICAN JOURNAL OF ROENTGENOLOGY Vol.142: 603-608. 1984.

Rimm 1993. E.B Rimm + M.J. Stampfer + A. Ascherio + E. Giovannucci + G.A. Colditz + W.C. Willett, "Vitamin-E Consumption and the Risk of Coronary Heart Disease in Men," NEW ENGLAND JOURNAL OF MEDICINE Vol.323: 1450-1456. 1993.

Roback 1990, 1994.
    Please see AMA 1990, 1994.

Roberts 1972. William C. Roberts + L.M. Buja, "The Frequency and Significance of Coronary Arterial Thrombi and Other Observations in Fatal Acute Myocardial Infarction: A Study of 107 Necropsy Patients," AMERICAN JOURNAL OF MEDICINE Vol.52: 425+. 1972.

Roberts 1973. William C. Roberts, "Does Thrombosis Play a Major Role in the Development of Symptom-Producing Atherosclerotic Plaques?" CIRCULATION Vol.48: 1161+. 1973.

Roberts 1974. William C. Roberts, "Coronary Thrombosis and Fatal Myocardial Ischemia," CIRCULATION Vol.49: 1+. 1974.

Roberts 1975. William C. Roberts, "The Status of the Coronary Arteries in Fatal Ischemic Heart Disease," CARDIOVASC. CLIN. Vol.6. 1975.

Roberts 1978. W.C. Roberts, "The Autopsy: Its Decline and a Suggestion for Its Revival," NEW ENGLAND JOURNAL OF MEDICINE Vol.299: 332-338. 1978.

Roentgen 1895. Wilhelm Konrad Roentgen, "On a New Kind of Ray," PROCEEDINGS OF THE WURZBURG PHYSICAL-MEDICAL SOCIETY. December 28. 1985

Rohl 1976. A.N. Rohl + A.M. Langer + Irving J. Selikoff + 2 co-workers, "Consumer Talcums and Powders: Mineral and Chemical Characterization," JOURNAL OF TOXICOLOGY & ENVIR. HEALTH Vol.2: 255-284. 1976.

Ron 1991. Elaine Ron + Baruch Modan + Dale Preston + 3 co-workers, "Radiation-Induced Skin Carcinomas of the Head and Neck," RADIATION RESEARCH Vol.125: 318-325. 1991.

Ron 1995. Elaine Ron + F. Lennie Wong + Kiyohiko Mabuchi, "Incidence of Benign Gastrointestinal Tumors among Atomic Bomb Survivors," AMERICAN JOURNAL OF EPIDEMIOLOGY Vol.142, No.1: 68-75. 1995.

Rosano 1993. G.M.C. Rosano + P.M. Sarrel + P.A. Poole-Wilson + P. Collins, "Beneficial Effect of Oestrogen on Exercise-Induced Myocardial Ischemia in Women with Coronary Artery Disease," LANCET Vol.342: 133-136. 1993.

Rosenfeld 1987. M.E. Rosenfeld + T. Tsukada + A.M. Gown + R. Ross, ARTERIOSCLEROSIS Vol.7: 9-23. 1987.

Rosenfeld 1987-b. M.E. Rosenfeld et al, ARTERIOSCLEROSIS Vol.7: 24-34. 1987.

Rosenfeld 1990. M.E. Rosenfeld + W. Palinski + S. Yla-Herttuala + T.E. Carew, TOXICOL. PATH. Vol.18: 560-571. 1990.

Rosenson 1998. Robert S. Rosenson + Christine C. Tangney, "AntiAtherothrombotic Properties of Statins: Implications for Cardiovascular Event Reduction," J. AMER. MED. ASSN. Vol.279: 1643-1650. May 27, 1998.

Rosenstein 1979. Marvin Rosenstein + T.J. Beck + G.G. Warner, HANDBOOK OF SELECTED ORGAN DOSES FOR PROJECTIONS COMMON IN PEDIATRIC RADIOLOGY. (U.S. Department of Health, Education, and Welfare, Public Health Service, Bureau of Radiological Health, Rockville, Maryland, USA 20857.) 1979.

Ross 1973. Russell Ross + John A. Glomset, "Atherosclerosis and the Arterial Smooth Muscle Cell: Proliferation of Smooth Muscle Is a Key Event in the Genesis of the Lesions of Atherosclerosis," SCIENCE Vol.180: 1332-1339. 1973.

Ross 1976. Russell Ross + John A. Glomset, "The Pathogenesis of Atherosclerosis," NEW ENGLAND JOURNAL OF MEDICINE Vol.295: 369-377; 420-425. 1976.

Ross 1976-b. Russell Ross + Laurence Harker, "Hyperlipidemia and Atherosclerosis," SCIENCE Vol.193: 1094-1100. 1976.

Ross 1977. Russell Ross + John A. Glomset + Laurence Harker, "Response to Injury and Atherosclerosis," AMERICAN JOURNAL OF PATHOLOGY Vol.86, No.3: 675-684. March 1977.

Ross 1981. Russell Ross, ARTERIOSCLEROSIS Vol.1: 293-311. 1981.

Ross 1984. Russell Ross + T.N. Wight + E. Strandness + B. Thiele, "Human Atherosclerosis, I: Cell Constitution and Characteristics of Advanced Lesions of the Superficial Femoral Artery," AM. J. PATH. 114: 79-93. 1984.

Ross 1986. Russell Ross, "The Pathogenesis of Atherosclerosis: An Update," NEW ENGLAND JOURNAL OF MEDICINE Vol.314: 488-500. 1986.

Ross 1993. Russell Ross, "The Pathogenesis of Atherosclerosis: A Perspective for the 1990s," (review article), NATURE Vol.362: 801-809. April 29, 1993.

Rossouw 1990. J.E. Rossouw + B. Lewis + B.M. Rifkind, "The Value of Lowering Cholesterol after Myocardial Infarction," NEW ENGLAND JOURNAL OF MEDICINE Vol.323: 1112-1119. 1990.

Rothwell 1993. Norman V. Rothwell, UNDERSTANDING GENETICS: A MOLECULAR APPROACH. 656 pages. ISBN 0-471-58822-9. LCCN 92-36402. (New York: Wiley-Liss Publishers.) 1993.

Rowley 1987. K. Rowley + S. Hill + R. Watkins et al, "An Investigation into the Levels of Radiation Exposure in Diagnostic Examinations Involving Fluoroscopy," BRITISH JOURNAL OF RADIOLOGY Vol.60: 167-173. 1987.

Royal College 1983. Royal College of Physicians of London, HEALTH OR SMOKING? FOLLOW-UP REPORT OF THE ROYAL COLLEGE OF PHYSICIANS. Twelve chapters. ISBN 0-272-79745-6. (Pitman Publishing, Ltd., London.) 1983. UCSF Library: RA1242, T6, R68, 1983.

Rubin 1954. Leonard Rubin, "Serum Lipoproteins in Infectious Mononucleosis," AMERICAN JOURNAL OF MEDICINE Vol.17: 521+. 1954.

Rubin 1963. E. Rubin et al, "Radiation-Induced Cardiac Fibrosis," AMERICAN JOURNAL OF MEDICINE Vol.34: 71-75. January 1963.

Rubins 1995. H.B. Rubins + more than 9 co-workers, "Distribution of Lipids in 8,500 Men with Coronary Artery Disease. Dept. of Veterans Affairs HDL Intervention Trial Study Group," AM. J. CARDIOL 75: 1196-1201. 1995.

Rudel 1986. L. Rudel + C.A.Marzetta + F.L. Johnson, "Separation and Analysis of Lipoproteins by Gel Filtration," METHODS IN ENZYMOLOGY Vol.129: 45-57. 1986

Sacks 1996. (CARE Study.) F.M. Sacks + M.A. Pfeffer + L.A. Moye et al for the Cholesterol and Recurrent Events Trial Investigators, "The Effect of Pravastatin on Coronary Events after Myocardial Infarction in Patients with Average Cholesterol Levels," NEW ENGLAND JOURNAL OF MEDICINE Vol.335: 1001-1009. 1996.

Sandker 1993. G.N. Sandker + D. Kromhout + C. Aravanis, et at, "Serum Cholesteryl Ester Fatty Acids and Their Relation with Serum Lipids in Elderly Men in Crete and the Netherlands," EUROPEAN JOURNAL OF CLINICAL NUTRITION Vol.47: 201-208. 1993.

Sasaki 1975. Masao S. Sasaki, "A Comparison of Chromosomal Radiosensitivities of Somatic Cells of Mouse and Man," MUTATION RESEARCH Vol.29: 433-447. 1975.

Scandinavian 1994. (4S Study.) Scandinavian Simvastatin Survival Study Group (correspondence to Dr. Terje R. Pedersen), "Randomised Trial of Cholesterol Lowering in 4444 Patients with Coronary Heart Disease; the Scandinavian Simvastatin Survival Study (4S)," LANCET Vol.344: 1383-1389. November 19, 1994.

Scandinavian 1995. (4S Study.) Scandinavian Simvastatin Survival Study Group, "Baseline Serum Cholesterol and Treatment Effect in the Scandinavian Simvastatin Survival Study (4S)," LANCET 345: 1274-1275. 1995.

Schuler 1992. (Heidelberg Study.) G. Schuler + 10 co-workers, "Regular Physical Exercise and Low-Fat Diet: Effects on Progress of Coronary Artery Disease," CIRCULATION Vol.86: 1-11. 1992.

Seaborg 1972. Glenn T. Seaborg, NUCLEAR MILESTONES. 390 pages. ISBN 0-7167-0342-4. LCCN 72-3915. (W.H. Freeman & Company, San Francisco.) 1972.

Seaborg 1993. Glenn T. Seaborg with Benjamin S. Loeb, "THE ATOMIC ENERGY COMMISSION UNDER NIXON: ADJUSTING TO TROUBLED TIMES. 268 pages. ISBN 0-312-07899-4. LCCN 92-30137. (St. Martin's Press, New York City.) 1993.

SEER 1997. SEER CANCER STATISTICS REVIEW, 1973-1994. (Surveillance, Epidemiology, and End Results Program of the National Cancer Institute, USA.) Edited by Lynn A. Gloeckler Ries + 5 others. 479 pages. NIH Publication No.97-2789. (Cancer Statistics, National Cancer Institute, Executive Plaza North, Room 343-J, 6130 Executive Blvd., MSC-7352, Bethesda MD 20892-7352, USA.) 1997.

Segal 1982. A.J. Segal + H.D. Maille + J.A. Lemkin, "Uroradiographic Dosimetry Using a Rare Earth Screen Film System," AMERICAN JOURNAL OF ROENTGENOLOGY Vol.139: 923-936. 1982.

Segaloff 1971. Albert Segaloff + William S. Maxfield, "The Synergism between Radiation and Estrogen in the Production of Mammary Cancer in the Rat," CANCER RESEARCH Vol.31: 166-168. February 1971.

Shah 1997. Keerti V. Shah, "Human PapillomaViruses and AnoGenital Cancers," (editorial), NEW ENGLAND J. OF MEDICINE Vol.337, No.19: 1386-1388. November 6, 1997.

Shapiro 1990. Jacob Shapiro, RADIATION PROTECTION: A GUIDE FOR SCIENTISTS AND PHYSICIANS, Third Edition. 494 pages. ISBN 0-674-74586-8. (Harvard University Press, Cambridge, Massachusetts, USA.) 1990. UCSF Library: RA569 S5 1990. See Blatz 1970.

Shepherd 1995. James Shepherd + 7 co-workers for the West of Scotland Coronary Prevention Study Group, "Prevention of Coronary Heart Disease with Pravastatin in Men with Hypercholesterolemia," NEW ENGLAND JOURNAL OF MEDICINE Vol.333, No.20: 1301-1307. November 16, 1995.

Shimizu 1987. Yukiko Shimizu + Hiroo Kato + William J. Schull + Dale L. Preston + 2 co-workers, LIFE SPAN STUDY REPORT 11, PART 1. COMPARISON OF RISK COEFFICIENTS FOR SITE-SPECIFIC CANCER MORTALITY BASED ON THE DS86 AND T65DR SHIELDED KERMA AND ORGAN DOSES. 56 pages. Technical Report RERF TR-12-87. (Radiation Effects Research Foundation, Hiroshima.) 1987. www.rerf.or.jp

Shimizu 1988. Yukiko Shimizu + Hiroo Kato + William J. Schull, LIFE SPAN STUDY REPORT 11, PART 2. CANCER MORTALITY IN THE YEARS 1950-85 BASED ON THE RECENTLY REVISED DOSES (DS86). 102 pages. Technical Report RERF TR-5-88. (Radiation Effects Research Foundation, Hiroshima.) December 1988.

Shimizu 1992. Yukiko Shimizu + Hiroo Kato + William J. Schull + David G. Hoel, "Studies of the Mortality of A-Bomb Survivors. 9. Mortality, 1950-1985: Part 3. Noncancer Mortality Based on the Revised Doses (DS86)," RADIATION RESEARCH Vol.130: 249-266. 1992. See also: Kodama 1993. And Ueda 1995. And Wong 1993.

Shleien 1977. B. Shleien + T.T. Tucker + D.W. Johnson, THE MEAN ACTIVE BONE-MARROW DOSE TO THE ADULT POPULATION OF THE U.S. FROM DIAGNOSTIC RADIOLOGY. U.S. Dept. of Health, Education, and Welfare (HEW), Food and Drug Admin.(FDA), Rockville, Maryland 20857. HEW-FDA Publication 77-8013. 1977.

Shope 1996. Thomas B. Shope, "Radiation-Induced Skin Injuries from Fluoroscopy," RADIOGRAPHICS Vol.16, No.5: 1195-1199. 1996.

Shope 1997. Thomas B. Shope, "Proposed Fluoroscopic Amendments," memo & letter March 18, 1997 to "Fluoroscopic Xray System Manufacturers, Users, and Other Interested Parties," from T.B. Shope, U.S. Public Health Service, Center for Devices and Radiological Health, 5600 Fishers Lane, HFZ-140, Rockville MD 20857 USA. 1997.

Shore 1993. Roy E. Shore + Nancy Hildreth + Philip Dvoretsky + Bernard Pasternack + Elena Andresen, "Benign Thyroid Adenomas among Persons X-Irradiated in Infancy for Enlarged Thymus Glands," RADIATION RESEARCH Vol.134: 217-223. 1993.

Shrivastava 1980. P.N. Shrivastava, "Model to Analyze Radiographic Factors in Mammography," MEDICAL PHYSICS Vol.76: 222-225. 1980.

Simonton 1951. John H. Simonton + John W. Gofman, "Macrophage Migration in Experimental Atherosclerosis," CIRCULATION Vol. IV: 557-562. 1951.

Simopoulos 1998. Artemis P. Simopoulos + Jo Robinson, THE OMEGA PLAN: THE MEDICALLY PROVEN DIET THAT GIVES YOU THE ESSENTIAL NUTRIENTS YOU NEED. 380 pages. ISBN 0-06-018281-4. LCCN 97-35344. (Harper-Collins Publishers, New York City.) 1998.

Singh 1992-a. R.B. Singh + R.R. Rastogi + R. Verma + B. Laxmi + S. Ghosh + M.A. Niaz, "Randomised Controlled Trial of CardioProtective Diet in Patients with Recent Acute Myocardial Infarction: Results of One-Year Follow-Up," BRITISH MEDICAL JOURNAL Vol.304: 1015-1019. 1992.

Singh 1992-b. R.B. Singh et al, "An Indian Experiment with Nutritional Modulation in Acute Myocardial Infarction," AMERICAN JOURNAL OF CARDIOLOGY Vol.69: 879-885. 1992.

Slattery 1989. M.L. Slattery et al, "Cigarette Smoking and Exposure to Passive Smoke Are Risk Factors for Cervical Cancer," JOURNAL OF THE AMERICAN MEDICAL ASSN. Vol.261: 1593-1598. 1989.

Slyper 1994. A.H. Slyper, "Low-Density Lipoprotein Density and Atherosclerosis: Unraveling the Connection," JOURNAL OF THE AMERICAN MEDICAL ASSN. Vol.272: 305-308. 1994.

Smith 1967. Elspeth B. Smith + P.H. Evans + M.D. Pownham, "Lipid in the Aortic Intima: The Correlation of Morphological and Chemical Characteristics," J. ATHEROSCLER. RES. Vol.7: 171-186. 1967.

Smith 1972. Elspeth B. Smith + R.S. Slater, "Relationship between Low Density Lipoprotein in Aortic Intima and Serum Lipid Levels," LANCET 1: 463-469. 1972.

Smith 1974. Elspeth B. Smith, "The Relationship between Plasma and Tissue Lipids in Human Atherosclerosis," ADV. LIPID RESEARCH Vol.12: 1-49. 1974.

Smith 1975. Elspeth B. Smith + D.C. Crothers, "Interaction between Plasma Proteins and the Intercellular Matrix in Human Aortic Intima," in PROTIDES OF THE BIOLOGICAL FLUIDS Vol.22: 315-318. Edited by H. Peeters. (Pergamon Press, New York City.) 1975.

Smith 1976. Elspeth B. Smith + K.A. Alexander + I.B. Massie, "Insoluble 'Fibrin' in Human Aortic Intima: Quantitative Studies on the Relationship between Insoluble 'Fibrin,' Soluble Fibrinogen and Low Density Lipoprotein," ATHEROSCLEROSIS Vol.23: 19–39. 1976.

Smith 1977. Elspeth B. Smith, "Molecular Interactions in Human Atherosclerotic Plaques," AMERICAN JOURNAL OF PATHOLOGY Vol.86, No.3: 665–674. March 1977.

Smith 1992. G. Davey Smith + J. Pekkanen, "Should There Be a Moratorium on the Use of Cholesterol Lowering Drugs?" BRITISH MEDICAL JOURNAL Vol.304: 431–434. 1992.

Smith 1997. Karen Smith + Julie Parsonnet, "Association between Prior Cytomegalovirus Infections and the Risk of Restenosis...", (letter), NEW ENGLAND JOURNAL OF MEDICINE Vol.336, No.8: 587. Feb. 20, 1997.

Speir 1994. Edith Speir + six co-workers, "Potential Role of Human Cytomegalovirus and p53 Interaction in Coronary Restenosis," SCIENCE Vol.265: 391–394. July 15, 1994.

Spodick 1959. D.H. Spodick, p.140 in ACUTE PERICARDITIS. (Published by Grune & Stratton Inc.) 1959.

Srinivasan 1972. S.R. Srinivasan et al, "Isolation of Lipoprotein–Acid Mucopolysaccharide Complexes from Fatty Streaks of Human Aortas," ATHEROSCLEROSIS Vol.16: 95–104. 1972.

Stamler 1987. J.S. Stamler, "Epidemiology, Established Major Risk Factors, and the Primary Prevention of Coronary Heart Disease," in CARDIOLOGY, edited by W.W. Parmley + K. Chatterjee. (J.B. Lippincott, Philadelphia, PA.) 1987.

Stamler 1996. Jeremiah S. Stamler + A. Slikva, "Biological Chemistry of Thiols in the Vasculature and in Vascular–Related Disease," NUTRITION REVIEWS Vol.54: 1–30. 1996.

Stampfer 1992. Meir J. Stampfer + M.R. Malinow + Walter C. Willett et al, "A Prospective Study of Plasma Homocyst(e)ine and Risk of Myocardial Infarction in U.S. Physicians," J. AMER. MED. ASSN. 268: 877–881. 1992.

Stampfer 1993. M.J. Stampfer + C.H. Hennekens + J.E. Manson + G.A. Colditz + B. Rosner + W.C. Willett, "Vitamin–E Consumption and the Risk of Coronary Disease in Women," NEW ENGL. J. MED. 328: 1444–1449. 1993.

Stampfer 1996. Meir J. Stampfer + Ronald M. Krauss + Jing Ma + Patricia J. Blanche + Laura G. Halt + Frank M. Sacks + Charles H. Hennekins, "A Prospective Study of Triglyceride Level, Low–Density Lipoprotein Particle Diameter, and Risk of Myocardial Infarction," J. AMER. MEDICAL ASSN. Vol.276: 882–888. Sept. 18, 1996.

Stanton 1983. R. Stanton + O. Tretiak, "Dose Reduction through Variable Dose CT Scanning: Optimality of the Filtered Backprojection Algorithm," JOURNAL OF COMPUTER ASSISTED TOMOGRAPHY Vol.7: 1054–1061. 1983.

Stary 1987. Herbert C. Stary, "Evolution and Progression of Atherosclerosis in the Coronary Arteries of Children and Adults," pp.56–63 in ATHEROGENESIS AND AGING, edited by S.R. Bates + E.C. Gangloff. (Springer Verlag.) 1987.

Stary 1992. Herbert C. Stary + David H. Blankenhorn + A.B. Chandler + S. Glagov + W. Insull, Jr. + M. Richardson + M.E. Rosenfeld + S.A. Schaffer + C.J. Schwartz + W.D. Wagner + R.W. Wissler, "A Definition of the Intima of Human Arteries and of Its Atherosclerosis–Prone Regions: A Report from the Committee on Vascular Lesions of the Council on Arteriosclerosis, American Heart Assn.," (special report), CIRCULATION Vol.85: 391–405. 1992.

Stary 1994. Herbert C. Stary + 9 co-authors, "A Definition of Initial, Fatty Streak, and Intermediate Lesions of Atherosclerosis: A Report from the Committee on Vascular Lesions of the Council on Arteriosclerosis, American Heart Association," an AHA Medical/Scientific Statement (approved October 20, 1992 by the AHA SAC/Steering Committee) Special Report in ARTERIOSCLEROSIS, THROMBOSIS Vol.14, No.5: 840–856. May 1994.

Stary 1995. Herbert C. Stary + 9 co-authors, "A Definition of Advanced Types of Atherosclerotic Lesions and a Histological Classification of Atherosclerosis: A Report from the Committee on Vascular Lesions of the Council on Arteriosclerosis, American Heart Association," an AHA Medical/Scientific Statement. CIRCULATION Vol.92, No.5: 1355–1374. 1995.

Steer 1973. A. Steer + I.M.Moriyamaa + K. Shimizu, "ABCC–JNIH PATHOLOGY STUDIES, HIROSHIMA AND NAGA-SAKI. REPORT 3. THE AUTOPSY PROGRAM AND THE LSS STUDY: January 1951–December 1970. Atomic Bomb Casualty Commission Report TR–16–73. (Radiation Effects Research Fdn, Hiroshima). 1973.

Steinberg 1987. D. Steinberg, "Current Theories of the Pathogenesis of Atherosclerosis," in HYPERCHOLESTEROLEMIA AND ATHEROSCLEROSIS: PATHOGENESIS AND PREVENTION, edited by D. Steinberg + J.M. Olefsky. (Churchill/Livingston, New York City.) 1987.

Steinberg 1991. Daniel Steinberg, "Anti–Oxidants and Atherosclerosis: A Current Assessment," (editorial), CIRCULATION Vol.84, No.3: 1420–1425. Sept.1991.

Steinbrecher 1990. U.P. Steinbrecher + H.F. Zhang + M. Lougheed, FREE RADICAL BIOL. MED. Vol.9: 155–168. 1990.

Stemerman 1972. M.B. Stemerman + R. Ross, "Experimental Arteriosclerosis. I. Fibrous Plaque Formation in Primates; an Electron Microscope Study," JOURNAL OF EXP. MEDICINE Vol.136: 769–789. 1972.

Stewart 1956. Alice M. Stewart + J.W. Webb + B.D. Giles + D. Hewitt, "Preliminary Communication: Malignant Disease in Childhood, and Diagnostic Irradiation In–Utero," LANCET 2: 447. 1956.

Stewart 1958. Alice M. Stewart + J.W. Webb + D. Hewitt, "A Survey of Childhood Malignancies," BRITISH MEDICAL JOURNAL Vol.2: 1495–1508. 1958.

Stewart 1960. William H. Stewart + Maryland Y. Pennell, HEALTH MANPOWER SOURCE BOOK 10. PHYSICIANS' AGE, TYPE OF PRACTICE, AND LOCATION. 199 pages. Public Health Service Publication No. 263, Section 10. 1960. Produced by U.S. Dept. of Health, Education, and Welfare, Public Health Service, Division of Public Health Methods. Printed by U.S. Government Printing Office. UCSF Library: RA410.7, A1, A36, No.10, 1960.

Stewart 1967. J. Robert Stewart + Keith E. Cohn + Luis F. Fajardo + E. William Hancock + Henry S. Kaplan, "Radiation-Induced Heart Disease: A Study of 25 Patients," RADIOLOGY Vol.89: 302–310. August 1967.

Stewart 1970. Alice M. Stewart + George W. Kneale, "Radiation Dose Effects in Relation to Obstetric Xrays and Childhood Cancers," LANCET 1: 1185–1188. 1970.

Stewart 1978. J. Robert Stewart + Luis F. Fajardo, "Cancer and Coronary Artery Disease," (editorial), INTERNATL. J. RADIATION ONCOLOGY BIOL. PHYS. Vol.4: 915–916. 1978.

Stewart 1984. J. Robert Stewart + Luis F. Fajardo, "Radiation-Induced Heart Disease: An Update," PROGRESS IN CARDIOVASCULAR DISEASES Vol.27, No.3: 173–194. Nov./Dec. 1984.

Storey 1998. Alan Storey + 9 co-workers, "Role of a p53 Polymorphism in the Development of Human PapillomaVirus-Associated Cancer," NATURE Vol.393: 229–234. May 21, 1998.

Straume 1995. Tore Straume, "High-Energy Gamma Rays in Hiroshima and Nagasaki: Implications for Risk and Weighting Factor," HEALTH PHYSICS Vol.69, No.6: 954–956. December 1995.

Strong 1966. J.P. Strong + L.S. Solberg + C. Restrepo, "The International Atherosclerosis Project," CIRCULATION Vol.33–34 (Suppl.3): 31+. 1966. Greater detail in GEOGRAPHIC PATHOLOGY OF ATHEROSCLEROSIS, edited by H. McGill and published in 1968 (Williams & Wilkins, Baltimore, MD).

Suleiman 1992. Orhan H. Suleiman + Barton J. Conway + Fred G. Rueter + Robert J. Slayton, "Automatic Film Processing: Analysis of 9 Years of Observations," RADIOLOGY Vol.185: 25–28. 1992.

Sullivan 1986. (Please compare with next entry.) Chuck Sullivan, "Linear Regression, the Easy Way," LOTUS JOURNAL, pp.85–88. June 1986.

Sullivan 1986. (Please compare with preceding entry.) D.R. Sullivan + T.A.B. Sanders + I.M.Trayner + G.R. Thompson, "Paradoxical Elevation of LDL Apoprotein B Levels in Hypertriglyceridaemia Patients and Normal Subjects Ingesting Fish Oil," ATHEROSCLEROSIS Vol.61: 129–134. 1986.

Sulzberger 1952. Marion B. Sulzberger + Rudolf L. Baer + Alexander Borota, "Do Roentgen-Ray Treatments as Given by Skin Specialists Produce Cancers or Other Sequelae?" AMERICAN MEDICAL ASSOCIATION ARCHIVES OF DERMATOLOGY AND SYPHILOLOGY Vol.65, No.6: 639–655. June 1952.

Suntharalingham 1982. N. Suntharalingham, "Medical Radiation Dosimetry," INTERNATIONAL JOURNAL OF APPLIED RADIATION AND ISOTOPES Vol.33: 991–1006. 1982.

SurgeonGen 1964. SMOKING AND HEALTH: REPORT of the ADVISORY COMMITEE TO THE SURGEON GENERAL of the PUBLIC HEALTH SERVICE. 387 pages. No index. Public Health Service Publication 1103. (U.S. Dept. of Health, Education, and Welfare --- now Dept. of Health and Human Services.) 1964.

Svedberg 1940. Thé Svedberg + Kai Pedersen, THE ULTRACENTRIFUGE. (Oxford University Press, London.) 1940.

TaskForce 1994. Recommendations of the Task Force of the European Society of Cardiology, European Atherosclerosis Society, and European Society of Hypertension, by K. Pyorala et al on behalf of the Task Force, "Prevention of Coronary Heart Disease in Clinical Practice," EUROPEAN HEART JOURNAL Vol.15: 1300–1331. 1994.

Taylor 1979. Kenneth W. Taylor + N.L. Patt + H.E. Johns, "Variations in Xray Exposures to Patients," JOURNAL OF THE CANADIAN ASSN. OF RADIOLOGISTS Vol.30: 6–11. 1979.

Taylor 1983. Kenneth W. Taylor, "Diagnostic Radiology," Chapter 16 in THE PHYSICS OF RADIOLOGY, Fourth Edition. Edited by H.E. Johns and J.R. Cunningham. (Charles C. Thomas, publisher, Springfield, Illinois, USA.) 1983.

Temin 1988. H. Temin, "Evolution of Cancer Genes as a Mutation-Driven Process," CANCER RESEARCH Vol.48: 1697+. 1988.

Terkel 1995. Studs Terkel, COMING OF AGE: THE STORY OF OUR CENTURY BY THOSE WHO'VE LIVED IT. 468 pages. ISBN 1-56584-284-7. LCCN 95-3806. (The New Press, New York City.) 1995.

Thom 1992. David H. Thom + 5 co-workers, "Association of Prior Infection with Chlamydia Pneumoniae and Angiographically Demonstrated Coronary Artery Disease," J. AMER. MED. ASSN. 268: 68–72. July 1, 1992.

Thomas 1963. W.A. Thomas + 5 co-workers, "Production of Early Atherosclerotic Lesions in Rats Characterized by Proliferation of 'Modified Smooth Muscle Cells'," EXPERIMENTAL MOL. PATHOL. Vol.2 (supplement 1): '40–60. 1963.

Thomas 1979. W.A. Thomas et al, "Population Dynamics of Arterial Cells during Atherogenesis: X. Study of Monotypism in Atherosclerotic Lesions...", EXPERIMENTAL MOL. PATHOLOGY Vol.31: 367–386. 1979.

Thomas 1983. W.A. Thomas + D.N. Kim, "Biology of Disease: Atherosclerosis as a Hyperplastic and/or Neoplastic Process," LABORATORY INVESTIGATION Vol.48, No.3: 245–255. 1983.

Thompson 1997. Paul D. Thompson, "More on Coronary-Plaque Rupture Triggered by Snow Shoveling," (letter), NEW ENGLAND JOURNAL OF MEDICINE Vol.336, No.23: 1678. June 5, 1997.

Tlsty 1993. T.D. Tlsty + 10 co-workers, "Loss of Chromosomal Integrity in Neoplasia," COLD SPRING HARBOR SYMPOSIA ON QUANTITATIVE BIOLOGY Vol. 58: 645–654. 1993.

Tobacco 1976. Tobacco Research Council Research Paper 1, STATISTICS OF SMOKING IN THE UNITED KINGDOM, 7th Edition, edited by P.N. Lee. 1976.

Tonomura 1983. Akira Tonomura + Kunikazu Kishi + Fumiko Saito, "Types and Frequencies of Chromosome Aberrations in Peripheral Lymphocytes of General Populations," Chapter 28 (pp.605–616) in RADIATION-INDUCED CHROMOSOME DAMAGE IN MAN, edited by Takaaki Ishihara + Masao S. Sasaki. ISBN 0-8451-2404-8. (Alan R. Liss, Inc., New York City.) 1983.

Tracy 1965. R.E. Tracy + K.R. Dzoga + R.W. Wissler, "Sequestration of Serum Low-Density Lipoproteins in the Arterial Intima by Complex Formation," PROC. SOC. EXP. BIOL. MED. Vol.118: 1095–1098. 1965.

Treasure 1995. C.B. Treasure + J.L. Klein + W.S. Weintraub et al, "Beneficial Effects of Cholesterol-Lowering Therapy on the Coronary Endothelium in Patients with Coronary Artery Disease," NEW ENGL. J. MED. 332: 481–487. 1995.

Trosko 1980. J.E. Trosko + C-c. Chang, "An Integrative Hypothesis Linking Cancer, Diabetes, and Atherosclerosis: The Role of Mutations and Epigenetic Changes," MEDICAL HYPOTHESES Vol.6: 455–468. 1980.

Tucker 1994. J.D. Tucker + 5 co-workers, "On the Frequency of Chromosomal Exchanges in a Control Population Measured by Chromosome Painting," MUTATION RESEARCH Vol.313: 193–202. 1994.

Tyndall 1987. D. Tyndall + D. Washburn, "The Effect of Rare Earth Filtration on Patient Exposure, Dose Reduction, and Image Quality in Oral Panoramic Radiology," HEALTH PHYSICS Vol.52: 17–26. 1987.

Ueda 1995. Hironori Ueda, "Arteriosclerosis in the Atomic-Bomb Survivors," RERF UPDATE Vol.7, No.2: 6–7. Summer 1995. (Rad'n Effects Research Fdn, Hiroshima). Preceded by Shimizu 1992.

● UNSCEAR 1958. United Nations Scientific Committee on the Effects of Atomic Radiation, REPORT TO THE GENERAL ASSEMBLY, Thirteenth Session, Supplement No.17 (A/3838). (United Nations, New York City.) 1958.

● UNSCEAR 1977. United Nations Scientific Committee on the Effects of Atomic Radiation, SOURCES AND EFFECTS OF IONIZING RADIATION, WITH ANNEXES. Report to the General Assembly. No index. (United Nations, New York.) 1977.

● UNSCEAR 1982. United Nations Scientific Committee on the Effects of Atomic Radiation, IONIZING RADIATION: SOURCES & BIOLOGICAL EFFECTS. Report A/37/45. 37th Session of General Assembly, Supplement No.45. (United Nations, New York City.) 1982.

● UNSCEAR 1986. United Nations Scientific Committee on the Effects of Atomic Radiation, GENETIC AND SOMATIC EFFECTS OF IONIZING RADIATION. No index. ISBN 92-1-142123-3. (United Nations, New York City. Sales No. E.86.IX.9.) 1986.

● UNSCEAR 1988. United Nations Scientific Committee on the Effects of Atomic Radiation, SOURCES, EFFECTS AND RISKS OF IONIZING RADIATION. 647 pages. No index. ISBN 92-1-142143-8. (United Nations, New York City. Sales No. E.88.IX.7.) 1988.

● UNSCEAR 1993. United Nations Scientific Committee on the Effects of Atomic Radiation, SOURCES AND EFFECTS OF IONIZING RADIATION: UNSCEAR 1993 REPORT TO THE GENERAL ASSEMBLY, WITH SCIENTIFIC ANNEXES. 922 pages. No index. ISBN 92-1-142200-0. (United Nations Sales No. E.94.IX.2. Available (US $90) from Bernan Associates, 4611-F Assembly Drive, Lanham, Maryland USA 20706-4391. 1993.

● UNSCEAR 1994. United Nations Scientific Committee on the Effects of Atomic Radiation, SOURCES AND EFFECTS OF IONIZING RADIATION: UNSCEAR 1994 REPORT TO THE GENERAL ASSEMBLY, WITH SCIENTIFIC ANNEXES. 272 pages. No index. ISBN 92-1-142211-6. U.N. Sales No. E.94.IX.11. 1994.

Van de Wal 1994. A.C. Van de Wal + A.E. Becker + C.M. Van der Loos + P.K Das, "Site of Intimal Rupture or Erosion of Thrombosed Coronary Atherosclerotic Plaques Is Characterized by an Inflammatory Process Irrespective of the Dominant Plaque Morphology," CIRCULATION Vol. 89:36–44. 1994.

Verhoef 1994. P. Verhoef + C.H. Hennekens + M.R. Malinow + F.J. Kok + W.C. Willett + M.J. Stampfer, "A Prospective Study of Plasma Homocyst(e)ine and Risk of Ischemic Stroke," STROKE Vol.25: 1924–1930. 1994.

Virchow 1856. Rudolph Virchow, PHLOGOSE und THROMBOSE IM GEFASSSYSTEM. GESAMMELTE ABHAND-LUNGEN ZUR WISSENSCHAFTLICHEN MEDICIN. (Meidinger Sohn and Co., Frankfurt.) 1856.

Vital Statistics USA. Please see specific entries: Census Bureau. Grove 1968. Linder 1947. MVS. NatCtrHS. PHS.

Wagner 1976. R.S. Wagner + K.E. Weaver, "Prospects for X-Ray Exposure Reduction Using Rare Earth Intensifying Screens," RADIOLOGY Vol.118: 183–188. 1976.

Ward 1985. John F. Ward, "Mammalian Cells Are Not Killed by DNA Single-Strand Breaks Caused by Hydroxyl Radicals from Hydrogen Peroxide," RADIATION RESEARCH Vol.103: 383–392. 1985.

Ward 1985-b. John F. Ward, "Biochemistry of DNA Lesions," RADIATION RESEARCH Vol.104, Supplement 8: S103–S111. 1985.

Ward 1988. John F. Ward, "DNA Damage Produced by Ionizing Radiation in Mammalian Cells: Identities, Mechanisms of Formation, and Reparability," pp.95–125 in PROGRESS IN NUCLEIC ACID RESEARCH AND MOLECULAR BIOLOGY Vol.35. (Academic Press.) 1988.

Ward 1990. John F. Ward, "The Yield of DNA Double-Strand Breaks Produced Intracellularly by Ionizing Radiation: A Review," INTERNATIONAL J. OF RADIATION BIOLOGY Vol.57, No.6: 1141-1150. 1990.

Ward 1991. John F. Ward, "Response to Commentary by D. Billen," (letter), RADIATION RESEARCH Vol.126: 385-387. 1991.

Ward 1991-b. John F. Ward, "DNA Damage and Repair," pp.403-421 in PHYSICAL AND CHEMICAL MECHANISMS IN MOLECULAR RADIATION BIOLOGY. W.A. Glass + M.N. Varma, editors. (Plenum Press, New York.) 1991.

Ward 1994. John F. Ward + J.R. Milligan + G.D.D. Jones, "Biological Consequences of Non-Homogeneous Energy Depositions by Ionizing Radiation," RADIATION PROTECTION DOSIMETRY Vol.52: 271-276. 1994.

Ward 1995. John F. Ward, "Radiation Mutagenesis: The Initial DNA Lesions Responsible," RADIATION RESEARCH Vol.142: 362-368. 1995. (Errata 1995: Rad'n Research Vol.143: 355.)

Warren 1942. Shields Warren, "Effects of Radiation on Normal Tissues; VI. Effects of Radiation on the Cardiovascular System," ARCHIVES OF PATHOLOGY Vol.34: 1070-1084. Dec.1942. UCSF Library: W1, AR, 468A. 1942. The series begins in the August 1942 issue, and continues into the January and February 1943 issues.

Waters 1994. (CCAIT Study.) D. Waters + L. Higginson + P. Gladstone + B. Kimball + M. Lemay + J. Lesperance, "Effects of Monotherapy with an HMG-CoA Reductase Inhibitor on the Progression of Coronary Atherosclerosis as Assessed by Serial Quantitative Coronary Arteriography: The Canadian Coronary Atherosclerosis Intervention Trial," CIRCULATION Vol.89: 959-968. 1994.

Watts 1992. (STARS Study.) G.F. Watts + B. Lewis + J.N.H. Brunt + E.S. Lewis + D.J. Coltart + L.D.R. Smith + J.I. Mann + A.V. Swan, "Effects on Coronary Artery Disease of Lipid-Lowering Diet, or Diet plus Cholestyramine, in the St. Thomas' Atherosclerosis Regression Study (STARS)," LANCET Vol.339: 563-569. 1992.

Webster 1954. WEBSTER'S NEW WORLD DICTIONARY OF THE AMERICAN LANGUAGE, College Edition. (New York: World Publishing Company.) 1954.

Webster 1990. M.W.I. Webster + 6 co-workers, "Myocardial Infarction and Coronary Artery Occlusion: A Prospective 5-Year Angiographic Study," (abstract), J. AMER. COLLEGE OF CARDIOL. Vol.15 (suppl.A): 218A. 1990.

Weiss 1998. Peter Weiss, "Another Face of Entropy: Particles Self-Organize to Make Room for Randomness," SCIENCE NEWS Vol.154: 108-109. August 15, 1998.

Wells 1923. H.G. Wells, "Relation of Clinical to Necropsy Diagnosis in Cancer and Value of Existing Cancer Statistics," JOURNAL of the AMERICAN MEDICAL ASSN. Vol.80: 737-740. 1923.

WHO 1958. World Health Organization (United Nations, Geneva), INTERNATIONAL STATISTICAL CLASSIFICATION OF DISEASES, INJURIES, AND CAUSES OF DEATH. The following list of the nine revisions (from PHS 1995, p.299) gives "year of conference by which adopted" and then "years of use in USA". First Revision, adopted 1900, in use USA 1900-1909. Second Revision, adopted 1909, in use USA 1910-1920. Third Revision, adopted 1920, in use USA 1921-1929. Fourth Revision, adopted 1929, in use USA 1930-1938. Fifth Revision, adopted 1938, in use USA 1939-1948. Sixth Revision, adopted 1948, in use USA 1949-1957. Seventh Revision, adopted 1955, in use USA 1958-1967. Eighth Revision, adopted 1965, in use USA 1968-1978. Ninth Revision, adopted 1975, in use USA 1979 to present.

WHO 1995. World Health Organization's international symposium, "Health Consequences of the Chernobyl and Other Radiological Accidents," held in Geneva on November 20-23, 1995. See coverage in "Chernobyl's Thyroid Cancer Toll," by Michael Balter, SCIENCE Vol.270: 1758-1759, Dec. 15, 1995.

WHO 1996. World Health Organization (United Nations, Geneva), Office of World Health Reporting, THE WORLD HEALTH REPORT 1996: FIGHTING DISEASE, FOSTERING DEVELOPMENT; REPORT OF THE DIRECTOR-GENERAL, WHO. 143 pages. Index. ISBN 92-4-156182-3. ISSN 1020-3311. WHO Office of World Health Reporting, 1211 Geneva 27, Switzerland. 1996. UCSF Library: RA8, A3, W67, 1996.

Wiatrowski 1983. W. A. Wiatrowski + D.T. Kopp + D.W. Jordan et al. "Factors Affecting Radiation Exposure and Radiographic Image Contrast in Urology," HEALTH PHYSICS Vol.45: 599-605. 1983.

Wilcox 1989. J.N. Wilcox + K.M. Smith + S.M. Schwartz + D. Gordon, "Localization of Tissue Factor in the Normal Vessel Wall and in the Atherosclerotic Plaque," PROC. NATL. ACADEMY SCI. (USA) 86: 2839-43. 1989.

Winkelstein 1990. W. Winkelstein Jr., "Smoking and Cervical Cancer --- Current Status: A Review," AMERICAN JOURNAL OF EPIDEMIOLOGY Vol.6: 945-957. 1990.

Winters 1982-a. T.H. Winters + J.R. DiFranza, "Radioactivity in Cigarette Smoke," (letter) NEW ENGLAND JOURNAL OF MEDICINE Vol.306: 364-365. February 11, 1982.

Winters 1982-b. T.H. Winters + J.R. DiFranza, "Radioactivity in Cigarette Smoke," (letter) NEW ENGLAND JOURNAL OF MEDICINE Vol.309: 312-313. July 29, 1982. Pages 309-312 of this same issue carry seven letters commenting on Winters 1982-a; combined, these letters cite most of the then-existing literature on this topic.

Witztum 1991. J.L. Witztum + D. Steinberg, "Role of Oxidized Low Density Lipoprotein in Atherogenesis," JOURNAL OF CLINICAL INVESTIGATION Vol.88: 1785-1792. 1991.

Wochos 1977. J.F. Wochos + J.R. Cameron, PATIENT EXPOSURE FROM DIAGNOSTIC XRAYS: AN ANALYSIS OF 1972-1974 NEXT DATA. HEW Publication (FDA) 77-8020. (U.S. Department of Health, Education, and Welfare, Public Health Service (FDA), Bureau of Radiological Health, Rockville, Maryland, USA 20857.) 1977.

Wochos 1979. J.F. Wochos + N. Detorie + J.R. Cameron, "Patient Exposure from Diagnostic Xrays: An Analysis of 1972-1975 NEXT Data," HEALTH PHYSICS Vol.36: 127-134. 1979.

Wolbach 1909. S.B. Wolbach, "The Pathologic History of Chronic X-Ray Dermatitis & Early X-Ray Carcinoma," JOURNAL MED. RESEARCH Vol.21: 415-449. 1909.

Wolfman 1986. L. Wolfman, WOLFMAN REPORT ON THE PHOTOGRAPHIC INDUSTRY IN THE UNITED STATES, 1963-1986. (ABC Leisure Publications, New York.) 1986.

Wong 1993. F. Lennie Wong + Michiko Yamada + Hideo Sasaki + Kazunori Kodama + 3 more, "Noncancer Disease Incidence in the Atomic Bomb Survivors: 1958-1986," RADIATION RESEARCH Vol.135: 418-430. 1993.

World Almanac 1991. WORLD ALMANAC AND BOOK OF FACTS, 1991, edited by Mark S. Hoffman. 960 pages. ISBN 0-88687-580-3. U.S. Population figures, 1790-1980, from U.S. Census, at pp.552-553. (Pharos Books, Scripps-Howard Company, New York City.) 1990.

World Health Organization (WHO).
        Please see WHO 1958, WHO 1995, and WHO 1996, above.

Wu 1999. Li-Jun Wu + Gerhard Randers-Pehrson + An Xu + Charles Waldren + Charles R. Geard + ZengLiang Yu + Tom K. Hei. "Targeted Cytoplasmic Irradiation with Alpha Particles Induces Mutations in Mammalian Cells," PROC. NATL. ACAD. SCI. USA Vol.96: 4959-4964. April 1999.

Yamashiroya 1988. H.M. Yamashiroya + L. Ghosh + R. Yang + A.L. Robertson Jr., "Herpesviridae in the Coronary Arteries and Aorta of Young Trauma Victims," AMERICAN JOURNAL OF PATHOLOGY Vol.130: 71-79. 1988.

Yew 1989. P. Renee Yew + Tripathi B. Rajavashisth + James Forrester + Peter Barath + Aldons J. Lusis, "NIH3T3 Transforming Gene Not a General Feature of Atherosclerotic Plaque DNA," BIOCHEMICAL AND BIOPHYSICAL RESEARCH COMMUNICATIONS Vol.165, No.3: 1067-1071. Dec. 29, 1989.

Yla-Herttuala 1986. S. Yla-Herttuala + 4 co-workers, "Glycosaminoglycans in Normal and Atherosclerotic Human Coronary Arteries," LABORATORY INVESTIGATION Vol.54: 402+. 1986.

Young 1960. Wei Young + John W. Gofman + Robert K. Tandy + Nathan Malamud + Eunice Waters, in AMERICAN JOURNAL OF CARDIOLOGY Vol.6: 294-308. 1960.

Young 1994. Stephen G. Young + Sampath Parthasarathy, "Why Are Low-Density Lipoproteins Atherogenic?" WESTERN JOURNAL OF MEDICINE Vol.160, No.2: 153-164. 1994

Zhou 1996. Yi Fu Zhou + M.B. Leon + Myron A. Waclawiw et al, "Association between Prior Cytomegalovirus Infection and the Risk of Restenosis after Coronary Atherectomy," NEW ENGL. J. MED. Vol.335: 624-630. Aug.29, 1996.

Zhou 1997. Yi Fu Zhou + Myron A. Waclawiw + Stephen E. Epstein, "Association between Prior Cytomegalovirus Infection and the Risk of Restenosis after Coronary Atherectomy," (letter), NEJM Vol.336: 588. Feb.20, 1997.

ZurHausen 1998. Harald zur Hausen, "Cervical Cancer: PapillomaVirus and p53," (News and Views), NATURE Vol.393: 217. May 21, 1998. See also: ZurHausen 1994 in LANCET Vol.343: 955-957.

>>>>>>>>>>

- The small letter "e" before a page-number identifies the page where the term is explained or defined. Example: e275

- The rows farthest to the left in this Index are alphabetized on the basis of letters exclusively. Spaces, hyphens, other punctuation marks, and numbers are treated as if they were absent.

- If a part of our text is hard to find again, it may be findable as a sub-entry under some of the big entries, such as:

A

C

E

G

H

I

J

M

N

O

Obesity: Heaviness & IHD risk, 350, 570-571, 580, 595-596

Occlusion (silent) of a vessel, 349

Occupationally-induced cancer, 3, 32, 495

Ockham's Razor, an important admonition in research, e215, e512, e514. Law of Minimum Hypotheses.

OConnor (Egan), co-author of Gofman 1985, 51

"Office building": Evacuation leaves the edifice, 347

Oleic acid (n-9), e562. And olive oil, 563

Oliver (M.F.) 1991+1992+1997: Warnings about lipid-lowering therapies, 319, 336

Omega-3 (n-3) & omega-6 (n-6) fatty acids, e562, e563. Health effects, 563-569

 Optimum ratio, n-6 over n-3 = 5:1, 568

Oncley (J.L.) 1947: Alpha & beta lipoproteins, 311

Ornish (Dean) 1990+1998: Lifestyle & IHD, 336

Orthovoltage xrays, e47, e48

Osler (Wm.) 1908: Infectious etiology, atheroscl, 306

Outlying datapoints, causes & effects, 96, 213, 399

 Visual demo with real-world data: Outliers move into line, 213, 220-221

 Outliers: A cause of negative Constants, 399

 Outliers: Persons w very high HDL-1 levels, 588-589

Ovarian cancer & genital talcum powder, 163-164

Oxidized lipoproteins, 307, 317, 345, 569

P

p-53.

 CMV product interacting with p53 protein, 308-309

Painless way to reduce cancer, IHD, 1

Palade (G.) 1956: Electron-microscope studies of endothelial vesicles, 317

Pappenheimer (J.R.) 1953: Capillary permeability, 317

"Paralytic" pessimism, 55

Parker (Herbert M.), co-author of Buschke 1942, 29

Parkes (J.L.) 1991, on Benditt hypothesis: Ref.List.

Parmley (Wiliam W.) 1997: Endothelial cells and atherogenesis, 307, 345. Plaque stabilization, 321, 345. ACE inhibitors, 351

Patel (P.) 1995: Infections and IHD, 309

Patterson (James T.) 1987: No cause assigned to 38% of deaths in year-1900, 498

Pauling (Linus) warned in 1950s about health effects of radioactive fallout, 31

"Paying the price" for weight-gain, 571

Peak year for cigarette smoking, USA, 363, 365

Peak year for IHD MortRate, USA, 287, 288

Pearson (Thomas A.) 1975+1977+1978b+1983a: Monoclonality in plaques, 326-328.

 1993: On LDL & endothel-derived relaxing factor, 345

 1998: Lipid-lowering in low-risk patients, 336

Pedersen (Terje R.) 1994 = Scandinavian 1994, 319.

 1995: NEJM editorial on lipid-lowering, 319

Peer review, viii, 8, 52, 381, 530. Appropriate, during peer-review, to challenge papers which uncritically incorporate the safe-dose fallacy, 529

Pelvimetry, pre-delivery, 17, e30, 34, 342, 524

Penicillin's introduction, 618

Penn (Arthur) 1981a+1986+1988+1989+1990+1991+1993 +1994+1996: Experimental work on mutagens in the etiology of atherosclerosis, 328, 330-331, 502.

 1990: Review paper in Mutation Research, 328

Pennell (M.Y.) 1952: PhysPop data & type, 57, 58

Per capita population dose from medical radn:

 Why "profound uncertainty among informed people," about magnitude of post-1960 decrement, is likely to be permanent, 37-38

 Number of exams (past, present) quite uncertain, 33, 609-611

 Doses per exam (past, present) not measured, 33, 36, 609, 610

 Downward & upward pressures on post-1960 per-capita population dose, explained, 34-35

  Net effect is probably downward, 37, 624

 Distinction between per-capita population dose and per-patient dose, 93-94

Perfect correlation, e12, e63, e95. Can persist while x-values rise, y values fall, 97-98

Perfect proportionality, e62, e63, e93, e97

Pericardium, e599

Per-patient dose, alike in 9 Census Divisions, 93-94, 623

Per-rad potency. Please see: Risk per rad.

Perry (I.J.) 1995: Plasma homocysteine & IHD, 331

Peterson (Ken M.) 1999: Co-author w. K.Z. Morgan, 50

Petrie (B.L.) 1987+1988: Viral etiology atheroscl, 308

Phillipson (B.E.), research on n-3 fatty acids, 563

Photofluorogram, e611

Photon, e38. The biological damage from xrays is due to the photons which never reach the film, 8

PHS 1959: Measures needed to limit radn exposure, 32

 Estimated annual per capita dose, 33

PHS 1992: Multiple reference years, age-adju MRs, 82

PHS 1995: State-distribution in the Census Divs, 57

 U.S. population in 1940 by age-bands, 82, 87

 PhysPop values 1975-1994, 59

 Breast cancer age-specific rates (1950-1993), 88

PhysPop = Physicians per 100,000 population, e5, e12.

 PhysPop values by Census Divisions reflect relative magnitude of per capita population dose from medical radiation, 53-54, 499

  NonCancer NonIHD analysis supports validity of of this surrogacy, 223, 514

 How PhysPop reflects accumulated per capita dose, 12, 56, 61, 65, 100

 Stability of PhysPop proportions & ranking over decades, by Census Divs, 61, 63-65, 356-358, 366

  Tabulated non-averaged, 66, 76. Averaged, 359, 360

 Merits of using PhysPops as dose surrogates, 11-12, 14-15, 54-55, 500

 PhysPop history permits this inquiry, 65, 356

Pickels (Edward G.) 1942+1943: The engineering/ physics genius who produced two remarkable instru- ments, the electrically driven analytical ultra- centrifuge & the preparative ultracentrifuge, 311

"Pie-charts" of estimated per-capita radn doses, 37

Pierce (Donald A.), Radn Effects Research Foundation:

 1996-a: A-Bomb Study really a LOW-dose study, 42

S

W

X, Y, Z

The guide to this index is located at page 677.

>>>>>>>>>>

Additional copies of this book are available for
purchase from the publisher, from online bookstores,
and from library distributors.

Library of Congress Catalog Number: 99-045096.
ISBN for the hardcover edition: 0-932682-97-9. USD $35
ISBN for the softcover edition: 0-932682-98-7. USD $27

The book's Table of Contents, Introduction, Abstract,
and Chapter 1 (the Executive Summary) are available
online at
    http://www.ratical.org/radiation/CNR/
Bound, printed copies of those same parts are
available from the publisher ($5 per copy, includes
first-class postage within the USA). Publisher:

Committee for Nuclear Responsibility Books
Post Office Box 421993, San Francisco CA 94142-1993.
Tel + Fax: 415-776-8299. E-mail: cnr123@webtv.net

Related books and documents by the author at:
www.x-raysandhealth.org
Also:
www.ratical.org/radiation/CNR/
Ditto   "   "   "   " /RMP/
Ditto   "   "   "   " /PBC/
Ditto   "   "   "   " /RIC/

• In 1972, Dr. Gofman shared the 1972 Stouffer Prize, one of the top awards for research in combatting arteriosclerosis. The 1972 Prize Committee was chaired by Professor Ulf S. von Euler, M.D., former chairman of the Nobel Prize Committee for Physiology and Medicine. The Committee's citation:

"The 1972 Stouffer Prize is awarded to Dr. John W. Gofman for pioneering work on the isolation, characterization and measurement of plasma lipoproteins, and on their relationship to arteriosclerosis. His methods and concepts have profoundly stimulated and influenced further research on the cause, treatment, and prevention of arteriosclerosis."

Radiation and Human Health. 1981. ISBN 0-87156-275-8.

• From the Journal of the American Medical Assn., March 19, 1982, p.1637, a review by Victor E. Archer, M.D.: "This remarkable and important book enables any intelligent person with a high school education to understand the complexities involved in assessing the risks to man from low levels of ionizing radiation. Gofman not only demonstrates his mastery of this complex subject but carefully explains the basic concepts of epidemiology, genetics, birth defects, carcinogenesis, radiobiology, physics, chemistry and even mathematics, which are necessary to an understanding of the subject."

Xrays: Health Effects of Common Exams. 1985. ISBN 0-87156-838.1. E.O'Connor, co-author.

• From the New England Journal of Medicine, Feb. 6, 1986, p.393, a review by Maurice M. Greenfield, M.D (radiologist): "This book is practical and important. It is destined to represent a watershed in the controversial field of low-dose radiobiology and will be of inestimable value to radiologists, other physicians, dentists, and patients."

• From the American Journal of Roentgenology, April 1986, p.774, a review by David S. Martin: "From a radiologist's point of view, this book represents a well organized and concise attempt to quantify the cancer risk from diagnostic xray exposures by age, gender, organ, and examination. As such, it is a useful starting point for comparisons."

Radiation-Induced Cancer from Low-Dose Exposure. 1990. ISBN 0-932682-89-8.

• From the New England Journal of Medicine, Feb. 14, 1991, p.497, a review by G. Theodore Davis, M.D., and Andre J. Bruwer, M.D. (radiologist) of two books jointly: The 1990 book by Gofman (above) and the 1990 BEIR-5 Report from the National Research Council, National Academy Press: "Both these works agree that previous assessments of the dangers of radiation underestimated the risk, but they reach substantially different conclusions about the magnitude of the risk, especially when the radiation is at lower doses (below 10 rem) and the doses are delivered slowly ... We strongly recommend both these excellent and timely books for physicians, engineers, and public health officials concerned with radiation, the environment, and public health."

Preventing Breast Cancer. 1995. ISBN 0-932682-96-0 (Second Edition).

• From the Journal of the American Medical Assn. "Medical News & Perspectives," August 2, 1995, a two-page feature (pp.367-368) by Andrew A. Skolnick about Gofman's book: "A respected authority on the biological effects of ionizing radiation has just published a book claiming that the vast majority of breast cancers in the United States were caused by ... medical xrays ..." Skolnick quotes from interviews with the author and with critics of the book.

• On August 3, 1995, Channel 3 in Britain telecast a report ("The Xray Effect") featuring the book's findings. The 1995 broadcast included these statements:

"John Gofman is a superb analyst and has always been at the cutting edge of medical science, particularly when it comes to protecting people." • - Mortimer Mendelsohn, M.D., Ph.D., then Assoc. Director of the Radiation Effects Research Foundation (the A-Bomb Survivor Study).

"Dr. Gofman is owed a debt of gratitude by the scientific community because he was one of the first people to raise the issue of cancer risks from radiation exposure." • - Edward P. Radford, M.D., epidemiologist and Chairman of the 1980 Committee on the Biological Effects of Ionizing Radiation (BEIR-3) of the National Academy of Sciences, National Research Council.